June 25–29, 2012
Low Wood Bay, Lake District, UK

Association for
Computing Machinery

Advancing Computing as a Science & Profession

MobiSys'12

Proceedings of the 10th International Conference on
Mobile Systems, Applications, and Services

Sponsored by:

ACM SIGMOBILE

In cooperation with:

ACM SIGOPS

Association for
Computing Machinery

Advancing Computing as a Science & Profession

The Association for Computing Machinery
2 Penn Plaza, Suite 701
New York, New York 10121-0701

Notice to Past Authors of ACM-Published Articles
ACM intends to create a complete electronic archive of all articles and/or other material previously published by ACM. If you have written a work that has been previously published by ACM in any journal or conference proceedings prior to 1978, or any SIG Newsletter at any time, and you do NOT want this work to appear in the ACM Digital Library, please inform permissions@acm.org, stating the title of the work, the author(s), and where and when published.

ISBN: 978-1-4503-1301-8 (Digital)

ISBN: 978-1-4503-1879-2 (Print)

Additional copies may be ordered prepaid from:

ACM Order Department
PO Box 30777
New York, NY 10087-0777, USA

Phone: 1-800-342-6626 (USA and Canada)
+1-212-626-0500 (Global)
Fax: +1-212-944-1318
E-mail: acmhelp@acm.org
Hours of Operation: 8:30 am – 4:30 pm ET

Printed in the USA

Welcome from the General Chair

On behalf of the entire organizing committee it is my pleasure to welcome you to MobiSys 2012. Situated on the banks of Lake Windermere in one of the most beautiful regions of England, I hope that you find the conference and this unique environment provides a though-provoking, enjoyable and rewarding experience.

MobSys 2012 represents the culmination of a year's work by many volunteers. Led by the PC-Chairs Srinivasan Seshan and Lin Zhong the entire Program Committee worked tirelessly to help produce a world-class scientific program. I would like to echo the PC Chair's thanks to all the authors who entrusted us with their work for review and publication.

In addition to a rich and varied program of technical paper presentations, MobiSys 2012 features an extensive poster and demonstration session that showcases new and exciting work in the area of mobile systems. The scale and quality of this session is a reflection of the efforts of our Demo/Poster Chairs David Chu and Albrecht Schmidt.

In common with previous years, MobiSys 2012 also includes a number of workshops. This year we had a substantial number of proposals and those that were selected have helped enrich the conference. I would particularly like to thank our Workshop Chair Moustafa Youssef who has essentially had to manage an entire series of mini-conferences – no small task and one for which I am extremely grateful.

I would also like to thank Christos Efstratiou our Publicity Chair who made sure that the news of MobiSys 2012 was broadcast far and wide. Finally, thanks to our Local Arrangements Chair Nick Race, ably assisted by Carol Airy, who together have helped make the conference such an enjoyable experience.

Finally, I would like to thank ACM SIGMOBILE and our generous corporate sponsors without which this event would not have been possible.

It has been a tremendous honor to serve as General Chair for this year's conference and I hope that you enjoy MobiSys 2012.

Nigel Davies
Mobisys 2012 General Chair
Lancaster University, UK

Welcome from the Program Co-Chairs

It is our great pleasure to welcome you to the *MobiSys 2012: The Tenth International Conference on Mobile Systems, Applications, and Services.* This year's conference continues the MobiSys tradition of being the premier forum for presentation of research in mobile and wireless systems.

The call for papers attracted 182 submissions. This was a new record for MobiSys and a strong sign that of the vibrancy of our community. Each paper was assigned to two Program Committee members for a first round of reviews. Each of the 155 papers with at least one review of "weak reject" was assigned an additional reviewer from the Program Committee in the second round. In the final round, we assigned an additional pair of reviewers or more to each of the top 88 papers. In all the Program Committee members and external reviewers generated 724 reviews and the top 88 papers received at least 5 reviews.

The Program Committee met for a full day at La Jolla, CA. The Committee discussed the top 60 papers and accepted 32 papers for the program. The 32 paper count is another record high for MobiSys and shows that there is a great deal of high quality work coming from our community. Each of these 32 papers was shepherded by a member of the Program Committee to help ensure that recommendations made by the reviewers and during the Program Committee meeting were addressed.

There were several new additions to the review process. First, we created an External Review Panel composed of people active in the mobile computing industry. In addition to getting the valuable opinions of these experts, a secondary goal of the panel was to increase the interaction between these industry leaders and the MobiSys community. Second, we had the Program Committee generate a single public review for each accepted paper. The public review highlights the key contributions of the paper that convinced the Program Committee to accept the paper and some of the remaining weaknesses in the paper. These public reviews were posted on an online forum (http://reviews.mobisys2012.rice.edu/) to help promote community discussion on each paper.

We owe thanks to many people who helped in putting together MobiSys 2012. First, we would like to thank all the authors of submitted papers. We are grateful to the Program Committee and the External Review Panel, who worked very hard in reviewing papers and providing feedback for authors.

We hope that you will find this program interesting and thought-provoking. In addition, we hope that the conference will provide you with a valuable opportunity to share ideas with other researchers and practitioners from institutions around the world. Welcome to MobiSys 2012.

Srinivasan Seshan & Lin Zhong
MobiSys 2012 Program Co-Chairs

Table of Contents

Session 7: Sensing

Session 8: More Networking

Demonstrations

Posters

MobiSys 2012 Organization

General Chair: Nigel Davies *(Lancaster University, UK)*

Program Co-Chairs: Srinivasan Seshan *(Carnegie Mellon University, USA)*
Lin Zhong *(Rice University, USA)*

Workshop Chair: Moustafa Youssef *(E-JUST, Egypt)*

Posters/Demos Co-Chairs: David Chu *(Microsoft Research, USA)*
Albrecht Schmidt *(University of Stuttgart, Germany)*

Local Arrangements Chair: Nick Race *(Lancaster University, UK)*

Publicity Chair: Christos Efstratiou *(University of Cambridge, UK)*

Student Travel Grants Chair: Marco Gruteser *(WINLAB / Rutgers University, USA)*

Steering Committee Chair: Victor Bahl *(Microsoft Research, USA)*

Program Committee:
Sharad Agarwal *(Microsoft Research, USA)*
Rajesh Krishna Balan *(Singapore Management University, SG)*
Lujo Bauer *(Carnegie Mellon University, USA)*
Gaetano Borriello *(University of Washington, USA)*
Kameswari Chebrolu *(IIT-Bombay, India)*
Landon Cox *(Duke University, USA)*
David Culler *(UC Berkeley, USA)*
Prabal Dutta *(University of Michigan, USA)*
Maria Ebling *(IBM Research, USA)*
Marco Gruteser *(WINLAB / Rutgers University, USA)*
Urs Hengartner *(University of Waterloo, Canada)*
Yih-Chun Hu *(University of Illinois at Urbana-Champaign, USA)*
Monica Lam *(Stanford University, USA)*
Anthony LaMarca *(Intel Labs, USA)*
Yang Li *(Google, USA)*
Z. Morley Mao *(University of Michigan, USA)*
Patrick McDaniel *(Penn State University, USA)*
Jeffrey Pang *(AT&T Labs, USA)*
Li-Shiuan Peh *(MIT, USA)*
Pablo Rodriguez *(Telefonica, Spain)*
Romit Roy Choudhury *(Duke University, USA)*
Mirjana Spasojevic *(Nokia Research, USA)*
Lakshminarayanan Subramanian *(New York University, USA)*
Doug Terry *(Microsoft Research, USA)*
Alexander Varshavsky *(AT&T Labs, USA)*

Program Committee (continued):

Arun Venkataramani *(University of Massachusetts Amherst, USA)*
Matt Welsh *(Google, USA)*
Guoliang Xing *(Michigan State University, USA)*

External Reviewer Panel:

Stuart Cheshire *(Apple)*
Rajarshi Gupta *(Qualcomm)*
Farooq Khan *(Samsung Electronics America)*
Erran Li *(Bell Labs, Alcatel-Lucent)*
Venu Vasudevan *(Motorola Mobility)*
Eran Yariv *(Microsoft)*
Hui Zang *(Sprint Labs)*

Additional reviewers:

Tico Ballagas
Ranveer Chandra
Kartikeya Chandrayana
Ben Greenstein
Dongsu Han
Mike Hazas
Phil Kortum
Balachander Krishnamurthy
Xiaozhu Lin
Graham Percival
Abhinav Srivastava
Jaideep Vaidya
Alec Wolman
Jianxin Wu

MobiSys 2012 Sponsor & Supporters

Sponsor: sigmobile

In cooperation with: SIGOPS ACM SIG on Operating Systems

Supporters: Microsoft Research

Google

TOYOTA InfoTechnology Center, U.S.A., Inc.

IBM Research

SwordFight: Enabling a New Class of Phone-to-Phone Action Games on Commodity Phones

Zengbin Zhang
UC Santa Barbara
zengbin@cs.ucsb.edu

David Chu
Microsoft Research
davidchu@microsoft.com

Xiaomeng Chen
University of Science and Technology of China
monachen@mail.ustc.edu.cn

Thomas Moscibroda
Microsoft Research Asia
moscitho@microsoft.com

ABSTRACT

Mobile gaming is a big driver of app marketplaces. However, few mobile games deliver truly distinctive gameplay experiences for ad hoc collocated users. As an example of such an experience, consider a sword fight dual between two users facing each other where each user's phone simulates a sword. With phone in hand, the users' thrusts and blocks translate to attacks and counterattacks in the game. Such Phone-to-Phone Mobile Motion Games (MMG) represent interesting and novel gameplay for ad hoc users in the same location. One enabler for an MMG game like sword fight is continuous, accurate distance ranging. Existing ranging schemes cannot meet the stringent requirements of MMG games: speed, accuracy and noise robustness. In this work, we design `FAR`, a new ranging scheme that can localize at 12Hz with 2cm median error while withstanding up to 0dB noise, multipath and Doppler effect issues. Our implementation runs on commodity smartphones and does not require any external infrastructure. Moreover, distance measurement accuracy is comparable to that of Kinect, a fixed-infrastructure motion capture system. Evaluation on users playing two prototype games indicate that `FAR` can fully support dynamic game motion in real-time.

Categories and Subject Descriptors

C.3 [**Special Purpose and Application-Based Systems**]: Real-time and embedded systems; C.5.3 [**Computer System Implementation**]: Microcomputers—*Portable devices*

Keywords

acoustic localization, mobile gaming, mobile motion games, smartphones

1. INTRODUCTION

Mobile gaming constitutes a large and fast growing industry. Estimates on the total worldwide market size differ, but typical numbers are in the order of $10 billion for 2011 [1]. Furthermore, continued growth in the underlying forces – smartphone and tablet sales, mobile Internet subscribers and app downloads – all point to a bright future for the industry.

The landscape of today's mobile games is rich and varied. However, one commonality among existing multiplayer games is that they invariably require the players to be physically passive, look at the screen, and conduct game action via interaction with the screen. The success of Nintendo Wii or Xbox Kinect in console gaming on the other hand, has demonstrated a demand for much more interactive games in which gameplay directly involves the user's physical activity. In this paper, we report on the development of a novel class of *Phone-to-Phone Mobile Motion Games* (MMG) that achieve a similar level of physical interactivity. These games are characterized by the fact that the *position, location, orientation, or movement of the phone is an integral part of game play.* In some cases, the phone may even serve as the equivalent of a Wii-stick, and players may not even need to look at the screen while playing. However, in contrast to Wii or Kinect, we do not rely on any external infrastructure such as a microphone array or cameras; MMG games are played purely phone-to-phone.

Consider for example the following *SwordFight* game. The rules are simple: Two players wield their phones, and try to attack each other. A player can attack the opponent by tapping the screen. If player A attacks, and her phone is within 20cm of the opponent's phone, she wins. However, attacking costs energy, and an attack can only be sustained for 4 seconds before the energy has completely depleted. Energy can be regained over time when the player remains in non-attack mode. Thus, one strategy to win is for a player to survive the opponent's attack by quickly moving his phone in such a way that it maintains sufficient distance from the opponent's phone; and then counter-attack while the opponent's energy is depleted.

The key enabling technology for a game like SwordFight is the ability to conduct *very fast*, *accurate*, and *robust distance measurements* between the phones, so that at any moment during play, both phones have precise distance information. Studies have shown that in order to sustain high-speed action games a lag of more than 100ms decreases user satisfaction and degrades player performance, and a lag of 200ms is unacceptable [2]. Therefore, we need to be able to conduct phone-to-phone distance measurements at a rate of at least 10Hz. The measurements also have to be accurate – with no more than a few centimeters of error – and robust in the face of mobility, noise, and networking issues. In the absence of any external infrastructure, the combination of these three requirements constitutes a significant technical barrier on commodity phones, and no existing solution is able to meet them simultaneously.

It is well-known that acoustic sound can be used for distance measurements. Works such as [16, 18] have demonstrated that under ideal circumstances (e.g., no mobility) and with sufficient computation time, accurate ranging can be achieved even on commodity phones. The problem is that in an motion game scenario, the circumstances are far from ideal and the requirements substantially more challenging. First, existing ranging protocols are based on the assumption that phone positions remain static during the process of taking a measurement. For a game like SwordFight, this is not a valid assumption as human hand speed can be as high as 2m/s. Furthermore, with two phones moving towards or apart from each other at high speed, aspects such as the Doppler effect need to be considered and dealt with. Secondly, acoustic ranging requires the use of expensive cross-correlation algorithms in order to determine the precise time-of-arrival of the sound signal. Cross-correlation algorithms are computationally intensive and cannot be run without modification on phones at sufficient speed to enable a SwordFight game. Third, not only *computation*, but also the *communication* (acoustic tone exchanges, protocol handshakes, etc) incurs a fundamental and significant delay.

In this paper, we address these challenges in a systematic manner. Our first contribution is to enable real time distance measurement by replacing the standard computationally intensive cross-correlation algorithm for finding sound peaks with a more sophisticated and efficient multi-stage algorithm. Our algorithm employs autocorrelation to fundamentally reduce computational complexity while preserving accuracy by targeting cross-correlation to a very narrow search window. Our second contribution is to employ a new pipelined streaming execution strategy that overlaps protocol communication and algorithm computation. While overlapping communication and computation is a well-known technique, in our case, this overlapping comes with a twist – we overlap the *sound waves* of the ranging protocol in addition to the typical networking data packets. It turns out that both pipelining and streaming are critical in order to realize real time measurements. Our third contribution is to understand and tackle the practical sources of measurement error during motion gaming for increased robustness. The first set of robustness optimizations addresses mobility and the effects of Doppler shift on the underlying ranging protocol. The second set of optimizations addresses environmental robustness from effects such as ambient noise, multipath and acoustic tone loss.

Combining these techniques, we develop `FAR` – a Fast, Accurate and Robust localization system that enables two potentially fast moving phones to keep accurate distance estimates to each other. Distance measurements can be taken at a rate of 12Hz with 2cm median error while withstanding up to 0dB noise (e.g., players or spectators talking while playing), multipath (e.g. as encountered in small rooms) and the Doppler effect. To practically demonstrate the system's ability to enable novel gaming concepts, we develop two prototype MMG games: SwordFight and ChaseCat.[1] In both isolated-player and in-situ gameplay experiments, we find that `FAR`'s distance measurements are comparable to Kinect, a dedicated fixed-infrastructure motion capture system. We have publicly tested our games at various locations with real players; and our experience shows that the games are fun and intuitive to play. We anticipate that with the high-speed ranging API we provide in this paper, many more MMG games can be developed.

2. ASSUMPTIONS & REQUIREMENTS

Enabling a game like SwordFight requires a distance ranging subroutine that allows two phones to keep accurate and up-to-date distance information between each other even in the face of noise and high mobility. This poses a unique set of challenges:

- **Phone-to-Phone:** MMG games should be playable everywhere at any time. We do not rely on any external infrastructure beyond the two phones.
- **Commodity phones:** We want MMG games to run on commodity hardware and OSs. This implies handling issues of on-phone sensor sampling rates, computation capacity, or audio playing and recording capabilities, whether they arise from the hardware, (typically closed source) audio driver, or OS.
- **Measurement Frequency:** To create the sensation of continuous real-time distance information, the ranging infrastructure must sample user movement as frequently as possible, at a rate of 10Hz or more [2].
- **High Accuracy:** Gameplay relies on the high accuracy of distance measurements even at a high degree of phone mobility. Specifically, we aim for a target accuracy of within 2-3cm up to a normal range of human social interaction of up to approximately 3m with line of sight.
- **User Mobility:** In a game like SwordFight, players should be able to play without artificial restrictions on their body or hand movement. The underlying ranging infrastructure must be capable of supporting the speed of natural human hand movement (up to 2m/s) [8].
- **Practical in Most Environments:** As it is natural for players and spectators to talk during a game, and because a game may be played indoors or close to walls, the underlying distance measurement framework must be robust against the impact of high ambient noise levels, multi-path effects and tone loss.

3. BACKGROUND

In this section, we review the well-known time-of-flight acoustic ranging principle that we share in common with prior phone-based acoustic localization work [16, 18], and we outline why these existing approaches are unsuited for enabling Mobile Motion Gaming.

[1]See `http://research.microsoft.com/mobile-motion-gaming` for a video illustrating the gameplay of SwordFight.

3.1 Acoustic Distance Measurement

The essential idea is to have two phones A and B play and record known audio tones one after another. These tones are often based on some form of pseudorandom sequences. Each phone records its own emitted tone as well as the tone originating from the remote phone. Suppose A emits a tone first and records this tone's arrival at its microphone at time t_{A1}. B records the arrival of this tone at time t_{B1}. Next, B emits a tone, which is recorded at the microphones of B and A at times t_{B2} and t_{A2}, respectively. The distance d between the phones can now be calculated as

$$d = \frac{1}{2} \cdot c \cdot [(t_{A2} - t_{A1}) - (t_{B2} - t_{B1})], \quad (1)$$

where c is the speed of sound.

Existing schemes [16, 18] follow a traditional execution strategy of *Play/Record then Compute and Exchange* (see Figure 2(a)). The figure shows that the two phones start the *Recording* step at the same time, and then send out the tones one after another (*Tone Playing* step). Guard periods are inserted prior to and following each tone to guarantee that the tones do not overlap, and that the *Recording* is complete.

Upon completion of the recording, the next step is for each phone to determine the exact local time when each tone was received. This is done by applying a *cross-correlation* algorithm to each of its recorded sound samples (*Computation* step). Cross-correlation is a standard signal processing technique that searches for the best match between a recorded sample and a reference signal (the best match is indicated by a sharp peak in the cross-correlation). It has been widely used in various ranging systems, e.g. [4, 6, 10, 16, 18, 19, 23] and is computed as

$$CC(t_0) = \frac{\sum_{t \in W}[X(t) - \overline{X(t)}][T(t-t_0) - \overline{T(t-t_0)}]}{\sqrt{\sum_{t \in W}[X(t) - \overline{X(t)}]^2 \sum_{t=1-L}^{0}[T(t) - \overline{T(t)}]^2}}$$

where $X(t)$ is the recorded sound sequence, $T(t), t \in [-L+1, 0]$ is the reference signal of length L, $\overline{X(t)} = \frac{\sum_{t \in W} X(t)}{L}$, $\overline{T(t)} = \frac{\sum_{t=1-L}^{0} T(t)}{L}$, $W = [t_0 - L + 1, t_0]$. For each t_0, the computation complexity is $O(|W|) = O(L)$, and thus $O(B * L)$ if we have B recorded sound samples. Finally, once the cross-correlation peaks are found at each phone, the corresponding tone arrival time-stamps are exchanged between the two phones e.g., via 3G or WiFi (*Measurement Exchange* step), and the distance can be computed.

3.2 Limitations

As pertains to MMG, existing phone-to-phone acoustic localization schemes all have the same drawbacks: they are too slow to serve the needs of MMG games due to the large measurement delay, and they are not designed for highly-mobile systems.

- The traditional execution strategy of *Play/Record then Compute and Exchange* is unsuited for high frequency measurements, as it takes too much time.
- Cross-correlation is well-known to be computationally expensive, as it grows super-linearly in the size of the reference signal L and the recorded sound sequence T. For example, although [18] applied an energy detection module to reduce the search space in the sound sequence T, the reported time for one measurement is still 800ms.

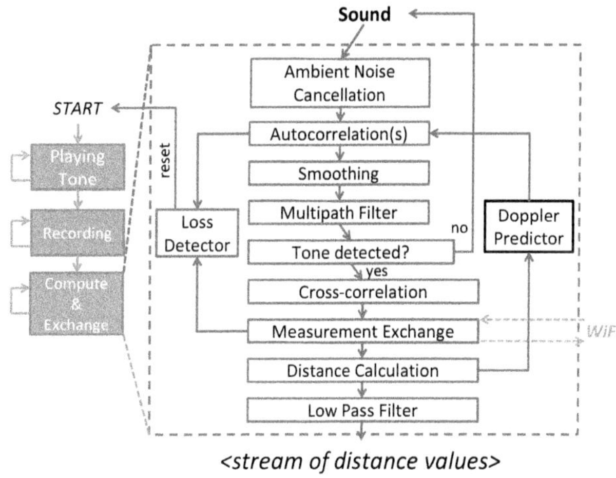

Figure 1: FAR Architecture

- Existing protocols are not designed for highly-mobile systems. They assume that during the course of a measurement, the phones remain stationary and if they are not, results will be erroneous.

Finally, in addition to these more fundamental reasons, there are also purely practical system design issues that prevent us from building MMG games on top of existing ranging infrastructure:

- Commodity phones do not offer tight timing guarantees on the operation of the playing and recording controls, and thus systematically incur large and unpredictable delays. Specifically, as we show in §8, there is a lag of high magnitude and high variance between the time a playback is initiated and the actual time the tone is played. For example, the lag is 60-90ms for both Nexus One phone running Android 2.3.4 and Samsung Focus phone running Windows Phone 7.5. This caps the frequency at which phones can emit tones. Based on our empirical observations, this lag occurs so commonly among commodity smartphones that we cannot simply discount it as a software- or device-specific phenomenon.

4. FAR SYSTEM DESIGN

In this section, we introduce FAR, our fast, accurate, and robust distance measurement system that serves as an API to phone-to-phone MMG games like SwordFight. Meeting the requirements in §2 in the face of the limitations in §3.2 is not easy and drives us to a novel system design as well as numerous optimizations at both the algorithm and systems level.

The key goal of the FAR system is to systematically improve upon existing phone-to-phone ranging schemes in two directions – i) by making them *faster* and reducing delays (See §5); and ii) by making them more *robust* in the face of mobility (See §6) – while at the same time keeping the required degree of accuracy.

Figure 1 illustrates the FAR architecture that achieves these goals. Upon initiation, each of the major stages *Playing*, *Recording*, *Computation* and *Exchange* happen continuously. §5 discusses the dynamic arrangement of these stages. *Computation* contains the main algorithmic contributions of our work. It is sequenced as follows.

1. An ambient noise filter mitigates environmental noise such as from shouting, talking and crowd noise common during gameplay.

2. One or more lightweight autocorrelators work in tandem to detect the presence of tone signatures.

3. Smoothing reduces the impact of anomalous local minima which are the result of sound distortion played and captured by commodity hardware.

4. Multipath readings (e.g. reflection from nearby objects) are filtered out to identify the actual tone received. If no tones are detected, expensive cross-correlation is not engaged.

5. The cross-correlator identifies the tone reception time stamp for accurate timing information.

6. The phones exchange their measured time stamps via WiFi. Failure to receive data here (due to signal jamming or tone misdetection in the autocorrelation step) over prolonged durations results in the loss detector automatically signaling a protocol reset.

7. The exchanged data is used in distance calculation. The distance calculation also informs the Doppler predictor, which adjusts the number of autocorrelators to use in the next round. This enhances measurement accuracy at high mobility.

8. A final low pass filter smooths the calculation before emitting the distance value to the game.

5. FAST DISTANCE MEASUREMENTS

In this section, we present the set of techniques that allows `FAR` to conduct distance measurements at high frequency and with low lag. To see how our techniques impact lag and frequency, it is useful to again consider the traditional execution strategy as illustrated in Figure 2(a). In the traditional execution strategy, the lag of each individual measurement is comprised of the following components:

- **Tone Length:** Time spent on sending out each tone. Since one measurement requires that both phones send out tones, the measurement lag includes two tone lengths.
- **Audio Playing Lag:** Delay between the time of calling the `Play()` API and the time when the tone is actually sent out from the speaker. This delay is observed across platforms on both Android and Windows Phone.
- **Sound Propagation Delay:** Time it takes for the sound to reach the other phone over the air.
- **Buffering Delay:** Time from when the tone has been fully recorded until when the audio driver passes the recorded buffer to the application layer. The reason for the delay is that the audio driver delivers a buffer only when the buffer is full, regardless of whether the tone has arrived.
- **Tone Detection Computation Time:** Time required to compute a recorded buffer and determine the exact time-stamp when the tone arrived.
- **Measurement Exchange & Distance Calculation:** Time it takes for the two phones to exchange their timestamps, and compute the final distance measurement result using Equation (1).

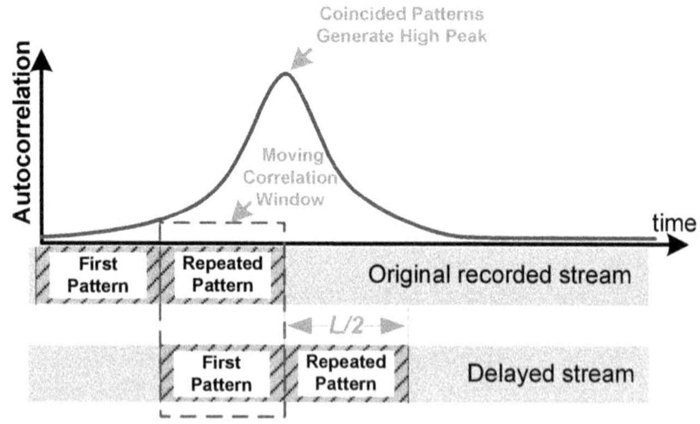

Figure 3: Autocorrelation-based Tone Detection. When sliding the correlation window, at a certain time, the repeated pattern in the original stream will coincide with the first pattern of the delayed stream, which generates a high correlation peak.

If we want to achieve measurement frequency above the 10Hz required for MMG games, we must address these delays across the board. `FAR` accomplishes this through an efficient system architecture and set of core acoustic signal processing algorithms. §5.1 discusses how we substantially reduce the tone-detection computation time by a very efficient algorithm. §5.2 introduces pipelining to further improve measurement frequency, and streaming to circumvent widespread system lags. Lastly, §5.3 summarizes how we reduce other components of delay.

5.1 Fast Tone Detection Algorithm

As discussed in §2, the problem of relying upon cross-correlation for tone detection is that its computation overhead is high. To find the start time of a tone of length L in a buffer of size B, a total of $O(B * L)$ multiplications are required. Our key observation is that instead of running the cross-correlation directly, we can first use a computationally much more efficient *autocorrelation* primitive. Autocorrelation's strength is that it is much more efficient than cross-correlation at finding *repeating patterns* in data sequences. However, autocorrelation does not give us an exact estimate of the tone's location because of its much flatter peak style. Therefore, after applying autocorrelation, we employ (as a second and third step) a simple smoothing algorithm and then a small-scale cross-correlation in a narrow search window centered around the smoothed autocorrelation peak. We now discuss these steps in detail.

Step 1: Autocorrelation-based Tone Detection. The basic methodology of autocorrelation we use is shown in Figure 3. Autocorrelation can only detect correlation of a signal to itself. Therefore, we let each phone send out a tone consisting of a pseudorandom sequence followed immediately by an exact copy of this sequence. To detect this tone, the receiver phone records the incoming stream $X(t)$, and then generates a second stream $Y(t)$ which is an identical copy of $X(t)$, except that it is delayed by half the tone length L, i.e, $Y(t) = X(t - L/2)$. The receiver now correlates $X(t)$ and $Y(t)$ (i.e., the incoming steam is correlated with a copied, delayed version of itself) in a moving correlation window of length $L/2$. This is done by incrementally computing for

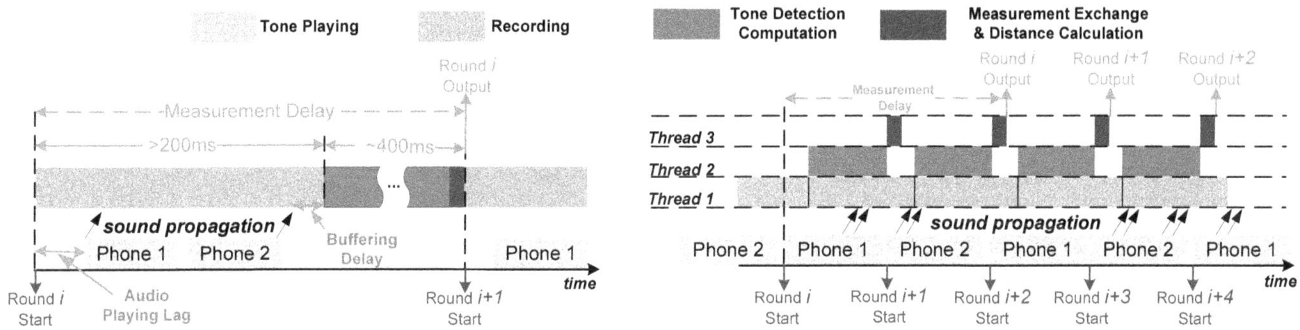

(a) Traditional Execution Strategy (b) FAR Execution Strategy

Figure 2: Execution Strategies

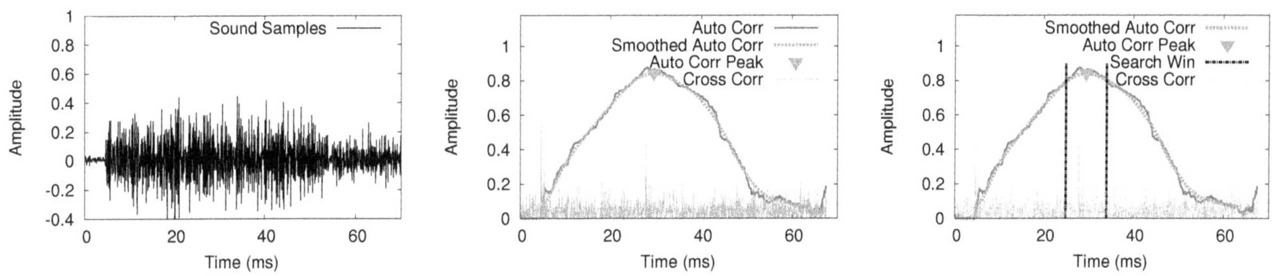

(a) Captured Sound Samples (b) Auto- and Cross-correlation Peaks (c) Peak Search in a Small Window

Figure 4: Fast Tone Detection Algorithm. (a) Recording of received tone with repeating patterns. (b) While cross-correlation peaks are sharp, the autocorrelation peak is flat and has an offset to the pattern's actual location. (c) Cross-correlation is used in a small window around the autocorrelation peak to find the precise location of the pattern.

each time step t_0 the function

$$R(L/2, t_0) = \frac{\sum_{t \in W}[X(t) - \overline{X(t)}][Y(t) - \overline{Y(t)}]}{\sqrt{\sum_{t \in W}[X(t) - \overline{X(t)}]^2 \sum_{t \in W}[Y(t) - \overline{Y(t)}]^2}}$$

where $\overline{X(t)} = \frac{\sum_{t \in W} X(t)}{L/2}$, $\overline{Y(t)} = \frac{\sum_{t \in W} Y(t)}{L/2}$, and $W = [t_0 - L/2 + 1, t_0]$. Higher autocorrelation values correspond to closer correlations of $X(t)$ and $Y(t)$ at t_0. The idea is that because phones send two identical sequences in each tone, at a certain time slot, one of the repeated patterns in $X(t)$ will match a first pattern in the delayed stream $Y(t)$, thus generating a large autocorrelation peak as shown in Figure 3.

The benefit of the above method is that we can compute $R(L/2, t_0)$ in $O(1)$ time. To see this, consider that the above expression can be rewritten as follows.

$$R(L/2, t_0) = \frac{\frac{L}{2}\widehat{XY} - \widehat{X}\widehat{Y}}{\sqrt{[\frac{L}{2}\widehat{XX} - \widehat{X}\widehat{X}][\frac{L}{2}\widehat{YY} - \widehat{Y}\widehat{Y}]}}$$

where

$$\widehat{XY} = \sum_{t \in W}[X(t)Y(t)],\ \widehat{XX} = \sum_{t \in W}[X(t)X(t)]$$

$$\widehat{YY} = \sum_{t \in W}[Y(t)Y(t)],\ \widehat{X} = \sum_{t \in W} X(t),\ \widehat{Y} = \sum_{t \in W} Y(t).$$

Note that all these variables can be computed in $O(1)$ time with a standard moving window method. Taking $\widehat{X}$ as an example, given $\widehat{X}(t_0)$, we can compute $\widehat{X}(t_0+1)$ as follows.

$$\widehat{X}(t_0+1) = \widehat{X}(t_0) + X(t_0+1) - X(t_0 - L/2 + 1).$$

Since $\widehat{X}(t_0)$, $X(t_0+1)$ and $X(t_0 - L/2 + 1)$ are all known, $\widehat{X}(t_0+1)$ is computed in $O(1)$. The other variables can also be updated in the same way. $R(L/2, t_0+1)$ can thus be obtained in $O(1)$ time based on the variables at t_0. For a buffer of size B, this autocorrelation solution has complexity $O(B)$, which is much faster than $O(B * L)$ via traditional cross-correlation.

Inaccuracy of Autocorrelation based Tone Detection. Unfortunately, we cannot rely solely on autocorrelation to find the tone peak as the result will be inaccurate. Figure 4(b) shows the output of the autocorrelation step. As can be seen, there is a noticeable offset between the autocorrelation peak and second cross-correlation peak (a sharp line). In an ideal scenario, these two peaks should perfectly align. The reason for this discrepancy is twofold. First, autocorrelation fundamentally has a much flatter peak. Intuitively, if a time slot's autocorrelation value is high, its neighboring time slots' values are also high. Second, the offset can arise due to signal distortion of the tone after propagation. This can result in up to 10cm error in the final result even in quiet environments as we show in the evaluation. For these reasons, we cannot rely exclusively on autocorrelation for tone detection. Rather, we seek to combine the respective benefits of cross-correlation and autocorrelation.

Step 2: Smoothing of Autocorrelation Peak. In order to obtain cleaner peaks, we smooth the autocorrelation curves within a 10ms window. Note that this smoothing process is necessary as the raw autocorrelation output can have many local maxima due to environmental noise.

Step 3: Small Scale Cross-correlation. We interpret the smoothed autocorrelation curve as an indicator for the location of the tone, then we search in a small window S around the autocorrelation peak using standard cross-correlation to detect the precise tone location (see Figure 4(c)). The computation overhead of the combined algorithm now is $O(B+|S|*L)$, which is still much smaller than $O(B*L)$, assuming that the autocorrelation indicator allows us to choose sufficiently small $|S|$.

Choosing the Search Window Size $|S|$. The above discussion reveals an interesting design trade-off concerning the selection of the window size $|S|$. A smaller $|S|$ yields faster cross-correlation computation, but risks mis-detecting a tone altogether because it is outside the searching window. This is especially true in noisy environments where the offset between cross-correlation and autocorrelation peaks can be large. Based on our empirical measurements, we set our window size to a very conservative size of 100 samples (offset by -80 to $+20$ around the detected peak), which is sufficient even at high noise. The fact that the search window is not symmetric around the detected peak is probably due to reflected signals, which may generate higher autocorrelation peaks after the direct signal's peak.

5.2 Pipelined Streaming Execution Strategy

With the preceding reductions in tone detection computation time, we next introduce FAR's *pipelined streaming execution strategy* for boosting measurement frequency. Its two key improvements upon the traditional execution strategy are that (1) it employs pipelining to increase measurement frequency by overlapping computation time with tone transmission, and (2) it employs streaming to minimize measurement lag by removing Audio Playing Lag. Figure 2(b) illustrates the FAR execution strategy.

Pipelined Execution to Mask Computation Time. We devise a pipelined execution strategy that uses separate threads to handle *Playing, Recording, Tone-Detection Computation*, and *Measurement Exchange* (Figure 2(b)). In the recording thread, the Audio Recorder periodically fills in a predefined audio buffer. For example, the minimum required buffer size for Nexus One running Android 2.3.4 is 1024 Bytes, which corresponds to 512 sound samples. Once the buffer is filled, the computation thread processes the buffered sound samples, while the recording thread waits for the next buffer. Once computation is finished, the measurement exchange thread exchanges the detected tone arrival times with the other phone using WiFi and outputs the final distance measurement. The net effect is that tone-detection computation is now fully overlapped with the exchange of sound waves (including recording delay, playing delay and signal propagation delay). Given that in our system, these two sets of operations take approximately equal

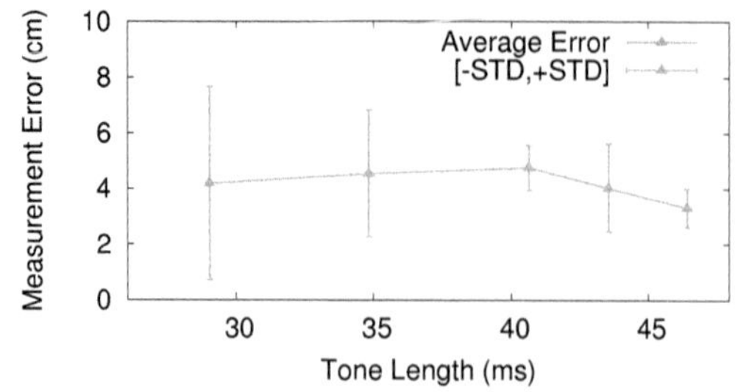

Figure 5: Impact of Different Tone Length. A tone length of 46.4ms has good accuracy and low variance.

amount of time, the pipelined execution strategy effectively doubles FAR's measurement frequency.

Note that when pipelining, it is of critical importance that the tone-detection computation time is tightly bounded. If the computation on one buffer is not finished before the next buffer arrives, the computation overhead will accumulate over time, resulting in buffer overflows and lost sound samples.

Streaming Execution to Mitigate Audio Playing Lag. Instead of blocking on computation completion as the traditional strategy does, FAR implements a streaming mode to continuously send and record tones. The `Play()` API of the Audio Driver is called at a frequency similar to the tone length, thus one tone will be sent right after another. This helps in eliminating the Audio Playing Lag from the overall measurement time. Similarly, we also implement the Audio Recorder to work in a streaming fashion, continuously recording sound samples for computation. As a result, if there is no lag between tones, measurements can theoretically be conducted at a frequency equal to $1/L$, provided tone detection is fast enough to support such a rate.

5.3 Further Optimizations

In addition to the techniques discussed in the previous sections, our system implements several additional optimizations to reduce delay.

Tone Length Minimization. Choosing the right Tone Length is important. Choosing a small value can negatively affect measurement accuracy, but choosing a large value not only adds lag to the overall execution, but also increases the complexity of the cross-correlation computation, thereby reducing the measurement frequency. We conducted extensive experiments over different tone lengths and found that using a tone length of 512 sound samples (corresponding to 46.4ms at a 11.025kHz sampling rate) achieves good accuracy and stability (see Figure 5), while still allowing us to keep the measurement frequency above 10Hz. The tone of 46.4ms is similar to that used in [16].

Buffering Delay Minimization. The buffer is pushed to the application layer whenever it is full. The buffering delay is therefore a random period in $(0, B)$. To minimize this delay in FAR, we use the phone's minimum buffer size (which is 46.4ms (1024Bytes) on Nexus One).

Networking Operation. We use UDP over WiFi, adding a delay of around 2-4ms per exchange.

Tone Overlapping. One attempted optimization that we were unable to execute in practice. Instead of spacing the two phone's tones over time, we tried to overlap them in time, but send them on different audio frequencies. Unfortunately, this approach failed because the local speaker's high power masks the remote phone's signal, which generates detection errors.

6. SYSTEM ROBUSTNESS

Besides measurement frequency and accuracy, another important challenge we need to tackle is the robustness of the system. A real-time motion game can not be fun if there are frequent measurement errors. In this section, we tackle the possible vulnerabilities a practical system could suffer in real game play environments.

6.1 Mobility Robustness

One unique challenge to distance estimation of motion gaming is the Doppler effect caused by player movement. Doppler effect happens when there is a relative movement between the sound player and recorder. Intuitively, when two phones are moving towards each other, the sound wave arrives at the recorder earlier than expected, so it appears *compressed*. However, since the recorder is recording the sound at a constant rate, the recorded sound samples are fewer than expected, which means in the recorded version of the tone, the repeating pattern has a shorter length. Thus, the offset used in the autocorrelation calculation needs to be shortened. Similarly, when the phones are moving further away, the tone is *diluted*, and thus a longer offset is needed.

We define the term *Doppler offset* D_{offset} as the number of samples between the two cross-correlation peaks i.e. the locations of the repeating patterns. In a scenario without Doppler effect, the Doppler offset should be equal to the repeating pattern's length, which is exactly half of the tone length, $D_{offset} = L/2$, and the autocorrelation of $X(t)$ and $X(t - L/2)$ should be calculated to detect the tone. If Doppler shift occurs, the repeating pattern will be separated by a D_{offset} not equal to $L/2$. In these cases, the autocorrelation of $X(t)$ and its delayed stream $X(t - D_{offset})$, which is $R(D_{offset}, t)$, should be calculated to find the tone.

In Figure 6(a) we show an example of a dilution, in which the tone does not show an autocorrelation peak. A careful diagnosis uncovers that the offset is $L/2+1$ rather than the expected $L/2$. This means that an extra sample has appeared in the tone because of Doppler sound dilution. Thus $D_{offset} = L/2 + 1$ is appropriate in this case.

Assume Phone 1 is sending the tone, Phone 1's velocity is v_{p1} (relative to earth), Phone 2's velocity is v_{p2}, Tone Length is L (so the repeated pattern length is $L/2$), and c is the speed of sound. The delay between the repeating pattern of received sound sequence D_{offset} can be expressed as

$$D_{offset} = \frac{c + v_{p2}}{c + v_{p1}} * L/2.$$

Figure 6(b) shows the relationship between D_{offset} and the relative velocity of the two phones. We assume the maximum speed of a player's hand is 2m/s, as we have observed from our experiments that a higher speed is extremely hard to reach. This is in accord with previous work [8] on human hand functions. Since this speed is not negligible compared to the speed of sound ($c = 343.2m/s$), we can see from the figure the possible range of the Doppler offset as $D_{offset} \in [L/2 - 3, L/2 + 3]$.

Recovery from Doppler Effect with parallel autocorrelators. To recover the correct autocorrelation peaks, a simple method is to calculate the autocorrelation with the appropriate offset. As an example, in Figure 6(c), we show that we can recover the peak by calculating $R(L/2 + 1, t)$, rather than $R(L/2, t)$. Given that $D_{offset} \in [L/2-3, L/2+3]$, we can set up a maximum of 7 parallel autocorrelators to catch any practical case of Doppler shift.

Parallel autocorrelators have both advantages and disadvantages. The advantage is that more autocorrelators increase the detection ratio. The disadvantage is that more autocorrelators lead to longer computation time. Therefore, one problem is to find the optimal number of autocorrelators $\hat{x}$ that maximizes the tones detected per unit time for a given velocity v, i.e.,

$$\hat{x} = \underset{x}{\text{argmin}} \left[D_v(x) \max(b, T(x)) \right],$$

where b is the communication bound, $D_v(x)$ is the detection ratio with x autocorrelators at velocity v, and $T(x)$ is the completion time for x parallel autocorrelators.

Unfortunately, each additional autocorrelator causes an additive increase in computation time. In response, we apply a *predictive Doppler estimate procedure*. The idea is to predict the likely Doppler shift based on the recent history of distance measurements, and only execute the autocorrelators likely to match the predicted velocity shift.

Figure 7 shows a fast movement case with a maximum player hand speed of 2m/s (and thus phone-to-phone relative speed reaches 4m/s). We show the two phones' distance change over time, and mark the tones affected by the Doppler effect. The Doppler-shifted tones are divided into two groups: diluted cases, and compressed cases. As evident from the figure, compressed cases occur when the distance is decreasing, and diluted cases happen when distance is increasing. Exceptions to this rule happen rarely.

Thus, we can use two groups of autocorrelators to deal with the two groups of Doppler effect cases: one group of autocorrelators $R(L/2-1, t), R(L/2-2, t), R(L/2-3, t)$ for compressed cases, and the other group of autocorrelators $R(L/2+1, t), R(L/2+2, t), R(L/2+3, t)$ for diluted cases. For each buffer of recorded samples, we always run the original autocorrelator $R(L/2, t)$. At the same time we predict the players' relative velocity based on the recent distance change history, then apply the appropriate group of autocorrelators, as shown in Figure 8. We can reduce the total overhead by nearly half with this method. In fact, our evaluation suggests that almost all of D_{offset} are within $[L/2 + 2, L/2 - 2]$, and thus only 3 parallel autocorrelators are needed to recover from the vast majority of Doppler shifts.

The prediction based method might not be correct when the player is changing direction. However, the user's speed naturally slows when changing direction, which will decrease the likelihood of experiencing Doppler shifts. As shown in Figure 7, almost all of the tones at the peaks or troughs are good cases.

6.2 Environmental Robustness

There are several environmental factors which can impair distance measurements. We address these presently.

Robustness to Ambient Noise. To study the character of ambient noise, we recorded sound from both indoor and

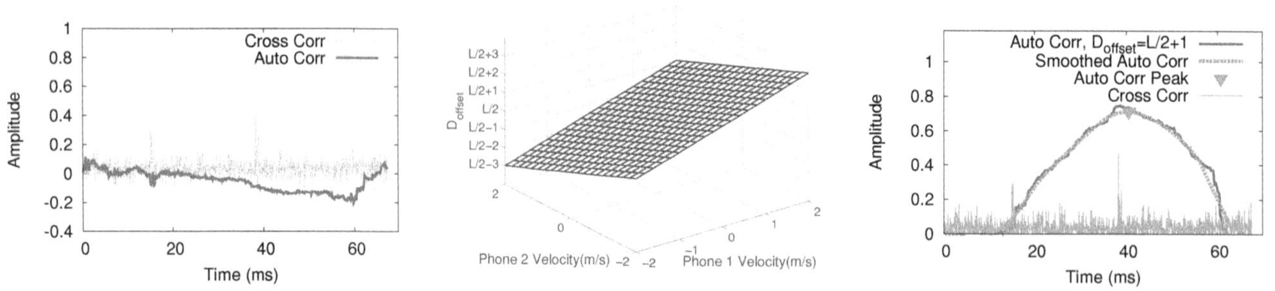

(a) Doppler Effect Example. (b) Phone Velocity *vs.* D_{offset}. (c) Peak Recovery by Changing Offset.

Figure 6: Doppler Effect and Recovery. (a) The autocorrelation peak is destroyed by the Doppler effect. The interval between the two cross-correlation peaks is $L/2+1$ rather than $L/2$. (b) Given player speed is limited, L_{recv} falls into a limited range. (c) Changing the offset of the autocorrelation to L/2+1 can detect the peak.

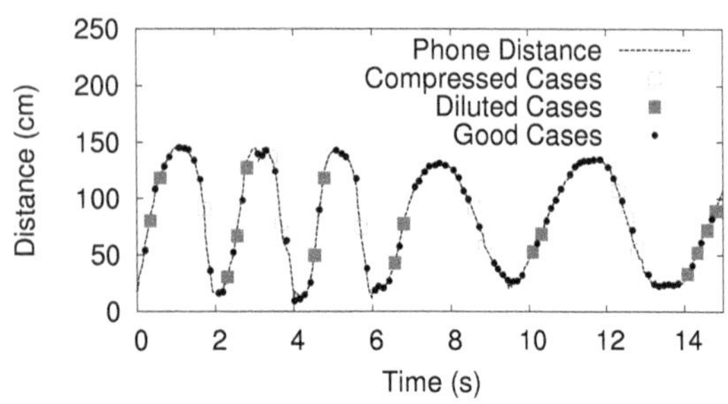

Figure 7: Doppler effect causes sound dilution and compression errors. However, we are able to predict these errors based on recent velocity.

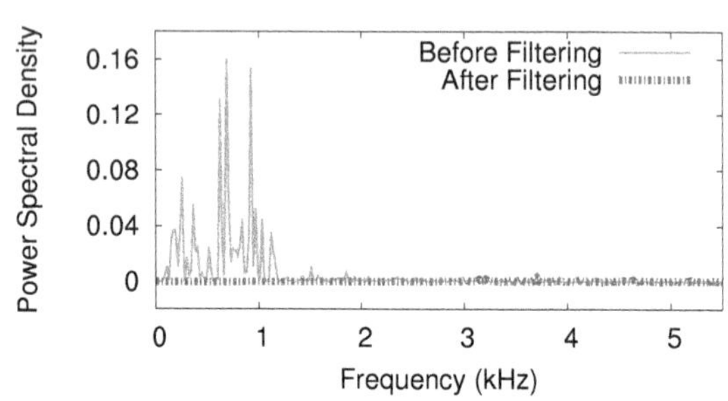

Figure 9: The frequency map of a recording of ambient sound from a crowded area before and after filtering.

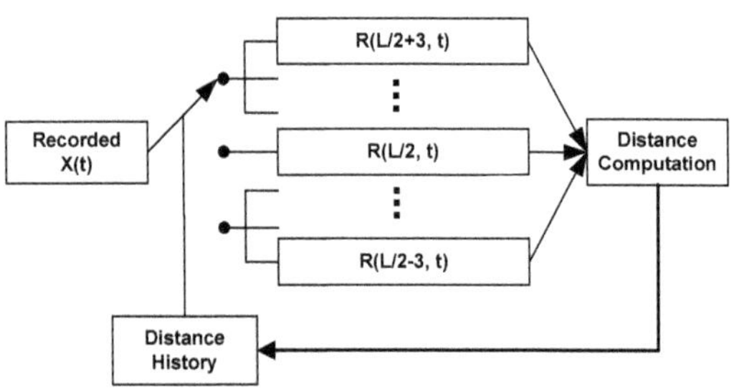

Figure 8: Predictive Parallel Autocorrelators. The distance history is leveraged to predict whether sound will be diluted or compressed, and subsequently which group of autocorrelators to use.

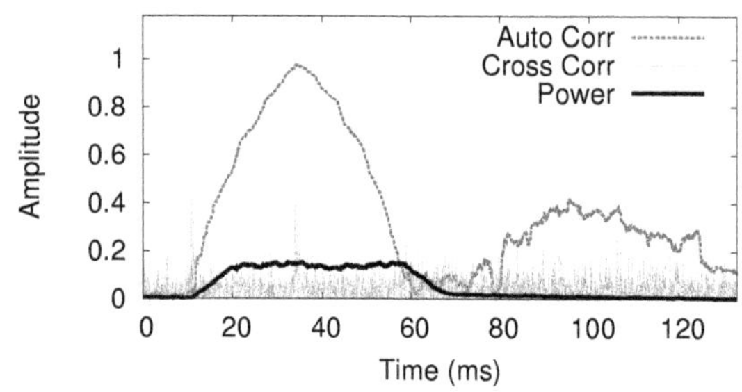

Figure 10: A multipath component generates a smaller autocorrelation peak (at around 100ms) after the correct peak, but with a much lower power level.

outdoor noisy scenarios. One finding is that the frequency for these sound usually is below 2kHz. Thus, in our system we apply a high pass filter (a 9-th order Butterworth filter) to filter out ambient noise. Figure 9 shows the spectrum density map of a 1-minute recording of a crowded place, before and after applying our filter respectively. It is clear that the high pass filter functions well in eliminating ambient noise.

Robustness to Multipath. Multipath effects can also impact the robustness of tone detection if the players are in a relatively small room or hallway. The reflected signal may generate correlation peaks or degrade the main component. Figure 10 shows a tone with an observable multipath effect. There is a lower autocorrelation peak at around 100ms which is caused by the multipath components. We observe that the multipath components usually have low power and signal quality after reflection or scattering. Note that the power threshold might not work well if there is high ambient sound. However, after applying the high pass filter, we can efficiently reduce the ambient sound power level. In our implementation, we empirically set the power threshold to 0.02 and peak level threshold to 0.4, which we show to be sufficient to filter out the multipath components experienced when playing in actual scenarios. In our evaluation section, we conduct experiment in a small room to show the performance.

Robustness to Occasional Tone Loss. FAR also provides failsafe methods to deal with occasional tone losses. Since the distance calculation requires the tone measurement from both phones' computation result, it is important to keep the two measurements synchronized. In particular, if a tone is not detected, the phone should be alerted of the situation. We achieve this by a *binary coding* method: we assign a unique pseudorandom sequence for each phone. If detection is flawless, each phone will see alternating codes. Thus if a phone receives any two consecutive tones whose codes are the same, the phone detects a missed tone condition and accounts for this in the distance calculation.

Coding beyond binary (e.g. ternary) is not necessary, because multiple consecutive missing tones are detected by simple threshold on the arrival interval between two consecutive tones. When such cases arise, the system will enter *Failsafe mode*. In Failsafe mode, one phone will send two tones separated by a much larger interval, e.g. 250ms. Both phone will then observe the large interval and reset their state.

7. GAME DESIGN

In this section, we briefly describe two novel games we have developed on FAR to date.

SwordFight. In SwordFight, the core gameplay works as described in the introduction. In order to ensure that attacks are oriented toward the opponent, the digital compass is used. At the start of a fight round, the players are instructed to point their phones toward one another and the phones' digital compass orientations are recorded. During the fight round, the digital compass is read to ensure that movements are actually oriented toward the opponent.

ChaseCat. ChaseCat is a fast round-based game. In each round, one player is randomly chosen as the Cat and the other player is chosen as the cat chaser, the Dog. The Dog player attempts to get her phone as close to the Cat player's phone as possible, while the Cat player tries to get her phone as far from the Dog player's phone as possible.[2] A random 15-25 seconds after the round starts, players are instructed to Freeze by an audio cue. If the Cat is within 30 centimeters of the Dog during Freeze, then the Dog catches the Cat and scores a point. Otherwise the Cat escapes and scores a point. Movement during Freeze is detected by distance and acceleration measurements.

Note that in the case of both games, distance measurements are expected to be accurate up to several centimeters. Our evaluation results in the next section show that FAR is accurate at this resolution. The maximum distance that can be supported is a function of the maximum volume of the phone speakers and the sensitivity of microphones. The current platforms we use happen to have an effective measurement range of 2 meters, so this is the current effective limit of gameplay. Out-of-range occurrences can be detected by sequences of dropped tones and low SNR.

8. IMPLEMENTATION

We implement both our games on two phones: Nexus One running Android 2.3.4 and Samsung Focus running Windows Phone 7.5.

System Configuration. We use two maximum length sequences (m-sequences) to generate unique tones for the two phones. Since the m-sequence is well-known for its distinctive peak when applying cross-correlation, it helps us to accurately identify the tone's location in the recorded sound. The sequence is comprised of 256 16-bit short integers. In the tone, the sequence is repeated once to support the autocorrelation-based detection method (§5.1). The frequency range of the tones are $[0, 11.025kHz]$ (Nexus One) and $[0, 16kHz]$ (Samsung Focus). The Audio Recorder samples the sound at a rate of $11.025kHz$ (Nexus One) and $16kHz$ (Samsung Focus), which is sufficient to support 1cm measurement granularity. As discussed, to achieve a good balance of robustness, accuracy, and speed, we choose a tone length of 46.4ms. As mentioned in §5, the buffer size of the Audio Recorder can incur a buffering delay. To minimize this delay, we use the minimum allowed buffer size in both phones: 1024 Bytes (46.4ms with 11.025kHz sampling rate) for Nexus One, and 2000 Bytes (62.5ms with 16kHz sampling rate) for Samsung Focus.

[2]We recommend players plant their feet during the game, but this has proven difficult to enforce.

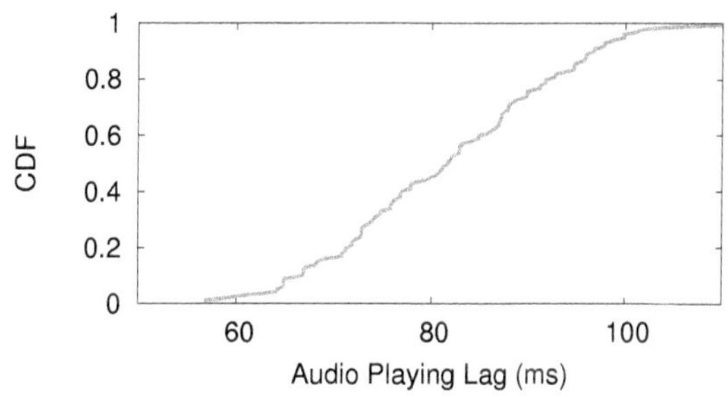

Figure 11: Audio Playing Lag Distribution

Randomness of Audio Playing Lag. As mentioned in §5.2, the Streaming Execution Strategy mitigates Audio Playing Lag. However, it turns out that the commodity devices we have encountered exhibit an additional random audio playing lag artifact. Figure 11 shows that the audio playing lag is randomly distributed between 64ms (5%-tile) and 99ms (95%-tile). Thus, to prevent two consecutive tones from overlapping with each other, we add a 35ms *guard period* when playing each tone. The interval between two consecutive `Play()` calls is then $L + 35ms = 46.4ms + 35ms = 81.4ms$. In fact, given our reductions in computation and communication overhead which are no longer the bottleneck, it is this random system artifact that currently prevents us from increasing the measurement frequency from $1/81.4ms \approx 12Hz$ to a theoretical $1/L = 1/46.4ms = 21.6Hz$.

9. EVALUATION

We demonstrate through our evaluation that FAR is fast, accurate and robust, and that it is suitable for MMG games. The primary metrics are as follows.

- *Measurement Frequency.* Is FAR efficient enough to support fast measurement? How fast can FAR make measurements?
- *Measurement Accuracy.* How accurate are the measurements?
- *Measurement Robustness.* How well does FAR handle noise and motion?
- *Gameplay.* Do games based on FAR work in practice?

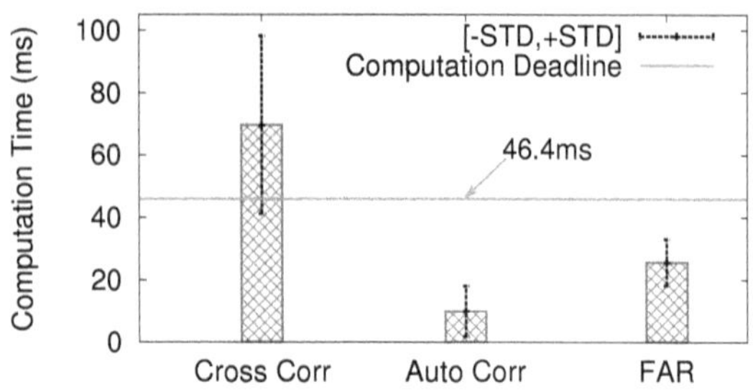

Figure 12: Compute time needed to process a 512-sample audio recording.

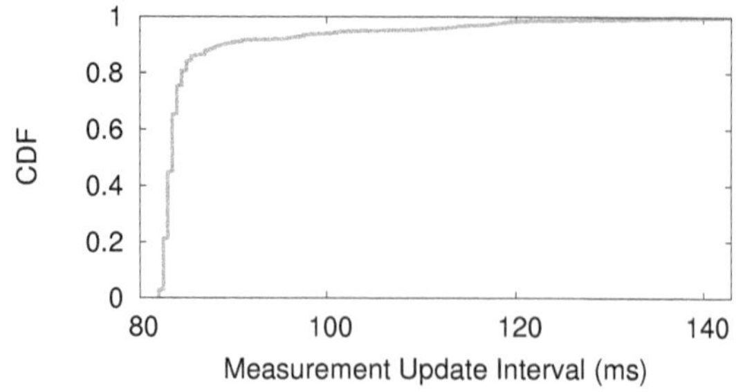

Figure 13: Measurement Update Distribution. More than 80% of the time, a new update arrives within 84ms of the previous update.

We conducted experiments with two Nexus Ones in a room measuring 14m by 7m by 2.5m that typically serves as a conference room. Static measurements are verified against manual ruler measurements. For dynamic measurements, the main comparable system used is the Xbox Kinect, a dedicated fixed-infrastructure motion tracking system for console gaming. Lastly, as real world validation, we invited public users to play SwordFight and ChaseCat on four separate occasions. The highlights of the evaluation are as follows.

- FAR achieves a measurement frequency of at least 12 Hz. In fact, FAR is fast enough to reach the limits imposed by the phone's audio driver.
- The median measurement error is 2cm. During times of rapid movement, FAR compare favorably to Kinect for both isolated player measurements and in-situ gameplay player measurements.
- Doppler effects, high noise and multipath and are effectively neutralized.
- Experiences of over fifty game players suggest that FAR is a good match for motion gaming, and that SwordFight and ChaseCat are fun to play.

9.1 Measurement Frequency

We show that FAR is efficient enough to take measurements as fast as the phone audio driver can support. We evaluate FAR's feasibility in supporting the pipelined streaming execution strategy, and then report on the achieved frequency.

Feasibility. The fundamental limitation of previous approaches is that the detection of tones requires intensive computation, and thus is not feasible within a pipelined streaming execution strategy bound by tight deadlines. Figure 12 illustrates that FAR is able to comfortably satisfy the strict deadline of 46.4ms corresponding to a recording of 512 samples. The traditional cross-correlation method is by far the slowest at 70ms.[3] Missing the computation deadline incurs serious buffer overflow problems, leading to loss of sound samples and thus measurements.

Measurement Frequency. Figure 13 shows the distribution of measurement updates for over 400 individual measurements as generated by FAR. The effective measurement frequency is 12Hz over 80% of the time. As noted in §8, a random audio playing delay of [64ms,99ms] inhibits us from achieving a theoretical frequency of 21.6Hz.

[3]Note that this is much smaller than as claimed in [18] because we only process a 512-sample recording, while [18] requires a much larger recording due to its traditional execution strategy.

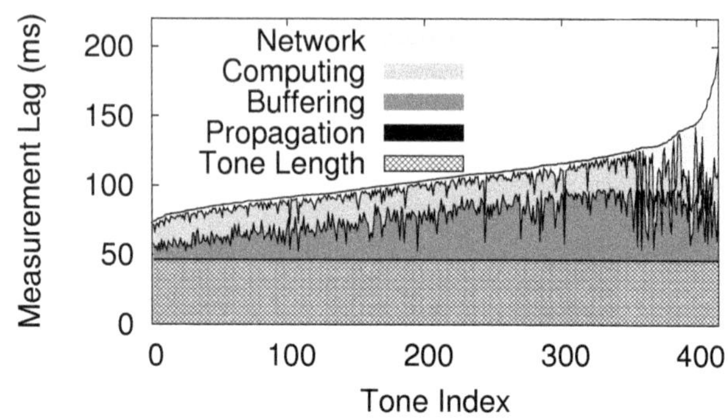

Figure 14: Measurement Lag Composition. Several individual delays contribute to the time between when a tone is played and when a distance measurement is completed.

Measurement Lag. For each distance measurement, the lag is comprised of the following components: *Tone Length, Propagation Delay, Buffering Delay, Computation Time* and *Measurement Exchange Delay*. For motion gaming, the distance between two phone does not exceed 2 meters. Therefore, the *Propagation Delay* is less than 6ms. Figure 14 shows the lag of 400 distance measurements when the phones are placed 50cm apart. The measurements are sorted by their total delay. With the exception of the last 5% of cases with large network delays, the majority of measurements exhibit a total lag of around 100ms. As expected, the delay due to the length of the tone is constant, and the network delay and propagation delay are very small. The buffering delay is uniformly distributed in (5ms, 46ms).

9.2 Measurement Accuracy

Next we juxtapose the measurement accuracy of autocorrelation (Auto Only) and FAR. We statically position the two phones in three typical phone-to-phone orientations at different distances. Figure 15 illustrates the measurement accuracy of the two methods, each computed from over 400 experimental cases. It is clear that autocorrelation exhibits measurement error that is severe, whereas FAR's error is very low with a median of 2cm and standard deviation of 1cm. The figure also highlights that FAR is robust across varying phone orientations and distances.

9.3 Impact of Player Movement

We next investigate whether our system is capable of accurate measurements while players are moving. After ex-

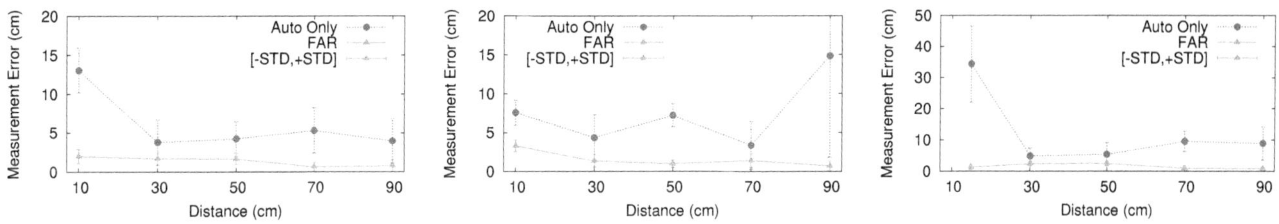

(a) Back To Back (b) Head To Back (c) Head To Head

Figure 15: Measurement Accuracy in a Large Room

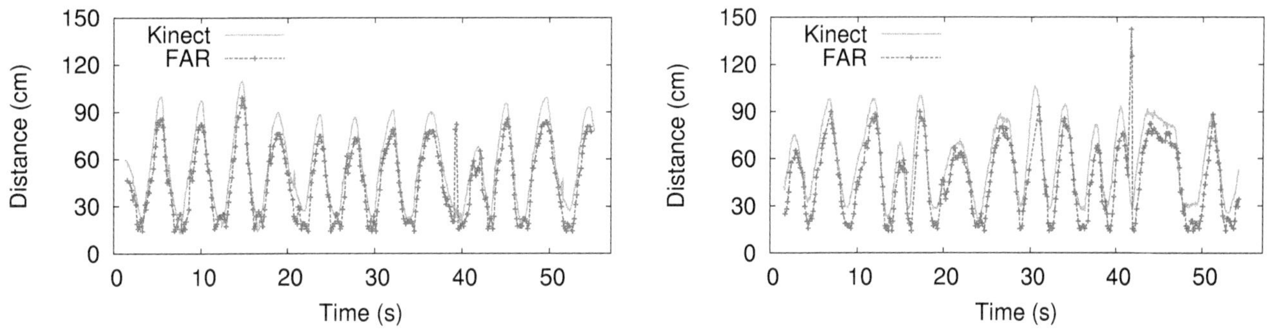

(a) Experiment Case 1 (b) Experiment Case 2

Figure 16: Isolated Player Measurements. Simultaneous distance measurements of FAR and Kinect when a single player is moving her hands freely with a phone in each hand.

ploring several possibilities including a studio motion capture system, we settled upon Kinect, an off-the-shelf motion capture system for the Xbox console gaming system, as the candidate comparable system. The reason Kinect is appealing is because it is commercially available, practical to use, and has already proven successful for infrastructure-based console gaming.

Kinect measurements were obtained with the Kinect SDK which is a Kinect-to-computer interface that allows programmatic collection of the human skeleton coordinates as seen by the Kinect camera. We employed both isolated player experiments and in-situ real player experiments.

Isolated Player Movement. In isolated player experiments, we positioned the Kinect camera statically, and placed a subject in front of the camera with a phone running FAR in each hand. The subject was then asked to vary her hand position at her discretion. FAR performed phone-to-phone distance measurements while Kinect independently tracked the hands.

Figure 16 illustrates the distance measurements simultaneously measured by Kinect and FAR for two representative traces. For the most part, FAR matches the Kinect motion path closely. In both traces, there are only 5 measurements out of 700 that deviate significantly from the Kinect estimate. In some instances, Kinect and FAR measurements can be offset from one another by up to 12cm. The reason for this is actually a reflection of the Kinect SDK, which can exhibit 5-15cm variance of the reported hand coordinate because it is ambiguous as to whether it measures distance from the base or tip of the hand. While an ideal ground truth would not have such issues, the results are encouraging enough to suggest that FAR compares favorably to a commercially successful motion gaming system.

In-Situ Gameplay Player Movement. In the in-situ gameplay experiments, six players were invited from the UC Santa Barbara Link Lab to conduct nine in-situ rounds of measurement. Players were each given a phone and asked to try out the SwordFight MMG game. During the game, FAR provided the in-game distance measurements while Kinect independently recorded their skeletal hand positions.

Figure 17 compares the corresponding FAR and Kinect measurements for two representative traces. There are occasionally slightly more mismatches between FAR and Kinect during in-situ gameplay than during isolated player experimentation. This is due to the fact that players – in the heat of competition – tend to grip or orient the phones such that mics or speakers may be temporarily blocked.

Interestingly, we have found that our system can in fact be more accurate than Kinect because Kinect cannot accurately track the hand when the player's hand is obstructed from the camera's view (e.g. by her own body or the opponent's body). These situations happen frequently in SwordFight and similar games. For example, two right-handed players facing off will often mean one player is poorly oriented toward Kinect. In these situations, FAR measures distance better because there are typically no obstructions between the two phones in the players' hands for games in which opponents are facing each other. In the in-situ experiments presented in Figure 17, we asked that the players compensate for this Kinect limitation for the sake of experimentation: one player was instructed to play with the left hand, the other player was instructed to play with the right hand,

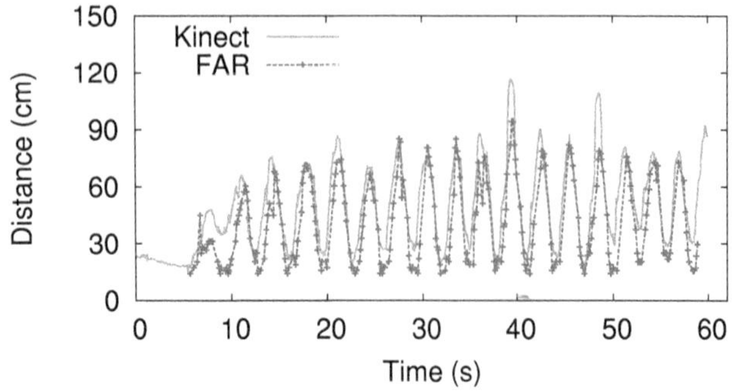

(a) Experiment Case 1

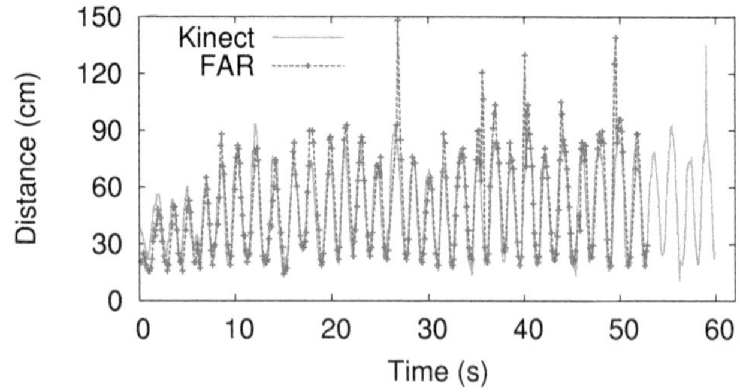

(b) Experiment Case 2

Figure 17: In-Situ Gameplay Measurements. Simultaneous distance measurements with FAR and Kinect of players engaged in SwordFight.

and both players were instructed to not obstruct the line-of-sight between the Kinect and their phones. These patently artificial constraints on gameplay highlight why FAR may be potentially even more suitable than fixed infrastructure measurement systems for certain styles of motion gaming, regardless of infrastructure availability.

9.4 Measurement Robustness

We quantify FAR's effectiveness at handling three sources of measurement error: Doppler effect, multipath and noise.

Robustness to Doppler Effect. FAR detects Doppler shifts with the use of parallel autocorrelators. Our experimental setup consisted of recording FAR and Kinect measurements simultaneously while instructing the players to swing hands at varying rates, including as fast as possible. We collected fifteen motion traces, each consisting of over 400 distance measurements. We calculated a phone-to-phone relative velocity from the Kinect data (which is not affected by Doppler shift). We group motion traces according to the maximum velocity reported by Kinect.

Figure 18(a) and Figure 18(b) show the detection performance of parallel autocorrelators (without prediction and with prediction respectively) for representative traces with motion speeds of 0.5m/s, 1m/s and 2m/s where the Doppler shift does occur. As expected, even one autocorrelator performs reasonably for 0.5m/s but more autocorrelators are needed as the velocity increases. Most of the tones affected by Doppler shift can be recovered by five parallel autocorrelators as shown in Figure 18(a). This means the Doppler offset is usually within $[L/2-2, L/2+2]$.

Figure 18(b) shows that after applying prediction, only three parallel autocorrelators can recover most of the errors, and achieve 85+% detection ratio even in a high speed scenario of 2m/s. The detection ratio gain for more than three autocorrelators is marginal. Figure 18(c) shows the computation time required for increasing numbers of parallel autocorrelators. Three correlators can be run simultaneously while maintaining or just barely exceeding the computation deadline required for the fastest measurement frequency. Our empirical observations indicate that sustained velocities of 2m/s or even 1m/s are very hard for players to maintain due to physical human limits; there is a natural opportunity to detect and complete measurements after

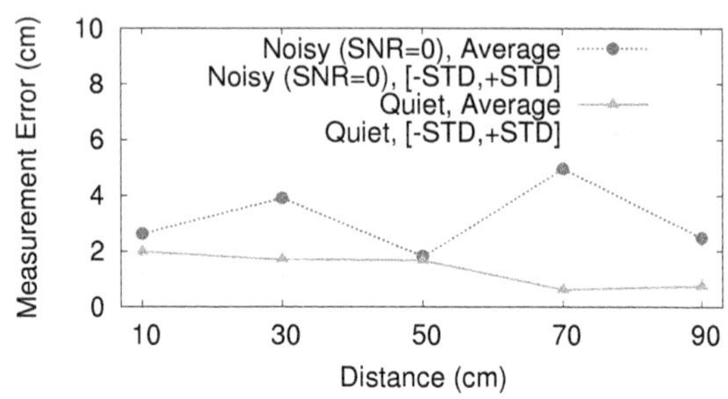

Figure 20: Measurement accuracy when injecting heavy noise (SNR=0). The accuracy drops a little but can still achieve an accuracy of 5cm in average.

sudden bursts of fast movement. Therefore, FAR uses three parallel autocorrelators which work well in practice.

Robustness to Multipath and Ambient Noise. To test robustness to reflected sounds, we ran FAR in the smallest contained space we could find, a private room measuring 3m by 2m. Figure 19 illustrates the performance in the small room with various phone orientations and distances. While the variance of measurement error is 5cm and larger than in the original room, the measurement error is again very small, with a median error of 2cm.

We injected an ambient sound to evaluate the anti-noise property of FAR. In order to ensure repeatability, the ambient sound is a 1-minute recording from a very crowded area, with people talking and laughing loudly. We define the SNR of the signal as the energy of the sound recorded by the remote phone over the energy of the ambient sound. While we tested over different scenarios, we report only the performance of the back to back position result with SNR=0 here in Figure 20 due to the similarity of the results. As expected, the ambient sound degrades the accuracy, but the impact is manageable. Recall that SNR=0 means a noise energy level equal to the tones', which is high.

9.5 Deployment Experiences

We have prepared SwordFight and ChaseCat for public play on multiple occasions: at the SenSys 2011 conference for any conference participant [24]; in Beijing for interns working at the Microsoft Research Asia building late at night; in Redmond for office colleagues taking a work break,

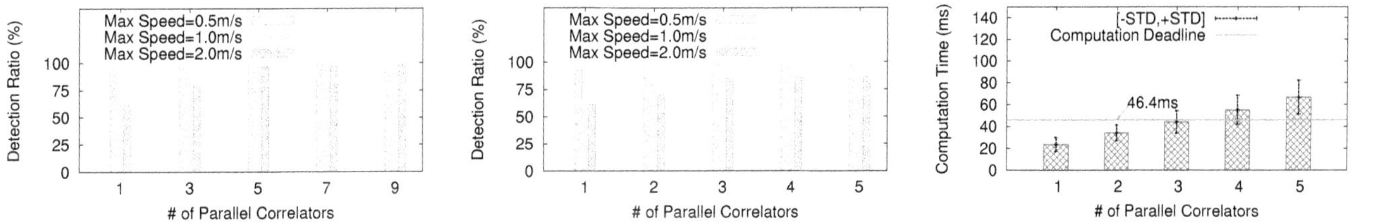

(a) Detection Ratio w/o Prediction (b) Detection Ratio w/ Prediction (c) Computation Time of Autocorrelators

Figure 18: Parallel Autocorrelators. (a) 90+% of the tones can be detected by five parallel autocorrelators, which corresponds to offset range $[L/2-2, L/2+2]$ (b) With predictive parallel autocorrelators, only three autocorelators are needed. 85+% of the tones can be detected even at 2.0m/s hand movement speed. (c) Three parallel autocorrelators can still meet the computation deadline.

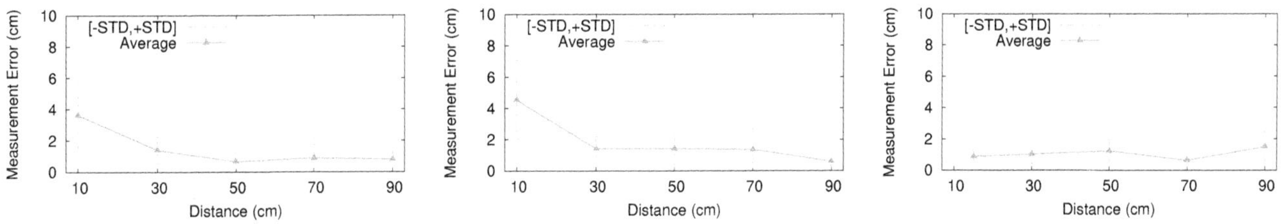

(a) Back To Back (b) Head To Back (c) Head To Head

Figure 19: Accuracy In A Small Room. The impact of the reflection is not noticeable.

and; at UC Santa Barbara for officemates, friends and family. Qualitatively, the public reception has been positive. Just as the Wii and Kinect ushered in a new class of console gameplay, SwordFight and ChaseCat introduce an element of physical action to mobile gaming that players seem to enjoy.

10. RELATED WORK

Motion Gaming. Recently, non-mobile console gaming systems have embraced player motion-based gameplay, developing a variety of schemes for player localization. The Wii remote [15] uses an infrared camera to track the relative position of the TV-mounted sensor bar which emits two infrared light sources. The PS3 Move controller [22] has a light-emitting ball affixed which is tracked by a TV-mounted video camera. The TV-mounted Kinect [14] projects an infrared mesh and tracks its time-of-flight back to the Kinect camera. These console gaming systems demonstrate the compelling nature of motion gaming, but are unfortunately tethered to fixed console-based infrastructure.

The idea of phone-to-phone motion gaming was first mentioned in the recent work of Qiu et al. [18], described in that paper's introduction as high-speed, locational, phone-to-phone (HLPP)-gaming. [18] was the first to propose phone-to-phone localization as a primitive for mobile gaming. The authors developed a 3D acoustic localization system that utilizes each device's two microphones, one speaker, 3-axis accelerometers and 3-axis magnetometers to calculate coordinates of one phone in three dimensions relative to the other. However, the scheme proposed in [18] is not actually applicable to high-speed motion games such as SwordFight since the reported update interval is over 800ms. Nevertheless, their 3D positioning algorithm may in the future serve as a complement to the high-speed ranging techniques presented in this work.

Acoustic Ranging. The authors of [16] developed an earlier method for acoustic ranging between two phones but its update rate can be arbitrarily slow since the two devices are not synchronized, and the applications are unclear.

Much research exists on infrastructure-based localization algorithms (see [11] for a survey). ActiveBat [5] and Cricket [17] are two representative infrastructure-based systems. Each is capable of achieving centimeter resolution, but dense ultrasonic infrastructure requirements are not compatible with ad hoc mobile gaming. Infrastructure-based acoustic localization includes [3, 12, 21]. These systems achieve resolution in meters, which is insufficient for action gaming. Several systems have reported achieving centimeter resolution [4, 7, 9] with custom-built hardwares which makes them less appealing in light of the pervasiveness of commodity mobile phones.

Components of our algorithm's signal processing techniques have been used previously in the communication systems but not in the context of localization. [20] proposed using autocorrelation to reduce the complexity of frame and carrier synchronization. [13] uses a technique similar to running multiple parallel detectors for handling Doppler effects underwater, but does not consider predicting Doppler shifts.

11. DISCUSSION AND CONCLUSION

In this paper, we report on the design and implementation of a new class of phone-to-phone games: Mobile Motion Games (MMG). The key underlying technology is a new highly-accurate and real-time phone-to-phone distance measurement substrate, which provides the API on top of which games such as SwordFight and ChaseCat are built.

Our experience with having people play these games is positive. They are intuitive and fun to play. However, our work here is a starting point. There remain open challenges that could affect the game experience. One issue is that continuous acoustic tones are noticeable, since mics and speakers on commodity phones only support the audible frequency range. A possible approach is to embed these tones in game music. A second issue is that line-of-sight blockage between the phones degrades the measurement accuracy, and players can accidentally or purposefully block the mics and speakers. Future work includes creation of a blockage warning or cheat detection protocol. In addition, we are interested in extending the API to permit simultaneous ranging between more than two phones. Lastly, we are working on designing and prototyping additional MMG games on top of our phone-to-phone gaming API.

12. ACKNOWLEDGEMENTS

We extend our thanks to our shepherd Monica Lam and the anonymous reviewers for their feedback and suggestions. We would also like to thank Jian Qiu for consulting advice, and members of the UC Santa Barbara Link Lab for volunteering their time to help us collect in-situ gameplay player data.

13. REFERENCES

[1] Market trends: Gaming ecosystem, Gartner 2011.

[2] T. Beigbeder, R. Coughlan, C. Lusher, J. Plunkett, E. Agu, and M. Claypool. The effects of loss and latency on user performance in unreal tournament 2003. In *Proceedings of the 3rd ACM SIGCOMM Workshop on Network and System Support for Games (NetGames)*, pages 144–151, 2004.

[3] X. Bian, G. Abowd, and J. Rehg. Using sound source localization in a home environment. *Pervasive Computing*, pages 281–291, 2005.

[4] L. Girod, M. Lukac, V. Trifa, and D. Estrin. The design and implementation of a self-calibrating distributed acoustic sensing platform. In *Proceedings of the 4th International Conference on Embedded Networked Sensor Systems (SenSys)*, pages 71–84, 2006.

[5] A. Harter, A. Hopper, P. Steggles, A. Ward, and P. Webster. The anatomy of a context-aware application. *Wireless Networks*, 8(2):187–197, 2002.

[6] M. Hazas and A. Hopper. Broadband ultrasonic location systems for improved indoor positioning. *IEEE Transactions on Mobile Computing*, 5(5):536–547, 2006.

[7] M. Hazas, C. Kray, H. Gellersen, H. Agbota, G. Kortuem, and A. Krohn. A relative positioning system for co-located mobile devices. In *Proceedings of the 3rd International Conference on Mobile Systems, Applications, and Services (MobiSys)*, pages 177–190, 2005.

[8] L. Jones and S. Lederman. *Human hand function.* Oxford University Press, USA, 2006.

[9] G. Kortuem, C. Kray, and H. Gellersen. Sensing and visualizing spatial relations of mobile devices. In *Proceedings of the 18th Annual ACM Symposium on User interface Software and Technology (UIST)*, pages 93–102, 2005.

[10] M. Kushwaha, K. Molnár, J. Sallai, P. Volgyesi, M. Maróti, and A. Lédeczi. Sensor node localization using mobile acoustic beacons. In *Proceedings of the IEEE International Conference on Mobile Adhoc and Sensor Systems (MASS)*, pages 491–500, 2005.

[11] A. LaMarca and E. de Lara. Location systems: An introduction to the technology behind location. *Synthesis Lectures on Mobile and Pervasive Computing*, 3(1):1–122, 2008.

[12] C. Lopes, A. Haghighat, A. Mandal, T. Givargis, and P. Baldi. Localization of off-the-shelf mobile devices using audible sound: architectures, protocols and performance assessment. *ACM SIGMOBILE Mobile Computing and Communications Review*, 10(2):38–50, 2006.

[13] S. Mason, C. Berger, S. Zhou, and P. Willett. Detection, synchronization, and doppler scale estimation with multicarrier waveforms in underwater acoustic communication. *IEEE Journal on Selected Areas in Communications*, 26(9):1638–1649, 2008.

[14] Microsoft. Xbox Kinect. `http://www.xbox.com/kinect`.

[15] Nintendo. Nintendo Wii. `http://www.nintendo.com/wii`.

[16] C. Peng, G. Shen, Y. Zhang, Y. Li, and K. Tan. Beepbeep: a high accuracy acoustic ranging system using cots mobile devices. In *Proceedings of the 5th International Conference on Embedded Networked Sensor Systems (SenSys)*, pages 1–14, 2007.

[17] N. B. Priyantha, A. Chakraborty, and H. Balakrishnan. The Cricket Location-Support System. In *Proceedings of the 6th Annual International Conference on Mobile Computing and Networking (MobiCom)*, pages 32–43, 2000.

[18] J. Qiu, D. Chu, X. Meng, and T. Moscibroda. On the feasibility of real-time phone-to-phone 3d localization. In *Proceedings of the 9th ACM Conference on Embedded Networked Sensor Systems (SenSys)*, pages 190–203, 2011.

[19] J. Sallai, G. Balogh, M. Maroti, A. Ledeczi, and B. Kusy. Acoustic ranging in resource constrained sensor networks. In *Proceedings of the International Conference on Wireless Networks (ICWN)*, pages 467–474, 2004.

[20] T. Schmidl and D. Cox. Robust frequency and timing synchronization for ofdm. *IEEE Transactions on Communications*, 45(12):1613–1621, 1997.

[21] J. Scott and B. Dragovic. Audio location: Accurate low-cost location sensing. *Pervasive Computing*, pages 307–311, 2005.

[22] Sony. Playstation Move. `http://us.playstation.com/ps3/playstation-move`.

[23] K. Whitehouse and D. Culler. Calibration as parameter estimation in sensor networks. In *Proceedings of the 1st ACM International Workshop on Wireless Sensor Networks and Applications (WSNA)*, pages 59–67, 2002.

[24] Z. Zhang, D. Chu, J. Qiu, and T. Moscibroda. Demo: Sword fight with smartphones. In *Proceedings of the 9th ACM Conference on Embedded Networked Sensor Systems (SenSys)*, pages 403–404, 2011.

ExerLink - Enabling Pervasive Social Exergames with Heterogeneous Exercise Devices

Taiwoo Park, Inseok Hwang, Uichin Lee[†], Sunghoon Ivan Lee[‡], Chungkuk Yoo, Youngki Lee, Hyukjae Jang, Sungwon Peter Choe, Souneil Park and Junehwa Song

Department of Computer Science, KAIST, [†]Department of Knowledge Service Engineering, KAIST and [‡]Computer Science Department, UCLA

{twpark, inseok, chungkuk, youngki, hjjang, sungwon, spark, junesong}@nclab.kaist.ac.kr, [†]uclee@kaist.edu, [‡]silee@cs.ucla.edu

ABSTRACT

We envision that diverse social exercising games, or exergames, will emerge, featuring much richer interactivity with immersive game play experiences. Further, the recent advances of mobile devices and wireless networking will make such social engagement more pervasive—people carry portable exergame devices (e.g., jump ropes) and interact with remote users anytime, anywhere. Towards this goal, we explore the potential of using heterogeneous exercise devices as game controllers for a multi-player social exergame; e.g., playing a boat paddling game with two remote exercisers (one with a jump rope, and the other with a treadmill). In this paper, we propose a novel platform called ExerLink that converts exercise intensity to game inputs and intelligently balances intensity/delay variations for fair game play experiences. We report the design considerations and guidelines obtained from the design and development processes of game controllers. We validate the efficacy of game controllers and demonstrate the feasibility of social exergames with heterogeneous exercise devices via extensive human subject studies.

Categories and Subject Descriptors

C.3 [**Special-purpose and Application-based Systems**]: Real-time and embedded systems; H.5.2 [**Information Interfaces and Presentation**]: User Interfaces—*evaluation/ methodology, input devices and strategies, prototyping*

General Terms

Design, Human Factors, Experimentation

Keywords

Exergame, Exercise, Social, Pervasive, Heterogeneous

MobiSys'12, June 25–29, 2012, Low Wood Bay, Lake District, UK.

1. INTRODUCTION

Applying diverse pervasive devices to game interfaces opens up many opportunities for game developers to interweave real world activities into the virtual world fabric of games. One promising application domain is exergaming that combines exercise and gaming to provide positive and engaging fitness experiences [6, 10, 4, 45].

More importantly, we envision that *pervasive* social exergames supporting multiple exercise modalities will enable true ubiquitous social interactions, fostering social bonds and friendships. Further, such exergames will allow people with different exercising capabilities and preferences to gather from diverse situations and to play/exercise together. For instance, a homemaker running on a treadmill at home, a son who is a college student cycling in his school gym, the husband on a business trip using his jump rope, and a young daughter using a hula hoop at her friend's house can all meet virtually and exercise together over a multi-exercise game. They will be able to choose the right exercise appropriate for their given situation or preference, enjoying exercising together via social exergames.

In this paper, we explore the potential of using multiple exercise devices as game controllers and incorporating multiple heterogeneous controllers into a game. Specifically, we consider a class of exercise equipment used for repetitive, individual, and aerobic (RIA) exercises such as treadmill running, stationary cycling, hula hooping, and jump roping. RIA exercises are widely performed by many people, largely because RIA exercises are simple and easy to learn and bring positive effects to physical health [21].

However, realizing an interactive multi-player exergame using RIA exercise devices as game controllers gives rise to important challenges resulting from the differences between exercise modalities, e.g., rotating ropes and taking steps. First, converting exercise devices into game controllers requires exercise-specific considerations of sensor hardware and software support to acquire the exercise intensity metrics to be used as game inputs, e.g., exercise speeds. Traditionally such support has been provided only for specific types of exercise equipment such as stationary bikes. Second, it may cause significant imbalances in game performance among game players who simultaneously utilize different types of controllers designed for different exercises. While imbalance due to network delay/jitter has been well investigated in the literature [16, 29, 18], little study has been done to understand the performance difference with heterogeneous exercise devices and personal preferences/capabilities which is a unique problem in pervasive social exergames.

In this paper, we propose the ExerLink platform featuring conversion and balancing mechanisms that are carefully tuned through an iterative design process including a series of field trials. ExerLink aims to deliver fair game play experiences by considering the unique aspects of social exergames. The main contributions of this paper are summarized as follows. First, we explore the possibility of incorporating different exercises into an interactive exergame and propose novel balancing mechanisms that consider heterogeneity of exercise devices and personal differences in capabilities and preferences. Second, we provide design considerations and guidelines obtained from the design and development processes for four game controllers. The conversions have been thoughtfully crafted over eight months of iterations incorporating three preliminary user tests. Third, we use an interactive game called Swan Boat [6] over ExerLink to demonstrate the feasibility of a pervasive multi-exercise game in co-located and remote environments. We also report the unique patterns of social interactions in this multi-exercise game.

The rest of this paper is organized as follows. In Section 2, we provide a brief summary of background and motivation. In Section 3, we present the key design issues and overview the platform architecture. In Section 4, we illustrate the conversion and balancing mechanisms that aim to deliver fair game play experiences. In Section 5, we detail how we build exercise-game controller prototypes using existing equipment (e.g., hula hoops, interactive treadmills). In Section 6, we perform a human subject study to evaluate the performance of exercise-game controllers. In Section 7, we conduct two case studies to evaluate the proposed balancing mechanisms and to understand social interaction patterns in both co-located and remote play. In Section 8, we discuss the limitations of the current work and describe future research directions. Finally, we present the related work in Section 9 and conclude the paper in Section 10.

2. BACKGROUND AND MOTIVATION

2.1 Why Social Exergames?

One of the main goals of exergaming is to encourage physical activity as it can transform tedious exercise into an enjoyable experience. This approach complements existing persuasive techniques that rely on peer pressure and social recognition. For instance, in Fish'n'Steps [28], an exercise logging application using a pedometer, a user raises a virtual fish that grows according to the number of steps its owner takes. In Houston [14], users can check each other's exercise records and exchange comments using text messages. Kukini, an extended version of Nike+iPod [10] supports social interactions by allowing users to share exercise results.

Beyond encouraging physical activities, recent literature also showed that due to their rich interactive nature, exergames can be used to foster social bonds and friendships. For instance, Wii Sports [5] provides social game play experiences mostly in a co-located place. Gamebike [3] allows individuals riding stationary bikes to compete with each other in a virtual race. Muller et al. [39] explored the relationship between social interactions and physical activities through their own work *Table Tennis for Three* and further showed that remote exergames bring similar social benefits based on their remote jogging game called *Jogging over a Distance* [39]. While the benefits of social exergames are well documented, existing approaches cannot be fully extended to *pervasive* social exergame scenarios because players must visit gyms and use designated equipment regardless of their preferences or physical abilities. Our goal is to relax this restriction such that users with portable exercise devices can join social exergames anytime, anywhere and those users with heterogeneous devices can enjoy the same social exergames with a fair game experience.

2.2 Motivating Scenario

Jessica, a 31-year-old woman working at a sales department, has gradually gained weight over the past couple years. She tried a few different forms of exercises, but she failed to lose weight because she felt that exercising is not so enjoyable. Moreover, due to the nature of her career, she often travels across town for sales, and it is very difficult to consistently visit exercising facilities. She found that two of her close friends, Christine and Nicole, had the same problem, and they decided to use ExerLink to make exercising more enjoyable. Jessica wants to exercise on a treadmill or jump rope because her major objective is to burn calories. On the other hand, Nicole prefers hula hooping because her objective is to strengthen her abdominal muscles and Christine prefers to ride an exercise bike due to her knee problem. At 6 PM everyday, they either meet up at the gym or remotely join the group at their own houses to exercise. One day Jessica has to visit the other side of the town. Fortunately, she doesn't have to worry because she can bring a jump rope and join the group exercise remotely using her smartphone. As a result, Jessica and her colleagues are able to consistently participate in the group exercise, and continue to lose weight for the next several months.

2.3 Repetitive-Individual-Aerobic (RIA) Exercises

We consider repetitive, individual, and aerobic exercises such as treadmill running, stationary cycling, hula hooping, jump roping, etc. These repetitive-individual-aerobic (RIA) exercises are suitable for use in multiplayer exergames due to their intrinsic characteristics as follows:

- **Popularity**: They are popular among many people of all different age groups, largely due to positive effects on physical health [21].
- **Ease of access**: They are usually simple and easy to learn and perform.
- **Monotony**: The solitary and monotonous nature of RIA exercises provides strong motivation to create games to relieve the monotony.
- **Long-lasting**: Aerobic exercises are usually performed for a long enough time to play a game as compared to anaerobic exercises.
- **Measurability**: By measuring the degree of repetitive intensity while exercising, we can easily employ exercise metrics as game input. Namely, the RIA exercise equipment can be viewed as 1-dimensional game controllers.

(a) (b)

Figure 1: (a) Team playing a game on ExerLink (b) Architecture overview of the system

2.4 Target Social Exergames

As RIA exercise devices can be regarded as 1-dimensional game controllers, we search for some popular types of games whose core mechanics largely depend on 1-dimensional control, namely racing and vertical position-oriented games. In racing games, players' exercise intensities can be easily mapped to the avatar's speed, as instantiated in several exergames [6, 19, 45]. In vertical position-oriented games such as horizontally-scrolled shooters and paddle games, it is natural and intuitive to control the vertical position of a game avatar using players' exercise speeds.

3. EXERLINK PLATFORM DESIGN

3.1 Design Considerations

Exercise- and Player-specific differences: In our scenario, users could play a game using heterogeneous types of exercise devices at the same time. Moreover, users have different preferences on the range of exercise intensity, according to personal physical ability or purpose of exercise. For example, Jessica may want to walk within 7 km/h as she has a cardiovascular problem or aims to lose her weight more effectively, whereas her friend George wants to run faster than 9 km/h. To cope with diverse needs, the platform should be designed to support different types of exercise devices to be used in one game, as well as to capture users' preferred ranges of exercise intensity. Most importantly, in such cases the game (and its supporting platform) should provide a fair game experience to all users.

Network latency: High network latency would result in a poor game experience, particularly when users are playing over cellular networks [36]. Social exergames should provide effective ways to mitigate the delay/jitter effects on the game play experience in terms of smoothness and fairness.

Interaction modalities: As one of the goals of social exergames is to foster social bonds, the platform should provide rich interaction modalities among users. Building social bonds is one of the strengths of co-located play, as participants can talk to each other or even just glance at each other. In remote game play scenarios, the platform should support some means of computer-supported communications (e.g., voice, video) to effectively facilitate interaction among players.

Exercise history information: It is important to provide users with a summary of exercise performed, such as duration and total calories burned, because it can be used in many useful ways; e.g., logged personally, shared with others, and utilized for setting up further exercise plans. The platform should collect exercise information during game play and provide summary information to the exercisers.

3.2 Architecture Overview

ExerLink mediates social exergames and diverse RIA exercise devices (see Figure 1) and supports mobile devices including smartphones, tablets and Mobile Internet Devices (MIDs). ExerLink provides a run-time environment for game applications, and it is capable of running multiple exergames at the same time. The ExerLink platform processes exercise intensity information from devices and converts it into game input values. The conversion process considers the heterogeneity of exercise devices and personal differences in physical abilities and preferences. The platform mediates communication between game applications and a game server considering the different network delays among remote game players. It also provides a voice communication channel between game players to support social interactions in a remote play situation. Lastly, it tracks and summarizes the amount of exercise performed by players to help maintain the motivation of the players.

ExerLink consists of four key components: the *Game Input Converter*, the *Voice Channel Manager*, the *Network Manager*, and the *Exercise Information Manager*. The Game Input Converter is responsible for translating exercising speeds from exercise devices into game inputs. Specifically, it considers the heterogeneity of exercise devices and users' preferred exercise intensities to provide fair game play experiences. The Voice Channel Manager supports voice communications to facilitate social interactions during game play. It selectively supports various modes of communications on-demand, namely peer-to-peer, team, and all-player communications. The Network Manager supports communication between game applications and a game server while effectively minimizing unfairness in game play due to network delay variation among players, as shown later in Section 4.3. The Exercise Information Manager summarizes exercise statistics. It automatically calculates the energy cost of physical activities from exercising speed in units of metabolic equivalent (MET), utilizing speed-MET mappings such as those described by the Compendium of Physical Activities (CPA) [7].[1] This makes more sense than direct measurement of energy costs, as carrying and wearing measurement devices such as heartbeat sensors can additionally burden users.

4. PROVIDING FAIR GAMEPLAY

To provide fair game experiences, ExerLink considers the following unique aspects of social exergames: (1) significant intensity variation among different exercise devices (e.g., hula hoop vs. jump rope), (2) differences in users' preferred exercise intensities (e.g., preferring slow jumping to fast jumping), and (3) differences in network delay due to heterogeneous network conditions while playing social exergames over the Internet. In this section, we present the conversion and balancing mechanisms which were carefully tuned through an iterative design process including preliminary field trials.

4.1 Converting Exercise Intensity into Game Input

As shown below, a user's exercise intensity using a certain device can be measured using standard metrics, e.g., rotations per minute (RPM) for hula hoops, jump ropes, and stationary bikes, or speed (km/h) for treadmills. We propose a model that converts exercise-specific intensity to game-specific input data. Establishing such a model is non-trivial and requires careful consideration of the exercise type and the individual's preference. In the following, we discuss standard strategies of converting exercise intensity values to game values and then propose a mapping method for social exergames that considers the heterogeneity of exercise devices and individuals' preferences.

4.1.1 Converting Strategies: Throttling and Adjusting

In the field of game design, two distinct strategies are widely used to convert exercise intensity to game input data, namely throttling and adjusting. The concept of throttling is analogous to a throttle lever; each position of the throttle lever uniquely corresponds to some value. Players can easily relate exercise intensity to game value because exercise intensity is proportional to the game input value. For example, when the throttling strategy is adopted in a game using a treadmill as the primary controller, the game input values would be proportional to the player's running speed.

In the adjusting strategy, the player's input acts like a first-derivative that changes the game state, e.g., an acceleration or deceleration control button. For instance, a vehicle keeps increasing its speed as long as the acceleration button is on, and it maintains its current speed when the button is off. In the example of the treadmill-controlled game as above, the game would maintain the current game state if the player keeps running within a predefined speed range which is considered as 'cruise,' similarly to the case of 'no button is on' in the previous vehicle example. If the player's running speed exceeds that range, the game state changes to the acceleration mode; if it falls below that range, it changes to the deceleration mode. The adjusting strategy is very different from the throttling strategy in that running at a given speed does not uniquely determine the game state; it only determines whether the game state changes or is maintained. Note that both strategies work well in practice, and the choice of strategy depends on the game type. In our prototype implementation, we design a multi-player social exergame where throttling is used as the default converting strategy.

4.1.2 Personalized Mapping Mechanism

To enable exercise- and player-specific intensities to take effect in games, we design a personalized mapping function for each exercise device, which converts the intensities into game input values. Those mapping functions are designed in thoughtful consideration of the following unique aspects of social exergames, which we learned over iterative studies on users and different exercise devices.

Ease of Control: Players expect ease of control for game play. For example, suppose that the RPMs of the hula hoop are mapped to the vertical position of a game avatar. The major range of the avatar's positions required by the game should be reachable by the RPM ranges in which the player can hula-hoop naturally, without putting too much effort. If the player is required to change the speed of hula-hoop to either too high or too low value frequently, the player will blame the controller for its lack of controllability.

Consistent Movement: Interestingly, players are quite sensitive to inflection points of movement patterns. An inflection point is a certain point on the intensity curve of an exercise; crossing the point naturally involves the change of the exercise mode. In the treadmill, for instance, an inflection point is around 7 km/h where the mode of exercise is switched between walking and running. Raising the speed of jump-roping over 170 RPM requires players to change their jumping method to alternating their feet or doing double-unders. During our iterative system design, we identified the existence of inflection points in several types of exercises. More importantly, we observed that controlling a game across such an inflection point incurs poor gaming experiences like unintended abrupt motions of the avatar in a game.

Marginal Range: From the two considerations previously discussed, we can define a preferred range for a player, in which the player can change the exercise intensity easily and smoothly. Another important consideration is that, even if the player's input goes out of the one of the two ends of the preferred range, it is desirable for the game to respond to the player's input to some extent. It may be analogous to the "bouncing effect" in iPhones or iPads, which is a gentle implication to let the user know that he/she has scrolled the page to the very end. If the game is designed to suddenly ignore the user's input as soon as it goes just above or below the preferred range, the player would complain that the game controller is unresponsive. To this reason, we need to design "marginal ranges" at the both ends of the preferred range where the responsiveness gradually becomes blunt, so that the players can naturally feel that they are getting out of the preferred range.

Given these observations, we introduce two types of ranges, namely preferred and marginal ranges. The pre-

[1]MET represents the energy cost of physical activities. It is approximately equal to the energy cost of the activity, expressed as kilocalories per hour per kilogram of body weight.

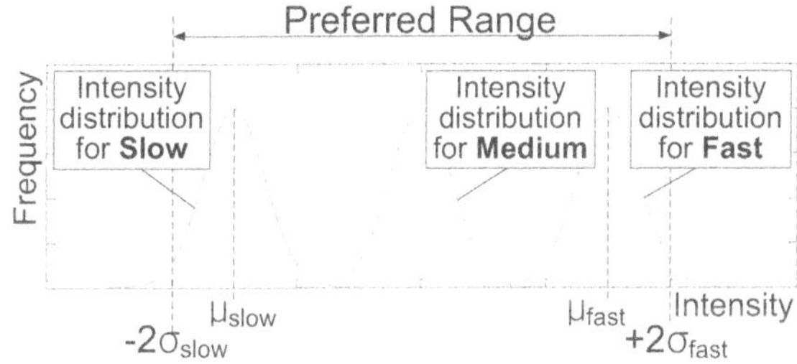

Figure 2: Conceptual overview of personalized mapping

ferred range is determined during the training phase on a per-player, per-exercise basis.

1. The system instructs the user to use an exercise device in three different representative intensities, namely slow, medium, and fast modes. We let the player to determine the preferred intensities for slow, medium, or fast, as we intend the mapping functions to be tuned subjectively to each player.

2. We determined the minimum and the maximum values of the preferred range based on the measured intensity distributions of the players. To be specific, we first observed that the players' intensity distributions roughly follow the normal distributions. Then we define the preferred range to include two-time standard-deviation intervals around the means of each mode, as shown in Figure 2. Accordingly, the preferred range for player i is defined by its minimum and maximum values as the following.

$$min_i = \mu_{slow} - 2\sigma_{slow}$$
$$max_i = \mu_{fast} + 2\sigma_{fast}$$

   where σ_{slow} and σ_{fast} are the standard deviations of intensity samples from slow and fast modes, respectively and μ_{slow} and μ_{fast} are the means of the samples from the modes, respectively.

3. Importantly, the preferred range should be defined not to include an inflection point in the middle, so as not to negatively affect the gaming experience as described above.

To summarize, for a given exercise, a player's preferred range captures a continuous, inflection-free range of the exercise intensities over which the player can change their exercise intensities without much difficulty. For instance, if a user decides to use a jump rope as a game controller, she simply goes through the training phase, and the preferred range is configured accordingly (e.g., from 120 to 140 RPM).

Besides, the marginal ranges are placed at both ends of the preferred range, namely lower and upper marginal ranges, respectively. In the case of a jump rope, the lower marginal range corresponds to [0, 120), and the upper marginal range corresponds to (140, R^{jr}_{max}) where R^{jr}_{max} is the maximum achievable RPM of jump-roping. We then need to determine the game value range for both preferred and marginal ranges. In our design, we empirically assign the preferred range to be mapped to the range of 20% to 80% of the game value (i.e., total 60% of the game value). The lower marginal range takes the lower 20% of the game value, and the upper range takes the upper 20% of the game value. As illustrated in Figure 3, we construct a resulting mapping function using a Bezier curve connecting the following points, namely (0, 0), (X_{min}, 20%), (X_{max}, 80%), and (R^k_{max}, 100%). Here, a player's preferred range is [X_{min}, X_{max}], and the maximum achievable exercise intensity of device k is R^k_{max}.

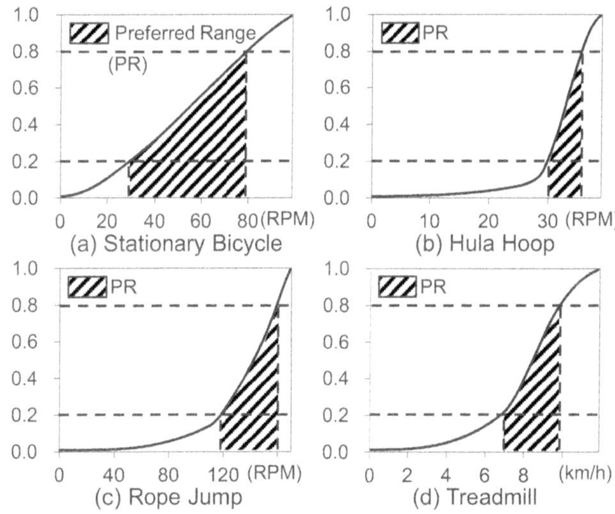

Figure 3: Examples of mapping functions for RIA controllers. The horizontal axis of each figure represents the exercise intensity of the exercise. A player has a unique Preferred Range (PR) for each device. The game input value range of 20-80% is used to map PR, and the remaining ranges (0-20% and 80-100%) are used to map Marginal Ranges (MRs). A mapping function is defined by connecting the four points shown in the graph.

4.2 Balancing Heterogeneity

We present a scheme for balancing heterogeneity for fair and collaborative game play. For the sake of illustration, we use a multi-player game called Swan Boat [6]. In Swan Boat, we have a team of two players exercising in close harmony with one another to collaboratively control the direction of a boat.[2] The difference in exercise intensity between team members determines the direction of their boat (either left or right). If the intensity levels of both users are the same, the boat moves straight; otherwise, it slants towards the side with higher intensity, with the angle proportional to the intensity difference. As illustrated earlier, our target games are 1-dimensional, and thus, they should not require any additional controllers apart from its core mechanism; i.e., an object is steered using exercise intensity in one direction. To use both directions, the game must restrict a team of two users to control the object or incorporate additional features (e.g., automatically changing direction after reaching at a certain point in 1-dimensional space). In this example game, players should closely collaborate in a synchronized manner, and thus, we expect that they can feel imbalances, particularly when heterogeneous devices are used, or/and their physical capabilities or preferences are different.

How fast one player can steer the boat toward one direction is related to the *movement time*, i.e., the rate of exercise intensity changes, with the exercise device that the

[2]As shown later, two teams can compete with one another to win a boat race.

```
Function : AccelerationBalancing
Input    : Vin is the original game input, t is current time, β is the target rate
Output   : Vout is the converted game input

1.  // detect acceleration
2.  if ((Vin − Vin_old) / (t − t_old) > acceleration_threshold)
3.      if (current_state != ACCELERATING)
4.          current_state ← ACCELERATING
5.          t_accel_begin ← t_old
6.          Vin_accel_begin ← Vin_old
7.      // convert acceleration rate from the beginning of the current acceleration
8.      current_accel ← (Vin − Vin_accel_begin) / (t − t_accel_begin)
9.      converted_accel ← β · current_accel
10. else
11.     if (current_state == ACCELERATING)
12.         current_state ← ACCEL_END_WAIT
13. if (current_state == ACCELERATING || current_state == ACCEL_END_WAIT)
14.     Vout ← Vout_old + converted_accel × (t − t_old)
15. Vin_old ← Vin
16. Vout_old ← Vout
17. t_old ← t
```

Figure 4: Pseudo code of the acceleration balancing. The deceleration balancing can be done similarly.

player is manipulating. Thus, the heterogeneity of devices or player's capability/preference is directly related to the movement time, and this imbalance directly affects the fairness of a game. For instance, a player on the bike can quickly ramp up its speed, whereas a player with the hula hoop can only slowly increase its speed. Given that a player's exercise intensity distribution falls mostly within the preferred range, we consider this range for balancing heterogeneity. The movement time is mainly dependent on a player's preferred range and minimum/maximum acceleration, which are measured during the training phase. For a given set of players and their exercise devices currently in use, we can set the target movement rate through which all players' movement rates are normalized. In our design, we set the target rates by averaging the rates of all players. Here, the acceleration and deceleration rates are inversely proportional to the movement times of upward and downward directions, respectively. We then adjust the acceleration and deceleration rates of each user to match the target rates. In other words, we artificially raise the rates of a player which has relatively lower rates than the target rates (e.g., the players with the hula hoop and treadmill), whereas we decrease the rates of the other users with higher rates than the target rates. Thus, each user u_i has a linear scaling factor β_{u_i} which is used to normalize a player's current speed. As shown in Figure 4, we also propose a simple heuristic which detects whether a player is currently accelerating or decelerating if the rate of change is above some threshold value (line 2). If so, the current state is changed to ACCELERATING or DECELERATING, and the output is scaled accordingly after a fixed amount of time (line 9).

4.3 Balancing Delay Variation

ExerLink uses a client-server model where all players interact with an authoritative game server to play the game, and the server maintains the global state. For a given action that a player takes, the action message is sent to the server. After collecting all the action messages, the server then changes the global state and notifies the resulting state to the players using an update message. In general, the tolerable latency of games with avatars is about 200ms [13], and yet, due to the unique characteristics of game controllers (i.e., changing intensity takes some time), exergames are more lenient to the delay variations. However, large delay variation would result in a poor game experience, particularly when users are playing over cellular networks [36]. To mitigate the problem we employ dead reckoning technique that allows a local client to predict the state of another player using the last known vectors in between update messages from the server [16].

The global state updates at the game server are propagated through all the remote players whose network delay may be different. Thus, the update messages may be delivered at different times. This means that a player who is close to the game server will see the update message earlier and may be able to react faster than the rest of users, which leads to an unfair advantage for the players with lower delays. At the server side, the action messages must be properly ordered such that players' responses are ordered based on their reaction times (i.e., the time duration between update message arrival and action message departure). Several earlier work addressed this issue of fair-ordering of action messages such as Sync-MS [29] and Fair-Ordering [18]. In the Sync-MS service, the game server can process action messages in a fair order based on the following assumptions; i.e., the clocks of all the participants (including the server) are synchronized, and the one-way delay from the server to each player is measured. In contrast, the Fair-Ordering service does not have such requirements and yet provides fair-ordering by simply processing the action message in the order of increasing *reaction time* that is the time between the reception of an update message and the departure of an action message by a player. The server processes all the action messages received during a waiting window and handles the delayed and out-of-order action messages. In ExerLink, we adopt the Fair-Ordering service to balance delay variation in social exergames.

5. EXERCISE-GAME CONTROLLER PROTOTYPE

We discuss our design considerations and technical explorations in redesigning RIA exercise devices as game controllers. There are a number of exercises that belong to the RIA category such as running, biking, hula-hooping, rowing, and jump-roping. We can largely classify these exercises based on the types of exercise equipment (either stationary machines or portable devices). For the purpose of prototyping, we chose treadmill running and stationary cycling for the exercises on machines, and hula-hooping and jump-roping for exercises with handheld equipment.

We modified the existing RIA exercise devices to measure the exercise intensities in real-time. To this end, we implanted sensors and micro-controllers onto an off-the-shelf hula hoop and a jump rope. For the treadmill and stationary cycle, their built-in sensing capabilities are exploited. We carefully augmented existing devices so as not to disrupt the players' natural exercise activities.

The entire prototyping procedure consists of a series of iterations. Importantly, we revised our technical designs of the prototypes until we resolved most of dissatisfaction reports during game plays. The three tests were carried out over eight months. Total 21 paid participants were recruited, and their ages range from 20 to 35. The participants were allowed to freely use all kinds of RIA controllers developed for that test. We interviewed them and collected comments to guide the revision of the controllers which in most cases inspired us with ideas for improvements. Below, we report

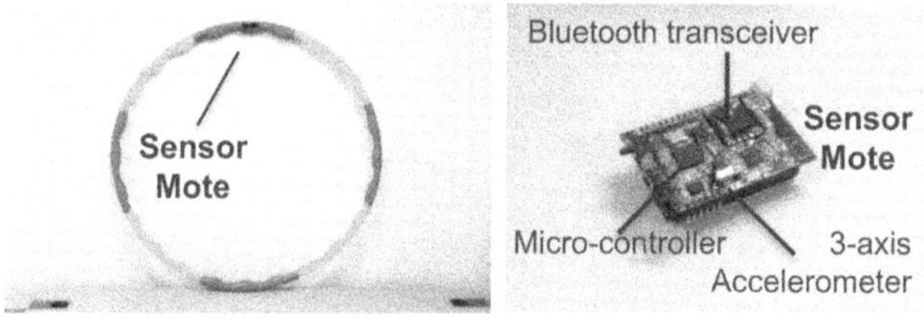

Figure 5: The first hula hoop controller prototype measures rotation speed of the hoop by analyzing a waveform of centrifugal force.

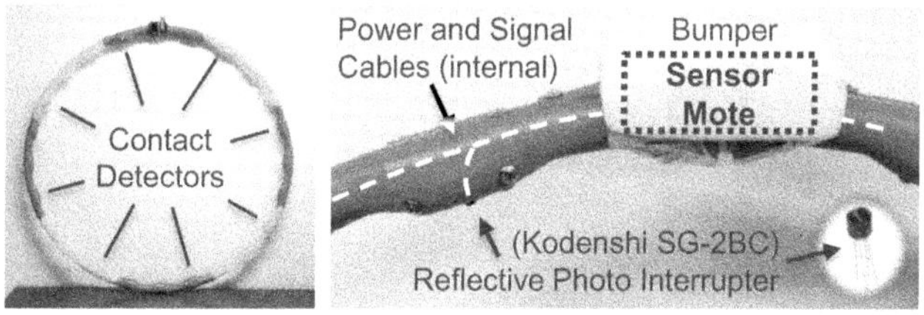

Figure 6: The revised hula hoop, measuring rotation speed eight times faster than the previous one

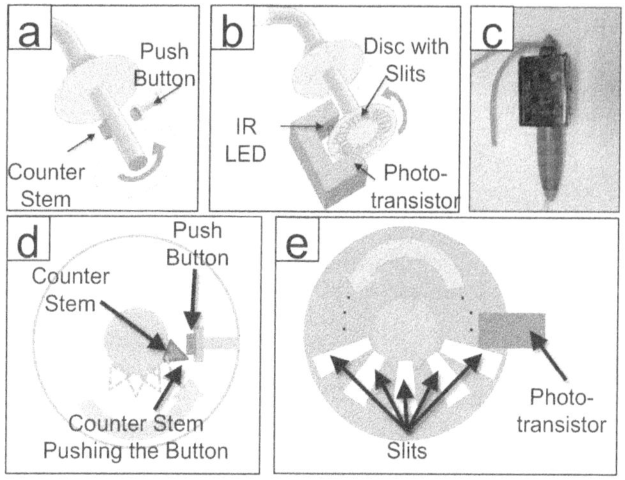

Figure 7: Conceptual diagram of measuring components for (a) the first prototype and (d) the illustration of its functionality. The improved version and its functionality is illustrated in (b) and (e), respectively. The exterior of the improved version is provided in (c)

our design process and key implementation details for each controller.

5.1 Hula Hoop

Our initial attempt was to use minimal sensor hardware and to largely rely on subsequent signal processing. We modified an off-the-shelf hula hoop by integrating a single sensor mote with a built-in 3-axis accelerometer and a Bluetooth transceiver. Figure 5 shows our original hula hoop design. We embedded a sensor mote inside the hula hoop tube, placing the X-axis of the accelerometer facing away from the center of the hoop. Our original strategy with the sensor mote was to measure the centrifugal force exerted when the player revolves the hoop around the waist. Note that the centrifugal force shows periodic characteristics, as the radius of rotation, the distance between the player's body and the sensor mote varies periodically while the hoop rotates. Then we measured the period and evaluated the hoop's rate of rotation. However, our tests showed that the duration of the period was typically too long when we tested the hoop as a game input. Many testers reported that the game value was updated too slowly even when they changed their hula-hooping rates quickly. It makes the game play unresponsive and frustrating.

To address these problems, we revised the design of a hula hoop to detect the contacts between the hoop's inner surface and the player's waist. Then the rate of rotation is calculated from the number of contacts per unit time. This revised design can update changes of the hoop's RPM much more frequently. To this end, we integrated eight reflective photo interrupters (Kodenshi SG-2BC) along the eight equi-spaced points of the inner circumference of the hoop as shown in Figure 6. These interrupters are wired to the sensor mote. This revision allowed us to increase the sampling rate by eight times over the original design, providing sufficient responsiveness to the game experiences.

5.2 Jump Rope

Similar to the initial approach of a hula hoop, we adopted the angular velocity of the rope measured at the grip as the game input parameter. Our original design was to install a counter stem and a button inside of the grip as shown in Figure 7 (a), along with an Atmega128L micro-controller and a Bluetooth transceiver. This design is simple to implement; the counter stem rotates together with the rope, pushing the button once every revolution of the rope as shown in 7 (d). The micro-controller computes the period of a revolution and converts it to angular speed. Lastly, the transceiver reports the value to the local game system.

We identified two major drawbacks of this mechanical counter. First, it was not sufficiently durable; the button-stem tends to loose after a few weeks and eventually failed after a few months in our field trials. Second, and more importantly, updating the rotation frequency once per revolution was not frequent enough in terms of the sampling rate. Its negative impact on the game experience was worse than in the hula hoop case, largely because the jump-ropers were generally able to slow down more abruptly than the hula hoopers. Note that the update interval gets longer as the jump-roping slows down. Many players complained that the game was not responsive enough especially when they intended to slow down abruptly.

To address these drawbacks, we devised an alternative scheme inspired by an opto-mechanical trackball [25]. We installed a small disc with 60 closely spaced radial slits, an LED, and a photo-transistor as shown in Figure 7 (b) and (e). As illustrated in Figure 7 (e), the photo-transistor can detect the presence the IR light if a slit lies upon the location of the transistor (i.e., logical one); otherwise, the IR light will be physically blocked (i.e., logical zero). Based on this method, we can measure the angular speed of the rope 60 times per revolution of the rope, or equivalently, at a sampling rate of 60 times higher than that in the original button-based design. Eliminating a mechanical button also improved the durability; the revised rope survived the entire prototyping phase.

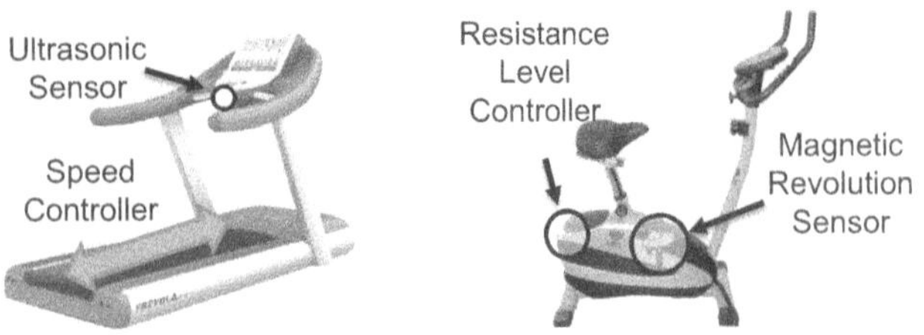

Figure 8: Interactive treadmill (left) and Stationary bicycle (right)

5.3 Stationary Bike

We built two types of bikes that differed in the location where the magnetic rotation sensor is installed: one that measures cadence on the flywheel, which controls the level of resistance, and the another that measures the cadence on the pedal shaft directly. We chose the latter ones, as shown in Figure 8, because the flywheel still rotates even when the player stops pedaling. As we intend to employ the player's current cadence as the game input, measuring flywheel RPM would provide an inaccurate cadence information. In our prototype, we use a mote to collect the data which are wirelessly delivered to the local game client via Bluetooth. The player's cadence is automatically measured and reported periodically in RPM.

5.4 Interactive Treadmill

In order to enhance a treadmill with interactive features required for game inputs, we designed and developed an interactive treadmill that builds upon our earlier work [6] (see Figure 8). In the case of a conventional treadmill, the speed of a treadmill is usually fixed by players or a predefined workout plan. In contrast, the interactive treadmill naturally adjusts its speed to the player's running pace. In detail, the built-in ultrasonic sensors monitor the player-to-console distance, automatically matching the treadmill's speed to the player's pace. Therefore, the interactive treadmill enables players to play games simply by changing their running pace, i.e., when the player increases running speed, she gets closer to the front of the treadmill, and subsequently, the treadmill speeds up; similarly, when she decreases her running speed, the treadmill slows down. Note that unlike other devices, players should be familiarized with the controlling mechanism of an interactive treadmill. Hence, it may take a slightly longer time for players to learn how to control the speed of a treadmill when compared to the cases of other exercise devices. In the following section, we take into account this learning effect in the experiments.

6. EVALUATION OF EXERCISE-GAME CONTROLLERS

We evaluate the performance of the prototype controllers by conducting an experiment concerning a basic and essential pointing activity—most of the core mechanics of the target exergames can be implemented on the basis of this pointing activity. For comparison, we basically measured how rapidly and accurately users can change a game value to a desired level.

6.1 Method

Participants: We recruited 20 participants (10 males and 10 females) via announcements posted in a local university campus. We advertised that the experiments may require sufficient physical abilities in four kinds of exercises as well as stamina. The participants were undergraduate or graduate students in the age range of 18 to 34. The average age was 22.35 (SD: 4.60).

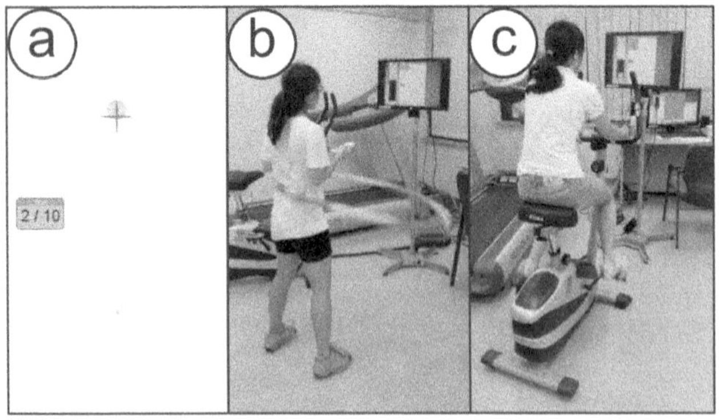

Figure 9: (a) Screenshot of the experiment. (b)(c) Participants performing the experiment using a hula hoop and a stationary cycle

Apparatus: The usability test software was developed in Unity, a casual game development framework. Output was presented on a 24" monitor. Input was via the aforementioned four exercise equipments. A wireless handheld button was given to each participant to click the targets.

Procedure and Design: We randomly assigned participants to one of the four groups (i.e., 5 participants per group). Each participant performed the tests with all four devices. The order of devices in each group was shuffled by following a balanced Latin square experiment design.

To measure the pointing performance, we designed a simplified 1-dimensional version of the multi-directional point-select task as defined in ISO 9241-9 (also known as Fitts' law tests) [22]. As shown in Figure 9 (a), there are two targets at the high intensity position (80% point of the converted game value) and the low intensity position (20%). The distance between the targets and the diameter of the targets are 480 pixels (130 mm) and 48 pixels (13 mm), respectively. The nominal index of difficulty [35] was 3.46 bits.

For each trial, participants moved the cross pointer to the current target using the exercise equipment and then clicked a button of the wireless handheld device to mark the target. After the click, the current target disappears and the next target appears in the opposing position. Participants repeated this task for 10 times (called a block), i.e., five times for acceleration and five times for deceleration. For each equipment, a series of 10 blocks were repeatedly performed. To evaluate the pointing performance, we collected movement times between the clicks and errors on distance between the targets and the clicked positions. From these data, we derived the throughput according to the ISO guidelines [22]. Before the beginning of the experiment, the participants were instructed using a demonstration and given a sequence of warm-up trials. During the warm-up trials, equipment settings were personalized, including adjusting the preferred ranges and controlling the resistance level of the bike.

Because the performance can be influenced by a participant's physical capability and stamina, we periodically reminded the participants' physical condition and asked them to take a rest if they look/feel tired in order to avoid the

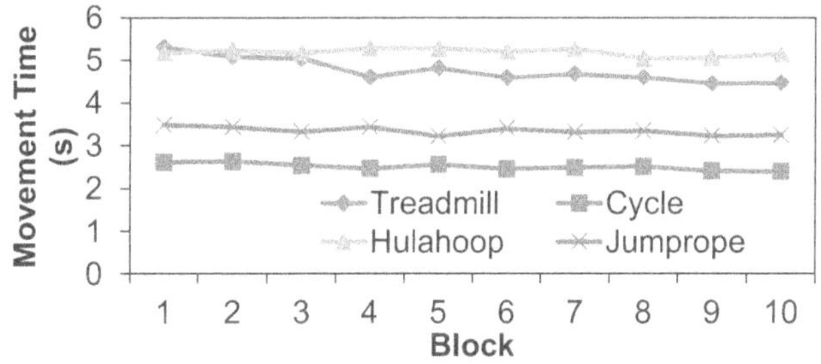

Figure 10: Movement time by equipment and block

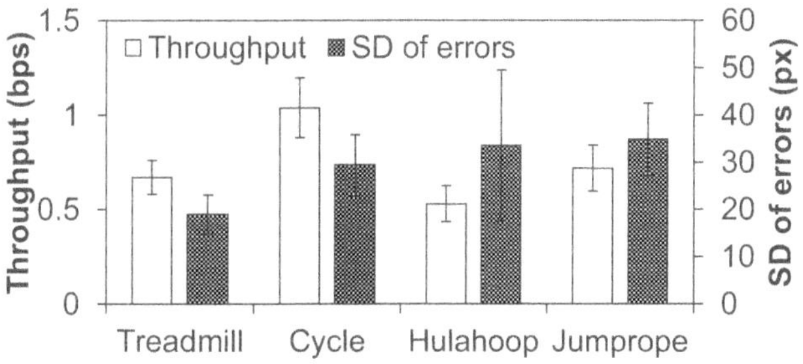

Figure 11: Throughput and SD of errors with 95% confidence intervals

physical conditions that could potentially degrade overall performance. Also, during the intermissions between blocks, the participants were given time for a rest, and beverages were supplied. The total duration of each experiment was around two hours. During the experiment, if participants loose control of the jump rope or the hula hoop, the corresponding trial was discarded and retried. The experiment was a 4×10 within-subjects design. The participants performed a total of 8000 clicks (20 participants $\times$ 10 clicks $\times$ 4 equipments $\times$ 10 blocks).

6.2 Results

6.2.1 Movement Time and the Learning Effect

As mentioned above, the interactive treadmill is equipped with a novel speed controlling mechanism. Given that all the participants had no previous experience with such an interactive treadmill, participants had to get familiar with the tests. Therefore we tried to check whether a learning effect exists. Figure 10 shows the effects of learning (i.e., block) and exercise equipment on movement time. The main effect was significant for equipment ($F_{4,760} = 407.8$, $p < 0.001$). The main effect for block and the equipment by block interaction were not significant.

However, still we can find that there seems to be the learning effect in the case of the treadmill, as shown in Figure 10. We further tested the learning effects for each equipment, and found significant differences on movement time between blocks in the case of the treadmill only ($F_{9,190} = 2.66$, $p < 0.01$). We analyzed the Helmert contrasts by following the work [35], and it showed that the block effect was not significant after block three. Therefore, in the subsequent analysis we used the mean of blocks ranging from four to ten.

The average movement time was $4.60s$ for the treadmill, $2.47s$ for the cycle, $5.19s$ for the hula hoop, and $3.31s$ for the jump rope. Paired t-tests revealed significant differences in movement time across all equipment combinations except between treadmill and hula hoop. The stationary bike was the fastest equipment, and the treadmill and the hula hoop were among the slowest.

6.2.2 Throughput and SD of errors

The throughput was 0.67 for the treadmill, 1.04 for the bike, 0.53 for the hula hoop, and 0.72 for the jump rope. The main effect on the type of equipment was clearly significant ($F_{3,76} = 14.22$, $p < 0.001$). Paired t-tests revealed significant differences in throughput across all equipment combinations except between treadmill and jump rope. However, the characteristics between these two types of equipment are somewhat different; the treadmill shows slower movement time and a lower standard deviation (SD) of errors, whereas the jump rope shows faster movement time and a higher SD of errors.

6.2.3 Movement Direction Effects

We found that there are differences in the measures for the two movement directions: up and down. We tested differences between directions as well as between exercise types using a two-way ANOVA test on the throughput, the movement time and the SD of errors. In overall, a main effect of direction was found, indicating that the throughput of *down* was greater than that of *up* ($F_{1,152} = 19.6$, $p < 0.001$) at the lower SD of errors ($F_{1,152} = 14.79$, $p < 0.001$) and the shorter movement time ($F_{1,152} = 24.87$, $p < 0.001$). This result indicates that increasing exercise speeds is harder than decreasing speeds. For example, in the case of the hula hoop, players should rotate their waist faster to increase the hooping speed, whereas they can easily decrease the speed by simply relaxing. Our balancing scheme on the device heterogeneity copes with these performance differences in movement directions by treating acceleration and deceleration rates separately, as mentioned earlier in Section 4.2.

6.2.4 Gender Effects

We observe the performance difference between male and female participants. We tested whether there exist a gender effect as well as an interaction effect between gender and exercise type using a two-way ANOVA test on the throughput, the movement time and the SD of errors. We found a main effect of gender, which indicates that the female participants achieved significantly greater throughput ($F_{1,72} = 66.56$, $p < 0.001$) at the fairly low errors ($F_{1,72} = 26.71$, $p < 0.001$) and the faster movement time ($F_{1,72} = 6.39$, $p < 0.05$).

Interestingly, there was an interaction between gender and exercise type only on the movement time ($F_{3,72} = 2.22$, $p < 0.1$). Simple effects tests showed that the movement time of the female participants did not show significant differences between exercise types ($F_{3,36} = 2.07$, $p > 0.1$), whereas the male participants did ($F_{3,36} = 3.00$, $p < 0.05$). Actually, we observed that the male participants tend to have larger preferred ranges than those of the female participants. We conjecture that these differences on preferred ranges affect the performance of controlling the game values.

One more interesting observation is that the errors of the female participants on hula hoops are quite lower than those of the male participants. From the open-ended exit interview, we found that about half of the male participants reported that it was unusual to change the speed of hula-hooping, whereas only one female participant reported the same. Recalling gender differences on physical abilities such

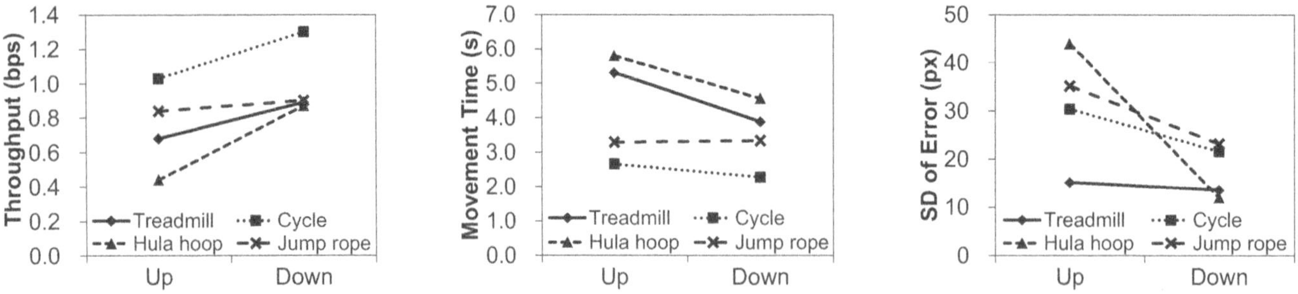

Figure 12: Interaction effects of movement direction and exercise type

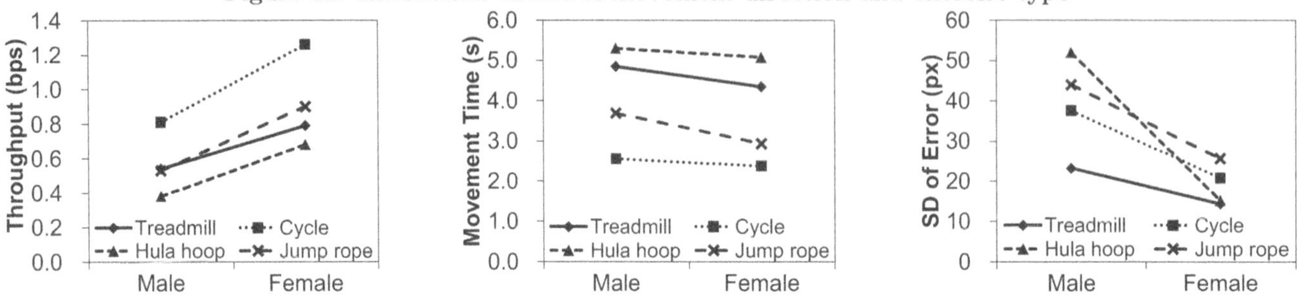

Figure 13: Interaction effects of gender and exercise type

as strength and flexibility [9], it is required to give careful attention on the gender differences. In some cases, it is possible to actively utilize these differences, as the traditional role playing games usually do in their character design, to motivate players to build their own strategy to cope with the differences.

7. CASE STUDIES WITH SWAN BOAT

We conducted two case studies to observe (1) whether our balancing scheme results in a fair game play of Swan Boat (Section 7.1), and (2) patterns of social interactions among remote game players (Section 7.2). In both case studies, we use Swan Boat with the four types of RIA exercise equipment as game controllers. For the remote exercise experiments, we use KT WiBro portable wireless routers (Egg KWI-B2200) for the players with a hula hoop and jump rope.[3]

7.1 Case Study 1: Balancing and Fairness

7.1.1 Method

For this study, we recruited 20 paid participants from the age group 20 to 29, via announcements posted in a local university campus. We assured that none of the participants joined the earlier experiments. The selected participants went through a training session to get familiarized themselves with the exercise equipment. The participants were given an hour for training. The training session was separately held a day before the experiments, to ensure enough time to recover from fatigue. The 20 participants were randomly divided into five groups with the same size, and each group was divided into two teams (each team with two members). The two teams of the same group competed with one another in the boat race.

Experiment setting and procedure: We asked the participants to compete in a race; i.e., the goal is to reach the finish line faster than the opponent team to win the race. The game was played over the 'rally map,' designed for a short fun race and close collaboration between the teammates. It has a winding path, and the path includes many obstacles. In addition, two types of items are randomly distributed along the path; they are boosters which increase the boat speed for 3 seconds, and traps which decreases the boat speed for 3 seconds. The length of a race was set to be about 2 minutes. To see the collaboration and competition effect altogether, we had one team use a heterogeneous pair of the exercise devices and had the other team use a homogeneous pair of exercise devices in order to compare the game performance. We also ensured that all exercise types were used in the games. Thus, we had the competing two teams play a total of four games.

We collected data from questionnaires, interviews, and game statistics including win/loss records, finishing times, etc. We also examined the participants' conversations during the experiment. The participants were asked to answer a short questionnaire at the end of a session. The questionnaire included a set of statements (e.g., the game is fun), and the participants were asked to rate each statement using a five-point Likert scale (5=highly agree). The exit interview was conducted to better understand questionnaire responses.

7.1.2 Results

Table 1 shows the game records of the 20 matches. The results show that players with different RIA controllers can collaborate and compete well while playing the game. A team composed of heterogeneous RIA controllers shows comparable game performance to a team composed of homogeneous RIA controllers. The game statistics clearly show that the game is playable with heterogeneous RIA controllers. The game performance was also comparable between the homogeneous exercise pair and the heterogeneous exercise pair.

The responses from the participants did not explicitly demonstrate the fairness issues between parings of exercise

[3]Egg KWI-B220 supports nation-wide WiBro connectivity (whose performance is comparable with a 4G LTE cellular network).

Table 1: Game records of the 20 matches

	Homogeneous team	Heterogeneous team
Playing time (sec)	86.30	86.35
# of acquired items	3.35	3.40
# of obstacle collisions	3.30	3.30
# of wins	11	9

types. Among the participants, some answered that the exercise type influenced the outcome whereas some did not. 45% agreed to the statement 'it is possible to win the game regardless of exercise types,' and 30% (strongly) disagreed. As for the statement 'other factors, such as exercise or game capability, influence the outcome of the race more than the exercise type,' 55% agreed whereas 40% disagreed.

However, contrary to the reports, we speculate that the exercise type is not a major factor of affecting the outcome. Considering those who disagreed to these statements, we tried to identify whether certain exercise types are commonly pointed out to be advantageous or disadvantageous. Interestingly, we found that the selection of such exercise types varied depending on the participants. It seemed their perception of unfairness is largely induced by the fact that their personal skills in the exercise types vary as well. For example, some participants said the jump rope appeared to be the most advantageous exercise type because it was easy to control the speed. In contrast, there was another participant gave an opposite explanation that it was hard to control the jump roping speed. Conflicting descriptions were made for the hula hoop and stationary cycle as well.

In addition, a majority agreed that teammates' exercise type is not a major factor for the outcome. 65% agreed to the statement 'other factors, such as teammate's exercise capability, have a greater influence than the partners' exercise type,' whereas 20% disagreed. The participants who agreed brought up many factors: for example, the teammate's skill in the exercise type, collaboration ability, and understanding of their exercise type.

7.2 Case Study 2: Social Interactions

7.2.1 Method

For this case study, a total of 16 paid students were recruited on campus from the age group of 18 to 33. Participants were carefully selected such that there is no participant overlapping with those in the previous case study. In our advertisements, we encouraged the participants to join the study with people whom they were familiar or friends with, because the objective of this case study is to observe social interactions among participants. We had four groups, and each group was divided into two teams of two members to allow competition. The participants went through a training session for an hour in order to familiarized themselves with the exercise equipment, and the actual experiment was conducted the day after the training session.

Experiment setting and procedure: The game setup was similar to that of Section 7.1. This case study involves two sets of experiments representing two different configuration in terms of players' location (i.e., co-located vs. remote play). In the co-located configuration, four exercise devices were placed at the same place, where participants can freely speak to each other. In the remote configuration, each exercise device was placed at an isolated place in the same building such that participants can only communicate via VoIP with Bluetooth headsets. In each set of experiments, participants played a total of four games (i.e., overall eight games were played). Each group performed two set of experiments in two distinct days, and the sequence of the experiments are randomized to mitigate the learning effect. We performed content analysis of participants' conversations during the game play, inspired by previous work on social interactions in online game play [17], and exit interview results to investigate the patterns of social interactions under different experiment configurations.

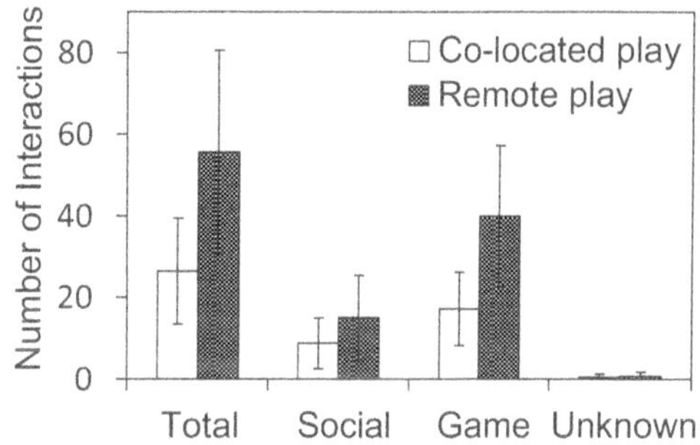

Figure 14: Number of comments between two configurations with 95% confidence intervals. *Total* represents average number of total comments conducted. *Social* and *Game* represent the average number of social and game-instructing comments, respectively. Note that *Undefined* indicates undefined comments that we could not interpret due to background noise.

7.2.2 Results

Our findings show that ExerLink can effectively facilitate social interactions in a remote environment similar to that in a co-located environment. Participants commented on their experiences with our remote social exergame compared to the co-located exergame as follows: *"I didn't feel any inconvenience in communication among people"* [P8], *"When played remotely, my teammate and I could control the boat as if we were in a same room. I could even hear what other teams were saying and use it for our game strategy"* [P4]. In summary, 14 out of 16 participants commented that they did not feel uncomfortable when communicating each other in a remote environment. The results of content analysis on the participants' conversations during the experiments, shown in Figure 14, also support these comments. It even shows that our system increases the average number of conversational exchanges in a remote environment, i.e, the difference on the total number of comments between two configuration is statistically significant ($t(21) = -3.98$, $p < 0.001$). Note that these numbers only include those comments that bring something meaningful during the game. For example, in Figure 14, *Game* represents the average number of comments affecting the game performances controlling the speed or direction of the swan boat (e.g., "go left," "slow down," etc). *Social* includes emotional expression, encouragements or other social conversation that does not affect the game.

It was also interesting to observe that participants tend to focus more on the game in a remote environment: *"When we played remotely, I could concentrate more on the game. I actually felt like I was playing Mario Cart"* [P1], *"When we were playing the game in the same room, I felt that I was disturbed when other participants created noise during the game. In contrast, I could fully concentrate on the game at a separate room"* [P11]. Figure 14 shows that the number of the game-instructing comments among the total number of comments increased when participants were playing the game remotely. The difference on the number of the game-instructing comments between two configuration is statistically significant ($t(21) = -4.49$, $p < 0.001$). It supports the above comments that subjects participated in the game more actively.

7.2.3 Potential of ExerLink as a social medium

We found that our system has potential to be a socializing medium. After each experiments, we conducted an exit survey, and an open-ended interview to understand the individual's willingness to use our system and its reason. Most of the participants appreciated the system as an excellent conduit for social experiences while 'playing together,' even in a remote play situation. 14 out of 16 participants (strongly) agreed to the statement 'I am willing to use this system to socialize with acquaintances and friends.' One participant stated: *"I think I can use this system to socialize with others, have fun, and exercise at the same time"* [P5].

Also, the participants liked the socializing feature of the system; they strongly agreed that the game play improved forming bonds between players. One participant stated: *"Before the game play, I and my teammate were not so familiar with each other. However, after playing just two games, I realized that we were chatting together as if we were close friends"* [P8].

In addition, we found that game play with strangers is acceptable. 8 out of 16 participants agreed to the statement 'I will virtually meet and play the game with strangers online,' whereas 3 disagreed. One participant stated: *"I think that it will be not so different to play StarCraft or League of Legends with strangers. If none of my friends are available to play together, I will find my teammates online"* [P13].

8. DISCUSSION

The focus of this study has been on inter-controller differences rather than inter-player differences. To that end, we conducted our study with a tester population of males and females mostly in their 20s. However, we note that there will be differences across age groups. We need to respect these differences and carefully study them to further develop games for broader populations.

Our evaluation results on controller performance are mostly obtained through short-term studies, leaving further studies on human factors to be conducted in a longitudinal study in the near future. We expect that it would be an interesting topic to study the effects of the controller performances as the players get accustomed to using the RIA controllers.

We have suggested guidelines on balanced game design and reported our experiences from case studies to evaluate the effectiveness of the balanced game design. We expect that a certain pattern would be derivable after multiple balancing cases have been accumulated.

This work has studied the effects of our custom-designed exercise equipment in the viewpoint of controlling exergames, yet leaving the effects of actual exercise amount unexplored. While the players enjoy the game as we suggested, it is an important issue to study the amount of exercise involved and the effectiveness of that exercise. Over-exertion, a potential problem which might occur due to the game, is also a concern, requiring careful observation in terms of exercise intensity and duration.

We did not thoroughly investigate the impact of network delay in this work since our focus has been demonstrating the feasibility of multi-exercise social games. During our field trials, we also tried to play Swan Boat over a 3G celluar networks and fortunately did not find any noticeable disruptions thanks to the proposed mechanisms in ExerLink. We are currently investigating the user experience under different network delay/jitter conditions using Dummynet [1].

9. RELATED WORK

As sensor and device technology advances, creating new kinds of exergames has been an active area of research in recent years. Yet existing exergames mostly focused on a single exercise modality and did not consider combining heterogeneous exercises into the same game. Exercise equipment has been revised or newly developed as an exergame such as an arm ergometer [19], an interactive treadmill [6], a spirometer [27], a tangible ball [23], and playful gadgets [11, 12, 30]. Exergame Fitness Co. provides several types of exergame controllers including game cycles, interactive floor and wall systems [3]. Several exergames [15, 40, 43, 45] used heart rate signals for a multi-player game. Mueller et al. recently developed exergames called "sports over distance" that directly use players' physical actions for a remote sport play (e.g., Jogging Over a Distance, Remote Impact, Table Tennis for Three) [37, 39, 40]. Readers can find more information about recent advances in exergames and their design principles and guidelines in the following articles [10, 38, 44]. Note that cellular phones can be also used as game controllers to play a remote game. AirPlay gamifed live TV sports games and shows such that the TV audience can participate in the game using their cell phones while watching TVs [31]; e.g., in "Call the Play Football," people predict how upcoming plays will unfold and those participants who correctly guessed are rewarded. Beyond a means of communications, we expect that recent smartphones with various internal sensors will create new opportunities for developing exergame controllers (e.g., monitoring physical activity levels with accelerometers).

So far little work has been done on assessing the performance or usability of exergame controllers on actual game play. In user interface literature [34, 33, 35], a pointing device's performance (e.g., mouse) is typically measured using the metric called throughput [22]. In addition, some works assessed the performance of game controllers in terms of pointing accuracy and rapidness [26, 41]. In the viewpoint of games, Natapov et al. compared the performance of computer-games when different game controllers are used, namely an Xbox gamepad and a standard PC mouse [42]. The main departure from existing work is that we consider exergame controllers and aim to address heterogeneity of exergame controllers in a multi-player social exergame.

A number of techniques have been proposed for balancing game play between players with disparate abilities, as summarized in the work by Stach et al. [45]. Traditional sports and games (e.g., golf, bowling and chess) used techniques such as handicapping, ladders, and asymmetric roles. These methods are also widely used in computer games in the form of prior simulations, player monitoring [8], and autonomous advantages [20]. Balancing enables inter-generational game play in which players from different generational groups can interact with each other in the same game [46, 23, 24, 44, 47]. For exergames, several works attempted to balance the game play among players with different physical abilities, by sharing heart rate information [15, 40, 43, 45].

There exist several cases that employ different types of game controllers. Rock Band and Guitar Hero successfully incorporate several types of musical-instrument-like controllers to create a metaphor of playing a rock concert. Time Crisis 2 permits the use of the game pad in PlayStation 2 as well as a custom-designed gun-type controller. Age Invaders [24] incorporates heterogeneous types of game controllers, i.e., the floor platform and conventional PC interfaces. However, these games do not address the balancing issues when heterogeneous controllers are used for collaboration and competition.

To our knowledge, there are only a few cases that incorporate multiple types of game controllers in a balanced manner. One impressive effort—in terms of emphasizing the importance of balancing between game controllers in the game development process—is Shadowrun, a multiplayer tactical first-person shooter that provides two different controlling methods, i.e., an Xbox gamepad and a mouse with a keyboard. The developers of this game balanced the disparities by providing game-side advantages or penalties for each controller [2]. In fact, this work gave us one important and practical lesson when creating and evaluating game controllers: *'If the player is blaming the controls for their failures, then we have done something wrong.'*

Exercise heterogeneity in exergames can be addressed by using the exercise intensity information, which can be predicted via the player's physiological status or can be directly measured via modified equipment. Notably, the heart rate is a good physiology metric to use, as it is typically proportional to the exercise intensities regardless of an exercise type [15, 32, 40, 43, 45]. However, a game controller harnessing such a mediated metric is limited in its responsiveness, compared to the one that directly measures the intensity level from exercise equipment. In practice, it takes tens of seconds for a change in the player's exercise intensity to be reflected in her heart rate [32, 45]. Accordingly, heart rate-based controllers cannot effectively deal with fast-paced games in which the players have to respond to changing game conditions within a matter of seconds. In our work, we revamped existing exercise devices to directly measure the intensity level and addressed the heterogeneity problem with a personalized inter-device balancing scheme.

Our framework allows the game designers and developers to easily employ multiple types of controllers at their discretion, thereby delivering more fun—players can choose their favorite equipment through which collaboration and competition can happen in a fair manner. We envision this kind of game development will soon be established in the near future.

10. CONCLUSION

We proposed ExerLink to explore the possibility of using heterogeneous exercise devices as game controllers and to realize pervasive social exergames that transforms isolated disparate exercise activities into a fun collaborative activity. When converting exercise intensity to game input, ExerLink aims to provide fair game play experiences by carefully considering the unique aspects of social exergames such as heterogeneity of exercise devices and personal differences of capabilities and preferences. We designed and implemented four exergame controllers by significantly augmenting existing exercise devices and performed preliminary human subject studies to evaluate the performance of exergame controllers and the user experience of an exergame. Our study results show that players with different exercises can effectively collaborate and compete well while playing the same game. In remote exergames, we found that missing visual cues of other players did not make any significant impact on the game play experiences, and yet, players were more engaged in both conversation and exercise as opposed to co-located exergames.

11. ACKNOWLEDGMENTS

We thank all of the subjects who participated in this study. Special thanks to our shepherd, Dr. Mirjana Spasojevic, and all the anonymous reviewers for their great reviews. This work was supported by the National Research Foundation of Korea grant (No. 2011-0018120) funded by the Korean Government (MEST), and the SW Computing R&D Program of KEIT (2011-10041313, UX-oriented Mobile SW Platform) funded by the Ministry of Knowledge Economy of Korea.

12. REFERENCES

[1] Dummynet, http://info.iet.unipi.it/~luigi/dummynet, 2009.

[2] Controller Balancing, http://replay.web.archive.org/20071011062350/http://shadowrun.com/backstage/controller_balancing.htm, 2012.

[3] Exergame Fitness, http://www.exergamefitness.com, 2012.

[4] Nike + iPod, http://www.apple.com/ipod/nike, 2012.

[5] Wii Sports, http://wiisports.nintendo.com, 2012.

[6] M. Ahn, S. Kwon, B. Park, K. Cho, S. Choe, I. Hwang, H. Jang, J. Park, Y. Rhee, and J. Song. Running or gaming. In *Proceedings of the International Conference on Advances in Computer Enterntainment Technology*, pages 345–348. ACM, 2009.

[7] B. Ainsworth, W. Haskell, M. Whitt, M. Irwin, A. Swartz, S. Strath, W. O'BRIEN, D. Bassett, K. Schmitz, P. Emplaincourt, et al. Compendium of physical activities: an update of activity codes and met intensities. *Medicine & Science in Sports & Exercise*, 32(9):S498, 2000.

[8] T. Alexander. *Massively Multiplayer Game Development.* Charles River Media, 2003.

[9] M. Alter. *Science of flexibility.* Human Kinetics Publishers, 2004.

[10] T. Campbell, B. Ngo, and J. Fogarty. Game design principles in everyday fitness applications. In *Proceedings of the ACM conference on Computer supported cooperative work*, pages 249–252. ACM, 2008.

[11] Y. Chang, J. Lo, C. Huang, N. Hsu, H. Chu, H. Wang, P. Chi, and Y. Hsieh. Playful toothbrush: ubicomp technology for teaching tooth brushing to kindergarten children. In *Proceedings of the 26th annual sigchi conference on human factors in computing systems*, pages 363–372. ACM, 2008.

[12] M. Chiu, S. Chang, Y. Chang, H. Chu, C. Chen, F. Hsiao, and J. Ko. Playful bottle: a mobile social persuasion system to motivate healthy water intake. In *Proceedings of the 11th international conference on Ubiquitous computing*, pages 185–194. ACM, 2009.

[13] M. Claypool and K. Claypool. Latency and player actions in online games. *Communications of the ACM*, 49(11):40–45, 2006.
[14] S. Consolvo, K. Everitt, I. Smith, and J. Landay. Design requirements for technologies that encourage physical activity. In *Proceedings of the SIGCHI conference on Human Factors in computing systems*, pages 457–466. ACM, 2006.
[15] R. de Oliveira and N. Oliver. Triplebeat: enhancing exercise performance with persuasion. In *Proceedings of the 10th international conference on Human computer interaction with mobile devices and services*, pages 255–264. ACM, 2008.
[16] C. Diot and L. Gautier. A distributed architecture for multiplayer interactive applications on the internet. *Network, IEEE*, 13(4):6–15, 1999.
[17] N. Ducheneaut and R. Moore. The social side of gaming: a study of interaction patterns in a massively multiplayer online game. In *Proceedings of the 2004 ACM conference on Computer supported cooperative work*, pages 360–369. ACM, 2004.
[18] K. Guo, S. Mukherjee, S. Rangarajan, and S. Paul. A fair message exchange framework for distributed multi-player games. In *Proceedings of the 2nd workshop on Network and system support for games*, pages 29–41. ACM, 2003.
[19] S. Guo, G. Grindle, E. Authier, R. Cooper, S. Fitzgerald, A. Kelleher, and R. Cooper. Development and qualitative assessment of the gamecycle exercise system. *IEEE Transactions on Neural Systems and Rehabilitation Engineering*, 14(1):83–90, 2006.
[20] J. Habgood and M. Overmars. *The Game Maker's Apprentice: Game Development for Beginnerst*. Apress, 2006.
[21] W. Haskell, I. Lee, R. Pate, K. Powell, S. Blair, B. Franklin, C. Macera, G. Heath, P. Thompson, and A. Bauman. Physical activity and public health: updated recommendation for adults from the american college of sports medicine and the american heart association. *Medicine & Science in Sports & Exercise*, 39(8):1423, 2007.
[22] ISO 9241-1. *Ergonomic requirements for office work with visual display terminals (VDTs) - Part 9: Requirements for non-keyboard input devices*. International Organization for Standardization, 2000.
[23] D. Kern, M. Stringer, G. Fitzpatrick, and A. Schmidt. Curball–a prototype tangible game for inter-generational play. In *Workshops on Enabling Technologies: Infrastructure for Collaborative Enterprises*, pages 412–418. IEEE, 2006.
[24] E. Khoo, T. Merritt, and A. Cheok. Designing physical and social intergenerational family entertainment. *Interacting with computers*, 21(1-2):76–87, 2009.
[25] S. Kim. Trackball device. *US Patent*, 4404865, 1983.
[26] C. Klochek and I. MacKenzie. Performance measures of game controllers in a three-dimensional environment. In *Proceedings of Graphics Interface 2006*, pages 73–79. Canadian Information Processing Society, 2006.
[27] B. Lange, S. Flynn, A. Rizzo, M. Bolas, M. Silverman, and A. Huerta. Breath: a game to motivate the compliance of postoperative breathing exercises. In *International Conference on Virtual Rehabilitation, 2009*, pages 94–97. IEEE, 2009.
[28] J. Lin, L. Mamykina, S. Lindtner, G. Delajoux, and H. Strub. Fish'n'steps: Encouraging physical activity with an interactive computer game. *UbiComp 2006: Ubiquitous Computing*, pages 261–278, 2006.
[29] Y. Lin, K. Guo, and S. Paul. Sync-ms: Synchronized messaging service for real-time multi-player distributed games. In *10th IEEE International Conference on Network Protocols*, pages 155–164. IEEE, 2002.
[30] J. Lo, T. Lin, H. Chu, H. Chou, J. Chen, J. Hsu, and P. Huang. Playful tray: adopting ubicomp and persuasive techniques into play-based occupational therapy for reducing poor eating behavior in young children. *UbiComp 2007: Ubiquitous Computing*, pages 38–55, 2007.
[31] D. B. Lockton. Live television show utilizing real-time input from a viewing audience. *US Patent*, 0028272, 2007.
[32] B. Maciel, L. Gallo, J. Marin Neto, and L. Martins. Autonomic nervous control of the heart rate during isometric exercise in normal man. *Pflügers Archiv European Journal of Physiology*, 408(2):173–177, 1987.
[33] I. MacKenzie. Fitts' law as a research and design tool in human-computer interaction. *Human-computer interaction*, 7(1):91–139, 1992.
[34] I. MacKenzie and P. Isokoski. Fitts' throughput and the speed-accuracy tradeoff. In *Proceedings of the 26th annual SIGCHI conference on Human factors in computing systems*, pages 1633–1636. ACM, 2008.
[35] I. MacKenzie, T. Kauppinen, and M. Silfverberg. Accuracy measures for evaluating computer pointing devices. In *Proceedings of the SIGCHI conference on Human factors in computing systems*, pages 9–16. ACM, 2001.
[36] J. Manweiler, S. Agarwal, M. Zhang, R. Roy Choudhury, and P. Bahl. Switchboard: a matchmaking system for multiplayer mobile games. In *Proceedings of the 9th international conference on Mobile systems, applications, and services*, pages 71–84. ACM, 2011.
[37] F. Mueller and S. Agamanolis. Sports over a distance. *Computers in Entertainment (CIE)*, 3(3):4–4, 2005.
[38] F. Mueller, M. Gibbs, and F. Vetere. Taxonomy of exertion games. In *Proceedings of the 20th Australasian Conference on Computer-Human Interaction: Designing for Habitus and Habitat*, pages 263–266. ACM, 2008.
[39] F. Mueller, M. Gibbs, and F. Vetere. Design influence on social play in distributed exertion games. In *Proceedings of the 27th international conference on Human factors in computing systems*, pages 1539–1548. ACM, 2009.
[40] F. Mueller, F. Vetere, M. Gibbs, D. Edge, S. Agamanolis, and J. Sheridan. Jogging over a distance between europe and australia. In *Proceedings of the 23nd annual ACM symposium on User interface software and technology*, pages 189–198. ACM, 2010.
[41] D. Natapov, S. Castellucci, and I. MacKenzie. Iso 9241-9 evaluation of video game controllers. In *Proceedings of Graphics Interface 2009*, pages 223–230. Canadian Information Processing Society, 2009.
[42] D. Natapov and I. MacKenzie. Gameplay evaluation of the trackball controller. In *Proceedings of the Conference on the Future of Game Design and Technology*, pages 167–174. ACM, 2010.
[43] V. Nenonen, A. Lindblad, V. Häkkinen, T. Laitinen, M. Jouhtio, and P. Hämäläinen. Using heart rate to control an interactive game. In *Proceedings of the SIGCHI conference on Human factors in computing systems*, pages 853–856. ACM, 2007.
[44] J. Sinclair, P. Hingston, and M. Masek. Considerations for the design of exergames. In *Proceedings of the 5th international conference on Computer graphics and interactive techniques in Australia and Southeast Asia*, pages 289–295. ACM, 2007.
[45] T. Stach, T. Graham, J. Yim, and R. Rhodes. Heart rate control of exercise video games. In *Proceedings of Graphics interface 2009*, pages 125–132. Canadian Information Processing Society, 2009.
[46] V. Vanden Abeele and B. Schutter. Designing intergenerational play via enactive interaction, competition and acceleration. *Personal and Ubiquitous Computing*, 14(5):425–433, 2010.
[47] F. Vetere, H. Davis, M. Gibbs, P. Francis, and S. Howard. A magic box for understanding intergenerational play. In *CHI'06 extended abstracts on Human factors in computing systems*, pages 1475–1480. ACM, 2006.

ACE: Exploiting Correlation for Energy-Efficient and Continuous Context Sensing

Suman Nath
Microsoft Research
sumann@microsoft.com

ABSTRACT

We propose ACE (Acquisitional Context Engine), a middleware that supports continuous context-aware applications while mitigating sensing costs for inferring contexts. ACE provides user's current context to applications running on it. In addition, it dynamically learns relationships among various context attributes (e.g., whenever the user is `Driving`, he is not `AtHome`). ACE exploits these automatically learned relationships for two powerful optimizations. The first is *inference caching* that allows ACE to opportunistically infer one context attribute (`AtHome`) from another already-known attribute (`Driving`), without acquiring any sensor data. The second optimization is *speculative sensing* that enables ACE to occasionally infer the value of an expensive attribute (e.g., `AtHome`) by sensing cheaper attributes (e.g., `Driving`). Our experiments with two real context traces of 105 people and a Windows Phone prototype show that ACE can reduce sensing costs of three context-aware applications by about 4.2×, compared to a raw sensor data cache shared across applications, with a very small memory and processing overhead.

Categories and Subject Descriptors

C.5.3 [**Microcomputers**]: Portable devices; K.8 [**Personal Computing**]: General

General Terms

Design, Experimentation, Measurement, Performance

Keywords

Continuous context-aware applications, Sensor-rich mobile environment, Rule-based optimizations

1. INTRODUCTION

The increasing availability of sensors integrated in smartphones provides new opportunities for continuous context-aware applications that react based on the operating conditions of users and the surrounding environment [12, 14, 17, 18, 19, 22]. Examples of such applications include a *location-based reminder* that reminds a user of something when he is at a predefined location (e.g., GeoNote for Android phones), a *jogging tracker* that keeps track of how much exercise the user does in a day (e.g., JogALot for iPhones), and a *phone buddy* that mutes the phone when the user is in a meeting (e.g., RingWise for Android phones) or plays a customized greeting message when he is driving. These applications use context information inferred from various sensors such as GPS, accelerometer, and microphone of the mobile phone.

A big overhead of running context-aware applications is their high energy consumption [15, 17]. Many of these applications require monitoring user's context *continuously*. Even some applications that require context information occasionally may need continuous or frequent context monitoring due to relatively high context detection latency. For example, the phone buddy application may need to know if a user is in a meeting to mute an incoming call. However, without continuously (or frequently) monitoring user's context, it may not be able to quickly infer whether the user is in a meeting when a call arrives. If not designed carefully, the cost of continuous context sensing can be prohibitively high, especially when applications require expensive attributes such as location and multiple such applications run concurrently.

Prior work has proposed various optimizations for energy-efficient continuous context sensing [12, 13, 15, 17, 18]. These are orthogonal to, and can be used with, our work. Prior work also proposed exploiting redundancy *across* applications—by sharing sensor data and inferred context attributes among applications through a shared cache [10]. However, as we will show, this alone can still be costly if applications involve expensive sensors such as GPS and microphone.

Our Contributions. We propose ACE (Acquisitional Context Engine), a middleware that supports continuous context-aware applications while mitigating sensing cost for acquisition of necessary context attributes. ACE exposes an extensible collection of context attributes such as `AtHome` and `IsDriving` (more examples later). Applications interact with ACE by requesting the current values of these attributes. In providing context information to applications, ACE can achieve significant energy savings—in our experiments with three applications and 105 real users' context traces, ACE reduces sensing costs by about 4.2× compared to a shared cache.

How does ACE get such a high energy savings? The key observation we exploit is that *human contexts are limited by physical constraints* and hence values of various context

MobiSys'12, June 25–29, 2012, Low Wood Bay, Lake District, UK.

attributes can be highly correlated. If we can find perfect relationships among various context attributes (e.g., there exists a one-to-one mapping from one set of attribute values to another), we can infer an unknown context attribute from a known context attribute or from an unknown but cheaper context attribute, reducing the total energy consumption of context acquisition. We illustrate this with an example.

Suppose Bob is running multiple applications on ACE including the following two: *App1* that uses accelerometer to monitor Bob's transportation mode (`Walking` vs. `Running` vs. `Driving`) and *App2* that uses location sensors (GPS, WiFi, and Bluetooth) to monitor whether he is `AtHome` or `InOffice`. Suppose ACE, by analyzing Bob's context history, discovers various *context rules* such as "When `Driving` is `True`, `AtHome` is `False`."[1] We will give more examples of context rules later; for now we assume rules that naturally make sense. ACE exploits these context rules in two ways.

Inference-based Intelligent Context Caching. Suppose at 10am, App1 asks ACE for the value of `Driving`. ACE determines the value of the attribute and puts it in cache. If, within a short time window, another application asks for the value of `Driving`, ACE can return the value like a traditional cache. In addition, if App2 asks for the value of `AtHome` within a short time window, and if the cached value of `Driving` is `True`, ACE can also return the value `AtHome` = `False`. ACE can do this by exploiting the negative correlation between the attributes `Driving` and `AtHome` and *without acquiring any sensor data.* Thus, ACE can reuse the cached value of `Driving` not just for an exact lookup, but also as a *proxy* for a *semantically related* lookup.

Speculation-based Cheaper Sensing Plan. Context rules allow finding the value of an attribute in many different ways. For example, the value of `InOffice` can be determined by acquiring necessary sensor data (e.g., GPS, WiFi signatures). Moreover, by exploiting context rules, it can also be answered `False` if `Driving` = `True` or `Running` = `True` or `AtHome` = `True`; and can be answered `True` if `InMeeting` = `True`. Each of these proxy attributes can further be determined through additional proxy attributes. This gives ACE the flexibility to choose, and speculatively sense values of, proxies that yield the value of the target attribute with the lowest cost. For example, if `Running` is the cheapest attribute and if there is a high probability that the user is jogging now, ACE will use `Running` as a proxy, and speculatively determine its value, to determine the value of `InOffice` (by using the rule that whenever `Running` is `True`, `InOffice` is `False`). If the value of `Running` is indeed `True`, ACE can infer the value of `InOffice` to be `False`, without using expensive location sensors. If `Running` is found to be `False`, however, ACE cannot make such a conclusion and needs to proceed with the next best proxy attribute.

The benefit of the above *speculative sensing* idea may seem counter-intuitive— evaluating an attribute can become cheaper by evaluating additional attributes. However, as we will show later, if such additional attributes are low-cost and are most likely to uniquely determine the value of the target attribute, such additional attribute can reduce the total *expected* cost of the system.

The above ideas are powerful in practice and can be generally applicable to a broader class of applications. However, realizing them requires addressing several challenges. First, how can the system automatically and efficiently learn context rules that can help finding cheap proxy attributes? Second, how can the system find suitable proxy attributes in the cache that can determine a value of the target attribute? Third, how can it determine the best sequence of proxy attributes to speculatively sense in order to determine the value of a target attribute? All these challenges need to be addressed without much overhead on the phone and with the goal of reducing the overall energy consumption. In the rest of the paper, we describe how ACE achieves this goal.

Intuitively, the energy savings in ACE comes from sensing redundancy across applications and context attributes, well-defined relationship among attributes, and the possibility of inferring expensive attributes through cheaper attributes. The savings depend on application mix and user habits. *How common is such redundancy in practice?* To understand this, we have evaluated ACE with three applications and two real datasets containing context traces of total 105 users. Our results show that ACE consumes around 4.2× less energy than consumed by a raw sensor data cache shared across applications. Our ACE prototype on a Samsung Focus phone running Windows Phone 7.5 OS shows that ACE imposes little overhead ($<$ 0.1 ms latency and $<$ 15 MB RAM footprint).

Note that inference caching and speculative sensing based on rules can sometimes produce results that are (a) incorrect, if user's behavior deviates from context rules, or (b) stale, if results are inferred from stale values in cache, and hence incorrect if cached values no longer hold. ACE employs several mechanisms to reduce such inaccuracy. Experiments show that such inaccuracy is $<$ 4% for our datasets and is small compared to classification errors in typical context inference algorithms [13, 15, 16]. If an application cannot tolerate such small additional inaccuracy, it can request ACE to get true context values by collecting sensor data (and paying high energy cost).

In summary, we make the following contributions.

- We describe design and implementation of ACE, a middleware for efficient and continuous sensing of user's context in a mobile phone (§ 2).
- We describe how ACE efficiently learns rules from user's context history (§ 3), and use the rules for intelligent context caching (§ 4) and speculative sensing (§ 5).
- By using a prototype on a real phone and two datasets containing continuous context traces of 105 users, we show that ACE can significantly reduce sensing cost with a minimal runtime overhead (§ 7).

2. ACE OVERVIEW

2.1 ACE Architecture

Figure 1 shows a high level architecture of ACE. It works as a middleware library between context-aware applications and sensors. An application interacts with ACE with a single API *Get*(X), asking for the current value of a context attribute X. ACE consists of the following components.

(1) Contexters. This is a collection of modules, each of which determines the current value of a context attribute (e.g., `IsWalking`) by acquiring data from necessary sensors and by using the necessary inference algorithm. An appli-

[1]Currently ACE uses simple association rules. However, the core ideas of ACE can be used with more general rules such as temporal or probabilistic rules.

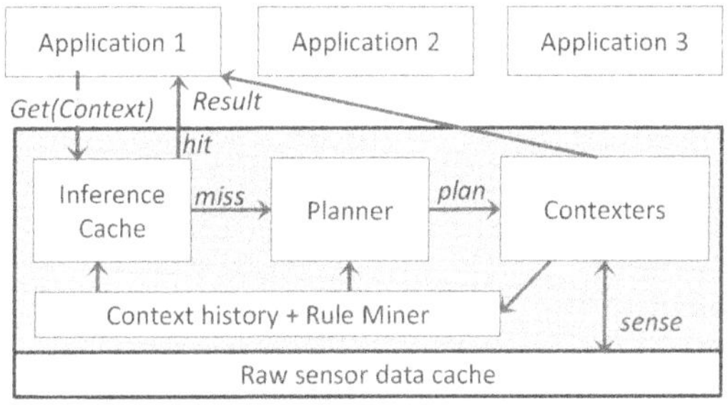

Figure 1: ACE Architecture

cation can extend it by implementing additional contexters. The set of attributes sensed by various contexters is exposed to applications for using with the *Get*() call.[2]

(2) Raw sensor data cache. This is a standard cache.

(3) Rule Miner. This module maintains user's context history and automatically learns *context rules*—relationships among various context attributes—from the history.

(4) Inference Cache. This implements the intelligent caching behavior mentioned before. It provides the same Get/Put interface as a traditional cache. However, on $Get(X)$, it returns the value of X not only if the value of X is in the cache, but also if it can be inferred by using context rules and cached values of other attributes.

(5) Sensing Planner. On a cache miss, this finds the sequence of proxy attributes to speculatively sense to determine the value of the target attribute in the cheapest way.

The last three components form the core of ACE. We will describe them in more detail later in the paper.

Importance of Rule Miner. One might argue that relationships among various context attributes can be hard coded within applications. Rule Miner automates this process. Moreover, it can learn additional rules that are not-so-obvious to developers, can change over time, are specific to individual users, and involve attributes from different applications (and hence developer of one application may not find them). Our experiments show that the energy savings of ACE would have reduced by more than half had it used a set of static rules for all users, instead of dynamically learning personalized rules for each user.

For simplicity, we assume Boolean context attributes. Even though our techniques can be generalized to real-valued attributes, we believe that Boolean attributes are sufficient for a large collection of context-aware applications as they react based on Boolean states of arbitrarily complex context attributes (e.g., mute the phone when in a meeting or in a movie theater).

2.2 Contexters

We have implemented a number of contexters for context attributes shown in Table 1. The table shows attribute names and the shorthand we use for them for brevity. We denote an attribute and its value such as `IsDriving` = `True` as a *tuple*, and often write it in shorthand as `D` = `T`.

Table 1 also shows the sensors various contexters use. The contexters for attributes `W`, `D`, `J`, and `S` collect accelerome-

[2]ACE also exposes raw sensor data for applications that need additional context attributes and more accuracy than is provided by ACE.

Table 1: Context attributes implemented in ACE

Attribute	Short	Sensors used (sample length)	Energy (mJ)
IsWalking	W	Accelerometer (10 sec)	259
IsDriving	D	Accelerometer (10 sec)	259
IsJogging	J	Accelerometer (10 sec)	259
IsSitting	S	Accelerometer (10 sec)	259
AtHome	H	WiFi	605
InOffice	O	WiFi	605
IsIndoor	I	GPS + WiFi	1985
IsAlone	A	Microphone (10 sec)	2895
InMeeting	M	WiFi + Microphone (10 sec)	3505
IsWorking	R	WiFi + Microphone (10 sec)	3505

ter data over a 10-second window and derive values of the attributes based on a decision tree learned from training data [13]. The values of **AtHome** and **InOffice** are determined by matching the current WiFi signature with a pre-learned dictionary. The contexter for **IsIndoor** uses GPS location and WiFi signature; i.e., the user is assumed to be indoor if GPS signal is not available or current WiFi access points are in a pre-learned dictionary. The value of **IsAlone** is determined by collecting an audio sample of 10 seconds and comparing the average signal strength with a pre-learned threshold [16]. Finally, the values of **InMeeting** and **IsWorking** are determined by using signal strength of surrounding sound and fine-grained location based on WiFi signature and acoustic background signature [25].

We have measured energy consumed by various contexters on an Android HTC desire phone (shown in Table 1). The key observation, which we expect to hold on other platforms and contexter implementations as well, is that energy consumptions of various contexters vary by an order of magnitude. This alludes to the potential savings in ACE when an expensive attribute can be inferred from a cheap attribute.

Extensibility. Note that implementation of various contexters is not at the core of this work, and one can think of better implementations. ACE treats contexters as black boxes; the only two pieces of information a contexter needs to expose to ACE are (1) the name of the attribute that it determines, and (2) its energy cost. Therefore, implementation details of contexters do not affect the rest of ACE. One can implement a new contexter or replace an existing contexter with a more accurate and energy-efficient one, and can expose the attribute name and energy cost to ACE through a configuration file; ACE will seamlessly incorporate the new contexter.

This highlights another useful feature of ACE: it does not require applications to agree on semantic meaning of context labels. Applications can simply register new contexters that other applications are unaware of, but they still benefit indirectly through correlations that ACE might detect and exploit.

2.3 Work flow of ACE

Figure 2 shows the next level of details of ACE's architecture to show how ACE works end-to-end. Applications interact with ACE by requesting the current value of a context attribute X, by calling the API $Get(X)$. If an unexpired value of X (computed previously for the same or a different application) is found in Inference Cache, ACE returns the value. Otherwise, ACE uses the current set of context rules (learned by Rule Miner) and currently unexpired tuples in the cache to see if a value of X can be inferred. If so, ACE returns the inferred value of X. Otherwise, ACE needs to

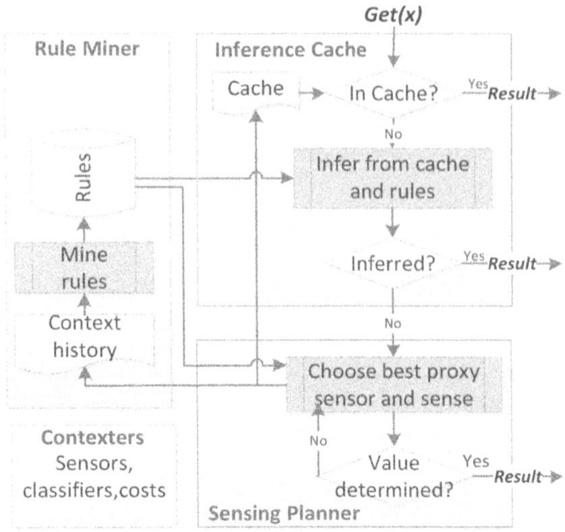

Figure 2: Workflow in ACE

invoke necessary contexters to determine the value of X. To explore the possibility of determining the value of X through cheaper proxy attributes, ACE invokes the Sensing Planner module. It repeatedly chooses the next best proxy attribute and speculatively runs its contexter until the value of X is determined. The value of X is finally put in the cache (with an expiration time), added to context history (for being analyzed by Rule Miner), and returned to application.

Reducing Inaccuracies. ACE uses several mechanisms to reduce possible inaccuracies in results, as mentioned in Section 1. First, ACE conservatively uses rules that almost always hold for a user. Second, the rules are dynamically updated, to capture any change in user's habit. Third, it uses a small (5 minutes) expiration time of tuples in the Inference Cache so that inference is done only based on fresh tuples. Fourth, it occasionally cross-validates results returned by Inference Cache and Sensing Planner with true contexts determined by running contexters. The default cross-validation probability is 0.05—our experiments show that this gives a good balance between energy and accuracy. Finally, an application can optionally provide feedback (e.g., got from the user) to ACE about inaccuracies in ACE results. ACE conservatively removes any rule inconstant with ground truths.

2.4 Applications

We target context-aware applications that need to monitor user's context continuously or frequently. Different applications may require different context attributes exposed by ACE such as user's location (home/office), transportation mode (walking/driving), work state (working/meeting), and group state (alone/in group). For concreteness, we consider the following three applications we have built in ACE.

GeoReminder. Like GeoNote for Android, it continuously monitors a user's current location and displays custom messages when he is at a pre-defined location. The location is computed based on GPS data and WiFi access point signature (using similar techniques as [15]).

JogTracker. Like JogALot for iPhone, it monitors a user's transportation mode (e.g., `IsWalking`, `IsRunning`, `IsDriving`) and keeps track of how much calories he burns in a day from walking and running.

PhoneBuddy. Like RingWise on Android, it monitors whether a user is in a meeting (`InMeeting`) or is driving (`IsDriving`); it mutes the phone if he is in a meeting and plays a custom greeting message if he is driving.

2.5 Datasets

Apart from a real prototype, we use two real datasets to evaluate various techniques in ACE. Using real traces is important because effectiveness of ACE depends on its ability to infer context rules from a user's context history. Each dataset contains continuous context trace of a number of users, where each trace contains a sequence of lines of the following format:`timestamp`, $\texttt{attrib}_1 = \texttt{value}_1, \texttt{attrib}_2 = \texttt{value}_2, \cdots$. There is a line in the trace every time any attribute changes its value. Thus, a user is assumed to be in the same state between two consecutive timestamps in the trace.

►**Reality Mining Dataset.** This dataset,[3] collected part of the Reality Mining project at MIT [7], contains continuous data on daily activities of 100 students and staff at MIT, recorded by Nokia 6600 smartphones over the 2004-2005 academic year. The trace contains various continuous information such as a user's location (at a granularity of cell towers), proximity to others (through Bluetooth), activities (e.g., making calls, using phone apps), transportation mode (driving, walking, stationary),[4]etc. over different times of the day. The cell-tower id is mapped back to latitude-longitude, based on the dataset in [8]. We consider 95 users who have at least 2 weeks of data. The total length of all users' traces combined is 266,200 hours. The average, minimum, and maximum trace length of a user is 122 days, 14 days, and 269 days, respectively.

The dataset allows us to infer the following context attributes about a user at any given time: `IsDriving`[5] (`D` in short), `IsBiking` (`B`), `IsWalking` (`W`), `IsAlone` (`A`), `AtHome` (`H`), `InOffice` (`O`), `IsUsingApp` (`P`), and `IsCalling` (`C`).

►ACE **Dataset.** This dataset, similar to the Reality Mining dataset, is collected with our continuous data collection software running on Android phones. The subjects in the dataset, 9 male and 1 female, worked at Microsoft Research Redmond. They all drove to work and had usual 9am-5pm work schedule. The dataset has all the context attributes in Table 1. The maximum, minimum, and average trace length of a user is 30 days, 5 days, and 14 days respectively.

Even though the dataset is smaller than the Reality Mining dataset, its ground truths for context attributes are more accurate. They come from a combination of three sources: (1) labels manually entered by a user during data collection time (e.g., when the user starts driving, he manually inputs `Driving` = `True` through a GUI), (2) labels manually provided by the user during post processing, and (3) outputs of various contexters.

3. RULE MINER

The Rule Miner component of ACE maintains a history of timestamped tuples (e.g., `AtHome` = `True`), sensed for various applications. It then incrementally derives rules regarding relationships among various context attributes.

[3]`http://reality.media.mit.edu/download.php`

[4]This is inferred from a user's speed and survey data.

[5]Denotes if the user is driving is own car or is taking a bus.

Table 2: A few example rules learned for one user

{IsDriving = True} ⇒ {Indoor = False}
{Indoor = T, AtHome = F, IsAlone = T} ⇒ {InOffice = T}
{IsWalking = T} ⇒ {InMeeting = F}
{IsDriving = F, IsWalking = F} ⇒ {Indoor = T}
{AtHome = F, IsDriving = F, IsUsingApp = T} ⇒ {InOffice = T}
{IsJogging = T} ⇒ {AtHome = T}

Discovering interesting relationships among variables in large databases has been studied extensively in the data mining research, with many efficient algorithms developed (see [26] for a survey). In ACE, we would like to generate rules that are easy to compute and use, and are compact to represent. We choose the well-known Apriori algorithm for association rule mining [1]. The algorithm can efficiently discover association rules from databases containing transactions (for example, collections of items bought together by customers). In ACE, we are interested in association rules that have the following general form: $\{l_1, l_2, \cdots, l_n\} \Rightarrow r$, which implies that whenever all the tuples $l_1, \cdots, l_n$ hold, r holds as well. Thus, the left side of a rule is basically a conjunction of tuples $l_1, l_2, \cdots, l_n$. For example, the rule {O = T, W = F, A = F} ⇒ M = T implies that if a user is in the office (O = T) and not walking (W = F) and is not alone (A = F), then he is in a meeting (M = T).

Apart from simplicity, association rules naturally support ACE's desired inference process: if for any rule $A \Rightarrow b$, *all* the tuples in A exist (and not expired) in ACE's cache, then we can immediately infer the tuples b, without additional sensor data acquisition.

The Apriori algorithm takes as input a collection of *transactions*, where each transaction is a collection of co-occurring tuples. Each association rule has a support and a confidence. For example, if a context history has 1000 transactions, out of which 200 include both items A and B and 80 of these include item C, the association rule $\{A, B\} \Rightarrow C$ (read as "If A and B are true then C is true") has a support of $8\%(= 80/1000)$ and a confidence of $40\%(= 80/200)$. The algorithm takes two input parameters: $minSup$ and $minConf$. It then produces all the rules with support $\geq minSup$ and confidence $\geq minConf$.

The original Apriori algorithm works in two steps. The first step is to discover frequent itemsets. In this step, the algorithm counts the number of occurrences (called *support*) of each distinct tuple (e.g., W = T) in the dataset and discards infrequent tuples with support $< minSup$. The remaining tuples are frequent patterns of length 1, called *frequent 1-itemset* S_1. The algorithm then iteratively takes the combinations of S_{k-1} to generate frequent k-itemset candidates. This is efficiently done by exploiting the anti-monotonicity property: any subset of a frequent k-itemset must be frequent, which can be used to effectively prune a candidate k-itemset if any of its $(k-1)$-itemset is infrequent [1].

The second step of the Apriori algorithm is to derive association rules. In this step, based on the frequent itemsets discovered in the first step, the association rules with confidence $\geq minConf$ are derived.

Parameters. The Apriori algorithm takes two parameters: $minSup$ and $minConf$. Many real-world applications care only about frequent rules, and hence use a large $minSup$ (e.g., $> 75\%$). We, however, are also interested in rules that hold relatively infrequently. For example, if a user drives < 1 hour in a day, Apriori will require a very small support of $< 4\%$ to find rules involving the context attribute IsDriving. We are interested in such rare rules because they can sometimes result in big energy savings (e.g., when InMeeting is inferred from IsDriving). Setting the $minSup$ close to zero, however, can produce many infrequent and useless rules. We found that the value $minSup = 4\%$ works well in practice.

Ideally, we would like to use $minConf = 100\%$, so that ACE uses rules that always hold and hence ACE does not introduce any inference errors. However, real contexters use imperfect classification algorithms on noisy data, and hence they can occasionally produce inconsistent context attributes. E.g., two contexters can mistakenly conclude that the user is atHome and isDriving at the same time. Our experiments show that a $minSup$ value of 99% strikes a good balance between tolerating such occasional inconsistencies and capturing good rules.

Example Rules. Table 2 shows a few example rules automatically learned by ACE for a single user in the ACE Dataset. Some rules are not so obvious; e.g., consider the fifth rule in Table 2: if the user is not at home and not riding a bus but is using phone apps, he is in the office. Apparently, this particular user uses apps only at home, in bus, and in office. Some rules can be specific to a single user and may not generally apply to all users. For example, the last rule in Table 2 applies to a user who jogs in a treadmill at home, and it may not apply to someone who runs outside.

3.1 Challenges

Even though many association rule mining algorithms exist, we need to address several challenges to use them in ACE.

3.1.1 Defining Transactions

Association rule mining algorithms take *transactions* of tuples: each transaction defines which tuples occur together. For a rule $A \Rightarrow B$ to be derived, both A and B need to appear together in the same transaction. In ACE, various contexters may be invoked for different applications asynchronously. Thus, co-occurring context attributes may not be inferred exactly at the same time, rather within a short window of time. Yet, for mining useful rules, we need to batch them as a single transaction of co-occurring tuples.

One straightforward way to address this is to use a time window: all attributes sensed within a window is considered to be co-occurring and hence are batched together in a single transaction. Unfortunately, deciding a good window size is nontrivial. A small window will generate small (or even singleton) transactions that will miss many rules involving multiple tuples. On the other hand, a large window may put conflicting tuples (e.g., AtHome = True and AtHome = False) in the same transaction when the user changes his context in the middle of the window. Then, a rule mining algorithm can produce invalid itemsets such as {AtHome = True, AtHome = False} and meaningless rules such as {AtHome = True} ⇒ {AtHome = False}.

To understand the impact of window size, we run Rule Miner on ACE Dataset as follows. We assume that a hypothetical application wakes up once every 30 seconds, and determines the value of a randomly chosen attributes in Table 1. (This simulates the combined behavior of a random collection of continuous context-aware applications.) Thus, we get a sequence of tuples, with two consecutive tuples 30 seconds apart from each other. Then, for a given window size W, we batch all tuples within W as a single transaction. Note that a batch may contain the same tuple multiple times, in which case we keep only one. Moreover, a batch may also contain conflicting tuples, in which case we drop all tuples involving the corresponding attribute.

Figure 3 shows the effect of various window sizes. The y-axis in the graph shows the fraction of rules learned, normalized to the maximum number of manually-verified "good" rules we could learn for various parameters and averaged over all users in ACE Dataset. The graph shows the trade-off introduced by various window sizes. A 5-minute window gives the maximum number of good rules.

Dynamic Window Size. Even with a carefully chosen window size, we do not find the maximum number of rules. This is because many transactions contain a small number of tuples, due to removal of conflicting tuples within the same window. Since a user can change his context any time within a window, such conflicting tuples occur very often. To avoid this, we use *dynamic windowing*, where we use a default window size (say 5 minutes), but *trim* a window into a shorter one whenever the value of any context attribute changes (e.g., when the user's `AtHome` context is changed from `True` to `False`). After such a trimmed window, we restore the default window size. Thus, a window never contains conflicting tuples. We also ignore the windows that have only a single tuple. Figure 3 shows that dynamic windowing with 5-minute default window size produces 18% more "good" rules than static windowing with 5-minute window.

Increasing default window size can reduce the number of rules produced, as shown in Figure 3. This is because increasing window size can reduce relative frequencies of popular tuples since multiple occurrences of a popular tuple are treated as a single occurrence within a transaction. If the relative frequency of an itemset containing a popular item goes below the $minSup$ threshold, APriori algorithm will ignore the itemset and no rules containing the itemsets will be produced.

Based on the discussion above, ACE uses dynamic window with 5 minutes default window.

3.1.2 Dealing with Low Support

As mentioned before, ACE uses a small $minSup$ value (e.g., 4%). With such a small support value, Apriori needs to enumerate a very large number of itemsets and becomes extremely expensive. This is because with a small support, many itemsets will be considered frequent and each of the frequent itemsets will contain a combinatorial number of shorter sub-itemsets, each of which will be frequent as well.

Figure 4 shows the time required by ACE's rule mining algorithm run with the Reality Mining dataset on a Samsung Focus phone with 1 GHz Scorpion CPU and 512MB RAM. The $minSup$ and $minConf$ are set to be 4% and 99% respectively. Depending on the length of the history, the process can take more than a few hours. Such a high overhead is not acceptable in practice.

Offloading. ACE addresses this problem by offloading the task of rule mining to a remote, powerful server (similar to many systems such as Maui [5]). ACE locally logs all context tuples, periodically uploads them to a server when it has connectivity and time, and gets the rules back. By default, this is done once a day. Computing rules in a server takes a few minutes. With a 2.7GHz dual-core server, the maximum time required for offloading context history, computing rules, and downloading them to the phone is around 15 minutes for a 269-day trace in the Reality Mining Dataset. For traces smaller than 90 days, the time is around 3 minutes.

Incremental Update. Between periodic downloads of rules from the remote server, Rule Miner incrementally updates its rules in the phone as follows. With rules, ACE also downloads various statistics required to compute supports and confidence of various rules. As new context tuples are added to context history, Rule Miner incrementally updates support and confidence statistics of existing rules and deletes the rules whose supports and confidence go below Rule Miner thresholds. Note that this only removes existing rules if they no longer hold, but does not produce new rules. However, this affects ACE's efficiency only (as ACE may miss opportunities to exploit potential rules), not the correctness of its inference. The inefficiency is addressed when new rules are downloaded from remote server.

3.1.3 Dealing with Inaccuracies

Since Inference Cache and Speculative Sensing are based on context rules, it is important than the rules are as accurate as possible. ACE conservatively uses rules that almost always hold (e.g., $minConf$ = 99% in Apriori). A user, however, may change his habits, which may invalidate some rules. For example, he may buy a treadmill and start running at home instead of outside. Then, the rule {`IsRunning` = `True`} ⇒ {`AtHome` = `False`} needs to be replaced with a new rule. To achieve this, ACE cross validates its inferred results with ground truths (collected by occasionally running contexters or getting feedback from users/applications, as mentioned in Section 2). ACE conservatively drops a rule if its support and confidence go below $minSup$ and $minConf$ due to the ground truths. New rules emerged due to change in habits are eventually learned by the Rule Miner as their itemsets become frequent.

3.1.4 Suppressing Redundant Rules

Rule mining algorithms often produce a large number of rules, especially with a small support. The cost of using the rules for Inference Cache and Sensing Planner increases linearly with the number of rules. Fortunately, many rules generated by standard rule mining algorithms are redundant; for example, the standard APriori algorithm can produce rules such as $\{A \Rightarrow B, B \Rightarrow C, A \Rightarrow C, AD \Rightarrow B\}$. However, given the two rules $\{A \Rightarrow B, B \Rightarrow C\}$, other two rules can be generated and hence they are redundant.

In order to reduce the inference overhead, ACE uses an algorithm for mining a compact, non-redundant set of association rules [28]. We have used both this algorithm and the Apriori algorithm on both our datasets. The basic Apriori algorithm generates around 750 rules on average on both datasets; eliminating redundant rules brings the number of rules below 40, on average, for both the datasets.

3.1.5 Bootstrapping

It takes a few days of context history for ACE to learn meaningful rules. Thus, in the beginning, ACE does not have any context history or rules that it can exploit. To deal with that, ACE starts with a set of seed rules that apply to almost all the users. Such *universal* rules are learned by the remote server where users offload their rule mining task.

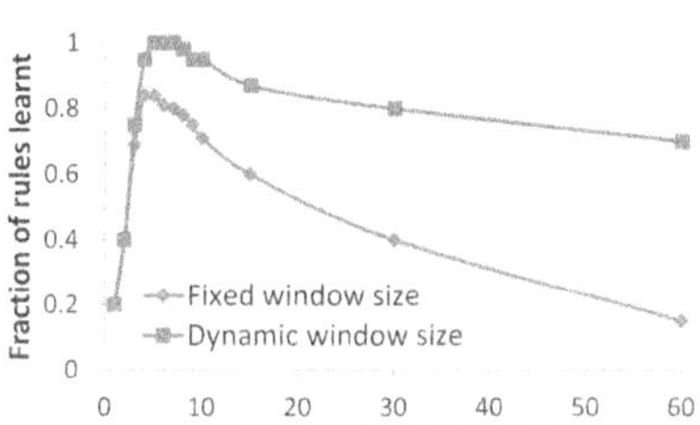

Figure 3: Effect of window size in batching

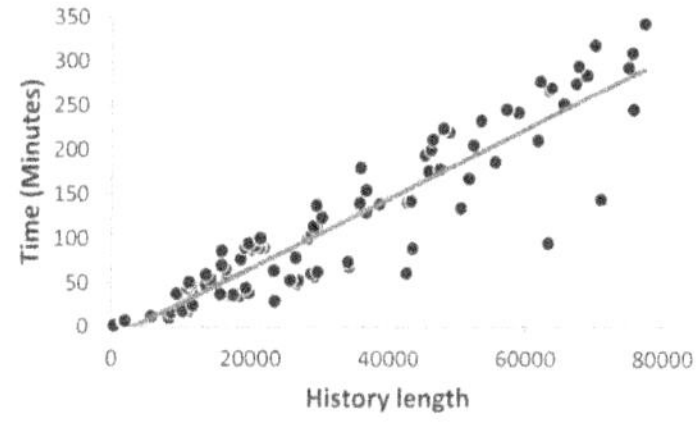

Figure 4: Time required to learn rules on a Samsung Focus phone

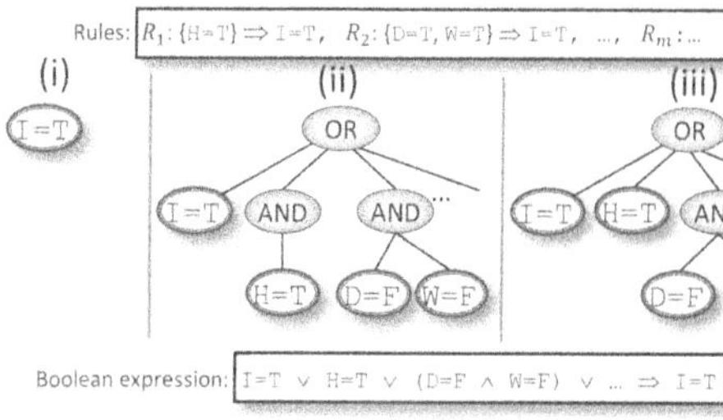

Figure 5: Expanding a node

To learn such rules, the server runs the same rule mining algorithm, but on all users' context history. In the Reality Mining Dataset, more than 60% of the rules of each user are such universal rules. As new context tuples are added to context history, ACE updates its rule set by learning new personalized rules and dropping universal rules that do not apply to the user.

4. INTELLIGENT CONTEXT CACHING

The Inference Cache in ACE, like traditional cache, provides a Get/Put interface. $Put(X, v)$ puts the tuple $X = v$ in the cache, with a predefined expiration time (default is 5 minutes). On a $Get(X)$ request for a context attribute X, it returns the value of X not only if it is in the cache, but also *if a value of X can be inferred from unexpired tuples in the cache by using context rules learned by the Rule Miner.*

To illustrate the operation of Inference Cache, assume that an application requires the value of `D` (`IsDriving`). Suppose, ACE determines its value to be `True` by using the corresponding contexter. The tuple `D` = `T` is put in the cache with an expiration time of, e.g., 5 minutes. Suppose, before this tuple expires, another application requires the value of `D`. Like traditional cache, Context Cache returns the value `True`. Now, suppose a third application requires the value of `I` (`IsIndoor`). Even though there is no tuple with `I` in the cache, Context Cache infers that the user cannot be indoor, thanks to the rule {`D` = `T`} ⇒ `I` = `F` , and hence it returns the value `False` as the answer.

Note that Inference Cache can sometimes return results that are (a) incorrect, if user's behavior deviates from context rules, or (b) stale, if results are inferred from unexpired but stale values in cache, and hence incorrect if cached values no longer hold. As mentioned before, the severity of (a) is reduced by the Rule Miner conservatively choosing rules that almost always hold. Similarly, to reduce the effect of (b), we choose a small cache expiry time of 5 minutes. This helps us keep the overall inaccuracy of ACE within $< 4\%$ for our dataset. (Using an expiry time of 10 minutes and 15 minutes make the inaccuracy 6% and 10% respectively.)

A key challenge in enabling Inference Cache is to efficiently exploit rules. One straightforward way to do this is to go over all the rules to find the ones whose right hand sides contain the target attribute and all the tuples in their left hand sides are satisfied by currently unexpired tuples in the cache. Then, the cache can output the tuples on the right hand sides of those rules. However, this simple approach fails to exploit transitive relationship among rules. For example, consider two rules: {`M` = `T`} ⇒ `O` = `T` and {`O` = `T`} ⇒ `I` = `T` Without using transitivity of the rules, the cache will fail to infer the state of `I` = `T` even if `M` = `T` is in the cache. Such transitivity is common among the rules, especially since our Rule Miner produces a succinct set of non-redundant rules.

To address this, ACE constructs *expression trees* over rules and performs inference on the trees.

4.1 Expression Trees

Informally, an expression tree for a tuple `a` = `v` is a tree representation of the Boolean expression that implies the value `v` for the attribute `a`. It is generated by combining various rules that imply `a` = `v` directly, or indirectly via transitive relationships with other rules. More formally, an expression tree is a Boolean AND-OR tree, where (a) a non-leaf node represents an AND or OR operation on the value of its child nodes, and (b) a leaf node represents a tuple. An AND node (or, OR node) evaluates to `True` if all (or any, respectively) of its child nodes recursively evaluate to `True`. The expression tree for the tuple `a` = `v` is the one that evaluates to `True` if and only if the tuple holds according to context rules. Figure 6(b) shows an expression tree for the tuple `Indoor` = `True`.

The expression tree for a given tuple `a` = `v` is constructed as follows. We start from a tree with a single leaf node `a` = `v` (Figure 5(i)). Then, we continue *expanding* each leaf node in the current tree by using context rules until no leaf nodes can be further expanded or all the rules have been used for expansion. To expand a leaf node `a` = `v`, we find all the rules $R_1, R_2, \cdots, R_m$ that have `a` = `v` on the right hand side. We then replace the leaf node with a subtree rooted at an OR node. The OR node has child nodes `a` = `v`, $X_1, X_2, \cdots, X_m$, where X_i is an AND node with all the tuples on the left hand side of R_i as child nodes. An example of this process is shown in Figure 5(ii), where the node `I` = `T` is expanded using the rules shown at the top of the figure.

After a node is expanded, all the leaf nodes in the tree are recursively expanded (to deal with transitive relationship). Note that we do not need to expand the same tuple more than once; for example, in Figure 5, if the `I` : `T` node is expanded again, we will unnecessarily replicate a portion of the tree, without changing the semantics of the tree. We thus expand each tuple only once—only if it has not already been expanded and there are rules with it on the right hand side. This ensures termination of the expansion process.

4.2 Inference Cache

The Inference Cache maintains one expression tree for each tuple; e.g., one for the tuple `D` = `T`, one for `D` = `F`, and so on. On a Get request on attribute `X`, the cache evaluates two expression trees—one for `X` = `T` and one for `X` = `F`—and see if any of these trees is satisfied (explained shortly) by unexpired tuples in the cache. If the tree for `X` = `T` is satisfied, the Get call returns the value `True`, if the tree for `X` = `F`

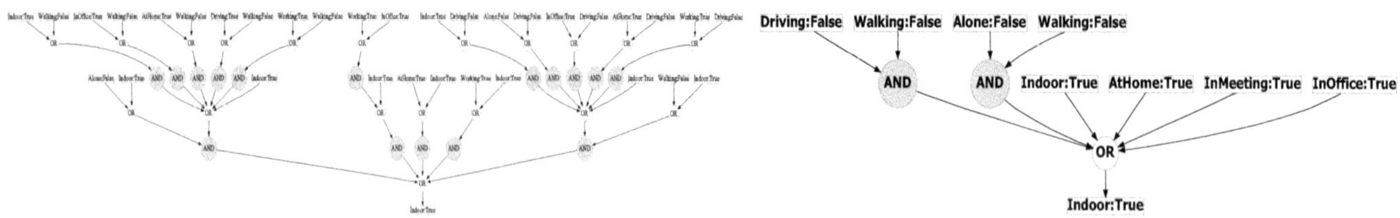

(a) Initial expanded tree (labels can be ignored)　　(b) Equivalent reduced tree

Figure 6: An expression tree for `Indoor` = `True`, shown upside down

is satisfied, `False` is returned, otherwise, the cache invokes the Sensing Planner module to invoke necessary contexters.

The procedure to check if an expression tree is satisfied with the tuples currently available in the cache is recursive. A leaf node is satisfied if and only if the corresponding tuple is in the cache. A non-leaf OR node is satisfied if *any* of its child nodes are satisfied (recursively). Similarly, an AND node is satisfied if *all* of its child nodes are satisfied. Finally, the whole tree is satisfied if the root node is satisfied.

4.3 Minimal Expression Tree

Expression trees obtained by expanding rules can be large. Since Inference Cache needs to maintain one tree for each tuple and needs to evaluate two trees on every Get request, larger trees impose memory and computation overhead. To avoid this, ACE normalizes the trees to shorter ones, by using standard Boolean logic to eliminate their redundancy.

►Alternating AND-OR level: An AND-OR tree can always be converted so that (a) a level has either all AND nodes or all OR nodes, and (b) AND levels and OR levels alternate. This can be done as follows: if an OR node (or an AND node) u has an OR child node (or an AND child node, respectively) v, child nodes of v can be assigned to u and v can be deleted. This compacts the tree.

►Absorption: If a non-leaf node N has two child nodes A and B such that the set of child nodes of A is a subset of the set of child nodes of B, we can remove B and its subtrees. In a tree with alternating AND-OR levels, if N is an OR node, then A and B will be AND nodes. Suppose $A = a \wedge b$ and $B = a \wedge b \wedge c$. Then, $N = A \vee B = (a \wedge b) \vee (a \wedge b \wedge c) = a \wedge b = A$. Similarly, if N is an AND node, $(a \vee b) \wedge (a \vee b \vee c) = (a \vee b)$, and hence the longer subexpression $B = (a \vee b \vee c)$ can be removed.

►Collapse: If a node N has one child, the child can be assigned to N's parent node and N can be removed (Fig. 6(iii)).

Such transformation significantly reduces the size of expression trees without changing their semantics. Figure 6 shows the result of such transformation: (a) shows the tree for the tuple `Indoor` = `True` after initial expansion, and (b) shows the result of compacting the tree. (In (a), we intend to show the large size of the tree; the labels inside nodes are not expected to be read.)

5. SENSING PLANNER

When Inference Cache fails to determine the value of X on a $Get(X)$ call, ACE needs to call necessary contexters to determine the value of X. The straightforward way to find the value is to invoke the contexter that directly produces value of X (e.g., the `AtHome` contexter that uses GPS and WiFi signature). However, by using context rules, ACE may

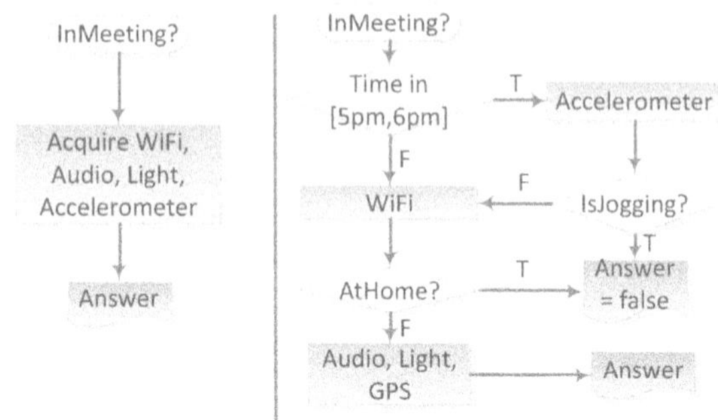

(a) Traditional plan　　(b) Conditional plan

Figure 7: A conditional acquisition plan that acquires different sensor data depending on various conditions

be able to find the value of X in an indirect and cheaper way, by *speculatively sensing* additional proxy attributes.

5.1 Conditional Plan for Speculative Sensing

We define a *sensing plan* as the order in which various attributes are checked in order to determine the value of a target attribute. An optimal ordering may depend on various conditions such as the cost c_i of checking an attribute i (i.e., the cost of running the corresponding contexter) and the likelihood p_i of the attribute returning a `True` value. A plan that orders attributes to check based on such conditions is called a *conditional plan*. We illustrate this with an example.

Suppose an application wants to find if the user is `InMeeting`. A traditional system will determine the value of the attribute by acquiring necessary sensor data, e.g., audio, GPS, and WiFi signature and by using necessary classifiers. However, ACE's conditional planning exploits context rules to opportunistically infer the value of `InMeeting` by acquiring attributes that are cheaper (in terms of sensing cost) and are likely to conclude a value of `InMeeting`. For example, if the current time is 5.30pm and the user is likely to be jogging between 5pm and 6pm, ACE can speculatively acquire the value of `IsJogging`, and if it is found to be `True`, ACE can infer `InMeeting` = `False` according to the rule `IsJogging` = `True` ⇒ `InMeeting` = `False`. Similarly, if checking `AtHome` is cheaper (by sensing WiFi only) and the user is likely to be at home, ACE can check if the user is actually at home and if so, can conclude `InMeeting` = `False`. If no other cheaper ways provide the value of the target attribute, ACE acquires the necessary sensor data to directly determine the value of the target attribute. Figure 7 shows both traditional and conditional plans for `InMeeting`.

Given a conditional plan, the true sensing cost depends on the exact path ACE takes to generate a result. In the previous example, if the user is indeed jogging, then the cost equals to the cost of acquiring accelerometer data only. The expected sensing cost of a plan can be computed by using the values of c_i and p_i of various attributes in the plan. ACE knows the value of c_i from the configuration file for contexter for attribute i. The value of p_i is computed from context history—ACE incrementally maintains one p_i for every hour of the day.

Figure 7 shows only one conditional sensing plan for the attribute `InMeeting`. However, there are many other plans—another valid plan is to first check whether both `IsDriving` and `IsWalking` are `False`, and if not, to check other attributes. In general, the number of valid plans can be exponential in the number of attributes. Our goal in ACE is to choose the plan with the *minimum expected cost.*

Note that, even when using the best conditional plan, ACE's speculation may occasionally end up spending more energy than directly acquiring the target attribute. In the previous example, `IsJogging` may return `False`, in which case ACE needs to acquire `IsIndoor`, making the cost of acquiring `IsJogging` a waste. However, hopefully such mistakes are not made often (e.g., `IsJogging` is acquired only if it is likely to return `True`), and the expected sensing cost is less than the cost of directly acquiring target attributes.

Hardness of the Problem. The value of an attribute `A` is `True` if the Boolean expression given by the expression tree for the tuple `A` = `True` is satisfied. How can we evaluate a Boolean expression with minimum cost? Note that value of a Boolean expression may be determined by evaluating only a subset of variables in the expression. Thus the evaluation cost depends on the order various attributes are checked. Suppose A_1 = `False`, A_2 = `True`, A_3 = `False`. Then the value of the Boolean expression $A_1 \lor A_2 \lor A_3$ can be determined by checking A_2 alone, without checking the values of A_1 and A_3. On the other hand, first checking A_1 does not yield any value of the expression, and requires checking other variables. An optimal algorithm would evaluate variables with small cost but high likelihood to determine the value of the whole expression before other variables.

The problem outlined above is an instance of the *probabilistic And-Or tree resolution* (PAOTR) problem. The problem is shown to be NP-Hard in general [9]. Efficient algorithms can be found for special cases when (a) each attribute appears at most once in the expression, (b) attributes are independent, and (c) the expression tree has a height of at most 2 [9]. The assumptions do not hold for our expression trees—they may violate any of these conditions.

5.2 Speculative Sensing API

Before describing how ACE chooses the best sensing plan, we describe how it programmatically uses a plan. One option is to enumerate the entire plan as a graph as shown in Figure 7. However, given such a plan, ACE will traverse only part of the graph, based on the outcomes of conditional nodes. Therefore, instead of enumerating the whole plan, ACE incrementally enumerates only the next node to traverse in the plan. More specifically, it implements following four methods, the details of which we will describe later.

- `Init(X)` : Initializes for a target attribute `X`.
- `attribute` ← `Next()` : Returns the attribute whose value needs to be determined next, or `null` if the target attribute value has already been determined.
- `Update(a, v)` : Updates the planner's internal state with the tuple `a` = `v`. The planner moves to the conditional branch of the plan based on this value of `a`.
- `value` ← `Result()`: Returns the value of attribute `X`.

Algorithm 1 Exhaustive search for the optimal sensing plan

```
procedure INIT(X)
    target ← X
    trace, next, result, dpCache ← φ

procedure RESULT
    return result

procedure UPDATE(attrib, value)
    trace ← trace ∪ [attrib = value]
    if attrib = target then
        result ← value

procedure NEXT
    if result = φ then
        return φ
    (attrib, cost) ← NextHelper(trace)
    return attrib

procedure NEXTHELPER(trace)
    if trace is in dpCache then
        [next, cost] ← dpCache[trace]
        return [next, cost]
    minCost ← ∞
    bestAttrib ← φ
    for all State s ∉ trace do
        traceT ← trace ∪ {s = true}
        if traceT satisfies expressionTree(target = T) then
            CostT ← 0
        else
            [next, CostT] ← NextHelper(traceT)
        traceF ← trace ∪ {s = false}
        if traceF satisfies expressionTree(target = F) then
            CostF ← 0
        else
            [next, CostF] ← NextHelper(traceF)
        ExpctedCost ← Cost(s) + Prob(s = true) · CostT +
Prob(s = false) · CostF
        if ExpectedCost < minCost then
            minCost ← ExpectedCost
            bestAttrib ← s
    dpCache[trace] ← [bestAttrib, minCost]
    return [bestAttrib, minCost]
```

When the Inference Cache requests the Sensing Planner to determine the value of the attribute X, it first calls `Init(X)`. Then, it repeatedly calls `a` ← `Next()`, determines the value `v` of `a` by using necessary contexter, and calls `Update(a, v)`. The process continues until the `Next()` call returns a `null`, at which point it returns the output of the `Result()` method to the Inference Cache.

5.3 Optimal Speculative Sensing Algorithm

The hardness result in the previous section indicates that a polynomial time algorithm for obtaining an optimal plan is unlikely to exist. However, when the number of attributes is relatively small (fewer than 10), it may be feasible to exhaustively search for all possible ordering of the attributes and then to choose the one that minimizes the expected cost. With n attributes, there are $n!$ such ordering. Fortunately, we can reduce the computational overhead with a dynamic programming algorithm that prunes the search space by early evaluation and caching.

The search space of our dynamic programming algorithm is defined over *traces*, where a trace is a sequence of tuples already sensed by various contexters. For example, suppose three invocations of the `GetNext()` method returned attributes a, b, and c and in response, their contexters have sensed their values to be True, False, and True respectively. Then the trace at that point is the sequence a = T, b = F, c = T. A trace represents a sensing plan as it defines the attributes to be sensed and the order of sensing. The key observation that allows us to employ dynamic programming is that once an attribute X is sensed, the original problem of finding the optimal plan is decomposed into two independent subproblems: one for finding the best trace having X = True and the other for finding the best trace having X = False. Each subproblem covers a disjoint subspace of the attribute-value space covered by the original problem, and hence can be solved independently.

Algorithm 1 shows the pseudocode of four methods in Sensing Planner. Given a target attribute, the methods start with an empty trace. Given the current trace T, the `Next()` method finds the optimal next attribute as follows. For each attribute A $\notin T$, it considers its True and False value. If the current trace plus A = T or A = F satisfies the expression tree, the search is pruned on additional attributes. Otherwise, expected cost of each possibility is recursively computed and added together, with weights equal to the probability of A = T and A = F, to compute the overall expected cost (Line 32). Finally, the attribute with the minimum expected cost is returned. For efficiency, values of various traces are cached in *dpCache* for memoization.

5.4 Heuristic Algorithm

When the number of attributes is large, the exhaustive algorithm can become prohibitively expensive. In such cases, ACE uses an efficient heuristic algorithm. The algorithm is *sequential* in nature—it does not use any conditioning predicate to dynamically choose the next best attribute, rather it ranks the attributes once based on their c_i and p_i values, and the attributes are sensed in that order regardless of observed attribute values. One ordering of attributes is computed for each target attribute. The nice aspect of this algorithm is that the ordering of attributes can be determined offline once (e.g., by a remote server). `Next()` method simply returns attributes according to the optimal order.

We now describe how the ordering for a target attribute X is computed; pseudocode for four methods of Sensing Planner is omitted for brevity. Intuitively, the algorithm traverses an expression tree in a depth-first manner and ranks attributes with small cost but high probability of producing a conclusion before other attributes.

1. Convert expression trees for X = T and X = F to their disjunctive normal forms. Combine them into one tree T by merging roots of both trees into one OR node.

2. Initialize a queue Q to be empty. At the end of the procedure, Q will contain attributes in the same order they need to be sensed. I.e., `Next()` simply needs to dequeue attributes from Q.

3. For each leaf node i in T, let $\mathcal{C}_i = c_i$ and $\mathcal{P}_i = p_i$.

4. Use the Procedure AssignCostAndProb shown in Algorithm 2 to assign costs and probabilities of all non-leaf nodes of T. The procedure simply traverses the tree in a depth-first manner, first considering nodes that have lower cost and higher probability to conclude a value for its parent node. If node i's parent is an OR node, then the probability $\mathcal{P}_i$ that i can conclude (a True value of) its parent is $\mathcal{P}_i$, the same as the probability of i being True. If the parent is an AND node, then the probability $\mathcal{P}_i$ that i can conclude (a False value of) its parent is $1 - \mathcal{P}_i$.

Algorithm 2 Generating attribute sensing order

```
procedure AssignCostAndProb(And-OR Tree Node T)
    if T is a leaf node then
        (C_T, P_T) ← (c_T, p_T)
        return (C_T, P_T)
    for all child node N_i of T, 1 ≤ i ≤ k do
        (C_N, P_N) ← AssignCostAndProb(N)
        if T is an AND node then
            P_N ← 1 − P_N
    Sort the child nodes of T in decreasing order of their values of P_{N_i}/C_{N_i}, 1 ≤ i ≤ k
    C_T ← C_{N_1} + Σ_{i=2}^{k} C_{N_i} Π_{j=1}^{i-1} (1 − P_{N_j})
    if T is an AND node then
        P_T ← Π_{i=1}^{k} P_{N_i}
    else
        P_T ← 1 − Π_{i=1}^{k} (1 − P_{N_i})
    return (C_T, P_T)

procedure FindOrdering(And-Or Tree Node T)
    if T is a leaf node then
        Q.Enqueue(T)
    for all child N_i of T in ascending order of P_{N_i}/C_{N_i} do
        FindOrdering(N_i)
```

5. Finally, use the procedure FindOrdering in Algorithm 2 to enqueue attributes to Q in the order they should be sensed. The ordering prefers nodes with higher values of $\mathcal{P}_i/\mathcal{C}_i$; i.e., with higher $\mathcal{P}_i$ and lower $\mathcal{C}_i$.

The heuristic above is a generalization of the 4-approximate algorithm proposed by Mungala et al. [20] for finding the optimal sequential plan for conjunctive queries. The generalization is done in order to support arbitrary AND-OR expression. The depth-first nature and sorting child nodes based on $\mathcal{P}/\mathcal{C}$ values are similar to an algorithm in [9], which is optimal for a depth-2 tree with independent and non-repeated attributes. Our evaluation in Section 7 shows that the heuristic algorithm provides < 10% worse plans than the optimal algorithm, but with very little overhead.

6. DISCUSSION

As mentioned before, energy savings in ACE come at the cost of occasional inaccuracy in context inference. This limitation is fundamental to ACE because the absolute true value of a context attribute can be found by sensing the attribute, which ACE tries to avoid as much as possible. There are several important parameters that can affect this inaccuracy: rule mining parameters (support and confidence), cache expiry time, and cross-validation frequency (i.e., how often outputs of ACE are compared with outputs of contexters, in order to drop invalid rules). There parameters need to be carefully set based on target application scenarios.

There are few other limitations in the current version of ACE. These limitations are not fundamental and can be addressed in future work. First, currently ACE supports Boolean attributes only. Supporting non-Boolean attributes (e.g., a user's location) would introduce complexity in various components of ACE. One possible approach would be to make the Rule Miner learn more complex rules in the form of decision trees, the Inference Cache to employ decision tree-based inference algorithm, and the Sensing Planner to use planning techniques such as the ones used in [6].

Second, ACE currently does not exploit temporal correlation across attributes. We believe that capturing and exploiting rules such as "if a user is `inOffice` now, he cannot be `atHome` for the next 10 minutes" can be very useful in practice. Such rules can also be useful in efficient support for context-aware triggers (e.g., a location-based reminder) as the rules would enable applications to check contexts less frequently (similar to SeeMon [12]).

Third, the Inference Cache uses the same expiry time for all context attributes. In practice, expiry time can depend on attribute semantics. ACE can support this by letting each contexter define a suitable expiry time of its attribute—ACE can easily use this expiry time without knowing the semantics of the attribute.

Fourth, the cost function in Sensing Planner considers only the energy cost of acquiring a context attribute. However, ACE could use a more complex cost function (e.g., a weighted linear combination) of various parameters such as energy, latency, and accuracy of relevant contexters. How these various parameters are combined may depend on application scenarios. Once such a cost function is defined, the algorithms described in Section 5 will work seamlessly.

Fifth, the Inference Cache currently returns only the value of an attribute. It can be easily extended to return the confidence of a returned value. Note that the Rule Miner computes confidence of each rule (to keep the ones with confidence $\geq minConf$). Thus, the Inference Cache can compute the confidence of a result from the confidence of all rules used to infer the result.

7. EVALUATION

In this section we evaluate various components of ACE and find that (1) the Inference Cache alone can reduce the sensing energy consumption by a factor of 2.2× (for the Reality Mining Dataset), compared to a baseline cache that shares sensor data and context attribute across applications. (2) Sensing Planner can further reduce the energy consumption by an additional factor of $\approx 2\times$. Overall, applications running on ACE consume 4.2× less sensing energy on average compared to what they consume when running on a baseline cache. The savings would have been less than half had ACE used a set of predefined rules instead of dynamically-learned personalized rules for each user. (3) the latency and memory overhead introduced by ACE is negligible ($<$ 0.1ms and $<$ 15MB respectively).

Our performance related experiments are done with a prototype of ACE running on a Samsung Focus phone running Windows Phone 7.5 OS. The phone has 1 GHz Scorpion CPU, 512MB RAM, and a Li-Ion 1500 mAh battery. On the other hand, since effectiveness of various ACE components depends on individual user's context history, we measure the effectiveness with an ACE prototype running on a laptop with the Reality Mining Dataset and the ACE Dataset.

Prototype Implementation. We have implemented ACE in Windows Phone 7.5 (Mango) platform. It is implemented as a C# library shared by context-aware applications. (We use the same prototype on a laptop having .Net platform for our trace-driven evaluation.)

Windows Phone OS currently imposes several restrictions on its apps. First, it does not allow cross-application sharing or communication within the phone; all sharing/communication happen through the Cloud. To avoid such expensive communication, we build our context-aware "applications" as separate tabs (i.e., *pivots*) within a single Windows Phone app. This allows ACE to share sensor data and context attributes among multiple "applications." This restriction, however, does not affect the overhead numbers we report here. We are currently porting our prototype to Android that does not pose the above restriction and hence independent applications will be able to use ACE as a library.

Trace-driven Experiments. We use three applications and two datasets described in Section 2 for evaluating ACE's effectiveness in reducing sensing energy of the applications. The applications use various context attributes mentioned in Table 1. Specifically, the JogTracker application requires the attributes `IsWalking`, `isJogging`, and `IsBiking` (Reality Mining dataset only), the GeoReminder application requires `AtHome`, `InOffice`, `Indoor` (ACE dataset only), and PhoneBuddy requires `IsDriving`,`InMeeting` (ACE dataset only), `IsAlone`, and `InOffice`.

To work with the datasets, we port our ACE prototype to run on a laptop and replace its contexters with fake ones that return a user's current context from a given trace, instead of by acquiring necessary sensor data. The energy costs of various contexters are modeled according our energy measurements of real contexters, as shown in Table 1. We process each user's trace in discrete ticks, where each tick represents 5 minutes in wall clock time.[6] At each tick, we run all three applications and a contexter invoked by an application returns the true context of a user at corresponding wall clock time from his trace. To avoid any bias in the caching results, at each tick we run the applications in a random order, and each application requests for required attributes in a random order. Sensor data and context attributes have an expiration time of 1 tick; i.e., applications can share them within the same tick, but not across ticks. Increasing the expiration time will further increase the benefit of ACE.

For each user, we start our experiment with rules learned from first 1 week of data (and then the rules are dynamically updated). We report the sensing energy only; i.e., the energy consumed by contexters executed in order to satisfy requests of various applications.

7.1 Effectiveness of Inference Cache

We first show that Inference Cache of ACE gives a significant energy-savings over a baseline cache that shares sensor data and context attributes across applications.

Figure 8 shows the average and 95% confidence interval of hit rates of various applications under two datasets. The applications have some overlap in terms of required sensor data and context attributes, and hence the baseline cache gives nonzero hit rates (5 − 30%). The hit rate is the highest for PhoneBuddy because it has the maximum overlap with other applications. Compared to the baseline cache, Inference Cache consistently provides significantly higher hit

[6]A shorter tick will increase the absolute energy savings by ACE.

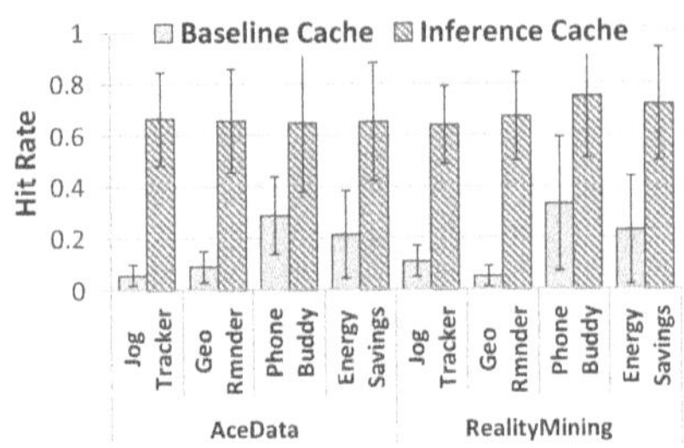

Figure 8: Hit rates and energy savings of Inference Cache

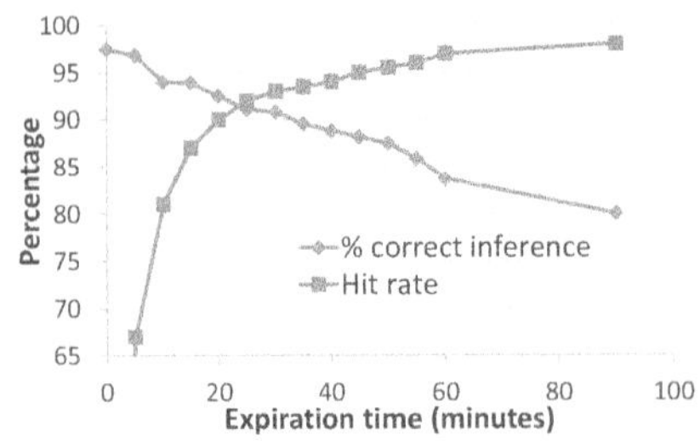

Figure 9: Effect of cache expiration time on hit rate and accuracy

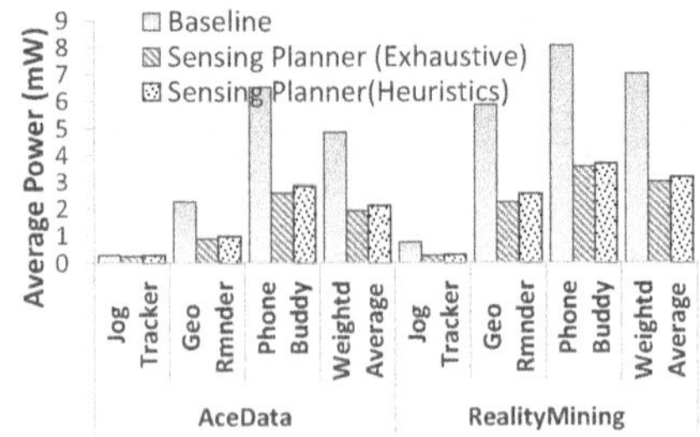

Figure 10: Energy savings by Sensing Planner

rates (65 – 75%), for all applications and datasets. The additional benefit comes from its ability to return the value of an attribute even if it is not in cache, yet can be inferred from existing attributes in the cache.

Since various attributes have different energy costs, the overall energy savings due to a cache hit differs across attributes (and applications). We therefore plot the fraction of energy savings, compared to a *No Sharing* scheme that does not share sensor data or context attribute among applications, in the same Figure 8. Both baseline cache and Inference Cache provides a non-zero energy savings; but the energy savings with Inference Cache is $> 3\times$ more than that with the baseline cache.

7.2 Accuracy

Figure 9 shows the effect of expiry time on cache hit rates and accuracy of results returned by *Get*(). The effect is shown for the Reality Mining dataset only. As expected, increasing expiration times increases cache hit rates. However, with large expiry times, Inference Cache can contain stale tuples, and any inference based on them can be stale or incorrect. Figure 9 shows inaccuracy due to such staleness as well as due to users deviating from automatically learned context rules. As shown, the inaccuracy is negligible with a small expiration time (e.g., $< 4\%$ with ACE's default expiration time of 5 minutes).

7.3 Effectiveness of Sensing Planner

In Figure 10, we show average power consumed for context sensing by various applications. We compare with a baseline planner that determines the value of an attribute by directly executing corresponding contexter. Both schemes are run with the Inference Cache enabled; i.e., the value of an attribute is determined only on a cache miss.

Figure 10 shows that, compared to the baseline planner, the heuristics-based Sensing Planner reduces energy consumption by 5 – 60% for various applications and datasets. The savings is more significant in applications requiring expensive attributes (e.g., PhoneBuddy), because often those expensive attributes are computed by executing cheaper contexters. Since Jog Tracker already uses cheap attributes, the Planner does not provide significant savings beyond the savings due to Inference Cache.) Overall, the savings is around 55% on average for both datasets.

The above savings can be slightly higher if Sensing Planner uses the dynamic programming-based exhaustive algorithm, instead of the heuristic algorithm. However, as shown in Figure 10, the heuristic algorithm performs close to the exhaustive algorithm—the difference is less than 10% for both datasets. On the other hand, the heuristic algorithm runs significantly faster than the exhaustive algorithm, as we will show later. Therefore, ACE uses the heuristic algorithm as default.

As mentioned in Section 5, speculative sensing may occasionally be useless and consume more energy than the baseline planner. Figure 11 shows the fraction of times Sensing Planner consumes no more energy than the baseline planner. As shown, Sensing Planner consumes less or equal energy as the baseline planner most of the time. The only case Sensing Planner makes a lot of "mistakes" ($\approx 40\%$) is with the GeoReminder application and Reality Mining dataset. This is due to the fact that probability distributions of various attributes in this dataset have high variances for a few users (because the data is not precise enough; e.g., location is based only at the granularity of a cell tower). Since Sensing Planner chooses an attribute a_i based on its *average* probability p_i of having `True`, it makes more mistakes for p_i with high a variance. The possibility of such high variance decreases if the user's behavior is repetitive and the prior probability distribution is tight. This is generally the case, as we found in the ACE Dataset. Note that even though Sensing Planner consumes more energy sometimes, on average, it consumes less energy than the baseline planner.

7.4 Overhead of ACE

Figure 12 shows the average latency of a *Get*() request when it is answered by the Inference Cache (i.e., a cache hit). The latency depends on complexity of expression trees, and we report both average and 95% confidence interval of the latency on the Samsung phone. The latency is negligible.

On an Inference Cache miss, ACE needs to use Sensing Planner to determine the value of the target attribute. Figure 13 shows the time spent by Sensing Planner on such a miss. (The latency does not include the sensing latency.) We show both the average and maximum latency for the exhaustive algorithm and the heuristic algorithm. The exhaustive algorithm can be slow and take up to a second. Planning time of exhaustive algorithm depends on the number of attributes. The ACE Dataset has one more attribute than the Reality Mining Dataset, and this results in an average latency increase of $\approx 2\times$ for the ACE Dataset. This shows that the exhaustive algorithm does not scale well as the number of attributes increases. In contrast, the heuristic algorithm, which is the default in ACE, runs very fast (< 0.1 ms) and remains comparable across various datasets and applications.

Overall, the overhead introduced by ACE on a *Get*() request (hit or miss) is less than a millisecond. This is negligible compared to the time needed to acquire sensor data and to infer user contexts (can be up to many seconds). The memory footprint is $< 15MB$ for all users in both datasets.

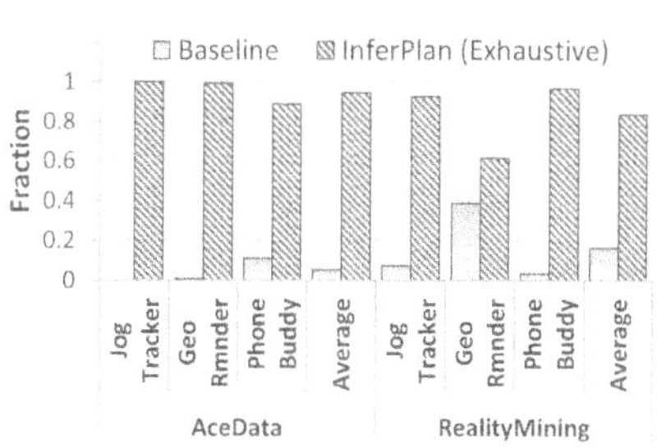

Figure 11: Fraction of times Sensing Planner does better

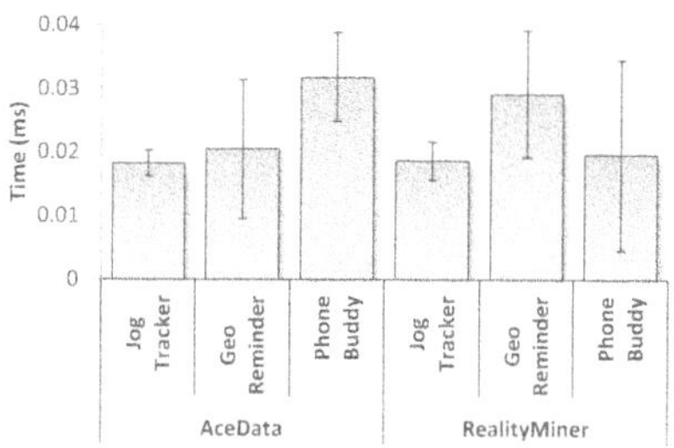

Figure 12: Latency of a Get() request on Inference Cache hit

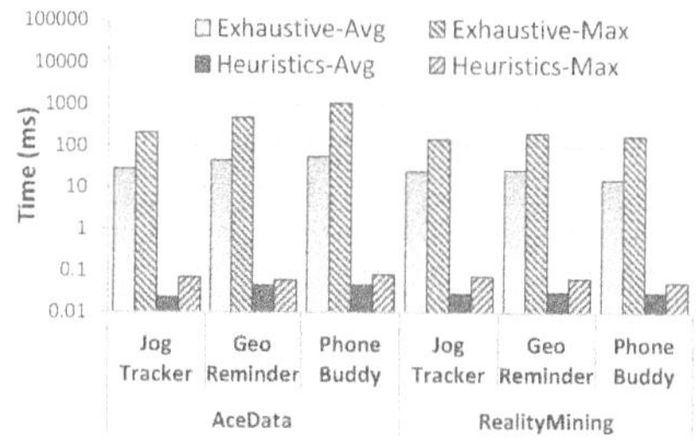

Figure 13: Time required to generate a plan on a Samsung Focus phone

7.5 End-to-end Energy Savings

Figure 14 shows the average end to end energy savings of ACE for the Reality Mining dataset. "No sharing" refers to a scheme where applications do not share any sensor data or context attributes with each other (e.g., if two applications require `IsDriving` attribute, they both invoke its contexter, which collects accelerometer data independently). As shown, a baseline cache reduces the energy consumption by 24%. In comparison, Inference Cache and the combination of Inference Cache and Sensing Planner reduce energy consumption by 64% and 82% respectively. Overall, ACE consumes 4.23× less sensing energy than the combination of baseline cache and baseline planner.

Clearly, the overall benefit of ACE depends on individual users as the context rules vary across users. Figure 15 shows the savings of various schemes for all 95 users in the Reality Mining dataset. User IDs are sorted according to energy consumed by ACE. As shown, for most users, both Inference Cache and Sensing Planner provide significant savings.

7.6 Importance of Dynamic Rule Learning

Finally we ask *how much value do dynamically learned rules add compared to a set of predefined rules?* To answer this, we compare two versions of ACE: **V1:** ACE with its default Rule Miner and **V2:** ACE with a set of predefined static rules.

The performance of **V2** will depend on the set of rules chosen a priori. For concreteness, we select universal rules from the ones discovered by Rule Miner. Ideally, we would like to select universal rules that hold for *all* the users. However, Rule Miner may miss a few of such rules for some users if their context traces do not have enough evidence of necessary patterns. We therefore choose k most frequent rules among all users. A caveat of this approach is that some of the selected rules may not hold for some users, and hence the value of k needs to be carefully chosen.

We measure the end-to-end energy of **V1** and **V2** for the Reality Mining dataset. **V1** consumes around 3.5mW (shown in Figure 14). In contrast, **V2** consumes around 10mW and 7.5mW with $k = 10$ and $k = 15$, respectively. (Energy consumption increases for larger k as some rules do not hold for some users). This $> 2\times$ savings in **V1** shows the significance of dynamically learned personalized rules.

8. RELATED WORK

The idea of context-aware computing is not new [23]; however it has gained a lot of interest recently, mostly due to the increasing availability of rich sensors on today's mobile devices. iOS, Android, and Windows Phone marketplaces contain a large number of context-aware applications. Most of these applications aim to enable new scenarios and to provide new features, with no or very basic optimizations for reducing energy overhead. Research community has also proposed many techniques to infer various context attributes of a user. They include techniques for inferring a user's activity [4, 13], location [2, 15], group state [27], proximity to others [3], surrounding sound events [16], etc. techniques have also proposed for inferring context attributes in energy-efficient ways (e.g., a-Loc [15] for energy-efficient location and [13] for energy-efficient activity). All these works are orthogonal to ACE as context inference is done within its black-box contexters. ACE can use any of these techniques in its contexter implementation.

SeeMon [12], like ACE, focuses on efficient and continuous context monitoring. It uses three optimizations: detecting change in context at earlier stages of the processing pipeline, exploiting temporal continuity of contexts, and selecting a small subset of sensors sufficient to answer a query. These optimizations can be used inside ACE's black-box contexters. The second optimization can also be used in ACE to produce richer rules such as "if a user is `inOffice` now, he cannot be `atHome` in next 10 minutes." Exploring such rich rules is part of our future work.

Our idea of sharing context attributes across applications is similar to using middleware cache in a mobile system [10]. However, these systems support exact lookup only—unlike Inference Cache, they do not support returning tuples that are not in the cache but can be inferred based on what is in the cache. Sensor substitution techniques substitute one sensor with another semantically related sensor for inferring a target attribute. SENST* [24] and a-Loc [15] have shown this for inferring location by substituting one expensive sensor with another cheap sensor. Inference cache is more powerful in that 1) it supports arbitrary context attributes, 2) attributes related to each other by a set of general association rules, and 3) the rules are learned automatically.

Our idea of speculative sensing is analogous to speculative execution used in pipelined processor [11] and other system architecture [21]. In contrast to these works, we use speculation for sensing and with conditional planning. Prior work has used conditional planning for acquisitional query processing [6], however this prior work 1) exploits correlation of various attributes, instead of association rules, 2) uses plans at sensor-level rather than context attribute-level, and 3) uses conditional plan for query processing rather than general context-aware applications.

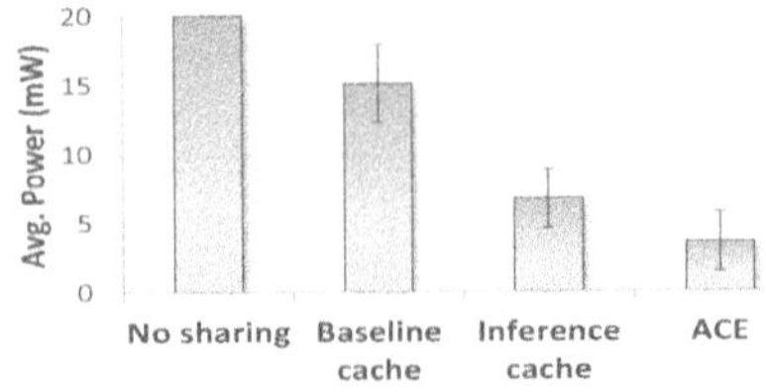

Figure 14: End-to-end energy savings

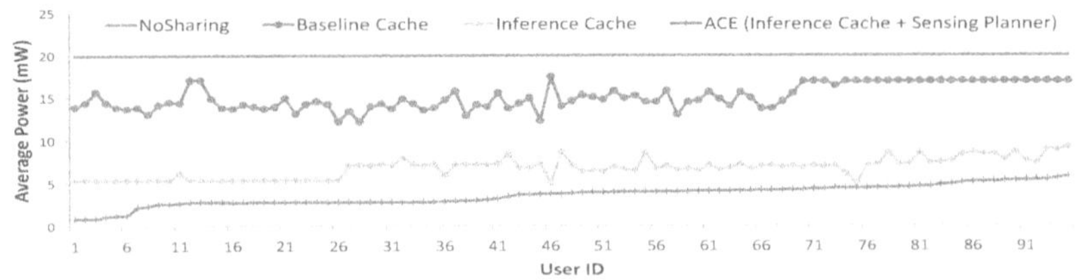

Figure 15: Per user power consumption (users sorted by ACE power)

9. CONCLUSION

We have presented ACE, a middleware for efficient and continuous sensing of user's context in a mobile phone. ACE automatically learns relationships among various context attributes and use them for two novel optimizations: inference caching and speculative sensing. Experiments with two real context traces of total 105 people and a Windows Phone prototype show that, compared to a traditional cache, ACE can reduce sensing costs of three context-aware applications by around 4.2×, with a very small memory and processing overhead.

Acknowledgements. We thank our shepherd, Alexander Varshavsky, and anonymous reviewers for their feedback. We thank Tarek Abdelzaher, Jie Liu, and Oriana Riva for feedback on earlier drafts; Miguel Palomera for writing an earlier version of the ACE Data collection tool; and various members of the Sensing and Energy Research Group at Microsoft Research for collecting context data.

10. REFERENCES

[1] AGRAWAL, R., IMIELIŃSKI, T., AND SWAMI, A. Mining association rules between sets of items in large databases. In *ACM SIGMOD* (1993).

[2] AZIZYAN, M., CONSTANDACHE, I., AND ROY CHOUDHURY, R. Surroundsense: mobile phone localization via ambience fingerprinting. In *ACM MobiCom* (2009).

[3] BANERJEE, N., AGARWAL, S., BAHL, P., CHANDRA, R., WOLMAN, A., AND CORNER, M. Virtual compass: Relative positioning to sense mobile social interactions. In *Pervasive* (2010).

[4] CHOUJAA, D., AND DULAY, N. TRAcME: Temporal activity recognition using mobile phone data. In *IEEE/IFIP International Conference on Embedded and Ubiquitous Computing* (2008).

[5] CUERVO, E., BALASUBRAMANIAN, A., CHO, D., WOLMAN, A., SAROIU, S., CHANDRA, R., AND BAHL, P. Maui: making smartphones last longer with code offload. In *MobiSys* (2010).

[6] DESHPANDE, A., GUESTRIN, C., HONG, W., AND MADDEN, S. Exploiting correlated attributes in acquisitional query processing. In *ICDE* (2005).

[7] EAGLE, N., PENTLAND, A., AND LAZER, D. Inferring social network structure using mobile phone data. In *Proceedings of the National Academy of Sciences (PNAS)* (2009), vol. 106, pp. 15274–15278.

[8] FICEK, M., AND KENCL, L. Spatial extension of the reality mining dataset. In *International Conference on Mobile Adhoc and Sensor Systems (MASS)* (2010).

[9] GREINER, R., HAYWARD, R., JANKOWSKA, M., AND MOLLOY, M. Finding optimal satisficing strategies for and-or trees. *Artificial Intelligence 170* (January 2006), 19–58.

[10] H. HÖPFNER, K. S. Cache-supported processing of queries in mobile dbs. *Database Mechanisms for Mobile Applications* (2003), 106–121.

[11] HAMMOND, L., WILLEY, M., AND OLUKOTUN, K. Data speculation support for a chip multiprocessor. In *ASPLOS* (1998).

[12] KANG, S., LEE, J., JANG, H., LEE, H., LEE, Y., PARK, S., PARK, T., AND SONG, J. SeeMon: scalable and energy-efficient context monitoring framework for sensor-rich mobile environments. In *ACM MobiSys* (2008).

[13] KWAPISZ, J. R., WEISS, G. M., AND MOORE, S. A. Activity recognition using cell phone accelerometers. *SIGKDD Explor. Newsl. 12* (March 2011), 74–82.

[14] LANE, N. D., MILUZZO, E., LU, H., PEEBLES, D., CHOUDHURY, T., AND CAMPBELL, A. T. A survey of mobile phone sensing. *Comm. Mag. 48* (September 2010), 140–150.

[15] LIN, K., KANSAL, A., LYMBEROPOULOS, D., AND ZHAO, F. Energy-accuracy trade-off for continuous mobile device location. In *ACM MobiSys* (2010).

[16] LU, H., PAN, W., LANE, N. D., CHOUDHURY, T., AND CAMPBELL, A. T. SoundSense: scalable sound sensing for people-centric applications on mobile phones. In *MobiSys* (2009).

[17] LU, H., YANG, J., LIU, Z., LANE, N. D., CHOUDHURY, T., AND CAMPBELL, A. T. The jigsaw continuous sensing engine for mobile phone applications. In *ACM SenSys* (2010).

[18] MILUZZO, E., LANE, N. D., FODOR, K., PETERSON, R., LU, H., MUSOLESI, M., EISENMAN, S. B., ZHENG, X., AND CAMPBELL, A. T. Sensing meets mobile social networks: the design, implementation and evaluation of the CenceMe application. In *ACM SenSys* (2008).

[19] MUN, M., REDDY, S., SHILTON, K., YAU, N., BURKE, J., ESTRIN, D., HANSEN, M., HOWARD, E., WEST, R., AND BODA, P. PEIR, the personal environmental impact report, as a platform for participatory sensing systems research. In *ACM MobiSys* (2009).

[20] MUNAGALA, K., BABU, S., MOTWANI, R., AND WIDOM, J. The pipelined set cover problem. In *ICDT* (2005).

[21] NIGHTINGALE, E. B., CHEN, P. M., AND FLINN, J. Speculative execution in a distributed file system. *ACM Trans. Comput. Syst. 24* (November 2006), 361–392.

[22] QIN, C., BAO, X., CHOUDHURY, R. R., AND NELAKUDITI, S. TagSense: A smartphone based approach to automatic image tagging. In *ACM MobiSys* (2011).

[23] SCHILIT, B., ADAMS, N., AND WANT, R. Context-aware computing applications. In *Proceedings of the 1994 First Workshop on Mobile Computing Systems and Applications* (1994).

[24] SCHIRMER, M., AND HÖPFNER, H. SENST*: approaches for reducing the energy consumption of smartphone-based context recognition. In *CONTEXT* (2011).

[25] TARZIA, S. P., DINDA, P. A., DICK, R. P., AND MEMIK, G. Indoor localization without infrastructure using the acoustic background spectrum. In *ACM MobiSys* (2011).

[26] ULLMAN, J. D. A survey of association-rule mining. In *Third International Conference on Discovery Science* (2000).

[27] WIRZ, M., ROGGEN, D., AND TROSTER, G. Decentralized detection of group formations from wearable acceleration sensors. In *Intl. Conf. on Computational Science and Engineering - Volume 04* (2009).

[28] ZAKI, M. J. Mining non-redundant association rules. *Data Mining and Knowledge Discovery 9*, 3 (Nov 2004), 223–248.

CoMon: Cooperative Ambience Monitoring Platform with Continuity and Benefit Awareness

Youngki Lee, Younghyun Ju, Chulhong Min, Seungwoo Kang, Inseok Hwang, Junehwa Song
Computer Science Department, KAIST, Daejeon, Republic of Korea
{youngki, yhju, chulhong, swkang, inseok, junesong}@nclab.kaist.ac.kr

ABSTRACT
Mobile applications that sense continuously, such as location monitoring, are emerging. Despite their usefulness, their adoption in real-world deployment situations has been extremely slow. Many smartphone users are turned away by the drastic battery drain caused by continuous sensing and processing. Also, the extractable contexts from the phone are quite limited due to its position and sensing modalities. In this paper, we propose CoMon, a novel cooperative ambience monitoring platform, which newly addresses the energy problem through opportunistic cooperation among nearby mobile users. To maximize the benefit of cooperation, we develop two key techniques, (1) continuity-aware cooperator detection and (2) benefit-aware negotiation. The former employs heuristics to detect cooperators who will remain in the vicinity for a long period of time, while the latter automatically devises a cooperation plan that provides mutual benefit to cooperators, while considering running applications, available devices, and user policies. Through continuity- and benefit-aware operation, CoMon enables applications to monitor the environment at much lower energy consumption. We implement and deploy a CoMon prototype and show that it provides significant benefit for mobile sensing applications.

Categories and Subject Descriptors
K.8 [**Personal Computing**]: General; C.3 [**Special-Purpose and Application-based Systems**]: Real-time and embedded systems

Keywords
Cooperation, Ambience, Context, Sensing, Energy, CoMon

1. INTRODUCTION
As continuous mobile sensing applications have been increasingly emerging, mobile users are starting to recognize smartphones as a personal sensing platform. Location monitoring services have been popularly deployed, for instance, spatial alarm and trajectory logging services [14][35]. New applications are also appearing; some notify a user about imperceptible information such as UV or dust levels [18], and others provide meaningful statistics from continuous activity observation [22]. These applications provide useful services to mobile users while running in the background, not requiring any explicit user intervention. However, many users are still reluctant to run such applications; they incur significant energy consumption and take up computational resources, potentially disrupting other common uses of the smartphones. This significantly compromises the spread of continuous sensing applications, depriving users of the benefits of running multiple such applications concurrently on their smartphones.

We approach the problem from a novel perspective, by utilizing *in-situ cooperation* of mobile users. We note that, for most people, their daily lives are highly social; they spend a significant portion of their time with others, e.g., family members, friends, colleagues, or even some strangers. According to our study, a user is co-located with acquaintances about 8.5 hours out of 15 active hours of a day, and even more, when accounting for co-location with strangers. In addition, 65% of meetings last for more than 30 minutes and 47% continue for more than an hour, allowing significant opportunities for stable cooperation in continuous sensing. Further, mobile users and their applications often share common interests in many situational contexts related to *ambience* such as locations, atmosphere, and social activities. These contexts can potentially be shared by nearby users, e.g., friends in a social gathering or people traveling in a bus. Thus, users can avoid repetitive sensing and processing redundantly performed by individual users that consume precious energy. This sharing becomes more practical due to the probable cost savings of the sharing. The power consumption for sensing and processing often exceeds the overhead to obtain context data from nearby users; for instance, directly performing location sensing every 10 seconds consumes 410 mW on a Nexus One phone while it consumes only 34 mW to receive the same data indirectly from others through Bluetooth communication (see Section 7 for detailed setup).

As an initial attempt to realize this approach, we propose CoMon, a novel cooperative ambience monitoring platform. CoMon automatically finds cooperators in situ and initiates the cooperation in a way that either enhances its energy capacity or extends its sensing modalities. Applications simply delegate their monitoring requests to CoMon and fully exploit cooperators' resources if available. By employing cooperation, CoMon significantly mitigates the quick battery depletion of devices, or overcomes the absence of specific sensing modalities.

A key challenge in the design of CoMon is how to construct cooperation groups and build network channels for continuous cooperation. We employ a continuity-aware cooperator detection method, which enables CoMon to maintain stable cooperation channels and reduce the complexity in cooperation network management. It leverages heuristic predictors based on social relationships and encounter history, and estimates the potential cooperation duration for candidate cooperators. The system is further designed to support pair-localized cooperation for efficient group management. This makes CoMon more robust against dynamic changes in group membership, by localizing the events such as the departure of a cooperator to a small number of pairs.

Table 1. Context examples and their categories

Context Category		Context Types
Ambience Context	*Spatial Context*	location, ambient sound, place, temperature, humidity, UV, dust-level, noise-level, mood, pollution (CO_2, O_3, …), crowdedness, …
	Social Context	discussion, meeting, conversation, lecture, group exercise, …
Personal Context		activity (walking, standing, …), gesture, health (heartbeat, gait, …), emotion, …

Another important challenge is to provide incentives to cooperating participants. Without benefits, a mobile user would be reluctant to actively participate in cooperation and share her resources for ambience monitoring. However, it is not a straightforward problem to guarantee mutual benefits to all cooperators. Cooperators often run different sets of applications, and possess different sensing devices. Also, they have their own preferences and policies in the use of energy of their devices. Such differences complicate the negotiation to guarantee fair and mutual benefit for cooperators. We propose a benefit-aware negotiation mechanism, which addresses the challenges and builds a mutually beneficial cooperation contract.

CoMon opens a new dimension to address the resource problem for continuous ambience monitoring. Many research efforts have been made to reduce energy consumption for location and context sensing and processing [16][26][27][30], taking an *intra-device optimization* approach, e.g., deactivating GPS based on mobility pattern [30], or optimizing a sound-data processing pipeline applying a noise-level filter [26]. Our *cooperation* approach complements such intra-device optimization techniques, providing further reduction in energy consumption. This additional dimension of benefit is significant, considering continuous and background operation of concurrent mobile sensing applications.

The contributions of this paper are as follows. First, we propose a novel cooperative ambience monitoring platform, CoMon; it significantly improves energy efficiency of smartphones and newly adopts unavailable sensing modalities. Second, we support the practicality of our cooperation approach through motivational studies on ATUS data [1] and Bluetooth-based encounter data [9]. These studies highlight the prevalence of continuous cooperation opportunities as the co-location duration of people comprises a significant portion of a day and social events like meetings continue stably in many cases. Third, as core techniques, we develop continuity-aware cooperator detection and benefit-aware negotiation mechanisms, which enables CoMon to obtain resource benefits from cooperation. Finally, we perform extensive experimental studies based on our prototype implemented over Android phones and custom-designed sensor motes, with diverse sensing capabilities. We show the resource benefits and overheads for diverse cooperation scenarios and applications.

In the rest of the paper, we first motivate CoMon in Section 2 with scenarios and studies on mobility data revealing many opportunities for cooperation. Section 3 describes the model of cooperation benefits and the CoMon architecture. Section 4 and Section 5 describes the two core techniques. We then present our implementation in Section 6 and Section 7 shows experimental results. In Section 8, we discuss other potential issues for CoMon and Section 9 introduces related work. Finally, we conclude the paper in Section 10.

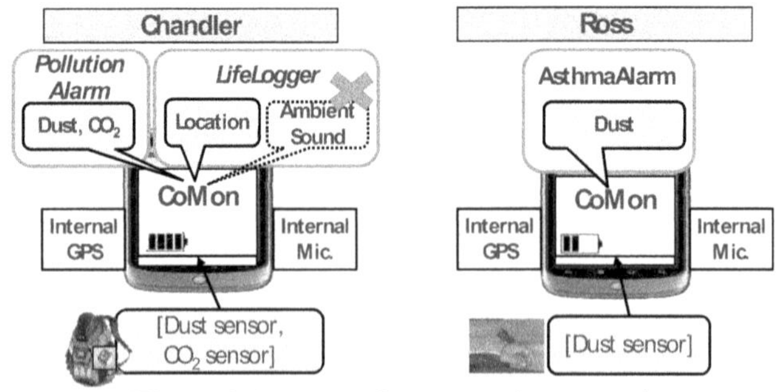

Figure 1. An example cooperation scenario

2. OPPORTUNITY FOR COOPERATIVE CONTEXT MONITORING

Mobile sensing applications have high potential to leverage cooperation between nearby people. As they become popular, many of them will run concurrently, actively utilizing diverse user contexts. Table 1 shows example contexts used by emerging mobile sensing applications [2][18][22][26][27][31][32]. Among them, a number of *ambience contexts* including *spatial* and *social contexts* would be potentially shareable with nearby people.

Along with increasing demands on monitoring diverse contexts, smartphones will incorporate more sensors and people will carry external sensors in diverse form factors, e.g., a watch-type health sensor for everyday precautions against heart attacks, a backpack-attached air-quality sensor to alert the user to possible fine-granule pollutant exposure in everyday life spaces. Considering the globalized nature of consumer product manufacturing, many people will likely have devices with similar sensing capabilities and thus have a high potential for sharing sensing resources.

Understanding that there will be many sharable contexts, two key questions are raised: (1) Does the cooperation result in actual energy benefits for context monitoring? (2) Are there enough cooperation opportunities in the everyday life of mobile users?

We first demonstrate an interesting scenario showing the expected cooperation cases and their benefits in Section 2.1. Note that the energy-related figures used in our scenario are presented based on our actual measurements (See Section 7 for detailed settings including sensor specifications). Then, we show that the opportunities for cooperation are actually prevalent in everyday life through our analysis of human activity and mobility datasets in Section 2.2. The cooperation benefits are further elaborated through comprehensive experiments in Section 7.

2.1 Cooperative Context Monitoring Scenario

Chandler, Ross, and Joey are friends in Manhattan. On Saturday, Chandler plans to meet Ross for shopping in the SoHo area. Chandler always runs two apps, *PollutionAlarm* and *LifeLogger* as in Figure 1. He runs *PollutionAlarm* to avoid exposure to air pollution such as dust and exhaust fumes, and *LifeLogger* to record his route (using GPS) and optionally ambient sound contexts (using the microphone to record music genres, meetings, etc.) [26][27]. Today, he turns off the ambient sound monitoring to extend the phone's battery life. Ross runs *AsthmaAlarm* due to his asthma problem. It monitors the dust levels in the air, a major allergen for asthmatics. While Ross is on his way to SoHo, he discovers that his dust sensor blinks notifying him that there are 'fewer than 3 hours of battery remaining'. Ross gets anxious, regretting that he forgot to recharge the sensor last night.

Table 2. Summary statistics of activity and mobility datasets

Dataset	ATUS	MIT/BT
Data source	American time use study (interview)	Bluetooth scanning trace (period: 5 min)
Participants	13,258	100
Start time	01/01/2010	09/08/2004
Duration	1 year	3 months
# of events	257,193 activities	285,512 encounters

When Ross meets Chandler, Ross's CoMon starts cooperation with Chandler's to monitor the dust level in turn. This reduces the net power-on duration of each sensor by half; the average power consumption by Ross's dust sensor decreases to almost half, from 848 to 487mW, and the estimated sensor lifetime is increased from 3 to 5.2 hours. Note that Ross's smartphone requires a slight additional power expenditure of 25mW to update the dust level to and from Chandler's phone during the cooperation.

Joey was walking in a park near SoHo for his daily exercise; he runs the *CalorieMonitor* application which uses his movement speeds for calorimetry calculation. He also uses *LifeLogger*. On his way home, Joey happens to meet Chandler and Ross and they decide to go to a café. Detecting Joey's devices, Chandler's CoMon system entrusts sound monitoring to Joey's CoMon while supporting location monitoring for Joey instead. This cooperation enables Chandler's *LifeLogger* to again be fully functional by reactivating the disabled sound monitoring. Now Joey's phone turns off energy-intensive GPS sensing which has consumed 440mW; instead, it needs only 11mW to receive location context from Chandler. The additional cost to Joey's device to provide the ambient sound context is marginal (12mW), since he has been monitoring this context for his own purpose. Through the cooperation, the total power consumption of Joey's phone is reduced from 570 to 365mW, increasing its lifetime by about 56%.

Note that CoMon achieves savings on other resource types such as CPU and memory as well. For instance, for ambient sound processing, over 20 operators for feature extraction and context classification such as FFT and GMM run continuously, requiring over 10% CPU cycles on a Nexus One phone. In the above scenario, CoMon allows Chandler to monitor the sound-driven context without such a cost through the cooperation with Joey. Furthermore, the benefit of CoMon is not limited to applications utilizing ambience contexts. Since smartphones are shared by multiple concurrently running applications, the amount of resources saved by cooperation can be leveraged to monitor other personal contexts or for other common uses.

2.2 Study on Cooperation Opportunity

To study cooperation opportunities in the daily life of mobile users, we analyze two public datasets on human activity and mobility behaviors. Table 2 shows their summaries. Note that while there have been many efforts on mobility data analysis, few focused on revealing cooperation opportunities; most of them focuses on revealing human mobility patterns,

ATUS: The American Time Use Survey (ATUS) dataset [1] includes the list of all activities of American participants over a 24 hour period and the acquaintances who were present during each activity. We use the dataset collected in 2010 from 13,258 interviewees over wide age, sex, and occupation distributions. We analyze this data to find the cooperation opportunities in everyday activities, especially in terms of the acquaintances being together.

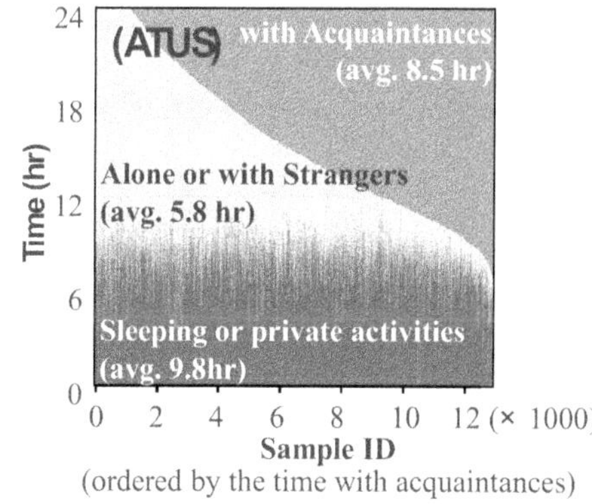

Figure 2. Distribution of the time together

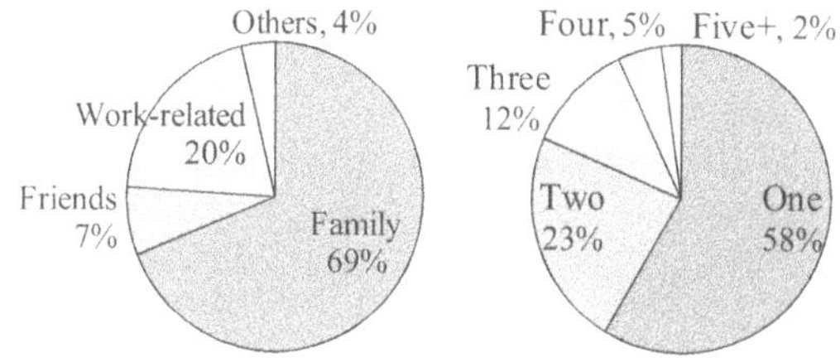

(a) Acquaintance type (b) Number of acquaintances

Figure 3. Detailed analysis of the time with acquaintances

MIT/BT: The MIT/BT dataset is the mobility dataset collected from 100 mobile phones of MIT students and staffs [9]. It is collected by Bluetooth scanning performed every 5 minutes. We analyze the encounters between the phones, i.e., the encounters with nearby people including strangers as well as acquaintances. We use the three-month dataset from the fall semester of 2004.

2.2.1 How Many Opportunities for Cooperation?

The longer people are together with others, the more opportunities for cooperative context monitoring we can exploit. To quantify the amount of such time in everyday life, we analyze the ATUS dataset. We do not use the MIT/BT dataset here since it is limited to the devices only discoverable by Bluetooth scanning.

Figure 2 shows the daily amount of time in terms of the presence of acquaintances for every participant. The average time with one or more acquaintances is 8.5 hours. We can confirm that people have lots of cooperation opportunities with acquaintances, i.e., more than one-third of a day. Specifically, 78% of the participants have more than 4 hours of co-located time with others, and 50% have more than 9.3 hours.

Figure 3(a) elaborates on with whom and how long participants spent time with acquaintances, i.e., family (average 5.9 hours), work-related people (1.7 hours), friends (0.6 hours), etc. Figure 3(b) shows the number of acquaintances a user is together with; it gives an intuition on the number of cooperator candidates at a time. For 42% of the time, people are with more than one acquaintance, giving more chances of cooperation.

Note that the opportunities for cooperation are not limited to those with acquaintances but can include those with strangers, e.g., a user in a bus can cooperate to monitor the route of the bus with other passengers. However, the ATUS dataset does not contain encounters with strangers.

2.2.2 Continuity of Cooperative Monitoring

We study the continuity of meetings, i.e. how long people are together during an encounter. Once cooperative monitoring has been

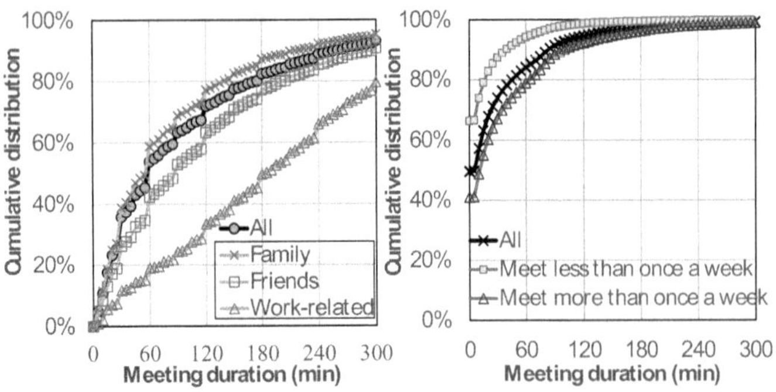

(a) ATUS (b) MIT/BT

Figure 4. CDFs for meeting duration between people/devices

established when they are together, this monitoring could continue as long as they remain together. Long-lasting cooperation would enable prolonged support for applications.

We study the ATUS dataset to identify how long the cooperation with acquaintances could last. Figure 4(a) shows the distribution of meeting durations. We define a meeting as consecutive activities with the same people, e.g., shopping at a department store with the spouse, then travelling together to a restaurant and having dinner. The figure shows the distributions for different acquaintance types. For 'all', it shows that 65% of meetings with acquaintances last for more than 30 minutes and 47% last for more than one hour. From these results, we can obtain typical durations of cooperative monitoring between acquaintances.

Figure 4(a) also shows that the meeting duration largely depends on the type of acquaintance; meetings with work-related people usually continue for the longest time, followed by friends and family. When there are multiple acquaintances nearby, we can consider their social relationship to select better promising cooperators in terms of continuity.

We study the MIT/BT data to determine the duration of meetings between two arbitrary mobile users without distinguishing between strangers and acquaintances. Figure 4(b) shows the distribution of encounter durations. We define the duration of an encounter as the time during which a user's mobile device is consecutively scanned by the other users'. Overall, the encounter duration is shorter than that in ATUS; 50% of people pass by within a duration of 5 minutes (a single detection only) and 26% stay together for more than 30 minutes. This is mainly because it includes strangers and people passing by. Also, if two people temporarily move apart from each other for a few minutes, e.g., one goes to the restroom, the encounter is segmented and treated as two separate meetings. Importantly, we found that the meetings between people who meet frequently, namely familiars, last longer. For people who meet once a week or more, 36% of the meetings between them last for more than 30 minutes while only 13% of the meetings with the others do. This implies that meetings with familiars would yield longer cooperative monitoring opportunities.

3. COMON DESIGN

3.1 Benefit-aware Cooperation Approach

A key goal of CoMon is to maximize the energy benefit from the opportunistic collaboration with nearby users. For effective design of CoMon, we first attempt to model the energy benefits obtainable from the cooperation with a cooperator, as shown in Figure 5. In the model, we divide the operation time into two periods, i.e., discovery period and cooperation period.

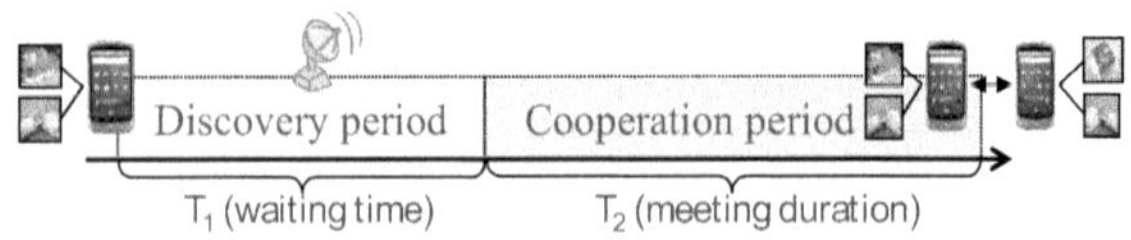

(a) Cooperation model for a cooperator

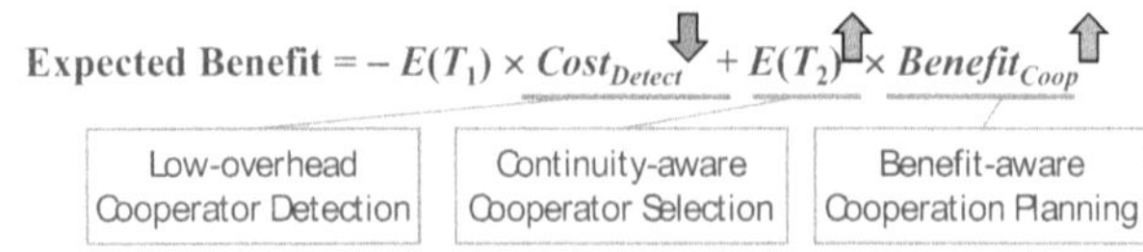

(b) Cooperation benefit and approaches

Figure 5. Cooperation benefit model

In the discovery period, the system attempts to detect nearby cooperator candidates. This incurs cost, which can be represented as $Cost_{Detect} \times E(T_1)$; $Cost_{Detect}$ is the average discovery cost per unit time, T_1 is the random variable of the waiting time until meeting a cooperator, and $E(T_1)$ is the expected waiting time. Once the cooperation starts with a cooperator, it will produce benefit. We model the benefit for the cooperation period as $Benefit_{Coop} \times E(T_2)$, where $Benefit_{Coop}$ is the average benefits from cooperation per unit time and $E(T_2)$ is the expected duration of cooperation. Taking all the cost and the benefits into account, the expected total benefit can be specified as follows:

$$Expected\ Benefit = Benefit_{Coop} \times E(T_2)\ -\ Cost_{Detect} \times E(T_1)$$

To increase the expected energy benefit, we devise the steps of cooperation as below.

Cooperator detection: The first step is to continuously detect nearby users as potential cooperators. The discovery period should be carefully chosen; a long interval reduces the discovery cost, i.e., $Cost_{Detect}$, but might decrease the potential cooperation duration, $E(T_2)$. (See Section 4)

Cooperator selection: Increasing the cooperation period T_2 is crucial for higher benefit $E(T_2)$. We explore the potential to predict the meeting durations upon discovery of a cooperator candidate. The negotiation for real cooperation starts only with the candidates who can lead to long enough cooperation to provide benefits. We develop the continuity-aware cooperator selection method (Section 4.1).

Cooperation planning: To increase the benefit per unit time, it is important to carefully determine a cooperation plan, e.g. selection of contexts to share or distribution of tasks to different devices. Different plans could significantly influence the benefit from the cooperation. We develop a planning method, which carefully decides the cooperation plan for higher $Benefit_{Coop}$. It ensures mutual benefits to both cooperators, since either of them would be reluctant to participate if benefits are not mutual (Section 5).

3.2 Architecture Overview

We carefully designed the architecture of CoMon as shown in Figure 6 applying the benefit-aware cooperation approach. It runs as a middleware on top of a smartphone OS and additionally supports external sensing devices [17][21]. CoMon provides mobile sensing applications with intuitive APIs, allowing the applications to specify the contexts of interest (e.g. location, activity) in a declarative query [16]. Consider a pollution monitor application that wants to monitor CO_2 level with 90% of accuracy every 30 seconds. Then, it specifies the query as follows:

CONTEXT CO_2 level **ACCURACY** 90%
PERIOD 30 Seconds **DURATION** Always

CoMon processes registered queries by leveraging cooperation opportunities with nearby users; it handles the query with the user's own devices if it finds no potentially beneficial cooperator. CoMon takes charge of all the underlying tasks for opportunistic cooperation, which are transparent to the applications. In terms of applications, the quality of service (QoS) provided by CoMon might vary from time to time due to the heterogeneity of devices or dynamic system situations. We believe that the slight QoS difference caused by cooperation does not cause severe problem for many applications; current Android sensing APIs do not also guarantee fine-granule QoS for GPS and accelerometer sensing. For the applications that have hard QoS requirements, CoMon may not initiate cooperation or can check QoS condition further while planning.

The benefit-aware cooperation approach is realized by two key components: *cooperator detector* and *cooperation planner*. The *cooperator detector* dynamically discovers nearby devices by periodic Bluetooth scans at a small overhead, and selects candidates that will potentially stay in the vicinity for a long period (long stayers) (Section 4). The *cooperation planner* negotiates with the selected one, and then decides the best cooperation plan (Section 5). Note that when the cooperator leaves, CoMon instantly processes the contexts provided by the cooperator with its own devices.

According to the cooperation plan, the *context processors* on the smartphone and sensors continuously process the requests and deliver the processing results to the applications and cooperators (via Bluetooth in current implementation). It incorporates a variety of processing modules for sensing, feature extraction, and context classification to support diverse types of contexts.

The *device manager* provides the cooperation planner with up-to-date energy information, required to make a proper plan. As a basic support of privacy, CoMon employs *access controllers*, which restrict unauthorized accesses to certain contexts. CoMon allows users to specify the access rules about what context information can be shared with whom. We further discuss the privacy issues and access rules in Section 8.

We employ a smartphone-centered architecture; external sensor devices and their data are exposed to cooperators only through smartphones. We suppose that external sensors would be designed in a small size with minimal computational resources. It would be hard for such sensors to perform as an independent participant for cooperation since they are limited in supporting multi-user connections or processing raw sensor data with complex modules.

3.3 Design Considerations and Choices

We below present key considerations and our choices for the design of practically working cooperation systems like CoMon.

Long-term cooperation: Dynamic changes of cooperators due to their mobility could incur overheads for frequent discovery, negotiation, and connection management. To minimize such overheads, CoMon targets the cooperation with long stayers only. Even when a user walks around in crowded Manhattan, CoMon selects the cooperators only among acquaintances doing the activity together, or familiar strangers who stay together for more than a certain amount of time. We find that even with only long-term cooperation, there are sufficient opportunities.

Pair-wise negotiation: When there are multiple cooperator candidates, it is important to determine how to organize the group for cooperation planning and execution. Our key idea is to localize

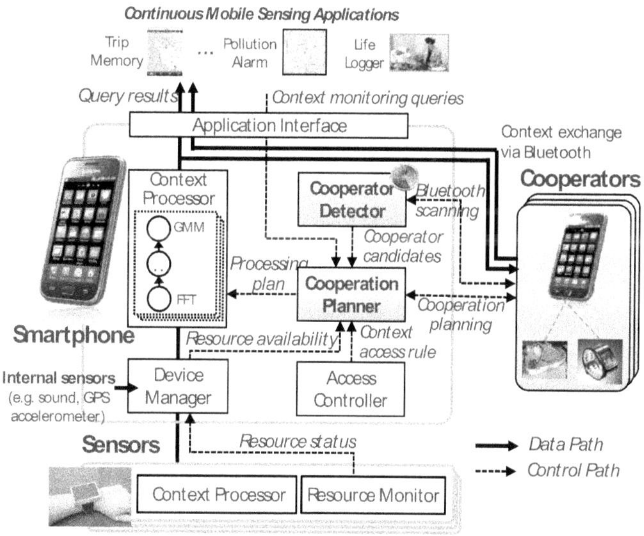

Figure 6. CoMon architecture

the effect of membership changes. CoMon performs the cooperation in the unit of a pair to localize the effect within some pairs. It negotiates with the cooperator candidates one at a time and incrementally continues the negotiation. An alternative approach would consider the whole group as a single cooperation unit, and perform a group-wide negotiation at once. Although this approach would lead to the group-wide resource optimum, it is relatively vulnerable to the mobility of users. Whenever a single cooperator joins or leaves, all the remaining cooperators should re-perform the negotiation process. Also, the group-wide negotiation significantly escalates the complexity of cooperation planning due to numerous options and policy conflicts.

Context-level service as cooperation interface: For negotiation, an important design choice is the appropriate abstraction level in exposing a mobile user's resources to cooperators. CoMon exposes underlying resources of the user's devices as context-level services. A context hereby means high-level information extracted by classifying or aggregating raw sensing data, e.g., a place category extracted from ambience sound or a 30-second averaged dust count in the air. The context-level service hides heterogeneity and dynamics of other cooperators' resources. Also, context-level exchanges could greatly save energy which might be high if high-rate raw data are exchanged. We assume that there would be consensus on a common context model as in [30], which could help extend the scope of the cooperation. Based on such model, different applications running over heterogeneous devices can share and exchange context information. Even with the common consensus on a context, different applications may require different level of accuracies and sampling and processing intervals. CoMon can evaluate such condition in the planning process but we do not handle such cases here for simplicity.

4. COOPERATOR DETECTION

A first step of CoMon operation is to find out the cooperators who will provide benefits through cooperation. The benefits will be provided when 1) cooperators stay longer enough to breakeven the overhead to start cooperation, and 2) there exists sharable contexts for which local execution takes more resources. A good detection method should locate such cooperators with low overhead. However, this is a significant challenge. First, it is difficult to accurately

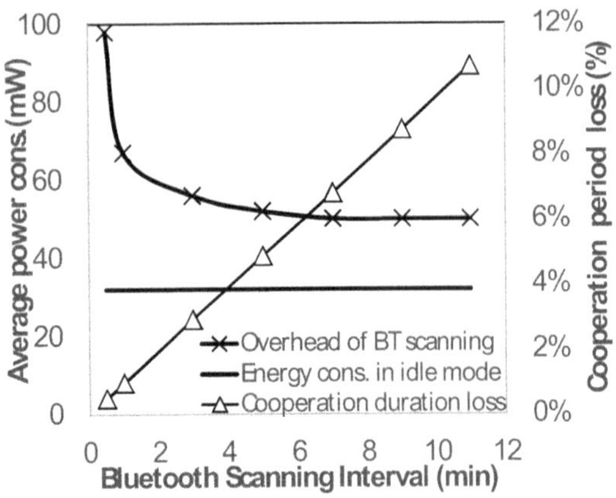

Figure 7. Tradeoff between energy overhead and potential loss in cooperation time

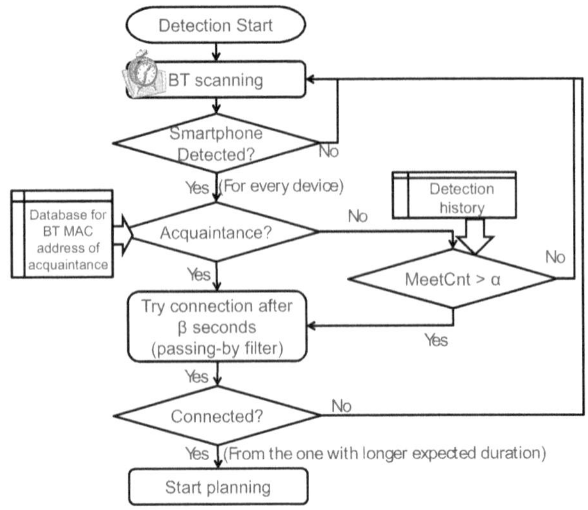

Figure 8. Continuity-aware cooperator selection

predict the length of encounters in advance. Moreover, it is costly to check with all candidates if they have been monitoring contexts that are sharable and beneficial.

4.1 Bluetooth-based Cooperator Detection

An early decision to build CoMon lies in selecting networking interfaces for power-efficient communication among nearby users. As candidates, we consider several popularly used interfaces such as ZigBee, Bluetooth, and WiFi-Adhoc. ZigBee achieves low power but at low bandwidth. WiFi-Adhoc makes opposite tradeoffs and Bluetooth is in the middle. Also, each protocol has its own characteristics, such as broadcast supportability and limited number of connections. The current CoMon implementation adopts Bluetooth, mainly because the interface is pervasive in most commodity smartphones. Also, it provides reasonable coverage, i.e., about 10 meters, for context sharing among the users who do the same activities. Note that for context sharing, protocols like ZigBee might be more adequate once available; it consumes an order of magnitude less power than Bluetooth with reasonable bandwidth (<250kbps).

To detect cooperator candidates, CoMon performs periodic Bluetooth scans, a common method for identifying devices in proximity [23]. A key issue for the discovery lies in deciding an appropriate scan interval T. More frequent scans incur higher energy overheads but enables quicker discovery of potential cooperators. The probable loss in cooperation duration would be $0.5T$ on average. We study the potential tradeoff by measuring power consumption for the scanning on a Nexus One phone, and estimating the potential loss in cooperation duration from ATUS dataset. Specifically, the loss is estimated by subtracting $0.5T$ from all the activity duration with acquaintances, assuming the acquaintances as potential cooperators. Figure 7 shows the results. A good value of T can be carefully determined leveraging the results. For instance, CoMon can use 5-minute interval, where the power consumption starts to saturate. At this interval, the average loss in potential cooperation duration would not be significant, i.e., 4.8% out of expected cooperation time per person; note that many acquaintances are likely to stay for a long period of time.

In addition to the energy overhead, the available network bandwidth is reduced during scanning [12]. It may make the connections with cooperators unstable, even disconnected. Against such instability, CoMon buffers the messages and defers the exchange during the scanning for 10 seconds. Also, CoMon stops scanning when the current cooperation benefit exceeds a certain threshold.

4.2 Continuity-aware Cooperator Selection

CoMon needs to distinguish cooperator candidates that are likely to stay together among the ones discovered. There would be a broad spectrum of candidates in terms of potential meeting durations, from those quickly passing by to those staying for hours. It would clearly be useful to know the potential meeting durations with the candidates.

A most interesting observation to automate estimation was made with respect to the social relationship between the potential collaborators. From our study on the ATUS dataset, meetings with acquainted people, e.g., family members, friends, and co-workers, are likely to continue long enough; this can be expected since the acquainted people usually share common purposes for meetings, e.g., doing an activity together. Also, the expected meeting durations differ according to the types of acquaintances or their activities, e.g., socializing or working.

For unknown relationships, the system can further utilize previous encounter or meeting histories. One may expect that the duration of a meeting with a person can be estimated from the distribution of the previous meeting durations with that person. However, our studies on MIT/BT dataset saw little correlation between the durations of previous meetings and the current one. Interestingly instead, we could see that number of previous encounters affects the current meeting duration.

Based on the observation, we develop a *continuity-aware cooperator selection* method. It distinguishes the candidates into multiple categories with different expected cooperation duration, e.g., passing-by, acquaintances, and strangers with multiple encounters. CoMon preferentially starts to negotiate with those in the category of longer expected duration. Figure 8 shows the overall flow of the selection process.

Acquaintance selection: Following the observation, the method first selects the devices owned by acquaintances. According to the ATUS data, the average expected continuation time is 51 minutes for acquaintances. The method further utilizes the type of acquaintances if the information is known, e.g., the expected cooperation time with a friend can be estimated as 128 minutes (see

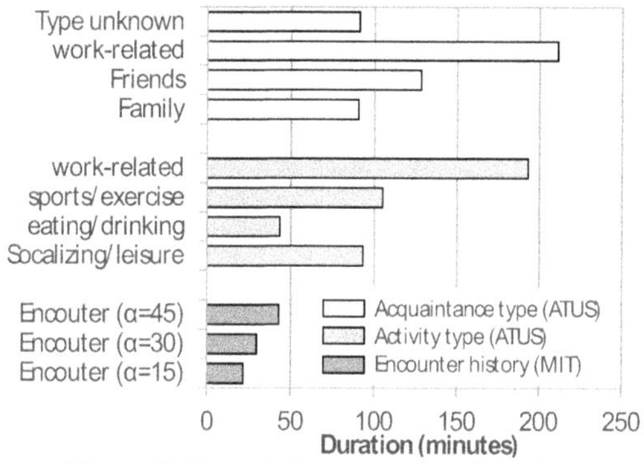

Figure 9. Expected cooperation duration

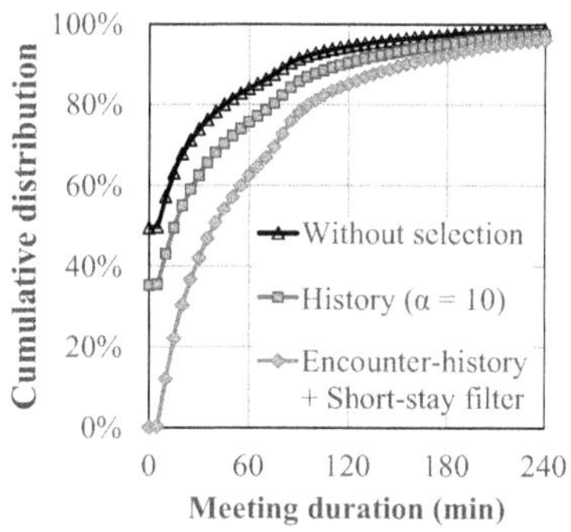

Figure 10. The effect of the cooperator selection (MIT/BT)

Figure 9). It can further consider activity types once it becomes available to the smartphones. To identify if a device is owned by acquaintances, the detector maintains a small database to match acquaintances' devices with their Bluetooth MAC addresses. To obtain this information, CoMon asks the user to manually designate the owner of a device from the phone book upon the first pairing; CoMon allows the user to disable the new pairing to avoid frequent new pairings with strangers.

Encounter history-based selection: When the cooperation with acquaintances are not possible, however, the method further finds other opportunities with the devices that have been frequently discovered more than a certain threshold, α within a time period, e.g., a month. Even when they are not in the contact list, they are highly likely to be a familiar stranger [28], e.g., a person in the same commuter bus. Figure 9 also shows the expected cooperation duration for different α values estimated from MIT/BT dataset.

Short-stay filter: Lastly, the method applies a *short-stay filter* to exclude passing-by devices out of the devices recommended by the two predictors. This is important since not every encounter by acquaintances or frequently meeting people leads to a longer meeting. The key idea is that the devices that already stayed for a certain time are likely to stay longer. We can find an analogy with a least-recently-used (LRU) paging algorithm in OSs. The method attempts to connect to the devices selected by predictors after a certain time, β seconds, not instantly after the discovery.

We validate the effectiveness of the encounter history-based selection and short-stay filter by emulating them over the MIT/BT dataset (α=10 in a two-week window, β=300 seconds). We skip acquaintance selection as we do not know about the relationships among the participants with the dataset. Figure 10 shows our method effectively selects the users who are likely to stay longer. For example, whereas the probability that a meeting lasts over 30 minutes is 26%, it increases to 38% when the encounter-history selection is applied, and further increases to 58% with the short-stay filter. The short-stay filter seems to be quite useful while the effect of encounter-history selection is not very significant. We thus apply the encounter-history based predictor only when cooperation is not available with acquaintances. We acknowledge that a large-scale deployment will be more realistic than our emulation with the MIT/BT dataset, and are working towards that.

5. COOPERATION PLANNING

With the candidates recommended by the detector, CoMon conducts cooperation planning, to decide which contexts to share and trade. For each user, the main goal of the planning is to maximize her benefit. For the system, the goal is to provide mutual and fair benefits for cooperators.

However, providing such maximized and mutual benefits is not a simple problem. An initial solution is that cooperators take turns to monitor the common contexts when the cooperation leads to benefits. For this, the system just needs to identify common requests of cooperators. For each context, it compares the energy demands to process it locally with the potential demands upon cooperation. Note that the cooperation is often beneficial for the contexts requiring energy-intensive sensing like GPS or dust sensing, or heavy processing over high-rate data like ambient sound context. For the beneficial contexts, a cooperator could monitor them for certain time duration, e.g., ten minutes and then the other monitors them for the next ten minutes.

Such a solution works in simple cases, but it needs to be further improved to deal with complex system environments. A key challenge results from the complexity in benefit estimation. The cooperation benefit cannot be statically determined in advance; even for the same cooperation pattern, the benefit could vary depending on resource availability, operational applications, and user policies. To be specific, first, the energy demand to monitor a context might be differently accounted considering other concurrently monitored contexts. CoMon processes multiple contexts in a shared way; it figures out the overlapping tasks among contexts, e.g., sensing, processing and communication tasks, and eliminates redundancy. Accordingly, the cooperation benefit for a context needs be evaluated, taking such shared evaluation into account. Moreover, cooperators expect different types of benefits and have different policies on energy use. A user who will be outside quite a while would want to save energy as much as possible, but the one who will soon go home would not mind consuming energy if he can benefit from new contexts.

The planning becomes even more challenging as we target future environments where a user carries a number of wearable sensors together with a smartphone. In this case, monitor-able contexts among users vary quite much, and sharing common contexts only provides limited benefits. Also, the user policies could be more complex reflecting the in-situ availability of sensing devices and their remaining energy in combination. For example, in Section 2, Joey obtains significant benefit from location sharing, saving battery of his power-hungry smartphone. However, the benefit would be relatively less if he uses a powerful external GPS; in this case, sharing through the smartphone might be even a loss.

5.1 Cooperation Planning Problem

To understand the problem in depth, we first clarify the problem and cooperation benefit. According to our context-level sharing principle, we describe a cooperator, u, as follows:

Input: $\{ctx_d\}$, a set of contexts to monitor
Output: cost to execute $\{ctx_d\}$
1. ***EDVector*** **←** ***GetEDVector***($\{ctx_d\}$)
2. *totalEC* ← 0 // init total energy consumption
3. for ∀ d_j where d_j is a device in *EDVector*
totalEC ← *totalEC* + $weight_j$ · ***EDVector***(d_j)
4. Return *totalEC*

Figure 11. A cost function for a policy 2

Def 1. A cooperator, *u*, is specified as: ***u*** **= <*D*, *S*, *P*>**, where

- *D* is a set of demanding contexts, $\{ctx_d\}$, by applications; the set is obtained from the registered queries in CoMon.
- *S* is a set of supply-able contexts, $\{ctx_s\}$, which CoMon can monitor and provide to cooperators; CoMon identifies the set based on available devices and their resource availability.
- *P* is a policy that denotes the desirable benefit from the cooperation. It is given by the user based on his preference or resource situation. The policy is substantialized as a cost function, $cost_P$, within the system. If $cost_P$ is reduced as a result of cooperation, the cooperation is considered beneficial.

Now, given two cooperators u_1=<D_1, S_1, P_1>, and u_2=<D_2, S_2, P_2>, the cooperation planning problem is to find *a cooperation plan, CP*, as its output for the estimated cooperation duration, where

- CP = {(ctx_c, u_i, t) | ctx_c is a context to cooperatively monitor, u_i is a cooperator in charge, either u_1 or u_2, t is a time duration to take charge},

such that $cost_{p1}$ and $cost_{p2}$ should decrease by applying CP. In each person's viewpoint, the purpose is to minimize her cost function but it could conflict with the purpose of the cooperator.

5.2 Cooperation Benefits and Policies

A user can apply diverse policies to describe his preferential benefits from cooperation. We first introduce useful example policies which are described in terms of device resources and application supportability.

Policy 1: A basic policy is to save the battery consumption of a smartphone for context monitoring. Since a phone is a generic personal computing platform utilized for diverse applications, it is better to save the battery power by default.

Policy 2: When a smartphone works together with external sensing devices, a user might want to consider the battery status of other devices as well. According to the importance and use cases of devices, a policy can be defined to reduce the weighted sum of power consumption over distributed sensing devices.

Policy 3: In terms of application supportability, a policy can be defined as to increase the number of supported queries. CoMon may not continue to support some requests due to shutdown or low battery level of corresponding devices. The policy attempts to resume the support for such requests through cooperation.

Policy 4: Sometimes, it is expected that a user will recharge the devices after a certain time, *T*, e.g., 3 hours. In this case, a policy is specified to increase the running time up to 3 hours for all applications if some cannot be satisfied only with local resources.

CoMon provides several system functions to enforce policies as cost functions. The key primitives are *getEDVector*($\{ctx\}$) and *getEAVector*(). *getEAVector()* returns the remaining energy of all sensing devices. Given a set of contexts to monitor, $\{ctx\}$, *getEDVector*($\{ctx\}$) returns the expected power consumptions on relevant sensing devices. For example, *getEDVector*($\{Dust\}$) returns an energy demand vector, (28.5mW, 720.7mW), where the elements represent the energy demands on the smartphone and dust sensor, respectively. Figure 11 shows an example cost function, $Cost_{P2}$, realizing policy 2 based on the primitives.

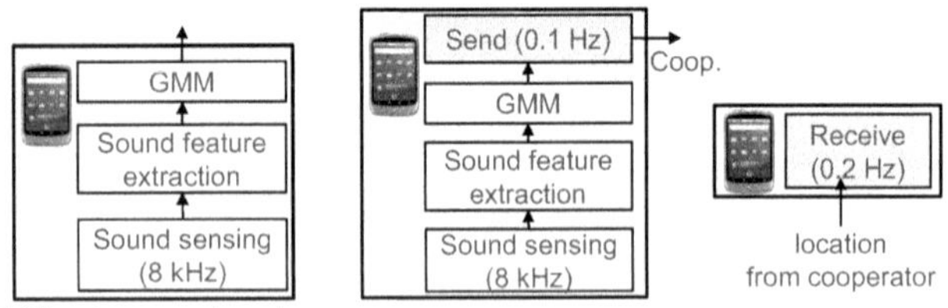

(a) local processing plan for ambient sound (b) $cplan_{out}$(ambient *sound*) (c) $cplan_{in}$(*location*)

Figure 12. (a) a local plan for ambient sound context (simplified), (b)(c) *cplans* for *case_ex*(*sound*, *location*) when Joey cooperates with Chandler in the scenario in Section 2.

To compute the energy demands, CoMon manages energy use profiles for the sensing, processing and communication tasks required to monitor contexts; currently, energy use are profiled offline while on-line profiling can be applied further. In CoMon, the processing of a context is represented as a graph of tasks, denoted as a *processing plan*. Figure 12(a) shows the example plan for the ambient sound context (see [15] for a detailed graph). CoMon estimates the energy demand to execute a plan by adding the energy demands for all tasks constituting the plan. Note that CoMon properly reflects the effect of the shared processing; the energy demands for redundant tasks between multiple contexts are accounted only once.

We build the system functions extending our previous systems [17][21]; they leverage such information for the coordination of multiple applications' resource use over personal devices. While they utilize multiple alternative plans for a context, we suppose that CoMon has a single plan for each context to focus on cooperation.

5.3 Benefit-aware Negotiation Mechanism

Careful cooperation planning is essential to provide cooperation benefits under complications in running applications, available sensing devices and their energy, and user policies. It becomes more complicated to provide mutual and fair benefits for both participants as they may have conflicts in planning decisions.

To address such challenges, we develop a benefit-aware negotiation mechanism. As a key idea, the mechanism pursues the fairness of opportunity to make beneficial cooperation decisions by themselves, rather than guaranteeing mutually identical benefits to cooperators; the identical benefit is not even possible due to participants' different policies and energy availability. In this principle, the mechanism utilizes one-to-one context exchange as a first-stage negotiation unit, providing each cooperator a chance to weigh up the unit by its own cost function. For each unit, it estimates the benefit reflecting in-situ resource availability and concurrent requests. Then, the benefit is cross-validated by each cooperator to ensure mutual benefits; the cases beneficial to only one side are excluded in advance, so that the planning results ensure the mutual benefit. Finally, the mechanism allows the cooperators to take turns to select the unit of exchange, providing each participant with fair opportunities to maximize its benefit.

In more detail, the mechanism introduces a *cooperation case*, as an atomic unit of cooperation planning; we describe the mechanism in perspective of a cooperator, u_1. We identify two representative types of cooperation cases as follows. Note that cooperation cases are built on a context level, hiding the low-level resource details of a cooperator.

- **Exchange of two contexts, ctx_{out} and ctx_{in},** denoted as **$case\_ex(ctx_{out}, ctx_{in})$**, is a case that u_1 obtains a context ctx_{in} from u_2 in exchange of providing ctx_{out}. This case enables the participants to save the energy by delegating the costly monitoring of a context or obtain an unavailable context.
- **Co-monitoring of a context, ctx_{co}, $case\_co(ctx_{co})$**, is a case that u_1 and u_2 monitor ctx_{co} in rotation. This case enables the participants to save the energy by halving the monitoring duration of the context.

With the cooperation cases, our planning method is performed in following three steps.

Step 1. Cooperation case generation: First, participants generate applicable cooperation cases by exchanging their demanding and supply-able contexts, i.e., D and S, with each other. The generated cases include a set of exchange cases, EX, and a set of co-monitoring cases, CM, where

- $EX = \{case\_ex(ctx_{out}, ctx_{in}) \mid ctx_{out} \in (S_1 \cap D_2), ctx_{in} \in (D_1 \cap S_2), ctx_{out} \neq ctx_{in}\}$, and
- $CM = \{case\_co(ctx_{co}) \mid ctx_{co} \in (S_1 \cap D_1 \cap S_2 \cap D_2)\}$.

For an exchange case, ctx_{out} is the one that u_1 provides and u_2 demands. ctx_{in} is vice versa. Second, a co-monitoring case is generated for a context that u_1 and u_2 both can provide and demand at the same time. If a cooperator has been already cooperating with another one u_3, it excludes the contexts involved in the cooperation with u_3 from its S and D for the case generation, following our pair-localized negotiation design.

Step 2. In-situ benefit estimation and cross-validation: The second step is to estimate the benefit of each generated cooperation case and exclude the cases that provide only one-side benefit. Since the benefit of a case can be differently estimated depending on each participant's policy and energy availability, the benefit estimation is separately done by each participant based on its *cost* function. Based on the estimated benefit, each participant excludes the cases that are not beneficial to the participant. Then, they exchange the list of the cases to exclude the cases that are not beneficial to the other participant as well. The cross-validation results in only mutually beneficial cases.

Details on benefit estimation. The key of this step is to estimate the benefits for a cooperation case. In detail, the benefit is calculated in two sub-steps: 1) introducing *a cooperation plan*, and 2) policy-based benefit calculation applying the new plan.

Internally, a cooperation case introduces a new processing plan to monitor the corresponding context. We denote such newly introduced plan as *a cooperation plan*, *cplan*, while denoting the original local plan as *lplan*. The new cooperation plans are differently introduced for exchange cases and co-monitoring cases. For an exchange case, $case\_ex(ctx_{out}, ctx_{in})$, a new $cplan_{in}(ctx_{in})$ is created for ctx_{in}. The $cplan_{in}(ctx_{in})$ simply consists of a task to receive the results for ctx_{in} from the cooperator. For ctx_{out}, a new $cplan_{out}(ctx_{out})$ is built by inserting a task to send results at the end of its original local processing plan. Figures 12(b) and (c) show example *cplans* created by the *case_ex(sound, location)* for Joey in Section 2.1. For a co-monitoring case, $case\_co(ctx_{co})$, a cooperation plan $cplan_{in}(ctx_{co})$ is used for every first half of rotation epoch to receive ctx_{co} and $cplan_{out}(ctx_{co})$ is used for the second half to provide ctx_{co}.

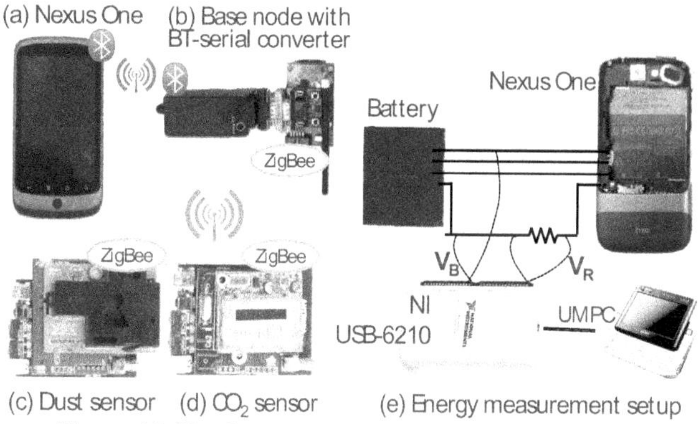

Figure 13. Hardware and energy measurement setup

With the new *cplans*, CoMon recalculates the cost using the *GetEDVector*() and *GetEAVector()* function. Then, the benefit is calculated by subtracting the new cost from the previous cost before applying the *cplans*, i.e., only with local plans.

Step 3. Turn-by-turn case selection: The final step is to select the validated cooperation cases one-by-one in turn. A participant who has a turn selects the case of the maximum estimated benefit and notifies it to the other. After the selection, the participants delete the cases associated with the contexts in the selected case. For example, if a participant selects the co-monitoring case of location context, both participants delete the cases that exchange the location context with another context. The selection process continues until there is no case to select. Such turn-based selection provides each cooperator with fair opportunity to maximize its benefit. After the selection, the cooperation planner applies and executes the cooperation plan for the selected cases.

6. IMPLEMENTATION

We prototyped CoMon on Android phones and various types of sensor devices. Figure 13 shows our hardware setup. We used Google Nexus One with 1GHz CPU, 512MB RAM. We connect a base sensor node to Nexus One via Bluetooth-to-serial converter to support ZigBee communication between Nexus One and sensor devices. We used commercially available ZigbeX sensor motes running TinyOS 1.1.11. They are equipped with Atmega 128L MCU, CC2420 RF transceiver supporting ZigBee protocols, and an additional extension board of dust and CO_2 sensors.

We developed mobile-side CoMon architecture as a background service on the Android platform. It provides applications with a service interface for query registration. For the registered queries, CoMon continuously delivers context monitoring results via a callback message handler. The major components implement the Android handler interface and interact with each other through message exchanges. On the sensors, we implemented the sensor-side architecture in NesC. For efficient utilization of limited sensor resources, the architecture identifies the common tasks in multiple application queries, and utilizes shared buffers to avoid duplicate data sensing and reduce the transmission by packing messages for different requests into a single packet.

7. EXPERIMENT

To demonstrate the effectiveness of CoMon, we evaluate the system based on the prototype described in Section 6. First, we present the

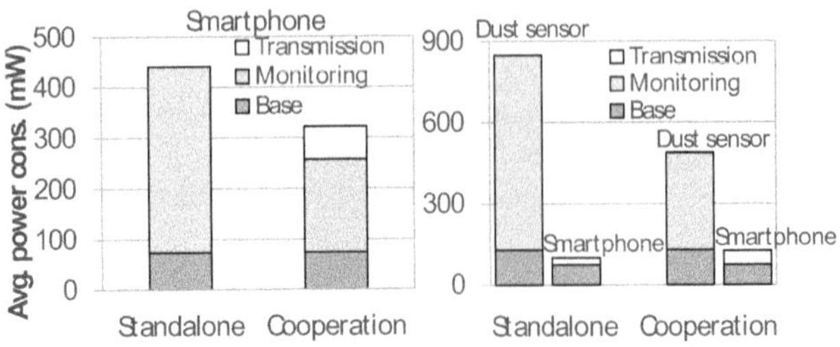

(a) *case_co(location)* (b) *case_co(dust)*

Figure 14. Power consumptions for co-monitoring case

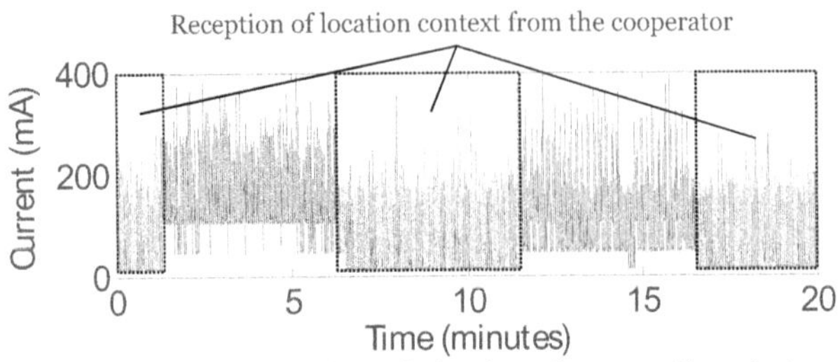

Figure 15. Current draw behavior of *case_co(location)*

energy benefit achieved in diverse cooperation cases. Second, we show that our cooperation planning method effectively provides mutual benefit. Third, we investigate the energy overhead for the cooperation. Lastly, we deploy CoMon on 12 people's devices and present interesting results and valuable lessons. For the power measurements besides the deployment experiments, we used a data acquisition tool, NI USB-6210, as shown in Figure 13(e). The tool measures the voltage (V_B) of the battery and the voltage (V_R) across a register at a sampling rate of 125 kHz. We measure the energy consumption over a certain period of time, 10 minutes by default, as follows:

$$Energy = \sum Power \times Time = \sum_t (\frac{V_R}{R} \times V_B) \times (t_i - t_{i-1})$$

For each sample, the power is obtained by multiplying V_B by the current that is calculated by dividing V_R by the resistance value, R (in our setting, 100mΩ). As a metric for energy, we use the average power consumption (mW), which is obtained by dividing the total energy use for the whole time period by the time.

7.1 Energy Benefits of Cooperation

We evaluate the energy benefits achieved by cooperative context monitoring. We measure and analyze the energy saving of the smartphone and the sensor devices for basic cooperation cases, i.e., co-monitoring cases and exchange cases. We compare the power consumption after applying the cooperation cases against the non-cooperative, standalone setting.

For the detailed analysis, we break down the power consumption into three states; *base*, *monitoring*, and *communication*. The *base* represents the power consumption for the primitive operations of the smartphones and the sensor devices. The *monitoring* includes the power consumed by data sensing and processing. The *communication* is the energy to send and receive the data. This includes both cases to communicate with a person's own external sensor devices and the cooperators; note that both communications are done via Bluetooth.

Co-monitoring cases: Figures 14(a) and (b) show the average power consumption for two co-monitoring cases, *case_co(location)* and *case_co(dust)*, respectively. Figure 14(a) shows that CoMon

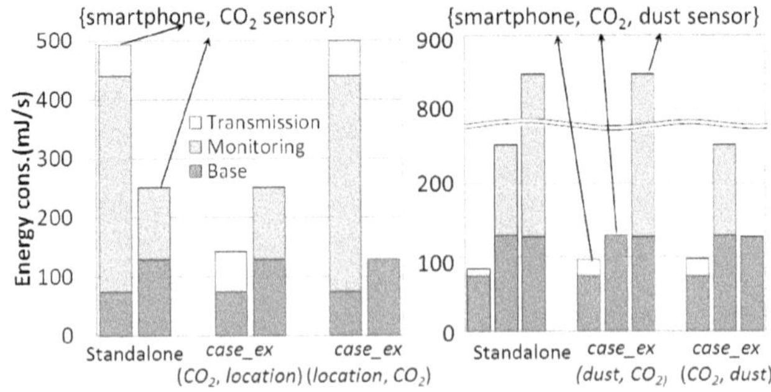

(a) {*location*, CO_2} (b) {*dust*, CO_2}

Figure 16. Power consumptions for exchange cases

achieves 27% of power saving on the smartphone through the co-monitoring of the location context, i.e., from 440 to 321mW. We expect that the energy saving extends the lifetime of the smartphone by 37%. The major contribution comes from halving the total duration of GPS activation. Note that the amount of power saving is less than the exact half, because of the 'base' consumption and the Bluetooth transmission for context exchange. Figure 15 shows the current drawn by the smartphone for 20 minutes for *case_co(location)*. It clearly shows the periodic activation of the GPS, and the lower current draw when the location context is being provided by the cooperator.

Figure 14(b) shows the results for *case_co(dust)*, which employs an external dust sensor. CoMon reduces the power consumption of the dust sensor by 43% (from 848 to 487mW), since it is turned off for a half of the monitoring duration. On the other hand, the power consumption of the smartphone slightly increases by 26mW as it transmits the monitoring results to the cooperator during its monitoring turn. This overhead is marginal in most cases; this is because, even in the standalone setting, the smartphone consumes the energy for the '*transmission*', to receive the data from the external sensor device. Taking such overheads or not is governed by the user's policy.

Exchange cases: Figure 16(a) shows the average power consumption for two exchange cases: when the user takes charge of CO_2 in return of location (*case_ex*(CO_2, *location*)), and vice versa (*case_ex*(*location*, CO_2)). For *case_ex*(CO_2, *location*), CoMon significantly reduces the power consumption of the smartphone (492 to 142mW) by deactivating its GPS; the additional cost to deliver its CO_2 context is insignificant, i.e., 7mW. The consumption of the CO_2 sensor remains the same at 251mW. In contrast, for *case_ex*(*location*, CO_2), the power consumption of CO_2 sensor is largely reduced from 251 to 129mW, whereas the smartphone slightly consumes 9mW of more power to transmit the location context. Figure 16(b) shows the exchange cases of CO_2 and *dust* contexts. These cases provide energy benefits similarly as shown in Figure 16(a).

7.2 Cooperation Planning for Mutual Benefit

We validate our cooperation planning mechanism and its effectiveness in terms of mutual benefits. We conducted an experiment with three users, u_A, u_B, and u_C, each having different devices and monitoring queries (see Figure 17). We investigate how cooperation planning is performed when the users come across, stay with, and leave each other. Figure 18(a) depicts four phases separated by the users' meeting and parting events. We show only u_A's viewpoint for concise description and verify the actual energy benefits. We set different cooperation policies for each user as: u_A

	u_A	u_B	u_C
Queries	Location, Dust, Ambient Sound	Ambient Sound, Dust	Location, Ambient Sound
Devices	Smartphone, Dust	Smartphone, Dust	Smartphone

Figure 17. Experimental setup

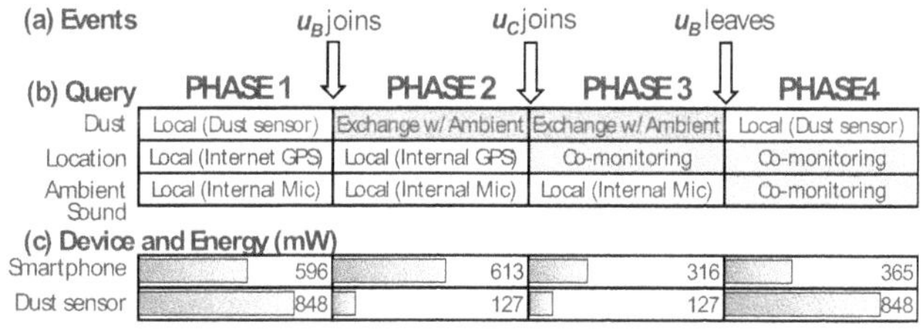

Figure 18. Experiment results for u_A

Table 3. Cooperation cases and expected benefits

Cooperation case		u_A's saving (mW)			u_B's saving(mW)	
ID	Description	Total	Phone	Dust	Phone	Dust
1	*case_co(ambient)*	30	30	0	12	0
2	*case_co(dust)*	336	-25	361	-24	361
3	*case_ex(u_A:dust, u_B:ambient)*	131	131	0	-5	127
4	*case_ex(u_A:ambient, u_B:dust)*	**704**	-17	721	**225**	0
5	*case_ex(u_A:ambient, u_B:location)*	231	231	0	-191	0
6	*case_ex(u_A:dust,* u_B:location)	454	454	0	-270	721

wants to maximize the total energy saving of the smartphone and the sensor devices, whereas u_B and u_C want to maximize the energy saving only for their smartphones.

Phase 1: u_A registers her location, ambient sound, and dust monitoring queries. As there is no cooperator, all those queries are processed by u_A's own resources. Figure 18(b) shows the power consumption of u_A's smartphone and dust sensor in Phase 1, i.e., 596 and 858mW, respectively.

Phase 2: Phase 2 begins when u_A meets u_B. Upon meeting each other, their CoMons start the cooperation planning process. By exchanging their demanding and suppliable contexts, both CoMons generate six cooperation cases and estimate the benefits for each case as in Table 3. u_A's CoMon determines that all cases are beneficial while u_B's determines that only *Case1* and *Case4* are so. Among the mutually beneficial cases, i.e., *Case1* and *Case4*, u_A selects *Case4*, *case_ex*(*ambient*, *dust*), as it is the most beneficial according to her policy. Now that u_A's monitoring queries are processed cooperatively, u_A's CoMon reduces the total power consumption from 1,444 to 740mW.

Phase 3: While u_A is being together with u_B, u_C comes across them. u_A's CoMon starts the cooperation planning with u_C as well, generating one cooperation case, i.e., *case_co*(*location*). Note that cooperation cases regarding the ambient sound context are not generated as it is already under the cooperation with u_B. u_A's CoMon begins additional cooperation with u_C by selecting and applying *case_co*(*location*); the total power consumption of u_A's devices has further reduced from 740 to 443mW.

Phase 4: u_B has just left u_A. Detecting the event, u_A's CoMon restores the dust monitoring to be processed by her own devices. Accordingly, u_A's total power consumption increases to 1,294mW. u_A begins additional planning with u_C on the dust and ambient sound contexts which u_A has cooperatively monitored with u_B. They generate and select a cooperation case, *case_co*(*ambient*), which is

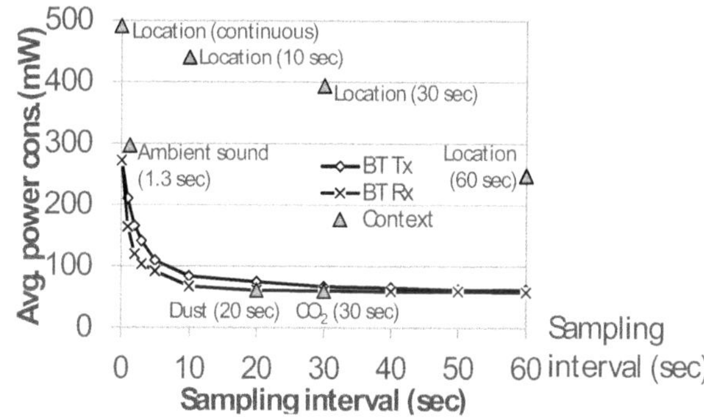

Figure 19. Exchange overheads and monitoring costs; the dust and the CO_2 sensor consumes 848 and 252mW, respectively

mutually beneficial. This new cooperation reduces the total power consumption of u_A's devices from 1,294 to 1,213mW.

Cooperation between u_B and u_C: In phase 3, u_B and u_C also meet. However, any cooperation cases are not applicable because u_B had been already performing the cooperation with u_A for all the requested contexts, i.e., ambient sound and dust.

7.3 Cooperation Overhead

We examine the energy overhead for cooperative monitoring. We observe two major causes of overheads: (1) to discover nearby cooperator candidates, and (2) to exchange the monitoring results. We measure and analyze the overheads, showing that those are insignificant compared to the expected benefits.

Discovery overhead: CoMon conducts periodic Bluetooth scans for discovery, consuming additional energy. We measure the overheads for the scanning intervals in Section 4. The overhead is 20mW in our default interval of 5 minutes. This is relatively small compared to the expected benefits of many cooperation cases in Section 7.1. For example, if CoMon has been looking for cooperators for 60 minutes, it just needs 6 minutes to break even after starting the cooperation of *case_ex*(*ambient sound*, *location*).

Context Exchange Overhead: To identify the overhead for context exchange, we measure the smartphone's power consumptions for Bluetooth message exchanges as shown in Figure 19. To figure out the relative amount of the overhead, we also plot the smartphone's energy cost for monitoring several example contexts. For the contexts requiring power-hungry sensing or heavy computation, the energy overhead to exchange a context is much smaller than the cost to monitor the context. For instance, receiving the location context consumes only 60mW of the smartphone, whereas monitoring the context using GPS costs 333mW. In the case of *dust* and *CO_2* contexts, the smartphone does not benefit from the cooperation, but the sensor devices significantly save their energy as in Section 7.1. Note that some sensors such as accelerometers and illuminometer hardly consume power (<35mW) (less than the communication overhead for cooperation), and thus the cooperation might not be beneficial. These cases are filtered out through cooperation planning process.

7.4 Deployment

On top of the CoMon platform, we prototyped a proof-of-concept application, *TripMemory*. It is an Android application that tracks the user's travelling path and logs her surrounding events extracted from ambient sound [26]. We aim at presenting the effectiveness of cooperative context monitoring rather than showing the novel

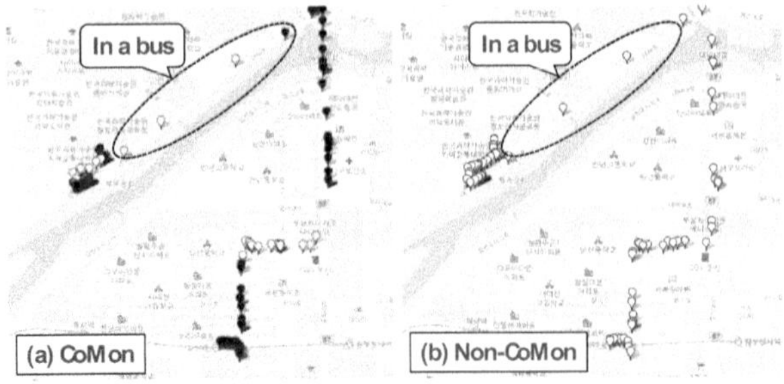

Figure 20. Location tracking from KAIST to the downtown

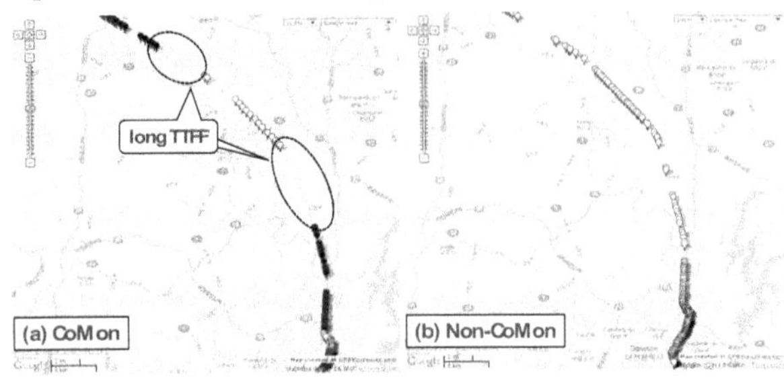

Figure 21. Location tracking in a high-speed rail

concept of the application. Upon the start of TripMemory, it registers either one or both of the following queries to CoMon requesting for user preferences; the query registration is performed once a day only.

CONTEXT location	**CONTEXT** ambient sound
PERIOD 5 Seconds	**PERIOD** 10 Seconds
DURATION Always	**DURATION** Always

CoMon notifies the application of monitoring results through the Android service interface. We recruited 12 participants consisting of 6 pairs of friends via the bulletin board of our school. Each participant was given a Nexus One phone with the CoMon platform and TripMemory installed. For comparison, each was given another phone with the same setting but deactivating the cooperation functionality of CoMon (named Non-CoMon). For fair comparison of the energy consumption, we used new batteries for the smartphones. The battery level of the smartphone is logged using Android library. We also required the participants to fully charge every night and not to run any application other than TripMemory. They roamed freely for a week.

According to our data, each participant runs TripMemory 6.2 days on average and 8.6 hours per day; some forgot to run for a day. On average, a pair cooperated 5.9 hours per day across 6.8 times of meetings. The average meeting duration is longer than we expected; we guess that this is because the participants are mostly close friends who are roommates or attending classes together. CoMon's average battery consumption is 19.7 % less than those of non-CoMon phones; this means that about 19.7% battery remains for a CoMon phone at the moment that the corresponding non-CoMon phone runs out of battery. Looking into the data, the cooperation benefits vary largely depending on the cooperation patterns. Only accounting for when a user turns on location monitoring, the benefits are 31.1 % in average. When both users turn on location and ambient sound monitoring, the benefits differ for the ones providing locations and the sound contexts. For the location providers, the average benefit is 6.9% only while the average benefit is 22% for the sound providers; note that location monitoring consumes a lot more energy. We expect that the benefit of CoMon will increase as CoMon is deployed by more people and more

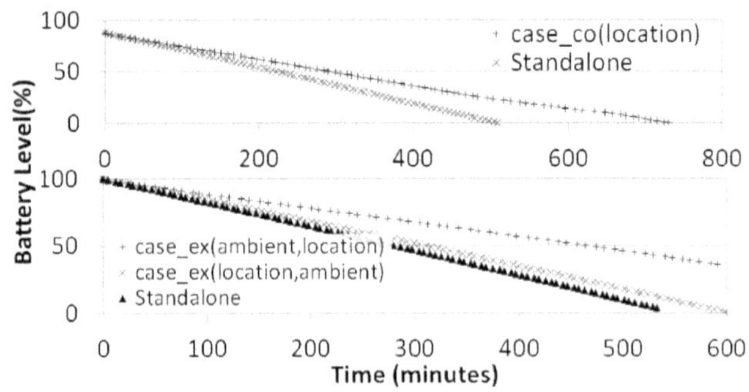

Figure 22. Battery level over time

Table 4. Stability results

Number of situations	Average duration (minutes)	Number of disconnection	Average disconnection time (minutes)
10	206	2.2	5.06

energy-intensive context processing is performed. In the rest of this section, we present interesting results and lessons from our deployment.

7.4.1 Quality of Application

To study the effect of the cooperation on the quality of the application, we compare the monitoring results from CoMon and Non-CoMon. We show a representative case in which a pair of participants, u_1 and u_2, trips together from KAIST to the downtown. During the trip, their CoMons performed *case_co*(*location*) and *case_co*(*ambient*) at the co-monitoring interval of 5 minutes. Figures 20(a) and (b) depict parts of the trip paths logged by TripMemory on u_1's CoMon and Non-CoMon, respectively; white dots represent the logs obtained from u_1's smartphone and black dots are those provided by u_2's. In the figures, the trip path recorded through the cooperation has very similar patterns to the path of Non-CoMon. It shows that CoMon does not compromise the quality of context monitoring much while achieving resource benefits. Note that different densities of the dots result from different transportation modes and constant monitoring period; they took a bus around the sparse part.

Figure 21 shows a rare but noteworthy case in which the monitoring quality was compromised. In the case, a pair of participants had an inter-city trip from Daejeon to Seoul by high-speed rail, running at a speed of up to 305km/h (190mph). Unlike the previous result shown in Figure 20, some location data are missing when switching the turns of co-monitoring. This is because it mostly takes a longer TTFF (time to first fix) as the device travels longer distance from the last detected location in its previous activation. We figured out that this problem happened when the user was moving extremely fast. It addresses two considerations that: (1) CoMon design needs to be further polished including a consideration of the delay to reactivate sensors, and (2) those delays of some sensors might be dependent on the some situational parameters, e.g., GPS sensor and the moving speed. We leave this issue for future work.

7.4.2 Long-term Observation of Battery

We present long-term observations of smartphone's battery level on TripMemory. We examine the battery behavior over time with the exemplary cases that the cooperation continues stably until the battery runs out. Figure 22(a) shows the battery level over time for co-monitoring of location information. CoMon extends the lifetime of the smartphone by 44% (from 508 to 732 minutes). Note that our benefit function and the offline profiling estimated the lifetime extension as 37% (see Section 7.1). We believe that the error comes from the fluctuations of the GPS and 3G signal strength, which affect the actual power consumption. Figure 22(b) describes the battery level when devices exchange the ambient sound context in return of the location and vice versa. *case_ex*(*location*, *ambient*) extends the lifetime from 532 to 598 minutes. *case_ex*(*ambient*, *location*) is more beneficial; the smartphone still shows 36% of the battery level at 598 minutes.

7.4.3 Cooperation Stability

We also investigate the stability of cooperation on top of CoMon using the TripMemory application. This shows that the stability of the connection is maintained while the users are cooperating, being physically close to each other. For this additional part of the experiments, we further recruited ten pairs of participants, where the participants in each pair are acquainted with each other. As a baseline, we have the users annotate the presence of their cooperators manually in case of disconnection and re-connection; we provide the users with vibrating and auditory alarms to notify the connection events. We collect the connection logs and the annotated data in diverse situations including dating, working in an office, and promenading along a river.

We look into the disconnection events and analyze them based on the participants' manual annotation. Table 4 shows the overall results. We first verified that there was no disconnection when they had been actually together. The intermittent disconnections occurred in our deployment situations were reconnected within 5 minutes on average, for example, the case of going to a restroom. This implies that, upon a disconnection from an acquaintance, waiting for 5 minutes might be worth trying, rather than immediately establishing a new cooperation with a stranger.

8. DISCUSSION

Coverage of Context Sharing. In the current design, we simply assume the range of Bluetooth (< 10m) as the coverage of context sharing. This works quite well in our deployment, where a pair of cooperators stays closeby during most of their meeting time. However, simply being within Bluetooth range does not ensure that two users have common contexts. For example, a user may detect another in the next room but may not have many common contexts. We believe this issue can be addressed in several ways. Exploiting Bluetooth RSSI [23] may deliver fine-grained clues on the inter-user proximity or the presence of obstacles separating them. Exchanging some contextual signature prior to cooperation may help to determine if they are in the same place. The place detection techniques such as SurroudSense [2] could be adopted for this purpose.

Privacy. Letting others know my context inherently raises privacy concerns. To be optimistic, we believe that such concerns might be relatively mitigated in the target environments of CoMon, where the users are physically in the same contexts. A study on location sharing supports that people are less conscious of sharing their locations when they are closeby [6]. A study on phone sharing shows that sharing is more acceptable with those in close social relations such as families, friends, or colleagues [24].

To be conservative, privacy concerns largely depend on different users and how the sensed contexts are to be used [19]. A user study implicates that people would be highly selective during their private time depending on their context and activities [5]. In this light, CoMon aims to provide users with the controllability and visibility on the sharing of their contexts. First, CoMon allows users to specify their sharing policies, i.e., the rules governing the access to their contexts from other cooperators as follows:

SHARE	Context
WITH	Target [EXCEPT Exception target]
RESOURCE	High | Middle | Low

CoMon provides simple UI showing the currently shared contexts and the cooperator information. We acknowledge that, the rules and UI address only some basic concerns on the privacy; it is still an open research question requiring further in-depth studies.

9. RELATED WORK

Collaborative Applications and Techniques. Opportunistic collaboration among smartphones has drawn attention in many domains, e.g., video playback and recording [3][33] and context inference [29]. CoMon takes collaboration opportunities for different purposes, e.g., saving energies for continuous context monitoring or obtaining new sensing modalities. Also, CoMon is the first to incorporate personal sensing devices into cooperation beyond smartphones.

Collaborative sensing techniques has been proposed to incorporate new sensing modalities and enhance data fidelity [10][20]. They share a high-level goal with CoMon in that it aims to increase the capability of individual mobile users through the collaboration. CoMon conducts its in-depth study on the cooperation opportunity and resource benefits of cooperation for continuous context monitoring.

There has been a research thread to develop middleware to construct mobile ad-hoc networks (MANETs) [4]. They group multiple mobile devices in an ad-hoc manner and enable applications such as file sharing and instant messaging. A major focus has been constructing network topologies through efficient discovery and advertisement of resources, such as mobile JXTA libraries. CoMon opens a new application domain of context monitoring, which clearly suits for the concept of in-situ sharing. In addition, different from conventional dynamic ad-hoc cooperation, we target continuity-lasting group of users to reduce the complexity and overheads for cooperation.

Participatory Sensing. This concept has been proposed to exploit the widely distributed mobile devices for urban-scale sensing applications. It has been adopted by many applications, e.g. pothole patrol [11], and has evolved into common platforms, e.g. PRISM [8] and Bubble Sensing [25]. These applications extend the spatio-temporal sensing coverage of a mobile user. Different from such works, CoMon aims to reduce the monitoring redundancies among the users in close proximity to resolve resource scarcity. CoMon is not a competing technology with participatory sensing but can complement each other. CoMon can serve as a client of participatory sensing, providing the contexts in greater energy efficiencies. In the other way, CoMon could utilize participatory sensing to extend its spatial context coverage.

Energy Optimization. There have been huge research efforts to reduce energy consumption for continuous sensing and data processing [16][30][34]. They focus on optimizing energy use within a single device whereas CoMon newly attempts to optimize resource use in consideration of multiple users and devices. A variety of techniques have been proposed, such as an energy-fidelity tradeoff [34], user interest-based sensor management [16], hierarchical sensor management [31], and low-power hardware mode utilization. Recent studies have delivered the energy efficiency for location monitoring [30]. CoMon complements such techniques for a single device, by further improving resource efficiency through active cooperation with nearby mobile users.

Task offloading as in MAUI [7] and Odessa [32] reduces resource consumption of smartphones; heavy back-end tasks in a processing pipeline are offloaded to servers. However, CoMon takes cooperation approach distributing tasks over nearby devices, having benefits not provided by the offloading approach. Many sensing tasks are not transferrable to remote servers since the sensing itself can be performed only where the context exists. Even for processing tasks, the overhead to transfer high-rate data could often overwhelm the resource benefit from offloading.

10. CONCLUSION

In this paper, we present the design and implementation of CoMon, a novel cooperative context monitoring system. We built CoMon by exploiting the prevailing cooperation opportunities among mobile users. CoMon allows every participant to take benefits from cooperation, through the continuity-aware cooperator selection and benefit-aware negotiation. We deployed CoMon prototype on off-the-shelf smartphones and diverse sensor devices. We evaluated that CoMon significantly improves resource efficiency for continuous mobile sensing and processing. Moreover, it extends the available contexts beyond those from one's own devices.

As people spend significant portion of time for social activities in their daily lives, smartphone applications and systems needs to be further evolved to support and leverage such situation. CoMon takes an initial step toward this direction, introducing social awareness into smartphones. It opens a broad spectrum of technical issues, e.g., networking for in-situ cooperation, co-activity group detection and managements, resource sharing and coordination, privacy. Based on CoMon and its underlying techniques, we are working on building a new platform to facilitate diverse in-situ social activities.

11. ACKNOWLEDGEMENTS

Our special thanks go to our shepherd, Professor Li-Shiuan Peh for her valuable suggestions to improve the quality of this paper. This research was supported by the NRF of Korea under grants No. 2011-0018120, SW computing R&D program of KEIT (2011, 10041313) funded by Ministry of Knowledge Economy of Korea, and partially by Microsoft Research Asia.

12. REFERENCES

[1] American Time Use Survey. http://www.bls.gov/tus

[2] Azizyan, M., Constandache, I., and Choudhury, R. R. SurroundSense: Mobile Phone Localization via Ambience Fingerprinting. In *MobiCom,* 2009.

[3] Bao, X. and Choudhury, R. R. MoVi: Mobile phone based video highlights via collaborative sensing. In *MobiSys,* 2010.

[4] Bisignano, M., Modica, G. D., and Tomarchio, O. JMobiPeer: a middleware for mobile peer-to-peer computing in MANETs. In *ICDCSW*, 2005.

[5] Choe, E. K., Consolvo, S., Jung, J., Harrison, B., and Kientz, J. A. Living in a Glass House: A Survey of Private Moments in the Home. In *UbiComp*, 2011.

[6] Consolvo, S., Smith, I. E., Matthews, T., LaMarch, A., Tabert, J., and Powledge, P. Location Disclosure to Social Relations: Why, When, & What People Want to Share. In *CHI,* 2005.

[7] Cuervo, E., Balasubramanian, A., Cho, D., Wolman, A., Saroiu, S., Chandra, R., and Bahl, P. MAUI: making smartphones last longer with code offload. In *MobiSys*, 2010.

[8] Das, T., Mohan, P., Padmanabhan, V. N., Ramjee, R., and Sharma, A. PRISM: Platform for Remote Sensing Using Smartphones. In *Mobisys,* 2010.

[9] Eable, N. and Pentlan, A. S. Reality Mining: Sensing Complex Social Systems. *Journal of Personal and Ubiquitous Computing*, Vol. 10, Issue 4, Mar. 2006.

[10] Eisenman, S. B. People-Centric Mobile Sensing Networks. Ph.D. Dissertation, 2008.

[11] Eriksson, J., Girod, L., and Hull, B. The Pothole Patrol: Using a Mobile Sensor Network for Road Surface Monitoring. In *MobiSys,* 2008.

[12] Google Android API. http://developer.android.com/guide/topics/wireless/bluetooth.html

[13] Gu, T., Pung, H. K., and Zhang, D. Q. A service-oriented middleware for building context-aware services. Journal of Network and Computer Applications, Vol. 28, Issue 1, 2004.

[14] iWake. http://itunes.apple.com/us/app/iwake-location-alarm/id406866935?mt=8

[15] Ju, Y., Min, C., Lee, Y., Yu, J., and Song, J. An Efficient Dataflow Execution Method for Mobile Context Monitoring Applications, in *PerCom*, 2012.

[16] Kang, S., Lee, J., Jang, H., Lee, H., Lee, Y., Park, S., Park, T., and Song, J. SeeMon: Scalable and Energy-efficient Context Monitoring Framework for Sensor-rich Mobile Environments. In *MobiSys*, 2008.

[17] Kang, S., Lee, Y., Min, C., Ju, Y., Park, T., Lee, J., Rhee, Y., and Song, J. Orchestrator: An Active Resource Orchestration Framework for Mobile Context Monitoring in Sensor-rich Mobile Environments. In *PerCom*, 2010.

[18] Kim, S. and Paulos, E. inAir: Measuring and Visualizing Indoor Air Quality. In Ubicomp, 2009.

[19] Klasnja, P., Consolvo, S., Choudhury, T., Beckwith, R., and Hightower, J. Exploring Privacy Concerns about Personal Sensing. In *Pervasive*, 2009.

[20] Lane, N., Lu, H., Eisenman, S. B., and Campbell, A. T. Cooperative Techniques Supporting Sensor-Based People-Centric Inferencing. In *Pervasive*, 2008.

[21] Lee, Y., Iyengar, S. S., Min, C., Ju, Y., Kang, S., Park, T., Lee, J., Rhee, Y., and Song, J. MobiCon: Mobile Context-Monitoring Platform. *Communications of the ACM*, 2012.

[22] Lester, J., Hartung, C., Pina, L., Libby, R., Borriello, G., and Duncan, G. Validated caloric expenditure estimation using a single body-worn sensor. In Ubicomp, 2009.

[23] Liu, S. and Striegel, A. Accurate extraction of face-to-face proximity using smartphones and Bluetooth. In *ICCCN,* 2011.

[24] Liu, Y., Rahmati, A., Huang, Y., Jang, H., Zhong, L., Zhang, Y., and Zhang, S. xShare: Supporting Impromptu Sharing of Mobile Phones. In *MobiSys*, 2009.

[25] Lu, H., Lane, N. D., Eisenman, S. B., and Campbell, A. T. Bubble-sensing: Binding SensingTasks to the Physical World. *Pervasive and Mobile Computing*, Vol. 6, Issue 1, Feb. 2010.

[26] Lu, H., Pan, W., Lane, N. D., Choudhury, T., and Campbell, A. T. SoundSense: Scalable Sound Sensing for People-Centric Applications on Mobile Phones. In *MobiSys*, 2009.

[27] Lu, H., Yang, J., Liu, Z., Lane, N. D., Choudhury, T., and Campbell, A. T. The Jigsaw continuous sensing engine for mobile phone applications. In *SenSys*, 2010.

[28] Milgram, S. The Familiar Stranger: An Aspect of Urban Anonymity. Addison-Wesley, Reading, 1977.

[29] Miluzzo, E., Cornelius, C. T., Ramaswamy, A., Choudhury, T., Liu, Z., and Campbell, A. T. Darwin phones: the evolution of sensing and inference on mobile phones. In *MobiSys,* 2010.

[30] Paek, J., Kim, J., and Govindan, R. Energy-Efficient Rate-Adaptive GPS-Based Positioning for Smartphones. In *MobiSys*, 2010.

[31] Park, T., Lee, J., Hwang, I., Yoo, C., Nachman, L., and Song, J. E-Gesture: a collaborative architecture for energy-efficient gesture recognition with hand-worn sensor and mobile devices. In *SenSys*, 2011.

[32] Ra, M-R., Sheth, A., Mummert, L., Pillai, P., Wetherall, D., and Govindan, R. Odessa: Enabling Interactive Perception Applications on Mobile Devices. In *MobiSys*, 2011.

[33] Shen, G., Li, Y., and Zhang, Y. MobiUS: Enable Together-Viewing Video Experience across Two Mobile Devices. In *MobiSys,* 2007.

[34] Sorber, J., Kostadinov, A., Garber, M., Brennan, M., Corner, M. D., and Berger, E. D. Eon: A Language and Runtime System for Perpetual Systems. In *SenSys*, 2007.

[35] Walkmeter. http://itunes.apple.com/us/app/walkmeter-gps-walking-stopwatch/id330594424?mt=8

MicroCast: Cooperative Video Streaming on Smartphones

Lorenzo Keller
School of I&C
EPFL, Lausanne, CH
lorenzo.keller@epfl.ch

Anh Le
CS Dept & CalIT2
UC Irvine
anh.le@uci.edu

Blerim Cici
Networked Systems & CalIT2
UC Irvine
bcici@uci.edu

Hulya Seferoglu
LIDS
MIT
hseferog@mit.edu

Christina Fragouli
School of I&C
EPFL, Lausanne, CH
christina.fragouli@epfl.ch

Athina Markopoulou
EECS Dept. & CalIT2
UC Irvine
athina@uci.edu

ABSTRACT

Video streaming is one of the increasingly popular, as well as demanding, applications on smartphones today. In this paper, we consider a group of smartphone users, within proximity of each other, who are interested in watching the same video from the Internet at the same time. The common practice today is that each user downloads the video independently using her own cellular connection, which often leads to poor quality.

We design, implement, and evaluate a novel system, MicroCast, that uses the resources on all smartphones of the group in a cooperative way so as to improve the streaming experience. Each phone uses simultaneously two network interfaces: the cellular to connect to the video server and the WiFi to connect to the rest of the group. Key ingredients of our design include the following. First, we propose a scheduling algorithm, MicroDownload, that decides which parts of the video each phone should download from the server, based on the phones' download rate. Second, we propose a novel all-to-all local dissemination scheme, MicroNC-P2, for sharing content among group members, which outperforms state-of-the-art peer-to-peer schemes in our setting. MicroNC-P2 is designed to exploit WiFi overhearing and network coding, based on a local packet broadcast framework, MicroBroadcast, which we developed specifically for Android phones. We evaluate MicroCast on a testbed consisting of seven Android phones, and we show that it brings significant performance benefits without battery penalty.

Categories and Subject Descriptors

C.2.1 [**Computer-Communication Networks**]: Network Architecture and Design—*Wireless communication*; C.2.2 [**Computer-Communication Networks**]: Network Protocols—*Applications*

Keywords

Video Streaming, Wireless Networks, Smartphones, Network Coding

MobiSys'12, June 25–29, 2012, Low Wood Bay, Lake District, UK.

1. INTRODUCTION

Today's smartphones are equipped with significant processing, storage and sensing capabilities, as well as wireless connectivity through cellular, WiFi and Bluetooth. They provide ubiquitous Internet access, primarily through their cellular connection and secondarily through WiFi, and enable a plethora of new applications. Among those applications, video (including consuming and creating/posting video content) is increasingly popular. However, meeting the growing demand for high quality video is currently a challenge in cellular networks.

Indeed, cellular traffic is growing exponentially (tripling every year), with a share of video traffic increasing from 50% now to an expected 66% by 2015 [7]. Credit Suisse reported in [13] that 23% of base stations globally have utilization rates of more than 80 to 85% in busy hours, up from 20% last year. This dramatic increase in demand poses a challenge for 3G networks, which is likely to remain in 4G networks as well. Furthermore, the data rate of the cellular connection may fluctuate over time (*e.g.,* throughout the day); the service loss rate can be as high as 50% [36]; and coverage can be spotty depending on the location and user mobility.

In this paper, we are interested in the scenario where a group of smartphone users, within proximity of each other, are interested in watching the same video at the same time. Watching the video on one phone screen is not comfortable for more than two people; users may also prefer to watch the video on their own screens. The default operation today is that each user with a cellular connection downloads the video independently from the server. However, each phone's individual cellular connection may not be sufficient for providing high video quality.

Consider, for example, that a user wants to show to her friends a YouTube videoclip while being in a bar; a group of friends who want to watch a live soccer match together on their phones while at a remote location (*e.g.,* a camping or skiing site); or a family who wants to watch a movie on their phones in the train or in the car; or a group of co-workers who want to watch a lecture using WiFi at a busy hotspot, such as the company cafeteria or a conference room. In all these cases, some or all of the users may have poor or intermittent cellular connectivity, depending on the coverage of their providers, or may face congestion in the local network (*e.g.,* when they use WiFi to download).

Fortunately, because the users engage in a group activity, this scenario naturally lends itself to cooperation. Furthermore, when every phone has multiple parallel connections (*e.g.,* 3G, WiFi, and Bluetooth), there are even more available resources that, if properly used, can further improve the user experience.

A special case of the general scenario is when the video content is stored locally on one of the phones and the user wants to share it

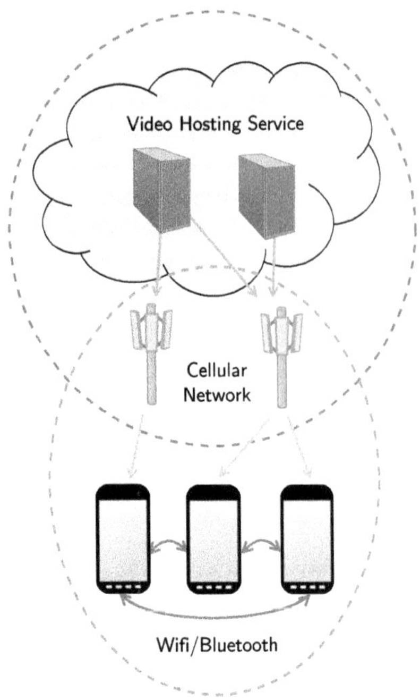

Figure 1: *MicroCast scenario.* A group of smartphone users, within proximity of each other, are interested in watching the same video at the same time. Each smartphone connects to the video source (*e.g.,* UStream, YouTube, or Netflix) using its cellular (3G or 4G) connection. The base station may be the same or different for different users, depending on the provider they use. Each smartphone can receive packets from the source as well as from other smartphones in the neighborhood through device-to-device (*e.g.,* WiFi) links.

with the other members of the local group. For example, a teacher wants to share a video with the students in a class; or a group of friends want to watch a video recorded of a group activity stored on one of the phones. In these cases, cooperation can also help.

We propose a novel cooperative scheme, called *MicroCast*[1], for video streaming to a group of smartphones within proximity of each other. Each phone utilizes simultaneously two network interfaces: one (cellular) to connect to the video server and download parts of the video; and the other (WiFi) to connect to the rest of the group and exchange downloaded parts. Our scheme optimizes the usage of the cellular links to ensure that all the available bandwidth is used when channel conditions are good so as to compensate for potential long periods of bad channel conditions. It also optimizes the dissemination in the local area so as to ensure that even in presence of heavy interference from other networks, the phones can still collaborate. In our scheme, each phone downloads the video much faster than it is played out when the conditions are good, in order to reduce the likelihood of buffering during the playback when the radio conditions significantly deteriorate, *e.g.,* it downloads at 2 Mbps or more on current cellular networks videos that are typically encoded at 750 Kbps.

Our system consists of the following key components.

I. *MicroDownload:* This is a simple yet effective scheduler that decides which parts of the video each phone should download from the server and distribute on the local network. The decision is based on feedback from each phone and takes into account the cellular download rate. Compared to prior work, *e.g.,* [20], [29], [25], [26], [38], where each phone tries to download all the data and the local links are used for error recovery, our scheme *jointly* utilizes two interfaces for data delivery. Compared to cooperative mobile systems for offloading cellular networks, *e.g.,* [17], [14], [42], our goal is to improve the 3G downlink utilization of a group of smartphones. Compared to [35], our scheme tries to maximize the rate at which the content is downloaded to prevent buffer underflow, which leads to freezing of the video.

II. *MicroNC-P2:* This is a novel all-to-all dissemination scheme for locally sharing content among group members within proximity of each other. It leverages the combination of network coding and WiFi overhearing (provided by the MicroBroadcast mechanism, described below) to efficiently share the downloaded parts of the video locally. We show that, in our setting, MicroNC-P2 significantly outperforms state-of-the-art P2P schemes, including BitTorrent [8] and the network coding-based R2 [41]. This is because MicroNC-P2 is explicitly designed to exploit the capabilities of the underlying MicroBroadcast to introduce less redundancy than the previous algorithms [41]. MicroNC-P2 is not only a key component of our MicroCast system but also a standalone contribution that can be used independently, *e.g.,* for local all-to-all dissemination.

III. *MicroBroadcast:* We develop a component that enables high rate packet broadcast, building on WiFi of Android phones. Although the wireless medium has inherently the broadcast capability, this cannot be fully exploited using standard WiFi broadcast, due to various problems, including low rates. lack of reliability etc. Pseudo-broadcast on WiFi ad-hoc mode, which was previously used in [18], cannot be used on Android phones, as we discuss later. To the best of our knowledge, MicroBroadcast is the first Android-based system to fully exploit the potential of broadcast. MicroBroadcast is a contribution on its own right and also a necessary component for the high performance of the entire MicroCast system: *e.g.,* MicroNC-P2 exploits its capabilities.

We implement a prototype of our system as an application for Android phones. An attractive feature of our architecture is that it is modular: one can change each component (*e.g.,* the link layer, the algorithms for local distribution or download, the streaming protocol, etc.) independently of the others. We plan to make the application and its related libraries publicly available [1]. In this paper, we present first the system design and then the performance evaluation results from experiments on a testbed consisting of seven Android phones. We evaluate each of the three components separately as well as the combined system as a whole and we demonstrate that there are significant performance benefits (in terms of decrease download time and increase per-user download rate) compared to alternative approaches, without significant battery cost. A video demonstration and supporting materials can be found in [1].

The rest of the paper is structured as follows. Section 2 presents related work. Section 3 describes the MicroCast architecture. Section 4 presents the MicroDownload and MicroNC-P2 algorithms and their interaction. Section 5 discusses implementation details, including challenges specific to Android phones and design choices

[1] "Micro" indicates locality: there is a small number of users and they are all within proximity of each other. "Cast" indicates a multicast scenario: all users in the group are interested in the same content sent by a single source, and that we use local broadcast.

made to address them. Section 6 presents performance evaluation results on our Android testbed for a range of scenarios. Section 7 discusses limitations and extensions. Section 8 concludes the paper.

2. RELATED WORK

Cooperative Mobile/Wireless Systems. When several users are interested in the same content and they are in proximity of each other, some of them may be able to use device-to-device connections, *e.g.,* through WiFi or Bluetooth, to get the content in a cooperative and/or opportunistic way. Opportunistic device-to-device communication is often used for the purpose of offloading the cellular network. For instance, [17], [14], and [42] consider a scenario in which device-to-device and cellular connections are used to disseminate the content, considering the social ties and geographical proximity. Instead of offloading cellular networks, our goal in this paper is to use cellular and local connectivity so as to essentially allow each user to enjoy the aggregated downlink rate.

Furthermore, cooperation between mobile devices for content dissemination or in delay tolerant networking, possibly taking into account social ties [16, 3], has extensively been studied. However, dissemination of content stored on a mobile device is only a special case of our framework, which uses only the local links, but not the downlinks. More importantly, we exploit single-hop broadcast transmissions, as opposed to multi-hop peer-to-peer communication that exploits mobility (at the expense of delay, which is crucial in our setting), but ignores broadcast.

The idea of using multiple interfaces of mobile devices has been used before but not in the same way as in this paper. For example, [37] effectively exploits cellular and WiFi interfaces simultaneously to create multiple paths to mobile devices. [34] improves the mobile networking experience using concurrent WiFi connections from multiple WiFi hot-spots. [28, 5] exploit the diversity of multiple interfaces on the same device to achieve better connectivity. Instead, we use the cellular connections of multiple smartphones to improve the download rate and jointly utilize local connections. [2, 35] address the same problem as we do, and use a similar strategy. However, they use only unicast communication among peers, as opposed to our (micro)broadcast. Therefore, they are more sensitive to local bandwidth fluctuations.

The recently announced Android Ice Cream Sandwich (Android 4.0) devices provide peer-to-peer (P2P) connectivity using WiFi Direct [43]. WiFi Direct is an industrial standard that prompts WiFi-equipped devices to establish ad-hoc peer-to-peer connections, while maintaining the infrastructure mode connection to the Internet. This is achieved by having one of the devices acts as a *virtual AP* to provide infrastructure to the rest. In MicroCast the mobile devices create P2P connections in a similar way, but in our case, we try to reduce the number of frames (in the channel) required for each data packet exchanged between the peers. This is done by "pseudo-adhoc", which is part of the MicroBroadcast component. The core idea of MicroBroadcast is to exploit the broadcast nature of the wireless channel, while the core idea of WiFi Direct is to offer ad-hoc connectivity to mobile devices.

Network Coding in Cooperative/Wireless Systems. Cellular links (3G or 4G) as well as WiFi suffer from packet loss due to noise and interference. One possible solution to this problem is to have several devices in a close proximity help each other with retransmissions of lost packets. Network coding is particularly beneficial as it can make each retransmission maximally useful to all nodes. Rate-distortion optimized network coding for cooperative video system repair in wireless peer-to-peer networks is considered in [21]. Wireless video broadcasting with P2P error recovery is proposed in [20]. An efficient scheduling approach with network coding for wireless local repair is introduced in [29]. [38], [26], and [25] propose a system, in which a group of smartphone users (which are connected to the Internet via cellular links) help each other for error correction, while base stations *broadcast* packets, which is not implemented in current systems; in contrast, we consider unicast (between a base station and a mobile device). A cooperative video streaming system is implemented over mobile devices (PDAs) in [27]. Compared to prior work, *e.g.,* [20], [29], [25], [26], [38, 15], where each phone tries to download all the data, and the local links are used for *error recovery*, our scheme *jointly* utilizes two interfaces, *i.e.,* 3G and WiFi, for data delivery.

Network Coding Implementation. There are WiFi testbeds that implement network coding, such as the COPE testbed [18]. COPE is a practical scheme for one-hop network coding across unicast sessions in wireless mesh networks. [24] proposes a cooperative IPTV system with pseudo-broadcast to improve reliability. Medusa [31] considers a scheme in which multiple unicast flows (video streams) are transmitted from a base station to clients with network coding. This scheme considers rate adaptation and video packet scheduling jointly.

The practicality of random network coding over iPhones is discussed in [32]. A toolkit to make network coding practical for system devices from servers to smartphones is introduced in [33]. [25] implements network coding on mobile devices and presents the performance in terms of throughput, delay and energy consumption. [26] extends [25] for picture transmission. A gesture broadcast protocol is designed for concurrent gesture streams in multiple broadcast sessions in [10] over smartphones using inter-session network coding. [38] proposes a system, in which a group of smartphone users (which are connected to the Internet via cellular links) help each other for error correction. A cooperative video streaming system is implemented over mobile devices (PDAs) in [27].

To the best of our knowledge, our work is the first to implement network coding and overhearing on the Android platform. These tasks are much more challenging on Androids than on laptops [18]. Network coding has been implemented before on other types of phones, [32, 25, 26], although not on Androids, but it has not been combined with overhearing, which is key in our setting.

Network coding for peer-to-peer systems. Network coding has been applied to P2P networks for content distribution [12, 39, 40] and live streaming [41, 22] (an excellent review is presented in [19]). We will show that, in our "micro"-setting, MicroNC-P2 significantly outperforms state-of-the-art P2P schemes, including the widely used BitTorrent [8] as well as the network coding-based R^2 [41]. This is because MicroNC-P2 is explicitly designed to exploit WiFi overhearing and network coding, and to avoid fundamental weaknesses of previous algorithms in our setting, such as the redundancy in R^2.

Network Optimization. Our system design is loosely inspired by the network utility maximization of the problem and its distributed solution [30]. For example, our preliminary analysis showed the benefit of cooperation and broadcast, which motivated the design of MicroNC-P2 and MicroBroadcast. It is worth noting that our experimental results closely match the trend obtained by the theoretical analysis, as explained in Section 6. Our MicroDownload is also inspired by the distributed rate control at [30].

Contributions. The contributions of this work lie (i) in the individual components, namely MicroDownload, MicroNC-P2, MicroBroadcast; each of which is novel, outperforms baselines, and can be used standalone; and (ii) in their combination into the overall MicroCast system. Our MicroCast architecture is modular, thus allowing for swapping various components (*e.g.,* different wireless technologies, streaming protocols, scheduling and dissemination algorithms). We optimized the mechanisms used by each component, in our particular setting, so as to outperform alternatives and work in a synergetic way. For example, MicroNC-P2 exploits network coding and high-rate broadcast that is made available at the system-level by the MicroBroadcast; MicroBroadcast makes local WiFi broadcast at high rates possible, for the first time, on Android devices; MicroNC-P2 reduces the congestion at the local network, thus simplifying the design of MicroDownload; and cooperation is used jointly on the downlink and local links.

3. MICROCAST ARCHITECTURE

3.1 Setup and Assumptions

We consider the scenario presented in Fig. 1: a group of smartphone users, within proximity of each other, are all interested in downloading and watching the same video at the same time. To achieve this goal, we use cooperation among the users. Furthermore, each phone simultaneously uses two interfaces: the cellular interface (3G) to connect to the server, and the local interface (WiFi) to connect to all other users in the group. Each phone downloads segments of the video from the server and shares these with the rest of the group locally.

We optimized our system under certain assumptions that hold for our setup of interest: First, there is a small number of users (up to 6-7). Second, these users know and trust each other, as it is the case in the motivating examples we provided in the introduction. Third, all users are within proximity of each other and all local links have similar rates on average. This is important as we need to use only single-hop, broadcast transmissions (although we do not need multihop transmissions due to close proximity, we could still incorporate them in our system – as we do in our Bluetooth-based implementation that supports multihop; see Section 7). Fourth, in every phone, we use the cellular connection for the downlink, and WiFi to establish local, device-to-device links. (Alternatives, including WiFi for the downlink or Bluetooth for the local link, are discussed in Section 7.) This has the implication that we can use the two connections (cellular and WiFi) on each phone simultaneously and independently; this would not be possible otherwise, as we discuss later. Note that cellular and WiFi connections are used in parallel, but one connects directly to the server (3G), while the other connects to the other users (WiFi); we do not use both of these connections to connect a user *directly* to the server through two different paths.

3.2 Architecture

We implement a prototype of our system as an application on Android phones, with the architecture shown in Fig. 2. In the remaining of this section, we give an overview of the building blocks and the interaction between them. More specifically, MicroCast consists of the following main components.

MicroDownload runs only on one of the phones (in a group) that initiates the download. It instructs the requesters of all phones which segments to download from the server. The download algorithm is described in Section 4.1.

MicroNC-P2 is responsible for distributing segments using the local wireless network. (We describe the algorithm in detail in Section 4.2.) MicroNC-P2 specifically exploits the broadcast capability provided by its lower layer MicroBroadcast to distribute segments quickly and efficiently.

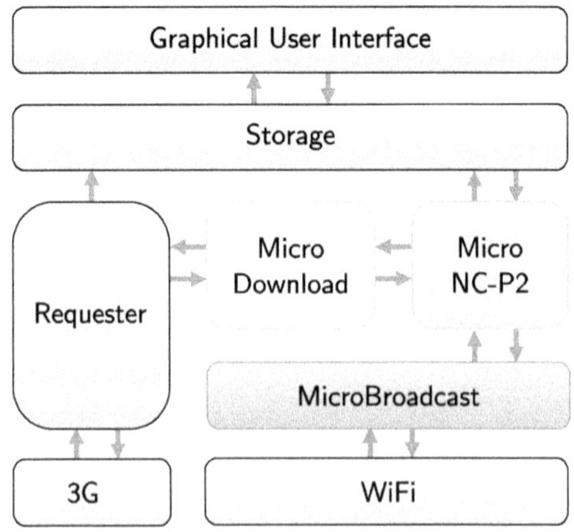

Figure 2: Architecture of MicroCast.

MicroBroadcast implements a comprehensive networking stack, which currently operates on wireless technologies including WiFi 802.11 and RFCOMM Bluetooth. The most important functionality that MicroBroadcast provides is the ability to pseudo-broadcast over WiFi. Nonetheless, MicroBroadcast also supports unicast, reliable and un-reliable message exchange between the network participants over both WiFi and Bluetooth. It also includes multi-hop routing, network-wide flooding, and peer discovery.

Requester retrieves segments of the video from the video source. Notice that depending on the video source location, only a subset of the phones might be able to request segments. For instance, if the video is locally available on one of the phones, only the phone having the file requests segments (from its local memory where the file is stored). If the video is on a remote HTTP server, only the phones with cellular connections request segments.

Storage is used to cache the received segments for successive playback.

Graphical User Interface (GUI) provides an interface to users to create video streams, share local videos, start/stop downloading video files, join existing video streams, and play/pause video. In addition to these basic features, the GUI lets users discover and connect to other devices, specify the wireless interface that should be used for the local dissemination, and decide whether the phone should collaborate for video downloading. The GUI also allows users to select system modules as well as run-time parameters for experimental purposes. The GUI provides live feedback during the experiments, and provides detailed statistics after the experiments. These functionalities of the GUI facilitate field tests. We provide some screen shots of our GUI in Fig. 3.

4. MICROCAST ALGORITHMS

In this section, we present the algorithms that we chose to implement in each component.

4.1 MicroDownload Algorithm

The video is divided into segments of fixed size. MicroDownload handles the scheduling of which phone should download which segment. Its main idea is that the next segment to be downloaded is assigned to a phone which has the smallest backlog (*i.e.*, the set of segments a phone has to download from its cellular connection). The MicroDownload algorithm is summarized in Alg. 1 and in the following.

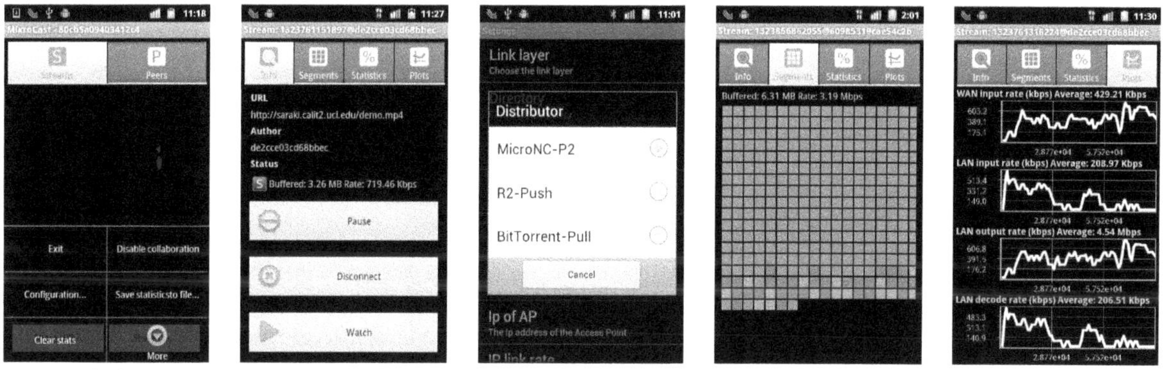

(a) Main Options (b) Information (c) Distributors (d) Downloading (e) Statistics

Figure 3: MicroCast Graphical User Interface.

Algorithm 1 MicroDownload Algorithm

```
while there are segments to assign do
    Find the phone with the smallest backlog
    if the backlog of the phone is smaller than K then
        Schedule the phone to download the next segment
    else
        Sleep until new feedback is received
    end if
    if feedback from phone indicates a failure then
        Schedule the phone to download another segment
        Add the segment that failed to the list of segments
    end if
end while
```

MicroDownload has a list of segments that should be assigned to the phones. Initially, it assigns a fixed number (K) of segments to each phone. The phones try to download one after the other the segments that are assigned to them. If a phone downloads a segment successfully, it notifies MicroDownload. Otherwise, it reports failure. MicroDownload re-assigns the failed segments (based on the backlogs). This mechanism ensures that no segment remains trapped in phones which have bad cellular connectivity. Also, this mechanism adapts segment download to the varying rates and conditions of cellular links. For example, if a phone has a bad cellular connection, the requests being handled by it will be re-scheduled by assigning them to other phones, which hopefully have better connections. However, MicroDownload will still assign some segments to the phone with the bad cellular connection so that it can start downloading immediately as soon as its channel quality improves.

4.2 MicroNC-P2 Algorithm

Each segment downloaded by a phone, is divided into packets, and distributed to the remaining phones using the local network. To do so, we design a custom dissemination scheme that exploits the benefits of overhearing and network coding. At a high level, our scheme takes advantage of pseudo-broadcast, *i.e.*, unicast and overhearing, to reduce the number of transmissions. Furthermore, instead of disseminating regular packets, our scheme disseminates random linear combinations of packets of the same segment, *i.e.*, dimensions of subspaces or network coded packets. This is to maximize the usefulness of overheard packets [11]. We desribe in detail how we use network coding in Section 5.3. As we will show in Section 6.2, existing distribution protocols, such as BitTorrent [8] and R2 [41], do not exploit overhearing and thus have worse performance.

We term our dissemination scheme MicroNC-P2, where P2 refers to an *initial Push* and *subsequent Pulls*. Fundamentally, MicroNC-P2 is built based on traditional pull-based P2P dissemination schemes. In MicroNC-P2, a phone, A, periodically advertises the segments that it currently has to its neighbors. Then, a neighbor, say phone B, requests segments that it does not have based on the advertisement. Upon receiving the request, phone A sends the requested segments to phone B. More specifically,

- When phone B requests a segment s from phone A, it takes into account previously overheard dimensions of the subspace representing segment s. In particular, it explicitly indicates in the request how many additional dimensions it needs to receive to decode s. This reduces the number of dimensions to be sent.

- When phone A is about to serve a segment s requested by phone B, it first checks if there are pending requests for the same segments from other neighbors. If there are, it finds the maximum number of dimensions requested among these requests. Denote this maximum dimension by d. If there is none, d is the number of dimensions requested by B. Afterwards, A serves d dimensions of segment s to B. The other phones, which need up to d dimensions of s, should be able to get the dimensions through overhearing.

After serving B, A notifies all phones that requested some dimensions of segment s. Upon receiving the notification, these phones check if they received all the necessary dimensions to decode s. If not, they send requests for additional dimensions. This is necessary, because overhearing is not guaranteed for all dimensions sent by A and for all phones. Finally, the scheme gives higher priority to requests that are closer to the playback time when serving them. Overhearing and unicasts effectively allow for pseudo-broadcast. As described, the amount of traffic saved by pseudo-broadcasting segment s depends not only on the quality of the overhearing but also on the number of requests of segments s from other phones that A processes at the time of broadcasting.

To be concrete, consider a local network consisting of four phones: A, B, C, and D. After finishing downloading segment s using 3G, A advertises it to B, C, and D. B, C, and D then send requests for this segment to A. For simplicity, assume perfect overhearing. First, consider the case where all requests arrive at A at a similar time. In this case, when A serves, *e.g.*, B's request for segment

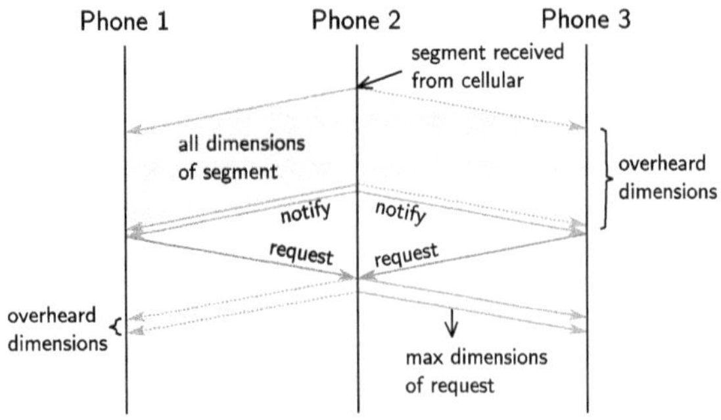

(a) MicroNC-P2

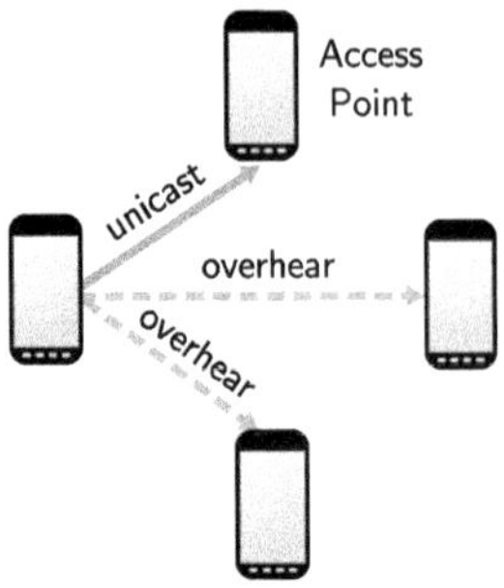

(b) MicroBroadcast

Figure 4: Space-time Diagram of MicroNC-P2 and MicroBroadcast.

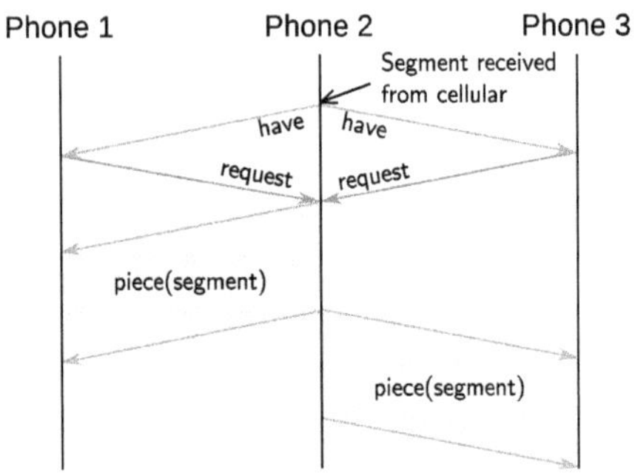

(a) BitTorrent-Pull

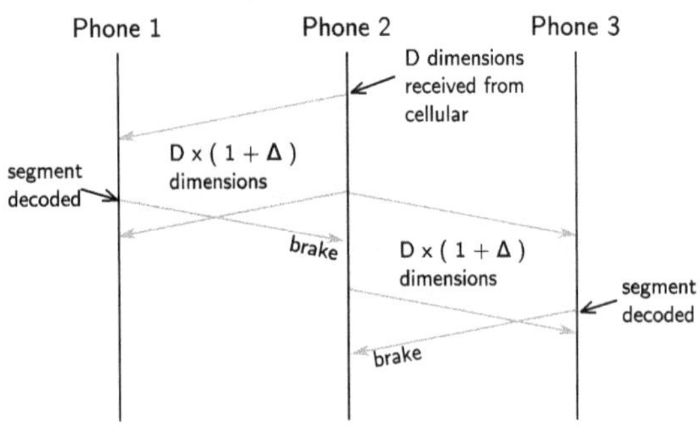

(b) R2-Push

Figure 5: Space-time Diagrams of BitTorrent-Pull and R2-Push.

s, A removes C and D's requests for s. Effectively, A serves all B, C, and D using a single transmission of s. Now, consider the case where the request from D arrives later than the time A initially serves s to B and C. The late arrival of D's request could be due to various reasons, such as large receiving and sending queues

Algorithm 2 MicroNC-P2 Algorithm

when a new segment s is received
  if s is received by the requester **then**
    // initial push
    Send all dimensions of s to a neighbor
  end if
  Add s to the list of segments to be advertised
end when

when a packet p is received from A
  if p is an advertisement
    or notification containing s **then**
    // subsequent pulls exploit overhearing
    Request A for the missing dimensions of s
  else if p is a request for d dimensions of s **then**
    Add this request to the request queue
  else if p is a dimension of s **then**
    Progressively decode s using p
  end if
end when

when there is a request for d dim. of s from A
  // pseudo-broadcast
  if there are other similar requests **then**
    Let d be largest requested dimension
    Remove these requests from the request queue
  end if
  Send d dimensions of s to A
end when

of D. A now has to serve D all dimensions of s even though D may overheard some dimensions initially sent by A to B (or C). Apparently, A needs to send more than needed.

To address this issue, we propose an initial push of segment s. Specifically, when A finishes downloading segment s, it sends all dimensions of s to a randomly selected neighbor before advertising the segment. By doing so, A ensures that the initial dissemination of segment s is taken into account in subsequent requests of segment s (if any) of A's neighbors. This effectively creates a *perfect synchronization* of the reception of the initial requests of segment s. We provide the pseudocode of MicroNC-P2 distribution algorithm in Alg. 2 and the space-time diagram of MicroNC-P2 in Fig. 4(a).

Last but not least, in order to address loss of request and notification packets, which could lead to missing segments, MicroNC-P2 includes a *recovery thread*. This thread periodically re-requests segments that were requested after a certain amount of time but never received.

4.3 Reception Rate

Our system aims to allow each phone to receive at a rate equal to the sum of the 3G/4G download rates of all phones in the cooperative group. This is the best rate we can hope to achieve since this is the maximum rate at which our network gets new information from the server. This rate may be higher than the playback rate of the video; this is useful to reduce the probability of a buffer underflow during playback. Downloading at a rate higher than the playback rate requires caching of the stream locally. This is not a problem on modern phones which come with large storage space. For instance, a Nexus S can store up to 16 GB of data. Assume for simplicity that each phone can receive rate R_c from the cellular network and rate R_l from the local network. If we have N phones, this implies that we need to maintain $R_l \geq (N-1)R_c$, since each phone is expected to receive through the local network the segments that the other $N-1$ phones have downloaded. In our set-up of interest, this was possible thanks to our efficient use of the local network through network coding and overhearing: our bottleneck (min-cut) has been between the phones and the cellular network.

5. IMPLEMENTATION DETAILS

In this section, we describe the implementation details, the major challenges we faced specifically on Android phones, and the design choices we made to address them.

5.1 Architecture Details

Our software is developed mostly in Java, with some minor parts in C, and runs on Android 2.3 and 4.0 and Java 2 SE. Both the Android and Java versions share the same structure and code, except for the graphical user interface (GUI) and the code that uses the local network wireless API. In this paper, we restrict our attention to the Android version.

Requester. It internally uses components called producers to retrieve segments of the video from the source of the stream. Our current implementation contains producers for three types of sources: HTTP, file, and content. The first one (HTTP) loads segments from an HTTP server using range requests. The second one (file) loads video from locally available files. Finally, the third one (content) retrieves data using the ContentProvider API of Android (*e.g.,* videos can be captured with the phone camera). The implementation allows to easily add new producers. The HTTP producer is agnostic to the actual networking technology used to access the server. Therefore, it can work not only when the phones use a 3G network to access the video server, but also when they use an infrastructure mode WiFi network. The overhead of using range requests measured in bytes is relatively small, around 3% when using segments of 22500 bytes. The download of segments is affected by round-trip time, but if the HTTP server hosting the video supports persistent connections, all requests can be carried out in a single TCP connection, thus saving some overhead. To further reduce the impact of round-trip time, our code supports the usage of multiple parallel TCP connections on each phone.

Storage. The segments are stored in the internal flash memory of the phone to keep the application memory requirements low. It is possible to access the segments from the storage either using a Java API, as done by the requester and MicroNC-P2, or via an embedded HTTP server that we have developed. This second interface allows us to play the video stream using the native Android media API. In order to support playback of non-streamable video, for instance, MP4 files with `moov` atom at the end of the file, our embedded HTTP server supports range requests; thus, the Android Media Player can perform random access of the video stream. If a range that has not yet been downloaded is requested, the HTTP server waits until it receives the full range before answering.

Graphical User Interface (GUI). The GUI automatically displays the locally reachable peers to form a group, and which streams (if any) is being downloaded in the group. The user only needs to select the desired stream and join it. The phones can play the video in a synchronized manner or at their own pace. Playback and download are decoupled thanks to our storage mechanism: a phone could be participating in the download while its video player is paused. To render the video, the application uses the media playback API included in Android, which supports various containers and video formats, such as H.264 in a MP4 container. The video can be displayed while MicroCast is still downloading; therefore, live streaming is supported.

MicroBroadcast. In order to facilitate porting of the application to different wireless technologies, MicroBroadcast contains an application layer implementation of a networking stack. Depending on the wireless technology used, features of MicroBroadcast are either implemented using a native mechanism or emulated. For instance, the Bluetooth implementation re-uses the native peer discovery mechanism while WiFi nodes run a custom peer discovery protocol. We give the implementation details for pseudo-broadcast in Section 5.4.

5.2 Multiple Network Interfaces

Each phone needs to use an interface as downlink (*e.g.,* 3G or WiFi) and another interface (*e.g.,* WiFi or Bluetooth) for the local cooperation. For the connection to the server, we chose 3G over WiFi mainly because it provides ubiquitous Internet access. A second reason is that Bluetooth and WiFi share partially overlapping parts of the spectrum and are often implemented in the same chip, while 3G is usually implemented on a different chip and uses a different part of the spectrum. This suggests that using WiFi (for the connection to the server) together with Bluetooth (for the local network) may noticeably decrease the transmission rate, which was indeed the case during our initial experiments. 3G is independent from both Bluetooth and WiFi, so the combination of 3G and either WiFi or Bluetooth does not reduce the transmission rate. For the local connection, we use WiFi instead of Bluetooth because it can support a larger number of connections at higher rates.

We also note that the Android connectivity manager imposes additional challenges when 3G is utilized simultaneously with WiFi. In particular, in order to improve the battery life, every time the WiFi interface is activated, Android turns the 3G data connection off. We solve this problem using an undocumented API that forces routing of packets for the HTTP server through the 3G interface, and therefore preventing the interface from being shut down.

5.3 Network Coding

Network Coding Scheme. We use generation-based network coding [6] over the field GF(2^8). Each segment is broken down into m packets $\hat{\mathbf{b}}_i$, which together form one generation (or segment), where m is the segment or generation size. Each packet contains n bytes, and we treat each byte as a symbol in GF(2^8). We also augment each packet with the m coding coefficients, each of which is selected uniformly at random from GF(2^8). Thus, each packet can be seen as a vector of length $n+m$ symbols from GF(2^8)

Phone A sends to phone B linear combinations of packets of the same segment, where the coding coefficients used to create linear combinations are selected uniformly at random from GF(2^8). Phone B can decode a segment upon receiving m linearly independent combinations of packets of the segment. Let M denote the matrix formed by m linearly independent packets: $M = [E \mid C]$, where E is of size $m \times n$ and C is the coefficient matrix of size $m \times m$. Original packets $\hat{\mathbf{b}}_i$ can be recovered by finding the inverse of C. In particular, $C^{-1} \cdot [E \mid C] = [B \mid I]$, where B is the matrix of size $m \times n$ whose row i is $\hat{\mathbf{b}}_i$ and I is the $m \times m$ identity matrix. Inverting C takes $\Theta(m^3)$ and multiplying C^{-1} with $[E \mid C]$ takes $\Theta(m^2(n+m))$ in terms of finite field multiplication. Thus, the decoding takes $\Theta(m^3+nm^2)$ in total. Generating m randomly encoded packets can be done by generating a random coefficient matrix R of size $m \times m$ and multiplying R with $[B \mid I]$. Thus, the encoding of a segment also takes $\Theta(m^3+nm^2)$ in total.

CPU Limitations. Network coding is a CPU intensive operation. In MicroCast, encoding and decoding must be performed efficiently, at a rate matching that of the local network dissemination; otherwise, CPU risks to become the bottleneck of the video distribution. Therefore, in our implementation, we explored several ways to optimize the coding speed.

The first method to reduce the CPU usage is to limit the size of the coding generation. The smaller the number of packets in each segment, the smaller the coding complexity. Using smaller segment sizes, however, reduces the diversity of encoded packets, *i.e.,* packets are less likely to bring innovative information to their recipients. In Section 6.5, we give encoding and decoding rates as a function of segment size.

Second, we seek to optimize our implementation of network coding. In particular, we test two implementation approaches: pure Java and native code. In the first implementation, the encoding and decoding operations are performed by code that runs in the Dalvik virtual machine. In the second approach, the code runs natively on the phone CPU and is invoked through the Java Native Interface. The Java implementation has the advantage of being portable across different hardware platforms but is less efficient than the native version. In both implementations, we use table lookups to perform finite field multiplication and division, and we use the bit-by-bit XOR operation to perform addition and subtraction.

For packet length equal 900 (bytes), segment length equal $22,500$ (bytes), and (resulting) generation size equal 25, inverting the coefficient matrix C takes roughly 8% of the decoding time (measured on the native implementation). The rest of the time is used to recover the original vector by linearly combining the received packets (multiplying C^{-1} with $[E \mid C]$). The Java implementation can encode at 2.9 Mbps and decode at 4.3 Mbps (the significant difference in rate is due to a different memory usage pattern), while the native implementation can both encode and decode at 24 Mbps. The Java implementation is sufficient for low bit-rate videos while the native implementation can support even high-quality video streaming (indeed, with the native implementation, our experiments show that MicroCast can support 2.5 Mbps stream to a group of 7 phones). Both implementations, although sufficient for our needs, are not fully optimized and thus the rate could potentially be further improved.

5.4 Implementing High-Rate WiFi Broadcast

MicroBroadcast provides to to MicroNC-P2 an interface for high-rate local broadcast. To the best of our knowledge, this is the first system that provides this capability on top of WiFi on Android phones.

Although phones within proximity of each other can, in principle, overhear all transmissions, high-rate broadcast was not possible with the existing modes. The *unicast mode of 802.11* does not exploit broadcast: it (redundantly) transmits the same packets to each receiver separately. The *broadcast mode of 802.11* has its own disadvantages, [31]: (i) it lacks a back-off mechanism, which may harm the performance of other flows; (ii) its transmission rate is limited to the minimum (base rate, 1 Mbps); (iii) finally, unlike laptops, it is not always possible to adapt the broadcast transmission rate on Android phones due to wireless driver and firmware limitations.

A possible solution is to use *pseudo-broadcast, i.e., overhearing*, which combines the benefits of unicast and broadcast. Unicast is used as the transmission mode, but the phones overhear all transmissions in their neighborhood. Therefore, pseudo-broadcast combines the desirable properties of unicast (high rate, back-off) with overhearing, which makes it attractive. Although it has been implemented in several frameworks [18, 31, 4], when implementing pseudo-broadcast, we faced several challenges that are specific to Android phones.

First, the phones we used do not readily support the *promiscuous mode* due to the constraints imposed by the WiFi firmware and drivers. Therefore, we needed to update the WiFi drivers of all the Android 2.3 phones we used, and the firmware in some of them. In particular, we updated the WiFi driver and firmware by installing CyanogenMod 7 ROM [9] (a custom Android firmware) on the phones after testing various possible firmwares. With the CyanogenMod 7 ROM, promiscuous mode is available, but only in infrastructure mode.

Second, even with the promiscuous mode enabled, Android does not support pseudo-broadcast mode natively, *i.e.,* does not pass the overheard packets up to the application layer. We had to develop our own *overhearing API* for that purpose "under the hood of Android framework" by developing our own C library and a C binary executable program that runs as a daemon. This involved filtering out irrelevant overheard packets (*i.e.,* packets which do not belong to video data that the phones are interested in) so as not to overload the CPU.

Third, as we mentioned above, since overhearing is not available in ad-hoc mode, this pseudo-broadcast implementation works only in infrastructure mode. Using infrastructure mode has a major disadvantage compared to ad-hoc mode: when a phone transmits a packet to another phone, the packet has to be relayed by the access point, which results in double amount of traffic.

To avoid this disadvantage of infrastructure mode as well as to exploit the benefits of overhearing, we implemented a *pseudo-ad-hoc mode*, which is shown in Fig. 4(b). In this mode, one of the phones acts as an access point (AP), and all other phones transmit data to it. These transmissions are overheard opportunistically by all other phones. When the AP phone receives a packet, it *does not* forward it (as it would normally do in the infrastructure mode), since the other phones should already have received it via overhearing. In this manner, we are able to enable overhearing (which is only possible in infrastructure mode), while ensuring only one transmission per packet (which is the case in ad-hoc mode), thus the term *pseudo-adhoc mode*.

6. PERFORMANCE EVALUATION

In this section, we first evaluate the performance of MicroDownload and MicroNC-P2 and compare their performance to baseline, popular alternatives. We then evaluate the entire MicroCast system as a whole. We show that our schemes significantly improve the streaming experience in terms of download time and video rate, without introducing significant battery and CPU penalty.

In our evaluation, we perform experiments on an Android testbed consisting of seven smartphones: four Samsung Captivate and three Nexus S. All smartphones have a 1 Ghz Cortex-A8 CPU and 512 MB RAM. Six of them use Android Gingerbread (2.3) and one (Nexus S) uses Android Ice Scream Sandwich (4.0) as their operating systems.

6.1 Evaluation of MicroDownload

In this section, we present experimental results that motivate the necessity of implementing MicroDownload and we show its effectiveness. The setup is the following: we used three Nexus S connected to the same cellular network provider, and placed them within proximity of each other (the distances among them are approximately 2 cm) in an indoor environment. The phones were placed in their positions 5 minutes before the experiment started to eliminate any possible positive or negative bias due to mobility.

In our experiment, we disabled MicroNC-P2 and we measured

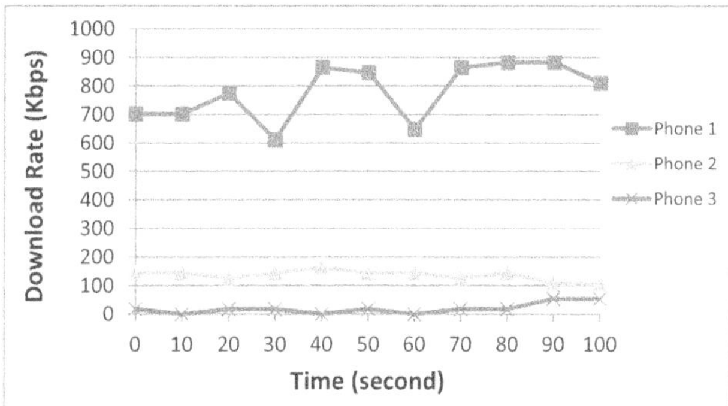

Figure 6: The cellular link rates of three smartphones in the same geographical area.

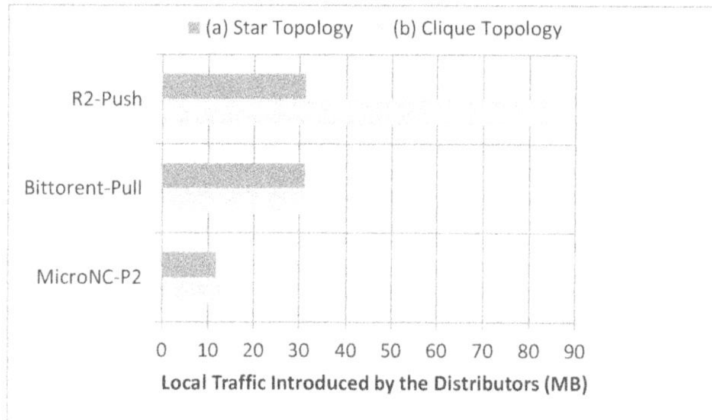

Figure 7: The amount of local traffic introduced by the phones when using different distributors. The file being downloaded is 9.93 MB. Bandwidth of the local network is sufficient to support the local dissemination. All phones receive the file at a similar rate 550 Kbps. MicroNC-P2 manages to introduce the least amount of traffic thanks to network coding and overhearing.

the download rates of the smartphones over 100 seconds. The results are presented in Fig. 6. The figure shows that despite being in close proximity and being connected to the same operator the phones experience significantly different average download rates. Phone 3 has a very low rate because it uses EDGE. The other two phones use the same 3G network but still have significantly different download rates. Moreover, phone 1 experiences significant rate variations. This variability in time and across phones is our motivation to develop MicroDownload to adaptively request data according to the downlink rates of cellular links, instead of making static decisions, such as splitting the requests equally among the phones.

Using these measurements, we can compare the effectiveness of MicroDownload algorithm to a simpler static strategy. We consider a scenario where the three phones download a 750 kB file, and MicroDownload makes a static decision, *i.e.,* each phone requests one third of the file. In this case, phone 3 (considering the same channel realization as in Fig. 6) is the bottleneck for downloading the file, and the total download duration is 80 seconds. However, if MicroDownload makes adaptive requests, as proposed in Section 4.1, then phone 3 is not a bottleneck anymore, and the total download duration is less than 10 seconds. This shows the importance of the adaptive request mechanism of MicroDownload.

6.2 Evaluation of MicroNC-P2

In this section, we compare the performance of MicroNC-P2 to a BitTorrent-based distributor [8] and an R2-based distributor [41]. We refer to these two distributors as BitTorrent-Pull and R2-Push, respectively. The performance metric of interest is the amount of local network traffic introduced by the phones when using different distributors to disseminate the same amount of information. We consider a clique and a star overlay topologies for local connectivity. Packets are exchanged locally using UDP.

We implemented the BitTorrent-Pull scheme based on the description of the BitTorrent protocol [8]. In particular, our implementation of the protocol supports three main types of messages: (i) *bitfield* and *have* messages, which are used by a phone to advertise the segments to its neighbors; (ii) *request* messages, which are used by a phone to request specific segments from its neighbors; and (iii) *piece* messages, which contain the actual data. The space-time diagram of BitTorrent-Pull is provided in Fig. 5(a). Fundamentally, BitTorrent is a pull-based P2P protocol: when a phone has downloaded a new segment, it advertises this segment to its neighbors. The neighbors then explicitly request the segments that they are missing. To account for the wireless loss rate (when using UDP), we implemented a recovery thread which periodically re-requests missing segments.

We implemented the R2-Push scheme based on the description in [41]. The R2 protocol was introduced to exploit the benefit of random network coding and random push. Following [41], our implementation of R2-Push supports two main types of messages: (i) *data* messages, which are random linear combinations of packets belonging to the same segment; and (ii) *brake* messages, which are used by phones to inform their neighbors that they successfully received and decoded specific segments. The purpose of brake messages is to ensure that the neighbors would stop pushing (unnecessary) linear combinations of the decoded data segments.

The space-time diagram of R2-Push is provided in Fig. 5(b). In contrast to BitTorrent-Pull, with R2-Push, the phones start pushing linear combinations of segments as soon as they receive them from either the cellular network or their neighbors. In our implementation, for a particular segment a phone is downloading, we limit the number of linear combinations that it can push to its neighbors to the rank of the matrix formed by the received packets plus a fixed amount of redundancy, Δ, to account for the wireless loss rate.

Fig. 7 shows the total amount of traffic introduced to the local network by four phones to disseminate a file when the phones are connected using star and clique topologies. Note that the difference between these topologies are that in the star topology, a hub is used as an access point (AP), and in the clique topology a random phone is chosen to serve as an AP. In both topologies, all phones overhear all the transmissions in the group, *e.g.,* from phones to the AP which is either a hub or a phone. MicroNC-P2 utilizes pseudo-adhoc as described in Section 5. The file size is 9.93 MB and is downloaded by a single phone using its 3G connection. The average rate of the 3G connection is measured at 550 Kbps. The

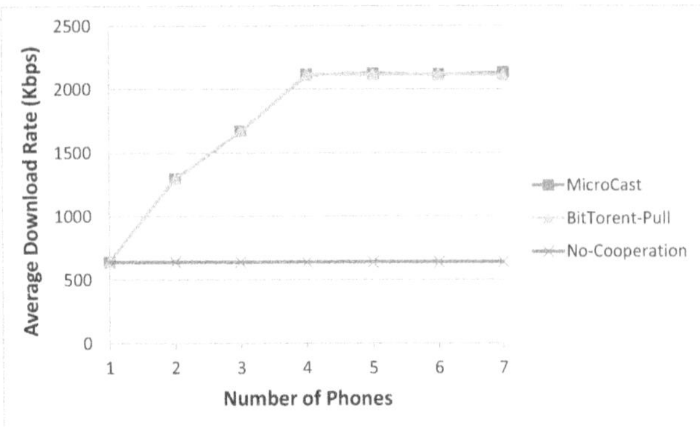

Figure 8: Average download rate as a function of number of phones when the local network bandwidth is 20 Mbps. Performance of MicroCast and BitTorrent-Pull almost coincide on the plots.

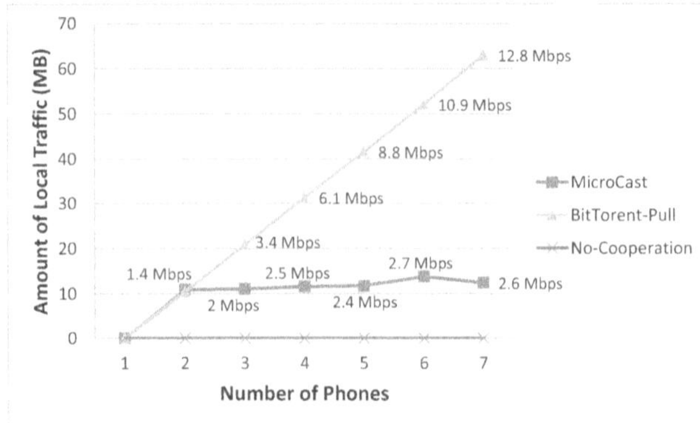

Figure 9: The amount of local traffic introduced by all phones.

phone that downloads the file is chosen at random. For R2-Push, we choose $\Delta = 3\%$. The local network can support 20 Mbps UDP traffic, measured using *iperf* [23]. This bandwidth is much larger than what is needed to support the traffic introduced by the phones in both topologies. Since the local network bandwidth is sufficient, each of the phone receives at the rate similar to the phone which downloads the file through 3G. Each reported number is averaged over three experiments.

We first observe from Fig. 7 that the amount of traffic introduced by both BitTorrent-Pull and R2-Push are more than three times higher than that of MicroNC-P2. Intuitively, this is due to the fact that when using MicroNC-P2, a packet sent by a phone may be beneficial to three phones instead of one thanks to network coding and overhearing.

Fig. 7 also shows that in a clique topology, R2-Push introduces much more traffic as compared to the star topology, while BitTorrent-Pull and MicroNC-P2 introduce similar amount of traffic in both clique and star topologies. This is due to the fact that in a clique topology, a phone may simultaneously receive linear combinations of the same segment from multiple neighbors. When this happens, it is critical that the neighbors which are sending to this phone stop pushing linear combinations in a timely manner. This could only be achieved with a timely arrival of the brake (stop) messages, which is not always possible in the clique topology, or in a setup where additional traffic is very high. The authors of R2 also observed the problem and reported it in [22]. In their setting [22], the amount of redundancy could be reduced by using larger segment sizes. However, due to limited computational power of mobile devices, we cannot afford having large segment sizes as discussed in Section 5.

To summarize, the set of experimental results presented in this section clearly show that by exploiting the broadcast nature of the wireless medium, MicroNC-P2 manages to introduce less amount of traffic into the local network as compared to BitTorrent-Pull and R2-Push.

6.3 Evaluation of the MicroCast System

In this section, we present the performance evaluation of the entire MicroCast system. We compare the average download rates of

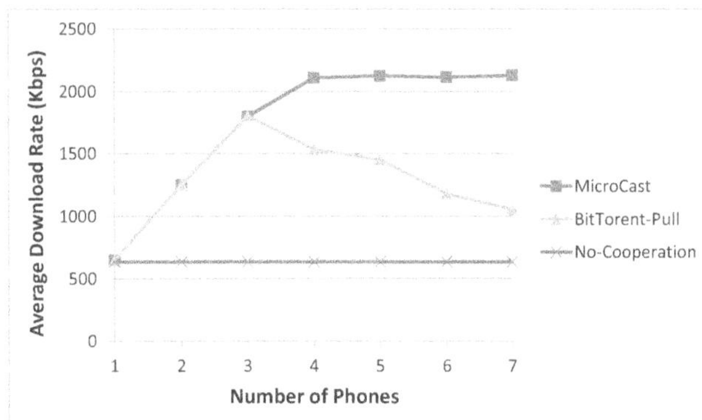

Figure 10: Average download rate as a function of number of phones when the local network bandwidth is 4 Mbps.

MicroCast to two other schemes: no cooperation, which we will refer to as No-Cooperation, and the combination of MicroDownload and BitTorrent-Pull, which we will simply refer to as BitTorrent-Pull. Note that we do not include R2-Push as a baseline in this section due to its inefficiency in our setup, as explained in Section 6.2.

In our experimental setup, we used up to seven phones, the first four of which had 3G rates varying from 480 Kbps to 670 Kbps. The rest of the phones did not have 3G connections. Packets are exchanged locally using UDP. The local network can support up to 20 Mbps UDP traffic. We use the star topology as explained in Section 6.2. We also use pseudo-adhoc. The size of the video file is 9.93 MB. Each value reported in this section is averaged over three experiments.

Fig. 8 shows the average download rate versus the number of phones. We observe that both MicroCast and BitTorrent-Pull are able to improve the average download rate up to the sum capacity of the 3G links. Note that MicroCast and BitTorrent-Pull do not provide any improvement for more than four phones because only four phones have 3G connections in our setup, and the average download rate is limited by the sum capacity of the four corresponding 3G links. Fig. 9 shows the amount of local traffic versus the number of phones. Although in Fig. 8 we see similar average download rates for both MicroCast and BitTorrent-Pull, Fig. 9 shows that BitTorrent-Pull introduces a larger amount of local traffic (which increases linearly in the number of phones) as compared to MicroCast. This behavior of BitTorrent-Pull is detrimental in terms of the average download rate in congested networks. An important observation is that, as the number of phones increases, MicroCast rate does not increase. This indicates that, even with many nodes, overheard packets are lost very rarely.

We updated our experimental setup to evaluate the performance of MicroCast and BitTorrent-Pull in a congested network. In our new setup, the congested network is generated by introducing 16 Mbps background UDP traffic on the same 802.11 channel (note that there is also interference from other sources in the environment which contributes to the background traffic). Since the local network can support up to 20 Mbps traffic, the leftover traffic is less than 4 Mbps. Fig. 10 presents the average download rate versus the number of phones in this setup. We see that the average download rate of BitTorrent-Pull reduces when we have more than three phones. This is because BitTorrent-Pull introduces a large amount of local traffic (as illustrated in Fig. 9), which leads to congestion.

Note that the addition of the 5th, 6th, and 7th phones also increases the local traffic in BitTorrent-Pull, even though they do not have 3G connection. This is because they still need to receive the file in the local area, which contributes to local area traffic.

On the other hand, Fig. 10 shows that MicroCast still improves the average download rate up to the total capacity of 3G links (of four phones) in a congested network. This is because it introduces only a small amount of local traffic (*e.g.*, even for seven phones, MicroCast only introduces 2.6 Mbps traffic to the local network). It can be observed from Fig. 10 that the average download rate of MicroCast is more than three times higher than that of No-Cooperation. Also, the improvement of MicroCast over BitTorrent-Pull in terms of average download rate is as high as 75% (we observed even more improvement for different setups, *e.g.*, with 18 Mbps background traffic), which is significant. We also note that our experimental results are consistent with the theoretical findings in our earlier work [30], *e.g.*, Fig. 8 and Fig. 10 are consistent with Fig. 3 in [30].

6.4 Evaluation of Energy Consumption

In this section, we evaluate the energy consumption of MicroCast when compared to our baselines: BitTorrent-Pull and No-Cooperation. We consider a setup similar to the one in Section 6.3, *i.e.*, three phones are connected to a fourth one that acts as AP , the 3G rates vary from 450 Kbps to 700 Kbps, and the size of the video file is 95.4 MB. We used the *BatteryManager* class of the Android SDK for the power consumption measurements. Before the experiment, all four phones are fully charged. During the experiment, the battery states are recorded every 10 seconds. In the experiments that do not use the local network, the wireless interface is turned off. The experiments are repeated three times, and their average is reported.

Fig. 11 presents the battery state (100% corresponds to the fully charged battery) versus time. Note that "MicroNC-P2 Access Point" and "BitTorrent-Pull Access Point" show the battery consumption levels of the phone which is selected to act as an AP in MicroCast and BitTorrent-Pull schemes, respectively. On the other hand, "MicroNC-P2 Normal" and "BitTorrent-Pull Normal" show the battery consumption levels of a phone which is not an AP. We observe from Fig. 11 that the No-Cooperation scheme has less battery con-

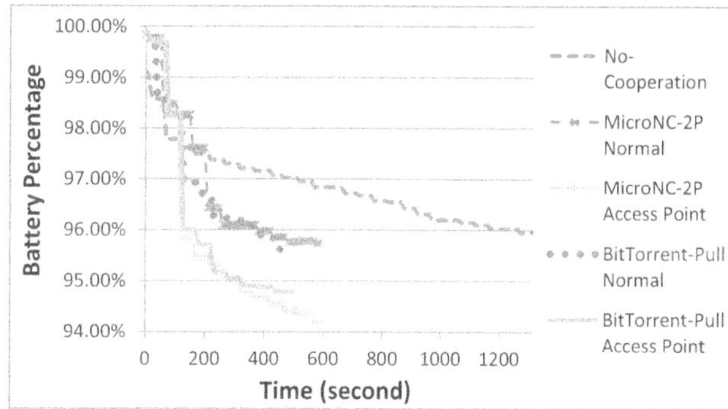

Figure 11: Battery drain when downloading the same file of size 95.4 MB using different schemes.

sumption compared to MicroCast and BitTorrent-Pull at a given time, *e.g.*, at 400 second. However, the time required to download the video file of No-Cooperation is very high (more than two times) as compared to MicroCast and BitTorrent-Pull.

If we look at the battery levels of all schemes when the file transmission is completed, we see that the battery consumption levels of No-Cooperation, MicroCast, and BitTorrent-Pull are similar. This demonstrates that employing cooperation in the long term (to download a video file) does not bring any significant battery cost. Finally, the phones which act as APs consume more battery as compared to the other phones: this is expected because the AP phone has additional tasks. However, even in the worst case (for the AP phone), MicroCast consumed approximately 6% of the battery to download the whole file. Considering that the phones are downloading a large file (*i.e.*, 95.4 MB), this battery consumption level is reasonable. These considerations show that the rate benefit of MicroCast comes at no significant battery cost.

6.5 Evaluation of NC Implementations

In this section, we present the performance of the two implementations of network coding we developed: native (written in C) and Java-based coding. The two implementations are described in Section 5.3. Fig. 12 shows the highest achievable decoding and encoding data rates. The slowest encoding rate for the Java implementation is 1 Mbps, while for the Native implementation, it is 8 Mpbs, which is significantly higher than what is needed for video streaming applications considered in this paper. Fig. 12 also shows that the Java implementation is more than sufficient to stream today's typical Internet videos when using generation size equal 25.

7. LIMITATIONS AND EXTENSIONS

The system presented in this paper builds on the assumptions described in Section 3: there is a small number of trusted users within proximity of each other, using the cellular connection to download and WiFi (with broadcast and single-hop transmission) for local sharing. However, our architecture is highly modular and allows for easily swapping several components, transparently to the rest of the system.

First, we have already implemented and successfully tested a version of the system that uses Bluetooth in the local area, in-

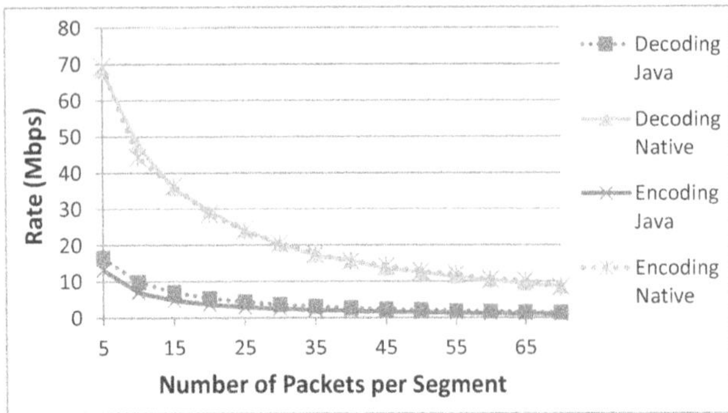

Figure 12: Coding and decoding throughput as a function of the generation length. Each packet is 900-byte long.

stead of WiFi. An attractive feature of Bluetooth is that it can be used without rooting the phone. Furthermore, it supports both single-hop and multi-hop transmission via piconets and scatternets respectively. On the downside, Bluetooth does not support broadcast, which was a necessary ingredient for achieving MicroCast's benefits. It would be interesting to study potential benefits of Bluetooth in multi-hop scenarios, *e.g.,* when not all phones can overhear each other sufficiently well. Second, one could use WiFi (instead of cellular) for downloading the content from the server: our architecture already supports that. However, in that case, the downlink and the local transmissions are no longer independent as they both use WiFi. We plan to explore this interaction in future work; we have already analyzed this problem theoretically (in a network utility maximization framework) in [30].

Our current implementation of MicroDownload uses a simple algorithm that could be improved in multiple ways. It currently does not take into account the local link quality between the nodes and does not try to download in parallel blocks on multiple devices when spare capacity is available (for instance, at the end of the stream). The current implementation of MicroDownload also cannot exploit broadcast capabilities of the WAN link. This is typically not available on cellular networks but could be used if, for instance, the phones are connected to an 802.11 AP. Notice that even when broadcast is available on the WAN link, it is still useful to perform collaboration: local dissemination can help correcting uncorrelated erasures experienced by the nodes.

Finally, the current implementation cannot support more than 7 concurrent devices (when an Android 4.0 device acts as the AP) or 6 devices (when an Android 2.3 device acts as the AP). This is due to the limitation of the softAP currently implemented in Android. In order to further increase the number of users, one could use more than one phones as softAPs, but this requires modification of the system. In particular, promiscuous mode does not allow for overhearing of packets associated with a different softAP. We will investigate such extensions as part of future work.

8. CONCLUSION

In this paper, we designed, implemented, and evaluated a novel system, MicroCast, that enables a group of smartphone users within proximity of each other to watch the same video from the Internet at the same time. MicroCast cooperatively uses the resources on all smartphones of the group, such as cellular links and WiFi connections, to improve the streaming experience. The system consists of three key components, namely: MicroDownload, which determines which parts of the video each phone should download from the server; MicroNC-P2, which exploits overhearing and network coding over WiFi; and MicroBroadcast, which provides, for the first time, high-rate broadcast over WiFi on Android phones. Experimental results demonstrate significant performance benefits in terms of per-user video download rate without battery penalty.

9. ACKNOWLEDGEMENTS

This work was funded by the following grants: ArmaSuisse Wissenschaft+Technologie (W+T) Project no. 8003413832, ERC Project NOWIRE (ERC-2009-StG-240317), AFOSR MURI award FA9550-09-0643, and NSF CAREER award 0747110.

10. REFERENCES

[1] Wireless network coding: from theory to practice, project wiki page. http://odysseas.calit2.uci.edu/doku.php/public:muri09.

[2] G. Ananthanarayanan, V. N. Padmanabhan, L. Ravindranath, and C. A. Thekkath. COMBINE: leveraging the power of wireless peers through collaborative downloading. In *Proceedings of the 5th International Conference on Mobile Systems, Applications and Services (MobiSys)*, pages 286–298, 2007.

[3] C. Boldrini, M. Conti, and A. Passarella. Exploiting users' social relations to forward data in opportunistic networks: The HiBOp solution. *Pervasive and Mobile Computing*, 4(5):633–657, Oct. 2008.

[4] R. Chandra, S. Karanth, T. Moscibroda, V. Navda, J. Padhye, R. Ramjee, and L. Ravindranath. DirCast: a practical and efficient Wi-Fi multicast system. In *Proceedings of the 17th IEEE International Conference on Network Protocols (ICNP)*, pages 161–170, Oct. 2009.

[5] J. Chesterfield, R. Chakravorty, I. Pratt, S. Banerjee, and P. Rodriguez. Exploiting diversity to enhance multimedia streaming over cellular links. In *Proceedings of the 2005 IEEE INFOCOM*, volume 3, pages 2020–2031, Mar. 2005.

[6] P. Chou and Y. Wu. Network coding for the internet and wireless networks. *IEEE Signal Processing Magazine*, 24(5):77–85, Sept. 2007.

[7] Cisco Systems. Cisco visual networking index: Global mobile data traffic forecast update, 2011-2016. http://www.cisco.com.

[8] B. Cohen. Incentives build robustness in BitTorrent. In *Proceedings of the First Workshop on Economics of Peer-to-Peer Systems*, Berkeley, CA, USA, 2003.

[9] CyanogenMod Team. Cyanogenmod. http://www.cyanogenmod.com.

[10] Y. Feng, Z. Liu, and B. Li. GestureFlow: streaming gestures to an audience. In *Proceedings of the 2011 IEEE INFOCOM*, pages 748–756, Apr. 2011.

[11] C. Fragouli and E. Soljanin. *Network Coding Fundamentals*. Now Publishers Inc, Delft, The Netherlands, June 2007.

[12] C. Gkantsidis and P. Rodriguez. Network coding for large scale content distribution. In *Proceedings of the 2005 IEEE INFOCOM*, volume 4, pages 2235–2245, Mar. 2005.

[13] P. Goldstein. Credit suisse report: U.S. wireless networks running at 80% of total capacity. *FierceWireless.com*, July 2011.

[14] B. Han, P. Hui, V. A. Kumar, M. V. Marathe, G. Pei, and A. Srinivasan. Cellular traffic offloading through opportunistic communications: a case study. In *Proceedings of the 5th ACM Workshop on Challenged Networks (CHANTS)*, pages 31–38, 2010.

[15] S. Hua, Y. Guo, Y. Liu, H. Liu, and S. Panwar. Scalable video multicast in hybrid 3G/Ad-Hoc networks. *IEEE Transactions on Multimedia*, 13(2):402–413, Apr. 2011.

[16] P. Hui, J. Crowcroft, and E. Yoneki. Bubble rap: social-based forwarding in delay tolerant networks. In *Proceedings of the 9th ACM International Symposium on Mobile Ad Hoc Networking and Computing (MobiHoc)*, pages 241–250, 2008.

[17] S. Ioannidis, A. Chaintreau, and L. Massoulie. Optimal and scalable distribution of content updates over a mobile social network. In *Proceedings of the 2009 IEEE INFOCOM*, pages 1422–1430, Apr. 2009.

[18] S. Katti, H. Rahul, W. Hu, D. Katabi, M. Medard, and J. Crowcroft. XORs in the air: Practical wireless network coding. *IEEE/ACM Transactions on Networking*, 16(3):497–510, June 2008.

[19] B. Li and D. Niu. Random network coding in Peer-to-Peer networks: From theory to practice. *Proceedings of the IEEE*, 99(3):513–523, Mar. 2011.

[20] S. Li and S. Chan. BOPPER: wireless video broadcasting with Peer-to-Peer error recovery. In *Proceedings of the 2007 IEEE International Conference on Multimedia and Expo*, pages 392–395, July 2007.

[21] X. Liu, G. Cheung, and C. Chuah. Rate-distortion optimized network coding for cooperative video stream repair in wireless peer-to-peer networks. In *Proceedings of the 2008 International Symposium on a World of Wireless, Mobile and Multimedia Networks (WoWMoM)*, pages 1–6, June 2008.

[22] Z. Liu, C. Wu, B. Li, and S. Zhao. UUSee: Large-Scale operational On-Demand streaming with random network coding. In *Proceedings of the 2010 IEEE INFOCOM*, pages 1–9, Mar. 2010.

[23] NLANR/DAST. IPerf. http://sourceforge.net/projects/iperf.

[24] Y. Park, C. Jo, S. Yun, and H. Kim. Multi-Room IPTV delivery through Pseudo-Broadcast over IEEE 802.11 links. In *Proceedings of the IEEE 71st Vehicular Technology Conference (VTC)*, pages 1–5, May 2010.

[25] M. Pedersen and F. Fitzek. Implementation and performance evaluation of network coding for cooperative mobile devices. In *Proceedings of the 2008 IEEE International Conference on Communications (ICC) Workshops*, pages 91–96, May 2008.

[26] M. Pedersen, J. Heide, F. Fitzek, and T. Larsen. PictureViewer - a mobile application using network coding. In *Proceedings of the 2009 European Wireless Conference*, pages 151–156, May 2009.

[27] M. Ramadan, L. El Zein, and Z. Dawy. Implementation and evaluation of cooperative video streaming for mobile devices. In *Proceedings of the 19th IEEE International Symposium on Personal, Indoor and Mobile Radio Communications (PIMRC)*, pages 1–5, Sept. 2008.

[28] P. Rodriguez, R. Chakravorty, J. Chesterfield, I. Pratt, and S. Banerjee. MAR: a commuter router infrastructure for the mobile internet. In *Proceedings of the 2nd International Conference on Mobile Systems, Applications, and Services (MobiSys)*, pages 217–230, 2004.

[29] J. B. Saleh, D. Qiu, and A. K. Elhakeem. Performance of an efficient scheduling approach to network coding for wireless local repair. *Cyber Journals: Multidisciplinary Journals in Science and Technology, Journal of Selected Areas in Telecommunications*, Jan. 2011.

[30] H. Seferoglu, L. Keller, B. Cici, A. Le, and A. Markopoulou. Cooperative video streaming on smartphones. In *Proceedings of the 49th Annual Allerton Conference on Communication, Control, and Computing (Allerton)*, pages 220–227, Sept. 2011.

[31] S. Sen, N. K. Madabhushi, and S. Banerjee. Scalable WiFi media delivery through adaptive broadcasts. In *Proceedings of the 7th USENIX Symposium on Networked Systems Design and Implementation (NSDI)*, 2010.

[32] H. Shojania and B. Li. Random network coding on the iPhone: fact or fiction? In *Proceedings of the 18th International Workshop on Network and Operating Systems Support for Digital Audio and Video (NOSSDAV)*, pages 37–42, 2009.

[33] H. Shojania and B. Li. Tenor: making coding practical from servers to smartphones. In *Proceedings of the 2010 International Conference on Multimedia (MM)*, pages 45–54, 2010.

[34] H. Soroush, P. Gilbert, N. Banerjee, M. D. Corner, B. N. Levine, and L. Cox. Spider: improving mobile networking with concurrent wi-fi connections. *SIGCOMM Computer Communication Review*, 41(4):402–403, Aug. 2011.

[35] M. Stiemerling and S. Kiesel. A system for peer-to-peer video streaming in resource constrained mobile environments. In *Proceedings of the 1st ACM Workshop on User-provided Networking: Challenges and Opportunities (U-NET)*, pages 25–30, 2009.

[36] M. Sullivan. A day in the life of 3G. *PCWorld.com*, June 2009.

[37] C. Tsao and R. Sivakumar. On effectively exploiting multiple wireless interfaces in mobile hosts. In *Proceedings of the 5th International Conference on Emerging Networking Experiments and Technologies (CoNEXT)*, pages 337–348, 2009.

[38] P. Vingelmann, M. Pedersen, F. Fitzek, and J. Heide. On-the-Fly packet error recovery in a cooperative cluster of mobile devices. In *Proceedings of the 2011 IEEE Global Telecommunications Conference (GLOBECOM)*, pages 1–6, Dec. 2011.

[39] M. Wang and B. Li. How practical is network coding? In *Proceedings of the 14th IEEE International Workshop on Quality of Service (IWQoS)*, pages 274–278, June 2006.

[40] M. Wang and B. Li. Lava: A reality check of network coding in Peer-to-Peer live streaming. In *Proceedings of the 2007 IEEE INFOCOM*, pages 1082–1090, May 2007.

[41] M. Wang and B. Li. R2: Random push with random network coding in live Peer-to-Peer streaming. *IEEE Journal on Selected Areas in Communications*, 25(9):1655–1666, Dec. 2007.

[42] J. Whitbeck, M. Amorim, Y. Lopez, J. Leguay, and V. Conan. Relieving the wireless infrastructure: When opportunistic networks meet guaranteed delays. In *Proceedings of the 2011 IEEE International Symposium on a World of Wireless, Mobile and Multimedia Networks (WoWMoM)*, pages 1 –10, June 2011.

[43] WiFi Alliance. Wi-Fi direct. http://www.wi-fi.org.

Flit: A Bulk Transmission Protocol for RFID-Scale Sensors

Jeremy Gummeson, Pengyu Zhang, Deepak Ganesan
Department of Computer Science
University of Massachusetts, Amherst, MA 01003
{gummeson, pyzhang, dganesan}@cs.umass.edu

ABSTRACT

RFID-scale sensors present a new frontier for distributed sensing. In contrast to existing sensor deployments that rely on battery-powered sensors, RFID-scale sensors rely solely on harvested energy. These devices sense and store data when not in contact with a reader, and use backscatter communication to upload data when a reader is in range. Unlike conventional RFID tags that only transmit identifiers, RFID sensors need to transfer potentially large amounts of data to a reader during each contact event. In this paper, we propose several optimizations to the RFID network stack to support efficient bulk transfer while remaining compatible with existing Gen 2 readers. Our key contribution is the design of a coordinated bulk transfer protocol for RFID-scale sensors that maximizes channel utilization and minimizes energy lost due to idle listening while also minimizing collisions. We present an implementation of the protocol for the Intel WISP, and describe several parameters that are tuned using empirical measurements that characterize the wireless channel. Our results show that the burst protocol improves goodput in comparison to vanilla EPC Gen 2 tags, improves energy-efficiency, allows multiple RFID sensors to share the channel, and also coexists with passive, non-sensor tags.

Categories and Subject Descriptors

C.2.2 [**Network Architecture and Design**]: Network Protocols

General Terms

Design, Experimentation, Measurement, Performance

Keywords

RFID, Bulk Transfer, Sensing

1. INTRODUCTION

A wide range of sensing applications require miniature, ultra-low power sensors in urban and indoor areas. This has led to an increased interest in the design of small, cheap, harvesting-based devices that are attached to common everyday objects (e.g., books, furniture, walls, doors, produce, etc), which can be used for tracking these items [9]. One example of such devices are RFID-scale sensors that exploit ambient light or RF for energy, and use backscatter communication with an RFID reader for data transfer.

RFID-scale sensors, also referred to as Computational RFIDs or CRFIDs, present a new frontier for distributed sensing [4]. These devices are distinct from existing battery-powered sensor platforms as well as commercial RFIDs. They are designed for continuous sensing, however, unlike existing continuous sensing devices (e.g. Motes), they use small capacitor buffers, rely solely on energy harvesting, and use more power-efficient backscatter communication for data transfer. CRFIDs are also distinct from commercial RFIDs in that they use hybrid harvesting to enable continuous sensing, computation, and storage rather than just vanilla identification.

In this paper, we investigate how to efficiently utilize the energy buffer of an energy harvesting CRFID node for burst message exchange. We focus our attention on mobile CRFIDs [21], whose movements result in two communication states: connected and disconnected. While tags traverse their environment they perform a series of sensing and computation operations. As time elapses, these devices buffer some amount of data during disconnected operation; occasionally they encounter a reader, resulting in a variable connection interval during which a tag may offload buffered data.

Our goal is to optimize the bulk transfer of this buffered data from the CRFID sensor to an RFID reader. Because CRFID sensors have small energy buffers, it is imperative that communications maximize goodput while minimizing the amount of energy per unit data. This presents several challenges. First, commercial RFID readers follow the EPC Gen 2 protocol which is optimized for large numbers of tags that each transfer a small amount of data (tag identifier). This protocol is inefficient when considering sparsely deployed CRFID sensors that each potentially need to transfer large amounts of buffered data to a reader. EPC Gen 2 also makes duty-cycling for CRFIDs very difficult to implement because they must listen to a potentially large number of messages while waiting to transmit; this is unacceptable as CRFIDs treat energy as a precious commodity. While a complete re-design of the protocol stack is possible, this would mean that CRFIDs could not take advantage of existing commercial RFID readers, making them far less attractive for widespread use. Rather, we seek to support efficient bulk data transfer while still being compatible with commercially available EPC Gen 2 RFID readers. Second, when compared with other sensing platforms, CRFIDs have different hardware components, use different energy sources, use different energy buffers, and follow a different communication protocol. Thus, designing an energy-optimized bulk transfer protocol for RFID sensors requires an entirely new set of bandwidth and energy-optimization mechanisms. Third, CRFIDs present different usage scenarios since they are largely deployed indoors, and often on mobile objects or people. Thus, any data transfer protocol should operate effectively under scenarios where there are short contact durations with readers, and considerable changes in link characteristics during mobility.

Our protocol, Flit, provides a fast and efficient alternative to the existing EPC Gen 2 protocol for bulk data transfer from sensors, while still remaining compatible with existing RFID readers. Flit makes three fundamental changes to the protocol stack. First, it enables each sensor to transfer data in a burst by responding to all slots in a query round rather than just its assigned slot. This design choice improves goodput and energy-efficiency by reducing wasted slots, and takes advantage of extended query rounds with less control overhead. Second, Flit coordinates across sensors by using explicit burst notifiers that are echoed by RFID readers, rather than devices randomly picking a slot in which to transmit. This approach serializes burst transfers across nodes, thereby allowing greater goodput while reducing potential for collisions. Third, Flit improves energy-efficiency by duty-cycling the RFID sensor when another CRFID is in the middle of a burst. This avoids wasted energy due to overhearing of reader messages during the burst, thereby enabling better use of a small buffer of stored energy on the CRFID sensor.

Our results show that:

- Flit achieves 60% greater goodput than EPC Gen 2 for a single tag at different distances from the reader, and for different mobility conditions. A breakdown shows that much of these gains are due to the use of larger query rounds, and avoiding wasted slots in the round.
- Flit achieves 4.5x more goodput than an EPC Gen 2 tag when three tags are transferring data concurrently, and 9.2x goodput when five tags are transferring simultaneously. In addition, Flit has considerably higher fairness than EPC Gen 2, which is skewed towards the node with highest SNR to the reader. This allows sensors to take advantage of shorter contact durations with readers.
- Duty-cycling in Flit achieves up 6.04x better average power efficiency than a non-duty cycled implementation. These power savings are acheived by minimizing energy consumpted from listening to other CRFIDs' transmissions.

2. AN EPC CLASS 1 GEN 2 PRIMER

The Gen 2 protocol for RFID tags is designed to inventory large tag populations over a number of communication rounds. To realize this protocol, an RFID must traverse a simple state machine and respond appropriately to a set of reader commands. Throughout this discussion, refer to Figure 1 to understand how a sequence of reader commands and tag responses are used to transmit data to a reader. The critical subset of EPC commands a CRFID must implement are:

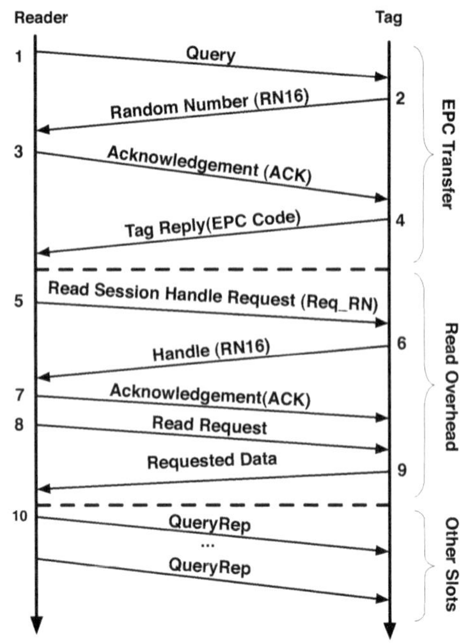

Figure 1: A series of message exchanges are required between a reader and tag to read the tag's EPC code or user memory.

Query, QueryRep, and QueryAdjust: A *Query* message (1) initiates a round of communication. This message specifies several round parameters. The most critical of these parameters is Q, which defines the number of slots in a round to be $2^Q - 1$ where $0 \leq Q \leq 15$. Tags generate a non-negative, random slot counter within the range specified by Q. The reader chooses Q such that collisions between tags are minimized.

Slots after a *Query* are occupied by *QueryRep* (10) and *QueryAdjust* messages. *QueryRep* messages indicate a successive slot for this round; *QueryAdjust* messages indicate a successive slot and additionally adjust the current Q value by +/- 1. After receiving either message, the tag decrements its slot counter; when the counter reaches 0, the tag proceeds.

Ack(RN16): To disambiguate tags in the event of collision, an *RN16* (2) message is used. The RN16 is a 16-bit value randomly generated by the tag. Upon decoding an RN16 from tag(s), the reader will echo one of the RN16s it received as an ACK

EPC: A tag knows it was chosen for communication if the received ACK matches the sent RN16; if this is not the case, the tag gives up on this round of communication to avoid further collision. After receiving its own RN16, the tag may backscatters its *EPC* (4) code to the reader. After sending its EPC, the tag will not respond to subsequent QueryReps or QueryAdjusts during this round of communication.

Req_RN: A reader that wants to further investigate a tag's state may send a *Req_RN* message (5). This message establishes a 16-bit handle to be use for subsequent communication. The tag echoes the handle back to the reader as an ACK (7).

Read: After establishing a session handle, the reader may send a *read* command (8) to request a segment of the tag's memory; after receiving a Read command, the tag responds with the requested data (9). The tag appends the session handle and a two byte CRC computed across the payload.

3. LIMITATIONS OF GEN 2 FOR CRFIDS

In this section, we discuss several limitations the EPC Gen 2 protocol has on designing an energy-efficient bulk data transfer protocol that transfers data from a CRFID sensor to a reader. To understand these limits, we conducted a benchmark study that quantifies the timing and energy requirements of relevant Gen 2 messages for the Intel WISP 4.1; the results of this study are presented in Table 1. For each message type, we report the time required to finish sending or receiving a particular message in the *Active* column. The amount of time between a message and subsequent message is reported in the *Idle* column. We compute the energy for a particular operation by multiplying the platform power consumption by the sum of the *Idle* and *active* durations. For each message type, we also note whether WISP sends(TX) or receives(RX) the message, as the WISP consumes more power when receiving a message because it increases its clock frequency.

3.1 Singulation Inefficiency

The Gen 2 RFID protocol is designed around inventorying large numbers of tags that need only report a static identifier. It is therefore primarily focused on collision avoidance for a large number of passive tags. In this section, we show that EPC Gen 2 is inefficient for bulk transfer both in terms of throughput and energy-efficiency.

A key parameter that controls the efficiency of the EPC Gen 2 protocol is the window size, Q. The window size is a parameter that is set by a reader based on the tag population that it observes, as described in §2; during a round a number of slots are chosen such that the probability of collision between two tag responses is negligible. The EPC Gen 2 standard provides some general guidelines as opposed to a specific algorithm for how to set Q, so the implementation of the algorithm is vendor specific and is typically unavailable to the customer. In addition, there is often no way to control the Q values set by a reader since modern RFID readers are designed for ease of use and hide low-level protocol parameters from the operator. In particular, the Impinj Speedway reader we use in this paper offers no visbility into the chosen Q value; the resulting window size is completely decided by the reader's proprietary algorithm.

The efficiency of the EPC protocol depends on the value of Q set by the reader in each round. For example, if a reader picks $Q = 3$ and there is only one tag present, then there are 8 slots in this round, including the Query, one of which is utilized by the tag. In addition to the obvious throughput inefficiency, this is also inefficient energy-wise as a tag incurs the energy overhead of listening to the QueryRep or QueryAdj messages for the slots that it does not respond to.

To understand the practical inefficiencies of Gen 2 singulation, we looked at the round lengths selected by an Impinj Speedway reader when a single WISP tag is placed in front of it. While the reader does not provide an interface to obtain

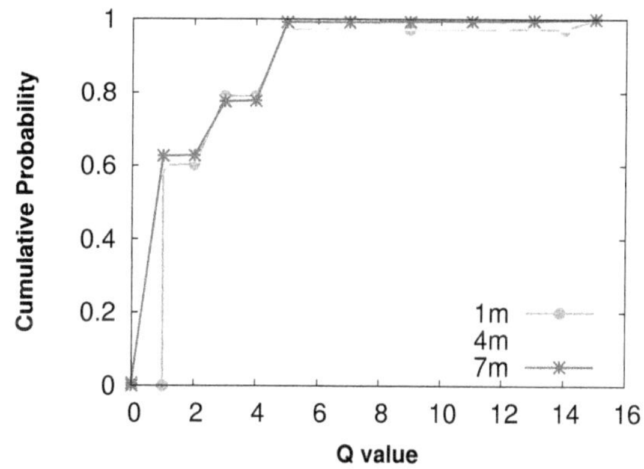

Figure 2: This plot shows what Q value a reader actually chooses when non-burst communication is used

Operation	# bits	Time Active	Time Idle	Energy
Query(RX)	22	983 μs	52 μs	648 nJ
QueryRep(RX)	4	273 μs	50 μs	210 nJ
QueryAdj(RX)	9	415 μs	51 μs	319 nJ
Read(RX)	52	2100 μs	50 μs	1615 nJ
RN16(TX)	16	641 μs	2390 μs	422 nJ
Ack(RX)	18	660 μs	36 μs	508 nJ
Req_RN(RX)	40	1616 μs	51 μs	1241 nJ
EPC(TX)	128	2450 μs	2360 μs	1615 nJ
CRC16	-	452 μs	-	307 nJ

Table 1: A CRFID emulates Gen 2 in software leading to widely varying amounts of energy consumption depending on the command.

the chosen Q value, the WISP is programmable, therefore we were able to obtain the numbers by transmitting this information in place of the EPC code. Figure 2 shows that the reader typically chose a Q value between 1 and 6, with a mean of 2.5; this behavior held for distance up to 7 m. These Q values indicate that the number of slots in a round varies between 2 and 64 slots despite only a single tag being present, clearly a major source of inefficiency.

To further drive this point, we refer to Table 1. Based on a mean Q value that varies between 2 and 6 as in Figure 2, the extra communication slots result in degradation throughput that varies between 9.7 - 294.2% and energy consumption that increases by between 34.9 - 546.4% as compared to a single tag communication during a single slot round. It is also important to note that this is a lower bound on the amount of energy required as CRFIDs will likely remain in an active state between received messages. These performance penalties change as a function of Q, which is in turn a function of the number of tags present.

3.2 Inefficiency of Read Messages

Gen 2 supports tag user memory operations in addition to simple EPC queries. Of particular interest is the Read command, which allows a reader to request a region of the tag's user memory. While at first glance, Read message seem to ideal for transmitting sensor data from a tag, we show that they are inefficient in terms of channel utilization and energy consumption.

Read messages are attractive because they support variable response lengths; a long read message could potentially overcome the singulation inefficiencies we previously highlighted, in addition to allowing CRFIDs to transmit large amounts of data to a reader. In theory, large Read messages are possible since Gen 2 specifies that an upper limit of 255 bytes on their size, but in practice, the size of read messages is limited by factors such as bit error rate, hardware limitations, and timing drift. For example, we found that the Impinj Speedway reader supports Read requests of lengths upto 60 bytes. For the Intel WISP, we found that read error rates sharply approached 100%, when 16 bytes of data were requested via Reads. Our hypothesis is that these practical limitations stem from three reasons: a) long messages are vulnerable to high bit-error rate (BER) at longer distances, particularly since the path loss on a backscatter link drops as the fourth power of distance [20], b) longer messages incur more timing drift, and RFID-scale devices often do not have real-time clocks to adjust for these, and c) large messages incur high memory overhead, which is limiting for RFID-scale devices. On the Intel WISP, both BER at higher distances and the timing drift were issues that made it difficult to get longer Read messages across to the reader.

Read messages also incur significant control overhead, which results in considerable throughput and energy inefficiency. As seen in Figure 1, a tag needs to be singulated prior to handling a read request, and depending on the Q value chosen for the round, may need to listen to several slots before it can set up a Read with the reader. This design clearly outlines the priorities of EPC Gen 2 — it is designed for obtaining identifiers from tags, and Reads are a second-class citizen that is intended to be used sparingly. The overhead is compounded by the fact that long Read messages are not practical, and is inefficient energy-wise since the tag is forced to listen to a long series of messages before its turn.

3.3 Lack of Duty-Cycling Support

Another major limitation of EPC Gen 2 is its lack of support for duty-cycling. While duty-cycling of a CRFID may seem unimportant for communication since the device receives power from the reader, this is not entirely true. The distance at which a CRFID can communicate with a reader is far more than the distance at which a tag can receive power from a reader. It is for this reason that passive tags have operating distances of a few feet from a reader, whereas a hybrid-powered CRFID (RF + ambient harvesting) or a battery-powered active tag can have communication ranges of 50-70 feet [10]. At longer distances, a CRFID needs to duty-cycle and leverage low-power states since they are using precious reserves of stored energy or are operating on small amounts of ambient power.

Wireless MAC protocols designed with duty-cycling in mind typically use a number of mechanisms to synchronize senders and receivers, and buffer packets while waiting for synchronization to occur. For example, the 802.15.4 MAC layers uses preambles to synchronize sleeping senders and receivers, the 802.11 power save mode (PSM) relies on the access point buffering, and TDMA MACs have fixed slots, allowing a device to sleep for a fixed duration without the risk of missing messages. In contrast, EPC Gen 2 has non-deterministic arrival times of messages and variable round lengths. While Q determines the length of a round, this length is often not set at the beginning of a round. Instead, Q can be dynamically changed using QueryAdjust messages (based on estimated tag density), and a tag would not know the current value of Q if it misses a QueryAdjust message. In addition, slot lengths can be different since slots can terminate at different times due to timeouts after different steps of the protocol. The consequence of lack of duty-cycling support is that Gen 2 can cause CRFIDs to waste excessive energy on idle listening while waiting for their communication slot.

4. FLIT DESIGN

There are a number of factors to consider when designing a bulk data transfer mechanism for Gen 2. Such a mechanism must strive to: 1) maximize data transfer rates so that sensor tags can transfer their data quickly and efficiently to a reader during short contact events, 2) minimize power consumption so that a CRFID can maximize the amount of data transferred using its small energy buffer, and 3) interoperate with standard commercial RFID readers, so that CRFIDs can leverage existing RFID reader infrastructure.

To realize these goals, we present the design of a burst protocol for CRFID sensors. First, we discuss the design tradeoffs in using an EPC Query versus the Read command as the data transfer primitive. Next, we demonstrate how sensors can achieve high levels of goodput using burst-mode data transfer that leverages unused slots in the EPC Gen 2 protocol. Third, we show a coordination mechanism that uses burst notifiers to avoid collisions among bursting tags. Fourth, we present a duty-cycling mechanism that minimizes the energy lost to idle listening. Finally, we discuss implications of the design choices that we make when there are a mix of sensor tags that are bursting and standard EPC Gen 2 tags that are only transmitting their identifier.

4.1 Read vs EPC for Burst Transfer

The first question in designing a burst data transfer protocol is which EPC Gen 2 message primitive to use as the building block for transferring data. Two options present themselves in terms of adapting the EPC Gen 2 protocol for bulk data transfer from the sensor to the reader. The first option is to use EPC Read command which allows a variable amount of data to be transmitted from a tag to reader, but has several inefficiencies as described above. The second is to use the *EPC* message, and send application data instead of the 12 byte static identifier within this message.

We first look at the energy efficiency of Reads vs EPC messages using the set of energy benchmarks in Table 1. The energy efficiency of the read command varies with the length of the data sent in response to the read request, while an EPC message is always 12 bytes. These benchmarks were captured using the Intel WISP 4.1 [16] (more details in §5).

From this breakdown, we compute the amount of energy consumed per byte of data transfer. Each Read command incurs energy overhead for steps 1–8 in Figure 1 that precede the read payload. The energy consumed for each EPC command varies a small degree based on whether it is in response to a Query, QueryRep or QueryAdjust since they have different sizes.

Suppose that EPC codes are used to transfer data to the RFID reader. An analysis based on a round with 4 slots (Q

= 2) will result in 12 bytes of data arriving at the reader per slot, for a total of 48 bytes of data. This process takes 39.77 μs.

Now, suppose that a Read message is used to request data from a tag's user memory. Since we must now use the EPC data to singulate an individual tag for the subsequent Read command, a larger Read message is required to compensate for this additional overhead. When considering the same 4 slots as in the EPC based approach, the Read command incurs the previously computed time delay as overhead, as well as steps 5 - 8 from Figure 1. In order to match the throughput of a pure EPC-based approach under this scenario, a Read request of at least 116 bytes is required. This result indicates that Reads are a poor choice for high throughput tag to reader communications based on the reasons outlined in §3.2. A similar analysis of the energy required per byte of transmitted data shows that a Read with 143 byte request size is required to match the energy efficiency of the pure EPC approach. A similar conclusion applies if energy efficient tag to reader communication is needed.

One advantage of Gen 2 Reads, is that they have built in options for security. After a tag receives the handle message depicted in Figure 1, the tag may optionally be sent an *access* message that contains a password; reception of a correct password moves the tag into a logical state called *Secured*. This state may be utilized to protect portions of tag memory targeted by a subsequent Read command. However, since computational RFIDs can implement cryptography, they could instead encrypt data locally if a particular application requires it.

4.2 Burst-Mode EPC Transfer

Having selected the 12 byte EPC message as the building block for bulk transfer, we turn to the question of improving efficiency when several hundreds of bytes of data need to be transferred using this message primitive. If the Gen 2 protocol were followed, data transfer would need to be over several tens or hundreds of rounds, and a CRFID would receive only one slot in each round. As described in §3.1, this would be extremely inefficient due to poor choices of Q at the reader.

The central idea in burst transfer is to ignore Gen 2 semantics of rounds, and to treat the protocol simply as a sequence of unassigned request/response slots. Each of these slots can be initiated by a Query, QueryRep, or QueryAdjust, but the burst protocol does not treat them differently. Instead, a CRFID sensor assumes that every slot is available to it for burst transfer, and just transfers its data in a sequence of consecutive slots. Before discussing issues of coordination across multiple tags (§4.4), we look at the benefits that this offers to a single tag.

The key benefit of burst transfer from a single tag is that we are no longer limited by poor selection of Q by a reader (§3.1). In fact, we turn a drawback into an advantage. To obtain the full benefits of burst transfer, we want the reader to choose a large Q. As previously shown, a round is initiated by a Query message for the first slot, and the other $2^Q - 1$ slots are initiated by QueryReps. A few slots are initiated by QueryAdjust messages, whose purpose is to increment or decrement Q in the middle of a round. A subtle benefit of QueryReps and QueryAdjs, as opposed to Queries, is their brevity. Based on protocol specs, Query, QueryRep and QueryAdjust messages have lengths of 22, 4, and 9 bits respectively. As Q grows, the energy expended during a round of communication becomes dominated by round trips involving reps and adjusts.

The energy and throughput benefits of using longer Q are quantified in the following equations:

$$\mathrm{E_{round}} = \mathrm{E_{query}} + \left(2^{Q+A} - 1\right) \cdot \mathrm{E_{rep}} + n \cdot \mathrm{E_{adjust}} \quad (1)$$

$$\mathrm{Goodput_{rnd}} = \frac{12}{\mathrm{T_{query}}} + \frac{12 \cdot (2^{Q+A} - 1)}{\mathrm{T_{rep}}} + \frac{12 \cdot n}{\mathrm{T_{adj}}} \quad (2)$$

Equation 1 shows the total energy spent on a round of communication, which includes the energy spent on listening and replying to the first query slot(Equery), the energy spent listening/replying to subsequent slots initiated by QueryReps (Q is the initial value assigned by the reader, and A indicates how it was adjusted during the round), and finally the energy spent on n QueryAdj messages that were sent during the round. Equation 2 shows the goodput counterpart, which takes into account the length of an EPC message (12 bytes), and the time for the three types of queries.

Using numbers from our microbenchmarks in Table 1, we see that a long round with $Q = 15$ can give about 10% benefit in both energy and goodput over a short round with $Q = 0$.

In summary, treating the Gen 2 protocol as a sequence of unassigned slots enables us to a) limit inefficiency due to empty slots caused by poor selection of Q, and b) improve efficiency by taking advantage of shorter slots initiated by QueryRep messages.

4.3 Coordination via Burst Notifier

Responding in every slot has a severe limitation: if multiple CRFIDs are present, they will suffer from collisions and see reduced energy efficiency and goodput instead of the improvements. An active radio system could solve this problem using control messages such as RTS/CTS or an overhearing-based approach such as CSMA to coordinate transfers between peer nodes. These approaches are not suitable for backscatter communication circuits because they are unable to decode messages transmitted by peers. An alternative would be for the reader to explicitly select a tag in the Query or QueryRep message, and all other tags that receive the message can ignore the slot. However, as mentioned earlier, QueryReps are only *4 bits* long, and leaves no room for such addressing. Besides there is the limitation that readers do not allow modifications of Query messages, making any such approach impractical. Thus, we ask the question: *How can CRFID sensors use the existing EPC protocol to efficiently coordinate bursts?*

A closer look at the Gen 2 Query/EPC exchange reveals that there is a two-way handshake being performed, which presents a solution to this problem. As shown in Step 3 of Figure 1, the reader echoes the RN16 of the RFID it chooses to occupy a given communication slot. Our strategy is to overload the RN16 to signify that a particular CRFID is currently bursting. We accomplish this by providing a special interpretation of a segment of reserved RN16s; we partition the space of RN16s as $0 < n < 2^{16}$, where n is the number of CRFID sensors deployed and values less than n are considered burst notifiers. The value of n is statically selected at compile time and is chosen based on the maximum number of CRFIDs envisioned for a particular application. A sensor that wishes to send a burst of EPCs will use its statically selected burst notifier chosen from the available pool, instead of a random value. Note that the sensor selects a notifier just once for an entire burst, rather than once per slot as is done by a standard tag.

The RN16 burst notifier is used in the following way: prior to initiating a burst, a CRFID sensor listens to the channel after decoding a query, rep, or adjust message. If the sensor observes an Ack within the range of burst RN16s, it should remain silent to avoid colliding with an ongoing burst. If the slot contains an RN16 outside of this range, it can go ahead and start a burst transfer after the current slot using its own burst notifier, as non-burst EPC messages occupy only one slot.

It is, of course, possible that another CRFID sensor is in the middle of its burst and either the reader might have missed the burst notifier or the listening sensor may not have received the notifier echoed by the reader due to channel error. Both cases would lead the listening sensor to conclude that the channel is free and start to burst, resulting in collisions at the reader. A collision at the reader typically results in the reader receiving the stronger signal among the colliding tags due to capture effect. The reader echoes the burst notifier that it receives, which results in only the sensor with stronger signal continuing to burst. While a collision could also result in neither signal being received by the reader, we handle this case by assigning a random back-off interval after hearing no Ack when one is expected.

To prevent the sensor from holding the channel indefinitely, the burst will terminate after a small, fixed amount of time that is large enough to amortize coordination overheads, but small enough to allow mobile tags with limited communication opportunities a chance to offload a burst of data to the reader.

4.4 Duty-cycled Coordination

Burst transfer is a natural fit for CRFID duty-cycling for two reasons: a) transfer is in large chunks of consecutive slots, enabling other nodes to sleep for longer durations and re-charge while waiting for a burst to end, and b) inefficiencies incurred due to duty-cycling such as wasted slots because a tag is asleep or wasted energy for listening because it is awake too early can be amortized over the longer sleep durations.

While bursts are convenient for duty-cycling, the lack of enough bits in the Query/QueryRep messages impacts duty-cycling efficiency as well. If a waiting tag knew precisely how much longer a burst from another tag would last, it could sleep for exactly that duration. However, this information is unavailable since a sensor tag relies on the burst notifier from the reader to detect a burst, which provides no information on the time remaining for the burst. Thus, a tag needs to periodically wakeup to check the channel and detect if a burst has ended. Thus, a key challenge for a duty-cycling strategy is to efficiently find the end of a burst so that sensors can capture the channel from another sensor between bursts and react quickly to mobility dynamics while avoiding most of the energy wastage caused by overhearing.

Thus, there are two questions that remain regarding how to duty-cycle an CRFID sensor: a) the amount of time a sensor should probe the channel and b) how much time a sensor should sleep. Since tag-to-tag communication is impossible,

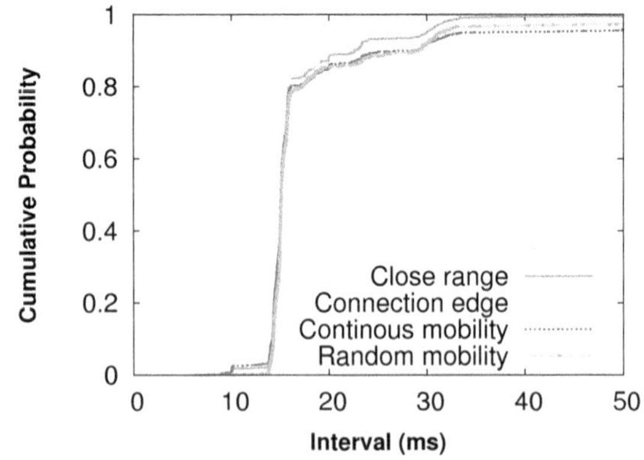

Figure 3: Most Query, QueryRep, and QueryAdj messages have inter-arrival times of less than 20 ms.

we do not consider adaptive policies for determining these intervals since the new probe and sleep intervals would need to be shared with all tags. We now describe how these intervals should be statically selected.

Probe Duration: The probe duration should be long enough such that a tag can detect whether another tag is continuing to burst. This duration is equivalent to a single slot in a query round. A sensor tag wakes up, listens to the first Query, QueryRep, or QueryAdjust slot, and sees whether a burst notifier is echoed by the reader during this slot. If so, it concludes that another tag is bursting and goes to sleep; if not, it concludes that it can initiate its own burst and starts transmission in the next slot. In the middle of a burst, if a tag detects that the reader has echoed a different burst notifier it concludes that another CRFID sensor is bursting and goes to sleep to save energy.

A potential issue here is that the duration of a slot can vary because a) Query, QueryRep, and QueryAdjust messages are of different lengths, b) a slot can terminate at different times depending on whether the reader times out after the RN16, Ack or EPC steps in its state machine and c) mobility can introduce additional dynamics. To address this, we look at the probe duration empirically by measuring the inter-arrival time of Query, QueryRep or QueryAdjust messages for a continuous exchange between an Intel WISP programmed with the EPC Gen 2 protocol and a reader. We look at this distribution for different distances from the reader, as well as for different mobility patterns. Figure 3 shows the CDF of the inter-message duration. The results show that the inter-query intervals do not depend significantly on the distance, and are impacted a little ,but not a lot, by mobility. The knee of the curves is in the 15-20 ms range, thus we select 20 ms as our probe duration; this probe duration is short and provides a reasonable guarantee that a query will be heard by the WISP during the period it is awake.

Sleep Interval: There are several considerations in determining the sleep interval. First, the sleep interval must be long enough that we get significant energy benefits from duty-cycling. Second, it should be short enough that a tag can quickly react to mobility-induced channel dynamics. Third, it should have sufficient randomization so that

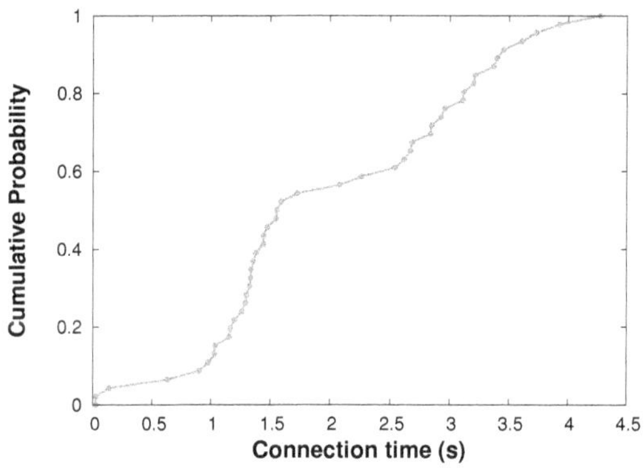

Figure 4: For human-scale mobility rates, the connection time between a tag and reader typically lasts several seconds.

we avoid unwanted synchronization issues that can result from multiple tags waking up at the same time.

In terms of the energy consumption, we want a duty-cycle of lower than 10%, hence the sleep duration should be at least 200 ms when the probe duration is 20 ms. To understand the typical contact duration at walking speed, we use 1 reader in a corridor, and walk in circles around it. We found that a typical contact duration is a few seconds in duration (see Figure 4), hence the sleep duration should be much smaller than this number. To prevent synchronization issues, the tag can randomize the sleep time within a tolerable range that provides a desired amount of energy savings while maintaining reactivity to expected mobility patterns for a given deployment.

4.5 Coexistence with Non-burst Tags

While our discussion thus far has assumed the tag population comprises solely of sensor tags that have to transfer data in a burst, we now look at the implications when a mix of sensor tags and standard EPC Gen 2 tags are communicating with the same reader infrastructure. Not surprisingly, the net effect is that standard EPC Gen 2 tags incur more delay in communicating with a reader infrastructure. However, there are mitigating factors that can enable better coordination across tags.

The burst mode transfer mechanism that fills up all slots of a round impacts standard Gen 2 tags in two ways. First, a standard tag which picks a slot within a round will collide with a burst tag, resulting in loss of one of the messages. In practice, we find that because of reader sensitivity, the reader gets one of the CRFID's messages with high probability (due to capture effect), hence it is still possible that the standard tag gets its data through. However, if the burst tag has the stronger signal, the standard tag suffers. Second, the burst transfer approach results in large Q values, which makes a round long; since standard tags only respond in one slot within a round, this makes their response slow. This effect is mitigated by the fact that we limit bursts to a relatively short duration of time (1sec in our implementation), after which a sensor goes to sleep for a short window of time before bursting again. The duration between bursts is sufficient for a few short communication rounds, enabling standard tags to get their data through.

The use of burst notifier facilitates co-ordination across

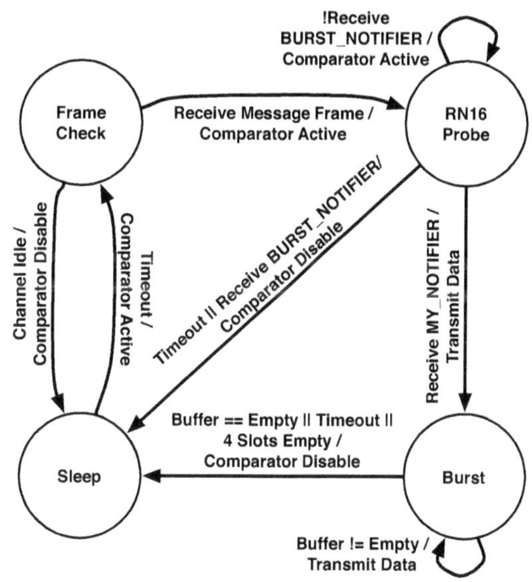

Figure 5: A coordinated bursting protocol for CRFID sensors can be implemented as a state machine. The protocol uses sleep states to both avoid contention and achieve energy efficiency.

CRFIDs, however, passive tags are free to choose any value from 0 to 2^{16} as its RN16, so it is possible that a passive tag could choose an RN16 that conflicts with a burst notifier. However, we choose a small part of the space for burst notifiers since we expect the number of sensors in the vicinity of a reader to be in the tens (equivalent to the number of objects in the vicinity of a reader) as opposed to thousands. Thus, the probability of collision is low. In addition, the RN16s are chosen anew in each round, hence a standard tag would likely choose a non-colliding RN16 in the next round.

5. IMPLEMENTATION

Our bulk transmission protocol is well suited for implementation on CRFID sensors because it is a modification of the EPC Gen 2 protocol they already support. CRFID sensors that want to implement the protocol need only modify their state machine to properly handle burst-mode transmission, burst notifers, and duty cycle appropriately in response to received message frames from a reader. In this section, we show the state machine we used to implement our Bulk Transmission Protocol for the Intel WISP. Next, we describe how state machine parameters can be defined based on results from channel measurements and mobility experiments. Finally, we give some details that describe the evaluation methodology that drove the design of Flit.

Figure 5 shows the state machine used to implement our Bulk Transmission Protocol for the Intel WISP. Sensors that have data to send initialize a timer interrupt and begin operation in the *Sleep* state. After this timer expires, the sensor activates its comparator and enters state *Frame Check*; after initializing another timeout value, the microcontroller enters a low-power mode, only waking up to handle an incoming message frame. Upon receiving a valid message frame, the sensor will enter state *RN16 Probe*; else, if the sensor does not hear a valid delimiter from a reader, it goes back to state *Sleep* after timing out. While in state *RN16 Probe*, the sensor initializes its timer with another timeout value; after hearing at least one empty frame from the reader, in which the sensor does not hear another sensor's BURST_NOTIFIER, it will send its own BURST_NOTIFIER in response to a Query, QueryRep, or QueryAdjust message. If the sensor hears its own BURST_NOTIFIER, it enters state *Burst*; if another sensor's BURST_NOTIFIER is heard, instead of an empty slot or if the timeout value is reached, the sensor enters state *Sleep*. Upon entering state *Burst*, the sensor will again initialize a timer, then begin transferring the contents of its buffered data as an EPC message in response to Query,QueryRep, or QueryAdjust messages; the sensor uses its BURST_NOTIFIER to send every message within the burst. Upon completion, timeout, or detecting 4 slots during which it finds no acknowledgement, the sensor returns to state *Sleep*.

5.1 Parameter Selection

While the state machine we previously described is a useful framework to constuct our protocol, implementation of the state machine was not straightforward, and needed several parameters to be carefully chosen and implementation aspects to be carefully addressed. We describe a few of these challenges in this section.

Timeouts: The timeout values used in our state machine are chosen based on Figures 3 and 4 in §4 that give good insight into expected connection intervals and message interarrival times respectively. In practice, these timeout values are used as comparison values for Timer A_1 on the WISP's MSP430 microcontroller. When considering hardware constraints and initialization overheads, one must also be careful to not choose a set of timeout values that generate too many interrupts that interfere with the WISP's ability to timely respond to reader messages. In practice, timeouts > 2 ms give the WISP sufficient time to listen for messages, while also providing the time needed to for timer initialization.

Duty-cycling: To implement the state machine, we also need to understand how the RF subsystem operates, and how to achieve maximal duty-cycling benefits. The RF subsystem comprises two components: a) the analog comparator that senses the channel to detect the presence of a bit, and b) the microcontroller that wakes up upon each interrupt from the comparator to process the bit and check if a valid message is present. The duty-cycling strategy is straightforward — shutting off the comparator avoids any energy lost from responding to interrupts and idle listening.

Burst Length: We choose one second as the length of a burst since it is long enough to obtain substantial duty-cycling benefits. After using up a burst, a tag pause 250 ms before trying to capture the channel for another burst. This duration provides a window for other tags to capture the channel or passive tags to transmit their identifier.

Burst Notifiers: The final parameter we consider is the burst notifier used by tags to coordinate their burst transfers. When modifying the WISP firmware, we found it can be difficult to get the state machine to stay within the tight timing constraints specified by EPC Gen 2 protocol. Complex operations in the firmware diminish responsiveness and in the end manifest as a reduction in goodput. For example, choosing a poor ordering of comparisons while looking for a burst notifier can lead to a 30% reduction in goodput. We also found that messages sent from the reader to the WISP can contain bit errors; one example is the RN16 field, which in actuality contains only 15 bits of consistent data. Thus, a careful implementation was needed to make burst notifiers operate correctly.

5.2 Debugging and Evaluation Methodology

Implementing and evaluating our system on the Intel WISP was particularly challenging due to the limited visibility, extremely low-power nature, and tight timing constraints. Any logging on the device would dramatically change performance and the nature of bugs, therefore all our evaluation had to be performed external to the device. We highlight how we overcame these challenges below:

Visibility: To overcome visibility issues, a WISP can be tethered to a JTAG debugging tool to observe its internal state. In many mobile scenarios, JTAG tethering is not a viable option; in these cases, software state of interest can be transmitted to a reader as an EPC code. To understand timing-related phenomena, Timer A1, which is unused by the firmware, can be used to capture the intervals which again are sent as EPC messages. In many cases, we found initializing a timer took long enough to violate the timing requirements of EPC Gen 2 and resulted in the reader being unable to decode tag message frames. To solve this problem, we use a Telos mote [2] to count the time between two GPIO interrupts generated by the WISP and log the resulting timing information TinyOS-2.x [13] application. The Telos platform is well-suited for this scale of timings; the platform itself does not significantly impact mobility because of its small size.

Energy Limitations: Evaluating a burst mode transfer mechanism with a limited energy supply is difficult. Because our work assumes tags have some amount of buffered energy when initiating contact with a reader, it would be difficult to use a capacitor-based buffer while allowing for repeatability. To remove measurement dependencies on harvested power, we power tags directly with batteries; this setup assumes they have a plentiful store of energy while in contact with a reader. Separately, we collected energy benchmarks that measured the energy required for sending and receiving protocol messages and other computation overhead. Combining these two approaches in our evaluation, we were able to evaluate the duty-cycling aspects of our protocol.

Channel Measurements: Another evaluation challenge lies in being unable to measure the channel directly. The protocol used to communicate with the Impinj reader, LLRP, does not provide an interface to directly obtain information about the PHY layer. This makes it impossible to get a completely accurate picture of channel characteristics and bit-error rates. One way to capture low-level channel measurements would be to use a passive EPC Gen 2 packet sniffer that listens to message exchanges between a tag and reader. A USRP-based Gen 2 channel monitor [7] has been implemented for this very purpose, but we found that the channel conditions observed at the monitor were much different than those observed by the tag.

To overcome these obstacles, we embed counters into EPC messages that indicate the number and type of messages received. These message counters assist in determining the messages sent in the forward link, as well as their contribution towards data delivery. Because an EPC code transfer happens directly after receiving ACK messages from the reader, the ratio of ACKs to EPC codes received at the reader gives a measure of the losses in the backscatter link. Additionally, the counter information we send allows us to align messages with GPIO timing triggers on the Telos mote.

6. EVALUATION

In this section, we evaluate the implementation of our bulk transmission protocol. The evaluation consists of four parts: 1) quantifying the goodput achievable by burst-mode EPC transfer, 2) showing that our burst notifier based coordination mechanism retains most of the goodput achieved by bursts by avoiding collisions, 3) demonstrating the energy benefits of duty-cycling, and 4) evaluating the interaction between bursting tags and standard EPC tags.

All evaluation results are obtained empirically; we feel this is important because of the complex RF environment that affects tag performance. We present results for small numbers of CRFIDs due to the limited number of prototypes available; the class of applications we consider (described in §1) are well aligned to the number of tags we use. In all of our experiments we use an Impinj Speedway reader with a fixed set of configuration parameters. The reader paramters used were: 1) Type A reference interval (Tari) = 25.0 μs, 2) Pulse Interval encoding (PIE) = 2.0:1, 3) Forward link = PR-ASK, 4) Pulse width = 0.5 5) Link Frequency = 256 KHz 6) Reverse Modulation = Miller 4, 6) Transmit power = 30.00 dBm, 7) channel = frequency hopping.

6.1 Burst mode transmission

The burst mode transmission protocol that we described in Section 4 ensures that all slots created for a round of communication are utilized by CRFIDs. In this section, we evaluate our burst transmission protocol in three ways: 1) We measure the window size allocated by Impinj reader when burst mode transmission is utilized, 2) We evaluate the goodput benefit gained by burst mode transmission, and 3) We provide a breakdown to show where the goodput benefits comes from.

Window size: In §4.2, we argued that the burst mode transmission strategy results in an RFID reader choosing a large window size (Q value) within a round of communication, leading to greater opportunities for using shorter messages, such as QueryRep or QueryAdjust, for data delivery. To validate this argument, we design an experiment that compares the Q value chosen by an Impinj reader for standard EPC transfer vs burst transfer. Our experiment setup places an Intel WISP in line-of-sight of an Impinj reader's antenna. The WISP piggybacks the Q value in place of the EPC code that it backscatters to the reader. We found that the Impinj reader consistently selects Q $\approx$ 10 for burst transmission independently of distance. As described in §4.2, a large Q value is good for burst transmission. In contrast, the Impinj reader selects $Q \approx 2$ when the standard EPC Gen 2 protocol is used. Any Q value larger than 0 leads to un-utilized slots, therefore, this choice results in a significant fraction of wasted slots.

Goodput: To quantify how burst-mode transfer of EPC codes better utilizes slots, we design an experiment that compares the goodput of a burst-optimized version where a single tag responds within every slot versus EPC Gen 2. In this experiment, we measure the goodput of an Intel WISP tag programmed to act as a conventional EPC Gen 2 tag vs

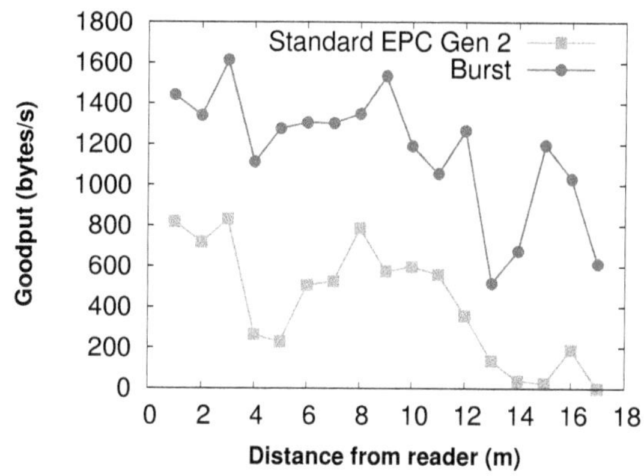

Figure 6: A CRFID sensor can achieve a 60% improvement in goodput by utilizing all slots in a communication round.

a WISP programmed for burst mode operation. We log the average throughput observed by the reader at different distances for several minutes and compare the results. Figure 6 shows that burst mode communication achieves 60% higher goodput than standard EPC Gen 2, and that these benefits are sustained across different distances. It is also notable that, in practice, the benefits of burst transfer are considerably larger than those that we predicted in §4.2. Our hypothesis is that additional gains are a result of the reader over-allocating slots for passive tags, resulting in comparatively low goodput.

Goodput breakdown: The increase in goodput for burst transmissions stems from two factors. First, we utilize every slot rather than one slot in each inventory round to transmit data. As shown in Figure 7, a standard EPC Gen 2 protocol only utilize only 64.8% of slots in each inventory round at 1m and 68.0% at 7m since it responds only to either a Query message or a QueryRep message, but not both – the other slots will go unutilized. In contrast, bursts will utilizes 100% of the slots. Second, in burst mode, we get more opportunities to exploit the shorter inventory messages: QueryReps and QueryAdjusts. Shorter messages in the forward link have lower loss rates; this leads to tags successfully capturing slots for data transfer with higher probability. For large Q, communication is dominated by slots that are initiated by QueryRep and QueryAdjust messages; as shown in Figure 7, the percentage of QueryReps at 1 m while bursting increases to 76.6%, as compared to the 34.2% slots in standard EPC Gen 2. This result holds true at a distance of 7 m as well. By exploiting QueryRep and QueryAdjust messages, tags get more opportunities for data transfer that contribute to higher goodput.

6.2 Coordinating bursts

While burst-mode transfer provides significant improvements to goodput, it also introduces problems when multiple RFID sensors wish to use the channel simultaneously. In this experiment, we evaluate how much the co-ordination mechanism benefits goodput when multiple tags are transmitting.

We consider a baseline case when a single CRFID is present, a low-contention case when three CRFIDs are simultaneously transferring data and a high-contention case when five

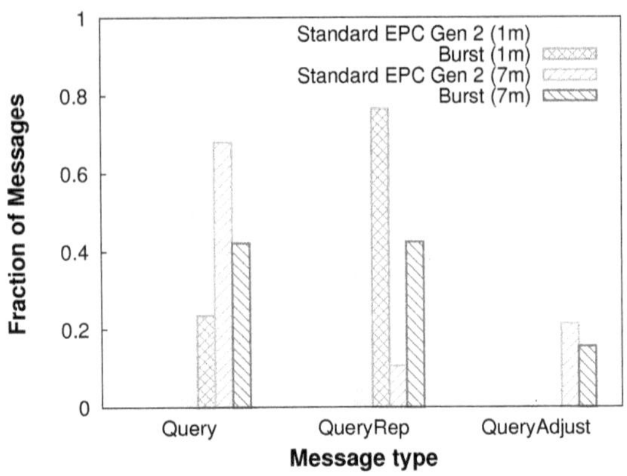

Figure 7: Bursting CRFIDs cause the number of round slots to increase. We observe this as an increase in the fraction of QueryReps received.

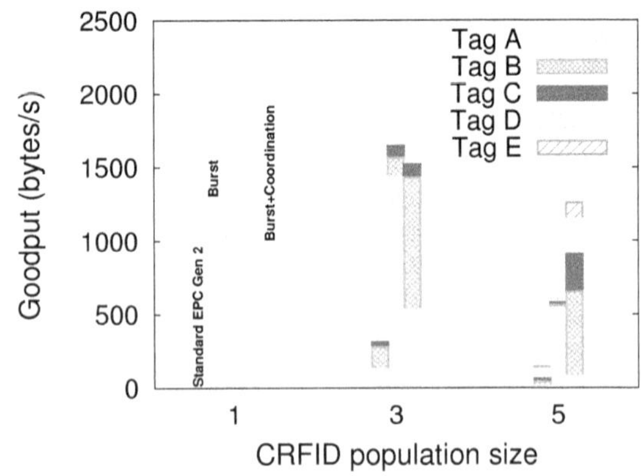

Figure 8: Coordination improves throughput and fairness as more bursting CRFIDs compete for the channel. For small tag populations, coordination incurs a small protocol overhead.

tags are simultaneously transferring data. The number of tags in our experiment is limited by the WISPs that we have available, however, we expect that the number of sensors transferring simultaneously will be a relatively small number. All tags are placed in a line with their antennas placed parallel to the Impinj reader's antenna within line-of-sight at a distance of 1 meter. Figure 8 shows a breakdown of the goodput for three protocols: a) standard EPC Gen 2, b) Burst mode without coordination, and c) Burst mode transfer with cooordination.

First, we look at the case where there is a single tag. We see that the standard EPC Gen 2 tag performs significantly worse than the bursting sensor tags, as expected. We also see that the co-ordination scheme performs about 9.2% worse than the case without co-ordination. This is because of the overhead required for coordination. After finishing a burst transmission, a coordinated tag releases the channel for a 50 ms to allow other waiting tags to acquire the channel, which reduces goodput. Next, we increase the tag population to three; we see a similar behavior in terms of overall goodput, but the split across nodes is very different. Overall, we see that burst mode transfer with coordination is still a little lower (7.7%) in total goodput than the uncoordinated case. However, the un-coordinated case gives out more than 87.3% of the channel to one of the three tags. The burst protocol with co-ordination is considerably fairer — all three tags receive a good chunk of the overall goodput. Finally, when the tag population reaches five, the impact of collisions becomes significant and adversely impacts the performance of uncoordinated sensors. For bursts without coordination, the total goodput reduces by 65.7% as compared to the single tag and the three tag cases. In addition, the uncoordinated scheme is highly unfair and allocates 90.0% of the goodput to one of the five tags. In contrast, when the five tags coordinate, they acheive similar goodput as observed in the one and three node cases.

Table 2 shows the Received Signal Strength Indicator (RSSI) values logged at the reader for the backscattered data from the five tags and their respective goodputs. Despite having similar distance from the reader, the tags observe different RSSI values as a result of differences in antenna orientation. As a result the tag(Tag A) with strongest RSSI(-39 dBm) captures the channel entirely and achieves a goodput of 540.5 bytes/second. The other four tags cumulatively get 10.0% of the total goodput, each achieving less than 30 bytes/second. In contrast, coordination results in a more even partitioning of goodput. Table 2 shows the coordination mechanism doesn't necessarily favor the tag with highest RSSI — in fact although Tag D's RSSI is lower than Tag A, it gets a large fraction of the goodput.

ID	Coordination		No Coordination	
	RSSI (dBm)	Goodput	RSSI (dBm)	Goodput
A	-41.7	78.9 Bps	-39.6	540.5 Bps
B	-37.7	571.1 Bps	-45.1	21.5 Bps
C	-39.9	258.8 Bps	-47.2	16.9 Bps
D	-46.0	238.2 Bps	-45.9	17.8 Bps
E	-50.1	112.8 Bps	-54.9	4.0 Bps

Table 2: A breakdown of the goodput from Fig 8 for the 5 tag case, shows that the amount of goodput a tag achieves is related to the average RSSI value at the reader.

6.3 Coordination-Aware Duty-Cycling

Coordination allows tags to avoid collisions between each others bursts, but does nothing to prevent energy consumed due to idle listening. We now evaluate our duty cycling mechanism which duty-cycles the comparator to reduce energy consumed due to idle listening. Our evaluation answers three questions: a) how much power is saved due to duty-cycling?, B) does duty-cycling result in degradation in goodput? and C) how does duty cycling affect the distribution of burst length among tags?

To quantify duty-cycling benefits, five tags continuously transmit EPCs for 10 minutes while deployed in a linear topology several feet from the reader with their antennas perpendicular to the reader's antenna. We set the probe and slew durations as defined in §4.4 as 20 and 200 ms respectively and look at the power consumed by each tag.

Figure 9 shows that duty-cycling reduces the average power consumption by 3.24x to 6.04x across the five tags when considering the amount of time the MCU spends in sleep vs

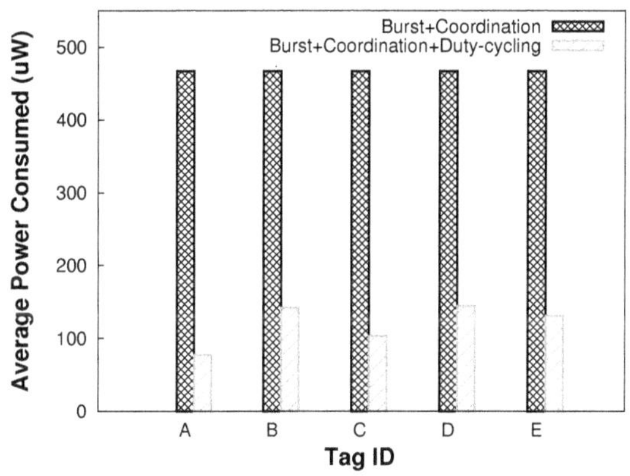

Figure 9: The Average power consumption of a CRFID is reduced considerably by reducing the energy lost to idle listening.

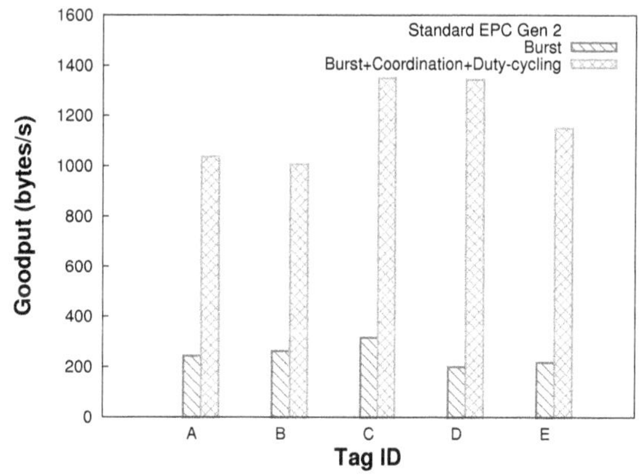

Figure 10: Nodes that duty cycle achieve considerably higher goodput. This is a result of reduced channel contention while non-bursting tags sleep.

active modes. The benefits of duty cycling differ depending on how often an individual tag gets access to the channel.

While duty-cycling improves overall energy consumption, a natural question is whether these gains come at the cost of goodput. Figure 10 shows that duty cycle based coordinated bursting tags have the highest goodput. This demonstrates that our duty-cycling mechanism does not sacrifice goodput to achieve its energy gains because of reduced channel contention that makes burst notifiers even more effective.

Finally, to better understand the dynamics of coordination and how frequently nodes switch among each other, we look at a breakdown of the duration that each tag holds the channel for the same dataset. We define a burst period as a contiguous segment when a single tag is bursting; at the end of the period, some other tag acquires the channel and starts bursting. These results are plotted in Figure 11. The results show that bursts are variable in length, even though they are allowed a maximum length of 1 second.

Fragmentation of bursts can result in unfair channel sharing because tags with better placement will see fewer spurious losses and hold the channel longer more frequently. We see this phenomena manifest prominently in Figure 11, as Tag A was placed in an unintentionally poor orientation. Although Tag A's bursts are shorter on average, they maintain throughput (Figure 10 – when the channel becomes poor according to tag A, another node grabs the channel. Upon hearing another tag's burst notifier, tag A immediately goes to sleep, resulting in very high power efficiency.

6.4 Coexistence with passive tags

In this section, we investigate the interaction between our burst protocol and standard EPC Gen 2. We answer two questions when both bursting tags and standard EPC tags are transmitting to an Impinj reader: 1) How does the goodput of CRFIDs change when they compete for the channel with multiple standard EPC tags? 2) What is the time between inventorying standard EPC tag when CRFIDs are bursting? We setup experiments where both bursting tags and passive tags are collectively within the field of an Impinj Reader's antenna. We measure the goodput achieved by the tags that run the Flit protocol, and the interval required to inventory passive Gen 2 tags.

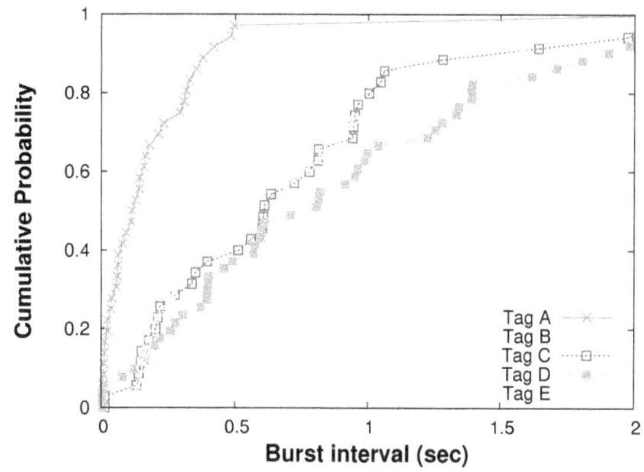

Figure 11: When tags use coordination, they achieve similar burst lengths; in this case Tag A is an exception because of particularly poor antenna orientation.

Goodput: Our goal in this experiment is to understand how increasing passive tag populations impact the goodput of bursting CRFIDs. Since we did not have enough Intel WISPs to use as passive tags, we use commercial passive tags for this experiment. We deploy five CRFIDs that are continually bursting to a reader, and increase the commercial passive tag population from 5 to 30. All commercial passive tags and CRFIDs are placed 1 meter from reader and spread in a line parallel with the reader antenna. Table 3 shows the average goodput as the passive tag population increases. The results show that there is only a small effect until about 20 tags (the average goodput is 253.3 bytes/second which is only a bit lower than 264.4 bytes/second when there are no passive commercial tags). For larger populations of passive tags, there are more collisions in slots which impacts goodput of burst CRFIDs. However, we see that despite a relatively large passive tag population, CRFIDs continue to perform well while bursting.

Time between inventory: In this experiment, we look at how the time to inventory a population of passive tags is impacted by the presence of a burst CRFID. We consider two variants of Flit — one with the standard 50 ms sleep

# Passive Tags	Avg Goodput (B/s)
0	264.4
5	222.1
10	238.0
15	270.7
20	253.3
25	184.2
30	184.9

Table 3: Deploying passive tags alongside a bursting CRFID causes the goodput of the CRFID to degrade as the number of passive tags increases.

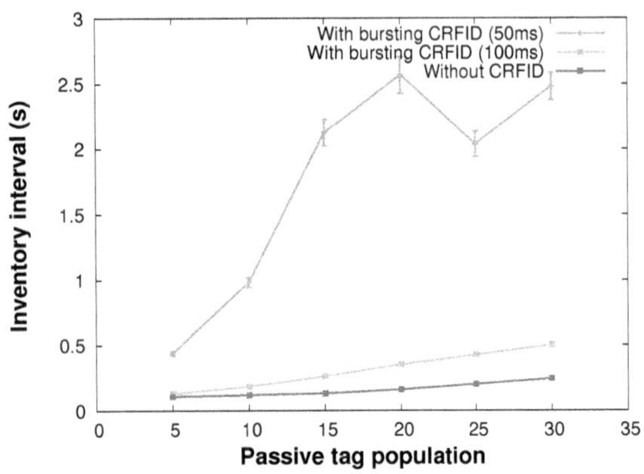

Figure 12: The average interval for a passive tag to be read increases with the tag population.

period between bursts, and one where this duration is increased to 100 ms. Figure 12 shows the average inventory time and associated 95% confidence intervals when there's a population solely comprising commercial passive tags vs a mix of passive tags and bursting CRFIDs. While the average inventory time increases with increasing population, the increase is greater when a bursting CRFID is in the mix. However, the total time is still of the order of a few seconds, showing that passive tags still see opportunities to get data through. In addition, increasing the sleep duration between bursts to 100 ms dramatically reduces the inventory time to be only slightly larger than the case when there are only passive tags. This provides a simple knob in deployments where inventorying time for passive tags needs to be low.

7. RELATED WORK

Empirical Wireless Measurements: In recent years there has been significant work in measuring the characteristics of wireless communication channels for a variety of communication mechanisms (e.g. [18]). Perhaps most relevant to our work is [6], which quantifies wireless performance of Gen 2 through an empirical study that looks only at messages sent by a reader. In contrast, our focus is on improving tag to reader communications for CRFIDs. Additionly, our measurements provide visibility into the backscatter link by piggybacking statistics in EPC codes.

Bulk Data Transfer: The bulk transmission of data through a wireless channel has been studied in a variety of contexts including hardware and software systems. In an 802.11 setting [14] describes several mechanisms that together provide a transport layer optimized for bulk data transmission. Our work looks at RFID backscatter as opposed to 802.11; specifically, we focus on optimizations at the MAC layer for improving the throughput of bulk data transfer. Also of interest is [12], which provides a bulk transport protocol for 802.15.4 based wireless sensor networks. This work uses an end-to-end acknowledgement-based protocol to provide reliability, a rate control mechanism to minimize transfer time and a metric derived from combined signal strengths to avoiding hidden terminal issues. Instead, we focus on detecting other CRFID bursts using reader messages rather than explicitly avoiding them with a collision avoidance metric. Most similar to our work is [8], which proposes the use of persistent read handles in EPC Gen 2 to offload large amounts of data from an individual tag. While more efficient than requiring singulation for each read command, this approach is limited to data transfer from a single tag and cannot take advantage of the relatively shorter QueryRep and QueryAdjust commands.

While Flit provides bulk data transfer at the MAC layer, it does not preclude performance improvements at the PHY layer as defined by EPC Gen 2. In [22], the authors present a system that optimizes the channel and bit-rate selected by a reader, such that tag goodput is maximized. This approach is completely compatible with Flit; a CRFID communicating with an optimized reader would only see further goodput gains.

The elimination of idle slots introduced by coordination is not a new idea and has been looked at in wired contexts, where a common data bus is arbitrated for use by multiple entities. One good example of this is the bus parking mechanism used in the PowerPC 60x [3]. Here, the bus arbiter speculatively grants a bus master before receiving a master request; this grant message is coordinated by broadcasting to all attached devices. This is a very similar concept to the coordination mechanism we use in Flit, where the reader broadcasts a burst notifier to all tags within range.

Energy Management: There has been some recent work on energy management for CRFID systems. The work presented in [15] instruments code at compile time to enable checkpointing software state to non-volatile storage; the goal is to avoid energy wasted on work that was lost to power outages. In [5], a run-time system is presented that adaptively schedules a task to avoid energy wastage. Wastage is defined as work that does not complete due to of energy limitations, as well as harvested energy that cannot be stored. [10] looks at tradeoffs when ambient harvesting is used with RF harvesting, and explores the choice of hardware components to satisfy application requirements given an anticipated amount of harvested energy. Our work is complementary to these efforts, as we improve the bandwidth and energy-efficiency of the communication stack and efficiencies would only further improve with a run-time management strategy.

Looking beyond CRFID research, there has been substantial work on energy management in harvesting-based sensors. For example, [17] achieves perpetual operation by scheduling tasks to match predicted energy harvesting rates, [11] and [19] looks at adaptive duty-cycling strategies for harvesting-based systems, [9] looks at using efficient solar harvesting in combination with an ultra-wideband impulse radio to balance energy usage at an even smaller scale, and commercial efforts have looked at micro-energy harvesting from miniature solar panels, thermal differences, vibrations, and wireless power-over-distance technology (e.g. Powercast [1]). Such techniques may be useful to ensure a suitable amount of buffered energy for bursts is available during reader contact periods, and is complementary to this paper.

8. CONCLUSION

In this paper we presented the design, implementation, and evaluation of Flit, a bulk transmission protocol for RFID-scale sensors. Through a careful analysis of the EPC Gen 2 protocol for passive RFIDs, we identified several opportunities for improvements to both goodput and energy efficiency when considering small numbers of CRFIDs that have large amounts of data to send. Through empirical evaluation, we demonstrated that significant gains in goodput are possible over a variety of distances when compared to a tag that implements vanilla EPC Gen 2. To enable the simultaneous bulk transfer of data from multiple CRFIDs, we designed a simple coordination mechanism that works well in practice and through an experimental evaluation, showed the complete system retains most of the performance improvements we observed for a single, uncoordinated CRFID. Finally, we demonstrated that our protocol can coexist with passive RFIDs can, but are inventoried with increased latency.

While flit is an unreliable MAC protocol, Flit provides an excellent building block for a reliable transfer protocol. A simple window-based reliable protocol over Flit would work as follows: After each burst of messages from the sensor to the reader, the reader sends a read command that has a bitmap of the messages that were received by the reader. In the next burst, the sensor can re-transmit these missing frame in addition to adding other data. We are currently implementing this protocol over Flit.

9. ACKNOWLEDGMENTS

This work was funded by NSF grants CNS-0916577, CNS-0855128, and CNS-0923313 . We thank Intel Research Seattle for providing several WISPs; Our shepherd Guoliang Xing for providing guidance in preparing our final draft; and the anonymous reviewers for their insightful comments.

10. REFERENCES

[1] Powercast website. `http://www.powercastco.com`.

[2] TelosB Datasheet. `http://www.xbow.com/Products/Product_pdf_files/Wireless_pdf/TelosB_Datasheet.pdf`.

[3] M. Allen, M. Alexander, C. Wright, and J. Chang. Designing the powerpc 60x bus. *Micro, IEEE*, 14(5):42, oct. 1994.

[4] M. Buettner, B. Greenstein, A. Sample, J. R. Smith, and D. Wetherall. Revisiting Smart Dust with RFID Sensor Networks. In *HotNets*, October 2008.

[5] M. Buettner, B. Greenstein, and D. Wetherall. Dewdrop: An energy-aware runtime for computational rfid. In *NSDI*, 2011.

[6] M. Buettner and D. Wetherall. An empirical study of uhf rfid performance. In *MobiCom*, pages 223–234, 2008.

[7] M. Buettner and D. Wetherall. A "gen 2" rfid monitor based on the usrp. *Computer Communication Review*, 40(3):41–47, 2010.

[8] M. Buettner and D. Wetherall. A software radio-based uhf rfid reader for phy/mac experimentation. In *RFID (RFID), 2011 IEEE International Conference on*, pages 134–141. IEEE, 2011.

[9] M. Gorlatova, P. Kinget, I. Kymissis, D. Rubenstein, X. Wang, and G. Zussman. Challenge: Ultra-low-power Energy-harvesting Active Networked Tags (EnHANTs). In *MobiCom*, 2009.

[10] J. Gummeson, S. S. Clark, K. Fu, and D. Ganesan. On the limits of effective hybrid micro-energy harvesting on mobile crfid sensors. In *MobiSys*, pages 195–208, 2010.

[11] A. Kansal, J. Hsu, S. Zahedi, and M. B. Srivastava. Power Management in Energy Harvesting Sensor Networks. *ACM Transactions Embedded Computing Systems*, 6(4):32, 2007.

[12] S. Kim, R. Fonseca, P. Dutta, A. Tavakoli, D. E. Culler, P. Levis, S. Shenker, and I. Stoica. Flush: a reliable bulk transport protocol for multihop wireless networks. In *SenSys*, pages 351–365, 2007.

[13] P. Levis, S. Madden, J. Polastre, R. Szewczyk, K. Whitehouse, A. Woo, D. Gay, J. Hill, M. Welsh, E. Brewer, and D. Culler. TinyOS: An Operating System for Sensor Networks. In *Ambient Intelligence*. Springer Verlag, 2004.

[14] M. Li, D. Agrawal, D. Ganesan, and A. Venkataramani. Block-switched networks: A new paradigm for wireless transport. In *NSDI*, pages 423–436, 2009.

[15] B. Ransford, J. Sorber, and K. Fu. Mementos: system support for long-running computation on rfid-scale devices. In *ASPLOS*, pages 159–170, 2011.

[16] J. R. Smith, A. P. Sample, P. S. Powledge, S. Roy, and A. Mamishev. A Wirelessly-Powered Platform for Sensing and Computation. In *UbiComp*, 2006.

[17] J. Sorber, A. Kostadinov, M. Garber, M. Brennan, M. D. Corner, and E. D. Berger. Eon: A Language and Runtime System for Perpetual Systems. In *SenSys*, November 2007.

[18] K. Srinivasan, P. Dutta, A. Tavakoli, and P. Levis. An empirical study of low-power wireless. *TOSN*, 6(2), 2010.

[19] C. M. Vigorito, D. Ganesan, and A. G. Barto. Adaptive Control of Duty Cycling in Energy-harvesting Wireless Sensor Networks. In *SECON*, pages 21–30, 2007.

[20] R. Want. RFID Explained. In *Synthesis Lectures on Mobile and Pervasive Computing*. Morgan & Claypool Publishers, 2006.

[21] E. Welbourne, K. Koscher, E. Soroush, M. Balazinska, and G. Borriello. Longitudinal study of a building-scale rfid ecosystem. In *Proceedings of the MobiSys*, MobiSys '09, pages 69–82, New York, NY, USA, 2009. ACM.

[22] P. Zhang, J. Gummeson, , and D. Ganesan. Blink: A high throughput link layer for backscatter communication. In *MobiSys*, 2012.

COBRA: Color Barcode Streaming for Smartphone Systems

Tian Hao, Ruogu Zhou, Guoliang Xing
Department of Computer Science and Engineering, Michigan State University, USA
{haotian, zhouruog, glxing}@cse.msu.edu

ABSTRACT

This paper presents COBRA – a visible light communication (VLC) system for off-the-shelf smartphones. COBRA encodes information into specially designed 2D color barcodes and streams them between screen and camera of smartphones. Due to the directionality and short range of visible light, COBRA can preserve user privacy and security in many near field communication scenarios such as opportunistic data exchange between smartphones. We develop a new 2D color barcode for COBRA that is optimized for streaming between small-size screen and low-speed camera of smartphones. COBRA adapts the size and layout of code blocks in streamed barcodes to deal with the significant image blur in mobile environments, and adopts new image processing techniques to achieve real-time barcode stream decoding. Our approach is evaluated through extensive experiments on Android smartphones.

Categories and Subject Descriptors

C.2.1 [**COMPUTER-COMMUNICATION NETWORKS**]: Network Architecture and Design—*Wireless communication*

Keywords

Visible Light Communication, Color Barcode, Smartphone

1. INTRODUCTION

As a potential alternative for radio frequency (RF) technology, visible light communication (VLC) has received significant attention recently. Due to the high directionality of narrow light beams, VLC can enable secure and interference-free wireless links. These characteristics make VLC an attractive solution for high-speed, short-range wireless communications in many scenarios such as indoor WLANs and vehicular networks [11] [13] [7].

This work explores a special form of VLC for off-the-shelf smartphone systems. Specifically, information can be encoded as a stream of images and played on smartphone/computer screens while a smartphone uses camera to record and then decodes the video stream. In contrast to RF technologies, the direction and distance of screen-camera link can be easily controlled, preserving the communication privacy and security. As a result, the troublesome authentication process, which is typically required for radios such as Bluetooth and Wi-Fi, can potentially be simplified or even avoided.

We have designed and implemented a novel VLC communication system described above called COBRA: COlor Barcode stReaming for smArtphones. COBRA encodes information into specially designed 2D color barcodes and uses several new techniques to deal with the significant image blur in dynamic environments and achieve real-time barcode stream decoding between smartphones. COBRA can enable a range of interesting applications for mobile users. By leveraging the abundance of LCD displays in retail stores and museums, COBRA can be easily implemented to provide patrons significantly more information (e.g., coupon brochures and maps) than popular static 2D barcodes. Moreover, COBRA users can use screens and cameras to opportunistically exchange small pieces of information such as contacts, photos, short video clips etc. In these application scenarios, screen and camera offer a more intuitive, easy-to-use communication interface without the reliance on additional wireless interfaces or the Internet infrastructure.

Near Field communication (NFC) is a short-range wireless communication technology that is increasingly popular for smart payments and data exchanges between mobile users. However, currently, NFC is supported by only a few smartphone platforms. In contrast, COBRA can be implemented on almost all smartphone platforms that are shipped with a camera, and also leverage the abundance of public LCD displays for applications like advertisements. Moreover, while NFC limits the communication to a short, fixed range (about 10 cm), COBRA can offer more flexible communication ranges by adapting the screen brightness, block size of barcodes, and the angle between screen and camera etc. Such flexibility does not compromise security because of the highly directional nature of visible light.

Recent works [15] [5] have explored the LCD-camera pair to build wireless links. These approaches can achieve high wireless transmission rate between large LCD monitors and high-speed digital cameras using complex communication encoding (e.g., OFDM) and computer vision techniques. However, they are not designed to handle the challenges of off-the-shelf smartphone platforms such as limited computation/power budget, small screen size, and poor image quality in mobile environments. Specifically, we make the following contributions in this paper.

1. We propose a new 2D color barcode design that is particularly optimized for real-time streaming between small-size screens and low-speed cameras of smartphones. Our barcode features a *smart border* composed of color corner trackers and timing reference which enable the code blocks to be quickly localized and extracted from low-quality images with significant blur and distorted perspective.

2. We develop several novel techniques to maximize the throughput of screen-camera link between two smartphones. COBRA sender adaptively changes the size and layout of code blocks in streamed barcodes to mitigate the impact of image blur in mobile environments. COBRA receiver adopts a new metric called degree of blur (DOB) to quickly assess the quality of captured images and restores the colors of pixels based on the hue saturation value (HSV) color model. Moreover, by using fast corner detection and code scan algorithms, COBRA can achieve real-time decoding of barcode stream.

3. We implement COBRA on Android smartphones and evaluate its performance through extensive experiments under a range of settings on system parameters and environmental factors such as view angle, distance, orientation, ambient lighting, and level of mobility. Our results show that COBRA can achieve robust high-throughput communication while incurring moderate computational overhead.

The rest of the paper is organized as follows. Section 2 reviews related work. Section 3 describes the challenges of system design and provides an overview of COBRA architecture. Section 4 and 5 present the design and implementation of COBRA, respectively. Section 6 presents the testbed experiment results. Section 7 concludes the paper and discusses future work.

2. RELATED WORK

As a potential alternative to radio frequency (RF) communication, visible light communication (VLC) has recently received significant attention. It has several advantages over RF technology such as security against eavesdropping and not generating RF interference. The VLC technology has been first explored by VLCC (visible light communication consortium) in Japan [11]. In [13], the authors propose to use LEDs for digital communication. In [7], the authors use white LEDs and orthogonal frequency division multiplexing (OFDM) to implement a VLC system. In their prototype, DSP hardware boards (32 stereo codec and 96 *kHz* sampling rate) are connected to the sender (9 LEDs) and the receiver (a single photodiode) for encoding and decoding, respectively. They conclude that supported by extra DSP hardware, low-cost off-the-shelf LEDs and photodiodes can easily achieve a satisfactory communication speed for most Internet applications. In [5], the authors discuss the challenges of building a visual MIMO system based on LED and CMOS by leveraging computer vision approaches. In [4], the authors propose a rate adaption protocol VMRA (Visual MIMO Rate Adaptation).

Instead of using LEDs and photodiodes, recent works such as pixelated wireless optical channel [8] and PixNet [15] employ the LCD-camera pair to build wireless links. These approaches can achieve high throughput over a long distance based on OFDM and complex computer vision algorithms. Several challenges in designing a Visual MIMO system are discussed in [16]. An initial system is built using LCD monitor displaying black/white checkboard patterns. Unfortunately, the existing approaches for LCD-camera communication are not suitable for smartphone platforms. First, they rely on high-end digital cameras to capture images. However, images captured by cameras of off-the-shelf smartphones have significantly poorer quality than those of high-end cameras. Moreover, the maximum capturing rate of the smartphone camera is typically lower than 30 frame per second (*fps*), which is less than half of the capturing rate of high-end digital cameras. Second, compared with LCD monitors, smartphone screens have smaller size and lower resolution. Third, the OFDM-based encoding and decoding methods used by PixNet are not suitable for smartphones because of high computational overhead.

Both 1- and 2-dimensional barcodes (e.g., UPC code [1], QR code [3] and Data Matrix [2]) are widely used to item tagging in retail and logistics. Recently, due to the popularity of smartphones, many applications have been developed to read barcodes using smartphone cameras. Unlike traditional laser scanning devices, camera is able to detect more colors [12]. By taking this advantage, new barcodes with higher capacity are developed to use more colors than just black and white. One of the most popular color barcodes is the High Capacity Color Barcode (HCCB) [14]. Compared with QR Code and Data Matrix, HCCB enables higher density storage of data on analog printed media. However, due to the limited capacity of a single barcode, its applications are limited to storing a small piece of information like website address. 4D barcode [9] is designed to achieve robust data transmission using off-the-shelf phones. Because of the computation overhead of operations like border detection and barcode rectification, the throughput on smartphone platforms is significantly limited[1]. Therefore, they are not suitable for real-time barcode streaming on smartphones.

3. CHALLENGES & SYSTEM OVERVIEW

The major challenges in building a real-time VLC system for smartphones are brought by the poor image quality and the limited computational resource of smartphones. First, smartphones have small size screen (typically around 4 inches) and low resolution. As a result, the neighboring pixels displaying different colors often blur with each other. In this work, the blur effect refers to the fact that the color sampled by the camera deviates from the actual color displayed on the screen of the sender. Second, the sender and receiver smartphones often have relative movement in mobile scenarios. This causes more severe blur effect in the captured images, making it challenging to achieve a high decoding rate. Finally, since the power consumption of the screen and camera on smartphones is high, it's desirable to maximize the communication throughput. The cameras on smartphones are typically capable of capturing images in 30

[1]For instance, the barcode designed in [9] can only achieve a streaming rate of about 100 bits/s, which is three orders of magnitude lower than the typical throughput of our system

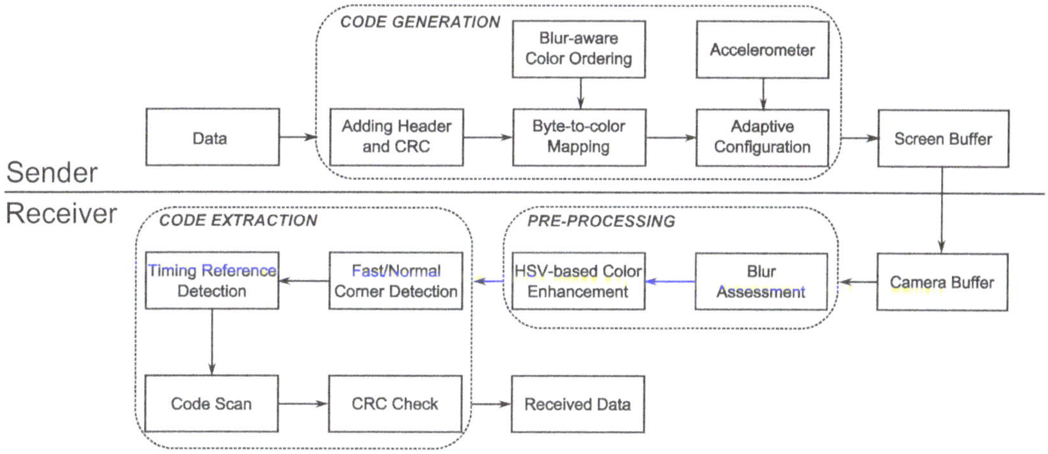

Figure 1: System architecture of COBRA

fps. As a result, it requires the system to capture and process a barcode image within around 30 *ms*. This imposes a challenge on many main-stream smartphones.

Keeping the above challenges in mind, we aim to achieve high throughput communication between screen and camera of smartphones in dynamic environments. Fig. 1 illustrates the architecture of the proposed COBRA system. In the sender-end of COBRA, the main objective is to encode data into color barcodes and display on the screen at certain frequency (15 *fps* in our implementation). COBRA adopts a new color barcode design that features a smart border that can significantly accelerate the decoding process. In order to generate color code from data, first we add header and CRC checksum to the data and form a data block with fixed size, then map each byte in the data block to certain color. The mapped colors are then used to generate an image of color barcode to be displayed on the screen. COBRA adopts two novel approaches, namely *adaptive configuration* and *blur-aware color ordering*, to mitigate the blur effect. First, as smaller block size causes more blur in the captured image, COBRA adaptively changes the block size according to the current level of mobility indicated by accelerometer readings. Second, motivated by the observation that blur usually occurs along the border of blocks of different colors, COBRA optimizes the order of color blocks in a barcode to minimizes the total border length of differen colors.

The receiver-end of COBRA consists of two main components, namely *pre-processing* and the *code extraction*. The pre-processing component is responsible for selecting and enhancing the images for the subsequent code extraction. Since the capturing rate of the receiver is twice of the refreshing rate of the sender's screen, more than one images containing the same barcode can be captured. To avoid unnecessary computation, COBRA will select the image with the best quality for each barcode for further processing. We develop a simple blur assessment metric called *degree of blur* (DOB) which can be computed efficiently by sampling a small number of pixels in the image. If an image has acceptable DOB, its colors will be enhanced based on the HSV (Hue, Saturation and Value) color space. Compared with the original RGB (Red Green Blue) color space, HSV is much more effective in recovering colors from a blurred image. After the pre-processing, the enhanced images undergo a sequence of operations for code exaction. *Corner detection* and *timing reference detection* are used to locate the color code area. Previous corner detection methods [14] for color barcodes usually involve computation intensive operations such as Hough transform. We develop a method called *fast corner detection* where corner locations of previous images can be reused to significantly accelerate the corner detection of the current image. This design leverages on the fact that the corner locations only have limited displacement between adjacent images captured in high frequency. After the corners are localized, the code is scanned to extract data. Using the *timing reference blocks* on the border of a barcode, COBRA is able to directly locate color blocks in code area and un-map them into bytes. A common challenge in code scan is perspective distortion of the barcode. Traditional approaches [14] address this challenge by transforming the barcode into a rectangle using complex geometric operations. The timing reference blocks on our color code allow COBRA to accurately and quickly locate blocks in code area.

COBRA is designed to achieve only one-way communication while two-way communication is often required by higher layer protocols to provide feedback to the sender. Many mainstream smartphones are equipped with front-facing cameras. A duplex two-way VLC communication can be realized between two smartphones by using front-facing cameras as transmitters. However, we leave this to the future work.

4. COBRA DESIGN

4.1 2D Color Code

Our goal is to design a color barcode with high capacity and fast decoding speed. As shown in Fig. 2, the color barcode is formed by square *color blocks* of the same size. The color blocks are arranged back-to-back in rows and columns. There are three types of areas in our color barcode, which are *corner trackers*, *timing reference blocks* and *code area*. The corner trackers are used to quickly locate the color barcode corners. Based on the location of corners, timing reference blocks on the border of the color barcode can be detected and then used as references to locate all color blocks in code

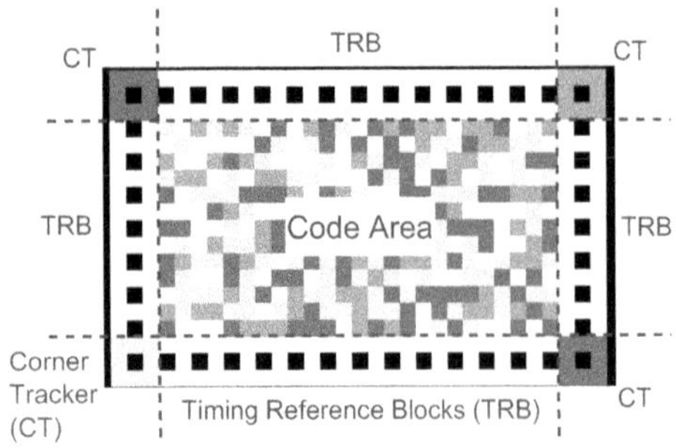

Figure 2: The design of the 2D color code. Different code areas are divided by dotted lines.

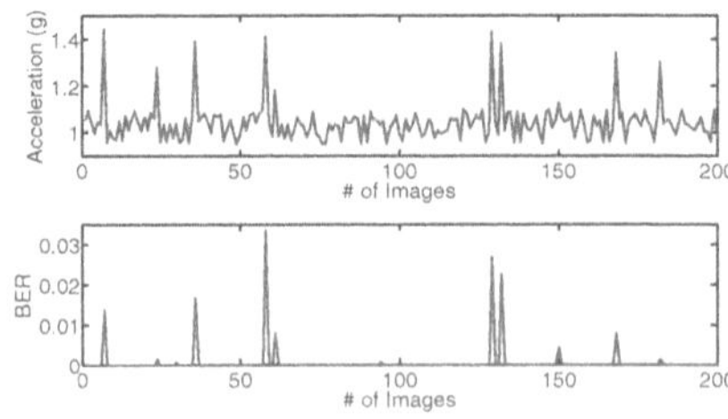

Figure 3: The BERs of a sequence of barcode images (bottom) and the corresponding real-time acceleration (top) measured on a Google Nexus S phone.

area. The code area stores the information encoded in the barcode. An important parameter of our code design is the number of colors used to encode data in a color block. In our implementation, 5 colors (black, white, red, green and blue) are used in the color barcode. Depending on the image quality, another 3 colors (magenta, cyan and yellow) can be used to increase the capacity. However, because of the limited quality of smartphone cameras, using more colors will cause more decoding errors. Therefore, it may not lead to improved system throughput.

The core feature of our color barcode design is the *smart border* consisting of corner trackers and timing reference blocks. Corner trackers are widely used by traditional barcodes (e.g., QR Code) to detect corner locations. Our corner tracker design features small-size color blocks that facilitate continuous localizations in a stream of images. Black blocks are only used in smart border for fast identification. The unique pattern where a black block is surrounded by 8 blocks of the same color (red, green or blue) is used to indicate the corner. In order to determine the barcode orientation in the image, we place corner trackers with green and red blocks as the top-left and bottom-right corners, respectively. The other two corners share the blue color. The timing reference block is a black block surrounded by 8 white color blocks. They significantly accelerate the decoding process by allowing direct location of color blocks in code area. Data (header, payload and CRC checksum) is encoded using a sequence of color blocks in code area. When 4 colors are used for encoding, each color block in code area represents 2 bits of data. For example, in our implementation, red, green, blue and white colors represent 00, 01, 10 and 11 respectively. In this way, a byte with value of 11100100 can be encoded as [*white blue green red*]. On a typical 4-inch phone screen with 800×480 resolution, a single color barcode with 6-pixel block size contains 18.8K bits. This capacity can be increased by taking advantage of the larger screens on new-generation smartphones. For example, with a 4.65-inch screen and 1280×720 resolution, the latest Galaxy Nexus can encode 47K bits in a single barcode.

4.2 Adaptive Code Generation

Theoretically, a color barcode formed by smaller color blocks leads to higher transmission rate. However, this does not always increase the communication throughput in reality due to the blur effect. Blur often leads to inaccurate color detection and hence reduces the decoding rate. In the captured image, blur usually occurs along the borders of two color blocks when pixels of each block display different colors. As a result, using smaller color blocks will increase the total length of border and consequently leads to more severe blur effect. Moreover, the relative movement between the screen of sender and camera of receiver is also a major factor that may cause blur of images. Therefore, it is critical to choose a proper block size to maximize the barcode capacity while maintaining a high decoding rate. In our design, COBRA adjusts the block size adaptively based on the phone's level of mobility. As most smartphone platforms are equipped with accelerometers, the level of mobility can be estimated based on the real-time acceleration of the device. As COBRA is designed to achieve one-way communication, it only accounts for the acceleration of sender. When the acceleration level of receiver is available, COBRA can estimate the relative movement between sender and receiver more accurately. However, this is left for future work. We also note that this acceleration-based adaptation mechanism only deals with the dynamic blur caused by minor displacements between two phones (e.g., due to shaking on a driving vehicle). When two phones experience significant displacements causing camera focus loss or shallow view angel, several components of COBRA, e.g., blur assessment and corner detection, can detect the resulted anomalies and discard the low-quality/incomplete images.

Fig. 3 shows the Bit Error Rate (BER) of a sequence of barcodes and the corresponding acceleration of a Google Nexus S phone. The details of barcode decoding are discussed later this section. We can observe that, when the acceleration is blow $1.2g$, nearly 99% images have zero BER. When the acceleration is above $1.2g$, however, the BERs of most images exceed 1%. This is because high level of movement leads to severe blur in the captured images leading to significant color distortion. Our further experiments show that, given a short distance (e.g., < 10 inches) between two phones, this threshold varies for different phones but are affected only slightly by environmental conditions. It is thus possible to determine this threshold through offline experiments.

Motivated by the empirical observations, we develop a simple algorithm to adaptively adjust the block size based on the mobility level. First, the COBRA sender is initialized with a threshold T_{acc} and the default block size. It samples the accelerometer continually at 60 *Hz* and computes the average acceleration in a moving time window. Once the average acceleration exceeds T_{acc} for 1 second, the sender enters the *mobile mode* by increasing the block size to a preset value ($1.5\times$ default block size in our implementation).

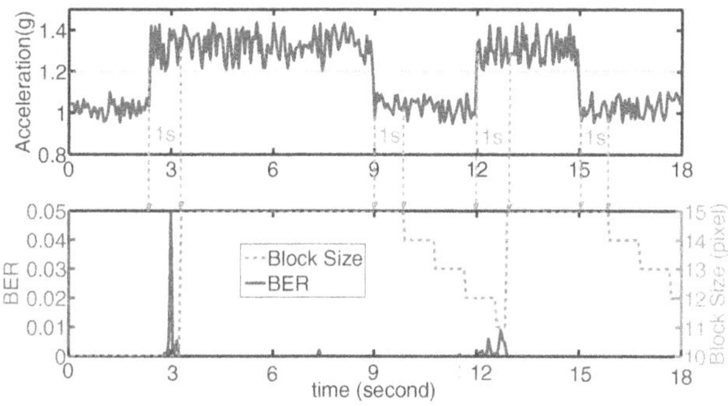

Figure 4: Top: The average acceleration of sender. Bottom: The dynamic block size adapted based on the acceleration and BER of each barcode at receiver.

When the acceleration is detected to be smaller than T_{acc} for 1 second, the sender switches to the *transition mode* by decreasing the block size at a rate of 1 pixel per second. In highly mobile scenarios, the linear block size decrease prevents the system from recovering from the mobile mode prematurely.

Fig. 4 illustrates a typical process of adaptive code generation. We can observe that at around 2.5 seconds, the BER of the receiver increases due to the acceleration of the sender. Then the sender enters the mobile mode by increasing the block size from 10 to 15 pixels. At around 10 seconds, the sender is relatively stable. It takes COBRA 1 second to confirm the stability before it starts to gradually decrease the block size. At 13 seconds, the sudden acceleration triggers the increase of block size to 15. Finally, at 15 seconds, due to the smaller acceleration, the block size starts to drop gradually again.

4.3 Blur-aware Color Ordering

Our analysis of blurred color barcode images shows that blur usually occurs along the border of blocks of different colors. Therefore, it is desirable to reduce the length of border by optimizing the layout of color blocks (referred to as *color ordering* hereafter) at the sender. This idea is analogous to modulation in communication systems where digital signals are modulated to mitigate the impact of channel noise. However, this is nontrivial for our problem because of two reasons. On the one hand, the optimal color ordering strategy is highly dependent on the color blocks (i.e., data payload) contained in a barcode. On the other hand, the ordering strategy must be relatively simple and consistent across different images in order to be encoded in a barcode so that the decoder of receiver knows how to restore the color clocks to the original order.

To balance the efficiency and computational overhead, COBRA adopts a simple color ordering strategy called *random stride*. As shown in Fig. 5, first, the sender divides the color blocks of a barcode into groups of equal size. The color blocks with the same index in all groups are then arranged together. The group size in this algorithm can be viewed as *stride* in the walk through the original color block sequence. The value of stride is searched within in a random chosen range to minimize the total length of border between differ-

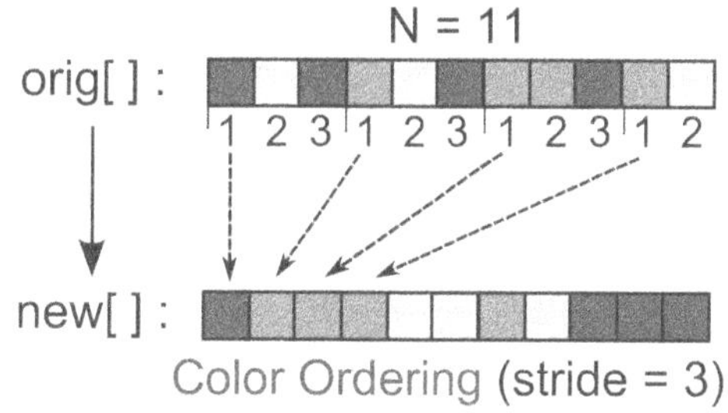

Figure 5: Illustration of the blur-aware color ordering.

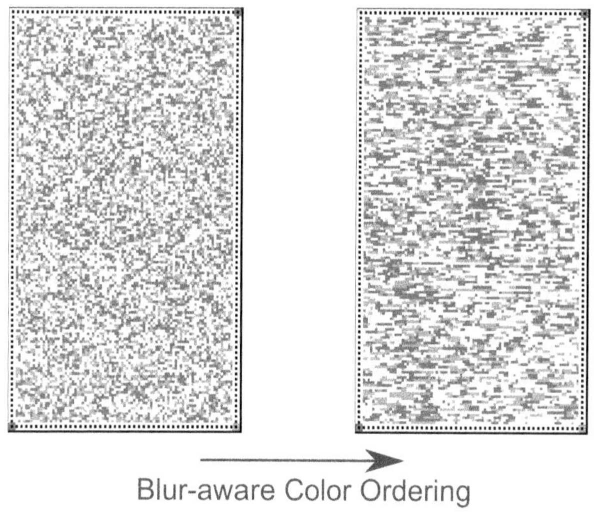

Figure 6: Left: the original barcode with block size = 6; Right: the barcode with the least border length among 20 barcodes re-ordered with randomly chosen strides.

ent colors. The reason of using a random stride search space is to reduce the incurred computational overhead. Formally, the random stride algorithm can be described as follows. Suppose array $orig[0...N-1]$ represents the original color sequence of length N. First, COBRA randomly chooses a stride length s in the range $[1, N/2]$. Then it re-orders the original color sequence using the following mapping function:

$$new\left[(n\%s)\cdot\left\lceil\frac{N}{s}\right\rceil+\left\lfloor\frac{n}{s}\right\rfloor\right]=orig[n] \qquad (1)$$

where $n \in [0, N-1]$ and array $new[0...N-1]$ represents the color sequence after re-ordering. In our implementation, COBRA randomly chooses 20 strides and keeps the one that produces the minimum border length. The stride length used for re-ordering the original color sequence is carried in the header of the barcode. As the sender only applies the color re-ordering operation on payload, it is easy for the receiver to recover the original color sequence using Eqn. 1 and the stride length contained in the header. Fig. 6 shows the original and reordered barcodes under a typical setting. It can be seen that the blocks of same color are often clustered together in the reordered barcode, reducing the blur along the borders of different colors.

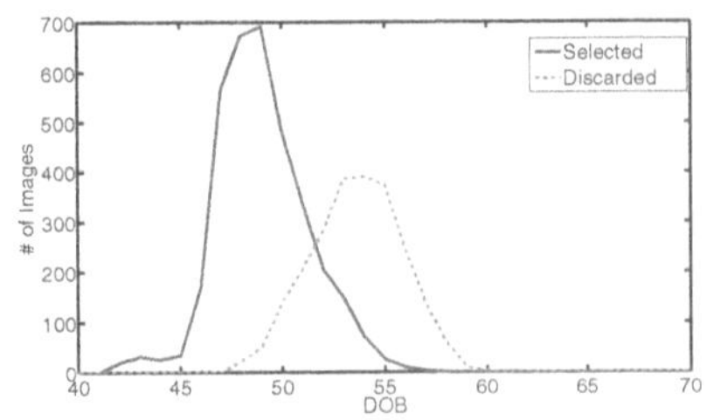

Figure 7: Distribution of DOB of over 9000 barcode images captured by a Google Nexus S. The blue solid line indicates the distribution of DOB of selected barcode images. The two curves represent the distributions of DOBs of selected and discarded barcode images, respectively.

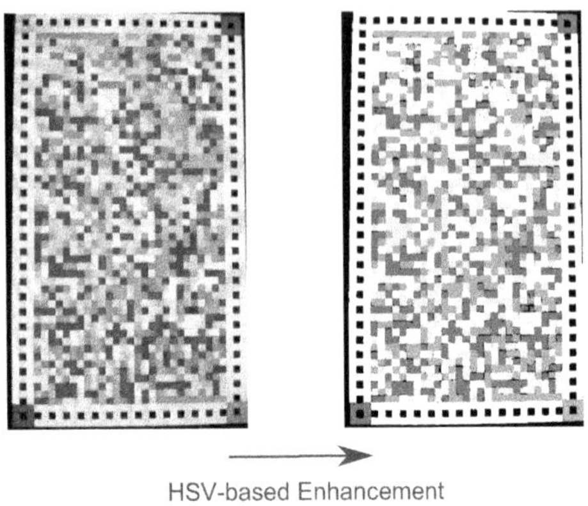

Figure 8: Left: raw image with $DOB = 45$. Right: the image after color enhancement. The major blur area is labeled by black dash line.

4.4 Blur Assessment and Color Enhancement

As discussed earlier, the blur of captured image directly affects the decoding rate of our system. In order to ensure that the camera is able to capture all the color barcodes displaying on the screen, the capturing rate of the camera has to be at least twice of the refreshing rate of the screen. For example, in our implementation on Google Nexus S, the capturing rate is set to be 30 *fps* while the sender refreshes 15 color barcodes per second. As a result, the receiver is likely to capture more than one image with different degrees of blur containing the same color barcode. In addition, when the camera captures the image while the screen is in the middle of refreshing the screen, the captured image usually contains the mixture of two color barcodes. Although a severely blurred image may still contain decodable areas, processing such images often yield little gain on system throughput because of the substantially higher overhead. Therefore, among the images of the same color barcode, COBRA chooses the one with the best quality for further processing. In order to reduce computation overhead, it is important for COBRA to filter out the poor-quality images. We develop a simple and efficient metric to measure the degree of blur effect. The key idea is to evaluate the deviation in pixel values of an image to the real pixel values. First, we divide the raw image into 4 equal regions (top left, top right, bottom left and bottom right). In each region, we randomly sample

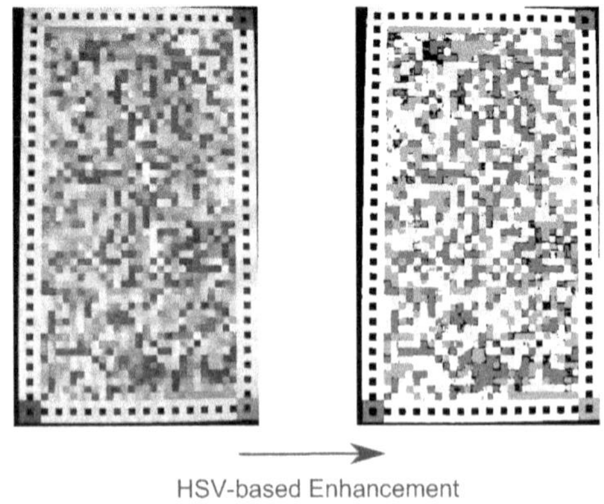

Figure 9: Left: raw image with DOB=52. Right: the image after color enhancement. The major blur area is labeled by black dash line.

N pixels. In our implementation, we use black, white, red , green and blue in barcode. The original pixel value in its three RGB components are either 0 or 255. Thus, for each pixel p, we define its degree of blur (DOB) as follows,

$$DOB_p = \sum_{Y \in \{R,G,B\}} \frac{min(|255 - Y|, Y)}{3} \quad (2)$$

where R,G and B indicate the RGB values of the target pixel. For example, the DOB_p of a typical blurred red pixel with RGB value $(210, 45, 45)$ is 45, while the DOB of the pure red pixel is 0. The DOB of the whole image is defined as

$$DOB_{img} = \frac{\sum_p DOB_p}{4N} \quad (3)$$

where DOB_{img} is simply the average DOB of randomly sampled pixels. The range of DOB_{img} is from 0 to 255. According to the definition, lower DOB_{img} indicates smaller blur effect. Fig. 8 and 9 show the images with DOB_{img} of 45 and 52, respectively. We can observe that the raw image with DOB_{img} of 45 has substantially much less blur. The difference is even more evident after the colors of two images are enhanced (discussed later in this section). Fig. 7 shows the distributions of DOB_{img} for the selected and discarded images among totally over 9000 continuously captured images on Google Nexus S. The two distributions are clearly distinctive, which suggests the effectiveness of the DOB metric. In summary, COBRA is able to use DOB as a metric to quantify the blur effect and select images with less blur for further processing.

Due to the poor quality caused by the blur effect, the pixel values in the captured image may not correctly reflect the actual color used to encode data. This will create difficulty in the following steps in code extraction such as corner detection. Although the blur assessment can filter out images of poor quality, the colors of pixels of selected images still deviate from the original ones. Therefore, after blur assessment, the selected images with less blur are then processed by the color enhancement component. The objective is to restore the original color of the image through analyzing its pixel values. As the colors used to transmit data is known (e.g., red, green, blue and white etc.), COBRA is able to compute the most likely actual color based on the proximity in HSV (Hue Saturation Value) color space [6]. In the HSV model, hue represents the color, saturation refers to the dominance of hue in the color, and value indicates the lightness of the color. Instead of using three components (R, G and B) to represent colors, HSV only has one component (hue) indicating different colors other than white and black. For this reason, the HSV model is more effective in detecting and restoring colors from a blurred image. An additional advantage of color enhancement is the pixel compression. The OS of smartphones like Android typically uses a 32-bit integer to represent the pixel value. After color enhancement, COBRA converts the colors into black, white, red, green and blue which can be represent by 3 bits. This pixel compression not only reduces memory usage but also accelerates the speed of further processing. As shown in Fig. 8, after HSV-based color enhancement, colors of all the pixels are enhanced according to the real colors used for encoding.

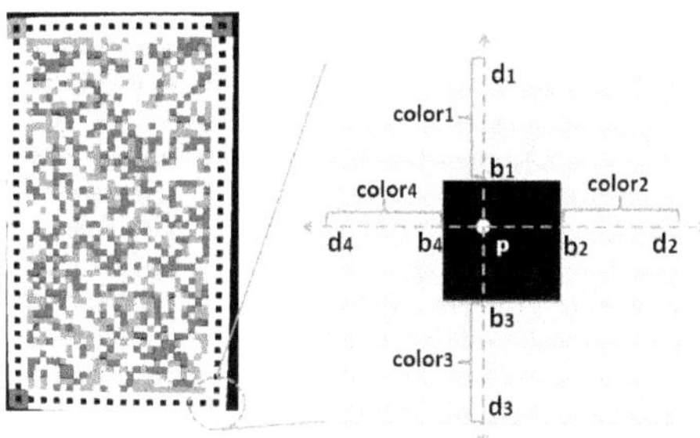

Figure 10: Illustration of the method of detecting pixels black block within corner tracker

4.5 Fast Corner Detection

Taking the enhanced images as input, COBRA will first detect the corner locations of the barcode. Then the locations of the timing reference blocks between the barcode corners are determined. Finally, the blocks in code area can be read according to the timing reference locations.

The task of corner detection is to find the location of barcode corners in the captured image. Corner detection is not as easy as it seems on smartphones because the captured image is subject to severe blur effect. Moreover, it is possible that some blurred pixels in code area are mistakenly recognized as part of a corner, making the corner detection more challenging. In order to capture and process high resolution (e.g., 1280 × 720) barcode stream (30 *fps*) in real-time, COBRA needs to minimize the processing time. The methods for decoding existing 2D barcodes often involve computation intensive operations such as hough transform [14]. Corner detection can also be implemented using object tracking algorithms in computer vision. However, the common algorithms used for object tracking are not suitable for COBRA due to their high computational complexity. For example, object tracking approaches such as [10] involve convolving the image with Gaussian filters at different scales, which incurs high barcode processing delay. Our design is based on a special structure called smart border. A unique pattern is placed in each corner of the smart border in order to accelerate corner detection. Specifically, the top left corner can be recognized by detecting the unique pattern of a black block surrounded by 8 green blocks. The patterns of the other three corners differ from each other in the color of surrounding blocks, so that the direction of the color code can be determined.

We now describe the process of corner detection in COBRA. Suppose the pixels of the input image are stored in the buffer $pixBuf[\ ]$, where each element in the $pixBuf$ represents the color (after color enhancement) of the pixel. After reading each element p from $pixBuf$, COBRA will conduct the following operations. First, if p is not black, COBRA will do nothing and continue reading the next element. Otherwise, COBRA will analyze the color of pixels in four directions starting from p (as shown in Fig. 10). Specifically, when searching upward, COBRA records the point b_1 where the pixel changes its color from black to another color $color_1$, and continues searching upward until the color of the pixel changes again at point d_1. After finishing searches in four directions, COBRA determines whether point p is inside the corner block based on the following 3 conditions: (1)$color_1$, $color_2$, $color_3$ and $color_4$ are the same color; (2)Both b_1b_3 and b_2b_4 are longer than $minBS$ (minimum block size, which is usually 5 pixels in our implementation); (3)b_1d_1, b_2d_2, b_3d_3 and b_4d_4 are all longer than $minBS$. The purpose of condition (1) is to check if the block containing p is surrounded by blocks of the same color. Condition (2) makes sure that point p is located in a black block, filtering out the case where p is just a pixel in code area misclassified as black pixel. Condition (3) checks the width of the surrounding color blocks, filtering out the case where black blocks for timing are surrounded by color pixels due to blur effect. Pixel p is confirmed as a pixel inside the corner block only if the searching result satisfies all the three conditions above. Once p is confirmed, the coordination of the corner (center of the black block) can be easily computed. Note that our corner detection method works regardless of the color code position and orientation in the image.

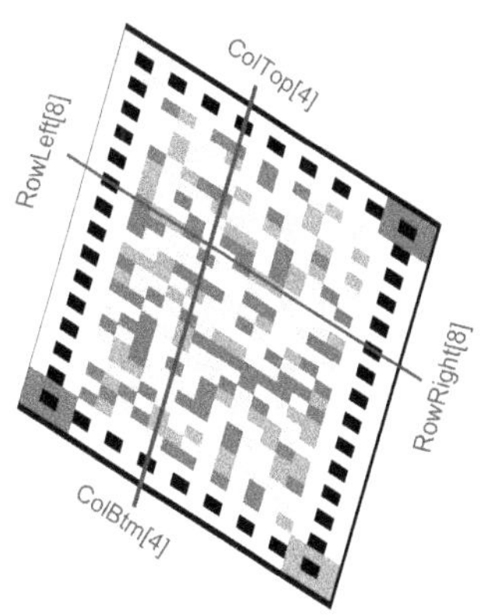

Figure 11: Locating blocks in code area using TR blocks in barcode with perspective distortion

In order to further minimize the processing time, we develop a method called fast corner detection. The idea takes the advantage of the high image capturing rate (e.g., 30 *fps* on typical smartphones). A high capturing rate implies that the displacement of the barcodes in adjacent captured images is limited in a very short range (typically shorter than a few pixels). Thus, instead of running the corner detection algorithm for every captured image, COBRA is able to infer possible corner positions of the current image using those of the previous captured images. For example, if the top left corner position is (x_0, y_0) in the previous image, COBRA only needs to search for the top left corner of the current image in a small area $(x_0 - r < x < x_0 + r$, $y_0 - r < y < y_0 + r)$ where r is the maximum dislocation range (in pixels) between adjacent images. Therefore, COBRA reduces the overhead by searching corners in only 4 small areas surrounding the previous corners.

4.6 Code Extraction

After obtaining corner locations, COBRA is finally able to access the code area. The challenges of code scan include possible prospective distortion and code segmentation. When the camera of receiver is capturing the screen of sender from an angle, the rectangular barcode on the screen becomes a trapezoid in the captured image (as shown in Fig. 11). That is, some blocks of the barcode expand in the image, while others shrink in size. For this reason, the system can not locate and read blocks in order. To deal with the prospective distortion, existing methods rely on transforming the code area into a rectangular area so that each block can be located without the impact of perspective distortion [14]. However, code area transformation requires large amount of buffer scanning and computation. In order to achieve fast code extraction, we add special blocks surrounding the code area which are used as timing references (TR) to locate each block in code area faster and more accurately. With the TR blocks, COBRA is able to directly access the blocks in code area without transforming.

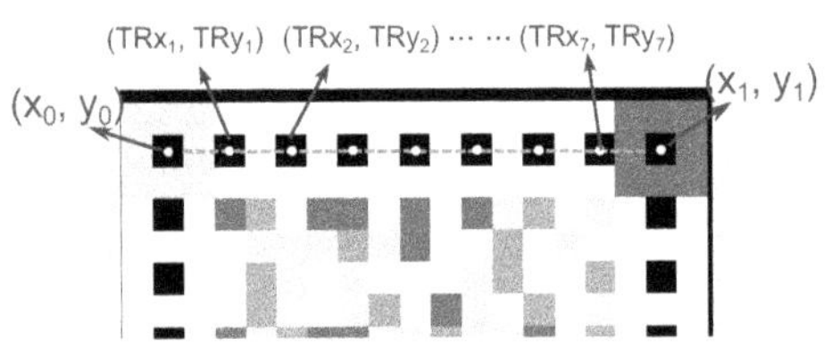

Figure 12: Extracting timing reference blocks between two corners. The yellow dot represents the center of the block. The red dotted line indicates the scanning path.

The TR blocks can be easily detected based on corner locations. Fig. 12 shows an example of detecting TR blocks between top left (at (x_0, y_0)) and top right (at (x_1, y_1)) corners. The process starts with scanning the pixels from (x_0, y_0) to (x_1, y_1). For the i-th consecutive black pixels, compute the center location (TRx_i, TRy_i) of the corresponding TR blocks. Given the corner location, detecting the TR blocks of the color code only requires scanning pixels between 4 corner points hence incurs little computation overhead. Since the TR blocks are placed in every other column or row on the border, before scanning code area, COBRA also needs to compute the center of the white blocks between adjacent TR blocks by simply taking their middle point.

The objective of code scan is to extract the color of the blocks in code area using TR blocks and un-map them in data. First, we mark the pixels between corresponding top and bottom TR blocks. Then COBRA will horizontally scan all the pixels between corresponding left and right TR blocks from top to bottom. Whenever COBRA encounters a marked pixel in horizontal scan, it stores the pixel color for later un-mapping operation. Suppose we store the

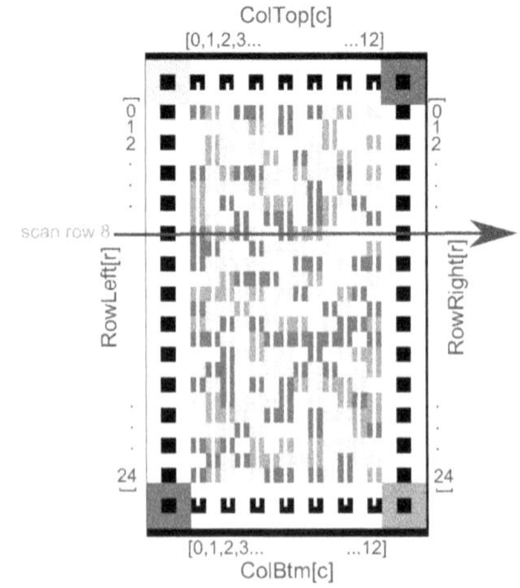

Figure 13: Extracting code blocks from code area. The black line with arrow indicates the scanning path and direction.

coordinates referencing the columns of the code area in arrays $ColTop[\,]$ and $ColBtm[\,]$. Likewise, coordinates referencing the rows of the code area in arrays $RowLeft[\,]$ and $RowLeft[\,]$. For example, as shown in Fig. 13, the center of the leftmost TR block on the top border is ($ColTop[0].x$, $ColTop[0].y$), and the center of the next white block on its right is ($ColTop[1].x$, $ColTop[1].y$).

The process of code scan is described in the following pseudo-code. Line 2 to 4 describe that the system marks all pixels between TR blocks for each column. Line 6 to 10 explains that in the process of a row scan between corresponding left and right TR blocks, the system stores the color of each marked pixel in order. Line 5 indicates that the row scan will be performed in a top-down order for all rows.

1: create array $colorBuf[ColNum \times RowNum]$ {/*For storing extracted colors*/}
2: **for all** c such that $0 \le i < ColNum$ **do**
3: Mark all the pixels between $ColTop[c]$ and $ColBtm[c]$
4: **end for**
5: **for all** r such that $0 \le i < RowNum$ **do**
6: **for all** p from $RowLeft[r]$ to $RowRight[r]$ **do**
7: **if** p is marked **then**
8: Store the color of p into $colorBuf[\,]$
9: **end if**
10: **end for**
11: **end for**

5. IMPLEMENTATION

COBRA is implemented on Android smartphone platforms running Android 2.3.3 Gingerbread. System evaluation is conducted mainly based on two handsets, Google Nexus S and HTC Inspire. In the sender end of COBRA, we implement an Android application for encoding and displaying barcodes. The size of the application file is about 24 *KB* and takes 56 *KB* storage on the phone after installed, requiring about 5 *MB* RAM allocation while running. In order to make sure each barcode is displayed on time, the encoding and displaying functions are implemented in separate threads. First, as described in Section 4.2, the encoding thread determines the block size based on feedback

Figure 14: A typical experimental setup of two COBRA phones. The sender (Google Nexus S) on the right is sending data to the receiver (HTC Inspire) on the left.

from accelerometer. Then, it determines the data segment size based on the block size and the screen size and maps every 2 bits to a pre-defined color (red, green, blue and white). The length of the header is 6 bytes. The sequence number of the barcode is encoded in the first 3 bytes, which can index over 1.67×10^7 barcodes. Lastly, as described in Section 4.3, the generated color sequence is re-ordered to minimize the border length and then passed to the displaying thread. In Google Nexus S, the typical time taken to refresh a barcode on screen is around 25 *ms*.

In the receiver end of COBRA, we implement an Android application for capturing and processing the barcode stream. The size of the application file is 30 *KB* and takes 72 *KB* storage on the phone after installed. It requires 3.5 to 12 *MB* RAM allocation depending on the captured image resolution (600×480 to 1280×720). COBRA provides both *real-time* and *buffered* decoding modes. In the real-time mode, the receiver retrieves captured images in real time from the callback function of camera preview, which is provided in package android.hardware.Camera since Android 2.2.x (FROYO). This function returns a buffer containing captured images in 30 *fps*. In the buffered mode, a video file on flash serves as the buffer between the receiving and decoding threads. The buffered mode incurs lower instantaneous CPU overhead while introducing an initial decoding delay. After getting a captured image, the receiver first computes its DOB as described in Section 4.4. Then, the receiver detects the corner trackers and timing reference blocks and use them to extract the sequence number (encoded in the first 12 blocks) from the barcode. The other blocks are not scanned until all the images containing the same sequence number are received. As discussed in 4.4, the receiver selects the image with the lowest DOB and discards other images that have the same sequence number. Lastly, all the blocks in the code area of the selected image are scanned and decoded into data stream.

6. EVALUATION

6.1 Experimental Methodology

We conduct several sets of experiments to evaluate the system performance of COBRA. Section 6.2 to 6.5 present the experimental results under a variety of settings on COBRA configurations and environmental factors including block size, screen refreshing rate, level of mobility, sender-receiver view angle/orientation/distance, ambient lighting, screen brightness, and camera resolution. Section 6.6 compares the overhead and throughput of COBRA and PixNet [15] that is a VLC system primarily designed for communication between large LCDs and high-end digital cameras.

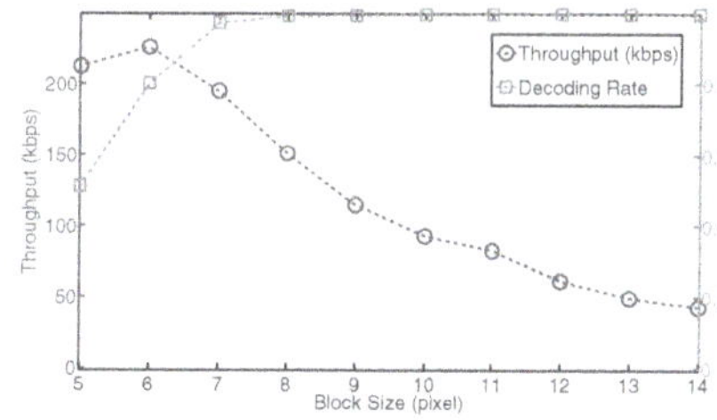

Figure 15: The throughput and decoding rate under different block sizes.

Our evaluation is focused on two metrics: 1) *decoding rate*, which is defined as the percentage of correctly decoded data in the total amount of data contained in a barcode; 2) *throughput*, which is the average amount of data received per second in the correctly decoded barcodes (i.e., 100% decoding rate). We note that the system throughput can be potentially improved by recovering partial data from faulty barcodes based on error correction codes. However, this is left for future work. For the purpose of evaluation, we measure the overhead of real-time receiving mode while using the buffered mode for evaluating the system decoding rate and throughput. Note that these performance metrics are independent of the receiving mode. The original data sent by the sender is used as the ground truth.

Unless otherwise specified, the experiments are carried out using Google Nexus S as sender and HTC Inspire as receiver, and the resolution of the captured images is 1280×720. The Google Nexus S has a 4-inch screen with 800×480 pixels and the HTC inspire has an 8 *MP* camera with autofocus. Fig. 14 shows a typical setup using two phones. When the phones' positions need to be precisely controlled (e.g., for measuring the view angle and distance between two phones), we mount them on two car phone holders.

6.2 Block Size and Refreshing Rate

We first investigate the decoding rate and throughput with respect to the block size. This set of experiments is conducted under normal indoor lighting condition, the distance between two phones fixed to 5 inches, and view angle fixed to 0 degree. The block size is changed from 5 to 15 pixels. Fig. 15 shows the throughput and decoding rate under different settings of block size. We can observe that the throughput achieves the maximum of 225 *Kbps* with block size of 6 pixels and gradually drops as the block size is decreased. A block size of smaller than 6 reduces the throughput because the resulted blur in captured images counteracts the increased barcode capacity. This is evident from the decoding rates under different block sizes. The decoding rate is only 51.1% when block size is 5.

We next examine the impact of barcode refreshing rate. The refreshing rate affects the blur in the captured images because lower refreshing rate leads to longer display time of the same barcode, allowing COBRA receiver to capture more images of the same barcode. As a result, the quality of selected image of a certain barcode is increased. In the experiment, we change the refreshing rate from 10 to 15 *fps* when the resolution of the captured image is 1280×720 and fix the other COBRA settings to examine its relationship

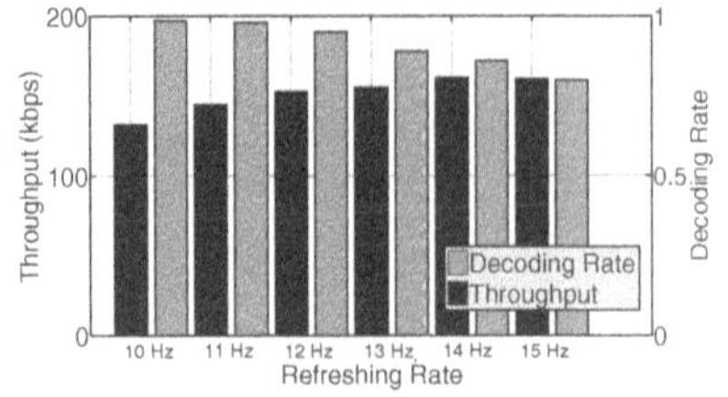

Figure 16: The decoding rate and throughput under different barcode refreshing rates. Barcodes are displayed on Google Nexus S using block size = 7.

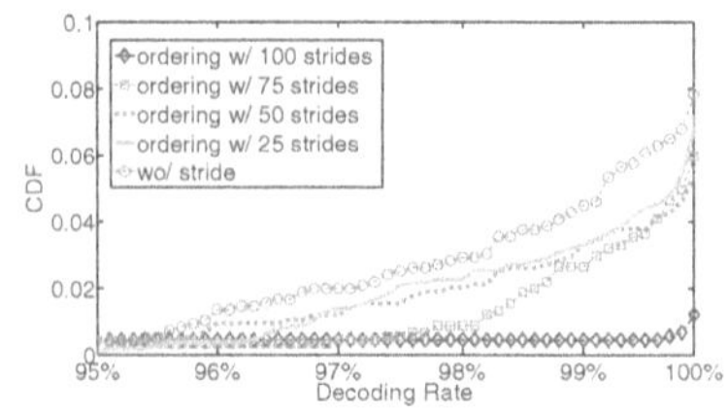

Figure 17: The CDF of the decoding rate with different stride settings.

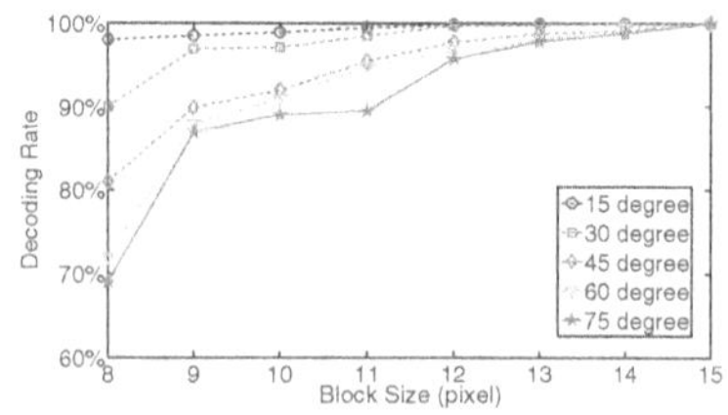

Figure 18: Decoding rates with different view angles.

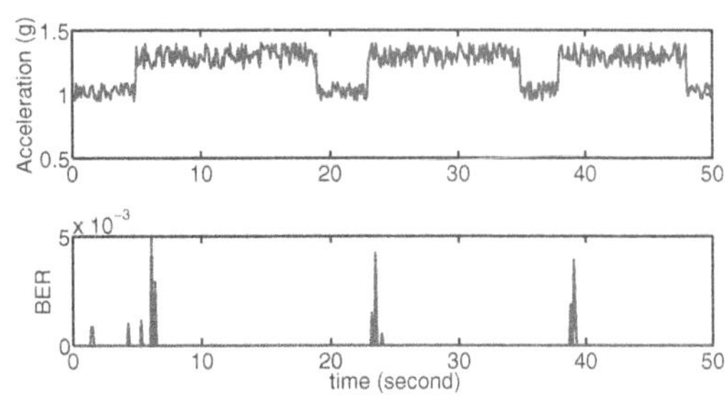

Figure 19: BER and acceleration (mean=1.25g) under adaptive configuration

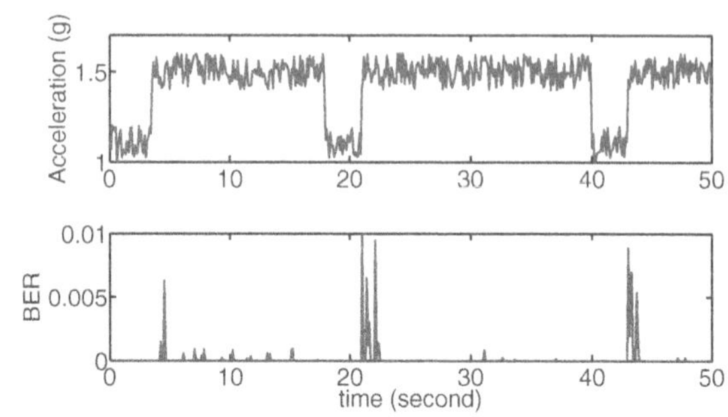

Figure 20: BER and acceleration (mean=1.42g) under adaptive configuration

with decoding rate. As shown in Fig. 16, lower refreshing rate leads to higher decoding rate and lower throughput. For example, when refreshing rate is 10 *Hz*, over 98.5% receiced barcodes have 100% decoding rate, while only 80.1% receiced barcodes have 100% decoding rate when refreshing rate is 15 *Hz*.

6.3 Mobility and Color Ordering

To examine the impact of mobility, we manually simulate different levels of phone movement and measure the mean and max acceleration. The block size is fixed to 7 pixels in this set of experiments. Fig. 19 and 20 shows the CDF of decoding rate under different movement levels. The scenario of $mean = 1.25g$ roughly corresponds to a slow walking person while $mean = 1.42g$ a vehicle driving at moderate speed. We can observe that a sudden acceleration always results in a spile in BER. However, once COBRA adapts the block size, the BER drops to zero quickly. When the level of mobility is increased (from average 1.25g to 1.42g acceleration), COBRA receiver suffers a higher BER. The overall results show that COBRA can maintain robust communication performance while suffering sporadic BER increase in mobile scenarios. We also evaluate the effectiveness of blur-aware color ordering described in Section 4.3 when the block size is fixed to 7 pixels. Fig. 17 shows the CDF of decoding rate for different numbers of strides used to search for the best ordering result. We can observe that 75 and 100 strides significantly reduce the number of barcodes with decoding rate lower than 98% due to the effective mitigation of blur. However, compared with 25 and 50 strides, 75 strides do not significantly improve the overall system performance. For a typical barcode of 8-pixel block size on 800 × 480 screen, re-ordering its 5000 blocks with 100 strides (see Section 4.3) takes less than 50 *ms*.

6.4 View Angle, Orientation and Distance

In real scenarios, the sending and receiving phones of COBRA may have various view angles, orientations and distances. These factors substantially affect the quality of communication channel. First, we investigate the impact of view angle that is defined as the angel between the two lines perpendicular to phone screens. A smaller view angel indicates a better alignment between two screens, leading to higher communication quality. When the images are captured from a view angle, the distances between the pixels on screen and the focal point of the camera are different. As a result, the size of the blocks far from the focal point expands in the captured image, while the size of those near the focal point shrinks, making it more difficult to extract the barcode. As the sender and receiver of COBRA are in the proximity to each other, we expect a typical view angle to be smaller than 45 degree in real scenarios. As shown in Fig. 18, when the view angle is larger than 30 degree, the minimum decoding rate falls below 90%. When the view angle is larger than 45 degree, it causes highly variable decoding rate. An interesting observation is that, larger block size mitigates the impact of view angle. For instance, when block size is 8, decoding rate ranges from 68.3% to 98.1%, compared with 89.5% to 99.5% with a block size of 11. The similar results are observed for system throughput in Fig. 21.

Next, we evaluate the impact of screen orientation on the captured images. The orientation directly affects the size and effective resolution of barcode in the captured image. A 90 degree orientation can reduce the size of captured barcode image by up to 50%. As shown in Fig. 22, the decoding

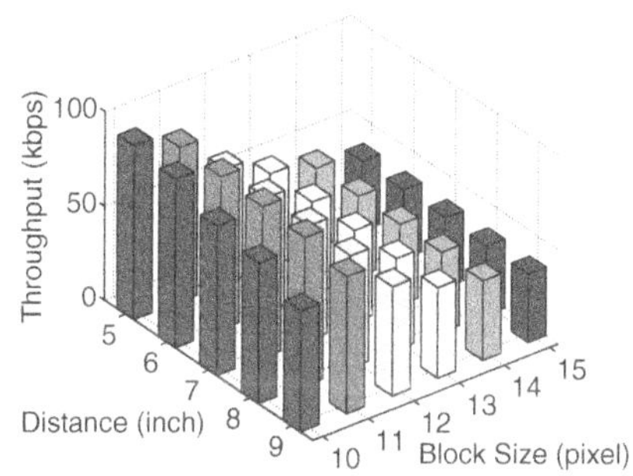

Figure 27: The throughput for different block sizes and distances

rate decreases as the angle of orientation increases. We can observe that when the orientation angle is larger than 30 degree, the barcodes with relatively small block size yield very low (less than 70%) decoding rate. Fig. 23 shows the throughput over different orientation angles. We can see that the throughput decreases as the orientation angle increases, and COBRA provides the optimal throughput when the block size is 11.

Lastly, we examine the impact of distance between sender and receiver for different block sizes. COBRA is designed for scenarios where sender and receiver are in proximity (< 1 foot). As shown in Fig. 27, 10-pixel block size provides the highest throughput (92.2 and 89.3 *Kbps*) when the distance is 5 and 6 inches, respectively. However, when the distance is longer than 6 inches, the throughput with 10-pixel block size is lower than that with 11-pixel block size. This is mainly because longer distance leads to smaller barcode size in the captured images. These results show that COBRA can control the effective communication range by varying the block size. This feature is desirable for near field communication scenarios with stringent security requirements.

6.5 Ambient Lighting, Screen Brightness, and Camera Resolution

As COBRA employs visible light as communication medium, its performance is naturally affected by ambient lighting condition and screen brightness. To evaluate the impact of ambient light, we choose 5 different lighting conditions ranging from completely dark to normal indoor office environments. For each condition, the light intensity is measured by a light sensor on TeloB mote. The experiments are carried out under sender screen brightness of 48 *lux* with block size of 7. The CDF of the decoding rate is shown in Fig. 24. We can observe that the decoding rate with the ambient light intensity of 1 and 20 *lux* is significantly lower than those with 50 *lux* or above. We also investigate the impact of screen brightness of Google Nexus S, which ranges from minimum 10 to maximum 115 *lux*. The experiments are carried out at night (20 *lux*) using block size of 7. Fig. 25 shows less than 97% of the total barcodes have 100% decoding rate when the screen brightness is lower than 30 *lux*. Another observation is that when the screen brightness is higher than 48 *lux* (similar to the default Android setting), it does not obviously affects the decoding rate.

The resolution of camera directly affects the quality of the captured images. We choose three image resolutions and evaluate their impact on the decoding rate using block size of 11. The CDF of decoding rate is shown in Fig. 26. As expected, the decoding rate increases with camera resolution, because each block in the barcode contains more pixels in higher resolution images making it easier to extract. Moreover, the 1280 (High Definition) resolution yields significant performance gain over 800 resolution. As more mainstream smartphones are shipped with HD cameras, COBRA can take advantage of the higher resolutions to improve system throughput.

	Component	Execution time
COBRA	Blur assessment	0.05 ms
	HSV enhancement	12 ms
	Corner detection (normal)	4 ms
	Corner detection (fast)	0.1 ms
	Timing reference detection	0.05 ms
	Code scan	2 ms
PixNet	Demodulation	**10 ms** (native)
		194 ms (Java)

Table 1: Running time of COBRA and PixNet.

6.6 Comparison with PixNet

In this section, we compare the performance of COBRA and PixNet [15]. PixNet is a VLC system primarily designed for communication between large LCDs and high-end digital cameras. PixNet encodes data in the frequency domain by adopting a 2D OFDM modulation scheme. The main components of Pixnet include a perspective correction algorithm, blur-adaptive OFDM coding, and ambient light filter.

As we were not able to obtain the source code from authors of PixNet, we implement a simplified Android version of PixNet that contains only OFDM modulator and demodulator while the perspective correction algorithm and ambient light filter are not implemented. For a fair comparison, we use two perfectly aligned Google Nexus phones that are 5-inch apart (i.e., zero view angle), and conduct experiments under mild ambient lighting conditions. Without the impact of perspective distortion, the decoding errors at receiver are thus mainly attributed to the blur of captured images. Nevertheless, our experimental results identify fundamental design principles of the OFDM algorithm that make PixNet ill-suited for screen-camera communication on smartphones.

Due to the small screen pixel size of and poor image quality of inexpensive phone camera, we have to increase the OFDM symbol size of PixNet in order to achieve acceptable link quality. In our implementation, we adopt a 97 by 97 OFDM symbol, and map it to a 388 by 388 block on the screen. Two OFDM symbols can be transmitted in a frame of 800 by 480 pixels. The maximum capacity of a single frame is 18.8 K bits. For both PixNet and COBRA, the frames are transmitted at 15 *fps* and recorded at 30 *fps*.

Our initial tests show that PixNet experiences excessive delay on Android Java VM due to the expensive 2D-FFT operations in OFDM modulation and demodulation. We then optimize the code using native Android C functions that contain a fast FFT library called KissFFT. It achieves 10X speedup over the pure Java FFT library in our experiments. As COBRA requires no expensive operations, it is implemented in Java.

We first compare the CPU overhead of COBRA and PixNet on Google Nexus S that has a 1 GHz CPU and 512M memory. The CPU execution time of a code module is measured as

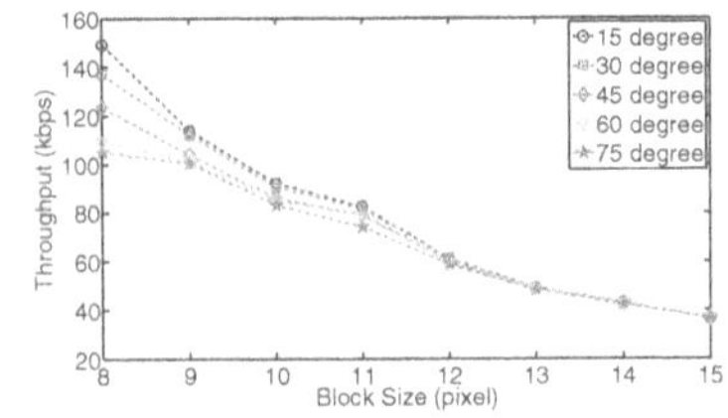

Figure 21: Throughput with different view angles.

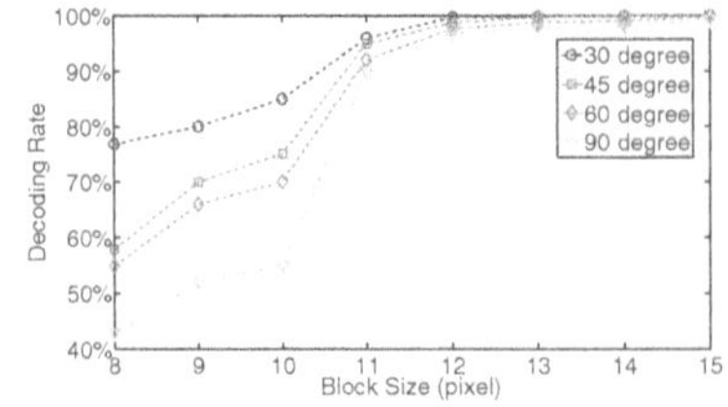

Figure 22: Decoding rate of barcode with different block size in different orientation

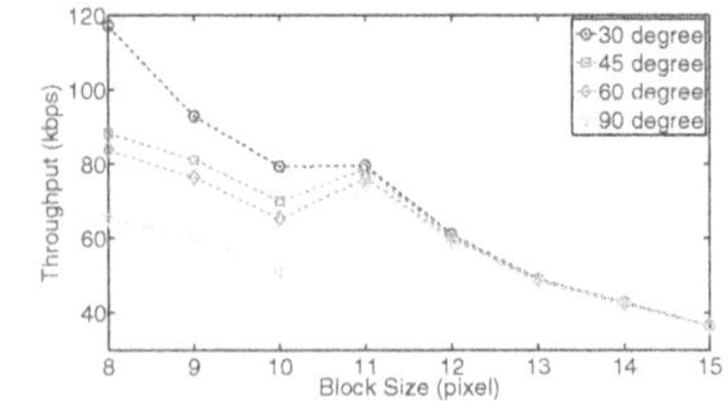

Figure 23: Throughput of barcode with different block size in different orientation

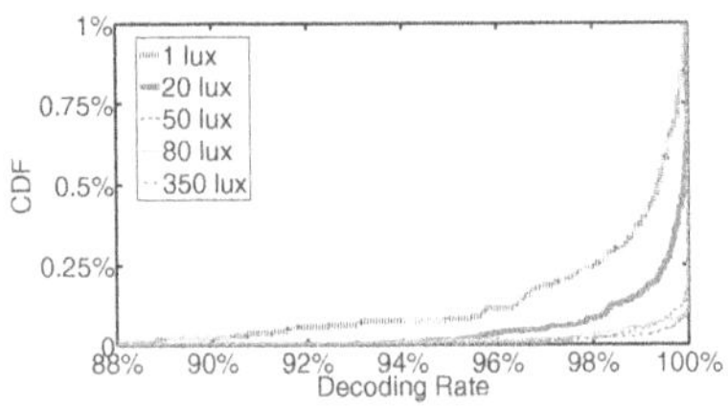

Figure 24: CDF of decoding rate under different ambient lighting conditions that range from completely dark indoor to daylight outdoor environments.

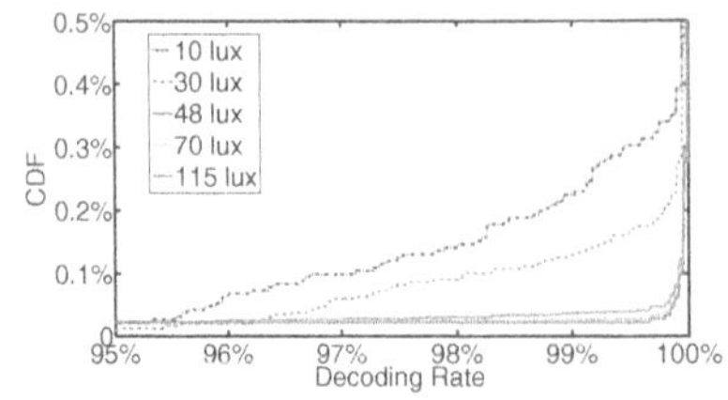

Figure 25: CDF of decoding rate with different screen brightness.

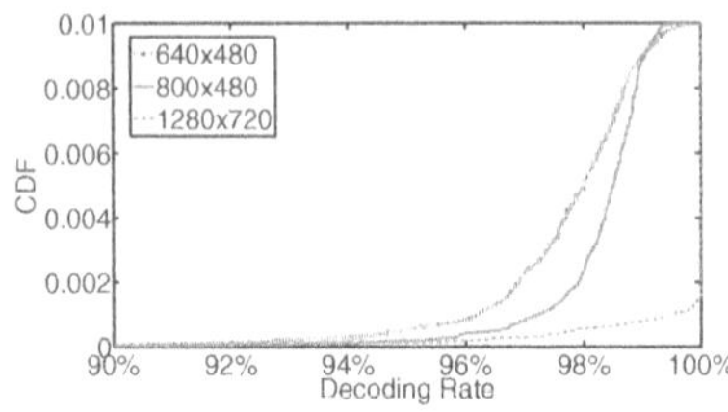

Figure 26: CDF of decoding rate with different image resolutions.

Setting		Correct Frames	Throughput
COBRA	block size 5 to 10	64% to 98%	91 to 172 Kbps
PixNet	3/7 coding	12.4%	6.4 Kbps

Table 2: Communication throughput of COBRA and PixNet

the elapsed time using function System.nanoTime()[2]. Note that we only measure the overhead of decoding of PixNet. The preprocessing procedures such as perspective correction would introduce additional overhead. Table 1 summaries the results. As the CPU overhead of sender is substantially smaller than receiver, we only show the results of receiver. The demodulation times of C and Java implementations of PixNet are 10 and 194 *ms*, respectively. This result shows that demodulating PixNet frames is very computational intensive on smartphones. On low-end smartphones with lower CPU speed, the frame decoding time can easily exceed 30 *ms*, causing frame loss on 30 *fps* camera. In contrast, although implemented using standard Java functions, COBRA can finish a frame decoding in less than 20 *ms*.

We also evaluate the throughput of COBRA and PixNet. Our initial tests show that PixNet suffers significant error rate as it is designed for unreliable one-way communication. To improve communication reliability, we extend the design of PixNet by adding reed-solomon error correction code in each frame. We tested various coding rates and found the 3/7 coding rate yields the best performance. Nevertheless, Table. 2 shows that, only 12.4% of the total frames can be correctly recovered by PixNet while all other frames fail the CRC check. The resulted throughput is 6.4 Kbps. In contrast, even without using any error correction code, COBRA correctly decodes 64% to 98% of total frames, leading to 91 to 172 Kbps throughput.

[2]The measured time may be higher than the actual execution time due to interruptions of Android system processes.

We believe the significant performance gap between PixNet and COBRA is attributed to the fundamental differences in their design principles. PixNet solely relies on the brightness information of LCD screen to encode and transmit data. The OFDM modulation method generates symbols composed of pixels with multi-level brightness. Due to the small pixel size and low resolution of smartphone screen, two adjacent pixels in a OFDM symbol may have very similar brightness in a blurred image, resulting in significant decoding errors. COBRA adopts fundamentally different techniques specially geared toward smartphones, including using distinctive colors to code data, optimizing the size/layout of code blocks, and restoring distorted colors using HSV color space. Because of the high error rates, PixNet is only practical for scenarios that require very limited bandwidth, such as wireless payments, authentication and key exchange before activating WiFi/Bluetooth radios. In contrast, COBRA achieves significantly higher throughput and reliability, enabling a wider range of applications such as mobile advertisements and data exchange between smartphone users.

7. CONCLUSION AND FUTURE WORK

This paper presents the design and implementation of a novel VLC communication system called COBRA for off-the-shelf smartphones. COBRA encodes information into specially designed 2D color barcodes and streams them between screen and camera of smartphones. We develop a new 2D color barcode that is optimized for streaming between small-size screen and low-speed camera of smartphones. COBRA adapts the size and layout of code blocks in streamed barcodes to deal with the significant image blur in mobile environments, and adopts new image processing techniques to achieve real-time barcode stream decoding. Our experimental results on Android smartphones validate the effectiveness of our approach.

COBRA is mainly designed for "asymmetric" communication scenarios such as advertisements where most or all the data flows along a particular direction. Without the real-time feedback from receiver, it is challenging to realize communication reliability mechanisms such as selective retransmissions. Such feedback, however, typically does not require high communication bandwidth and hence may be implemented using front-facing cameras or even ambient light sensors. In the future, we will extend COBRA to support feedback from the receiver. A challenge identified in our preliminary experiments is that, when two phones are placed face to face, a refreshing screen may cause reflection from the other screen causing significant noise to the front-facing camera. We will investigate noise canceling techniques that can address this issue.

The primary design objective of COBRA is to maximize the throughput of short-duration communications between two smartphones. Although high throughput brings energy benefits, energy efficiency is not explicitly considered in this work. We plan to optimize the energy efficiency of COBRA based on two approaches. Many organic light-emitting diode (OLED) smartphone screens are inefficient for displaying bright colors. We will account for such inefficiency by optimizing the color mapping in our barcode design. Moreover, we will investigate time synchronization techniques that allow two phones to synchronize screen and camera operations. As a result, power consumption can be reduced due to the lower requirements on screen refreshing and camera recording rates.

8. ACKNOWLEDGEMENTS

This research was supported in part by the U.S. National Science Foundation under grants CNS-0954039 (CAREER). We thank the anonymous reviewers and the paper shepherd Marco Gruteser for their valuable feedbacks.

9. REFERENCES

[1] I. 15420:2009. *Information technology - Automatic identification and data capture techniques - EAN/UPC bar code symbology specification.*

[2] I. 16022:2006. *Automatic identification and data capture techniques - Data Matrix bar code symbology specification.*

[3] I. 18004:2006. *Automatic identification and data capture techniques - QR code 2005 bar code symbology specification.*

[4] A. Ashok, M. Gruteser, N. Mandayam, T. Kwon, W. Yuan, M. Varga, and K. Dana. Rate adaptation in visual mimo. In *Communications Society Conference on Sensor, Mesh, and Ad Hoc Communications and Networks*, 2011.

[5] A. Ashok, M. Gruteser, N. Mandayam, J. Silva, M. Varga, and K. Dana. Challenge: Mobile optical networks through visual mimo. In *Proceedings of the sixteenth annual international conference on Mobile computing and networking*. ACM, 2010.

[6] D. Cardani. Adventures in hsv space. *Laboratorio de Robótica, Instituto Tecnológico Autónomo de México*, 2001.

[7] H. Elgala, R. Mesleh, and H. Haas. Indoor broadcasting via white leds and ofdm. *IEEE Transactions on Consumer Electronics*, 2009.

[8] S. Hranilovic and F. Kschischang. A pixelated mimo wireless optical communication system. *Selected Topics in Quantum Electronics, IEEE Journal of*, 2006.

[9] T. Langlotz and O. Bimber. Unsynchronized 4d barcodes: coding and decoding time-multiplexed 2d colorcodes. In *Proceedings of the 3rd international conference on Advances in visual computing-Volume Part I*. Springer-Verlag, 2007.

[10] D. Lowe. Distinctive image features from scale-invariant keypoints. *International journal of computer vision*, 60(2):91–110, 2004.

[11] M. Nakagawa. *The world of visible optical communication - It opens with LED - "Light Communications"*. Kogyo Chosakai Publishing, Inc.

[12] E. Ohbuchi, H. Hanaizumi, and L. Hock. Barcode readers using the camera device in mobile phones. In *Cyberworlds, 2004 International Conference on*. IEEE, 2004.

[13] G. Pang, T. Kwan, H. Liu, and C. Chan. Led wireless. *Industry Applications Magazine, IEEE*, 2002.

[14] D. Parikh and G. Jancke. Localization and segmentation of a 2d high capacity color barcode. In *Applications of Computer Vision, 2008*. IEEE, 2008.

[15] S. Perli, N. Ahmed, and D. Katabi. Pixnet: interference-free wireless links using lcd-camera pairs. In *Proceedings of the sixteenth annual international conference on Mobile computing and networking*. ACM, 2010.

[16] W. Yuan, K. Dana, M. Varga, A. Ashok, M. Gruteser, and N. Mandayam. Computer vision methods for visual mimo optical system. In *Computer Vision and Pattern Recognition Workshops (CVPRW), 2011 IEEE Computer Society Conference on*, 2011.

BLINK: A High Throughput Link Layer for Backscatter Communication

Pengyu Zhang, Jeremy Gummeson, Deepak Ganesan
Department of Computer Science
University of Massachusetts, Amherst, MA 01003
{pyzhang, gummeson, dganesan}@cs.umass.edu

ABSTRACT

Backscatter communication offers an ultra-low power alternative to active radios in urban sensing deployments — communication is powered by a reader, thereby making it virtually "free". While backscatter communication has largely been used for extremely small amounts of data transfer (e.g. a 12 byte EPC identifier from an RFID tag), sensors need to use backscatter for continuous and high-volume sensor data transfer. To address this need, we describe a novel link layer that exploits unique characteristics of backscatter communication to optimize throughput. Our system offers several optimizations including 1) understanding of multi-path self-interference characteristics and link metrics that capture these characteristics, 2) design of novel mobility-aware probing techniques that use backscatter link signatures to determine when to probe the channel, 3) bitrate selection algorithms that use link metrics to determine the optimal bitrate, and 4) channel selection mechanism that optimize throughput while remaining compliant within FCC regulations. Our results show upto 3× increase in goodput over other mechanisms across a wide range of channel conditions, scales, and mobility scenarios.

Categories and Subject Descriptors

C.2.1 [**Network Architecture and Design**]: Wireless communication

General Terms

Design, Experiment, Measurement, Performance

Keywords

Backscatter communication, Mobility detection, Rate adaptation, Channel switching

MobiSys'12, June 25–29, 2012, Low Wood Bay, Lake District, UK.

1. INTRODUCTION

An emerging class of applications require small form-factor, ultra-low power sensors that are ubiquitously deployed in ways presently difficult to achieve with conventional wireless sensor technologies. These devices may be deployed in areas where small form-factor is critical (e.g. on-body sensors for gesture and activity recognition), maintenance is difficult or impossible (e.g. embedded monitoring of bridges and planes), or on a host of small everyday objects (Internet of Things).

An acute problem for this class of sensor devices is operating under extremely low energy budgets. The sizes of batteries or capacitors have to be tiny, necessitating new designs that can operate at orders of magnitude lower power than typical mote-class sensor platforms. While micro-controllers have become increasingly efficient in terms of their active and sleep-mode power consumption, active radios such as 802.15.4 still consume too much power. There is therefore a dire need for an ultra-low power communication mechanism for this class of sensors.

One paradigm that promises to dramatically reduce energy cost is backscatter communication. A backscatter base-station provides a carrier wave to the sensor node, which can decode these transmissions using a simple analog comparator circuit. To transmit data, a sensor toggles the state of a transistor to detune its antenna and "reflects" the carrier wave back to the reader with its own information bits. Since a sensor does not need to actively transmit a radio signal, this can lead to extremely low-power communication. In addition, backscatter requires low-complexity circuits which can be implemented at low-cost. Despite these benefits, widespread use of backscatter communication for sensors has been stymied by two drawbacks: a) it is restricted to short-range, single-hop communication, and b) RFID readers are bulky and difficult to deploy in a dense manner.

Recent developments, however, have mitigated these drawbacks and renewed the case for wider use of backscatter communication on sensors. The first development is dramatic increases in communication range by enabling tags to take advantage of alternate energy sources, for example tiny solar panels [15, 5]. The second development is miniaturization in RFID reader hardware leading to small portable readers that can be attached to a mobile phone [2]. Thus, backscatter communication offers a considerably more energy-efficient and increasingly practical alternative to active radio circuits on existing sensor systems.

Despite the possibilities, we have limited understanding of how a backscatter link layer should be optimized for data transfer from sensor devices. Our focus in this work is on two central elements in such a link layer — bit-rate adaptation and channel selection. While these topics have been widely studied in systems that employ active radios (e.g. 802.11 [9, 21, 23, 27, 28] and 802.15.4 [24]), the applicability of prior results to our problem is limited. Unlike a conventional active radio link, a backscatter link comprises both the forward link from reader to sensor and the backward link from the sensor to reader, since the sensor is essentially "reflecting" the carrier wave from the reader while modulating with data. This difference results in different path loss for the forward vs backward link, and unique multipath self-interference behavior. In addition, the highly asymmetric hardware capabilities of the reader and the sensor tag imply that modulation and coding schemes for the forward and reverse link are very different.

In this paper, we explore the design of a high-throughput **link layer** for sensors that use backscatter communication. Our work provides novel insights into self-interference in backscatter communication and shows that a combination of RSSI and packetloss metrics are required to capture link characteristics. We use these metrics to design a novel link layer with three fundamental contributions. First, we design a mobility-aware link layer that reduces channel probing overhead by comparing link signatures over time to detect sensor mobility. Second, we present a classifier that uses RSSI and packetloss information to select the optimal bitrate for data transfer. Third, we design channel selection mechanism that optimizes the use of channels for communication while remaining within FCC specifications. Our results show that:

- We achieve less than 10% false positive rate and less than 1% false negative rate for detecting sensor tag mobility across a range of bitrates.
- Our bitrate classifier selects the optimal bitrate with over 80% accuracy under most conditions, and achieves over 88% of the optimal goodput even under heavy multipath scenarios.
- We show that channel selection and switching improve goodput by over 2× in comparison with default schemes used by readers.
- BLINK scales well to large numbers of static and mobile tags. In a deployment with 20 static tags, BLINK is 1.6× better than the default scheme used by Impinj readers (AutoSet) and 1.3× better than a backscatter-optimized version of SampleRate. In a deployment with 10 mobile tags, BLINK is 2.4× better than AutoSet and 2× better than SampleRate.

2. BACKGROUND

This section provides an overview of the physical layer defined by EPC Gen 2, a dominant standard for backscatter based communication. Our focus is on mechanisms that are applicable to commercial readers, since they are readily available and in widespread use.

2.1 RF Harvesting

Backscatter communication is designed to both provide power to a sensor tag as well as to enable communication. As shown in Figure 1, the RFID reader provides a carrier

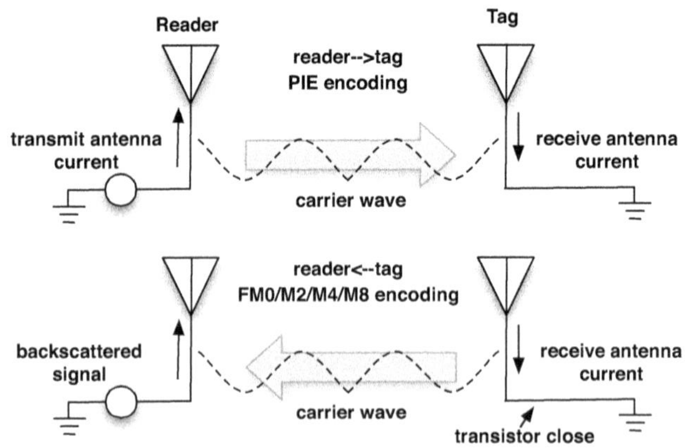

Figure 1: Backscatter signaling at PHY [3]

wave, which can be rectified by the sensor to produce DC voltage. This voltage is boosted to an appropriate level by a charge pump at the sensor, and accumulated in a small storage capacitor until the voltage reaches an appropriate threshold before any computation (or sensing) can begin. Once voltage is sufficient to power the device, it can begin to receive and transmit data, both of which are done by modulating the same carrier wave. The RF harvester at the sensor includes an analog comparator circuit that can decode reader transmissions from the reader. To transmit data, a sensor toggles the state of a transistor to detune its antenna and reflect the carrier wave back to the reader with its own information bits. The Intel WISP [6] and UMass Moo [1] are examples of sensor platforms that rely on backscatter communication.

2.2 Forward vs Backward link

The forward and backward links in backscatter communication differ in several ways. First, the path loss is very different for the two links. The signal to noise ratio (SNR) for typical backscatter communication decays with the square of distance for the forward link and to the fourth power of distance for the backscatter link. Second, the encoding schemes for the links are different — reader to sensor communications use pulse-interval encoding (PIE), which allows easy decoding, whereas sensor to reader communication uses more complex encodings (FM0, Miller2, Miller4, Miller8). Third, the antenna sensitivity at the sensor and reader are vastly different. A typical RFID reader (e.g. Impinj [4]) uses a mono-static antenna for sending and receiving data, which has a sensitivity of -80 dBm. In contrast, an RFID-scale sensor (e.g. the Intel WISP [6]) uses a simple dipole antenna for data transfer, which is significantly less sensitive than the reader antenna. These factors contribute to different link qualities in the two directions. The forward link uses weaker encoding and is received by a less sensitive antenna, but has lower path loss. The backward link uses stronger encoding and is received by a highly sensitive antenna, but has much higher path loss.

Our focus in this paper is on optimizing the backward link from the sensor tag to the reader. We make this decision for two reasons: 1) the forward link offers almost no choices - only one encoding scheme (PIE) is supported and no baud rate option is clearly proposed, whereas sensor to reader communication uses more complex encodings (FM0, Miller2, Miller4, Miller8) and several baud rates (32 kbps to 640 kbps) since the reader has more computational resources for decoding, and 2) the backward link is crucial since it has greater path loss and unique multipath self-interference behavior, and is therefore very sensitive to sensor placement relative to the reader (discussed in greater detail in §4.1).

2.3 Backscatter Channels

RFID readers use frequency hopping to avoid interference across readers when reading sensors in the same area. A typical UHF reader hops between 50 channels in the 902MHz ~ 928MHz ISM band. FCC regulations specify that a reader can have a maximum channel dwell time of 0.4 seconds in any ten second period to reduce interference in a channel [11]. Commercial reader implementations address this by spending an equal amount of time in each of the 50 channels (0.2 secs) and hopping in sequence from the first to last channel. As we show in §4.4, this is inefficient, and more intelligent choice of channels while remaining within FCC specifications can improve throughput.

3. SYSTEM OVERVIEW

Figure 2 shows the BLINK link layer architecture. At the core of BLINK are link metrics that capture path loss and multipath fading characteristics of a backscatter link. We show that backscatter communication has unique characteristics: RSSI is a better measure of path loss and packet loss rates are a better measure of multipath fading. These metrics are used by a mobility detector that detects changes in the mobility patterns of a sensor tag and triggers bitrate and channel adaptation. The central idea of the mobility detector is to compare link signatures over time to identify whether a sensor has moved from one location to another, or whether it is in continuous motion.

When mobility is detected, the rate adaptation module needs to choose a new bitrate that best suits the channel characteristics. This is done by using a classifier that maps from the link metrics (RSSI and packetloss vectors) to the appropriate bitrate. An additional challenge in rate adaptation is that the link metrics need to be obtained at the lowest bitrate to ensure that sensors are not missed, but obtaining these metrics across 50 channels is exceedingly slow and reduces goodput. The rate adaptation module therefore uses a fast probe technique that exploits loss patterns on backscatter channels, as well as knowledge of the feature needed for the classifier to optimize probing duration.

Mobility also triggers channel selection; this module is responsible for selecting channels that maximize throughput. We take advantage of flexibility in FCC regulations that allow the use of fewer than 50 channels while ensuring per-channel dwell times are under 0.4s. This module measures channel characteristics such as the burstiness and sharpness of transitions between good to bad channels, and uses this information to decide whether to select the best channels a priori, or whether to dynamically switch between them to take advantage of bursts of good throughput.

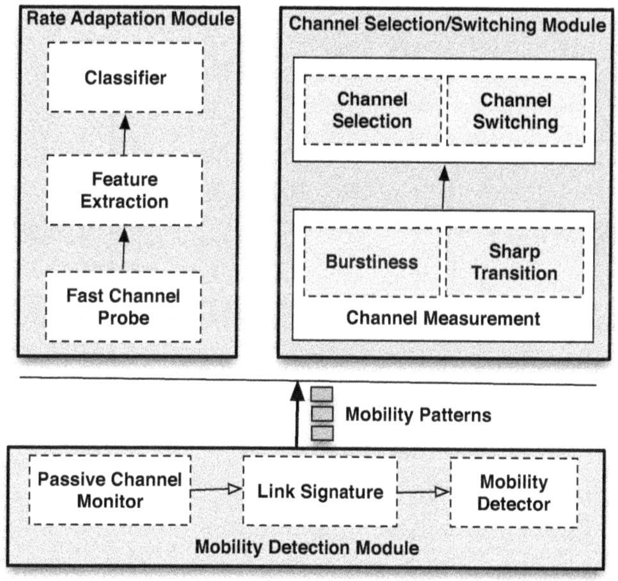

Figure 2: System overview.

4. BLINK DESIGN

This section describes the key design elements of BLINK. We first present the link metrics that are the foundation of our link layer, and then discuss how each module in our system takes advantage of the link metric to detect mobility, select optimal bitrate, and select the best channels to use.

4.1 Backscatter Link Metrics

Commercial RFID readers expose two link metrics: a) the RSSI value for each query response from a sensor tag, and b) the aggregate per-channel loss rate for each dwell time interval. In a traditional wireless networks that use active radios such as 802.11 or 802.15.4, RSSI and loss rate are strongly correlated. As a consequence, most link-layer metrics rely on either fine-grained RSSI information or coarse-grained packet loss rates (e.g. [21, 23]). However, a unique feature of backscatter communication is that packet loss and RSSI provide complementary information about path-loss and self-interference, and therefore need to be used in conjunction. To understand why, let us look at the nature of multipath interference in backscatter systems.

Backscatter multipath: Multi-path phenomena is often caused by the transmitted signal being reflected by objects like walls and buildings. As a result, the receiver will detect multiple copies of the same signal which traverse different paths. The summation of these signals distorts the original shape of transmitted signal resulting in pockets of constructive interference where RSSI is high and destructive interference where RSSI is low. In addition, the summation of signals is sensitive to frequency and leads to frequency selective signal distortion.

Multipath effects are a bit different for backscatter communication. Consider the case shown in Figure 3 where the reader sends out a signal S1 and the sensor tag backscatters with signal S2 on the same carrier wave. In addition to the backscattered signal from the sensor, a wall-reflected multipath signal, $S1^{'}$, is also received at the reader. Thus, the backscatter signal received by the reader is the summation of two signals: wall-reflected multipath signal $S1^{'}$ and sensor-backscattered signal S2. $S1^{'}$ could be seen as carrier wave modulated by information from the reader and S2 is carrier wave modulated by information from the sensor. As a re-

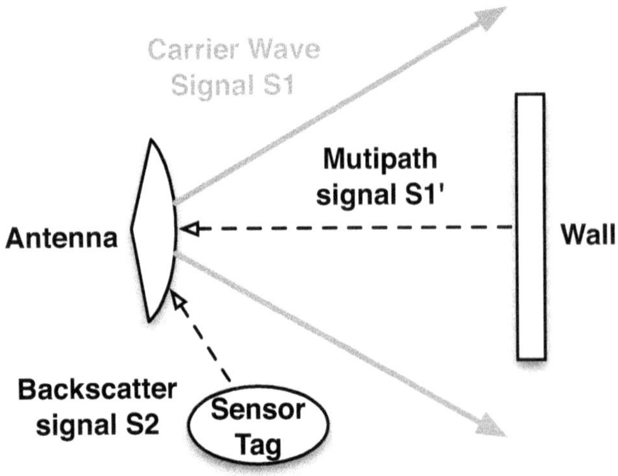

Figure 3: Multipath self-interference when tag is placed at the edge of reader beamformer direction.

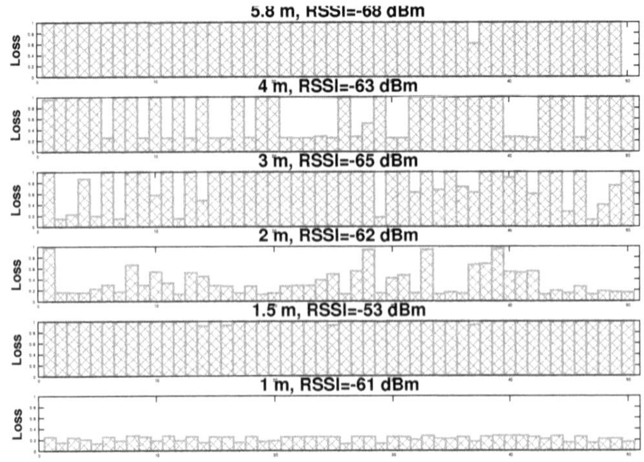

Figure 4: Empirical results about multipath self-interference.

sult, even if constructive interference happens between the two signals , it may not be decoded correctly at the reader since $S1'$ and S2 have different information bits. In other words, RSSI may be high due to constructive interference but packet loss would be high as well.

To contrast against a traditional communication scenario, assume that the sensor tag in Figure 3 is equipped with an active radio. In this case, its transmission signal, S2, to the base-station might interfere with a multipath version of the signal $S2'$ caused by reflection from the wall. Unlike backscatter multipath, copies of the same signal (S2 and $S2'$) interfere with each other, and techniques such as equalization can handle the offset between the signals.

This phenomena is exacerbated by RFID placement relative to the reader's antenna beamformer. When a sensor tag is placed within the beamformer direction of the reader antenna, then the signal strength of the backscattered signal (S2) is typically high enough that it can be decoded despite multipath effects. However, if the sensor is placed outside the primary beamformer direction, then the signal that it receives is weak and the backscattered signal (S2) may not be strong enough to overcome the interference from $S1'$.

Empirical evidence: To provide empirical evidence of the above behavior, we use a passive tag to measure the packet loss rate on each channel for a tag placed at the edge of the antenna beamformer. Figure 4 shows RSSI and corresponding packet loss rate across the 50 channels when the tag is placed at different locations. When the tag is 1 meter from reader, it is within the beam of the reader antenna. As a result, the reader obtains high RSSI and low loss rate. Moving the tag to 1.5m results in a sharp improvement in RSSI, but packet losses become severe. This point clearly shows the effect of multi-path self interference — RSSI is high, which would suggest excellent channel quality, but packet loss rate is high as well due to multipath. The multipath effect reduces when the tag moves a bit more to 2m. Beyond this point, we do not observe self-interference and link quality degrades predictably with distance.

Implication: Why does this distinction between RSSI and packet loss matter? The result shows that we need to consider both RSSI and packet loss to optimize communication on backscatter channels. Together, they capture the effect of path loss and multipath self-interference.

4.2 Mobility Detection

The mobility detection module generates triggers whenever it believes that the sensor tag's location or mobility pattern has changed. Detecting the mobility pattern of a sensor is important for a reader to decide how to interact with the sensor. When a sensor has moved, a reader may need to change the encoding, baudrate, or channels to maximize throughput [22]. In this section, we introduce a zero-overhead approach to determine mobility behavior of sensor tags.

At the core of mobility detection is the notion of a backscatter link signature. The link signature is defined as the distance between the RSSI vectors and lossrate vectors across successive scans of all the channels. Here, the RSSI vector comprises a vector of the RSSI values on each of the N channels that an RFID reader uses to communicate with a sensor, and the lossrate vector is the loss rate across the same channels. We denote the RSSI vector for location A to be $(a_1, a_2, ..., a_N)$ and the packet loss vector to be $(a_1^*, a_2^*, ..., a_N^*)$

We use a simple euclidean distance metric to measure the distance between the vectors for successive scans. The distance d between successive RSSI vectors, $(a_1, a_2, ..., a_N)$ and $(b_1, b_2, ..., b_N)$ is defined as:

$$d = \frac{\sqrt{\sum_{i=1}^{N}(a_i - b_i)^2}}{N}$$

The distance between two packet loss vectors d^* can also be calculated in the above manner. Given the two distances, the mobility detector identifies if a sensor is static or mobile as follows:

$$\text{Output} = \begin{cases} \text{Static} & \text{if } d < d_T \ \ \& \ \ d^* < d_T^* \\ \text{Mobile} & \text{if } d \geq d_T \ \ || \ \ d^* \geq d_T^* \end{cases}$$

where d_T and d_T^* are empirically measured thresholds (see §5). The rationale for generating mobility triggers when either distance was larger than the threshold was because

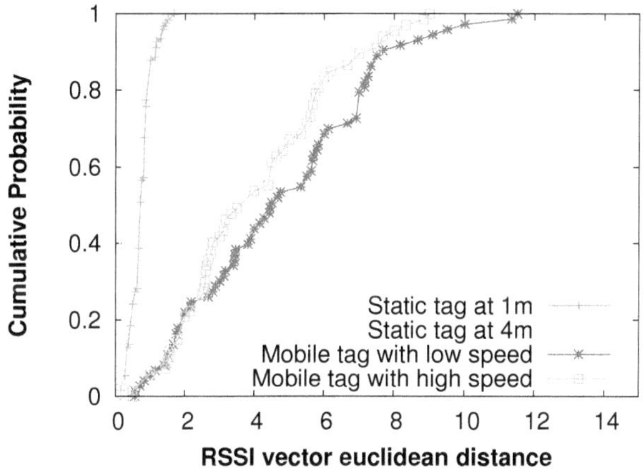

Figure 5: RSSI euclidean distance when tag is stationary and mobile

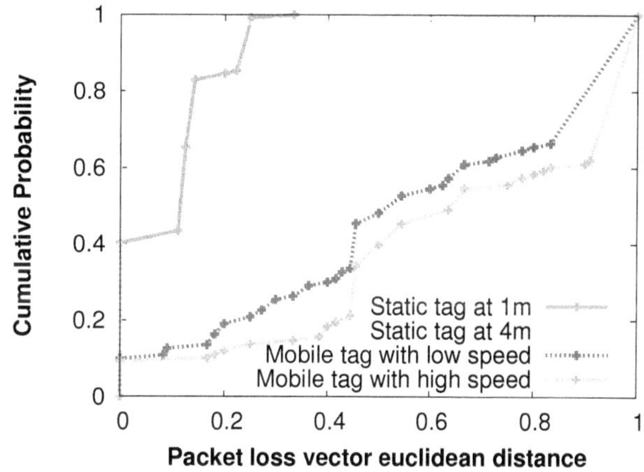

Figure 6: Packet loss euclidean distance when tag is stationary and mobile

we were willing to sacrifice a few false alarms, but wanted to ensure that mobility scenarios were not missed.

Note that changes in the environment such as movement of metal objects in the vicinity of the sensor could be misinterpreted by the mobility detector as a change in location of the sensor since the link signature might have changed. These triggers are, in fact, useful since environmental changes can impact link characteristics and therefore require adjusting communication parameters between the reader and the sensor.

Empirical evidence: To provide an empirical comparison of link signatures for stationary and mobile sensor tags, we use a passive tag to measure link signatures in two cases. The first is when the tag is stationary and placed at two spots that are at different distances from the reader. We use an Impinj reader and the fastest bitrate, FM0/640, for capturing link signatures. The second case is when a tag is mobile at different speeds. To ensure repeatable mobility experiments, we use a LEGO toy train on an oval track which was mounted with a tag. The train was run at different speeds to measure the link signatures under mobility. In each of these experiments, we get a sequence of link signatures, one for each consecutive pair of reader scans.

The distance between RSSI vectors are shown in Figure 5 and the distance between packet loss vectors are shown in Figure 6. The graphs clearly show that it is straightforward to distinguish between the mobile and stationary case since they have vastly different signatures. Even while considering the cases within one of these groups, there is sufficient difference in the signature that we can detect changes in location, or mobility speeds. In §5, we provide a more in-depth breakdown to show that this is indeed possible.

4.3 Rate adaptation

The rate adaptation module exploits the link metrics (RSSI and packet loss vectors) that it obtains from the probes to determine the bitrate that would achieve highest goodput for communication with a reader. As shown in Table 1, an EPC Gen 2 reader specifies six bitrates that are a result of different encoding/baudrate combinations. Our goal is to select the best of these options.

Bitrate (symbol/s)	Throughput (kbps)
FM0/640	640
FM0/160	160
Miller4/640	160
Miller4/256	64
FM0/40	40
Miller8/256	32

Table 1: Ranked encoding and baud rate combinations based on throughput.

4.3.1 *RSSI/Packet loss → Encoding/Baudrate*

RSSI and packet loss rate are two dominant factors that affect the performance of backscatter system. This intuition for how these link metrics impact bitrate is shown in the RSSI-lossrate map in Figure 7. Consider the first row where RSSI goes from high to low and packet loss is low. As we showed earlier, RSSI in backscatter communication is primarily impacted by pathloss and not multipath effects. Thus, as RSSI reduces, the maximum throughput of the channel reduces and a bitrate that leads to a lower throughout should be chosen. The optimal choice of bitrate follows the order shown in Table 1 i.e. order of reducing throughput.

Now lets look at the case where packet loss goes from good to bad and RSSI is fixed at high. We know that packet loss rate shows the effect of multipath self-interference. Therefore, for a fixed RSSI value, increase in loss rate should lead to a choice of more advanced encoding schemes. Using a more advanced encoding would mean that more symbols are employed to transmit the same information, which in turn would mean more tolerance for signal distortion in the time domain. This would reduce the negative impact of multipath interference. This is shown in the first column in Figure 7 — loss rates increase, the choice of encoding shifts from FM0, which employs one high-low pulse to encode 1 bit of information, to Miller-4, which uses four high-low combinations, and finally Miller-8, which uses eight.

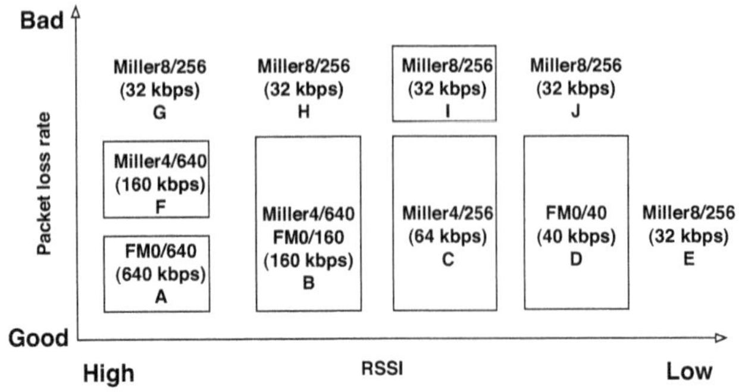

Figure 7: The link metrics map of encoding and baud rate for bit rate adaptation. When RSSI decreases, we should choose encoding/baud rate with lower throughput. When packet loss rate increases, we should increase the complexity of encoding to overcome interference.

4.3.2 *Classifier design*

The RSSI/packet loss to bitrate map shown in Figure 7 provides the intuition for the operation of a classifier, but doesn't provide the boundaries between classes. We learn these boundaries using training data collected from various indoor settings, and use a classifier to map from the current link metrics to the appropriate bitrate.

The classifier uses two features, the sorted list of RSSI across the channels and the sorted list of packet loss across the channels. We use the sorted list since we find that the shape of RSSI and packet loss is a better indicator of the optimal bitrate than using specific channel information. Given the RSSI and loss vectors, the classifier chooses the closest bitrate cluster among those shown in Figure 7. In other words, it chooses cluster i using the following expression:

$$\forall i \in \{A, B, ..., I\} \quad i : \min(|R_i - R|_{l_2} + |L_i - L|_{l_2})$$

where R and L are the current sorted RSSI vector and loss rate vectors, and R_i and L_i are the empirically-determined centroids of the cluster i from Figure 7.

4.3.3 *Mobility-aware channel probing*

An important question when designing a rate adaptation algorithm is the cost of probing in-order to obtain the link metrics across different channels. Unlike the mobility detector, which can passively monitor the RSSI and loss rate metrics at any bitrate that has been currently chosen by upper layers, the channel probe needs to use the simplest encoding and slowest baud rate. This is because of two reasons: 1) using a complex encoding and fast baudrate can mask channel errors, making it harder to estimate the channel accurately, and 2) higher bitrates reduce communication range, and can lead to misses of corner and far-away sensor tags. To give an estimate of the time taken for a channel probe, consider an Impinj reader at the slow FM0/40 setting. By default, the reader spends 100ms in each channel to transmit 7 queries. Since there are 50 channels, the total time for a channel probe is 5 seconds. This is clearly a large number, and results in significant loss of goodput and responsiveness. This section presents a fast probing mechanism that uses two key ideas: a) one-query channel probing, and b) random channel probing.

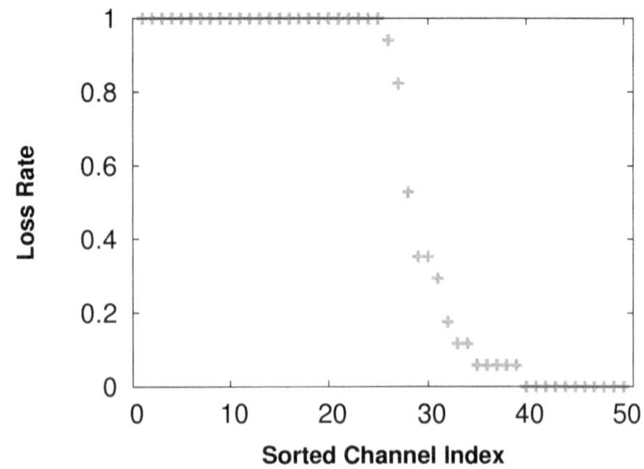

Figure 8: Short transition between "good" and "bad" channels. We plot sorted packet loss rate on 50 channels. Only 10% channels have loss rate between 80% and 20%. 90% channels are either consistently good (zero loss) or consistently bad (over 90% loss).

One-Query Channel Probing: How many queries in each channel are required to estimate its loss rate? The more queries it takes, the more expensive a channel probe. The key idea in this approach is that one query per channel may be enough since backscatter links have a sharp phase transition behavior (from "good" to "bad" channel quality).

To validate our approach, we look at the loss rate patterns across channels on the Impinj reader. Figure 8 shows the packet reception rates sorted from the highest to lowest across the 50 channels. We observe that there is a fairly sharp transition and that channels tend to be either consistently good (zero loss) or consistently bad (over 90% loss). Only 10% of the channels have loss rates between 20% and 80%. This behavior is typical of wireless channels — studies have shown that the transition between low to high loss rates is sharp across distance [31], and we observe that this is true across frequency as well.

The sharp transition makes probing more efficient — a successful packet on a channel is likely be followed by several continuous successful packets, and vice-versa. As result, we can use a single query probe in each channel to estimate channel loss rate. This simple change reduces the overall cost of probing by 7× from 5 seconds to 0.7 seconds.

Random channel probe: A single query per channel reduces probe time but 0.7 seconds is still a long duration, so we look at reducing this even further. This approach exploits the fact that the features used by the classifier are the *shape* of the RSSI and loss rate vectors. Thus, it is not essential to probe all the 50 channels, only to probe as many channels as are needed to accurately estimate the RSSI and loss vector shape. We exploit this idea to perform a random sampling of a subset of the 50 channels (in our case, 10), which reduces the channel probe time by another 5×.

4.4 Channel selection/switching module

The RFID reader hops among 50 channels in the $902MHz \sim 928MHz$ range for communicating with sensor tags. While FCC regulations prevent the reader from dwelling on a single channel for too long, there is some flexibility in how chan-

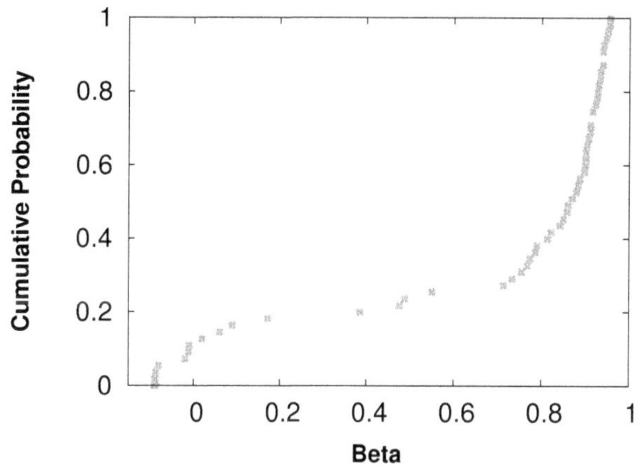

Figure 9: Distribution of β values for the intermediate links. Most links are highly bursty ($\beta > 0.8$) making the case for a fast switching algorithm

nels are chosen. FCC allows 0.4s channel usage within 10s, therefore, the minimal number of channels can be used is 25 rather than 50. When 25 channels are chosen, an RFID reader can stay on each channel for the maximum-allowed 0.4s duration.

Our channel selection module takes advantage of two characteristics of backscatter channels. The first is the sharp transition between "good" and "bad" channels as shown in Figure 8. Since a large number of channels tend to be consistently good or bad, a simple channel selection approach would be to select only the 25 channels that have least packet loss rate, and dwell in each for the maximum allowed duration, 0.4s.

The second characteristic of channels that we take advantage of is burstiness. Wireless channels are known to exhibit bursty loss patterns [24], and this behavior impacts performance of wireless protocols [7]. Channel burstiness has been shown to be mainly caused by RSSI and interference, but can also result from hardware or protocol choices, for example, an RFID reader might scan the field by changing beamformer orientation, resulting in bursty loss characteristics. Figure 9 plots the β metric defined in [24] for channels where we observe intermediate loss rates (i.e. neither 100% nor 0%) across several placements, where $\beta = 1$ means that link is bursty while $\beta = 0$ tells us that independent packet loss is observed. We clearly see that channels with intermediate loss rates tend to be bursty, with success and losses occurring in clusters.

The bursty nature of channels presents an alternate approach to selecting channels. Instead of choosing the top 25 channels, we can use *fast switching* to switch across channels. If a query transmission in a channel fails, the reader can immediately switch to another next channel. This mechanism also helps us deal with occasional loss bursts that we observe due to external interference, and lets us take advantage of intermediate links during bursts of good activity. To ensure that we meet FCC regulations, we keep a counter on the amount of time spent per channel and mark the channel as "used" when it completes the 0.4s dwell time quota.

The transition characteristics and burstiness can depend on the deployment, external interference, and also on the choice of reader hardware. To handle such variability, we use an online approach to measure the sharpness of transitions (fraction of intermediate channels) and the extent of burstiness (beta value). If the channel is not sufficiently bursty, we pick the top 25 channels but if there is high burstiness and sharp transitions, we use the fast switching approach.

5. IMPLEMENTATION

In this section, we describe key implementation details that are not covered in previous sections. Our implementation is entirely done on the reader-side and requires no modifications to the tags.

Gen 2 Reader: Our mobility-aware bitrate adaptation and channel selection protocol is designed to operate on commercial readers that support the EPC Gen 2 protocol. However, despite being compliant with EPC Gen 2, commercial readers often do not expose all the parameters that we need to tune bitrate and channels. On the Impinj reader, we experienced two limitations: a) changing the bitrate was only possible at the beginning of a round, and it took ~30ms to finish the configuration, which sacrificed some goodput, and b) the reader does not expose hooks that would allow us to select specific channels or change the order of switching between channels. To evaluate aspects of our link layer such as channel selection and switching that are not possible to implement on the Impinj reader, we use trace driven simulations using traces captured by a commercial reader.

USRP Reader: One option that we considered during our implementation is the use of a USRP software radio reader developed by Buettner et al [13] to evaluate our link layer. Clearly, the USRP reader gives us much greater flexibility in terms of evaluating our techniques, but, it also has several hardware limitations that make it hard to fully evaluate BLINK. For example, the USRP has a maximum range of 2.5 meters, making it difficult to observe the multipath interference behavior. In addition, compared to the commercial reader, the achievable goodput with a sensor tag was lower, transitions between good and bad channels were much less sharp, and burstiness was less evident. To avoid hardware artifacts from influencing our results, we use the commercial Impinj reader for almost all our experiments, and sparingly use the USRP reader in cases where the commercial reader does not expose necessary hooks.

Baseline: We compare BLINK against two baseline schemes: a) the default Impinj reader configuration, called "AutoSet", and b) a backscatter-optimized version of SampleRate [9], a widely used WiFi-based rate adaptation algorithm. AutoSet is a rate adaptation algorithm used by default on the Impinj Speedway RFID reader [12]. Although six bit rate configurations are available on the Impinj reader, AutoSet only utilizes three of these — FM0/320 kbps, Miller4/68 kbps, and Miller8/20 kbps. Among these, the Impinj reader uses Miller4/68 when the tag has good connectivity to the reader (close range), and FM0/320 when the tag has bad connectivity (long range). Miller8/20 kbps is used for the very last query to pick up any stragglers that have very poor connectivity to the reader.

We also compare BLINK with SampleRate [9] a commonly used rate adaptation algorithm for WiFi communication. SampleRate maintains a ranked list of bit-rates, based on the average per-packet transmission times observed at each rate. The highest rate on this list is used to transmit data, except every tenth data packet, which is transmitted at one of the other bit-rates and used as a channel probe. SampleRate stops using a bit rate if it experiences four successive packet losses.

A direct implementation of SampleRate on the Impinj reader turned out to be inefficient since the Impinj reader incurs significant overhead for switching across bit-rates. For example, a single query packet at FM0/640 takes about 6ms whereas switching bit-rates incurs a latency of 30ms. As a result, SampleRate with default parameters performs poorly and can only achieve around 10 reads/s goodput even at close ranges. We therefore optimized SampleRate parameters, and found that it performs best when probes are done for 0.5 secs after every 5 secs of transmission at the best rate. We use these parameters for SampleRate in our evaluation.

Thresholds for link signatures: Our mobility detection protocol uses two distance thresholds, d_T and d_T^*, to determine if a sensor tag is mobile. We now describe how we set these thresholds. Figure 10(a) shows the link signatures of a sensor tag in two cases: a) a stationary tag placed at several different locations in a room, and b) a mobile tag that is placed on a toy train moving along a oval train track at different speeds.

First, we look at the distances for the static case (tag at a single location) vs the mobile case. Figure 10(a) shows a significant difference between the static and mobile cases, and the choice of thresholds ($d_T = 0.3$ and $d_T^* = 0.15$) is relatively straightforward. Second, we peer more deeply into the static cases to see whether we can distinguish between a tag placed at one location vs another. Figure 10(b) shows the RSSI and packet loss distances between one of the locations (referred to as A) and several other locations where the tag was placed. Again, its clear that there is a substantial difference across locations, and the same threshold that we used for the mobile case works in this case as well.

Note that by using the same threshold in the two cases, we are not distinguishing between a tag that has moved to another location vs a tag that is in continuous motion. While a simple extension of looking at a window of distances could address this issue, distinguishing the cases is not important for us since we only care about generating appropriate triggers for rate adaptation and channel selection.

6. EVALUATION

In this section, we evaluate the implementation of our link layer using commercial Impinj readers and passive Alien RFID tags. The evaluation consists of four parts: 1) benchmarking the accuracy of our mobility detection algorithm, 2) validating the link metrics to bitrate map, and evaluating the benefit of our classifier-based bitrate selection algorithm, 3) demonstrating the goodput benefit of using channel selection/switching, and 4) evaluating the overall performance of our high throughput backscatter link layer.

6.1 Mobility detection

The mobility detection module enables a reader to be aware of tag mobility patterns. In this section, we evaluate our mobility detection algorithm in two steps: 1) we benchmark accuracy when we use RSSI and packet loss distance exclusively, and a combination of both to detect mobility, and 2) we compare accuracy when we detect mobility under different choices of bitrate (since a reader can be communicating with tags at different bitrates).

	RSSI	LossRate	RSSI+LossRate
False positive	1.01%	8.04%	9.05%
False negative	5.63%	3.03%	0.87%

Table 2: False positive/negative rate when different channel features are utilized. FM0/640 is used to obtain link signature.

	False positive	False negative
FM0/160	0%	1.67%
Miller4/640	4.08%	10%
Miller4/256	0%	2.38%
Miller8/256	0%	0%
FM0/40	10%	3.92%

Table 3: False positive/negative rate when different bitrates are used to obtain link signature.

Detection accuracy under different link metrics: To evaluate mobility detection accuracy, we attach an RFID tag to a toy train that moves along a 1m × 2.5m oval track. The train follows a stop and move pattern; it moves along the oval track for two mins, stops for three mins, and follows this pattern eight times.

We evaluate the accuracy of mobility detection by measuring the false positive and false negative rate, where the null hypothesis is defined as the tag not being mobile. Given this hypothesis, the false positive rate can be defined as the ratio of being notified mobile when tag is actually stationary, and false negative rate is the ratio of being identified as stationary when tag is in fact mobile.

Table 2 shows that by combining RSSI and lossrate features, we can reduce false negatives to under 1%, which is 3× lower than the rate when only loss rate were used, and 5× lower than the rate if only RSSI were used. We lose a bit on false positives since when either RSSI or lossrate distances suggest mobility, we consider the sensor to have moved. This causes more false alarms but our goal was to ensure that actual mobility scenarios were not missed, which is shown to be the case.

Detection accuracy under different bitrates: Since the reader could be operating at different bitrates at different times, one question is how the accuracy of mobility detection is impacted by bitrate. To understand this, we replicate the above experiment, but do this across all bitrates. To reduce the time to run the trace, we use two stops for each. We then measure the false positive/negative rate at different bitrates.

The results are shown in Table 3. We observe that both false positive and negative rates are lower than 10% for all other five bitrates. The results show that no matter which bitrate is selected for obtaining the link signature, our algorithm can achieve similar mobility detection accuracy.

6.2 Rate adaptation

We now turn to the rate adaptation algorithm, which uses a classifier to select the optimal bitrate. We evaluate our rate adaptation algorithm in four steps: 1) we verify the correctness of link metrics to bitrate map, 2) we benchmark the accuracy of the classifier in choosing the best bitrate when full channel probing is done, 3) we evaluate the accuracy

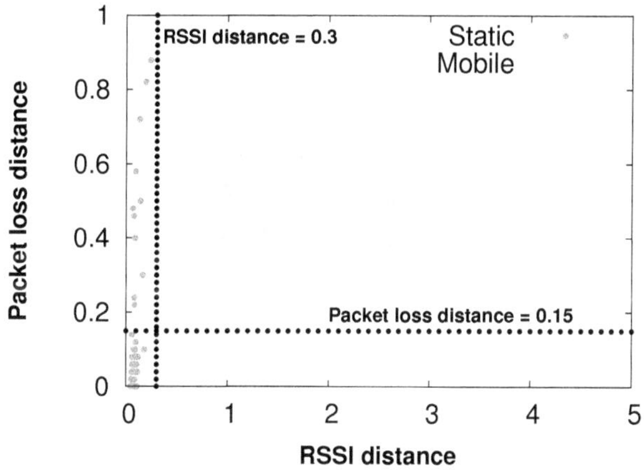

(a) Link signature of static and mobile tag.

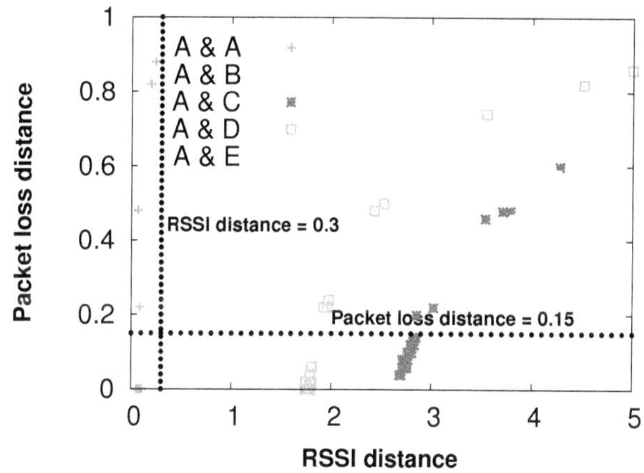

(b) Link signature for tag move from location A (1m from reader) to B(4m), C(7m), D(10m), E(12m).

Figure 10: Link signature of static and mobile tag.

of the classifier when fast probing is done, and 4) we compare the goodput achieved by the rate adaptation algorithm against the optimal goodput as well as the one achieved by fast probing.

Correctness of link metrics to bitrate map: We now provide micro-benchmarks that validate the link metrics to bitrate map that we presented in §4.3. We look at two cases — the tag is placed within the beamformer direction, and the tag is placed at the edge of the beamformer direction (i.e. placed close to reader but far from the beamformer direction). In each case, we look at how the optimal bitrate as well as the second-based bitrate changes as the tag is moved away from the reader.

Figure 11(a) shows the results for the case when the tag is placed within the beamformer direction. Multipath self-interference is less likely in this setting, and therefore the optimal bitrate follows a more predictable pattern. As the tag moves away from the reader, the optimal bitrate goes from Block A to B to D, roughly following the order in the lowest row of the map in Figure 7.

Figure 11(b) shows the case when the tag is placed at the edge of the beamformer. As discussed in §4.1, we expect multipath self-interference to be more severe in this case. The results here are more unpredictable. When the tag is 1 meter from reader, the highest bitrate (Block A) is chosen. When the tag moves just a bit further to 1.5m, the optimal bitrate moves from Block A to Block F (as predicted by the first column of Figure 7). This point is clearly caused by multi-path self-interference as described in Figure 3 — RSSI is high, which would suggest excellent channel quality, but packet loss is high as well. The multipath effects reduce when the tag moves a bit more to 2m, and the optimal bitrate changes to Block B. After that, channel quality progressively degrades and the optimal selection moves to Block C and finally to Block I.

The two figures also show the goodput obtained by the second-highest bitrate (green circles). There are cases where the second-highest bitrate has considerably lower throughput than the optimal one, where errors in classification will be expensive. But there are also several cases where the difference is small, and classifier errors will not lead to significant reduction in goodput.

Bitrate boundaries: While Figure 7 provides an intuition for how the combination of link metrics can be used to predict bitrate, it is an idealized model and the actual boundaries between the different clusters in real-world data are likely to be less regular. Understanding these block boundaries can provide a better idea for how well a classifier would work and where classification errors might occur.

Figure 12 shows the empirically measured optimal goodput mode and their corresponding RSSI and lossrate across several reader-tag distances, placements and times of day. Each circle shows the case when a particular bitrate was selected as the optimal bitrate, and maps to one of the blocks in Figure 7. We did not observe points for some of the clusters, and do not plot them in the figure. As can be expected, the boundaries between clusters are not as regular as shown in the idealized bitrate map, however similar trends can be observed. Also, it can be seen that the classifier might have errors in boundary regions — for example, some points in Block A fall into the Circle B (red points in the green circle), so these might lead to mis-classification. However, we also notice that in these boundary regions, the goodput of both choice A or B is similar, hence both are good choices.

Classifier Accuracy: We turn to evaluating the accuracy of our classifier. To train the classifier, we use 158 sample points from a room over a day, where each sample involves placing the tag at a random location relative the reader, and measuring the optimal bitrate, the RSSI across channels and the packet loss across channels. After the training process, the classifier has an empirically measured RSSI/packet loss to bitrate map.

Table 4 describe the three settings that we use for testing, each of which stresses the classifier in a different way. Group 1 is a dataset from the same room as the training set, but on a different day, Group 2 is a dataset on a corridor with different multipath propagation characteristics, and Group 3 is a dataset in a corridor where we expected substantial multipath since there were a significant number of metal objects including servers nearby. For each test set, we compare

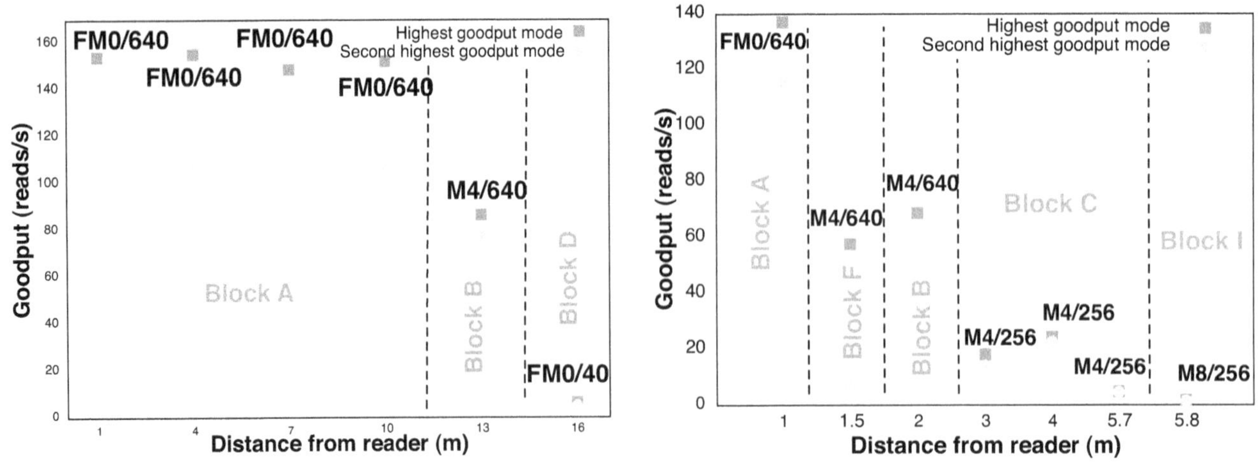

(a) Top two goodput modes when tag is placed within reader beamformer direction

(b) Top two goodput modes when tag is placed at the edge of reader beamformer direction

Figure 11: Top two goodput modes

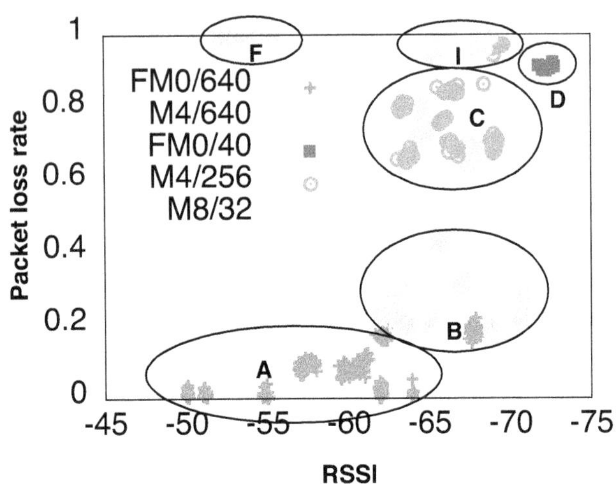

Figure 12: Empirical measured link metric map. It shows the optimal bitrate under different RSSI and packetloss.

	Group 1	Group 2	Group 3
Training data	room/day 1	room/day 1	room/day 1
Testing data	room/day 2	corridor1/day 3	corridor2/day 4
Training size	158	158	158
Test size	347	161	162

Table 4: Three groups of experiments for verifying the accuracy of classifier.

Training data	Testing data	Accuracy	% optimal goodput
room/day1	room/day2	83.95%	98.18%
room/day1	corridor2/day3	86.47%	89.75%
room/day1	corridor2/day4	51.38%	88.95%

Table 5: The accuracy for classifier to choose optimal goodput mode. It also shows the percentage of goodput achieved compared with optimal goodput.

the selected bitrate given by classifier to the ground truth optimal bitrate to evaluate the accuracy of the classifier.

The accuracy of classifier in the three groups of experiments is shown in Table 5. We observe that for the first and second groups, the classifier picks up the optimal goodput mode with over 83% accuracy and achieves over 89% of the optimal goodput. For the third group, the accuracy of selecting the optimal bitrate drops dramatically to around 50%, which seems poor. However, a large number of these cases fall into the boundaries between clusters, hence despite low accuracy, the classifier achieves 88% of the optimal goodput.

To summarize, these results show that a) our classifier can achieve approximately 89% of the goodput achieved by optimal bitrate selection under varying conditions, and b) the classifier can be trained in one region and used in several other regions with vastly different propagation and multipath effects without significant effect on goodput.

Comparison against reader bitrate control: In this evaluation, we ask two questions: a) how does BLINK's selected bit-rate compare to the optimal bit-rate, and b) how does our classifier perform if it only uses RSSI or loss-rate as feature. We use data from all four settings in Table 4 in this evaluation.

Figure 13 shows a CDF of goodput from an optimal scheme that always selects the best bit-rate, the goodput achieved by BLINK, and the goodput obtained by our classifier when utilizing only RSSI/loss rate as feature. We find that BLINK largely follows the optimal case in terms of goodput, and picks a sub-optimal rate only 18% of the time. The loss-rate based classifier is the next best mechanism and picks a sub-optimal bit-rate 30% of the time, although the difference in goodput is only about 3%. The RSSI-based classifier performs the worst, and picks a sub-optimal bitrate 62% of the time, and has on average 15% lower throughput than BLINK. This shows that using a combination of loss-rate and RSSI performs better than using only one of these metrics, and that BLINK performs close to an optimal scheme.

Effect of fast channel probing on bitrate: The results so far assumed that the classifier was given the full channel

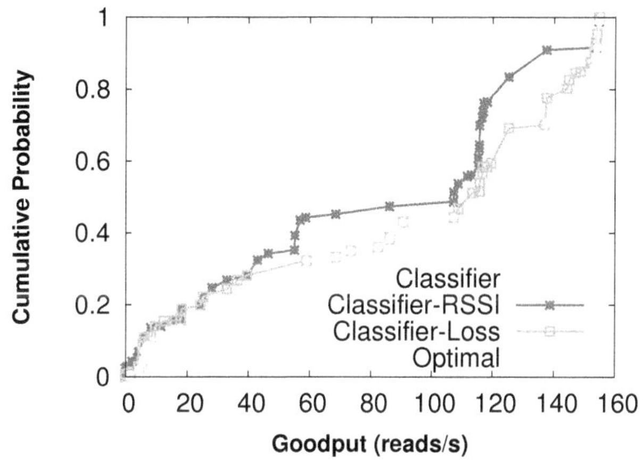

Figure 13: The CDF goodput of optimal mode, mode chosen by classifier, mode chosen by classifier with only RSSI/loss rate feature and default Impinj algorithm.

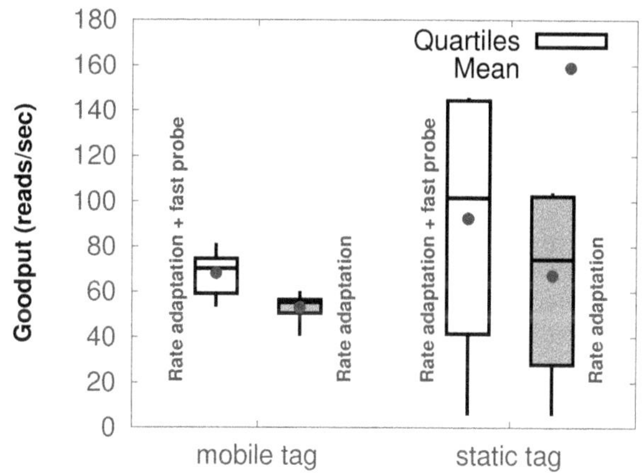

Figure 14: The goodput of rate adaptation and rate adaptation under fast probing for 1 static and mobile tag.

	Accuracy	% optimal goodput
Regular probe	85.45%	97.08%
One Query probe	88.72%	98.06%
10 random probe	73.44%	92.75%
One Query probe + 10 random probe	72.34%	91.48%

Table 6: The accuracy for classifier to choose optimal goodput mode under fast channel probing. It also shows the percentage of optimal goodput retained.

probe *i.e.* full information across 50 channels. As described in §4.3.3, this is very time consuming, and fast channel probing can reduce channel probe time by 35×. So, we ask: how does fast probing impact classification accuracy? Our results are based on a dataset collected in the high-multipath corridor (same environment as corridor2/day4), since it represents the worst case scenario. We use a trace-driven simulation to identify how different schemes would perform.

Table 6 summarizes our results. The second row shows that using one query per channel rather than the default seven queries does as well, or in fact a bit better in terms of classifier accuracy and fraction of optimal goodput. The third row shows that cutting down on the number of channels to sample from 50 to 10 reduces accuracy to 73% but still achieves 93% of the goodput of the optimal scheme. Finally, combining the two metrics has a similar effect. This result shows that on readers where the probe time is very long, a combination of one query per channel and random subset of channels can achieve 35× reduction in probe times without significantly impacting goodput.

Effect of fast channel probing: We now look at the benefits of fast channel probing on performance both for a static tag and mobile tag. In the mobile case, we attach a tag to a LEGO toy train which moves along a 1m×2.5m oval track. The mobility detector triggers rate adaptation roughly every five seconds in this experiment. Figure 14 shows the goodput achieved by the two schemes. Fast channel probing increases goodput by 29% for one mobile tag and 38% for one static tag.

6.3 Channel selection/switching

We now turn to an evaluation of the goodput achieved by the channel selection/switching algorithm. Even though FCC regulations and EPC Gen 2 provides flexibility for channel selection, commercial readers do not provide user-level control over these parameters. As a result, we evaluate the algorithm in two ways. First, we use a trace-driven simulation, where the trace is a repeated 50-channel scan of a tag by an Impinj reader. We stitch the dwell times for each channel together to form a single continuous trace on top of which we emulate the channel selection algorithm. While this trace does not entirely reflect reality, it can give a sufficient picture of the performance of our algorithm. Second, we implement the channel selection algorithm on the USRP software radio reader. The downside of this study is that the USRP reader behaves quite differently from the commercial reader — packet losses are much higher, loss transitions are less steep and burstiness is less evident. Despite this, the USRP reader evaluation provides a baseline for how our techniques would perform when the channel behaves differently from expected.

We evaluate both channel selection and channel switching — the former assumes that lossrate information is available for all channels, whereas the latter switches dynamically and doesn't require prior knowledge of which are good or bad channels. Figure 15 compares the performance of channel selection and switching against the default scheme used by commercial readers.

The results on trace-driven simulations show that channel switching is 2.16× better than the default scheme and channel selection is marginally worse and is 2.06× better than the default. This shows that switching is effective due to burstiness and sharp transitions. For the USRP reader, the results are a bit different — switching is 1.12× better than default and selection is 1.31× better.

6.4 Overall system performance

We now look at the performance of an integrated system on a commercial reader. Since channel selection or switching cannot be enabled on these readers, we do not include the module in our combined evaluation. Our goals are two-

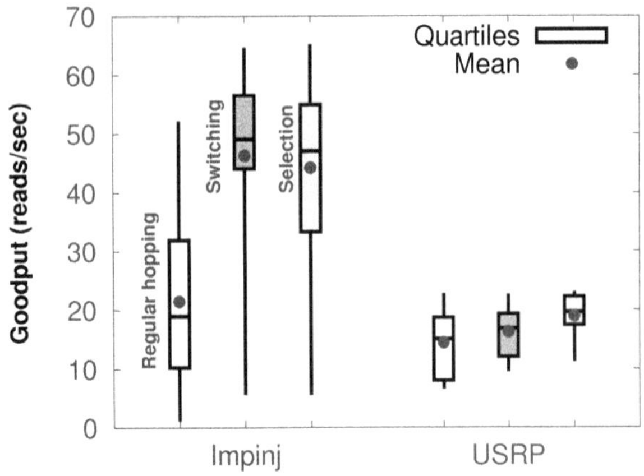

Figure 15: The goodput achieved with channel selection and switching.

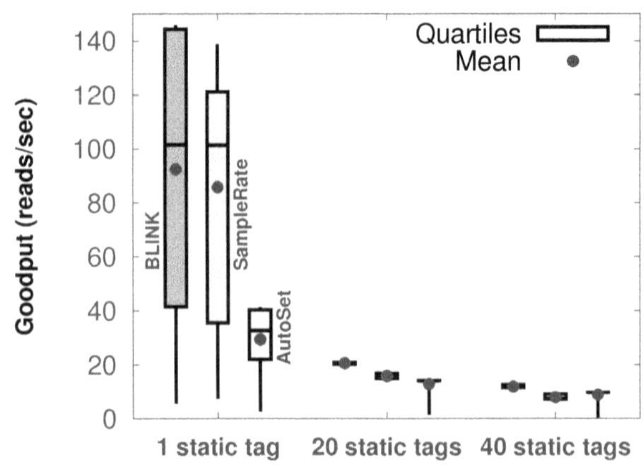

Figure 16: The goodput of AutoSet, SampleRate, and BLINK for 1, 20 and 40 static tags.

fold: a) to provide a comparison between BLINK, AutoSet, and SampleRate, and b) to understand the benefits of our schemes at scale when there are large numbers of static or mobile tags.

First, we look at the static case, where we evaluate performance as we increase the number of static tags placed in front of a reader from 1 to 40. These tags are placed at random locations. To obtain statistically significant results, the results are aggregated over several runs and several placements. Figure 16 shows that for one static tag, BLINK is 3.15× better than AutoSet. SampleRate achieves 2.92× better goodput than AutoSet which is just slightly lower than BLINK. When the tag population scales to 20, BLINK and SampleRate are 1.6× and 1.24× better than AutoSet respectively. When the tag population is further increased to 40, the goodput increase is 1.34× for BLINK but SampleRate degrades and has a throughput of only 0.9× that of AutoSet. This is likely because of the increase in MAC layer collisions leading to higher variability in packet loss rates and consequently poor decision-making by SampleRate. Thus, BLINK has significant benefits as deployment density increases even under static settings.

Second, we look at the mobile case. We place one, five, and ten RFID tags at fixed locations on a person who moves continuously in a 15m×15m room. The movement pattern involves walking towards and away from the reader, and walking in a circle along the edges of the room. We repeat the same mobility trace at roughly the same speed to obtain a fair comparison among BLINK, SampleRate and AutoSet. Figure 17 show the summary statistics across 10 mobility traces for each scheme. We see that for one mobile tag, BLINK is 2.29× better than AutoSet, and SampleRate is 1.77× than AutoSet. However, as the number of mobile tags grow, the benefits of BLINK increases. When 5 mobile tags are employed, the gains of BLINK and SampleRate over AutoSet are 1.96× and 1.42× respectively. When tag population scales to 10, BLINK is 2.44× better than AutoSet, whereas SampleRate is only 1.21× better than AutoSet. These results clearly demonstrate the benefits of BLINK across a wide range of scenarios and scales for a realistic number of RFID sensors.

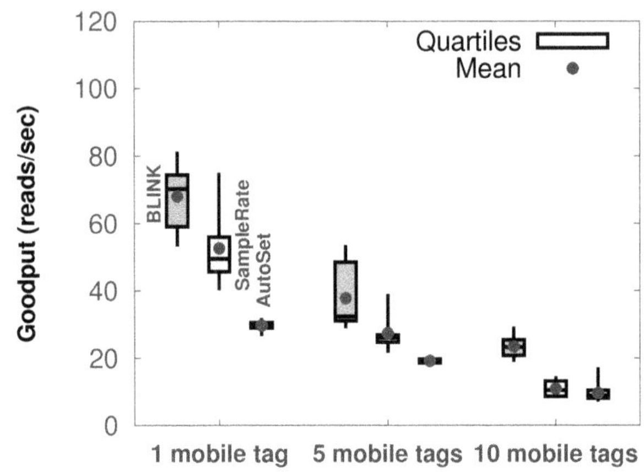

Figure 17: The goodput of AutoSet, SampleRate, and BLINK for 1, 5, and 10 mobile tags.

7. RELATED WORK

There has been substantial prior work on wireless channel characteristics and link layers. Our primary contribution is to design a link layer that is optimized to the unique characteristics of backscatter communication.

Backscatter system: Much of the work on backscatter communication is specific to optimizing communication from EPC Gen 2 tags, for example, better tag density estimation [26], better search protocols to reduce inventorying time [19], better tag collision avoidance [20], more accurate tag identification [29], better recovery from tag collisions [8]. Some work has also looked at how physical layer parameters account for the major propagation characteristics of the backscatter channel [14]. Other work has looked at more efficient use of harvested energy, for example Dewdrop [10] balances task demands with available energy, and our prior work uses hybrid energy harvesting to increase communication range to many tens of feet [15]. None of these tackle link-layer adaptation mechanisms to improve goodput.

ECP Gen 2 optimization: Recent work with Computational RFID devices such as the Intel WISP includes Buettner et al [11], who explore how the RFID PHY layer interacts with EPC Gen 2 and identify physical layer parameters that degrade overall performance and reliability. [11] also suggests adapting the link layer to improve goodput by using FM0/640 kbps to read tags until only low SNR tags are left (because they can't be read at the high rate), and then changing to Miller-8/256 kbps to read the rest of the tags. Our work addresses backscatter bit-rate selection in a much more rigorous manner.

In an effort complementary to this paper, we have designed Flit, a high-throughput and low-power MAC layer that addresses inefficiencies of EPC Gen 2 for bulk data transfer from sensors [16]. BLINK is complementary to this effort and differs in two ways: 1) Flit involves optimizations at the protocol level of the EPC Gen 2 MAC, whereas BLINK is entirely at the link layer and agnostic of how the protocol works, b) Flit involves optimizations on the tag-side and therefore requires programmable RFID tags (e.g. WISP) whereas BLINK is entirely at the reader-side and can operate with any commercial RFID tag.

Link signatures: Recent work has also explored the problem of location distinction using link signatures [30]. The main idea is to form a signature of the channel impulse response from physical layer data at a receiver, and use this to detect location changes. This is tested for WiFi software radios where PHY information is available. At a high level, our mobility detector uses similar ideas, but we differ in our focus on backscatter communication and link metrics that are relevant to capturing the channel signature for RFIDs. Link signatures have also been used in localization — for example, SkyLoc [25] exploits GSM signal strength fingerprinting for identifying the current floor of a user in tall multi-floor buildings. While BLINK exploits radio fingerprints as well, it does so for location distinction, which is simpler and does not involve an extensive training phase.

Estimating channel quality: Link layer mechanisms for active radios rely on different link metrics including SNR, loss rate and PHY-layer hints. SNR is used by AccuRate [23], which uses it to select the appropriate constellation map, FARA [21], which uses per-frequency SNR to evaluate the depth of fading in each channel of 802.11, and [17], which proposes an effective SNR metric to predict the highest link bit rate. Packet loss rates are widely used, for example, [28] uses short-term loss ratio to opportunistically guide its rate change decisions. More recently, PHY-layer hints have been used, for example SoftRate [27] uses "SoftPHY" or confidence values conveyed by PHY to learn the bit error rate (BER). Our work addresses unique aspects of backscatter communication where we find RSSI and packet loss rate are needed in conjunction to capture path loss and multipath fading characteristics.

Channel selection: [24] investigates β-factor metric to measure intermediate link burstiness and proposes opportune transmissions to improve bursty link packet reception ratio. In contrast, our work exploits burstiness for fast channel switching in search for better transmission opportunities. Our work also differs from CSMA-based techniques for channel selection (e.g. [18]) since we deal with backscatter-specific considerations.

8. CONCLUSION

Backscatter communication offers an attractive alternative to existing active-radio based sensor systems. However, backscatter has predominantly been used for communication with RFID tags which have limited data to transfer, and there has been limited prior work on understanding how backscatter can be used for high-throughput communication between readers and sensors. Our work bridges the gap by providing an in-depth exploration of the physical layer behavior of backscatter links, what metrics are needed to capture this behavior, and how bitrate adaptation and channel selection can exploit these channel characteristics to improve throughput while remaining within FCC guidelines for the UHF RFID frequency range. Results show that our link layer, BLINK, provides substantial throughput benefits for a range of channel conditions, scales, and mobile conditions.

9. ACKNOWLEDGEMENT

We thank Prof. Lakshminarayanan Subramanian for shepherding and the anonymous reviewers for their valuable comments. This research was supported by NSF grants 0855128, 0916577 and 0923313.

10. REFERENCES

[1] A Batteryless Programmable RFID-Scale Sensor Device. `http://spqr.cs.umass.edu/moo/`.

[2] How to NFC. `http://developer.android.com/videos/index.html#v=49L7z3rxz4Q`.

[3] RFID Basics: Backscatter Radio Links and Link Budgets. `http://rfidtribe.com/index.php?option=com_content&view=article&id=423&Itemid=102`.

[4] The Impinj UHF Gen 2 Speedway RFID Reader. `http://www.impinj.com`.

[5] Ultracapacitor Offers 75-Foot Read Range for Passive Tags. `http://www.rfidjournal.com/article/articleview/8565`.

[6] WISP: Wireless Identification and Sensing Platform. `http://seattle.intel-research.net/wisp/`.

[7] D. Aguayo, J. Bicket, S. Biswas, G. Judd, and R. Morris. Link-level measurements from an 802.11 b mesh network. In *ACM SIGCOMM*, 2004.

[8] C. Angerer, R. Langwieser, and M. Rupp. Rfid reader receivers for physical layer collision recovery. *IEEE Transactions on Communications*, 2010.

[9] J. Bicket. *Bit-rate selection in wireless networks*. PhD thesis, Massachusetts Institute of Technology, 2005.

[10] M. Buettner, B. Greenstein, and D. Wetherall. Dewdrop: an energy-aware runtime for computational rfid. In *USENIX NSDI*, 2011.

[11] M. Buettner and D. Wetherall. An empirical study of uhf rfid performance. In *ACM MobiCom*, 2008.

[12] M. Buettner and D. Wetherall. A gen 2 rfid monitor based on the usrp. *ACM SIGCOMM CCR*, 2010.

[13] M. Buettner and D. Wetherall. A software radio-based uhf rfid reader for phy/mac experimentation. In *2011 IEEE International Conference on RFID*, pages 134–141, 2011.

[14] J. Griffin and G. Durgin. Complete link budgets for backscatter-radio and rfid systems. *Antennas and Propagation Magazine, IEEE*, 51(2):11–25, 2009.

[15] J. Gummeson, S. Clark, K. Fu, and D. Ganesan. On the limits of effective hybrid micro-energy harvesting on mobile crfid sensors. In *Proceedings of the 8th international conference on Mobile systems, applications, and services*, pages 195–208. ACM, 2010.

[16] J. Gummeson, P. Zhang, and D. Ganesan. Flit: A bulk transmission protocol for rfid-scale sensors. In *MobiSys*, 2012.

[17] D. Halperin, W. Hu, A. Sheth, and D. Wetherall. Predictable 802.11 packet delivery from wireless channel measurements. *ACM SIGCOMM*, 2010.

[18] N. Jain, S. Das, and A. Nasipuri. A multichannel csma mac protocol with receiver-based channel selection for multihop wireless networks. In *IEEE ICCCN*, 2001.

[19] C. Law, K. Lee, and K. Siu. Efficient memoryless protocol for tag identification. In *Proceedings of the 4th international workshop on Discrete algorithms and methods for mobile computing and communications*, pages 75–84. ACM, 2000.

[20] V. Namboodiri and L. Gao. Energy-aware tag anticollision protocols for rfid systems. *Mobile Computing, IEEE Transactions on*, 9(1):44–59, 2010.

[21] H. Rahul, F. Edalat, D. Katabi, and C. Sodini. Frequency-aware rate adaptation and mac protocols. In *ACM MobiCom*, 2009.

[22] L. Ravindranath, C. Newport, H. Balakrishnan, and S. Madden. Improving wireless network performance using sensor hints. In *USENIX NSDI*, 2011.

[23] S. Sen, N. Santhapuri, R. Choudhury, and S. Nelakuditi. Accurate: Constellation based rate estimation in wireless networks. In *Proceedings of the 7th USENIX conference on Networked systems design and implementation*, pages 12–12. USENIX Association, 2010.

[24] K. Srinivasan, M. Kazandjieva, S. Agarwal, and P. Levis. The β-factor: measuring wireless link burstiness. In *ACM SenSys*, 2008.

[25] A. Varshavsky, A. LaMarca, J. Hightower, and E. de Lara. The skyloc floor localization system. In *Pervasive Computing and Communications, 2007. PerCom'07. Fifth Annual IEEE International Conference on*, pages 125–134. IEEE, 2007.

[26] H. Vogt. Efficient object identification with passive rfid tags. *Pervasive Computing*, pages 98–113, 2002.

[27] M. Vutukuru, H. Balakrishnan, and K. Jamieson. Cross-layer wireless bit rate adaptation. In *ACM SIGCOMM CCR*, 2009.

[28] S. Wong, H. Yang, S. Lu, and V. Bharghavan. Robust rate adaptation for 802.11 wireless networks. In *ACM Mobicom*, 2006.

[29] D. Zanetti et al. Physical-layer identification of uhf rfid tags. In *ACM MobiCom*, 2010.

[30] J. Zhang, M. Firooz, N. Patwari, and S. Kasera. Advancing wireless link signatures for location distinction. In *ACM MobiCom*, 2008.

[31] J. Zhao and R. Govindan. Understanding packet delivery performance in dense wireless sensor networks. In *ACM SenSys*, 2003.

Fast App Launching for Mobile Devices Using Predictive User Context

Tingxin Yan[1], David Chu[2], Deepak Ganesan[1], Aman Kansal[2], Jie Liu[2]
[1]University of Massachusetts Amherst, [2]Microsoft Research
{yan, ganesan}@cs.umass.edu, {davidchu, kansal, jie.liu}@microsoft.com

ABSTRACT

As mobile apps become more closely integrated into our everyday lives, mobile app interactions ought to be rapid and responsive. Unfortunately, even the basic primitive of launching a mobile app is sorrowfully sluggish: 20 seconds of delay is not uncommon even for very popular apps.

We have designed and built FALCON to remedy slow app launch. FALCON uses contexts such as user location and temporal access patterns to predict app launches before they occur. FALCON then provides systems support for effective app-specific prelaunching, which can dramatically reduce perceived delay.

FALCON uses novel features derived through extensive data analysis, and a novel cost-benefit learning algorithm that has strong predictive performance and low runtime overhead. Trace-based analysis shows that an average user saves around 6 seconds per app startup time with daily energy cost of no more than 2% battery life, and on average gets content that is only 3 minutes old at launch without needing to wait for content to update. FALCON is implemented as an OS modification to the Windows Phone OS.

Categories and Subject Descriptors

C.5.3 [**Computer System Implementation**]: Microcomputers—*Portable devices*; D.4.8 [**Operating Systems**]: Performance—*Modeling and prediction*

Keywords

Mobile devices, Context, Location, Application prediction

1. INTRODUCTION

Mobile apps have blossomed in popularity and ubiquity. With each generation, the multitude and diversity of apps continues to grow. A recent tally revealed 380,000 iOS apps, 250,000 Android apps and 35,000 Windows Phone (WP) apps. With so many apps out there, systems support that can improve our daily app interaction experience is poised to be widely beneficial.

Early pioneers in mobile computing recognized that during mobile interactions, human attention is a scarce resource [5, 12]. Our analysis of a long term user study [13] supports this view: 50% of mobile phone engagements last less than 30 seconds, and 90% of engagements last less than 4 minutes. With such brief periods of interaction, it is all the more important that interaction is rapid and responsive.

Unfortunately, interacting with apps can be surprisingly sluggish. Applications can take dozens of seconds to launch, negating any sense of agility. A particularly troubling class of apps are games, some of which can take upwards of 20 seconds to fully launch past splash screens, and even more time to reach a playable state. Our investigation shows that even relatively simple apps such as weather reporters can experience launch delays of 10 seconds or more.

Some applications, such as email clients, are optimized for quick launch within seconds. However, these applications mask actual latency by showing stale content while asynchronously fetching new content over the network. While appropriate for a restricted set of interactions, stale content fails to engage the user in the same way that the latest fresh content can. When considering the full launch time starting from touch until the latest content is displayed, apps with asynchronously loaded content can also exhibit severe load times of up to a dozen seconds.

One possible approach to decrease launch time is to cache apps in memory. This approach is implemented by iOS, Android and WP, yet suffers from several drawbacks. First, some apps, especially games, demand large amounts of memory, overwhelming memory real estate for other apps. Therefore, naïve caching schemes can exhibit low benefit. Second, caching does not address the issue that content may become stale by the time the user interacts with the app.

Another possible approach that addresses stale content is the use of push notifications. While push notifications can ensure content freshness, the energy cost of push communication can be prohibitively high [11]. Not surprisingly then, four out of the top five battery saving suggestions offered by Apple concern disabling push notifications [2].

Ideally, we would like to take advantage of both available cache and proactive content updates without incurring their drawbacks. It so happens that mobile devices are equipped with an increasingly sophisticated array of sensors such as IMUs, mics, cameras and geolocation units. Each can provide insightful context for the OS [3]. For a particular user, Angry Birds usage may be much more likely

MobiSys'12, June 25–29, 2012, Low Wood Bay, Lake District, UK.

at home, whereas Email and Calendar usage may be much more likely at work. Or, a user may habitually check Facebook after receiving an SMS when at school.

We have built the FALCON[1] system to predictively *prelaunch* apps based on context signals and user access patterns. Prelaunch entails both preloading apps in memory and app-specific fresh content preparation.

At the foundation of FALCON is a data-driven analysis that reveals surprising insights about how and when we use mobile apps. We find an intrinsic bundling behavior in how mobile apps are used — once a user starts using the phone, say for a phone call or SMS, they tend to use several other applications before turning away from the phone to another task. Bundling behavior manifests not just at the timescale of single user session with the phone, but across longer durations of several weeks. We find that games are often used in bursts where users tend to obsessively use a game for a few weeks, and then ignore it entirely for several months. Similarly, other contexts such as location and time-of-day plays a central role in app usage, perhaps indicative of our tendency toward structured behavior.

While such behavioral insights can be distilled into a set of preload prediction rules, a key challenge is that users are not alike in how these behaviors apply to them. Some users are more likely to open several apps after a phone call whereas others do so after opening the browser; some users have strong tendencies at a core set of tightly defined locations such as home and work whereas others exhibit more diversity. To address this, FALCON personalizes the context triggers to the individual to maximize prelaunch accuracy.

Applications also differ widely in terms of the benefits from correct prelaunch versus the cost of incorrect prelaunch. As an example comparing two popular games, *Need for Speed* takes over 2× more energy than *Plants vs. Zombies* even though both load equally slowly. FALCON uses a novel cost-benefit adaptive prediction algorithm that accounts for both the expected latency benefit of correct predictions and the expected energy cost incurred for wrong predictions. The algorithm adapts to budget constraints and jointly optimizes the applications to prelaunch as well as the temporal and location contexts under which such prelaunching is enabled. Both features and learning algorithm are designed to achieve very low runtime overhead.

We have implemented FALCON as an operating system modification to the Windows Phone 7.5 operating system and show that it has minimal energy, memory and processing overhead. In addition to preloading apps from local storage, FALCON ensures content freshness by prefetching content before app launch. This is achieved with a WP event handler that FALCON invokes whenever it predicts the app is about to be launched. For example, a Twitter app can fetch new tweets during this event. As a result, a user's time to first meaningful app interaction is no longer bound by network latency.

Our major contributions are as follows.

- From extensive data analysis, we design spatial and temporal features that are highly indicative of mobile app access patterns.
- We design a cost-benefit learning algorithm whose predictions lead to launch latency reduction of nearly 6 seconds per app usage with the daily energy cost of no more than 2% battery lifetime. Predictions also lead to getting content that is only 3 minutes old at launch without waiting for content to update at launch.
- We prototype FALCON on a Windows Phone, and demonstrate that it has minimal runtime energy overhead of a few μAh and a small memory footprint of under 2 MB.

[1]Fast App Launching with Context. The peregrine falcon can swoop dive at more than 200 mph – making it the fastest animal on earth.

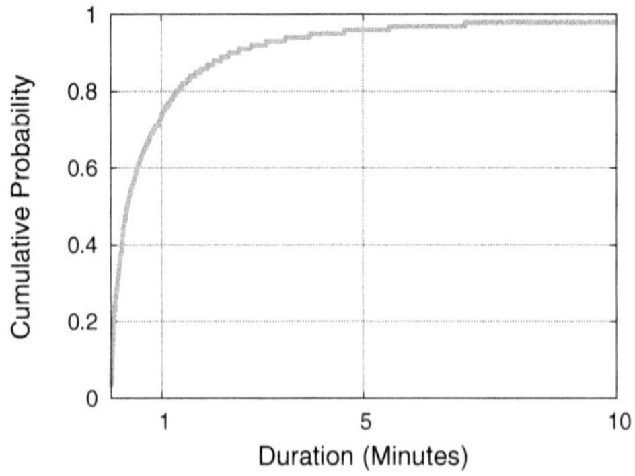

Figure 1: App usage durations for a representative individual over a one year long trace.

Taken as a whole, FALCON effectively replaces a subset of the policy decisions traditionally reserved for the memory manager and scheduler. Context permits FALCON to make much more informed app decisions which lead to substantial launch time reductions at minimal overhead.

The rest of this paper is organized as follows. §2 provides quantitative background of the problem. §3 provides an overview of the system. §4 discusses the launch predictor, features and learner. §5 details implementation aspects. §6 evaluates FALCON. §7 discusses related work, and §8 offers a discussion and conclusion.

2. BACKGROUND AND PROBLEM SCOPE

In this section, we take a closer look at the issue of app launch latency. Our investigation is based on trace data collected from the Rice LiveLab user study [13], and our own internal study. In the Rice LiveLab's study, thirty four volunteers were given iPhones for a period of fourteen months from 2010-2011. During this time, their device usage was monitored. Our own internal double-blind study consisted of three WP users monitored over one month. We focus on the following elements which were recorded in both traces: *app usage, location, datetime*. The LiveLab's trace is much more extensive; it will be the main focus of investigation unless otherwise noted.

We first examine the thesis that mobile interactions are inherently brief [5]. Figure 1 illustrates the CDF of app usage durations across all users in the Rice trace. Indeed, most interactions with the phone are short with 80% of the apps being used for less than two minutes. This indicates that short engagements are the norm and the proportional effect of slow app startup can be expected to be even more severe for mobile users.

Qualitative reports suggest that slow app launch is a substantial drag on user experience. As an example, a recent iOS upgrade caused an outcry from iPhone owners that were

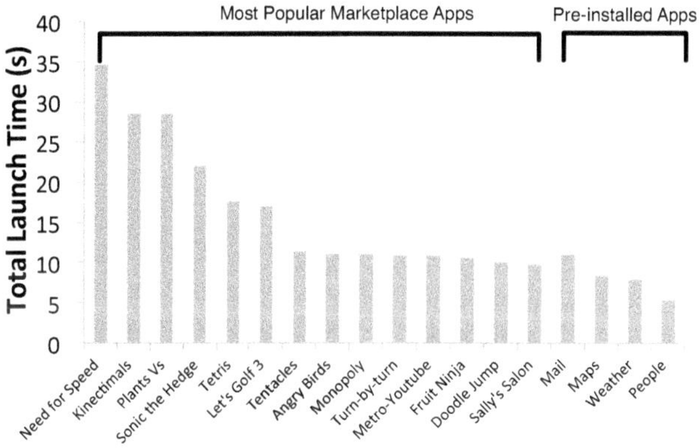

Figure 2: Total launch times of popular apps on Windows Phone. The first 14 are most popular apps in the Marketplace, and the last four are first-party apps that are preinstalled.

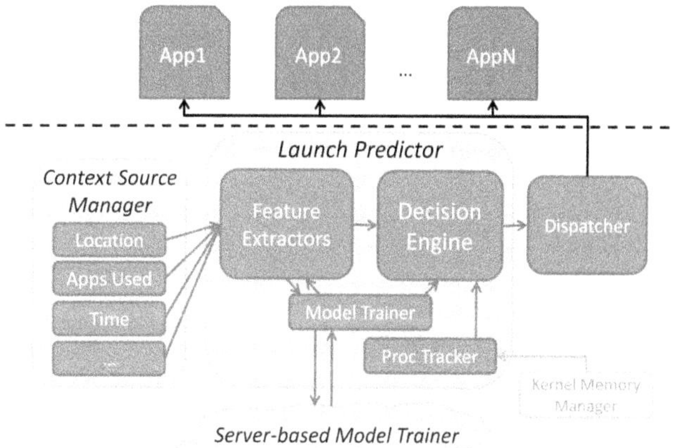

Figure 3: The FALCON Architecture.

hit with slower speeds [6]. Popular blogs suggest that users are willing to try involved system hacks to improve app launch times by ten seconds [4].

To quantify the extent of the issue, we studied the top 20 popular apps in the WP marketplace. *Launch time* represents the elapsed time from when a user taps on an app's icon to the first time a user is able to interact with the application. Figure 2 shows that launch times are fairly long, with a median launch time of 11.2 seconds and an average launch time of 16.7 seconds. Surprisingly, no app launched in under 5 seconds. The trend for iPhone was similar.

The use of asynchronous launch operations is a popular coping mechanism for slow launch times. Two typical classes of asynchronous operations are fetching content from the network while displaying stale content, and loading content from local storage while showing a splash screen. For example, when the native WP Email app is launched, it displays old messages while asynchronously fetching new messages. As a result, the launch time is 1.3-1.7 seconds. However, when the time for asynchronous message fetch is also accounted for, the elapsed time is 11 seconds, even on WiFi with no new emails present. Our anecdotal evidence suggests that checking for new email quickly is much more valuable than reviewing old email. Similarly, checking social networking apps for the latests posts is much more valuable than rereading old posts. For these network-bound apps, fast completion of asynchronous launch operations is important for much of the app's utility. Similarly, games commonly employ splash screens while asynchronously loading graphics and art objects into memory from local storage. To count total launch time in these cases, we disable or skip any splash screens. Therefore, in this work we focus on *Total Launch Time (TLT)* which we define as the base launch time plus completion of asynchronous launch operations. Figure 2 also shows the TLT of several first-party apps: Email, People (a Facebook and Twitter news reader), Maps and Weather. The TLT range is 5-11 seconds.

Slow launch times paired with brief usage durations are a troubling combination. It suggests that users are experiencing far too much overhead waiting for their apps to launch rather than engaging their apps. Fortunately, as §4 will show, our data-driven exploration indicates that user behavior is highly habitual, and linkable to observable context such as time and location.

3. SYSTEM OVERVIEW

`FALCON` offers an architecture in which to systematically observe and utilize context and execute predictive app launch actions. Figure 3 shows the `FALCON` system architecture. The central component is the launch predictor. Its role is to use context signals to predict app launches. The launch predictor's feature extractors convert raw data from context sources into features for the decision engine and model trainer. The decision engine performs *inference* to determine which features to use and what applications to prelaunch. The prediction is then passed on to the dispatcher, which loads apps into memory and executes the prelaunch routine of the selected app(s). The model trainer records observed features in order to periodically *train* and update the decision engine and feature extractors with new parametrizations.[2] Note that training occurs infrequently and is only critical when access patterns change significantly, whereas inference occurs frequently and is required whenever a launch prediction is made. The process tracker handles communication with the kernel's memory manager in order to ensure that the decision engine's view of the apps running on the system is in sync with what is actually running.

Supporting the launch predictor, the context source manager is a lifecycle manager and container for various context sources. It also helps shepherd raw data from individual context sources to the launch predictor.

A benefit of the `FALCON` architecture is its modularity. We have been able to use several context sources (location, time, ambient light level, accelerometer and gyroscope), and three separate decision engines within the architecture. In the following sections of this paper, we focus on the instantiation of the `FALCON` architecture which we found most suitable for launch prediction.

4. LAUNCH PREDICTOR DESIGN

The Launch Predictor is a machine learner with several feature extractors and a decision engine. This section first discusses the important features we identified, and then de-

[2]Unlike standard features, ours involve some training.

feature	description
f_t	triggers
f_{dt}	dynamic triggers
f_λ	location cluster
$f_{dt,\lambda}$	dynamic triggers per location
f_{ToD}	time of day
f_β	burst behavior indicator

Table 1: Feature Summary

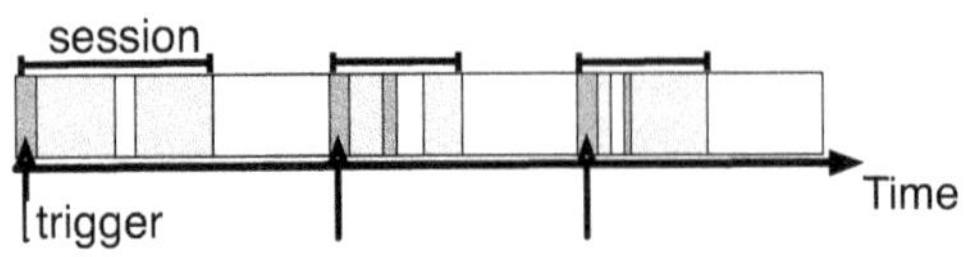

Figure 4: During this 2 hour timeline, SMS (in red) triggers other follower apps (in green). Arrows indicate start of triggered sessions.

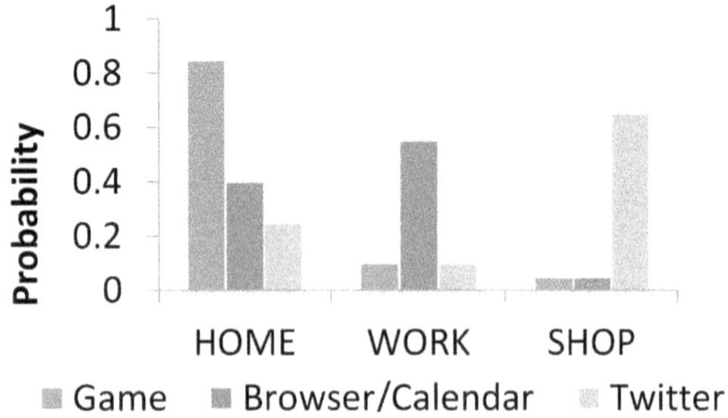

Figure 5: App usage is correlated with location.

scribes a novel cost-benefit learner and decision engine well-suited to launch prediction.

4.1 Personalized Features

Our primary objective for feature design is to identify a set of features that give strong insight into the next app to launch and that are very inexpensive to compute at inference time. As is typical of feature design, we take a data-driven approach. This investigation is heavily guided by our data exploration of the LiveLab and internal user studies. For each feature, we present a motivating vignette uncovered from the traces, followed by a formal definition of the feature. Table 1 summarizes the features considered.

4.1.1 Triggers and Followers

A novel feature in FALCON is session triggers and followers. Our data analysis reveals that once the user starts using the phone – for example to answer a phone call or view an SMS message – they have a tendency to open a sequence of other applications within relatively short durations of each other. We refer to such a sequence of application launches as a *session*.

The data traces reveal that certain apps are more likely than others to be the *trigger apps* that start sessions, whereas other apps are more likely to be *follower apps* that are used in the same session as the trigger apps. Figure 4 shows a two hour extract of one user's trace. During this time SMS was the trigger for four separate sessions. Analysis of aggregate user behavior revealed that SMS, Phone, Email, Facebook, and Web Browser are among the most popular triggers across all users. This is intuitive since trigger apps are ones that grab the users' attention (e.g. SMS and Phone) or satisfy immediate information needs (e.g. Browser).

Let f_t denote a trigger feature. Intuitively, the trigger features identify the occurrence of a trigger app weighted with the likelihood of a follower app. For random variable follower app F, trigger app T, and candidate launch app α, the trigger feature is calculated as follows.

$$f_t = \Pr(F = \alpha \mid T \in \tau_t)\,\mathrm{I}(\tau_t)$$

where τ_t is a set of triggers, for example $\{SMS, Phone\}$, and $\mathrm{I}(x)$ is an indicator function that is 1 if (any member of) x is observed, and 0 otherwise. The term $\Pr(\mathrm{F} = \alpha \mid \mathrm{T} \in \tau_{\mathrm{st}})$ expresses the probability that the follower app is the candidate launch app α given that the trigger app is one of SMS or Phone. Note that probabilities are based on the data distribution observed during training.

Interestingly, we found that the best triggers were different for different applications, even for a single individual. For example, the best three triggers for Email are different from those for AngryBirds. There is also significant variability across users — a heavy user of games might have a game as a trigger whereas others may not. These differences lead us to propose dynamic triggers that are calculated on a per-launch-candidate basis as the set of top-k triggers most likely to lead to the launch candidate as a follower.

$$\hat{\tau}_k = \operatorname*{argmax}_{\tau_k} \Pr(F = \alpha \mid T \in \tau_k)$$
$$f_{dt,k} = \Pr(F = \alpha \mid T \in \hat{\tau}_k)\,\mathrm{I}(\hat{\tau}_k)$$

where $\hat{\tau}_k$ represents the k best triggers for α. §4.2 discusses the procedure for choosing k to arrive at a fixed set of dynamic triggers per app f_{dt}.

4.1.2 Location Clustering

Empirical observations suggest that location is correlated with app usage. Figure 5 illustrates the effect of location on app usage for a brief two week time window for a representative user from our internal user study. For this user, there is a surprisingly strong tendency for game usage at home, Browser and Calendar usage at work, and Twitter usage at a frequently visited shopping center.[3]

The location feature f_λ is computed by assigning the user's current location to the nearest neighbor cluster.

$$f_\lambda = \operatorname*{argmin}_{c \in Clusters} \mathrm{DIST}(\lambda, l_c)$$

where λ is the current location and l_c is the central location of cluster c. Clusters are computed from the user's historical data during the training phase. The literature offers many geospatial clustering algorithms [10], and after experimenting with several, we chose k-means clustering which incurs low computation cost while performing comparably to the others.

We can use location in conjunction with triggers as a joint feature $f_{dt,\lambda}$. For each location, we compute the best set of

[3]The labels Home, Office and Shop are educated guesses according to the recorded geocoordinates.

Figure 6: This particular user launched the app Angry Birds frequently from mid-Jan to mid-Feb only. Both yearly and monthly timelines are shown.

dynamic triggers as follows.

$$\hat{\tau}_{\lambda,k} = \underset{\tau_{\lambda,k}}{\text{argmax}} \Pr(F = \alpha \mid T \in \tau_{\lambda,k}, L = \lambda)$$

$$f_{dt,\lambda,k} = \Pr(F = \alpha \mid T \in \hat{\tau}_{\lambda,k}, L = \lambda)\, \mathrm{I}(\hat{\tau}_{\lambda,k})\, \mathrm{I}(\lambda)$$

This feature captures dynamic trigger variation across locations.

4.1.3 Temporal Bursts

We also find that a user's app usage changes significantly over longer time scales. Figure 6 illustrates an app, Angry Birds, becoming popular during a month-long period, then ceasing to be used at all afterward. Perhaps not surprisingly, this behavior turns out to be very common for games and entertainment apps. However, our data analysis reveals that such *burst* usage behavior is dependent on the type of application, for example, games have shorter but more intense bursts in comparison with other applications. Therefore, a key challenge is to determine the frequency and duration of these bursts for each application in an efficient manner.

To address this issue, we develop a novel *burst predictor*. The burst predictor uses multiple sliding windows, each of a different resolution, to detect and react to sudden upticks in app usage. Each app has its own burst predictor to track its bursty behavior.

As a basis for our burst predictor, we use a wavelet-based burst detection algorithm [15] as a subprocedure. Given a stream of data, the detector's goal is to find all windows of any length during which the aggregate statistic of the window exceeds a threshold. In our instantiation, this corresponds to detecting any sequence of days during which an app is launched more than a threshold per day on average. We currently set the threshold to one.

The wavelet detector uses a unique modified wavelet data structure in order to bound the computational complexity of detection by $O(n)$ where n is the number of days, whereas the brute force solution is $O(n^2)$. The output is an indicator for each window $w_{i,j}$ where i indicates the offset from the start of the stream and j indicates the length of the window. Let $b_{i,j}$ be a burst indicator which corresponds to a sliding window starting from day i of length j days. $b_{i,j}$ is true if the corresponding sliding window is a burst, and false otherwise.

Provided the burst detector, our burst predictor must decide whether the next time step is part of a burst or not. To make the decision, our predictor measures the predictive power of each burst window size j in being able to successfully predict whether the app is used in the day following the window. Let w_j represent the predictive power of window size j.

$$w_j = \frac{1}{n-j} \sum_{i}^{n} [b_{i,j} b_{i+j,1} + (1 - b_{i,j})(1 - b_{i+j,1})]$$

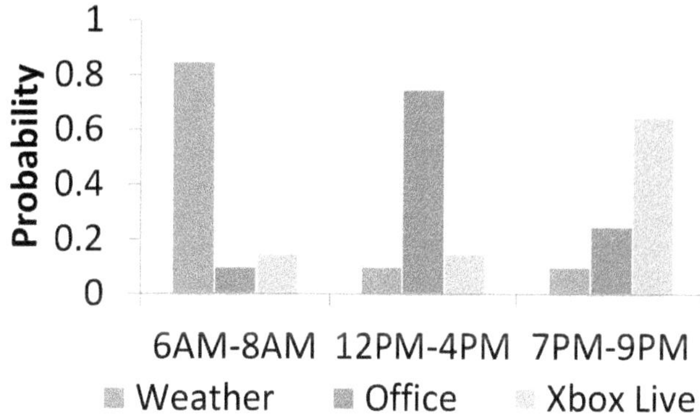

Figure 7: Time of day is correlated with app usage.

By analyzing historical data, we can now determine the best window size as the one with the highest score, and use it to detect a burst. This window size can be used to compute f_b, which is set to 1 if we are currently within a burst window and 0 otherwise.

$$\hat{j} = \underset{j}{\text{argmax}}(w_j)$$

$$f_\beta = b_{n-\hat{j},\hat{j}}$$

Time of Day Feature: Besides multi-day temporal bursts, app usage likelihood changes throughout the day as well. Figure 7 illustrates this for a typical user in our internal user study. A simple Time of Day (ToD) feature f_{ToD} represents the time of the day, segmented by hour. Our experiments with finer and coarser granularities did not yield much difference.

4.1.4 Additional Context Signals

We also built context sources mapping to accelerometer, gyroscope and ambient light level sensors. These proved sufficiently straightforward to incorporate into the `FALCON` architecture. However, we do not yet have sufficient user trace data to analyze their effectiveness at predicting app launches.

4.1.5 Multi-feature Decision Engine

The different features that we have described can be used as the core inputs of a decision engine that makes application prelaunch decisions. Once training phase has provided the trigger features location clusters, and burst predictions, the decision engine can decide whether or not to launch an application α when a trigger T is met. The decision procedure takes three steps.

1. f_β is checked to see whether a burst of α is underway.
2. f_λ is checked to see whether the current location suggests use of α.
3. $f_{dt,\lambda}$ is checked to see whether a trigger for α started the current session.

The three steps are executed in a sequential order. The launch decision for α can be summarized as follows.

$$decision(\alpha) = \begin{cases} true & \textbf{if} \quad f_\beta\, f_\lambda\, f_{dt,\lambda} > 0 \\ false & \textbf{else} \end{cases}$$

Our description of the decision engine, however, has a major omission — it does not take into account the actual cost and benefit of a correct preload versus an incorrect preload. In the following section, we describe how we can incorporate these cost-benefit considerations into our decision engine.

4.2 Cost-Benefit Learner

The Cost-Benefit Learner (CBL) combines the output of the decision engine for each application together with knowledge of the cost-benefit tradeoffs of launching the application to decide whether to preload an application. We first provide an intuition for how the CBL operates before launching into the specifics of its mechanics and the details of the cost-benefit trade-off.

4.2.1 CBL Overview

To decide whether to prelaunch or not, CBL takes three steps: First, it computes the precision and recall metrics for an application from the decision engine (§4.1.5). Second, it converts the precision and recall number to cost and benefit. Third, it determines the best followers for each trigger by maximizing benefit for a fixed cost budget.

Precision and Recall: The precision and recall for a trigger and follower pair $\{T, F\}$ at location cluster λ are defined from observed user data as follows.

$$precision(T, F, \lambda) = \frac{\text{launches of } F \text{ after } T \text{ at } \lambda}{\text{sessions with trigger } T \text{ at } \lambda}$$
$$recall(T, F, \lambda) = \frac{\text{launches of } F \text{ after } T \text{ at } \lambda}{\text{sessions with follower } F \text{ at } \lambda}$$

Cost and Benefit: We measured the energy cost and loading time of all apps apriori. We can use the measurement results to determine the cost and benefit of prelaunching by combining the precision and recall results.

Precision describes the expected number of prelaunchs that leads to an actual usage. Multiplying by energy cost of a prelaunch, we can derive the energy cost of a prelaunch that leads to an actual usage. Recall describes the probability that a follower is preloaded. Multiplying by the launch time of the follower, we can derive the loading time benefit of a prelaunch across all launches. We will quantitatively analyze the cost and benefit in the next section.

Trading-off cost and benefit: We assume that the CBL is provided a daily energy budget. The energy budget is a predefined percentage of the total battery capacity that the CBL engine can use for its prelaunches within a day. (For example, the CBL may only be able to use 1% of the battery each day across all prelaunches.) Given this constraint, the goal of the CBL is to pick, for a given user, the best set of followers to prelaunch. Intuitively, we should prelaunch the items that have least energy cost and highest loading time benefit. The goal of our optimization framework is to choose the best option given the energy budget.

4.2.2 CBL Optimization Framework

We first describe the optimization procedure which runs during training. For ease of exposition, we cover the CBL's method for finding dynamic triggers on a per location- and per app-basis. In our implementation, the technique is extended in the straightforward way to additionally include time-of-day.

Let c_α and b_α represent the energy cost and latency reduction benefit respectively for a single prelaunch of α. Then the expected cost c' and benefit b' of selecting to prelaunch α for all encounters of t and λ is calculated as follows.

$$b'_{t,\lambda,\alpha} = b_\alpha \times recall(t, \alpha, \lambda)$$
$$c'_{t,\lambda,\alpha} = c_\alpha \times (1 - precision(t, \alpha, \lambda))$$

Note that while the benefit b' is scaled by the proportion of triggers leading to the follower, the cost c' is scaled by the proportion of triggers *not* leading to the follower. This is because mistaken predictions incur a cost c_α whereas energy for correct predictions can be accounted toward the user's actual launch.

We map our formulation to the 0-1 Knapsack problem to model the cost and benefit trade-off. Consider all tuples $< t, \lambda, \alpha >$ as candidates for the Knapsack. The predefined energy budget corresponds to the max allowable Knapsack weight, the expected reduced launch time $b'_{t,\lambda,\alpha}$ corresponds to the benefit of inclusion in the Knapsack, and the expected energy cost $c'_{t,\lambda,\alpha}$ corresponds to the weight in the Knapsack. It is a 0-1 Knapsack since a tuple can be included or excluded in its entirety.

The 0-1 Knapsack model for the cost-benefit trade-off is formulated as follows.

$$\text{maximize} \sum_{t,\lambda,\alpha} b'_{t,\lambda,\alpha} \mathrm{I}(< t, \lambda, \alpha >)$$
$$\text{subject to} \sum_{t,\lambda,\alpha} c'_{t,\lambda,\alpha} \mathrm{I}(< t, \lambda, \alpha >) \leq W$$

As before, the function I() is either zero or one, indicating whether or not to include the tuple $< t, \lambda, \alpha >$. W is the energy budget.

One possibility to solve the Knapsack problem is with dynamic programming. However, a dynamic programming solution requires pseudo-polynomial time. Instead, we use a standard greedy approximation to solve the Knapsack problem. The greedy approximation sorts the tuples in decreasing order of their benefit per unit of cost. It then proceeds to pick the tuples according to the ranked order, starting with the top-ranked one, until the energy budget is reached. The output of the greedy approximation is K, a list of tuples sorted by benefit per unit cost within the energy budget.

The final step of the optimization updates the dynamic trigger feature $f_{dt,\lambda}$ to use the appropriate top-k triggers. Recall from §4.1.1 that for a given α, the value of k is used by $\hat{\tau}_{\lambda,k}$ for the feature $f_{dt,\lambda,k}$. k is set according to the number of appearances of α in K. The decision engine, which looks at $f_{dt,\lambda,k}$ for the chosen k, performs inference in time linear to the length of K.

Dynamic Energy Constraints

There are several ways to define an energy budget for the Knapsack optimization framework. For example, one could budget a fixed percentage of the battery per day, pick a percentage of the current battery level or have an finer-grained budget at an hourly level. While our optimization can work with different definitions of the energy budget, we use the representative example of a budget that is chosen based on the residual energy. The Knapsack model is adapted to the dynamic energy constraint W' as follows.

$$W' = (W - U)^\theta$$

where U is the energy consumed so far by prelaunching,

and θ is a tuning parameter in the range $(0, 1]$ that can be chosen to discourage aggressive prelaunching when energy is abundant. We currently set $\theta = 1$. Note that meeting the dynamic budget constraint does not require recomputation of Knapsack. Instead, we maintain a pointer to the position in K which represents the dynamic energy budget cutoff.

Eviction Policy: Most phone OSs use an LRU policy for app eviction.[4] While the CBL is primarily used for launch prediction, the sorted tuple list K is also used to evict the app least likely to be used. As part of launch prediction, apps already resident in memory are assigned very low cost scores. Nonetheless, such an app may still be evicted due to its meager benefit b' according to the current trigger t and location λ. This conveniently integrates launch prediction with eviction policies.

Server-based Training If the user app launch behavior changes (e.g. a new location is visited or a new app is installed), retraining is recommended. Even though the greedy Knapsack approximation is efficient, training involves scanning through all the logged context data. Therefore, we offload training to a server. We currently initiate retraining manually. The log with data records since the last training are sent to the server, and new parameters are sent back to the phone.

Memory Constraints In the preceding discussion, we treated all preloads equally in terms of memory cost. The FALCON implementation and its associated evaluation also make this simplifying assumption. In practice, apps vary in memory usage. A logical extension of our Knapsack model can handle multiple-constraints where both energy and memory are treated as constraints.

5. IMPLEMENTATION

We have implemented FALCON as an OS modification to Windows Phone (WP) 7.5. We first discuss the elements of WP's app model relevant to our implementation before describing the implementation details.

5.1 Background on Windows Phone Apps

WP apps are similar to webapps in organization. Each app is collection of pages with links pointing across pages. An app is referenced by its unique ID (UID), and a page within an app is referenced by its page name. Pages can take arguments as parameters. The following is an example of these elements: *app://<app-id>/ShowContactInfo?id=8*. Each app has a default page *Default* that is loaded at startup. Apps can also support multiple entry points by supporting "deep links" to specific pages. For example, the standard Address Book app supports pinning of contact cards to the home screen via deep linking. Links are not functional cross-app; WP enforces strict app isolation.

The WP app lifecycle consists of four states: Not Launched, Active, Deactivated and Tombstoned. Apps transition from Not Launched to Active when tapped from the home screen, and from Active to Deactivated whenthe user navigates away to the home screen or another app. An app undergoing deactivation has a 10 second grace period to perform cleanup activity before receiving no further cycles. A deactivated app is still resident in memory until it is evicted by the memory manager which then changes its state to Tombstoned. For Tombstoned apps, the OS retains some minimal tombstones of bookkeeping information, such as the last page of the app the user was on. In very low memory conditions, even tombstones may be dropped. To limit battery drain, background apps are only permitted 15 seconds of CPU time every 30 minutes.

Figure 8: FALCON ambient display widget is shown at the top of the home screen. Current predictions are *Monopoly*, *FindSomeone* and *Music+Video*.

5.2 Implementation Description

FALCON is implemented in three parts. The first is an OS service. The second is a new application event upcall. The third is an ambient display widget.

The FALCON OS services is a privileged daemon written in native code. As such, it has access to the registry, a persistent structured datastore, and can call privileged library and system calls. For example, starting a new process is a privileged call. FALCON replaces a subset of the standard policies of the system's memory manager and scheduler with its prelaunch decisions.

The service manages a thread for each context source. Currently, three context sources are used corresponding to the features described in §4.1. Context sources post updates at different rates, depending upon the rate of change of the underlying signal. In order to support variable update rates, the main service shares a *model* data structure with each source and the launch predictor. Each source independently updates the model with observations for the context for which it is responsible.

The launch predictor uses the model's observations in its inference process. The launch predictor is executed whenever one of several defined events occurs: (1) the phone receives an SMS or phone call, (2) the user unlocks the phone, or (3) the user launches a trigger app. These events already activate the phone's application processor from a low to high energy state so it is inexpensive to piggyback the launch predictor execution.

Upon generating an inference, the launch predictor passes the candidate app(s) to the *dispatcher*, which vectors the launch signal through the appropriate prelaunch event up-

[4]Without paging, apps are evicted in their entirety.

call. For third party apps which are written in managed C#, this process involves translating native calls to managed code calls.

Apps implement the event handler as a custom page named *Prelaunch*, e.g. *app://<app-id>/Prelaunch*. The dispatcher jumps directly to the Prelaunch deep link. An app can implement any custom prelaunch logic in its Prelaunch page. For example, our modified Email app fetches new emails during prelaunch, and our modified game loads the last level the user played during prelaunch. In general, network content fetch or local storage load operations are good candidates for Prelaunch. For apps that do not implement the Prelaunch link, the dispatcher opens its Default page instead. This is sufficient for us to reduce launch time for many apps without requiring app modification.

In WP, pages are meant to show a display to the user, which is not appropriate for prelaunching. Once the dispatcher opens an app to either the Prelaunch or Default page, it immediately deactivates it. With the existing WP deactivation grace period, this gives the app an opportunity to complete its app-specific prelaunch operation. For prelaunch deactivation, the grace period is modified to 25 seconds, which is high enough to cover the total app launch times of most apps. The deactivation event also gives apps an opportunity to adjust for any side-effects (e.g. unregistering server state) that the Default page may have initiated. Conveniently, developers already implement the deactivation event as part of the regular app lifecycle management protocol.

Lastly, FALCON includes an ambient display widget using the WP live tiles mechanism. Live tiles are simply home screen icons that can be updated programmatically. Figure 8 shows a screenshot of the FALCON widget. The launch predictor updates the live tile to show the highest likelihood candidate apps.

6. EVALUATION

In this section, we evaluate the performance of FALCON through extensive data analysis as well as measurements of our prototyped system implementation.

6.1 Dataset

The primary dataset in our analysis is the Rice LiveLab user study [13]. In this study, thirty four volunteers were given iPhone 3GSs for a period of up to fourteen months. We use two traces from the LiveLab users study: the app usage trace which records the start time and duration of every app usage, and the location trace which records the user's location with an interval of five minutes when GPS is available. By joining the app usage table and location table, we have a trace of the time and location information corresponding to each app usage. Note that since location is obtained at a relatively coarse granularity compared to app usage data, there are several instances where we do not have data at precisely the times that an app was used. To address this, we interpolate location points using a nearest neighbor method. In total, the interpolated trace of time and location information contains over 400 thousand samples across fourteen months. The majority of our evaluation is conducted based on the LiveLab dataset, except for microbenchmarks which are measured on the Windows Phone platform.

6.2 Microbenchmarks

Before measuring the performance of FALCON, we first benchmark three critical parameters for each app: a) the time taken for launching an app, b) the amount of energy consumed during app launch, and c) the memory consumed by the app at launch time. Note that we focus on these parameters at launch time as opposed to how they vary during app usage.

Our measurements are conducted on an LG Mazza phone running Windows Phone 7.5. We measured the top twenty most popular apps in the Windows Phone Marketplace, as well as the first party apps that are preinstalled on the phone, including People, Email, and Maps.

Figure 9(a) shows the CDF of app launch time with and without prelaunching. The launch time includes both the OS loading time and network fetch time. The network fetch time is the average over a number of measurements under a good WiFi connection, which is an optimistic estimate since a cellular data connection is likely to be slower. Without preloading, half of the apps require more than 10 seconds to be fully launched, and around 10% of apps have a launch time of more than 20 seconds. With preloading, the app launch time can be reduced dramatically to close to zero.

Figure 9(b) shows the CDF of energy cost to launch an app. The energy cost as measured with a Monsoon power meter ranges from 500 to 2500 μAh, with a median energy consumption of 750 μAh. For reference, a typical LiOn battery on a mobile phone has a capacity of 1500 mAh, therefore the energy cost ranges from 0.03% to 0.17% of the battery capacity for a *single launch*. Clearly, when app usage is high, the total energy cost of launching applications can be a significant fraction of the overall budget. A benefit of FALCON is that we can cap the total amount of energy for preloading to ensure that its effect on battery life is bounded.

Figure 9(c) shows the CDF of memory cost for launching an app. The memory consumption varies from 10 MB to 90 MB, with a median memory consumption of 30MB. This is a significant fraction of available memory on smartphones, which typically have memory capacity between a few tens to hundreds of MB. (Our latest WP phone has about 150MB available for user apps.)

6.3 Benefits of Individual Features

At the core of FALCON are three features: session triggers and followers, temporal bursts, and location clusters. We now look at the benefits of using each of these features in isolation.

6.3.1 *Session Triggers and Followers*

Our goal in this section is to understand trigger and follower behavior across different users and applications. Specifically, we ask three questions: a) what applications are the best triggers across different users, b) are there differences between individuals in their trigger and follower behavior, and c) for a single user, are there differences in triggers for different follower apps?

Figure 10(a) shows the top 5 session triggers across all users. Not surprisingly, this list includes Email, Facebook and Safari — all popular apps. What is surprising, however, is that SMS and Phone calls are excellent triggers as well. In other words, receiving an SMS or a phone call creates an interruption into an individual's routine, and once inter-

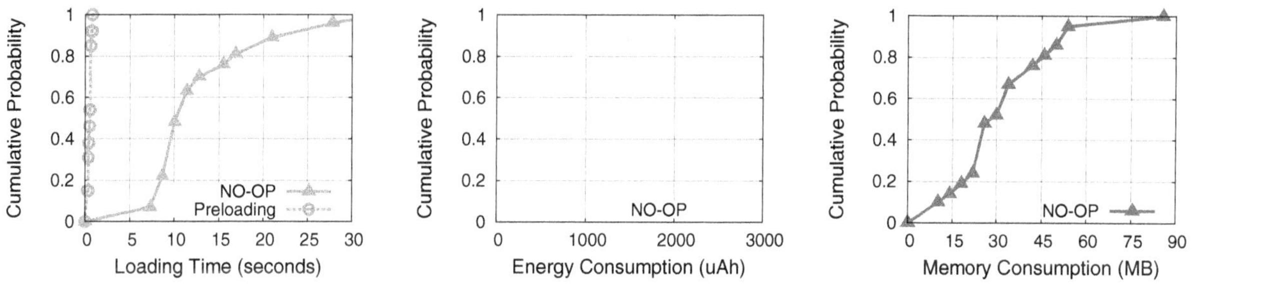

(a) CDF of launch time (b) CDF of energy consumption (c) CDF of memory consumption

Figure 9: Microbenchmark of launch time, energy cost, and memory cost. Apps measured include the top 20 most popular ones in Windows Phone Marketplace, and preinstalled first party apps.

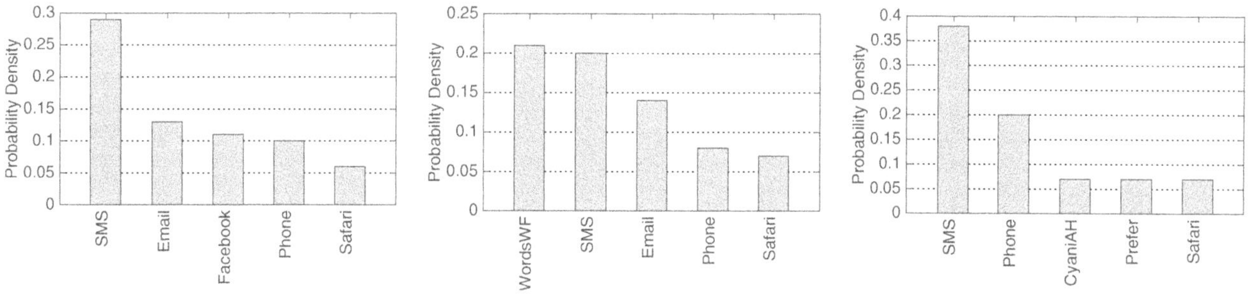

(a) Top 5 triggers across all users. (b) Top 5 triggers for a specific user. (c) Top 5 triggers for "Angry Birds".

Figure 10: Different set of session triggers for (a) aggregation across multiple users, (b) a specific user, and (c) a specific app across different users. Y-axis shows the probability that each app acts as a trigger. App name abbreviations in the graphs are: "WordsWF" = "WordsWithFriendsFree," "CyaniAH" = "CyanideAnd-Happiness."

rupted the users tend to use other applications before they close the phone.

Figure 10(b) shows the top 5 triggers across all users are not necessarily the best triggers for an individual. This graph shows the best triggers for a user who is heavy user of games — in this case, the best trigger is in fact the *WordsWithFriends* game. This indicates that personalization is needed for the user, and we cannot assume that triggers are the same across everyone.

To understand if different apps have different triggers, we look at the triggers for a specific application across all users. We choose *Angry Birds* as our target app in this evaluation since it is very popular across multiple users and its loading time is representative for games. Figure 10(c) shows that triggers for *AngryBirds* are also different from those across all applications, suggesting that triggers should be chosen differently for each application.

To conclude, this evaluation shows that triggers need to be both user- and app-specific, and it makes the case for dynamic personalized triggers that we will evaluate in §6.4.2.

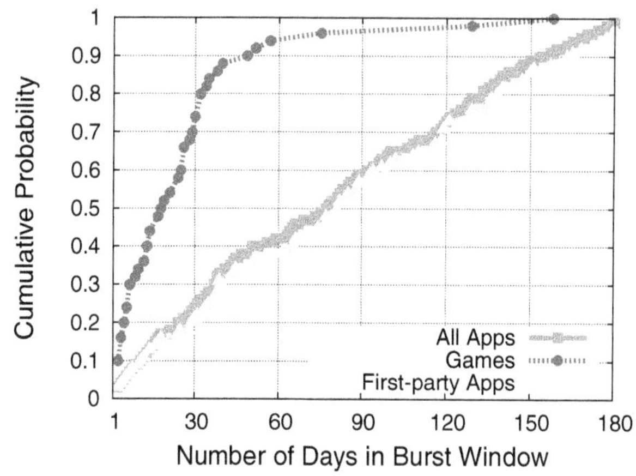

Figure 11: CDF of number of days that involve bursty usage for each application. 80% of games have less than 30 bursty days over 14 months usage.

6.3.2 Temporal Bursts

We now turn to evaluate the performance of the temporal feature, namely bursty app usage. Temporal bursts can be an effective feature for improving prelaunch accuracy since it indicates when an application is heavily used during a relatively short window of time, such as a span of several days. Our evaluation looks at which applications are likely to be used in a bursty manner and which are not. Intuitively, one would expect that games are likely to be used in a bursty manner whereas First-party apps such as Email or People (WP Facebook client) should have a more steady usage pattern. Therefore, we explore the burstiness for three categories of apps — All applications, Games, and First-party apps.

Figure 11 shows the CDF of number of bursty days for each app category. Each point in this graph corresponds to a {user, app} pair. From this graph, we can conclude that games have stronger bursty behavior than apps such as Email and Facebook. Around 80% of games are used within 30 days out of 14 months trace length, while only 20% of the first party apps are used within the same time frame. This observation suggests that bursty behavior is a discriminating feature that we can exploit to improve the prelaunching performance, especially for games.

6.3.3 *Location Clusters*

After temporal burst feature, we now look at the performance of location clusters. Similar with temporal features, we look for applications that are heavily used within certain location clusters to improve prelaunch accuracy. We evaluate the distribution of number of location clusters per application (determined via k-means clustering). We use the same categories as the previous section — All applications, Games, and First-party apps. Our intuition is that games tend to be used in fewer locations than other apps.

Figure 12 shows the CDF of number of clusters where each app category is used. From this graph, we can see that games have a higher locality than other apps — 90% of game apps are used in less than 25 clusters, and more than half of games are used in less than 10 clusters. Thus, we conclude that location clustering is also a discriminating feature, particularly for games.

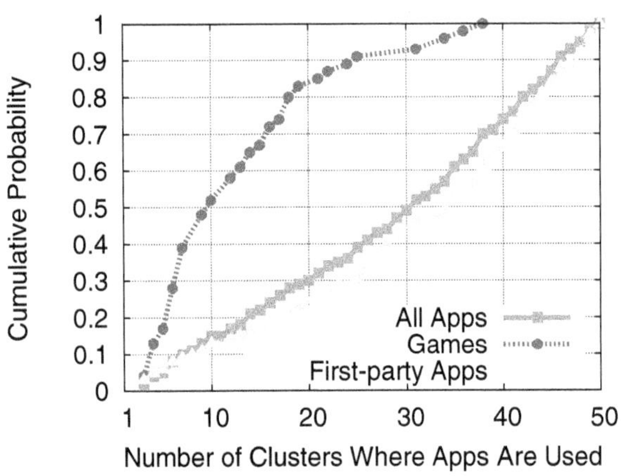

Figure 12: CDF of number of location clusters per app. Over 50% of games are used within less than ten clusters.

6.4 Combining Features

We now look at the benefits of combining the features of temporal burst and location clusters on prediction accuracy. Our evaluation consists of two steps. First, we look at how adding temporal features and location clusters improves accuracy when the set of triggers are fixed across all users and apps. Second, we evaluate the benefits of making the triggers dynamic, in which case triggers can be modified within a cluster and burst window based on the individual and target application for each user.

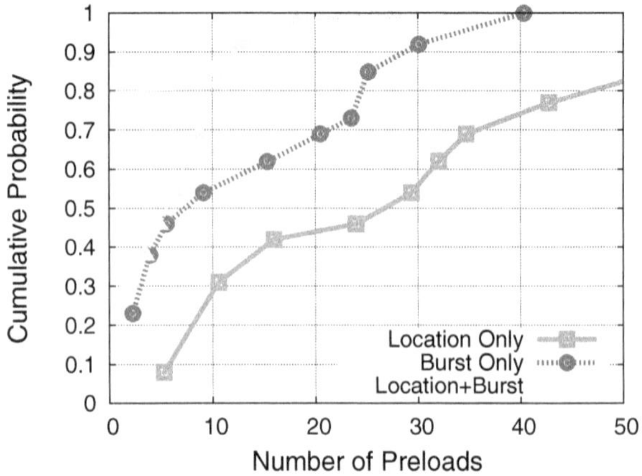

Figure 13: Benefit of location and temporal burst features, measured in terms of number of preloads per correct usage.

6.4.1 *Benefits of location + temporal features*

In this experiment, we look at the benefits of combining location and temporal features. We fix trigger apps to include the most common triggers across users, Phone, SMS, and Safari. We then explore the benefit of using only location clusters, only temporal bursts, and finally a combination of the two for each follower app. For each setting, we look at improvement in prelaunch accuracy, specifically, the number of preloads per app usage for each user.

Figure 13 shows the CDF of number of preloads across all users. Note that all curves have the same set of triggers. As seen from the figure, location clustering by itself performs poorly — the median number of preloads that lead to an actual app usage is around 27. If we use only temporal bursts, the median preload number is reduced to only six. Combining the two features gives even better performance — median number of preloads reduces to around five.

6.4.2 *Benefits of dynamic triggers*

While combining location and temporal features improves preload accuracy, it still incurs the cost of several preloads per app usage. We now look at how dynamic triggers improve performance over static triggers. This experiment evaluates improvement in precision and recall by personalizing triggers for each user based on both location and temporal burst contexts. We focus on "Angry Birds", a popular game that provides a representative example of an application that exhibits both location and spatial selectivity.

We compare the precision and recall of preloading under three trigger settings: static triggers for all users and location clusters, dynamic triggers for each user but the same for all location clusters, and dynamic triggers for each user at each location clusters. Figure 14(a) and Figure 14(b) shows the precision and recall for two different users when the follower app is *AngryBirds*. For both users, dynamic triggers for each location cluster achieves the best precision and recall, and clearly provides dramatic improvements over the other two cases. In addition, the performance of dynamic triggers for each location cluster is also stable for different users, while the performance of the other two schemes varies significantly for different users. This observation clearly in-

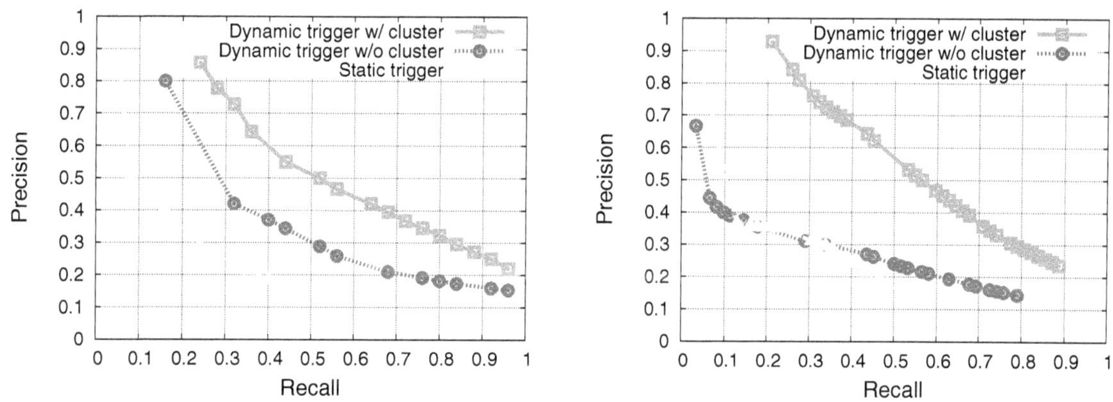

(a) User 1 (b) User 2

Figure 14: Precision and recall of prelaunching a follower app for two individual users under static and dynamic trigger settings. In this graph, "Angry Birds" is used as a representative follower app that exhibits both temporal and spatial selectivity.

dicates that we should use dynamic personalization of triggers at the location cluster level.

6.5 Evaluation of cost-benefit learner

Having comprehensively evaluated how features can enable better preloading, we turn to evaluate the performance of Cost-Benefit Learner (CBL). Our cost-benefit analysis includes information about the cost of a preload in terms of energy, and the benefits of the preload in terms of app loading latency, or in the case of network prefetching, in terms of the freshness of fetched data. We use the energy/latency benchmarks from §6.2 in this evaluation. We assume for simplicity that each app has equal memory consumption.

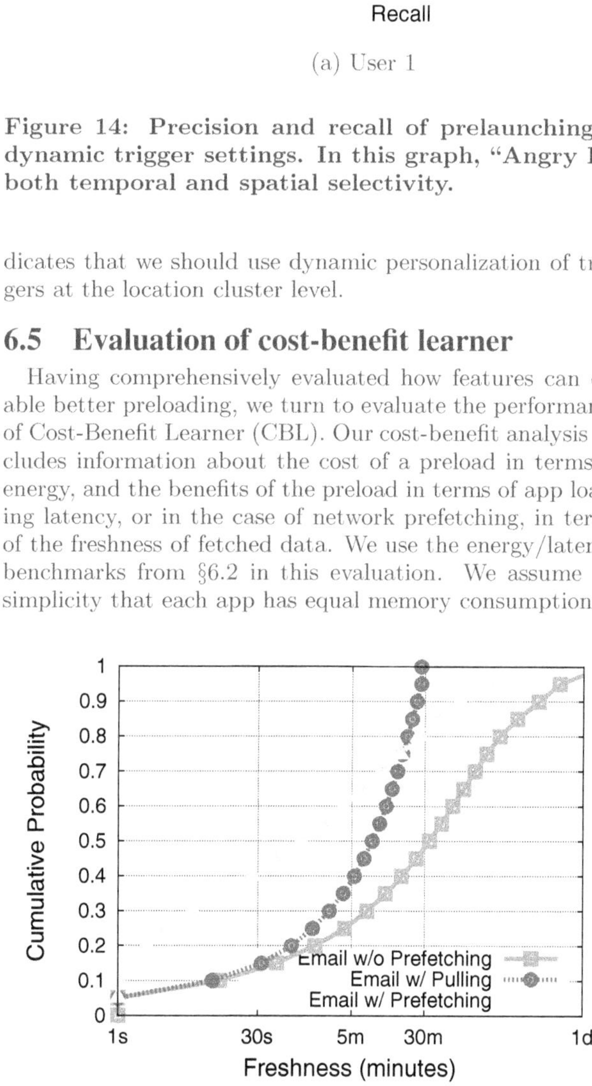

Figure 15: Benefit of prelaunching for app freshness. Email is used as a representative application that is both heavily used and has high network content fetch time.

6.5.1 *Performance of Prefetching*

In this experiment, we look at the benefit for apps that require fetching new content during the launch process. We take the Email application as a representative app, which is both heavily used and incurs significant delay for fetching emails from servers.

First, we compare CBL (noted as Email w/ Prefetching) against two baseline schemes: a) no prefetching (noted as Email w/o Prefetching), and b) active pull with an interval of 30 minutes(noted as Email Pulling). Our evaluation metric is freshness of the app, which is defined as the time that elapses between retrieval of content from the server and when the app is opened by the user.

Figure 15 shows the freshness of rhe Email app using our scheme (note that the x-axis is in log scale). Compared with no prefetching, we improve freshness significantly by reducing the median freshness from 30 minutes to around 3 minutes. Our scheme also outperforms the pulling-based scheme, whose median freshness is about 10 minutes.

6.5.2 *Performance of Preloading*

In this experiment, we look at how app loading time can be improved by our CBL learner. We compare our scheme against LRU caching, which is the standard policy on most smartphones including Windows Phone. In this experiment, we select 24 popular apps (including games, social apps, and First-party apps) as follower to cover the most common app usage cases on WP platform. We assume that the memory manager allows up to five apps, as is the case on the current Windows Phone platform. We also assume that every app has equal memory consumption.

While both CBL and LRU take advantages of bursty behavior of apps, there is a subtle but important difference in the way FALCON vs. LRU work for bursty apps. It might seem like LRU would work well when apps are bursty, but the burstiness of apps (e.g. games) is at the timescale of days, not minutes. In other words, a "bursty game" might be used once or twice a day. LRU is designed to work well when burstiness is at short timescales, where apps are so frequently used that they are not evicted. CBL is able to take advantage of burstiness at the timescale of days because we are looking at longer term contextual and session behavior.

Our CBL outperforms LRU by up to 100% in terms of loading time. Figure 16 shows the median loading time is around 1 second and 2 seconds for CBL and LRU, respec-

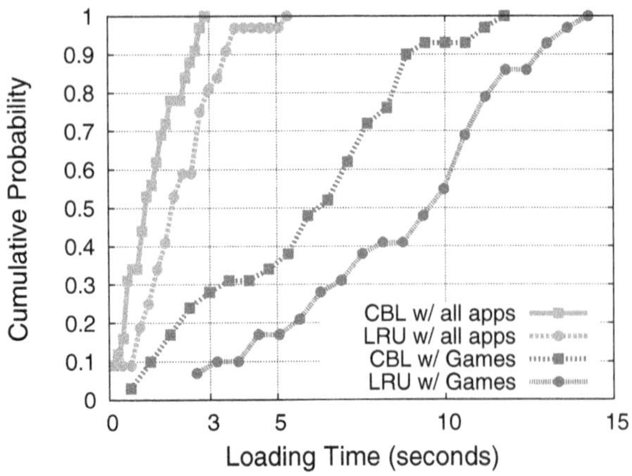

Figure 16: Benefit of loading time with prefetching. CBL is compared against LRU, both when all applications are used, and when only games are used. We look at games separately since these apps exhibit both temporal burstiness and spatially correlated usage patterns.

tively. Even when LRU already performs well in terms of loading time, our scheme still improves the loading time by half. For games, the median loading time is also reduced from 9 seconds with LRU, to 6 second with CBL.

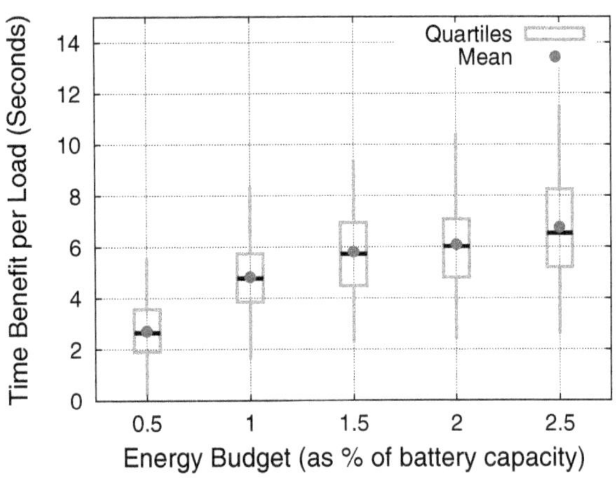

Figure 17: Energy cost and launching time benefit of using the combination of dynamic triggers, temporal bursts, and location clustering for prelaunch.

6.5.3 Overall benefits

Figure 17 shows the aggregated cost and benefit of using CBL. The x-axis of this graph represents the energy budget provided to CBL, and is normalized to the percentage of a fully charged phone battery(1500 mAh) as the daily budget. The y-axis represents the benefit of loading time which includes both app loading time and content fetching time. Our estimate for content fetching time is conservative since we used our micro-benchmark results under good WiFi connections. Our target metric is the reduced loading time per app usage. Figure 17 shows that with CBL, the average loading time for an app is reduced to 5.5 seconds with the energy cost of less than 1% battery power, and less than 6 seconds with 2% battery power.

Binary Size	129 KB
Memory (stable state)	1840 KB
Processor utilization (stable state)	<1%
Processor utilization per prediction	<3%
Energy cost per prediction	< 3 μAh

Table 2: FALCON Overhead Profile

6.6 Bootstrapping FALCON

Having evaluated the overall performance of cost-benefit learner of FALCON, we now ask how fast the cost-benefit learner can learn from history of app usage to make accurate prelaunch decisions. In this experiment, we tune the training dataset size by progressively adding new app usage data to the set, and study the performance of CBL in terms of precision and recall aggregated across all follower candidates.

We split the 14-month long app usage trace as follows in this experiment: we use the data from day one to day N as training data, and we tune the number N such that the training dataset size is set from 1% of total dataset size to 60%. We then use the 20% of data immediately following the training dataset as the testing dataset. For each setting, both location clustering and temporal burst windows are re-calculated to reflect temporal dynamics in features. To simplify the analysis, we do not consider energy constraint in this experiment but only study precision and recall of up to the top five follower candidates.

From Figure 18(a), we conclude that the topmost candidate from FALCON begins to reach over 50% prediction precision with around 35 days of training data. When the training dataset size grows to 84 days, or 2.8 months, prediction precision is close to 70%. The precision climbs to over 80% when the training dataset size is over 252 days, or 8 months. Figure18(b) shows that the recall of FALCON ins improved as the fraction of training set increases. If we prelaunch the top 5 follower candidates, over 70% of the targeted apps can be successfully preloaded by FALCON when the training dataset is 42 days long. The recall increases to over 80% when the training dataset size is more than 252 days, or 8 months long.

In conclusion, the performance of FALCON grows as the training data size increases. Even with a relatively small training dataset of two months, FALCON can make prelaunch decisions with good accuracy, demonstrating that good performance can be achieved with a few months of training.

6.7 System Overhead

The system overhead of our FALCON implementation is very modest. Table 2 shows its resource consumption profile. Memory usage during stable state is less than 2MB, and does not increase measurably during prediction. Processor utilization is low during stable state, and reaches up to 3% during prediction. We also measured the energy overhead of FALCON predictions with a Monsoon power meter, and we observed that the extra cost FALCON incurs for each launch prediction is only a few μAh.

Note that we have not accounted for the periodic geolocation sampling costs as a component of the overhead. This is

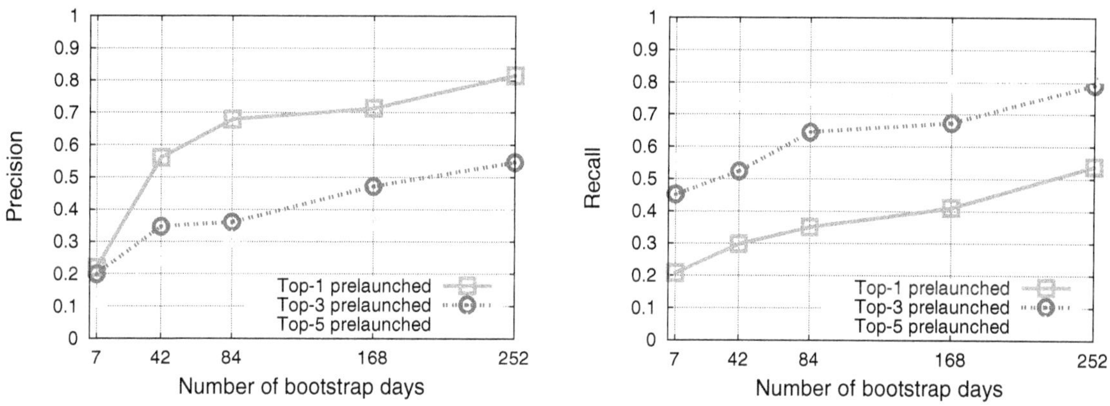

(a) Precision

(b) Recall

Figure 18: Precision and recall of bootstrapping.

for two reasons: a) we don't need GPS geo-location since we only need cluster-level information, hence cell tower triangulation will suffice, and b) we can be opportunistic about obtaining GPS readings by leveraging cached geo-location results generated from other app's requests. In this manner, geo-location information can be obtained with low cost.

Another component of our system is online feature extraction. Our current implementation of location clustering is relatively expensive to perform on a mobile phone, and therefore needs to be done in the cloud. In contrast, the burst detection can be done online with a slightly modified version of the wavelet-based burst detection algorithm described in §4. We implemented a light-weight online burst window detection algorithm that adapts the sliding window size based on historical data, and uses the window to detect bursts. Our measurements indicate that our online burst detection has a small performance loss compared to the wavelet based scheme as presented in [15], with 4% less precision, and 10% less recall.

7. RELATED WORK

Systems optimizations for faster application launching has a long history on non-mobile computer systems. One area that has received significant treatment is the arrangement of code and data on disk for faster loading. Windows sorts code and data blocks on disk to minimize disk head movement during launch, and defragments files on disk for faster launching. Intel TurboMemory supports pinning selected applications' blocks to high performance cache to speed launch times. Joo et al. [7] investigate interleaving I/O fetch and code execution to increase app launch times. These schemes are complementary to our use of context for making prelaunch decisions.

SuperFetch, a part of Windows OS, is an extension to the memory manager that tries to predict future code and data accesses and preload them. The scheme works by learning the user's favorite apps based on time of day and day of the week, and loading these when there are free pages available. Unlike FALCON, SuperFetch does not use predictive mobile context such as location nor energy budgets in its loading decisions. Whereas both FALCON and SuperFetch interface with the memory manager to load apps, FALCON further reduces the total launch time by calling an app's prelaunch routine.

Recently, the authors of [3] proposed the use of user context information from sensors to improve mobile operating system performance, and suggested that OS-managed context could benefit user privacy, energy, system performance. The authors speculated that core system services – among others, memory management and process scheduling – could be beneficiaries of sensor context signals.

Browsers and other web systems have long known how to employ content prefetching and caching to decrease page load times [1]. A draft HTML specification includes `prefetch` attributes as hints to the browser [8]. These prefetch mechanisms generally do not guide prefetch decision making, as our launch predictor does, and are not cognizant of mobile context nor energy concerns. Recently in [9], the authors studied mobile web usage data and built a predictive page loading mechanism based on dividing pages into regularly visited sites and spontaneously visited sites.

Several platform-specific tweaking programs allow mobile users to configure app loading according to preset rules [14]. The shortcomings of these programs are that they stop short of addressing total launch time, and manually maintaining a rule-based system for the typical user is challenging.

8. DISCUSSION AND CONCLUSION

Mobile device usage is inherently brief, competing with myriad other real demands on human attention. As such, it is important that app interactions are rapid and responsive. Despite the fact that mobile devices are becoming increasingly resource rich, launch times are and will remain a significant challenge. This is due to two reasons. First, network-bound apps will continue to be limited by network fetch latency, which is constrained by cellular technology. Second, increasingly graphics- and media-rich apps targeting up-market phones will perpetually leave previous-generation and budget smartphones feeling sluggish to launch apps.

Our work addresses this important problem through the design of FALCON, a system that uses location and temporal contexts to predictively launch applications, thereby reducing perceived delay for users. We designed a set of features to capture the essential characteristics of app access patterns, and a launch predictor to adaptively balances latency reduction benefit with energy launch costs. We have shown that launch prediction using context can yield latency reductions of around 50% vs. LRU at a manageable 2% energy cost, and present fresh content at launch within the last 3 minutes. The FALCON modification to the Windows Phone OS consists of a context source manager, embedded launch predictor and prelaunch event upcall dispatcher. These elements are sufficiently low overhead to require only 2MB memory and negligible energy and processing.

Although our results show dramatic benefits of FALCON, there are several possible improvements and opportunities that we have not explored. One area is eliminating reliance on external servers or cloud services for model training in FALCON, thereby mitigating the need for sharing personal data with external service providers. An additional benefit of reducing reliance on the cloud is faster adaptation to underlying changes in app usage habits. Another open question is how users' expectations will change as the OS predictively prelaunches apps on their behalf. Will the fact that prelaunching can result in variable launch time hinder usability of mobile applications or will users appreciate the quicker interactions that are made possible by such techniques? We will address these questions in future studies.

9. ACKNOWLEDGEMENTS

We extend our thanks to Lin Zhong, Ahmad Rahmati, Clayton Shepard, and the Rice LiveLab team for making available their extensive mobile user study dataset. We would also like to thank Alec Wolman for helpful discussions and draft feedback, Alexander Varshavsky for shepherding, and the anonymous reviewers for their comments. This research was supported by NSF grants CNS-0546177 and CNS-0910900.

10. REFERENCES

[1] C. C. Aggarwal, J. L. Wolf, and P. S. Yu. Caching on the world wide web. *IEEE Transactions on Knowledge and Data Engineering*, 11(1):95–107, 1999.

[2] Apple - batteries - iphone. `http://www.apple.com/batteries/iphone.html`, 2011.11.28.

[3] D. Chu, A. Kansal, J. Liu, and F. Zhao. Mobile apps: it's time to move up to condos. In *13th USENIX conference on Hot topics in Operating Systems (HotOS)*, pages 16–16, Napa, California, 2011.

[4] Daily tip: How to make your iphone camera launch instantly [jailbreak]. `http://www.tipb.com/2011/04/20/daily-tip-iphone-camera-launch-instantly-jailbreak/` 2011.4.20.

[5] D. Garlan, D. P. Siewiorek, and P. Steenkiste. Project aura: Toward distraction-free pervasive computing. *IEEE Pervasive Computing*, 1:22–31, 2002.

[6] iOS 5 slowing iPhone 4 and 4S complaints. `http://www.phonesreview.co.uk/2011/10/25/ios-5-slowing-iphone-4-and-4s-complaints/`, 2011.10.25.

[7] Y. Joo, J. Ryu, S. Park, and K. G. Shin. Fast: Quick application launch on solid-state drives. In *9th USENIX Conference on File and Storage Technologies, San Jose, CA, USA*, pages 259–272. USENIX, 2011.

[8] Link prefetching. `https://developer.mozilla.org/en/link_prefetching_faq`.

[9] D. Lymberopoulos, O. Riva, K. Strauss, A. Mittal, and A. Ntoulas. Pocketweb: instant web browsing for mobile devices. In *Proceedings of the 17th International Conference on Architectural Support for Programming Languages and Operating Systems (ASPLOS)*, pages 1–12, London, UK, 2012.

[10] H. J. Miller and J. Han. *Geographic Data Mining and Knowledge Discovery*. Taylor & Francis, Inc., Bristol, PA, USA, 2001.

[11] F. Qian, Z. Wang, A. Gerber, Z. M. Mao, S. Sen, and O. Spatscheck. Profiling resource usage for mobile applications: a cross-layer approach. In *Proceedings of the 9th International Conference on Mobile Systems, Applications, and Services (MobiSys 2011)*, pages 321–334, Bethesda, MD, 2011.

[12] M. Satyanarayanan. Pervasive computing: Vision and challenges. *IEEE Personal Communications*, 8:10–17, 2001.

[13] C. Shepard, A. Rahmati, C. Tossell, L. Zhong, and P. Kortum. Livelab: measuring wireless networks and smartphone users in the field. *SIGMETRICS Performance Evaluation Review*, (3):15–20, Jan. 2011.

[14] Tasker. `http://tasker.dinglisch.net/`.

[15] Y. Zhu and D. Shasha. Efficient elastic burst detection in data streams. In *Proceedings of the ninth ACM international conference on Knowledge discovery and data mining (SIGKDD)*, pages 336–345, New York, NY, USA, 2003.

Web Caching on Smartphones: Ideal vs. Reality

Feng Qian
University of Michigan

Kee Shen Quah
University of Michigan

Junxian Huang
University of Michigan

Jeffrey Erman
AT&T Labs Research

Alexandre Gerber
AT&T Labs Research

Z. Morley Mao
University of Michigan

Subhabrata Sen
AT&T Labs Research

Oliver Spatscheck
AT&T Labs Research

ABSTRACT

Web caching in mobile networks is critical due to the unprecedented cellular traffic growth that far exceeds the deployment of cellular infrastructures. Caching on handsets is particularly important as it eliminates all network-related overheads. We perform the first network-wide study of the redundant transfers caused by inefficient web caching on handsets, using a dataset collected from 3 million smartphone users of a large commercial cellular carrier, as well as another five-month-long trace contributed by 20 smartphone users. Our findings suggest that redundant transfers contribute 18% and 20% of the total HTTP traffic volume in the two datasets. Also they are responsible for 17% of the bytes, 7% of the radio energy consumption, 6% of the signaling load, and 9% of the radio resource utilization of *all* cellular data traffic in the second dataset. Most of such redundant transfers are caused by the smartphone web caching implementation that does not fully support or strictly follow the protocol specification, or by developers not fully utilizing the caching support provided by the libraries. This is further confirmed by our caching tests of 10 popular HTTP libraries and mobile browsers. Improving the cache implementation will bring considerable reduction of network traffic volume, cellular resource consumption, handset energy consumption, and user-perceived latency, benefiting both cellular carriers and customers.

Categories and Subject Descriptors

C.2.2 [**Computer-Communication Networks**]: Network Protocols – Applications; C.4 [**Performance of Systems**]: Performance attributes

General Terms

Measurement, Performance

Keywords

HTTP Caching, Redundant Traffic, Cellular Networks, Redundancy Elimination, Smartphone Applications

MobiSys'12, June 25–29, 2012, Low Wood Bay, Lake District, UK.

1. INTRODUCTION

Caching plays a vital role in HTTP that dominates the Internet traffic usage [21]. Web caching can effectively decrease network traffic volume, lessen server workload, and reduce the latency perceived by end users.

Web caching in cellular networks is even more critical. HTTP traffic generated by mobile browsers and smartphone applications far exceeds any other type of traffic. HTTP accounts for 82% of the overall downstream traffic based on a recent study of a large cellular ISP [20]. From network carriers' perspective, cellular networks operate under severe resource constraints [31] due to the explosive growth of cellular data traffic (a growth of 5000% of its data traffic over 3 years reported by a major U.S. carrier [20]). Therefore even a small reduction of the total traffic volume by 1% leads to savings of tens of millions of dollars for carriers which are expected to spend $40.3 billion on cellular infrastructures in 2011 [4] (for all U.S. carriers).

The benefits of caching are also significant from customers' perspective. *(i)* Fewer network data transfers cut cellular bills since most carriers impose a monthly limit on data plan usage. *(ii)* The user experience improvement brought by caching is more notable in cellular networks whose latency is usually higher than those in Wi-Fi and wired networks. *(iii)* Transferring less data also improves handset battery life, as the power consumed by the cellular radio interface contributes 1/3 to 1/2 of the total device power during the normal workload [32].

The location of a cache can be at one or more places on a path from a handset (*i.e.*, a mobile device) to a web server. Although most of prior works focus on caching proxies placed in the network (*e.g.*, hierarchical caching proxies [18], caching proxy placement [21], and caching within 3G networks [20]), we emphasize that in cellular networks, *caching on handsets* is particularly important because it does not incur any network-related overhead. In particular, it eliminates data transmitted over the last-mile, *i.e.*, the radio access network, which is known as the performance and resource bottleneck of a cellular network [31]. In contrast, a network cache does not bring the aforementioned benefits but can cut the wide-area latency and enable sharing across users. In this study we focus on handset caches, which can coexist with network caches.

The rules of web caching are defined in HTTP 1.1 protocol (RFC 2616 [23]), although there exist numerous caching proposals [17, 26, 27, 29] that were never widely deployed. As specified in RFC 2616, HTTP employs *expiration* and *revalidation* to ensure cache consistency: the server sets for each cacheable file an expiration time before which the client should safely assume the freshness of the cached file. After the cache entry expires, the client must send a small revalidation message to the server to query the freshness of the cache entry. We detail the procedure in §2.

Redundant network transfers due to cache implementation. Let an *ideal* handset cache be a cache that has unlimited size and strictly follows the aforementioned caching rules. Any practically implemented handset cache is not ideal due to one or more reasons below: *(i)* it has limited size, *(ii)* the implemented caching logic does not satisfy RFC 2616, and *(iii)* the cache is not persistent (*i.e.,* it does not survive a process restart or a device reboot). A non-ideal cache potentially incurs *redundant transfers* that do not occur if an ideal cache were used. Also, redundant transfers can be caused by developers not utilizing the caching support even if it is provided by the HTTP library.

In this paper, we present to our knowledge the first network-wide study of the redundant transfers (defined above) by investigating the following important characteristics:

- The prevalence of redundant transfers within today's smartphone traffic, in terms of both the traffic volume contribution and the resource impact;
- The root cause for the redundant transfers;
- The handset caching logic, identified to be the main reason of redundant transfers, of HTTP libraries and browsers on popular smartphone systems.

We summarize our main findings as follows.

- We leveraged a large dataset containing 695M HTTP transaction records collected from 2.9M customers of a commercial cellular network in the U.S. To complement this data set which has a short duration of 24 hours, we collected another five-month trace from 20 smartphone users using smartphones instrumented by us. We measured the prevalence of redundant transfers for both datasets using an accurate caching simulation algorithm, assuming an ideal cache. To our surprise, redundant transfers due to caching implementation issues contribute 18% and 20% of the total HTTP traffic volume, for both datasets, respectively. The fraction slightly decreases to 17% at the scope of *all* traffic for the user study trace. Almost all redundant transfers are caused by the handsets instead of the server. Further, the redundancy ratio varies among applications. Some popular smartphone apps have unacceptably high fractions (93% to 100%) of redundant transfers.
- By strategically changing the simulation algorithm and analyzing the cache access patterns, we found the impact of limited cache size and non-persistent cache on the redundancy to be limited. For example, the fraction of redundant bytes is at least 13% (compared to 18% for an ideal cache) within all HTTP traffic even when our simulation uses a small cache of 4 MB. This implies the problematic caching logic is the main reason for redundant transfers.
- We investigated the resource overhead (handset radio energy, radio resources, and signaling load) incurred by redundant transfers, by leveraging our previously designed Radio Resource Control (RRC) simulation tool for 3G networks [32]. The results indicate that the impact on resource consumption is smaller than their impact on the traffic volume (yet still significant, *i.e.,* 7% for radio energy, 9% for radio resources, and 6% for signaling load, based on our analysis of the user study trace), due to reasons explained in §5.
- We performed comprehensive tests of ten state-of-art HTTP libraries and mobile browsers, many of which were found to not fully support or strictly follow the HTTP 1.1 caching specifications. In particular, among the eight HTTP libraries, four (three for Android and one for iOS) do not support caching at all. Smartphone apps using these libraries thus cannot benefit from caching. By exposing the shortcomings of existing implementations, our work helps encourage library and platform developers to improve the state of the art, and helps application developers choose the right libraries to use for better performance.

Overall, our findings suggest that for web caching, *there exists a huge gap between the protocol specification and the protocol implementation on today's mobile devices*, leading to significant amount of redundant network traffic, most of which could be eliminated if the caching logic of HTTP libraries and browsers fully supports and strictly follows the specification, and developers fully utilize the caching feature provided by the libraries. Fixing this cache implementation issue can bring considerable reduction of network traffic volume, cellular resource consumption, handset energy consumption, and user-perceived latency, benefiting both cellular carriers and customers.

Paper organization. §2 provides background of HTTP caching. §3 describes the measurement goal, data, and methodology. Then we measure the traffic volume impact and the resource impact of redundant transfers in §4 and §5, respectively. In §6 we perform caching tests for popular HTTP libraries and mobile browsers. We summarize related work in §7 before discussing our future work and concluding the paper in §8.

2. BACKGROUND: CACHING IN HTTP

This section provides background of web caching defined in HTTP 1.1 [23]. As described in §1, our study focuses on caching on handsets instead of within the network. In the remainder of the paper, a *cache* refers to an HTTP cache on a handset unless otherwise specified. We refer to a web object (*e.g.,* an HTML document or an image) carried by an HTTP response as a *file*.

Caching consistency (*i.e.,* keeping cached copies fresh) is the key issue. To realize that, the server sets the expiration time for each file by specifying either `Expires` or `Cache-Control:max-age` header directive. Then before a cache entry expires, a handset should safely assume the freshness of the file by serving the request using the cached copy without generating any network traffic. After a cache entry expires, a handset should perform cache *revalidation*, *i.e.,* asking the origin server whether the file has changed, by issuing conditional requests using `If-Modified-Since:<time>` or `If-None-Match:<eTag>` directive (or both). The former directive instructs the server to send a new copy if the file has changed since a specified date, which is usually the last modified time indicated by the `Last-Modified` directive in the previous response. The latter allows the server to determine the freshness using a file "version identifier" called eTag (entity tag), which is a quoted string attached to the file in the previous response. An eTag might be implemented using a version name or a checksum of the file content. In either case, if the file has changed, the server sends a new copy to the handset. Otherwise, a small `304 Not Modified` response is returned to the handset, without a document body, for efficiency.

A handset determines a cache entry has expired if and only if $t_{\text{arrive\_age}} + t_{\text{cache\_age}} \geq t_{\text{fresh\_life}}$. $t_{\text{arrive\_age}}$ is the age of the file when it arrives at the cache. It is computed from the `Date` or `Age` directive of a response, plus an estimated round-trip time compensating for the network delay. $t_{\text{cache\_age}}$, which can be trivially calculated, indicates how long the file has been in the cache. $t_{\text{fresh\_life}}$ is the cached copy's freshness lifetime (similar to the shelf life of food in grocery

Table 1: Our measurement datasets.

Dataset	ISP	UMICH
Data collection period	May 20 2011 0:00 GMT ~ May 20 2011 23:59 GMT	May 12 2011 ~ Oct 12 2011
Collection point	Commercial cellular core network	Directly on users' handsets
Number of users	2.92 million (estimated)	20
Dataset size	271 GB	119 GB
Traffic volume	24.3 TB	118 GB
Platforms	Multiple (mainly iOS and Android)	Android 2.2
Data format	695 million records of HTTP transactions	Full packet trace (including payload) of all traffic

stores) derived from the `Expires` or `Cache-Control:max-age` directive where the latter one overrides the former one if both exist. $t_{\text{fresh\_life}}$ is usually specified by the server while a handset can also specify `Cache-Control` request directives (`max-age`, `max-stale`, and `min-fresh`) to tighten or loosen expiration constraints although they are rarely used in practice.

Non-storable and must-revalidate files. A *non-storable* file (marked by `Cache-Control:no-store`) forbids a cache from storing (*i.e.*, caching) the file. A *must-revalidate* file (marked by `Cache-Control:must-revalidate`, `Pragma:no-cache`, or `Cache-Control:no-cache`) can be stored in a cache. However, the cache must bypass the freshness calculation mechanism and always revalidate with the origin server before serving it ("`no-cache`" is misleading because the file actually can be cached). A handset can also explicitly set the above caching directives in a request, indicating it will not store or will always revalidate the file.

Clearly, well-behaved caching logic requires correct implementation at both the handset and the server side. Our analysis described in §3 helps identify inefficient caching behaviors and which side is responsible for any of them.

3. MEASUREMENT GOAL, DATA, AND METHODOLOGY

This section highlights our measurement goal (§3.1), describes the measurement data (§3.2), and then details our analysis approach for redundant data transfers (§3.3).

3.1 The Measurement Goal

We define *inefficient caching* as caching behaviors that *(i)* lead to redundant data transfers, whose negative impact on performance, resource consumption, and billing is particularly high for mobile users, and *(ii)* relate to the *implementation* (as opposed to the *semantics*, as described below) of the HTTP caching mechanism. Inefficient caching can be caused by multiple reasons including the following factors covering the most important aspects of cache implementation (we discuss less significant factors in §3.3.2).

- **Problematic caching logic** due to any of the following reasons. *(i)* The handset does not fully support or strictly follow the protocol specification[1]. *(ii)* The server does not properly follow the caching specification. *(iii)* In order to leverage the caching support provided by an HTTP library, a developer still needs to configure it. Many developers may skip that for simplicity, or simply be unaware of it, therefore missing the opportunity of caching even if it is supported.
- **A limited cache size** causing the same content to be fetched twice if the first cached copy is evicted from the cache.
- **A *non-persistent* cache** whose cached data does not survive a process restart or a device reboot (unlike a *persistent* cache that survives both).

As highlighted in §1, we aim at understanding the impact of redundant transfers caused by the above inefficient caching factors, for commercial cellular networks (§4 and §5), as well as how the caching logic of popular HTTP libraries and mobile browsers deviates from the specification (§6).

Caching implementation vs. semantics. All the factors listed above relate to caching *implementation*. Redundant transfers may also be attributed to the misconfiguration of caching parameters, which ideally should be properly set by a server according to the *semantics* of files. For example, for a news website, conservatively marking all news articles and images as non-storable or setting for them very short freshness lifetime values may lead to redundant transfers while bringing negligible benefits given that the article contents rarely change. A thorough study of the file semantics and caching parameter settings is out of the scope of this paper, although we show examples where caching parameter settings are obviously too conservative (§8).

From this point on, unless otherwise specified, *redundant transfers* refer to redundant transfers caused by caching implementation.

3.2 The Smartphone Measurement Data

We collected two diverse datasets summarized in Table 1 to measure redundant transfers caused by inefficient caching.

3.2.1 The ISP *Dataset*

The ISP dataset was collected from a large U.S. based cellular carrier at a national data center on May 20, 2011, on the interface between the GGSN (Gateway GPRS Support Node) and SGSNs (Serving GPRS Support Node) without any sampling. Each record in the dataset corresponds to one HTTP transaction, containing three pieces of information: *(i)* a 64-bit timestamp, *(ii)* summaries of header fields in the request and the response, and *(iii)* the actual amount of data transferred based on the TCP data associated to each HTTP transaction. To preserve subscribers' privacy, the URLs were anonymized using a 128-bit hash function and also all cookie information was removed from the HTTP headers[2].

Subscriber identification. Since our cache simulation is performed at a per-user basis (§3.3), we need to identify the subscriber ID for each record. Instead of using MSISDN (the phone number) or IMEI (the device ID) information that was not collected due to privacy concern, we used anonymized session-level information to correlate multiple HTTP transaction records with a single subscriber. One disadvantage of our approach is that one real

[1]Not following the specification can also cause consumption of expired contents. But this was never observed in our tests in §6.

[2]In HTTP, caching and cookies are decoupled, and a server is responsible for explicitly disabling caching when appropriate [16].

subscriber may be identified with multiple subscriber IDs (but one subscriber ID never maps to multiple real subscribers). This leads to an underestimation of the amount of redundant data due to increased cold start cache misses [34] (explained in §3.3.2).

3.2.2 *The* UMICH *Dataset*

The ISP dataset is representative due to its large user base, however it is limited in terms of the trace duration and recorded content, as only summarized HTTP transaction records were captured. This is complemented by our second dataset called UMICH, collected from 20 smartphone users for five months, allowing us to keep detailed track of each individual user's web cache for a much longer period. These participants consisted of students from 8 departments at University of Michigan[3]. The 20 participants were given Motorola Atrix (11 of them) or Samsung Galaxy S smartphones (9 of them) with unlimited voice, text and data plans of the same cellular carrier from which we obtained the ISP dataset. All smartphones use Android 2.2. The participants were encouraged to take advantage of all the features and services of the phones. We kept collected data and users' identities strictly confidential.

We developed custom data collection software and deployed it on the 20 smartphones. It continuously runs in the background and collects two types of data: *(i)* full packet traces in **tcpdump** format including both headers and payload, and *(ii)* the process name responsible for sending or receiving each packet, using the method described in [32] by efficiently correlating the socket, the inode, and the process ID in Android OS in realtime. Both cellular and Wi-Fi traces were collected without any sampling performed. The data collector incurs no more than 15% of CPU overhead although the overhead is much lower when the throughput is low (*e.g.,* less than 200 kbps).

We also built a data uploader that uploads the captured data (stored on the SD card) to our server when the phone is idle. The data collection is paused when the data is being uploaded so the uploading traffic is not recorded. Also the upload is suspended (and the data collection is resumed) by any detected network activity of user applications. The entire data collection and uploading process is transparent to the users, although we do advise the users to keep their phones powered on as often as possible.

3.3 Analyzing Redundant Transfers

We explain our data analysis approach. We feed each user's HTTP transactions in the order of their arrival time to a web cache simulator developed by us. The simulator behaves like a cache that strictly follows the HTTP 1.1 caching mechanism specified in RFC 2616 (§2). Redundant transfers can be identified through the simulation process. Our approach differs from previous trace-driven cache simulations [17, 34, 21, 20] in two ways. *(i)* Our simulation is performed at a *per-user* basis to capture redundant transfers for each handset, while previous ones consider aggregated HTTP transfers of all users to compute, for example, the cache hit ratio, for a network cache. *(ii)* Ours is more fine-grained in that it distinguishes various causes of the redundancy (and also non-redundancy).

In the remainder of the paper, a *handset cache* and the *simulated cache* refer to the cache on a real handset and the cache maintained by our simulator, respectively.

The basic simulation algorithm is illustrated in Figure 1. It assigns to each HTTP transaction a label indicating its caching status. A file can be **NOT_STORABLE** due to its **Cache-Control:no-store** directive or causes **CACHE_ENTRY_NOT_EXIST** because it has not yet been cached. **NOT_EXPIRED_DUP** is an undesired case where a handset issues a request for a file cached in the simulator *before* it expires, resulting in redundant transfers ("**DUP**" means duplication).

Then Line 10 to 17 deal with the scenario where a cached file has expired. If the file has changed, then the entire file needs to be transferred again (**FILE_CHANGED**). If the file remains unchanged, the ideal way to handle it is that the handset performs cache revalidation and the server sends back an **HTTP_304** response. Otherwise, the problem either comes from the handset side, which does not issue a conditional request (**EXPIRED_DUP**), or from the server side, which does not recognize a conditional request (**EXPIRED_DUP_SVR**). Among the aforementioned labels, **NOT_EXPIRED_DUP**, **EXPIRED_DUP** and **EXPIRED_DUP_SVR** correspond to redundant transfers.

Is our simulated cache complete? Let us first assume our simulated cache is persistent with unlimited size (*i.e.,* an ideal cache defined in §1). Consider the Venn diagram shown in Figure 2. Let U be all HTTP transactions carried out by applications. What we observe in the data, D, is a subset of U since requests of $U \setminus D$ (the relative complement of D in U) are already served by the handset cache so $U \setminus D$ does not appear on the network. However, remember that in order for a file f to be cached, it must be transferred over the network (*i.e.,* $f \in D$) at least once. Therefore our simulated cache will not miss any file that is in a handset cache, if the simulated cache is ideal (we discuss the only exception in §3.3.2). More importantly, as explained in §3.1, our goal is to study the redundant data transfers that always belong to D instead of $U \setminus D$.

```
01 foreach HTTP transaction r
02     if (file is not storable) then
           //the file contains "Cache-Control: no-store"
03         assign_label(r, NOT_STORABLE);
04         continue;
05     else if (cache entry not exists) then
           //cache entry not found
06         assign_label(r, CACHE_ENTRY_NOT_EXIST);
07     else if (cache entry not expired) then
           //a request is issued before the file expires
08         assign_label(r, NOT_EXPIRED_DUP);
           //the response is ignored by the simulator because the
           //request should not have been generated
09         continue;
10     else if (file changed) then
           //the file has changed after the cache entry expires
11         assign_label(r, FILE_CHANGED);
12     else if (HTTP 304 used) then
           //the file has not changed after the cache entry expires,
           //and a cache revalidation is properly performed
13         assign_label(r, HTTP_304);
14     else if (revalidation not performed) then
           //the file has not changed after the cache entry expires,
           //but the handset does not perform cache revalidation
15         assign_label(r, EXPIRED_DUP);
16     else
           //the file has not changed after the cache entry expires,
           //but the server does not recognize the cache revalidation
17         assign_label(r, EXPIRED_DUP_SVR);
18     update_cache_entry(r);//update the simulated cache
19 endfor
```

Figure 1: The basic caching simulation algorithm.

[3]This user study has been approved by the University of Michigan IRB-HSBS #HUM00044666.

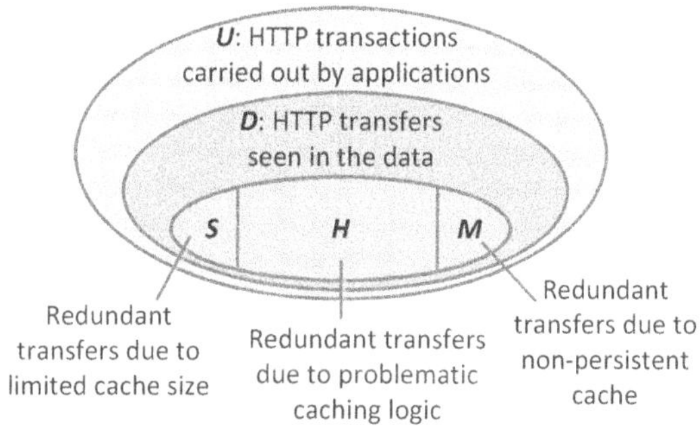

Figure 2: A Venn diagram showing HTTP transactions observed by applications and by our simulator, as well as the redundant transfers.

On the other hand, a file in our simulated cache may be missing in a handset cache because of its problematic caching logic (the set H in Figure 2), the limited size (the set S), or a lack of persistent storage (the set M) of the handset cache. Ideally we want to distinguish the three cases. However, the redundant transfers identified by our simulator are in fact $H \cup S \cup M$ where H, S, and M are indistinguishable, given that the simulated cache is ideal.

3.3.1 *Algorithm Details*

Figure 1 sketches the basic simulation algorithm. We provide details of the algorithm below.

The key of a cache entry (our simulated cache was implemented using a hash map) consists of three parts: *(i)* the host name indicated by the **Host** request directive, *(ii)* the file name followed by the **GET** command, and *(iii)* the eTag. Part *(i)* and *(ii)* must exist otherwise the HTTP transaction is assigned a special label "**OTHER**" (Table 2) while the eTag part is optional. Also the file name includes the entire **GET** string containing query strings so **/a.php?para=1** and **/a.php?para=2** have different cache entries. Empty or error responses (*e.g.*, **404 Not Found**) are also counted as **OTHER**, which only accounts for 0.5% of all HTTP traffic in both datasets.

The change of a file is identified through a different **Last-Modified**, **Content-Length** or **Content-MD5** value for the same cache entry.

A heuristic freshness lifetime can be used when neither **Expires** nor **Cache-Control:max-age** exists in a response, according to RFC 2616. We use a heuristic lifetime of 24 hours and later we show our analysis results are not sensitive to this value (§4.1).

A partially cached file is caused either by a byte-range request (using the **Range** directive) for a subrange of the origin file, or by a prematurely broken connection. Our simulator supports partial caching by allowing a cache entry to contain one or more subranges of a file. We implemented the following logic according to RFC 2616. Assume one or more subranges of a file have been cached, and an incoming response transfers another subrange R. The new subrange R is then combined with the existing range(s) if both the existing and the new range have the same eTag value (the eTag value must exist). Otherwise, all previously cached range(s) are removed before R is put into the cache entry.

If a file is partially cached, then a single transfer of the whole or a part of the file may contain both redundant and non-redundant bytes. To handle such a case, for each of the ***EXPIRED_DUP*** labels, we use two new labels, ***EXPIRED_DUP*_CACHED**, and ***EXPIRED_DUP*_UNCACHED** (not shown in Figure 1), to distinguish the redundant and the non-redundant ranges, respectively.

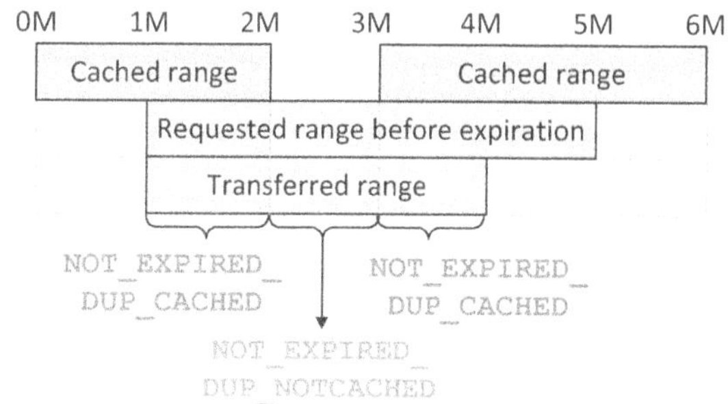

Figure 3: A partially cached file and a partial cache hit.

Consider a file of 6 MB shown in Figure 3. Assume two ranges [0, 2M) and [3M, 6M) are already cached by the simulator. The handset makes a byte-range request of [1M, 5M) before the cache entry expires. However the user cancels the transfer in the middle so only [1M, 4M) is actually transferred, as observed in our dataset. In this example, ideally the HTTP library should only request for [2M, 3M), which is not in the cache, using the **Range** and the **If-Range** directives[4]. We therefore label [2M, 3M) as **NOT_EXPIRED_DUP_NOTCACHED**, the non-redundant range, and label [1M, 2M) and [3M, 4M) as **NOT_EXPIRED_DUP_CACHED**, the redundant ranges. The two labels substitute for the original **NOT_EXPIRED_DUP** label which is not used for partially cached files. We introduce similar labels for **EXPIRED_DUP_SVR** and **EXPIRED_DUP**. Table 2 summarizes all labels.

3.3.2 *Limitations*

We discuss five limitations of our simulation approach.

- Inefficient caching and hence redundant transfers exist in both unencrypted HTTP and encrypted HTTPS traffic. However, the ISP trace contains only HTTP records. For the UMICH trace, the simulator cannot parse the HTTPS traffic that was collected by **tcpdump** running below the SSL library. HTTPS accounts for only 11.2% of the total traffic volume, compared to 85.4% for HTTP transfers.
- As mentioned before, the simulator cannot precisely distinguish S, H, and M shown in Figure 2. We qualitatively address such indistinguishability in §4.2 based on robust heuristics.
- A file may already be cached in a handset cache before the data collection started but the simulator does not know that. As an inherent problem (called cold start cache miss [34]) of any trace-driven cache simulation algorithm, it leads to an underestimation of the cache hit ratio (or the redundancy ratio in our case). But both our traces (in particular the UMICH dataset) are sufficiently long so the impact of cold start cache miss is expected to be small.
- Redundant transfers can also be caused by users explicitly reloading a file (*e.g.*, refreshing a web page). In that case, the application may override the default caching behavior by, for example, requesting for a file before its cached copy expires, but the simulator has no way to identify such manually triggered redundant transfers.

[4]The **Range** directive is often used together with an **If-Range:<eTag>** conditional request. It means "if the file is unchanged, send me the range that I am missing; otherwise, send me the entire new file".

Table 2: Detailed breakdown of caching entry status.

Label	Cache hit or miss?	Fully or partially cached?	Redundant due to caching?	% HTTP bytes ISP (GC)*	% HTTP bytes UMICH (GC)	% HTTP bytes UMICH (PC)*
1. **NOT_STORABLE**	-	-	-	15.4%	19.7%	19.7%
2. **CACHE_ENTRY_NOT_EXIST**	Miss	-	-	47.5%	42.0%	42.3%
3. **FILE_CHANGED**	Miss	-	-	1.9%	0.5%	0.5%
4. **HTTP_304**	Hit	Either	-	0.1%	0.0%	0.0%
5. **NOT_EXPIRED_DUP**	Hit	Full	Yes	13.6%	15.0%	14.7%
6. **NOT_EXPIRED_DUP_CACHED**	Hit	}Partial	Yes	2.3%	1.3%	1.3%
7. **NOT_EXPIRED_DUP_NOTCACHED**	Miss		-	16.0%	17.0%	17.0%
8. **EXPIRED_DUP**	Hit	Full	Yes	1.7%	4.0%	4.0%
9. **EXPIRED_DUP_CACHED**	Hit	}Partial	Yes	0.1%	0.0%	0.0%
10. **EXPIRED_DUP_NOTCACHED**	Miss		-	0.9%	0.0%	0.0%
11. **EXPIRED_DUP_SVR**	Hit	Full	-	0.0%	0.0%	0.0%
12. **EXPIRED_DUP_SVR_CACHED**	Hit	}Partial	-	0.0%	0.0%	0.0%
13. **EXPIRED_DUP_SVR_NOTCACHED**	Miss		-	0.0%	0.0%	0.0%
14. **OTHER**	-	-	-	0.5%	0.5%	0.5%

* GC: all processes on a handset share one single global cache; PC: each process has its own cache.

Table 3: Statistics of file cacheability.

Count by	Dataset	Normally Cacheable	Must-revalidate	Non-storable	Other
Bytes	ISP	69.8%	14.3%	15.4%	0.5%
	UMICH	78.2%	1.6%	19.7%	0.5%
Files	ISP	72.4%	12.4%	14.9%	0.4%
	UMICH	65.6%	6.8%	25.4%	2.1%

- Similarly, our simulator cannot identify redundant transfers caused by users manually clearing the cache after browsing sessions. The amount of such redundant data is expected to be small due to the observed strong temporal locality of accessing the same cache entry (described in §4.2).

4. THE TRAFFIC VOLUME IMPACT

In this section, we investigate the *traffic volume* impact of redundant transfers caused by inefficient caching behaviors.

4.1 Basic Characterization

We first assume our simulated cache is ideal (*i.e.*, it is persistent with unlimited cache size). Thus all redundant transfers caused by the three factors described in §3.1 can be identified.

File cacheability. Table 3 breaks down all HTTP bytes (files) transferred over the network into four categories: normally cacheable (*i.e.*, following the standard expiration and freshness calculation mechanism), must-revalidate (§2), non-storable, and other HTTP transfers (§3.3.1). For both datasets, most bytes (70% to 78%) and most files (66% to 72%) are normally cacheable, indicating the potential benefits of caching if handled properly by a handset.

A detailed breakdown of caching entry status is shown in Table 2, which lists the 14 labels described in §3.3. We show two simulation scenarios for the UMICH dataset: *(i)* all processes on a handset share one single global cache, and *(ii)* each process has its own cache. They correspond to "GC" and "PC" in Table 2, respectively. The per-process cache simulation is feasible because the UMICH trace contains packet-process correspondence for each packet. We summarize our findings as follows.

- **NOT_EXPIRED_DUP** contributes most bytes (77% for ISP and 74% for UMICH) among the four labels (5, 6, 8, 9) incurring redundant transfers. In other words, redundant transfers are usually caused by a handset issuing unnecessary requests *before* received files expire. For the ISP (UMICH) trace, 14% (31%) of all HTTP transactions (not shown), corresponding to 14% (15%) of all HTTP bytes (Row 5 in Table 2), are unnecessary because if handsets properly cache previous responses, no request needs to be sent out over the network and the responses can be served from local caches. On the other hand, the contribution of **EXPIRED_DUP** is much less. For the ISP (UMICH) trace, for 4.5% (10.2%) of all HTTP transactions (not shown), conditional requests to check the freshness of the cached data need to be sent, so that their responses, corresponding to 1.7% (4.0%) of all HTTP bytes (Row 8 in Table 2) will end up being served from local caches (if handsets properly cache previous responses) because they do not change.
- When **HTTP_304** is not used, it is almost always attributed to the handset instead of the server which properly handles cache revalidation, as indicated by the negligible contribution of **EXPIRED_DUP_SVR***.
- Partially cached files incur limited redundant transfers in that the traffic volume contribution of ***_DUP_CACHED** is small. In contrast, ***_DUP_NOTCACHED** (illustrated in Figure 3) account for considerable amount of non-redundant bytes. We found most (95%)[5] of ***_DUP_NOTCACHED** bytes originate from servers using byte-range responses for streaming large multimedia files. Note that a recent measurement study [22] showed that 98% of multimedia streaming traffic for a commercial cellular network is delivered over HTTP.
- The results of UMICH (GC) and UMICH (PC) are almost identical, indicating small overlap among files requested by different applications.
- Cellular and Wi-Fi traffic exhibit similar breakdown (not shown in Table 2), indicating the caching strategies on both the server and the handset side are independent of the network interface.

[5]Identified by their **User-Agent** strings. See §4.3 for details.

Table 4: The overall traffic volume impact of redundant transfers when different heuristic freshness lifetime values are used.

Dataset	% of redundant bytes of all HTTP traffic [% of redundant bytes of all traffic (HTTP and non-HTTP)] under different values of heuristic freshness lifetime				
	1 hour	6 hours	1 day*	1 week	1 month
ISP (GC)	17.74% [16%]**	17.74% [16%]	17.74% [16%]	17.74% [16%]	17.74% [16%]
UMICH (GC)	20.24% [17.29%]	20.25% [17.30%]	20.28% [17.33%]	20.32% [17.36%]	20.34% [17.38%]
UMICH (PC)	19.94% [17.03%]	19.95% [17.04%]	19.98% [17.07%]	20.02% [17.10%]	20.04% [17.12%]

* 1 day is the heuristic freshness lifetime used for other results in the paper.
** Assume the fraction of HTTP traffic is 90%, based on a recent large-scale measurement study for a commercial cellular data network [36].

The overall traffic volume impact is summarized in Table 4. We highlight key observations below.

- The first number in each grid of Table 4 is the fraction of redundant bytes within all HTTP traffic. Recall the redundant bytes come from label 5, 6, 8, 9 in Table 2, and they correspond to $H \cup S \cup M$ in Figure 2. By eliminating redundant transfers, the reduction of HTTP traffic is as high as 17.7% and 20.3% for ISP and UMICH, respectively.
- Even at the scope of *all* traffic (HTTP and non-HTTP), the redundancy ratio is also significant (17.3% for UMICH) as indicated by the second number. Note that this is an underestimation because we did not consider the redundancy of HTTPS traffic accounting for 11.2% of the total traffic volume of UMICH.

  We do not have the number for the ISP dataset that only contains HTTP records. As reported by a recent measurement study [36], HTTP accounts for at least 90%[6] of the total traffic volume of an aggregated one-week dataset involving 600K cellular subscribers collected in August 2010. If we assume that fraction is representative and apply it to our ISP dataset, then its overall redundancy ratio at the scope of all traffic is at least 16%.
- Recall that our simulator introduced a heuristic freshness lifetime when a response contains no expiration information. Table 4 shows this parameter has negligible impact on the amount of redundant data.
- Although the duration of the ISP trace is much shorter, its redundancy ratio is only marginally smaller than that of UMICH, implying the usage duration has limited impact on the redundancy ratio. This is partly explained by the strong temporal locality of accessing the same cache entry (Figure 5 in §4.2).
- For the UMICH dataset, the difference between redundancy ratios of per-process caches (PC) and a single global cache (GC) is as small as 0.3%.

4.2 The Impact of the Cache Size

Now we discard the assumption of unlimited cache size and consider a *finite* cache for the simulator. This helps quantify the impact of limited cache size on redundant transfers. We implemented LRU (Least Recently Used) algorithm for our simulator since all HTTP libraries and browsers we tested in §6 use LRU as the replacement algorithm, if they support caching. LRU discards the least recently accessed files first when the cache is full. With a finite cache size, the simulated cache may not capture all redundant transfers in the trace. We refer to those captured ones as *detected* redundant transfers.

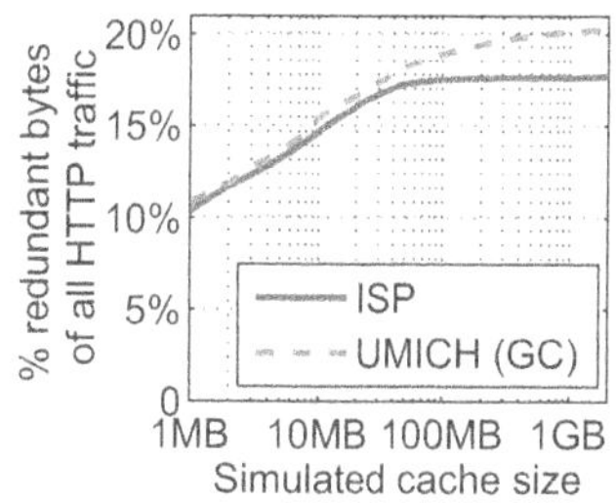

Figure 4: Relationship between the simulated cache size and the fraction of detected HTTP redundant bytes.

Consider Figure 2 again. Let us decrease the simulated cache size. Then we see fewer detected redundant transfers (each of $|S|$, $|H|$, and $|M|$ decreases) since more previously detected redundant transfers are classified as `CACHE_ENTRY_NOT_EXIST` due to cache misses as the simulated cache becomes smaller. In particular, when the simulated cache size is smaller than the handset cache size, redundant transfers due to limited size of the handset cache will be eliminated (*i.e.*, $|S|$ becomes 0) because if a cache entry is evicted from the handset cache, it must have been evicted from the simulated cache (assuming both use LRU). Therefore, if the detected redundant bytes decrease to $x\%$ when the simulated cache size goes below the handset cache size, then the traffic volume impact of $H \cup M$ is at least $x\%$. Note this is a very loose lower bound in that $|H \cup M|$ also decreases as the simulated cache becomes smaller.

Our measurement results are shown in Figure 4 where we vary the simulated cache size from 1 MB to 2 GB for both datasets (note that the X-axis is in log scale). Figure 4 shows that even when the cache has a very small size of, for example, 4 MB[7], the detected redundant bytes is still as high as 12.8% and 13.2% (compared to 17.7% and 20.3% when the simulated cache has unlimited size),

[6] See Figure 1(a) of the paper [36]. The following categories use HTTP: `web_browsing`, `smartphone_apps`, `market`, and `streaming`.

[7] In comparison, our caching tests in Table 9 (§6.2) show that the cache sizes for the Android 2.2 browser and the Safari browser of iOS 4.3.4 / iPhone 4 are 8 MB and 100 MB, respectively.

Table 5: Measuring redundant transfers for top applications in both datasets.

The UMICH dataset (HTTP Bytes: 101 GB)			The ISP dataset (HTTP Bytes: 24.3 TB)		
Android process*	% HTTP bytes	% redundant HTTP bytes	`User-Agent` regex (device/OS name)*	% HTTP bytes	% redundant HTTP bytes
Streaming Service 1	29.8%	8.8%	Streaming Service 1 (A)**	37.8%	20.1%
Streaming Service 2	12.4%	0.5%	Internet Radio (A)	11.6%	1.9%
Web Browser 1	11.5%	20.4%	Web Browser (A)	11.3%	14.6%
Entertainment	6.6%	12.0%	Media player (A)	8.9%	3.1%
News and Weather	6.3%	55.3%	Map (A)	3.1%	0.0%
Lifestyle	3.9%	99.4%	HTTP Library (B)	2.4%	86.6%
Music and Audio 1	3.4%	0.0%	Web Browser (C)	2.1%	1.6%
Music and Audio 2	2.9%	0.1%	Weather (A)	2.0%	93.0%
Web Browser 2	1.8%	8.6%	Social Networking (A)	1.6%	7.3%
Social Network Manager	1.7%	99.3%	Streaming Service 2 (D)	1.5%	19.1%
Media and Video	1.7%	0.8%	Web Browser (B)	1.4%	13.5%
Web Browser 3	1.4%	18.3%	Ad library (B)	1.0%	100.0%
(Total or average)	83.4%	18.3%	(Total or average)	84.7%	18.2%

* The process names, `User-Agent` regular expressions, and device/OS names have been anonymized.
** For the ISP dataset, A, B, C, D refer to four different device/OS names after anonymization.

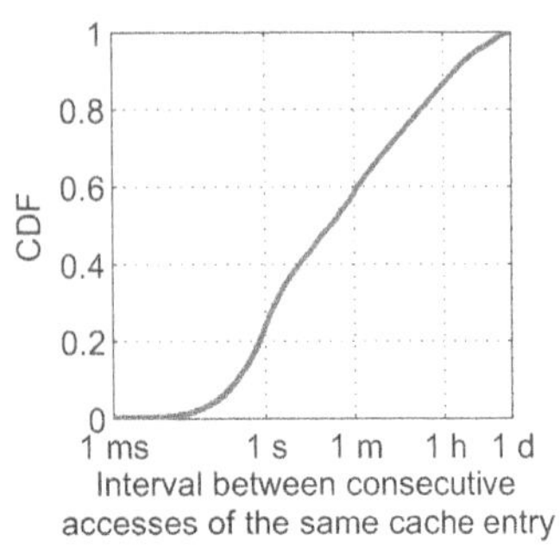

Figure 5: Distribution of intervals between consecutive accesses of the same simulated cache entry (for the ISP trace).

which are the aforementioned (loose) lower bounds of $|H \cup M|$, for ISP and UMICH, respectively. We therefore conclude that the problematic caching logic (instead of the limited cache size) takes the major responsibility for redundant transfers.

Figure 4 also suggests how to set the cache size, which can be small enough to entirely fit into today's smartphone memory with very limited additional cache misses incurred. For example, reducing the simulated cache size from infinity to 50 MB causes additional cache misses for only 2.0% and 0.4% of HTTP bytes (they are the loose upper bounds of the reduction of $|S|$) for UMICH and ISP, respectively, as indicated by the decrease of the detected redundant bytes shown in Figure 4.

Persistent vs. non-persistent cache. As described in §3.1, a non-persistent cache does not survive a process restart or a device reboot while a persistent cache does survive both. Based on our caching tests of 6 HTTP libraries and mobile browsers that support caching (§6.2), only one library for iOS (`NSURLRequest`) uses a non-persistent cache (Table 9). All Android libraries as well as both iPhone and Android browsers use persistent caches.

Our simulation assumes a persistent cache, which is consistent with the UMICH trace involving only Android handsets. For the ISP trace involving iOS devices, redundant transfers can also be caused by the non-persistent cache, which cannot be simulated since we do not know when a user restarts a process or reboots a handset. However, we expect the fraction of redundant transfers caused by the non-persistent cache is small due to two reasons. *(i)* Restarting a process happens infrequently in iOS [11]. On iPhone and iPad, pressing the "home" button simply puts an application to background. *(ii)* More quantitatively, Figure 5 plots the CDF of the intervals between consecutive accesses of the same simulated cache entry for the ISP trace. It is generated during the cache simulation. As shown in Figure 5, 59% of the intervals are less than 1 minute and 87% are less than 1 hour. We expect such strong temporal locality of cache entry access makes a non-persistent cache comparable to a persistent cache in terms of efficiency.

4.3 Diversity Among Applications

This subsection investigates the caching efficiency of individual smartphone applications.

Identifying smartphone applications is trivial for the UMICH trace, which contains the process name for each packet and hence for each HTTP transaction. 741 unique processes were observed from the UMICH dataset.

For the ISP dataset whose application identification is less straightforward, we used the `User-Agent` field in HTTP requests to distinguish different applications. First, from the 48,214 unique `User-Agent` strings appeared in the dataset, we picked the top 500 strings with the highest HTTP traffic coverage, yielding an overall traffic coverage ratio of 95.5%. We found many `User-Agent` strings belong to the same application. They are only slightly different in OS versions, hardware specifications, and languages, *etc.* For example, Apple iTunes has the following `User-Agent` format: `iTunes-`*Device*`/`*iOS version* `(`*device version*`;` *memory size*`)`, such as `iTunes-iPhone/4.3.3(4;16GB)` or `iTunes-iPhone/4.2.1 (2;8GB)`. To avoid duplicated applications, we generated 95 regular expressions, each corresponding to an app for a specific device and/or OS, that cover all 500 `User-Agent` strings. All regular expressions follow a simple pattern of *app_name★device/OS_name★*, such as `iTunes★iPhone★` and `Pandora★iOS★`, by ignoring other less significant fields such as the OS version number.

Redundant transfers for top applications are measured in Table 5 (assuming an ideal cache for simulation). For the UMICH (ISP) trace, we show the top 12 process names (`User-Agent` regular expressions), their contributions of HTTP traffic, and their fractions of redundant bytes (their names have been anonymized). Due to the heavy-tail distribution of the smartphone application

Table 6: **A summary of our findings regarding to the traffic volume impact of redundant transfers.**

Question	Our finding based on the two datasets	§
File cacheability	Most bytes (70% to 78%) and most files (66% to 72%) are cacheable.	4.1
Traffic volume impact of redundant transfers	They account for 18% to 20% of HTTP traffic, and about 17% of the overall traffic for the UMICH trace.	4.1
Main reason for redundant transfers	Problematic caching logic of the handsets (instead of the sever).	4.2
Impact of cache size on redundant transfers	Limited. The detected redundant bytes are at least 13% even for a simulated cache of as small as 4 MB.	4.2
Suggested handset cache size	Reducing the cache size from infinity to 50 (100) MB causes cache misses for at most 2.0% (1.4%) of HTTP bytes.	4.2
Difference between persistent cache and non-persistent cache	Very small. Both have similar caching efficiency due to strong temporal locality of accessing the same cache entry.	4.2
Caching efficiency of individual mobile apps	Some popular apps have unacceptably high fractions (93% to 100%) of redundant transfers.	4.3

usage [36], these top apps are responsible for more than 83% of all HTTP traffic. We found that while some apps incur small fractions of redundant data, some have unacceptably high redundancy ratios.

To validate the cache simulation results, we further studied the four apps with high redundancy ratios greater than 90% as shown in Table 5. We used them locally by exploring their common application usage scenarios, whose packet traces were simultaneously collected by `tcpdump` running on our handsets. By analyzing the traces, we found that all four apps use HTTP as the application-layer protocol but none of them performs caching. For example, for the "Weather (A)" app, HTTP responses of the weather information contain the `Expires` and the `Cache-Control:max-age` directives both specifying a freshness lifetime of 5 minutes. But when we checked the weather for the same location again (we verified from the trace that the URLs were identical), the handset always issued a non-conditional request regardless of the freshness of the downloaded file.

Inefficiency caused by HTTP POST. Table 5 indicates that the "Map (A)" app incurs negligible redundant transfers, because almost all its bytes are not cacheable (not shown). Specifically, we found that instead of employing HTTP GET, the application heavily uses HTTP POST requests that point to a single static file name by including the parameters in the body of a POST request, making it impossible for HTTP to cache the responses. A more caching-friendly approach is to attach query strings to the URLs to make them cacheable.

We summarize important findings regarding to the traffic volume impact of redundant transfers in Table 6.

5. THE RESOURCE IMPACT

§4 reveals the traffic volume impact of redundant transfers due to inefficient caching. In cellular networks, resources such as handset battery life, radio resources, and signaling load could also become critical bottlenecks. We now focus on understanding the resource impact of redundant transfers.

5.1 Cellular Resource Management Policy

To efficiently utilize the limited radio resources, cellular networks employ a radio resource management policy distinguishing them from wired and Wi-Fi networks. In particular, there exists a radio resource control (RRC) state machine [31] that determines the radio resource usage based on application traffic patterns, affecting the handset energy consumption and the user experience.

The RRC state machine of the 3G UMTS cellular carrier used by the 20 participants generating the UMICH trace is depicted in Figure 6, which was inferred by our prior work [31]. The UMTS RRC state machine consists of three power states: `IDLE` (no connection), `CELL_FACH` (low-power, low bandwidth state), and `CELL_DCH` (high-power, high-bandwidth state). Similar RRC state machines exist in other types of cellular radio access networks such as GPRS/EDGE [12], 3G EvDO [19], and 4G LTE [25].

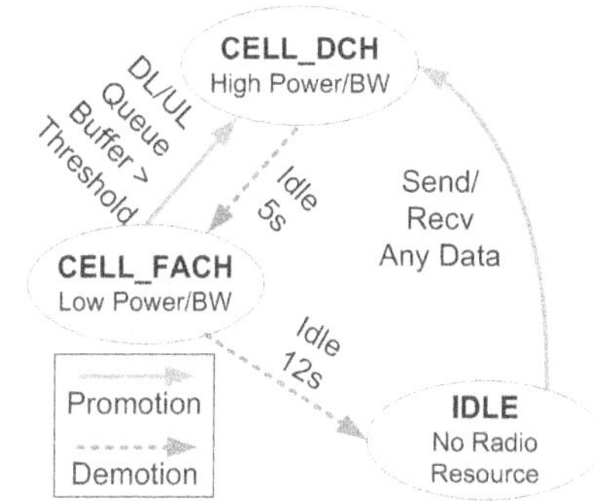

Figure 6: The RRC state machine of the 3G UMTS cellular carrier used by the participants generating the UMICH trace.

A state promotion (going from low to high power state) is triggered by a relatively high data transmission rate. It incurs a long delay up to several seconds during which tens of control messages are exchanged between a handset and the radio access network for resource allocation. A large number of state promotions increase the signaling overhead and worsen the user experience [31].

A state demotion (going from high to low power state) takes negligible time. However it is controlled by an inactivity timer that may cause significant waste of resources, because when waiting for timeout, a handset still occupies the transmission channel and the WCDMA codes, and its radio power consumption is kept at the corresponding level of the state[8]. If the inactivity timer, which is reset by any uplink or downlink packet, finally expires, we say the RRC state is demoted after a *tail*, which is the idle time period matching the inactivity timer value before a state demotion. The negative resource impact of tails in cellular networks was investigated in prior work [13, 31].

[8]In typical UMTS networks, each handset is allocated dedicated high-speed channel on `CELL_DCH`. For HSDPA [24], a UMTS extension with higher downlink speed, a high-speed channel may be shared by a limited number of handsets. Occupying it during the tail time can potentially prevent other handsets from using it.

5.2 Computing the Resource Impact

Three metrics are used to quantify the resource impact. We leveraged the RRC state machine simulator (configured with the state transition model shown in Figure 6) and the handset power model used by the ARO tool [32] for computing the metrics below.

- ***E***: the handset radio energy consumption. It is the energy consumed by the cellular radio interface whose radio power contributes 1/3 to 1/2 of the overall handset power during the normal workload [32]. E is calculated by associating each RRC state or state promotion an average power value measured from a particular phone [31, 32]. The exact radio power values may also depend on other factors such as the signal strength, but our qualitative conclusions in terms of the energy impact should hold.
- ***S***: the signaling overhead quantified by the total state promotion delay.
- ***D***: the radio resources. It is estimated by the total `CELL_DCH` (the high-speed dedicated channel) occupation time (including the tail time).

We only studied the UMICH trace, because accurate RRC state reconstruction, a prerequisite for computing the above metrics, requires for each incoming and outgoing packet its precise timing and size, which is not available in the ISP trace. Also, we only considered the 3G traffic within the UMICH trace since the resource management policy of Wi-Fi is much more efficient due to its short-range nature[9].

To quantify the resource impact of redundant transfers, we *take the difference* of the computed metrics for the original trace and the modified trace with all redundant transfers removed. Specifically, the radio energy impact is computed as $\Delta E = (E_O - E_R)/E_O$ where E_O and E_R correspond to the radio energy consumed by the original and the redundant-transfer-free trace, respectively. ΔE is positive as removing redundant transfers reduces energy consumption. The radio resource impact ΔD and the signaling impact ΔS are computed in similar ways. Note this general method [32] can be used to compute the resource impact of any transfers.

Two factors may lead to underestimation of the resource impact. *(i)* We conservatively consider all HTTPS traffic non-redundant. *(ii)* In the above approach, by removing redundant transfers (or any transfers of our interest), we assume that any of the remaining traffic is *unaffected* in terms of its schedule and occurrence. This may not be true in reality: if some redundant transfer is eliminated, the subsequent transfer may happen sooner, or some other traffic may not occur at all (*e.g.*, a DNS lookup). Ignoring such dependency leads to an underestimation of the resource impact because in reality the resultant redundant-transfer-free trace has shorter duration and/or less traffic than our simulated one. Although it is difficult to handle all such cases, we do address a common case where a DNS lookup would not occur if its corresponding HTTP transfer were eliminated. Specifically, when removing a redundant transfer X, we also try to eliminate a DNS lookup D right before X, using a time window δ tolerating the handset processing delay. We empirically choose δ=300 ms but varying it from 100 to 500 ms has negligible impact on the results.

5.3 Measurement Results

Table 7 measures the resource impact of redundant transfers in two scenarios: *(i)* consider only 3G HTTP traffic and exclude 3G non-HTTP traffic, and *(ii)* consider all 3G traffic in the UMICH trace. The "ΔE (HTC)" and "ΔE (Nexus)" columns refer to the radio energy impact using power parameters of an HTC TyTn II smartphone and a Google Nexus One smartphone [32], respectively. The presented results also assume an ideal cache. We found that similar to our findings in §4.2, as long as the simulated cache size is reasonably large (*e.g.*, greater than 10 MB), its impact on the measured resource impact of redundant transfers is small (results not shown).

[9] Wi-Fi has very short tail time and negligible promotion delay[13].

Table 7: Resource impact of redundant transfers (the UMICH trace), under the scopes of HTTP traffic and all traffic.

	ΔS	ΔE (HTC)	ΔE (Nexus)	ΔD
HTTP only	26.9%	26.1%	25.9%	27.1%
All traffic	6.1%	7.0%	6.7%	9.0%

As shown in Table 7, by considering non-HTTP traffic, the resource impact of redundant transfers decreases sharply from more than 25% to less than 10%, although non-HTTP traffic accounts for only 13% of the total 3G traffic volume. This is attributed to two reasons explained below.

First, the resource impact of non-HTTP traffic can be significant although their traffic volume contribution is small. Due to the tail effect explained in §5.1, intermittently transmitting very small amount of data may utilize much more resources than transmitting large amount of data in one burst [32]. One representative example of such an inefficient traffic pattern identified in the UMICH trace is Android push notification (identified by TCP port 5228) and XMPP (Extensible Messaging and Presence Protocol, a popular instant messaging protocol using TCP port 5222 [1]) traffic. They account for 1% of the total 3G traffic volume while their resource impact in terms of E (HTC) is 18%[10]. On the other hand, for all HTTP traffic dominating the overall 3G traffic volume (87%), the resource impact in terms of E (HTC) is only 20%. Note that the traffic patterns of push notifications (and in general, delay-tolerant transfers) can be optimized to be more resource efficient [13, 30], resulting in higher resource impact of redundant transfers.

The second reason is *resource sharing*. Unlike the traffic volume measured in §4, resources in cellular networks can be shared by multiple HTTP sessions, or be shared by HTTP and non-HTTP transfers. Recall that in §5.1, the "radio-on" period of a transfer consists of a state promotion, its data transmission period and the following tail. If the radio-on periods of two transfers are fully or partially overlapped, then D and hence E are shared during the overlapped period. The signaling load S is also shared in that only one state promotion is triggered by the two transfers. Resource sharing significantly reduces the resource impact of redundant transfers, many of which do not incur additional resource overhead because their channel occupation periods overlap with those of other transfers. For S, E (HTC), E (Nexus), and D, the fractions of resource reduction due to resource sharing among redundant transfers and other transfers are 70%, 61%, 63%, and 50%, respectively[11].

[10] It is computed using the method described in §5.2 by taking the difference of the radio energy for the entire trace and the modified trace where push notification and XMPP transfers are removed.

[11] It is computed as $1 - (U(T) - U(T_2))/U(T_1)$ where $U(\cdot)$ computes the resource consumption for a certain trace. Trace T is the original trace of all traffic. T_1 consists of only redundant transfers. T_2 corresponds to T with T_1 removed.

Table 8: Our tested HTTP libraries and smartphone browsers.

Name	HTTP library or browser	Platform	Handset
UC	`java.net.URLConnection`	Android 2.3	Samsung Galaxy S
HUC	`java.net.HttpURLConnection`	Android 2.3	Samsung Galaxy S
HC	`org.apache.http.client.HttpClient`	Android 2.3	Samsung Galaxy S
WV	`android.webkit.WebView`	Android 2.3	Samsung Galaxy S
HRC	`android.net.http.HttpResponseCache`	Android 4.0.2	Samsung Galaxy Nexus
T20	`Three20` (Version 1.0.6.2)	iOS 4.3.4	iPhone 4
NSUR	`NSURLRequest`	iOS 5.0.1	iPhone 4S
ASIHR	`ASIHTTPRequest` (Version 1.8.1)	iOS 4.3.4	iPhone 4
AB	The Android web browser	Android 2.3	Samsung Galaxy S
SB	The Safari web browser on iPhone	iOS 4.3.4 and 5.0.1	iPhone 4 and 4S

6. FINDING THE ROOT CAUSE

We learn from §4.2 that the main reason for redundant transfers is the problematic caching logic. We verify this by performing comprehensive caching tests for state-of-art HTTP libraries and browsers of Android and iOS.

Previously, professional developers also spent efforts investigating HTTP cache implementation issues leading to poor performance, focusing on mobile browsers [5, 7, 8, 6, 9]. Our tests go beyond them in two aspects. *(i)* Our tests are much more complete, covering all important aspects of caching implementation. To our knowledge, only three (Test 7, 10, 11) out of the thirteen tests described in §6.1 were performed before. *(ii)* Prior efforts only investigated browsers but we further examined HTTP libraries that are heavily used by today's smartphone applications.

6.1 Test Methodology

We examined eight HTTP libraries listed in Table 8. To the best of our knowledge, they cover all publicly available HTTP libraries for Android and iOS. To test them, we wrote small applications using these libraries as HTTP clients. We also investigated the default browsers on Android and iPhone, using strategically generated HTML pages embedding multiple web objects to perform tests involving multiple files (for Test 11 and 12).

We performed all tests on real handset devices: Samsung Galaxy S with Android 2.3, Samsung Galaxy Nexus with Android 4.0.2, iPhone 4 with iOS 4.3.4, and iPhone 4S with iOS 5.0.1. Each handset has non-volatile storage of at least 10 GB. In each test, a client only requested files, whose caching directives were properly configured, from our controlled HTTP server running Apache 2.2. We ran `tcpdump` on the server to monitor incoming HTTP requests to tell whether a request we made was served by the handset cache or by the server.

We took the following measures to further eliminate external factors that may affect the accuracy. *(i)* Before launching each test, the handset cache (if existed) was always cleared either manually (for browsers) or by calling the corresponding APIs (for libraries). *(ii)* We verified that the caching behaviors of the server in all tests were correct by analyzing the traces collected at the server[12]. The clocks of both the server and the handset were synchronized as well. *(iii)* We also ran `tcpdump` on the handset and compared HTTP requests and responses observed at the handset and at the server. We found both to be identical in all tests, implying that the cellular middleboxes did not change any caching directives. This is further confirmed by the fact that using cellular and Wi-Fi yielded the same testing results.

[12]The server was also tested by `http://redbot.org`, an online tool for checking HTTP caching implementation of web servers.

We designed 13 black-box tests to understand how caching was implemented in the state-of-art HTTP libraries and mobile browsers. Test 1 to Test 7 verify whether key features (*e.g.*, revalidation) were supported. Failure to support any of them may lead to redundant transfers. Test 8 to Test 13 determine important attributes of a cache (*e.g.*, its size) whose implications on redundant transfers cannot be overlooked either. Note the testing methodology is generally applicable to HTTP libraries and browsers on any platform. We detail the tests below.

Test 1 (Basic caching). The handset requests for a small cacheable file f. The server transfers f with a proper `Expires` directive. Then the client requests for f again before it expires. If the basic caching is supported, the second request should not incur any network traffic.

Test 2 (Revalidation). It is similar to Test 1 except that the client requests for f after it expires. For the second request, the client should issue a conditional request with an `If-Modified-Since` directive. Then the server will respond with an HTTP 304 indicating the file has not changed.

Test 3 (Non-caching directives). It tests whether the following directives are correctly followed: `Cache-Control:no-store`, `Pragma:no-cache`, and `Cache-Control:no-cache`. Their caching logic is described in §2.

Test 4 (Expiration directives). It is similar to Test 1 except that the server uses other expiration directives: *(i)* `Cache-control:max-age`, *(ii)* `Expires` and `max-age`, and *(iii)* `max-age` and `age`. In *(ii)*, `max-age` should override `Expires`. In *(iii)*, the file should always expire if `age` is greater than `max-age`.

Test 5 (URL with query string). It is similar to Test 1 except that the URL contains a query string (*e.g.*, `query.php?p=123`).

Test 6 (Partial caching). The handset performs a byte-range request $[p_1, q_1]$ for a small cacheable file f. The server transfers the corresponding range of f with a proper `Expires` directive and an eTag. Before the response expires, the client performs another byte-range request $[p_2, q_2]$ for f. If $p_2 \geq p_1$ and $q_2 \leq q_1$, then we should not observe the second request over the network.

Test 7 (Redirection). We test whether a handset caches two types of responses: `301 Moved Permanently` and `302 Found`. They are common ways of performing a permanent and a temporary redirection, respectively. A 301 response is always cacheable unless indicated otherwise (*e.g.*, by `Cache-Control:no-store`). A 302 response is only cacheable if indicated by a `Cache-Control` or `Expires` header field. The testing procedure is similar to Test 1 except that the response of the server is HTTP 301 or 302, redirecting the request to another small file.

Test 8 (Shared or non-shared cache). We run two applications A and B using the same library to be tested on the same phone. A requests for a small cacheable file f. Then B requests for f again before f expires. If the cache is shared by both applications, the second request should not incur any network traffic. Otherwise we will see two requests over the network. We did not find any publicly available API that allows one to read or modify cache entries of the default Android or iPhone browser, whose caches are thus assumed to be non-shared.

Test 9 (Persistent or non-persistent cache). We perform a test similar to Test 1 except that we reboot the phone after receiving the first response. A persistent cache must survive a device reboot (and therefore a process restart).

Test 10 (Cache entry size limit). A large file may not be cached by a given cache implementation with a cache entry size limit. We perform binary search for this limit by fetching cacheable files of varying sizes from the server. For a file of size s_i, we perform Test 1 after clearing the cache to see whether s_i is above the cache entry size limit. Previous measurement [8] reported that HTML pages and external web objects (*e.g.*, JavaScript and CSS) may have different cache entry size limits, which are measured separately.

Test 11 (Total cache size). We perform binary search for the total cache size. To test whether it exceeds a particular value z, we clear the cache and download n cacheable files $f_1, ..., f_n$ each with a size of z/n (smaller than the cache entry size limit inferred by Test 10). We then request for the n files again. z exceeds the total cache size if and only if any file is transferred over the network in the second pass.

Test 12 (Replacement policy). We test for five cache replacement algorithms known to be commonly implemented [35]: *(i)* LRU (Least Recently Used), *(ii)* LFU (Least Frequently Used), *(iii)* evicting the oldest cache entry, *(iv)* evicting the cache entry with the nearest expiration time, and *(v)* evicting the cache entry of the largest size.

To test whether the replacement policy is LRU, we first fill up the cache using n cacheable files $f_1, ..., f_n$ such that $\sum_{i=1}^{n} f_i < z$ where z is the total cache size inferred by Test 11. Next, we randomly generate an n-permutation $p_1, p_2, ..., p_n$, and then request for $f_{p_1}, ..., f_{p_n}$ again. Subsequently the handset downloads a new file f_{n+1} such that $\sum_{i=1}^{n+1} f_i > z$, thus triggering a cache entry eviction. If f_{p_1} is evicted (based on Test 1), then we know LRU is the replacement policy since f_{p_1} is the least recently accessed file. Other replacement algorithms are tested in similar ways.

Test 13 (Heuristic freshness lifetime). The HTTP server is configured in a way that it puts neither `Cache-Control:max-age` nor `Expires` in a response. Then we test whether a small file can be cached by performing Test 1 in which the two requests are sent back to back. If it can, we do binary search for the heuristic freshness lifetime (§3.3.1) by varying the interval between the two requests.

6.2 Test Results

Table 9 summarizes the results (refer to Table 8 for acronyms of the libraries and browsers). Each feature in Test 1 to Test 7 can be fully supported (indicated by a "●" symbol), not supported at all ("○") or partially supported ("◑" with the reason explained). For each of the attribute tests (Test 8 to Test 13), a "✕" symbol means the test was not performed since the corresponding API does not support HTTP caching. We highlight key findings as follows.

- To our surprise, among the eight HTTP libraries, four (three for Android and one for iOS) do not support caching at all. Smartphone apps using these libraries thus cannot gain any benefit from caching.
- For libraries and browsers that do support caching, they may not strictly follow RFC 2616, as detailed by footnotes d, h, j, k in Table 9. To our knowledge, only observations h and j were reported by previous measurements [9, 6] (for browsers only). All such cases of non-compliance potentially incur redundant transfers.
- The Android browser uses a small cache of 8 MB. As described in §4.2, increasing the cache size brings non-trivial reduction of cache misses.
- No library or browser supports partial caching, although its impact on redundant transfers is limited (§4.1).
- We found that for all libraries that support caching, in order to leverage the caching support, a developer still needs to configure the library. However, developers can easily skip that for simplicity, or they can simply be unaware of it, therefore missing the opportunity of caching and incurring redundant transfers.

By exposing the shortcomings of existing implementations, our work helps encourage library and platform developers to improve the state of the art, and helps application developers choose the right libraries for better performance.

7. RELATED WORK

We describe related work in four categories.

Extensive research on web caching has been done since the World Wide Web was in its nascent state. We summarize the important topics. *(i)* Web server workload modeling and characterization, focusing on the implication on caching [14, 16]. *(ii)* Efficient cache replacement algorithms [17, 35]. *(iii)* Efficient cache validation and invalidation techniques for strong consistency [26, 27]. Note a *validation* is initiated by a client which verifies the validity of its cached files (as used in HTTP), while an *invalidation* is performed by the origin server which notifies clients which of its cached files have been modified. *(iv)* Cooperative proxy caching [18, 34] where individual proxies share their cached files with each other's clients. *(v)* Caching-friendly content representation such as delta encoding [29].

Caching in mobile networks. Recent study [20] explored the potential benefits of HTTP caching in 3G cellular networks by analyzing traffic traces collected from a large 3G cellular carrier. They found that the cache hit ratio is 33% when caching at the Internet gateway. Another study [22] investigated the potential for caching video content in cellular networks, indicating that 24% of the bytes for progressive video download can be served from a network cache located at the Internet gateway. By comparison, our study investigates caching efficiencies from the perspective of individual handsets.

Another recent study [33] examined three client-only solutions: caching, prefetching, and speculative loading, using web usage data collected from 24 iPhone users over one year. The authors focus on improving the smartphone browsing speed instead of saving the bandwidth. They found that 40% of resource requests can be served by a local browser cache of 6 MB, implying the necessity of HTTP caching. However, its effectiveness of reducing the latency is found to be not as good as that of reducing the traffic volume, mainly because revalidation cannot hide network RTT, which is an important factor affecting mobile browser performance.

HTTP cache implementation on browsers. Professional developers spent efforts investigating HTTP cache implementation

Table 9: Testing results for smartphone HTTP libraries and browsers. ●: fully supported ○: not supported ◐: partially supported ×: not applicable. Refer to Table 8 for acronyms of the libraries and browsers.

Test Name	UC	HUC	HC	WV	HRC	T20	NSUR	ASIHR	AB	SB
1. Basic caching	○	○[b]	○[a]	●	●	○	◐[j]	●	◐[j]	◐[j]
2. Revalidation	○	○	○	●	●	○	●	●	●	●
3. Non-caching directives	○	○	○	●	●	○	●	◐[d]	●	●
4. Expiration directives	○	○	○	●	●	○	●	●	●	●
5. URL with query string	○	○	○	●	●	○	●	●	●	●
6. Partial caching	○	○	○	○	○	○	○	○	○	○
7. Redirection	○	○	○	○	○	○	◐[h]	○	◐[h]	◐[h]
8. Shared or non-shared cache	×	×	×	Non-shared	Non-shared	×	Shared	Shared by default[e]	Non-shared	Non-shared
9. Persistent or non-persistent cache	×	×	×	Persistent	Persistent	×	Hybrid[l]	Persistent	Persistent	Persistent
10. Cache entry size limit	×	×	×	2 MB	No limit	×	NP: 50 KB P: 2 MB[l]	No limit / 512 KB[i]	2 MB / 4 MB[i]	250 KB[g] / 4 MB[i]
11. Total cache size	×	×	×	Storage Capacity	Configurable[c]	×	NP: 1 MB P: 40 MB[f]	Storage Capacity	8 MB	100 MB
12. Replacement policy	×	×	×	LRU	LRU	×	LRU	LRU	LRU	LRU
13. Heuristic fresh lifetime	×	×	×	17.5 hrs	30 mins	×	48 hrs	0[k]	48 hrs	48 hrs

[a] Including subclasses `AbstractHttpClient`, `AndroidHttpClient`, and `DefaultHttpClient`. None supports basic caching.
[b] The class provides caching interfaces through the abstract classes `ResponseCache`, `CacheRequest`, and `CacheResponse`. But developers need to implement them by themselves.
[c] The cache size must be specified by developers.
[d] The class does not cache responses with `Pragma:no-cache` or `Cache-Control:no-cache`.
[e] Developers can make it non-shared by specifying a private cache storage path.
[f] The default sizes are 1 MB for the non-persistent cache and 40 MB for the persistent cache. But they are also configurable by developers.
[g] Safari on iOS 5 has a larger cache entry size limit of 2 MB for an HTML page.
[h] They do not cache a cacheable HTTP 302 response.
[i] The first and the second numbers are the cache entry size limits for an HTML page and an external web object (*e.g.*, JavaScript), respectively.
[j] When loading the same URL back-to-back, the second load is treated as a reload without using a cached copy or issuing a conditional request.
[k] Revalidation is always performed when neither `Cache-Control:max-age` nor `Expires` exists in a response.
[l] Both a persistent and a non-persistent cache are used. Given a file whose size is s, it is stored in the non-persistent cache if s <50KB, or stored in the persistent cache if 50KB$\leq s <$2 MB, or not stored if $s \geq$ 2MB.

issues leading to poor performance, focusing on mobile browsers. The following measurement studies were reported on various technical blogs. Early measurement [5] in 2008 shows that the iPhone 3G browser has a non-persistent cache with an entry size limit of 25 KB (for HTML files) and a total size of 500 KB, implying potential performance issue for large web pages. The experiments were revisited in 2010 [7], and larger cache sizes of iOS 4 on iPhone 4 and Android 2.1 were observed. Similar tests of caching sizes were performed in [11]. Blog entry [8] further pinpoints that for iPhone and Android browsers, the cache entry size limit differs depending on the file type (we considered this in our tests). Blog entry [6] revealed an implementation bug of Safari on iOS 4 shown in footnote *j* in Table 9. We confirmed this and found this problem also exists in the Android 2.3 browser and the `NSURLRequest` library. Blog entry [9] discovered that most desktop and mobile browsers do not cache HTTP redirections properly. Our caching tests cover all aforementioned aspects, and are much more complete as described at the beginning of §6.

Data compression. Besides caching, another orthogonal approach for redundancy elimination is data compression, which can be performed at each single object (*e.g.*, gzip [10]), across multiple objects (*e.g.*, shared dictionary compression over HTTP [15]), or for packet streams (*e.g.*, MODP [28]). Compression can be jointly applied with caching to further save the network bandwidth.

8. DISCUSSION AND CONCLUSION

Web caching in mobile networks is critical due to the unprecedented cellular traffic growth that far exceeds the deployment of cellular infrastructures. Caching on handsets is particularly important as it eliminates all network-related overheads. We have performed the first network-wide study of the redundant transfers caused by inefficient web caching on handsets. We found that redundant transfers account for 17% and 20% of the HTTP traffic, for the two large datasets, respectively. Further analysis on the UMICH trace suggests that redundant transfers are responsible for 17% of the bytes, 6% of the signaling load, 7% of the radio energy consumption, and 9% of the radio resource utilization of *all* cellular data traffic. Most of the redundancy can be eliminated by making the caching logic fully support and strictly follow the protocol specification, and making developers fully utilize the caching support provided by the HTTP libraries.

We plan to pursue three potential directions for further optimizing web caching for mobile networks.

The offline application cache is a new feature in HTML5, the latest HTML standard [3]. It differs from HTTP caching in two ways. *(i)* Caching information of all objects embedded in an HTML page is specified in a small *cache manifest* file associated with the HTML page. *(ii)* There is no explicit expiration, which is instead indicated by a new version of the manifest file that is always downloaded whenever the HTML page is fetched over the network. Although the usage of HTML5 caching is very unpopular in our datasets (only one app in the UMICH trace used it), analysts envision it will eventually be widely used as almost all smartphones are expected to support HTML5 by 2013 [2]. We expect strategically employing this coarse-grained caching mechanism with the traditional per-file-based HTTP caching can achieve more reduction of revalidation traffic causing non-trivial resource consumption despite their small sizes [33].

Optimizing caching parameter settings based on the file semantics is not addressed in this paper, as described in §3.1. However we do observe from our datasets examples where caching parameter settings are obviously too conservative. For example, in the UMICH trace, for the built-in weather app of Motorola Atrix, 95% of its bytes are marked by server as non-storable. We are conducting a more in-depth investigation on optimizing caching parameter configurations.

Previous caching proposals such as delta encoding [29] and piggyback cache validation [26] may provide additional benefits in cellular networks. For example, batching multiple validation requests into a single message [26] potentially reduces the resource overhead as otherwise each individual validation request may incur a separate tail. We plan to revisit both studies in our future work.

9. ACKNOWLEDGEMENTS

We thank Mark Gordon and Zhiyun Qian for their contribution to the data collection software for collecting the UMICH dataset, as well as the 20 students at University of Michigan who participated in our user study. We appreciate the valuable comments from Thomas Andrews, Yudong Gao, Mark Gordon, Sanae Rosen, Zhiyun Qian, and Qiang Xu. We would also like to thank the Mobisys reviewers and especially Matt Welsh for shepherding the paper. This work is partly funded by NSF grants CNS-1059372, CNS-1050157, CNS-1039657 and Navy grant N00014-09-1-0705.

10. REFERENCES

[1] Extensible Messaging and Presence Protocol. http://xmpp.org/xmpp-protocols/.

[2] HTML5-enabled phones to hit 1 billion in sales in 2013. http://news.cnet.com/8301-1023_3-57339156-93/html5-enabled-phones-to-hit-1-billion-in-sales-in-2013/.

[3] HTML5 (W3C working draft). http://www.w3.org/TR/html5/.

[4] Invest in Cell Phone Infrastructure for Growth in 2010. http://pennysleuth.com/invest-in-cell-phone-infrastructure-for-growth-in-2010/.

[5] iPhone Cacheability - Making it Stick. http://www.yuiblog.com/blog/2008/02/06/iphone-cacheability/.

[6] (lack of) Caching for iPhone Home Screen Apps. http://www.stevesouders.com/blog/2011/06/28/lack-of-caching-for-iphone-home-screen-apps/.

[7] Mobile Browser Cache Limits: Android, iOS, and webOS. http://www.yuiblog.com/blog/2010/06/28/mobile-browser-cache-limits/.

[8] Mobile cache file sizes. http://www.stevesouders.com/blog/2010/07/12/mobile-cache-file-sizes/.

[9] Redirect caching deep dive. http://www.stevesouders.com/blog/2010/07/23/redirect-caching-deep-dive/.

[10] The gzip home page. http://www.gzip.org/.

[11] Understanding Mobile Cache Sizes. http://www.blaze.io/mobile/understanding-mobile-cache-sizes/.

[12] GERAN RRC State Mchine. 3GPP GAHW-000027, 2000.

[13] N. Balasubramanian, A. Balasubramanian, and A. Venkataramani. Energy Consumption in Mobile Phones: A Measurement Study and Implications for Network Applications. In *IMC*, 2009.

[14] L. Breslau, P. Cao, L. Fan, G. Phillips, and S. Shenker. Web Caching and Zipf-like Distributions: Evidence and Implications. In *INFOCOM*, 1999.

[15] J. Butler, W.-H. Lee, B. McQuade, and K. Mixter. A Proposal for Shared Dictionary Compression over HTTP. http://groups.google.com/group/SDCH.

[16] R. Caceres, F. Douglis, A. Feldmann, G. Glass, and M. Rabinovich. Web proxy caching: the devil is in the details. *SIGMETRICS Perf. Eval. Rev.*, 26(3), 1998.

[17] P. Cao and S. Irani. Cost-aware WWW proxy caching algorithms. In *USITS*, 1997.

[18] A. Chankhunthod, P. Danzig, C. Neerdaels, M. Schwartz, and K. Worrell. A hierarchical internet object cache. In *USENIX ATC*, 1996.

[19] M. Chatterjee and S. K. Das. Optimal MAC State Switching for CDMA2000 Networks. In *INFOCOM*, 2002.

[20] J. Erman, A. Gerber, M. Hajiaghayi, D. Pei, S. Sen, and O. Spatscheck. To Cache or not to Cache: The 3G case. *IEEE Internet Computing*, 2011.

[21] J. Erman, A. Gerber, M. Hajiaghayi, D. Pei, and O. Spatscheck. Network-Aware Forward Caching. In *WWW*, 2009.

[22] J. Erman, A. Gerber, K. Ramakrishnan, S. Sen, and O. Spatscheck. Over The Top Video: the Gorilla in Cellular Networks. In *IMC*, 2011.

[23] R. Fielding, J. Gettys, J. Mogul, H. F. L. Masinter, P. Leach, and T. Berners-Lee. Hypertext Transfer Protocol - HTTP/1.1 . RFC 2616, 1999.

[24] H. Holma and A. Toskala. HSDPA/HSUPA for UMTS: High Speed Radio Access for Mobile Communications. John Wiley and Sons, Inc., 2006.

[25] J. Huang, F. Qian, A. Gerber, Z. M. Mao, S. Sen, and O. Spatscheck. A Close Examination of Performance and Power Characteristics of 4G LTE Networks. In *Mobisys*, 2012.

[26] B. Krishnamurthy and C. Wills. Study of Piggyback Cache Validation for Proxy Caches in the World Wide Web. In *USITS*, 1997.

[27] C. Liu and P. Cao. Maintaining Strong Cache Consistency in the World-Wide Web. In *ICDCS*, 1997.

[28] C. Lumezanu, K. Guo, N. Spring, and B. Bhattacharjee. The Effect of Packet Loss on Redundancy Elimination in Cellular Wireless Networks. In *IMC*, 2010.

[29] J. Mogul, F. Douglis, A. Feldmann, and B. Krishnamurthy. Potential benefits of delta encoding and data compression for HTTP. In *SIGCOMM*, 1997.

[30] F. Qian, Z. Wang, Y. Gao, J. Huang, A. Gerber, Z. M. Mao, S. Sen, and O. Spatscheck. Periodic Transfers in Mobile Applications: Network-wide Origin, Impact, and Optimization. In *WWW*, 2012.

[31] F. Qian, Z. Wang, A. Gerber, Z. M. Mao, S. Sen, and O. Spatscheck. Characterizing Radio Resource Allocation for 3G Networks. In *IMC*, 2010.

[32] F. Qian, Z. Wang, A. Gerber, Z. M. Mao, S. Sen, and O. Spatscheck. Profiling Resource Usage for Mobile Applications: a Cross-layer Approach. In *Mobisys*, 2011.

[33] Z. Wang, F. X. Lin, L. Zhong, and M. Chishtie. How Far Can Client-Only Solutions Go for Mobile Browser Speed? In *WWW*, 2012.

[34] A. Wolman, G. Voelker, N. Sharma, N. Cardwell, A. Karlin, and H. Levy. On the scale and performance of cooperative Web proxy caching. In *SOSP*, 1999.

[35] K.-Y. Wong. Web Cache Replacement Policies: a Pragmatic Approach. *IEEE Network*, January/February, 2006.

[36] Q. Xu, J. Erman, A. Gerber, Z. M. Mao, J. Pang, and S. Venkataraman. Identifying Diverse Usage Behaviors of Smartphone Apps. In *IMC*, 2011.

Mobius: Unified Messaging and Data Serving for Mobile Apps

Byung-Gon Chun
Yahoo! Research
Santa Clara, CA
bgchun@yahoo-inc.com

Carlo Curino
Yahoo! Research
Santa Clara, CA
krl@yahoo-inc.com

Russell Sears
Yahoo! Research
Santa Clara, CA
sears@yahoo-inc.com

Alexander Shraer
Yahoo! Research
Santa Clara, CA
shralex@yahoo-inc.com

Samuel Madden
MIT
Cambridge, MA
madden@csail.mit.edu

Raghu Ramakrishnan
Yahoo! Research
Santa Clara, CA
ramakris@yahoo-inc.com

ABSTRACT

Mobile application development is challenging for several reasons: intermittent and limited network connectivity, tight power constraints, server-side scalability concerns, and a number of fault-tolerance issues. Developers handcraft complex solutions that include client-side caching, conflict resolution, disconnection tolerance, and backend database sharding. To simplify mobile app development, we present Mobius, a system that addresses the messaging and data management challenges of mobile application development. Mobius introduces MUD (Messaging Unified with Data). MUD presents the programming abstraction of a logical table of data that spans devices and clouds. Applications using Mobius can asynchronously read from/write to MUD tables, and also receive notifications when tables change via continuous queries on the tables. The system combines dynamic client-side caching (with intelligent policies *chosen on the server-side*, based on usage patterns across multiple applications), notification services, flexible query processing, and a scalable and highly available cloud storage system. We present an initial prototype to demonstrate the feasibility of our design. Even in our initial prototype, remote read and write latency overhead is less than 52% when compared to a hand-tuned solution. Our dynamic caching reduces the number of messages by a factor of 4 to 8.5 when compared to fixed strategies, thus reducing latency, bandwidth, power, and server load costs, while also reducing data staleness.

Categories and Subject Descriptors

C.2.4 [**Computer-Communication Networks**]: Distributed Systems

General Terms

Design, Algorithms, Experimentation, Performance

MobiSys'12, June 25–29, 2012, Low Wood Bay, Lake District, UK.

Keywords

Mobile cloud computing, data management, messaging, push notification, caching, mobile apps

1. INTRODUCTION

Increasingly, applications for mobile devices create and publish content that must be shared with other users in a timely fashion, or retrieve data personalized to a user's location or preferences from a cloud. Examples include Yahoo! applications like Yahoo! News, Sports, Finance, Messenger, Mail, Local, and Livestand, shop review applications like Yelp, crowd-sourced traffic collection applications like Waze, and social networking applications like Facebook, Foursquare and Twitter. Developing such applications atop create/read/update/delete (CRUD) APIs is challenging for a number of reasons: 1) cellular networks have high and variable network latencies and costly bandwidth, 2) mobile phones have limited battery-power, 3) network disconnections are common, and 4) server-side scalability and fault-tolerance are a requirement.

This forces developers to implement complex networking and data management functionality and optimizations, including: asynchronous interactions (to hide the cost of remote operations), client-side read caching (to reduce bandwidth and latency, while bounding data staleness), batching of write operations (to reduce power costs), protocols to arbitrate concurrent and disconnected operations on shared data, timeouts and reconnection strategies (to handle network coverage issues), notification mechanisms (to allow data to be pushed from the server side), and manual sharding of data and replication (to guarantee server side scalability and fault-tolerance).

We believe that developers should not have to worry about these issues when writing mobile applications. Unfortunately, existing frameworks provide limited support for addressing these challenges. On the one hand, RESTful backend storage services focus on well-connected hosts [16, 12] and provide scalability, but provide little support for data sharing across users, caching, write batching, and so on. For example, iCloud [5] only provides mechanisms for applications to backup a user's private state on servers; developers are left to implement the rest themselves. On the other hand, replication systems for mobile computing such as Coda, Ficus, Bayou, PRACTI, and Cimbiosys [37, 34, 49, 22, 41] support replication of data for sharing and disconnected operations, but they are not designed to manage data and notifications for a large number of users.

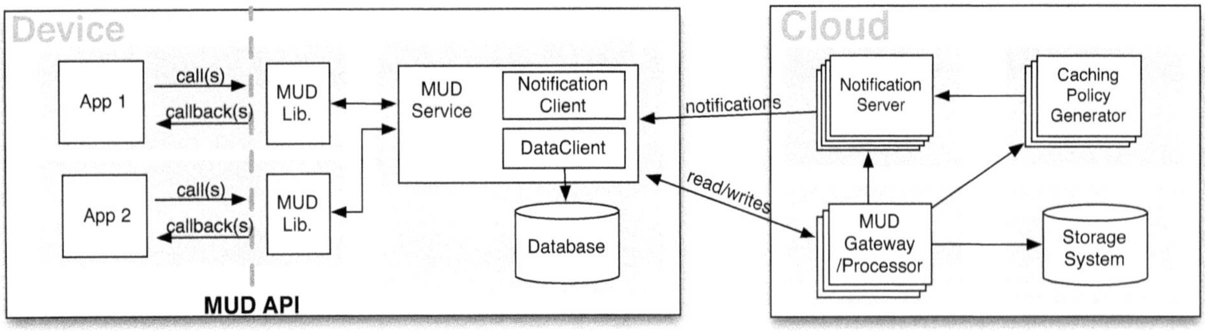

Figure 1: Mobius architecture.

In this paper, we bridge this gap with Mobius, a reusable data management and messaging platform that provides a simple, unified abstraction for mobile apps and addresses the challenges transparently underneath the abstraction.

Mobius employs a number of features that are motivated by data consumption, creation, sharing, and messaging requirements of Yahoo! mobile applications/platforms, which can be applicable to a broad set of applications. The requested features are:

- Reading/writing data and messaging
- Flexible predicate-specified real-time notification
- Writer's control on the visibility of its data
- Disconnected operation handling for reading and writing
- Transparent, intelligent caching/prefetching
- Hosted, scalable backend service

The first novel feature of Mobius is the *MUD* programming abstraction, short for "Messaging Unified with Data". MUD presents a table-oriented interface that supports asynchronous, low-latency access for messaging as well as for predicate-based lookups for data-intensive applications. Applications can read/write data or send/receive messages by reading and writing a shared logical table through the unified read/write API. Mobius implements the features that are needed by both networking and data stacks around this abstraction: disconnection-tolerance, support for changing IP addresses, caching, push notifications, and so on.

MUD tables are partitioned across mobile nodes and one or more server back-ends. From the perspective of the mobile app developer, accesses are done to local tables. These accesses may cause data to be fetched on an as-needed basis from servers, depending on the application's tolerance to stale data and the previous records that have been fetched. We support several kinds of read operations with a range of freshness guarantees.

The MUD API provides two less-conventional means for accessing data. First, it provides "writer predicates" where the visibility of writes in the table is constrained to readers who satisfy a particular predicate. For example, a user's writes may only be visible to other users in his family, or to users in a certain geographical region. This provides a form of predicate-based communication or late binding, where the exact recipients of a message are not known when the message is created. Second, the API provides a notification interface that sends applications "push notifications" when a particular record or range of a table is updated. This is useful both for inter-device messaging and for providing continuous queries, where the result of a query is updated as the records that satisfy it change. Our design and implementation of Mobius unifies backend storage and processing for event notification and data serving, which can scale to a large number of concurrent push clients.

The second novel aspect of Mobius is support for *server-directed* caching and prefetching. To express preferences about what data should be cached, and to enable prefetching of data the application is likely to need in the future, the MUD API allows applications to specify "utility functions" that expose to our system the tradeoffs between various characteristics of a query answer, e.g., freshness, completeness, latency, bandwidth cost, and power cost.

Using this preference information, and knowledge of workload and system status, a server component can automatically devise intelligent caching strategies that maximize utility for each user. For example, knowing the frequency of updates, a caching policy can reduce power costs by suggesting that when asking for traffic data from the middle of the Nevada desert, it is best to serve data from the local cache since the chance of missing an update is minimal. Mobius provides several off-the-shelf utility functions that are designed to make it easy for programmers to implement common caching and prefetching values; e.g., programmers can provide a simple freshness threshold to ensure that content is more recent than the threshold.

We evaluate the performance of Mobius on several microbenchmarks to demonstrate that it does not incur substantial overhead over conventional APIs for data access and messaging, and to illustrate the potential for significant performance optimizations. In our initial prototype, remote read and write latency overhead is less than 52% when compared to a hand-tuned solution. We also show that our server-controlled client-side caching, based on simple machine learning based strategies, can reduce the number of queries pushed to the server by $4\times$ to $8.5\times$ for reads and $3\times$ to $6\times$ for writes, on real-world user traces from an iPhone crowd-sourcing application deployed to over 600 cars.

The rest of the paper is organized as follows. Section 2 presents the Mobius system architecture. Section 3 introduces the MUD abstraction, API, and usage. Section 4 presents Mobius write/read protocols, Section 5 presents Mobius continuous read/notification protocols, and Section 6 discusses server-directed, dynamic, context-based caching. We describe our implementation and experimental evaluation of the prototype in Section 7. Finally, we survey related work in Section 8, and conclude in Section 9.

2. MOBIUS ARCHITECTURE

Mobius is a unified messaging and data serving system for mobile apps that addresses the challenges discussed in Section 1. Mobius is motivated by data consumption, creation, sharing, and messaging requirements of current and future mobile apps; in particular, our design is guided by discussions with Yahoo! mobile app/platform developers. Many of them use their own custom layer

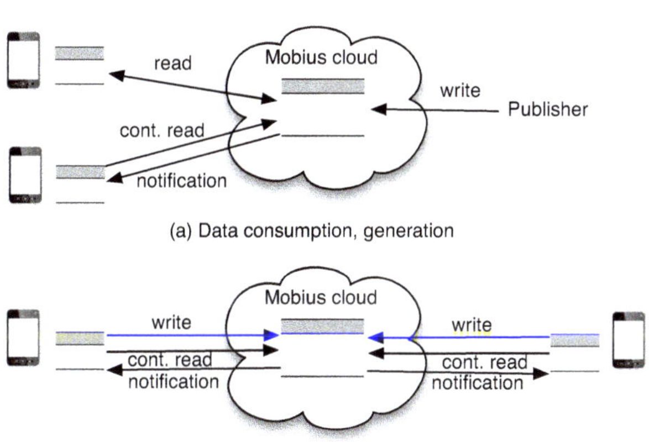

(a) Data consumption, generation

(b) Data synchronization, messaging

Figure 2: MUD usage scenarios.

to deal with caching, disconnection handling, push notification, or messaging in a limited fashion. Mobius provides all these features through the MUD abstraction.

Mobius consists of both client-side and back-end (cloud) infrastructure, as shown in Figure 1.

Client-side infrastructure: Services on the client include the client-side storage system, the MUD Service and application-level library code. The MUD Service is composed of a *notification client* and a *data client*. The notification client receives push notifications from the Mobius back-end infrastructure and forwards them to the appropriate MUD Library instance. The data client is responsible for sending data and queries to the back-end and for processing data sent in response. It is also responsible for query and update processing using client-side storage and as a runtime for the caching and prefetching policies devised by the server.

Back-end infrastructure: Services on the back-end include the back-end storage system, notification servers, gateways, MUD processors and caching policy generators.[1] Each *notification server* maintains a network connection to multiple client devices and is responsible for pushing notification messages to these clients. Device identifiers are used to partition clients among notification servers. The *gateways* accept client requests, forward them to MUD processors, and relay responses back to the clients. The *processor* performs query and update processing, including continuous queries, which are maintained as soft-state. Continuous query results are routed back to clients via the notification server. The back-end storage system stores records and provides scan operations that enable clients to re-sync when they connect. The storage system is partitioned by primary key ranges; each partition is handled by a single MUD processor, although each processor may be responsible for multiple partitions. Furthermore, storage is geo-replicated across multiple data-centers, each of which may run instances of notification servers and gateways. Our prototype targets PNUTS [30] as the back-end store, but other standalone storage systems may be used. The *caching policy generator* creates caching and prefetching policies for *each client* by mining data access patterns of *all clients* (this information is obtained from MUD processors). Clients use these policies to decide what to cache or prefetch and when. Next, we describe each component in detail, beginning with a description of the MUD data model and API.

[1]We also assume the existence of a naming service, which could be implemented atop the underlying storage service, or be provided using a pre-existing standalone system, such as ZooKeeper [17].

Key	Value				Predicate
	C1	C2	C3	C4	
k1	2	CA			F1 == 1
k2	3		Red		
k3		MI	Blue	100	F1> 100

Table 1: MUD table example.

3. THE MUD ABSTRACTION

MUD relieves mobile app developers from dealing with messaging and data management issues by providing them with a unified programming interface for both messaging and data. This interface is based on the simple abstraction of shared logical tables that can span devices and back-end clouds. Apps can read/write data or send/receive messages by reading and writing a shared logical table. In addition, apps can register continuous read operations and get notified when relevant updates are made to a table. Figure 2 shows different MUD usage scenarios. We also mediate access to tables stored solely on the device local storage, but this is trivial and not discussed further in this paper.

In the MUD data model, each logical table is a collection of records. Each record contains a (required) primary key, a sequence of attribute and value pairs, a predicate, and metadata (such as the record's version). Each attribute corresponds to a column in the logical table. Schemas are flexible: a record can have values for any subset of table columns. Mobius supports dynamically adding and removing columns from a table. An example of a table is shown in Table 1.

A unique feature of the abstraction is the symmetric nature of read and write operations. As usual, reads include a predicate specifying the attributes of records to be read. In addition, in MUD a writer (or sender) includes a *writer predicate* with the data. A reader (or receiver) includes *reader data* with its queries, such as authentication tokens, current location, time, or other context information. Writer predicates are then evaluated against the reader data. Hence, writers can control which readers may observe the written data. For example, an advertiser can send special coupons only to its premium users, and use user location and preferences to filter out uninterested users.

In its simplest form, a writer predicate does not require the *authenticity* of reader data. A writer predicate may be simply used to limit which readers can receive the data as an optimization (e.g., discount coupons for users in a particular location). These context-based predicates allow writers to *late bind* to readers that can see their updates, such as when the list of users in a particular location is not known at update time, or when adding a user to a group for advertisements. In more complex cases, a writer predicate can be an access control list (ACL) specifying clients that may read the record (note that, as mentioned in Section 3.1, each table has an ACL as well) when reader data's authenticity is verifiable. For example, only clients with proper identities can read the record (e.g., user messages in the writer's friend list).

We provide simple and yet well-defined consistency semantics (per-record *sequential* [39] and *fork-sequential* [40] consistency), and let app developers choose the appropriate consistency for their application. We achieve this by exposing read operations with different consistency levels. Applications can gracefully degrade consistency whenever latency or service availability are more important. Informally, our consistency guarantees define the order in which clients see updates, while allowing clients to return cached updates. In Section 4 we describe the different types of read and write operations.

Table manipulation
create(tb, properties)
drop(tb)
alter(tb, properties)
Table update
write(tb, key, type, data, predicate, utility)
delete(tb, key, type, utility)
Table query
read(tb, predicate, type, data, utility)

Table 2: MUD API.

Obviously, the back-end store must itself provide some level of consistency so that the system as a whole may be consistent. Interestingly, for simple read and write operations the back-end store need only guarantee eventual propagation of updates. That is, all non-faulty back-end storage replicas should eventually receive all record updates made to their partition, but the updates do not need to be sequentially consistent. In practice, we use PNUTS, Yahoo!'s geographically distributed data serving platform [30]. PNUTS provides *timeline consistency*: each PNUTS replica applies increasingly newer updates. We note that other systems, such as Amazon's SimpleDB, also provide the required back-end store functionality [12], while others, such as Microsoft's SQL Azure provide even stronger primitives, such as single-partition transactions [25].

3.1 MUD API

Operations in MUD are categorized into three groups: table manipulation, update, and query. Table 2 lists the supported operations (for brevity, in a simplified form). All operations are asynchronous and their results are returned through a provided callback interface.

Table manipulations include three methods: *create*, *drop*, and *alter*. *create* makes a logical table *tb* with the supplied *properties*, which may include an access control list (ACL), an optional schema, etc. An app that creates a table may later invoke *drop* to delete it. Finally, *alter* is used to modify the table's properties.

The ACL defines a security level of the MUD table, which can be public, app, or user. Our authorization follows OAuth [9], a well-known authorization protocol for publishing and interacting with protected data. For a user to access the table, the user must be authorized at the level of the table or higher. The public security level does not require any authorization. The app security level requires at least app authentication, and the user security level requires user authentication.

The *write* and *delete* operations get as a parameter the table identifier *tb*, a primary key of the record to be written or deleted, the type of the update operation (blind or conditional, described in Section 4) and a utility function. (We discuss utility in Section 6.) In addition, a *write* operation is passed the data to be written and a writer (or sender) *predicate*, that constrains the visibility of the written data to readers whose reader-data satisfies the predicate. The predicate follows the same structural rules as a read's predicate discussed below. If no record exists in *tb* with the given primary key, the write inserts it and otherwise updates the record.

A read operation passes a *predicate*, which is a boolean expression over the rows of a logical table *tb*, similar to a SQL SELECT statement. Specifically, the predicate may be a conjunction or a disjunction of conditions. Each condition imposes a boolean constraint on record key and attributes using binary operators: $=$, $\neq$, $<$, $>$, $\leq$, $\geq$, and a limited number of boolean functions and regular expressions on strings. In addition, a *predicate* may include user-defined functions (UDFs), such as aggregation or limited joins (for example, a UDF can summarize the values of a given attribute for all matching records). Such functions help conserve network bandwidth in cases where transferring records from back-end to the client may be easily avoided. The language is similar to Yahoo! Query Language (YQL) [16].

A *read* additionally takes *data* as parameter which typically contains (but is not limited to) information about the end-user or device. For example, *data* may include the reader's location, identifier and so on. A read retrieves all tuples that match its predicate, and whose write-predicate matches the read's *data*. Similar to updates, a *read* additionally includes a *utility* and a *type*. The read utility is analogous to the write utility, and is used for caching.

We support three different "one-time" read types (committed_wio, committed and uncommitted), described in Section 4, which guarantee different consistency semantics. A fourth type of read is the continuous read, which sets up a continuous query. A continuous read gets an optional *sync-timestamp* parameter (which does not appear in Table 2), that instructs it to perform a one-time read returning records no older than *sync-timestamp*, in addition to registering the continuous query. This may be used if currently stored records are of interest in addition to future updates or to restablish a continuous query after a failure or device disconnection. A continuous read always returns a *handle*, which may be later used to unregister or modify the subscription. To this end, we provide two auxiliary methods:

unregister(handle)
update-data(handle, data)

For example, consider a client interested in restaurants within 1 mile of its location. Such a client may register a continuous query, including its current location in the *data* of the read. As it moves and its location changes, the client invokes *update-data* to update its subscription with its latest location (although a subscription may be updated by unregistering and re-submitting it, this would require more messages as well as querying for any updates missed in the process).

For the convenience of app developers, higher level APIs are provided as a layer on top of the basic MUD API we presented above, as indicated in Figure 1. For example, we provide blocking operation APIs as well as a simple pub/sub API. In the future, even richer interfaces could be considered, such as an SQL-like declarative query language.

3.2 Using MUD

To showcase the benefits of Mobius and illustrate how to use the MUD API, we present three sample applications, social news reading, experience sharing and continuous journal editing, that combine messaging and data interactions and need late binding and caching.

Social News Reading: This application is a social news-reader application supporting instant social sharing and updates. The application supports reading news articles of users' interests, being instantly notified of any urgent news, writing user-generated content, and exchanging interesting news articles and messages with friends in real time. The social news-reader application can be easily implemented with the MUD API. The app developer first creates two tables, *news (topic, article)* and *chat (group, msg)*, using *create*. Suppose that Jane uses this app for sharing articles with her friends. Jane's application requests news articles on sports by issuing *read(news, topic='sports', committed, ...)*. It also registers Jane's interest in art articles by invoking a continuous read *read(news, topic='arts', continuous, ...)*. Whenever a new article becomes available (e.g., when the application's cloud service discovers a new article online and makes it available to the app clients by invoking *write(news, unique article id, blind, [topic='arts', article='new exhibition'], ...))*. Jane's application is instantly notified. When Jane requests to share an article with her friends, the app performs *write(chat, unique message id, blind, [group='Jane's friends',msg=article id], ...)*, assuming her friends have already registered continuous reads on the *chat* table. The same table can be used for exchanging discussion messages. When Jane wants to send a local event message to her friends located within 1 mile from her current location, the app can perform *write(chat, unique message id, blind, [group='Jane's friends', msg=event url],* $|receiver.loc - this.loc| \leq 1$ *mile, ...)* where *this.loc* is Jane's location and *receiver.loc* is the location of a friend who sets up a continuous read to receive Jane's messages.

Experience Sharing: This application is aimed at enhancing the experience at large events, such as fairs or concerts. The first set of data consists of map overlays, schedules and other information provided by the organizers. These are queried by the users and can be cached, since their content is likely to be stable. This is achieved by issuing reads such as: *read(overlaymaps,lon>y1 and lon<y2 and lat>x1 and lat<x2, ..., freshness<24h, ...)* where *freshness<24h* represents a simple utility function that allows for heavy caching. Organizers also push timely news about schedule changes, and other unpredictable events to the users. They can do this by issuing simple writes. For example to notify every person over 21 years old of a free beer event one could write: *write(flashnews, pk1, blind, "free beer at south kiosk", receiver.age≥21,...)* to a table over which user devices have posed subscriptions, i.e., continuous reads. In this context predicates can be used to avoid broadcasting to the entire crowd, and can be applied to user profiles and context. Users contribute by messaging each other, observing friends location, sharing comments, tweets and pictures. All these use cases are easily supported with some combination of reads, writes and continuous reads. Furthermore, "relive the event" user experiences can be supported by issuing rich historical queries such as: "What was the name of the band playing when I sent that comment?," or, upon reading a geo-tagged tweet: "Tell me if anyone uploaded pictures taken at this time and place." Users can operate seamlessly on local data under disconnections, and the system syncs as soon as a connection is available.

Continuous Journal Editing: This application supports keeping context-tagged journals for the users. The application is implemented on multiple platforms — smartphones, tablets, and desktop machines. The users can update journals in any of the devices they use. For example, a user edits a journal on a tablet disconnected from the network. She then edits the journal with her smartphone on the way home and writes it to the back-end. When the user reads the journal at the tablet at home, she should be able to see both changes; i.e., the edits made on the smartphone should not be overwritten by the updates locally queued on the tablet when the tablet comes back online. This is achieved by using conditional writes such as: *write(journal, pk1, conditional, [content="about cafe A,..."], ...)*. Unlike blind writes, which simply overwrite the content, a conditional write succeeds only if the version of the record at the back-end is identical to the one locally stored on the device at the time the update is performed. Thus, whenever two conditional writes are made concurrently to the same shared journal, they conflict and one of the updates (in our case the one performed on the tablet) fails. The application issuing the failed update would then provide the user with information about the conflict so that she can reconcile them.

Given this high-level description of Mobius and the MUD API, we proceed to describe Mobius design in more detail. We start with write/one-time read operations, and then discuss continuous reads and caching. After that, we present the results of some initial experiments with our Mobius prototype.

4. WRITE/READ PROTOCOLS

Mobius provides app developers with the freedom to choose the appropriate consistency semantics for their application. The weakest consistency model our system provides is *fork-sequential consistency* [40], which is the guarantee provided when mobile devices execute operations without blocking under disconnection. As we restrict the types of records the client can read, we support stricter consistency semantics. Table 3 summarizes the consistency semantics Mobius provides. In the following, we describe different types of write and read operations, their consistency semantics, and protocols to implement them.

4.1 Operations and Consistency Semantics

Writes: Our system supports two record write operations: a "blind" write, which simply overwrites a record, and a "conditional" write, which requires the version of the record at the back-end to match the version in the client's local store. As demonstrated by the Continuous Journal Editing application in Section 3.2, conditional writes are especially useful for mobile applications.

Each record in Mobius has monotonically increasing versions, as described in detail later in this section. If the record being updated is not in the local store, a conditional write succeeds only if the record does not exist at the back-end store. When the record is written successfully at the back-end store, the record is *committed.*

The type of the write affects its liveness guarantee: A blind write invoked by a non-faulty client (that eventually becomes connected) is guaranteed to succeed regardless of the operations of other clients. Conditional writes have weaker liveness; while system-wide progress is guaranteed (some client's write will succeed), progress is not guaranteed for each individual client, as other clients may simultaneously update the same record causing a conditional write to fail.

Reads: We support three different one-time *read* operations: uncommitted, committed, and committed_wio.

A committed read returns a "recently" committed update; the notion of recent is briefly explained below. This read type provides sequential consistency [39]. Sequential consistency is equivalent to the concept of one-copy serializability found in database literature [23]. It guarantees that the real execution is equivalent to a sequential one, where the program order of each client is preserved.

With sequential consistency, the view of each client must preserve the precedence order of its own operations. However, as clients may execute operations asynchronously, without waiting for previous operations to complete, precedence order is only a partial order. Hence, the definition of sequential consistency does not restrict the order in which writes executed concurrently by a client may take effect.[2] We thus additionally support a slightly stronger notion of consistency, which requires the *invocation order* of writes to be respected. A committed_wio (committed with write invocation order) read thus requires that the returned committed record reflects all non-failed writes previously invoked by the client (and is therefore executed after all such writes complete, which may require blocking).

Intuitively, an uncommitted read is allowed to return uncommitted writes, even writes that may later fail (if some global constraints are not met). For example, a client reading and writing data while

[2]In fact, all definitions of sequential consistency we are aware of, assume that each client executes one operation at a time.

Read type	Semantics	Write visibility
uncommitted	fork-sequential consistency	pending writes
committed	sequential consistency	committed writes
committed_wio	sequential consistency with write invocation order	committed writes, blocks till client's previously invoked writes complete

Table 3: Consistency semantics

disconnected, or a client accessing data for which it is the only writer, may be fine with seeing data that it has previously written before it has been committed. Clients may also use uncommitted reads if they value latency more than consistency. This read type provides fork-sequential consistency [40]. Fork-sequential consistency allows client views to diverge when they execute operations locally, and yet guarantees that clients see writes in the same order. Specifically, it guarantees that if a client observes a write issued by another client, then their views are identical up to the write.[3] Formal definitions of the properties are presented in Appendix A.

All read types preserve the "timeline" of updates at the client, that is, a client is never returned an old version of a record after having seen a newer one. Note that this requires a client to maintain the last version identifier of each record it has seen, even if the record itself is deleted from its local store. This information may be garbage-collected when the client is sure that older versions of the record may not be returned by the back-end (for example, when the back-end indicates that all replicas are up to date). We note that while each read type returns increasingly newer versions of each record, this is not guaranteed across read types (for example, an uncommitted read may return a newer update than a subsequent committed read).

Stronger semantics: As we have already mentioned, we target PNUTS as the back-end store. PNUTS provides *timeline consistency* [30] on a per-tuple basis. Timeline consistency is a weaker variant of sequential consistency; it guarantees that each tuple will undergo the same sequence of transformations at each back-end store replica.[4] With some additional support from the back-end, it is possible to guarantee stronger consistency with only minor changes to our protocols. Specifically, if the back-end supports queries of the *latest* value of each record, the protocols guarantee linearizability [35] instead of sequential-consistency and fork-linearizability [24] instead of fork-sequential consistency. The only change required on the client-side is that committed and committed_wio reads can no longer be served locally; linearizability requires a read to return the latest complete update. Note that even if the back-end supports querying for the latest update, such queries usually require contacting the record master or multiple back-end replicas, and are therefore more expensive than queries that do not have to return the latest update.

4.2 Protocols

Next, we describe the protocol for executing reads and writes. In the text below, we denote client C's local store by DB_C.

Writes: A write is executed in two stages at the client:

1. insert R = *(key, data, predicate, ver, pver, ts, committed)* to table *tb* in DB_C, where *ver* is a version of the update, *pver* is the previous version of this record in DB_C (if any), *ts* is the local time of this update and *committed* is *false*. Once insert completes, the operation returns.
2. send the record R to a MUD gateway. Upon receiving a response *callback* is invoked and, if the update was successful, *R.committed* is set to *true*, *R.ts* to the current local time, which is used for local freshness checking, and *R.ver* is set to the global version assigned to this update by the back-end store.

When an update is submitted, *ver* is assigned a local version. Committed updates have versions assigned by the back-end, which we call *global* versions (for example, in PNUTS, global versions are simply update counters). Note that a local version is itself a triple *(base, netid, seq)*, where *base* is a global version of the last committed update to the record in DB_C (or 0 if *key* is not in DB_C), *netid* is the client identifier, and *seq* is the number of pending (uncommitted) updates issued by the client. *pver* is the version (whether global or local) of the previous update to the record in DB_C. *pver* of an update is *ver* of the preceding update to the same record stored in DB_C (0 if *key* is not in DB_C—in this case the conditional write is treated as a conditional insert).

For a conditional write to succeed, the writer must have the necessary permissions,[5] and *pver* must be equal to the current version of the record at the back-end. This check is performed atomically and therefore requires back-end support for compare-and-swap updates as provided, e.g., in SimpleDB, Windows Azure Storage or PNUTS.

While a client is disconnected from the system, its updates are stored locally and then sent to the back-end when the client reconnects. Sending updates to the back-end can be done for every update separately, or periodically, batching multiple updates together. Updates are sent to one of the MUD gateways, which forwards each request to the appropriate MUD processor responsible for the partition to which the updated record belongs. Each MUD processor is responsible for many partitions. The MUD processor forwards the write and read requests to the back-end store, maintaining any ordering constraints required for the write and committed_wio read requests. Once a write is executed against the store, the completion status is sent back to the MUD gateway which in turn returns a response to the invoking client.

In case a conditional write fails because of a conflicting update (the current version of the record at the back-end is different than *pver*), the client is given a chance to resolve the conflict. Specifically, the client is sent (a) the conflicting update found at the back-end; (b) last update to the record in DB_C and (c) the previously committed update in DB_C (*base* when the update was invoked). Given this information the client can resolve the conflict and issue

[3]Originally, "forking" consistency conditions were used in the Byzantine system model, where a malicious server may "fork" the views of two clients. We are not aware of previous work that used the conditions to model disconnected client operations in a non-malicious setting.

[4]PNUTS does not guarantee that the "timeline" of updates is preserved *at the client*, as a client may connect to different back-end replicas.

[5]Interactions between conditional writes and writer predicates have security implications, and must be handled with care.

a new update, or simply discard its local changes. This method is akin to the conflict resolution employed by software revision control systems such as SVN. Note that when such conflicts occur, the pending update is deleted from DB_C and the conflicting update returned from the back-end is stored locally.

Deletes are a special type of update. Locally, they are treated just like any other write, except that the data being written is a special *tombstone*, which marks the record for deletion. When such updates are executed on the back-end, they are translated to blind or conditional deletes.

Reads: Like updates, read operations involve two stages. First, the client attempts to answer the read locally. It evaluates the query against DB_C and identifies the primary keys of matching records. If rules to decide local serving do not allow serving some of the matching keys locally, or if it is possible that there are records which may match the query but are not stored in DB_C, the query is sent to a MUD gateway.

To illustrate the flow of read operations with the rules, we use a maximal acceptable record age (expressed as a parameter *max-age*) as a simple utility function (in Section 6 we discuss more general utility functions). The following auxiliary procedure is used to decide whether a record R in DB_C is "fresh" enough (*time* is the current local time):

---

procedure *isFresh*(R)
  if R.*committed*=*false* **or** *time*−R.*ts*<*max-age* **then**
    return *true*
  else return *false*

As an extension, one can use utility functions such as the ones employed in bounded inconsistency models (e.g., numerical error, order error, staleness) to define what records are acceptable in an application-specific fashion [51].

For each such key k of matching records, let R be the latest update to k in DB_C (a pending update is always considered more recent than a committed update for the choice of R) and R_{base} the last *committed* update to k in DB_C. For each of the read types, we now specify the conditions for serving k locally:

- *uncommitted*: return R if *isFresh*(R).
- *committed*: return R_{base} if *isFresh*(R_{base}).
- *committed_wio*: return R_{base} if *isFresh*(R_{base}) and R=R_{base}.

When the MUD gateway receives the read, the gateway breaks the read into multiple sub-queries, each corresponding to a different partition, and forwards each sub-query to the MUD processor responsible for that partition. The read type determines how the processor handles the read: *committed_wio* reads are queued after writes and other committed_wio reads previously received by the processor. Uncommitted and committed reads are maintained separately—such reads may be executed right away without waiting for writes to complete. When processing a client's read, the processor queries the back-end store and sends back all matching records to the MUD gateway through which this read was submitted. The gateway then forwards any returned data to the invoking client. When all processors to which sub-queries corresponding to a read have returned a response, the gateway sends a special completion message to the client.

Clients filter received records corresponding to updates older than they have already seen. This check is performed using the version *ver* of a record. If a newer record is received, it is both returned to the application through the MUD Library, and stored in DB_C (in practice, if the data is large it does not have to be stored locally—we may store just its version information). In addition to any pending updates, at most one committed version of each record is kept in DB_C (the one having the highest observed version).

5. CONTINUOUS READ PROTOCOL

In this section, we describe the processing of continuous reads and push notifications sent to clients. The notification flow is exercised when a registered continuous read for a logical table matches changes to the table's records.

As with a normal read, a continuous read is submitted to one of the MUD gateways, which splits it into one or more sub-queries, each of which corresponds to a different partition and is sent to the corresponding MUD processor. A continuous read is first executed against the back-end store just like any committed or uncommitted read. A minor difference is that, unlike normal reads, a continuous read includes an optional *sync-timestamp* parameter, which determines the maximal age of existing records which is of interest to the client (if *sync-timestamp* is set to zero, no existing records are queried). These records are returned through the response flow of a normal read. Continuous reads are then maintained as soft state at MUD processors.

When a write operation updates a record, the MUD processor of the corresponding partition is notified. It then matches the updated record and all relevant continuous reads. Finally, for each such continuous read, it looks up the notification server responsible for push notifications to the client that registered the read, using a unique client identifier. The MUD processor then forwards the updated record to all relevant notification servers, including both the client identifier, *netid*, and the registering app identifier, *appid*. Each notification server in turn forwards the write to notification clients. When a client is not reachable (e.g., it is turned off or disconnected from the network), the notification is stored in a message queue for a limited amount of time. Eventually, the continuous read is terminated by the server. Upon reconnecting, the client detects this, and re-submits the read if needed.

A *netid* is a unique and fixed client device identifier (unrelated to a client's IP address, which may change over time). Since multiple apps may share the same device (and the notification channel), the pair (*netid*, *appid*) is used to uniquely identify the origin of a continuous read. We call such pair an *endpoint*. A notification server exposes a low-level API to other servers, allowing them to send messages to endpoints. Each server is responsible for a range of *netid*s. This mapping is maintained by the name service and cached at each notification server.

Each client maintains a persistent connection to a notification server for push-notification messaging. The client locates a notification server by performing a DNS resolution of a publicly known notification service domain name. It is possible that this server is not responsible for the client's *netid*, in which case the client is redirected to the appropriate notification server. After establishing a channel, the notification client sends periodic heartbeat messages (e.g., once every 20 minutes). Such heartbeats are used by the notification server to detect a client's failure and garbage-collect its registered continuous reads. Whenever the IP address of the client device changes, the client re-establishes a persistent connection to the notification server. When the notification server pushes a message to the client, the notification client routes the message to the appropriate MUD Library using the *appid* included in the message.

6. CACHING

Our consistency semantics restrict the allowed ordering of updates seen by clients, and yet are amenable to caching, allowing our system to choose when to serve a record using the device's local store and when to fetch a record from the back-end, and similarly when to propagate writes to the server. Hence, appropriate caching policies need to be devised to complement the consistency semantics with proper control over the tradeoffs between performance and answer quality.

We model this tradeoff as a utility maximization problem and exploit a developer-provided *utility function* in order to choose the appropriate execution plan for each query (e.g., whether to serve from client-side storage or fetch remotely), in a way that maximizes utility.

Given a query Q, and an execution plan P, a utility function is defined as: $Utility(Q,P) = Value(Q,P) - Cost(Q,P)$ where $Value$ is a function that assigns a numerical value to various quality characteristics (e.g., freshness, completeness, bias) of the answer to query Q as produced by plan P, and $Cost$ is a function assigning a numerical value to the costs (e.g., latency, bandwidth, power) of executing query Q according to plan P.

We aim to provide developers with a library of common utility functions to choose from, while allowing advanced developers to specify their own. An example of a common utility function is the classic notion of cache *freshness*, where $Value$ is defined as a step function—assuming value zero when Q is answered using locally cached data older than some threshold time t_0 and a large constant value otherwise—and $Cost$ is defined as a linear function of query latency. Under this utility function, answering from the client-side store yields higher utility whenever locally stored data is "fresh enough", while fetching data remotely is only beneficial when the required information does not exist locally or is too stale (otherwise we incur unnecessary costs). More complex utility functions accounting, for example, for battery and communication bandwidth can be easily devised.

Caching policies are most often implemented as fixed heuristics, hard-coded in the application logic at design time. Our system does not follow this convention and instead supports dynamic, context-aware, and globally-selected caching policies. Our architecture is designed so that the client runtime (specifically, the caching decision logic in the MUD data client) is a simple executor of cache policies that are determined at the server-side back-end. A server-side *caching policy generator* has a unique vantage point to determine the most efficient caching strategies since it has access to the usage patterns of all users (and can potentially collect such information across applications). Usage or access patterns include (but are not limited to) read/write ratios, temporal access correlations, current system conditions (e.g., network congestion, bandwidth cost, etc.), and user context (e.g., location). Taking such information into account allows for more informed caching decisions. For example, for a check-in application such as FourSquare, the server-side policy generator can use the density of check-in reads and updates in a certain area to determine how frequently a user's cache should expire and how often a user should push updates to the back-end. For example, when traveling to a destination where the user has no friends, a user's device could push his check-ins lazily and answer mostly from its local store, significantly reducing battery and bandwidth costs with no significant impact on utility.

Note that MUD notifications may be leveraged for pushing caching policies *dynamically* to clients. For example, in a map/-traffic application, traffic may increase dramatically following an accident, in which case it is important that users increase their frequency of traffic reports and get updated frequently on the developing conditions. A caching policy generator can react by increasing the poll and update rate for all devices in the vicinity of the accident.

Dynamically and globally coordinated caching policies have the potential to significantly outperform hard-coded heuristics (typically used today). In our current prototype, we cast the read cache policy selection problem as a classification problem: the features include the query predicates and user context, and the classification labels indicate whether a query should be served from the local cache or be sent to the server. In this initial implementation, the utility function is provided in the form of a cost-matrix, that indicates to the system the relative impact on utility of stale data versus the cost of unnecessary queries sent to the server. In our experiments, we use a real trace of user GPS locations and queries, derived from the CarTel application [36], to train a cost-sensitive classifier based on a J48 decision tree, and demonstrate that even our simple initial solution leads to up to a $4\times$ reduction in the number of queries sent to the server (providing a significant improvement in read latency, and reducing bandwidth), and reduces by $3.5\times$ the number of queries serving stale data (thus increasing the perceived data quality), when compared to a rather common baseline caching strategy.

The output of the policy selector is a trained decision tree model that is shipped to the client through our notification channel. For each query the client-side runtime tests the decision tree, providing the current user context (e.g., location and time) and the query predicates (e.g., geographical area of interest and time window). The decision tree determines whether to answer locally or remotely.

We applied similar techniques to automatically devise a write caching strategy. The goal is to batch as many write operations as possible in a single network message, reducing the cost of communication [21], while trying to guarantee that most reads will observe all previously generated writes. Again, we can cast this as a classification problem, using cost-sensitive decision tree classifiers, where we try to estimate the maximum delay acceptable for each write operation if it is to be delivered to the server in time to be observed by the earliest subsequent read. The utility function provided here specifies the relative cost of delaying a write for too long (thus impacting quality of reads) versus delaying it insufficiently, thus missing batching opportunities. Section 7.3 shows experimentally that this leads to a $3\times$ reduction of the number of messages sent, while maintaining the number of queries receiving partially stale answers unchanged.

The fact that these techniques do not require application-specific models (e.g., we do not model user mobility), but instead use read-/write ratios and their correlation to predicates and contextual variables suggests it is possible to build general purpose cache policy selectors.

In future work, our architecture will allow us to explore many other interesting optimization strategies, such as:

- cross-app synchronization of recurring reads and writes
- cross-app prefetching and piggybacking
- dynamic cache eviction policies
- gracefully degrading answer quality when disconnected
- push-based cache maintenance
- dynamically changing caching strategies
- compaction of outstanding updates stored locally

Addressing such challenges at the platform level provides much more than simple factoring of the functionalities in a common layer. We have a vantage point that is unique, and allows us to exploit machine learning and statistical techniques to improve the performance of our system in a much more globally-optimal, context-aware, and dynamic way. The simple strategies we implemented in our prototype are only used to demonstrate some of the opportunities in this space, and a more complete investigation is part of our future research agenda.

7. EVALUATION

In this section, we describe a series of experiments we ran to evaluate our Mobius prototype. The goal is to demonstrate that:

1. The Mobius architecture and MUD API do not introduce significant additional client-side latency when sending messages or accessing local storage.
2. Utility-based caching using a server-generated caching policy can offer substantial reductions in access latency and total amount of network I/O versus simpler, local-only caching policies.
3. The Mobius back-end scales to large numbers of partitions, allowing developers to finely partition (and, by extension, parallelize) workloads.

7.1 Implementation and Experiment Setup

We implemented our prototype of Mobius using an event-driven, asynchronous architecture. We built Android and Java MUD client libraries and created an Android service that handles notification and data serving, which is shared by multiple apps running on the device. The app-side library talks to the service through Binder. The service stores data locally in SQLite [14]. Requests are handled by asynchronous Futures, and responses are returned by local callbacks. Requests from clients to gateways are encoded as JSON and are sent over HTTP. The MUD gateway, processor, and notification server are all built on JBoss Netty [6]. The protocol between notification client and server is a custom protocol optimized for this purpose.

Our prototype stores application data in JSON format, and supports queries written in a subset of the MVEL [8] expression language. We use a bLSM index for server-side storage [43]. Rather than attempting to optimize read queries using conventional database optimization techniques, we assume the underlying data (and associated queries) are finely partitioned, and chose an implementation designed to scale well with the number of partitions. This prototype comprises approximately 9,000 lines of Java code.

We ran end-to-end experiments by running servers on a machine located at UC Berkeley. The machine is equipped with two Xeon X5560 CPUs and 12GB RAM, running Ubuntu server 11.04. We ran our mobile apps for benchmarks on a Samsung Galaxy Nexus running Ice Cream Sandwich with a Verizon 4G LTE connection and a Nexus S phone running Gingerbread with a T-mobile 3G connection. We ran the storage backend experiments on a machine with a single Intel Core i7 2600 (3.4GHz) CPU, 16GB of RAM, and Ubuntu 11.04. This machine is equipped with a two-disk RAID-0 consisting of 10K RPM SATA disks.

7.2 Latency

To understand the latency overheads, we compared Mobius with baseline systems that directly execute local or remote SQL statements without considering consistency, scalability or fault tolerance. We ran microbenchmarks to measure latencies of local/remote writes, and local/remote reads. The baseline system for local writes/reads performs operations directly on local SQLite. For remote writes/reads, the baseline system sends requests to an HTTP server that executes the operations on a MySQL server at the backend. Note that the baseline systems are representative of typical developer choices. In contrast, Mobius also maintains a local copy of both read and written data. This allows local query serving.

Table 4 shows the latency overhead of Mobius relative to the baseline systems. We measured the mean latency of 500 requests.

Galaxy Nexus	Mobius Latency (ms)	Baseline Latency (ms)
Local write	94.10	64.99
Local read (cache)	40.58	2.89
Remote write (4G)	281.74	202.66
Remote read (4G)	301.94	198.08
Nexus S		
Local write	141.69	102.02
Local read (cache)	51.47	10.54
Remote write (3G)	540.07	1364.45
Remote read (3G)	364.02	1276.38

Table 4: Latency of MUD operations.

With a 4G connection, for remote writes, Mobius has up to 39% overhead, whereas for remote reads, Mobius has up to 52% overhead. While we plan to optimize our prototype to further reduce latencies, we believe that even the current overhead is acceptable, especially given that many applications are designed to deal with much higher latencies, and higher variances of latencies that occur on wireless connections or as users move from one location to another. Local writes and reads are slower with Mobius since Mobius performs these operations with asynchronous Futures unlike synchronous baseline SQLite systems. With the 3G connection, Mobius's remote writes and reads are much faster than the baseline systems. This demonstrates that simple baseline systems are hard to get right when environments change. The reason Mobius does better is that it maintains persistent HTTP connections, but the baseline system does not, re-establishing a connection each time a request is made. On 3G the additional round-trips for connection establishment incur more latency than the additional latency Mobius itself adds.

7.3 Intelligent Server-directed Caching

In Section 6 we argued in favor of server-generated caching strategies. We now present some initial results that demonstrate the benefits that one can obtain from this approach—a simple initial solution based on decision tree classifiers. Exploring more sophisticated and robust machine-learning and statistical techniques is part of our research agenda.

For these experiments, we use data traces from the CarTel [36] iPhone application. The traces consist of vehicles location reports and geo-spatial queries (in this case, for traffic data in a region). Specifically, the data comprises

- *Writes*: GPS locations of about 600 cars, reported at regular intervals (approximately 3.7M data points).
- *Reads*: User-initiated queries for data points within a bounding box (20x20km) and a two hour window.

We first report experiments on read caching strategies. We compare: 1) a *baseline* based on the actual policy implemented in the CarTel application, which answers queries from the local store (cache) whenever the query is no further than 1km from a previously cached query not older than 10 min, and 2) a *decision-tree-based* policy, which analyzes the execution trace and devises a policy based on the current query and cache contents.

As input, the decision tree takes the timestamp and GPS coordinates of the current and previous query. We generate the decision tree using Weka's cost-sensitive J48 implementation. The classifier is trained to predict whether a query should be answered from the local cache or sent to the server. In certain geographical regions our data is very sparse, and the classifier learns that issuing a query

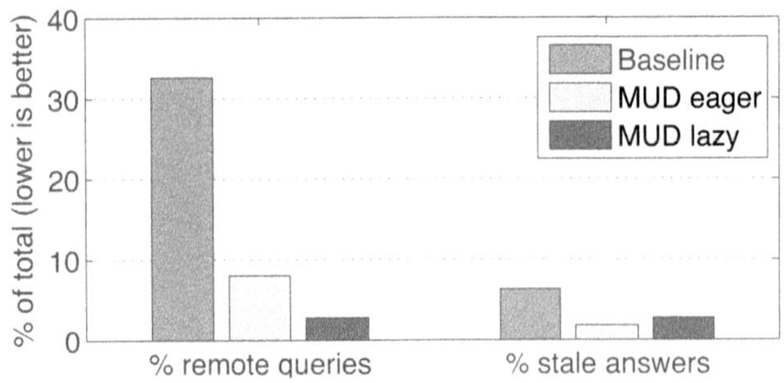

Figure 3: Mobius server-determined read caching vs Baseline on CarTel dataset

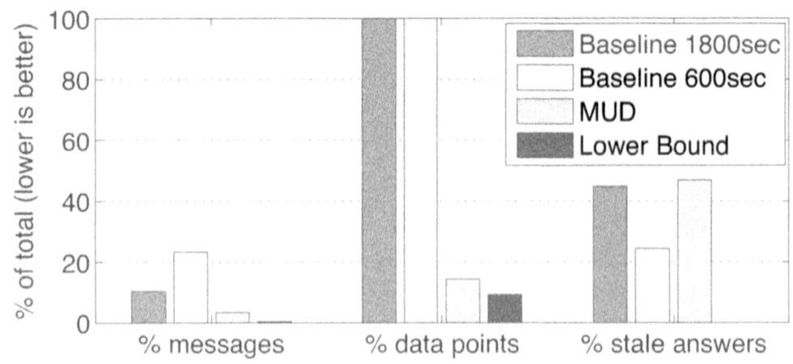

Figure 4: Mobius server-determined write batching vs Baseline on CarTel dataset

to the server if the client is interested in sparse areas is inconsequential even if the local cache is empty. We use standard 10-fold cross-validation in this experiment.

Figure 3 shows the results comparing the baseline described above and the decision tree classifier, for two different utility functions provided by the developer (one more sensitive to latency of the answer, and the other more sensitive to freshness of the data). In both cases, the decision tree significantly reduces the number of remote queries, improving perceived latency, bandwidth usage, power consumption and sever load, as well as result quality—fewer queries return stale data to the end user.

The second experiment focuses on write caching (or write batching). We compare two baseline policies that delay writes up to 10min and 30min respectively, our decision-tree based approach and a lower-bound computed *a posteriori* on the dataset.

We train the decision tree classifier to predict the maximum delay we can accept for a write without impacting a query. The classifier receives as input the GPS location (content of the write) and the time at which the write was performed in the trace. During training, we also include the time of the first subsequent read of the modified data. We quantize the delay into 10min buckets from 0 to 120min, and include an "infinite delay" bucket that accounts for unobserved writes. During training, we heavily penalize overestimates of maximum delay (which cause queries to receive stale data), and lightly penalize underestimates, increasing the cost progressively with increased error. For this experiment, we sample 10% of the data for training and use 90% for testing.

Figure 4 compares our approach with the baseline according to three metrics: the percentage of messages sent, the percentage of data points sent (this is to demonstrate the opportunity to not send data if we are confident the data will never be read), and the percentage of queries impacted by at least one missing data point. We show that our system's query results are comparable to the 30min fixed heuristic in exchange for a 7× reduction in the number data points sent (this translates in bandwidth savings and server side load reduction), and a 3× reduction in number of messages (which mostly translates into battery power reduction). When compared to a more aggressive 10min baseline we produced more stale results but up to 6× fewer messages.

While the presented approach is not completely general, it is also a somewhat naive use of classification. Even so, it is a significant improvement over reasonable baselines, showing that cloud-based caching policies can be very effective. In the context of the CarTel application, the policy selector determined that in certain regions, the density of writes is so low that pushing queries to the server would return no new results. Thus, based on location and time of the day the system is able to suggest that the client be more or less eager at pushing queries to the server.

7.4 Backend Scalability

For this experiment, we use the backend server to persist messages to disk and forward them to a single subscriber. The goal is to show impact of partitioning for a one-to-one messaging application. To stress test predicate evaluation, we group subscriptions into partitions, forcing the query processor to evaluate multiple subscription predicates for each incoming message. We focus on the gateway and query processor performance here, notification server scalability is well-studied, e.g., Google Thialfi [18].

The query processor processes subscriptions as it would in a real deployment: it decides correct recipient(s) for each message, and forwards the message to the notification server(s) for delivery. Each time a new tuple arrives, it is evaluated against each active subscription in its partition. Scenarios with only a few subscriptions per partition perform a small number of such evaluations per published tuple, while scenarios with many subscriptions per partition must perform many predicate evaluations for each published tuple.

The results of our experiment are shown in Figure 5. In the ideal case, each subscription is assigned to a separate partition and the query processor forwards approximately 5,000 messages per second. The cost of forwarding messages increases linearly as the number of subscriptions per partition increases. In our simple experiment, each message will be delivered to one subscriber; the work performed by the other predicate evaluations is wasted. Increasing the number of predicate evaluations while holding the other costs constant gives us a good idea of the performance of predicate evaluation: our simple prototype can evaluate approximately 70,000 predicates per second.

The query evaluation strategy performs well when most predicate evaluations succeed, but is unable to efficiently "prune" irrelevant queries as it processes incoming messages. We could avoid this linear slowdown in the number of subscriptions by implementing an index over query predicates, as in database continuous query processing systems like NiagaraCQ [28] or the ARIEL trigger processing system [29]. However, we observe that our architecture presents an interesting tradeoff to application developers: In order to improve performance, one can increase the sophistication of the predicate evaluation system, or one can resort to increasingly sophisticated partitioning schemes, reducing the number of subscriptions that must be considered when a tuple arrives. On the one hand, static partitioning schemes are unwieldy, but can be trivially parallelized and have good (and predictable) worst case performance. On the other hand, sophisticated query processing schemes are complex, but automatically improve the performance of a wide range of applications. We suspect that both techniques will be useful in practice.

Regardless of the underlying query processing scheme, subscription patterns that cause certain messages to be broadcast to large

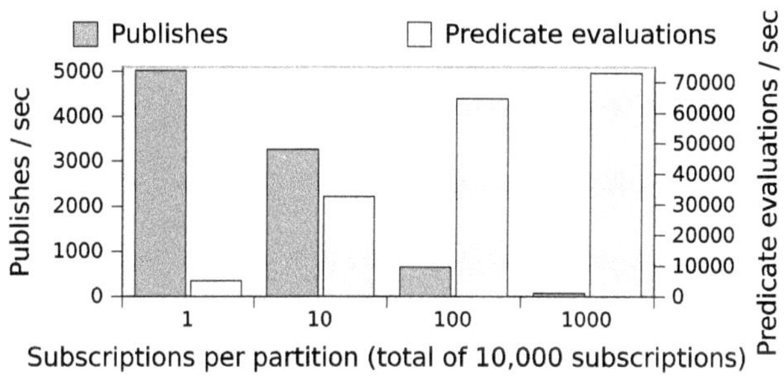

Figure 5: Impact of partitioning on one-to-one messaging via notifications. With small partitions, network and I/O overheads dominate. With large partitions, our stress test becomes CPU bound due to unnecessary predicate evaluations.

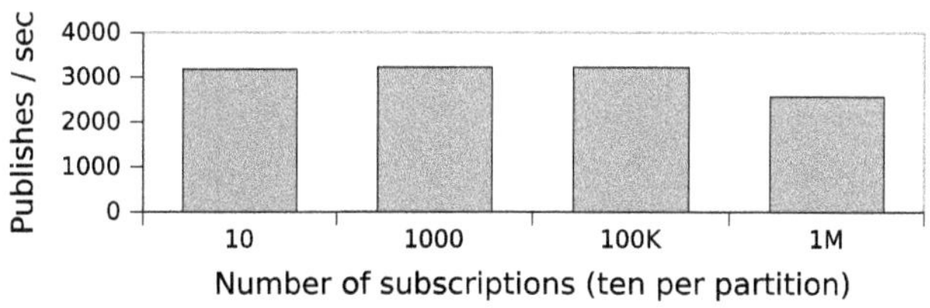

Figure 6: Notification throughput as subscriber count varies

numbers of users are likely to lead to hotspots, overwhelming some of the backend machines. Two common solutions to this problem are replication of partitions and read caching. Both approaches have the same effect: the computational and network load associated with hot partitions is spread across multiple backend servers.

Our simple prototype ignores such issues, and focuses on scaling up to applications with large numbers of small subscription groups. To this end, we ran the same publish / subscribe benchmark as above, holding the number of subscriptions per partition constant (Figure 6). With ten subscriptions per partition, the prototype is able to establish approximately one million subscriptions before running out of RAM to store active subscriptions. Although we suspect that other performance issues will manifest before we hit this limitation in practice, it is worth noting that we could spill the query cache to the backing bLSM index.

This design scales linearly by partitioning data and subscriptions across multiple servers. Partitioning schemes can be provided by application developers, but we also plan to further explore these issues by adapting our previous work on automatic and fine-grained partitioning [31, 46] to this context.

8. RELATED WORK

Mobius and MUD build on prior work from many fields, including mobile systems, pub-sub systems, messaging, continuous query processing and scalable "key-value" storage, but differ from existing work in several ways. First, the MUD API unifies messaging with data serving through the abstraction of a logical table that spans devices and clouds. Another novel aspect of MUD is that it introduces "writer predicates" to control writes, in addition to "reader predicates", in contrast to conventional pub-sub systems and database continuous queries. Second, Mobius allows utility functions to decide caching/prefetching decisions whose policies are derived by mining access patterns across clients at backend servers. Finally, Mobius's design integrates multiple components in a non-trivial way to provide app developers with a unified messaging and data platform, and also provides cloud operators with linear scaling in the number of applications, users and in the amount of data.

Existing cloud services are exposed to mobile apps through REST APIs or wrapper libraries of the APIs. Amazon Mobile SDKs for Android and iPhone [1] provide a wrapper library over the REST APIs of Amazon cloud services such as storage (S3), database (SimpleDB), and simple notifications service (SNS). SimpleDB is a non-relational data store that provides eventual or strong consistency, and SNS is a topic-based pub/sub system. SQL Azure database [13] and Google CloudSQL [3] provide cloud RDBMSs. In contrast, Mobius provides an integrated messaging and data serving platform that addresses challenges of mobile apps such as disconnection, mobility and intelligent caching for mobile devices.

Much research has been done on replication protocols for mobile computing [47]. The protocols can be categorized based on communication topologies (client-server vs. peer-to-peer), partial or full replication, and consistency guarantees. Coda and iCloud are example systems with client-server replication protocols; clients cannot directly share updates each other. Coda [37] is a distributed file system with which clients can perform disconnected operations. Coda caches a set of files specified by a hoarding profile. iCloud [5] provides synchronization of files among a set of apple devices (iPhone and iPad) and clouds, which supports a narrow app interface and semantics. Several projects have explored peer-to-peer replication protocols. Ficus [34] is a peer-to-peer replicated file system. Bayou [49] is a fully replicated database system that provides eventual consistency with session guarantees for choice of consistency [48]. PRACTI [22] supports topology independence, partial replication, and arbitrary consistency. To support partial replication, it separates invalidation messages from body messages with updates. CouchDB [2] is a schema-free "document store" with eventual consistency guarantees. Cimbiosys [41] and Perspective [42] use filters to selectively replicate content at mobile devices. Cimbiosys supports eventual filter consistency. With regard to data replication, Mobius has a client-server model since our main applications consume data from cloud, write data to cloud, and share data through cloud. Mobius provides partial replication since its cache contains partial records of tables. By avoiding direct peer-to-peer communications between devices, Mobius provides stronger consistency guarantees, and requires a simpler architecture. More importantly, it inherits all of the advantages of existing cloud storage architectures, such as analytics, multi-tenancy, elasticity, fault-tolerance, scalability, automatic application upgrades (to avoid legacy support issues), and so on.

Cloud to device push notification services such as Apple APNS [20], Google C2DM[4], Microsoft MPNS [11] provide a network channel to send small messages from cloud to device by maintaining a persistent connection between device and cloud. Similarly, our notification client/server provides channels for push notification. Thialfi [18] is a recent notification service that reliably propagates object version changes from cloud to client. In contrast, Mobius provides a high-level abstraction for data management and messaging. Unlike Mobius, Thialfi, does not deal with client-side caching, disconnected operations, consistency semantics, and so on. In addition, Mobius supports writer predicates which allow late binding of subscribers or readers to published content.

Communication service abstractions for mobile apps have also been explored. Project Hawaii [10] provides a collection of services and libraries, including a message relay service and a rendezvous service. Contrail [44] is a communication platform for decentralized social networks that uses the cloud to relay encrypted data. Junction [7] provides a switchboard service with which devices talk each other, especially in peer-to-peer settings. These services are low-level communication services. Mobius provides higher-level abstraction for messaging and data management.

Various systems provide asynchronous, distributed write/read primitives. Pub/sub systems provide communication between publishers which produce information and subscribers which are interested in receiving it. Depending on the expressive power of subscriptions, pub/sub systems are classified into topic-based or content-based ones [27, 32]. Continuous query databases such as NiagaraCQ [28] and YFilter [33] evaluate queries when data is written to databases. Tuplespaces such as Linda [26] and TSpace [15] provide distributed associative memory spaces; applications write tuples to the tuple space and read tuples from the tuple space based on a select predicate. Mobius's write/continuous read is built on the same principle, but Mobius has both reader and writer predicates.

Caching/prefetching has been studied extensively in distributed systems. We briefly describe a few relevant mobile computing projects. SEER [38] provides automated, dynamic hoarding decisions for distributed file systems that support disconnected operations. The system observes file access patterns, clusters files with small semantic distances, and colocates them; this pattern mining occurs in each client machine. WhereStore [45] is a device store that uses filtered replication to create partial replicas based on device location histories. Mobius provides a generalization: arbitrary caching/prefetching policies based on intelligent *backend* policy inference from *all* clients. Cedar [50] uses content addressable storage to reduce data transmission by detecting commonality between client and server query results of relational databases. Query containment checking [19] extracts matching queries with ranges of exiting tuples. Both approaches rewrite queries to exploit locally available data. Mobius employs similar rewriting techniques at the device. ICEDB [52] uses global feedback from the server in deciding what tuples to upload. Mobius covers a wide variety of push/pull primitives between device and server based on global and local policies.

9. CONCLUSIONS AND FUTURE WORK

In this paper, we introduced MUD, a unified messaging and data serving abstraction for mobile apps, and presented a system, Mobius, that supports it. Mobius provides several key features: automated caching, writer-/reader-predicate based reads and notifications, server-directed caching policies, and protocols for handling disconnected operation. In addition to simplifying mobile application development, we believe that Mobius will ultimately make more efficient use of hardware resources than is possible with adhoc application storage and messaging stacks.

Early experiments with our prototype implementation of Mobius are promising. Our MUD processing engine scales up to approximately one million subscriptions per machine, and is designed to linearly scale out to multiple back-end servers. Similarly, our client-side caching policies significantly reduce network bandwidth use. Finally, the overhead we impose upon client requests is reasonable. In conclusion, we believe that the MUD API, combined with our network and caching technologies, will significantly simplify the lives of mobile application developers.

Acknowledgments

We would like to thank Tyson Condie for his comments and contributions. We also would like to thank the anonymous reviewers for their comments and our shepherd, Doug Terry, for his guidance.

10. REFERENCES

[1] AWS SDK. aws.amazon.com/mobile.

[2] CouchDB. couchdb.apache.org.

[3] Google Cloud SQL. code.google.com/apis/sql.

[4] Google cloud to device messaging. code.google.com/android/c2dm/index.html.

[5] iCloud. www.icloud.com.

[6] JBoss Netty. www.jboss.org/netty.

[7] Junction. openjunction.org.

[8] MVEL. mvel.codehaus.org.

[9] OAuth. oauth.net.

[10] Project Hawaii. http://research.microsoft.com/en-us/um/redmond/projects/hawaii/.

[11] Push notifications for windows phone. http://msdn.microsoft.com/en-us/library/ff402537(v=VS.92).aspx.

[12] SimpleDB. aws.amazon.com/articles/3572?_encoding=UTF8&jiveRedirect=1.

[13] SQL Azure Database. msdn.microsoft.com/en-us/library/windowsazure/ee336279.aspx.

[14] SQLite. www.sqlite.org.

[15] TSpace. www.almaden.ibm.com/cs/tspaces.

[16] Yahoo! Query Language. developer.yahoo.com/yql.

[17] ZooKeeper. zookeeper.apache.org.

[18] A. Adya, G. Cooper, D. Myers, and M. Piatek. Thialfi: A client notification service for internet-scale applications. In *SOSP*, 2011.

[19] K. Amiri, S. Park, R. Tewari, and S. Padmanabhan. Scalable template-based query containment checking for web semantic caches. In *ICDE*, 2003.

[20] Apple push notification service. In *iOS Developer Library*. 2011.

[21] N. Balasubramanian, A. Balasubramanian, and A. Venkataramani. Energy consumption in mobile phones: a measurement study and implications for network applications. IMC, 2009.

[22] N. Belaramani, M. Dahlin, L. Gao, A. Nayate, A. Venkataramani, P. Yalagandula, and J. Zheng. PRACTI replication. In *NSDI*, 2006.

[23] P. A. Bernstein, V. Hadzilacos, and N. Goodman. *Concurrency Control and Recovery in Database Systems*. Addison-Wesley, 1987.

[24] C. Cachin, A. Shelat, and A. Shraer. Efficient fork-linearizable access to untrusted shared memory. In *PODC*, 2007.

[25] B. Calder et al. Windows azure storage: A highly available cloud storage service with strong consistency. In *SOSP*, 2011.

[26] N. Carriero and D. Gelernter. Linda in context. *Communications of the ACM*, 1989.

[27] A. Carzaniga, D. S. Rosenblum, and A. L. Wolf. Design and evaluation of a wide-area event notification service. *TOCS*, 2001.

[28] J. Chen, D. J. DeWitt, F. Tian, and Y. Wang. NiagaraCQ: A scalable continuous query system for internet databases. In *SIGMOD*, 2000.

[29] E. H. Chris, C. Carnes, L. Huang, M. Konyala, L. Noronha, S. Parthasarathy, J. B. Park, and A. Vernon. Scalable trigger processing. In *ICDE*, 1999.

[30] B. Cooper, R. Ramakrishnan, U. Srivastava, A. Silberstein, P. Bohannon, H.-A. Jacobsen, N. Puz, D. Weaver, and R. Yerneni. PNUTS: Yahoo!'s hosted data serving platform. In *VLDB*, 2008.

[31] C. Curino, Y. Zhang, E. P. C. Jones, and S. Madden. Schism: a workload-driven approach to database replication and partitioning. In *VLDB*, 2010.

[32] P. Eugster, P. Felber, R. Guerraoui, and A.-M. Kermarrec. The many faces of publish/subscribe. *ACM Computing Surveys*, 35(2), 2003.

[33] Y. D. P. Fischer, M. J. Franklin, and R. To. YFilter: Efficient and scalable filtering of XML documents. In *ICDE*, 2002.

[34] R. G. Guy, J. S. Heidemann, W.-K. Mak, T. W. P. Jr., G. J. Popek, and D. Rothmeier. Implementation of the ficus replicated file system. In *USENIX Summer*, 1990.

[35] M. P. Herlihy and J. M. Wing. Linearizability: A correctness condition for concurrent objects. *TOPLAS*, 12(3), 1990.

[36] B. Hull, V. Bychkovsky, Y. Zhang, K. Chen, M. Goraczko, A. Miu, E. Shih, H. Balakrishnan, and S. Madden. CarTel: A distributed mobile sensor computing system. In *SenSys*, 2006.

[37] J. J. Kistler and M. Satyanarayanan. Disconnected operation in the coda file system. *TOCS*, 1992.

[38] G. H. Kuenning and G. J. Popek. Automated hoarding for mobile computers. In *SOSP*, 1997.

[39] L. Lamport. How to make a multiprocessor computer that correctly executes multiprocess programs. *IEEE Trans. Comput.*, 28, 1979.

[40] A. Oprea and M. K. Reiter. On consistency of encrypted files. In *DISC*, pages 254–268, 2006.

[41] V. Ramasubramanian, T. L. Rodeheffer, D. B. Terry, M. Walraed-Sullivan, T. Wobber, C. C. Marshall, and A. Vahdat. Cimbiosys: A platform for content-based partial replication. In *NSDI*, 2009.

[42] B. Salmon, S. W. Schlosser, L. F. Cranor, and G. R. Ganger. Perspective: Semantic data management for the home. In *USENIX FAST*, 2009.

[43] R. Sears and R. Ramakrishnan. bLSM: A general purpose log structured merge tree. In *SIGMOD*, 2012.

[44] P. Stuedi, I. Mohomed, M. Balakrishnan, T. Wobber, D. Terry, and M. Mao. Contrail: Enabling decentralized social networks on smartphones. In *Middleware*, 2011.

[45] P. Stuedi, I. Mohomed, and D. Terry. WhereStore: Location-based data storage for mobile devices interacting with the cloud. In *MCS*, 2010.

[46] A. Tatarowicz, C. Curino, E. P. C. Jones, and S. Madden. Lookup tables: Fine-grained partitioning for distributed databases. In *ICDE*, 2012.

[47] D. B. Terry. *Replication Data Management for Mobile Computing*. Morgan-Claypool, 2008.

[48] D. B. Terry, A. J. Demers, K. Petersen, M. J. Spreitzer, M. M. Theimer, and B. B. Welch. Session guarantees for weakly consistent replicated data. In *PDIS*, 1994.

[49] D. B. Terry, M. M. Theimer, K. Petersen, A. J. Demers, M. J. Spreitzer, and C. Hauser. Managing update conflicts in bayou, a weakly connected replicated storage system. In *SOSP*, 1995.

[50] N. Tolia, M. Satyanarayanan, and A. Wolbach. Improving mobile database access over wide-area networks without degrading consistency. In *MobiSys*, 2007.

[51] H. Yu and A. Vahdat. Design and evaluation of a continuous consistency model for replicated services. In *OSDI*, 2000.

[52] Y. Zhang, B. Hull, H. Balakrishnan, and S. Madden. ICEDB: Intermittently-connected continuous query processing. In *ICDE*, 2007.

APPENDIX

A. FORMAL CONSISTENCY SEMANTICS

We describe the consistency semantics that Mobius provides to clients accessing its tables formally. As operations take time, they are represented by two events occurring at the client, an *invocation* and a *response*. A *history* is the sequence of invocations and responses that occurred in an execution. An operation o *precedes* another operation o' in the history if o completes before o' is invoked. Two operations are *concurrent* if neither one of them precedes the other. Finally, we say that a sequence of events is *sequential* if it does not contain concurrent operations.

Sequential views: We define consistency in terms of the possible *views* clients have of the execution (as previously mentioned, each record is treated as a separate consistency domain). Intuitively, a view is a sequence of invocations and responses which is sequential and equivalent (from the client's point of view) to the execution history. The view must contain the client's own operations in addition to any operation apparent from the client's interaction with the shared storage. The view must be *legal* according to the "sequential specification" of the shared object (i.e., it could have resulted from a sequential interaction with the storage). Finally, the view must respect the operation precedence order of every client's program: for every two operations executed by the same client o and o' in a view, if o completes before o' is invoked then it appears before o' in the view. Note that there are usually multiple views possible at a client.

Sequential Consistency: The first consistency property provided by the system is *sequential consistency* [39]. Here, we formalize the definition using the notion of views:

DEFINITION 1 (SEQUENTIAL CONSISTENCY). *A history σ is* sequentially consistent *if there exists a sequence of events π which is a view of σ for all clients.*

Sequential consistency with write invocation order: We say that a client's view *preserves write invocation order* of a history σ if for every two operations o and o' of the same client in the view, if o is invoked before o' in σ and at least one of them is a write, then o precedes o' in the view. We call the resulting consistency condition *sequential consistency with WIO* (write invocation order):

DEFINITION 2 (SEQUENTIAL CONSISTENCY WITH WIO). *A history σ is* sequentially consistent with WIO *if there exists a sequence of events π which is a view of σ for all clients, and which preserves the invocation order of writes in σ.*

Fork sequential consistency: We formalize the possibility of view divergence using the notion of *fork-sequential consistency* [40].

DEFINITION 3 (FORK-SEQUENTIAL CONSISTENCY). *A history σ is* fork-sequentially consistent *if for each client C_i there exists a sequence of events π_i such that (1) π_i is a view of σ at C_i; and (2) For every client C_j and every operation o in both π_i and π_j, the prefix of π_i up to o is identical to the prefix of π_j up to o (called the* No-join *property).*

Informed Mobile Prefetching

Brett D. Higgins[*] Jason Flinn[*] T.J. Giuli[†]
Brian Noble[*] Christopher Peplin[†] David Watson[†‡]
University of Michigan[*] Ford Motor Company[†] Arbor Networks[‡]

ABSTRACT

Prefetching is a double-edged sword. It can hide the latency of data transfers over poor and intermittently connected wireless networks, but the costs of prefetching in terms of increased energy and cellular data usage are potentially substantial, particularly for data prefetched incorrectly. Weighing the costs and benefits of prefetching is complex, and consequently most mobile applications employ simple but sub-optimal strategies.

Rather than leave the job to applications, we argue that the underlying mobile system should provide explicit prefetching support. Our prototype, IMP, presents a simple interface that hides the complexity of the prefetching decision. IMP uses a cost-benefit analysis to decide when to prefetch data. It employs goal-directed adaptation to try to minimize application response time while meeting budgets for battery lifetime and cellular data usage. IMP opportunistically uses available networks while ensuring that prefetches do not degrade network performance for foreground activity. It tracks hit rates for past prefetches and accounts for network-specific costs in order to dynamically adapt its prefetching strategy to both the network conditions and the accuracy of application prefetch disclosures. Experiments with email and news reader applications show that IMP provides predictable usage of budgeted resources, while lowering application response time compared to the oblivious strategies used by current applications.

Categories and Subject Descriptors

D.4.4 [**Operating Systems**]: Communications Management—*network communication*; D.4.8 [**Operating Systems**]: Performance

Keywords

Mobile prefetching, budgeted resources, adaptation

MobiSys'12, June 25–29, 2012, Low Wood Bay, Lake District, UK.

1. INTRODUCTION

Prefetching is a fundamental technique for improving application performance. It is employed in numerous domains, including computer architecture [28], databases [22], file systems [13], and distributed systems [12]. In essence, a prefetching system predicts what data a higher system layer, such as an application, will request and speculatively retrieves and caches that data in anticipation of future need. If the prediction is correct, the cached data is provided on request — this improves performance by eliminating the fetching of the data from the critical path of servicing the request. However, if the prediction is incorrect, prefetching consumes resources that could be used for other activities. Most prefetching systems use heuristics to balance these concerns; such heuristics decide when and how much data to prefetch.

In many ways, mobile computing is an ideal domain for prefetching. Applications that run on smart phones and tablets frequently fetch data from the cloud. Yet, mobile devices must often rely on wireless networks, and such networks can exhibit low bandwidth, high latency, and intermittent connectivity. By prefetching and caching data, a mobile application can avoid on-demand use of poor and unreliable wireless networks, leading to a substantial improvement in interactive response time. Thus, the potential rewards of prefetching are high.

Unfortunately, the potential costs of mobile prefetching are also high. Since wireless network bandwidth is limited, care must be taken that prefetching does not interfere with interactive traffic and degrade application response time. Further, prefetching data from a remote server can consume battery energy and cellular data allotments, so profligate prefetching may result in a mobile device's battery expiring too soon or in overage charges on a cellular plan.

Unfortunately, balancing costs and benefits is complex. The prefetching system must consider at least three different concerns (performance, energy usage, and wireless data consumption), no two of which are measured by a common metric. It must understand how the costs of fetching data from a remote server vary as wireless network conditions change, and it must take into account network-specific oddities such as 3G tail time [24]. It must account for differences in hit rates caused by variation in behavior across users, applications, and even among different classes of data within the same application. Finally, it must ensure that prefetching does not substantially degrade foreground network activity.

Given this host of complexities and the limited resources of many mobile application developers, it is understandable that no current mobile application addresses all (or even most) of the above concerns when deciding when to prefetch data. Instead, mobile applications employ simple, yet often suboptimal, heuristics such as prefetch nothing, prefetch everything, or prefetch data subject to an arbitrary constraint (such as only prefetching data items with size less than 32 KB). Our evaluation shows that use of such heuristics represents a substantial missed opportunity.

Instead, we propose that the mobile computer system provide explicit prefetching support to all applications. Our prototype system, which we call IMP (Informed Mobile Prefetching), is structured as a library to which any mobile application may link. The application interface is unobtrusive; applications must specify the items that could potentially benefit from prefetching, and they must inform IMP when those items are consumed or are no longer needed. The IMP library manages the complex mobile prefetching process, guiding applications to prefetch the right amount of data at the right time in a manner that improves performance while still meeting wireless data and battery lifetime goals. IMP both helps overzealous applications rein in prefetching to meet user budgets and allows tentative applications to make better use of available resources to improve the user's experience.

A key observation behind the design of IMP is that many of the prefetching complexities detailed above have been addressed individually, albeit not always within the domain of mobile computing. We view the major contributions of IMP as unifying these disparate ideas, applying them specifically to the domain of mobile prefetching, and providing prefetching services through a simple, easy-to-use interface that is well-suited for rapid mobile application development.

The design of IMP is inspired by Transparent Informed Prefetching (TIP), which employs a cost-benefit analysis to determine when to fetch data from an array of disks into a file cache [23]. Like TIP, IMP uses a shadowing strategy to predict hit rates for different prefetch depths. IMP uses a substantially different algorithm than TIP's to address unique features of mobile prefetching. IMP considers not just performance but also energy and wireless data usage. IMP also considers how aggregation of multiple prefetches can reduce energy usage. Finally, whereas cache space is the bottleneck resource for disk prefetching, IMP considers wireless network usage and its associated resource costs to be the bottleneck in mobile prefetching.

IMP leverages lower network layers to prioritize foreground network traffic over prefetching activity and to discover, characterize, and effectively match traffic to multiple available wireless networks. While IMP could be built on any layer that provides such services, it currently uses Intentional Networking [10] for these purposes.

Finally, IMP proposes a unified approach for dealing with *budgeted resources*. A budgeted resource has a fixed limit on the amount of the resource that can be consumed within a specified time period. One example is cellular data allotments: many cellular plans allow usage of up to a specified amount of data for a fixed charge. Another example is battery energy: for predictable behavior, it is reasonable to expect the battery of a smart phone to last for an entire day so that it can be recharged at night. While the negative consequences of exceeding a budget have been widely observed, the negative consequences of under-utilizing a budget have received little attention. Yet, we observe that unused cellular data and leftover battery energy represent a substantial wasted opportunity when the mobile system has degraded user experience in order to conserve those resources. In response to this observation, IMP takes a "Price is Right" approach to budgeted resources — it tries to come as close as possible to the budget as possible without exceeding it. IMP continually measures how well it is doing in meeting its budgets and uses a control loop to adjust the cost of battery energy or cellular data usage relative to performance accordingly. This is a variation of Odyssey's goal-directed adaptation [8], which IMP modifies to manage multiple resources and applies to the domain of mobile prefetching.

We have implemented IMP as a Java library and modified two applications, the K9 e-mail client and the OpenIntents news reader, to use IMP. Our experimental results show that IMP is able to meet all specified energy and cellular data usage goals. Additionally, in most cases, whenever simple heuristic-based prefetching strategies also meet the goals, IMP outperforms those strategies on interactive fetch time, often by a factor of two or more.

2. DESIGN CONSIDERATIONS

In this section, we examine the tradeoffs involved in mobile prefetching, and discuss why the prefetching decision is more complex in the mobile domain than in many other areas.

2.1 Why is mobile prefetching so complex?

Most current prefetching systems have the sole goal of maximizing performance, subject to constraints such as cache space and network bandwidth. Prefetching is beneficial when a data item that was prefetched speculatively is later requested by a higher layer of the system. The latency of servicing the request is reduced by the time that would have been required to fetch the data. However, prefetching consumes resources. For instance, because prefetch requests are interwoven with demand fetches from the higher layer, a prefetch request might degrade the performance of those foreground requests. This interference mitigates the performance benefit of a successful prefetch. Further, when prefetched data is never consumed by the application, the interference with foreground activity degrades overall performance. Thus, prefetching systems employ heuristics to try to maximize performance by only prefetching data when the benefits of doing so exceed the costs.

Similar concerns arise in the domain of mobile prefetching when data is fetched from remote servers. The performance benefit of prefetching can be substantial, especially when wireless networks offer poor or intermittent connectivity. However, because wireless bandwidth is limited, the potential for prefetch requests to degrade other traffic is magnified.

In mobile computing, the prefetch decision must consider resources other than performance. Battery lifetime is often a critical concern. While the performance impact of prefetching can be mitigated by scheduling prefetches for intervals with little or no demand requests for data, the energy cost of prefetching data is difficult to mask. Yet, prefetching can still have substantial energy benefits. For instance, consider that the energy required to fetch data over a WiFi network is often less than that required to fetch data over a cellular network and that the energy required to fetch data over networks varies in proportion to the quality of the wireless connection. Thus, by prefetching data during periods of WiFi or good cellular connectivity, the mobile computer uses less energy than if it were to service demand fetches for the same data during periods of poorer connectivity. Further, prefetching allows the mobile computer to batch multiple requests. This saves energy by amortizing transition costs across multiple requests — for instance, the energy costs of 3G tail time can be ameliorated by sending multiple requests back-to-back.

Since it is increasingly common to cap cellular data consumption, prefetching systems must also consider cellular data usage when deciding whether to prefetch. Like energy usage, prefetching can impact cellular data usage either positively or negatively. Prefetching data over a cellular network will increase data usage if the prefetched item is never requested. On the other hand, the mobile computer can prefetch data over a WiFi network and thereby reduce data usage by avoiding a subsequent demand fetch over a cellular network. One could imagine setting a monthly budget for all prefetching and switching prefetching off if the budget is exceeded. However, it is unclear how to set such a budget accurately because non-prefetch data consumption can vary widely from month to month and from user to user.

In summary, mobile prefetching is complex because one must consider at least three metrics (performance, energy usage, and cellular data usage), whereas current prefetching algorithms often consider just performance. Further, it is difficult to compare the three metrics considered in mobile prefetching because no two are expressed in a single currency. For instance, if a potential prefetch would improve performance and save energy, but would also require additional cellular data transmission, should the mobile computer perform that prefetch? The next section describes our solution to this dilemma.

2.2 Balancing multiple concerns

Battery energy and cellular data allotments are examples of a resource class that we term *budgeted resources*. For such resources, there exists a fixed amount of the resource that must last for a period of time. For instance, a cellular data plan may provide 4 GB of data for a month. The consequence of exceeding the budget is severe (e.g., additional cost or degradation of quality of service). On the other hand, any resources not consumed during the budgeted period are wasted (unused data allotments do not roll over to the next month).

Battery energy is also best thought of as a budgeted resource. In this case, the budget is the amount of energy in the battery, and the budgeted period is the time until the battery is recharged. Current energy management strategies correctly worry about the negative consequences of exceeding the budget (running out of battery energy). However, they do not focus nearly enough on the negative consequences of undershooting the budget. Since mobile computers increasingly degrade the user experience to preserve battery energy, any time the battery is recharged with a substantial amount of energy remaining represents a substantial wasted opportunity — the user experience was degraded for no purpose.

The inherent challenge is that the importance of a budgeted resource may change dramatically over time. Near the end of the month when there is substantial data left in a cellular plan, or near the end of the day, when the user has a mostly-full battery and plans to recharge soon, conserving a budgeted resource such as a data allotment or energy is relatively unimportant. Instead, performance should be maximized. On the other hand, when the battery is low or most of a data allotment is consumed, then the relative importance of the budgeted resource is very high.

Strategies that assign a fixed conversion rate for budgeted resources (e.g., saving a Joule of battery energy is worth a 10% degradation in application performance) are doomed to be incorrect as the relative importance of the budgeted resource changes. A fixed conversion rate will be too high when the budgeted resource is unimportant and too low when the resource is precious.

Instead, we argue that the management of the budgeted resource should be adaptive. IMP uses a control loop to adjust the conversion rate used to equate budgeted resources with performance. When the system is projected to exceed its budget based on measurements of current supply and demand, the conversion rate is increased to make the budgeted resource more precious. This causes future decisions (prefetch decisions in the case of IMP) to place more weight on conserving the budgeted resource. Thus, demand is reduced to match supply. On the other hand, when the system is projected to use significantly less than the budgeted amount, the conversion rate is decreased and the system becomes more aggressive about using the resource to improve performance (or reduce the use of other budgeted resources).

An additional benefit of the control approach is that the user experience becomes more predictable. For instance, users can come to expect that their phone battery will just last all day in most circumstances, rather than having to constantly monitor the battery level and adjust their own behavior. Further, the budget itself is a simple knob through which the user can adjust system behavior. For instance, if one knows that one will recharge in an hour, one can just change the budget to reflect that decision rather than adjust numerous settings like screen brightness, processor speed, etc.

3. DESIGN

We next discuss the design of Informed Mobile Prefetching. We first give some background on Intentional Networking, a multi-network abstraction leveraged by IMP. We then describe how IMP decides whether to prefetch data in response to application hints.

3.1 Background: Intentional Networking

Intentional Networking [10] is an abstraction for multiplexing traffic over multiple wireless networks. Intentional Networking discovers and simultaneously connects to multiple networks such as cellular and WiFi. Higher layers of the system (such as IMP) annotate network sends with qualitative labels, which Intentional Networking uses to route traffic over the most appropriate network interface.

IMP uses Intentional Networking to differentiate prefetch traffic from other application network activity. By designating prefetches as background requests, IMP causes Intentional Networking to prioritize other data over prefetches. Further, Intentional Networking throttles low-priority traffic slightly to ensure that queuing delays due to excessive in-flight background data do not adversely affect foreground requests. Experimental results have shown that foreground requests can be serviced with low latency with only a few percent degradation in background throughput. When appropriate, Intentional Networking provides greater throughput for large requests by striping them across multiple networks.

IMP also gathers passive and active measurements of network quality from the Intentional Networking layer. Intentional Networking provides notifications when new networks are available and when existing networks are disconnected.

3.2 Prefetch decision algorithm

IMP decides when and how much data to prefetch using a cost/benefit analysis inspired by Transparent Informed Prefetching (TIP) [23]. The TIP algorithm regulates prefetching of file system data from an array of disk drives in a server environment. The core idea behind TIP is that the application should disclose hints that describe opportunities to prefetch data — these hints are predictions of future accesses. TIP dynamically decides when to prefetch data corresponding to the disclosed hints based on a cost/benefit analysis that estimates how prefetching will impact application performance. One of the nice features of the TIP design is that applications do not have to know any details about the environment in which they are operating (e.g., resource constraints), nor do applications have to specifically estimate how effective prefetch hints will be at predicting future accesses. These details are managed by the prefetching system.

Since the use case that TIP targets (prefetching file system data from disk arrays) is considerably different than prefetching data to mobile computers, IMP retains the structure of the TIP cost/benefit model but changes most of the details. For instance, TIP considers cache buffers to be the bottleneck resource, and it attempts to optimize only performance. On the other hand, in mobile prefetching, the wireless network is the bottleneck resource and one must consider both battery energy and data usage in addition to performance.

We next describe how IMP estimates the potential cost and benefit of prefetching a data item for each of its three metrics: performance, energy use, and data consumption. IMP separately considers the impact of prefetching over each currently available network (e.g., it calculates the impact for cellular and WiFi when both are available).

3.2.1 Performance

The performance benefit of prefetching comes from decreasing the time to service a subsequent request for the prefetched data. The precise amount of benefit depends on the size of the data item and the network conditions that will exist at the time the data is requested. If the item is not prefetched, a demand fetch of the data will take time T_{fetch}, where

$$T_{fetch} = \frac{S}{BW_{future}} + L_{future}$$

as given by the size of the data item (S) and the future network bandwidth and latency at the time the demand fetch occurs (BW_{future} and L_{future}, respectively).

Although it may sometimes be feasible to predict future network conditions [19], for many applications it is quite difficult to predict when the user will request data. For this reason, IMP uses the average network conditions observed in the past for the mobile computer as a reasonable approximation of future conditions.

For each type of network, IMP tracks the average availability, latency, and bandwidth of the network. To calculate averages, it obtains active network measurements and supplements those observations with passive measurements of network behavior that occur when data is prefetched or fetched on demand by applications. IMP uses network availability to merge the costs of performing the demand fetch over different network types. It assumes that if the lowest cost network is available, that network will be used. If not, the next lowest cost network will be used if available, etc. In practice, since IMP currently considers just cellular and WiFi transmission and since WiFi almost always dominates 3G when all costs (performance, energy, and data) are considered, this works out to: $T_{fetch-WiFi} \times Availability_{WiFi} + T_{fetch-cellular} \times (1 - Availability_{WiFi})$ or the cost of fetching data over the two network types weighted by the observed average availability of WiFi.

IMP allows applications to specify the size of each data item when a prefetch is requested. For the applications we have modified to use IMP, it has been trivial for the application to retrieve the size of each data item from a remote server at the same instance that it learns of that item's existence. For example, when an email client requests a listing of all new messages, it can ask for the size of each email to be returned in the same request. If an application declines to supply the size of an item to be prefetched, IMP instead uses the average size of items it has fetched for that application in the past.

The calculation as described so far assumes that prefetch hints are always accurate; i.e., that the prefetched data will be requested at some time in the future. However, many prefetched items are never consumed by the application, so IMP must also consider the estimated *accuracy* of a prefetch hint in its calculations. This is calculated on a per-application basis as the number of prefetch hints for which the prefetched data was later consumed divided by the total number of prefetch hints issued. IMP increments the count of total hints when a new hint is issued, and it increments the count of consumed hints when an application first requests prefetched data. It uses shadow caching to determine the number of hints for which it did not prefetch data but for which the application later requested the data. Specifically, IMP remembers all prefetch hints issued, regardless of whether or not it has yet decided to prefetch the hinted data. Applications inform IMP when a hinted request that has not yet been prefetched must be fetched on demand — this information is necessary since prefetching is no longer useful. IMP increments the count of consumed hints when this happens.

IMP allows applications to optionally differentiate among prefetch hints by associating each hint with a *class*. The choice of how to apply classes is left up to the application; in general, applications will benefit most if their prefetch classes separate prefetches by some application-specific or user-specific quality that makes some prefetches more likely to be requested than others. For instance, an email application might use one class for messages in a priority inbox, and another class for the remaining messages. A news reader application might associate a different class with each news feed. IMP maintains a separate estimate of accuracy for each class and uses the appropriate class estimator when new prefetch hints are issued. In this way, the abstraction we provide simplifies the creation of application-specific prefetch strategies; the application need only divide its hints into classes, and IMP takes care of discovering which classes have the most valuable hints.

In summary, IMP calculates the performance benefit of prefetching as the product of the average time to fetch the data on demand and the application-specific or class-specific accuracy estimate. The accuracy estimates allow IMP to help the application focus on prefetching the items that are most likely to be used. Note that since IMP relies on Intentional Networking to ensure that prefetch traffic does not interfere much with foreground application activity, it need not assign a performance cost to each prefetch.

3.2.2 Energy usage

IMP calculates the effect of prefetching on battery energy by comparing the energy that would be used to prefetch the data immediately ($E_{prefetch}$) with the expected energy cost of later fetching the item on demand (E_{fetch}). The previous section describes how IMP calculates T_{fetch}, the expected time to perform a future demand fetch of an item, as well as how it calculates the expected size of the data item (if this is not specified by the application). IMP calculates $T_{prefetch}$, the time to perform a prefetch, the same way it calculates the time for a demand fetch, except that it uses the current estimates of bandwidth and latency for the network over which prefetching is being considered (recall that IMP considers prefetching over WiFi and cellular networks independently). IMP then uses models developed by PowerTutor [30] to calculate the energy impact of both an immediate prefetch and a possible future demand fetch. These power models are specific to the mobile device and cellular carrier. They are generated with a suite of automated tests that isolate and exercise different components of the device in turn. Once a model is derived, it can be re-used for many different purposes (as we are reusing models developed by others in our work).

For prefetching on WiFi, the power model provides a power coefficient $P_{WiFi-xmit}$ that encapsulates the average power usage of the WiFi interface when actively sending and receiving; thus, $E_{prefetch} = P_{WiFi-xmit} \times T_{prefetch}$. When prefetching on a 3G cellular network, IMP also includes the transition costs that may be incurred due to network activation. For instance, tail energy [24] has been shown to be a significant phenomenon in 3G cellular networks. 3G networks use inactivity timers to avoid the delays that can result from frequent channel release and reacquisition. This results in the network interface spending substantial time in a high-power state after the last transmission completes.

If a network transmission is predicted to cause a promotion to a high-power state, IMP includes the transition costs to and from the high-power state (which includes the tail energy cost for 3G networks) in the energy cost of the prefetch. However, if the radio is predicted to already be in a high-power state, only the additional time spent in the high-power state is included. Like PowerTutor, IMP monitors the number of bytes queued on the cellular interface to infer when power state transitions occur. When calculating the energy cost of an immediate prefetch, IMP queries the state transition model to learn whether or not the cellular radio is in a high power state. When calculating the energy cost of a future demand fetch, it uses the average distribution of power states observed for that interface to determine the likelihood of being in each power state. Thus, the energy cost of a 3G prefetch is $(P_{3G-xmit} \times T_{prefetch}) + E_{tail}$, where

$$E_{tail} = \begin{cases} P_{tail} \times T_{tail} & \text{if transition occurs and tail begins} \\ P_{tail} \times T_{inactivity} & \text{if tail time is extended} \end{cases}$$

For simplicity, we here omit details such as the difference between DCH and FACH power states, though we include them in our calculations.

Finally, the net energy cost (or benefit) of a prefetch is $E_{prefetch} - (E_{fetch} \times Accuracy)$, since the energy cost of the future demand fetch is only paid if the user actually requests the item. Note that a prefetch can have an expected energy benefit if current network connectivity is strong (so less energy than average is required to fetch data) or a lower-power network is currently available, and the estimated accuracy of a prefetch is high. Batching (described in Section 3.2.5) can also lead to energy benefits for prefetching by amortizing transition costs across multiple requests.

3.2.3 *Data consumption*

IMP also considers the effect of prefetching on cellular data consumption. The estimated future data cost of a demand fetch is calculated as $D_{fetch} = S \times (1 - Availability_{WiFi})$. The benefit of prefetching over WiFi is D_{fetch}, since the cellular data cost on WiFi is zero. The cost of prefetching over a cellular network is simply S. As in the case of energy cost, when referring to the data cost of prefetching below, we denote it as $D_{prefetch} - Accuracy \times D_{fetch}$, where $D_{prefetch}$ is zero or S depending on which network is used.

3.2.4 *Putting it all together*

As explained in the last three sections, IMP separately calculates the cost or benefit of prefetching over each type of available network in terms of performance, energy, and data usage. Unfortunately, these three metrics are expressed in different units (seconds, Joules, and bytes), making it difficult to determine what to do when the metrics disagree about whether or not prefetching would be beneficial.

A common solution to this dilemma is to employ a conversion rate or utility function to equate metrics expressed in different units (e.g., one could make 10 Joules of energy expenditure equal to 1 second of interactive delay). However, as argued in Section 2.2, the importance of each metric can change substantially over time, making static conversion rates arbitrary and dooming them to be incorrect at least some of the time.

Instead, IMP dynamically adjusts conversion rates using a feedback loop that takes as input how well the system is meeting its resource budgets. Specifically, IMP maintains two values, c_{energy} and c_{data}, that represent the current conversion rate between each respective budgeted resource and performance. For instance, if c_{energy} is 0.2, then 5 Joules is considered equal to 1 second of interactive delay.

IMP uses Odyssey's goal-directed adaptation to adjust these conversion rates. Since we did not modify this algorithm and it is described elsewhere in detail [7], we provide a simple overview of its design here. Note that Odyssey applied goal-directed adaptation only to energy; one of the contributions of this work is the observation that this strategy can also be used to regulate cellular data usage. Additionally, Odyssey used application-specific fidelity adjustments to regulate the amount of resource spending over its budget period, whereas we regulate spending by adjusting the amount of prefetching.

Once a second, IMP measures the remaining supply of a resource. Goal-directed adaptation uses an exponential smoothing algorithm to calculate the rate at which the resource is being consumed from recent observations of supply. It multiplies the estimated rate of consumption by the time remaining in the budgeted period to estimate the remaining demand for the resource.

IMP relies on having an estimate of the future time at which each budgeted resource will be replenished. Many cellular data plans have predictable periods coinciding with a monthly billing cycle, making it simple to estimate the budget period. Additionally, recent work [25] has identified personal patterns in different users' battery recharge behaviors, which suggests it may be possible to predict the budget period for the available battery energy. The accuracy of such predictions is important, however; a too-long or too-short prediction of the budget period will increase the likelihood that IMP underspends or overspends the budget, respectively.

Next, the goal-directed adaptation algorithm subtracts from the latest supply measurement 5% of the current remaining supply plus 1% of the original supply. This provides some insurance against future changes in the resource budgets, which can occur if the user downloads a new application that consumes a lot of data, or if they arrive home later than usual to recharge their phone. Finally, it computes a ratio that expresses how much resource consumption is currently over or under budget:

$$c_{adjustment} = \frac{estimated\_demand}{reduced\_supply}$$

IMP then updates the current conversion rate with this ratio: $c_{new} = c_{old} \times c_{adjustment}$. Thus, if energy consumption is over budget, a Joule becomes more precious compared to other resources. This feedback process leads IMP to bias further against prefetch decisions that cost energy, which will eventually bring the energy consumption within budget.

In summary, IMP calculates the net benefit of prefetching an item over a particular type of network as benefit minus cost, or:

$$T_{fetch} \times Accuracy - (c_{energy} \times (E_{prefetch} - Accuracy \times E_{fetch}) + c_{data} \times (D_{prefetch} - Accuracy \times D_{fetch}))$$

If this value is positive for a single network type, IMP prefetches the item over that network. If the value is positive for multiple network types, IMP tells the Intentional Networking layer that the data item may be prefetched over all such networks. Intentional Networking stripes the item over all such networks if the item is large enough that striping makes sense.

3.2.5 *Batching*

Because some costs (most notably 3G tail time) can be amortized across multiple prefetches, prefetching may be beneficial if several items are fetched in a batch, even though prefetching each item individually is estimated to have a negative impact. IMP checks for possible batching benefits by considering the impact of fetching one item, two items, etc. up to the number of currently outstanding

Application → IMP		IMP → Application	
prefetch(Fetcher)	→ Future	Fetcher.fetch(Labels)	→ Result
Future.get()	→ Result		
Future.cancel()			
Fetcher.setSize(int)			
Fetcher.setPrefetchClass(int)			

This figure shows the IMP API used for hinting, executing, consuming, and canceling prefetches. Applications create a `Fetcher` that implements the `fetch` callback to execute a (pre)fetch. The `fetch` method takes as an argument the set of Intentional Networking *labels* to be used for the invoked fetch. The `prefetch` function takes a `Fetcher` and returns a `Future`, a handle that the application uses to retrieve the result of a fetch. The type of `Result` that a fetch returns is specified by the application. The application may optionally call `setSize` and `setPrefetchClass` to pass additional information to IMP.

Table 1: Informed Mobile Prefetching API

application prefetch requests. If it predicts a positive impact for any batch size, it initiates prefetching.

3.2.6 *Network disconnection*

If a network becomes unavailable while a prefetch is being performed using that network, IMP recalculates the potential impact of prefetching over the remaining networks as above. If a potential benefit is found, it resumes the prefetch over the selected remaining networks. If not, the prefetch is delayed and resumed when an appropriate network is found. For simplicity, IMP currently relies on application-level support for resuming prefetch requests without retransmitting the data already received, though it could potentially provide this feature transparently with additional buffering.

3.2.7 *Discussion*

We have designed IMP to manage resources that have a fixed budget to be spent over a predictable time period. However, there are also cellular data plans where the subscriber pays for the amount of data actually consumed, rather than paying for a monthly budget. Though IMP does not handle such cases in its current form, we could imagine extending it to do so. Rather than targeting a single budget over a single time period, we could treat the pay-as-you-go data plan as a series of small budgets, in the amount of the data purchase increments. Once purchased, a unit of data can be treated in the same way as the budgets IMP currently considers, since there are no refunds for unused data and partially spent units are generally rounded up to the next full unit for billing. However, we would also need to consider the monetary cost of each data increment and also incorporate some notion of how much the user is willing to spend, which would require direct input from the user.

The concerns in this scenario differ from what we consider in IMP's current long-term budget approach, because of the added goal of saving the user money, the smaller units of data for which money is spent, and the fact that less money spent is always favorable. However, this scenario can still benefit from IMP's fixed-budget adaptation strategy for each allotted unit of data. In addition, the same strategy can be applied for each extra gigabyte that the user effectively purchases if they exceed their data cap on a pay-per-month plan.

Caching systems typically have a limited amount of storage available for cached data, along with an eviction policy. We observe that battery energy and cellular data are usually far scarcer than storage on today's mobile devices, so we chose to focus on improving the use of those resources. We could also apply a TIP-like strategy to include the value of cache space in our cost-benefit analysis.

Finally, we note that some users have unlimited data plans, and some users are frequently able to charge their phones throughout the day so that they are rarely in danger of running out of battery. In these situations, IMP will prefetch more aggressively, since it is given larger resource budgets. In the case of unlimited data, the throttling that some carriers perform after the user has crossed a certain data limit can be viewed as a budget in itself, since the reduced bandwidth is undesirable.

4. IMPLEMENTATION

This section describes the IMP's Application Programming Interface and the applications we have modified to use IMP.

4.1 API

IMP is implemented as a Java library targeted at the Android platform. Its prefetching API is inspired by Java's ExecutorService and Future [9], which simplify programming tasks with arbitrary asynchronous computation. The interface, which is shown in Table 1, is designed so that IMP need not replicate the application-specific details of (pre)fetching data.

4.1.1 *Prefetch hinting and execution*

Applications use the `prefetch` method to provide a prefetch hint to IMP. As described in Section 3.2, IMP uses a cost/benefit analysis to decide when and if to act on the hint. The application supplies a `Fetcher` object that both identifies the data item to be fetched and implements the application-specific `fetch` method that retrieves it from the server. This method is used for both prefetches and on-demand fetches. In the former case, IMP initiates the prefetch by calling `fetch`. The application can optionally supply the size of the item to be fetched, as well as a *prefetch class*. Applications can use classes to indicate prefetch priority. IMP will prefetch items in classes with higher observed accuracy in preference to items in other classes. Within a class, IMP prefetches objects in the order that prefetch hints are supplied.

4.1.2 *Prefetch consumption*

Like the `submit` method in Java's ExecutorService, IMP's `prefetch` method returns a Future — this is a handle that allows the application to later retrieve the result of the fetch operation by calling its `get` method. If the prefetch has started but not yet completed, the `get` method blocks until the prefetch completes or until an optional timeout expires. If the prefetch has not yet started, calling `get` triggers an immediate demand fetch of the item (which is effected through the `fetch` callback).

4.1.3 *Prefetch cancellation*

If the application no longer needs a item for which it has issued a prefetch hint, it cancels the prefetch by either calling the `cancel` method on the Future object or by simply removing the last reference to the Future object, in which case the prefetch is implicitly canceled when the Future object is garbage-collected.

4.2 Applications

We have modified the K9 [11] open-source email client and the OpenIntents [20] news reader to use IMP.

4.2.1 *K9 email*

K9 [11] is is an open-source email application for Android, originally based on the stock Android email client. By default, K9 prefetches messages that are less than 32 KB (the threshold can be adjusted from the application). We added IMP support by using an IMAP proxy to intercept traffic between K9 and our IMAP server.

On application start, the proxy downloads the user's email headers (including the size of each message), decides which emails to prefetch, and issues prefetch hints for those messages (including their attachments), in order from newest to oldest.

4.2.2 *News reader*

We also added prefetching to OpenIntents, an open-source Android Atom/RSS feed reader. OpenIntents stores the content available in a feed but does not prefetch linked articles. Since the application uses HTTP for its network communication, we modified Apache HttpComponents [3], a Java HTTP library included in the Android SDK, to add support for article prefetching.

Frequently, Atom and RSS feeds only contain a summary and link to a given article, rather than the full contents of the article. Our modified news reader therefore grabs the full-article links from each feed and decides which articles to prefetch. The application-specific fetcher specified via the `Fetcher` object issues an HTTP request for the full content of the article plus any embedded images, and it then stores the fetched data persistently for later retrieval.

Our news reader associates a prefetch class with each feed and uses the `setPrefetchClass` method to specify the class for each hint. To evaluate the impact of using classes, we also created a version of the news reader that does not specify prefetch classes.

4.3 Discussion

IMP needs some information from the application in order to make good prefetching decisions, but we must avoid making the developer's task onerous, or else applications will not use our system. Thus, we have only added features when our applications required them (*e.g.* prefetch classes). We could imagine applications that would desire additional features, such as the ability to retrieve a partial prefetch result instead of waiting for all data to arrive. Our API design would make such an addition straightforward.

In general, applications must be modified to work with IMP. We can support unmodified email clients via an email proxy, as features of the IMAP protocol help us discover which items the user might request and when items are requested on-demand or deleted. For other applications such as OpenIntents, it is difficult to gather this information without modifying the application.

5. EVALUATION

Our evaluation compares IMP with the prefetch strategies most commonly employed by current mobile applications. We compare results across three metrics: application performance, energy usage, and data consumption. In addition, we examine how well IMP meets budgets for energy and data usage. Finally, we quantify the benefit of using prefetch classes.

5.1 Experimental Setup

5.1.1 *Testbed*

We run K9 and OpenIntents on the AT&T version of the Nexus One phone, running Android version 2.3.4. We modified Android system software to allow the simultaneous use of WiFi and cellular networks and added Intentional Networking support. To generate repeatable experiments, the phone connects to an isolated WiFi access point and a private Cisco MicroCell that is connected to AT&T's network. Since the MicroCell acts as a miniature cellular tower, our evaluation captures the effects of the cellular wireless medium on network bandwidth and latency. We emulate network conditions by passing all traffic through a computer that inserts delays with the netem [16] network emulator and throttles throughput using the Linux Advanced Routing and Traffic Control tools [15]. We run servers for each application on a Dell Optiplex GX270 desktop with a 2.8 GHz Pentium 4 processor and 1GB DRAM.

We measure energy usage by applying the Nexus One power model used in the implementation of PowerTutor [30]. We measure cellular data usage by reading the number of bytes sent and received through the Linux sysfs interface.

When we compare with strategies that do not use IMP, we allow those strategies to use Intentional Networking to isolate prefetching traffic from other foreground traffic and to take advantage of both cellular and WiFi networks. Thus, the results reported in this paper show only the *additional* benefit that IMP provides on top of the benefit already provided by Intentional Networking. All reported results are the mean of five trials — graph error bars are 95% confidence intervals. Where there is a resource budget set for IMP, the budget is indicated by a horizontal line above the bar.

To quantify the benefit of different prefetching strategies, we report the average fetch time over all emails or news articles in a run. We report the average rather than the median because the median does not distinguish between two prefetch strategies that produce more than a 50% cache hit rate, and it is the long fetch delays that dominate the user's experience when not enough data is prefetched. Though the individual fetch times can vary considerably due to several factors – data size, network bandwidth at the time of the fetch – lower average response time generally indicates more cache hits and thus more successful prefetching.

5.1.2 *Trace-driven evaluation*

To evaluate how our system performs in realistic mobile networking conditions, we use a vehicular network trace gathered in Ypsilanti, MI [10]. Trace-driven evaluation provides experimental repeatability and allows for meaningful comparison between different prefetch strategies.

We gathered this trace by driving a vehicle equipped with WiFi and Sprint 3G network interfaces and continuously measuring the downlink and uplink bandwidth and latency via active probing to a server at the University of Michigan. The median 3G bandwidth is 368 Kb/s downlink and 40 Kb/s uplink, with maximum bandwidth of 1.2 Mb/s downlink and 74 Kb/s uplink. WiFi is available 27% of the time, with a median session length of 7 seconds; the longest WiFi session length is 131 seconds.

We collected a second trace by walking in downtown Ann Arbor and across the University of Michigan campus, carrying a Nexus One phone and measuring the available open and campus WiFi and the AT&T cellular network. The median 3G bandwidth is 695 Kb/s downlink and 216 Kb/s uplink, with maximum bandwidth of 1.3 Mb/s downlink and 358 Kb/s uplink. WiFi is available 18% of the time, with a median session length of 56 seconds; the longest WiFi session length is 99 seconds.

When running benchmarks, we replay the traces on the emulation computer, which throttles bandwidth and delays packets for each network according to the conditions observed. When no WiFi or cellular coverage is observed in a trace, the throttling computer causes the connection to drop — the Android OS typically discov-

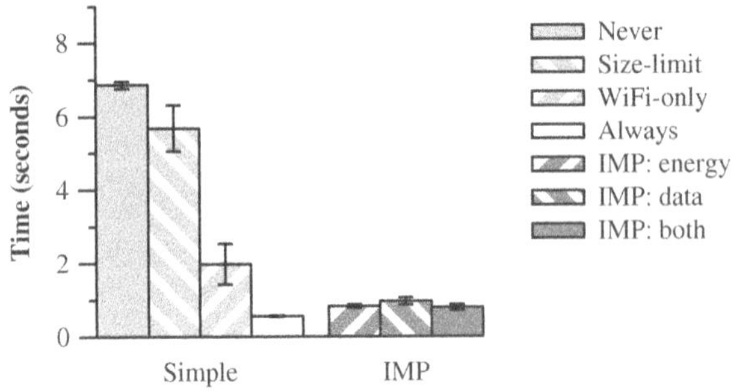

Average fetch time for 28 emails fetched over 20 minutes. The left set of bars shows results for simple heuristic-based prefetching strategies; the right set of bars shows results for IMP when energy, cellular data, or both are constrained. Each bar is the mean of 5 trials. The error bars are 95% confidence intervals.

Figure 1: Average email fetch times, driving trace

ers the network disconnection after several seconds. Since the collected traces are longer than our experiments, we use only the first portion of each trace.

5.1.3 Comparison strategies

For each application, we compare IMP's performance to four prefetch strategies commonly used by mobile applications today. Unlike IMP, none of these strategies explicitly considers the cost and benefit of prefetching. The strategies are: never prefetch anything, prefetch items with size less than an application-specific threshold, prefetch over WiFi when it is available but never over the cellular network, and always prefetch everything.

5.1.4 IMP scenarios

IMP attempts to maximize performance subject to one or more resource constraints. For each application, we consider three scenarios: one where energy usage is constrained, one where cellular data usage is constrained, and one where both resources are constrained. Although we do not show the results, we did verify that when neither budgeted resource is constrained, IMP (correctly) emulates the always-prefetch strategy because that strategy maximizes performance.

To run a large number of experiments, we limit execution time to 20 minutes. The goals for energy and data usage are scaled proportionately. Note that even though longer experiments would likely show a wider variety of network conditions and usage behaviors, shorter experiments like these are much more challenging for a feedback strategy like the one employed by IMP because there is little time to react to under-usage or over-usage of a budgeted resource. For instance, if IMP is over-using the cellular data budget after half the budgeted time, it has 15 days to react during a month-long experiment, but only 10 minutes to bring the usage back under budget in these experiments. Similarly, IMP would have more time to react to unexpected changes in a user's mobility patterns over the course of a long experiment – for example, taking a different route to the office and encountering far fewer open WiFi APs. Thus, we expect that if IMP can meet budgets in these shorter experiments, it should do quite well in longer ones.

5.2 Results

5.2.1 Email

We constructed an email workload from day-long traces of email activity within our department collected during prior projects. We

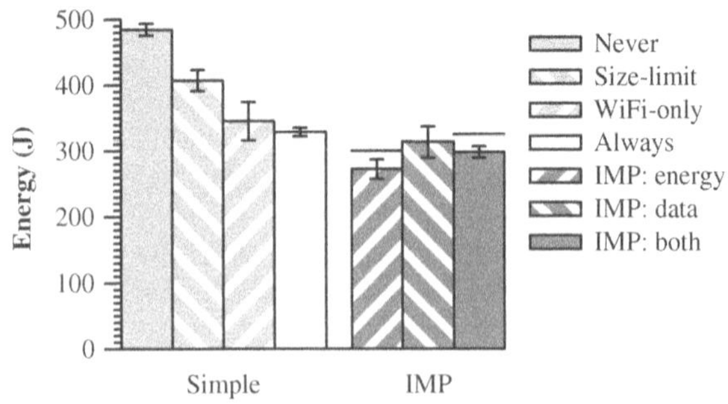

Total energy usage for the email benchmark. The left set of bars shows results for simple heuristic-based prefetching strategies; the right set of bars shows results for IMP when energy, cellular data, or both are constrained. Where there is an energy budget set for IMP, it appears as a solid line above the bar. Each bar is the mean of 5 trials. The error bars are 95% confidence intervals.

Figure 2: Email energy usage, driving trace

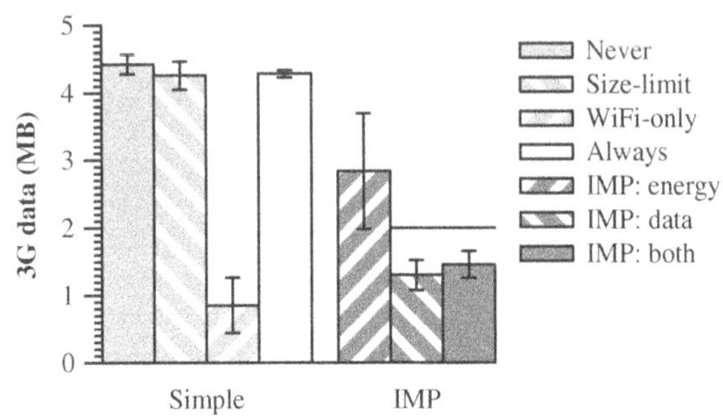

Total cellular data usage for the email benchmark. The left set of bars shows results for simple heuristic-based prefetching strategies; the right set of bars shows results for IMP when energy, cellular data, or both are constrained. Where there is a data budget set for IMP, it appears as a solid line above the bar. Each bar is the mean of 5 trials. The error bars are 95% confidence intervals.

Figure 3: Email data usage, driving trace

use a trace of an email user fetching messages from a server [2] to derive the timing of email fetches and the interleaving user think time. The size of each email is randomly sampled from a distribution of email sizes in another trace [27].

At the start of the benchmark, the proxy fetches the list of 35 emails from the server along with their sizes. It issues prefetch hints for all emails. The reported accuracy of Gmail's Priority Inbox is 80% [1]. Since this seems a reasonable mechanism for deciding which email should be hinted, we assume that 28 of the emails are read during the experiment while the other 7 are deleted before they are read. This models a scenario in which a user flips through the inbox and decides, based on the subject line and message preview text, whether or not to read the email. Thus, the accuracy of prefetch hints is 80%. Of course, IMP is not told which hinted emails will be read and which will be discarded.

K9's default settings specify that email messages less than 32 KB will be fetched automatically upon inbox update. We use that value for the size-limit strategy.

Figure 1 shows the average email fetch time for each strategy. As expected, the always-prefetch strategy produces the lowest fetch time, since every fetch is a cache hit. However, as shown in Figures 2 and 3, this strategy uses more energy and data than any IMP strategy.

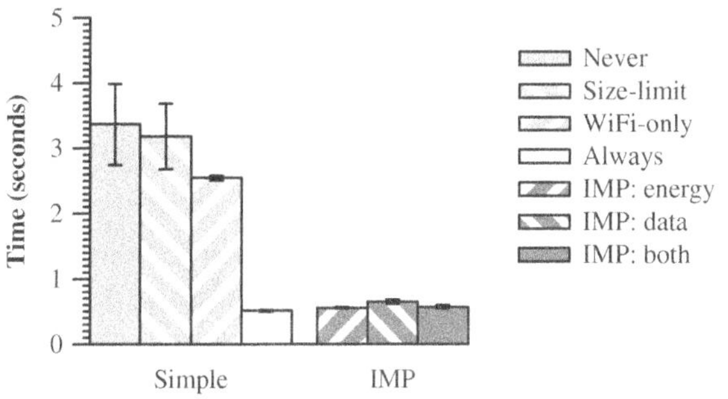

Average fetch time for 28 emails fetched over 20 minutes. The left set of bars shows results for simple heuristic-based prefetching strategies; the right set of bars shows results for IMP when energy, cellular data, or both are constrained. Each bar is the mean of 5 trials. The error bars are 95% confidence intervals.

Figure 4: Average email fetch times, walking trace

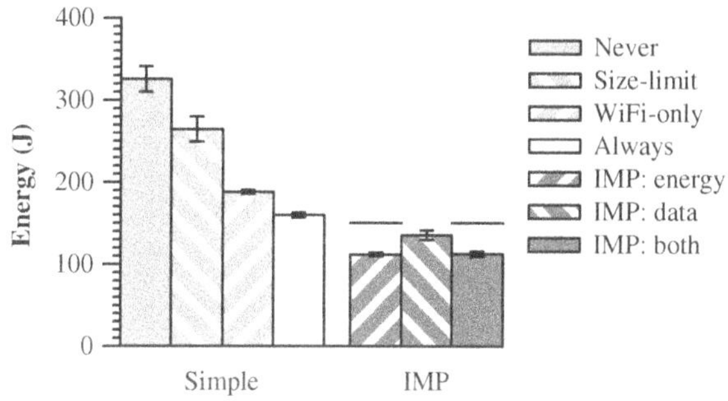

Total energy usage for the email benchmark. The left set of bars shows results for simple heuristic-based prefetching strategies; the right set of bars shows results for IMP when energy, cellular data, or both are constrained. Where there is an energy budget set for IMP, it appears as a solid line above the bar. Each bar is the mean of 5 trials. The error bars are 95% confidence intervals.

Figure 5: Email energy usage, walking trace

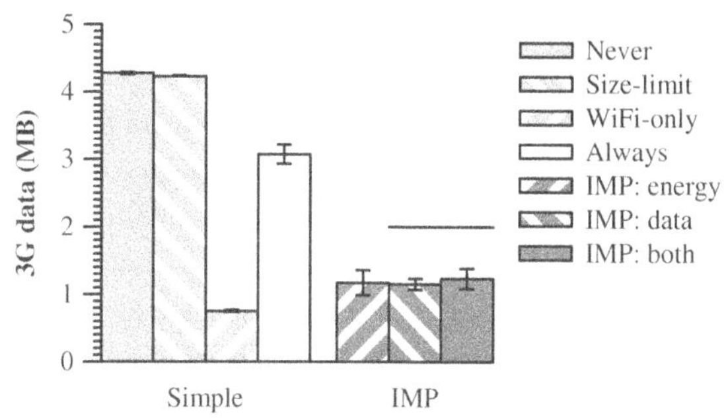

Total cellular data usage for the email benchmark. The left set of bars shows results for simple heuristic-based prefetching strategies; the right set of bars shows results for IMP when energy, cellular data, or both are constrained. Where there is a data budget set for IMP, it appears as a solid line above the bar. Each bar is the mean of 5 trials. The error bars are 95% confidence intervals.

Figure 6: Email data usage, walking trace

Interestingly, the never-prefetch strategy uses more energy than any other strategy. This is due to a combination of the tail energy incurred with each demand fetch and poorer 3G bandwidth in later portions of the trace causing the 3G interface to remain in the high-power state longer.

When resources are scarce, IMP is able to meet resource budgets that are set lower than the amounts used by most simple strategies. In the constrained-energy scenario, we set the energy budget to use no more than 300 Joules over 20 minutes (this is indicated by the solid line over the `IMP: energy` bar in Figure 2). Despite this goal being lower than the energy usage of any simple strategy, IMP comes in under budget in every trial. Further, IMP provides average fetch time within 300 ms of the (performance-optimal) always-prefetch strategy. Compared to the other strategies, IMP improves average fetch time by 2–8x and reduces energy usage by 21–43%. The fact that IMP produces a lower average fetch time than the common WiFi-only prefetching strategy lends credence to our assertion that prefetching on 3G is often a wise way to spend resources when the budget allows for it, since doing so provides a clear benefit to the user.

It is perhaps surprising that IMP uses less energy than the WiFi-only strategy in this scenario, especially since WiFi transmission is usually more energy-efficient than cellular transmission. Our analysis shows that because the WiFi-only prefetching strategy does not prefetch over the cellular network, it cannot prefetch some items before the user requests them. This leads to demand fetches later in the experiment that must be serviced over cellular because WiFi is unavailable at the time. Further, the cellular network quality at the time is poor, causing even more energy usage than normal. IMP avoids these demand fetches through more aggressive prefetching. Additionally, although IMP sends more data over cellular than the WiFi-only strategy, it reduces energy usage due to 3G tail time by batching multiple requests together. In contrast, the demand fetches cannot be batched due to user think time.

In the constrained-data scenario, we set the budget to use no more than 2 MB of data. Figure 3 shows the 3G data usage for each prefetching strategy. IMP meets this goal in all trials, whereas WiFi-only is the only simple strategy to use as little data. However, the WiFi-only strategy more than doubles average fetch time compared to IMP. Meanwhile, IMP is still able to provide average fetch time within 410 ms of that of the always-prefetch strategy and 2–7x less than that of the never-prefetch and size-limit strategies.

To constrain both resources, we set an energy budget of 325 Joules and a data budget of 2 MB. The results are similar to the single-budget scenarios. IMP provides average fetch time within 240 ms of the always-prefetch strategy and 2–8x lower than the other strategies, while reducing energy usage by 9–38% compared to the simple strategies. Compared to the always-prefetch, never-prefetch, and size-limit strategies, IMP reduces cellular data consumption by 3x.

IMP sometimes undershoots its budgets because it is hedging against future spikes in demand. In this experiment, however, undershoot occurs simply because IMP runs out of items to prefetch.

Figures 4, 5, and 6 show the results for the email application on the walking trace. In the walking scenario, the best strategy is to always prefetch, because the best WiFi is available at the beginning of the trace, and the little available WiFi that comes later is poor. In all scenarios, IMP emulates always-prefetch closely, achieving average fetch time within 40–150 ms of that of the always-prefetch strategy and 4–6x lower than the other strategies.

For the energy-constrained experiments using the walking trace, we set an energy budget of 150 Joules. IMP meets the goal in all trials, reducing energy usage by 30–65% compared to the simple strategies. For the data-constrained experiments, we set a 3G data budget of 2 MB. IMP meets this goal as well in all trials, reducing

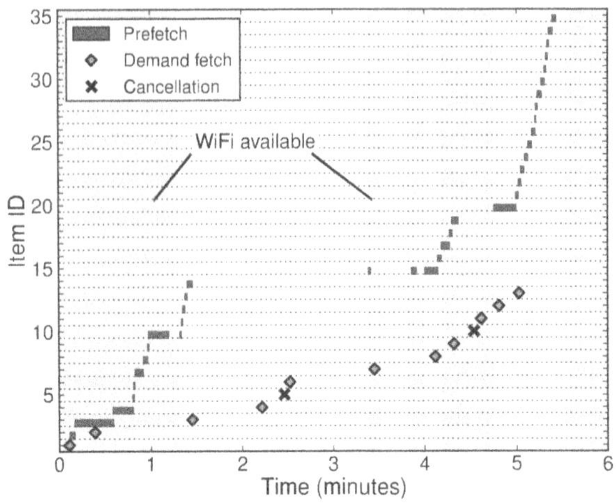

Email prefetch and demand fetch activity for one run of IMP, energy-constrained, on the driving trace. The horizontal bars show issued prefetches. The diamonds show demand fetches. The shaded regions show periods when WiFi is available. Each canceled prefetch hint is marked with an X. Since prefetching finishes early in the experiment, only the first six minutes are shown.

Figure 7: Email + IMP prefetch activity

3G data usage by 2–4x compared to the never-prefetch, size-limit, and always-prefetch strategies. Only the WiFi-only strategy meets the data goal, but it increases average fetch time by over 4x compared to IMP.

To provide some insight into the decisions that IMP makes over the course of an experiment, we show in Figure 7 the activity of the email application in one run of the driving trace, with IMP running with the energy budget only. IMP starts prefetching on WiFi, until the WiFi fades shortly after the 1-minute mark. At this point, the 3G interface is already active due to an Intentional Networking control message, so IMP starts prefetching on 3G. It completes a few prefetches before deciding that the next prefetch is too large to send on 3G, so it subsides. A few minutes later, it begins encountering some spotty WiFi, which it uses to attempt prefetching the next message. Soon after it starts making forward progress on WiFi, the WiFi network fails again. IMP then decides that it has enough budget to send a large batch of prefetches over 3G (taking the tail time savings into account), so it finishes prefetching the rest of the messages.

5.2.2 *News reader*

We constructed a news reader workload based on one of the authors' Google Reader subscriptions. Google Reader provides statistics for a user's feeds for the past 30 days, including information such as the number of articles published and the number of articles read, along with the calculated read percentage. We choose five feeds with articles widely varying in size (primarily due to attached images). We used the relative feed volume over the past 30 days to choose a subset of each feed's articles for the benchmark, starting with the most recent articles. There are 25 articles across the five feeds.

Our simulated user selects a feed and begins iterating through the articles in that feed, opening and reading some of them and marking others as read without actually opening them. The fraction of articles that the user actually reads is determined by the read rate

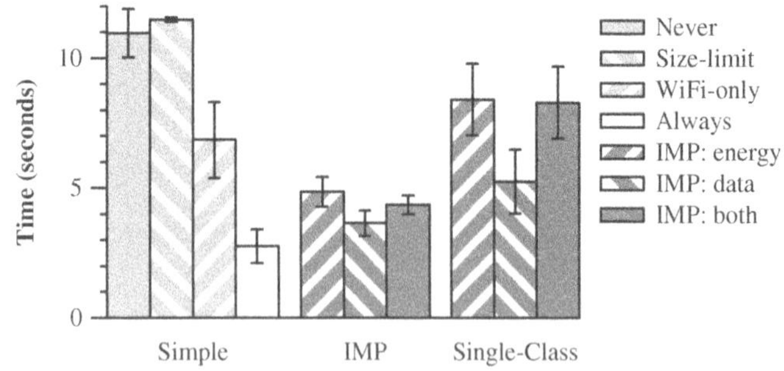

Average news article fetch time for 16 articles fetched over 20 minutes. The left set of bars shows results for simple heuristic-based prefetching strategies. The middle set of bars shows results for IMP when energy, cellular data, or both are constrained. The right set of bars shows results for IMP with the same constraints when IMP does not separate prefetch hints by class. Each bar is the mean of 5 trials. The error bars are 95% confidence intervals.

Figure 8: Average news article fetch times, driving trace

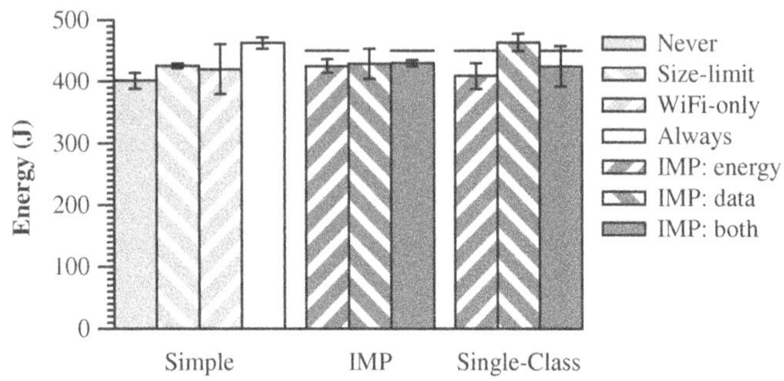

Total energy usage for the news reader benchmark. The left set of bars shows results for simple heuristic-based prefetching strategies. The middle set of bars shows results for IMP when energy, cellular data, or both are constrained. The right set of bars shows results for IMP with the same constraints when IMP does not separate prefetch hints by class. Where there is an energy budget set for IMP, it appears as a solid line above the bar. Each bar is the mean of 5 trials. The error bars are 95% confidence intervals.

Figure 9: News reader energy usage, driving trace

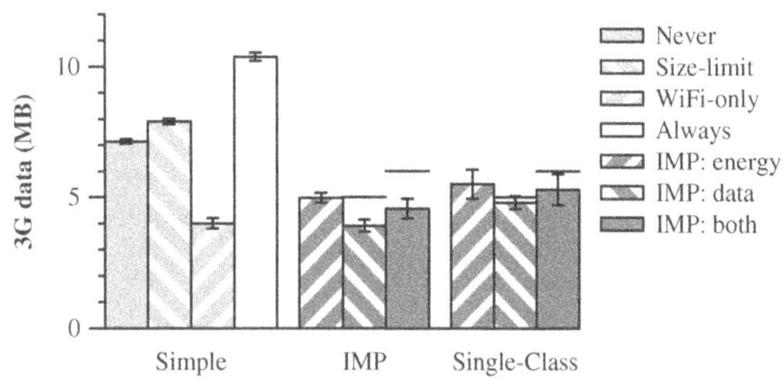

Total cellular data usage for the news reader benchmark. The left set of bars shows results for simple heuristic-based prefetching strategies. The middle set of bars shows results for IMP when energy, cellular data, or both are constrained. The right set of bars shows results for IMP with the same constraints when IMP does not separate prefetch hints by class. Where there is a data budget set for IMP, it appears as a solid line above the bar. Each bar is the mean of 5 trials. The error bars are 95% confidence intervals.

Figure 10: News reader data usage, driving trace

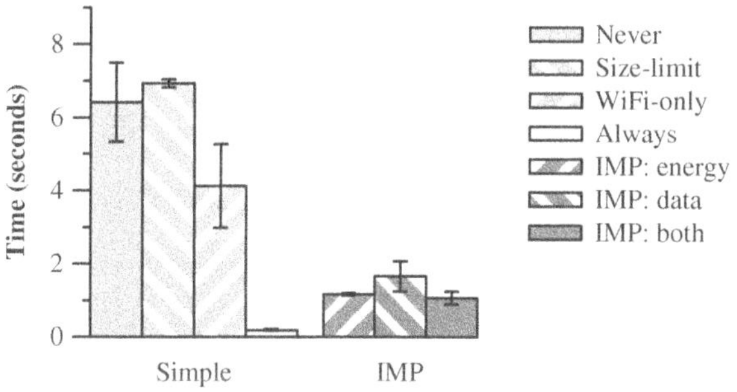

Average news article fetch time for 16 articles fetched over 20 minutes. The left set of bars shows results for simple heuristic-based prefetching strategies. The right set of bars shows results for IMP when energy, cellular data, or both are constrained. Each bar is the mean of 5 trials. The error bars are 95% confidence intervals.

Figure 11: Average news article fetch times, walking trace

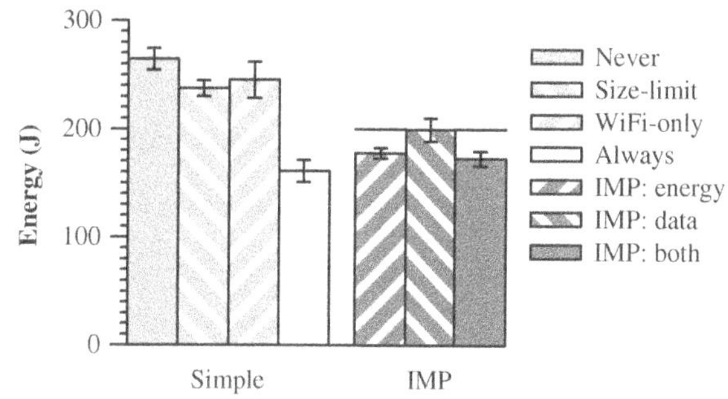

Total energy usage for the news reader benchmark. The left set of bars shows results for simple heuristic-based prefetching strategies. The right set of bars shows results for IMP when energy, cellular data, or both are constrained. Where there is an energy budget set for IMP, it appears as a solid line above the bar. Each bar is the mean of 5 trials. The error bars are 95% confidence intervals.

Figure 12: News reader energy usage, walking trace

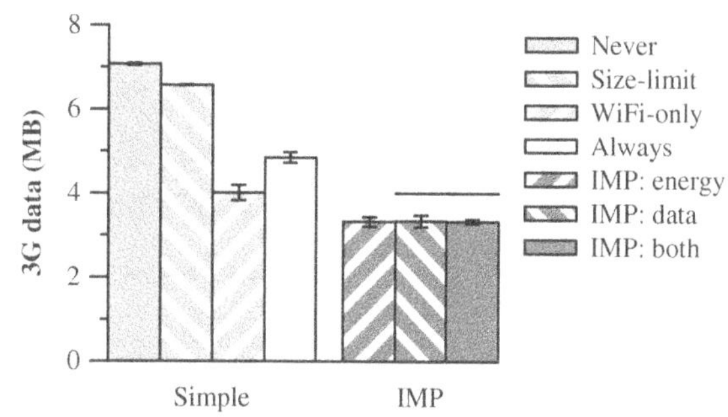

Total cellular data usage for the news reader benchmark. The left set of bars shows results for simple heuristic-based prefetching strategies. The right set of bars shows results for IMP when energy, cellular data, or both are constrained. Where there is a data budget set for IMP, it appears as a solid line above the bar. Each bar is the mean of 5 trials. The error bars are 95% confidence intervals.

Figure 13: News reader data usage, walking trace

for that particular feed. For the chosen feeds, the read rate ranges from 25% to 100%, and the total read rate across all feeds is 64%.

Lacking a trace for newsreader user behavior, we chose to model a user's think time between article fetches as a random variable. Recent work in modeling web user dwell times observes that the majority of page visits last one minute or less [17]. Thus, after the user reads an article, the benchmark pauses for a random think time uniformly distributed between 30 and 60 seconds, then moves on to the next article in the feed, continuing until the feed is exhausted. The user does not pause after choosing to skip an article. As with the email application, the benchmark lasts 20 minutes.

Since the news reader does not prefetch article contents by default, we set the value used in the threshold strategy to 128 KB, which is just above the median article size of the 25 articles.

Our news reader associates a different class with each feed. The results of this evaluation are shown by the middle set of bars in Figures 8, 9, and 10. In the next section, we evaluate the benefit of prefetch classes by using only a single class. For ease of comparison, these results are shown by the rightmost set of bars in each figure.

As with the email benchmark, the always-prefetch strategy produces the best average fetch time, though the greater inaccuracy of the prefetch hints in this experiment reduces the benefit of prefetching compared to the email experiment.

In contrast to the email benchmark, aggressive prefetching has an adverse affect on energy usage. This is primarily due to the lower prefetch hint accuracy and the large size of some articles. Nevertheless, when given an energy budget of 450 Joules, IMP performs enough prefetching to reduce average fetch time by 29–58% compared to the never-prefetch, size-limit, and WiFi-only strategies, while also meeting the energy goal. As we discuss in the next section, IMP benefits from separating prefetch hints into classes and first prefetching articles most likely to be read. While the always-prefetch strategy has better performance than IMP, it does not meet the energy goal.

When we constrain data usage to 5 MB, IMP has more freedom in scheduling prefetches than in the energy-constrained scenario and reduces average fetch time by 47–68% compared to the never-prefetch, WiFi-only, and size-limit strategies. IMP meets the cellular data goal in all runs, reducing 3G data usage by 45–62%.

When constraining both resources, we kept the 450 Joule energy budget and set a 6 MB data budget. IMP's behavior is similar to its behavior in the energy-constrained scenario: it meets each goal in all runs and reduces average fetch time by 36–62% compared to the never-prefetch, size-limit, and WiFi-only strategies.

IMP uses less than the given cellular data budget, for the same reasons as above: IMP finishes prefetching all the hinted articles, after which no more articles are sent on the cellular network.

Figures 11, 12, and 13 show the results for the newsreader application on the walking trace. As in the case of the email application, the best strategy for this trace is to always prefetch. In this application, IMP comes closer to the average fetch time of always-prefetch than any of the other simple strategies, achieving average fetch time 2–6x lower than those strategies.

For the energy-constrained experiments using the walking trace, we set an energy budget of 200 Joules. IMP meets the goal in all trials, reducing energy usage by 25–35% compared to the never-prefetch, size-limit, and WiFi-only strategies.

IMP uses 7–10% more energy than the always-prefetch strategy on average. The reason for this is that IMP decides to pause prefetching when it encounters a particularly large prefetch early in the benchmark. At this time in the experiment, IMP has sent a few prefetches on 3G to amortize the tail time, but it decides the potential savings of the tail time is not worth the energy that will be spent to fetch the large article, especially since the article is less likely to be read than the articles that IMP has already prefetched.

This results in two demand fetches on 3G later in the benchmark, since IMP does not resume prefetching until it has ceased spending energy for a while to catch up to its target rate. Thus, IMP incurs additional tail time compared to the always-prefetch strategy.

For the data-constrained experiments, we set a 3G data budget of 4 MB. IMP meets this goal as well in all trials, reducing 3G data usage by 17–53% compared to the simple strategies. Interestingly, IMP reduces 3G data usage even compared to the WiFi-only strategy. The reason for this is that IMP uses prefetch classes, whereas WiFi-only does not, which causes IMP to prioritize articles most likely to be read. Since WiFi is limited in this trace, the WiFi-only strategy fails to prefetch some particularly large articles that later must be demand-fetched over 3G.

5.2.3 *Benefit of prefetch classes*

To evaluate the benefit of prefetch classes, we executed the same benchmark, using the driving trace, without having the news reader use classes to differentiate prefetch hints of articles in different feeds.

The benefit of prefetch classes is greatest in the scenarios in which energy is constrained; in these two scenarios, the multi-class news reader reduces average fetch time by 42–47% compared to the single-class news reader scenarios. Though the energy constraint causes both versions of the news reader to prefetch cautiously, the multi-class news reader benefits greatly by first prefetching the articles with the highest likelihood of being read.

When the application uses prefetch classes, IMP meets all goals in every trial. However, without prefetch classes, IMP meets only 87% of the resource goals, failing to meet the data goal in 1 of 5 trials of the data-constrained scenario and the energy goal in 1 of 5 trials of the dual-constraint scenario. In the two trials in which IMP missed the goal, it overshot by less than 5.3% due to a late demand fetch in the benchmark. While no adaptive system will ever be perfect, we expect that a longer experiment with larger goals (typical of mobile phone usage today) would give IMP substantially more room for error and probably eliminate these overshoots.

Nevertheless, prefetch classes provide an added margin of safety by allowing IMP to make more informed decisions. IMP is able to target its prefetch effort to articles that are more likely to be read. Further, IMP's per-item predictions are more accurate since they are based on per-feed statistics rather than overall hit rates.

5.2.4 *Discussion*

From the above results, we conclude that IMP does an excellent job of meeting budgets for constrained resources. IMP meets all budgets in every trial for both the email and news reader applications (as long as the latter uses prefetch classes to differentiate feeds).

In the newsreader experiment on the walking trace, when energy is the only constraint, the always-prefetch strategy is best for both fetch time and energy usage, as discussed above. However, in all other cases, when one or more budgetary constraints are specified, IMP outperforms all heuristic-based prefetching strategies we examined that also meet the budgetary constraints. Often, the performance improvement is a factor of two or more.

The benefit of prefetch classes is apparent not only in improving the percentage of budgets met for the news reader from 87% to 100%, but also in substantially improved response time. There is clearly a tradeoff between complexity and accuracy in the application interface. IMP can make better decisions if the application provides more information, but each additional specification places more burden on the application developer. In the case of prefetch classes, our results indicate that the small effort of differentiating hints by class (in the case of the news reader) can substantially improve results. This indicates that other, more contextual methods for differentiating prefetch accuracy could be beneficial.

6. RELATED WORK

One of the primary contributions of IMP is the unification of prior solutions to a seemingly disparate collection of problems to meet the unique challenges of mobile prefetching. In this section, we describe prior work in prefetching and in managing limited resources in mobile computing, and we discuss how IMP unifies and builds upon these ideas.

6.1 Prefetching

Prefetching is a long-studied technique used in a wide variety of computing domains. Processors predict which instructions or data will be needed and populate cache lines in advance [28]. File systems and databases attempt to infer application access patterns and fetch data items before they are needed [13, 22]. Distributed file systems fetch files from servers in advance of their use, both to improve performance and to maintain availability in the face of unreliable network connections [12].

The design of IMP is inspired by Transparent Informed Prefetching (TIP), which uses cost/benefit analysis to manage allocation of disk cache buffers between competing consumers: prefetched data and the LRU cache [23]. Such work recognizes that prefetching must be done carefully, lest it harm more than it helps. We observe the same high-risk/high-reward nature of prefetching in mobile computing and construct a cost/benefit analysis based on user-perceived performance improvement and the cost of prefetching in terms of battery energy and cellular data.

In the domain of mobile computing, prefetching has long been recommended as a primary technique for improving application performance [29]. Prior theoretical work in prefetch algorithm modeling and analysis observes, as we do, the tension between improving data access performance and spending limited bandwidth and energy resources to achieve that end. Persone et al. develop a prefetch cost model by which to numerically evaluate prefetch strategies based on mobility patterns [6]. However, their work considers different costs of prefetching in isolation from each other and from the benefit of prefetching. In contrast, IMP explicitly considers how different costs, such as energy usage and 3G data consumption, may be meaningfully combined and weighed against the benefits that prefetching can provide.

Lee et al. seek to improve the efficacy of prefetching and counteract its inherent uncertainty by improving the accuracy of predictions, using location-specific or context-specific information [14]. Web prefetching has long used spatial locality to predict what data users will request next [21]. Such efforts are beneficial and complementary to IMP. IMP allows the application to decide which data should be prefetched and instead addresses the decision of when to prefetch given limited resources and changing network conditions.

As we have demonstrated, having more accurate prefetch hints from the application allows IMP to deliver better results. Thus, IMP might benefit from prediction of network availability and quality, as done systems such as BreadCrumbs [19]. By knowing with greater accuracy what future bandwidth and WiFi coverage to expect, IMP can make better-informed decisions about whether prefetching now or waiting until later would be more cost-effective.

6.2 Limited resources in mobile networking

Several recent projects have proposed techniques for dealing with the different strengths and weaknesses of WiFi and cellular networks. Wiffler [4] explored the use of intermittently-available WiFi to reduce 3G data usage and ease pressure on cellular networks, subject to application-specific delay tolerance. Bartendr [26] observed that the energy cost of sending data on a cellular network increases significantly as signal strength drops and that energy savings can be realized by predicting periods of good signal strength and, when possible, delaying transmissions to target those periods.

Additionally, much recent work targets the poor interaction between naive applications' cellular data usage patterns and the energy cost of 3G tail time. Balasubramanian et al. measured the significant amount of energy wasted in 3G tail time and developed TailEnder to amortize tail energy cost with batching and prefetching [5]. The Tail Optimization Protocol [24] predicts long idle periods and direct the cellular radio to release radio resources and enter the low-power idle mode without waiting for the tail time. TailTheft [18] "steals" parts of the tail time for small transmissions without extending the tail time.

We share these systems' goal of reducing energy and cellular data costs, and indeed, many of the techniques they describe are applicable to IMP as well. However, we also observe that resource conservation, while a worthy goal, is not always the most important goal for a mobile user, and that often, energy and cellular data can and should be spent more freely in order to improve the user's experience. We also observe that prefetching is not always beneficial, and that aggressively prefetching data that will not be used is unnecessarily wasteful. Hence, IMP explicitly considers the changing relative importance of these goals to tune its decisions over time, and it also considers the observed accuracy of prefetch hints to determine the value of prefetching.

IMP uses Intentional Networking [10] to simplify the use of multiple wireless networks and to prevent prefetch traffic from penalizing the performance of interactive traffic. IMP also benefits from the power modeling work of the PowerTutor project [30], which enabled automatic derivation of power models through a series of tests that exercise different components of a mobile device in turn and isolate the power consumption of each. Finally, IMP adopts goal-directed adaptation techniques developed in Odyssey for energy savings [8], which we additionally apply to the cellular data resource. IMP applies these ideas to the new domain of mobile prefetching.

7. CONCLUSION

Prefetching is an invaluable tool for improving the performance of mobile applications. However, prefetching can be costly in terms of the energy and cellular data consumed, and these costs are exacerbated by the possibility of prefetching data that the user never requests. As a result, many mobile applications employ simple but suboptimal heuristics as a means of grasping some benefit while avoiding undue complexity.

We propose that mobile computing systems provide explicit prefetching support to applications. IMP shifts the burden of complex prefetching decisions from the application to the system. It assumes responsibility for estimating the cost and benefit of prefetching, tracking the importance of energy and cellular data over time, and scheduling prefetches to improve interactive performance while meeting specified resource usage goals.

Our evaluation shows that IMP is able to meet specified energy and cellular data usage goals. Additionally, we show that in most cases, whenever simple heuristic-based prefetching strategies also meet the goals, IMP outperforms those strategies on interactive fetch time, often by a factor of two or more.

Acknowledgments

We thank the anonymous reviewers and our shepherd, Arun Venkataramani, for comments that improved this paper. The work reported in this paper was supported in part by DARPA under grant HR0011-10-1-0081. The views and conclusions contained in this document are those of the authors and should not be interpreted as representing the official policies, either expressed or implied, of DARPA, the University of Michigan, Ford, Arbor Networks, or the U.S. government.

8. REFERENCES

[1] ABERDEEN, D., PACOVSKY, O., AND SLATER, A. The learning behind Gmail priority inbox. In *NIPS 2010 Workshop on Learning on Cores, Clusters and Clouds* (Mount Currie, British Columbia, Canada, December 2010).

[2] ANAND, M., NIGHTINGALE, E. B., AND FLINN, J. Ghosts in the machine: Interfaces for better power management. In *Proceedings of the 2nd International Conference on Mobile Systems, Applications and Services* (Boston, MA, June 2004), pp. 23–35.

[3] APACHE HTTPCOMPONENTS. `http://hc.apache.org/`.

[4] BALASUBRAMANIAN, A., MAHAJAN, R., AND VENKATARAMANI, A. Augmenting mobile 3G using WiFi. In *Proceedings of the 8th International Conference on Mobile Systems, Applications and Services* (San Francisco, CA, June 2010), pp. 209–221.

[5] BALASUBRAMANIAN, N., BALASUBRAMANIAN, A., AND VENKATARAMANI, A. Energy consumption in mobile phones: A measurement study and implications for network applications. In *Proceedings of the 9th ACM SIGCOMM Conference on Internet Measurement* (Chicago, IL, USA, November 2009), pp. 280–293.

[6] DE NITTO PERSONE, V., GRASSI, V., AND MORLUPI, A. Modeling and evaluation of prefetching policies for context-aware information services. In *Proceedings of the 4th International Conference on Mobile Computing and Networking* (Dallas, TX, USA, October 1998), pp. 55–65.

[7] FLINN, J. *Extending Mobile Computer Battery Life through Energy-Aware Adaptation*. PhD thesis, Department of Computer Science, Carnegie Mellon University, December 2001.

[8] FLINN, J., AND SATYANARAYANAN, M. Managing battery lifetime with energy-aware adaptation. *ACM Transactions on Computer Systems (TOCS) 22*, 2 (May 2004), 137–179.

[9] FUTURE (JAVA 2 PLATFORM SE 5.0). `http://docs.oracle.com/javase/1.5.0/docs/api/java/util/concurrent/Future.html`.

[10] HIGGINS, B. D., REDA, A., ALPEROVICH, T., FLINN, J., GIULI, T. J., NOBLE, B., AND WATSON, D. Intentional networking: Opportunistic exploitation of mobile network diversity. In *Proceedings of the 16th International Conference on Mobile Computing and Networking* (Chicago, IL, September 2010), pp. 73–84.

[11] K-9 MAIL. `http://code.google.com/p/k9mail/`.

[12] KISTLER, J. J., AND SATYANARAYANAN, M. Disconnected operation in the Coda file system. *ACM Transactions on Computer Systems 10*, 1 (February 1992), 3–25.

[13] KOTZ, D., AND ELLIS, C. S. Practical prefetching techniques for parallel file systems. In *Proceedings of the First International Conference on Parallel and Distributed Information Systems* (Miami Beach, FL, USA, December 1991), pp. 182–189.

[14] LEE, D. L., XU, J., ZHENG, B., AND LEE, W.-C. Data management in location-dependent information services. *IEEE Pervasive Computing 1*, 3 (2002), 65–72.

[15] LINUX ADVANCED ROUTING AND TRAFFIC CONTROL. `http://lartc.org/`.

[16] THE LINUX FOUNDATION. *netem*. `http://www.linuxfoundation.org/collaborate/workgroups/networking/netem`.

[17] LIU, C., WHITE, R. W., AND DUMAIS, S. Understanding web browsing behaviors through weibull analysis of dwell time. In *Proceedings of the 33rd International ACM SIGIR Conference on Research and Development in Information Retrieval* (Geneva, Switzerland, July 2010), pp. 379–386.

[18] LIU, H., ZHANG, Y., AND ZHOU, Y. TailTheft: Leveraging the wasted time for saving energy in cellular communications. In *Proceedings of the 6th ACM International Workshop on Mobility in the Evolving Internet Architecture* (Washington, D.C., USA, June 2011), pp. 31–36.

[19] NICHOLSON, A. J., AND NOBLE, B. D. BreadCrumbs: Forecasting mobile connectivity. In *Proceedings of the 14th International Conference on Mobile Computing and Networking* (San Francisco, CA, September 2008), pp. 46–57.

[20] OPENINTENTS NEWS READER. `http://www.openintents.org/en/newsreader`.

[21] PADMANABHAN, V. N., AND MOGUL, J. C. Using predictive prefetching to improve world wide web latency. *Computer Communication Review 26*, 3 (1996), 22–36.

[22] PALMER, M., AND ZDONIK, S. B. Fido: A cache that learns to fetch. In *Proceedings of the 17th International Conference on Very Large Data Bases* (Barcelona, Spain, September 1991), pp. 255–264.

[23] PATTERSON, R. H., GIBSON, G. A., GINTING, E., STODOLSKY, D., AND ZELENKA, J. Informed prefetching and caching. In *Proceedings of the 15th ACM Symposium on Operating Systems Principles* (Copper Mountain, CO, December 1995), pp. 79–95.

[24] QIAN, F., WANG, Z., GERBER, A., MAO, Z. M., SEN, S., AND SPATSCHECK, O. TOP: Tail optimization protocol for cellular radio resource allocation. In *Proceedings of the 18th IEEE International Conference on Network Protocols* (Kyoto, Japan, October 2010), pp. 285–294.

[25] RAHMATI, A., QIAN, A., AND ZHONG, L. Understanding human-battery interaction on mobile phones. In *Proceedings of the 9th ACM International Conference on Human Computer Interaction with Mobile Devices and Services* (Singapore, September 2007), pp. 265–272.

[26] SCHULMAN, A., NAVDA, V., RAMJEE, R., SPRING, N., DESHPANDE, P., GRUNEWALD, C., JAIN, K., AND PADMANABHAN, V. N. Bartendr: A practical approach to energy-aware cellular data scheduling. In *Proceedings of the 16th International Conference on Mobile Computing and Networking* (Chicago, IL, USA, September 2010), pp. 85–96.

[27] SHAH, S., AND NOBLE, B. D. A study of e-mail patterns. *Software: Practice and Experience 37*, 14 (2007), 1515–1538.

[28] SMITH, A. J. Cache memories. *ACM Computing Surveys 14*, 3 (September 1982), 473–530.

[29] WATSON, T. Application design for wireless computing. In *Proceedings of the First Workshop on Mobile Computing Systems and Applications (HotMobile)* (Santa Cruz, CA, USA, December 1994), pp. 91–94.

[30] ZHANG, L., TIWANA, B., QIAN, Z., WANG, Z., DICK, R. P., MAO, Z. M., AND YANG, L. Accurate online power estimation and automatic battery behavior based power model generation for smartphones. In *Proceedings of the Eighth IEEE/ACM/IFIP International Conference on Hardware/Software Codesign and System Synthesis* (Scottsdale, AZ, USA, October 2010), pp. 105–114.

FM-based Indoor Localization

Yin Chen[†], Dimitrios Lymberopoulos[‡], Jie Liu[‡], Bodhi Priyantha[‡]
[†]Department of Computer Science
Johns Hopkins University
Baltimore, MD
yinchen@cs.jhu.edu
[‡]Microsoft Research
One Microsoft Way
Redmond, WA
{dlymper,liuj,bodhip}@microsoft.com

ABSTRACT

The major challenge for accurate fingerprint-based indoor localization is the design of robust and discriminative wireless signatures. Even though WiFi RSSI signatures are widely available indoors, they vary significantly over time and are susceptible to human presence, multipath, and fading due to the high operating frequency. To overcome these limitations, we propose to use FM broadcast radio signals for robust indoor fingerprinting. Because of the lower frequency, FM signals are less susceptible to human presence, multipath and fading, they exhibit exceptional indoor penetration, and according to our experimental study they vary less over time when compared to WiFi signals. In this work, we demonstrate through a detailed experimental study in 3 different buildings across the US, that FM radio signal RSSI values can be used to achieve room-level indoor localization with similar or better accuracy to the one achieved by WiFi signals. Furthermore, we propose to use additional signal quality indicators at the physical layer (i.e., SNR, multipath etc.) to augment the wireless signature, and show that localization accuracy can be further improved by more than 5%. More importantly, we experimentally demonstrate that the localization errors of FM and WiFi signals are *independent*. When FM and WiFi signals are combined to generate wireless fingerprints, the localization accuracy increases as much as 83% (when accounting for wireless signal temporal variations) compared to when WiFi RSSI only is used as a signature.

Categories and Subject Descriptors

H.4 [**Information Systems Applications**]: Miscellaneous; C.2.1 [**Network Architecture and Design**]: Wireless communication

General Terms

Experimentation, Measurement, Performance

Keywords

FM, localization, mobile systems, fingerprinting, wireless

1. INTRODUCTION

Accurate indoor positioning information has the potential to revolutionize the way people search, locate and navigate to points of interest inside buildings in a similar way that GPS revolutionized the way people navigate outdoors. For instance, a user in a mall could leverage his mobile device, equipped with accurate indoor positioning technology, to instantly search, locate and navigate with real-time turn-by-turn directions to any store in the mall. When entering a store, the user's mobile device could automatically provide directions to the exact aisle or section where the desired product is located. At the same time, businesses and advertisers could push coupons and offers to the user in real time based on his current position within the mall or the store, maximizing customer targeting effectiveness.

Enabling these scenarios has been challenging mainly due to the unavailability of GPS signals in indoor environments. In the absence of GPS, fingerprint-based indoor localization techniques have been the most accurate approach to indoor localization [1, 9, 27]. The major challenge for fingerprint-based approaches is the design of robust and discriminative signatures. The most popular approach, that does not require any hardware deployment, has been to leverage already available wireless signals (e.g., WiFi, cellular) to profile a location, usually in the form of received signal strength indicator (RSSI) values [1, 24]. In previous work, RSSI values of WiFi signals have been primarily used for this purpose, as WiFi access points are widely deployed indoors, and every mobile device is equipped with a WiFi receiver.

Even though this approach has been successful in localizing people at a coarser grain (e.g., at the building level [21]), it exhibits several limitations when considering indoor environments where a person needs to be localized at the room level. First, the operating frequency range of WiFi signals makes them susceptible to human presence and orientation as well as to the presence of small objects in a room. This introduces variability in the recorded fingerprints that can lead to localization errors. Second, several of the deployed WiFi access points are commercial in nature and employ optimizations, such as frequency hopping, to improve network's throughput. These optimizations can result in significant variations in the observed received signal strength (i.e., RSSI values change across WiFi channels), and therefore in the localization process. Third, WiFi RSSI values exhibit high variation over time that, as we show in this work, can adversely impact localization accuracy. Fourth, the area of coverage of a WiFi access point is significantly reduced in indoor environments due to the presence of walls and metallic objects, easily creating blind spots (i.e., basement, parking lots, corner rooms in a building, etc.).

To address these limitations, we study the feasibility of leveraging alternative wireless signals to *augment* or even *replace* WiFi

Signal	Frequency	Range	RX Power
WiFi	2.4 GHz, 5GHz	30 m indoor	800 mW
FM	88-108 MHz	300 km outdoor	40 mW

Table 1: Basic properties of WiFi and FM broadcast signals.

signals for fingerprinting. In particular, we propose to use FM broadcast radio signals for fingerprinting indoor environments. FM signals operate at the frequency range of 88-108MHz in the US, which makes them less susceptible to the presence and orientation of humans and small objects. Furthermore, FM signals are significantly stronger than WiFi signals in the sense that they can easily cover areas of hundreds of kilometers, while achieving good indoors penetration (Table 1). From the infrastructure point of view, there are thousands of commercial and amateur FM signals being broadcasted continuously across the world, eliminating the need for deploying any custom infrastructure. Also, most mobile devices, even the lower-end ones, are equipped with FM radio receivers that are lower power and less costly compared to the WiFi receivers (Table 1).

However, in the case of FM radio signals, the access points (FM towers) are located up to several hundred of kilometers far away from the user and transmit signals at a very high power. As a result, the recorded FM RSSI signatures might not exhibit significant variation across nearby locations, and therefore fine grain localization might not be feasible. Previous work that has already examined the use of FM radio signals for localization in *outdoor* environments has verified this intuition by demonstrating coarse-grained localization accuracies (e.g., zip code level [10] or tens of meters [8]).

In this work, we demonstrate through a detailed experimental study that FM broadcast radio signals can be used to achieve room-level indoor localization with similar or better accuracy to the one achieved by WiFi signals. Even though FM radio reception may not vary significantly across nearby *outdoor* locations, in the case of *indoor* environments the internal structure of the building can significantly affect the propagation of FM radio signals, providing enough resolution in the FM signal signatures to accurately localize mobile devices.

This paper makes the following contributions:

- We demonstrate through detailed experiments in 3 representative buildings across the US (a residential building, an office building, and a shopping mall) that FM radio signals can achieve similar room-level accuracy in indoor environments when compared to WiFi signals.
- We propose to exploit additional information at the physical layer, such as multipath or frequency offset information, to create more reliable fingerprinting of indoor spaces, and demonstrate through real world experiments, that this approach can improve the accuracy of FM-radio based indoor localization by more than 5% when compared to the accuracy achieved by FM or WiFi RSSI-only signatures.
- We study in detail the effect of wireless signal temporal variation and demonstrate that WiFi RSSI values exhibit significantly higher variation over time compared to FM RSSI values. This enables FM-based indoor localization to achieve approximately 57% higher room-level localization accuracy when considering temporal variations of wireless signals.
- We experimentally demonstrate that FM and WiFi signals are complementary in the sense that their localization errors are independent. Our experimental results indicate that when FM and WiFi signals are combined to generate fingerprints, the localization accuracy increases by 11% (without accounting for temporal variation) or up to 83% (when accounting for wireless signal temporal variation) compared to when WiFi RSSI only is used as a signature.

2. ARCHITECTURE OVERVIEW

Figure 1 provides an overview of the proposed indoor localization approach. As in most fingerprinting approaches, there is a training and a positioning stage. The training stage is responsible for collecting location-annotated wireless signal fingerprints that form the fingerprint database. The fingerprint database can be automatically crowdsourced from real mobile users as they check-in to different businesses or it can be manually created through detailed profiling. Every time a business check-in takes place, the wireless fingerprint is recorded on the mobile device and the business location is retrieved from freely available web services. The recorded wireless fingerprint is properly annotated with the business' location information and stored in the database. At the positioning stage, the mobile device records its wireless signal fingerprint and compares it against the available fingerprints in the database. The location associated to the fingerprint in the database that is the closest to the fingerprint recorded on the mobile device, in terms of a distance metric, such as euclidean distance, is assumed to be the current location of the device.

The most challenging task in fingerprint-based localization is the engineering of the fingerprint itself. To enable accurate localization, fingerprints need to be carefully engineered so that even nearby locations have sufficiently different fingerprints. Most previous approaches have adopted the received signal strength (RSSI) of nearby WiFi access points as the wireless signal fingerprint. In this work we extend this approach in two fundamental ways. First, we augment the wireless fingerprint to include the RSSI information obtained by FM radio signals. As Figure 1(b) shows, mobile devices record RSSI information from several FM radio station signals that are broadcasted from one or more radio towers. The RSSI value for each FM frequency can be used along with the WiFi RSSI values to form the wireless fingerprint. By combining WiFi RSSI with RSSI values from another wireless signal that is less susceptible to human presence and orientation, small objects, and multipath and fading due to its lower wavelength, we manage to encode a more robust profile of the location into the wireless signal fingerprint that, as it will be shown later, can lead to better localization accuracy.

Second, to enable unique fingerprints even for nearby locations, we propose to extract more detailed information, that goes beyond RSSI, at the physical layer (Figure 1(b)). Even though RSSI has been proven to be a good high-level signal indicator, it does not provide the necessary granularity to enable robust fine grain localization. For instance, RSSI values at different rooms inside a building might be identical due to different reasons such as human presence and multipath. However, lower level information at the physical layer, such as signal-to-noise ratio (SNR), and multipath indicators, can provide enough insight on *how* an RSSI value was generated. For instance, the way wireless signals are reflected in a room is unique and it depends on the room's setup and location in the building. As a result, multipath indicators might be different across rooms even though the RSSI values for these rooms might be identical. By augmenting wireless signal fingerprints with additional low level signal indicators, we enable fingerprints to capture more robust information about the wireless signal transmission *and the way it is affected from the current room's structure*. This infor-

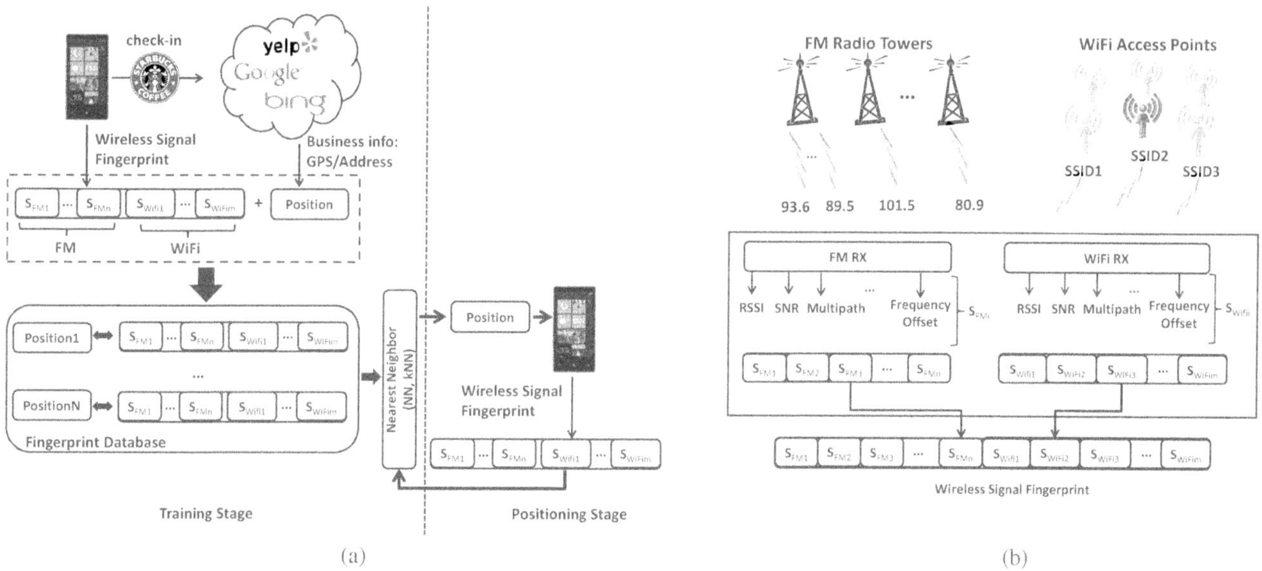

Figure 1: (a) Training and positioning stages for indoor localization (b) Signal fingerprinting using FM and WiFi radios.

mation could eventually be used to differentiate two rooms with the same RSSI values.

Note that this information could be leveraged for any wireless signal (e.g., WiFi) as long as it is exposed through the software driver. In fact, most recently Sen et al. [18, 19] exploit 802.11n PHY layer (OFDM) impulse responses and report notable localization accuracy gains. In this work, we focus on FM radio signals and therefore leverage the additional signal indicators for FM radio signals only. Nevertheless, we believe that in addition to WiFi and FM, getting additional signal indicators from the PHY layer can improve the localization accuracy of other wireless signals too.

3. EXPERIMENTAL SETUP

To evaluate the proposed approach we conducted detailed experiments in three typical building environments (Figure 2): an office building in a major corporate campus, a mall consisting of various restaurants and retail shops, and an apartment in a typical US residential building. The office and mall buildings are part of a major corporate campus located in the west coast of the US. Both buildings have a steel skeleton, and their perimeter is covered by large windows. The office building consisted of 3 different floors, with each floor containing approximately 40 rooms of $9ft \times 9ft$ size each. The mall building consisted of a single floor with a total number of 13 large rooms (i.e. $100ft \times 30ft$) of varying size and shape (Figure 2(b)). The residential building is located in a major city on the east coast of the US, and is built using steel-reinforced concrete. The particular apartment profiled in this study consisted of 5 different rooms as shown in Figure 2(c).

All buildings had exceptional WiFi and FM signal coverage. During data collection we recorded 434, 379, and 117 unique WiFi access points at the office, shopping mall, and residential buildings respectively. In every room of all 3 buildings, the FM receiver was able to tune to more than 32 FM radio stations.

3.1 Hardware

FM radio signatures were collected using the SI-4735 FM radio receiver from Silicon Labs [20] (Figure 3). The particular receiver was chosen for two reasons. First, Silicon Labs' FM radio receivers are very popular among a wide variety of consumer products such as cars, cell phones, portable media players and more. This enabled us to experiment with one of the most widely used FM receivers in the market today. Second, the particular receiver is among the few ones that expose low-level reception signal information to the application layer. In particular, besides reporting RSSI values, it provides 3 additional indicators[1] of signal reception: signal-to-noise ratio (SNR), multipath (MULTIPATH), and frequency offset (FREQOFF). SNR takes values between 0 and $128db$, and indicates how strong the received signal is compared to noise (i.e., interference, signal reflections, etc.). MULTIPATH takes values between 0 and 100, and indicates the severity of the multipath effect (i.e. the number and power of wireless signal reflections) in the current signal reception. FREQOFF mostly takes values between -10 and 10, and quantifies the difference in offset between the actual received signal and its different reflections. In general, the higher the value of FREQOFF the higher the effect of multipath and fading on the received signal. The combination of RSSI, SNR, MULTIPATH, and FREQOFF indicators represents the FM signature collected at each location during data collection (Figure 1(b)).

For embedded designs, such as cell phones, where space and size is important, the FM receiver can be connected to a patch antenna that can be directly implemented on the main PCB of the device. This would enable the device to acquire reliable signal readings, eliminating the unpredictability of loose earphone wires that are currently used as the FM antenna on most cell phones. Unfortunately, we were not able to find any patch antennas that meet the design criteria of Silicon Labs for the current FM receiver. As a result, we opted to use the typical FM antenna provided in the evaluation board. To approximate the behavior of a patch antenna, we folded the antenna as much as possible as shown in Figure 3. In our measurement setup the length of the antenna has twice the length of a typical smartphone.

WiFi signatures were collected using an 802.11a/b/g/n compatible WiFi Link 5300 card from Intel. All WiFi signatures consisted of RSSI values only. Both WiFi and FM receivers were connected to a Lenovo T61p laptop that simultaneously recorded WiFi and FM wireless signal fingerprints for a given location (Figure 3).

[1]The exact algorithm used to compute these indicators is sensitive, and is not provided by the manufacturer.

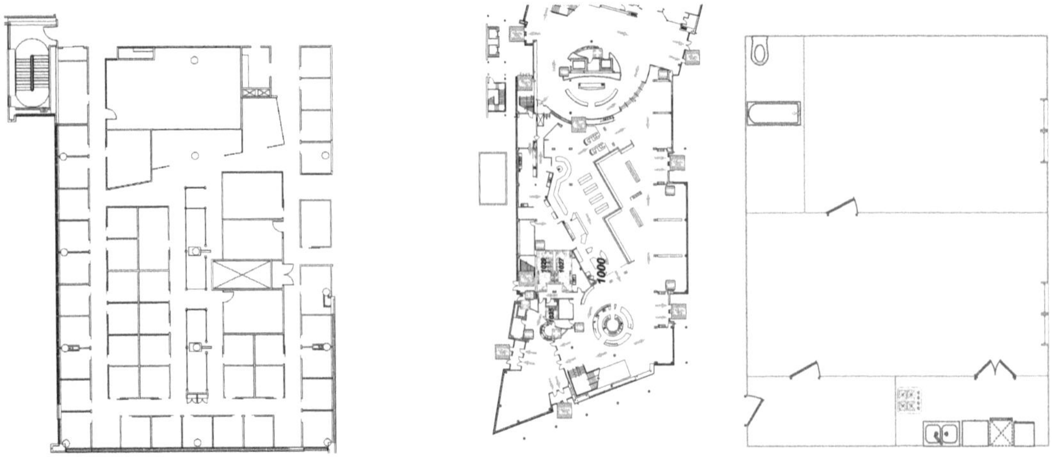

(a) Office (1 out of 3 floors) (b) Mall (c) Apartment

Figure 2: Maps of the different buildings used in our experimental study.

3.2 Data Collection and Evaluation

The goal of our experiments was to achieve accurate room-level localization[2]. In other words, given a wireless signal fingerprint from a mobile device, provide the room number where the device is located ("Position" in Figure 1 is actually a room number). To achieve this, we profiled each room in the building using both WiFi and FM signatures as described in Figure 1(b). For every room we collect the WiFi and FM signatures for 3 random points inside the room. We chose to only profile a small number of points within the room so that we can evaluate the ability of FM and WiFi signals to achieve high localization accuracy with sparse profiling, i.e., a small fingerprint database. We explore the benefit of having larger databases in Section 5.2. Depending on the experiment, multiple data collections were performed for each room over different time windows.

Specifically, at each location we record the FM signatures for 32[3] FM broadcast stations and scan the signal strengths of all available Wi-Fi access points, as described in Section 3.1. We denote $r_i, s_i, m_i, f_i \in \mathbf{R}^{32}$ as the RSSI, SNR, MULTIPATH, and FREQOFF values for the 32 FM broadcast stations at the i-th location, and similarly $w_i \in \mathbf{R}^M$ as the RSSI values for the Wi-Fi access points. Here M is the total number of Wi-Fi access points in the building, and i is the profiled location. We concatenate the signature vectors of the i-th location and denote as $a_i \in R^{128+M}$. In total, for every profiled location i within a room there are $128 + M$ values corresponding to 4 values for every one of the 32 FM channels and M values, one for every WiFi access point. As a result, the whole dataset for a given building can be written as $A = \{a_i : i = 1, 2, \ldots, 3 \times R\}$, where R is the number of rooms.

A typical fingerprinting-based localization scheme involves an offline phase to construct the database and an online phase for positioning unknown locations. In order to evaluate the localization performance, we emulate this two-phase process by separating the signatures into training and test sets. We first partition the whole dataset into three complimentary subsets of equal size: A_1, A_2 and A_3, where each set A_j contains one and only one location from each room. Next, we can group two subsets together as the training set, i.e., the fingerprint database, and use the third subset as the test set. We repeat this process using each of the three subsets as the test set and correspondingly the other two subsets as the fingerprint database, and report the average localization accuracy across all combinations. For each test location in the test set, we compare its signature vectors against the fingerprint database and return the location of the nearest neighbor in signal space as the localization result.

In all experiments, we report the localization accuracy when Euclidean and Manhattan distance metrics are used to compute the distance between wireless signal signatures. Even though more distance metrics have been evaluated, we opted to show only these two, as they consistently provided the highest localization accuracy across all experiments.

[2]We focus on room-level resolution as this is the resolution at which data can be crowdsourced in a robust way from business check-in events. In Section 7, we evaluate FM's capability to achieve fine-grain indoor localization.

[3]The seek/tune time for the SI4735 FM receiver is 60 ms per channel, and the power-up time is 110 ms. It takes approximately 2 seconds to scan 32 FM radio stations.

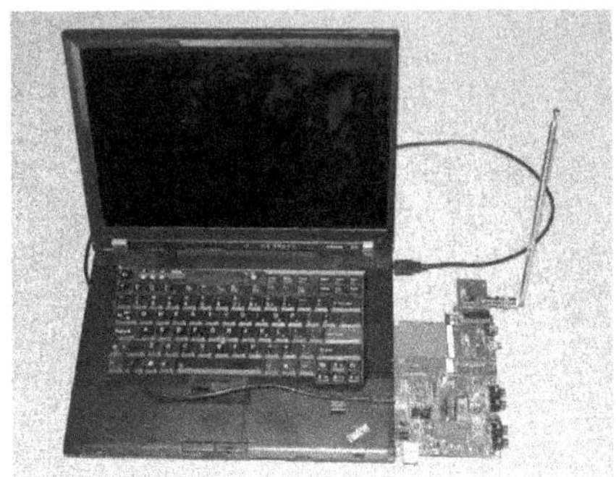

Figure 3: Data collection setup based on the SI-4735 FM radio receiver from Silicon Labs and the Intel WiFi Link 5300 wireless card connected to a Lenovo T61p laptop.

4. FM-BASED INDOOR LOCALIZATION

In this section, we focus on the office building environment consisting of 3 different floors and 119 rooms in total. We first investi-

Signature Type	Distance Metric	
	Euclidean	Manhattan
FM RSSI	85%	87%
Wi-Fi RSSI	76%	88%

Table 2: Room level localization accuracy for 119 rooms on 3 floors using FM and WiFi RSSI values as signatures.

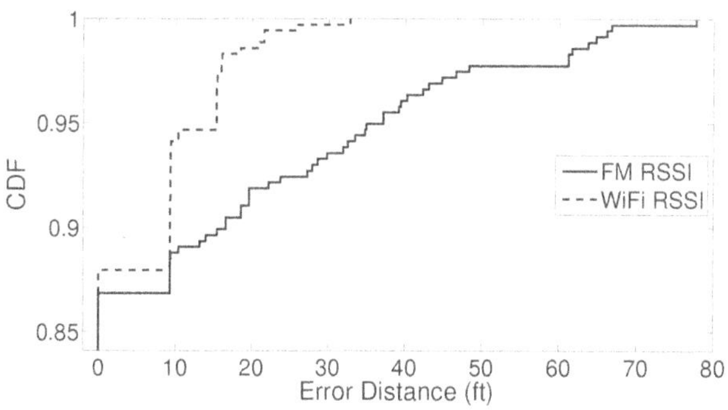

Figure 4: Distribution of the localization errors for FM and Wi-Fi RSSI. The nearest neighbors in signal space are determined using the Manhattan distance between the RSSI vectors.

gate the performance of using RSSI values alone for both FM (i.e., r_i) and Wi-Fi signals (i.e., w_i). Next, we use more signal quality indicators for FM (i.e., s_i, m_i, f_i) to see whether extracting more information from the physical layer can improve the localization accuracy. Last, we combine the FM and Wi-Fi vectors to investigate the effect on the localization accuracy and perform sensitivity analysis on the number of FM radio stations and WiFi access points used for fingerprinting.

4.1 RSSI-based Indoor Localization

Table 2 lists the room level localization accuracy results when signature vectors consist of FM or WiFi RSSI values only. It is clear that FM and WiFi RSSI values achieve similarly high accuracies that are close to 90%. Of the two distance metrics, Manhattan distance (i.e., the L_1 norm) yields slightly higher accuracy than Euclidean distance (i.e., the L_2 norm).

Figure 4 shows the distribution of the localization errors in terms of physical distance when using FM and WiFi RSSI signatures. Although both signals exhibit similar room level accuracies (Table 2), the localization errors are lower in the case of WiFi. In other words, when WiFi localization erroneously predicts rooms, those rooms are closer to ground truth compared to FM-based localization. This is expected given that there are orders of magnitude difference between WiFi and FM signals in terms of both deployment density and communication range. In general, a WiFi access point is only visible in a subset of the rooms in the building. This significantly limits the search space, and therefore the localization error that is generally lower than $30ft$. Conversely, in the case of FM signals, there are only a handful of radio towers at a given region that might be tens or even hundreds of kilometers away from the building. These FM signals can be received throughout the whole building, making every room in the building a possible candidate location.

This effect is better illustrated in Figures 5(a) and 5(b), where the Manhattan distance between every pair of profiled locations is shown when FM RSSI and WiFi RSSI signatures are used respectively. In the case of WiFi signatures (Figure 5(b)), errors are usually constrained within the vicinity of the diagonal (3 squares along the diagonal, where each square corresponds to one of the 3 floors profiled) mainly due to the communication range of access points. Very rarely the distance between WiFi RSSI vectors is low for distant locations in the building. On the other hand, the error profile of FM RSSI signature is the exact opposite (Figure 5(a)). The effect of the three squares shown in Figure 5(b) has disappeared, but now distant locations in the building can generate low distance values. As a result, even though FM and WiFi achieve similar localization accuracy overall, the localization error of FM signals is higher in terms of absolute physical distance.

Signature Type	Distance Metric	
	Euclidean	Manhattan
FM RSSI	85%	87%
FM SNR	80%	84%
FM Multipath	68%	76%
FM Frequency Offset	53%	53%
FM All	81%	91%
FM All Normalized	90%	93%

Table 3: Room level localization accuracy for 119 rooms on 3 floors using additional FM signal signatures. Combining multiple signal indicators in a single signature provides more accurate localization.

4.2 Robust Fingerprinting by Exploiting the Physical Layer

To further increase localization accuracy and to constraint errors, in this section, we leverage additional information at the physical layer to generate more robust signatures. The SI4735 FM receiver provides three additional signal quality indicators (SNR, MULTIPATH, and FREQOFF) as described in Section 3. Each of these signal indicators could be used as an individual signature, or they could all be combined with RSSI to form a single more detailed signature. The additional signal indicators can enhance the resolution of FM signatures by providing more insight about signal reception. Signal-to-noise ratio, multipath, and frequency offset, all capture detailed information about the wireless signal reception and *the way it is affected from the current room's structure and position in the building*.

When combining multiple signal indicators into the same signature, calculating the distance between signatures becomes more challenging. Different signal indicators have different value ranges that could result into biasing the distance calculation (i.e. higher value/range indicators become more important). For example, the multipath value range is between 0 and 100 whereas the frequency offset value is usually in the range of -10 to 10. Therefore, we normalize the value of each signal indicator using the standard deviation of the signal indicator's values in the fingerprint database. For example, we can compute the standard deviation for the RSSI signatures as

$$\delta_r = \left[\frac{1}{N|D| - 1} \sum_{i \in D} \sum_{j=1}^{N} (r_{ij} - \overline{r}) \right]^{\frac{1}{2}} \quad (1)$$

where D represents the set of location indices that are in the fingerprint database, and r_{ij} is the RSSI value of the j-th FM broadcast station at the i-th location. $\overline{r}$ is the average RSSI value in the database, $|D|$ is the cardinality of the set D, and $N = 32$ is the number of FM broadcast stations. Using the standard deviation, we

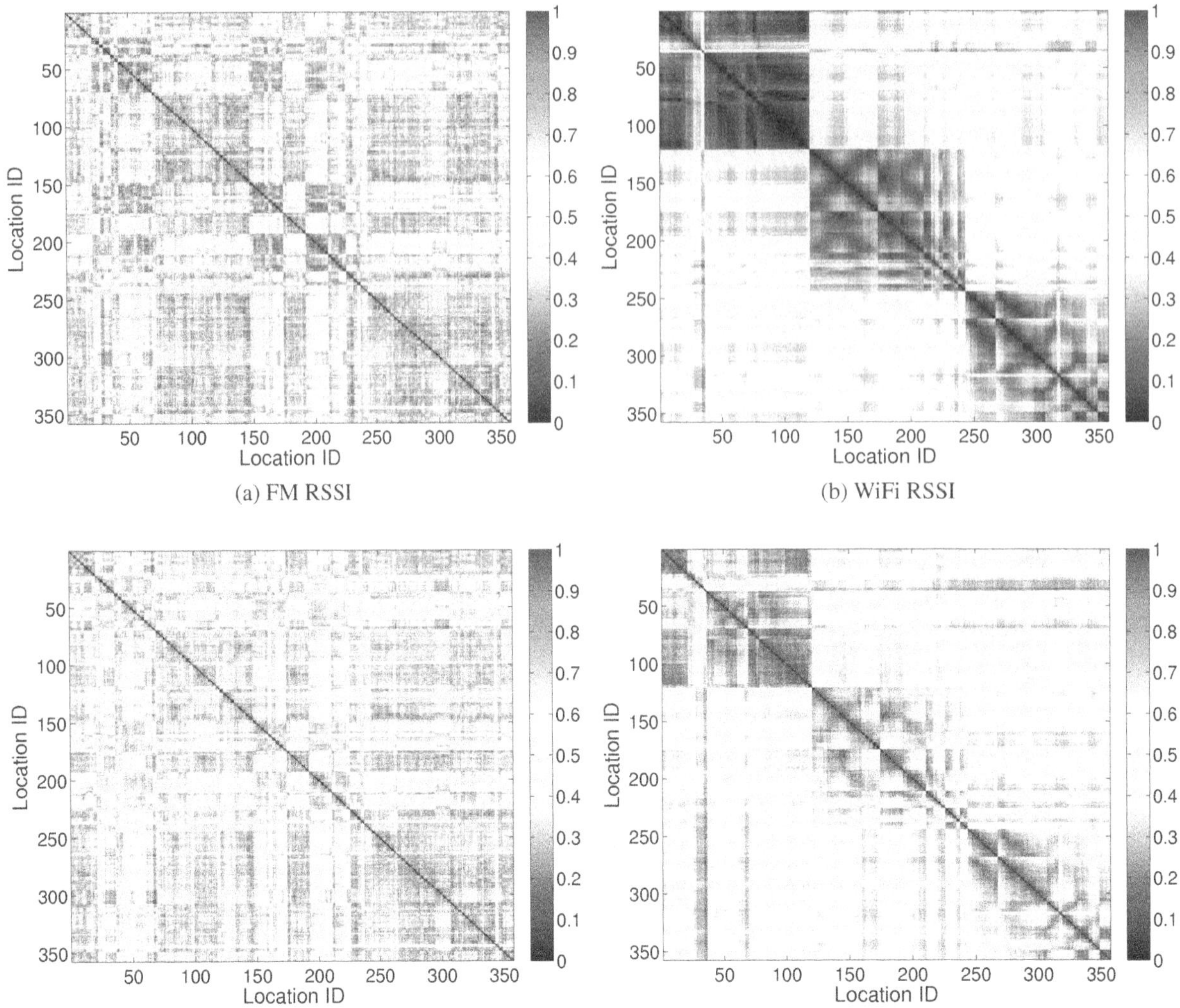

(a) FM RSSI
(b) WiFi RSSI
(c) FM ALL Normalized
(d) FM & WiFi ALL Normalized

Figure 5: The Manhattan distance of signature vectors between all pairs of profiled locations in the office building and for 4 different signature types. The distances are normalized by the maximum pairwise distance in each figure to have the same range of $[0, 1]$ across all figures. Diagonal values are 0 as they are the distances between identical vectors that correspond to the same locations.

can normalize the RSSI signatures as

$$\hat{r}_i = \frac{r_i}{\delta_r}, \quad \forall i \in D \cup T, \tag{2}$$

where T represents the set of location indices that are in the test set. All signal indicators' values are normalized in the same way, enabling us to compute un-biased distance values between signatures.

Table 3 lists the room level localization accuracy when each signal indicator is used as a single signature, and when all signal indicators are combined together into a single signature. The "FM All" signature corresponds to combining all *raw* signal indicator values into a single signature. The "FM All Normalized" signature corresponds into combining all *normalized* signal indicator values into a single signature.

It is clear that among all individual signal indicators, RSSI achieves the best accuracy. On the other hand, multipath and frequency offset indicators seem to not be able to provide the necessary resolution to achieve accurate localization on their own. However, com-

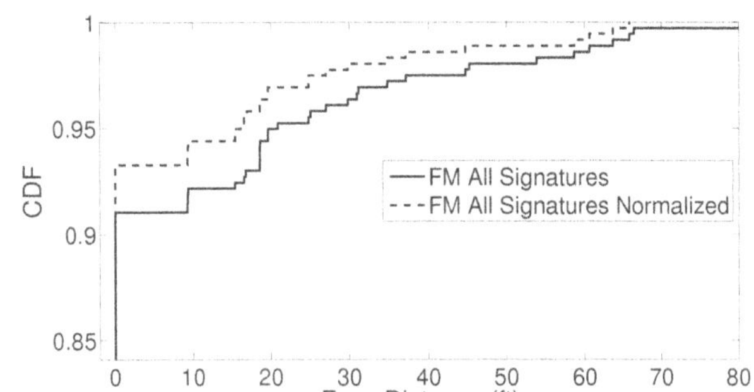

Figure 6: Distribution of the localization errors when combining all available signal indicators from the FM signals into a signature. The nearest neighbors in signal space are determined using the Manhattan distance between the signature vectors.

Signature Type	Distance Metric	
	Euclidean	Manhattan
FM All	81%	91%
FM All Normalized	90%	93%
Wi-Fi RSSI	76%	88%
FM & Wi-Fi All	93%	98%
FM & Wi-Fi All Normalized	94%	98%

(a) Room level localization accuracy

True	Floor #	2	2	2	2	2	3	3
	Room #	3	12	16	19	35	21	23
Predicted	Floor #	2	2	2	2	2	3	3
	Room #	5	11	17	20	36	19	24

(b) Error location list for FM & Wi-Fi All Normalized

Table 4: Room level localization accuracy for 119 rooms on 3 floors. Table (a) shows the localization accuracy achieved across different signature types. Table (b) shows a complete list for the room and floor numbers of the incorrectly identified locations when using the normalized FM and WiFi signatures with the Manhattan distance metric.

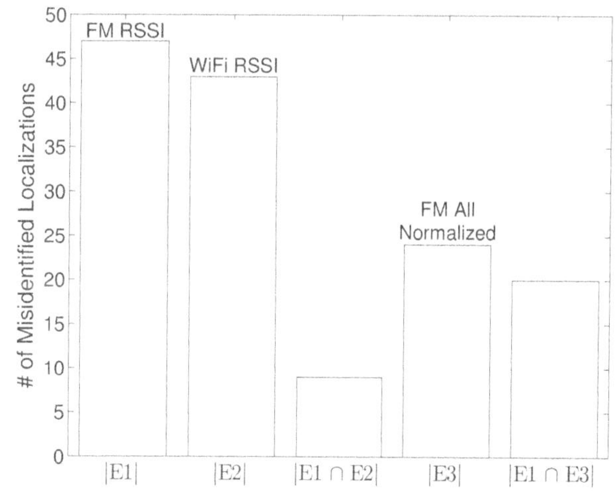

Figure 7: E1 and E2 represent the set of locations that are misidentified by FM RSSI only and WiFi RSSI only respectively. E1 $\cap$ E2 are the set of locations that misidentified by both FM and WiFi RSSI signatures. |E1 $\cap$ E2| $\ll$ |E2|, suggesting that FM and WiFi positioning errors are not correlated. As a comparison, E3 denotes the set of misidentified locations using the FM All normalized signature.

bining all signal indicators into a single signature achieves higher accuracy than any individual signal indicator. This highlights the benefit of extracting more information from the physical layer and reflects the intuition that each type of signal indicator can capture a unique set of interplays between the propagating radio wave and its surrounding environment.

The impact of the additional signal indicators on the localization accuracy of FM signature is better illustrated in Figures 5(a) and 5(c), where the Manhattan distance between every pair of profiled locations is shown when the FM signature ignores or takes into account the additional signal indicators. By comparing Figures 5(a) and 5(c), it is obvious that when all signal indicators are leveraged in the FM signature, the distance matrix appears to be significantly less noisy, in the sense that the distances between non-neighboring locations in the matrix are significantly higher compared to the FM RSSI matrix. As a result, higher localization accuracy is achieved when all FM signal indicators are combined into a single signature.

Furthermore, Table 3 shows that normalizing the signatures, as described above, can further improve localization accuracy. When compared to Table 2, FM-based localization achieves 5.7% higher localization accuracy compared to the WiFI RSSI signatures.

Normalization not only improves accuracy, but also constraints the error when wrong predictions are made. Figure 6 shows the distribution of the localization errors for FM-based fingerprinting. Normalization increases the percentage of correctly identified locations, and also reduces the errors for the incorrectly identified locations.

4.3 Combining FM and Wi-Fi

In this section we investigate whether the FM and WiFi signal indicators could be combined into a single signature to further improve indoor localization accuracy.

Table 4(a) lists the room level localization accuracy when WiFi and all FM signal indicators are combined into a single signature. The combination of WiFi and FM signals can eliminate almost all localization errors, achieving 98% accuracy; an 11.3% increase compared to WiFi RSSI fingerprinting (Table 2). This suggests that the localization errors generated by the FM signatures are not correlated with the errors generated by WiFi signatures. To further investigate the correlations, we collect and count the number of locations misidentified by each signature, as shown in Figure 7. E1 and E2 are the set of locations misidentified by FM RSSI and WiFi RSSI respectively. The fact that |E1 $\cap$ E2| $\ll$ |E2| indicates the set of locations misidentified by FM RSSI rarely overlap with those by WiFi RSSI. As a comparison, we also collect the locations misidentified by all normalized signatures of FM and denote as set E3. One can see that |E1 $\cap$ E3| $\approx$ |E3|, and therefore the locations misidentified by FM all normalized signatures highly overlap with those misidentified by FM RSSI. Overall, one can see that FM localization errors are not correlated with the WiFi errors. On the other hand, using more FM signatures removes many of the localization errors by FM RSSI.

The complementary nature of FM and WiFi signals is clearly illustrated in Figure 5. Initially, FM RSSI signatures (Figure 5(a)) provide high localization accuracy, which further increases when the additional signal indicators are leveraged (cleaner distance matrix in Figure 5(c)). However, errors are still distributed throughout the building, in the sense that signature distances even for distant locations in the building are low. On the other hand, localization errors in WiFi RSSI signatures are mostly constrained to only nearby locations as demonstrated by the 3 dark squares in Figure 5(b). When FM and WiFi signal indicators are combined into a single signature (Figure 5(d)), the benefits of both FM and WiFi signals show up in the resulting distance matrix. FM signatures significantly reduce the dark square effect which is the main source of errors in the case of WiFi signals (Figure 5(b)). At the same time, WiFi signals reduce the number of cases where distant locations have low distance values, the major source for errors in the case of FM signals. As a result, a mobile device can leverage its ability to receive both signals to effectively enhance localization accuracy with marginal overhead.

Table 4(b) lists the floor and room numbers of the locations that are identified incorrectly when combining the normalized FM and WiFi signatures (i.e., the last row of Table 4(a)) with the Manhattan distance. A total of seven test locations (2% of the 357 locations)

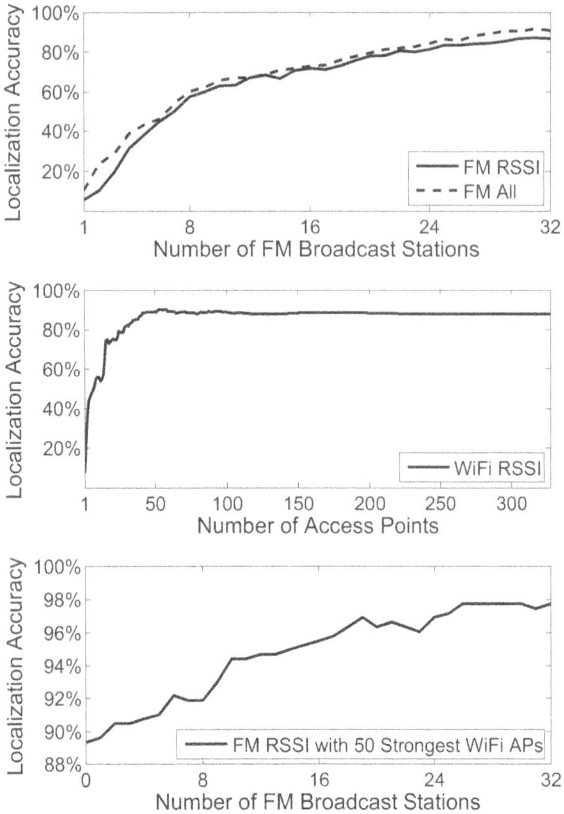

Figure 8: The sensitivity of localization accuracy on the number of FM broadcast stations and WiFi access points. Stations and access points are added in descending order of their average signal strength. In the last graph we employ the 50 strongest WiFi access points and only vary the number of FM radio stations.

are misidentified. However, all the erroneously predicted rooms are on the same floor and nearby the true rooms.

4.4 Sensitivity Analysis on the Number of FM Stations and WiFi Access Points

So far we have been using the signatures from all 32 FM broadcast stations and 434 visible WiFi access points in the office building for indoor localization. However, it is unclear how many FM radio stations and WiFi access points are actually needed to provide accurate localization. To answer this question, we perform a sensitivity analysis where we withhold a certain percentage of FM radio stations and WiFi access points, and rerun the nearest neighbor based localization algorithms. Specifically, we sort the FM stations and WiFi access points in descending order of their RSSI values averaged over all locations. At each step, we incrementally add one station/access point at a time, and rerun the localization algorithm. As a result, when we evaluate the localization performance with n stations, we use the n strongest stations in terms of their average RSSI values.

Figure 8 shows the localization accuracy achieved when different number of FM radio and WiFi access points are used. It is clear that for both signals, additional infrastructure leads to higher accuracy. To achieve the maximum localization accuracy (i.e., accuracy when all radio stations or access points are used), 30 FM radio stations and approximately 50 WiFi access points are required. In the case of FM, the accuracy increases very fast till around 8 stations, but it does not saturate afterwards. Instead it keeps rising, but at a lower pace, indicating that a large number of radio stations is required to achieve high localization accuracy. In the case of WiFi, given the high density of access points and their limited communication range, the number of useful access points saturates relatively fast at around 50 access points.

When we start by using the 50 strongest WiFi access points as the base line, the localization accuracy increases as we are incrementally adding FM stations, and seems to saturate at the point where 25 radio stations are used (bottom graph of Figure 8).

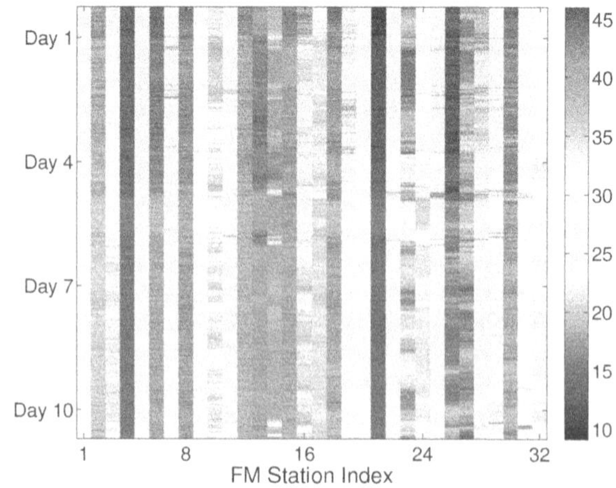

Figure 9: Raw RSSI values in dBμV of the 32 FM broadcast stations over the course of 10 days. Results for the rest 3 FM signal indicators are not shown in the interest of space.

5. TEMPORAL VARIATIONS

The results in the previous section are derived without considering the temporal variations of FM and WiFi signals. However, it is known that signal signatures are likely to change overtime. For example, Haeberlen et al. [9] achieve 95% room level localization accuracy using WiFi RSSI signatures when the test and training data are collected in close time proximity. With data from different time and day, however, the localization accuracy drops to 70%, as pointed out in [22].

In this section we explore the temporal variations of the broadcasted FM signals and the impact on localization accuracy. First, we continuously monitor the FM signals for ten days at a fixed location in one room to gain intuition on how signatures vary over time. To quantify the impact of temporal variations on localization accuracy, we collect fingerprints for the 40 rooms on the 2nd floor on different days and run the localization algorithm against fingerprint databases that were recorded at different points in time.

5.1 Continuous Monitoring of FM Signals Over Ten Days

Figure 9 shows the raw RSSI values of the 32 FM stations at a fixed location in one room over the course of ten days. This room is a regular office and therefore its door and furniture could change states due to the presence of humans. The room is at the perimeter of the office building and has a window that faces a busy street. We note that there were rainy, cloudy, and also sunny days during the experiment. We configured the receiver to record all signal in-

Signature Type	Distance Metric	
	Euclidean	Manhattan
FM RSSI	100%	100%
FM SNR	100%	100%
FM Multipath	100%	100%
FM Frequency Offset	99%	77%
FM All	100%	100%
FM All Normalized	100%	100%

Table 5: Room level localization accuracy using the data collected at a fixed location over the course of ten days as the test set, against the fingerprint database collected in Section 4. The first three measurements taken at the fixed location were inserted into the fingerprint database.

dicators for the 32 FM broadcast stations once per three minutes. Therefore, Figure 9 includes more than 4000 rows of data.

Figure 9 shows that, overall, RSSI values at a given frequency do not change drastically over time. However, all FM stations seem to exhibit fluctuations in RSSI values, but to a different extent. For instance, FM station 27 seems to exhibit way larger fluctuations in RSSI values when compared to FM station 21. We believe that this is due to the fact that different radio stations are broadcasted from different radio towers and at different transmission power levels.

In order to quantify whether these fluctuations would impact this room's localization result, we use the first three rows of data in conjunction with the fingerprints collected in section 4 as the fingerprint database, and use the rest of the rows in Figure 9 as the test set. Table 5 lists the percentage of rows that are identified correctly. The localization accuracies are consistently high for all signature types except frequency offset. This suggests that as long as a location has been profiled before, the temporal variations should not cause this location to be mislabeled in the future.

5.2 Collecting Fingerprints on Different Days

In this section, we extend the temporal variation analysis to multiple locations. Specifically, we collect four additional sets of fingerprint measurements for the 40 rooms on the second floor in exactly the same way as before, but on different days. We chose to study the second floor because this floor exhibits most localization errors as shown in Table 4(b).

We first study the pairwise localization performance between two datasets, where one dataset is chosen as the test set and a different dataset is chosen as the fingerprint database. Note that this way the size of the test set is the same as that of the fingerprint database. We run the nearest neighbor localization algorithm on all combinations of pairs of datasets (i.e., 20 pairs across 5 datasets) and present the average room level localization accuracy in Table 6(a). Compared to the results in Table 4, it is obvious that temporal variations of the signal signatures can lead to noticeable degradation of localization accuracy. WiFi RSSI signatures seem to be affected the most by temporal variations as the localization accuracy decreases by 44% (from 88% to 49%) in the presence of temporal variations. On the other hand, FM signatures seem to be less susceptible to temporal variations, as the localization accuracy decreases by only 13% (from 93% to 81%), and the achieved accuracy remains above 80%. In general, because of the differences in frequency and wavelengths between FM and WiFi signals, WiFi signals are more susceptible to human presence/orientation, and to the presence of even small objects in the room. However, both human presence and

Signature Type	Distance Metric	
	Euclidean	Manhattan
FM RSSI	78%	77%
FM All	79%	82%
FM All Normalized	75%	81%
Wi-Fi RSSI	43%	49%
FM & Wi-Fi All	85%	89%
FM & Wi-Fi All Normalized	80%	90%

(a) Pairwise: same size test and training sets

Signature Type	Distance Metric	
	Euclidean	Manhattan
FM RSSI	91%	92%
FM All	91%	92%
FM All Normalized	88%	91%
Wi-Fi RSSI	57%	61%
FM & Wi-Fi All	94%	96%
FM & Wi-Fi All Normalized	91%	95%

(b) One versus many: four datasets in the fingerprint database

Table 6: Room level localization accuracy for 40 rooms on the second floor. Table (a) lists the localization accuracy where one dataset is chosen as the test set and a different dataset as the fingerprint database. Table (b) includes four datasets in the fingerprint database.

setup of objects in a room changes over time, significantly lowering the localization accuracy of WiFi RSSI-based fingerprinting.

The signature that is affected the least by temporal variations is the one that combines WiFi RSSI and the 4 FM signal indicators. In particular, localization accuracy decreases by only 8%, and absolute accuracy is 90%. Note that these observations are consistent with the observations in Section 4.3, indicating that WiFi and FM errors are uncorrelated.

As more data is crowdourced or manually collected over time, the quality of the fingerprint database improves. The rationale is that more datasets that are collected across different days can potentially capture more patterns of the signal signatures in the temporal domain. Table 6(b) quantifies the impact of the size of the fingerprint database on the localization accuracy. It lists the results of using four datasets as the fingerprint database and one dataset as the test set. It loops through the five different combinations and reports the average numbers. Clearly, adding more datasets into the database can lead to notable gains in the localization accuracy, indicating that a bigger fingerprint database can better cope with temporal variations.

We note that during the course of our experiments, most of the changes in the environment were the movements of people, chairs, doors, and other smaller objects. Our experimental results indicated that both WiFi and FM signals are susceptible to these changes, however, FM signals are affected significantly less compared to WiFi. More dramatic changes such as moving big metal shelves should affect the signatures to a larger extent for both WiFi and FM signals, but have not been studied in this work.

6. DIFFERENT TYPES OF BUILDINGS

In this section, we investigate whether the results obtained in office environments can apply to other types of buildings and geographic regions in the US.

Signature Type	Distance Metric	
	Euclidean	Manhattan
FM RSSI	77%	80%
FM All	65%	72%
FM All Normalized	67%	71%
Wi-Fi RSSI	94%	97%
FM & Wi-Fi All	94%	98%
FM & Wi-Fi All Normalized	96%	98%

(a) Pairwise: same size test and training sets

Signature Type	Distance Metric	
	Euclidean	Manhattan
FM RSSI	87%	90%
FM All	79%	82%
FM All Normalized	92%	87%
Wi-Fi RSSI	97%	100%
FM & Wi-Fi All	97%	100%
FM & Wi-Fi All Normalized	97%	100%

(b) One versus many: four datasets in the fingerprint database

Table 7: Table (a) lists the localization accuracy where one dataset is chosen as the test set and a different dataset as the fingerprint database. Table (b) includes four datasets in the fingerprint database.

6.1 Shopping Mall

Shopping malls are different from office buildings in many aspects. The ceilings are taller and the rooms are sparser and bigger, which makes malls resemble outdoor environments more than office buildings do. Given that FM-based indoor localization depends on the internal structure of the building to achieve high localization accuracy, it is unclear whether FM signatures can be used in this type of environments.

We opted to collect fingerprint measurements at the first floor of a two-story mall building that hosts various restaurants and retail shops (Figure 2(c)). We collected five fingerprint datasets on three different days. For each dataset, we take measurements in 13 rooms and at 3 random locations for each room. Therefore, every dataset includes a total of 39 locations. The first 4 datasets are collected during a weekend, and the 5th dataset is collected on a Wednesday afternoon. Table 7(a) shows the average pairwise localization accuracy across all possible combinations of database and test datasets. Interestingly, FM signatures perform slightly worse compared to the office building (cf. Table 6), but WiFi signatures perform significantly better. The degradation of FM signatures' accuracy can be attributed to the fact that mall buildings resemble more of outdoor environments as the impact of the internal structure of the building on signal propagation is lower compared to office buildings. On the other hand, WiFi signatures can more reliably distinguish the 13 rooms because of the large size and clear spatial separation of these rooms, to the extent that some of them are covered by completely different sets of access points. Nevertheless, combining WiFi and FM signatures still provides the highest positioning accuracy.

Table 7(b) shows the localization accuracy when using only one dataset as the test set and the remaining four datasets altogether as the database. Similar to the case of the office building environment, having more fingerprints in the database increases localization accuracy.

Signature Type	Localization Accuracy
FM RSSI	100%
WiFi RSSI	90%
FM & WiFi All	97%
FM & WiFi All Normalized	100%

Table 8: Localization accuracy for 5 rooms in a residential building when the Manhattan distance metric is used.

6.2 Residential Building

Residential buildings differ from office and mall buildings in size, shape, structure and often building materials. In this section, we study the performance of FM based indoor localization in an apartment unit that has multiple rooms of various sizes (Figure 2(b)). This apartment is located on the east coast of the US, and therefore the list of FM stations are completely different from the ones used in the office and mall buildings which are on the west coast. Nevertheless, we still chose 32 audible stations such that the number of stations remains consistent.

We collected two datasets on two different days. Each dataset gathered measurements in 5 rooms and at 3 random locations per room. We chose each of the two datasets in turn as the test set and fingerprint database to evaluate the positioning accuracy. Table 8 shows the average room level localization accuracy for FM and WiFi signatures. Overall, both FM and WiFi signatures demonstrate above 90% accuracy in this environment, with FM signatures achieving perfect room-level localization accuracy. These results are comparable to the results at the office and mall buildings, suggesting that: (1) the achieved localization accuracies are independent of the building type, and (2) the FM based indoor localization approach is applicable to other geographic regions with different FM broadcast infrastructure.

7. FINE-GRAIN LOCALIZATION AND DEVICE VARIATIONS

So far, we have been focusing on room-level localization as this is the resolution at which data can be crowdsourced reliably from current business check-in events. However, it is still feasible to collect more fine-grained, location annotated indoor fingerprints through detailed profiling. To study the feasibility of using FM radio signals to perform fine-grain location estimation, we performed experiments along the hallway on the second floor of the office building.

Specifically, we gathered FM and WiFi signatures at 100 locations that formed a straight line along the hallway, with the distance between every two adjacent locations being approximately one foot (± 0.06 ft). We collected a total of three datasets in different days to capture the temporal variation of signal signatures. The exact same locations were profiled in all three datasets. The coordinates of each location were measured accurately using a laser distance meter.

7.1 Leave One Out Evaluation

For each one of the three datasets, we perform the leave-one-out evaluation. In particular, we remove one and only one location at a time from the dataset and compare its signature against the other 99 signatures in the dataset. The position of the closest signature is assumed to be the position of the test location. In such an evaluation, a robust signature scheme should always return one of the two neighboring locations to the test location.

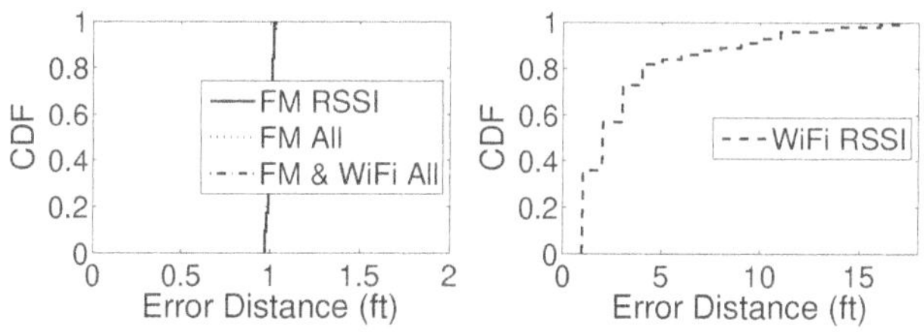

Figure 10: Distribution of the leave-one-out localization errors for FM and WiFi RSSI signals. The nearest neighbors in signal space are determined using the Manhattan distance between the RSSI vectors.

Datasets	Signature Type	Localization Error (ft)	
		50%	90%
1 vs 2	FM RSSI	0.02	2.02
1 vs 2	WiFi RSSI	4.97	15.00
3 vs 1&2	FM RSSI	0.01	1.00

Table 9: Localization errors using FM and WiFi RSSI signatures in the hallway using the Manhattan distance to determine the nearest neighbor in signal space. All datasets were taken at different days. Datasets 1 and 2 were taken using the same FM receiver, while dataset 3 leveraged a different FM receiver of the exact same type.

Figure 10 shows the distribution of the localization errors for FM and WiFi RSSI signatures across all 3 datasets. FM RSSI signatures provide highly accurate positioning with errors around 1 ft. Recall that the 100 locations formed a line with the distance between two neighboring positions set to one foot (± 0.06 ft). This indicates that each location is identified as one of its two neighbors on the line. On the other hand, WiFi RSSI signatures exhibit significantly larger errors, with the 90 percentile error approximating 10 ft.

Figure 11 visualizes the spatial resolution of the FM and WiFi RSSI signatures by plotting the Manhattan distances between all pairs of locations for one of the three datasets collected. The distances are linearly mapped to the [0,1] range for each signature such that the graphs are directly comparable. The diagonals are always zero because the vectors are identical for the same location. By comparing Figures 11(a) and 11(b), it becomes apparent that not only FM RSSI signatures have the necessary spatial resolution for accurate fingerprinting, but they provide significantly finer-grain resolution compared to WiFI RSSI signatures. In the case of FM signals, the RSSI distances are low only for the same or neighboring locations. In most other cases, the FM RSSI signatures differ significantly. Conversely, in the case of WiFi signals, RSSI distances are low even for locations that can be more than $15ft$ far away from each other, leading to high localization errors as shown in Figure 10.

7.2 Temporal and Device Variation

All 3 datasets were collected at different days, and the third dataset was collected using a different FM receiver chip of the same model and manufacturer. This allowed us to study how temporal and device variations affect the accuracy of fine-grain localization.

First, we examine temporal variations using datasets 1 and 2. We let one of them be the test set and the other be the fingerprint database, and then flip their roles to compute the average localization error. The 50th and 90th percentile errors are shown in Table 9. Both FM and WiFi RSSI signatures show higher localization errors compared to Figure 10, due to temporal variations. Nevertheless, FM still outperforms WiFi significantly.

To study the effect of device variations, we choose dataset 3 as the test set (fingerprint database) and one of the other two datasets in turn as the fingerprint database (test set), and compute the average CDF of localization errors. The results are summarized in the last row of Table 9. Note that the results are subject to both temporal and device variations since dataset 3 was collected in a different day too. Even under device variations, the localization error does not increase significantly as compared to temporal-only variations. In fact, the localization error seem to be slightly lower with device variations. These variations are most likely due to the random changes in the environment, indicating that device variations do not notably affect the localization performance.

8. RELATED WORK

Previous approaches to fingerprint-based indoor localization can be roughly classified into two categories: *infrastructure-based* and *infrastructure-less* approaches. *Infrastructure-based* approaches rely on the deployment of customized RF-beacons, such as RFID [13], infrared [25], ultrasound [17], Bluetooth [3], and short-range FM transmitters [11]. The advantage of these approaches lies on the fact that the deployed beacons can be carefully engineered/optimized for indoor localization, and of course they can be deployed at the necessary density to provide accurate indoor positioning. However, the high overhead of deploying custom hardware usually prohibits the feasibility of infrastructure-based approaches.

Infrastructure-less approaches, on the other hand, do not require any hardware to be deployed, as they leverage already available wireless signals to profile a location, usually in the form of RSSI values. The state-of-the-art approach to signal fingerprinting relies on WiFi signals as WiFi access points are widely deployed indoors, and every mobile device is equipped with a WiFi receiver. The early RADAR [1] indoor localization system demonstrated the effectiveness of WiFi fingerprinting by achieving localization accuracies in the range of 2 meters. More recent work [9, 27] reported higher accuracy by statistically modeling the signal strength as Gaussian distributions. WiFi signals, however, operate at the 2.4GHz or 5Ghz range that makes them particularly susceptible to multipath, fading, small objects and most importantly to human presence. In particular, human body and its orientation can drastically impact WiFi RSSI values. As a result, profiling of locations becomes extremely tedious as for every location signatures need to be recorded for different body orientations [1]. Also, in agreement with our findings, the temporal variations of WiFi RSSI values tend to be high due to the sensitivity of the signal to the presence of humans and small objects. Most recently, Sen et al. [18, 19] exploit 802.11n PHY layer (OFDM) impulse responses and report more robust localization performance. On the other hand, commercial WiFi-based localization systems (e.g., Skyhook [21]) focus more on coarse-grained localization (e.g., 70 meters in Skyhook [6]) to alleviate the need for war-driving inside buildings.

Varshavsky et al. developed GSM fingerprint-based indoor localization systems that can achieve slightly worse accuracy to that of WiFi based systems [14, 24]. The key concept of their system is to record fingerprints from not only the six surrounding GSM stations, but also farther away stations whose signal can still be heard by the mobile phones.

Recently, Tarzia et al. explored the possibility of using acoustic background spectrum for room-level localization [22]. Their localization system takes advantage of the observation that each room tends to have its own unique background noise. Using currently available smartphones and their embedded microphones, they demon-

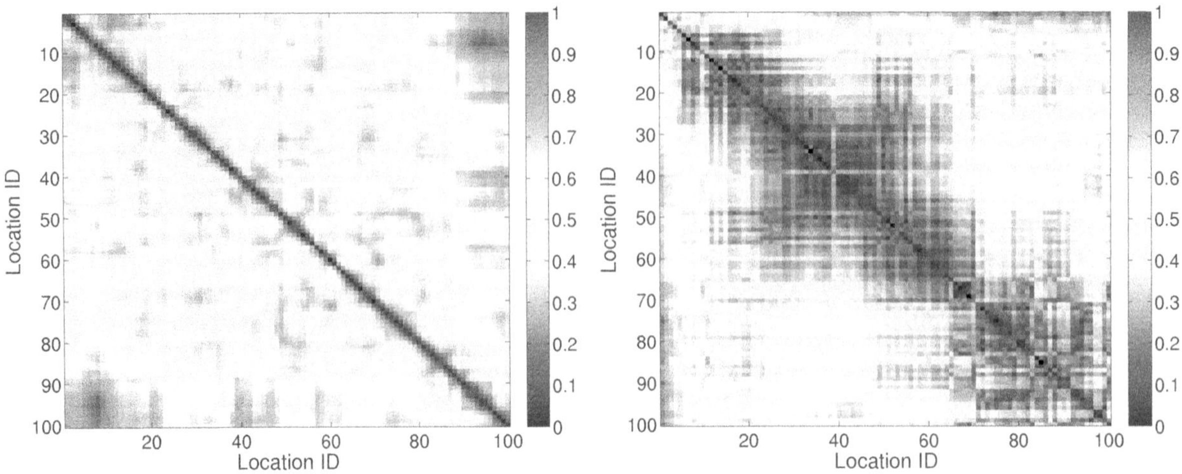

(a) FM RSSI (b) Wi-Fi RSSI

Figure 11: The Manhattan distance between the RSSI vectors measured at 100 evenly spaced ($1ft$) locations in the hallway of an office building. The distances are normalized by the maximum pairwise distance in each figure to have the same range $[0, 1]$ across the two figures. Diagonal values are 0 as the RSSI values for the same location are identical.

strate room level accuracies around 70% from an experiment involving 33 rooms.

Chung et al. investigated indoor localization with geo-magnetic sensors [5]. Their system is based on the observation that the steel and concrete skeletons of buildings can distort the geomagnetic field, such that fingerprinting is feasible. The localization error of this system is within 1 meter 88% of the time. However, the values of the geo-magnetic sensors change drastically even for nearby locations, requiring enormous profiling. Furthermore, the proposed system uses custom-made geo-magnetic sensors that is unclear how they can be reliably integrated into current phones.

In this paper, we consider FM broadcast radio signals for fingerprint based indoor localization. Because of the lower frequency, FM signals are less susceptible to human presence, multipath and fading, they exhibit exceptional indoor penetration, and according to our experimental study they vary less over time when compared to WiFi signals. From the infrastructure point of view, there are thousands of commercial and amateur FM signals being broadcasted continuously across the world, eliminating the need for deploying any custom infrastructure. Also, most mobile devices, even the lower-end ones, are equipped with FM radio receivers.

FM radio based localization systems have been studied before in the context of *outdoor* localization. Krumm et al. measured the signal strength from a set of FM broadcasting stations in outdoor environments and used the strength rankings to distinguish 6 suburbs [10, 26]. Also in outdoor environments, Fang et al. compared the performance of FM radio signals and GSM signals using a spectrum analyzer [8]. They report that both signals achieve similar accuracy in the order of tens of meters using the fingerprinting approach. In this work, we consider FM radio signals for *indoor* localization, and we experimentally demonstrate that higher accuracies can be achieved indoors. This is due to the impact that the internal structure of the building has on the FM radio signal propagation. The internal walls, floors, and ceilings affect FM radio signal propagation enabling higher spatial resolution in the recorded RSSI signatures.

In parallel with our work, Andrei et al. [16], as well as Moghtadaiee et al. [12] explored FM broadcast signals for indoor localization. Andrei et al. reported median localization error of 0.91 meters when leveraging FM RSSI signatures [16]. However, this is preliminary work that only included profiling a single room in a building. More recently, Moghtadaiee et al. measured RSSI values of FM broadcast radios in several rooms with USRP2, and reported a mean localization error of 2.96 meters using fingerprinting[12].

Our work differs from these approaches in four fundamental ways. First, to the best of our knowledge, this is the first large scale study of using FM radio signals for the purpose of indoor localization. We collected measurements in more than 100 rooms on multiple floors, buildings and regions in the US. The volume of experimental data enabled us to investigate room level as well as fine-grain localization performance in the 2D, as well as in the 3D space. Previous work has only evaluated 2D errors, yet at a much smaller scale. Second, we go beyond RSSI-based fingerprinting. We propose and evaluate the use of additional signal strength indicators at the physical layer to create more robust and discriminative signatures. Third, we study in detail the impact of temporal and device variations on the localization accuracy for both WiFi and FM signals. Fourth, we experimentally demonstrate that WiFi and FM errors are independent, and that the two signals can be combined to generate signatures that can achieve up to 83% higher localization accuracy when accounting for wireless signal temporal variations.

Last but not least, energy efficiency is an important consideration when designing localization systems [6, 7, 15]. FM receiver consumes significantly less power compared to the receivers for WiFi and GPS, and thereby can be integrated into general localization systems to tradeoff accuracy and efficiency by duty-cycling various components.

9. CONCLUSIONS

We have presented and evaluated a new approach to fingerprint-based indoor localization that leverages FM broadcast radio signals. Our experimental results show that when FM RSSI values are combined with additional information about signal reception at the physical layer to form the wireless signal signature, localization is 5.7% more accurate than WiFi-based techniques. Furthermore, we experimentally demonstrated that due to differences in operating frequency and wavelength, FM signals are more robust to temporal variations when compared to WiFi signals. More importantly, we have shown that FM and WiFi signals exhibit uncorrelated errors. The combination of FM and WiFi signal indicators into a single signature provides up to 83% higher localization accuracy than when WiFi only signals are used.

FM infrastructure is already widely deployed across the world, and many of the mobile devices include FM receivers. With minor modifications in the hardware of commercial devices and the inclusion of patch FM antennas, the proposed approach could be immediately deployed in the wild. In fact, on most smartphones today there is a single chip that houses all wireless infrastructure (WiFi, Bluetooth, FM radio etc.) such as the BCM4329 from Broadcom [2], WiLink 7.0 from Texas Instruments [23], and CSR9000 from Cambridge Silicon Radio [4]. Most, if not all, of the chip vendors include FM radio components that in general constitute of only a small fraction of the overall chip. This makes the cost impact of FM radio minimal. RSSI information is already available on most radio chips, and localization using RSSI can already achieve satisfactory accuracy. On the other hand, from the FM chip vendor's perspective, there are no major obstacles or limitations in exposing this information to the application layer. We believe that accurate indoor localization can justify and motivate chip vendors to easily expose this information to the application layer.

Based on our intuition and the experimental results for FM signals, we believe other signals such as AM and TV broadcast radios can also be used for localization. Also, combining more signals into a richer fingerprint should further increase the localization performance.

10. ACKNOWLEDGMENTS

We sincerely thank Dr. Romit Roy Choudhury for shepherding this paper, as well as all the anonymous reviewers for their valuable feedback.

11. REFERENCES

[1] P. Bahl and V. N. Padmanabhan. RADAR: An In-Building RF-based User Location and Tracking System. In *Proceedings of INFOCOM*, 2000.

[2] Broadcom. Low-Power 802.11n with Bluetooth 2.1 + EDR and FM (Tx and Rx), 2008.

[3] R. Bruno and F. Delmastro. Design and analysis of a bluetooth-based indoor localization system. In *Personal Wireless Communications*, Lecture Notes in Computer Science. 2003.

[4] Cambridge Silicon Radio. CSR9000.

[5] J. Chung, M. Donahoe, C. Schmandt, I.-J. Kim, P. Razavai, and M. Wiseman. Indoor location sensing using geo-magnetism. In *MobiSys '11*, 2011.

[6] I. Constandache, R. R. Choudhury, and I. Rhee. Towards mobile phone localization without war-driving. In *Proceedings of the 29th conference on Information communications*, INFOCOM'10, pages 2321–2329, Piscataway, NJ, USA, 2010. IEEE Press.

[7] I. Constandache, S. Gaonkar, M. Sayler, R. R. Choudhury, and L. P. Cox. Enloc: Energy-efficient localization for mobile phones. In *INFOCOM*, pages 2716–2720, 2009.

[8] S.-H. Fang, J.-C. Chen, H.-R. Huang, and T.-N. Lin. Indoor localization using fm radio signals: A fingerprinting approach. In *IWCMC '08*, 2008.

[9] A. Haeberlen, E. Flannery, A. M. Ladd, A. Rudys, D. S. Wallach, and L. E. Kavraki. Practical robust localization over large-scale 802.11 wireless networks. In *MobiCom '04*, 2004.

[10] J. Krumm, G. Cermak, and E. Horvitz. Rightspot: A novel sense of location for a smart personal object. In *UBICOMP '03*, 2003.

[11] A. Matic, A. Popleteev, V. Osmani, and O. Mayora-Ibarra. Fm radio for indoor localization with spontaneous recalibration. *Pervasive Mob. Comput.*, 6, December 2010.

[12] V. Moghtadaiee, A. G. Dempster, and S. Lim. Indoor localization using fm radio signals: A fingerprinting approach. In *2011 International Conference on Indoor Positioning and Indoor Navigation (IPIN)*, September 2011.

[13] L. M. Ni, Y. Liu, Y. C. Lau, and A. P. Patil. Landmarc: indoor location sensing using active rfid. *Wirel. Netw.*, 10, November 2004.

[14] V. Otsason, A. Varshavsky, A. L. Marca, and E. de Lara. Accurate gsm indoor localization. In *UbiComp'05*, September 2005.

[15] J. Paek, K.-H. Kim, J. P. Singh, and R. Govindan. Energy-efficient positioning for smartphones using cell-id sequence matching. In *Proceedings of the 9th international conference on Mobile systems, applications, and services*, MobiSys '11, pages 293–306, New York, NY, USA, 2011. ACM.

[16] A. Popleteev. *Indoor positioning using FM radio signals*. PhD thesis, University of Trento, April 2011.

[17] N. B. Priyantha, A. Chakraborty, and H. Balakrishnan. The cricket location-support system. In *Mobicom '00*, 2000.

[18] S. Sen, B. Radunovic, R. R. Choudhury, and T. Minka. Precise indoor localization using phy layer information. In *Proceedings of the 10th ACM Workshop on Hot Topics in Networks*, HotNets '11, pages 18:1–18:6, New York, NY, USA, 2011. ACM.

[19] S. Sen, B. Radunovic, R. R. Choudhury, and T. Minka. Spot Localization using PHY Layer Information. In *MobiSys '12*, 2012.

[20] Silicon Laboratories. Si4735 Broadcast AM/FM/SW/LW Radio Receiver IC Rev D60, 2011.

[21] Skyhook, Inc. http://www.skyhookwireless.com/.

[22] S. P. Tarzia, P. A. Dinda, R. P. Dick, and G. Memik. Indoor localization without infrastructure using the acoustic background spectrum. In *MobiSys '11*, 2011.

[23] Texas Instruments. WiLink 7.0 single-chip WLAN, GPS, Bluetooth and FM solution, 2010.

[24] A. Varshavsky, E. de Lara, J. Hightower, A. LaMarca, and V. Otsason. Gsm indoor localization. *Pervasive Mob. Comput.*, 3, December 2007.

[25] R. Want, A. Hopper, V. Falcão, and J. Gibbons. The active badge location system. *ACM Trans. Inf. Syst.*, 10, January 1992.

[26] A. Youssef, J. Krumm, G. Cermak, and E. Horvitz. Computing location from ambient fm radio signals commercial radio station signals. In *IEEE Wireless Communications and Networking Conference, WCNC*, 2005.

[27] M. Youssef and A. Agrawala. The horus wlan location determination system. In *MobiSys '05*, 2005.

You are Facing the Mona Lisa: Spot Localization using PHY Layer Information

Souvik Sen
Duke University
souvik.sen@duke.edu

Božidar Radunović
Microsoft Research
bozidar@microsoft.com

Romit Roy Choudhury
Duke University
romit.rc@duke.edu

Tom Minka
Microsoft Research
minka@microsoft.com

ABSTRACT

This paper explores the viability of precise indoor localization using physical layer information in WiFi systems. We find evidence that channel responses from multiple OFDM subcarriers can be a promising location signature. While these signatures certainly vary over time and environmental mobility, we notice that their core structure preserves certain properties that are amenable to localization. We attempt to harness these opportunities through a functional system called *PinLoc*, implemented on off-the-shelf Intel 5300 cards. We evaluate the system in a busy engineering building, a crowded student center, a cafeteria, and at the Duke University museum, and demonstrate localization accuracies in the granularity of $1m$ x $1m$ boxes, called "*spots*". Results from 100 spots show that *PinLoc* is able to localize users to the correct spot with 89% mean accuracy, while incurring less than 6% false positives. We believe this is an important step forward, compared to the best indoor localization schemes of today, such as *Horus*.

Categories and Subject Descriptors

C.2.1 [**Network Architecture and Design**]: Wireless communication

General Terms

Design, Experimentation, Performance

Keywords

Wireless, Localization, Cross-Layer, Application

1. INTRODUCTION

Precise indoor localization has been a long standing problem. While the frontier of localization technology has advanced over time, new kinds of location based applications are raising the bar. For instance, the advertising industry is beginning to expect location accuracies at the granularity of an aisle in a grocery shop [1]. Museums are expecting user locations at the granularity of paintings [2], so tourists can automatically receive information about the paintings they stop at. In addition to such high accuracy demands, these applications are inherently intolerant to small errors. If a localization scheme incorrectly places a user in the adjacent aisle in the grocery store, or displays information about the adjacent painting, the purpose of localization is entirely defeated. This is unlike traditional applications – say GPS based driving directions – where small errors are tolerable. As a consequence, new localization schemes will need to meet strict quality standards, without incurring a heavy cost of installation and maintenance. We refer to this problem as *spot localization*, where a device in a specific $1m$ x $1m$ box needs to be accurately identified. Localizing the device outside the box will be useless, irrespective of whether the estimated location is close or far away from the box.

The state of the art in indoor localization is quite sophisticated. Different schemes optimize distinct objectives, including accuracy [3–5], computation [4, 6], ease of calibration [7, 8], energy [9], etc. While the literature is rich, we sample few of the representative schemes to outline the frontier of today's location technology. Cricket [10] achieves high accuracy using special (ultrasound-based) infrastructure installed on ceilings. Noting the difficulties of installing special hardware, RADAR, Place Labs and Horus [4, 6, 8] explored the feasibility of using signal strengths from existing WiFi APs. While RADAR and Horus both rely on signal calibration, EZ [7] recently demonstrated the ability to eliminate calibration at the expense of accuracy degradation. Summarizing all these schemes, we find that the state of the art achieves median location error of $4m$ and $7m$, with and without calibration, respectively [7]. While this accuracy can enable a variety of applications, there are others that need precision at the granularity of "$1m$ x $1m$". This paper targets such high accuracies while ensuring that the calibration complexity is no worse than RADAR or Horus. We call our proposal *PinLoc*, as an acronym for Precise indoor Localization.

PinLoc's main idea is simple. While most WiFi based localization schemes operate with signal strength based information at the MAC layer, we recognize the possibility of leveraging detailed physical (PHY) layer information. Briefly, the intuition is that the multipath signal components arrive at a given location with distinct values of phase and magnitude. When aggregated over multiple OFDM sub-carriers in 802.11 a/g/n, these rich data poses as a fingerprint of that location. Since we define each spot as a cluster of locations, war-driving each spot produces an array of location fingerprints. A training algorithm runs on each array of fingerprints to learn the statistical attributes of that spot. Later, when a mobile device arrives at a spot, it computes a fingerprint (from a sequence of overheard beacons), and classifies it to one of the spots by matching against the learnt attributes. We find that devices are reliably classified to the correct spot, despite movements of people and other objects in the environment. Our war-driving effort is comparable to RADAR or Horus – we mounted a laptop on a Roomba robot and programmed it to move randomly within each spot for around 4 minutes. Finally, where several other schemes are strongly reliant on multiple APs, PinLoc offers reasonable performance even in WiFi-sparse environments. In some cases, PinLoc is able to localize even with signals from a single AP.

At first glance, our findings seemed too good to be true. We expected the signal phases to be sensitive to the orientation of the laptop, human movements, and/or structural changes in the environment (such as repositioning of chairs, boxes, shelves). We suspected that frequent war-driving would be necessary to adapt to such environmental perturbations. While these concerns were natural, we were surprised to find that the fingerprints actually preserved statistical properties even under perturbations. For instance, although the channel response at a specific location varied with time and environmental dynamism, they could be consistently organized around a *set of few tight clusters*. When combined across 30 subcarriers and different APs (i.e., high-dimensional data), we found that even the sets of clusters could be reasonably unique. Further, since spots are composed of many "distinct locations", the fingerprint of a spot is a string of channel responses from multiple distinct locations inside that spot. Thus, even if the channel response from one distinct location is not unique, the probability that the string of channel responses appears in more than one spot is far lower. These and other factors (discussed later) together contribute to PinLoc's robustness. RSSI, on the other hand, is an average of the magnitudes on each sub-carrier, which hides fine-grained information about that location, ultimately limiting the accuracy of localization.

Harnessing the above opportunities into a working system (using off-the-shelf wireless cards) forms the core of PinLoc. The detailed PHY layer information is first extracted from the driver and sanitized using a phase correction operation. The sanitized parameters are then fed to a machine learning algorithm that models the channel response distribution. Later, during system tests, the channel parameters are extracted from received WiFi beacons, and classified to one of the war-driven spots. To address energy issues, PinLoc disables active scanning, and only uses beacons from APs in the same channel. Finally, the individual modules are combined into a full system, and tested over a variety of scenarios. The results are promising – with less than 4 minutes of wardriving per-spot, we observe 89% mean accuracy and false positives consistently below 6%. From the application's perspective, PinLoc was tested in the modern art wing of Duke University's museum. Spots in front of each of 10 paintings were localized with high accuracy.

To the best of our knowledge, no prior work has demonstrated PHY layer-based WiFi localization on off-the-shelf platforms. Zhang et. al. [11] used signal amplitudes and phases on USRP platforms to demonstrate *location distinction*. We note that location distinction detects when a node's location has changed (e.g., for security purposes), but does not need to establish uniqueness for each location. Localization is naturally a far stricter problem, especially when the target is sub-meter accuracies. PinLoc makes an early effort towards this goal – the main contributions may be summarized as follows.

- **We target the problem of *spot localization* where success is defined as the ability to place a device within a $1m$ x $1m$ area, called spots.** We break away from RSSI based schemes and explore the feasibility of using detailed PHY layer information.
- **We utilize the per-subcarrier frequency response as features of a location, and rely on machine learning algorithms to classify a device to one of the trained spots.** We use off-the-shelf Intel 5300 cards; the entire system relies on existing WiFi deployments, and requires no special installation.
- **We evaluate PinLoc at varying accuracy standards, namely, discriminating between seats in a lab, chairs in a cafeteria, and adjacent paintings in a museum.** We observe consistent accuracies under mobile/dynamic environments, outperforming Horus [4], the most accurate RSSI based localization.

The subsequent sections expand on each of these contributions, beginning with definition and applications, followed by measurement, design, and evaluation.

2. LOCATIONS, SPOTS, AND APPLICATIONS

The above section loosely used the terms "locations" and "spots" – we clearly define them here. Locations are like pixels that define the resolution of our localization system. Each location is a small area that has a unique fingerprint. As we show later, the "size" of a location is approximately $2cm \times 2cm$. Spots are larger boxes, say $1mx1m$, and composed of multiple locations. We will see later how the fingerprint of a spot is essentially a combination of fingerprints from all locations in that spot.

Apps for Spot Localization.
The ability to localize users to the granularity of a spot is obviously useful — for instance, precisely tracking a user's indoor location can empower numerous applications. What might be less obvious is that a reasonable number of applications may be enabled even if only a few spots are reliably identified. For instance: (1) Advertising agencies may post discounts on to the user's phone when she pauses in front of select products in the store. (2) Spot localization may be applicable to geofencing – students at different desks may see different sets of exam questions. (3) Logical locations [12] refer to places (Starbucks, Airport, public library) as opposed to geographic coordinates (latitude/longitude from GPS). Since such places

are often adjacent to each other, separated only by a wall, it has been difficult to tell the exact place in which a person is located. Spot localization may be able to detect when a user enters/exits through a door, thereby identifying the place of the user. In a related application, a blind person could come to an approximate location and be prompted with *"the door in front of you is Dr. Brown's eye clinic"*. This may also aid security applications, the content from a server may be downloadable only when a person is inside a restricted area [13]. (4) Even identifying which aisle a person is in could be achieved if the entry and exit points of an aisle are spot-localized. We argue that PinLoc is particularly amenable to these applications since the war-driving effort gets proportionally reduced with fewer spots. While the same reduction may apply to all war-driving based schemes in general, their error margins may still be an issue. Horus [4], for example, incurs $4m$ mean error, which may not discriminate between adjacent grocery aisles or between two adjacent wide-screen TVs at Best Buy. PinLoc is tasked to achieve this discrimination.

3. HYPOTHESES AND MEASUREMENTS

This section aims to show through experiments that PHY layer channel information from existing WiFi deployments can be an indicator of location. We identify the essential hypotheses that must hold, and verify them through measurements. The next section draws on the findings to design and implement the machine learning components in PinLoc. We begin our discussion with a brief background on channel frequency response.

3.1 Background

Most modern digital radios use OFDM communication, and transmit signals across orthogonal subcarriers at different frequencies. Each transmitted symbol $X(f)$ is modulated on a different subcarrier f, and the quality of the received symbol $Y(f)$ will depend on the channel $H(f)$:

$$Y(f) = H(f)X(f) \quad (1)$$

Vector $\mathbf{H} = (H(f))_{f=1,\cdots,F}$ is called *channel frequency response* (CFR), and it is a complex vector that describes the channel performance at each subcarrier. A 802.11 a/g/n receiver implements $F = 48$ data sub-carriers, and includes a channel estimation circuit as a part of the hardware. The Intel 5300 [14] card, released recently with a publicly downloadable driver, exposes the per-subcarrier CFR to the user. Figure 2(a,c) shows examples of some CFR vectors.

Two important properties of the CFR are of interest in PinLoc. (1) The CFR changes *entirely* once a transmitter or a receiver moves more than a fraction of a wavelength [15]. Since the WiFi wavelength is about $12\ cm$, the CFR offers a possibility to discriminate between two nearby locations. (2) Even when the device is static at a specific location, the CFR experiences *channel fading* due to changes in the environment at different time-scales. This introduces randomness in the CFRs, injecting ambiguity in signatures. However, it is unclear whether this randomness is completely unpredictable or whether it exhibits some statistical structure that lends itself to localization. The following hypotheses and measurements are designed to answer these and related questions.

3.2 Hypotheses

We present 3 main hypotheses that need to hold if CFRs are to be used for PinLoc.

1. *The CFRs at each location appear random, but actually exhibit a statistical structure.* This structure is preserved over time and environmental changes.
2. *The "size" of the location (over which the CFR structure is defined and preserved) is small.*
3. *The CFR structure of a given location is different from the structures of* all *other locations, with high probability.* The probability increases when aggregating the CFRs from multiple APs.

Towards verifying these hypotheses, we first describe our testbed environment and experiment methodology, followed by the measurements and findings.

3.3 Experiment Methodology

Our initial experiments were performed in a relatively busy engineering building (see floorplan in Figure 1). We consider a set of 15 different spots in our lab and the adjacent classroom. The center of these spots are approximately $2m$ apart from each other. We place a laptop equipped with the Intel 5300 WiFi card [14] at each of these spots, and associate them to existing WiFi APs (the same set of WiFi APs are audible from each of these spots). The laptops are made to download packets through each of the nearby APs (using iPerf) – the corresponding channel frequency responses (CFR) are recorded for each packet. The Intel 5300 card firmware exports a subset of the CFRs (30 out of 48 data subcarriers), and we only use these for our scheme.

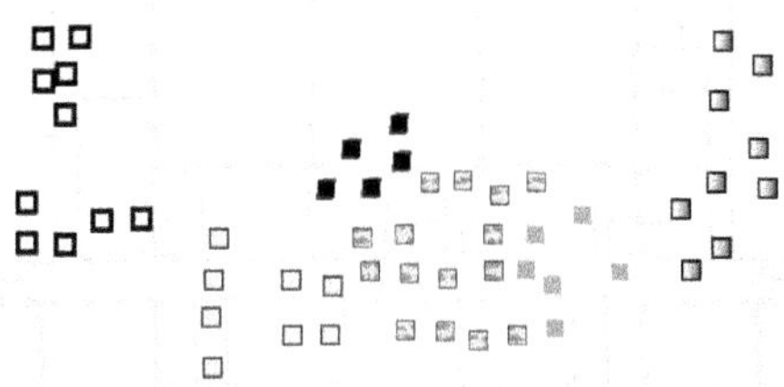

Figure 1: Engineering building floorplan. Different sets of spots shown in different colors – our initial measurements in this section uses only one set of 15 spots (shown in green).

For each location we perform 4-6 measurements at different times during busy office hours. During the measurements in the lab (a $10m$ x $10m$ area), there were between 3 to 5 people who frequently walked in and out. Classroom measurements were performed during and between classes (classroom capacity of 24 seats) – one measurement coincided with all students exiting the classroom at the end of a class.

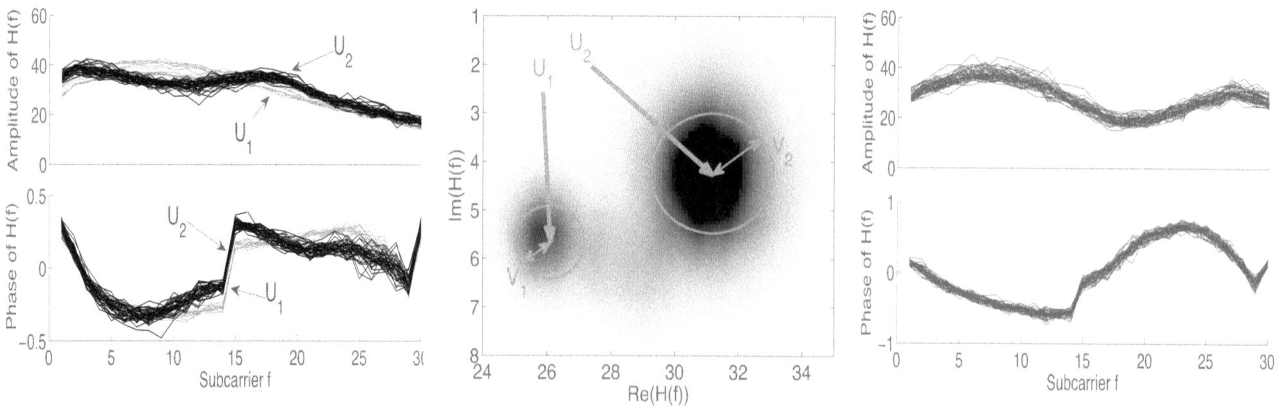

Figure 2: (a) The amplitudes and phases of the channel responses H of 50 (out of 20000) packets sent over the same link (we see two unique clusters, U_1 and U_2); (b) PDF of the complex value of the same 20000 channel responses $H(f)$ for a single subcarrier $f = 20$; (c) The amplitudes and phases of the channel responses H of 50 packets at a different client location.

3.4 Measurement and Verification

(Hypo. 1) *The CFRs at each location appear random but actually exhibit a statistical structure over time.*

Testing on a Single Location: Figure 2(a) shows the channel frequency responses (CFR) recorded on a laptop at a fixed location (the laptop received $20,000$ packets from a specific AP over a period of $100s$, but for visual clarity, we only show 50 CFRs from 50 randomly selected packets). We observe two emerging clusters, denoted with two vectors $\mathbf{U}_1$ and $\mathbf{U}_2$ – CFRs belonging to the same cluster are not identical but appear as noisy realizations of the cluster mean. This is an outcome of fading, caused by different electro-magnetic propagation effects and/or environmental changes.

We now take subcarrier $f = 20$, gather all its CFRs from all $20,000$ packets, and plot the empirical probability density function (PDF) in Figure 2(b). The CFRs are complex numbers, and hence we plot the *Real (Re)* and *Imaginary (Im)* values on X and Y axes – darker colors represent higher values of the PDF. We again see that two dominant clusters emerge, each cluster appearing as a complex Gaussian random variable, with means $U_1(f)$ and $U_2(f)$ and variances $V_1(f)$ and $V_2(f)$, respectively. Of course, this is only a visual indication – we will carefully model this later in section 4.2.

Figure 2(c) shows the outcome of the same experiment, but with the laptop placed at a different location. We find only a single cluster of CFRs, and the shapes of the CFRs differ distinctly from those in Figure 2(a). These few representative clusters hint at the possible existence of complex but invariant structures in per-location CFR, motivating further investigation. Figure 3 pictorially explains the cluster computation process.

Temporal stability of CFR Clusters: We now test whether the observations from these two locations generalize to a larger number of locations, under various environmental changes. Figure 4 shows the distribution of the number of representative CFR clusters from 30 distinct links in total. For each of the 30 links we aggregate all the available data, collected over 2 days – this adequately captures the links' temporal fluc-

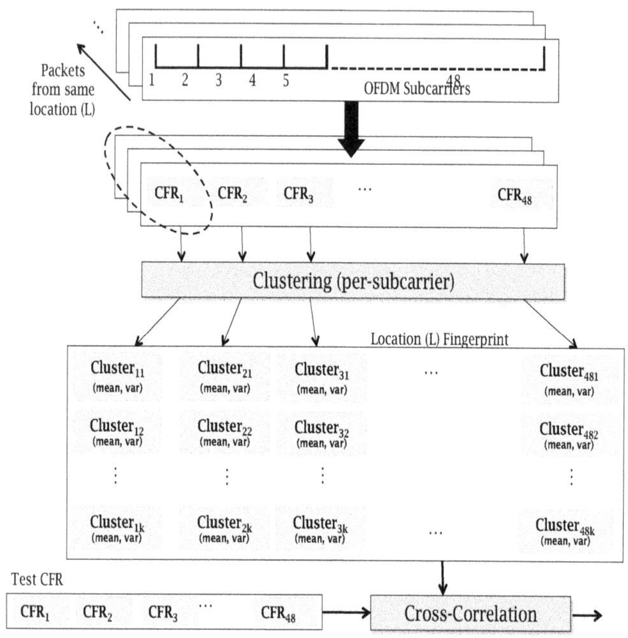

Figure 3: The process of creating clusters (which together form the location fingerprint), and how test CFRs are cross-correlated with the fingerprint.

tuations. We use the clustering algorithm, described later in section 4.3, to identify representative CFR clusters. Evident from Figure 4 (a), more than 80% of links experience 4 CFR clusters or less. However, we still see a substantial number of links with a large number of clusters, even up to 19. This could well suggest that the CFR structure is quite random in dynamic scenarios (e.g., in the classroom), and thus, PinLoc may only be applicable in very static environments.

To verify this, we next look at frequency of occurrence of different clusters. Figure 4(b) shows that the distribution is highly non-uniform, with a strong predominance of the more frequent clusters (i.e., the frequent clusters occur very frequently, and the vice versa). Evidently, the fourth-most frequent cluster occurs no more than 10% of the cases in any

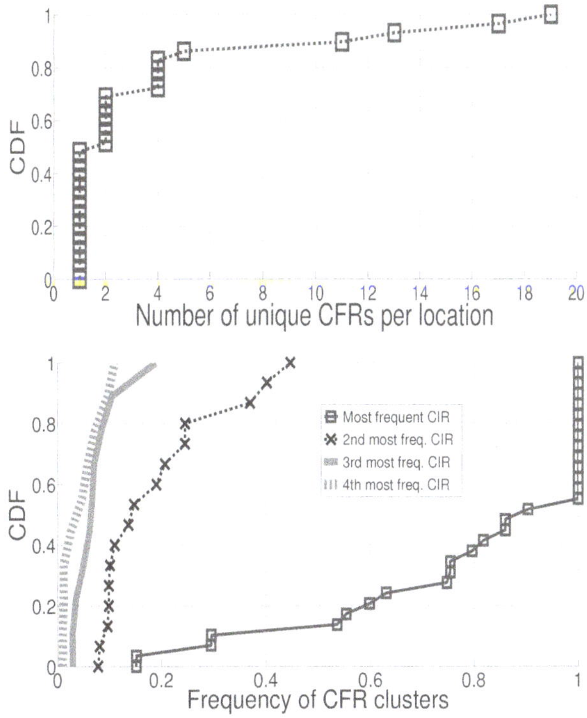

Figure 4: (a) CDF of the number of CFR clusters observed at 30 different client locations (i.e., 30 distinct links); (b) CDF of the probability of seeing the n-th most frequent CFR cluster.

spot, and the 5th, 6th, ... 19th clusters are almost rare. This suggests that even if a few spots experience large number of clusters, we are not very likely to see most of them, both during the training and localization.

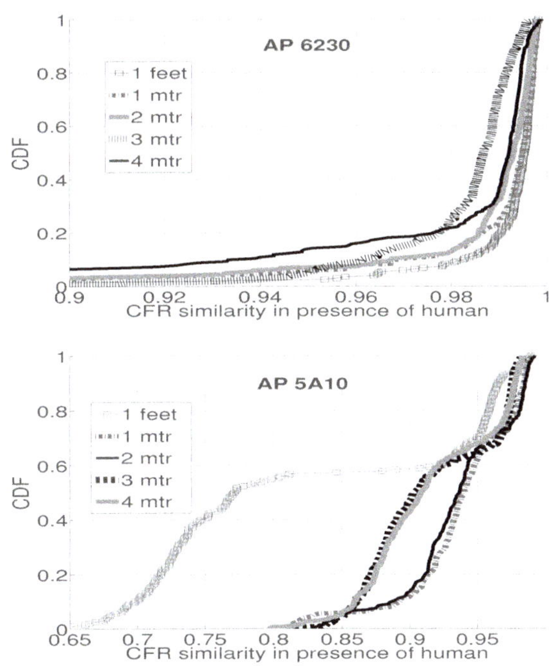

Figure 5: CFR cross correlation in presence of human beings at a location for 2 different APs at 2.4GHz

Impact of environmental changes: To understand the impact of environment changes on CFR clusters, we perform two controlled experiments. First, we study the effect of human mobility on CFR stability. We place a laptop at a fixed location and start gathering CFRs from two different APs. We run the experiment during night, and observe a single CFR cluster for both links. Then, we position a human at an increasing distance d from the laptop. We plot the CDF of cross-correlation of each received CFR with the CFR cluster observed without the human (Figure 3). We define cross-correlation[1] of two CFR cluster means $\mathbf{a}$ and $\mathbf{b}$ as

$$c(\mathbf{a}, \mathbf{b}) = \frac{\sum_{i=1}^{F} a_i b_i}{\sqrt{\sum_{i=1}^{F} a_i^2}\sqrt{\sum_{i=1}^{F} b_i^2}}. \qquad (2)$$

Figure 5(a) shows high correlation, suggesting that human obstructions may not create a significant change to the CFRs from $AP6230$. This is probably because the human does not alter any of the strong signal paths between the laptop and the AP. The link to $AP\ 5A10$, however, changes with human movement; nevertheless, the change is substantial only when the human is very close to the laptop (1 foot away). Even then the median cross-correlation is larger than 0.75. In all other cases, the cross-correlation remains high, suggesting that human movements around the device may not distort the CFRs too much from its core structure.

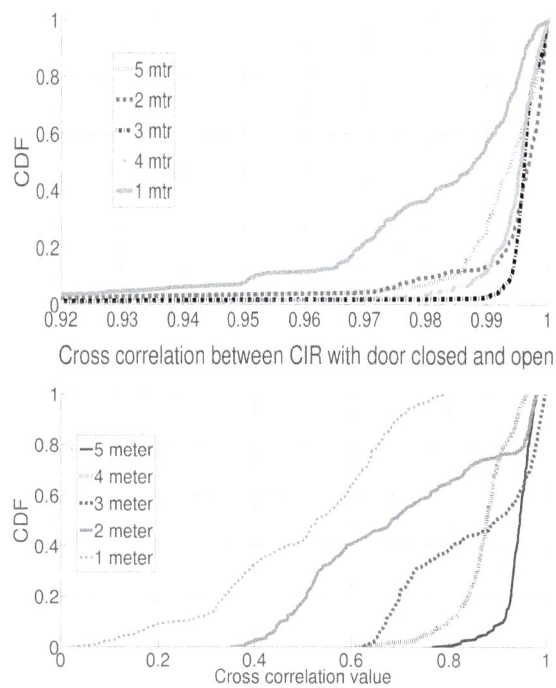

Figure 6: (a) CFR cross correlation for (a) door open vs. closed and (b) original metal shelf vs. moved; for various distances from the laptop.

Next, we study the effects of moving objects in the environment – doors, chairs, metal shelves, etc. We place a laptop at a fixed location and gather CFRs from different APs. Figure 6(a) shows that wooden door obstructions do not induce a significant change to the CFRs. Our measurements also

[1] We use cross-correlation as a metric here for illustrational purposes, and we design a more accurate metric in section 4.

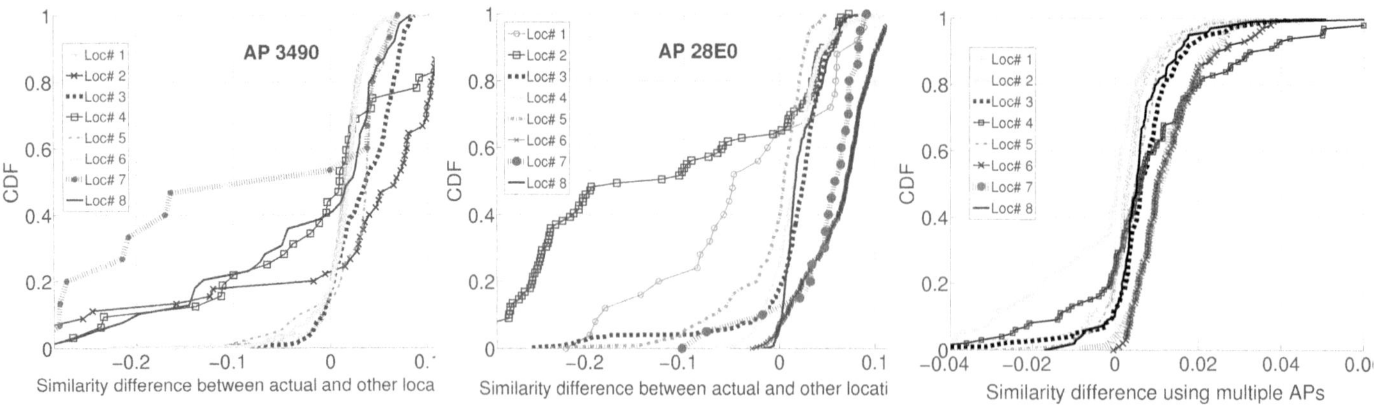

Figure 7: CDF of the difference in similarities $S_{own} - S_{others}$ observed at 8 locations, for two different access point: (a) AP 3490, (b) AP 28E0. (c) CDF of the maximum similarity difference ($S_{own} - S_{others}$) across all APs.

show similar results for other non-metallic objects. However, a repositioned metallic cupboard (approximately 3.5 feet high and 3 feet wide) altered the CFRs significantly – Figure 6(b) shows the impact. Importantly, however, these alterations are localized only around the shelf's original and final locations; spots more than 4 meters away from the shelf are much less perturbed, and need not be re-calibrated. While the above results are from controlled experiments, section 5 reports results from uncontrolled settings (student center, cafe, museum), with hundreds of mobile humans and shifting objects.

(Hypo. 2) The "size" of the location (over which the CFR structure is defined and stable) is small.

Precise localization will require the CFR structure to vary over space. If the structure varies in large granularities (say, multiple meters), PinLoc's accuracy will naturally be bounded by that granularity.
Thus to understand the "size" of each location, we move a test location increasingly further away from a reference location, and compute their respective CFR cross-correlations. Various existing channel measurements [11,15] show that the channel changes entirely once a receiver is moved more than a wavelength, which is $12cm$ in the case of WiFi. Figure 8 shows that the cross-correlation drifts apart with increasing distance, and is quite low even above $2cm$. While this may suggest that localization is feasible at $2cm$ resolution, we will see later that multiple locations in a room may exhibit matching fingerprints. This is why we will require PinLoc to collect multiple fingerprints from very nearby locations (within a $1m \times 1m$ spot). The combined fingerprint from a specific "spot" is much less likely to occur in other spots, and hence, we will attain reliable localization at the granularity of spots.

(Hypo. 3) The CFR structure of a given location is different from the structures of all other locations.

To evaluate the (dis)similarities of CFRs among different locations, we divide the measured data into a training and a test set. Each location has a set of CFR clusters pertaining to an AP, represented by their mean and variance and learnt from the training set. For a test CFR from a location L, we use correlation to find the best matching CFR cluster from L's training data. The correlation value, denoted as S_{own}, indicates *similarity* of the test CFR with the trained fingerprint at

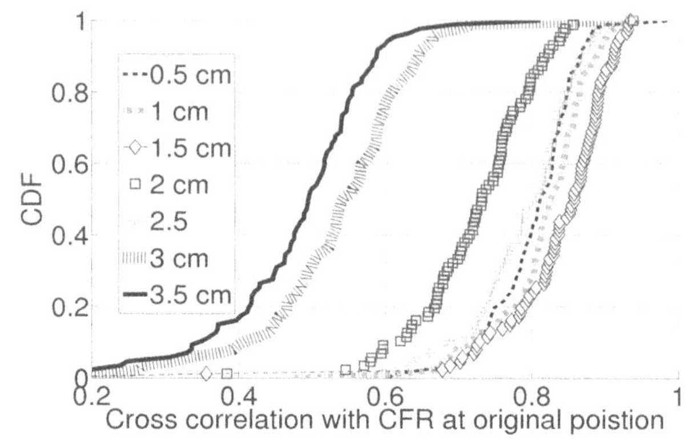

Figure 8: Cross-correlation drifts away for CFRs that are apart by 2cms or more.

the same location. Now, for all fingerprints of all other locations, we find the one that exhibits maximum *similarity* of this test CFR – we denote this similarity as S_{others}. If a device's measured CFR is more similar to a different location than its own, we will naturally misclassify the device's location.

Figure 7 (a) and (b) plot the CDF of the difference in similarities, $S_{own} - S_{others}$, for 8 different locations, for two distinct APs[2]. If the difference is negative, then the packet is likely to be misclassified. Also, the larger the difference, the greater the confidence in packet location. Figure 7 (a) and (b) show that the CFR *from a single AP* is often sufficient to correctly classify location. Of course, in some cases – such as $(Location\,7, AP3490)$ – more than 50% of the CFRs are more similar to other locations, implying misclassification. However, when considering the CFRs of location 7 to a different AP, the misclassification reduces significantly. This suggests that CFRs are diverse across different APs, and this diversity can be leveraged to improve localization. Figure 7 (c) shows the effect of exploiting AP diversity with 2 APs. Specifically, we now pick the AP with the highest similarity difference. Clearly, there is significantly less negative values in Figure 7 (c), implying that AP diversity can help create dissimilarity in location fingerprints.

[2]Note that S_{own} and S_{others} are computed per-AP.

One may ask: *Figure 7 shows that a packet may be classified to one out of 8 different spots. In reality, the system will need to discriminate between many more spots – will the system scale to such scenarios?* We note that PinLoc does not need to discriminate between all spots in a large area. Prior work has used WiFi SSIDs alone to localize devices to around $10m$ x $10m$ regions in indoor environments [8]. PinLoc will leverage such schemes to first compute a coarse-grained macro-location, and then discriminate only between the spots inside that macro-location. Having verified these hypotheses, we now design the full localization scheme, including CFR clustering and matching over multiple sub-carriers. Thereafter, we evaluate PinLoc's performance in section 5.

4. DESIGN AND IMPLEMENTATION

Figure 9 shows the architectural overview of PinLoc. During war-driving, a Roomba-mounted laptop moves randomly through a spot for around 4 minutes. Recall from sections 1 and 2, that a spot is composed of many "locations" (Figure 10), hence, the laptop measures the CFR of every location it hits. Of course, due to the Roomba's random mobility, the laptop may not be able to collect CFRs from all locations in a spot, making our war-driving process far less painstaking. The war-driven data is then sanitized through a phase correction operation and fed to a clustering algorithm, which outputs a few dominating CFR clusters per-location. As mentioned earlier, these clusters are expressed as cluster means and variances, and together form the training set. The training set from different locations in the same spot are stored under the corresponding spot database.

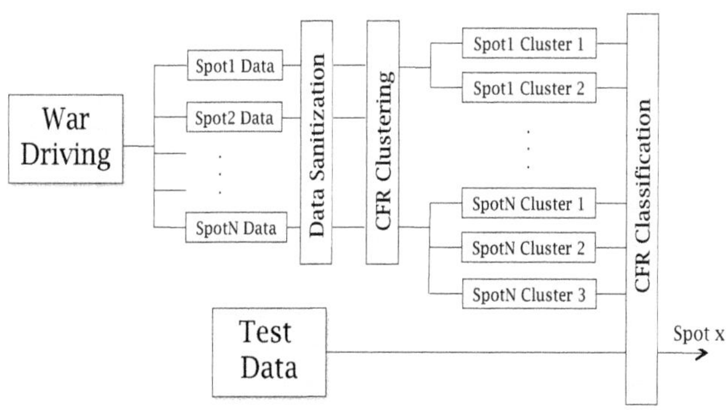

Figure 9: PinLoc architecture

During the real-time localization phase, each mobile node passively records *strings* of CFRs it receives from AP beacons (string length of 4 denoted with shaded squares in Figure 10). The mobile either sends these CFRs to a PinLoc server, or requests the spot databases for candidate spots in its (known) macro-location. The next step is matching. A single CFR reading will not match exactly the ones from the spot-databases due to random fluctuations, but on average, they are more likely to fall in the correct cluster than a wrong one. To improve the accuracy, instead of matching a single reading, we match the string of consecutive CFR readings. Each of these 4 CFRs may match well with a location from a random incorrect spot, but it is unlikely that all the CFRs from a string will match better with locations from the same wrong spot.

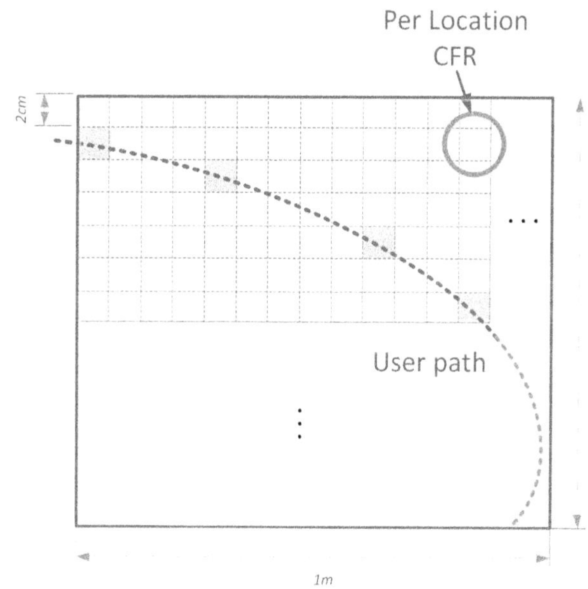

Figure 10: A device records multiple CFRs from a spot.

Results from the next section will confirm this robustness of spot localization. However, before presenting the results, we discuss the sanitizing, clustering, and matching modules in details.

4.1 Data Sanitization (Phase and Time Lag)

The CFRs received at a location cannot be directly used for calibration – an unknown phase β and time lag Δt (which also differ across subsequent packets) can distort the CFR. The sanitization module in PinLoc aims to correct for these offsets. The problem arises because the transmitter and the receiver do not attempt to precisely synchronize their timing and their phases (beyond symbol level) before transmitting a packet[3]. Hence, the phase of the channel response of subcarrier f will be $\hat{\phi}_f = \phi_f + 2\pi f_f \Delta t + \beta + Z_f$, where ϕ_f is the genuine channel response phase we are searching for and Z_f is some measurement noise. It is important to notice that we do not need to learn the exact values of Δt and β for each packet (which is probably impossible). Since we feed $\hat{\phi}_f$ to a classification algorithm, we need to make sure that whenever we measure $\hat{\phi}_f$, the measurement includes *the same* values of Δt and β.

We use a simple transform to achieve the goal. For every received channel response we calculate

$$\begin{aligned} a &= \frac{\hat{\phi}_F - \hat{\phi}_1}{2\pi F}, \\ b &= \frac{1}{F} \sum_{1 \le f \le F} \hat{\phi}_f. \end{aligned}$$

Intuitively, a is the slope of the received response's phase and b is the offset. It is then easy to verify that, if the measurement noise Z_f is small, $\hat{\phi}_f - af - b$ eliminates the random Δt and β time lags. We verify that this is indeed the case in our measured datasets and we use the post-processed phase value $\hat{\phi}_f - af - b$ in all our further calculations.

[3]This is not a problem for a conventional OFDM receiver that only needs to remove, but not learn the channel response.

4.2 Modeling the channel response

We see from a sample measurement, presented in section 3, that the channel responses look like noisy replicas of a few representative clusters. However, it may not be obvious how to identify clusters and the main challenge of the classification algorithm is how to deal with the measurement noise. We make a reasonable assumption to model the noise (also called fast-fading) as a complex Gaussian noise, which corresponds to Rayleigh fading [15]. We first verify this assumption visually by looking at the samples across subcarriers, such as the one illustrated in Figure 2 (b). We then take a few samples from the measurements and verify using QQ plots that the distribution fits well to a complex Gaussian and that the real and the imaginary parts are i.i.d. We also assume that the noise is independent across subcarriers.

Let us consider a link from a single location to a single AP. Recall that $\mathcal{U} = \{\mathbf{U}^1, \cdots, \mathbf{U}^u\}$ is the set of means of the representative CFR clusters of the link, as discussed in section 3, where each $\mathbf{U}^i = (U^i_f)_{f=1,\cdots,F}$. Let us suppose we observe packet $\mathbf{P} = (P_f)_{f=1,\cdots,F}$, where P_f is the complex channel response for subcarrier f.

Following the observations from the measurements, we propose to model the channel response as a random vector with a Gaussian mixture distribution. That is, the channel response is assumed to be drawn from one of the representative CFR clusters, chosen at random for each packet. The channel response is modeled as a combination of the representative CFR clusters. We model each representative CFR cluster as a complex Gaussian random vector with mean $\mathbf{U}^i$ and some variance $\mathbf{V}_i$ (since the real and imaginary parts are i.i.d, the variance is scalar). Assuming that the packet $\mathbf{P}$ belongs to the CFR cluster with the mean $\mathbf{U}^i$, we have that the probability of packet $\mathbf{P}$ is

$$\mathrm{P}(\mathbf{P}|\mathbf{U}^i, \mathbf{V}^i) = \prod_{f=1}^{F} \frac{1}{2\pi \left(V^i_f\right)^2} \exp\left(-\frac{||P_f - U^i_f||^2}{2\left(V^i_f\right)^2}\right). \quad (3)$$

We see that each representative CFR cluster is described with a pair of complex vectors $(\mathbf{U}^i, \mathbf{V}^i)$ representing the mean and the variance of the observations. In section 4.3 we discuss how to derive $(\mathbf{U}^i, \mathbf{V}^i)$ from the measured data set.

Furthermore, we can apply logarithm to (3) and remove constants to derive the log likelihood distance metric

$$d(\mathbf{P}, \mathbf{U}^i) = \sum_{f=1}^{F} \log(V^i_f) + \sum_{f=1}^{F} \left(\frac{||P_f - U^i_f||^2}{\left(V^i_f\right)^2}\right), \quad (4)$$

which we shall use as a distance metric in the classification algorithm. Note that d is indeed a distance metric, but it also has a probabilistic interpretation from (3), which we will use later while combining multiple readings to improve classification accuracy.

4.3 Clustering algorithm

Recall that we model the data at each location as a Gaussian mixture distribution, with K clusters with means and variances (U_k, V_k). We denote with w_k the probability that an observed packet belongs to a particular cluster k, which corresponds to how often we "see" cluster k in our training data. According to the Gaussian mixture model, this probability is independent across packets. Thus, the parameters of our model are $(w_k, U_k, V_k), k = 1, ..., K$. The classical approach to estimate these parameters is the expectation-maximization algorithm [16]. Instead, we estimate the parameters using variational Bayesian inference [16]. Variational inference is provided by the Infer.NET [17] framework that we use to implement the clustering algorithm. It is particularly convenient here because it tends to prune unneeded clusters. Instead of estimating the number of clusters K by running the algorithm multiple times with different values of K, we perform one run with $K = 10$ and drop the clusters with small weights w_k. Some locations may actually have more than 10 clusters but this is rare and discarding these extra clusters has little impact on performance.

We note that another potential clustering algorithm that could be used here is k-means clustering algorithm. However, our approach takes into account that the noise has a Gaussian distribution and hence can perform a more accurate clustering. For further discussion on the drawbacks of the k-means clustering, please see [16, Chapter 9].

4.4 Classification algorithm

Our classification algorithm is composed of two parts. First, PinLoc computes a macro-location based on WiFi SSIDs alone [8], and shortlists the spots within this macro-location; we call these spots the *candidate set*, C. The second task is to pick one spot from C, or to declare that the device is not in any of these spots. To this end, the WiFi device overhears beacons from the APs as it roams around (we discuss the energy implications in section 5.5). Let A be the set of all APs and let $AP(\mathbf{P})$ denote the AP which transmitted packet $\mathbf{P}$. We define the distance between a given packet $\mathbf{P}$ and a spot S_i as

$$d(\mathbf{P}, S_i) = \min_{\mathbf{U}^i \in Z_i, AP(\mathbf{U}^i) = AP(\mathbf{P})} d(\mathbf{P}, \mathbf{U}^i). \quad (5)$$

where Z_i is a set of representative CFR clusters learned from spot S_i. Then, for all values of i, we compute the minimum of $d(\mathbf{P}, S_i)$ – this outputs the most likely spot that the user is located at, based on the CFRs from packet $\mathbf{P}$. The operation repeats for every packet received within a short time window (typically 30 packets from 3 APs), and the spot that is picked most often (highest vote) is identified. PinLoc does not immediately declare the highest voted spot as the user's location. If the highest vote count is small, it suggests low confidence and the possibility that the user is not located at any of the trained spots[4]. Thus, PinLoc ensures that the number of votes is above a rejection threshold before announcing that spot as the user's location. If the number of votes is below the threshold, then PinLoc announces the location as "*not-a-spot*". The rejection threshold can be selected from the training data and a application-specified false positive rate. We use 15% of the number of possible votes as the threshold in our evaluation. E.g., if the maximum number of votes are 30, PinLoc announces a location as "*not-a-spot*", if the highest matching spot obtain less than 5 votes.

[4]We note that, in order to deal with the outliers, we use the majority voting scheme with the distance function (5) instead of the direct probabilistic interpretation (4).

5. EVALUATION

We evaluate PinLoc across 100 different spots using war driven training data and several test samples. Our wardriving approach is explained next.

5.1 War-Driving

The channel model and clustering algorithm from section 4.2 can be applied only to data from the same location. As we have seen in section 3, the size of a location that has the same representative CFR clusters is about 2cm × 2cm. It is important to get enough data from the same location to be able to accurately learn the channel responses. Also, it is important to collect responses from many (2cm × 2cm) locations within the 1m × 1m spot.

Our current war-driving procedure is in 2D. We transmit a packet from an AP every $1ms$. A Roomba robot, mounted with a laptop, moves at a programmed speed of $30cm/s$. Figure 11 shows an example of war-driving at the museum. We receive about 60 packets during a $2cm$ stretch. We divide all packets received in one spot into batches of 60 packets and run clustering algorithm on each batch.

Figure 11: PinLoc war-driving at different spots in the museum. The Roomba robot mounted with a laptop, and 4 virtual wall devices at the corners of the spot.

There are three important questions that arise about war-driving. The first question is *do we need to record every single representative CFR cluster during war-driving?* As we show later in section 5, PinLoc's performance is not too sensitive to war-driving accuracy – the accuracy degrades gracefully as the Roomba is made to war-drive for shorter durations.

The second question is *do we need to war-drive in a particular fashion?* During the localization we match each sample against all learned representative CFR clusters to find the best fit, thus it is not important in which order we store the clusters in the learning database. One can swipe through a spot in any direction, multiple times and in multiple rounds. Moreover, war-driving does not have to be done in any particular channel conditions (e.g. during busy or off-peak hours). In fact, the accuracy increases if we accumulate channel samples from diverse channel conditions.

The third question is *why not define a spot as a single location ($2cm \times 2cm$)?* Recall from Figure 10 that a string of CFRs improves the probability of spot localization. This is because a single test CFR may match with a different random spot; but it is unlikely that a string of test CFRs will all match better with the same wrong spot. In leveraging this diversity gain, we find that strings of length 5 are reasonable. Given that AP beacons are spaced $100ms$ apart in WiFi b/g/n, and human indoor walking speed is $1m/sec$, getting 10 beacons implies that the person moves almost $1m$. Hence, we fix our spot-size to $1m \times 1m$. Of course, even if a person is not walking, we assume that there will be inherent swaying motions of the body in the granularity of few cms – such motions will be sufficient for PinLoc.

5.2 Experiment Design

We evaluate PinLoc in 4 environments: (1) Hudson Engineering building with faculty offices and classrooms, (2) busier Bryan Student Center, (3) Twinnies cafe, and (4) Duke University Nasher Museum. Both training and test samples are taken from the laptop, mounted on the Roomba robot. In all the experiments, the laptop associates to all APs on the same frequency channel, and receives beacon packets from them for a duration of $1s$. This $1s$ duration ensures that a mobile user (walking at 1m/s) will remain inside the same spot while receiving all the beacons. For the engineering building scenario, we take test samples during daytime with more than 100 people around. The cafeteria experiments are performed during busy lunch hours, with more than 50 people present at any time, along with a high churn. For the museum measurements PinLoc was trained for 10 large paintings in one wing of the museum, and tested for these spots. The measurements were obtained on a slightly less busy day to minimize interference with real visitors. Both training and test samples are taken at the same height (we discuss the ramifications of height in section 6).

Metrics: Our goal is to show that we can accurately localize test samples to the corresponding spots, but also to detect when a sample does not belong to any spot. We use the following two metrics to evaluate PinLoc. (1) *Accuracy* – the fraction of cases in which the user was localized to the correct spot. (2) *False positives (FP)* – the fraction of cases in which the users were localized to an incorrect spot/non-spot. In other words, false positives also take into account the cases where PinLoc localizes the user to a trained spot, even though the user was not located at any of these spots.

Comparison with RSSI: We also evaluate whether we can use RSSI to achieve a similar accuracy with the same number of APs. For this, we compare PinLoc with a modified Horus algorithm. The original Horus algorithm [4] interpolates RSSI measurements to simplify war-driving. In order to provide a fair comparison, we modify Horus to use the same war-driven training set we use in PinLoc. We define the similarity analogously to (5), replacing correlation with a difference between two RSSI. This is consistent with equation (3) in Horus [4], where the localization metric is a joint probability of seeing different access points. We compare the modified Horus algorithm with PinLoc by using the same test data across the same number of APs.

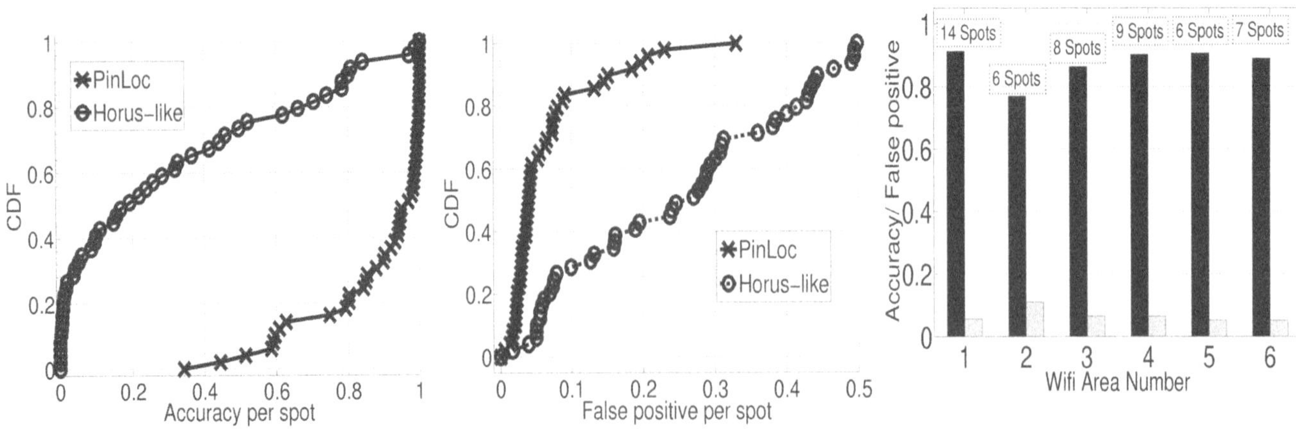

Figure 12: PinLoc in office environment: (a) Accuracy, (b) False positive against Horus (c) Per macro-location.

5.3 PinLoc accuracy and false positives

Figure 12 reports results from the engineering building experiments. PinLoc achieves nearly 90% mean accuracy across 50 spots (Figure 12(a)), consistently outperforming Horus. The false positives (FP) are also maintained to less than 7%, compared to more than 25% in Horus (Figure 12(b)). RSSI based algorithm is significantly worse than PinLoc, since it is represented with a single real number. CFR is represented with 30 complex numbers and contains much richer information. In this comparison, we were tuned to a single channel, and observed 1 to 4 APs on that channel. Ofcourse, PinLoc's, as well as Horus's, performance will improve if more APs are available. But this may require scanning across channels which may increase energy consumption.

Figure 12(c) zooms into the performance of PinLoc and shows the accuracy/FP on a per-WiFi macro location[5]. The number of spots per WiFi region is shown on top of the bars. Student center has a more dynamic environment with cafes and shops – we evaluate PinLoc at 34 spots across 3 WiFi macro locations. Figure 13(a) shows that even in such an active environment PinLoc can maintain low false positive(7.3%) and high accuracy(86%). An obvious question is whether such performance will degrade if adjacent spots need to be identified (i.e., center of spots are $1m$ apart). Figure 13(b) shows that PinLoc is able to discriminate in such settings with a reasonable average accuracy of 82%.

Similar accuracy/FP graphs are plotted for the cafeteria and museum in Figure 14. The mean accuracy for the cafeteria case is 90.07% and the mean FP is 4.5%. For the museum, the mean accuracy is 90.28%, and the mean FP, 4.1%. In all four scenarios, PinLoc achieves high accuracy/low FP for most of the spots, except around 20% where the performance drops. Careful investigation showed that these spots received packets at low SNR from many APs. To probe this further and understand their ramifications, we next perform an analysis across various system parameters.

[5]Recall that a macro-location is derived from WiFi SSIDs; PinLoc discriminates only between candidate spots inside the macro-location.

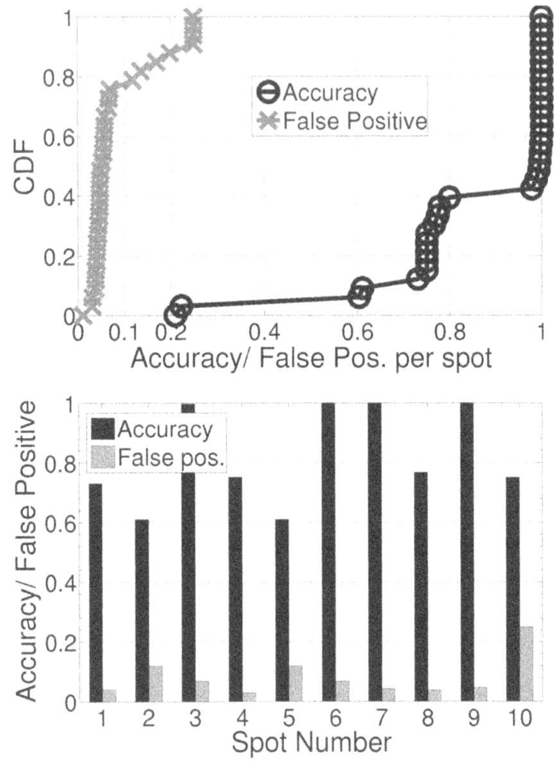

Figure 13: Pinloc performance in student center (a) Accuracy, false pos., (b) Performance of adjacent spots.

5.4 Impact of Parameters

Impact of number of test packets

A user extracts CFRs from beacon packets that are transmitted by APs every $100ms$. Thus, assuming up to $1m/s$ walking speeds, the user dwells for at least $1s$ inside a spot, thereby receiving 10 beacons per-AP. Figure 15 (a) shows the variation of accuracy and FP with the number of received beacons per-AP. With the typical size of 10 packets per AP, PinLoc achieves mean accuracy of 89% and false positives of 7% across 50 spots in the engineering building; 15 packets raise them to 91% and 5%, respectively. Figure 15 (a) also shows low accuracy (68%) and high FP (14%) if only 1 packet per AP is used. This is because one single reading may randomly match with

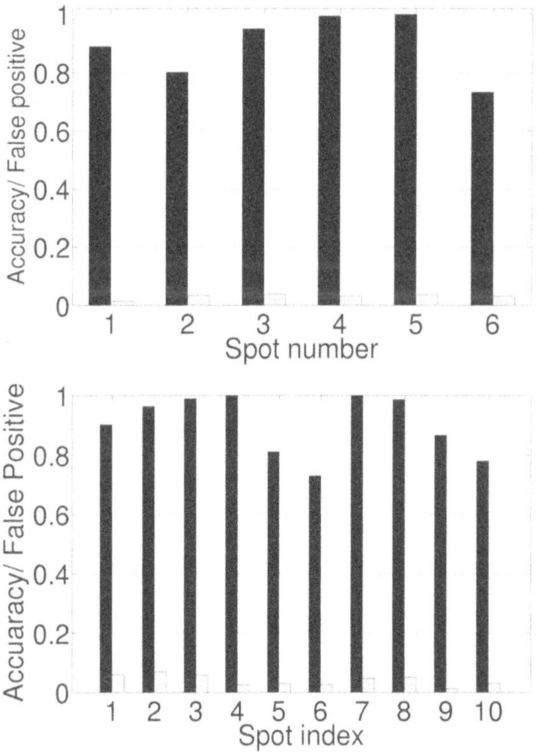

Figure 14: PinLoc performance in cafeteria and museum (a) Accuracy and FP per spot in cafeteria. (b) Accuracy and FP per-spot in the museum.

an incorrect spot. However, even with 5 packets, we find the accuracy above 82%, and FP less than 10%. This indicates that even at higher user mobility, or with failure in beacon reception, PinLoc can sustain a reasonably good performance.

Impact of the number of APs

PinLoc's performance in sparse WiFi environments is of interest. To this end, we analyze the performance for varying number of APs. While collecting the test data, we were tuned to a single channel, and observed 1 to 4 APs on that channel (we did not scan across channels to limit energy consumption). We divide the spots into different categories depending on how many APs are visible within each spot – Figure 15 (b) shows the results. An encouraging observation is that *even when only a single AP is visible, PinLoc can perform spot localization with accuracy of over 85% and false positives below 7%*. This is in contrast to other WiFi-based localization methods that need at least 3 APs to attain reasonable precision. Furthermore, as the number of visible APs increases, the performance improves quickly.

Impact of war-driving

It is important to understand how long we need to war-drive to achieve high localization accuracy. The tradeoff is that short war-driving will record fewer CFRs, incurring the possibility of overlooking an important CFR cluster. To understand the impact, we run PinLoc on different training sets, drawn from different war-driving durations. Figure 15 (c) plots the accuracy and the FP (per-spot) as a function of this duration. Evidently, a few minutes of war-driving per-spot suffices; we observe reasonable performance when using only 1 minute of war-driving data.

Impact of mobility

We turn to the cafeteria scenario to analyze the effect of the mobility on the accuracy of localization. We take one hour of test data for three spots in the cafeteria. We perform localization on each batch of 10 beacon packets and plot its success or failure in Figure 16. For all three spots, we see that the time instants when localization failed are short and uniformly spread over the measurement interval. The mean accuracy was 85% with 7% false positives. Thus, even in a very busy environment such as the cafeteria, we are able to provide localization without prolonged disruption.

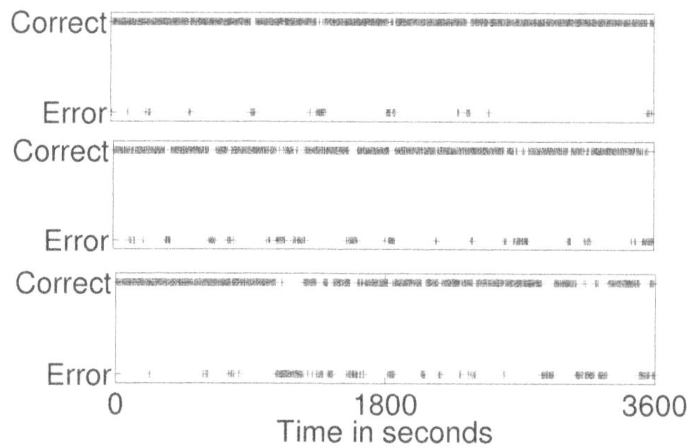

Figure 16: Success of PinLoc localization over time for three spots and over an interval of 1 hour.

Impact of old training data

We concede that PinLoc will need a fresh round of war-driving for spots that are affected by significant environmental changes (e.g., metallic shelves). But with "small" structural changes will PinLoc's war-driven training data remain valid over days and months? To evaluate this we tested 5 spots in our engineering building 7 months after wardriving them. Figure 17 shows a moderate median accuracy of 73% per spot in this scenario. Depending on application requirements, war-driving can be periodically scheduled to improve accuracy. This may not be hard since war-driving can be automatically be done using a Roomba robot.

5.5 Energy consumption

PinLoc is designed with energy efficiency in mind. Contrary to existing schemes that rely on power-hungry active scanning, PinLoc uses only beacons from APs in a single channel. For this, it synchronizes with the beacon-schedules of these APs and periodically wakes up to collect the CFRs. It sends this information to a central server for computing location. The amount of such data is low, approximately 1200 bytes per second for 2 APs and $1m/s$ speed, and can be easily batched with existing traffic that communicates to the location based service. Consequently, the data upload energy is marginal. Now, for the energy footprint of beacon reception, we performed measurements on Google Nexus One phones, using the Monsoon Power Meter. We found that receiving beacons from 2 APs every $100ms$ incurs an additional $5.28mW$ power on average. This may be negligible compared to $1326.72mW$ on

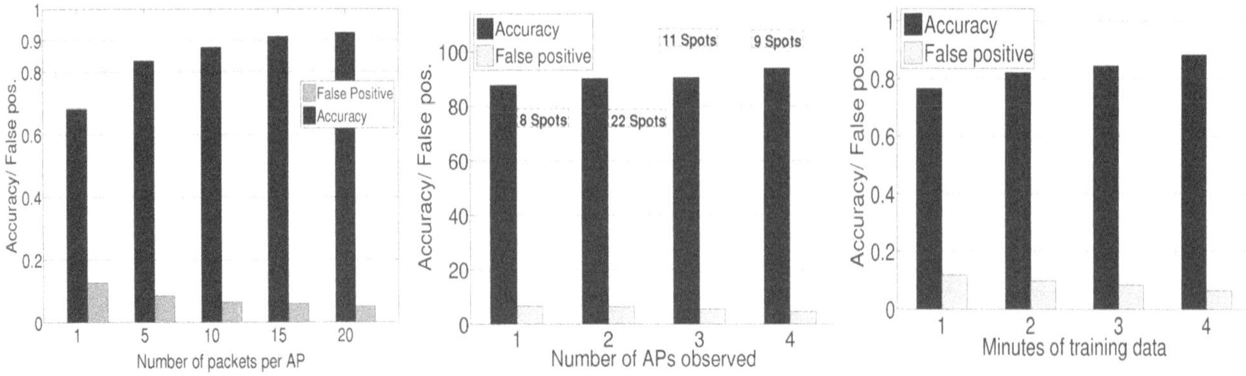

Figure 15: Accuracy of PinLoc against: (a) number of beacons, (b) number of APs and (c) duration of war-driving.

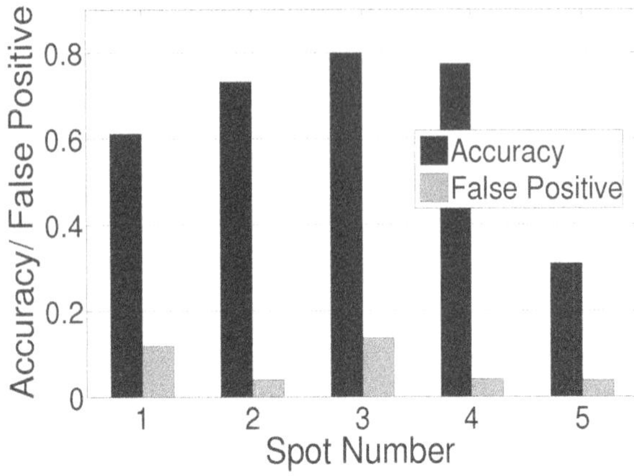

Figure 17: Accuracy of 5 spots tested 7 months after training.

average to stream YouTube video. We omit the details in the interest of space but argue that PinLoc's low energy overhead makes it a practical proposition.

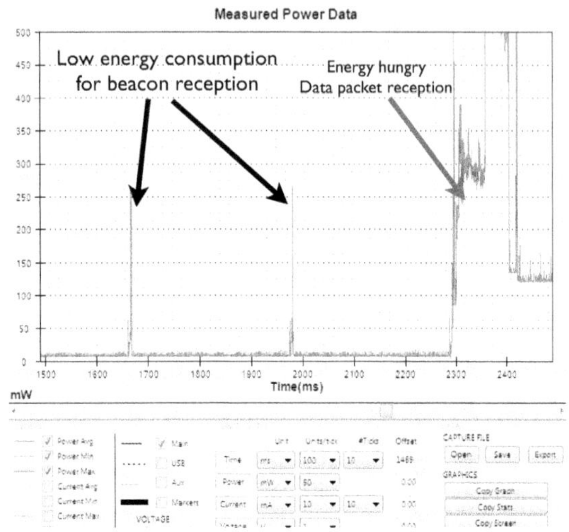

Figure 18: Beacon energy consumption from the Monsoon Power Meter tool.

6. LIMITATIONS AND DISCUSSION

• **Antenna's orientation:** PinLoc's war-driving and testing were all performed with a laptop on a Roomba robot, in a 2D plane. The antenna is placed in the laptop's lid and the lid was closed, hence the antenna was parallel to the testing plane. During wardriving and testing, Roomba robot took random turns, constantly pointing the antenna in different directions. Hence we can conclude our results are robust to the antenna's orientation on the 2D plane.

• **Height, 3D war-driving, and phone mobility:** In reality, users will carry their phones at different heights and 3D war-driving would be necessary. We discuss the effects of 3D wardriving with help of two simple micro-benchmarks.

(1) We manually war-drive a $1m \times 1m \times 50cm$ box for 15 minutes (a person holds a laptop in hand and moves it in random directions across the box), and we collect CFR samples from 4 APs. We then collect test data within the same box by moving the laptop around it and rotating the laptop. The person that holds the laptop also moves around it. We compare the test data with the manually war-driven 3D data and with the previously war-driven 2D data for 9 different locations. We plot the localization accuracy in time in Figure 19(a). The mean accuracy is 84%. We see that the accuracy is similar to the one reported in section 5 hence we can speculate that with suitable extensions, PinLoc might scale real-world scenarios.

(2) To understand the complexity of 3D wardriving, we measure the vertical "size" of a location (the z axis analogue of Figure 8). We see that cross-correlation between CFRs drifts apart at $10cm$ or more (Figure 19(b)). This suggests that 3D war-driving may be feasible, perhaps with a height-varying tripod on top of a Roomba. We were unable to construct a robot for 3D wardriving. We leave 3D localization to future work, as well as issues that may emerge from phone orientation, or users inserting their phones in pockets and bags.

• **Dependency on particular hardware cards:** We run the experiments reported in section 5 alternating among 4 laptops (each with an Intel 5300 card). Since Intel 5300 was the only available card that exposes CFR information, we were not able to test PinLoc with cards from different vendors. We speculate that cross-platform calibration would be necessary, similar to existing RSSI based schemes [7, 18].

• **Localization and MIMO:** We did not explore MIMO capabilities in our current test-bed. A MIMO receiver would provide as many CFR samples as the number of receive antennas,

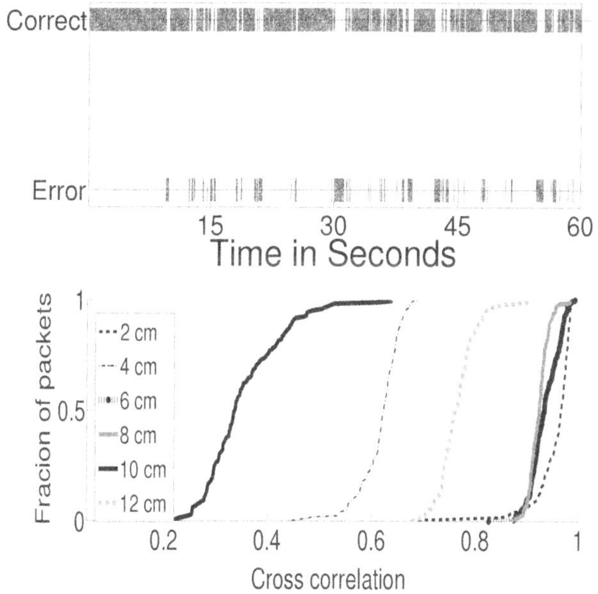

Figure 19: (a) Accuracy of a 3D spot over time. (b) Correlation drifts away with height differences of $10cm$ or more.

and could be highly valuable for localization. We leave this for future investigation.

7. RELATED WORK

The topic of indoor localization has seen a variety of approaches that may be broadly classified as active and passive, and sub-classified into RF and sensing based techniques. We sample some of the key ideas, and discriminate PinLoc from them.

RF signal strength-based localization has been the dominant theme of localization for more than a decade. RADAR [8] performs detailed site surveys a priori and then utilizes the SSIDs and received signal strength reported by wireless devices to generate location fingerprints. Horus [4] and LEASE [19] propose enhancements to RADAR by exploiting the structure of RSSI. Place Lab [6] and the Active Campus project [20] attempt to reduce the overhead of calibration – they show that collecting signal information from WiFi and GSM base stations through war-driving is adequate for reasonably accurate localization. Finally, Patwari and Kasera [11] have also recently explored the use of signal characteristics of a wireless channel to achieve location distinction – that is, they reliably identify when a device has moved from one location to another.

Time-based techniques utilize time delays in signal propagation to estimate distances between wireless transmit-receiver pairs. Examples include GPS [21], PinPoint [22], and work by Werb et. al. [23]. The TPS system uses difference in time of arrival of multiple RF signals from transmitters at known location [24]. Similarly, the PAL system [25,26] uses time difference of arrival between UWB signals at multiple receivers to determine location. The Cricket system [10,27] and AHLoS [28] utilize propagation delays between ultrasound and RF signals to estimate location of wireless devices. Such a solution requires installation of ultrasound detectors on wireless devices, limiting their applicability.

Angle-of-arrival based techniques utilize multiple antennas to estimate the angle at which signals are received, and then employ geometric or signal phase relationships to calculate bearings to other devices with respect to the device's own axis [29–31]. Besides positioning, such methods can also provide orientation capabilities. However, these techniques require extremely sophisticated antenna systems (4 to 8 antennas) and non-trivial signal processing capabilities, unforeseeable on mobile devices in the near future. PinLoc's reliance on WiFi alone, along with the ability to utilize PHY layer information from off-the-shelf interfaces [32], makes it a potential candidate for immediate deployment.

8. CONCLUSIONS

This paper shows that PHY layer information, exported by off-the-shelf Intel 5300 cards, offer new opportunities to localize WiFi devices in indoor environments. We leverage the observation that multipath signals exhibit stable patterns in the manner in which they combine at a given location, and these patterns can lend themselves to meter-scale localization. Evaluation results from the engineering building, cafeteria, student center, and the university museum, demonstrate a mean accuracy of 89% for 100 spots. From the application's perspective, PinLoc could enable product advertisements on a shopping aisle, or offer information on each exhibit at a museum. We believe this is a step forward in the area of indoor localization, even though some more work is necessary before it is ready for real-world deployment.

9. ACKNOWLEDGMENT

We sincerely thank Dr. Anthony Lamarca for shepherding our paper, as well as the anonymous reviewers for their valuable feedback. This work is supported in part by the NSF CNS-0747206 grant.

10. REFERENCES

[1] Placecast. Shopalerts. `http://placecast.net/shopalerts/index.html`.

[2] E. Bruns et al. Enabling mobile phones to support large-scale museum guidance. *Multimedia, IEEE*, 2007.

[3] N. Priyantha, A. Chakraborty, and H. Balakrishnan. The cricket location-support system. In *MOBICOM*, 2000.

[4] M. Youssef and A. Agrawala. The horus WLAN location determination system. In *MobiSys*, 2005.

[5] Chen Y. et al. Fm-based indoor localization. In *Mobisys*. ACM, 2012.

[6] Yu-Chung Cheng, Yatin Chawathe, Anthony LaMarca, and John Krumm. Accuracy characterization for metropolitan-scale wi-fi localization. In *MobiSys*, 2005.

[7] K. Chintalapudi, A. Iyer, and V. Padmanabhan. Indoor localization without the pain. In *MOBICOM*, 2010.

[8] V. Bahl et al. RADAR: An in-building rf-based user location and tracking system. In *INFOCOM*, 2000.

[9] I. Constandache, S. Gaonkar, M. Sayler, R. Roy Choudhury, and L. Cox. Enloc: Energy-efficient localization for mobile phones. In *IEEE Infocom Mini Conference*, 2009.

[10] N. B. Priyantha. *The cricket indoor location system*. PhD thesis, MIT, 2005.

[11] Junxing Zhang et al. Advancing wireless link signatures for location distinction. In *MobiCom*. ACM, 2008.

[12] M. Azizyan, I. Constandache, and R. R. Choudhury. SurroundSense: Mobile phone localization via ambience fingerprinting. In *MOBICOM*, 2009.

[13] A. Sheth, S. Seshan, and D. Wetherall. Geo-fencing: Confining Wi-Fi coverage to physical boundaries. *Pervasive Computing*, 2009.

[14] Intel Research. Intel 5300 mimo channel measurement tool. http://ils.intel-research.net/80211n-channel-measurement-tool.

[15] D. Tse and P. Viswanath. *Fundamentals of Wireless Communication*. Cambridge University Press, 2005.

[16] C. Bishop. *Pattern Recognition and Machine Learning*. Springer, 2006.

[17] T. Minka, J.M. Winn, J.P. Guiver, and D.A. Knowles. Infer.NET 2.4, 2010. Microsoft Research Cambridge. http://research.microsoft.com/infernet.

[18] A. Haeberlen et al. Practical robust localization over large-scale 802.11 wireless networks. In *MOBICOM*. ACM, 2004.

[19] P. Krishnan and others. A system for LEASE: Location estimation assisted by stationary emitters for indoor rf wireless networks. In *IEEE Infocom*, March 2004.

[20] William G. Griswold et al. Activecampus: Experiments in community-oriented ubiquitous computing. *Computer*, 2004.

[21] The global positioning systems (wikipedia entry). http://en.wikipedia.org/wiki/Global_Positioning_System.

[22] M. Youssef et al. Pinpoint: An asynchronous time-based location determination system. In *ACM Mobisys*, June 2006.

[23] J. Werb and C. Lanzl. Designing a positioning system for finding things and people indoors. *IEEE Spectrum*, 35(9), September 1998.

[24] X. Cheng et al. TPS: A time-based positioning scheme for outdoor sensor networks. In *IEEE Infocom*, March 2004.

[25] S.J. Fontana, R.J.and Gunderson. Ultra-wideband precision asset location system. In *IEEE Conference on Ultra Wideband Systems and Technologies*, May 2002.

[26] R.J. Fontana, E. Richley, and J. Barney. Commercialization of an ultra wideband precision asset location system. In *IEEE Conference on Ultra Wideband Systems and Technologies*, November 2003.

[27] A. Smith et al. Tracking moving devices with the cricket location system. In *ACM Mobisys*, June 2004.

[28] A. Savvides and M. Han, C.C. Srivastava. Dynamic fine-grained localization in ad-hoc networks of sensors. In *ACM MobiCom*, 2001.

[29] D. Niculescu and B. Nath. Ad hoc positioning system (APS) using AoA. In *IEEE Infocom*, 2003.

[30] D. Niculescu and B. Nath. VOR base stations for indoor 802.11 positioning. In *ACM MobiCom*, September 2004.

[31] J. Xiong and K. Jamieson. SecureAngle: improving wireless security using angle-of-arrival information. In *ACM HotNets*, 2010.

[32] S. Sen et al. Precise Indoor Localization using PHY Layer Information. In *ACM HotNets*, 2011.

No Need to War-Drive: Unsupervised Indoor Localization

He Wang
Duke University

Souvik Sen
Duke University

Ahmed Elgohary
EJUST, Egypt

Moustafa Farid
EJUST, Egypt

Moustafa Youssef
EJUST, Egypt

Romit Roy Choudhury
Duke University

ABSTRACT

We propose *UnLoc*, an unsupervised indoor localization scheme that bypasses the need for war-driving. Our key observation is that certain locations in an indoor environment present identifiable signatures on one or more sensing dimensions. An elevator, for instance, imposes a distinct pattern on a smartphone's accelerometer; a corridor-corner may overhear a unique set of WiFi access points; a specific spot may experience an unusual magnetic fluctuation. We hypothesize that these kind of signatures naturally exist in the environment, and can be envisioned as internal *landmarks* of a building. Mobile devices that "sense" these landmarks can recalibrate their locations, while dead-reckoning schemes can track them between landmarks. Results from 3 different indoor settings, including a shopping mall, demonstrate median location errors of $1.69m$. War-driving is not necessary, neither are floorplans – the system simultaneously computes the locations of users and landmarks, in a manner that they converge reasonably quickly. We believe this is an unconventional approach to indoor localization, holding promise for real-world deployment.

Categories and Subject Descriptors

H.3.4 [**Information Storage and Retrieval**]: Systems and Software; C.2.4 [**Computer-Comunication Networks**]: Distributed Systems

Keywords

Location, Mobile phones, Sensing, Landmarks, Recursion

1. INTRODUCTION

Despite innovative research [1–9, 9–16], indoor localization is still not in the mainstream. In trying to trace the reasons, we distilled two main messages: (1) Indoor spaces require fairly high location accuracy because the contexts vary at finer spatial granularity. For instance, $5m$ location error outdoors may still indicate the same street, but $5m$ indoors may mean two different aisles in a grocery store. An inventory-management application may very well require aisle-level precision. (2) While such precision is attainable with pervasive WiFi systems, they come at a prohibitively high cost, mostly in the form of meticulous (signal) calibration. Such calibration is not necessarily a one-time cost since RF fingerprints could change, perhaps due to changes in layout and objects in the environment. Attempts to simplify the calibration process have been successful, but at the expense of reduced location accuracy. This zero sum game between accuracy and calibration overhead has been an important hurdle to deploying indoor localization systems. This paper is tasked to break away from this tradeoff, and achieve *meter* level accuracy with zero calibration. Although a high bar, we believe this is feasible and make an attempt to do so.

Our scheme cuts across isolated ideas in mobile computing. We introduce these ideas first, and then describe the value in making them compatible.

(1) A few recent schemes have demonstrated the ability to compute the motion trajectory of a mobile phone, using its accelerometer and compasses [2, 17]. This has been called urban dead-reckoning. Due to noise in the mobile sensors, the dead-reckoned trajectories are accurate in the beginning, but diverge from the truth over time. Therefore, outdoor localization schemes like CompAcc [2] have triggered periodic GPS measurements to recalibrate the user's location. Unfortunately, GPS is unreliable indoors, rendering dead-reckoning based approaches useless. Nonetheless, if one can identify other means of recalibration, dead-reckoning could be applicable even indoors.

(2) Urban sensing and activity recognition literature have demonstrated the ability to recognize ambiences and user behavior. For instance, inertial sensors can detect when a user is walking, turning into a corridor, or climbing up the stairs [18]; microphones and magnetometers can detect ambient sounds and magnetic fluctuations [14, 19]. While these signatures have been primarily used for various forms of context-awareness, they lend themselves to localization as well. For example, these signatures can be treated as "landmarks", useful to recalibrate indoor dead-reckoning.

(3) Past work has mostly relied on signal calibration [1, 3] to develop WiFi based localization. We observe that WiFi can be

MobiSys'12, June 25–29, 2012, Low Wood Bay, Lake District, UK.

valuable even without calibration. For instance, one can use overheard WiFi APs to partition an indoor space into smaller sub-spaces. Thus, a landmark signature need not be unique in the entire building – so long as it is unique within a WiFi sub-space, it can be recognized without ambiguity.

UnLoc combines these 3 ideas (dead-reckoning, urban sensing, and WiFi-based partitioning) into a framework for unsupervised localization. The core idea is recursive but not complicated. Briefly, mobile users move naturally in the building collecting accelerometer, compass, gyroscope, and WiFi readings. By assimilating data from these devices, UnLoc detects sensory signatures (e.g., a corridor turn) that are unique within their respective WiFi sub-spaces. Now using the same collected data, UnLoc dead-reckons the devices starting from a known reference location, say the entrance of the building. Since dead-reckoning provides a rough location to the phone, it is also possible to roughly localize the signatures based on when the phone senses them.

Now, the locations of these signatures – also called *landmarks* – can be made more accurate by combining the rough estimates from multiple phones. These landmarks can then be used to improve dead-reckoning of subsequent phones, which in turn can refine the landmark locations. This recursive process continues to improve localization accuracy over time. Observe that, the system does not need calibration – the first few users may experience inferior location accuracy, but a little more data brings the system to convergence.

Of course, translating this idea-sketch into a functional system entails a variety of challenges: (1) Dead-reckoning is non-trivial in indoor environments where metals and electrical equipment significantly affect the compass bearing[1]. (2) Extracting the accurate location of the landmarks from multiple erroneous locations is problematic, since the errors may not be all equal. Some notion of confidence on the errors need to be built, so that the estimates can be suitably weighted before combination. (3) Identifying landmark signatures from the sensed data warrants unsupervised learning on sensor features. (4) Finally, the system needs to be optimized for energy, to avoid a significant battery drain during localization. This paper addresses these challenges one step at a time, prototypes on a testbed of Android Nexus series phones, and evaluates with 3 volunteers naturally walking in 2 university buildings and 1 shopping mall. Performance results show median localization accuracy of $1.69m$ when UnLoc runs online, and $0.89m$ when the locations are computed offline. The system quickly reaches convergence in less than 2 man-hours, and remains robust to dynamic changes in the environment. We believe this could be a promising direction, and with rigorous testing and tuning, a potential candidate for the real-world.

Our main contributions may be summarized as follows:

- **We identify an opportunity to simultaneously harness sensor-based dead-reckoning and environment sensing for localization.** Our approach does not require calibration or installation of additional infrastructure.
- **We design a practical scheme that employs unsupervised learning to extract unique sensor signatures – called landmarks.** We show that adequate landmarks exist inside buildings such that dead-reckoning is practical and reasonably accurate.
- **We develop UnLoc on the Android OS, and evaluate across 3 different indoor spaces, including the North Gate shopping mall in Durham.** We achieved less than $2m$ error without any pre-deployment effort; in fact, we used the map of the building only to compute ground truth for evaluation.

[1]Prior work demonstrated dead-reckoning mostly for outdoor environments.

The subsequent sections expand on each of these contributions, beginning with architectural overview and intuition, followed by measurement, design, and evaluation.

2. ARCHITECTURE AND INTUITION

We begin with a high level overview of UnLoc, focussing mainly on the core building blocks and intuitions (Figure 1). We will postpone the discussion on actual algorithms and engineering details to the next section. In fact, *we will even make a simplifying assumption that the building floorplan is available to UnLoc.*

Of course, this assumption need not hold in our final system – its sole purpose is ease of explanation. Once we have developed the core framework, we will discuss how the assumption can be relaxed.

Human Motion Traces

Consider the example where we intend to localize users in a shopping mall. When users visit the mall, the UnLoc app running on their smartphones collect time-stamped sensor readings. These readings (mainly from accelerometer, compass, gyroscope, and WiFi APs[2]) are assimilated in a central repository. UnLoc must operate on this data to track each user's location. For now, we perform this offline.

Seed Landmarks (SLMs)

As a first step, UnLoc looks into the floorplan of the building and identifies some *"seed landmarks"*. These seed landmarks are essentially certain structures in the building – stairs, elevators, entrances, escalators – that force users to behave in predictable ways. These predictable behaviors can be translated to sensor signatures. For instance, building entrances are characterized by a visible drop in the GPS confidence when the user moves from outdoors to indoors; elevators exhibit a distinct accelerometer signature, emerging from the start and stop of the elevator. If a user, Alice, used the elevator in the mall, UnLoc expects her trace to contain the elevator signature embedded in it. Let us assume that this signature occurs in Alice's trace at time t_i. Since the location of the elevator is known within the floorplan, UnLoc can precisely localize Alice at t_i. In similar ways, Alice can be precisely localized at other time-points when she passed through any of the other landmarks. The next step, then, is to localize her while she moved between these landmarks – for this, UnLoc adopts techniques from urban dead-reckoning.

Dead Reckoning

Urban dead-reckoning [2] is an established idea that uses the accelerometer and the compass to track a mobile user.

[2]Sound and light measurements could also be useful, but we do not activate them for energy-efficiency.

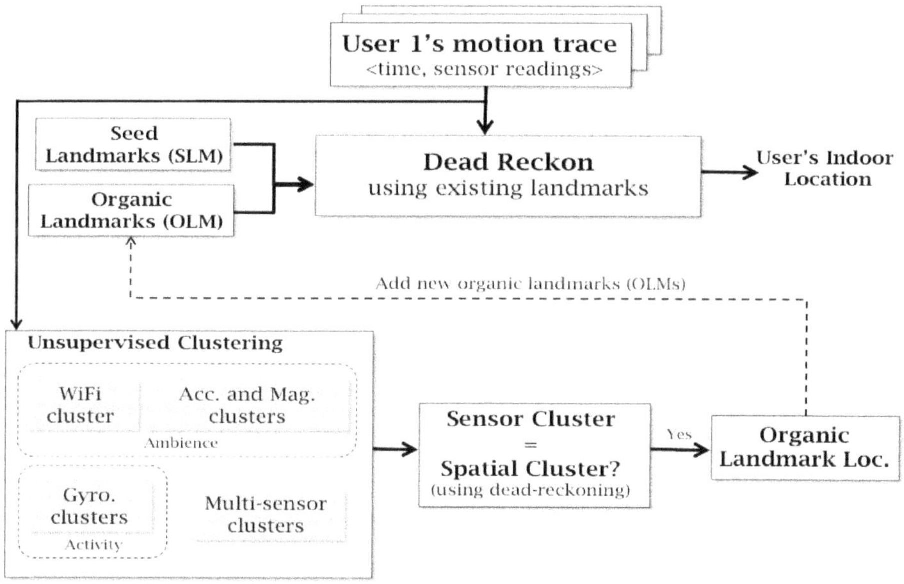

Figure 1: UnLoc architecture – motion vectors are computed from recorded sensor readings, and the user's location computed via dead-reckoning. The dead-reckoning error is periodically reset using landmarks in the environment. Further, the same sensor readings are mined to identify signatures of new landmarks. These landmarks help in improving localization accuracy for subsequent users.

The key idea is simple. Based on the accelerometer readings of the mobile phone, it is possible to count the number of steps a person has walked, and therefrom derive the displacement of the person. Based on the compass, the direction of each of these steps can be tracked[3]. Merging these, the <*displacement, direction, time*> tuple forms the human's motion vector. Pivoting these vectors at the seed landmarks, UnLoc tracks the location of Alice. Although the tracking operation is crude (because the noisy sensors accumulate error over time), the error gets reset whenever Alice crosses any of the landmarks. Thus, in the steady state, Alice's localization error exhibits a saw-tooth behavior – over time, the localization error grows and then sharply drops to zero at the landmark, and then grows again. Observe that increasing the density of landmarks will cause the error to reset frequently, thereby curbing the error growth. UnLoc will attempt to accomplish this by organically extracting additional landmarks from the indoor environment

Organic Landmarks (OLMs)

In addition to seed landmarks, UnLoc postulates that any indoor environment will offer some ambient signatures across one or many sensing dimensions. These signatures can be in the magnetic domain, wherein metals in a specific location may produce unique and reproducible fluctuations on the user's magnetometer, near that location. Signatures could also be WiFi-based – a spot may overhear a set of WiFi base stations, but the set may change at short distances away from that spot. A few (dead) spots inside a building may not overhear any WiFi or GSM/3G signals, which by itself is a signature. Further, even a water-fountain could be a signature – users that stop to drink water may exhibit some common patterns on the accelerometer and magnetometer domains. Whenever these pattern surface on Alice's trace, it could be an opportunity to recognize her location. Of course, even though these signatures can become useful landmarks, they cannot be known *a priori*, and will vary across different buildings. They have to be learnt dynamically.

Towards learning these landmarks, UnLoc subjects the sensor data (gathered from all phones) to a clustering algorithm (Figure 1). Actually, various features of this data are extracted and the clustering runs on this high dimensional space – detailed in the next section. Once the clustering operation has completed, each of the resulting clusters is expected to contain similar sensor patterns. Now, since each sensor reading is associated with time-stamps, it is possible to find their corresponding locations via dead-reckoning. UnLoc computes these locations to check whether all members of a cluster fall within a small area – Figure 2. If they do, then UnLoc deems it a new landmark. Since these landmarks were discovered automatically, without any external supervision, we call them *organic landmarks* (OLMs). If no OLMs are found, UnLoc waits for more traces and continues scanning for OLMs.

Simultaneous Localization and Mapping

Both SLMs and newly discovered OLMs are used to improve dead-reckoning for subsequent users, which in turn improves the location estimates of the SLMs/OLMs themselves. This circular process pushes the entire system to better accuracy and UnLoc continues to improve over time. The evaluation section will quantify this behavior, for different users and buildings.

2.1 Intuitions and Supporting Measurements

The success of UnLoc hinges on at least 3 performance-related expectations: (1) Dead reckoning can attain desired levels of accuracy, if periodically recalibrated by landmarks. (2) Indoor environments indeed offer the requisite number of landmarks. (3) The locations of the landmarks can be computed

[3]However, as we will find in the next section, magnetic fluctuations in the environment will derail the compass readings, forcing us to shift to gyroscopes.

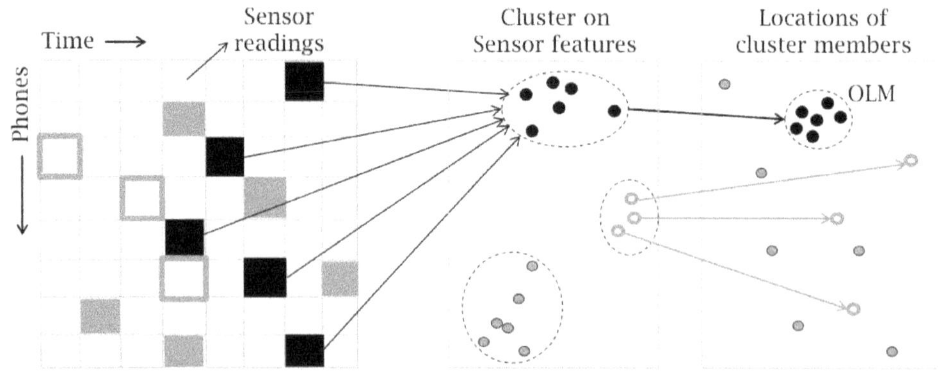

Figure 2: Matrix showing sensor readings collected by devices across time. Readings are clustered and location of cluster members computed. If all cluster members fall within a small region, UnLoc deems it an OLM.

from rough estimates of multiple devices (i.e., the dead-reckoning errors are indeed independent). We discuss our intuitions in each of these, and present supporting measurements.

(1) Dead-Reckoning Accuracy

We performed some initial experiments in the computer engineering building at Duke University. Volunteers were asked to carry NexusS phones in their pockets, and walk naturally in one wing of the building. The UnLoc app running on the phones records the accelerometer and compass readings, and extracts from them the <*displacement, direction, time*> tuples. To record ground truth, we marked different doors and windows in the buildings with a distinct number[4] – when a user passed by that number, she entered it into her phone. Since we know the mapping between the number and the door/window, we are able to extract the ground truth (almost accurately). We gathered 10 traces, each starting from the entrance of the building. We performed trace-based analysis to understand how the dead-reckoned path diverges from the true path, with varying number of landmarks.

Figure 3 shows the accumulated error over time (light gray curve) when using pure dead-reckoning with zero landmarks. Evidently, the error accumulates dramatically fast, and is completely unusable. Hence, we simulate landmarks by periodically reseting the user's location to the correct location – the black curve shows the results. While the performance improves, the mean localization error is still $11.7m$, greater than our target. On analyzing the data, we observed that the magnetic field in indoor environments is heavily distorted by metallic and electrical equipments in the (engineering) building. Thus, although these magnetic fluctuations are beneficial for finding landmarks, they derail dead-reckoning by injecting heavy error in the compass. UnLoc appeared impractical at this point.

Fortunately, newer smartphones are embedded with gyroscopes that measure the angular velocity of the phone in 3 dimensions and are not affected by the magnetic field. We carefully processed the gyroscope readings to compute the angular changes during walking, i.e., how the user turned. However, gyroscope readings are relative and we needed to combine it opportunistically with the compass to estimate the user's absolute walking direction. We will report several technical challenges of this operation in section 3, including compass offsets and gyroscope drifts. However, once we harnessed the gyroscope and the compass in tandem, the dead-reckoning error reduced appreciably (blue curve in Figure 3). When re-calibrated by periodic landmarks, the average error dropped further to $1.2m$, offering us confidence to build a fuller system.

[4]Recall that GPS is unavailable indoors.

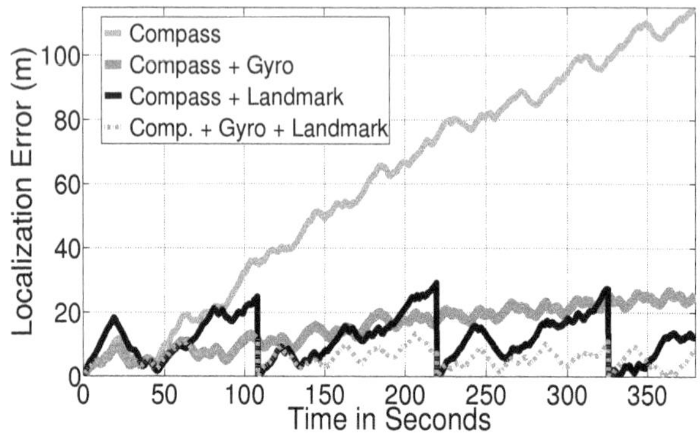

Figure 3: Error from dead-reckoning reduces with gyroscope, and further with landmarks.

In addition, once a landmark is encountered, the user's path can be retraced and corrected between the last two landmarks. Although this does not help in realtime tracking of the user, it helps in offline analysis, and more importantly, for improving the location estimate of organic landmarks.

(2) Landmark Density

It is natural to question *why indoor environments would exhibit sensor signatures to be used as landmarks?* While there is of course no guarantee, our observation is that an indoor environment is rich with ambient signals, like sound, light, magnetic field, temperature, WiFi, 3G, etc. Moreover, different building structures (e.g., stairs, doors, elevators) force humans to behave in specific ways. If one "combs" through all these sensor signals and their high-dimensional combinations, we postulate that some signatures are likely to emerge. The intuition is essentially rooted in diversity, i.e., the chances that all of the signals are similar and no pattern "shows up as different", seems unlikely. Further, these signatures need not be unique in the entire building – so long as they are unique within the WiFi sub-space, they can be valid landmarks. We performed a small scale measurement study to verify this intuition – the results follow.

• **WiFi Landmarks**:
Consider an area within which all locations overhear a distinct set of WiFi APs. An indoor space is likely to have many such areas of varying sizes. Figure 4 shows the distribution of the sizes of these areas. While most of the areas are quite large – explaining why simple WiFi based localization is not very accurate – there are a few areas (at the left side of the X axis) that are very small. UnLoc aims to exploit these small areas as a landmark. If one of this small areas overhears a set of WiFi APs, denoted W, then a mobile phone overhearing the same set can be assumed to be within that area. Since the area is small, the localization error of the phone will be small too, enabling a location recalibration. Our measurements show that in two floors of the engineering building, we find 8 and 5 such WiFi landmarks, each of area less than $4m^2$. Thus WiFi APs can offer landmarks to enhance dead-reckoning.

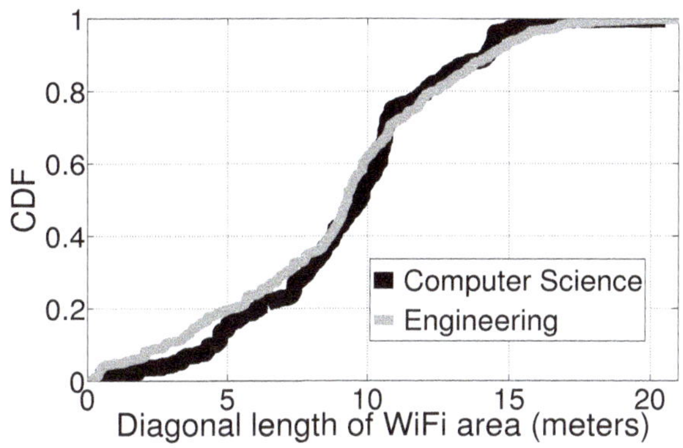

Figure 4: Some WiFi areas are very small (tail of distribution), and hence, an ideal landmark.

• **Magnetic/Accelerometer Landmarks**:
In search of signatures in other sensing domains, we executed K-means clustering on accelerometer and compass measurements (we present the algorithm and parameter details later). For each cluster, we mapped their members to their corresponding physical locations (using ground truth). For most clusters, we found that their member-locations were widely scattered in space, and hence, were unusable as a landmark. However, members of a few clusters proved to be tightly collocated in space as well. For example, we discovered a unique/stable magnetic fluctuation near our networking lab. We found another spot with a distinct accelerometer signature – a pair of symmetric bumps in opposite directions (Figure 5).

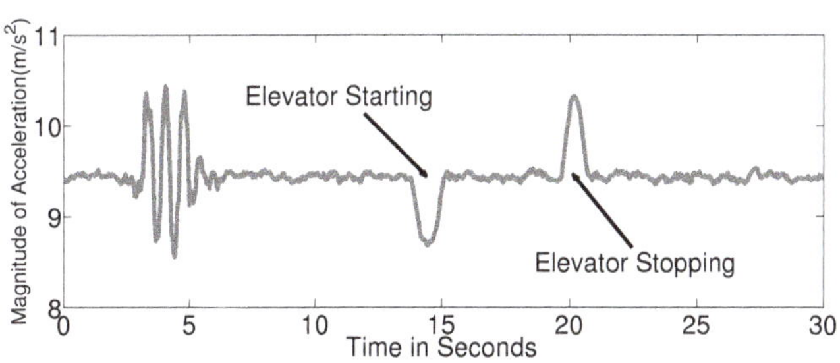

Figure 5: Accelerometer signature inside the elevator (caused by the elevator starting and stopping).

Although UnLoc need not understand the semantic meaning of these signatures, out of curiosity we analyzed the data, and discovered that they were caused by the elevator starting and stopping. In fact, the direction of the bumps (upward or downward) even indicated whether the user went upstairs or downstairs. These spatially collocated patterns were natural landmarks, and when we assimilated all sensing dimensions, the number of landmarks proved to be 6 and 8 for each floor. In fact, when we combined accelerometer and compass together – a higher dimensional signature – we found even more landmarks due to turns in the building. In sum, the organic and seed landmarks together seemed to offer the needed density to support indoor dead reckoning.

(3) Computing Landmark Locations

Perhaps an important pitfall in UnLoc lies in computing the locations of the landmarks. Figure 6(a) is intended to explain the problem. In this example, assume that UnLoc has combined the user's sensor data, clustered on them, and discovered 3 sensor signatures (in distinct WiFi areas), that can be used as landmarks. Now the locations of these landmarks need to be computed, but there is no ground truth to learn that. One way to estimate the location of a landmark is to use dead-reckoning, but that will not be accurate since dead-reckoning itself is erroneous. Our approach is to compute the landmark location by combining all the (dead-reckoned) estimates of a given landmark. The intuition is that dead-reckoning errors have been observed to be random and independent, due to the noise in hardware sensors and human step sizes [2]. By combining these errors from adequate measurements, one could expect the estimated mean to converge to the actual landmark location. Figure 6(b) illustrates the opportunity through a simple centroid calculation.

Figure 8 visualizes data from real measurements to verify independence in dead-reckoning error – each line joins the estimated landmark location to the actual location. Visually, the errors appear uncorrelated. Of course, this is a preliminary overview of the inner workings of the system – we later discuss the algorithmic details of this error combining process, and measure performance in greater detail.

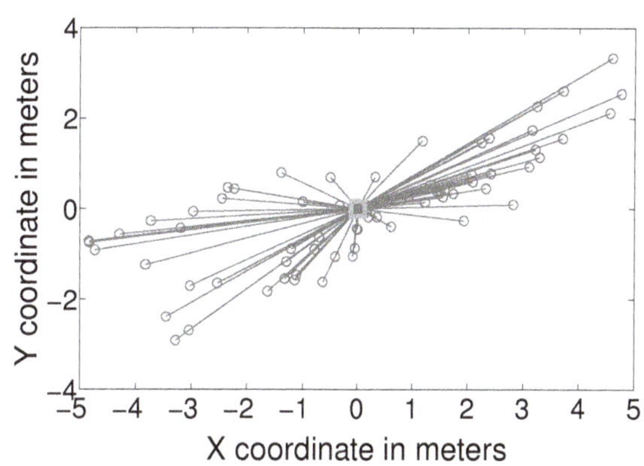

Figure 8: Uncorrelated errors from multiple dead-reckoning estimates.

3. DESIGN DETAILS

This section describes the algorithms and engineering details underlying UnLoc.

3.1 Seed Landmarks

If the building's floorplan is known (which is often necessary to visualize the user's location), then UnLoc can infer the lo-

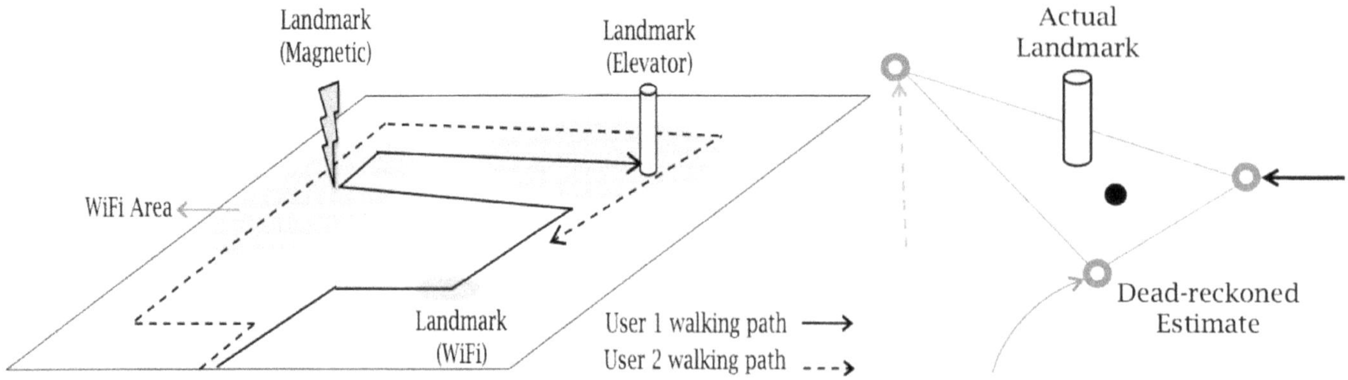

Figure 6: (a) UnLoc users walk and periodically encounter landmarks – refines landmark locations, corrects own location. (b) The solid circle showing the centroid of the dead-reckoned estimates. Multiple erroneous estimates leading to a better approximation of the landmark location.

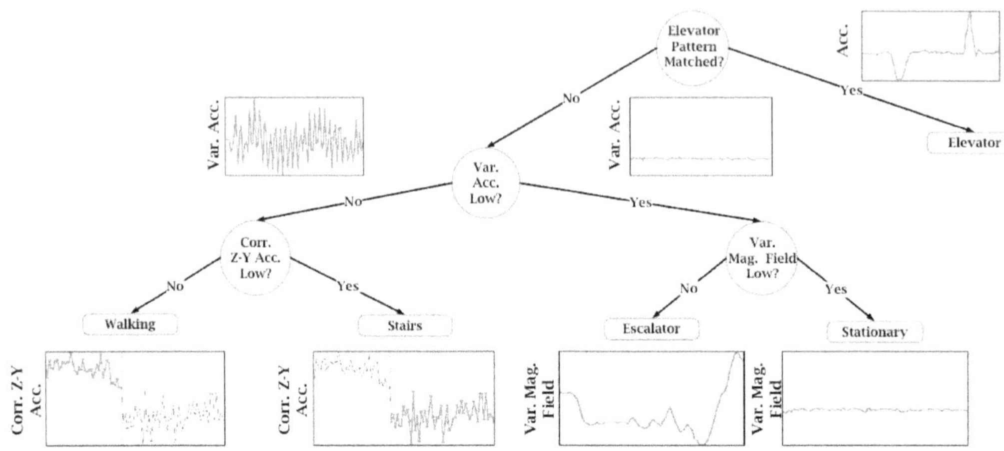

Figure 7: Decision tree for detecting Seed LandMarks (SLMs). The top level separates the elevator based on its unique acceleration pattern. The second level separates the constant velocity classes (stairs and escalator) from the other two classes (walking and stairs) based on the variance of the acceleration. The third level uses the variance of magnetic field to separate the escalator from the stationary case and the correlation between the Z and Y acceleration components to separate between the stairs and walking cases.

cations of doors, elevators, staircases, escalators, etc. This implies that the location of seed landmarks (SLMs) are immediately known. As long as the smartphone can detect these SLMs while passing through them, it can recalibrate its location. Thus, the goal of the SLM detection module is to define sensor patterns that are global across all buildings.

Elevators, Staircases, and Escalators

This class of SLMs are based on using the inertial sensors. These sensors have the advantage of being ubiquitously installed on a large class of smart phones, having a low-energy footprint, and being always on during the phone operation (to detect the change of screen orientation). We focus on three particular examples that are common in indoor environments: elevators, escalators, and stairs. Figure 7 shows a classification tree for detecting the three classes of interest and separating them from walking and being stationary. We note that a false positive leads to errors in estimating the location of the SLM while a false negative leads to missing an opportunity for recalibration. Therefore, high detection accuracy with low false positive/negative rates are highly desired.

Elevator: A typical elevator usage trace consists of normal walking period, followed by waiting for the elevator for sometime, walking into the elevator, standing inside for a short time, an over-weight/weightloss occurs (depending on the direction of the elevator), then a stationary period which depends on the number of the floors the elevator moved, another weight-loss/over-weight, and finally a walk-out. To recognize the elevator motion pattern, we developed a Finite State Machine (FSM) that depends on the observed state transitions. Different thresholds are used to move between the states. Evaluation over 22 traces show that the thresholds are robust to changes in the environment and can achieve 0.6% and 0% false positive and negative rates, respectively.

Escalator: Once the elevator has been separated, it is easy to separate the classes with constant velocity (escalator and stationary) from the other classes (walking and stairs) using the variance of acceleration. Now, to separate the escalator from stationarity, we found that the variance of the magnetic field can be a reliable discriminator. Of course, a user may sometimes climb up the escalator – we find that magnetic variance also differentiates between this and an actual staircase-climb (see Figure 9).

Stairs: Once the scenario with constant speed is separated,

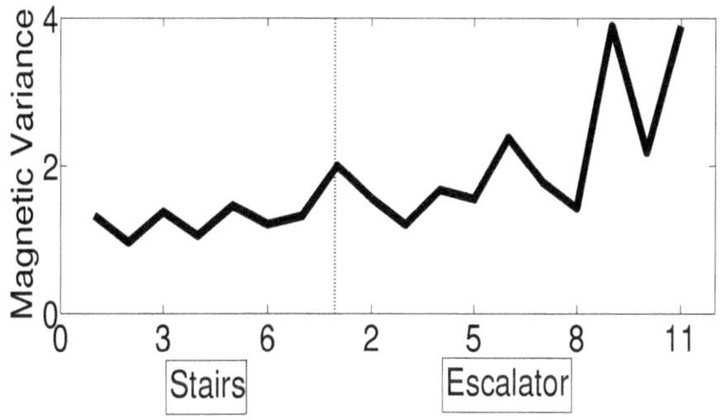

Figure 9: Magnetic variance when a user is climbing stairs versus escalator.

we need to differentiate between the stair and walking case. The main observation here is that when the user is using the stairs, her speed increases or decreases based on whether the gravity is helping her or not. This creates a higher correlation between the acceleration in the direction of motion and direction of gravity as compared to walking. As reported later, staircases can sometimes lead to false negatives (1.8%).

3.2 Dead Reckoning

The two sub-tasks in dead reckoning are (1) computing the user's displacement from accelerometer and (2) continuously tracking the direction of movement.

Displacement from accelerometer:
One possible solution is to double-integrate the accelerometer readings. Figure 10(b) shows the unacceptable results – the difference between the estimated and actual displacement reaches more than $100m$ only after $30m$ of actual displacement. This is an attribute of a noisy accelerometer, low sampling rate (24Hz), as well as jerky movements of the phone when carried by the user. A better approach [2,20] is to identify a human-walking signature as in Figure 10(a). This signature arises from the natural up/down bounce of the human body for each step taken. To capture this, we pass the signal through a low pass filter, and identify two consecutive local minima. Between these local minima, we search for a local maxima, and check whether the difference between the maxima and minima is greater than a threshold. If so, we increment the step count.

The physical displacement can be computed by multiplying step count with the user's *step size*, a function of the user's weight and height [21][5]. Employing a fixed step size across all users can clearly be erroneous. However, UnLoc has the opportunity to infer the step size by counting the number of steps for a known displacement (i.e., between two landmarks). Figure 10(c) shows the error accumulated by UnLoc using these techniques in place – the results are encouraging. The step count accuracy (verified with 10 users) was also 98%.

Orientation using compass/gyroscope:
Past work has demonstrated the feasibility of dead-reckoning using the smartphone compass. However, to the best of our knowledge, all these results are for outdoor environments. Indoors, the magnetic field due to ferromagnetic material and electrical objects in the vicinity, completely derailed our dead

[5]A more accurate approach is to estimate the user step size based on her gait [22].

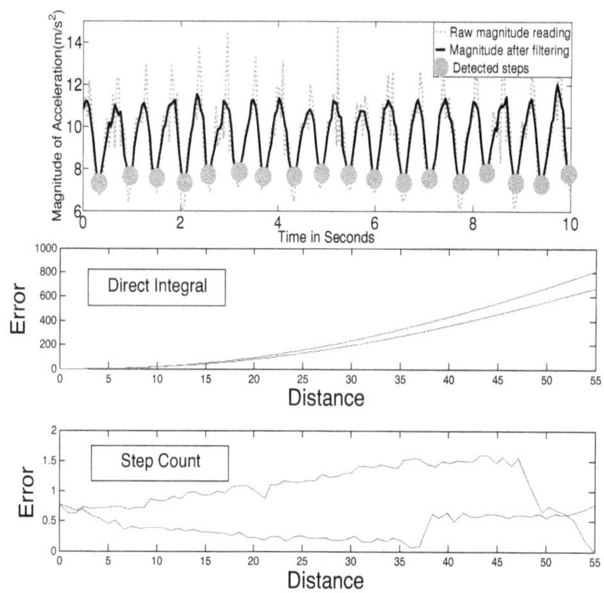

Figure 10: (a) Accelerometer readings (smoothened) from a walking user. (b) Displacement error with double integration for two users. (c) Error with step count method.

reckoning attempts[6]. In response to this, we explored the feasibility of using gyroscope to *infer* the user's movement directions. Our intuition was that the gyroscope is decoupled from the magnetometer sensor, and hence, could be insensitive to ambient magnetic fields.

While this insensitivity proved true, i.e., the gyroscope indeed remained unaffected by changing magnetic fields, the tradeoff was that the gyroscope offered *relative angular velocity*. This is in the form of a 3D rotation matrix which when multiplied by a time interval, yields the relative angular displacement (RAD) of the device. Unfortunately, the RAD is with respect to a direction that is not necessarily the absolute direction. Thus, while we could track the structure of the user's motion path using the gyroscope, these paths were biased by the error in their initial direction. Thus all the estimated paths appeared as rotated versions of the true path, shown in Figure 11(a).

We observe that encountering landmarks can help infer this bias. Figure 11(b) explains the opportunity. Consider a user encountering a known landmark L_1 at time t_1, and later another landmark L_2 at time t_2. UnLoc identifies that the user encountered these two landmarks because the signatures matched, and hence, regardless of the dead-reckoned estimates, UnLoc "pins down" the user's path at these locations. Now, let θ denote the angle between the line joining L_1, L_2 and the line joining L_1, X_2, where X_2 is the dead-reckoned estimate at time t_2. We observe that θ is the initial bias, and therefore, UnLoc rotates the entire motion segment by θ. Importantly, the same θ can be used to track the user for the subsequent motion segment – say until the user encounters landmark L_3 – at which point the bias can again be updated. This process of learning and updating the bias at every landmark leads to stable and consistent results. The only remaining problem lies in tracking the user until she has encountered the second

[6]We attempted at least 5 different techniques to learn and correct for these fluctuations.

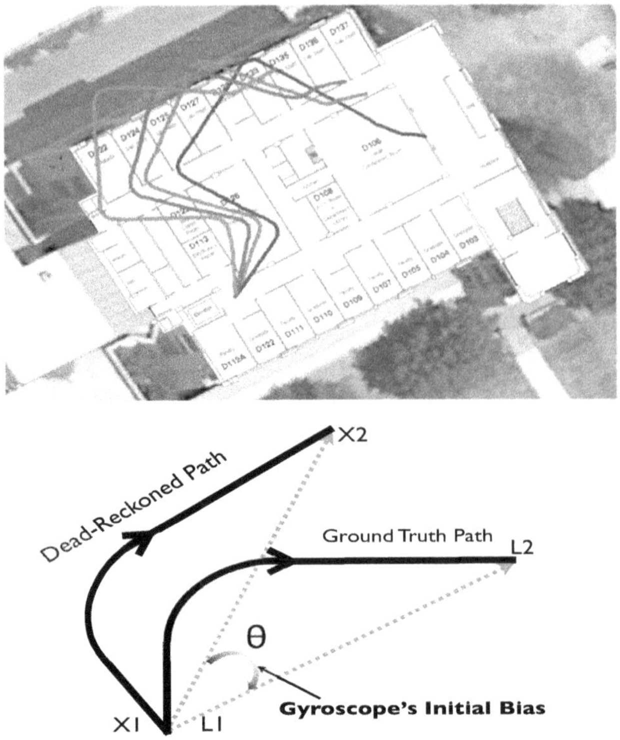

Figure 11: (a) Rotated walking trail due to gyroscope's initial bias. (b) Correcting gyroscope's bias using landmarks.

landmark – during this phase, her dead reckoning error can be arbitrarily high.

UnLoc turns to the compass during the initial phase when the gyroscope bias is still unknown. Clearly, the compass value cannot be used blindly – ambient magnetic field fluctuations can actually degrade the results. Thus, UnLoc juxtaposes the gyroscope and compass readings – Figure 12(a) – and whenever the trends are correlated, the compass value is selected as the direction of motion. The gyroscope's bias is now inferred, and used thereafter. The intuition here is that correlated trends in compass and gyroscope is an indicator of proper compass readings; and if the compass is not reflecting the gyroscope's trend, it is probably affected by other factors. Figure 12(b) shows the eventual outcomes in angular direction estimation. The compass helps with the initial phase, while gyroscope based dead reckoning proves to be effective. In fact, to the best of our knowledge, UnLoc is the first to leverage smartphone gyroscopes for indoor dead reckoning.

3.3 Organic Landmarks

The task of discovering organic landmarks (OLMs) is rooted in (1) recognizing distinct patterns from many sensed signals, (2) and testing whether a given pattern is spatially confined to a small area. Recall that Figure 2 illustrates the flow of operations. All the sensor readings are gathered in a matrix: element <i,j> of the matrix contains sensor readings from phone i at time j. These sensor readings are essentially *features* of the raw sensed values (from the accelerometer, compass, gyroscope, magnetometer, and WiFi). Features for the magnetic and inertial sensors include *mean, max, min, variance, mean-crossings,* while for WiFi, they are *MAC ID, RSSI.*

UnLoc normalizes these features between $[-1, 1]$ and feeds

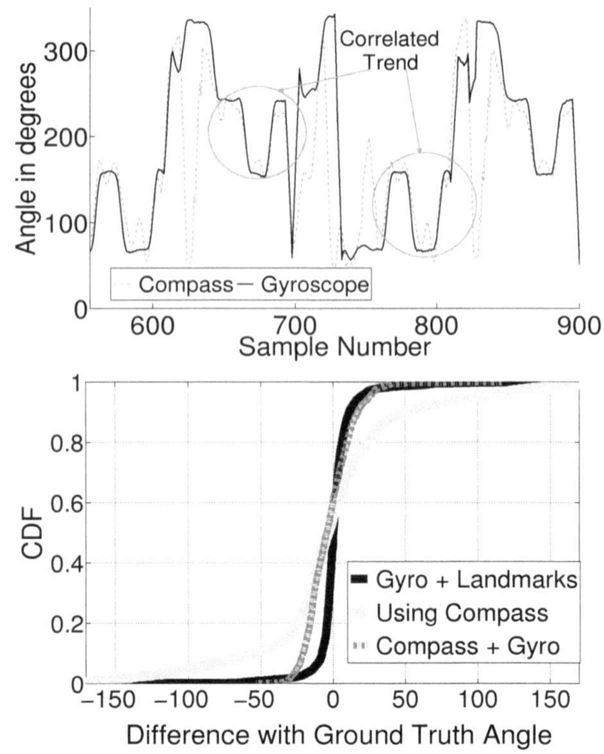

Figure 12: (a) Compass orientation is more reliable when it is correlated with the gyroscope. (b) CDF of orientation estimation error.

them to a $K - means$ clustering algorithm. (we tested with *Expectation Maximization* (EM) clustering as well, but it was not consistently better than K-means). The clustering is executed for each individual sensing dimensions, as well as their combinations (such as accelerometer and compass together). Figure 13, for example, shows the clusters from the magnetometer readings for $K = 3$. We varied the value of K and recorded the clusters in each case. Our goal is to identify clusters that have *low similarity* with all other clusters; this will suggest a good signature. For this, we compute the correlation between a given cluster and all other clusters – if the maximum correlation is less than a *similarity threshold,* we consider this cluster as a candidate for landmark.

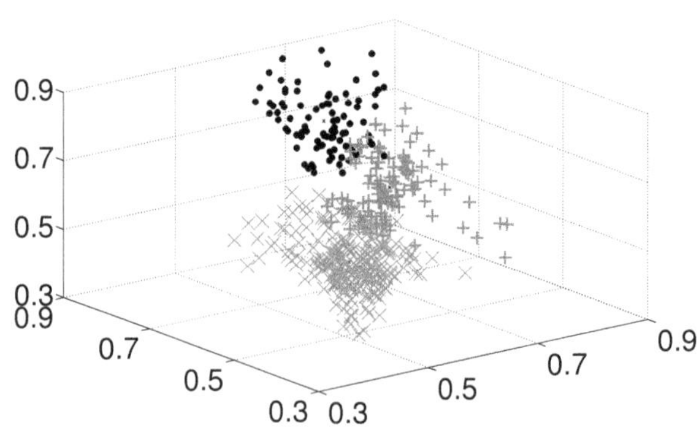

Figure 13: Clusters identified by kmeans algorithm

To qualify as an OLM, the candidate cluster must also be confined to a small geographical area. For this, we first test whether the members of a cluster are within the same WiFi area (i.e., they overhear the same WiFi APs). While this is necessary, it is not sufficient because many WiFi areas are large. Therefore, for clusters within a WiFi area, we compute the dead-reckoned locations for each of their members. If locations of all cluster-members are indeed within a small area – we use $4m^2$ – then we recruit this cluster as an OLM. As an anecdote, we found that using accelerometer, one of the sensor clusters were scattered all over the indoor space; upon investigation, we detected that this cluster roughly captured walking patterns. On the other hand, another cluster that proved to be within the $4m^2$ area was from a magnetic signature near an electrical service room in the building. UnLoc announces the centroid of the cluster as the OLM's location.

While the above describes the generalized version of the OLM detection algorithm, the different sensing dimensions require some customization, discussed next.

3.3.1 WiFi Landmarks

We use MAC addresses of WiFi APs and their corresponding RSSI values as features. To remain robust to signal variations (which alters the set of overheard APs), we only consider APs that are stronger than a threshold RSSI. Now, applying K-means clustering, we identify small areas ($4m^2$) that have low similarity with all locations outside that area. We compute similarity of two locations, l_1 and l_2, as follows. Let us denote the sets of WiFi APs overheard at locations l_1 and l_2 as A_1 and A_2, respectively. Also, let $A = A_1 U A_2$. Let $f_i(a)$ denote the RSSI of AP a, $a \in A$, overheard at location l_i; if a is not overheard at l_i, then $f_i(a) = 0$. We now define similarity $S \in [0, 1]$, between locations l_1 and l_2 as:

$$S = \frac{1}{|A|} \sum_{\forall a \in A} \frac{min(f_1(a), f_2(a))}{max(f_1(a), f_2(a))}$$

The rationale for this equation is to add proportionally large weights to S when an AP's signals are similarly strong at both locations, and vice versa. We threshold on S to define a WiFi landmark. We choose a threshold of 0.4 in our system, indicating that all locations within the WiFi landmark needs to exhibit less than 0.4 similarity with any other location outside the landmark. Of course, this rather strict threshold ensures that landmarks are quite distinct, but also reduces the number of possible landmarks. Figure 14 shows this tradeoff using traces from two Duke University buildings. We observed that 0.4 was a reasonable cut-off point, balancing quality and quantity of WiFi OLMs.

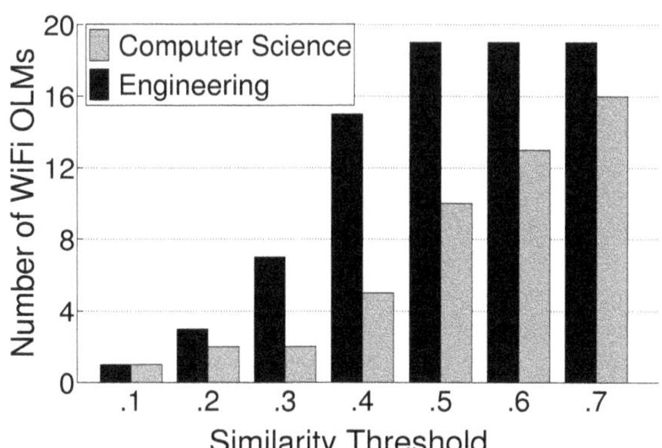

Figure 14: Tradeoff between similarity threshold and number of WiFi landmarks.

3.3.2 Magnetic and Inertial Sensor Landmarks

Indoor environments are characterized by at least a few turns (at the end of corridors, into offices, classrooms, stairs, etc.). Since the gyroscope offers reliable angular displacements, we recognize the opportunity to use them as organic landmarks. We design a special feature called the *bending coefficient*. Essentially, the coefficient captures the notion of path curvature, computed as the length of the perpendicular from the center of a walking segment to the straight line joining the end-points of the segment. We compute the bending coefficient over a sliding window on the user's walking path, and use them as a separate feature. Later, when we cluster on bending coefficient and WiFi together as features, similar turns within a WiFi area gather in the same cluster. The turns in the cluster could still be doors of adjacent classrooms in a corridor – these turns may very well lie within the same WiFi area. To avoid coalescing all these turns into the same landmark, UnLoc checks if the cluster is confined to within a $4m^2$ area; only then is the cluster declared a landmark.

Magnetic landmarks are also derived through similar techniques. So long as the magnetic signature is unique within one WiFi area, and the sensed locations are spatially confined within $4m^2$, we deem it as a magnetic OLM. Figure 15 shows an anecdotal example where the magnetic field near our networking lab demonstrates a unique distortion.

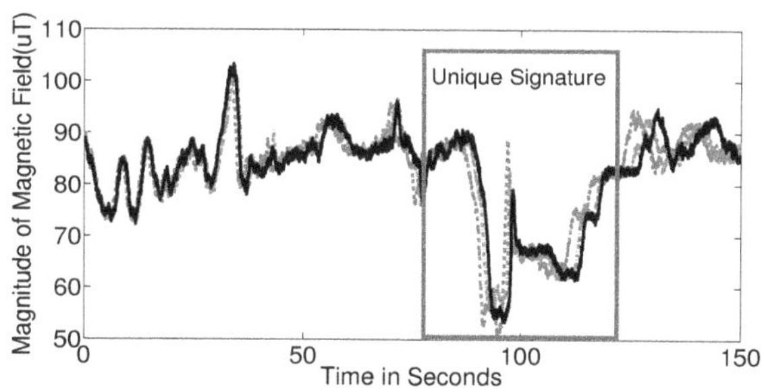

Figure 15: Magnetic signature near networking lab.

3.4 Simultaneous Localization and Mapping

UnLoc uses the SLMs and OLMs to reset dead reckoning error and track the user. The improved dead-reckoned paths help in refining the landmark locations because different paths offer *independently* erroneous estimates of a specific landmark. Combining these independent errors produces the refinement (the intuition derived from law of large numbers, where the sampled mean converges to the true mean for a large number of samples). We combine the estimates of a landmark, say L_i, as follows. While the obvious approach would be to compute the centroid, we actually take advantage of the observation that all estimated locations may not be equally incorrect. Consider 2 users who arrive at L_i from landmarks L_j and L_k, respectively. If L_j is closer to L_i than L_k, then the user that walks from L_j is likely to have incurred less error. This is because pure dead reckoning is known to accumulate error over time. Thus, accounting for this confidence in landmark estimates, UnLoc computes a weighted centroid. The result is declared as the location of the landmark.

3.5 Points of Discussion

• **While we assumed the knowledge of a floorplan, we can relax that assumption now.** Observe that in reality, we need just one ground truth location of any seed landmark. This could be the location of the building's entrance, staircase, elevator. Once we know the GPS coordinate of one SLM, the rest of the SLMs and OLMs can be organically grown, using this known coordinate as the origin. Importantly, the location of this origin SLM – say the building's entrance – needs to be learnt *only once* when UnLoc bootstraps for the first time. In the steady state, even if users do not know when they are passing through the entrance, their locations will become known once they encounter a landmark. The same is true when a user enters through a different entrance of the building, or turns on her phone at some time after she is inside the building. The localization service will activate once they pass through the first landmark.

In this paper's implementation, we extracted the GPS location of the building entrance from Google Satellite View. During bootstrap, a user activated UnLoc when entering through this entrance. Alternatively, the entrance location could have been collected by other means, or even estimated from locations where GPS fixes are lost (after entering a building). We firmly believe that obtaining the GPS coordinate of one SLM, just one time, will not be difficult in real life.

• **Activity based landmarks are feasible too – a busy cafe may invariably have a queue, or visiting a restroom may have a unique signature.** These activities can very well be landmarks as long as their patterns surface upon clustering. Even temporary landmarks can be learnt (i.e., queue exists between noon and 2:00pm only), and even unlearnt if the queueing behavior disappears during, say, winter vacations. Our current implementation has not explored these opportunities – we have only used signatures that are stable over time.

• **The early adopters of UnLoc will help with localizing the OLMs and bring the system to convergence – is this not war-driving?** We argue that this is certainly not war-driving because the early adopters behave naturally and do not collect ground truth (since they do not need GPS). In fact, they collect exactly the same sensor readings as all other users. The only difference is that the early users may experience less localization accuracy (Figure 19). The process of war-driving, on the contrary, is associated with the notion of (ground truth) calibration, which naturally requires additional equipment.

The next section discusses the implementation, experiment methodology, and performance of UnLoc.

4. EVALUATION

Prototype Implementation

UnLoc is implemented on Google NexusS phones using JAVA as the programming platform. The phone samples the 4 sensors (magnetometer, compass, and accelerometer at $24Hz$; gyroscope at highest permissible rate) and WiFi at $1Hz$. Various features derived from these measurements are sent to the UnLoc server.[7] The server side code is written using C# and MATLAB, and implements the dead reckoning, clustering, and landmark signature-matching algorithms. Whenever a new landmark is detected from clustering, the server updates the OLM list.

[7]In a real-life deployment of UnLoc, communication to the server may not be necessary. The landmarks of a building can be downloaded *a priori*; clustering and landmark matching algorithms can be executed locally on the phone.

Methodology

We design real-life experiments with 3 different users in 3 different university buildings – (1) Computer Science (2) Engineering, and (3) North Gate shopping mall. Approximately, we covered $1750m^2$, $3000m^2$, and $4000m^2$ respectively, in these buildings. Each user walked around arbitrarily in the building for 1.5 hours, covering multiple floors; they carried 2 phones, one in the pocket and another in the hand with the screen facing up. We made separate arrangements to collect ground truth (recall that GPS is not available inside any of these buildings). Briefly, we pasted markers on the grounds at precisely known locations, such as the center of a classroom door, the first step in a staircase, the entry-point to the elevator, in front of a window, etc. Each of these markers had a number on them; as a user walked through a marker, she spoke out the number on the marker, and the phone recorded it. By superimposing the map of the building on Google Earth, and identifying the corresponding locations of the markers, we extracted their GPS locations. This offered us ground truth at these markers. Between two markers (separate by $5m$ on average), we interpolated using step-count. Of course, UnLoc did not rely on any of the ground truth markers to compute its location. Thus, at any given time, the difference between ground truth and the UnLoc-estimated location, is UnLoc's instantaneous localization error.

4.1 Evaluation Results

We intend to concentrate on the following questions:

- How many landmarks are detected in different buildings? Are they well scattered? (Figure 16))
- Do real users encounter these landmarks (i.e., is the matching between the sensor reading and established landmark signatures, reliable)? (Figure 17)
- What is the localization accuracy of a user, in real-time, and offline? (Figure 18, 19, 20)

SLM Detection Performance

Table 1 shows the confusion matrix for the detection of all SLMs (using traces from 2 malls in Egypt). The matrix shows that some SLMs are easier to detect than others due to their unique patterns. This leads to zero false positive and negative rates for the elevators and walking cases. However, even with the difficult SLMs, the UnLoc's template signatures can still achieve a high accuracy, with an overall 0.2% false positive and 1.1% false negative rates.

Detecting Organic Landmarks (OLMs)

Figure 16(a) shows the number of landmarks detected inside different buildings. For the engineering building, the breakup of the landmarks is: 9 magnetic, 8 turns, and 15 WiFi OLMs. For the computer science building, the breakup is: 9 magnetic, 10 turns, and 10 WiFi OLMs. Perhaps more importantly, Figure 16(b) shows how these landmarks are quite homogeneously scattered inside the buildings. We observe that with these numbers of well-scattered landmarks, a user's dead reckoning error is not likely to grow excessively, in turn

Table 1: Confusion matrix for classifying different seed landmarks

	Elevator	Stationary	Escalator	Walking	Stairs	FP	FN	Traces
Elevator	24	0	0	0	0	0%	0%	24
Stationary	0	31	1	0	0	0%	3.1%	32
Escalator	0	0	22	0	0	0.6%	0%	22
Walking	0	0	0	39	0	0%	0%	39
Stairs	0	0	0	1	52	0%	1.8%	53
Overall						0.2%	1.1%	170

helping landmark localization. This well matches our core intuition and expectation.

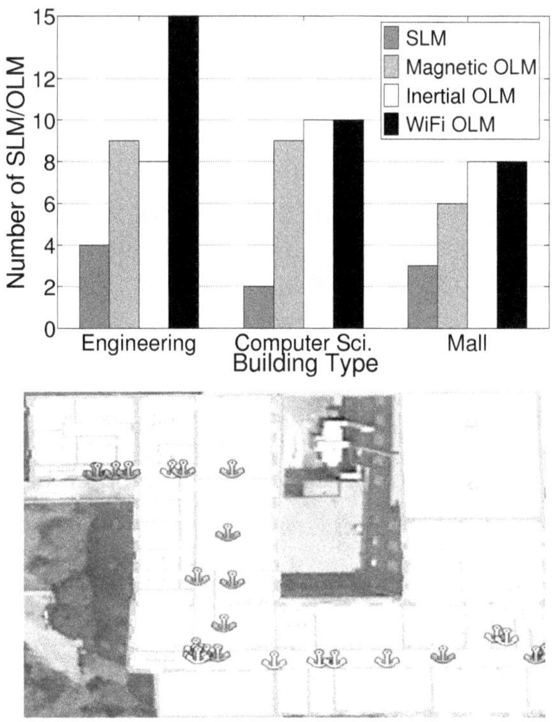

Figure 16: (a) Number of SLMs and OLMs located in different buildings (b)Location of different types of OLMs in the engineering building

Figure 17(a) reports the accuracy of the landmark locations, as computed by UnLoc. Observe that aligned with our intuition, the number of landmarks increase over time, as more users explore the space (Figure 17(b)). Moreover, the accuracy of these landmarks also increase, since different paths bring different independent estimates. In fact, our data sets are limited in diversity of paths since volunteers could not walk around into any rooms or auditoriums in these buildings - many were research offices, faculty offices, or classrooms. In reality, the diversity of different independent paths may be expected to augment the accuracy.

Landmark Signature Matching

If landmark signatures fluctuate quickly over time, then OLMs will be unstable. Thus, users may never encounter the established OLMs because their signatures are changing faster than they can be learnt. Furthermore, it is entirely possible that users at a different location sense a signature that matches a far-away landmark. In such a case, the user's location will be repositioned to the (highly erroneous) landmark. To verify if such variations occur in our buildings, we collected sensor readings on multiple days. We found sound consistency in the signatures. This is not surprising because all our signatures are designed to be stable, particularly WiFi and accelerometer/gyroscope based turns. While the magnetic signatures can change, in our cases we did not observe anything appreciable.

Nonetheless, recall that UnLoc embraces a conservative approach by using a low "similarity threshold" while declaring a landmark. In other words, the signature of the landmark should be very dissimilar with other signatures to qualify as a landmark. This ensures that when a test user matches her sensed readings with existing OLMs, the false positive (FP) rate is low. Figure 17(c) quantifies false positives - evidently FP is less than 1%. As a tradeoff for choosing very distinct signatures, it is possible that a test user may not match it well. Figure 17(c) shows that the matching accuracy is reasonably high, although not perfect. We believe that this is an acceptable tradeoff - given that the number of landmarks are high, missing a few will affect performance much less than matching to an incorrect landmark. In other words, UnLoc is in favor of trading off matching accuracy to maintain low false positives.

Of course, changes in the ambience - say relocation of major electrical equipment to a different room, or deactivated WiFi APs - will affect existing landmarks, and UnLoc will not be able to match them. However, UnLoc will learn these changes changes over time, both the disappearance of the landmark from its original location, as well as the emergence of a landmark at a different location. So long as these changes are not all at the same time, UnLoc should be able to remain resilient to landmark churn. We leave the quantification of this claim to future work.

Localization Performance

In offline localization, whenever a user encounters a landmark, UnLoc learns her errors, and therefore can track back and partly correct her past trail. Online, real-time localization does not offer this benefit. Figure 18(a) shows the comparison of these two forms of localization - evidently, the advantage of offline is appreciable. This implies that for applications which do not need online tracking, localization error can be within $1.15m$ on average, even with a few landmarks.

Beyond the bootstrapping phase, as UnLoc identifies several organic landmarks, error correction opportunities will increase. How much accuracy can we expect in steady state when many OLMs have already been recognized? Figure 18(b) demon-

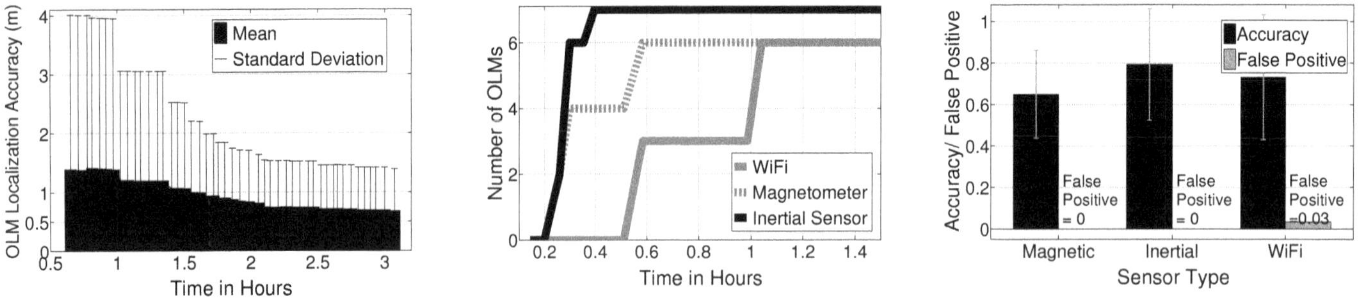

Figure 17: (a) OLM localization accuracy improves over time. (b) Also more number of OLMs are detected as more users walk around the building. (c) OLM matching accuracy and false positives

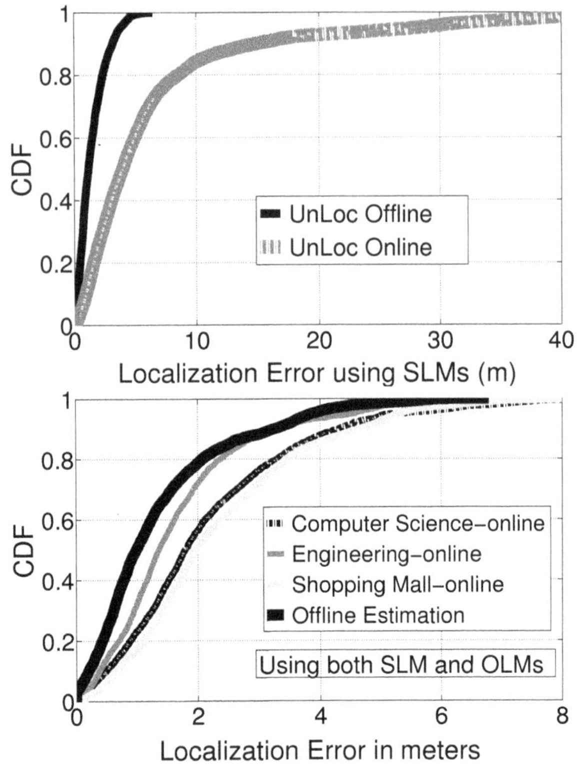

Figure 18: CDF of localization accuracy using (a) only SLMs (b) both SLMs and OLMs

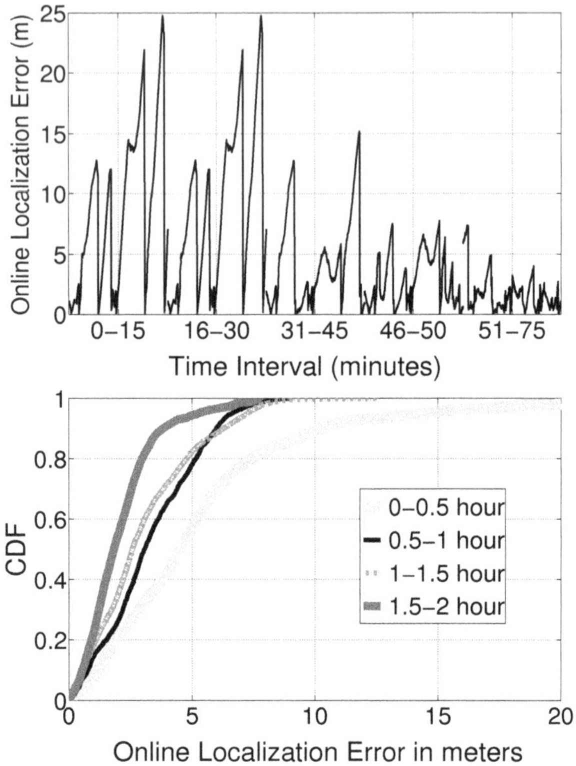

Figure 19: Localization error over time for (a) a single user. (b) averaged over all users.

strates the result; mean instantaneous localization error is within $1.69m$. However, the performance will only improve over time as more OLMs are detected. Figure 19(a) quantifies the intuition. As a user moves away from a landmark, her location error grows and eventually gets reset at the next landmark. Initially, UnLoc only has a few landmarks and hence the error between two landmarks is high. As more landmarks are identified and added to the system, the error growth is curbed frequently. We aggregate the error over time across all the users walking on multiple routes, and present in Figure 19(b). The benefits of additional OLMs are evident.

The above results are from experiments across 3 different buildings using several landmarks. It is difficult to predict how many such landmarks exist in other buildings. Thus, it is natural to ask if *localization performance will get derailed if only a few landmarks exist?* Figure 20 presents location error for varying landmarks. Even with 10 out of the 28 landmarks, average instantaneous location error is within $1.9m$. We expect a few landmarks to be available in most buildings – when in doubt, UnLoc could even turn on the microphone to expand to ambient acoustic signatures. While far more rigorous experimentation is necessary (across more buildings, people, phone platforms, and time), we believe the results from these small scale tests are promising to justify moving forward.

Note on Performance Comparison

Table 2 shows a qualitative comparison of UnLoc with a number of other indoor localization schemes. The most relevant scheme for comparison is EZ [23], which to the best of our knowledge, was among the first to attempt calibration-free localization. However, we believe that EZ's reliance on "occasional GPS fixes" in indoor environments could be problematic. We believe UnLoc's requirement of only a door location is more dependable/scalable. Also, UnLoc is far less sensitive to RF signal fluctuations; the data volume needed to bootstrap is also less. Other schemes in Table 2 require war-driving or special infrastructure; UnLoc is free of both.

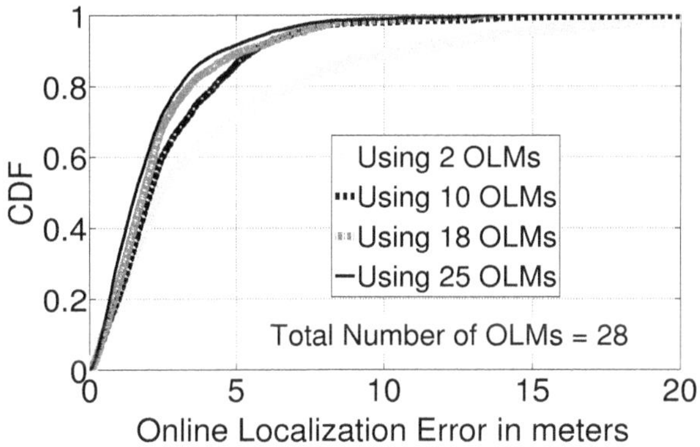

Figure 20: Tradeoff between number of landmarks and localization performance.

Table 2: Comparison with other Localization Systems

Name	**EZ**	**SLAM**	**Horus**	**UnLoc**
Accuracy	$2-7m$	$\sim 5m$	$\sim 1m$	$1-2m$
Pitfalls	GPS Lock?	Special Sensors (LIDAR, Odometer.)	RF Sensitivity	Door Location
Overhead	None	None	War-driving	None

5. LIMITATIONS AND FUTURE WORK

• **Heterogeneous Hardware:** It is possible that the landmark signatures will vary across smartphone platforms – UnLoc has not been tested for this. However, if data gathered from smartphones are indexed by the phone's make and model, it should be feasible to detect landmark signatures for each distinct model (as we have, for the Android Nexus series). Thus, in real-life, a phone would download landmark signatures that are specific to its hardware, and run UnLoc for localization. In other words, UnLoc would extend to all platforms from which initial sensor data has been gathered.

• **Phone Orientation Effect:** We have experimented with the phones in realistic orientation (in pocket and in hand). Handling arbitrary phone orientations and their effect on the reported sensor values is an important direction for future research. As gyroscopes are becoming more available on new phones, they can be leveraged to map arbitrary orientation to a specific frame of reference. Meanwhile, orientation-independent features, such as the magnitude of the acceleration, should be investigated.

• **Energy Footprint:** UnLoc avoids using sensors that have a high energy footprint, e.g., light and sound. This is expected to limit the energy consumed from sensing. Moreover, UnLoc can incrementally turn on sensors, perhaps depending on the density of landmarks available in the environment. For example, if a wing of a building has numerous landmarks, UnLoc could turn off the magnetometer, and use only the accelerometer and compass based landmarks for localization. This paper has not addressed such optimizations – these are part of our ongoing work.

• **Scalability Testing:** A complete UnLoc system needs to be tested over larger scale, in terms of the number of users and data size. However, as with other crowd-sourcing based systems, it is difficult for us to deploy a large scale testbed. This includes in part providing incentive systems for users to deploy the system and experiment with it.

6. RELATED WORK

Our goal and vision intersect with a number of projects in literature. In particular, the indoor localization literature is vast, ranging from theoretical models to simulations to implemented systems [1–9, 9–16]. In the interest of space, we heavily sub-sample this literature, focussing on systems related to UnLoc.

Calibration-free Localization: A recent work called *EZ* [23] was among the first to attempt indoor localization without war-driving. Their key intuition is that overheard WiFi APs and RSSI can together offer the user's location (via a genetic algorithm), provided that a mobile device occasionally gets a GPS fix. While EZ demonstrates accuracies between 2 to $7m$, the GPS locks inside a building can be erratic. Moreover, the precise RF signal propagation may vary over time due to environmental dynamism. UnLoc eliminates this reliance on periodic GPS fixes, and only relies on a one-time global truth information – e.g., the location of a door, or staircase, or elevator. The entire system can bootstrap from this.

WiFi based techniques: Other RF based techniques [1, 3, 5] also inspire UnLoc. Place Lab [5] is a highly successful project where signals from different WiFi and GSM base stations are utilized for localization. A wireless map is created by war-driving a region; the mobile device localizes itself by comparing overheard APs/cell towers against the wireless map. UCSD's Active Campus project [6] adopts similar techniques of localization, but assumes that the location of the WiFi access points are known *a priori*. RADAR [1] also operates on WiFi fingerprinting, and is capable of achieving up to $5m$ accuracy in indoor settings. As a tradeoff to accuracy, RADAR needs to carefully calibrate WiFi signal strengths at many physical locations in the building. High resolution calibration is time-consuming and may not scale over wide areas. Some techniques have bypasses the calibration effort through deployment of additional infrastructure, e.g., Cricket [10], Lease [4], PAL [8], Pinpoint [15]. UnLoc is designed to be an infrastructure-independent, calibration-free, system.

Dead-reckoning: Dead reckoning using inertial sensors is a well-studied research area. However, typical sensors used in such domains are expensive and high-quality ones – coping with noisy smartphone sensors in indoor environments is far less explored. The key problem is that dead-reckoning suffers from the accumulation of error, and can grow cubically even with foot mounted accelerometers [24]. To reduce such error, periodic recalibration with the GPS has been used in outdoor environments [2, 25]. However, GPS is not available indoors. UnLoc presents an option to replace GPS with indoor landmarks; the ability to identify these landmarks in an unsupervised way enables zero-calibration indoor localization.

Simultaneous Localization and Mapping (SLAM): A highly popular and successful technique in robotics, called SLAM, allows a robot to simultaneously discover landmarks and build a map-representation of an indoor environment [26]. However, SLAM typically depends on using explicit environment sensors, such as laser range finders and cameras. Moreover, the rotation of the robot wheels offer a precise computation of displacement. Recently, WiFi-SLAM [27] proposed the usage of the WiFi signal strength for SLAM based on a Gaussian Process Latent Variable Model (GP-LVM). However, the multipath propagation affects the accuracy of the model that captures the relation between the WiFi signal and distance. The paper also assumes that motion between multiple WiFi samples can be accurately determined. UnLoc is certainly reminiscent of SLAM, even inspired – however, it is completely different. Unlike SLAM, UnLoc uses smartphone sensors to compute the displacement and direction of users; the landmarks are essentially ambient signatures or user-activities. While ambient signatures and activity recognition have been utilized in the past [14, 28], its applicability towards dead-reckoning is, to the best of our knowledge, novel.

Sensor fusion: Multi-modal sensing has received strong interests – [29] uses a body worn sensory board (MSB) for human activity recognition. UnLoc, however, is an early attempt to apply activity recognition to indoor location tracking, using cheap mobile phone sensors.

7. CONCLUSION

We observe that inherent properties of indoor environments offer unique opportunities to perform localization. The core approach draws from existing ideas in literature, and combines them in an unconventional way. Essentially, mobile devices are dead-reckoned using their sensor measurements, but these same measurements are leveraged to detect unique environmental signatures within the building. These signatures are used to correct the dead-reckoning error, which in turn improves the location accuracy of these signatures – a recursive procedure. Performance results suggest promise, motivating us to pursue UnLoc to the point of real deployment.

8. ACKNOWLEDGMENT

We sincerely thank Dr. Jeffrey Pang for shepherding our paper, as well as the anonymous reviewers for their valuable feedback. This work is supported in part by the following grants – NSF CNS 0747206, IIS 0910846, and a Google Research Award.

9. REFERENCES

[1] V. Bahl et al. RADAR: An in-building rf-based user location and tracking system. In *INFOCOM*, 2000.

[2] I. Constandache, R.R. Choudhury, and I. Rhee. Towards mobile phone localization without war-driving. In *INFOCOM*. IEEE, 2010.

[3] M. Youssef and A. Agrawala. The horus WLAN location determination system. In *MobiSys*, 2005.

[4] P. Krishnan and others. A system for LEASE: Location estimation assisted by stationary emitters for indoor rf wireless networks. In *IEEE Infocom*, March 2004.

[5] Yu-Chung Cheng et al. Accuracy characterization for metropolitan-scale wi-fi localization. In *MobiSys*, 2005.

[6] William G. Griswold et al. Activecampus: Experiments in community-oriented ubiquitous computing. *Computer*, 2004.

[7] M. Youssef et al. Pinpoint: An asynchronous time-based location determination system. In *ACM Mobisys*, 2006.

[8] S.J. Fontana, R.J.and Gunderson. Ultra-wideband precision asset location system. In *IEEE Conference on Ultra Wideband Systems and Technologies*, May 2002.

[9] D. Niculescu and others. VOR base stations for indoor 802.11 positioning. In *ACM MobiCom*, 2004.

[10] N. B. Priyantha. *The cricket indoor location system*. PhD thesis, MIT, 2005.

[11] A. Savvides and M. Han, C.C. Srivastava. Dynamic fine-grained localization in ad-hoc networks of sensors. In *ACM MobiCom*, 2001.

[12] J. Xiong and K. Jamieson. SecureAngle: improving wireless security using angle-of-arrival information. In *ACM HotNets*, 2010.

[13] S. Sen et al. Precise Indoor Localization using PHY Layer Information. In *ACM HotNets*, 2011.

[14] M. Azizyan et al. SurroundSense: Mobile phone localization via ambience fingerprinting. In *MOBICOM*, 2009.

[15] M. Youssef, A. Youssef, C. Rieger, U. Shankar, and A. Agrawala. Pinpoint: An asynchronous time-based location determination system. In *ACM MobiSys*, 2006.

[16] S. Kumar and D. Shepherd. Sensit: Sensor information technology for the warfighter. In *Proc. 4th Int. Conf. on Information Fusion*, 2001.

[17] I. Constandache et al. Did you see bob?: human localization using mobile phones. In *ACM MobiCom*, 2010.

[18] L. Bao and S. Intille. Activity recognition from user-annotated acceleration data. *Pervasive Computing*, pages 1–17, 2004.

[19] J. Chung et al. Indoor location sensing using geo-magnetism. In *MobiSys*. ACM, 2011.

[20] F. Evennou and F. Marx. Advanced integration of wifi and inertial navigation systems for indoor mobile positioning. *EURASIP journal on applied signal processing*, 2006:164–164, 2006.

[21] Step size in pedometers. s.`http://walking.about.com/cs/pedometers/a/pedometerset.htm`.

[22] M. Alzantot and M. Youssef. Uptime: Ubiquitous pedestrian tracking using mobile phones. In *IEEE Wireless Communications and Networking Conference (WCNC 2012)*. IEEE.

[23] K. Chintalapudi, A. Iyer, and V. Padmanabhan. Indoor localization without the pain. In *MOBICOM*, 2010.

[24] O. Woodman and R. Harle. Pedestrian localisation for indoor environments. In *UbiComp*. ACM, 2008.

[25] M. Youssef et al. Gac: Energy-efficient hybrid gps-accelerometer-compass gsm localization. In *IEEE GLOBECOM 2010,*.

[26] M. Montemerlo et al. Fastslam: A factored solution to the simultaneous localization and mapping problem. In *Proceedings of the National conference on Artificial Intelligence*, 2002.

[27] B. Ferris, D. Fox, and N. Lawrence. Wifi-slam using gaussian process latent variable models. In *Proceedings of the 20th International Joint Conference on Artificial Intelligence*, pages 2480–2485, 2007.

[28] S.P. Tarzia et al. Indoor localization without infrastructure using the acoustic background spectrum. In *ACM MobiSys*, 2011.

[29] J. Lester, T. Choudhury, and G. Borriello. A practical approach to recognizing physical activities. *Pervasive Computing*, pages 1–16, 2006.

Satellites in Our Pockets: An Object Positioning System using Smartphones

Justin Manweiler
IBM T. J. Watson Research*
Hawthorne, NY, USA
jmanweiler@us.ibm.com

Puneet Jain
Duke University
Durham, NC, USA
puneet.jain@duke.edu

Romit Roy Choudhury
Duke University
Durham, NC, USA
romit.rc@duke.edu

ABSTRACT

This paper attempts to solve the following problem: *can a distant object be localized by looking at it through a smartphone*. As an example use-case, while driving on a highway entering New York, we want to look at one of the skyscrapers through the smartphone camera, and compute its GPS location. While the problem would have been far more difficult five years back, the growing number of sensors on smartphones, combined with advances in computer vision, have opened up important opportunities. We harness these opportunities through a system called *Object Positioning System (OPS)* that achieves reasonable localization accuracy. Our core technique uses computer vision to create an approximate 3D structure of the object and camera, and applies mobile phone sensors to scale and rotate the structure to its absolute configuration. Then, by solving (nonlinear) optimizations on the residual (scaling and rotation) error, we ultimately estimate the object's GPS position.

We have developed *OPS* on Android NexusS phones and experimented with localizing 50 objects in the Duke University campus. We believe that OPS shows promising results, enabling a variety of applications. Our ongoing work is focused on coping with large GPS errors, which proves to be the prime limitation of the current prototype.

Categories and Subject Descriptors

H.3.4 [**Information Storage and Retrieval**]: Systems and Software

General Terms

Algorithms, Design, Experimentation, Performance

Keywords

Augmented Reality, Localization, Structure from Motion

*This work was conducted at Duke University.

1. INTRODUCTION

Imagine the following scenario in the future. While leaving for office, Alice needs to ensure that the repairman comes to her home later in the day and fixes the leakage on the roof. Of course, the leak is small and Alice must point out the location of the leak. To this end, she walks across the road in front of her house, points her camera towards the leak, takes a few photos, and types in "`leaking from here`". Later, when the repairman comes to Alice's house, he points his camera towards the roof and scans – when the leak is inside the camera's view-finder, Alice's message pops-up. The repairman repairs the leak and leaves. Alice comes back home in the evening, points her camera towards the leak, and sees the repairman's tag: "`repaired, received payment, thanks!`". Before returning into her house, she cursorily scans the neighborhood with her phone to see if there was anything new. She finds a "`pool party Saturday evening`" tag at the community swimming pool, and another on a tall crane at a nearby construction site, that read "`too noisy: 13 votes`". Alice remembers how she has been frustrated as well, so points her camera at the crane and votes. She looks at the tag again to confirm, which now reads "`too noisy: 14 votes`".

While this may be an intriguing vision of the future, the core idea of tagging objects in the environment, and viewing them through a smartphone's viewfinder, is old. A variety of augmented reality applications have already built such frameworks – Wikitude and Enkin even offer them on the app store [16]. However, these applications implicitly assume that objects in the environment have been annotated out-of-band – that someone visited Google Earth, and entered a tag for the swimming pool. Later, when an Enkin user looks at the same pool through her camera viewfinder, tags of all the objects in her viewfinder pops up. We believe that out-of-band tagging is one of the impediments to augmented reality (AR) becoming mainstream. The ability to tag the environment spontaneously will be vital if users must embrace AR applications in their daily lives.

This project – *Object Positioning Systems (OPS)* – is tasked to address this "missing piece" in today's AR applications. Our ultimate goal is to offer a service that allows a lay user to point her smartphone to any object in the environment and annotate it with comments. While this is the front-end functionality of our system, the key challenge in the back-end pertains to object localization. Our system essentially needs to compute the GPS location of the desired object, and then trivially associate the user-generated tag to that location. Another user standing at a different location should be able to look at the same object, run our system to compute its location, and retrieve all tags associated to it. Ideally, the system should operate in real time, so the user can immediately view the tag she has created.

While translating this vision to reality warrants a long-term research effort, as a first step, we narrow down its scope as follows. We sidestep indoor environments due to their stringent requirements on object positioning accuracy – a tag for a chair cannot get attached to the table. Therefore, we focus on outdoor objects and assume desktop-type CPU capability (which if unavailable on today's phone, may be available through the cloud). Even under this narrowed scope, the challenges are multiple: (1) State-of-the-art in computer vision is capable of localizing objects from hundreds of pictures of the same object [19, 21]. In the case where a *few* pictures are available – such as those taken by Alice of her rooftop – computer vision becomes inapplicable. Our intuition suggests that sensor information from mobile devices should offer opportunities to compensate for the deficiencies in vision, but the techniques for such information fusion are non-trivial. (2) The smartphone sensors, such as GPS, accelerometer, compass, and gyroscope, are themselves noisy, precluding the ability to pivot the system on some ground truth. Hence, aligning sensor information with vision will become even more difficult, requiring us to formulate and solve a "mismatch minimization" problem. (3) OPS needs to identify the user's intention – different objects within the viewfinder may be at different depths/locations, and only the intended object's location is of interest. (4) Finally, the system needs to be reasonably lightweight in view of the eventual goal of on-phone, real-time operation.

The design of OPS has converged after many rounds of testing and modification. Our current prototype on Android NexusS phones has been used to localize 50 objects within the Duke University campus (e.g., buildings, towers, parking lot, cranes, trees). Performance evaluation shows that the system exhibits promising behavior. In some cases, however, our errors can be large, mainly stemming from excessively-high GPS errors. Nonetheless, OPS is able to identify and communicate such cases to the user – like a confidence metric – allowing them to re-attempt the operation. While not ready for real-world deployment, we believe OPS demonstrates an important first step towards a difficult problem with wide-ranging applications.

The key contributions in OPS are summarized as follows.

1. **Localization for distant objects within view:** We show opportunities in *multimodal sensing* to localize visible objects in outdoor environments, with core techniques rooted in mismatch optimization.
2. **System design and implementation on the Android NexusS platform:** Reasonably lightweight algorithms achieve promising location accuracy, with marked improvements over an optimized triangulation-based approach using GPS and compass.

The rest of the paper expands on these contributions, beginning with motivation and overview in Section 2 and primitives of OPS localization in Section 3. Next, in Section 4, we present the design of OPS. In Section 5, we address additional practical challenges for translating the core design into a complete system. We provide results from our testing experiences in Section 6 and our ongoing work to improve OPS in Section 7. We compare OPS with the state of the art in Section 8. Section 9 concludes with a brief summary.

2. MOTIVATION AND OVERVIEW

This section visits the motivation of the paper, with a generalization of OPS to other applications, and then presenting a functional overview of the system. The subsequent sections elaborate on the core technical challenges and solutions.

2.1 Applications beyond Tagging

An Object Positioning System (OPS) has natural applications in tagging the environment. While this was our initial motivation, we observed that the core capability to localize a distant object is probably a more general primitive. In contemplating on the possibilities, we envisioned a number of other applications that can overlay on OPS:

(1) Location-based queries have been generally interpreted as queries on the user's current location (e.g., "restaurants around me," "driving directions from here to the airport"). However, queries based on a distant object can be entirely natural, such as "how expensive are rooms in that nice hotel far away," or "is that cell tower I can see from my house too close for radiation effects?" While walking or driving up to the object location is one way to resolve the query, the ability to immediately look up the hotel price based on the hotel's location, is naturally easier. OPS could enable such "object-oriented queries."

(2) OPS could potentially be used to improve GPS, particularly where the GPS errors are large or erratic. This is true even though OPS actually depends on GPS. The intuition is that combination of multi-modal information – vision and GPS in this case – can together improve each of the individual dimensions. Thus, knowing the location of the object can help improve the location of the camera.

(3) High-end cars entering the market are embedded with a variety of safety features [13], such as adaptive cruise control, lane change detection, blind spot alerts, etc. Existing cars remain deprived of the capabilities since upgrades may be expensive, even if feasible. High accuracy OPS technologies on mobile smartphones may enable services that approximate these capabilities. Smartphones mounted near the car's windshield could estimate location of other objects in the surroundings, and trigger appropriate reactions.

To summarize, one may view OPS as somewhat analogous to GPS – GPS satellites help receivers estimate self-location, while OPS phones estimate other's-locations. It is this analog that motivates our metaphor – *satellites in our pockets.*

2.2 System Overview

We present an overview of OPS with the goal of introducing the functional components in the system, and their interactions. We expect it to help the transition to technical details.

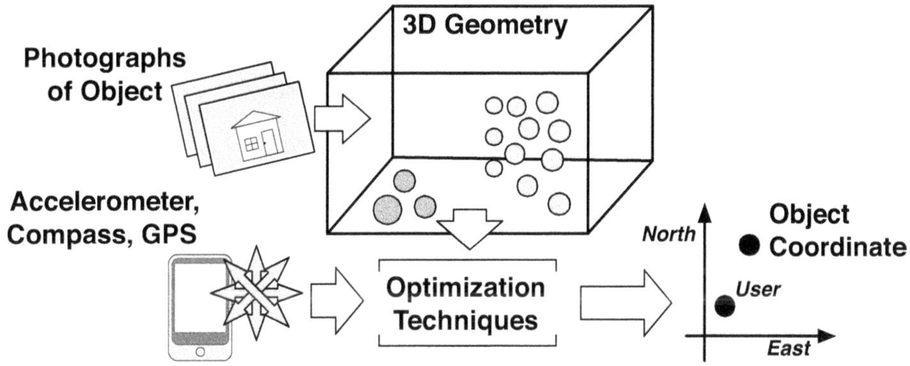

Figure 1: An architectural overview of the OPS system – inputs from computer vision combined with multi-modal sensor readings from the smartphone yield the object location.

When a user activates OPS on her smartphone, the camera is automatically turned on, along with the GPS, accelerometer, compass, and gyroscope. The user is expected to bring the object of interest near the center of her viewfinder, and take a few pictures from different positions. These positions can be separated by a few steps from each other in any direction – the goal is to get multiple views/angles of the same object. As few as 4 photos are adequate, however, more the better. Once completed, OPS displays the object's GPS coordinate.

While this is a simple front-end, Figure 1 shows the flow of operations at the back-end. The pictures taken by the user are accepted as inputs to the computer vision module, which implements a technique called *structure from motion* (SfM) [9]. Briefly, SfM is the process of extracting a 3D structure of an object from diverse views of a moving camera. As a part of this process, SfM first identifies *keypoints* in each picture – keypoints may be viewed as a set of points that together capture the defining aspects of the picture. The keypoints are matched across all the other pictures, and those that match offer insights into how the camera moved (or its angle changed) while the user clicked the different pictures. The final output of this process is a 3D structure, composed of the object and the camera locations.

Importantly, the 3D structure – also called the *point cloud* – is not in absolute scale. Rather, the point cloud offers information about the *relative* camera positions, as well as the *relative* distances between the cameras and the object. To be able to obtain the GPS location of the object, the point cloud needs to be "grounded" on the physical coordinate system. In an ideal scenario, where the point cloud and the GPS locations are both precise, it would be easy to scale the relative camera locations to match the GPS points. This will scale the object-distance as well, eventually yielding the absolute object location. Unfortunately, errors in the point cloud, and particularly in GPS readings, introduce a mismatch. Therefore, OPS uses the configuration of the camera-locations in the point cloud to first adjust the GPS positions. To this end, OPS formulates and solves an optimization problem to minimize the total adjustments.

The next goal is to use the compass readings from these (corrected) locations to triangulate the object of interest. Again, if all compass readings were accurate, any pairwise triangulation from the GPS points should yield the same object location. Unsurprisingly, compasses are noisy as well – therefore OPS executes another optimization that minimizes the total adjustments on all compasses, under the constraint that all triangulations result in the same object location. This corrects the compass readings, and also offers a rough estimate of the object's distance from the GPS locations. By applying the compass readings back on the 3D point cloud, and again solving an optimization problem (detailed later), OPS finally converges on the object location.

OPS also extracts the height of the object, by incorporating the *angular pitch* of the phone while taking the picture. Thus, the final output is a location in 3D space, represented as a GPS coordinate and a height above the ground. The following sections zoom into the details of each of these components, beginning with the primitives of object localization.

3. PRIMITIVES FOR OBJECT LOCALIZATION

Inferences of a distant location from a known point-of-origin is an old problem. Historically, the principles of *triangulation* date to Greek philosophers of the 6th Century BC. Land surveying applies the same basic techniques today at great precision. The related technique of *trilateration* (location determination through known distances, rather than angles) is the technical basis of GPS. As a starting point, we investigate the applicability of these techniques to object localization.

Why not use GPS/compass to triangulate?
Smartphones have embedded GPS and compass (magnetometer) sensors. The precise location and compass bearings from any two points determines a pair of lines[1]. The object-of-interest should fall at their unique intersection. We illustrate compass-based triangulation in Figure 2.

In principle, if a user points her phone at an object-of-interest from two distinct locations, we should be able to easily infer the object's location. Of course, to obtain these distinct locations, we cannot ask the user to walk too far, or using the system would be impractical. Instead, we can imagine the user walking just a few steps to infer the location of the object, say 40 meters away. When the scenario is

[1]The two points must not be collinear to the remote location.

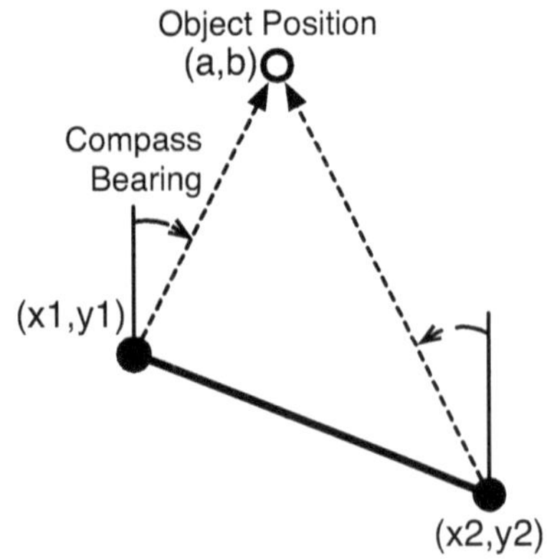

Figure 2: Compass-based triangulation from GPS locations (x_1, y_1), (x_2, y_2) to object position (a, b).

such (i.e., distance between camera views is much smaller than the distance from the camera to the object), compass precision becomes crucial. A few degrees of compass error can dramatically reduce the accuracy of triangulation. Similarly, if the distance between the camera views are slightly erroneous, the result can also be error-prone. Smartphone sensors are not nearly designed to support such a level of precision. GPS can be impacted by weather (due to atmospheric delay), clock errors, errors in estimated satellite ephemeris, multipath, and internal noise sources from receiver hardware. Compass magnetometer readings are imprecise and subject to bias, due to variation in the Earth's magnetic field and nearby ferromagnetic material. Triangulation, at least under these extreme conditions, does not apply immediately.

Can smartphones apply trilateration?

Trilateration requires estimating the distance to the object-of-interest (range), from two vantage points. GPS is a popular implementation of trilateration – the distances from multiple satellites are computed from the propagation delay of the corresponding signals. Unfortunately, a GPS-like scheme is inapplicable for object positioning, since the objects are not collocated with a wireless radio. The phone camera, however, may partially emulate this functionality without any requirement of infrastructure at the object. This can naturally be an appealing alternative.

So long as the object-of-interest remains clearly in the camera view, the size of an object in the picture is a function of the camera's distance to that picture. The size can be estimated by the *visual angle* needed for that object (Figure 3), which can be computed as $v = 2\arctan(s/d)$, where v is the visual angle, s is the size (or height) of the object, and d is the distance to the object. Since we do not know object size s, we cannot compute d. However, knowing two different visual angles from two distinct locations, it is possible to eliminate s and obtain a ratio of the distances to the object from these locations. Let σ denote this ratio; then σ can be computed as

$$\sigma := \frac{d'}{d} = \frac{\tan(v/2)}{\tan(v'/2)}$$

Thus, although visual trilateration cannot precisely localize the object, the value of σ can certainly offer hints about the

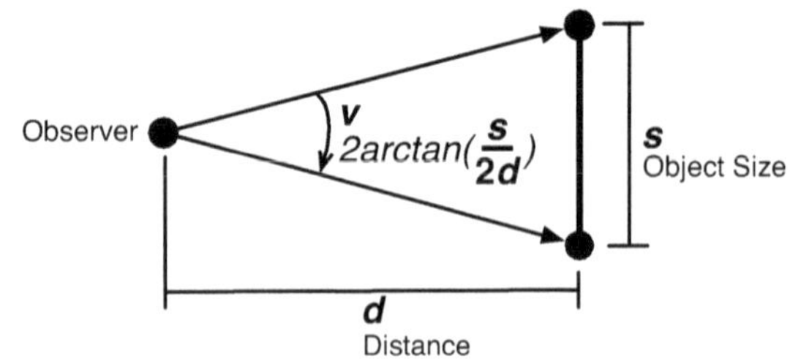

Figure 3: The visual angle v relates the apparent size s of an object to distance d from the observer.

object's position. If one plots all points in space that are away from two camera locations in the ratio of σ, one gets a curve as shown in Figure 4. The object will naturally lie at some location on this curve.

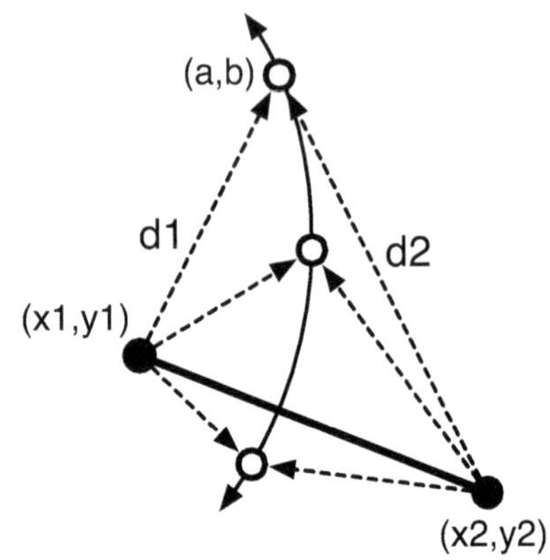

Figure 4: Visual Trilateration: unknown distances from GPS locations (x_1, y_1) and (x_2, y_2) to object position (a, b) are in a fixed ratio d_2/d_1.

Can phone cameras also triangulate?

Land surveying systems typically use optical sensing for precise triangulation. Possibly, the camera could be exploited to improve the accuracy of compass-based triangulation as well. Multiple views of an object from different angles, even if only slightly different, produce visual distortions, due to the phenomenon of *parallax*. Points in the foreground appear to change in relative position to points in the background. The properties of parallax, and visual perception in general, are well-understood. For example, stereo vision leverages parallax effects to invoke a three-dimensional perception from two-dimensional images. Thus, with a careful analysis of images taken from multiple nearby locations, it should be possible to invert these effects. In particular, it would be possible to infer the interior angle between a pair of GPS locations and the object position. However, knowing the interior angle is again not adequate to pinpoint the object location – instead it offers a curve and the object can be at any location on this cure. Figure 5 captures this efficacy of visual triangulation.

Combining Triangulation and Trilateration

While neither triangulation nor trilateration can pinpoint object location, observe that computing the intersection of the two curves (in Figure 4 and Figure 5) yields a small number of intersection points. Moreover, if compass triangulation is added, there is more than adequate information to

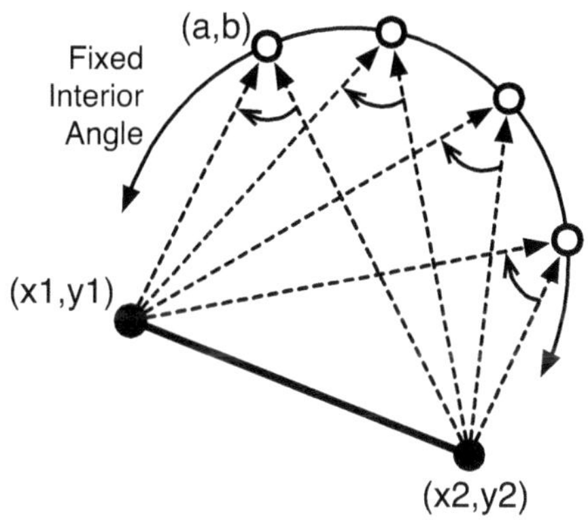

Figure 5: Visual Triangulation: fixed interior angle from known GPS location (x_1, y_1) to unknown object position (a, b) to known GPS position (x_2, y_2) .

uniquely identify the object position (a, b). Figure 6 shows the superimposition of all four curves – observe that this is an over-constrained system, meaning that there is more than sufficient information to compute an unique solution.

This excess of information will later form the basis for noise correction on imperfect sensors. This is necessary because, with errors from GPS, compass, and inaccurate parameter estimation from the visual dimensions, we do not obtain a single point of intersection across all curves. While increasing the number of camera views will help, it will also increase the number of curves (each with some error). Thus, ultimately, we are left with many points of intersection, many of which can be far away from the true object position. To find a single point of convergence, we will rely on optimization techniques, finding the most-likely true object point by minimizing estimates of sensor error.

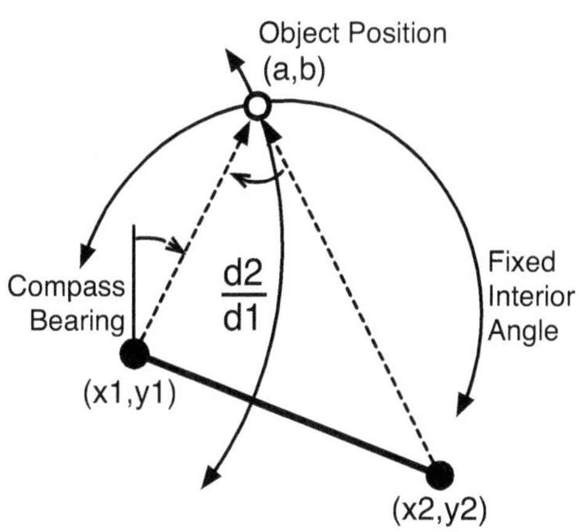

Figure 6: Intersection of the four triangulation curves for known points $(0, 0)$ and $(10, -4)$, localized point $(4, 8)$, distance ratio $\sigma = 6\sqrt(5)/4\sqrt(5) = 1.5$, and internal angle $\gamma = 2 \cdot \arctan(1/2) \approx 53°$.

Next, we describe the OPS system design, focusing mainly on how advanced computer vision techniques can be applied to implement visual trilateration and triangulation. In particular, vision will quantify relative distance and invert the effects of parallax to find the interior angle between a pair of photographs, which will guide the other sensors to ultimately yield object location.

4. OPS: SYSTEM DESIGN

In an ideal world, visual information should not be necessary – noise-free sensors should be able to triangulate the object position. Since real-world sensors are noisy, OPS uses visual information to combat the impact. However, visual information relies partly on sensors, and thus, the overall system needs to be optimized jointly, to marginalize the noise. For ease of explanation, we first describe the individual techniques in isolation (i.e., without considering the effect of noise). Then, we explain how noise forced many of our designs to fail, motivating our ultimate methods of "mismatch optimization." Figure 7 captures this flow of operations.

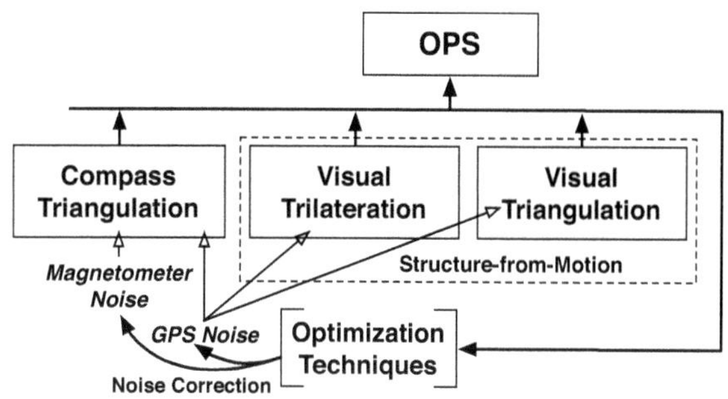

Figure 7: OPS builds on triangulation and trilateration, each underpinned by computer vision techniques, and multi-modal sensor information. Sensor noise affects the different techniques, and makes merging difficult.

4.1 Extracting a Visual Model

We begin with a discussion of the kind of information deducible from multiple photographs of the same object. Figure 8 shows how two observers each experience a different *perspective transformation* of the same object – a building. Note that the apparent *size* of the same building is different in each, as is the *shape*. The differences in scale[2] can be used to determine relative distances to the building. The shape of the transformation can reveal something about the difference in angle from each view to the building. As evident from our discussion of visual trilateration and triangulation, relative distances and angles from the cameras to the object can be valuable for estimating its location. OPS relies on a state-of-the-art computer vision technique, called *Structure from Motion* (SfM), to extract these distances and angles [9, 17]. As we will see next, SfM derives a 3D model of objects in the visible environment, including the object-of-interest and the camera, and computes these relations from them.

Structure from Motion

The SfM module functions as a multi-step process as follows. First, SfM accepts multiple photos from the user, and on each photo, runs an algorithm known as a *feature detector* [1, 11, 15]. The algorithm identifies various "interesting" points of the photo, called *keypoints*. Ideally, the keypoints

[2]After accounting for differences in camera focal length and lens distortions.

Figure 8: Two vantage points of the same object of interest. The "thought-bubbles" show the two different perspective transformations, each observing the same four feature corner points.

Figure 9: Example of a 3D point cloud overlaid on one of the images from which it was created.

should be those that are likely to be robust (stable) across different perspectives of the same object. Put differently, a robust keypoint is one that consistently reflects the same physical point on the object. For example, the peak of a pitched roof may be taken as a keypoint in each image of the same house. Of course, the keypoint detection algorithm is a heuristic, and is prone to inconsistencies across multiple photos. However, with a large number of keypoints per image, there is likely to be a substantial number of keypoints that derive from the same physical point in all photos.

The keypoints from each photo are updated to a server, which then executes a keypoint matching algorithm. The matching process entails comparison of *feature descriptors* associated with each keypoint. A feature descriptor can be thought of as a unique "fingerprint" of a photograph, taken from the pixels around the keypoint. If a pair of feature descriptors are a strong match (numerically), the corresponding pair of keypoints can be assumed to likely capture the same physical point in the real world. Once keypoints are linked across multiple photos, SfM now prepares to analyze the spatial relationship between the locations at which the photographs were taken.

For spatial reasoning, SfM applies algorithms that bear similarity to stereo vision. Perspective differences from multiple views of the same object (arising from parallax) can be used to reconstruct depth information. However, unlike stereo vision, SfM does not require a (known) fixed distance and relative orientation between a pair of views. Instead, SfM takes multiple sets of matched keypoints and attempts to reconstruct (1) a sparse 3D *point cloud* of the geometry captured by those keypoints, and (2) the relative positions and orientation of the camera when the original photographs were taken, known as *pose*. Figure 9 shows an example point-cloud for a building – observe that the points in the cloud are located on the surface of the building and other visible objects in the environment, as well as at the location of the camera.

SfM relies on *Bundle Adjustment* to perform a simultaneous refinement on the estimated 3D point cloud and parameters for each camera view (including, camera pose and lens distortions). Popular implementations of Bundle Adjustment use the Levenberg-Marquardt algorithm to perform a numerical nonlinear optimization on *reprojection error* between what is seen in each image (as described by the keypoints) and what is predicted by different parameterizations of camera pose and 3D geometry. Thus, in summary, the final output from SfM is a considerably-accurate 3D point cloud.

From 3D Point-Cloud to Physical Location

For OPS, we utilize SfM as a "black box" utility. As input, SfM takes the matched keypoints of the user's images. As output OPS receives a 3D *point cloud* of estimated <X, Y, Z> coordinates for each keypoint that was successfully matched across a sufficient number of images. We also have estimated <X, Y, Z> camera pose coordinates from where each photo was originally taken. This coordinate system, however, even if completely precise, exists at an unknown relative scaling, translation, roll, pitch, and tilt from the corresponding locations in the real-world. To compute the physical location of the object, the camera locations and orientations in the 3D model needs to be "aligned" with the GPS and compass readings from the smartphone. However, since the GPS/compass readings themselves will be noisy, this alignment will be non-trivial – the GPS/compass values will need to be adjusted to minimize the mismatch. Moving forward, OPS will focus on addressing these challenges.

4.2 Questions

Before we continue further into the challenges of mismatch minimization, we briefly discuss a few natural issues related to the system build-up.

(1) Capturing User Intent

The use of computer vision entails a second practical challenge – many objects may appear in the camera view. OPS must be able to infer, automatically, which object in view the user is most-likely interested in localizing. For example, a building may be partially occluded by trees. Thus, the point cloud may contain many keypoints that are not reflective of the specific object-of-interest. In general, we assume that the user positions the object-of-interest roughly at the center of the camera's viewfinder. Across multiple photographs, the intended object will become a "visual pivot." Near-foreground and distant-background points appear to shift away from this central point, due to parallax. More sophisticated techniques based on the computer vision techniques of *segmentation* are also relevant here. For example, in Section 5, we will consider an alternative approach for cases where we can assume the user is focused on a building. In our evaluation, however, we will avoid such assumptions.

(2) Privacy

We note that while OPS may offload computational tasks to a central server/cloud, users need not ever upload actual photographs of objects-of-interest (in fact, they can be discarded from the smartphone as well). Instead, they can only upload the keypoints and feature descriptors, that contain all the information needed by SfM. This serves to address any privacy concerns that a user may have with OPS.

4.3 Point Cloud to Location: Failed Attempts

In our original design, we expected that once a Structure-from-Motion point cloud is extracted, estimation of the real-world coordinate of the object would be reasonably straightforward. This would only require mapping vision <X, Y, Z> coordinates to real-world <latitude, longitude, altitude>. In practice, substantial sensor noise makes this mapping more difficult and error-prone than we had anticipated.

The point cloud reflects the structure of the object location relative to the user's locations, when the original photographs were taken, but at an unknown relative scaling, translation, roll, pitch, and tilt from the real-world. Importantly, the locations at which the photographs have been taken are known in both real-world coordinates (through GPS) as well as in the SfM-derived coordinate system. In principle, some *affine transformation* should exist to convert from one coordinate system to the other. We sought to apply state-of-the-art computer vision optimization techniques to estimate this affine transformation, and we describe three of our failed attempts, followed by the actual proposal.

Attempts using Point Cloud Registration
We applied the computer vision technique of Iterative Closest Point (ICP) to find the mapping. ICP is commonly used in the *registration* of one point cloud to another. Before applying ICP, we first eliminated what is typically a key challenge; we pre-defined the association of points from one point cloud to the other. We also eliminated as many degrees-of-freedom in the transformation as possible: we normalized the <X, Y, Z> coordinates to eliminate translation from the search space and constrained scaling to be uniform in all dimensions. We attempted three mechanisms for estimation of the affine transformation: first, an approach based on the Singular Value Decomposition (SVD); next, a nonlinear minimization of transformation error based on the Levenberg-Marquardt algorithm; third, we exploited knowledge of surface normals of camera pose (the 3D direction at which the camera points) in the SfM point cloud and attempted to match with a 3D phone rotation matrix from compass and accelerometer (to find orientation relative to the vector of gravity). In all cases, sensor noise (in GPS, compass, and accelerometer) resulted in a nonsensical transformation. With only a few camera locations to connect one coordinate system to the other, and a still-large number of remaining degrees-of-freedom in the transformation, there is simply insufficient data to overcome sensor noise.

Attempts Intersecting Triangulation/Trilateration
After many unsuccessful attempts at applying computer vision techniques to estimate the affine transformation between coordinate systems, we attempted to simplify the localization problem. We tried to directly apply our intuitions for (1) compass triangulation; (2) visual trilateration; and (3) visual triangulation. The parameters of relative distance and interior angles can be trivially estimated from a SfM point cloud. If all sensors and vision techniques were fully-precise, the true (a, b) object location should fall at the intersection of those equations. In practice, after numerically solving the roots of all equation pairs, sensor noise and bias create many intersection points. We applied a 2D hierarchical clustering on these intersection points, hoping that one of the clusters would be distinctly dense, and the centroid of that cluster would be the estimated object location. In many cases, this proved correct. However, in more cases, sensor error (especially GPS) was large. Intersection points became diffused, and no longer indicative of the true location. To be practical, OPS would need to apply a more robust approach.

Attempts Optimizing Across Error Sources
We were encouraged by the partial-success of our second approach, directly applying our equations of triangulation and trilateration. Next, we attempted to push this technique further, this time integrating an optimization approach. We formulated a minimization problem on the error terms for each sensor value and vision-derived parameter, and with the optimization constrained to find a single object location (a, b). This led to a complex nonlinear optimization with many error terms and constraints. While this would occasionally converge to the correct object location, more often it found a trivial and nonsensical solution. For example, GPS error terms would "correct" all GPS locations to the same point. Further, the complexity of the optimization led to an impractically-long running time.

4.4 The Converged Design of OPS

From our early attempts to build a real-world object location model, we learned two important lessons that influenced the final design of OPS. First, any optimization would need to be constrained to be limited in the number of degrees-of-freedom and avoid degenerate cases. Second, we would need a mechanism to reduce the impact of GPS error. The final design of OPS consists of two optimization steps, each designed to limit the potential for degenerate solutions.

Before continuing, it is helpful the simplify our notion of location. We do not consider latitude and longitude directly, as angular values are inconvenient for measuring distances. Instead, we apply a precise Mercator projection to the square UTM coordinate system. Therefore, we can now refer to a latitude/longitude position as a simple 2D (x, y) coordinate. Recovery of the final <latitude, longitude> coordinate of the object is a straightforward inversion of this projection.

"Triangulation" via Minimization on Compass Error
Before explaining the optimizations underlying OPS, it is instructive to consider a reasonable baseline comparison. Since we assume that the user will take more than two photographs when using OPS, it would be unfair to compare OPS to a triangulation with only two GPS readings $(G_1^x, G_1^y), (G_2^x, G_2^y)$ and two compass bearings θ_1, θ_2. Instead, we must generalize the notion of triangulation to support as many measurements as will be available for OPS.

In Table 1, we present a nonlinear optimization that represents a triangulation-like approach to object localization.

Unlike standard triangulation, this scales to support an arbitrary number of GPS (G_i^x, G_i^y) and compass heading θ_i pairs. In noise-free conditions, all lines of compass bearing originating at the corresponding GPS point would converge to a single point (a, b). Of course, due to sensor error, we can expect that all pairs $\binom{n}{2}$ of compass lines will result in $\binom{n}{2}$ different intersection points. The optimization that follows seeks to find the most-likely single point of intersection by rotating each compass bearing as little as possible until all converge at the same fixed point (a, b). We experimented with a number of other approaches for "generalized triangulation." For one, we considered joint optimizations on GPS and compass. For another, we considered the 2D median of all $\binom{n}{2}$ intersections points. After experimentation, we believe this is the most-effective technique, and thus the fairest baseline comparison method to OPS.

Minimize $\sum_{\forall i} |E_i^\theta|$

Subject to

$$\forall i \; : \; b - G_i^y = (a - G_i^x) \cdot \cot(\theta_i + E_i^\theta)$$

Solving for a, b

$\forall i \; : \; E_i^\theta$

With parameters

$\forall i \; : \; G_i^x, G_i^y, \theta_i$

Name	Parameter Sources
G_i^x, G_i^y	GPS position (of user at each photograph)
θ_i	Compass bearing (at each photograph)
Name	**Solved Variable Interpretation**
a, b	Estimated object location at (a, b)
E_i^θ	Estimated error for compass bearing θ_i

Table 1: Optimization for Triangulation

Now, we turn our attention to the two-step process employed in OPS. First, we apply the output of computer vision to correct for noise in GPS measurements. Second, we extend this baseline sensor-only optimization for triangulation to (1) use our corrected GPS points; and (2) exploit our intuitions for visual trilateration and triangulation.

Minimization of GPS Noise, Relative to Vision

From our earlier attempts, we realized that a direct application of our intuitions for visual trilateration and triangulation would be insufficient. Instead, we need a mechanism to reduce sensor noise before a object-positioning step can be effectively applied. Here, we rely on the output of structure from motion to correct for random GPS noise. Bias across all GPS measurements will remain. However, a consistent bias will be less damaging to the final localization result than imprecision in the relative GPS positions across multiple photographs. Structure from motion can help eliminate this noise between relative positions, as it tends to capture this relative structure with far greater precision than GPS. We design a nonlinear programing optimization that seeks to move the GPS points as little as possible, such that they match the corresponding relative structure known from vision.

We design an optimization that maps the original GPS points where photographs were taken $\{\forall i : (G_i^x, G_i^y)\}$ to a set of fixed GPS points $\{\forall i : (F_i^x, F_i^y) = (G_i^x + E_i^x, G_i^y + E_i^y)\}$. The optimization will also solve a scaling factor λ that proportionally shrinks or expands the point-to-point distances in the structure-from-motion point cloud to match the equivalent real-world distances measured in meters. The constraints simply enforce that the distance between any pair of GPS points, $\sqrt{(G_i^x - G_j^x)^2 + (G_i^y - G_j^y)^2}$, is equal to the distance between those same points in vision coordinates, $\sqrt{(V_i^h - V_j^h)^2 + (V_i^d - V_j^d)^2}$, after multiplying the vision distance by a constant factor λ.[3] Since we expect the GPS points to have some noise relative to the same points in vision, we introduce error terms for the GPS distance, $\sqrt{(E_i^x - E_j^x)^2 + (E_i^y - E_j^y)^2}$. With these error terms, the optimization is simply a minimization on the sum of squared error. Table 2 presents the complete optimization.

Minimize $\sum_{\forall i} \left({E_i^x}^2 + {E_i^y}^2 \right)$

Subject to

$$\forall i,j \; : \; \left[(G_i^x + E_i^x) - (G_j^x + E_j^x)\right]^2 + \left[(G_i^y + E_i^y) - (G_j^y + E_j^y)\right]^2 = \lambda^2 \cdot \left[(V_i^h - V_j^h)^2 + (V_i^d - V_j^d)^2\right]$$

Solving for σ

$\forall i \; : \; E_i^x, E_i^y$

With parameters

$\forall i \; : \; G_i^x, G_i^y, V_i^h, V_i^d$

Name	Parameter Sources
G_i^x, G_i^y	GPS position (of user at each photograph)
V_i^h, V_i^d	Vision position (arbitrary units)
Name	**Solved Variable Interpretation**
λ	Scaling factor, vision coordinates to GPS
E_i^x, E_i^y	Estimated camera GPS error

Table 2: OPS Optimization on GPS Error

OPS Optimization on Object Location

From the GPS-correction optimization (Table 2), we are left with a set of fixed GPS points, $\{\forall i : (F_i^x, F_i^y)\}$, and a scaling factor λ from vision to GPS coordinates. Now, we take these parameters to extend the baseline sensor-only optimization for triangulation (Table 1), along with additional context from visual trilateration and triangulation. We present this final optimization as Table 3.

The additional context for visual trilateration and triangulation is encoded as parameters $\{\forall i, j : \gamma_{ij}\}$, (C_x, C_y), and D. Each value γ_{ij} represents the angle from (F_i^x, F_i^y) to

[3] To avoid confusion with GPS x and y dimensions, we use h and d to represent the relevant two dimensions of the vision coordinate system. The h, or horizontal, dimension runs left/right of the object from the perspective of the user. The d, or depth, dimension runs towards/away from the object.

(a,b) to (F_j^x, F_j^y), estimated from vision. As represented in the notation, this is the positive acute angle between vectors $\vec{V_i}$ and $\vec{V_j}$. Thus, $\{\forall i,j : \gamma_{ij}\}$ directly encodes our original interpretation of visual triangulation. To avoid redundancies in the constraints (since triangulation and trilateration parameters are directly measured from the same vision point cloud), we only need to partially encode visual trilateration. Instead of encoding each of the relative distance from each camera point, we can simply enforce that the distance from the user's position to the object as a known, fixed value. We compute (C_x, C_y) as the 2D median across $\{\forall i : (F_i^x, F_i^y)\}$, by applying convex hull peeling. Next we enforce that the distance from (C_x, C_y) to the object at (a,b), $\sqrt{(a-C_x)^2+(b-C_y)^2}$, is equal to the distance D. We can compute D from the vision point cloud along with the vision-to-GPS scaling factor λ.

The minimization function must change to accommodate context from triangulation (visual trilateration is fully incorporated as hard constraints). The addition of γ_{ij} error terms $\{\forall i,j : E_{ij}^{\gamma}\}$ allow angular error to be traded between magnetometer-derived compass headings and vision-derived angles. The compass error scaling factor, $(n-1)/2$, balances for the lessor quantity of compass error terms relative to pairwise vision angles.

$$
\begin{aligned}
&\textbf{Minimize} && \frac{n-1}{2} \cdot \sum_{\forall i} |E_i^{\theta}| + \sum_{\forall i,j} |E_{ij}^{\gamma}| \\
&\textbf{Subject to} && (a-C_x)^2 + (b-C_y)^2 = D^2 \\
&\forall i && : \ b - F_i^y = (a - F_i^x) \cdot \cot(\theta_i + E_i^{\theta}) \\
&\forall i && : \ \vec{V_i} = (a - F_i^x)\hat{\imath} + (b - F_i^y)\hat{\jmath} \\
&\forall i,j && : \ \gamma_{ij} + E_{ij}^{\gamma} = \\
& && \arccos\left(\vec{V_i} \cdot \vec{V_j} / |\vec{V_i}||\vec{V_j}|\right) \\
&\textbf{Solving for} && a, b \\
&\forall i && : \ E_i^{\theta} \\
&\forall i,j && : \ E_{ij}^{\gamma} \\
&\textbf{With parameters} && C_x, C_y, D \\
&\forall i && : \ F_i^x, F_i^y, \theta_i \\
&\forall i,j && : \ \gamma_{ij}
\end{aligned}
$$

Name	Parameter Sources
F_i^x, F_i^y	Fixed GPS position (at each photograph)
θ_i	Compass bearing (at each photograph)
γ_{ij}	Vision estimate for vector angle $\angle \vec{V_i}\vec{V_j}$
C_x, C_y	2D Median of $\{\forall i : (F_i^x, F_i^y)\}$
D	Estimated distance from (C_x, C_y) to (a,b)
Name	**Intermediate Results**
$\vec{V_i}$	Vector from (F_i^x, F_i^y) to (a,b)
Name	**Solved Variable Interpretation**
a, b	Estimated object location at (a,b)
E_i^{θ}	Estimated error for compass bearing θ_i
E_{ij}^{γ}	Estimated error in vector angle $\angle \vec{V_i}\vec{V_j}$

Table 3: OPS Final Object Localization

5. DISCUSSION

Extending the Location Model to 3D

In Section 4.4, we described how OPS estimates the object-of-interest location in two dimensions, namely as the point (a,b). In some contexts, the 3D location of the object can also be useful. For example, we might want to localize a particular window of a multistory building. Ultimately, OPS should provide a $<$ latitude, longitude, altitude $>$ location tuple. However, the height dimension adds additional challenges not faced on the ground, following a plane tangential to the Earth's surface. First, GPS-estimated altitude is prone to greater inaccuracy than latitude and longitude. Second, while it is natural for a user to take multiple photos by walking a few steps in-between, a requirement to take photographs at multiple heights would become awkward. Thus, while vision provides three-dimensional geometry, GPS locations for photographs are roughly planar. Further, since the object localization task is already challenging in two dimensions, it is desirable to avoid integrating 3D into our location optimizations.

Instead, OPS finds the two-dimensional object location first. Next, it uses the now-known distance to the object, along with accelerometer and vision inputs, to estimate height. For each photograph, OPS records the raw three-axis accelerometer output. Since we can expect the phone to be roughly still while the photograph is taken, the accelerometer is anticipated to measure only gravitational force. This gravitational vector defines a unique orientation in terms of phone roll (rotational movement on the plane of the phone screen, relative to the ground) and pitch (rotational movement orthogonal to the plane of the phone screen, relative to the horizon). Pitch provides a rough estimate of how much higher (or lower) the user is focusing, relative to a plane parallel to the ground and intersecting the user at eye-level. To improve the accuracy of this measurement, we can "average" pitch measurements from every photograph. Importantly, the user might not align the object in every photo at exactly the same pitch. For example, the window might appear higher or lower on the screen. We can correct for this by leveraging vision once again. From our 3D point cloud, there is a unique mapping of every 3D point back to each original 2D image. We can now compute an adjustment value, measured in pixels, from the horizontal center line of the screen. We can convert this pixel value to an angle, given the known camera field-of-view. Next, the angular sum of pitch and adjustment, averaged across all photographs, can be used to estimate height when combined with the known two-dimensional distance.

$$h_{\text{object}} = h_{\text{observer}} + \frac{\text{2D distance}}{2 \cdot \tan\left(\frac{\text{pitch+adjustment}}{2}\right)}$$

Alternatives for Capturing User Intent

If we can make assumptions regarding the structure of the object-of-interest, computer vision techniques of segmentation can assist in isolation of the object from the structure-from-motion point cloud. For example, consider a typical multistory building with large flat sides. Images of an exterior wall will tend to yield many keypoints along a flat plane, roughly perpendicular to the ground plane. These points on the wall plane are often clearly distinct from points on the ground, smaller clusters of points in the nearest-foreground from occlusions, or sparse points in the distant background.

To determine where the object-of-interest lies within a point cloud, we attempt to segment the point cloud and find such a predominate plane. We apply Random Sample Consensus (RANSAC) [5], an iterative method to estimate a model for a plane, under an angular constraint that it must be roughly-perpendicular to the ground and parallel to the field-of-view. All points in the point cloud are then classified as either inliers or outliers to the plane. Next, we find the spatial centroid among inliers to the plane. This point is considered to be the object-of-interest.

6. EVALUATION

We take a systems-oriented approach in evaluating OPS, so as to capture real-world performance. Phone sensors are subject to noise and biases. GPS can be impacted by weather (due to atmospheric delay), clock errors, errors in estimated satellite ephemeris, multipath, and internal noise sources from receiver hardware. Compass magnetometers are affected by variations in the Earth's magnetic field and nearby ferromagnetic material. Computer vision techniques, such as structure from motion, can break down in a variety of scenarios. For example, keypoint extraction may fail if photographs have insufficient overlap, are blurred, are under or over-exposed, or are taken with too dark or bright conditions (such as when the sun is in the user's eyes). The primary goal of our evaluation is to consider how well OPS overcomes this naturally-challenging operational context.

6.1 Implementation

OPS is implemented in two parts, an OPS smartphone client and a back-end server application. We built and tested the OPS client on the Google NexusS phone, as a Java extension to the standard Android 2.4 camera program. Photographs may be pre-processed locally on the phone to extract keypoints and feature descriptors (reducing the required data transfer), or simply uploaded to our server for processing (faster with a high-performance WiFi connection). Along with photographs (or keypoints and descriptors), the phone uploads all available sensor data from when each photograph was taken, to include GPS, compass, and accelerometer. Our server is a Lenovo desktop running Ubuntu Linux 11.04.

Computer Vision

Both the client and server applications support the basic computer vision tasks of keypoint detection and extraction of feature descriptors. We use the SURF (Speeded Up Robust Feature) algorithm [1]. We choose SURF over the related SIFT (Scale Invariant Feature Transform) [11] as it is known to be considerably faster to compute and provide a greater robustness against image transformations. SURF detection and extraction are performed using OpenCV. For both the client and server-side applications, JavaCV provides java wrappers of native C++ calls into OpenCV.

For server-side structure from motion, we use Bundler, an open-source project written by Noah Snavely and the basis for the Microsoft Photosynth project [17]. Bundler operates on an unordered set of images to incrementally-build a 3D reconstruction of camera pose and scene geometry. As input, Bundler expects keypoints matched across multiple photos (on the bases of the corresponding feature descriptors). Bundler can operate on any keypoint type, expecting SIFT by default. We adapt the output of the OpenCV SURF detector to match the SIFT-based input expected by Bundler, substantially decreasing processing time per photo on the phone. As output, Bundler provides a sparse point cloud representation of the scene in view. OPS builds its real-world localization model on top of this point cloud, which exists at an unknown relative scaling, translation, roll, pitch, and tilt from the corresponding locations in the real-world.

Server-side Nonlinear Optimization

The OPS server is implemented primarily in Java. Mathematica is used through a Java API for solving nonlinear optimizations. We use differential evolution as the optimization metaheuristic for its strong robustness to local minima (simulated annealing proved to be similarly effective, with both outperforming the default Nelder-Mead) [18].

6.2 Accuracy of Object Localization

We tested OPS at more than 50 locations on or near the Duke University campus. We attempted to use OPS in the most natural way possible, focusing on localization tests that mirror how we would expect a real user would want to use the system. Primarily, we considered objects at distances between 30m and 150m away, for two reasons. First, the object should be far enough away that it makes sense to use the system, despite the hassle of taking photographs. Though it only takes about a minute to take the required photographs, a user should not be more willing to simply walk over to the object to get a GPS lock. Second, distances are limited by the user's ability to clearly see the object and focus a photograph. Building densities, building heights, and presence of significant occlusions (such as trees), constrain the distances at which photographs can be easily taken.

We compare OPS accuracy to "Optimization (Opt.) Triangulation." To provide the fairest comparison, Triangulation reflects the triangulation-like optimization described in Section 4.4, designed to scale to an arbitrary number of GPS, compass-heading pairs. For some graphs, we also show the distance from the position at which photographs were taken (centroid across all photographs) to the true object location. This is an important consideration; as sensor values are projected into the distance, noise and bias can be magnified.

Testing Parameters

OPS performance is dependent on the way OPS is used. For our evaluation, we took four photographs for each localization (the minimum required for structure from motion). Each successful photograph was taken between 0.75m and 1.5m from the location of the preceding photograph, in no specified direction. This flexibility is important, as it allows the user the compose the shot as is natural, given obstacles in the vicinity and visual occlusions. The distance (about one or two paces) is sufficiently far that GPS is able to estimate distance with sufficient (albeit still poor) accuracy. The distance is sufficiently small as to ensure a relatively small angular separation to the object of interest, enabling structure from motion to work well. Further, the distances are not too burdensome for the user; the required photographs can be taken in less than one minute. Once the user takes the required photos, processing requires approximately an additional 30-60 seconds, primarily attributable to structure from motion computation and depending on the

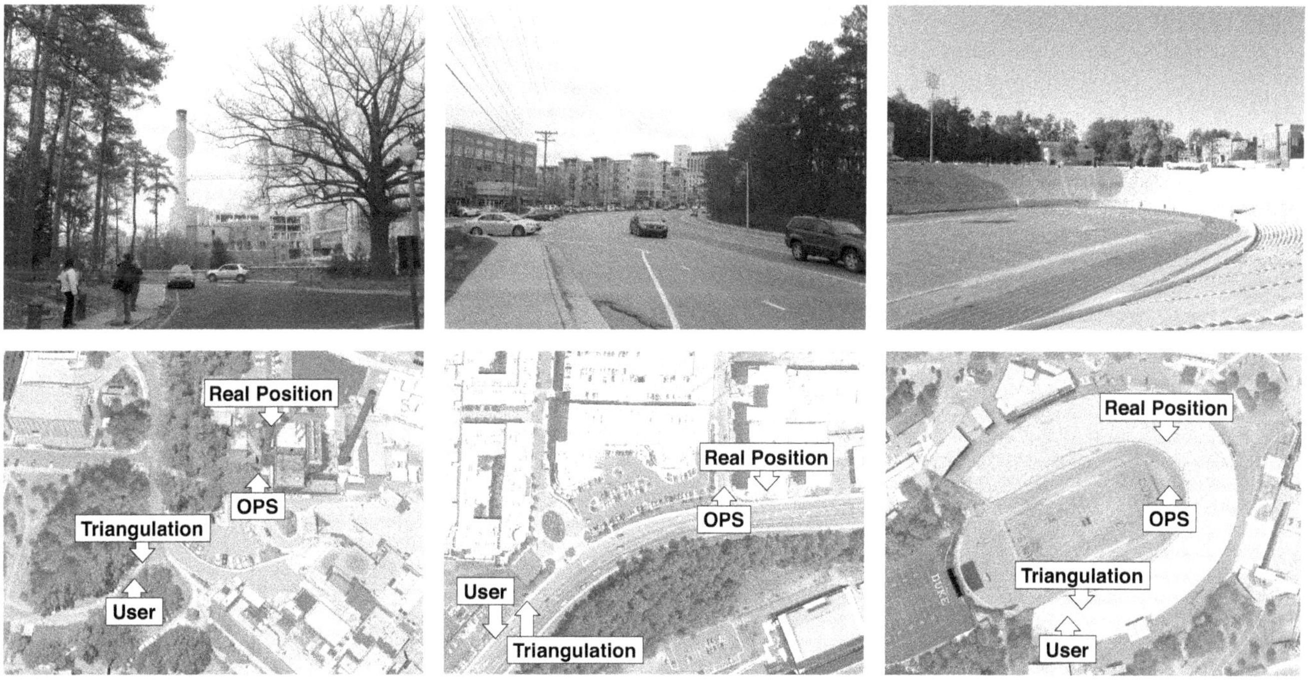

Figure 10: Sampled tests; circle denotes object-of-interest (top), Google Earth view (bottom): (a) smokestack of a coal power plant; (b) distant building with vehicles in foreground; (c) stadium seats near goal post.

number of keypoints detected in each photo. With additional photographs, both accuracy and processing time increase.

Structure from motion is sensitive to the quality of photographs. In poor lighting, keypoint detection becomes less robust. Thus, OPS is sensitive to the quality of lighting conditions. At night, there is unlikely to be sufficient light for high-quality photographs. Further, a camera flash is useless for the distances under consideration. At dawn and dusk, direct sun into the camera lens can also ruin photographs. We test OPS in daylight conditions, and avoid early morning and late evening. We are considering extensions to OPS to enhance performance in poor lighting, likely avoiding structure from motion in these cases.

Example Usage Scenarios

In Figure 10, we show three example photos, taken while using OPS. Below each photo, we show a screenshot taken from Google Earth with four annotations, (1) the location at which the user photographed the object-of-interest; (2) the true location of the intended object (positioned at the center of the screen in each photo; (3) the object position inferred by Opt. Triangulation; and (4) the object position inferred by OPS. We show these particular examples to highlight that OPS is a general approach, and can therefore localize for a wide variety of distant objects. Further, while the final result may still be a substantial distance from the precise true location, OPS is typically able to find a position that is far more representative of the object-in-view than triangulation.

Overall Performance

Figures 11 and 12 present overall performance of OPS across all locations. Figure 11 shows three CDF curves, (1) error in meters from the OPS-localized position to the true object position; (2) error in meters from the Opt. Triangulation position to the true object position; and (3) distance in meters from the position at which photographs were taken (centroid across all photographs) to the true object location. Figure 12 shows individual localization results, sorted by user-to-object distance. Overall, OPS provides a substantial performance improvement over triangulation.

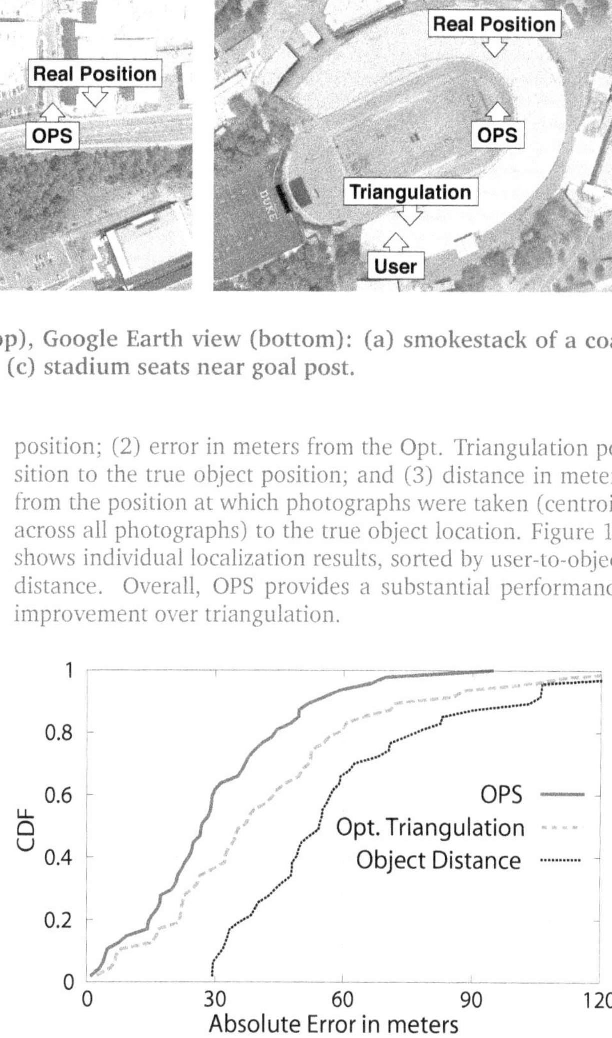

Figure 11: CDF of error across all locations. Graph reflects four photos taken per location. 50 locations.

Sensitivity to GPS and Compass Error

To better understand OPS's robustness to sensor noise, we wanted to carefully evaluate the impact of GPS and compass noise, in isolation. We took a set of four photographs of a Duke University Science Center building, from a mean distance of 87m. We ensured that this particular set of photographs provided a robust model from vision, through a manual inspection of the point cloud. For each photograph, we used Google Earth to mark the exact location at which

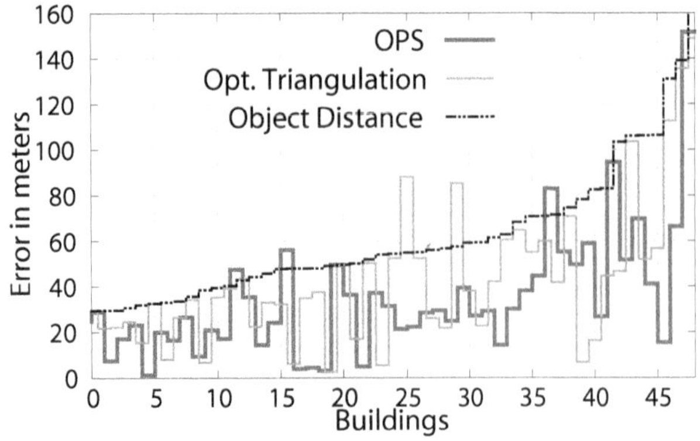

Figure 12: OPS and triangulation error at 50 locations. Graph reflects four photos taken per location.

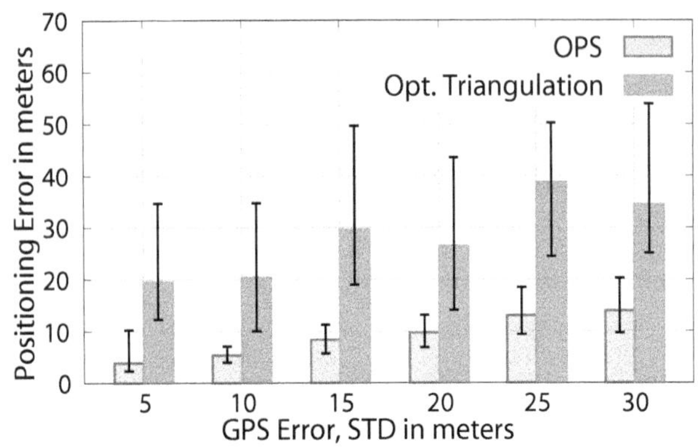

Figure 13: Error from ground-truth GPS camera locations. X-axis shows the standard deviation of introduced Gaussian GPS errors. Bars show median error; whiskers show first and third quartiles.

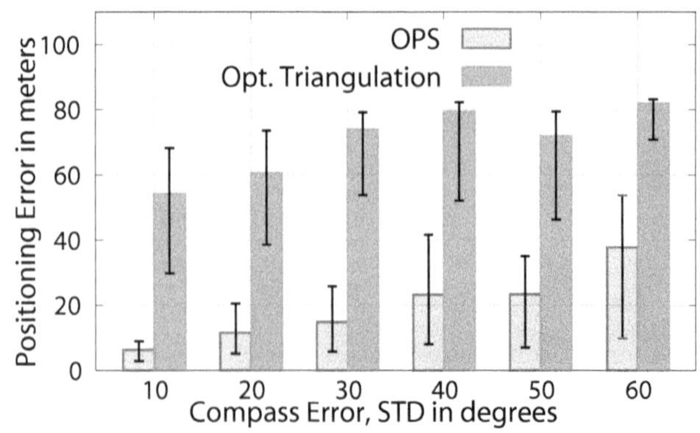

Figure 14: Error from ground-truth GPS camera locations. X-axis shows the standard deviation of introduced Gaussian compass errors. Bars show median error; whiskers show first and third quartiles.

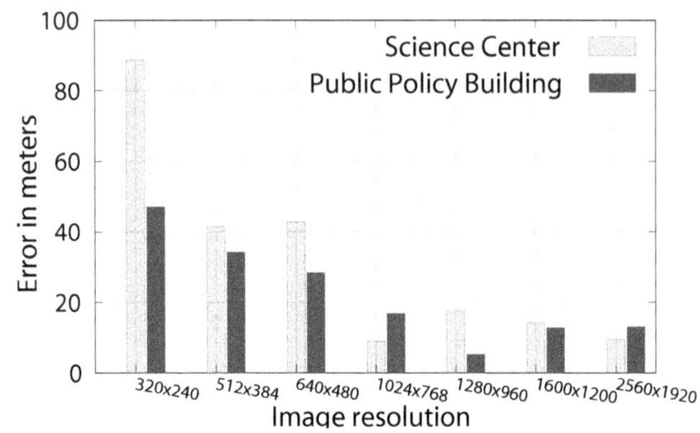

Figure 15: OPS error by photo resolution. Keypoint detection is less reliable below 1024x768 pixels.

it was taken, with less than one meter error. From these carefully-marked points, we mathematically computed the compass bearing to the object, assuring less than 5° error (attributable to error in marking the points). Next, we randomly perturbed each GPS point or compass angle according to random deviates of a Gaussian (Normal) distribution with mean 0 and a varied standard deviation. Figure 13 shows performance as a function of the standard deviation of injected GPS error (in meters). OPS is able to leverage vision to remove much of this GPS noise, providing strong robustness. Figure 14 shows performance as a function of the standard deviation of injected compass error (in degrees). Superior performance of OPS is especially reflective of how visual triangulation enforces the true relative angle between a pair camera locations and the object position. Bars reflect median of 50 trials with first and third quartiles on whiskers.

Sensitivity to Photograph Detail
Figure 15 shows OPS performance with varied levels of photographic detail. For this experiment, four photographs were used, along with ground-truth GPS and compass data. Full-resolution photos were downsampled, reducing the efficacy of keypoint detectors. With fewer keypoints, especially at resolutions below 1024x768 pixels, structure from motion suffers and can impact the object position.

7. ROOM FOR IMPROVEMENT

OPS remains an ongoing research project. We are exploring a variety of mechanisms to improve accuracy.

Live Feedback to Improve Photograph Quality
OPS is impacted by the quality of user photographs. Poor angular separation between two photographs (too small or large) can reduce the efficacy of structure from motion. We imagine a system of continuous feedback to the user, aiding in shot selection and framing. By combining an especially-lightweight keypoint detection heuristic, such as FAST [15], on the camera video with information from the phone gyroscope, it would be feasible to suggest when a "good" photograph can be taken. Otherwise, we would inform the user to take a corrective step (to the left or right).

Improving GPS Precision with Dead Reckoning
From Figure 13 and 14, it is clear that OPS is already relatively insensitive to GPS and magnetometer noise. Although OPS explicitly uses the output of structure from motion to cancel GPS noise, large GPS errors can still become disruptive – this is quite often the cause of poor performance. One possibility is to apply additional sensors to improve GPS precision. In particular, a constant bias is less damaging

than imprecise relative positions across camera locations. We are considering mechanisms to leverage gyroscope and accelerometer for dead reckoning between camera locations.

Continual Estimation of Relative Positions with Video
Continuous video could potentially be used to substantially augment or even replace the structure-from-motion point cloud. A frame-by-frame comparison of these keypoints, in fine-grained comparison with accelerometer, gyroscope, compass, and GPS, could provide a highly-detailed trace of how the smartphone (1) moved in 3D space and (2) viewed the object-of-interest in terms of relative angles/distances.

8. RELATED WORK

To the best of our knowledge, OPS is a first-of-kind system to find the precise position for objects of significant distance away, despite noisy sensors and without requiring a database of pre-existing photographs or maps. Nonetheless, there is a substantial body of related work, especially leveraging computer vision for recognition of well-known landmarks.

Localization through Large-Scale Visual Clustering
OPS is related to the problem of worldwide localization on the basis of visual classification [4, 6, 10, 19, 21]. Most closely related to OPS, [21] is the back-end system underlying Google Goggles: (1) online travel guides are mined for possible tourist landmarks; (2) photo-sharing databases of millions of GPS-tagged photos are searched using keywords from these travel guides; (3) unsupervised visual clustering on these search results provide a visual model of the landmark. From 20 million GPS-tagged photos, 5312 landmarks can be recognized by visual comparison of a photograph with these landmark models. OPS is designed to be more generic, able to localize objects which cannot be considered landmarks (without a pre-exiting photo library of these objects). We believe that a hybrid approach can be valuable, which leverages a photo database to enhance accuracy (where useful photos are available), but can also provide OPS functionality in the general case.

Aligning Structure from Motion to the Real World
Substantial computer vision literature has considered improvements and applications of structure from motion (SfM) [17]. However, our notion of "object positioning" should not be confused with the computer vision concept of object localization, which seeks to find the relative location of objects within an image or point cloud. For mobile systems, closely related to OPS, [7] uses SfM for a "landmark-based" directional navigation system. SfM, along with a preexisting database containing many photographs of the area in view, enable a augmented reality view with highlighted annotations for known landmarks. More-directly related to the goals of OPS, [8] seeks to align the output of structure from motion to real-world coordinates, using GPS coordinates to improve the scalability of structure from motion when using hundreds of photos. However, the techniques require the availability of overhead maps and are not suitable to the extremely small number of photos (typically four) expected for OPS.

Building Object Inventories
Our use of multimodal sensing inputs for estimating object position is related to the techniques in [3] for building inventories of books in a library. The authors project a rough indoor location for a book through WiFi and compass, then apply vision techniques to visually detect book spines.

Applying Computer Vision to Infer Context
CrowdSearch [20] combines human validation with location-specific image search (for objects such as buildings) through computer vision. By leveraging Amazon Mechanical Turk for human-in-the-loop operation, CrowdSearch enables precision in cases of poor image quality. TagSense [14] infers "tags" of human context for photographs, by combining computer vision with multimodal sensing. OPS is complementary to CrowdSearch and TagSense, by enabling awareness of precise location for objects in photographs.

Object Recognition and Location Inference
In [12], the authors try to estimate the 3D-orientation and location of an object in the real world scene using a polygonal 3D-model of a depicted object. The accuracy of OPS could be enhanced opportunistically by utilizing the size of known, recognizable objects when they happen to appear in the view.

Robotics Navigation and Obstacle Avoidance
Mobile robot navigation requires awareness of object position (localization) for navigation and obstacle avoidance. [2] surveys various techniques such as odometry, inertial navigation, active beacons and model matching for relative and absolute position measurement. These techniques typically require specialized hardware, OPS attempts to solve object localization using off-the-shelf smartphones.

9. CONCLUSION

Today's augmented reality applications are less useful than their potential. We believe this is due, at least in part, to an unfortunate usage asymmetry. Users can retrieve available localized content in a natural way, viewing pop-up annotations for the world through their smartphone, but there is no means to equivalently introduce new annotations of objects in the vicinity. By providing localization for objects in a user's view, this paper seeks to enable convenient content creation for augmented reality. Beyond augmented reality, we believe that a precise *Object Positioning System* can be widely enabling for a variety of applications, and is worthy of a significant research endeavor. In this light, we believe that our approach, OPS, takes a substantial first step.

10. REFERENCES

[1] H. Bay, T. Tuytelaars, and L. Van Gool. Surf: Speeded up robust features. In *ECCV*, 2006.

[2] J. Borenstein, H. Everett, L. Feng, and D. Wehe. Mobile robot positioning-sensors and techniques. Technical report, DTIC Document, 1997.

[3] D. M. Chen, S. S. Tsai, B. Girod, C.-H. Hsu, K.-H. Kim, and J. P. Singh. Building book inventories using smartphones. In *Proceedings of the international conference on Multimedia*, MM '10, pages 651–654, New York, NY, USA, 2010. ACM.

[4] G. Cuellar, D. Eckles, and M. Spasojevic. Photos for information: a field study of cameraphone computer vision interactions in tourism. In *CHI '08 extended abstracts on Human factors in computing systems*, CHI

EA '08, pages 3243–3248, New York, NY, USA, 2008. ACM.

[5] M. Fischler and R. Bolles. Random sample consensus: a paradigm for model fitting with applications to image analysis and automated cartography. *Communications of the ACM*, 24(6):381–395, 1981.

[6] J. Hays and A. Efros. Im2gps: estimating geographic information from a single image. In *Computer Vision and Pattern Recognition, 2008. CVPR 2008. IEEE Conference on*, pages 1–8, June 2008.

[7] H. Hile, A. Liu, G. Borriello, R. Grzeszczuk, R. Vedantham, and J. Kosecka. Visual navigation for mobile devices. *IEEE MultiMedia*, 17:16–25, April 2010.

[8] R. Kaminsky, N. Snavely, S. Seitz, and R. Szeliski. Alignment of 3d point clouds to overhead images. In *Computer Vision and Pattern Recognition Workshops, 2009. CVPR Workshops 2009. IEEE Computer Society Conference on*, pages 63–70, June 2009.

[9] J. Koenderink, A. Van Doorn, et al. Affine structure from motion. *JOSA A*, 8(2):377–385, 1991.

[10] Y. Li, D. Crandall, and D. Huttenlocher. Landmark classification in large-scale image collections. In *Computer Vision, 2009 IEEE 12th International Conference on*, pages 1957 -1964, 29 2009-oct. 2 2009.

[11] D. G. Lowe. Object recognition from local scale-invariant features. In *Proceedings of the International Conference on Computer Vision-Volume 2 - Volume 2*, ICCV '99, pages 1150–, Washington, DC, USA, 1999. IEEE Computer Society.

[12] M. Magnor. Geometry-based automatic object localization and 3-d pose detection. In *Image Analysis and Interpretation, 2002. Proceedings. Fifth IEEE Southwest Symposium on*, pages 144–147. IEEE, 2002.

[13] S. Moon, I. Moon, and K. Yi. Design, tuning, and evaluation of a full-range adaptive cruise control system with collision avoidance. *Control Engineering Practice*, 17(4):442–455, 2009.

[14] C. Qin, X. Bao, R. Roy Choudhury, and S. Nelakuditi. Tagsense: a smartphone-based approach to automatic image tagging. In *Proceedings of the 9th international conference on Mobile systems, applications, and services*, MobiSys '11, pages 1–14, New York, NY, USA, 2011. ACM.

[15] E. Rosten and T. Drummond. Machine learning for high-speed corner detection. In *European Conference on Computer Vision (ECCV)*, pages 430–443, 2006.

[16] G. Schall, J. Schöning, V. Paelke, and G. Gartner. *A survey on augmented maps and environments: Approaches, interactions and applications*. Taylor & Francis Group, 2011.

[17] N. Snavely, S. M. Seitz, and R. Szeliski. Photo tourism: exploring photo collections in 3d. In *ACM SIGGRAPH 2006 Papers*, SIGGRAPH '06, pages 835–846, New York, NY, USA, 2006. ACM.

[18] R. Storn and K. Price. Differential evolution–a simple and efficient heuristic for global optimization over continuous spaces. *Journal of global optimization*, 11(4):341–359, 1997.

[19] G. Takacs, V. Chandrasekhar, N. Gelfand, Y. Xiong, W. Chen, T. Bismpigiannis, R. Grzeszczuk, K. Pulli, and B. Girod. Outdoors augmented reality on mobile phone using loxel-based visual feature organization. In *Proceeding of the 1st ACM international conference on Multimedia information retrieval*, pages 427–434. ACM, 2008.

[20] T. Yan, V. Kumar, and D. Ganesan. Crowdsearch: exploiting crowds for accurate real-time image search on mobile phones. In *Proceedings of the 8th international conference on Mobile systems, applications, and services*, MobiSys '10, pages 77–90, New York, NY, USA, 2010. ACM.

[21] Y.-T. Zheng, M. Zhao, Y. Song, H. Adam, U. Buddemeier, A. Bissacco, F. Brucher, T.-S. Chua, and H. Neven. Tour the world: Building a web-scale landmark recognition engine. In *Computer Vision and Pattern Recognition, 2009. CVPR 2009. IEEE Conference on*, pages 1085–1092, June 2009.

Acknowledgments

We sincerely thank our shepherd, Gaetano Borriello, as well as the anonymous reviewers, for their invaluable feedback. We are grateful to NSF for partially funding this research through the following grants: CNS-0916995 and IIS-910846.

APPENDIX

We provide equations for curves of visual trilateration and triangulation. We assume at least two GPS points are known $(x_1, y_1), (x_2, y_2)$. The object position is inferred at (a, b).

(1) Equation of Visual Trilateration

Let $\sigma = d_2/d_1$ be the ratio of distance from (x_2, y_2) to (a, b) divided by the distance (x_1, y_1) to (a, b).

$$(a - x_2)^2 + (b - y_2)^2 = \sigma^2 \left[(a - x_1)^2 + (b - y_1)^2\right]$$

To derive, construct two right triangles with hypotenuses (x_1, y_1) to (a, b) and (x_2, y_2) to (a, b). Apply the Pythagorean theorem on each, substituting the first into the second.

(2) Equation of Visual Triangulation

Let γ be the interior angle from (x_1, y_1) to (a, b) to (x_2, y_2).

$$\left[(x_2 - x_1)^2 + (y_2 - y_1)^2\right] = \left[(a - x_1)^2 + (b - y_1)^2\right] + \left[(a - x_2)^2 + (b - y_2)^2\right] - 2\sqrt{(a - x_1)^2 + (b - y_1)^2}\sqrt{(a - x_2)^2 + (b - y_2)^2}\cos\gamma$$

To derive, apply the law of cosines $(C^2 = A^2 + B^2 - 2AB\cos\gamma)$ with C taken as the side $\sqrt{(x_2 - x_1)^2 + (y_2 - y_1)^2}$.

A Close Examination of Performance and Power Characteristics of 4G LTE Networks

Junxian Huang
University of Michigan

Feng Qian
University of Michigan

Alexandre Gerber
AT&T Labs - Research

Z. Morley Mao
University of Michigan

Subhabrata Sen
AT&T Labs - Research

Oliver Spatscheck
AT&T Labs - Research

ABSTRACT

With the recent advent of 4G LTE networks, there has been increasing interest to better understand the performance and power characteristics, compared with 3G/WiFi networks. In this paper, we take one of the first steps in this direction.

Using a publicly deployed tool we designed for Android called 4GTest attracting more than 3000 users within 2 months and extensive local experiments, we study the network performance of LTE networks and compare with other types of mobile networks. We observe LTE generally has significantly higher downlink and uplink throughput than 3G and even WiFi, with a median value of 13Mbps and 6Mbps, respectively. We develop the first empirically derived comprehensive power model of a commercial LTE network with less than 6% error rate and state transitions matching the specifications. Using a comprehensive data set consisting of 5-month traces of 20 smartphone users, we carefully investigate the energy usage in 3G, LTE, and WiFi networks and evaluate the impact of configuring LTE-related parameters. Despite several new power saving improvements, we find that LTE is as much as 23 times less power efficient compared with WiFi, and even less power efficient than 3G, based on the user traces and the long high power tail is found to be a key contributor. In addition, we perform case studies of several popular applications on Android in LTE and identify that the performance bottleneck for web-based applications lies less in the network, compared to our previous study in 3G [24]. Instead, the device's processing power, despite the significant improvement compared to our analysis two years ago, becomes more of a bottleneck.

Categories and Subject Descriptors

C.2.1 [**Network Architecture and Design**]: wireless communication; C.4 [**Performance of Systems**]: measurement techniques, modeling techniques, performance attributes; D.2.8 [**Metrics**]: Performance measures

Keywords

LTE, 4G, 4GTest, 3G, energy saving, power model simulation, network model simulation

1. INTRODUCTION

Initiated in 2004 by *3rd Generation Partnership Project* (3GPP), the *Long Term Evolution* (LTE), commonly referred to as a type of 4G wireless service, aims at enhancing the *Universal Terrestrial Radio Access Network* (UTRAN) and optimizing radio access architecture [1]. Since 2009, LTE starts entering the commercial markets and is available now in more than 10 countries, with an expectedly fast-growing user base. The targeted user throughput is 100Mbps for downlink and 50Mbps for uplink, significantly higher than the existing 3G networks, with less than 5ms user-plane latency [12]. Understanding the actual user-perceived network performance for LTE network and how it compares with its predecessor 3G and its competitors, *e.g.,* WiFi and WiMAX, is important, yet not straightforward. Our previous study [24] takes one of the first steps to measure 3G network performance directly from end users. In this study, we follow up with a new tool, 4GTest, with enhanced measurement design and global server support, allowing us to characterize network performance of LTE and other mobile networks.

Besides higher bit rate, lower latency and many other service offerings for LTE, *user equipment* (UE) power saving is an important issue. LTE employs *Orthogonal Frequency Division Multiplex* (OFDM [15]) technology, which suffers from poor power efficiency. To save power, LTE uplink uses an special implementation of OFDM called SC-FDMA for uplink, with improved power efficiency. *Discontinuous reception* (DRX) has been employed by existing wireless mobile networks to reduce UE energy consumption. In UMTS [14], during the idle state, UE periodically wakes up to check paging messages and sleeps for the remaining time. LTE supports DRX for both `RRC_CONNECTED` and `RRC_IDLE` modes [16], seeking more opportunities to conserve UE battery. DRX is configured at a per-UE basis and controlled by a list of parameters in Table 2. The configuration of these parameters incurs tradeoff among UE power saving, channel scheduling delay, and signaling overhead.

To understand this tradeoff, existing studies use either total on time to estimate UE power usage [36, 19], or a simplified LTE power model [25, 34], which ignores the impact of downlink/uplink data rates. In this paper, we develop the first empirically derived comprehensive power model of a commercial LTE network, which accurately models UE energy usage. Also, existing studies [36, 19, 25, 34] heavily rely on synthetic packet models instead of real user traces. Our study is the first that leverages a comprehensive real user data set, we call UMICH, consisting of 5-month traces of 20 smartphone users, to analyze the impact of LTE parameter configuration on realistic application usage patterns. UMICH data set only includes users traces in 3G and WiFi networks, but not in LTE network. We make our best efforts to overcome this limitation

§	Figures & Tables	Data source	Description
4.1	Figure 5, 6	Public deployment of 4GTest	Compare LTE network performance with other mobile networks
4.2	Figure 7	Local LTE Phone	Infer LTE parameters with network-based approach
4.3, 4.4	Figure 8	LTE Laptop and LTE Phone	Understand the impact of various factors on LTE performance
5.1, 5.2	Table 3, 4, Figure 9, 10, 11	Local LTE Phone	Derive LTE power model and compare with WiFi and 3G
5.3	Figure 12	Local LTE Phone	Compare bulk data transfer energy efficiency of LTE with WiFi and 3G
5.4	Table 5	Local LTE Phone	Validate LTE power model with real applications
6.1, 6.2	Figure 14	Simulation on UMICH data set	Analyze energy efficiency and the impact of different energy components
6.3	Figure 15, 19, 20, 16, 17, 18	Simulation on UMICH data set	Study the impact of LTE parameter configurations
7.1	Figure 22	Different local platforms	Compare JavaScript execution speed in 2009 and in 2011
7.2	Table 6, Figure 21	Local LTE Phone	Study performance bottleneck of web-based applications

Table 1: Summary and index of experimental results.

by feeding WiFi traces into our LTE model simulation framework, given that RTT of WiFi is close to LTE (§4.1) and throughput is less important than RTT for most web-based smartphone applications (§7.2).

Compared with 3G, LTE significantly improves the network performance. Meanwhile, device processing capability and software design have also improved remarkably over the last two years, comparing with our previous study [24]. To understand the potential performance bottleneck shift for smartphone applications, we perform case studies of several popular applications on Android. With the help of CPU, network and power traces, we devise a systematic methodology for application performance and power measurement and analysis.

We summarize the results of this paper in Table 1 and highlight our main contributions as follows.

(i) This paper is one of the first studies on commercial LTE networks. We characterize LTE network performance with both local experiments and global measurement by deploying a network performance measurement tool 4GTest with distributed server support globally. Generally, LTE is observed to have significantly higher downlink and uplink throughput than 3G and even WiFi, with a median value of 13Mbps and 6Mbps, respectively, which is even higher than the advertised throughput range of the LTE provider studied.

(ii) We develop the first empirically derived comprehensive power model of a commercial LTE network, considering both uplink and downlink data rates in addition to state transitions and DRX. It accurately quantifies the radio energy usage of UE with less than 6% error rate. Our power measurement also reveals that LTE is less power efficient than 3G and WiFi for small data transfers, *e.g.,*, for transferring one packet, the energy usage for LTE, 3G and WiFi is 12.76J, 7.38J and 0.04J, respectively. For bulk data transfer, LTE is more power efficient than 3G, yet still lags behind WiFi, *e.g.,* for downloading 10MB data, 3G and LTE require 34.77 and 1.62 times the energy of WiFi, respectively. Considering that LTE is more energy efficient if fully utilized, it is even more crucial to optimize applications using tools like ARO [28] in LTE than 3G to achieve the full LTE savings. The energy model derived in this paper will provide the necessary understanding to drive those optimizations.

(iii) With the aforementioned power model, we build a trace-driven LTE analysis modeling framework, which breaks down the total energy consumption into different components, to identify the key contributor for energy usage. It also quantifies the channel scheduling delay perceived by UE and signaling overhead of cell towers. This analysis framework is applied to a comprehensive data set UMICH consisting 5 months' data from 20 real users. By comparing the estimated energy usage of different networks, LTE is as much as 23 times less power efficient compared with WiFi, and even less power efficient than 3G, despite the presence of DRX designed for energy saving. We also study the impact of configuring various LTE parameters on radio energy, channel scheduling delay, and signaling overhead. Similar to 3G UMTS network, LTE tail timer T_{tail} remains a key parameter in determining UE energy usage and performance, as well as signaling overhead of cell towers.

(iv) We perform case studies of several popular applications on Android to understand the impact of improved LTE network performance and enhanced UE processing power on applications. The processing speed for contemporary smartphones have increased significantly compared with our analysis two years ago. However, we still identify that the performance bottleneck for web-based applications lies more in the device's processing power than in the network, indicated by the high average CPU usage of 79.3% in LTE network, as compared to 57.7% in 3G network, as well as the underutilized network capacity due to small object size in typical web transactions.

The remaining paper is organized as follows. We cover background on RRC and DRX for LTE in §2, followed by experimental setup and methodology in §3. We then present network characterization results, including 4GTest results in §4 and power measurements in §5. §6 summarizes the simulation analysis on UMICH data set, and §7 includes the case study of several popular Android applications. We discuss related work in §8 before concluding in §9.

2. BACKGROUND

We first cover the necessary background on LTE state machine behavior and corresponding power characteristics.

2.1 Radio Resource Control (RRC) and Discontinuous Reception (DRX) in LTE

LTE has two RRC states, **RRC_CONNECTED** and **RRC_IDLE** [10, 17], as shown in Figure 1. At **RRC_CONNECTED** state, UE can be in one of the three modes: Continuous Reception, Short DRX, and Long DRX. While at **RRC_IDLE** state, UE is only in DRX mode. Table 2 enumerates a list of important LTE parameters, which have significant impact on UE's radio energy consumption, user experience, and signaling overhead for cell towers. The terms in Table 2 are used consistently throughout this paper.

If UE is initially in **RRC_IDLE** state and receives/sends one packet, regardless of the packet size, a state promotion from **RRC_IDLE** to **RRC_CONNECTED** occurs with a relatively stable delay, similar to the promotion from IDLE to DCH/FACH in UTMS network [27]. We define the LTE promotion delay to be T_{pro}[1] consistently throughout this paper. During this period, radio resources are allocated to the UE.

After being promoted to **RRC_CONNECTED**, UE enters Continuous Reception by default and keeps monitoring the *Physical Down-*

[1]T_{pro} is a measured system property, different from the configurable LTE parameters in Table 2.

Symbol	Full name	Measured value	Description
T_i	DRX inactivity timer	100ms	UE stays in Continuous Reception for T_i before DRX starts when idling
T_{is}	Short DRX cycle timer	20ms	UE remains in Short DRX for T_{is} before entering Long DRX when idling
T_{tail}	RRC inactivity timer	11.576s	UE stays in **RRC_CONNECTED** for T_{tail} before demoting to **RRC_IDLE**
T_{on}	**RRC_CONNECTED** On Duration timer	1ms	The on duration of UE during each DRX cycle in **RRC_CONNECTED**
T_{oni}	**RRC_IDLE** On Duration timer	43ms	The on duration of UE during each DRX cycle in **RRC_IDLE**
T_{ps}	Short DRX cycle	20ms	The cycle period of Short DRX in **RRC_CONNECTED**
T_{pl}	Long DRX cycle	40ms	The cycle period of Long DRX in **RRC_CONNECTED**
T_{pi}	**RRC_IDLE** DRX cycle	1.28s	The cycle period of DRX in **RRC_IDLE**

Table 2: Important LTE RRC and DRX parameters.

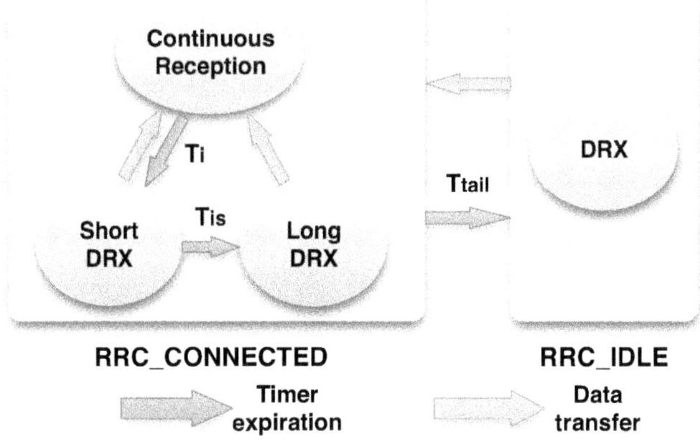

Figure 1: RRC state transitions in LTE network.

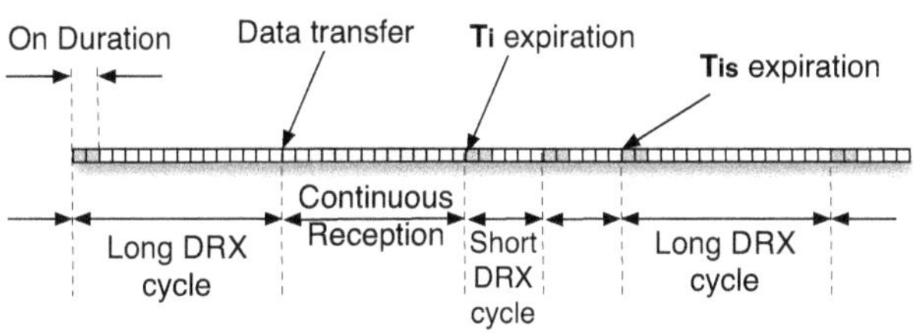

Figure 2: Illustration of the LTE DRX in RRC_CONNECTED.

link Control Channel (PDCCH), which delivers control messages to UE. UE also starts the DRX inactivity timer T_i, which is reset every time UE receives/sends a packet. Upon T_i's expiration without seeing any data activity, UE enters the Short DRX mode.

Discontinuous Reception (DRX) [16, 30], illustrated in Figure 2, is adopted by LTE for UE to "micro-sleep" to reduce power consumption while providing high QoS and connectivity. DRX in **RRC_CONNECTED** and **RRC_IDLE** have similar mechanisms, but different parameter settings. A DRX cycle includes an On Duration during which the UE monitors PDCCH. UE rests for the rest of the cycle to save energy. The tradeoff between battery saving and latency is the guideline for determining the parameterization of DRX cycle. With a fixed On Duration, a longer DRX cycle reduces energy consumption of UE while increasing user-perceived delay, and a shorter DRX cycle reduces the data response delay at the cost of more energy consumption. Short DRX and Long DRX modes, having the same On Duration and differing in cycle length, are to meet these conflicting requirements.

When UE enters Short DRX, Short Cycle Timer T_{is} is started. Upon T_{is}'s expiration, if there is no data activity, UE switches to Long DRX; otherwise, UE goes back into Continuous Reception.

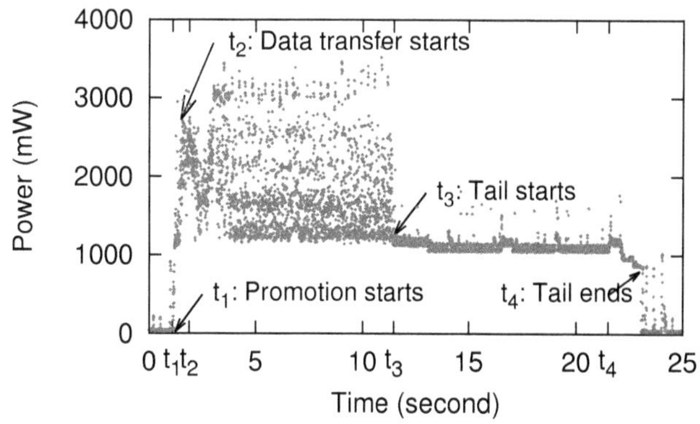

Figure 3: Power states of LTE.

For our measurement, T_{is} coincidentally equals T_{ps}, so only one cycle of Short DRX is expected to take place before T_{is} expires. Every time UE enters Continuous Reception when there is any data transfer, UE starts the tail timer T_{tail}, which is reset every time a packet is sent/received. When T_{tail} expires, UE demotes from **RRC_CONNECTED** to **RRC_IDLE** and the allocated radio resource is released. Notice that T_{tail} coexists with T_i and T_{is}.

2.2 Smartphone power model for LTE

Given the description of LTE state machine, we illustrate the power traces of an Android smartphone in a commercial LTE network based on local experiments described in §3.2. We observe that network activities match the corresponding state transitions indicated by different power levels.

Figure 3 shows the power trace of uploading at the speed of 1Mpbs for 10 seconds. With screen off, the energy is mostly consumed by the radio interfaces, as the power level is less than 20mW before t_1. At t_1, the application sends a TCP SYN packet triggering **RRC_IDLE** to **RRC_CONNECTED** promotion, and the application waits for T_{pro} until starting data transfer at t_2. Between t_2 and t_3, depending on the instant data rate, the power level fluctuates. We notice the power level during fast data transfer is significantly higher than the base power in **RRC_CONNECTED**, which motivates us to incorporate data rates into our LTE power model. After the data transfer completes at t_3, the device remains in **RRC_CONNECTED** for a fixed tail time T_{tail}, until t_4, when the device goes back to **RRC_IDLE**. The periodicity of DRX between t_3 and t_4 is not obvious due to limited sample rate.

In summary, this section provides necessary background information and some initial motivating observations for the following discussions on the network and power characterization of LTE network.

3. METHODOLOGY

In this section, we present the methodology for network and power measurement, as well as trace-driven simulation analysis and real application case studies.

3.1 Network measurement

In this section, we discuss the design of our publicly deployed tool 4GTest, followed by experimental setup and methodology for local network measurement.

3.1.1 *The design of* 4GTest

Based on our previous experiences of developing 3GTest [6, 24], we design a new mobile network measurement tool for Android devices, called 4GTest [2], with higher accuracy and improved user experience. As a new feature, 4GTest allows users to switch among different network types, *i.e.,* 3G, WiFi and LTE. We also improve the network measurement methodology leveraging the M-Lab [5] support. 4GTest server suite is deployed to 46 M-Lab nodes across the world, covering North America, Europe, Asia, and Australia. Each node has 4-core 2.00 GHz Intel Xeon CPU and our virtual slice has 4GB memory and 100Mbps Ethernet network access, which ensures that the network bottleneck is unlikely on the wired network path.

RTT and variation test LTE has significantly smaller latency compared with 3G [12], hence the network distance between users and the measurement servers for the wired Internet path becomes less negligible. Given that the GPS coordinates for M-Lab nodes are known, in 4GTest, a nearest M-Lab node is selected for a user based on the current GPS location if available, or the IP address otherwise, with the help of a IP address to GPS coordinates mapping [4]. Such a mapping is sufficient for our purpose of finding coarse-grained location estimates for server selection.

To measure RTT and variation, 4GTest repeatedly establishes a new TCP connection with the server and measures the delay between SYN and SYN-ACK packet. Both the median of these RTT measurements and the variation are reported to our central server.

Throughput test Since single-threaded TCP measurement is more sensitive to packet loss and hence less accurate [32], we use multi-threaded TCP measurement in 4GTest to estimate the *peak channel capacity*, *i.e.,* three nearest server nodes in M-Lab are selected for each user at runtime to start concurrent threads for throughput test. Despite the shared nature of M-Lab nodes with other test tools, it is unlikely that all three selected nodes are overloaded in terms of CPU and network.

A throughput test lasts for 20 seconds, to balance across bandwidth usage, user waiting time and measurement accuracy. The initial 5 seconds are ignored empirically due to TCP slow start. The remaining 15 seconds are separated into 15 1-second bins. The average throughput for each bin is calculated and the median of all bins is the measured throughput. Compared with using average throughput, median more accurately estimates the steady peak throughput by reducing the impact of abnormal bins, *e.g.,* a very high bin due to initial buffering or a low bin due to temporary signal problem. Uplink and downlink tests share the same methodology.

3.1.2 *Local network measurement*

Devices for accessing LTE networks in this study are named LTE Phone and LTE Laptop. LTE Phone is an HTC phone with LTE data plan from a cellular ISP. It has 768 MB RAM memory and 1 GHz Qualcomm MSM8655 CPU, running Android 2.2.1. LTE Laptop is a laptop equipped with LTE USB Modem. It has 8GB memory and 2.53 GHz Intel Core Duo CPU, running Mac OS X 10.7.2.

Packet and CPU trace collection In order to collect packet traces on LTE Phone, we cross-compile tcpdump [9]. With a small snapshot length, only to capture TCP/UDP and IP headers, the CPU overhead of the background tcpdump process is less than 5%. To capture CPU usage history, we write a simple C program to read `/proc/stat` in Android system every 25ms and the overhead is verified to be smaller than 5% on LTE Phone.

Network-based approach for LTE parameter inference is used to validate the power-based approach (§3.2). We use the following experiment to infer LTE state machine and `RRC_IDLE` DRX parameters. LTE Phone maintains a long lived TCP connection with a test server. For each experiment, server first sends a packet P_1 to trigger UE's promotion to `RRC_CONNECTED`, and after X seconds, server sends another packet P_2 to the phone to measure RTT. Assume the RTT at `RRC_CONNECTED` is RTT_b ignoring the minor impact of DRX in `RRC_CONNECTED`. When $X \leq T_{tail}$, $RTT(X) = RTT_b$. Otherwise, $RTT(X) = RTT_b + T_{pro} + T_{drx}$, where $T_{drx} = T_n - X$ and T_n is the start time of the next DRX on duration after P_2's arrival. If P_2's arrival is inside any DRX on duration, $T_{drx} = 0$.

To understand the **impact of packet size on one-way delay (OWD)**, uplink and downlink OWD is measured with varying packet size. For each packet size, 100 samples are measured, and we make sure that packets are not fragmented. OWD is measured between LTE Laptop and a test server. A GPS Receiver [3] is connected to LTE Laptop, which uses the received GPS messages for time synchronization. The test server is also time-synced with another GPS clock. The maximum error of our setup is less than 1ms.

We **compare LTE with 3G and WiFi** by local experiments on LTE Phone. When 4G is disabled in LTE Phone, it connects to 3G network and the network type is eHRPD. For WiFi, with the data network radio interface turned off, LTE Phone connects to a wireless router, Cisco-Linksys WRT54GL Wireless-G Broadband Router, 802.11g in channel 2.437GHz. The Internet ISP for WiFi is Comcast and the advertised downlink peak data rate is 6Mbps.

3.2 Power measurement

Similar to previous studies [27, 35], we use Monsoon power monitor [7] as power input for LTE Phone measuring power traces at the same time. The power trace contains two fields, timestamp and average instant power, and the sampling rate is 5000Hz. We share the same observation with previous study [35] that screen plays an important role in device power consumption, *i.e.,* with screen 100% on, the UE idle power is 847.15mW compared with 11.36mW with screen off. For all measurements, we keep the test application running in the background with screen completely off to minimize power noise, unless UI interactions are required and screen should be kept on, *i.e.,* measuring power for browser. In this case, we subtract screen power from the total, with slightly increased noise. All experiments are repeated at least 5 times to reduce measurement error.

To measure state transition power levels, UE keeps a long-lived TCP connection with the server and packet traces are collected to make sure there is no background traffic. In order to trigger state promotions, we make the device idle for sufficient time, *e.g.,* 30 seconds, and then send a packet from server to client. UE remains idle afterwards and demotes to idle state in the end, and the power trace covers the full tail.

3.3 Trace-driven analysis

To compare energy consumption for different networks using real user traces and evaluate the impact of setting LTE parameters, we devise a systematic methodology for trace-driven analysis, which is applied to a comprehensive user trace data set, named UMICH.

3.3.1 UMICH *data set for analysis*

UMICH data set is collected from 20 smartphone users for five months, 05/12/2011 ~ 10/12/2011, totaling 118 GB. These participants consist of undergraduate and graduate students from 8 departments at University of Michigan. The 20 participants are given Motorola Atrix (11 of them) or Samsung Galaxy S smartphones (9 of them) with unlimited voice, text and data plans, all running Android 2.2. As stated in our study consent form, we keep collected data and users' identities strictly confidential[2].

Custom data collection software developed by us are deployed on the 20 smartphones. It continuously runs in the background and collects full packet traces in tcpdump format including both headers and payload. Both cellular and Wi-Fi traces are collected without any sampling performed. The data collector incurs no more than 15% of CPU overhead, although the overhead is much lower when the throughput is low (*e.g.,* less than 200 kbps).

We also build a data uploader that uploads the data to our server when the phone is idle as indicated by low network throughput. The upload is suspended by detected network activity of user applications. The entire data collection and uploading process is transparent to the users.

3.3.2 *Trace-driven modeling methodology*

We build a network model and power model analysis framework for the trace-driven analysis, totaling about 8,000 lines of code in C++.

Network model simulator takes the binary packet trace files in libpcap [9] format and preprocesses the data following the same methodology as previous study [27], for the purpose of removing existing delays imposed by state promotions and extracting the actual traffic patterns, *e.g.,* for the purposing of simulating the trace in another network. It then applies a specific network model to the processed traces and adjusts timestamps for some packets, since different network models have different delay behavior due to state machine differences. There are two reasons for a packet's timestamp to be adjusted, promotion delay and DRX waiting time. We use the same methodology to adjust timestamp for promotion delay as previous study [27]. In LTE network, if there is no packet for over T_{tail} time, when the next packet P arrives, an extra delay T_{pro} is inserted before P. In terms of DRX waiting time, if a downlink packet P' comes at time $T_{P'}$ and UE is in DRX mode, either in `RRC_CONNECTED` or `RRC_IDLE`, there is no DRX waiting time if P' arrives inside any DRX on duration; otherwise, P' would experience an extra delay of $T_n - T_{P'}$, where T_n is the start time of the next DRX on duration after $T_{P'}$. The parameters for different network models are listed in Table 3, obtained from local experiments. We also empirically assume that UE goes into complete sleep with 0 power usage after being idle for T_{tail2} (1 minute). Notice that T_{tail2} is not actually observed in our experiments and is only used to bound the total idle energy and simplify our analysis.

The output of network model simulator is an array of packets with adjusted timestamps, in ascending order, as well as the RRC and DRX states of UE at any time.

Power model simulator takes the output of network model simulator and calculates the energy consumption based on the power model, detailed in §5. We break down the total energy into four components: *promotion, data transfer, tail,* and *idle*.

[2]This user study has been approved by the University of Michigan IRB-HSBS #HUM00044666.

For promotion energy, both promotion delay T_{pro} and average promotion power P_{pro} are fixed. If a trace includes N promotions, promotion energy is given by $NT_{pro}P_{pro}$.

For idle energy in `RRC_IDLE`, assume that DRX cycle is T_{pi} (with base power P_b) and on duration is T_{oni} (with on power P_{oni}), energy consumption of an duration of t_{idle} is given by

$$\lfloor \frac{t_{idle}}{T_{pi}} \rfloor \Big(T_{oni}P_{oni} + (T_{pi} - T_{oni})P_b \Big) + E_{res}$$

where the remaining energy E_{res} can be calculated as

$$E_{res} = \begin{cases} T_{res}P_{oni} & \text{if} \quad T_{res} \le T_{oni} \\ T_{oni}P_{oni} + (T_{res} - T_{oni})P_b & \text{if} \quad T_{res} > T_{oni} \end{cases}$$

with

$$T_{res} = t_{idle} - T_{pi}\lfloor \frac{t_{idle}}{T_{pi}} \rfloor$$

Tail energy is more complex for LTE since `RRC_CONNECTED` has three modes as in Figure 1, while 3G and WiFi do not have DRX in connected mode. We simulate the transitions among Continuous Reception, Short DRX and Long DRX to calculate tail energy, in a similar way of idle energy calculation. Note that we only count tail energy when the tail is complete, otherwise, the energy is considered as data transfer energy, *i.e.,* if one packet P is observed before UE demotes to `RRC_IDLE` at the end of the tail, the energy between P and the previous packet is still considered as data transfer energy.

For data transfer energy, we propose a novel model. Assume uplink throughput is t_u (Mbps) and downlink throughput is t_d (Mbps), the instant power level (mW) of UE is

$$P = \alpha_u t_u + \alpha_d t_d + \beta$$

Validation of this formula is in §5.2. We simply consider payload throughput in our power model, without considering header length and ACK packets due their small impact. Since TCP instant throughput changes frequently, in order to accurately estimate the instant power levels, we divide the trace into small time windows with size W seconds and within each window, throughput t_u and t_d are calculated, which determine the power P. W is set to be 1 second empirically in our setup and the total energy is not sensitive to the choice of W as long as W is larger than a few RTTs and smaller than most flow durations.

3.4 Application performance

As a case study, we select a few smartphone applications and collect both network and CPU traces on LTE Phone. Specifically, we select the default browser, YouTube, NPR News and Android Market as our sampled applications given their popularity. For the default browser, we choose two different usage scenarios, *i.e.,* visiting mobile version of `google.com` representing a simple website and visiting non-mobile version of `yahoo.com` representing a content-rich website. We name the two browser usage scenarios as Website G and Website Y, respectively. While for the other applications, we click the application icon until the initial page is fully presented to the user. We make sure the cache is cleared before each experiment.

Figure 4 shows co-located network and CPU trace of Website Y. At time 0, when GO button is clicked, loading starts. Before t_a, CPU usage stays low most of the time given that UE has not finished downloading the HTML or JavaScript objects. Starting from t_a, CPU usage jumps to 100% and remains a high average usage until t_c. We notice that network activity nearly stops after t_b, with only a few TCP FIN packets for closing TCP connections afterwards, some even come seconds later, *e.g.,* at time 10.5 seconds.

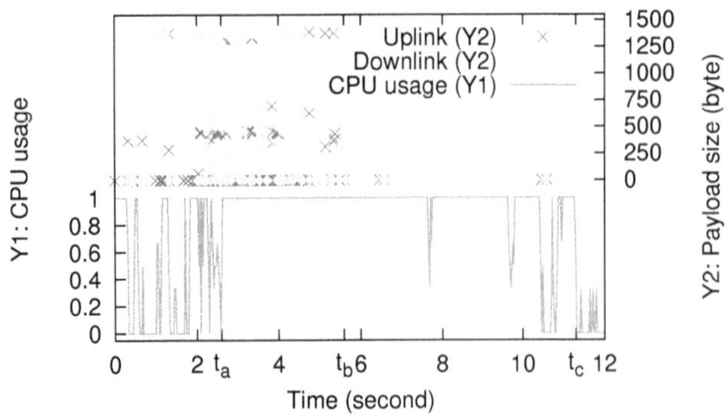

Figure 4: Network and CPU trace for Website Y **in LTE.**

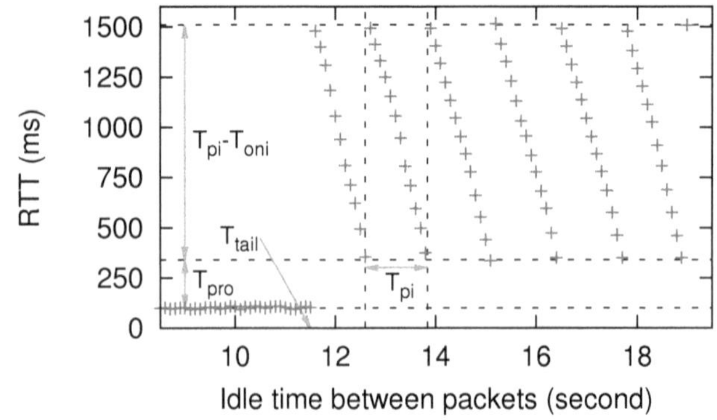

Figure 7: LTE RTT v.s. inter-packet idle time.

During the time between t_a and t_c, UE is rendering HTML pages or executing JavaScript, and during the time between t_a and t_b, UE is also downloading web objects in parallel.

We define t_c to be the *application loading time* instead of t_b, since at t_b, UE has not fully rendered the contents for the user, though the download process is complete. We validate this by collecting video traces with a camcorder facing the screen of the phone. We replay the video at normal speed and manually mark the start time and the end time, when the website gets fully loaded and all UI activity indicators stop. We verify that t_c is an accurate estimate for application loading time. We also define the average CPU usage between time 0 and t_c to be the *CPU usage* for this application.

Compared with a previous approach for measuring web browsing performance by modifying and instrumenting WebKit [33], our approach is more lightweight and easily extensible to other applications other then web browsers, with reasonable accuracy. One limitation of our approach is that it is not applicable for applications, *e.g.*, streaming video/audio and some game applications, whose CPU usage remains high after initial loading.

4. LTE NETWORK CHARACTERIZATION

We present network measurement results from publicly deployed 4GTest in this section. In addition, we carry out extensive local measurement on an LTE network to infer RRC and DRX parameters, and understand the impact of various factors on network performance.

4.1 Comparing LTE to other mobile networks

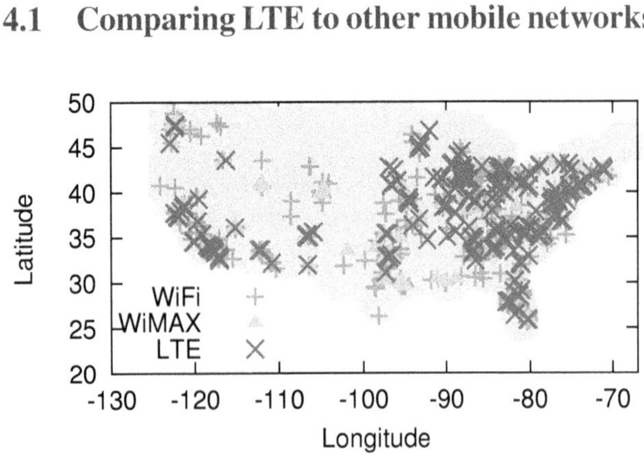

Figure 6: 4GTest **user coverage in the U.S.**

4GTest was initially released on Android Market on 10/15/2011. Till 12/15/2011, there has been altogether 3294 global users and 13768 runs. In the analysis, we focus on 2292 users and 7063 runs inside the U.S., among which 380 are LTE users with 982 runs. We check the distribution of 4GTest users for different network types. For example, in Figure 6, the coverage of LTE, WiMAX, and WiFi are mostly similar, covering 39, 37 and 44 states in the U.S., respectively. We also observe that other networks have similar coverage across the U.S. This indicates that 4GTest data set enables fair comparison on the distribution of performance metrics for different mobile networks in the U.S.

Figure 5 summarizes the performance comparison among various mobile networks based on 4GTest data set. We present anonymized data based on the technology type. The box plot for each metric shows the 95th, 75th, 50th, 25th and 5th percentile among all measurements for each network type. We observe that LTE network has a high downlink and uplink throughput, with the median to be 12.74Mbps and 5.64Mbps, respectively, which is much higher than WiFi's 4.12Mbps (downlink) and 0.94Mbps (uplink), as well as 4.67Mbps (downlink) and 1.16Mbps (uplink) for WiMAX. The 3G family, including eHRPD, EVDO_A and HSDPA clearly lag behind LTE. We also observe relatively high variation on LTE throughput for different users at different locations, and even for the same user at the same location across different runs.

In terms of RTT and RTT jitter, LTE with the median RTT 69.5ms and RTT jitter 5.6ms is comparable to WiFi's 64.5ms (RTT) and 7.9ms (RTT jitter). Similar to throughput comparison, WiMAX also lags behind LTE with median RTT to be 125.0ms and RTT jitter to be 21.0ms. For 3G networks, compared with our previous study [24], median RTT has been reduced from a median of 400ms to below 200ms, some even has a median of 78ms (HSDPA 2), close to LTE RTT. RTT jitter for 3G networks are all larger than 10ms.

To summarize, Figure 5 provides a fair comparison on performance distribution among different mobile networks, from which we observe that LTE significantly improves network throughput as well as RTT and jitter, making it comparable to or even better than WiFi for throughput. WiMAX lags behind LTE, followed by HSDPA, eHRPD and EVDO_A networks.

4.2 Network-based LTE parameter inference

Figure 7 summarizes the network-based approach for LTE state machine and DRX parameter measurement using LTE Phone, with the methodology discussed in §3.1.2. Each point is the median of five runs and for each run, RTT is measured after making LTE Phone idle for a specific time. The pattern of RTT is expected as inter-

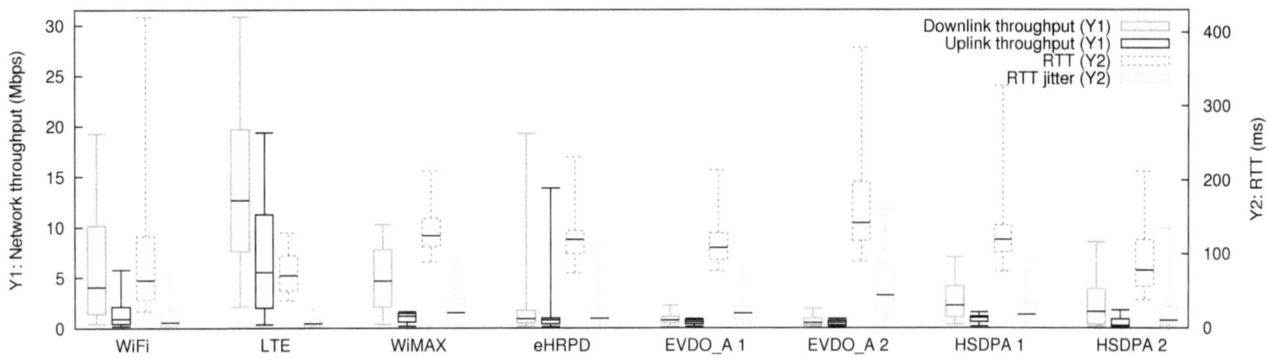

Figure 5: 4GTest **result analysis based on network type***.

*Each network type corresponds to the data of one cellular ISP, except for WiFi which has data of various ISPs.

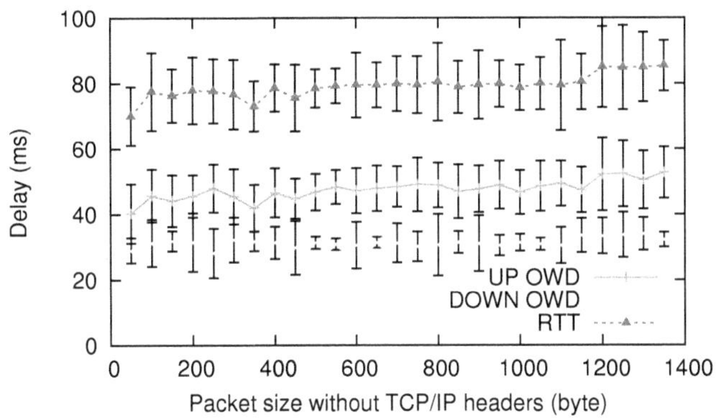

Figure 8: OWD/RTT v.s. packet size in LTE network.

packet idle time changes. Based on these results, we can infer the values of the following parameters in Table 2: T_{tail} is around 11.5s, T_{pi} is 1.28s and T_{oni} is close to 45ms. In addition, we observe T_{pro} is around 260ms. These measured values agree with those measured from the power-based approach in Table 3, cross-validating our measurement methodology.

4.3 One-way delay and impact of packet size

With the setup in §3.1.2, we measure uplink/downlink one-way delay (OWD) for both LTE and WiFi networks. Experiment in WiFi shows that both uplink and downlink OWD are around 30ms with little correlation with packet size, and RTT is stable around 60ms. However, for LTE, Figure 8 shows that uplink OWD is clearly larger than downlink OWD. We also observe that uplink OWD for LTE slightly increases as packet size grows, while the median of downlink OWD stays more stable around 30ms. The median RTT ranges from 70ms to 86ms as packet size increases. In summary, RTT in LTE is more sensitive to packet size than WiFi, mainly due to uplink OWD.

4.4 Mobility

Requirements for LTE [12] specifies that mobility across the cellular network should be supported and optimized for low mobile speed from 0∼15 km/h. Higher mobile speed of 15∼120 km/h should be supported with high performance. In this paper, we carry out one of the first studies to quantify the impact of mobile speed for a commercial LTE network.

We measure RTT and downlink/uplink throughput with 4GTest at three different mobile speeds: stationary, 35mph and 70mph, highway speed limit. All experiments are done in the same car for fair comparison. Given that 4GTest takes less than 2 minutes to finish, handover seldom happens during a single test and we filter out samples with cell ID changed during the test. Previous work [29] has shown that signal strength varies across locations and has a non-negligible impact on performance, so we only select sample points with the similar signal strength for fair comparison.

We observe that RTT remains stable at different speeds, with small variation. Uplink and downlink throughput both have high variation of 3∼8Mbps. Comparing throughput at different speeds, there is no clear difference, given the high variation. Our experiments show that at least at our test location, there is no major performance downgrade at high speed for LTE, satisfying the requirements in [12].

In addition, we also study the correlation between LTE performance and **time of day**. We periodically run 4GTest for 48 hours in two randomly selected days. RTT's median value remain stable at 68ms across different hours. Notice that we do not use the 4GTest results from real users to study this time of day pattern, because those results may also contain user-specific and location-wise factors which affect the performance and are hard to be excluded. For downlink and uplink throughput, variation across different hours is observed; however, there is no strong correlation with time of day. This may imply that, at our location, LTE currently does not have too many users and has well-provisioned resources available.

To summarize, with both global experiments, via 4GTest, and local experiments, we observe that LTE has significantly improved network performance over 3G, making it comparable to, if not better than, WiFi.

5. LTE POWER MODEL CONSTRUCTION

This section summarizes the construction of the new LTE power model, as well as the 3G and WiFi model measured from the same LTE Phone. We then compare energy efficiency in bulk data transfer for different networks and validate LTE power model in the end.

5.1 Power model for RRC and DRX

With the experimental setup described in §3.2, we measure power model for LTE, 3G, and WiFi on LTE Phone, summarized in Ta-

	Power* (mW)	Duration (ms)	Periodicity (ms)
Screen off (base)	11.4±0.4	N/A	N/A
Screen 100% on	847.2±2.7	N/A	N/A
LTE promotion	1210.7±85.6	T_{pro}: 260.1±15.8	N/A
LTE Short DRX On `RRC_CONNECTED`	1680.2±15.7	T_{on}: 1.0±0.1	T_{ps}: 20.0±0.1
LTE Long DRX On `RRC_CONNECTED`	1680.1±14.3	T_{on}: 1.0±0.1	T_{pl}: 40.1±0.1
LTE tail base	1060.0±3.3	T_{tail}: 11576.0±26.1	N/A
LTE DRX On `RRC_IDLE`	594.3±8.7	T_{oni}: 43.2±1.5	T_{pi}: 1280.2±7.1
3G promotion	659.4±40.4	582.1±79.5	N/A
3G DCH tail base	803.9±5.9	8088.2±149.6	N/A
3G FACH tail base	601.3±6.4	824.2±148.1	N/A
3G DRX (idle)	374.2±13.7	55.4±1.5	5112.4±37.7
WiFi promotion	124.4±2.6	79.1±15.1	N/A
WiFi tail base	119.3±2.5	238.1±9.2	N/A
WiFi beacon (idle)	77.2±1.1	7.6±0.1	308.2±1.0

Table 3: LTE, 3G, and WiFi power model.

*All power readings in this table include the base power (screen off), which has negligible impact on total energy.

ble 3. The LTE parameter values are validated by the network-based measurement in §4.2. For simplicity, we ignore the WiFi AP scanning and association, assuming UE is already connected with an AP.

First, we observe that LTE reduces the promotion delay (T_{pro}) from 3G's 582.06ms to 260.13ms. However, the power level is almost doubled, *i.e.*, 1210.74mW (LTE) v.s. 659.43mW (3G). WiFi has the most lightweight state promotion with smaller T_{pro} and much lower power level.

Secondly, LTE appears to have longest tail (11.576 seconds) with highest tail base power (1060.04 mW). Summing up DCH and FACH tail, 3G's total tail time (8.9 seconds) is smaller than LTE's T_{tail} of 11.6 seconds. Even 3G DCH's tail base power is 24.17% lower than LTE's tail base power, and the gap becomes 25.25% if we consider LTE DRX in `RRC_CONNECTED` with a high on duration power (1680.20mW). WiFi is much more power efficient, with shorter tail and much lower base power.

We also compare LTE DRX in `RRC_IDLE` with 3G DRX and WiFi beacon in the idle state. LTE has the highest on power and slightly smaller On Duration than 3G, while WiFi has smallest on power and On Duration. The cycle of LTE (1.28 seconds) is in between 3G and WiFi.

Based on these observations, LTE is less energy efficient during idle state and for transferring smaller amount of data. For example, if only one packet is transferred, the energy usage considering both promotion and tail energy for LTE, 3G and WiFi is 12.76J, 7.38J and 0.04J, respectively. One possible reason for LTE's higher power states is that devices must incorporate *multiple-input and multiple-output* (MIMO) to support LTE network, *e.g.*, the test device we use has 1 transmit antenna and 2 receive antennas, which contributes to higher power consumption.

5.2 Power model for data transfer

Previous work on 3G UMTS power modeling either treats DCH power state to have a fixed power value [35, 28], or assumes energy per bit to be the same constant for both uplink and downlink [18]. These assumptions might be reasonable given that 3G has relatively low data rates. However, for LTE, we observe that device power is much higher during high speed data transmission (up to 3300mW for uplink) relative to the base power (1060mW) in `RRC_CONNECTED`, and there is significant difference between downlink and uplink power levels at the same data rate. In this paper, we propose a new comprehensive power model for LTE empirically derived in a commercial LTE network.

	α_u (mW/Mbps)	α_d (mW/Mbps)	β (mW)	α_u/α_d
LTE	438.39	51.97	1288.04	8.44
3G	868.98	122.12	817.88	7.12
WiFi	283.17	137.01	132.86	2.07

Table 4: Data transfer power model.

We start with measuring device power states with controlled uplink or downlink throughput. The impact of TCP ACK packets, which are small in size, is minor, thus ignored.

Figures 9 and 10 present the power-throughput curve for LTE, 3G, and WiFi. The curves are limited by the peak data rate we can achieve at the test location. We observe that for all networks, a linear model fits well for both uplink and downlink. Assume uplink throughput is t_u (Mbps) and downlink throughput is t_d (Mbps), the power level (mW) for uplink is $P_u = \alpha_u t_u + \beta$ and for downlink $P_d = \alpha_d t_d + \beta$. The best fit parameters are listed in Table 4.

By looking at α_u/α_d, we notice that uplink power increases faster than downlink for all three networks types. This is expected because sending data requires more power than receiving data for wireless data access [21]. LTE has the largest gap of $\alpha_u/\alpha_d = 8.44$ among three network types. This is largely because α_d for LTE is quite small. For 3G, both α_u and α_d are larger than LTE. β is the base power when throughput is 0, with the ranking of LTE > 3G > WiFi. This is consistent with the tail base power comparison in Table 3. We notice that β is slightly higher than the tail base for all networks types. This is possibly because of the overhead of switching transmitters or receivers into high speed mode.

For simultaneous uplink and downlink transfers, given that transmitters and receivers are separate, we conjecture that the power level (mW) is given by the following formula:

$$P = \alpha_u t_u + \alpha_d t_d + \beta$$

To validate this conjecture, we measure the power levels for concurrent uplink and downlink transfers in Figure 11. Assume total throughput $t = t_u + t_d$ and the ratio of uplink throughput $\epsilon = t_u/t$:

$$P = \alpha_u t_u + \alpha_d t_d + \beta = (\alpha_u - \alpha_d)t\epsilon + \alpha_d t + \beta$$

When t is a constant, P grows linearly with ϵ and the slope is $(\alpha_u - \alpha_d)t$. Figure 11 shows two curves of t = 1Mbps and t = 2Mbps, both having a strong linear pattern and the slope is less than 5% off the expected value.

5.3 Energy efficiency for bulk data transfer

To compare the power efficiency of different networks in the wild, we use bulk data transfer experiments to measure energy per bit. Perrucci *et al.* [26] measure energy per bit for 3G and WiFi with a fixed bulk size. In addition to taking LTE into consideration, we vary the bulk size to cover more possible network usage scenarios. Figure 12 shows the measured energy per bit in $\mu J/bit$ ($10^{-6} Joule/bit$) with different bulk data size. All data is randomly generated so that there is no chance for caching. We do not include promotion or tail energy but instead focus on data transfer energy. Given that signal strength and peak data rate on wireless

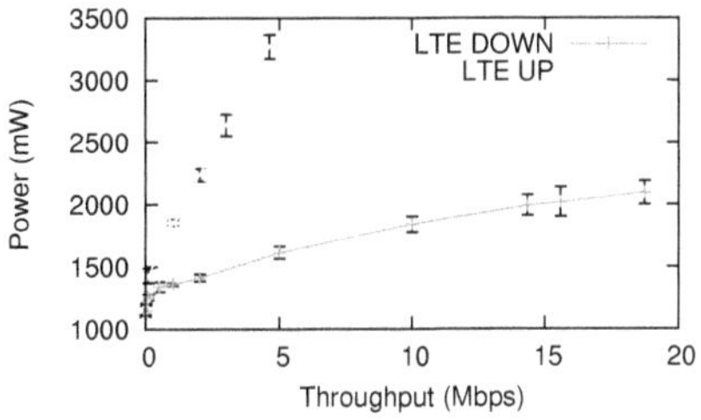

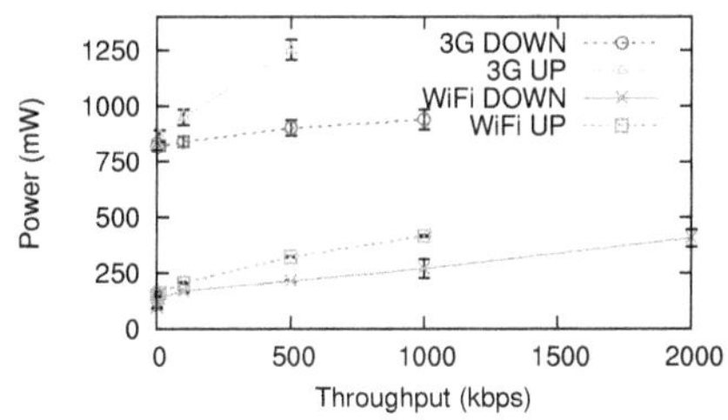

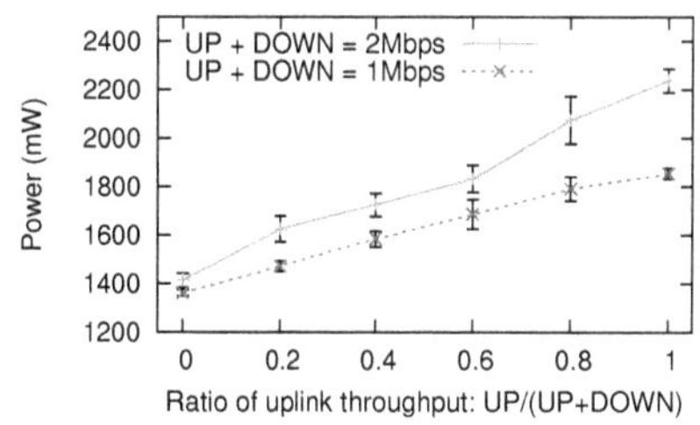

Figure 9: Power-throughput curve for LTE network.

Figure 10: Power-throughput curve for 3G and WiFi.

Figure 11: Power of simultaneous uplink and downlink transfers.

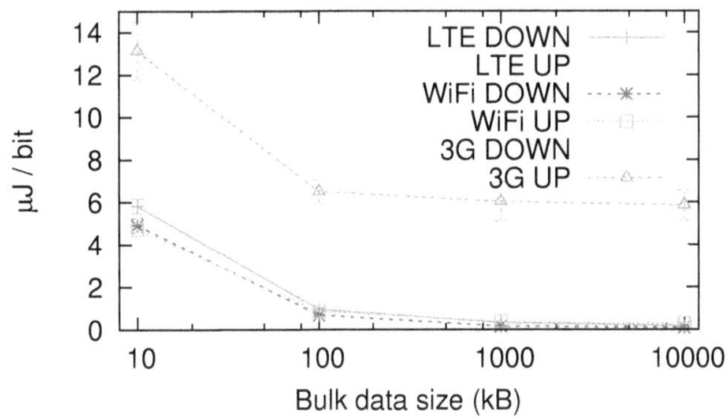

Figure 12: Energy per bit for bulk data transfers.

App	Measured energy (J)[1]	Simulated energy (J)[1]	Error
Website G[3]	24.77	24.37	-1.61% (-2.06%[2])
Website Y[3]	31.84	30.08	-5.53% (-7.04%)
YouTube	21.81	21.14	-3.07% (-4.17%)
NPR News	24.65	24.37	-1.12% (-1.70%)
Market	38.64	38.03	-1.58% (-3.03%)

[1]Both measured and simulated energy include tail energy
[2]This error is for simulated energy assuming $\alpha_u = \alpha_d = 0$
[3]Refer to §3.4 for the definition of Website G and Website Y

Table 5: LTE power model validation.

networks fluctuates, both affecting energy per bit, our measurement only serves as a sampled view for the energy efficiency of different networks.

First, energy per bit decreases as bulk data size increases, largely because with a small data size, throughput does not reach link capacity due to TCP slow start. We also observe that LTE's energy per bit in downlink is comparable with WiFi, although the absolute power level of LTE is much higher than WiFi. This is due to high downlink throughput for LTE at our test location, even compared with the WiFi network. Similarly, for LTE uplink, it drops from $10\mu J/bit$ to less than $1\mu J/bit$ as bulk data size increases. With bulk data size of 10MB, LTE consumes 1.62 times the energy of WiFi for downlink and 2.53 for uplink. With lowest throughput, 3G has the worst energy efficiency for large data transfer, *e.g.*, for downloading 10MB data, 3G requires 21.50 times the energy of LTE and 34.77 times the energy of WiFi, and for uplink, 7.97 times of LTE and 20.16 times of WiFi.

5.4 Power model validation

To validate the LTE power model and the trace-driven simulation (§3.3.2), we compare measured energy (measured from LTE Phone) with simulated energy for case study applications. Table 5 contains the sample application usage scenarios described in §3.4. The error rate is consistently less than 6%, with the largest error rate from Website Y, which includes heavy JavaScript execution and HTML rendering. Since our power model focuses on radio power and ignores the impact of CPU, for Website Y, the total energy usage is slightly underestimated.

The error rate is increased if the impact of downlink and uplink throughput is ignored, *i.e.*, assuming $\alpha_u = \alpha_d = 0$. However, the increase is not significant, at most 1.5%. This is because for these web-based applications, network throughput is low due to small object size (§7.2). For other applications, such as video/audio streaming and file download, we expect to see a larger gap in error rate if the impact of downlink/uplink is ignored.

In this section, in addition to comparing energy per bit in bulk data transfer for different networks, we construct a new LTE power model and validate its accuracy, which is the basis for the following analysis.

6. USER TRACE BASED TRADEOFF ANALYSIS

In this section, we apply the LTE power model to UMICH data set and compare energy efficiency with 3G and WiFi, with detailed break down of the total energy. We then study the tradeoff of configuring different LTE parameters via our analysis framework.

6.1 Energy efficiency comparison

We use the UMICH data set to simulate the LTE, WiFi and 3G model. Assume that the simulated energy usage for LTE, WiFi and 3G power model is E_{lte}, E_{wifi} and E_{3g}, respectively, the energy ratio of LTE/WiFi is defined as E_{lte}/E_{wifi}, and that for 3G/WiFi is calculated as E_{3g}/E_{wifi}. With the same traces, we can make fair comparison among different power models to understand their energy efficiency.

In Figure 13, we compare the energy efficiency of different networks for the 20 users both individually and in aggregate (summing up for all users). We first observe that LTE power model consumes significantly more energy than WiFi. The ratio of LTE/WiFi ranges from 16.9 to 28.9 and the aggregate ratio for all users is 23.0. Notice that the gap between LTE and WiFi is larger compared with the bulk data transfer experiments in §5.3. This is because, for bulk data transfer, LTE's high throughput could compensate the low energy efficiency, compared with real traces, which do not saturate the link capacity. Second, for 3G/WiFi ratio, the range is between 10.8

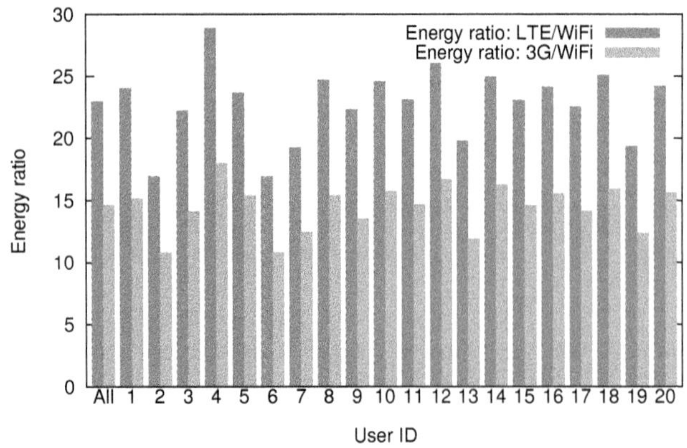

Figure 13: Power model simulation: energy ratio.

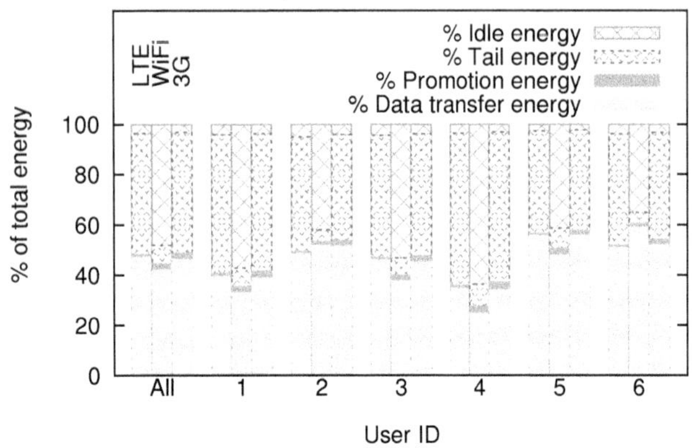

Figure 14: Break down of energy usage in analysis.

and 18.0, and in aggregate, 3G/WiFi ratio is 14.6, lower than LTE. In summary, the energy efficiency for LTE is lower than 3G, with WiFi having a much higher energy efficiency. Notice that in this work, we do not consider AP association energy for WiFi networks due to the lack of such information, so E_{wifi} is an underestimate. However, we believe that AP association does not happen very often and the impact is not significant.

6.2 Energy consumption break down

To further understand the energy consumption, with the methodology discussed in §3.3.2, it is decomposed into promotion energy, data transfer energy, tail energy, and idle energy in Figure 14 with results for aggregate analysis and 6 sample users. These energy values are calculated including the UE base power.

Promotion energy contributes to a small portion ($< 4\%$) of the total energy for all networks, *i.e.*, in aggregate, its contribution is only 1.2%, 2.5% and 2.6% for LTE, WiFi, and 3G, respectively. In terms of idle energy, WiFi has significantly higher percentage than LTE and 3G, despite small average power difference at idle state across networks, *i.e.*, 31.1mW [3] for LTE, 13.0mW for WiFi and 15.3mW for 3G. This is explained by WiFi's smaller total energy, making its idle energy contribution relatively higher.

Aggregate data transfer energy percentage is 47.1%, 42.1% and 46.2% for LTE, WiFi and 3G, respectively. The variation across users is high, *e.g.*, for LTE network, it ranges from 22.0% to 62.3%, due to traffic pattern differences across users.

Surprisingly, the biggest energy component for LTE network is tail, rather than data transfer. The average tail energy for LTE and 3G is 48.2% and 48.1% respectively compared to 7.2% for WiFi. Our observation for 3G is consistent with previous study [18]. Combined with high data transfer energy due to the higher power levels, tail energy lowers the energy efficiency of LTE and 3G compared to WiFi.

6.3 Impact of LTE parameters

Similar to §6.1, we use WiFi traces in the UMICH data set to study the impact of LTE parameter configuration on radio energy E, channel scheduling delay D, and signaling overhead S. E is the simulated total energy consumed by UE. Since most energy is consumed by radio interfaces, with display energy excluded, we approximate this total energy as radio energy. D is the sum of scheduling delay for all packets, resulting from two factors: waiting for the next DRX cycle's on duration for a downlink packet and waiting for T_{pro} during **RRC_IDLE** to **RRC_CONNECTED** promotion. S is the overhead of the LTE network for serving this UE. S has different definitions in our study given that different parameters studied affect signaling load in different ways, *e.g.*, T_{tail} affects mostly the state promotions, and T_{pi} affects the total on duration in **RRC_IDLE**. D and E directly affect end users via application delay and battery life time; S is of interest to cellular ISPs for supporting large user base at low cost.

6.3.1 *Impact of LTE tail timer* T_{tail}

In Figure 15, we study the impact of LTE tail timer T_{tail}, with S defined to be the total number of **RRC_IDLE** to **RRC_CONNECTED** promotions. T_{tail} varies from 1 second to 30 seconds. T_D is the default configuration of 11.58 seconds for T_{tail} in the measured LTE network with the corresponding values of E, D and S calculated as E_D, D_D and S_D, respectively. When T_{tail} is set to be T', similarly, we get E', D' and S'. The change relative to the default setting is calculated as $\Delta(E) = (E' - E_D)/E_D$, and similarly for $\Delta(D)$ and $\Delta(S)$. Δs for each T_{tail} values are plotted in Figure 15 and they are all 0 at T_D.

As expected, a larger T_{tail} value reduces both D and S, at the cost of E. We observe that $\Delta(S)$ and $\Delta(D)$ curves are almost identical, since most channel scheduling delay results from idle to active promotions. In addition, the impact of T_{tail} on the radio energy E is significant. For example, when T_{tail} is set to be 30 seconds, total radio energy E increases by 55%, while for $T_{tail} =$ 1 second, E decreases by 65%, at the cost of 143% $\Delta(D)$ and 169% $\Delta(S)$. This indicates that T_{tail} is a very important parameter for LTE network.

In previous study [28], traffic bursts attribute to low efficiencies of radio resource and energy utilization in 3G UMTS network, due to tail time. Given that LTE has similar tail time, we analyze the bursts in the UMICH data set. We follow the same definition of a burst as in [28], which is a sequence of consecutive packets whose inter-arrival time is less than a threshold δ, with δ set to be 1.5 seconds, since it is longer than common cellular RTTs [24]. From ISP's perspective, T_{tail} should not be smaller than majority of the inter-burst delays; otherwise, a larger number of promotions would be triggered.

We study the inter-burst delay in the 3G and WiFi traces of UMICH data set in Figure 19. For 3G traces, 67.3% inter-burst delays are smaller than the default value of T_{tail} (11.58 seconds), and 55.4% for WiFi traces. Compared with WiFi traces, 3G traces

[3]The idle power for LTE network is calculated as the average power of a DRX cycle in **RRC_IDLE**, and similarly for WiFi and 3G networks.

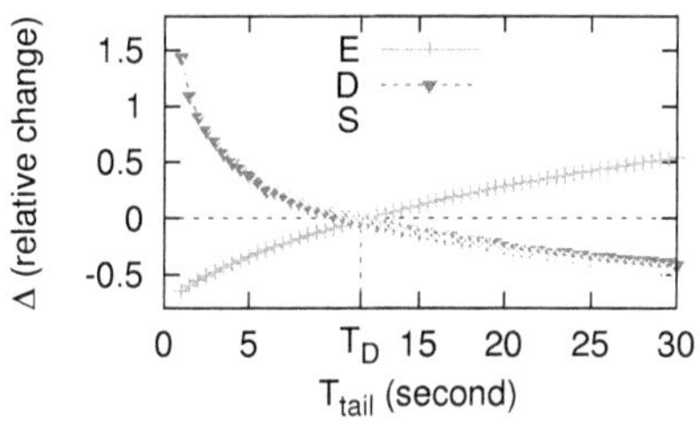

Figure 15: Impact of the LTE tail timer T_{tail}.

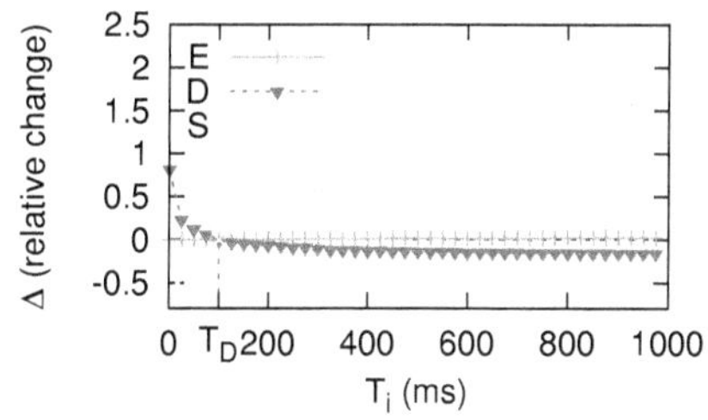

Figure 16: Impact of the DRX inactivity timer T_i.

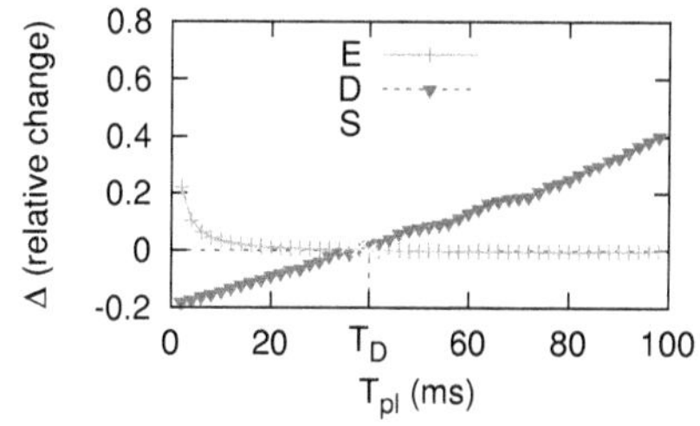

Figure 17: Impact of the DRX cycle in RRC_CONNECTED **(**T_{pl}**).**

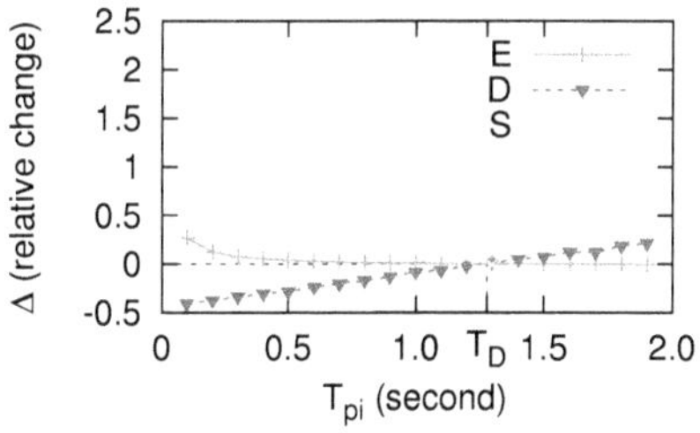

Figure 18: Impact of DRX cycle in RRC_IDLE **(**T_{pi}**).**

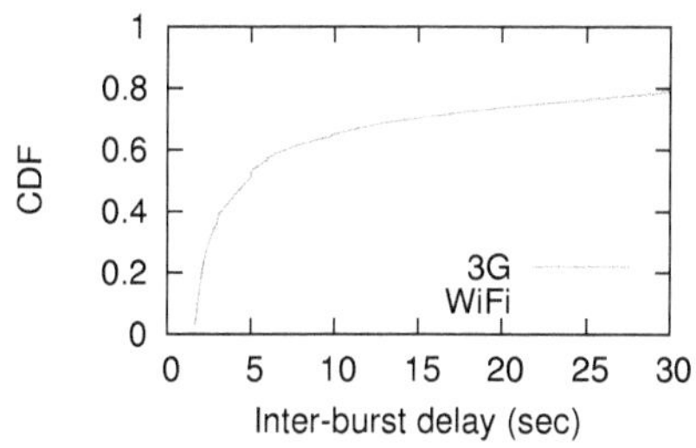

Figure 19: CDF of Inter-burst delay for UMICH **data set.**

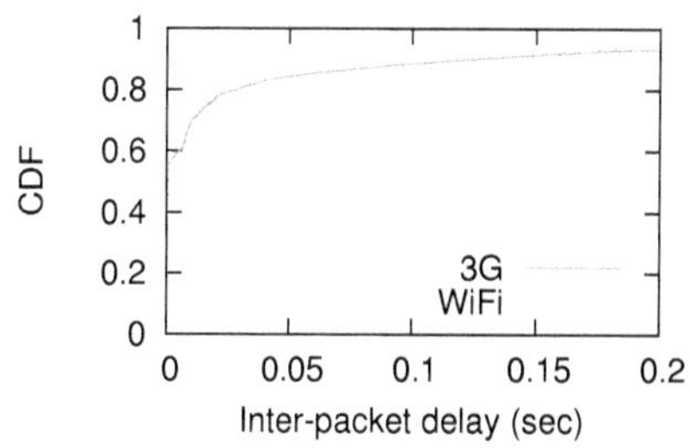

Figure 20: CDF of Inter-packet delay for UMICH **data set.**

have smaller inter-burst delays. We also observe the inter-burst delay distribution for different users have non-negligible differences. This makes us believe that a per-UE based dynamic T_{tail} configuration mechanism, adapting to traffic patterns, provides a better chance for optimizing radio resource efficiency and balancing energy utilization. Notice that this approach is inside the radio access network and transparent to UE. Previous study on 3G UMTS network also points out two possible approaches to deal with the tail effect with support from UE, *i.e.*, applications proactively alter traffic patterns based on the state machine behavior without interaction with the network, or cooperate with the radio access network in allocating radio resources by fast dormancy [11, 13]. These approaches remain to be applicable solutions for the tail problem in LTE.

6.3.2 Impact of DRX inactivity timer T_i

Figure 16 shows the tradeoff of setting DRX inactivity timer T_i. Signaling overhead S is defined as the sum of the continuous reception time and DRX on durations in **RRC_CONNECTED**, since during these time slots, UE exchanges control messages with eNB (base station of LTE network) [30]. A larger T_i keeps UE in continuous reception longer and reduces the scheduling delay for downlink packets, *i.e.*, no DRX waiting time, at the cost of higher energy usage, since continuous reception has higher power level than DRX idle state. T_D is the default value 100ms.

We observe that T_i has negligible impact on E, with only 2% $\Delta(E)$ as T_i is set to be 1 second. Different from E, S is significantly affected by T_i. For D, when T_i is set to be small than 100ms, $\Delta(D)$ grows up to 80%. However, even when T_i is set to be as large as 1 second, D is only reduced by 16%. This can be partly explained by Figure 20 that 88.6% packet pairs has $\leq$ 100ms inter-packet delay for 3G traces and 92.3% for WiFi traces in the UMICH data set. Decreasing T_i causes more packets arrive outside of continuous reception, which may experience DRX delay. Hence similar to T_{tail}, the setting of T_i is also affected by the traffic pattern, *i.e.*, inter-packet delay distribution, which may differ across users.

6.3.3 Impact of DRX cycle in RRC_CONNECTED

In **RRC_CONNECTED**, Short DRX cycle timer T_{is} determines the transition from Short DRX to Long DRX. Given that we already vary the DRX cycle to understand its impact, for simplicity, we assume T_{is} to be 0, *i.e.*, the DRX cycle in **RRC_CONNECTED** refers to Long DRX cycle T_{pl}.

In Figure 17, S is defined similarly as in Figure 16, *i.e.*, the sum of continuous reception time and DRX on durations. T_D is 40ms, the default setting for T_{pl}. First, we notice that compared with T_D, when T_{pl} is set to be 100ms, energy saving $\Delta(E)$ is minor (0.2%) with 13.0% reduction in S, at the cost of 40% increase in D. So T_{pl} is not recommended to be set much higher than T_D unless ISP prioritizes S reduction. When T_{pl} is set to be a small value, for example 2ms, E increases by 21.9% and D reduces by 18.1%, with a significant increase in S of 412.9%. Especially for ISPs with fast growing LTE user volume and congested radio access network, DRX cycle in **RRC_CONNECTED** is not recommend to be set to too small values. We omit the simulation results for Short DRX cycle T_{ps} and Short DRX timer T_{is}, since reducing T_{ps} or increasing T_{is} has similar effect of decreasing T_{pl}, *i.e.*, resulting in a shorter average DRX cycle, and their impact is similar as in Figure 17.

6.3.4 Impact of DRX cycle in RRC_IDLE

Similarly to **RRC_CONNECTED**, we study the impact of DRX cycle in **RRC_IDLE**. Under this setting, signaling overhead S is defined to be the total on duration of DRX cycles in **RRC_IDLE**, during which control messages are exchanged between UE and eNB [30]. In Figure 18, the default setting 1.28s is shown as T_D. The impact of T_{pi} on E is not significant when T_{pi} is larger than T_D, with 0.7% energy saving when T_{pi} set to be 2.0s, which also brings in 21.3% increase in D and 31.7% reduction in S. When T_{pi} is set to

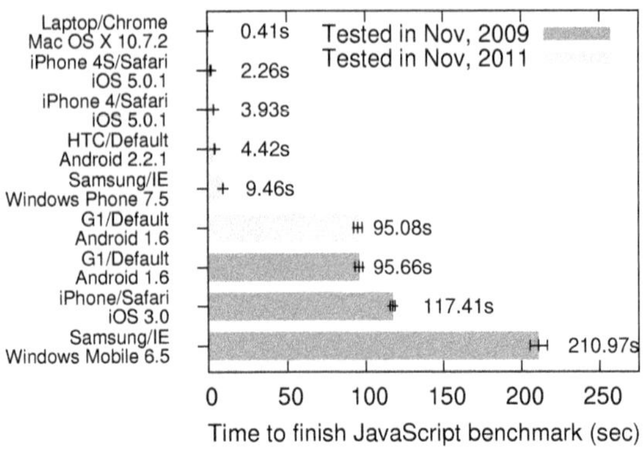

Figure 22: JavaScript execution speed comparison.

be 100ms, the energy overhead is 26.2%, with 40.7% reduction in D and up to 1147.2% increase in S.

Similar to T_{pl}, T_{pi} causes significant signaling overhead when set to be too small. Default values for both T_{pl} and T_{pi} are reasonable settings and adjustment around the default configurations results in moderate impact on channel scheduling delay and negligible impact on radio energy.

In this section, we observe that LTE consumes up to 23 times the total energy of WiFi in the simulation, and the tail problem in 3G UMTS network is still a key contributor for the low energy efficiency for LTE. Similar to 3G network, T_{tail} remains a key parameter for LTE. In addition to the UE-participated approaches including proactive traffic shaping and fast dormancy proposed by previous study [27], we discuss a possible approach to solve the tail problem in LTE network, which dynamically adapts T_{tail} to traffic pattern at a per-UE base, being transparent to UE. Similar observations are shared by the other LTE parameters and we leave the feasibility study of such dynamic parameter configuration framework for LTE as future work.

7. APPLICATION PERFORMANCE IMPACT

In this section, we try to identify the new bottleneck of web-based smartphone applications in LTE and start with comparing the processing power of contemporary smartphones.

7.1 JavaScript execution

JavaScript execution is an important component in web transactions and we use it as a sampled view of UE's processing power. In 2009, we found that the JavaScript execution speed for smartphone browsers could be up to 20∼80 times slower than desktop browsers [24]. We revisit the JavaScript execution experiments in this study, as shown in Figure 22. We consistently use the same benchmark [8] (version 0.9) to quantify the JavaScript execution speed. In Figure 22, the test results on the same G1 phone in 2009 and in 2011 are very close, validating that the benchmark of that version has not changed since 2009.

From 2009 to 2011, iOS has a speedup of 29.88 for iPhone 4 and 51.95 for iPhone 4S, while 21.64 for Android and 22.30 for Windows Phone. And the gap between smartphone and computer has dropped to 5.5∼23.1 times in JavaScript execution speed. Possible reasons for this improvement include fast CPU, larger memory, and better OS and application software for smartphones.

7.2 Application case study

Along with improvements in mobile client processing power, LTE technology significantly increased network speed. So we perform a case study of a few popular applications to understand the impact of network speed and processing power enhancement on user-perceived performance.

App	Total payload	Max obj size	# objects	Avg size
Website G	160.7KB	87.6KB	11	14.6KB
Website Y	509.9KB	124.2KB	59	8.6KB
YouTube	449.1KB	109.4KB	33	13.6KB
NPR News	306.2KB	169.5KB	9	34.0KB
Market	599.7KB	145.4KB	17	35.3KB

Table 6: HTTP object statistics.

With the measurement methodology discussed in §3.4, we compare loading time, CPU usage, and energy consumption for five application usage scenarios in Figure 21. These usage scenarios are representative for web browsing and other similar web-based applications, different from audio and video streaming applications.

First, loading time comparison shows that LTE and WiFi have comparable user-perceived delay for the studied scenarios, and 3G lags behind with 50%∼200% larger response time. The application loading time for LTE slightly lags behind WiFi. On one hand, given that promotion delay is counted into the loading time, LTE has larger T_{pro}. On the other hand, though LTE has faster throughput than WiFi with throughput tests, its RTT is slightly larger. To understand why RTT matters more than throughput for the usage scenarios in Figure 21, we analyze the object size statistics for different applications in Table 6.

In Table 6, across all scenarios, the maximum object size is 169.5KB for NPR News and the average object size is at most 35.3KB for Android market. Especially for Website Y, there are 59 HTTP objects with an average size of only 8.6KB. Due to TCP slow start, these objects are far from saturating the link capacity for either LTE or WiFi. Persistent TCP connections are observed to transfer multiple web objects; however, no HTTP pipelining is observed. We also observe up to 5 parallel TCP connections to different remote servers and even to the same remote server; however the average low throughput, *e.g.,* 364.2kbps for Website Y, indicates that the RTT is more critical than available bandwidth for these web-based applications in LTE and WiFi.

We observe that average CPU usage for 3G network is relatively low ranging from 35.5% to 70.8%, with an average of 57.7%. CPU usage for LTE is between 68.8% and 84.3%, averaging at 79.3%, compared with WiFi's 78.2%∼93.0% and 87.1% as average. This comparison implies that the gap between WiFi and cellular network has narrowed because of LTE's better network performance. For 3G network, network performance appears to be the bottleneck, resulting in underutilized processing power. Given that the average CPU usage for LTE and WiFi is already 80%∼90%, further performance optimization relies more on processing speed improvement and OS/application software enhancement, compared with network improvement. Work such as CloneCloud [20] on offloading computation into the cloud is one such promising direction.

In terms of energy usage, WiFi clearly has significantly higher energy efficiency for the usage scenarios studied, similar to previous discussions. 3G consumes 5.5∼7.8 times the energy of WiFi, with 20.9%∼55.6% of the total energy due to tail. While for LTE, energy usage is 5.4∼12.0 times of WiFi, with 32.2%∼62.2% tail energy. We notice that, most of these applications keep some of the existing TCP connections open even after the initial loading; so that if user generates subsequent requests, these connections could be reused. Similar to previous work [28], we observe that appli-

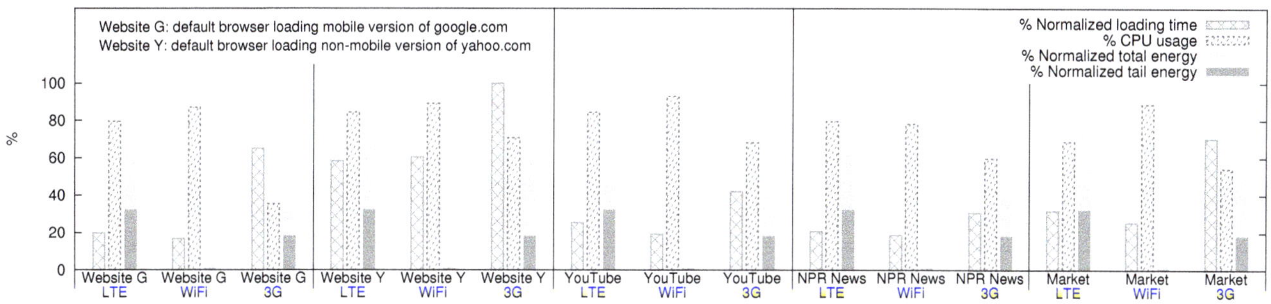

Figure 21: Loading time, CPU usage and energy consumption analysis on mobile applications.

*The loading time is normalized by the maximum loading time observed in this figure, and similarly for total energy.

cations remain idle after the initial loading and a few TCP control packets are generated by UE to close the existing applications a few seconds later, such as the packets seen at time 10.5 seconds in Figure 4, which reset T_{tail} and further increase the high-power tail length.

In this section, we observe that contemporary smartphones have reduced gap with desktop computers in terms of processing power. However, for web-based applications, downloading mostly small sized objects, the performance bottleneck is still at the UE processing side, given the significant improvement in network speed for LTE.

8. RELATED WORK

We summarize related work in three categories below.

Measuring 2G and 3G networks. Prior study [18] investigates the impact and the optimization of tail effects in 2G/3G networks. Previous work [27] characterizes the impact of RRC state machine on radio resources and energy by analyzing a dataset collected from a commercial UMTS network. Recent work [22] also investigates impact of traffic patterns on radio power management policy. Other measurement studies (*e.g.*, 3GTest [24], LiveLab [31], and [23]) collect traces from real smartphone users, focusing on characterization at only IP and higher layers. None of them investigate resource management policy in 4G LTE networks.

Smartphone power modeling has also been investigated by previous empirical measurements. The Bartendr project [29] studies the relationship between signal strength and smartphone radio power usage. PowerTutor [35] collects power traces for individual hardware components under controlled experiments then uses multi-variable regression to compute the coefficients for a linear power model considering all hardware components. ARO [28] employs an approach similar to [35] but focusing only on radio power usage. It performs more fine-grained simulation of transmission queues to capture state transitions. Our LTE power model is also empirically derived, but it differs from all aforementioned models in two aspects. First, it considers DRX in `RRC_CONNECTED`, a new power management mechanism in LTE networks. Second, it further takes into account both uplink and downlink data rates, resulting in a more accurate estimation of the radio power consumption when the throughput is high, as is a common case in LTE networks.

DRX in 4G LTE networks. Zhou *et al.* [36] model the DRX mechanism as a semi-Markov process, with which they analytically study the effects of DRX parameters on the performance, as well as the tradeoff between the power saving and wake-up delay. Kolding *et al.* [25] also investigate the balance between throughput and power saving by changing the configuration of DRX parameters, using a web-browsing traffic model. Bontu *et al.* [19] further considers the impact of DRX parameters on different applications with various delay sensitivities. Wigard *et al.* compares a long-DRX-only scenario and a scenario with both long and short DRX, in terms of throughput and power consumption. All above studies employing analytical models suffer from an inherent limitation: the expressiveness of an analytical model is quite limited and is unlikely to capture the characteristics of real-world traffic patterns using a statistical distribution with a few parameters. The existence of concurrent applications accessing the network further increases the difficulty of modeling the packet dynamics. In contrast, our work is the first empirical study to investigate the impact of the configurations of DRX parameters and tail timer. We overcome the above limitation by employing network traces from real smartphone users, thus more realistically revealing the complex tradeoffs incurred by various DRX parameters.

9. CONCLUSION

In this paper, we take one of the first steps in understanding performance and power characteristics of 4G LTE networks. With publicly deployed 4GTest, we observe that LTE has significantly higher downlink and uplink throughput, compared with 3G and even WiFi. To characterize power usage for LTE, we derive the first comprehensive power model of a commercial LTE network, with less than 6% error rate. By applying the power model to a comprehensive data set, UMICH, we identify that LTE is much less power efficient than WiFi, and the key contributor is the tail energy, controlled by T_{tail}. In addition, via a case study, we identify the UE processing to be the new bottleneck for web-based applications in LTE networks.

Our work opens new research opportunities for the new LTE network, such as optimizing smartphone applications with tools similar to ARO [28] with the energy model derived in this paper, and optimizing LTE parameter configuration with UE-based approaches (*e.g.*, fast dormancy) or network-based approaches to improve radio resource and energy efficiency.

10. ACKNOWLEDGEMENTS

This work is partly funded by NSF grants CNS-1059372, CNS-1050157, CNS-1039657 and Navy grant N00014-09-1-0705. We would like to thank Zhiyun Qian, Doug Sillars and the shepherd Sharad Agarwal for their valuable comments on the paper. We would also like to thank the participants in our user study and anonymous reviewers whose comments helped improve the final version.

11. REFERENCES

[1] 3GPP LTE. http://www.3gpp.org/LTE.

[2] 4GTest. http://mobiperf.com/4g.html.

[3] BU-353 USB GPS Receiver. http://www.usglobalsat.com/p-62-bu-353-w.aspx.

[4] MaxMind. http://www.maxmind.com.

[5] Measurement Lab. http://www.measurementlab.net/.

[6] MobiPerf (3GTest). http://mobiperf.com.

[7] Monsoon power monitor. http://www.msoon.com/LabEquipment/PowerMonitor.

[8] SunSpider JavaScript Benchmark 0.9. http://www.webkit.org/perf/sunspider/sunspider.html.

[9] TCPDUMP and LIBPCAP. http://www.tcpdump.org/.

[10] 3GPP TR 25.813: Radio interface protocol aspects (V7.1.0), 2006.

[11] UE "Fast Dormancy" behavior. 3GPP discussion and decision notes R2-075251, 2007.

[12] 3GPP TR 25.913: Requirements for Evolved UTRA and Evolved UTRAN (V9.0.0), 2009.

[13] Configuration of fast dormancy in release 8. 3GPP discussion and decision notes RP-090960, 2009.

[14] 3GPP TS 25.304: User Equipment (UE) procedures in idle mode and procedures for cell reselection in connected mode (V10.2.0), 2011.

[15] 3GPP TS 36.211: Physical Channels and Modulation (V10.3.0), 2011.

[16] 3GPP TS 36.321: Medium Access Control (MAC) protocol specification (V10.3.0), 2011.

[17] 3GPP TS 36.331: Radio Resource Control (RRC) (V10.3.0), 2011.

[18] N. Balasubramanian, A. Balasubramanian, and A. Venkataramani. Energy Consumption in Mobile Phones: A Measurement Study and Implications for Network Applications. In *IMC*, 2009.

[19] C. S. Bontu and E. Illidge. DRX Mechanism for Power Saving in LTE. In *IEEE Communications Magazine*, 2009.

[20] B.-G. Chun, S. Ihm, P. Maniatis, M. Naik, and A. Patt. CloneCloud : Elastic Execution between Mobile Device and Cloud. In *Proceedings of 6th European Conference on Computer Systems (EuroSys 2011)*, 2011.

[21] S. Cui, A. J. Goldsmith, and A. Bahai. Energy-Efficiency of MIMO and Cooperative MIMO Techniques in Sensor Networks. In *IEEE J. Sel. Areas Commun., vol. 22, no. 6, pp. 1089-1098*, 2004.

[22] H. Falaki, D. Lymberopoulos, R. Mahajan, S. Kandula, and D. Estrin. A First Look at Traffic on Smartphones. In *Proc. ACM SIGCOMM IMC*, 2010.

[23] H. Falaki, R. Mahajan, S. Kandula, D. Lymberopoulos, and R. G. D. Estrin. Diversity in Smartphone Usage. 2010.

[24] J. Huang, Q. Xu, B. Tiwana, Z. M. Mao, M. Zhang, and P. Bahl. Anatomizing Application Performance Differences on Smartphones. In *MobiSys*, 2010.

[25] T. Kolding, J. Wigard, and L. Dalsgaard. Balancing Power Saving and Single User Experience with Discontinuous Reception in LTE. In *ISWCS*, 2008.

[26] G. Perrucci, F. Fitzek, and J. Widmer. Survey on Energy Consumption Entities on Smartphone Platform. In *Vehicular Technology Conference*, 2009.

[27] F. Qian, Z. Wang, A. Gerber, Z. M. Mao, S. Sen, and O. Spatscheck. Characterizing Radio Resource Allocation for 3G Networks. In *IMC*, 2010.

[28] F. Qian, Z. Wang, A. Gerber, Z. M. Mao, S. Sen, and O. Spatscheck. Profiling Resource Usage for Mobile Applications: a Cross-layer Approach. In *MobiSys*, 2011.

[29] A. Schulman, V. Navda, R. Ramjee, N. Spring, P. Deshpande, C. Grunewald, K. Jain, and V. N. Padmanabhan. Bartendr: A Practical Approach to Energy-aware Cellular Data Scheduling. In *MobiCom*, 2010.

[30] S. Sesia, I. Toufik, and M. Baker. LTE: The UMTS Long Term Evolution From Theory to Practice. John Wiley and Sons, Inc., 2009.

[31] C. Shepard, A. Rahmati, C. Tossell, L. Zhong, and P. Kortum. LiveLab: Measuring Wireless Networks and Smartphone Users in the Field. In *HotMetrics*, 2010.

[32] S. Sundaresan, W. de Donato, R. Teixeira, S. Crawford, and A. PescapÃÍ. Broadband Internet Performance: A View From the Gateway. In *SIGCOMM*, 2011.

[33] Z. Wang, F. X. Lin, L. Zhong, and M. Chishti. Why are Web Browsers Slow on Smartphones? In *HotMobile*, 2011.

[34] J. Wigard, T. Kolding, L. Dalsgaard, and C. Coletti. On the User Performance of LTE UE Power Savings Schemes with Discontinuous Reception in LTE. In *ICC Workshops*, 2009.

[35] L. Zhang, B. Tiwana, Z. Qian, Z. Wang, R. P. Dick, Z. M. Mao, and L. Yang. Accurate Online Power Estimation and Automatic Battery Behavior Based Power Model Generation for Smartphones. In *CODES+ISSS*, 2010.

[36] L. Zhou, H. Xu, H. Tian, Y. Gao, L. Du, and L. Chen. Performance Analysis of Power Saving Mechanism with Adjustable DRX Cycles in 3GPP LTE. In *Vehicular Technology Conference*, 2008.

Human Mobility Modeling at Metropolitan Scales

Sibren Isaacman*, Richard Becker†, Ramón Cáceres†, Margaret Martonosi*,
James Rowland†, Alexander Varshavsky†, Walter Willinger†
*Princeton University, Princeton, NJ, USA †AT&T Labs, Florham Park, NJ, USA
isaacman@princeton.edu, {rab,ramon}@research.att.com, mrm@princeton.edu,
{jrr,varshavsky,walter}@research.att.com

ABSTRACT

Models of human mobility have broad applicability in fields such as mobile computing, urban planning, and ecology. This paper proposes and evaluates *WHERE*, a novel approach to modeling how large populations move within different metropolitan areas. WHERE takes as input spatial and temporal probability distributions drawn from empirical data, such as Call Detail Records (CDRs) from a cellular telephone network, and produces synthetic CDRs for a synthetic population. We have validated WHERE against billions of anonymous location samples for hundreds of thousands of phones in the New York and Los Angeles metropolitan areas. We found that WHERE offers significantly higher fidelity than other modeling approaches. For example, daily range of travel statistics fall within one mile of their true values, an improvement of more than 14 times over a Weighted Random Waypoint model. Our modeling techniques and synthetic CDRs can be applied to a wide range of problems while avoiding many of the privacy concerns surrounding real CDRs.

Categories and Subject Descriptors

I.6.5 [**Computing Methodologies**]: Simulation and Modeling—*Model Development*; K.4.0 [**Computing Milieux**]: Computers and Society—*General*

General Terms

Algorithms, Measurement, Human Factors

Keywords

Human mobility patterns, Call Detail Records

1. Introduction

Human mobility models have myriad uses in mobile computing research and other fields of study. Models that faithfully reproduce the movements of real people can help answer questions in areas as varied as mobile sensing, opportunistic networking, urban planning, ecology, and epidemiology. For example, a model of how people move around a city can help evaluate whether a sensing application running on mobile phones would be able to attain the desired geographic coverage.

Our work aims to produce accurate models of how large populations move within different metropolitan areas. In pursuit of this general aim, we have a number of more specific goals. Our first goal is to generate sequences of locations and associated times that capture how individuals move between important places in their lives, such as home and work. Previous work has shown that people spend most of their time at a few such places [13, 16, 31]. Our second goal is to aggregate the movements of many such individuals to reproduce human densities over time at the geographic scale of metropolitan areas. A model that operates at these scales can help address important societal issues such as the environmental impact of home-to-work commutes. Our third goal is to take into account how different metropolitan areas exhibit distinct mobility patterns due to differences in geographic distributions of homes and jobs, transportation infrastructures, and other factors. Previous work has shown significant differences between cities along metrics such as commute distances [16, 17, 18, 25].

Many human mobility models that fall short on one or more of these goals have been proposed in the past. Some models produce random motion that does not correspond to actual mobility patterns, e.g., [19, 24]. Their lack of memory about recurring movement patterns and of spatiotemporal realism about population densities results in unrealistic motion of modeled individuals. Some models are tailored to a small geographic area such as a university campus, e.g., [20]. They do not apply to larger geographic areas with more diverse populations. Some models aim to be universal, e.g., [13], and thus do not adapt to different geographic areas. There remains a need for a realistic model that matches empirical observations for large and distinct geographic areas.

This paper introduces a modeling approach that takes as input certain spatial and temporal probability distributions drawn from large populations of real people living across wide geographic areas. An especially good source of these distributions are the Call Detail Records (CDRs) maintained by cellular network operators. Billions of cellphone users worldwide keep their phones near them most of the time, and the networks need to know the rough location of all active phones to provide them with voice and data services. CDRs contain information such as the time a voice call was placed or a text message was received, as well as the identity of the cell tower with which the phone was associated at that time. When joined with information about the locations of those towers, CDRs can serve as sporadic samples of the approximate locations of the phone's owner. A growing body of work has shown that information derived from anonymized CDRs can accurately characterize many aspects of human mobility [3, 11, 13, 16, 17, 18, 31].

With cellular network data becoming more available, it is tempting to think that creating human mobility models from such data should be easy. This is, however, still not the case. For example, while CDRs readily yield insights into aggregate population densities, they do not convey whether their associated locations correspond to home, work, or other important places for particular cellphone users. Without such semantic information, it is difficult to abstract CDRs into models applicable to scenarios, regions, or populations that vary from those for which the real-life CDR data was collected. Furthermore, both the spatial and temporal granularity of CDR data is quite coarse. Spatially, CDRs are only accurate to the granularity of celltower spacings. Temporally, CDRs are only generated when phones are actively involved in a voice call or text message. Our work makes key contributions in overcoming the challenges stemming from lack of semantic information and coarse granularity, in order to produce usefully accurate models for arbitrary metropolitan regions.

Our modeling approach intelligently samples the spatial and temporal probability distributions from CDRs, or other population data, to generate sequences of locations and times for any number of synthetic people in any region for which the required distributions can be obtained. A generative model derived from CDRs has flexibility, compactness, and availability advantages over using CDRs directly. First, our models offer the option of perturbing the input distributions to evaluate what-if scenarios, for example to consider how the addition of a new residential or employment area might change traffic patterns. In contrast, the original CDRs are difficult to manipulate in meaningful ways. Second, our model for a metropolitan area with a 50-mile radius can be stored as a set of histograms that fit within 2 gigabytes. In contrast, an anonymized CDR dataset for the same area occupied approximately 100 gigabytes. Finally, our models can be made available to a larger research community because they do not to reproduce the mobility pattern of any individual real person. They thus avoid many of the privacy concerns associated with source CDRs.

The final stage of our modeling approach produces locations and times in the form of synthetic CDRs. These synthetic CDRs have the same format and call/text frequency characteristics of real CDRs. They are modeled to approximate the actual movement patterns of users. Increased model complexity results in more accurate movement patterns, which in turn produces higher-fidelity synthetic CDRs. We chose the CDR output format for several pragmatic reasons. One, we can compare this output directly against real CDRs, our best source of location information for large populations and regions. Two, this output can plug in directly into the growing body of analysis software that uses CDRs as input.

In this paper, we propose and evaluate WHERE ("Work and Home Extracted REgions"), a region-scale modeling approach. First, we identify the key properties of human movement, such as important locations and commute distances, that need to be represented as probability distributions. Then, we describe how these probability distributions can be used to generate synthetic CDRs for an arbitrary number of synthetic people.

We validate our approach by comparing the spatiotemporal dynamics of synthetic populations generated by WHERE to those of real populations. In particular, we compare the spatial population densities on an hourly basis for synthetic and real CDR sequences. Our validation begins with stylized examples that confirm our models' fidelity both quantitatively and visually. We validate both at the aggregate level, where simpler models may perform well, as well as at a finer granularity, which exposes the advantages of WHERE compared to other models considered. We then scale up our validation to large datasets containing real anonymized CDRs for the Los Angeles (LA) and New York City (NY) metropolitan areas. Our LA and NY datasets each span three months of activity for hundreds of thousands of phones, yielding billions of location samples.

Recognizing that real CDRs are not available to all researchers, we also evaluate models in which the same input distributions are derived from publicly available US Census data [32]. We show that models based on real CDRs closely approximate the real populations and movements of these cities. Models based on census data are also viable, but at a loss of significant accuracy.

Finally, we present example applications of our modeling approach. We create models for the LA and NY metropolitan areas and use the resulting synthetic CDRs to perform calculations that one may wish to perform on real CDRs. We show that calculations performed on the WHERE model produce far more accurate results than those performed on more naive models. For example, we can calculate daily ranges of travel that agree with real ranges, as well as perform more complex tasks such as investigating opportunistic message propagation in large urban environments.

The overall contributions of our work are:

- We introduce an approach to modeling human mobility patterns by generating fully synthetic CDRs from real-world probability distributions.
- Our approach works at the scale of large metropolitan areas and accounts for mobility differences between metropolitan areas.
- We show that our technique is extensible to greater levels of precision by providing it more complete input probability distributions (at the cost of increased model complexity).
- We validate our approach against large-scale location datasets drawn from two major US metropolitan areas. We compare our generated CDRs against real CDRs, and show that our location distributions achieve more than 4 times error reduction compared to a Random Waypoint model.
- As an example of how our models can help answer concrete questions about human mobility, we use our synthetic CDRs to compute daily ranges of travel. Our synthetic CDRs exhibit error at the median of less than 0.8 and 1 mile for NY and LA residents, respectively. This accuracy constitutes more than a 14 times improvement over that of a Weighted Random Waypoint model.

The rest of this paper is organized as follows. Section 2 presents the probability distributions needed to create our models. Section 3 explains the construction of synthetic CDRs. Section 4 describes our evaluation methodology. Section 5 validates our synthetic CDRs against stylized examples and Section 6 validates them against large datasets of real CDRs. Section 7 examines example uses of artificial CDRs. Finally, Section 8 discusses open issues and future directions, and Section 9 surveys related work.

2. Spatial and Temporal Parameters for Mobility Modeling

Modeling human mobility requires both spatial and temporal information about the places and times that humans move. Human mobility is tightly coupled to the geography of the city people live in [16, 17, 18, 25]. Therefore, any accurate mobility model should take into account both the area geography and individual user mobility patterns. In this section, we describe the probability distributions that serve as inputs to the WHERE method.

Figure 1 summarizes the overall flow of our approach, including key inputs and data structures. For each of the input distributions

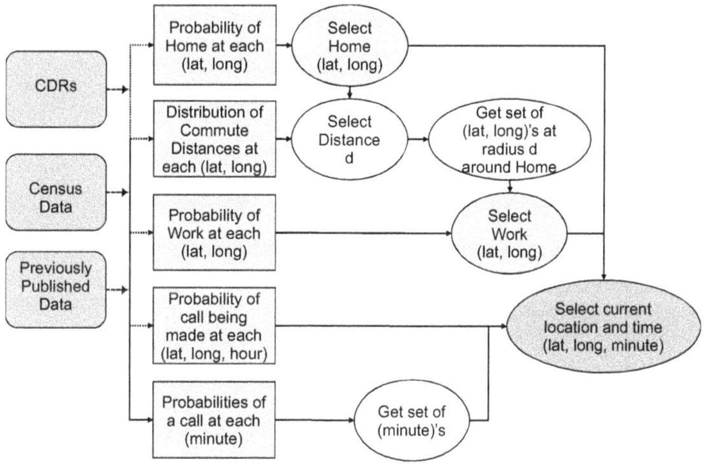

Figure 1: Overall view of the WHERE modeling approach. Five input probability distributions are important for our modeling technique, but their source can vary.

Distribution	Data Sources All Public	Hybrid	All CDR
Home	Census	Census	CDR
CommuteDistance	Census	Census	CDR
Work	Census	Census	CDR
Hourly	Census Home and Work data split by time of day	CDR	CDR
CallTime and PerUserCallsPerDay	Previously-published work [1, 5]	CDR	CDR

Table 1: For the five probability distributions used in our approach, the input data for these probabilities can be gathered from different sources. We have explored methods based entirely on public data, as well as methods using data from proprietary anonymized CDRs.

required, Table 1 summarizes different methods for generating the distributions. The subsections below describe the distributions in more detail, and discuss the different possible sources for the input data required.

2.1 Spatial Information: Important Locations

Capturing where people spend time is important for creating accurate human mobility models. Prior work has demonstrated that the majority of people's movement occurs between a small number of locations. A full 60% of mobility can be accounted for with just the top two cellphone towers with which a user is associated [31]. Therefore, in modeling mobility, good accuracy can be expected if we have probability distributions for the few locations in which users spend the majority of their time.

For most people, the two most important locations are "home" and "work." A design decision for the model is how to gather input information and how to express the probabilities of different home and work locations. If one simply pulled locations independently from two separate home and work distributions, the accuracy of the resulting model would be quite poor, since such an approach would ignore the strong correlation between where someone works and where they choose to live. We also explored methods that assumed commute distances independent of home location, but these were also too inaccurate to be useful.

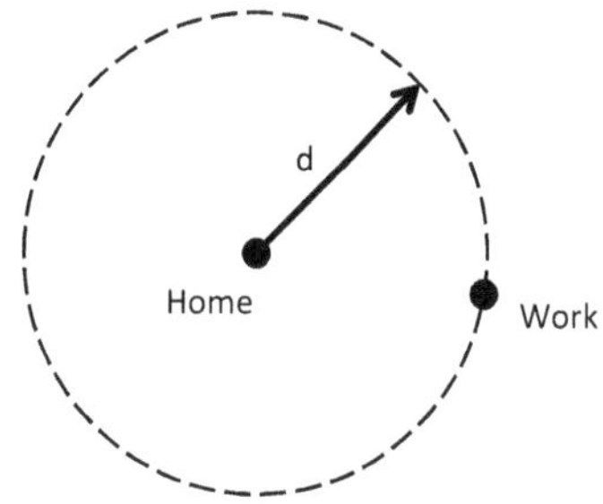

Figure 2: Selecting a user's home and work locations. Pick a home location, then select *d* from the distribution of commute distances at that home location. The work location is selected from a probability distribution for work locations on the resulting circle of radius *d*.

As shown in Figure 2, we use a different approach that estimates spatial home-work densities by relying on three distributions. First, we pull home locations randomly from a probability distribution across latitude and longitude expressing the likelihoods of where people live, i.e. *Home*. For each point in space, a second probability distribution *CommuteDistance* expresses the probability of having different commute distances, conditioned on that given home location. A commute distance, *d*, selected from this distribution can be envisioned as describing a circle of radius *d* around the selected home location. Next, our method selects a work location somewhere along this circle. To do so, a third distribution, *Work*, gives the probability of different work locations around a circle of commute distance *d* from the home location. These *Home* and *Work* locations are derived from population densities that correspond to the city we wish to synthesize. Such distributions could be computed from either real CDRs or the census data, which contains information about the number of people living or working in a particular area. We will discuss possible sources of each spatial distribution more thoroughly in a subsequent section.

2.2 Spatiotemporal Information: Hourly Population Densities

The distributions for *Home*, *Work*, and *CommuteDistance* give information about spatial probabilities, but do not link them to particular times of day. To improve the fidelity of our model, we need to include temporal information as well. For instance, a heavily

(a) Los Angeles

(b) New York

Figure 3: Logscale heatmaps of call densities in the LA and NY areas from 7 to 8 p.m. on weekdays over a 3-month period. (The apparent activity in offshore areas is only a byproduct of interpolation techniques used to produce these maps.)

residential area is likely to be more populated at night while a commercial district is likely to be more populated during the day.

We model the time-varying aspects of human mobility through the use of spatial population densities indexed by time t. Unlike the previous spatial distributions, which have specific meanings like home and work, this portion of our model is an aggregate distribution that simply reflects the probability of people being at a particular location at a particular time. For some it could be work, for others home, and for others neither.

Hourly population density distributions can be constructed in several ways. First, if available, CDR traces can be analyzed to estimate population at any point in space and time. This requires the assumption that the spatial densities of telephone calls is approximately equivalent to the spatial density of people. Prior work [17] has shown that cellular calls are an accurate representation of the locations of a user. In collections of actual CDRs, call locations are estimated as the location of the cell tower through which the call originated. Since cell towers vary in density (urban vs. rural) we interpolate the cell tower locations and call counts to a regular grid whose granularity is specified by the modeler. The hourly call counts per latitude-longitude grid area can be normalized to form a distribution across latitude and longitude called *Hourly*. Figure 3 shows instances of one-hour population density estimates made in this way for calling activity in the LA and NY regions.

If CDR data are not available, one could approximate the population densities from census data. For example, one could form an overall distribution by using census home location data to estimate population data during traditional non-work hours (e.g., 7pm to 7am) and census work location data to estimate population density during traditional work hours (e.g., 7am to 7pm). We evaluate this alternative in subsequent sections.

2.3 Temporal Information: Calling Patterns

The end goal of our modeling work is to create synthetic CDR traces that can be validated against real-world CDRs and that can be used in the same ways that previous research has used real-world CDRs. Thus, a final step in our model is to combine the spatial probabilities above with information about calling patterns, so that our model produces accurate synthetic CDRs for comparison and future use.

To incorporate realistic temporal information into the mobility model, we require realistic information about the distributions of user call rates. One can envision a user's daily call volume characteristics as being probabilistically chosen from a distribution of *PerUserCallsPerDay* indexed by different possible averages and standard deviation values.

Once a user's average and standard deviation of calls per day has been selected, the next question concerns the more detailed temporal patterns of when those calls are made. To this end, we separate users into similarity classes with separate temporal distribution functions for each similarity class. This method requires a detailed source of temporal input data, such as a real-world CDR trace. From this, we examine how many calls each user makes during each hour of the day, and this calling behavior gives us a 24-dimensional vector (i.e., one dimension for each hour). We normalize this vector so that it represents probabilities rather than call volumes, and then this normalized 24-dimensional vector is used to generate similarity scores between each pair of users. From this, we use a common clustering algorithm (X-means [27]) to cluster users. Given the option of one to 20 clusters of users, X-means determined 2 clusters of users to be the best choice for our data. Finally, we return to the CDR trace and record per-minute call probability distributions separately for each user class. From these we form a probability distribution *CallTime* for each UserClass and for each hour of the day. An example of empirically-collected call time probabilities is shown in Figure 4. While typical diurnal patterns are seen for both classes of users, one class clearly favors evening calls while the other favors afternoon calls.

Without access to CDR information, it is still possible to construct call distributions. One approach would be to assume that all users have similar temporal patterns, and simply use an overall

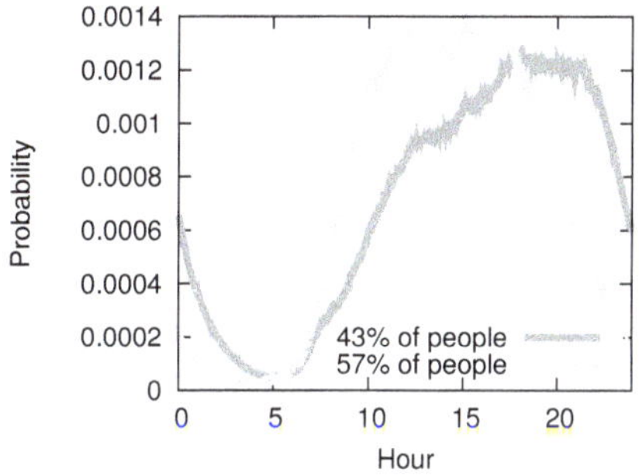

Figure 4: Distribution of call times for two classes of users as determined by X-means clustering.

Algorithm 1 Create

Ensure: *pop*[] is an N sized array of 4-element structures to be filled in with 4 properties for each of N synthetic users

for $user = 0 \rightarrow N$ **do**
  $pop[user].home \leftarrow$ location from *Home*
  $commute \leftarrow$ distance from *CommuteDistance* conditioned on $pop[user].home$
  $pop[user].work \leftarrow$ location from *Work* at distance $commute$ from $pop[user].home$
  $pop[user].callsbehavior \leftarrow$ user type from *CallTime*
  $pop[user].callsperday \leftarrow \mu$ and σ from *PerUserCallsPerDay* and independent of $pop[user].callsbehavior$
end for

Algorithm 2 Move

for $user = 0 \rightarrow N$ **do**
  for $day = 0 \rightarrow D$ **do**
    $callstoday \leftarrow$ normal random number with μ and σ from $pop[user].callsperday$ distribution
    for $call = 0 \rightarrow callstoday$ **do**
      $calltime \leftarrow$ time from $pop[user].callsbehavior$
      $location \leftarrow$ location using probabilities of $pop[user].work$ and $pop[user].home$ at time $calltime$ from the *Hourly*
      print $user, day, calltime, location$
    end for
  end for
end for

per-hour call probability distribution function as the guide for when calls are most or least likely to be made. Such probabilities can be drawn from prior work including [1, 5]. Our work validates the accuracy of approaches like these against real-world CDR collections.

To summarize, the sources of temporal input data on call patterns can either be published statistics [1, 5] or proprietary CDR data. Because these call patterns have been seen multiple times in multiple contexts, we assume that such a calling pattern is general enough to hold regardless of the spatial data to which it is applied.

3. Algorithms for Model Generation

The previous section argued for the importance of particular spatial and temporal information as inputs towards a stochastic mobility model. In this section we describe how to use such information as inputs for mobility models with different degrees of complexity and accuracy. We start in Section 3.1 with a two-place model that has synthetic people who alternate between home and work. We call this model WHERE2. Despite the simplicity of this model, the attention to spatial and temporal distributions gives it considerable accuracy. Section 3.2 shows how the technique can be expanded into a three-place model we call WHERE3, with even more accurate results.

3.1 Two-Place Model: Work and Home

WHERE2 makes use of the fact that most people spend the majority of their time either at home or at work. The process of generating a synthetic CDR trace embodying this model occurs in two stages. In the first stage, a synthetic user is created according to Algorithm 1, *Create*. First, a user is assigned a home from *Home*. Second, a commute distance is selected for the synthetic user. As illustrated in Figure 2, *CommuteDistance* is conditioned on a user's particular home location. In this way, we begin to tie users specifically to the geography of the area to be synthesized. It is not sufficient to select from a general probability distribution for commute distances, because this may unfairly bias toward commute distances for very dense areas. Intuitively, the likely commute distances for a person living in midtown Manhattan are quite different from those who live in outlying exurbs. Third, all possible locations that are the selected distance away from the chosen home location are considered as possible "work" locations, and a work location is chosen for the synthetic user. Possible work locations are weighted with probabilities given by *Work*, and again are conditioned on a particular home location and commute distance.

Finally, after having selected a user's home and work, the synthetic user is assigned a calling pattern (i.e., mean μ and standard deviation σ calls per day) according to the distributions from *CallTime* and *PerUserCallsPerDay*.

In the second stage, our synthetic users are moved between "home" and "work" according to Algorithm 2, *Move*. Movement "occurs" based on synthetic CDRs representing calls made at different locations at different times. For each day, a number of calls to make is selected from the user's average (μ) and standard deviation (σ) of the number of calls made per day. Each call is then made according to the calling distribution of the user's class, as given in *CallTime*. When a "call" is made, the location of the call is determined to be either home or work according to the probability of a person being in the location at that time of day (i.e., the probability is determined according to the distributions in *Hourly*).

3.2 Extension to Additional Places

This section discusses how WHERE2 can be augmented for greater realism by increasing the number of important locations that are considered for each synthetic user. The tradeoff of model complexity against the fidelity of the synthetic trace can be tuned to suit the model's purpose.

We can extend WHERE using Algorithm 3, *CreateExpand*, to create WHERE3. When the synthetic user is given locations, a third location is selected such that the distance of the additional location is selected from a distribution of distances from the "home" and "work" locations of users in the real city, the *AverageDistance* Distribution. Clearly, this must be conditioned on the commute distance because distances between the three locations (home, work, and the one to be added) are conditionally dependent on each other (i.e., they form three sides of a triangle). Probabilities for the additional locations can be drawn from a probability distribution of the user locations over time, the *AllLocations* Distribution. Similarly, we can create WHERE4, WHERE5, etc. by incrementally adding locations to the previous model that are conditionally dependent on

Algorithm 3 CreateExpand

Ensure: $pop[]$ is an N sized array of 5-element structures to be filled in with the 4 properties from $Create$ and 1 additional property for each of N users

for $user = 0 \rightarrow N$ **do**
  Perform $Create$ for user
  disttothirdloc $\leftarrow$ distance of third cluster to $pop[user].home$ and $pop[user].work$ from *AverageDistance*
  $pop[user].third$ $\leftarrow$ location from *AllLocations* at distance *disttothirdloc* from $pop[user].home$ and $pop[user].work$
end for

the locations before them. Such models will be increasingly precise but come at the expense of making the models increasingly complex.

In addition, when the synthetic CDRs are created, probabilities for a user "moving" to a given location must be modeled realistically. In our approach, they are conditioned on the location of the home cluster. This needs to be done so that home and work remain the most frequently visited locations instead of a generically more popular third location. This way, additional locations will be visited in a way that mimics the way those locations are visited by real people. This requires modification of the hourly probability distribution described in Section 2.2, a step not reflected in the *CreateExpand* algorithm. Specifically, we need a set of hourly probabilities for each possible home location. By conditioning the probabilities for Section 2.2 by the location of the synthetic user's home, we greatly increase the size of the required set of input probability distributions, but we are able to add realism to human movement patterns. Once the probabilities are modified, no modification needs to be done to the *Move* algorithm except that when selecting from the hourly probability distribution, the home location must be taken into consideration as well.

4. Evaluation Methodology

This section offers methodology and background information regarding our experimental evaluations. In particular, we discuss the metrics by which we gauge model accuracy, the other mobility models we compare against, and potential sources for the input data our model requires.

4.1 Earth Mover's Distance

Ultimately, the synthetic CDRs are intended to create movement patterns that mimic those seen in the real CDRs. Thus, in aggregate, a "good" synthetic trace has the synthetic user population distributed in a very similar way in space as the real trace, at any point during the day. To quantitatively assess the similarity or dissimilarity of two population density patterns at a given time, we need a measure for comparing two spatial probability distributions (i.e., the hourly locations of real and synthetic users). To this end, we rely on the Earth Mover's Distance (EMD) as our measure of choice [29, 30]. To calculate it efficiently, we use the Fast EMD code from [26].

EMD attempts to find the minimum amount of energy required to transform one probability distribution into another. If one visualizes a probability distribution as a hill of earth to be reshaped into the second probability distribution, this energy is given by the "amount" of probability to be moved and the "distance" to move it. Since different distance weightings lead to different EMD values, we provide Table 2 as a basic calibration for the meaning of EMD values in this work. Table 2 provides EMD calibration data for a location shift transform. The starting point is a probability distribution in which all of the probability is concentrated in a single location (e.g., delta function). The "Location Shift" transform maintains the concentrated probability spike, but shifts it linearly by the specified number of miles to a new geographic location.

Miles	EMD
1	8.67e+05
10	8.65e+06
20	1.74e+07
30	2.61e+07
40	3.48e+07
50	4.35e+07

Table 2: EMD values for linear shifts of the probability distribution.

Table 2 confirms that in the case of a simple location shift, the EMD changes linearly with the shift amount or distance. In particular, an error caused by a simple shift of 1 mile corresponds to an EMD of 8.67e+05, and larger shifts scale proportionately. For the remainder of this paper we normalize our EMD errors by this factor. This allows us to refer to them in terms of the more human readable "miles of error". To determine the average error between real and synthetic location density patterns, we generate location probability distributions for each hour of the day from our synthetic CDRs. Then we calculate the EMD between our synthetic distributiond and the reference probability distribution (one of the distributions from Section 4.2). To convert to miles of error, we simply divide the result by 8.67e+05.

Another benefit that comes with using EMD in this context is that since we deal with probability distributions and physical distances, EMD is a metric in the strict mathematical sense [2]. The ensuing properties regarding relative distances (e.g. triangle inequality) are useful when comparing the fidelity of different models or use cases.

4.2 Comparison Models

To demonstrate the value of our approach, we compare our synthetic models against two commonly-used mobility models: Random Waypoint [19] and Weighted Random Waypoint.

Random Waypoint: In the Random Waypoint (RWP) model, each user selects a random destination from all possible destinations in the area to be simulated [19]. Once a destination is selected, the user moves at a random velocity toward the destination. When the selected destination is reached, the user waits for a random amount of time and then selects a new destination and new velocity to begin the procedure anew. While known to be simplistic, RWP has nonetheless been used extensively by the mobile computing research community, because so few alternatives exist.

Weighted Random Waypoint: The Weighted Random Waypoint (WRWP) model behaves similarly to RWP in that a user moves at a random velocity towards a chosen waypoint and waits for a random amount of time. However, in this variant, the destinations are not chosen from a uniform distribution. Instead, a location probability distribution is used to weight possible waypoints. The distribution we use is obtained by combining all of the hourly probabilities from Section 2.2 into a distribution that gives the popularity of the location for making calls over the whole day. This results in waypoints being preferentially chosen based on their popularity throughout the day. Intuitively, this accounts for the importance of home and work locations, so it is interesting to note how our approach further improves on this.

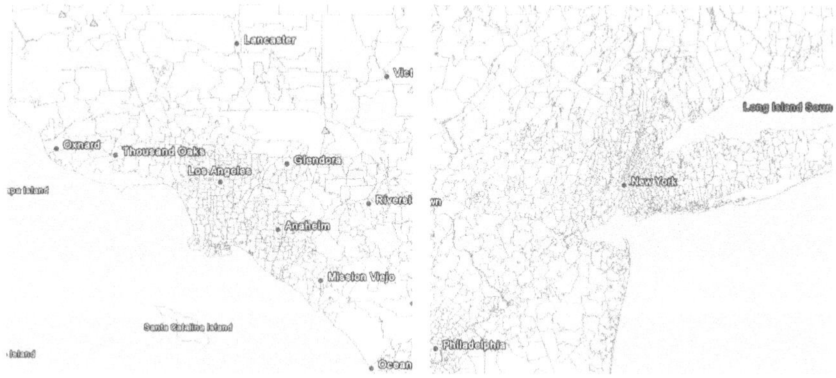

(a) Los Angeles (b) New York

Figure 5: ZIP codes in the LA and NY metropolitan areas used in our study. Note that NY and LA areas are drawn to the same scale.

	LA	NY
Total Unique Phones	318K	267K
Total Unique CDRs	1395M	1095M
Median CDRs/phone/day	18	18
Median calls/phone/day	6	7
Median texts/phone/day	6	5

Table 3: Characteristics of the CDR datasets for the LA and NY metropolitan areas. Each dataset spans 91 consecutive days from April 1 to June 30, 2011.

For either of the RWP and WRWP models, we generate synthetic agents that move on a grid and at random points in time, announce their location. In this way, the baseline models exhibit the same non-continuous behavior as the real CDRs. Once the baseline movement trace is generated, we calculate the EMD of a simulated hour of the baseline model against the reference CDRs.

4.3 Sources for Input Probability Distributions

As noted in Section 2, the probability distributions that form the input to WHERE can be obtained from a range of sources. We describe our sources here.

4.3.1 Real Call Detail Records

Call Detail Records maintained by cellular network operators provide up-to-date and low-cost information about human locations on a large scale. We have access to anonymized CDRs from a major US carrier, from which we extract the necessary spatial and temporal probability distributions.

Dataset Contents: We gathered location information for a random set of cellular phones whose billing addresses lie within the metropolitan regions of interest. First, we identified all ZIP codes within a 50-mile radius of the Los Angeles and New York city centers. These ZIP codes correspond to the darker colored regions in Figure 5. Second, we obtained anonymized and simplified CDRs for a random sample of phones registered to individuals with billing addresses in those ZIP codes. These CDRs contain the following information: a unique phone identifier in place of the telephone number, the starting time of the voice or text event, the duration of the event, and the locations of the starting and ending cell towers associated with the event. We excluded phones registered to businesses because their billing ZIP codes do not generally correspond to people's homes. We also excluded phones that appeared in their billing ZIP codes fewer than half the days they had voice or text activity, so as to exclude people who do not live in those ZIP codes.

Table 3 describes our CDR datasets for LA and NY. Each dataset contains more than a billion location samples for hundreds of thousands of phones over 3 months of activity, with 18 median location samples per day for each phone.

Data Validation: In previous work, we have verified that CDR datasets gathered using this same methodology accurately represent the mobility patterns of the population at large. More specifically, we have performed a number of comparisons against data from the US Census [32] and against ground truth provided by volunteers. One, we have confirmed that the number of sampled phones in each ZIP code is proportional to the population of that ZIP code [18]. Two, we have shown that the maximum pairwise distance between any two cell towers contacted by a phone in one day is a close approximation for how far the phone's owner traveled that day [17]. Three, we have shown that applying certain clustering and regression techniques produces accurate estimates of important locations in people's lives, in particular home and work [16].

Privacy Measures: Given the sensitivity of location information, we took several steps to preserve the privacy of individuals represented in our datasets. First, only anonymous records were used in this study. Personally identifying characteristics were removed by someone not involved in the data analysis. Second, we worked only with large collections of phones. We did not focus our analysis on any individual phone and we present only aggregate results.

In addition to these active steps, it is in the nature of CDRs to yield only temporally sparse and spatially coarse location information. A CDR is generated only when a phone is used for a call or text message—at all other times the phone is invisible to us. Furthermore, we only know phone locations to the granularity of a cell tower. Because a tower often covers an area greater than one square mile, our spatial resolution is limited.

4.3.2 Census Data

Clearly, not all researchers will have access to real CDRs. However, in the US and other countries, there is publicly available data regarding home and work locations, as well as commute distances, for large populations. To demonstrate the efficacy of using publicly available data, we used census information for the same regions defined in Section 4.3.1 to construct probability distributions for the locations of home and work.

Although the census can be used to determine densities of home and work locations, it provides little or no information about the hourly probabilities of a given location. Therefore, when we use WHERE on All Public data, we must make an assumption about the hourly distributions. To that end, we assume that the *Work* distri-

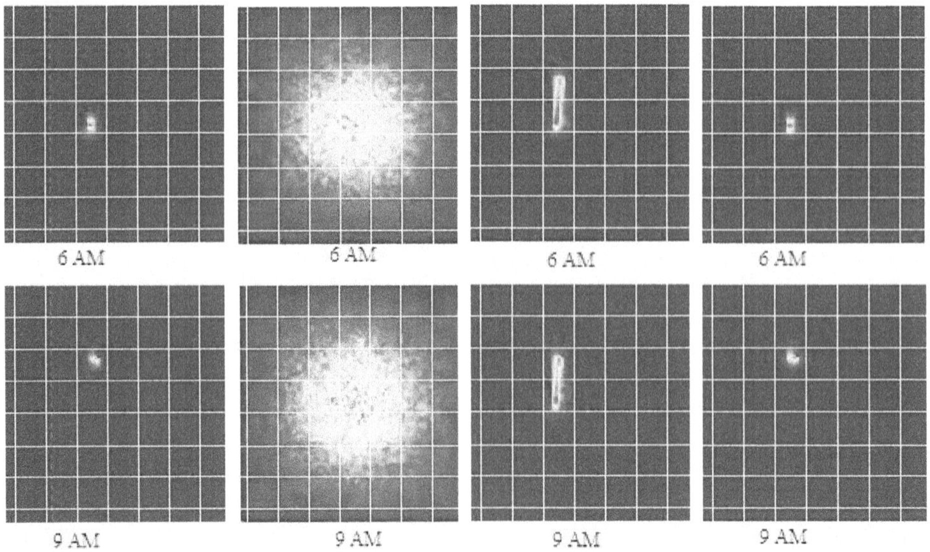

(a) Two Locations Ideal (b) RWP (c) WRWP (d) WHERE2

Figure 6: Heatmaps comparing the probability distributions of three models of mobility against the "Two Locations" test case during the "home" and "work" phase of the test case for 10,000 synthesized users. Hot (red/light) locations have more calls made at that time, while cold (blue/dark) locations have fewer.

bution doubles as the hourly distribution of all hours between 7am and 7pm on Weekdays and that the *Home* distribution provides the hourly distribution at all other times.

We also consider a Hybrid scheme in which we use a combination of public data and CDRs. In this scheme, home, work, and commute distances come from the census, but the other distributions are drawn from CDR data.

4.4 Artificial Test Cases

It is useful to construct some stylized examples to demonstrate the function of specific features of WHERE. By creating artificial test cases, we are better able to reason about the expected behavior of the model and its strengths and weaknesses. In this section, we explain how we construct three artificial reference probability distributions that exercise various features of WHERE.

Two Locations: Initially, we must determine that WHERE functions correctly. The model must be able to place synthetic users at home and work locations and move them according to time of day. Further, since we emulate CDRs with discrete call locations, the synthetic users must only exist at these locations.

We test these features with the "Two Locations" test case. The entire world is populated by users that move predictably between two locations at highly regimented intervals. From 7am to 7pm on weekdays, all of the probability is concentrated in a single "work" location. At all other times, the probability is clustered in a second "home" location.

Two Distinct Types: We then expand the test cases to include multiple possible home and work locations. WHERE must be able to correctly condition commute distances on home locations. If multiple possible works exist for a home, the model must select a realistic one.

We demonstrate this with the "Distinct Types" test case. The home distribution has half the probability at one location and half at a second location. The work distribution is also split between two locations. The commute distributions are constructed such that each home location uniquely identifies a work location.

5. Validation: Stylized Examples

In this section we evaluate the accuracy of WHERE2 in synthesizing roughly 10,000 people using input probability distributions for the stylized test cases described in Section 4.4. We compare, both graphically and quantitatively, errors in the WHERE2, RWP, and WRWP models relative to an ideal result for each stylized case.

5.1 Two Locations Test Case

Figure 6 illustrates the spatial density patterns of user locations created by the "Two Locations" test case. Figure 6(a) shows the ideal result, in which probability is a spike at the home location at 6am, and a spike at the work location at 9am. RWP (Figure 6(b)) clearly differs greatly, because it is given so little input information regarding spatial or temporal patterns to model. The WRWP model better concentrates probability near the home and work locations, but does not have sufficient temporal input to distinguish how 9am mobility probabilities should differ from the 6am ones. Finally, Figure 6(d) shows the heat maps for WHERE2, which are a strong visual match to the ideal case. Even though WHERE2 has no hardwired information about what "work hours" are, it is able to correctly model the call distribution information both during the "home" and "work" phases. In contrast, while the WRWP model is able to correctly move users from "home" to "work", the lack of timing information damages its functionality for generating realistic synthetic CDRs.

While Figure 6 allows for an easy visual differentiation between the different models, the EMD measure is better for quantifying the observed differences. In fact, the EMD for WHERE2 corresponds to a simple shift of the cumulative probability of less than 0.5 miles. This is a 10X improvement over WRWP, the next best model considered. The WHERE2 model correctly generates calls from the desired discrete locations, and among the considered models, it is the only one to recognize that not all locations are equally probable at all times. Because each user only has two locations, WHERE3 performs exactly like WHERE2.

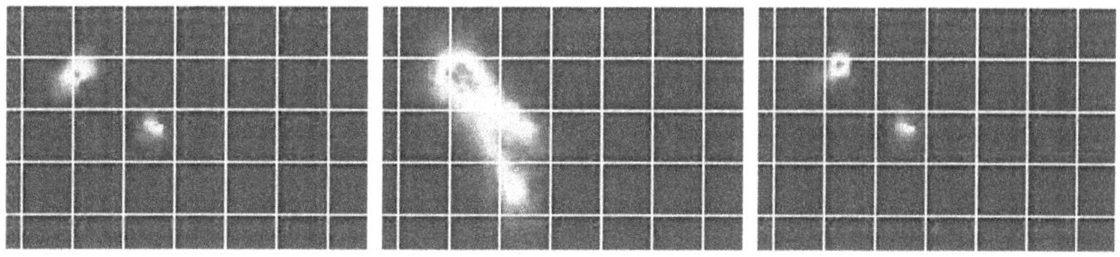

(a) Two Distinct Types Ideal (b) WRWP (c) WHERE2

Figure 7: Heatmaps comparing the WRWP model to WHERE2 for 10,000 synthesized users. The WRWP is unable to account for the two distinct patterns and thus sends synthetic users on paths never taken in the "real" trace.

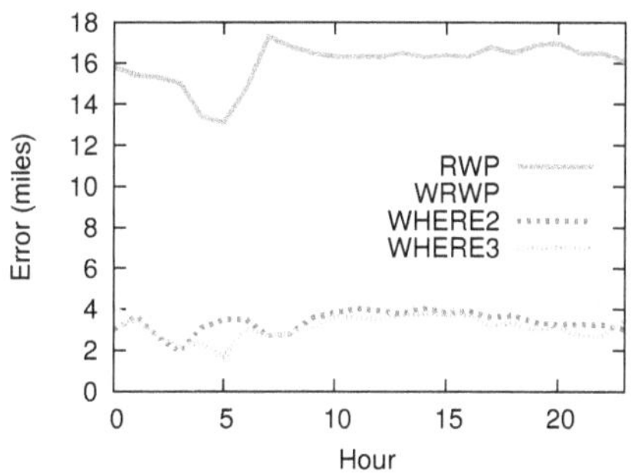

Figure 8: EMD error over time for different models of human movement in the NY area. Our expanded model shows average errors smaller than 3 miles. WHERE fits between WHERE3 and WRWP.

5.2 Two Distinct Types Test Case

Figure 7 shows the heatmaps of the "Two Distinct Types' ideal distribution, as well as WRWP and WHERE2. WHERE2 detects that users make calls only from one of two locations and thus correctly positions users. In contrast, the weighted model has a single user travel to each of the four possible locations, resulting in a significantly worse EMD. When comparing against the WRWP model, WHERE2 has an average of 5% improvement across all hours of the day. Because each user type only has two locations, WHERE3 again performs exactly like WHERE2.

6. Validation: Large-Scale Real Data

In this section, we evaluate the accuracy of WHERE in synthesizing more than 10,000 people using probability distributions from the datasets of real CDRs described in Section 4.3.1, and from the census data described in Section 4.3.2. We compare, both graphically and quantitatively, errors produced by the WHERE, RWP, and WRWP models relative to those two large-scale real-data sources.

6.1 Modeling Based on Real CDRs

Figure 8 shows EMD errors for the different possible models using distributions drawn from real CDRs. By including population density information, WRWP improves over original RWP by 4 times. WHERE, however, outperforms this WRWP model by an additional 20%. Furthermore, WHERE3 improves further still; it is able to recreate the NY probability with an accuracy of about 3 miles. WHERE and WHERE3, therefore, provide a powerful tool for obtaining accurate large-scale motion patterns that result from individual user movements.

Finally, Figure 9 gives an overall visual demonstration of WHERE's effectiveness. When this heatmap is compared to Figure 3(b), the two distributions are found to be visually very similar. The low EMD error combined with the visual confirmation of similarity

Figure 9: Logscale heatmap of the output of WHERE for the NY area.

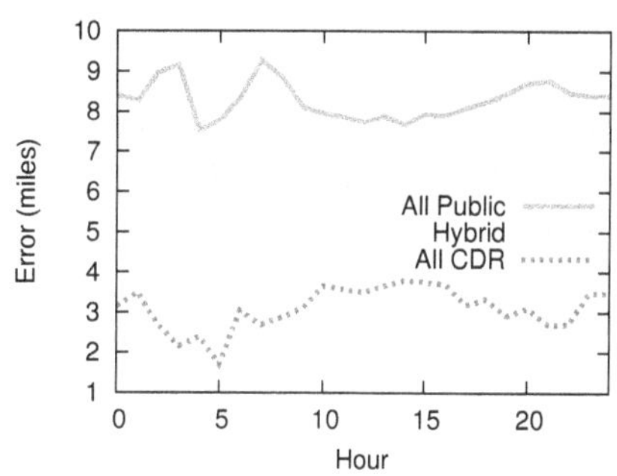

Figure 10: EMD error using different levels of available data. Even with fully public data, errors remain at an average of 8 miles.

shows that WHERE is an effective approach to modeling human mobility in a city.

6.2 Modeling Based on Census Data

Although the most accurate synthetic CDRs are derived from real CDRs, other data sources can be used at some loss in accuracy. For example, Figure 10 shows that using all-public data from the US Census, we are able to recreate location distributions in NY with an average error of 8 miles. Using a hybrid of census data with some CDR information regarding the call densities at different times of the day (not publicly available, but easily aggregated and thus potentially easier to obtain) reduces this error to an average of 6.8 miles. In contrast, using all-CDR information reduces the average error to some 3 miles in WHERE3, as shown earlier.

Although 8 miles of error seems large compared to the 3 miles of error possible with WHERE3, this error is aggregated across all users. Individually, user behavior remains qualitatively correct and

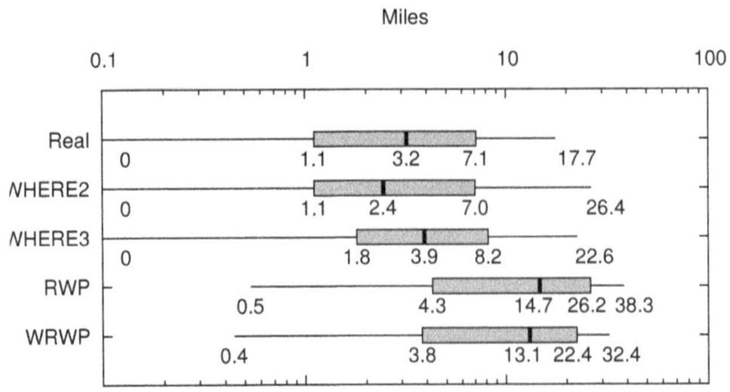

Figure 11: Daily Range from the real NY data as compared to WHERE and two variants of Random Waypoint. WHERE shows more realistic behavior than either RWP or WRWP.

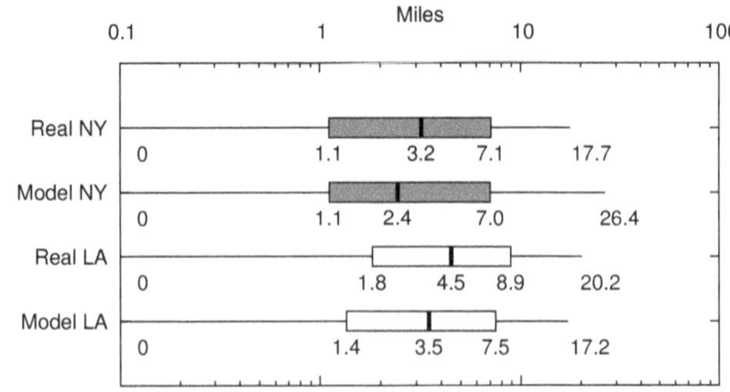

Figure 12: Comparing daily range statistics for NY and LA generated by WHERE2 and with real data. WHERE2 accurately recreates the characteristics of NY and LA mobility.

relative differences remain correct. Thus, though one could not rely on fully public data for an application such as placement of cellular towers, there remain many uses at such error levels. Notable examples include determining commute distances or experimenting with hypothetical population shifts (see Section 7.1 and Section 7.3), both of which depend on relative user behavior.

7. Example Uses

Creating synthetic CDRs has a wide range of implications for the scientific community. The inability to access real CDRs has frustrated many researchers. In generating artificial, yet realistic CDRs, we enable the greater community to perform a wide range of experiments, with an assurance that the results mimic those that would have been obtained with real CDRs.

This section presents three scenarios that highlight the value of our mobility models based on synthetic CDRs. We first present two usage experiments that we performed on synthetic CDR traces as well as on real CDRs. In Section 7.1 we replicate the daily range calculations reported in [18], which defined daily range as the maximum distance a person travels in one day. This comparison serves as an important, independent validation of WHERE because no daily range statistics were included in the input to our models. By performing well on the daily range metric, WHERE demonstrates an ability to model large-scale movement patterns while retaining realistic individual behavior. Second, in Section 7.2, we perform a simple message flooding experiment relevant in the context of opportunistic networking. Our third example highlights the major benefit of synthetic CDR traces in their ability to predict and visualize the impact of hypothetical changes to regional mobility patterns. Such changes might result, for example, from new employers moving into the region or new residential areas being developed. With this in mind, Section 7.3 presents an example of using WHERE to generate synthetic CDRs for a what-if scenario regarding mobility in the New York region.

The scenarios presented in this section constitute a set of example applications that demonstrate the broad utility of our models. These examples are far from exhaustive, but are important as illustrations of a wide range of possible uses.

7.1 Daily Range

Daily range, or the "diameter of a convex hull," has been shown to be a useful tool for characterizing human mobility patterns [18, 20]. We can therefore demonstrate the value of our modeling techniques by showing that synthetic CDRs they generate can be used to replicate daily range experiments and match the results obtained with real CDRs.

Additionally, daily range serves as an important metric for verifying correctness of the generated models. The EMD metric measures the aggregate behavior of the users, but daily range displays the results at a per-user granularity. Correctness at a metropolitan scale may or may not be correlated with correctness at the scale of individual users. The daily range metric tests whether individuals are modeled realistically.

We summarize our results with the help of boxplots in Figure 11. The boxplots depict five-number summaries of the complete empirical distributions of interest. The "box" represents the 25th, 50th, and 75th percentiles, while the "whiskers" indicate the 2nd and 98th percentiles. The horizontal axes show miles on a logarithmic scale. Nearly any difference between the medians is statistically significant due to the large sample sizes.

Figure 11 shows the daily range results for synthetic CDRs produced by WHERE2 and WHERE3, as well as those produced by RWP and WRWP. It is clear that the behavior of the random waypoint models differs greatly from reality. For example, WRWP exhibits qualitatively wrong results and errors greater than 300%. In contrast, WHERE2 is qualitatively correct, and its median daily range value is within 0.8 miles of the true value. WHERE3 closes this error further: 0.7 miles error at the median.

In other words, this daily range example shows that WHERE2 and WHERE3 capture aspects of mobility that are not captured by the random waypoint models. By exposing differences at the grain of individual synthetic users in this way, WHERE2 results in more than a 14X improvement over WRWP. This advantage was not visible under the more aggregate metric of EMD, when WHERE appeared to hold only a small advantage over WRWP.

This example also demonstrates another attractive feature of our model. Because much of human mobility is between two major points of interest, adding a third point to our model provides very little obvious benefit for experiments of this nature. A researcher interested in problems such as daily range can choose to use the simpler WHERE2 model, avoiding the effort of computing the more complex WHERE3 model with little or no loss of accuracy.

Comparing NY and LA: Though much of the evaluation thus far has focused on New York City, the technique is more broadly applicable. Figure 12 compares daily range statistics for both NY and LA, for both real CDR data as well as WHERE. As before WHERE generates very high-accuracy results for these statistical

	Real	WHERE2	WHERE3
Delivery Percentage	98%	53%	83%
Median Message Delay	17 hours	74 hours	18 hours

Table 4: Message passing properties for opportunistic flooding. For such a scenario an expanded model of human movement gives better results.

distributions. At the median value, WHERE computes daily range for the LA area with only 1 mile of error. Low errors are seen at the other percentiles as well. This demonstrates WHERE's applicability across cities with very different mobility patterns and geographic characteristics.

7.2 Message Propagation

Another use for CDRs is the investigation of human mobility for social contacts, epidemiology, and data carrying. For instance, in a delay-tolerant networking scenario [8], what sorts of routing algorithms work best? In this section we demonstrate that our artificial CDRs can be used to simulate inter-person contacts.

To that end, we developed a simulator for epidemic routing [33]. At random times in the CDR trace, a message is injected from a random source to a random destination. As users meet they exchange all messages. For the purposes of this simplified experiment, users "meet" if their last known positions are the same within one hundredth of a degree of latitude and longitude (roughly a circle with a radius of 0.5 miles). When the message reaches its intended recipient, the time is recorded. This allows us to examine properties of the CDRs, including delivery percentages and message delay rates. Note that only a small subset of all users is considered for this experiment, as our full user base would ensure message delivery regardless of movement patterns. Here, we limit the experiment to 5,000 messages passed between 5,000 users in the New York area over 30 days.

Table 4 displays statistics regarding message passing success in the real CDRs as well as in our synthetically constructed models. Although even WHERE3 is far from the actual values, it is clear that adding the third location provides significant improvement. This is consistent with the idea that opportunistic networking relies on highly mobile users [36].

We stress that these results are based on the most basic version of the model. This example is provided to demonstrate the extensibility of the model as more clusters are added.

7.3 Hypothetical Cities

A major advantage of mobility models over empirically-gathered real-world CDRs is the ability to create parameterized model of cities and user behavior patterns that cannot be observed in the real world. This allows researchers and city planners considerable power to experiment with the effects of modifications that are being considered.

As one example, Figure 13 shows a box plot of data that can be constructed from artificial CDRs if we imagine a scenario intended to study the impact of telecommuting. In this case, we adjusted the WHERE2 model for NY so that a randomly selected 10% of the people whose original work locations were in the borough of Manhattan were instead given work locations that matched their home location. Such a scenario might arise, for example, if tolls or subway fares were increased, or if the city enacted policies to encourage telecommuting. With only 10% of the Manhattan workers affected, the impact on daily range is subtle, but noticeable; the results do demonstrate the trend that daily ranges would be slightly

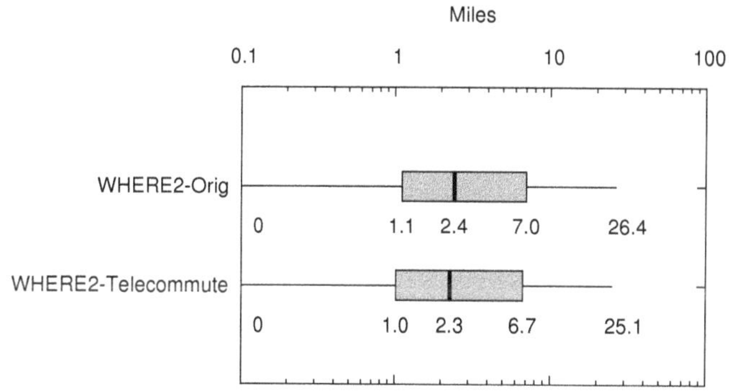

Figure 13: Boxplots of WHERE applied to a NY scenario in which 10% of Manhattan workers opt to telecommute.

diminished by this change. In general, the power of such examples is in allowing model users to explore the impact of different mobility and behavioral changes on their metric of interest.

This example demonstrates the flexibility of our modeling approach. Experiments that are not feasible in reality (e.g., convincing 10% of New Yorkers to telecommute) are simply a matter of adjusting some input probabilities.

8. Discussion and Future Work

As we have shown, our approach to human mobility modeling achieves its three main goals: capturing the motion of individuals among important places in their lives, aggregating that motion to reproduce human densities over time at the scale of a metropolitan area, and accounting for differences between metropolitan areas. However, there remain areas for refinement.

8.1 Travel Routes and Additional Locations

One such refinement would be to produce not only sequences of locations with associated times, but also routes taken between those locations. A tradeoff we incurred in obtaining location information for large populations in wide geographic regions is that both real CDRs and census tables are too coarse to provide route information. So far we have not attempted to improve on the spatial and temporal granularity of our source data, although other work [4] has had some success in identifying routes by making use of individual cellular antennas (not only towers that hold multiple antennas) and by analyzing longer sequences of antennas involved in the same call (not only the start and end towers).

In future work, we plan to infer such routes by using maps of transportation networks to interpolate between the locations produced by our current models. A similar idea has been applied to finer-grained location traces in smaller geographic regions, in particular traces of WiFi access-point associations in a university campus [34]. Working at the scale of metropolitan areas adds complexity because it becomes important to distinguish between modes of transportation beyond walking, e.g., driving a car, taking a train, or riding a bicycle. We can obtain some of the necessary information from public sources such as Table P30, Summary File 3, from the 2000 US Census: "Means of Transportation to Work for Workers 16+ Years" [32]. In addition, we can apply heuristics based on the distance and time between two location samples.

Once we select a mode of transportation between two locations, we can run proven routing algorithms such as those used by popular websites that provide driving, public transportation, biking, and walking directions [14]. Human mobility models that include route estimates would have broader applicability than those that produce only locations and times.

In addition, WHERE currently restricts users to a limited number of places. Future research directions may include incorporating heuristics about additional locations that a person may visit, perhaps by using WRWP as a rare, but possible additional destination. While it is true that the majority of movement occurs between important locations, there may be applications (such as the message passing experiment listed in Section 7.2) that could benefit from information about the long tail of other locations a person visits.

8.2 Differential Privacy

It is also worthwhile to discuss why we believe that our approach to human mobility modeling will preserve privacy. The intuitive reason is that we are careful not to reproduce the mobility pattern of any individual real person, for example someone represented in an input dataset of real CDRs. Instead, we create synthetic mobility patterns by sampling a sequence of probability distributions that each represents a large population. This approach makes it highly unlikely that any of our synthetic people exhibits a mobility pattern that identifies a single real person. However, the approach does not ensure that an adversary will be unable to reverse our algorithm to arrive at some portion of the source dataset, especially an adversary who brings to bear auxiliary datasets.

In future work, we plan to formalize our privacy argument by adjusting our algorithm to achieve *differential privacy* [7]. Informally, a differentially private algorithm is one that produces approximately the same output on two input datasets that differ only in the data for one individual. Our modeling approach naturally lends itself to a proven technique for achieving differential privacy without significantly affecting accuracy, namely to introduce controlled noise at key points in the algorithm. For example, we could introduce noise in the input CDRs, or when sampling one or more of our input probability distributions. Similar techniques have been successfully applied to network traffic traces [23].

We plan to show that our adjusted algorithm has the property that an individual's presence or absence in the input dataset will not alter the output by a significant amount. This property provides strong privacy guarantees in an information theoretic sense, regardless of how much auxiliary data an adversary applies.

9. Related Work

Characterizing human mobility based on cellular network data has recently received considerable attention. In our previous work [16], we developed an algorithm for identifying people's important locations based on anonymized cellular network data, and showed how to use these important locations to estimate home-to-work commute distances and commute carbon footprints for large populations in the New York and Los Angeles metropolitan areas. In other previous work [18, 17], we also characterized the daily range of travel of those same populations. Girardin et al. used cell phone usage within cities to determine locations of users in Rome [9], New York City [11], and Florence [10]. They were able to find where people clustered in these cities and the major paths people tended to take. The work presented in this paper goes beyond characterization to develop algorithms for constructing human mobility models from both cellular network data and census data.

A number of previous efforts attempted to model human mobility at various scales. Rhee et al. [28] studied statistical patterns of 44 participants carrying GPS devices for four months and concluded that people's movement has a resemblance to Levy flights, random walks where the step-length probability has a heavy-tailed distribution. They also proposed a Levy walk mobility model that can be used for network simulations. Kim et al. [20] developed an algorithm for extracting a human mobility model from wireless network traces collected from WiFi APs at Dartmouth College. Yoon et al. [34] present a trace-driven framework for generating realistic mobility models based on the association information between WiFi users and access points and maps of the area where WiFi traces were collected. Hsu et al. [15] used WLAN traces to extract a time-variant community mobility model that showed good performance via simulation. In contrast, we developed and validated mobility models based on CDRs for hundreds of thousands of people moving across large metropolitan areas.

González et al. [13] used cellular network data from an unnamed European country to create a universal model of how individuals move. Song et al. [31] studied similar cellular network data to predict an individual's movements. Specifically, the authors consider the towers associated with phone users and show that given sufficient past history, one could guess the current location of a given user with high accuracy. Other efforts have developed algorithms for predicting where a user will travel next [3, 6, 22]. To our knowledge, we are the first to develop and validate models that account for differences between geographic areas.

Several recent papers have looked at the privacy risks of releasing location traces. Krumm [21] showed that home addresses of individuals who shared their GPS traces could be accurately identified based on the combination of reverse geo-coding and white pages. Golle and Partridge [12] used the publicly available census data to show that knowing a person's home and work locations, even at a coarse resolution, is often enough to uniquely identify that person. Finally, Zang and Bolot [35] showed that releasing anonymized CDRs in its original format poses serious privacy risks, as a large fraction of the population could be re-identified from the anonymous data. These papers motivated us to look beyond a simple anonymization of location traces. In the future, we plan to show that WHERE preserves differential privacy.

10. Conclusions

Modeling human mobility is important for mobile computing research, urban planning, epidemiology, and ecology. In this paper, we proposed a method for generating realistic human mobility models for large metropolitan areas from real Call Detail Records (CDRs) and census data. Our models generate spatio-temporal data in the form of synthetic CDRs, which could then be processed by the same algorithms that operate on the real ones. We also demonstrated how our method can be extended to create more precise models of human mobility.

We validated our methodology by first showing that our synthetic CDRs maintain key properties of the real CDRs, in particular human densities over time. We then demonstrated that our WHERE model is more accurate than other models such as Random Waypoint and Weighted Random Waypoint. Finally, we gave examples of experiments enabled by our models, and demonstrated that our synthetic CDRs perform well compared to the real CDRs. Specifically, we showed that we can calculate median daily ranges with an error at the median of less than 1 mile, and good agreement across the rest of the distribution. In future work, we plan to show that our modeling approach preserves differential privacy, at which point we hope to release our models to the broader research community.

11. Acknowledgments

We thank our shepherd, Rajesh Balan, and the anonymous reviewers for their feedback. Parts of this work were supported by the National Science Foundation under Grant Nos. CNS- 0614949, CNS-0627650, and CNS-0916246. Isaacman also acknowledges support from a Princeton University Wallace Memorial Fellowship in Engineering and a research internship from AT&T Labs.

12. References

[1] S. Almeida, J. Queijo, and L. Correia. Spatial and temporal traffic distribution models for gsm. In *Vehicular Technology Conference*, Sept. 1999.

[2] D. Applegate, T. Dasu, S. Krishnan, and S. Urbanek. Unsupervised clustering of multidimensional distributions using earth mover distance. In *Proc. KDD'11*, 2011.

[3] M. A. Bayir, M. Demirbas, and N. Eagle. Discovering spatiotemporal mobility profiles of cellphone users. *World of Wireless, Mobile and Multimedia Networks and Workshops*, 2009.

[4] R. Becker, R. Cáceres, K. Hanson, J. M. Loh, S. Urbanek, A. Varshavsky, and C. Volinsky. Route classification using cellular handoff patterns. In *13th International Conference on Ubiquitous Computing (Ubicomp)*, Sept. 2011.

[5] J. Candia, M. C. González, P. Wang, T. Schoenharl, G.Madey, and A.-L. Barabási. Uncovering individual and collective human dynamics from mobile phone records. *MATH.THEOR.*, 41:224015, 2008.

[6] K. Dufková, J.-Y. Le Boudec, L. Kencl, and M. Bjelica. Predicting user-cell association in cellular networks from tracked data. *Intl. workshop on Mobile Entity Localization and Tracking in GPS-less Environnments*, 2009.

[7] C. Dwork. Differential privacy: A survey of results. In *Theory and Applications of Models of Computation (TAMC)*. Springer Verlag, April 2009.

[8] K. Fall. A delay-tolerant network architecture for challenged internets. In *SIGCOMM*, 2003.

[9] F. Girardin, F. Calabrese, F. Dal Fiorre, A. Biderman, C. Ratti, and J. Blat. Uncovering the presence and movements of tourists from user-generated content. In *Intn'l Forum on Tourism Statistics*, 2008.

[10] F. Girardin, F. Dal Fiore, J. Blat, and C. Ratti. Understanding of tourist dynamics from explicitly disclosed location information. *Symposium on LBS and Telecartography*, 2007.

[11] F. Girardin, A. Vaccari, A. Gerber, A. Biderman, and C. Ratti. Towards estimating the presence of visitors from the aggregate mobile phone network activity they generate. In *Intl. Conference on Computers in Urban Planning and Urban Management*, 2009.

[12] P. Golle and K. Partridge. On the anonymity of home/work location pairs. In *Seventh International Conference on Pervasive Computing (Pervasive 2009)*, 2009.

[13] M. C. González, C. A. Hidalgo, and A.-L. Barabási. Understanding individual human mobility patterns. *Nature*, 453, 2008.

[14] Google Maps. `http://www.census.gov`.

[15] W.-J. Hsu, T. Spyropoulos, K. Psounis, and A. Helmy. Modeling time-variant user mobility in wireless mobile networks. In *IEEE INFOCOM*, 2007.

[16] S. Isaacman, R. Becker, R. Cáceres, S. Kobourov, M. Martonosi, J. Rowland, and A. Varshavsky. Identifying important places in people's lives from cellular network data. In *9th International Conf. on Pervasive Computing*, 2011.

[17] S. Isaacman, R. Becker, R. Cáceres, S. Kobourov, M. Martonosi, J. Rowland, and A. Varshavsky. Ranges of human mobility in los angeles and new york. In *Eighth IEEE Workshop on Managing Ubiquitous Communications and Services*, 2011.

[18] S. Isaacman, R. Becker, R. Cáceres, S. Kobourov, J. Rowland, and A. Varshavsky. A tale of two cities. In *Workshop on Mobile Computing Systems and Applications (HotMobile)*, 2010.

[19] D. B. Johnson and D. A. Maltz. Dynamic source routing in ad hoc wireless networks. In *Mobile Computing*, pages 153–181. Kluwer Academic Publishers, 1996.

[20] M. Kim, D. Kotz, and S. Kim. Extracting a mobility model from real user traces. In *IEEE INFOCOM*, 2006.

[21] J. Krumm. Inference attacks on location tracks. In *Fifth International Conference on Pervasive Computing (Pervasive 2007)*, 2007.

[22] K. Laasonen. *Mining Cell Transition Data*. PhD thesis, University of Helsinki, Finland, 2009.

[23] F. McSherry and R. Mahajan. Differentially-private network trace analysis. In *Proc. ACM SIGCOMM*, 2010.

[24] W. Navidi and T. Camp. Stationary distributions for random waypoint models. *IEEE Transactions on Mobile Computing*, 3(1):99–108, 2004.

[25] A. Noulas, S. Scellato, R. Lambiotte, M. Pontil, and C. Mascolo. A tale of many cities: universal patterns in human urban mobility. *PLoS ONE*, 2012.

[26] O. Pele and M. Werman. Fast and robust earth mover's distances. In *ICCV*, 2009.

[27] D. Pelleg and A. Moore. X-means: Extending k-means with efficient estimation of the number of clusters. In *17th International Conf. on Machine Learning*, 2000.

[28] I. Rhee, M. Shin, S. Hong, K. Lee, and S. Chong. On the levy-walk nature of human mobility: Do humans walk like monkeys? In *IEEE INFOCOM*, 2008.

[29] Y. Rubner, C. Tomasi, and L. J. Guibas. A metric for distributions with applications to image databases. In *Proc. IEEE International Conference on Computer Vision*, 1998.

[30] Y. Rubner, C. Tomasi, and L. J. Guibas. The earth mover's distance as a metric for image retrieval. *International Journal of Computer Vision*, 40, 2000.

[31] C. Song, Z. Qu, N. Blumm, and A.-L. Barabási. Limits of predictability in human mobility. *Science*, 327, 2010.

[32] US Census Bureau. `http://www.census.gov`.

[33] A. Vahdat and D. Becker. Epidemic routing for partially connected ad hoc networks. Technical Report CS-200006, Duke University, 2000.

[34] J. Yoon, B. Noble, M. Liu, and M. Kim. Building realistic mobility models from coarse-grain traces. In *Proc. ACM MobiSys*, 2006.

[35] H. Zang and J. Bolot. Anonymization of location data does not work: A large-scale measurement study. In *Seventeenth Annual International Conference on Mobile Computing and Networking (MobiCom 2011)*, 2011.

[36] G. Zyba, G. M. Voelker, S. Ioannidis, and C. Diot. Dissemination in opportunistic mobile ad-hoc networks: The power of the crowd. In *IEEE INFOCOMM*, 2011.

Isolating and Analyzing Fraud Activities in a Large Cellular Network via Voice Call Graph Analysis

Nan Jiang*, Yu Jin, Ann Skudlark, Wen-Ling Hsu, Guy Jacobson, Siva Prakasam, Zhi-Li Zhang*

*Computer Science Dept., University of Minnesota, MN 55414 AT&T Labs - Research, NJ 07940

*{njiang,zhzhang}@cs.umn.edu {yjin,aes,hsu,guy,asp}@research.att.com

ABSTRACT

With widespread adoption and growing sophistication of mobile devices, fraudsters have turned their attention from landlines and wired networks to cellular networks. While security threats to wireless data channels and applications have attracted the most attention, *voice-related fraud activities* also represent a serious threat to mobile users. In particular, we have seen increasing numbers of incidents where fraudsters deploy malicious apps, e.g., disguised as gaming apps to entice users to download; when invoked, these apps automatically – and without users' knowledge – dial certain (international) phone numbers which charge exorbitantly high fees. Fraudsters also frequently utilize social engineering (e.g., SMS or email spam, Facebook postings) to trick users into dialing these exorbitant fee-charging numbers.

In this paper, we develop a novel methodology for detecting voice-related fraud activities using only call records. More specifically, we advance the notion of *voice call graphs* to represent voice calls from domestic callers to foreign recipients and propose a Markov Clustering based method for isolating dominant fraud activities from these international calls. Using data collected over a two year period from one of the largest cellular networks in the US, we evaluate the efficacy of the proposed fraud detection algorithm and conduct systematic analysis of the identified fraud activities. Our work sheds light on the unique characteristics and trends of fraud activities in cellular networks, and provides guidance on improving and securing hardware/software architecture to prevent these fraud activities.

Categories and Subject Descriptors

C.2.0 [**Computer Communication Networks**]: Security and Protection

General Terms

Measurement, Management, Security

Keywords

Cellular Network, Fraud, Malware, Mobile Apps

MobiSys'12, June 25–29, 2012, Low Wood Bay, Lake District, UK.

1. INTRODUCTION

The past decade has witnessed the rapid deployment and evolution of mobile cellular networks, which now support billions of users and a vast diverse array of mobile devices from smartphones, tablets, to e-readers and smart meters. It was reported [1] that in 2010 there were over 5 billion mobile phones in operation, in comparison to the total world population of 6.8 billion. Mobile phones and tablets are gradually replacing traditional wire-lines as well as personal computers, and are becoming an indispensable component in our daily life [2,3]. With breath-taking advances in smart mobile devices and the growing sophistication in the mobile applications (apps) and services (e.g., location services and cloud services) they spur, we are now entering in a new era of mobile computing.

With their wide adoption, smartphones, while providing valuable utility and convenience to mobile users, also bring with them new security threats. As smartphones function both as phones as well as mobile computers, smartphone users not only face the usual Internet security threats (e.g., malware and botnets [4]) through their web browsing and other data activities (albeit such data-related security threats are mitigated by the perhaps markedly improved hardware and software platforms compared to conventional desktop platforms); they may also encounter a variety of *voice-related* security threats, ranging from conventional voice scams similar to those on landlines, e.g., stealing customers privacy information or defrauding users of money through various social engineering techniques, to new forms of voice fraud that utilize the data functionality of smartphones for voice-related trickeries. For instance, we have seen increasing numbers of incidents where fraudsters deploy malicious apps, disguised as interesting games and other applications to entice users to download them; when invoked, these apps automatically – and without users' knowledge – dial certain (international) phone numbers which charge exorbitantly high fees. Fraudsters also frequently utilize other social engineering trickeries to deceive users, e.g., through SMS or email spam, Twitter tweets, or fake online postings to lure users into clicking on malicious URL links, resulting in automatic dialing of exorbitant fee-charging international numbers.

So far most research efforts have focused on applying and developing anomaly detection and prevention techniques to secure wireless data channels, mobile devices, and applications/services [5–9]. To the best of our knowledge, little work has been devoted to detecting and understanding various voice-related fraud activities targeting mobile users, especially when *international* phone numbers are involved.

Voice-related fraud activities can have a much wider impact in cellular networks, as potentially all mobile users can be victims of such activities. Furthermore, unlike data traffic, mobile international voice calls often follow the pay-per-call compensation model, and are far more expensive; hence voice-related fraud activities involving international phone numbers can bring direct and significant financial losses to both mobile users and cellular service providers. Detecting and rooting out such voice-related fraud activities, especially those that target users through the data plane triggered voice fraud, is not an easy task, due to the large user population, the vast phone number space and limited data.

Using voice call records collected over a two year period in one of the largest cellular networks in the US, the goal of our study is two-fold: 1) *to develop an effective approach to proactively isolate dominant fraud calls from a myriad of legitimate calls*; 2) and *to conduct a systematic analysis of the unique characteristics and trends of fraud activities in cellular networks*, e.g., techniques for soliciting fraud calls and social engineering. Achieving these goals can provide a means of alerting customers and cellular providers of potential fraud threats to avoid financial loss for both parties, and ultimately improve customers' satisfaction. Moreover, understanding different fraud activities can help gain useful insights in developing better hardware/software architecture for preventing future fraud activities.

Since voice call records only contain limited information, such as call time, originating/terminating numbers, country codes and call durations, we explore the relationship among parties participating in the calls (i.e., "who calls whom") for the fraud detection task. In particular, we advance the notion of *voice call graphs* for representing the call records. A voice call graph is a bi-partite graph, where two independent sets of nodes represent the groups of domestic originating numbers and foreign terminating numbers, respectively, and the edges stand for phone calls between these originating numbers and terminating numbers. By visualizing small scale voice graphs and characterizing large scale voice graphs with classic graph statistics, we find that fraud numbers and victims often exhibit very strong correlation, which results in community structures in the voice call graph. Therefore, the task of isolating fraud calls can be formulated as the problem of extracting dominant community structures from voice graphs. This serves as our basic heuristic for detecting voice-related fraud. Based on this heuristic, we propose a Markov Clustering (MCL) based algorithm to decompose voice call graphs in an iterative manner, which produces millions of disconnected subgraphs on a month-long voice graph. We further rely on the strength of community structures (measured by the number of cliques) and their popularity (measured by the number of callers) as a gauge to isolate fraud activities from these subgraphs.

We validate the proposed detection algorithm using two sources of ground truth: 1) a list of phone numbers that are reported by mobile users to the cellular service provider, which are then manually verified by fraud agents to be involved in international revenue sharing fraud (IRSF); 2) online reports from mobile users that are posted on forums, blogs or social media sites. By matching our detection results against the ground truth, we find that the proposed algorithm is able to isolate from millions of terminating numbers the most dominant IRSF fraud numbers. In particular, these IRSF numbers together have attracted more than 85% of the victims and resulted in 78% of the fraud calls. More importantly, in 60% of the cases, our method is able to detect fraud numbers at least 1 month prior to the earliest online user reports. Such an advantage in early fraud detection allows us to effectively reduce exposure to significant financial loss for both mobile users and cellular network providers. In addition to IRSF activities, our method also identifies a wide variety of other types of fraud, ranging from traditional voice scams to emerging fraud cases committed through mobile devices, smartphone apps and online social media sites. This enables us to gain a comprehensive view of voice-related fraud in today's large cellular networks.

Based on our detection results, we conduct extensive analysis of fraud activities in cellular networks. Our analysis unveils two major types of fraud: 1) IRSF fraud which brings direct revenue to the fraudsters through victims placing calls to premium rate international numbers; and 2) scams that rely on social engineering to defraud victims. For both types of fraud, we observe interesting characteristics that are unique to cellular networks. For example, we find malware apps, unlocked devices and online media sites can serve as new channels for carrying IRSF fraud, and smartphone users are more susceptible to many of these fraud activities. Also, personal information such as email contact lists and online transaction details are becoming popular components of social engineering techniques. In addition, we identify the *heteronym property* of fraud numbers, which take advantage of the fact that most mobile devices lack the ability of distinguishing foreign numbers from domestic ones to solicit calls to fraud numbers. Moreover, we find that the vetting process used by online app marketplace and online media sites plays an important role in effectively preventing fraud activities. All these observations provide us with useful insights in designing better and more secure hardware/software platforms to prevent future fraud activities.

We note that analyzing voice call graphs alone is not sufficient to detect fraud activities. Nonetheless, by analyzing voice graphs, we can winnow down and zero in on likely suspicious activities. Additional (expensive) approaches, e.g., which incorporate billing information and manual investigation, can then be applied to analyze the detection results to further confirm the fraud activities. In other words, our method serves as a "first-line" defense in alerting users and cellular providers of emerging fraud activities and isolating likely fraud activities proactively and quickly from a large number of call records. In addition to cellular networks, the proposed method can also be readily applied for combating fraud activities on landlines.

The remainder of this paper is organized as follows. The background of the UMTS network and the datasets studied are introduced in Section 2. In Section 3, we formally define voice call graphs and motivate the heuristics for fraud detection. We then propose a MCL based algorithm in Section 4 to decompose voice call graphs and isolate fraud related subgraphs. Using ground truth from two sources, we evaluate the detection results in Section 5 and conduct systematic analysis of detected fraud activities in Section 6. Section 7 discusses related work and Section 8 concludes the paper.

2. BACKGROUND AND DATASETS

In this section, we provide a quick overview of the cellular network under study, and briefly introduce the composition of telephone numbers. We then discuss the datasets and ground truth used in our fraud analysis.

2.1 UMTS Network Overview

The cellular network under study utilizes primarily UMTS (Universal Mobile Telecommunication System), a popular 3G mobile communication technology supporting both voice and data services. The key components of a typical UMTS network are illustrated in Fig. 1. When making a voice call or accessing a data service, a mobile device directly communicates with a cell tower (or node-B), which forwards the voice/data traffic to a Radio Network Controller (RNC). In the case of mobile voice (also including Short Message Service or SMS), the RNC delivers the voice traffic to the PSTN (Public Switched Telephone Network) or ISDN (Integrated Services Digital Network) telephone network, through a Mobile Switching Center (MSC) server. All voice call records, domestic or international, can be observed at MSCs.

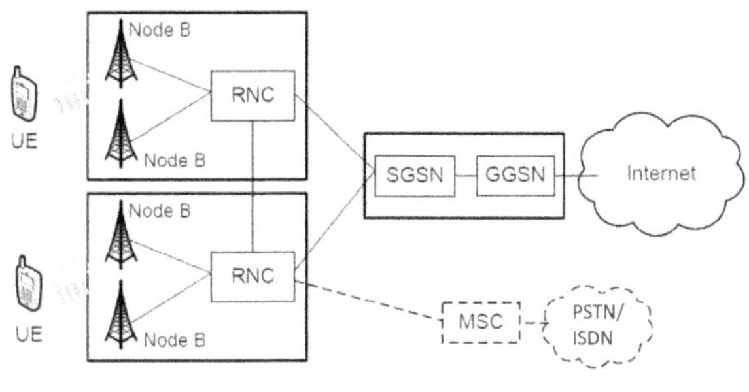

Figure 1: UMTS network architecture.

2.2 Primer on Phone Numbers

A telephone number consists of a sequence of digits for reaching a particular phone line in a public switched phone network. The phone line that initiates the call is associated with an *originating number* and the number of the targeting phone line is called the *terminating number*. In a UMTS network, the phone number is also referred to as a MSISDN (Mobile Subscriber Integrated Services Digital Network Number). A MSISDN comprises three components: a country code followed by an area code (also called a national destination code), then by the subscriber number. A majority of phone numbers within the same geographical area share the same country code and area code. Depending on the specific country, phone numbers vary in length. Under certain circumstances, phone numbers from two different countries can be exactly the same (see Section 6.4). An international dialing prefix (also referred to as an exit code) is attached in front of the country code to distinguish these numbers. The specific exit code is determined by both the originating country and the terminating country and is provided directly by the cellular service provider. Users need to explicitly dial the exit code in order to initiate an international phone call. However, when receiving a phone call from a foreign party, the exit code is already contained in the incoming foreign number. Therefore, when returning such a call, the exit code is often attached automatically by the mobile device without the user's knowledge.

2.3 Datasets

We use datasets consisting of a complete set of international voice calls collected at the MSCs of the UMTS network under study. Theses phone calls are initiated by mobile users in the cellular network (i.e., domestic users) to international terminating numbers. We emphasize here that no customer private information is used in our analysis and we have *anonymized* all customer identities (domestic originating numbers). In particular, the anonymization process keeps the area code intact and only anonymizes the remaining 7 digits in the originating numbers. More importantly, the distance between two originating numbers[1] is also preserved after anonymization. In addition to protecting users' privacy, this type of anonymization enables us to study the relationship among phone numbers that participate in the same fraud activities (see Section 6.2). Similarly, to adhere to the confidentiality under which we have access to the data, in places, we only present normalized views of our results while retaining the scientifically relevant magnitudes.

2.4 Obtaining Ground Truth

To evaluate the efficacy of the proposed fraud detection algorithm and to understand different fraud activities, we utilize two sources of user reports as our ground truth.

IRSF list: This list contains phone numbers associated with *international revenue share fraud* activities. In the scenarios of IRSF, an international revenue share provider designates a set of numbers as *premium rate service (PRS)* numbers, which are often priced much higher than normal calls terminating to the same foreign country[2]. The profit generated from calls to these PRS numbers are shared by the revenue share provider and the content provider. By serving as the content provider and attracting victims to call these PRS numbers, the attackers gain direct revenue from these IRSF calls. Cellular service providers are directly impacted by IRSF fraud because they will suffer from monetary loss when their customers refuse to pay for the cost due to IRSF calls. The IRSF list in our study is created from customer reports to the care center of a large cellular service provider regarding suspicious IRSF activities observed from their monthly bill statements. Manual validation then follows up by placing phone calls to the reported IRSF candidates before adding these numbers to the list. If such numbers are of premium rates, provider-specific strategies will be applied to prevent future customers' calls to these numbers. Some agents may even check the numbers that are adjacent to the reported numbers to see if they are also premium-rate. Even though these numbers may not be involved in any reported fraud activities so far, they can be acquired by fraudsters in future. We refer to the phone numbers identified in this way as *premium rate number ranges* in the rest of this paper.

Online feedback: This list contains phone numbers regarding which we have found customer complaints posted on forums, blogs and other online media sites, such as [10, 11], through popular search engines. This data source covers a wide variety of fraud activities, such as voice calls related to scams and malware, etc. We note that a small fraction of the IRSF numbers identified from the online feedback overlap with the IRSF list. However, there are also many IRSF numbers which are not covered by the IRSF list, possibly be-

[1]We treat two phone numbers on_i, on_j as two integers, and the distance between them defined as $|on_i - on_j|$.

[2]In a few countries, revenue sharing numbers can have similar rates as regular calls.

cause no one has complained about them yet. In addition, to understand different fraud activities and their associated social engineering techniques, we assign labels to the fraud numbers by distilling and summarizing keywords from user comments describing these numbers (see Section 6.3).

We note that since there is no guarantee that users will report all fraud activities they encounter, not to mention that many users lack the knowledge to identify fraud activities, these two data sources only cover a subset of all fraud activities in the network. Moreover, there is often a lag between fraud activities and user feedback. For example, users may only start to notice the IRSF activity when they observe unexpected charges on their monthly bills. As we shall see in Section 5.2, such a lag can last weeks to even months, rendering much less effective the widely used reactive fraud detection method based on user feedback.

3. VOICE CALL GRAPHS

In this section, we advance the notion of *voice call graphs* as a means to represent the communication patterns exhibited in the mobile voice channel. After characterizing voice graphs constructed from different time spans, we propose our key heuristic in identifying fraud activities.

3.1 Definition of Voice Call Graphs

In this paper, we only study a single direction of phone calls, i.e., the outbound phone calls placed by domestic phone numbers to international numbers. This choice not only helps reduce the volume of the data, but also enables us to observe both fraud calls initiated by domestic numbers and returned calls solicited by incoming fraud calls (e.g., random scanning, see Section 6).

In our dataset, each voice record contains the domestic originating number and the targeted international terminating number (note that we drop the terms "domestic" and "international" in the rest of the paper for simplicity). By depicting each record as an edge, all voice records can be readily captured by a bi-partite graph, which we refer to as a *voice call graph*, or a *voice graph* in short. Formally, we define a voice call graph $\mathcal{G} := \{\{\mathcal{ON}, \mathcal{TN}\}, \mathcal{E}\}$, where $\mathcal{ON}$ and $\mathcal{TN}$ stand for the set of originating numbers and the set of terminating numbers appearing within the observation time window T, respectively. An edge e_{ij} is drawn between $on_i \in \mathcal{ON}$ and $tn_j \in \mathcal{TN}$ if at least one voice call is made from on_i to tn_j within T. Note that since we only look at phone calls on one direction (i.e., from domestic numbers to international numbers), we treat edges as undirected[3].

3.2 Voice Call Graph Properties

Fig. 2 shows a call graph plotted using the Graphviz tool [12], which represents voice calls from 1,000 randomly sampled originating numbers in one single day, where the blue/red nodes represent originating/terminating numbers, respectively. At a glance, the voice graph in Fig. 2 is extremely sparse and contains a large number of disconnected components (subgraphs), with a majority of the subgraphs containing only one single edge. For those subgraphs with more than one edge, most of them exhibit a star structure centered on originating numbers, representing one originating number placing phone calls to a few terminating numbers. In comparison, most terminating numbers only have degree 1. As we extend the observation time period and the originating number population, the voice graph grows significantly and renders direct visualization inapplicable. Instead, we characterize larger voice graphs using popular graph statistics. Fig. 3[a] shows the log-log plot of the node degree distribution in a one-day voice graph. We observe that the degrees of both the originating numbers and terminating numbers display a power-law shape. The power-law shape of the originating numbers implies that a majority of domestic customers rarely call foreign numbers from their cell phones. In addition, the power-law shape of the terminating numbers indicates that, except for a few very popular terminating numbers (on the low end of the curve) associated with hotlines of popular hotels and resorts or foreign agencies like embassies, etc., most terminating numbers receive calls from a very small number of originating numbers.

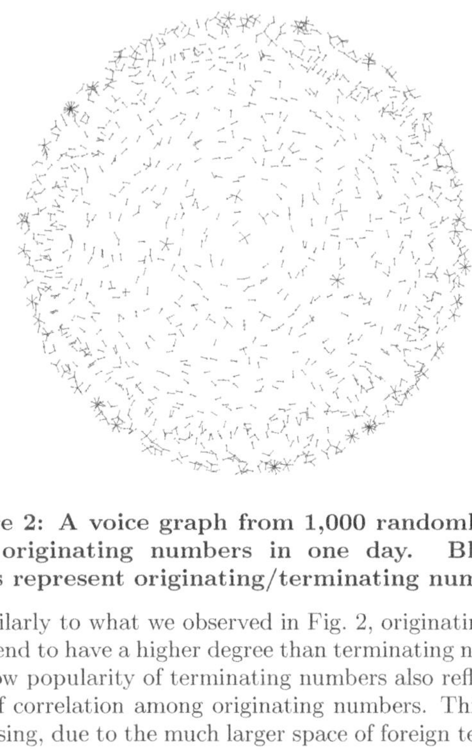

Figure 2: A voice graph from 1,000 randomly sampled originating numbers in one day. Blue/red nodes represent originating/terminating numbers.

Similarly to what we observed in Fig. 2, originating numbers tend to have a higher degree than terminating numbers. The low popularity of terminating numbers also reflects the lack of correlation among originating numbers. This is not surprising, due to the much larger space of foreign terminating numbers, in general, two mobile customers are unlikely to call the same international number(s). Therefore, voice graphs often consist of a large number of small disconnected subgraphs. Fig. 3[b] shows the increase in the number of subgraphs over time, which ranges from 0.3 million in a one-day graph to more than 3 million in a one-month graph. The growth is sublinear, plausibly due to the expansion of a giant connected component, which we will discuss in section 3.4. In addition, Fig. 3[c] demonstrates the distribution of the sizes of subgraphs from voice graphs spanning different time periods (in terms of the number of edges within each subgraph). The subgraph sizes display again a power-law shape, indicating the dominance of small subgraphs in voice graphs. The same observation holds for the voice graphs from one day, one week, and one month data, though we do observe that the number of subgraphs grows during a longer observation period. The extremely low connectivity in a voice graph

[3]The definition of voice call graphs can be easily extended to weighted graphs or directed graphs. For instance, the weight on an edge represents the number of calls associated with each edge.

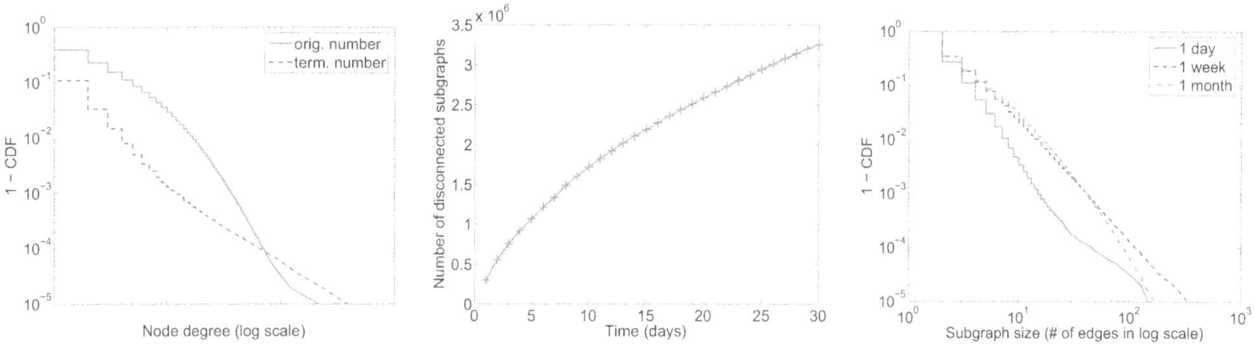

(a) Node degrees in a one-day graph (b) Number of subgraphs over time (c) Subgraph size distribution

Figure 3: Voice call graph properties

distinguishes it from other types of widely studied graphs, such as network traffic activity graphs [13] and online social network graphs [14], which often exhibit much stronger connectivity and correlations among nodes.

We note that, since the objective of this paper is to identify and analyze fraud activities in cellular networks, we want to identify as many fraud activities as possible. As we shall see in Section 6.4, fraud activities can take days to become noticeable. To cover such cases, we need to extend our observation period. Therefore, in the rest of this paper, *by default we choose one month as the observation window for constructing voice graphs.*

3.3 Heuristic for Detecting Voice-related Fraud

Based on our analysis of voice graphs, we propose our heuristic for detecting fraud activities from these graphs. We utilize two properties of voice graphs. First, in comparison to most legitimate terminating numbers that have low degrees, fraudsters intend to attract more victims to call fraud numbers and hence fraud numbers often appear to be much more popular (also referred to as heavy hitters). We note that detecting heavy hitters is a common approach used for identifying anomalous activities [15, 16]. However, the popularity of terminating numbers alone is not enough in this scenario. We find that the most popular terminating numbers are associated with hotel hotlines, traveling agencies, embassies, and so on. In comparison, as we shall see in Section 5, the most popular fraud activities only attract hundreds of victims during a month-long period, which are not among the top high degree terminating numbers. We need additional features to help identify fraud numbers.

The second property we utilize is the low connectivity of voice graphs. Based on our experience, fraudsters often employ several foreign numbers to increase the chance of reaching victims and to avoid detection and regulation. As we shall see in Section 6, these numbers can even come from different countries. Therefore, we are alerted with a potential fraud activity while observing many domestic users start placing phone calls to the same set of foreign numbers. The communication patterns exhibiting the above two properties are often referred to as *community structures* in voice call graphs, where a set of originating numbers place phone calls to a set of terminating numbers. This serves as our key heuristic for detecting voice fraud in cellular networks.

Based on this heuristic, we formulate the fraud detection problem as finding the community structures from voice graphs[4]. We note that a community structure can also originate to legitimate activities, for example, due to tourists calling hotlines in a popular resort or companies communicating with foreign branches. However, as we shall see in Section 5, this heuristic can help successfully isolate a large number of popular fraud activities from millions of phone calls. In addition, we shall discuss in Section 6 that certain types of fraud activities can be detected in a more accurate way by investigating properties of the community structures.

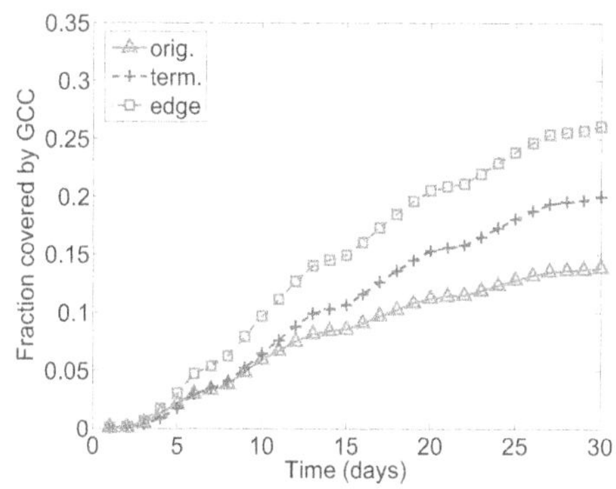

Figure 4: Evolution of GCC's in voice call graphs spanning different time intervals.

3.4 Challenges

Identifying all community structures is still a challenging task. This is mainly due to the appearance of random edges or weak connections which connect different communities, thereby forming large subgraphs mixed with different fraud activities. For example, Fig. 4 shows the evolution of the largest subgraph (a.k.a. giant connected component or GCC) in the voice call graph over a month-long time period. In Fig. 4, three curves display the coverage of GCC at a particular time in terms of originating numbers, terminating numbers and edges, respectively. We observe the GCC becomes significant from 10 days onwards and covers up to 25% edges in a 30-day voice graph. Since the voice graphs used in this paper are constructed using month-long

[4]We have also explored other call features such as call duration and call time. However, none of them exhibit significant difference between fraud numbers and legitimate numbers. In future work, we will investigate other features like user calling history to improve detection accuracy.

data, with such a long observation period, the GCC and other large subgraphs can grow to thousands or millions of edges. To separate different fraud activities within GCC and other large subgraphs, it is necessary to decompose them first. Therefore, we next propose a Markov clustering based method for decomposing large subgraphs and then identifying community structures from the decomposition results.

4. A MARKOV CLUSTERING BASED FRAUD DETECTION ALGORITHM

In this section, we propose a Markov Clustering (MCL) [17] based algorithm for decomposing voice graphs and identifying potential fraud activities.

4.1 Decomposing Voice Graphs using MCL

Alg. 1 shows the proposed algorithm, where we iteratively apply the MCL algorithm to large subgraphs which contain more than N edges ($N = 2,000$). For subgraphs with fewer than 2K edges, we can extract community structures with little cost (see Section 4.2).

Algorithm 1 Decomposing voice call graphs with MCL.

1: Input: $\mathcal{G}$, $N = 2,000$, $\beta = 2$;
2: Extract disconnected subgraphs $\mathbf{G} := \{\mathcal{G}_i\}$ from $\mathcal{G}$, where $\mathcal{ON} = \cup_i \mathcal{ON}_i$, $\mathcal{TN} = \cup_i \mathcal{TN}_i$ and $\mathcal{E} = \cup_i \mathcal{E}_i$;
3: **for** each $\mathcal{G}_i \in \mathbf{G}$ **do**
4:   **if** $\mathcal{E}_i > N$ **then**
5:     Construct symmetric adjacency matrix A from $\mathcal{G}_i$;
6:     **repeat**
7:       Normalize rows in A;
8:       $A := A^2$; //expansion
9:       $a_{ij} := a_{ij}^{\beta}$, for all entries in A;//inflation
10:     **until** A converges
11:     Extract disconnected subgraphs $\mathbf{G}_A$ from A;
12:     $\mathbf{G} = \mathbf{G} \cup \mathbf{G}_A - \{\mathcal{G}_i\}$;
13:   **end if**
14: **end for**

Introduction to MCL algorithm. The MCL algorithm is developed for graph partitioning, which is based on the assumption that random walks tend to stay within the same cluster for a longer time rather than traversing across clusters. MCL iterates two processes: expansion and inflation (line 8 and 9 in Algorithm 1). Expansion takes the power of the Markovian matrix using regular matrix product. For instance, taking the square of the matrix will compute random walks of length two. Since higher length paths are more common within clusters than between different clusters, expansion will increase the probabilities of intra-cluster walks. Inflation is the element-wise power to β followed by a diagonal scaling (to make the resulting matrix Markovian). Inflation changes the probabilities associated with the collection of random walks departing from one particular edge by favoring more probable walks. MCL terminates when the two processes converge. Cluster memberships can be identified by extracting connected components from the MCL result. We select MCL to decompose voice graphs for two reasons. First, in MCL, we do not need to specify the expected number of clusters. Second, MCL can scale up to large graphs consisting of millions of edges.

The standard MCL algorithm only takes regular (non bi-partite) undirected graphs as input. However, voice call graphs are bi-partite undirected graphs. Therefore, for each subgraph up to decomposition, before feeding it to MCL, we need to create its corresponding non bi-partite version. For example, let A_{asym} be the adjacency matrix corresponding to a voice graph $\mathcal{G}$, we construct a symmetric adjacency matrix A from A_{asym} as follows:

$$A = \begin{pmatrix} 0 & A_{asym} \\ A_{asym}^T & 0 \end{pmatrix}$$

The MCL algorithm then operates on A and finally decomposes G into a series of subgraphs after iterating the expansion and inflation steps. We have tested different selections of β and $\beta = 2$ yields the most stable and interpretable results, which is the default parameter setting that we use throughout this paper. By the end of the algorithm, all voice graphs larger than N will be decomposed and the remaining subgraphs are of less than N edges. We next isolate fraud activities from these subgraphs.

4.2 Isolating Fraud Activities

Recall that we have formulated the task of detecting fraud activities as the problem of identifying dense community structures in call (sub)graphs. In a bipartite graph, we refer to the atomic community structure as a *clique*, which represents the smallest complete (2-by-2 bi-partite) subgraph in the original graph. Fig. 5[a] shows the structure of a clique, which contains four edges, connecting two originating numbers and two terminating numbers. The total number of cliques varies across different voice graphs. For example, Fig. 5[b-d] shows three voice graphs corresponding to three different types of fraud activities identified using the proposed method (Please refer to Section 6 for detailed descriptions and analysis of these fraud activities). Fig. 5[d] exhibits a much denser community structure, and, as a consequence, is characterized by a large number of cliques[5]. In other words, the number of cliques serves as an indicator of the significance of the community structure in a subgraph. Therefore, in this paper, we only consider *subgraphs with at least one clique* as the candidates for fraud activities. After this filtering step, we isolate approximately 30K candidates from millions of call subgraphs every month. For the rest of the subgraphs which contain no clique, the originating numbers show little correlation and hence our heuristic does not apply. In Section 5.1, we will discuss those fraud activities that fall into subgraphs without any clique.

After filtering out subgraphs without cliques, we further reduce the number of candidates for investigation based on the graph properties. Based on the heuristic presented in Section 3.3, we differentiate subgraphs representing fraud activities from legitimate ones through two features: the popularity of terminating numbers (in terms of the number of originating numbers forming the community structure) and the connectivity of the community structure (in terms

[5]In fact, the community structure in Fig. 5[b] is formed by the return calls in response to three fraud numbers sequentially pinging the same set of victims. The community structure is not significant since victims may not necessarily respond to the same fraud numbers. In Fig. 5[c], the fraudster advertised fraud numbers through emails and the loose community structure is due to victims calling multiple fraud numbers. In comparison, Fig. 5[d] shows the fraud activity caused by a malware. A victim is programmed to call all the fraud numbers coded in the malware, and thereby showing a very strong community structure.

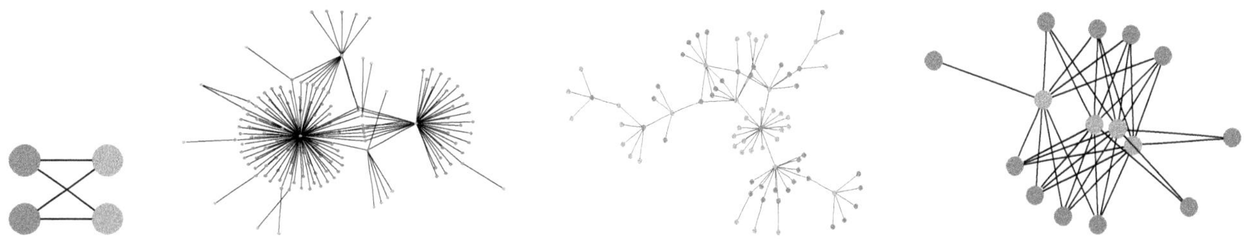

(a) A clique. (b) Random scanning (126 orig. nbrs, 18 term. nbrs, 74 cliques) (c) Email scam (40 orig. nbrs, 38 term. nbrs, 16 cliques) (d) Win mobile malware (11 orig. nbrs, 4 term. nbrs, 177 cliques)

Figure 5: Voice graphs corresponding to a single clique and three different fraud activities.

of the number of cliques in the subgraph). By matching the first month data against two sources of ground truth, Fig. 6 displays how the detection rate changes while we tune these two parameters. In Fig. 6, the x-axis represents the percentile of subgraphs extracted after we change each of the parameters, and the y-axis stands for the proportion of fraud activities that are *not* covered by the x percentile of subgraphs. We observe in Fig. 6 that 1) the top 5% subgraphs with no less than 5 cliques account for more than 90% of fraud activities; 2) the top 20% subgraphs with no less than 10 originating numbers also cover around 90% of fraud activities. These observations also validate our heuristic that fraud numbers are often associated with a higher popularity and fraud activities tend to exhibit dense community structures in their corresponding subgraphs. We note that in practice, we prefer to choose higher values for these two parameters. Though this limits the number of fraud activities identified, it can help reduce the workload of manual investigation by eliminating many false alarms. However, since one of our main objectives in this paper is to understand the diversity of voice fraud activities in cellular networks, we want to include as many fraud activities as possible. Therefore, in the rest of this paper we choose 5 cliques and 10 originating numbers as the thresholds, i.e., only subgraphs with at least 5 cliques and 10 originating numbers will be considered as containing potential fraud activities. Investigation based on the ground truth shows that such thresholds are able to help filter out 98% of these subgraphs, while still capturing more than 90% of the fraud activities.

We note that a limitation of the proposed algorithm is that it relies on strong community structures in voice call graphs to detect fraud activities. Therefore, fraudsters can potentially evade detection by eliminating the correlation among fraud numbers, e.g., by employing only one fraud number or by defrauding a different subset of victims with each individual fraud number. The proposed algorithm also fails to detect stealthy fraud activities that attract less than 10 victims in a month. However, as we shall see in Section 5.1, both strategies are not dominant and are only adopted in less than 10% of the observed fraud cases. Moreover, these strategies limit the impact of fraud activities and hence they attract far fewer victims than other fraud activities that employ multiple correlated phone numbers.

In the remainder of this paper, for each remaining subgraph, we call the terminating numbers within cliques *potential fraud numbers*. The originating numbers that phone these fraud numbers are referred to as *victims* and the associated calls are called *fraud calls*. Fraud numbers, victims, and fraud calls collectively form a *fraud activity*. The remaining nodes and edges in the subgraph are omitted in our studies. Next, we will evaluate our fraud detection results.

5. EVALUATION

In this section, we compare the detection results against the ground truth. The evaluation shows that proposed algorithm is able to detect most popular fraud activities and our detection results are usually months ahead of user reports.

5.1 Evaluation on IRSF Fraud Detection

We run the proposed fraud detection algorithm on five months of data (Jan 2011 to May 2011) Since there is usually an anticipated lag between the occurrence of fraud activities and user reports to the carrier, we use the IRSF list that contains IRSF numbers inserted between Jan 2011 and Oct 2011 by the service provider as the ground truth[6]. Moreover, in Section 6, we shall study the evolution of fraud activities using two years of data to track their entire life cycle.

Our first evaluation focuses on detecting IRSF numbers. After decomposing the monthly voice graphs from Jan 2011 to May 2011 and applying the filtering process, we isolate around 24K fraud number candidates in total. Any of the 24K candidates that are covered by the IRSF list is considered as a *true detection* (or a true positive). On the contrary, we consider a *missed detection* when a phone number from the IRSF list appears in the data but not among the 24K candidates. In addition, by searching for feedback on the Internet, we identify a separate set of IRSF numbers outside the IRSF list. Reports of these numbers are often associated with complaints regarding high charges. We refer to this new IRSF number set as *new detections*. We note that the above counting method only provides a lower bound on the detection performance, since some IRSF numbers may be reported after Oct 2011 given the potential lag of user reports (see Section 5.2).

To assess the severity and impact of fraud activities, we measure the number of victims and fraud calls attracted by each fraud number. We note that the lifetime of a fraud number determines the quantity of victims and fraud calls associated with that number, which can be affected significantly by user reports. For example, after being reported to the provider, the provider-specific strategies often restrict other users' access to the fraud number. The online user feedback can also prevent other users who read the post from making fraud calls. Therefore, to ensure a fair com-

[6]We also have access to the IRSF numbers inserted before Jan 2011. However, most of such numbers show no activity in 2011. This is likely due to the specific strategies exercised by the carrier to prevent users from calling these numbers.

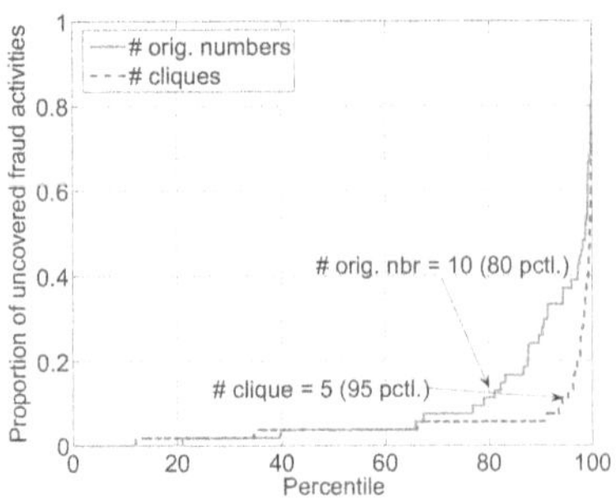

Figure 6: Determining thresholds for suspicious subgraphs

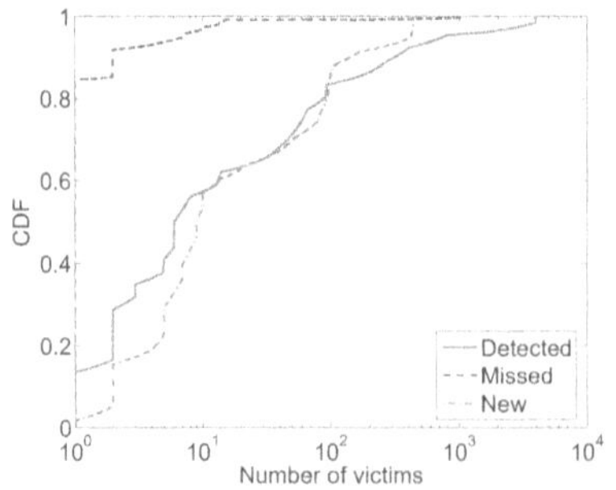

Figure 7: Popularity of fraud numbers

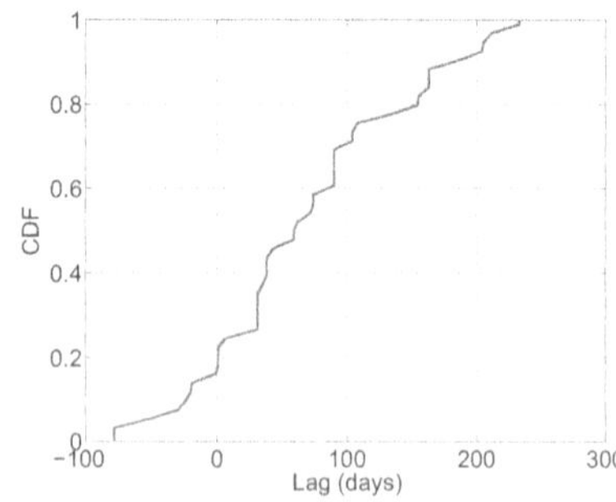

Figure 8: Lag of user reported IRSF numbers

parison and to capture the real impact of fraud activities, we count the number of victims and fraud calls of a fraud number only *within a 4-week time window prior to its first report time.* For *true detections* and *missed detections*, we consider the first report time as the time when the fraud numbers were inserted into the IRSF list. For *new detections*, we treat the time of the first online post regarding a fraud number as its first report time.

Table. 1 displays the detection results, in terms of the fraud numbers, victims and fraud calls associated with three IRSF number sets. We observe that though the detected and new fraud numbers only account for 11% of all the fraud numbers, together they attract 85% of all the victims and are the root cause of 78% of fraud calls. Fig. 7 shows the distribution of the number of victims associated with each IRSF number (note that the x-axis is in log scale). IRSF numbers belonging to the true detection and new detection groups are highly popular and attract a large number of victims, with mean values of 6 and 8, respectively. In comparison, most fraud numbers in the missed detection category are only associated with a single victim. This is plausibly due to two reasons. First, the high popularity of the detected IRSF numbers is actually due to the emergence of new IRSF activities, which utilize malware, unlocked devices, and online social media to draw attention from more victims and hence attract more fraud calls (we shall present case studies of these new IRSF activities in Section 6). Second, we find that a majority of the IRSF numbers (more than 90%) that our detection algorithm missed actually belong to the premium rate number ranges. These premium rate numbers were detected proactively by fraud agents exploring the adjacent numbers of confirmed IRSF numbers. However, since no user report is found complaining about these numbers, they may not yet be utilized for fraud activities or the fraudsters have not yet advertised these numbers. In other words, these numbers are *dormant IRSF numbers* in comparison to other *active IRSF numbers* reported by users. By excluding these dormant IRSF numbers, our detection rate on active IRSF numbers can exceed 50%.

5.2 Evaluation on Early Fraud Detection

We have demonstrated that our algorithm can successfully isolate many highly popular IRSF numbers. Taking prompt actions against these IRSF numbers can help both customers and cellular service providers avoid significant financial loss. In fact, one key advantage of our algorithm is the short response time to emerging fraud activities. For the IRSF numbers in the true detection category, we measure the gap between the detection time[7] and the time that they first appear in the IRSF list, which we refer to as the *lag of user reports.* A positive lag value means the detection time goes before the user report time. We demonstrate the cumulative distribution function (CDF) plot of the user report lag in Fig. 8. For more than 80% of the IRSF numbers, our detection result precedes user reports and in more than 60% of the cases, we can detect the IRSF numbers at least one month earlier than the user reports! Similar observations are made for the other types of fraud. Early detection of these IRSF numbers provides the cellular service provider with enough time to track and to take actions against these fraud activities to prevent other customers from calling these numbers. On top of our detection results, we are developing tools for alerting customers of potential fraud risk through mobile apps and SMS messages.

Table 1: IRSF number detection result

	Fraud nbrs.	Victims	Fraud calls
Detected	6.7%	66.9%	35.7%
Missed	89.0%	14.6%	22.5%
New	4.3%	18.5%	41.8%

5.3 Discussion

False positives. We argue that analyzing voice call graphs alone is not sufficient for identifying fraud activities with high accuracy. By matching the 24K fraud number candidates with the two sources of ground truth, we find that the proposed detection algorithm yields a precision of 9.3%, i.e., around 9.3% out of 24K phone numbers are associated with IRSF or other fraud activities. However, we believe that 9.3% only serves as a precision lower bound. Given the observation of significant delay of user reports in Section 5.2, we expect to see more numbers in our detection results to be reported as fraud numbers by mobile users in future.

Practical usefulness. Despite the relatively large number of false positives, the proposed algorithm still provides a valuable "first-line" defense in alerting users and cellular providers of emerging fraud activities. After isolating from millions of voice calls a small number of fraud candidates using our method, cellular service providers can allocate resources to track these suspicious numbers and combine with

[7]For IRSF numbers detected using data from one particular month, their detection time is defined as the first day in the following month, e.g., if an IRSF number is detected using the April data, its detection time is considered as May 1st.

other expensive data sources, like billing information and customer call history, to further confirm the fraud activities. For example, we can investigate the billing records associated with potential IRSF calls to look for exorbitant charges. IRSF numbers can also be confirmed by directly calling these numbers[8]. We believe many of these methods can be automated to reduce the manual cost, e.g., through automated query scripts or using auto-dialing tools, to make post-investigation of our automatic detection results more manageable. More importantly, our analysis using real voice call records shows that, by proactively identifying and restricting potential IRSF numbers based on our detection results, cellular service providers can benefit from a reduction of tens of thousands of fraud calls per month, and subsequently a huge saving in customer care cost and an implicit increase in customer satisfaction, which far outweighs the cost of investigating the detection results.

Other fraud. We note that, unlike IRSF activities, we are unable to assess the detection rate of other types of fraud, since we are not able to query all the numbers to find corresponding user feedback (if any) through online search engines. However, as we shall see in Section 6, our method has successfully identified many other types of voice fraud, including emerging fraud cases committed through mobile devices, smartphone apps and online social media sites. This enables us to gain a comprehensive view of voice-related fraud in today's large cellular networks. Understanding these fraud activities, e.g., how are they committed (Section 6.2) and what are their popular social engineering techniques (Section 6.3), can provide us with unique insights into how to design securer software/hardware architectures to prevent future fraud activities.

6. ANALYSIS OF FRAUD ACTIVITIES

In this section, we present detailed analysis of the fraud numbers detected using the proposed method.

6.1 Stability of Fraud Activities

Since we use whole month datasets for fraud detection, certain long lasting fraud activities that span several months can appear multiple times. In order to obtain an accurate count of different fraud activities, we first need to remove these duplicated cases. Our approach is as follows.

For any two detected fraud activities i and j, we calculate their similarity using the Jaccard similarity coefficient, defined as $J(i,j) := |\mathcal{FN}_i \cap \mathcal{FN}_j|/|\mathcal{FN}_i \cup \mathcal{FN}_j|$, where $\mathcal{FN}_i$ represents the set of fraud numbers associated with fraud activity i. $J(i,j)$ ranges from 0 to 1, with $J(i,j) = 1$ indicating that two fraud activities are caused by the same fraud numbers. For a fraud activity i, we find its most similar counterpart k from other fraud activities, where $k = argmax_j J(i,j)$. We display the Jaccard similarity between each fraud activity and its most similar counterpart in Fig. 10.

We observe that Fig. 10 displays a bi-modal shape, where a majority of the fraud activities do not have a counterpart with a similar set of fraud numbers. Based on the plot, we select 0.6 as the threshold and consider a fraud activity i is a duplicate if there exists another activity j such that $J(i,j) > 0.6$.

6.2 Categorization of Fraud Activities

After removing duplicated instances, we match the fraud number candidates from our five months of detection results against the online user feedback to categorize different types of fraud. In particular, we run queries for all fraud number candidates through popular online search engines and extract user comments regarding these numbers based on the search results. Most comments are from online forums, such as *whocallsme.com* [10] and *800notes.com* [11], etc. We manually investigate all the comments and identify different types of fraud activities based on the comments. We note that when we compare user comments regarding different phone numbers within the same fraud activity, we find consistency in terms of the user feedback describing these fraud numbers. This confirms the validity of our detection result, where phone numbers exhibiting strong correlation (e.g., appearing in the same subgraph) are employed for conducting the same fraud activity. In addition, this observation enables us to categorize voice fraud at the level of fraud activities instead of individual phone numbers.

We display the fraud categorization in Table 2. Based on whether the fraud activities can bring immediate monetary gain to the fraudsters, we classify them into two major categories: *IRSF* and *Scam* (column 1). Inside each major category, we further group fraud activities into subcategories according to the channels through which different fraud activities are committed (column 2). In the third column of Table 2, we present the details of each type of fraud activity. The dominance of each type of fraud in terms of their proportion out of all fraud activities is shown in the fourth column. In the fifth column, we display the *relative significance* (RS) of smartphone devices participating in each type of fraud, which we shall define formally at the end of this section. In the following, we discuss each fraud category in detail and present case studies.

IRSF. IRSF fraudsters devise various approaches to solicit users to make as many phone calls as possible to the fraud numbers to maximize their profit. Though IRSF activities are also common for landlines, we have discovered a number of IRSF activities unique to cellular networks, including:

1) Malware. The most dominant IRSF case, accounting for 13.2% of all fraud activities, is caused by malicious mobile apps. These mobile apps are designed to be able to initiate users' inadvertent phone calls to IRSF numbers. These IRSF numbers employed by the malware can be either static or dynamic. In the former case, we find a malware disguised as a gaming app on windows mobile devices (referred to as windows mobile malware), which has four IRSF numbers hard coded. At night, the app launches many calls to these IRSF numbers automatically. The corresponding voice graph corresponding is depicted in Fig. 5[d], which shows a large number of cliques since the app automatically calls all these 4 IRSF numbers.

In the latter case, we identify a series of apps (possibly with the same authorship) which are claimed to be able to spoof the caller's identity or changing the caller's voice (referred to as spoof call malware). However, when a user places a call (e.g., to a domestic number), the call will first pass through an IRSF number (i.e., as a proxy) before reach-

[8]Once accessing an IRSF number, the callers are often transferred to an interactive voice response, where they are kept on hold with background music, while their airtime is being used.

Table 2: Categorization of fraud activities

Revenue	Channel	Description	Pct.(%)	RS
IRSF (dir.)	Malware	Malicious mobile apps with automated dialer to IRSF numbers	13.2	ALL
	Random Scanning	Pinging a range of phone numbers sequentially to solicit return calls	3.3	1.37
	Online Media	Publishing IRSF numbers on blogs, forums and social networks	2.5	1.20
	Unlocked Device	Unlocked devices are pre-configured with IRSF numbers as SMS or MMS access points	2.5	0.53
	Device Exploit	Only certain brand and type of phones are involved	0.8	0.13
Scam (indir.)	Target Scam	Calls users who fraudsters already have information about and attempt to collect more information, to sell products, or to defraud them of money, etc.	55.4	1.04
	Online Media	Post phone numbers online and cheat for money from people who call	10.7	1.44
	Email	Hack somebody's email account, send phishing emails to all the contacts	8.3	1.46
	Malware	Lock users' PCs, force them to call certain numbers to remove the malware	3.3	1.01

ing the destination. More interestingly, we find that these apps are able to communicate with a remote server to download new lists of IRSF numbers periodically. Due to the employment of dynamic IRSF numbers, it is much harder to detect and disable these apps.

In Fig. 9[a-b], we compare the IRSF numbers associated with these two types of malware in terms of the number of victims attracted over time[9]. The x-axis is aligned to the first time when we observe the fraud number, and the y-axis represents the cumulative number of victims over time. In addition, for each terminating number, we display its length (i.e., number of digits) inside the parentheses.

Fig. 9[a] demonstrates an IRSF number employed by the Windows mobile malware. We observe that around 140 victims start calling the number within 2 weeks after its debut. However, the number of new victims quickly vanishes after 6 weeks when the malware is taken down from the online marketplace. In comparison, Fig. 9[b] shows a similar plot for the spoof call malware. After the malware appears, we observe a linear increase of victims to 50K in around 20 weeks. This is mainly because the malware targets unlocked devices and is hosted by an unofficial website, which lacks a formal process for identifying and removing malicious apps. Meanwhile, when `Term_A` became unavailable somehow at the 20th week, the fraudster acquired a new fraud number `Term_B` and remotely configured the malware to start calling this new number. Therefore, we can see a continuous increase of victims associated with the new number after 30 weeks, when `Term_A` stopped acquiring new victims.

2) Random Scanning. In this scenario, fraudsters employ automated calling system to ping a series of adjacent phone numbers and leave no message to attract users to call back. This type of fraud activity can be distinguished by measuring the distance of the originating numbers (recall our datasets contain only the outgoing calls).

In order to measure the relationship of originating numbers, we define the *closeness* as $median_i(\min_j |on_i - on_j|)$, where $on_i, on_j \in \mathcal{ON}$. Essentially, closeness measures the median distance of all originating numbers to their closest counterparts. Fig. 11 shows the distribution of closeness of all subgraphs after filtering (note that the x-axis is in log scale). We are able to identify successfully all such random scanning as the instances with a closeness less than 100 (to the left of the x-axis).

3) Online Media IRSF. We find several cases where fraudsters post IRSF numbers on popular online media, e.g., Facebook and Twitter, etc. For example, Fig. 9[c] shows the increase of victims corresponding to an IRSF case advertised through Twitter. Around the 12th week, the fraudster posted a tweet disclosing the contact numbers of a famous Puerto Rican singer, which had been retweeted thousands of times within a day. The advertising strategy is so successful that more than eight thousand victims were drawn to make phone calls to these IRSF numbers on the same day. Fortunately, these fraud calls soon disappeared on the following day when many tweets were posted and retweeted shortly to alert other people of this fraud activity.

As another example, we identify a website that advertises IRSF numbers for a public vote of popular scenic locations. Fig. 9[d] demonstrates the behaviors of the three IRSF numbers associated with this fraud case. During their lifetime, `Term_A` and `Term_B` steadily attracted new victims and number of victims was almost doubled close to the end of the vote. In comparison, the `Term_C` attracted many fewer victims, plausibly because it has 12 digits instead of 10 digits, which makes it much easier to be identified as an international number (see Section 6.4 for the heteronym property of fraud numbers).

4) Unlocked Device. Victims of this fraud activity are users of unlocked mobile devices. Unlocked mobile devices are cell phones that are not tied to a specific carrier. The advantage of purchasing such devices is that it offers the user more choices of voice/data plans and ease of switch between carriers, especially when a user is traveling abroad. There are many ways of acquiring an unlocked device. For example, after the contract with the carrier ends, a user can ask the service provider for a code to unlock her device. It is also possible to purchase unlocked devices through third party retailers. We have identified fraud activities on unlocked devices purchased from online third party retailers, where the fraudsters reconfigured the device and specified IRSF numbers as the access points for Short Message Service (SMS) or Multimedia Messaging Service (MMS). Consequently, sending SMS or MMS messages will trigger calls to IRSF numbers.

5) Device Exploit. This fraud activity only applies for one particular prepaid mobile device. Users complain that they will lose all the remaining credits in their accounts while making phone calls to these fraud numbers. We suspect this is caused by attackers taking advantage of certain vulnerabilities of the device or the operating system. Strictly speaking, this is not a regular IRSF case. However, since

[9]Because the curves corresponding to the fraud numbers of the same fraud activities in Fig. 9[a] and [c] look almost the same, we hence only show one example in each plot.

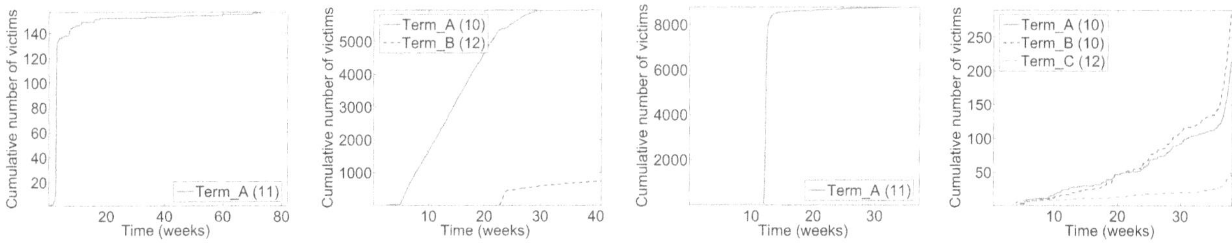

(a) Win mobile malware (b) Spoof call malware (c) Scam on Twitter (d) Voting scam

Figure 9: Cumulative number of victims attracted by different fraud activities over time

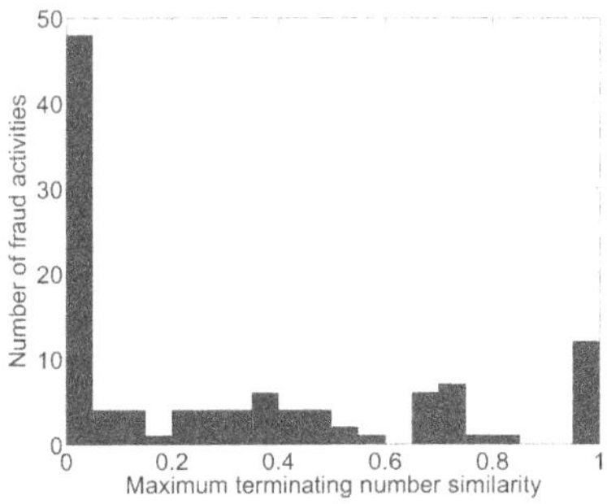

Figure 10: Stability of fraud activities measured by term. numbers.

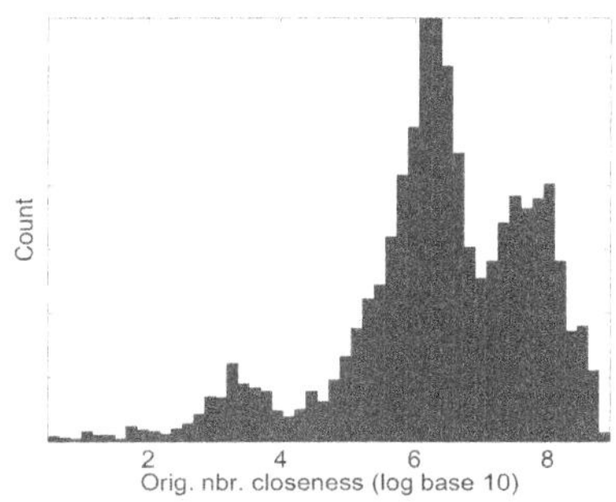

Figure 11: Closeness of originating numbers.

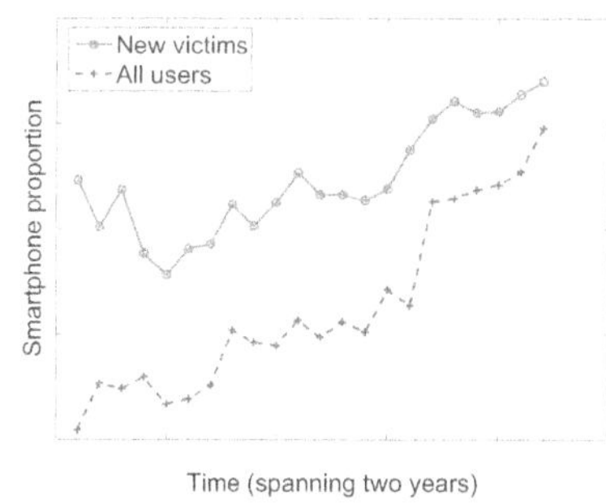

Figure 12: Change of smartphone proportion over time.

fraudsters are likely to gain profit from such activities, we include it here for completeness.

Scam. Scam, sometimes referred to as *phishing fraud*, often involves two stages. At the first stage, fraudsters solicit phone calls from mobile users. At the second stage, when victims call the fraud numbers, fraudsters apply various social engineering techniques to defraud victims of money or acquire private information from them. Scam includes:

1) Target Scam. This is the most common case which accounts for 55.4% of all fraud activities. In this scenario, fraudsters target victims for whom they have certain private information, e.g., their names, phone numbers, addresses, and medical prescription details.

2) Online Media Scam. In this scenario, fraudsters post advertisements on online forums, e.g., craigslist. The advertisements are often about inexpensive house rentals and real estate selling. Several phone numbers are often listed at the end of the posts to entrap victims baited by a low price.

3) Email. In this case, fraudsters hack into personal email accounts and send emails to the contact list disguised as the original account owners. In the email, fraudsters claim an emergency in the foreign country and request phone calls from people in the contact list.

4) Malware. The most interesting means of soliciting phone calls is using malware. We find a particular type of malware called *ransom malware* targeting the Windows operating system. Upon infected by such malware, users' personal computers are locked. The users are unable to execute any command and cannot even enter the safe mode to remove the malware. A dialog box often appears in the center of the screen, informing users to call a list of foreign numbers to obtain a code for unlocking their machines. Often the fraudsters will force the users to pay a ransom for the system restoration. We note that even though the ransom malware we have identified focused only on personal computers, it would not be surprising if such malware appear on mobile devices in the near future. Precaution against such malware (e.g., providing users with a scrutinized set of APIs) is highly suggested for mobile operating systems.

Discussion on Smartphone Penetration. Here we investigate different types of devices involved in the fraud activities above. We identify the model of an end-user device from the first 8-digit Type Allocation Code (TAC) in the associated International Mobile Equipment Identity (IMEI). The remaining 6-digit serial number has been anonymized to protect customers' privacy. Based on the functionality of devices, we further classify them into *smartphones* and *regular phones*. In comparison to regular phones, smartphones are characterized with high computational power, and support services like video streaming, web browsing, GPS and a variety of mobile apps.

For ease of exposition, we define $\alpha(g)$ as the proportion of victims involved in fraud activity g who are smartphone users. Similarly, we denote α_0 as fraction of smartphone users out of all mobile users in the network. Therefore, we define the *relative significance* of smartphones associated with fraud activity g as $RS(g) := \alpha(g)/\alpha_0$, which is displayed in the last column of Table 2. A higher value of RS (greater than 1) indicates the predominance of smartphones involved in that fraud activity.

Since *IRSF Malware* is propagated in the form of bogus mobile apps, all victims associated with this type of fraud are smartphones. Surprisingly, *Online Media IRSF*, *Online Media Scam* and *Email Scam* (line 3, 7 and 8) attract many more smartphone victims. We conjecture that smartphone users, who are capable of maintaining persistent access to online media sites and email services, are potentially exposed to more fraud activities. Moreover, many smart-

Table 3: Social engineering techniques applied by fraudsters.

Techniques	Description	Pct.(%)
Debt collection	claim the victim has an unpaid bill and threaten for legal actions if full payment is not received.	22.5
Medical	claim a non-existent prescription and ask for payment	17.5
Loan	offer low interest loans for education, auto, etc.	16.3
Credit card	claim to be credit card service representatives and try to collect personal information	12.5
Emergency	fraudsters send emails to the contact lists of hacked email accounts, claim loss of passport or wallet in a foreign country and ask to send money for help	10.0
Rental	post on craigslists or through emails, ask for deposit in order to get free/inexpensive housing	8.8
Computer support	claim to be computer expert who have discovered security issues on the victim's computer. Aim at collecting personal information or leading to downloading malware.	5.0
Fortune	claim to have certain amount of lucky money for you (e.g., winning a lottery, a tax return, or even a heritage).	3.8
Threatening	claim to have smuggled a family member, a relative or a friend and ask for money	2.5
Telemarketing	try to sell prescribed medicines, movies, etc.	1.3

phone devices support the "click to call" function which allows users to initiate phone calls by clicking on the phone numbers on the web page from the browser. This also increases their chance of responding to fraud activities. In comparison, *Unlocked Device* and *Device Exploit* (line 4 and 5), constrained by a few vulnerable devices, most of which are regular phones, attract a predominant number of regular phone victims. Moreover, since *Target Scam* and *Malware Scam* (line 6 and 9) are migrated unchangeably from landlines and Internet, they do not rely on specific mobile devices. Thus they have an RS value close to 1.

We further explore the fraction of new victims with smartphone devices over a two-year time period, which is depicted as the solid red curve in Fig. 12. For comparison, we also display the proportion of smartphone users out of all mobile users over the same time period (the blue dotted curve). The ascending trend of the dotted curve in Fig. 12 clearly indicates a sign of smartphone penetration, where the proportion contributed by smartphone users increases steadily, with a few jumps plausibly due to the release of popular smartphone devices and other social events. However, we observe that smartphone users always account for a much higher proportion of new victims (i.e., the solid curve stays above the dotted curve), indicating that smartphone users are more susceptible to fraud activities. In addition, accompanying the fast increase of smartphone popularity, we observe that the proportion of victims with smartphone devices is also increasing significantly, which is indicated by the ascending trend of the red solid curve in Fig. 12. This is possibly because of the emerging fraud activities targeting smartphone population, especially the ones carried by malware apps and online media sites, which can be much more effective in acquiring new victims than traditional fraud targeting landlines.

Implications: While mobile users enjoy the advance of smartphone devices, its rich functionality and strong computation power leave more room for fraudsters to carry out more sophisticated attacks on these devices, which makes smartphone devices more susceptible to fraudulent activities. We observe many fraud activities carried by malicious mobile apps targeting smartphones. These malware apps have integrated the functions of voice calls and data transmission and exhibit certain characteristics of modern botnets, e.g., employing dynamic fraud numbers to evade detection and regulation, maintaining connection with a command-and-control (C&C) server, etc. We expect to see in near future that new malware apps incorporate other channels, e.g., SMS and MMS, for fraud activities. Moreover, more techniques used by botnets today are expected to be adopted for developing malware apps, such as fast-flux, domain-flux and P2P network, etc. Lessons learned from combating botnets can be useful for preventing fraud in cellular networks.

Since all malware infections we have observed are triggered by users actively downloading malware from the online marketplace, detection and removal of malware at the online marketplace has proved to be an effective way of preventing fraud activities. In particular, our analysis suggests that the online marketplace should come up with more advanced vetting process for the apps, especially if it contains code segment involved with voice calls or SMS messages. Similarly, information posted on online media should also be carefully vetted (especially phone numbers appearing on online posts) to avert fraud activities.

6.3 Social Engineering Techniques

Given that many fraud activities adopt social engineering to defraud victims, we take a closer look at the social engineering techniques that fraudsters often employ to solicit victims to make phone calls, to disclose sensitive personal information, or even to inflict monetary loss. As mentioned in Section 6.2, our analysis of social engineering techniques also relies on descriptions from online feedback provided by the victims based on their conversations with the fraudsters.

We summarize the identified social engineering techniques in Table 3, where the first column shows the name of the technique and the second column provides explanation of each technique. The third column displays the popularity of each social engineering technique in terms of the proportion of fraud activities (i.e., subgraphs) that practice it. There are also fraud numbers (less than 10%) that either no one complaints about or the corresponding social engineering technique cannot be distilled from the online feedback. This may be due to the small victim population affected by this fraud activity or the lag in user reports (recall there can be several months of delay of user reports after the fraud activity is detected). We temporarily ignore these numbers and leave them for future investigation.

In Table 3, we see various techniques that fraudsters usually adopt. Though these techniques vary from case to case, they generally fall into several categories which are more likely to attract people's attentions. For example, people usually are responsive if they are told that they have an outstanding expense, which may affect their credit rating. Therefore, a very popular social engineering technique is regarding bill or debt, where fraudsters claim fictitious debts owed by the victims in order to acquire private information or defraud victims of money. Similarly, techniques regarding services that are crucial in our daily life, such as personal/car loan and credit cards, are also commonly used by fraudsters. An interesting observation shows that fraudsters applying these techniques often are aware of personal information of the victims, such as their names, addresses, bank account or former prescription information. For example, we observe reports regarding fraudsters pretending to be bank personnel, who mentioned victims' former bank account numbers. In fact, the bank has reported the leak of customer private information due to hacker attacks. Apparently, such information has somehow been acquired to conduct fraud activities. As another example, medical prescription information is adopted frequently by the fraudsters to defraud victims. With personal information, fraudsters often launch targeted scam to increase their success rate. In comparison, when lacking such information, fraudsters usually employ auto dialer to find victims or rely on social media sites, forums or other online media channels. The common social engineering techniques applied by these fraudsters are advertisement for discounted drugs, lottery defraud or even feigning an emergency regarding a family member. However, these fraud activities only account for less than 10%.

Implications: Personal information plays an important role in fraud activities. Based on user reports, fraud activities are more difficult to recognize when personal information is mentioned by fraudsters. We have identified cases where personal information is acquired by fraudsters from bank databases through security attacks. This suggests the importance of correlating fraud activities and network intrusions to understand and track the chain of fraudulent events. In addition, there are also cases where fraudsters are likely to acquire personal information from online sources, like social media sites or personal blogs. Vetting these online sources to avoid information leakage is crucial for combating fraud activities. Moreover, with the increasing functionality of mobile devices, users tend to store personal information, e.g., contact list, emails or even credit card information (for online purchase) in these devices. There is a demanding need for better security on these personal devices.

6.4 Originating Countries of Fraud Numbers

To understand the geographical distribution of fraud numbers, we investigate the top originating countries of these numbers. In fact, we find fraud numbers are distributed widely to 58 different countries, where the top countries, UK, Sao Tome and Principe and Russia, collectively contribute to less than 25% of fraud numbers, respectively. More interestingly, we find around 60% of the fraud activities contain phone numbers from multiple countries. All these make tracking and regulating fraud activities a much more challenging task. In addition, we have observed diversity of fraud activities from different countries. For example, the UK is related to a higher number of email related fraud activities, Nigeria is associated with more rental fraud activities, and Sao Tome and Principe has contributed to many malware related fraud numbers.

An interesting observation is made by measuring the length of the fraud numbers. Including country codes, around 60% of the fraud numbers contain 10 digits, the same number of digits as a typical US number excluding the country code "1". We refer to this observation as the *heteronym property* of fraud numbers. Having the same length as a US number, it is very hard for a user to distinguish such a foreign number, and hence increase chances of victims calling that number. In an extreme case, we find a scam activity employing three different foreign numbers looking very similar: 664-133xxxx in Montserrat, 66-4133xxxx in Thailand and 64-4133xxxx in New Zealand[10]. We have tried to initiate international phone calls to mobile devices equipped with four most popular mobile operating systems. Unfortunately, in the default firmware setting, none of the devices has successfully parsed the incoming number and displayed the right country name of the caller. This explains why many users complaining on the forums about high rate US numbers, which are essentially international fraud numbers.

Implications: Though certain mobile apps have started to offer the functionality for parsing incoming phone numbers, it is necessary for the mobile operating systems to provide direct support of such functionality. We believe it is a critical step for preventing fraud activities by presenting more information to the users to help them identify potential fraud calls. Our analysis also indicates that fraudsters often employ numbers from multiple countries to void tracking and regulation. Therefore, a forum for international telcom companies to share knowledge regarding premium rate numbers and known fraud numbers can be useful for detecting and preventing fraudulent activities.

7. RELATED WORK

Our work is related to many research topics as follows.

Fraud detection: There is a rich body of literature on detection of various Internet fraud activities. [18] studied the one click fraud by analyzing public reports of fraudulent websites. [19] designed NetProbe, a system to detect fraud in online auction networks. [20] detected fraud in web advertising networks. [21] used a spectrum based framework for fraud detection in social networks. [22] also studies fraud detection in wireless and landline networks. However, they focused on detecting subscription fraud (account holders who have no intention to pay for any bill) by assessing the similarity between accounts. In comparison, we focus on fraud that employ foreign fraud numbers and target mobile customers inside the cellular network. To the best of our knowledge, our work is the first to present a comprehensive study of voice fraud in large cellular networks and we have discovered many emerging attacks carried by mobile devices, smartphone apps and other channels in mobility networks.

Cellular network security: Due to the increasing popularity of cellular networks, security in cellular networks is becoming an important research area. There are many works which focus on detecting malware, botnets and other anomalous activities in cellular networks. For example, [5] studied 25 distinct families of mobile viruses and worms targeting the Symbian OS and developed a machine learning based malware detection algorithm using behavioral statistics of these malware. [6] designed a mobile botnet called *Andbot*

[10]Different international exit codes are used to differentiate calls to these numbers, e.g., 011 for Montserrat, 001 for Thailand, and 00 for New Zealand. However, such exit codes are often attached automatically by the device without informing the user.

which infects Android mobile devices and utilizes URL flux for the command and control channel. [7] proposed to use process state transitions and user operational patterns to differentiate behaviors of malware and human users. Most of the existing works focus on securing the mobile data channel. In comparison, our focus in this paper is on fraud activities through voice channels, which have potentially a much wider influence and can bring direct financial loss to both mobile users and cellular network providers.

Graph based anomaly detection: Many recent works focus on anomaly detection by representing the data as graphs and detecting suspicious activities by extracting community structures from these graphs. For example, [13] studied properties of community structures in different application traffic graphs, [8] proposed to utilize DNS failure graphs to identify suspicious network activities, [23] presented a data mining technique for detecting anomalies in various bi-partite social graphs, [9] detected malware by mining file relation graphs. [24] designed a method to capture graph evolution over time and applied the method to detect subscription fraud in telecom networks. Motivated by these works, we propose to use voice call graphs to isolate fraud activities in cellular networks.

8. CONCLUSION AND FUTURE WORK

In this paper, we proposed a graph based method to detect fraud activities of voice calls in a large cellular network, which enables us to efficiently isolate popular fraud numbers from millions of international phone numbers. Using call records spanning a two-year time period, we showed that our method successfully detected a great variety of voice call fraud activities. In addition, we identified unique characteristics of fraud activities in cellular networks, such as new fraud carried by malware apps, unlocked devices and online media sites, social engineering techniques and heteronym property of fraud numbers. Our analysis sheds light on the new characteristics and trends of voice-related fraud activities in current mobile networks.

Our future work will focus on investigating the potential of combining additional features, such as the user calling history, and instant user fraud reports to improve fraud detection accuracy. Meanwhile, we shall incorporate inbound phone calls for our analysis in order to detect more fraudulent activities. Our work also includes developing mobile apps for alerting users of potential fraud threats.

Acknowledgement

This work is supported in part by the NSF grants CNS-0905037, CNS-1117536 and the DTRA Grant HDTRA1-09-1-0050. We thank Cheri Kerstetter and Peter Coulter for their constructive comments. We also thank Colin Goodall and Puspalogan Sundaram for data collection and useful discussions. Finally, we thank our shepherd, Lujo Bauer, for his many suggestions on improving the paper.

9. REFERENCES

[1] Over 5 billion mobile phone connections worldwide. http://www.bbc.co.uk/news/10569081.

[2] Data capable cell phones replacing the pc. http://blogs.itbusiness.ca/2011/03/data-capable-cell-phones-replacing-the-pc/.

[3] Report: Cellphones replacing landlines. http://www.upi.com/Top_News/US/2009/11/19/Report-Cellphones-replacing-landlines/UPI-72441258651963/.

[4] iphone virus turns into a mobile botnet. http://www.redmondpie.com/iphone-virus-turns-into-a-mobile-botnet-9140130/.

[5] A. Bose, X Hu, K. Shin, and T. Park. Behavioral detection of malware on mobile handsets. In *MobiSys '08*, 2008.

[6] X. Cui, B. Fang, L. Yin, X. Liu, and T. Zang. Andbot: towards advanced mobile botnets. In *LEET'11*, 2011.

[7] L. Xie, X. Zhang, J. Seifert, and S. Zhu. pbmds: a behavior-based malware detection system for cellphone devices. In *WiSec '10*, 2010.

[8] N. Jiang, J. Cao, Y. Jin, L. Li, and Z.-L. Zhang. Identifying suspicious activities through dns failure graph analysis. In *ICNP '10*, 2010.

[9] Y. Ye, T. Li, S. Zhu, W. Zhuang, E. Tas, U. Gupta, and M. Abdulhayoglu. Combining file content and file relations for cloud based malware detection. In *KDD '11*, 2011.

[10] WhoCallsMe - Reverse phone number lookup. http://www.whocallsme.com.

[11] 800notes - Directory of unknown callers. http://www.800notes.com.

[12] Graphviz. http://www.graphviz.org/.

[13] Y. Jin, E. Sharafuddin, and Z.-L. Zhang. Unveiling core network-wide communication patterns through application traffic activity graph decomposition. In *Proc. of SIGMETRICS '09*, pages 49–60, 2009.

[14] A. Mislove, M. Marcon, K. Gummadi, P. Druschel, and B. Bhattacharjee. Measurement and analysis of online social networks. In *IMC '07*, 2007.

[15] Y. Zhang, S. Singh, S. Sen, N. Duffield, and C. Lund. Online identification of hierarchical heavy hitters: algorithms, evaluation, and applications. In *IMC '04*, 2004.

[16] A. Mahimkar, Z. Ge, A. Shaikh, J. Wang, J. Yates, Y. Zhang, and Q. Zhao. Towards automated performance diagnosis in a large iptv network. In *SIGCOMM '09*, 2009.

[17] S. Dongen. *Graph clustering by flow simulation*. PhD thesis, University of Utrecht, 2000.

[18] N. Christin, S. Yanagihara, and K. Kamataki. Dissecting one click frauds. In *CCS '10*, 2010.

[19] S. Pandit, D. Chau, S. Wang, and C. Faloutsos. Netprobe: a fast and scalable system for fraud detection in online auction networks. In *WWW '07*.

[20] A. Metwally, D. Agrawal, and A. Abbadi. Using association rules for fraud detection in web advertising networks. In *VLDB '05*, 2005.

[21] X. Ying, X. Wu, and D. Barbar á. Spectrum based fraud detection in social networks. In *CCS '10*, 2010.

[22] C. Cortes, D. Pregibon, and C. Volinsky. Communities of interest. In *IDA'01*, 2001.

[23] J. Sun, H. Qu, D. Chakrabarti, and C. Faloutsos. Neighborhood formation and anomaly detection in bipartite graphs. In *ICDM '05*, 2005.

[24] C. Cortes, D. Pregibon, and C. Volinsky. Computational methods for dynamic graphs. *Journal of Computational and Graphical Statistics*, 12, 2003.

What is keeping my phone awake? Characterizing and Detecting No-Sleep Energy Bugs in Smartphone Apps

Abhinav Pathak
Purdue University
pathaka@purdue.edu

Abhilash Jindal
Purdue University
jindal0@purdue.edu

Y. Charlie Hu
Purdue University
ychu@purdue.edu

Samuel P. Midkiff
Purdue University
smidkiff@ecn.purdue.edu

ABSTRACT

Despite their immense popularity in recent years, smartphones are and will remain severely limited by their battery life. Preserving this critical resource has driven smartphone OSes to undergo a paradigm shift in power management: by default every component, including the CPU, stays off or in an idle state, unless the app explicitly instructs the OS to keep it on! Such a policy encumbers app developers to explicitly juggle power control APIs exported by the OS to keep the components on, during their active use by the app and off otherwise. The resulting power-encumbered programming unavoidably gives rise to a new class of software energy bugs on smartphones called *no-sleep bugs*, which arise from mis-handling power control APIs by apps or the framework and result in significant and unexpected battery drainage.

This paper makes the first advances towards understanding and automatically detecting software energy bugs on smartphones. It makes the following three contributions: (1) we present the first comprehensive study of real world no-sleep energy bug characteristics; (2) we propose the first automatic solution to detect these bugs based on the classic reaching definitions dataflow analysis algorithm; (3) we provide experimental data showing that our tool accurately detected all 14 known instances of no-sleep bugs and found 30 new bugs in the 86 apps examined.

Categories and Subject Descriptors

D.2.5 [**Software Engineering**]: Testing and Debugging

General Terms

Design, Experimentation, Measurement

MobiSys'12, June 25–29, 2012, Low Wood Bay, Lake District, UK.

Keywords

Smartphones, Mobile, Energy, Energy-Bug, No-Sleep-Bug.

1. INTRODUCTION

1.1 Motivation

Smartphones have surpassed desktop machines in sales in 2011 to become the most prevalent computing platforms [1]. To enrich the user experience, modern day smartphones come with a host of hardware I/O components embedded in them. The list of components broadly fall into two categories: *traditional components* such as CPU, WiFi NIC, 3G radio, memory, screen and storage that are also found in desktop and laptop machines, and *exotic components* such as GPS, camera and various sensors. And they differ from their desktop/laptop counterparts in that power consumed by individual I/O components is often comparable to, or higher than, the power consumed by the CPU.

This, along with the fact that smartphones have limited battery life, dictates that energy has become the most critical resource of smartphones. Preserving this crucial resource has driven smartphone OSes to resort to a paradigm shift in component power management. On desktop machines, where the CPU accounts for a majority of the energy consumption, the default energy management policy is that the CPU stays on (or runs at a high frequency) unless an extended period of low load has been observed. The policy is consistent with the historical notion that energy management is a second class citizen since machines are plugged into a power source [2]. Smart phones, in sharp contrast, make power management policy a first class citizen. In fact, the power management policy on smartphones has gone to the other extreme: *the default power management policy is that every component, including the CPU, stays off or in an idle state, unless the app explicitly instructs the OS to keep it on!*

In particular, all smartphone OSes, *e.g.,* Android, IOS, and Windows Mobile, employ an aggressive *sleeping policy* which put the components of the phone to sleep, *i.e.,* puts them into an *idle* state immediately following a brief period of user inactivity. In the idle state, the smartphone as a whole draws near zero power, since nearly all the components, including CPU, are put to sleep. Such a sleeping policy is largely responsible for prolonged smartphone standby times – smartphones can last dozens of hours when idle.

The aggressive sleeping policy, however, severely impacts smartphone apps, since an app may be performing critical tasks by intermittently interacting with the external world using various sensors. For example, an app syncing with a remote server over the network may appear to perform no activity when waiting for the server to send its reply, and the system may be put to sleep by the aggressive sleeping policy, leaving the remote server with a view of lost connectivity.

To avoid such disruptions due to the aggressive sleeping policy, smartphone OSes provide a set of mechanisms for app developers to *explicitly* notify the OS of their intention to continue using each component. In particular, the OS exports explicit power management handles and APIs, typically in the form of power wakelocks and acquire and release APIs [3], for use by the app developer to specify when a particular component needs to stay on, or awake, until it is explicitly released from duty.

We argue such explicit management of smartphone components by app developers has presented to the app developer a profound paradigm shift in smartphone programming that we call as *power-encumbered programming*. This new programming paradigm places a significant burden on developers to explicitly manipulate the power control APIs (§4.1 details one such example of the burden placed on developer due to power encumbrance). This manipulation is *required* to ensure the correct operation of the apps. Consequently, power-encumbered programming unavoidably gives rise to a new class of software energy bugs on smartphones, called *no-sleep bugs*. No-sleep bugs are defined as energy bugs resulting from mis-handling power control APIs in an app or framework,[1] resulting in the smartphone components staying on for an unnecessarily long period of time. No-sleep bugs form one important category of the family of *smartphone energy bugs* which are defined in [4] as errors in the smartphone system (an app or the framework, the OS, or the hardware) that cause an unexpectedly high energy consumption by the system as a whole.

Discussions of energy bugs on numerous Internet forums have narrowed the causes to mis-handling of power control APIs by apps and the framework on smartphones OSes, including Android, IOS, and Windows Mobile. Our recent survey [4] has found that 70% of all energy problems in apps and frameworks reported by mobile users were due to no-sleep energy bugs. These and other types of energy bugs have caused a great deal of user frustrations. Despite their severity, *i.e.,* high battery drain, to the best of our knowledge there has been no study of any kind of smart phone energy bugs, much less no-sleep energy bugs. Drawing parallels with research on traditional software bugs (*e.g.,* concurrency bugs in concurrent programs [5]), a comprehensive treatment of energy bugs on smartphones will require (1) a good understanding of real world energy bug characteristics, learned from common mistakes programmers make in writing smartphone apps, to lead to effective debugging techniques; and (2) developing multi-faceted approaches to eliminating energy bugs, including avoiding energy bugs during app development (*e.g.,* by providing better programming language support for power management) and compile and runtime detection.

1.2 Our Contributions

This paper takes the first steps towards understanding and automatically detecting software energy bugs on smartphones. Specifically, our paper makes three concrete contributions:

(1) The first characterization study of no-sleep energy bugs in smartphone apps: We present the first comprehensive real world no-sleep energy bug characterization study. Our study is based on no-sleep energy bugs in real world apps and the Android framework, including popular apps (*e.g.,* Facebook) and built-in (*i.e.,* shipped with) apps and services (*e.g.,* the Android email app). The bugs are collected by crawling Internet mobile forums, bug repositories, commit logs of open source Android apps and by running our no-sleep bug detector developed in this paper. For each bug, we carefully examine its reported symptoms, corresponding source code and related patches (when available), and developer's discussions (when available), or the analysis performed by our bug detector. Our study reveals a taxonomy of three major causes of no-sleep energy bugs, which provide useful guidelines and hints to designing effective detection techniques. Our study also confirms the significant burden power-encumbered programming places on app developers. For example, making a single outgoing or incoming phone call in Android involves about 40 invocations of power control APIs in the Dialer app and the framework's Radio Interface Layer services to dynamically manage the power control of the CPU, screen, and other sensors!

(2) The first solution to automatically detect no-sleep energy bugs: We make the key observation that power control APIs are explicitly embedded in the app source code by the app developers, and two out of the three causes for no-sleep energy bugs from our characterization study are because a turn-on API call is missing a matching turn-off API call before the end of the program execution. We thus propose a compile-time solution based on the classic reaching definitions dataflow analysis problem [6] to automatically infer the possibility of a no-sleep bug in a given app. Our solution detects no-sleep bugs in single-threaded and multi-threaded apps, as well as event-based apps which have multiple entry points. Like all static analysis based tools, our detection tool can suffer false positives but has the tremendous advantage of *no* runtime overhead and *no* false negatives (to the best of our abilities of establishing the ground truth).

We further present the complete implementation of our static analysis detection tool for apps written for Android. The tool is capable of running directly on the app installers (`.apk` files) and hence source code is not required. Our implementation handles the specifics of event-driven mobile programming and of the Java language such as runtime null pointer exceptions and object references.

(3) Detecting new no-sleep bugs in Android apps and framework: We have run our no-sleep bug detection tool on 86 Android apps and the framework collected from the Android market. Experimental evaluation shows that our tool accurately detected all reported instances of no-sleep bugs, as well as 30 instances of new previously unreported no-sleep bugs. These include no-sleep bugs in many popular apps, *e.g.,* the default Android Email app. Our no-sleep bug detection incurred false positives in 13 out of the 55 apps it reported to contain a bug.

2. POWER ENCUMBERED PROGRAMMING

In this section, we first describe the energy management APIs and their semantics that are exposed to the developers in the Android smartphone OS, and discuss the burden they impose on the app developers. We first discuss programming APIs for traditional components (*e.g.,* screen, CPU) and then for exotic components (*e.g.,* GPS, Camera). We also discuss the issues arising from event based programming model of smartphone apps. We then introduce the prominent class of energy bugs studied in this paper: no-sleep bugs.

[1]In this paper, we use the term framework to refer to both the services in and the apps that are bundled with, the Android framework.

Table 1: Summary of power operations exported by Android APIs.

Component lock/manager name (API to start/stop)	Component(s)	Battery Drain up to (%/hr)	Comments
	Traditional Components		
PARTIAL_WAKE_LOCK (acquire/release)	CPU	5%	CPU runs despite any timers
SCREEN_DIM_WAKE_LOCK (acquire/release)	CPU and Screen (DIM)	12%	No illumination if shutdown, else illuminates till lock release (Flag ACQUIRE_CAUSES_WAKEUP forces illumination in all cases)
SCREEN_BRIGHT_WAKE_LOCK (acquire/release)	CPU and Screen (bright)	25%	
FULL_WAKE_LOCK (acquire/release)	CPU, Screen (bright) and Keyboard backlight	25%	
	Exotic Components		
PROXIMITY_SCREEN_OFF_WAKE_LOCK (acquire/release)	Screen, Proximity Sensor	25%	Screen shuts if sensor activates
LocationManager (requestLocationUpdate/removeUpdates)	GPS	15%	Tracks user location
SensorManager (registerListener/unregisterListener)	Accelerometer, Gyro, Proximity Magnetic Field, *etc.* [8]	10%	Sensormanager class controls various sensors on phone
MediaRecorder (start/stop)	Mic/Camera (for video)	20%	Usually stores media on sdcard
Camera (startPreview/stopPreview)	Camera (for still pictures)	20%	One app at a time registers camera

Listing 1: An example power wakelock usage.

```
PowerManager.WakeLock wl = pm.newWakeLock(PowerManager.
    PARTIAL_WAKE_LOCK);
wl.acquire();  //CPU should not go to sleep
net_sync();    //Perform critical task here
wl.release();  //CPU is free to sleep
```

2.1 Managing Traditional Components

The Android framework exports wakelock functionality through PowerManager.Wakelock [2] class, with 4 different options and associated APIs for managing several traditional components: CPU, screen, and the keyboard backlight. A wakelock is an *instance* (object in Java) of the wakelock class, instantiated using one of 4 options, and each option has a different effect on the hardware component, as summarized in Table 1. For example, option FULL_WAKE_LOCK instantiates a lock that when acquired both keeps the CPU and screen on at full brightness and turns on the keyboard backlight.

Listing 1 illustrates a basic wakelock usage: how to ensure that the CPU does not sleep during some critical phase. The app declares a wakelock (pm is an instance of PowerManager) and then *acquires* it, which instructs the OS not to put the CPU to sleep, irrespective of user activity, since it intends to perform some critical task. Once the critical task (in this case a remote network sync net_sync()) is completed the app *releases* the wakelock, indicating to the OS that CPU can now sleep according to its sleeping policies.

Semantics: The above simple usage of a wakelock is just like a conventional mutual exclusion lock, *i.e.*, an app explicitly *acquires* and *releases* it to instruct the OS to switch the component on and off, respectively. Like an object a wakelock can be shared among several threads of a process. The semantics of wakelocks, however, are quite different from those of conventional mutual exclusion locks.

First, as shown in Table 1, a single wakelock (instantiated with one of the four options) controls one or more components. Second, the power control effect of a single wakelock depends on the configuration of that wakelock: a wakelock can be configured to be reference counted [9]. In an almost peculiar sense, it is more like a condition variable when configured not to be reference counted, and a semaphore when configured to be reference counted.

We first consider non-reference counted wakelocks. An acquire() on a released or fresh instance of wakelock wakes up the corresponding component (the ACQUIRE_CAUSES_WAKEUP flag wakes up the screen), or keeps the component awake if it is already so. In other words, an acquire( ) called on an already acquired wakelock is treated as a nop. Similarly, a release() called on an acquired wakelock sets the component free to sleep as far as the perspective of this wakelock is concerned, irrespective of the number of times an acquire() has been called on the lock. In this sense, a non-reference counted wakelock is like a condition variable.

In contrast, reference counted wakelocks are like semaphores. Each acquire() of a wakelock increments the internal counter associated with the (instance of) wakelock, and a release() decrements the internal counter. It only lets the component sleep (from the perspective of this wakelock) if the internal counter value reaches 0. A *release* is a nop[3] in other cases.

To make matters even more complicated, an acquire can also be called with a timer [10], which instructs the system to release it automatically once the timeout interval expires.

Third, the above power control semantics are from the perspective of one wakelock. Unlike traditional *mutual exclusion* locks, different wakelocks (even instantiated with different options) on the same component can be held by multiple entities (*e.g.,* processes and threads) in the system at the same time. Even a single entity may hold multiple (instances of) wakelocks on the same component. The power control effect on a component must take into account the state of all wakelocks. The component is switched on when the first wakelock is held. Only when all the wakelocks from all the entities for the component are *released*, taking into account the reference counting semantics for each, can the component go to sleep, subject to higher level sleeping policies (administered by framework processes), *e.g.,* sleep after 5 seconds of user inactivity.

This demonstrates the new programming burden inflicted on app developers: power management is no longer just a transparent OS or driver task; the developers now need to perform explicit power management in the app layer.

2.2 Managing Exotic Components

In addition to traditional components, modern smartphones come with several "exotic" components embedded in them. These include GPS, camera, several sensors such as an accelerometer, proximity sensor, and gyroscope. Some of these components are the biggest energy consumers in smartphones, *e.g.,* GPS and camera, and drain the battery at a high rate [11, 12, 13].

[2] Apps need android.permission.WAKE_LOCK permission from users to use this class.

[3] Newer Android APIs throw an exception when *release()* is called on an unacquired wakelock. Hence, a release is usually called after testing if the lock is currently held (using API *wakelock.isHeld()*).

Unlike some of the traditional components, *e.g.,* WiFi NIC, the new exotic components are used in an explicit on-off fashion. For example, the GPS is explicitly turned on, using the OS exported API, to acquire the smartphone location and in this state it consumes battery at a high rate. Once the location is determined, the component is explicitly turned off, triggering the component to return to a low power state.

Like wakelocks, the explicit power management of exotic components places a significant programming burden on app developers. An incorrect or inefficient use of these APIs can easily lead to poor utilization of these components, wasting significant battery energy (see §2.4).

Semantics: Table 1 lists the APIs exported by Android for accessing the *exotic* components. Their semantics of power control of these components is similar to the plain wakelocks in §2.1, *i.e.,* no reference counting or timer-based release.

In summary, the developers are burdened with explicitly manipulating power control APIs for both traditional and exotic smartphone components to ensure the correct operations of the apps. We call this new smartphone programming paradigm as *power-encumbered programming*.

2.3 Issues from Event-based Programming

The complexity of power-encumbered programming is exacerbated by the event-based nature of smartphone apps. Compared to programming in desktop/server environments, smartphone programming is event-oriented because of the inherit interactive nature of phone apps. A typical user-facing smartphone app is written as a set of event handlers with events being user or external activities. The developer needs to keep track of each possible event and when it may be triggered, and manipulate the wakelocks accordingly.

We illustrate how the level of complexity introduced by power-encumbered programming is exacerbated by event-based programming through a concrete example from the Dialer app [14] that comes with the Android framework.

The Dialer app implements the dialing functionality of the phone. The app is triggered when the user receives an incoming call or when the user clicks the phone icon to make an outgoing call. To implement its functionality, the app explicitly maintains three wakelocks: `FULL_WAKE_LOCK` for keeping the screen on (*e.g.,* in situations like when the user is dialing the numbers to call), `PARTIAL_WAKE_LOCK` for keeping the CPU on (*e.g.,* in case of an incoming call when the phone is switched off), and `PROXIMITY_SCREEN_OFF_WAKE_LOCK` which switches the proximity sensor on and off (to detect user's proximity to the phone).

To manage the three wakelocks, the app explicitly maintains a state machine where the states represent the lock behavior, *i.e.,* which lock needs to be acquired and which needs to be released, and the "condition" of the phone represents the state transitions. The conditions are diverse and include events such as (a) if the phone gets a call, (b) if the phone is pressed against the user's ear in which case the proximity sensor triggers the screen to go off, (c) if the call ends, (d) if a wired or bluetooth headset is plugged in (*e.g.,* in the middle of a call), (e) if the phone speaker is turned on, (f) if the phone slider is opened in between calls, and (g) if the user clicked home button in the middle of a call. For each of these triggering events, the phone changes the state of wakelock state machine, acquiring one and releasing another.

In addition to wakelocks in the Dialer app, the Radio Interface Layer (RIL) in Android maintains additional 5 wakelocks to handle incoming and outgoing calls. Using explicit component access tracing through an instrumented Android framework running on a Google Nexus One handset, we observed that performing a single outgoing or incoming call in Android resulted in 30-40 distinct instances of wakelock acquires and releases!

2.4 No-Sleep Bugs

The new burden of explicit component power manipulation from power-encumbered programming, combined with the complexity of handling events in the app behavior, can easily overwhelm developers and lead to programming mistakes in manipulating the power control APIs. Incorrect or inefficient usage of such APIs can lead to an unexpected drain of the phone battery, known as no-sleep bugs [4].

A "no-sleep bug" is a condition where at least one component of the phone is woken up and is not put to sleep due to a mistake in manipulating power control APIs in an app. The component that is woken up continues to drain the battery for a prolonged period of time, resulting in *severe* and *unexpected* battery drain. Typically the battery drain continues until the app is forcefully killed[4] or the system is rebooted.

No-sleep bugs form one of the most important categories of software *energy bugs* in smartphone apps [4]. Unlike regular software bugs in apps, energy bugs do not lead to an app crash or OS blue screen of death [4]. An app hit by an energy bug continues to provide the intended functionality, with a single difference: the phone suffers a severe, unexpected battery drain. The severity of the energy drain due to the bug depends on the component that is not put to sleep. As shown in Table 1, for each of the 3 components (GPS, Screen with full or low brightness and camera), the impact of a no-sleep bug can be severe with the battery draining at a rate of 10-25% every hour[5] without any user interaction.

For other components, *e.g.,* the CPU and proximity sensor, the battery drains at a relatively low rate - up to 5% every hour. When a CPU wakelock (`PARTIAL_WAKE_LOCK`) is held, it prevents the CPU from 'freezing', a state where it would consume zero power (IDLE state). In a wakelock held state, the CPU draws minimal power (depending on the CPU specifications of the handset). However, as the CPU remains on, other activities continue to run, *e.g.,* WiFi NIC chatters, background periodic OS processes, hardware interrupts handling by OS, *etc.* These activities together consume, as measured on Google Nexus One, about 5% of the battery every hour. Any additional user activity is not accounted for in this. As a result, over a long period of time, say 12 hours, an only-CPU wakelock bug can drain about 50-60% of the battery without any user interaction or performing necessary activities.

3. METHODOLOGY

To characterize the root cause of no-sleep bugs observed in current mobile apps, we collected no-sleep bugs in smartphone apps in four ways. **(a) Mobile forums:** We crawled 4 popular mobile Internet forums (the same as in [4]): one general forum with discussions covering all mobile devices and OSes, and three OS/company specific mobile forums. In total we crawled 1.2M posts, from which we filtered out posts related to no-sleep bugs in smartphone apps. For each app reported by the user to contain a no-sleep bug, we downloaded the binary installers of the version of the app that was reported to contain the bug and the first version which had the problem solved. We then decompiled the app from binary installers

[4]In Android, apps (esp. background services) are not killed; they run in background once a user stops interacting with them. They are usually killed by the system only in case of memory pressure.

[5]These rates were calculated using HTC nexus and magic handsets running Android. A fully charged battery holds between 1100mAH to 1500mAH of charge.

to Java source code using ded[6] [16]. For apps that were successfully decompiled (*e.g.*, FaceBook), we studied the root cause of no-sleep bugs. **(b) Bug lists:** We crawled mobile bug repositories of open source mobile frameworks like Android [17] and Maemo [18]. We extracted bugs reported with no-sleep conditions and extracted the source code (open source) of the versions that actually contained the bugs (*e.g.*, no-sleep bug in Android SIP Service [19]) and its patch (if available). **(c) Open source code repositories:** We scraped the commit logs of open-source Android apps hosted on online code repositories like github [20]. We extracted the commit logs of no-sleep bug fixes and downloaded the versions both before and after the fix. **(d) Running our no-sleep bug detection tool:** Finally, we ran our solution of automatically detecting no-sleep bugs developed in this paper on 86 Android apps and the stock framework and discovered 42 apps with no-sleep code-paths as detailed in §6, §7 and §9 (labeled with "*" in Table 2). These apps are used in the characterization study in §4.

4. CHARACTERIZING NO-SLEEP BUGS

Using the bug-collection methodology described above, we now present a case-study of no-sleep bugs observed in smartphone apps. We characterize the root cause and impact of the bugs. Table 2 gives a summary of the three general categories of no-sleep bugs we have identified and their impact. Drain time shows the amount of time it will take to drain a fully charged battery under typical usage.[7] Without the bugs, it takes about 15 hours to drain a fully charged battery. The bug references in Table 2 refer to both the bug fix commit logs and user complaints about specific apps on Android bug repositories, all of which indicate the real impact and user frustration caused by the no-sleep bugs. The first two categories, *No-Sleep Code Paths* and *No-Sleep Race Condition*, exhibit typical symptoms of no-sleep bugs where a component is not put to sleep at all, whereas the third category, *no-sleep dilation*, represents the scenario where a component was held on much longer than the programmer's intention (on the order of hours). Below we present an in-depth analysis of these three categories of no-sleep bugs.

4.1 No-Sleep Code Path

The root cause for most of the observed no-sleep bugs in a single threaded activity was the existence of a code path in the app that wakes up a component, *e.g.*, by acquiring the wakelock for the component, but does not put the component back to sleep, *e.g.*, there is no release of the lock. This category captures a majority of the no-sleep bugs we have observed in our bug collection.

We observed three causes for the existence of a code path where the component was switched on but not put to sleep.

The first cause is that the programmer simply forgot to release the wakelock throughout the code, or the programmer released the lock in the `if` branch of a condition but not in the `else` branch. Although it seems like a simple mistake, this does happen in real apps. For example, a version of the Agenda widget [21] contained such a no-sleep bug.

The second cause is that the programmer did put code that releases the component wakelock on many code paths, but the code took an unanticipated code path during execution along which the component was not put to sleep. Listing 2 shows a code-snippet

[6]Not all app binaries could be transformed into (meaningful) source code, especially the ones that have been obfuscated during compilation (using tools like proguard [15]), *e.g.*, the NYTimes Android app.

[7]Typical usage assumes that the phone is used actively by a user for 20% of the total time. Drain times are calculated using the energy drain rate in the bug state measured on the HTC Nexus phone.

Listing 2: No-sleep bug: different code paths.

```
try{
    wl.acquire();   //CPU should not sleep
    net_sync();     //Throws Exception(s)
    wl.release();   //CPU is free to sleep
} catch(Exception e) {
    System.out.println(e); //Print the error
} finally {
} //End try-catch block
```

that represents a typical template of a no-sleep bug where an app takes a different, somewhat unanticipated code path after waking up a component and therefore does not put it back to sleep. As in Listing 1, the critical task in the app, `net_sync()`, is protected by acquiring and releasing the CPU wakelock instructing the CPU not to sleep during the remote syncing phase. However, routine `net_sync()` may throw exceptions [41], a Java language mechanism for notifying apps of some failure conditions, such as a connect to a remote end host failed, a string could not be parsed to integer, or a specific file to be opened does not exist. A thrown exception is explicitly *caught* by the `try`'s `catch` block which simply prints the exception for debugging purposes. Now the no-sleep bug can manifest itself in the following code path. First the `try` block executes and acquires the wakelock. Next a call is issued to the critical task, `net_sync()`. If an exception is raised inside `net_sync()`, the control directly jumps to the catch block, the debug output is printed, and the code exits the try catch block. Consequently, the code-path followed does not release the wakelock, keeping the CPU on indefinitely. To fix this problem, the wakelock should be released in the `finally` block so that it is always executed.

A large number of no-sleep bugs are caused by this second reason. These include popular apps such as FaceBook [22], Agenda widget [21] (another version), MyTrack [25] (no sleep of GPS), BabbleSink [26], CommonsWare and apps [27], as well as services that come with the Android framework, such as Android Telephony [28], Android Exchange [29, 30], and WifiService [31]. For example, in SIP service [19], the wakelock was not released since the objects containing the wakelocks were deleted and so the lock handlers were deleted along with them.

The third cause for a no-sleep code path is that a higher level condition (like an app level deadlock) prevented the execution from reaching the point where the wakelock was to be released. This is likely to happen in smartphone programming because event-based programming of smartphones can lead to many possible code branches (as in the example of the Android Email app) that makes it difficult for the programmer to anticipate all the possible code paths and keep track of the wakelock state. `LocationListener` [33, 34] in the Android framework contained such a no-sleep bug. The developer did release the wakelock, however, a higher level app deadlock prevented the code from entering the release phase of the app.

Finally, the most common pattern of no-sleep code-path bug is a result of the fact that developers do not properly understand the life-cycle of Android processes. In Android, an app activity once started is always alive. When the user exits any app, Android saves the state of the app and passes it back to the app if the user returns to it. The app is only completely killed when the phone is critically low on RAM or when the app kills itself. This methodology is used to reduce the startup time of the app and to maintain its state.

This essentially means that the app may not actually be destroyed for very long periods of time. But many app developers only release the wakelock in the `onDestroy()` call-back, instead of in

Table 2: No-sleep bug case study. Entries with (F) represent bugs in the Android framework, and with (*) represent new bugs found by our technique.

App	Description	Bug Description	Drain	Ref.
		No-Sleep Code Paths		
Agenda Widget	Popular Android widget managing news/calendar	Two bugs were reported in difference versions (a) not all branches release wakelocks in `AlarmService`; (b) programmer forgot to call wakelock release after acquiring it.	9 hrs	[21]
FaceBook	The default FaceBook App `v 1.3.0`	`facebook.katana.HomeActivity`, the central Activity, acquires wakelock to run `FaceBookService`. Not all possible branches in the service release wakelocks.	9 hrs	[22]
k9mail	One of the most popular email client for Android	Per-thread wakelock maintained. Wakelock acquired when IMAP `DONE` was sent, but was not released in `MessagingControllerPushReceiver` during IDLE stat	9 hrs	[23]
CheckinMaps	Visual stories on maps	GPS remains on, even when user closes the app (calling the `onPause()` handler).	5 hrs	[24]
(*) MyTrack	Track User Path online	GPS remains on, even after user navigates away from the app draining battery.	5 hrs	[25]
BabbleSink	Find phone's location	A `NullPointerException` causes the thread to exit without releasing wakelock.	9 hrs	[26]
CommonsWare	Android Training Book	Wakelock released without finalize.	9 hrs	[27]
Sip Service (F)	Std. voice protocol implementation in Android	A Sip handler (object) was deleted which had wakelocks acquired before releasing the wakelock. The deleted handlers can not perform `release()` to release the wakelocks.	9 hrs	[19]
Telephony (F)	Telephony Handler: RIL service in Android code	Android telephony does not release the partial wakelock right away if there is an error in sending the RIL request, preventing the phone to go in power collapse draining battery.	9 hrs	[28]
(*) Android Exchange (F)	The default email app in Android framework	During background syncing of mailboxes in an exchange account, the app acquires wakelock and does not release in all failure conditions, specifically in `IOExceptions`.	9 hrs	[29] [30]
WifiService (F)	Android WiFi Handlers	CPU does not go to sleep during a message removal and wakelock was held forever.	9 hrs	[31]
PowerManager(F)	PowerService Android	Two instances of wakelocks are not released in`PowerManagerService` in Android.	9 hrs	[32]
LocationListener (F)	GPS handling library in Android framework	A deadlock in `LocationManagerService` for releasing wakelocks after client notifications have been received prevented the release of the wakelocks draining battery.	9 hrs	[33] [34]
		No-Sleep Race Condition		
(*) Android Email app (F)	Default Android email app performing sync	Race condition between email synchronizing thread and the main thread which kills the synchronizing thread resulted in a shared wakelock to remain in acquired state after exit.	9 hrs	[35] [36]
		No-Sleep Dilation		
MyTrack	Track User Path online	Wakelock acquired and released much before and after the required functionality in app.	9 hrs	[25]
GoogleBackup (F)	Cloud backup [37]	Wakelock reported to be held for a long duration of time (up to an hour) in poor network.	11 hrs	[38]
GPS Driver (F)	Android GPS handler	Wakelocks are being held for longer than needed in low level GPS driver code.	15 hrs	[39]
Google Maps	Android Google maps	App was reported to hold wakelock for several hours even when it was not used.	10 hrs	[40]

Listing 3: Wakelock Complexity

```
@Override protected void finalize(){
    /**
     * It is understood that This finalizer is not
     * guaranteed to be called and the release lock
     * call is here just in case there is some path
     * that doesn't call onDisconnect and or
     * onConnectedInOrOut.
     */
    if (mPartialWakeLock.isHeld()) {
  Log.e(LOG_TAG, "[CdmaConn] UNEXPECTED; mPartialWakeLock
       is held when finalizing.");
    }
    releaseWakeLock();
}
```

`onPause()`. `onDestroy()` is only called when the app component is about to be destroyed. As a result, once an app with this bug is started, the phone will only sleep when it is running critically low on memory (which may take a long period of time).

Manually tracking all possible code paths for wakelock acquire/release appears to be a daunting task for app developers. Listing 3 shows an example of the complexity involved in power-encumbered programming. This is a code-snippet from the Android framework's class `CdmaConnection`. This class uses a PARTIAL_WAKE_LOCK for managing the connection and releases the wakelock when the connection is disconnected. However, there are many different possible program paths arising from different patterns of user interactions, hardware states dependent on the external environment, *etc.* The developer included `releaseWakeLock` in `finalize` as an additional safety measure, even though `finalize` is not guaranteed to be called. This example shows the need for an automated tool that can aid developers in checking all possible program paths for no-sleep bugs.

4.2 No-Sleep Race Condition

The second category of no-sleep bugs we observed was cause by race conditions in multi-threaded apps. Specifically, we observed that the power management of a particular component was carried out (*i.e.*, switched on and off) by different threads in the app. In the common case, one thread switches the component on, and some time later another thread switches the component off, resulting in the normal behavior of component utilization. However, in a corner case condition, it can happen that the thread that switches on the component gets scheduled to run after the thread that switches the component off, resulting in a no-sleep bug with the component left

Listing 4: No-sleep bug: race condition.

```
public void Main_Thread(){
    mKill = false;  //Unset kill flag
    wl.acquire();    //CPU should not sleep
    start(worker_thread); //Start worker
    //....Do Something
    mKill = true;     //Set kill flag
    stop(worker_thread); //Signal worker
    wl.release();     //CPU can sleep now
} //End Main_Thread();
public void Worker_Thread(){
    while(true){
      if(mKill) break; //Break if flagged
      net_sync();     //Critical task
      wl.release(); //Rel. wl before sleep
      sleep(180000);//Sleep for 3 minutes
      wl.acquire(); //CPU should not sleep
    } //End while loop;
} //End Worker_Thread();
```

on. Effectively there is a race condition between the manipulation of the wakelock by the two threads.

Listing 4 shows a code snippet of a no-sleep bug caused by a race condition. `Main_Thread` runs first, acquiring a wakelock (waking up a component, *e.g.,* the CPU), and then fires `Worker_Thread` which periodically executes a critical task, *e.g.,* syncing stock updates. After every synch `Worker_Thread` gives up the lock, sleeps[8] for 3 minutes (allowing the CPU to sleep), and re-acquires the lock after waking up. This process is repeated in an infinite loop until `Worker_Thread` is notified by `Main_Thread` using the `mKill` flag to break out of the loop. To initiate the termination of the app, `Main_Thread` sets the `mKill` flag, and calls the API `stop` to signal `Worker_Thread` to initiate the halt which wakes `Main_Thread` up if it is in sleep state. `Main_Thread` releases the wakelock after calling `stop`.

In the normal scheme of things, the code in Listing 4 executes without any energy bug. However, consider the following sequence of events. `Main_Thread` sets the `mKill` flag, signals `Worker_Thread` to stop, and releases the wakelock. Then `Worker_Thread` wakes up, acquires the lock and exits the loop because of the `mKill` flag. As a result, the wakelock remains held by the app and is never released. A key point to note here is that the semantics of the `stop()` API called by `Main_Thread` does not guarantee that the return from the call will be synchronized, *i.e.,* only after `Worker_Thread` exits. Had that been the case, there would have been no race condition and hence no no-sleep bug in the app.

Tracing no-sleep bugs in app source code caused by race conditions is particularly hard since it requires enumerating all the possible execution orderings of the threads. However, using our automatic techniques for detecting no-sleep bugs presented in §7, we were able to detect an instance of a no-sleep bug caused by a race condition in the Android Email App [35, 36, 42], which had a similar pattern as shown in Listing 4.

4.3 No-Sleep Dilation

This category of no-sleep bugs differs from the first two categories in a single aspect: the component woken up by the app is ultimately put to sleep by the app, but only after a *substantially* longer period of time than expected or necessary. For example, consider the code in Listing 1. Suppose routine `net_sync()` usually finishes in a few seconds, but during a particular run it hangs for an unexpected length of time before it returns, the the battery is drained for that prolonged period of time.

While it is arguable that keeping the system on during the execution of a critical task, no matter how long it takes, was indeed the intention by the app developer, we characterize such situations as the third category of no-sleep energy bugs, *no-sleep dilation*. These are considered no-sleep energy bugs for the following reasons: (a) such instances of prolonged component wakeup are usually unexpected, even by the app developer, as we found by reading the log of code commits; (b) the mobile programming API documentation strictly warn developers not to keep the components awake for prolonged periods of time unless it is actually required, *e.g.,* in the Skype app, where a user performing a video call requires the components (screen, CPU) to be switched on from the start till the end of the call irrespective of how long the call persists; (c) instances of no-sleep bugs in this category were observed to cause severe frustration among smartphone users since the energy drain was both severe and unexpected; and (d) the root cause of such prolonged completion time of critical tasks was usually because of a higher level bug (*i.e.,* programming mistake) in the code, which significantly inflated the running time of critical task.

We found two causes for no-sleep dilation in smartphone apps: app delay and app optimizations. We first discuss the dilation caused by app delay in the GPS driver [39] in Android. The driver held wakelocks for longer than needed. In some circumstances, after holding the wakelock, the driver issued a *wait*, waiting for an event. However, after being signaled, a second *wait* was issued causing another wait until the driver was signaled again. All this was done while holding a wakelock. As a result, a higher level bug in handling signals extended the time period the wakelock was being held.

Another cause of no-sleep dilation observed in the apps we studied results from poor placement of component wakeup code in the app. For example, consider the code in Listing 1. The dilation could happen if the app developer, instead of just protecting the critical part *net_sync()*, wrapped a large piece of code in wakelocks. We observed such a bug in the MyTrack app [25] where the developer acquired the CPU wakelock the moment the app was turned on and released it when the app completed. However, the critical part of the code was only the period where the user clicked the track button for location tracking.

5. DEBUGGING NO-SLEEP BUGS

Two general approaches exist to understanding program behavior: those done at compile-time and those done at run-time. Compile-time approaches incur no run-time overhead. While a run-time approach can gather perfect, or near perfect information about a given run, a compile-time approach will (conservatively) determine facts that may be true on any run. Because of the run-time overhead, compile-time approaches are preferred when they are sufficiently accurate, as is the case with our problem. In this paper, we present a static, compile-time solution for detecting no-sleep energy bugs in smartphone apps.

Our solution treats the acquire and release of a wakelock l as a definition of (assignment to) the variable v_l corresponding to l. A definition d of a variable v_l is said to reach some point p in a program if there exists a path from d to p that does not redefine v_l. Therefore, if a definition of v_l corresponding to acquiring a wakelock *reaches* the end of some code region there exists a no-sleep code path in the region. Thus, detecting no-sleep code paths corresponds exactly to the *reaching definitions (RD) dataflow problem* [6], which can be solved by a standard compile-time *dataflow analysis*.

[8] We assume that the `sleep` API call used in Listing 4 registers a wakeup timer with the CPU, which in case the phone freezes during the app sleep, wakes up the CPU at the end of the timeout.

We first present, in §6, our solution when only a single thread is being analyzed, and then, in §7, show how to apply the problem to multi-threaded smartphone apps to detect no-sleep bugs arising from races. We leave detecting and debugging no-sleep dilation bugs as future work.

6. NO-SLEEP CODE PATHS

We first give an overview of dataflow analysis, and then describe our solution as a dataflow analysis problem.

6.1 Dataflow Analysis: An Overview

Dataflow analysis refers to a set of techniques that ascertain facts about program properties by analyzing the effects of statements along different paths of a *control flow graph (CFG)* on those properties. There exist many useful dataflow analysis, *e.g.,* RD (discussed above), *live variable analysis* (which variable values are used after a block), and *available expressions* (which subexpressions have already been computed and are unchanged yet).

Each node in a CFG is a *basic block* of statements, *i.e.,* the block has exactly one entry point and one exit point. There exist a directed edge (B_i, B_j) in the CFG connecting every pair of blocks B_i, B_j such that block B_j can execute immediately after block B_i. There is also an edge from every exception to every `catch` that might catch it. Figure 1 shows an example of a dataflow graph. Two special blocks are added to the CFG: *ENTER* and *EXIT*. There exists an edge from *ENTER* to every block $B_i \neq ENTER$ with no predecessor, and an edge from *EXIT* to every block $B_j \neq EXIT$ with no successor. A *forward* dataflow analysis propagates facts about the program from the *ENTER* to the *EXIT* node while a *backward* analysis propagates information backwards through the graph from the *EXIT* to the *ENTER* node.

Each node in the CFG is annotated with two sets: *GEN* and *KILL*. The *KILL* set contains facts in the analysis that become false in this node, and the *GEN* set contains facts that become true. Each node also has an *IN* and an *OUT* set. The sets associated with a block B can be denoted as *IN[B]*, *OUT[B]*, *GEN[B]*, and *KILL[B]*. For a forward (backward) analysis the *IN* (*OUT*) set will contain facts that are true immediately before the node is visited, and the *OUT* (*IN*) set will contain facts that are true immediately after the node is visited. The *transfer* function describes how the *OUT* (*IN*) set is computed from the *OUT* (*IN*), *KILL* and *GEN* sets. For simplicity we only consider *forward* analysis from this point on.

CFGs with branches contain *join* points where multiple paths come together, *e.g.,* the block containing statements `d8` and `d9` in Figure 1. A *meet* operation decides how the values coming from the predecessor node are combined to form the value of the *IN* set. If the CFG contains cycles, an iterative algorithm that visit nodes repeatedly is used until the analysis converges to a fixed point such that revisiting all of the nodes does not change the values of any *IN* or *OUT* set. The algorithm works by adding the *ENTER* node to a work list. As nodes are processed, if their *OUT* set changes, their successors are added to the work list. When the work list is empty the algorithm has converged at a fixed-point.

We note that all dataflow schemas compute approximations to the actual ground truth. The actual problem being solved is undecidable (*e.g.,* constant propagation [43]). This is because it is undecidable, in general, if a particular path along a CFG will be taken during a program's execution. As a result, dataflow solutions return *conservative* or *safe* estimates to the actual problem. A *conservative* approach guarantees that the results obtained by the analysis will err on the side of safety. Thus while an RD analysis may say more definitions reach some point than actually do, it will never fail to find all definitions that do reach a program point.

6.1.1 *The Reaching Definitions Dataflow Problem*

The first task in applying these concepts to the RD problem is to construct a CFG. Next, we define the *GEN* and *KILL* sets for each block B. The last assignment to a variable v in the block creates a definition d of v that can reach other statements outside the block, and therefore the definition is placed in the *GEN* set. Thus for block *B1* of Figure 1, the definitions at `d1`, `d2`, and `d3` can reach the other blocks, and therefore are added to *B1*'s *GEN* set. The definition of the *KILL* set comes from the following observation: an assignment to some variable v in a block prevents any definition of the variable v outside from flowing through the block. Thus any definition outside the block become members of the *KILL* set. Thus in block *B5*, definition `d8` causes definition `d4` to be in the kill set.

We now define the *IN* set. Consider the set of predecessors of some block B. Any definition that is in the *OUT* set of one of these predecessors can reach B, and thus is in *IN[B]* set of *B*. Therefore, *IN[B]* is the union of the *OUT* sets of all of its predecessors, *e.g.,* $IN[B2]$ is $OUT[B5] \cup OUT[B1]$. Finally, the *OUT* set for a block B must be computed. The *OUT* set is simply the *IN* set with the effects of flowing through the block applied to it. The expression $IN[B] - KILL[B]$ gives those definitions that reached the block and can reach later blocks, and unioning this with the *GEN* set gives all definitions that can pass through this block and reach other blocks. Thus $f_B : OUT[B] = GEN[B] \cup (IN[B] - KILL[B])$ is the transfer equation for block B.

Interprocedural analysis, which incorporates the effects of routine calls and routine arguments, is beyond the scope of this discussion but is covered in detail in [44].

6.2 No-Sleep Code Path Dataflow Analysis

We now formulate the single-thread no-sleep code path problem as an RD problem, and show how to solve it using standard dataflow analysis [6]. We only analyze non-reference counted, no-timer wakelocks and exotic component power APIs. We leave the study of other categories as future work.

6.2.1 *No-Sleep Code Path to Reaching Definitions*

For no-sleep code path analysis, we are only interested in the points in the code path where the smartphone component power is managed, *e.g.,* the points in the CFG where wakelocks for the CPU or screen are acquired or released, or points where the camera is turned on and off. As a result, the *domain* of the dataflow problem is a set consisting of component wakelocks for traditional components and component power management assignments for exotic components. For brevity, from now on we use wakelocks to refer to the power control handles for both traditional and exotic components.

Once the transformation is completed, the no-sleep code path problem is reduced to finding the RD in the transformed CFG, *i.e.,* finding which definition of a wakelock reaches the `EXIT` node of the CFG. If only those definitions that declare all the variables as 0 (*i.e.,* the component can sleep) reach the `EXIT` node, the code is said to be free of no-sleep bugs, since all of the possible code paths put all accessed components to sleep before reaching the end of the CFG, and therefore the end of the code.

Solving the code path problem. We now show how to apply the standard iterative algorithm for dataflow analysis to solve our no-sleep code path problem.

For our no-sleep code path problem, the set of non-zero variable definitions reaching the `EXIT` node represents the no-sleep code path bugs in the app. Table 3 shows, for each block B, the $IN[B]$ and $OUT[B]$ sets at the end of three iterations. It shows the $IN[B]$ and $OUT[B]$ sets are the same at the end of the second and third

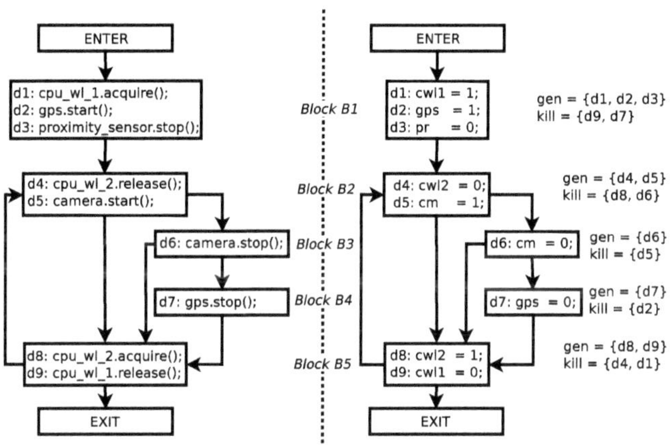

Figure 1: Transforming no-sleep code path into reaching definition dataflow problem, and the resulting *IN* and *OUT* sets.

iteration, and hence the algorithm has reached a fixed-point and converged in three iterations. The value of $OUT[EXIT]$ in the last iteration contains the reaching definitions at the end of the code: all definitions but d_1 and d_4 can reach the end of the code, including d_2, d_5 and d_8 which indicate the existence of a no-sleep bug. Their presence indicates no-sleep code paths along which a component (GPS, camera, and CPU wakelock_2, respectively) is woken up but not put to sleep.

6.2.2 *Handling Uncaught Runtime Exceptions*

Java runtime exceptions (RTE) [45] (*e.g.*, null pointer and array index out of bounds exceptions) can be thrown during normal Java Virtual Machine (JVM) operations. RTEs that are handled explicitly by a `try-catch` block in code are handled as before by adding a path from the source block to the handler block. However, RTEs are often not handled by a program and the thread raising the exception is terminated by the JVM when the exception is thrown.

Uncaught RTEs are a source of no-sleep bugs and must be handled by our analysis. Consider the code in Listing 5. A CPU wakelock is acquired, followed by a call to the critical routine from instance *b*. The wakelock is released after the call. If a RTE is raised (*e.g.*, a null pointer exception on line 3 caused by *b* being `null`), the thread is halted. This results in a no-sleep energy bug since the thread terminates before the wakelock is released. We identified an instance of this bug [26] in our characterization study. Listing 6 details the patch applied by the developer to fix a null pointer RTE (code lines appended with "-"or "+" indicate that these lines were removed from or added to the new version, respectively). The developer added handlers for the null pointer RTE and moved the lock release into a `finally` block to ensure that it is run regardless of any exceptions.

Our technique places an edge from each RTE that is not handled within a routine to the *EXIT* node for that routine. This creates a path for a lock acquire definition to reach the exit, and could lead to more false positives (although we have not seen that in our test cases). Techniques such as null pointer analysis [46], ABCD [47] for array bounds check, and RTE analysis techniques, *e.g.*, [48] can be used to make the analysis more precise and generate fewer false positives.

6.2.3 *Handling Event Based Entry Points*

Android app programming is primarily event-based programming. Unlike traditional code, where the `main()` routine starts the app with the app exiting when `main()` returns, Android app programming typically consists of several functions which are event handlers, one corresponding to each event the app handles. These events could be a button click, an incoming call, a notification from server, a response from a component (*e.g.*, GPS), *etc.* Each event handler is invoked when the event is fired and the handler may in turn invoke a tree of routines underneath it before exiting.

Listing 5: No-sleep code path due to runtime exceptions.

```
1 wake_lock_.acquire();//CPU should not sleep
2 Object b = xyz.getObject(); //b is a reference to an
      object
3 b.net_sync(); //Perform critical task here
4 wake_lock_.release();//CPU is free to sleep
```

Listing 6: Fixing no-sleep code path due to runtime exceptions.

```
      wake_lock_.acquire();
-     client = new AppengineClient(this);
      Log.d(TAG, "onHandleIntent");
+     try {
+       client = new AppengineClient(this);
        //....
+     } finally {
        wake_lock_.release();
+     }
```

Handling multiple entry points of an app creates a new challenge: each handler has its own CFG,[9] and a component may be turned on in one event handler and put to sleep under another (*e.g.*, start camera when start button is clicked and stop camera when stop button is clicked). However, the order of execution of the different events, which is needed to stitch together the CFGs of different handlers, may be unknown at compile time and depends on user interactions.

We handle this complication as follows. (1) For common event handlers (*e.g.*, `onCreate`, `onPause`) which have known invocation orders, we simply perform the RD analysis across them on the combined CFG obtained from stitching together individual CFGs following those invocation orders. For example, if a component is not put to sleep when the app is `paused` after being first `created`, it *usually* is a sign of a no-sleep bug. (2) For the remaining handlers, we ask the developers to specify all expected invocation orders, and then perform no-sleep bug RD analysis on the combined CFG for these orders.

7. NO-SLEEP RACE CONDITIONS

To statically detect possible no-sleep race conditions for multi-thread apps, we adapt the RD dataflow analysis previously developed for parallel programs [49, 50].

A multi-threaded program typically has a repeating pattern of sequential sections ending with a thread *fork*, interleaved execution of parallel threads followed by a thread *join*, followed by the next sequential section in the pattern. Execution is sequential within each thread and so a CFG can be built for the thread. CFGs for different threads can be stitched together by connecting the fork spawning the thread with the *ENTRY* node for the thread's CFG, and the *EXIT* node of the CFG with the join node using a *parallel control edge* [50].

The RD analysis is now modified for this new CFG. Three observations [50] motivate these modifications. (1) All threads in a parallel section are executed; (2) Any of the definitions $d_{ti}, d_{tj}, \ldots, d_{tk}$ to some variable v executing in different threads, and not ordered

[9]We extract the entry point of an event handler, *i.e.*, the root of its CFG, from various `.xml` files in the build tree (*e.g.*, `Manifest.xml`, `main.xml`).

Table 3: Computing IN and OUT for no-sleep code paths.

Block B	$OUT[B]^0$	$IN[B]^1$	$OUT[B]^1$	$IN[B]^2$ (=$IN[B]^3$)	$OUT[B]^2$ (=$OUT[B]^3$)
B_1	$\{\}$	$\{\}$	$\{d_1, d_2, d_3\}$	$\{\}$	$\{d_1, d_2, d_3\}$
B_2	$\{\}$	$\{d_1, d_2, d_3\}$	$\{d_1, d_2, d_3, d_4, d_5\}$	$\{d_1, d_2, d_3, d_5, d_6, d_7, d_8, d_9\}$	$\{d_1, d_2, d_3, d_4, d_5, d_7, d_9\}$
B_3	$\{\}$	$\{d_1, d_2, d_3, d_4, d_5\}$	$\{d_1, d_2, d_3, d_4, d_6\}$	$\{d_1, d_2, d_3, d_4, d_5, d_7, d_9\}$	$\{d_1, d_2, d_3, d_4, d_6, d_7, d_9\}$
B_4	$\{\}$	$\{d_1, d_2, d_3, d_4, d_6\}$	$\{d_1, d_3, d_4, d_6, d_7\}$	$\{d_1, d_2, d_3, d_4, d_6, d_7, d_9\}$	$\{d_1, d_3, d_4, d_6, d_7, d_9\}$
B_5	$\{\}$	$\{d_1, d_2, d_3, d_4, d_5, d_6, d_7\}$	$\{d_2, d_3, d_5, d_6, d_7, d_8, d_9\}$	$\{d_1, d_2, d_3, d_4, d_5, d_6, d_7, d_9\}$	$\{d_2, d_3, d_5, d_6, d_7, d_8, d_9\}$
$EXIT$	$\{\}$	$\{d_2, d_3, d_5, d_6, d_7, d_8, d_9\}$	$\{d_2, d_3, d_5, d_6, d_7, d_8, d_9\}$	$\{d_2, d_3, d_5, d_6, d_7, d_8, d_9\}$	$\{d_2, d_3, d_5, d_6, d_7, d_8, d_9\}$

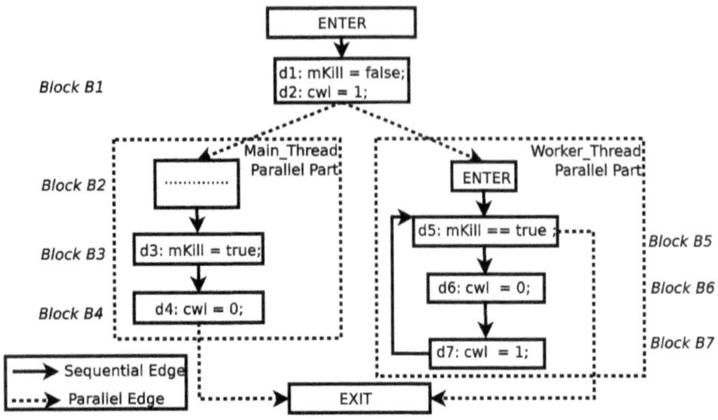

Figure 2: Tracing no-sleep race bug: Parallel Flow Graph (PFG) for code in Listing 4.

by synchronization, may be the last definition of v to execute, therefore no such definition t_{tp} can be said to kill another definition d_{tq}, $p, q \in \{i, j, \ldots, k\}$; (3) Any kill performed *unconditionally* by any thread in a parallel section (*i.e.*, along all sequential paths through the thread) kills all definitions that occur before the parallel section's fork; (4) Any definition d performed *conditionally* within a thread does not kill definitions d' before the thread's parallel section's fork since either d or d' may reach the parallel section's join, and following statements. The details of modifying the dataflow analysis to account for these additional constraints and ordering synchronization are discussed in detail in [50] and are not repeated here.

Figure 2 shows that the RD analysis on the PFG finds that definitions d_3, d_4 and d_7 can reach the `EXIT` node. Since d_7 turns on the component, this is a no-sleep race bug.

8. IMPLEMENTATION

ProGuard Extension: We implemented no-sleep bug tracing as a 1K-LOC extension to ProGuard [15]. The ProGuard tool is used to shrink, optimize and obfuscate Android code and helps to make smaller `.apk` installer files. It builds an intermediate Representation of the input source containing CFGs that we use. We chose ProGuard since it is integrated into the Android build system, it automatically runs when an Android app or the framework is compiled and does not require a separate, manual invocation. However, source code is not required to perform the analysis since ProGuard can run directly on the bytecode generated by the Java compiler. If we only have the `.apk` installer for an app we first use use `ded` [16] to decompile the embedded `.dex` files (Dalvik Executable [51]) and convert them to Java bytecode (`.class` files). We then run ProGuard and the no-sleep bug dataflow analysis on the `.class` bytecodes.

Handling Object References and Intent Resolution: Java object references and intent resolutions are the only indirect control transfer mechanisms in the Android framework and apps. An indirect

Listing 7: Handling simple code paths.

```
wl1.acquire(); wl2.acquire(); //wakeup
if(wl1 != null)     //if object is not null
    wl1.release(); //release the wakelock
if(wl2.isHeld())    //if wl2 is acquired
    wl2.release(); //release the wakelock
```

control transfer mechanism poses problems for static analysis since it is difficult to determine at compile time which class the object is an instance of, or which handler will service the intent, and hence which particular method (routine) will be called. We use a conservative approach by analyzing all routines' references that could possibly be referred to at runtime.

Handling Special Code Paths: To reduce the number of false positives, we handle the two special cases shown in Listing 7. Two wakelocks (`wl1, wl2`) are acquired before `if` conditions and are released under their respective `if`. We found these two usages to be common in mobile apps. The RD no-sleep bug analysis for code in Listing 7 would flag both lock acquires as reaching the end of the block since they are not released in the `else` branches. However, in both cases, it is evident that there is no bug since if the wakelock is either `null` or not held, it need not be released. We handle these two common usages specially, by inserting an `else` branch to the `if` condition which contains a definition of `release()`.

Runtime Exceptions: In our characterization study, we did not observe any occurrences of uncaught RTEs other than null pointer exceptions (NPEs). Hence, in our current implementation, we only handle NPEs. We leave handling other RTEs as future work. To handle NPEs, we trace the `null` reaching definitions at each access point of the object for every object declared in the program. If a `null` definition reaches an object access point, we add a path from that point to the `EXIT` node in the CFG.

Race Conditions: Our current implementation only implements analysis for no-sleep race condition for programs without synchronization points. We leave as future work extensions to handle synchronization points, which can be implemented by adopting the techniques proposed in [50].

9. EVALUATION

We now present experimental results of no-sleep bug detection using our dataflow analysis based no-sleep bug detection tool. We first present a summary of the detection results on 500 apps running on Android and then discuss false positives and the runtime of the scheme.

Methodology: We collected app installers (`.apk` files) for 500 apps, including popular apps like Facebook, Google apps such as gtalk and stock apps in the Android framework including Email and Dialer. These include all the apps listed in Table 2. Automatic analysis of the `manifest.xml` file for permissions reveals that 187 apps explicitly manipulate component wake/sleep cycles. We then decompiled the `.apk` installers using `ded` [16] and obtained

Table 4: Summary of detecting no-sleep code paths.

App type breakdown	#	Breakdown of 51 that contain no-sleep code paths	#
Total input set of apps	500	New bugs	30
Manipulated component	187	Incorrect event handling	26
Fully decompiled	86	`if,else` + exception paths	12
In the framework	6	Forgot release (incl. Services)	3
No-sleep code paths	42	Miscellaneous	1
True negatives	31	False positives	13

86 apps that were decompiled[10] to bytecode and Java source code. For these 86 apps, we ran our detection analysis tool directly on the bytecode, and then used the decompiled Java source code to identify false positives incurred by our analysis. Table 4 gives a breakdown of the apps (left) and causes of bugs in them (right).

No-Sleep Bug Analysis: Manually enumerating all possible paths in the program to verify the correctness of the tool along each of those paths is not humanly possible due to the exponential number of paths. We used the following approach. For each of the 86 apps, we manually[11] tally the no-sleep bugs (no-sleep code paths in single threads). We then check if the bugs had been reported by the tool. Using this approach we segregated the apps into four categories: (a) *True Positive (TP):* Apps in which we manually found the bug and it was reported by the tool; (b) *True Negative (TN):* Apps in which we could not manually find bugs and the tool reported no bug; (c) *False Positive (FP):* Apps in which we did not find bugs manually, but the tool reported bugs; and (d) *False Negative (FN):* Apps where we found a bug manually but the tool did not. The last category contained no apps. Table 4 summarizes the results. We found 31 apps in the TN set and 55 in the TP and FP sets. Below we break down these 55 apps according to the causes.

(1) Incorrect event handling (26): The largest category of bugs in these apps are bugs from inappropriate handling of events in Android apps, specifically the handlers in the default Android *activity* cycle [52]: `onPause()` and `onStop()`. An *activity* is a single focused window that a user interacts with. It is the foreground GUI part of an app that the user sees. The framework calls the `onPause()` event routine of the activity in the app whenever the activity is interrupted by another activity coming in front of it, *e.g.*, an incoming phone call displaying an "incoming call-box" while the user is playing a game. Once the activity is completely sent to the background, *i.e.*, the activity is no longer visible, the framework calls `onStop()`. The framework calls `onDestroy()` when the app finally exits.

The bug occurred in any app that wakes up a component when it is started or resumed in the default event handlers `onCreate()` or `onResume()`, respectively, and lets the component sleep only when the app finally exits, in handler `onDestroy()`. Basically, when such an app is interrupted during this interval by another activity started by the user (like clicking the home screen, or starting another app), the respective component is kept awake since the original app is frozen and may not run for a long time. We observed instances of this bug in 26 apps involving CPU wakelocks (in 23) and GPS (in 3).

(2) `if-else` + exceptions (12): In 12 apps, including Agenda Widget and Android Email App (Table 2), not all paths (using `if-else`) released the component. Also, no-sleep code paths akin to the code in Listing 2 were observed in apps including the

[10]Decompilation failed completely for the remaining apps, *i.e.*, not even a single source file was generated.

[11]These include leveraging bug reports when available for apps in Table 2.

Listing 8: No-sleep code path: false positives.

```
//Use a routine to manipulate component
void WakeUpCPU(boolean wakeup){
    if(wakeup) wl.acquire(); //wakeup
    else wl.release(); //release the lock
} //End WakeUpCPU
void CriticalTask(){
    WakeUpCPU(true);  //acquire the lock
    // Do critical task ...
    WakeUpCPU(false); //release the lock
} //End CriticalTask
```

Facebook and Android Email apps. The component (CPU in these cases) was not allowed to sleep in case of a thrown exception.

(3) Forgot to release a wakelock (3): In K9mail [23], Agenda Widget [21] and SIP service [19], the programmer forgot to release the CPU wakelock. In case of K9mail, this was due to using `ThreadLocal` variables for wakelocks [23].

(4) Miscellaneous (1): In Android WifiService [31], the wakelock is not released in all cases of received messages.

(5) False positives (13): 13 apps were reported to contain a no-sleep path, but upon further manual analysis, they turned out to *false positives*, as discussed next.

We detected a no-sleep bug caused by a race condition in the Android Email app, as was discussed in detail in §4.2. Analyzing multi-threaded Android apps involving explicit synchronizations is left as future work.

Reasons for False Positives: There were two major reasons for the false positives reported in the 13 apps: use of helper functions or variables for component access, and interference of higher level app logic in lock placement.

(a) Helper functions/variables (3): The code in Listing 8 describes an app using a helper function for wakelock manipulation. Routine `WakeUpCPU(boolean)` manipulates the wakelock depending on the boolean input variable. Routine `CriticalTask()` acquires and releases the wakelock using the helper routine by passing `true` and `false` as input arguments. RD analysis of routine `WakeUpCPU(boolean)` suggests that one code path (the `if` branch of the condition) reaches the end of the routine and hence there exists one path in routine `CriticalTask()` that reaches the end of the routine with the wakelock held.

Similarly, we found apps to use additional helper variables to track the liveness status of the component instead of relying on standard API routines (*e.g.*, `isHeld()`) that come bundled with the Android framework. The boolean helper variable is toggled each time the component is switched on or off and is checked before setting the component free. This use of multiple variables to achieve a single purpose not only results in false positives in our static analysis of no-sleep bugs, but also was previously found to be the root cause of multi-variable access bugs [7].

(b) Higher level App logic (10): Listing 9 demonstrates a false positive observed in the Android Dialer app due to higher level app logic. The incoming call handler acquires the wakelock if the caller is not blacklisted by the user and otherwise it immediately returns. Similarly, when the call is disconnected, the app releases the wakelock only when the caller was not blacklisted. This induces a false positive in static reaching definitions analysis since the `acquire()` in `HandleIncomingCall()` reaches the end of `DisconnectCall()`. We leave addressing this issue to future work.

Analysis Runtime and Wakelock Statistics: Table 5 presents detailed statistics of no-sleep bugs in 5 popular apps. For each app,

Table 5: Summary of no-sleep code paths for 5 popular apps.

App	KLOC (# classes) {# lib class}	# wakelock objs (# acq def.) {# rel. def.}	Analysis time (sec)
Facebook v1.3.0	93.5 (712) {710}	1 (256) {128}	408
Telephony	74.8 (326) {495}	7 (18) {29}	53
Exchange	17.0 (626) {952}	1 (19) {12}	51
SipService	3.8 (43) {366}	2 (6) {8}	33
CW	0.3 (8) {100}	1 (1) {1}	3

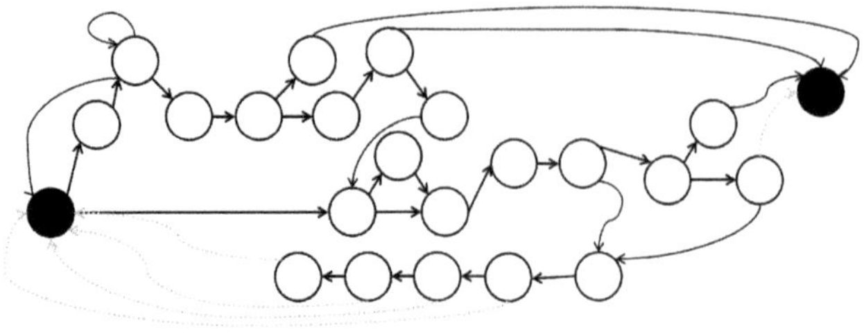

Figure 3: CFG of a routine in the Android Email app.

the table shows the LOC, the number of classes in the app and in the libraries, the number of wakelocks, and how many times they were defined, *i.e.,* acquired and released, and the time taken to run the no-sleep code path analysis.

From the table we see that the runtime of the analysis varied from 3 seconds (for CommonsWare (CW), 0.3 KLOC) to 408 seconds for FaceBook (93.5 KLOC decompiled). The 3 services that belong to the Android framework took about 50 seconds to perform the code analysis. Since the tool performs off-line static analysis, *i.e.,* on a desktop/server, there is no energy drain on the mobile devices. The table also shows that the Telephony service in Android holds 7 different wakelocks (all with option `PARTIAL_WAKE_LOCK`) with 18 acquire and 29 release definitions, for different utilities, including CDMA connection, GSM connection, SMS Dispatcher, Radio Interface Layer, *etc.* Other apps utilize one or two wakelocks. Although the FaceBook app manipulates only one wakelock, the wakelock APIs could be called from multiple locations, resulting in a total of 256 acquire and 128 release definitions.

Battery Drain of No-Sleep Bugs: Table 2 lists the amount of time it will take to drain a fully charged battery under typical usage on the Google Nexus One phone due to the corresponding no-sleep bugs.

Real App CFG (Android Email App): To illustrate the complexity of the apps, apart from the number of lines of code and number of wakelock objects shown in Table 5, we show in Figure 3 a trimmed down version of the CFG of one of the central routines in the Android Email app, `runPingLoop()`. The nodes in the graph show branch points. These include, but are not limited to, handling error conditions such as (a) PING returned empty, (b) PING returned with a login failure, (c) NAT failure, (d) synchronization aborted by the user, and (e) alarm event trigger. The left-most node (solid black) denotes the entry point and the right-most node denotes the exit point. The edges depicted in thick black lines are the ones along which the CPU wakelock is being held, while edges in dotted green lines are when wakelocks have been released. Out of the four edges reaching the exit node, only one reaches after releasing the wakelock.

10. RELATED WORK

Debugging in Mobile Environments: Diagnosing bugs in smartphone environment is a new domain, and there is little existing

Listing 9: No-sleep code path: false positives in the Dialer App.

```
void HandleIncomingCall(){
    if(caller != BLACKLISTED) wl.acquire();
    else return;
    //....Handle rest of incoming call
} //End HandleIncomingCall
void DisconnectCall(){
    if(caller != BLACKLISTED) wl.release();
    else return;
    //....Handle rest of disconnecting call
} //End DisconnectCall
```

work. MobiBug [53] is a framework for mobile debugging that focuses on how to perform lightweight logging on resource-limited smartphones. It proposes three ideas: spreading the logging task among many phones, building a conditional distribution model for the app behavior and its dependencies, and sampling what each phone logs. MobiBug is designed to be a runtime *traditional* bug tracing system, targeting bugs that usually result in app crashes. However, *energy bugs* [4] differ from traditional bugs in that they do not lead to any app crash; the apps continue to work normally except that the battery drains rapidly. [4] presents a taxonomy of many types of energy bugs in smartphones and estimates no-sleep bugs to constitute about 70% of all energy bugs in smartphone apps. Our paper is the first directed towards solving energy bugs in mobile phones.

Applications of Reaching Definitions: Reaching definitions analysis has many uses, and most are part of the compiling folklore (see [6]).

Debugging Software Concurrency Bugs: Debugging traditional software bugs on non-mobile devices is a well-studied topic. There have been several bug characterizing studies (*e.g.,* [5, 54]) which classify different categories of software bugs, *e.g.,* concurrency bugs, semantic bugs, and configuration bugs. The debugging solutions fall into three categories: static, dynamic, or hybrid analysis. Some examples of static analysis include Racerx [55] which uses a flow sensitive, inter-procedural analysis to detect data races and deadlocks, and MUVI [7] which applies data mining to infer patterns of multi-variable access correlations. Our solution for no-sleep energy bugs falls in the realm of static analysis, and is based on reaching definitions analysis.

11. CONCLUSION

This paper makes the first advances towards understanding and automatically detecting software energy bugs on smartphones. It makes three contributions. First, it presents the first comprehensive real world no-sleep energy bug characterization study. Our study reveals three major causes of no-sleep energy bugs and provides useful guidelines and hints to design effective detection schemes. Second, it proposes the first detection solution, based on the classic reaching definitions dataflow analysis, to automatically infer potential no-sleep bugs in an app. Third, evaluation of our tool on 86 Android apps and the Android framework shows that our tool accurately detected all reported instances of no-sleep bugs, as well as 30 instances of new no-sleep bugs.

Our work is only the beginning of smartphone energy bug research and opens a wide avenue for further studies. A sample list of future work includes: (1) develop solutions to no-sleep dilation bugs; (2) explore run-time and hybrid compile-time and run-time solutions to no-sleep bugs; (3) characterize and detect other types of energy bugs in smartphone apps; (4) understand energy bugs in the smartphone OSes; and (5) develop better programming language support to avoid no-sleep bugs at programming time.

Acknowledgments. We thank the anonymous reviewers, especially our shepherd, Maria Ebling, for their helpful comments. This work was supported in part by NSF grant 0916901. Abhinav Pathak was supported in part by a 2011 Intel PhD Fellowship.

12. REFERENCES

[1] "Smartphone sales overtake pcs for the first time." URL: http://mashable.com/2012/02/03/smartphone-sales-overtake-pcs/

[2] H. Zeng, C. S. Ellis, A. R. Lebeck, and A. Vahdat, "Ecosystem: Managing energy as a first class operating system resource," in *Proc. of ASPLOS*, 2002.

[3] "Android powermanager class." URL: http://developer.android.com/reference/android/os/PowerManager.html

[4] A. Pathak, Y. C. Hu, and M. Zhang, "Bootstrapping energy debugging for smartphones: A first look at energy bugs in mobile devices," in *Proc. of Hotnets*, 2011.

[5] S. Lu, S. Park, E. Seo, and Y. Zhou, "Learning from mistakes — a comprehensive study on real world concurrency bug characteristics," in *ASPLOS*, 2008.

[6] A. Aho, M. Lam, R. Sethi, and J. Ullman, *Compilers: principles, techniques, and tools*. Pearson/Addison Wesley.

[7] S. Lu, S. Park, C. Hu, X. Ma, W. Jiang, Z. Li, R. Popa, and Y. Zhou, "Muvi: automatically inferring multi-variable access correlations and detecting related semantic and concurrency bugs," in *SOSP*, 2007.

[8] "Android sensorevent class." URL: http://developer.android.com/reference/android/hardware/SensorEvent.html

[9] "Class powermanager.wakelock: Reference count." URL: http://developer.android.com/reference/android/os/PowerManager.WakeLock.html#setReferenceCounted(boolean)

[10] "Class powermanager.wakelock: Timer based." URL: http://developer.android.com/reference/android/os/PowerManager.WakeLock.html#acquire(long)

[11] A. Pathak, Y. C. Hu, M. Zhang, P. Bahl, and Y.-M. Wang, "Fine-grained power modeling for smartphones using system-call tracing," in *Proc. of EuroSys*, 2011.

[12] L. Zhang, B. Tiwana, Z. Qian, Z. Wang, R. Dick, Z. Mao, and L. Yang, "Accurate Online Power Estimation and Automatic Battery Behavior Based Power Model Generation for Smartphones," in *Proc. of CODES+ISSS*, 2010.

[13] A. Shye, B. Scholbrock, and G. Memik, "Into the wild: studying real user activity patterns to guide power optimizations for mobile architectures," in *MICRO*, 2009.

[14] "Dialer app." URL: http://www.java2s.com/Open-Source/Android/android-platform-apps/Phone/com/android/phone/PhoneApp.java.htm

[15] "Android proguard." URL: http://developer.android.com/guide/developing/tools/proguard.html

[16] "Decompiling apps." URL: http://siis.cse.psu.edu/ded/

[17] "Android - an open handset alliance project." URL: http://code.google.com/p/android/issues/list

[18] "Maemo community." URL: http://maemo.org/intro/

[19] "Sipservice: release wake lock for cancelled tasks." URL: https://github.com/android/platform_frameworks_base/commit/0c01e6e060d079b0a25a44c1159db63944afce17

[20] "Github: Social coding." URL: https://www.github.com/

[21] "Agenda." URL: http://www.androidagendawidget.com

[22] "Facebook 1.3 not releasing partial wake lock." URL: http://geekfor.me/news/facebook-1-3-wakelock/

[23] "K9mail: Simplifying wakelock." URL: http://code.google.com/p/k9mail/source/detail?r=1696

[24] "Checkinmap: Disable location updates when checkinmap is paused." URL: https://github.com/jmschanck/Ushahidi_Android/commit/337b48f5f2725f3e84796fab12947ffbec3c0357

[25] "My tracks android app." URL: http://mytracks.appspot.com

[26] "Babblesink: Move line inside try in case of npe before release of wake lock." URL: https://github.com/hatstand/babblesink/commit/9fbc6f01ce81ef4625a6bd62a3b4b787e6080e36

[27] "Ensuring that the wakelock is released during exception." URL: https://github.com/commonsguy/cwac-wakeful/commit/c7d440f1150887bb9a1a3c44015c7579d7ab1970

[28] "frameworks/base/telephony: Release wakelock on ril request send error." URL: https://gist.github.com/CyanogenMod/android_frameworks_base/commit/133d22d577aa86a8e4095e3af29851d1bd7f7b1b

[29] "Ensure wake lock is released when an ioexception is thrown during a sync." URL: https://github.com/mtuton/android_apps_email/commit/85fec873c4413ef86d40972cc3dbe925ee23e733

[30] "Android issue #9307 fixed - partial wake lock released." URL: https://github.com/CyanogenMod/android_packages_apps_Email/commit/f53bf8f178380ed882a0fa34e10c41f9e8242b93

[31] "Wakelock issue for driver stop." URL: https://github.com/buglabs/android-buglabs-frameworks-base/commit/3bf504df9fc1971078fdde7eed418a0dd8f601e2#wifi

[32] "Fix wakelock leak in powermanagerservice.sendnotificationlocked()." URL: http://gitorious.org/rowboat/frameworks-base/commit/93597ed1839de164c81f83832d4c2373ea32ac8f

[33] "Using a locationlistener is generally unsafe for leaving a permanent partial_wake_lock." URL: http://code.google.com/p/android/issues/detail?id=4333

[34] "Locationmanagerservice: Fix race when removing locationlistener." URL: https://gist.github.com/CyanogenMod/android_frameworks_base/commit/0528b9b26a9d64ba43acd0e334638303d514b8eb#location/java/android/location/ILocationProvider.aidl

[35] "Email application partial wake lock." URL: http://code.google.com/p/android/issues/detail?id=9307

[36] "E-mail app has a bug which causes a partial wake lock to be held until manually interrupted." URL: http://code.google.com/p/android/issues/detail?id=6811

[37] "Android backup service." URL: http://code.google.com/android/backup/index.html

[38] "Googlebackuptransport holds backup wake lock so long which leads to high current." URL: http://www.google.bg/support/forum/p/Google+Mobile/thread?tid=481ff31338a19536

[39] "Fix threading problem that resulted in the wakelock being held too long." URL: https://github.com/CyanogenMod/android_hardware_qcom_gps/commit/a162c4351926285892214b0726aaf07f0631dc72

[40] "Googlemaps holding wakelock for long." URL: http://www.google.com/support/forum/p/maps/thread?tid=016d2cec36d7410b

[41] "java.lang.exception class." URL: http://download.oracle.com/javase/1.4.2/docs/api/java/lang/Exception.html

[42] "Email 2.3 app keeps awake when no data connection is available." URL: http://www.google.com/support/forum/p/Google+Mobile/thread?tid=53bfe134321358e8

[43] J. B. Kam and J. D. Ullman, "Global data flow analysis and iterative algorithms," *J. ACM*, vol. 23, 1976.

[44] E. M. Myers, "A precise inter-procedural data flow algorithm," in *POPL*. ACM, 1981.

[45] "java.lang class runtimeexception." URL: http://docs.oracle.com/javase/1.4.2/docs/api/java/lang/RuntimeException.html

[46] F. Qian, L. Hendren, and C. Verbrugge, "A comprehensive approach to array bounds check elimination for java," in *Compiler Construction*, 2002.

[47] R. Bodik, R. Gupta, and V. Sarkar, "Abcd: Eliminating array bounds checks on demand," in *PLDI*, 2000.

[48] M. Bravenboer and Y. Smaragdakis, "Exception analysis and points-to analysis: better together," in *International symposium on Software testing and analysis*, 2009, pp. 1–12.

[49] J. Lee, D. Padua, and S. Midkiff, "Basic compiler algorithms for parallel programs," in *ACM SIGPLAN Notices*, 1999.

[50] D. Grunwald and H. Srinivasan, "Data flow equations for explicitly parallel programs," in *PPoPP*, 1993.

[51] ".dex: Dalvik executable format." URL: http://source.android.com/tech/dalvik/dex-format.html

[52] "Android activity." URL: http://developer.android.com/reference/android/app/Activity.html

[53] S. Agarwal, R. Mahajan, A. Zheng, and V. Bahl, "There's an app for that, but it doesn.t work. diagnosing mobile applications in the wild," in *Hotnets*, 2010.

[54] Z. Yin, X. Ma, J. Zheng, Y. Zhou, B. Lakshmi, and S. Pasupathy, "An empirical study on configuration errors in commercial and open source systems," in *SOSP*, 2011.

[55] D. Engler and K. Ashcraft, "Racerx: Effective, static detection of race conditions and deadlocks," *SOSP*, 2003.

RiskRanker: Scalable and Accurate Zero-day Android Malware Detection

Michael Grace [†*], Yajin Zhou [†], Qiang Zhang [*], Shihong Zou [*], Xuxian Jiang [†*]
[†]North Carolina State University
{mcgrace, yajin_zhou, xjiang4}@ncsu.edu
[*]NQ Mobile Security Research Center
{zhangqiang, zoushihong}@nq.com

ABSTRACT

Smartphone sales have recently experienced explosive growth. Their popularity also encourages malware authors to penetrate various mobile marketplaces with malicious applications (or apps). These malicious apps hide in the sheer number of other normal apps, which makes their detection challenging. Existing mobile anti-virus software are inadequate in their reactive nature by relying on known malware samples for signature extraction. In this paper, we propose a proactive scheme to spot zero-day Android malware. Without relying on malware samples and their signatures, our scheme is motivated to assess potential security risks posed by these untrusted apps. Specifically, we have developed an automated system called *RiskRanker* to scalably analyze whether a particular app exhibits dangerous behavior (e.g., launching a root exploit or sending background SMS messages). The output is then used to produce a prioritized list of reduced apps that merit further investigation. When applied to examine 118,318 total apps collected from various Android markets over September and October 2011, our system takes less than four days to process all of them and effectively reports 3,281 risky apps. Among these reported apps, we successfully uncovered 718 malware samples (in 29 families) and 322 of them are zero-day (in 11 families). These results demonstrate the efficacy and scalability of RiskRanker to police Android markets of all stripes.

Categories and Subject Descriptors

D.4.6 [**Operating Systems**]: Security and protection – *Invasive software*

General Terms

Security

Keywords

Android, Malware, RiskRanker

*The names of the first two authors are in alphabetical order.

MobiSys'12, June 25–29, 2012, Low Wood Bay, Lake District, UK.

1. INTRODUCTION

In recent years, smartphones have experienced explosive growth. Gartner [13] reports that worldwide smartphone sales in the third quarter of 2011 reached 115 million units – an increase of 42 percent from the third quarter of the previous year. CNN [28] similarly shows that smartphone shipments have tripled in the past three years. Not surprisingly, multiple smartphone platforms are vying for dominance on these mobile devices. At present, Google's Android platform has overtaken Symbian and iOS to become the most popular smartphone platform, being installed on more than half (52.5%) of all smartphones shipped [13].

The availability of feature-rich applications (or simply apps) is one of the key selling points that these mobile platforms advertise. By making it convenient for app developers to develop and publish apps, and easy for users to locate and install these apps, platform providers hope to set up a positive feedback loop in which apps will further attract users to their platforms, which in turn drive developers to develop more apps. Various organizations, therefore, have created app stores to facilitate this process. Platform providers tend to offer official distribution services such as Google's Android Market[1] or Apple's App Store. Cellular carriers also provide their own markets and stores, such as AT&T's AppCenter. Moreover, there are third-party markets altogether, ranging from publishing giant Amazon's Appstore to small, specialty markets like Freeware Lovers.

As platforms become more popular, it seems inevitable that they begin to attract developers of a different kind: malware authors. Moreover, the central role these markets play makes it possible for a tremendous number of mobile devices to be compromised in a very short time. For instance, the `DroidDream` malware infected more than 260,000 devices within 48 hours, before Google removed the related malicious apps from the official Android Market [1]. In light of these threats, there is a pressing need for market curators to examine or vet apps before accepting them for publication.

Unfortunately, the sheer number of new apps uploaded into these markets makes such examination challenging. Using the official Android Market as an example, in the first half of 2011 alone, 223,613 new apps were published [7], which translates to an average of 1,242 new apps each day. Examining such a large number of apps manually – in a timely fashion – is a daunting task. We could choose to deploy mobile anti-virus software to scan these uploaded apps before they are made available for download. However, the

[1]The official Android Market is now part of Google Play.

reactive nature of existing mobile anti-virus software makes it inadequate in identifying new or mutated malicious apps. Specifically, such software relies solely upon a priori knowledge of malware samples in order to extract and deploy signatures for subsequent detection. From another perspective, malware authors may produce new malware variants, or obfuscate existing ones, to evade detection. For instance, the `DroidKungFu` malware has at least 5 different variants; each variant was able to escape detection by existing anti-virus software when it was first reported.

In this paper, we propose a *proactive* scheme to spot zero-day Android malware by scalably and accurately sifting through the large number of untrusted apps in existing Android markets, including both official and alternative ones. Without relying on malware specimens (and their signatures), our scheme is motivated and accordingly designed to measure potential security risks posed by these untrusted apps. Specifically, we divide potential risks into three categories: *high-risk*, *medium-risk*, and *low-risk*. High-risk apps exploit platform-level software vulnerabilities to compromise the phone integrity without proper authorization from users. Medium-risk apps do not exploit software vulnerabilities, but can cause users financial loss or disclose their sensitive information. For example, these apps may illicitly subscribe to premium services unbeknownst to the user. Low-risk apps are similar, but milder; they may collect device-specific or generic, generally readily-available personal information.[2]

Based on this risk classification, we have accordingly developed an automated system called *RiskRanker* to assess the risks from existing (untrusted) apps for zero-day malware detection. The assessment performs a two-order risk analysis. In the first-order risk analysis, we aim to directly identify apps in high- and medium-risk categories. For example, if an app contains code designed to exploit platform-level vulnerabilities (Section 2), it will be flagged as a high-risk app. In the second-order risk analysis, we perform a further investigation to uncover suspicious app behavior. For example, some malicious apps may be designed to encrypt exploit code to evade our first-order analysis. With that in mind, we develop systematic ways to locate those apps and map them to corresponding risk categories. By focusing on these high- and medium-risk apps, we can substantially reduce the number of suspicious apps that require subsequent verification.

We have implemented a RiskRanker prototype and evaluated it using 118,318 apps (104,874 distinct apps)[3] collected over a two-month period, i.e., September and October 2011. We deploy our system and run it in parallel on a local cluster of five machines. The evaluation results are promising. In total, it takes about 30 hours to process all these apps and identify high- and medium-risk apps. Once this process finishes, the first-order risk analysis module reveals 2,461 suspicious apps and the second-order risk analysis reports 840 apps. In total, there are 3,301 suspicious apps (of which 3,281 are unique – as some apps may be flagged by both risk analyses). When such an app is reported, our system also automatically generates the related execution paths that may lead to security risks. With these detailed execution paths, it took a single co-author two more days to examine these apps. In other words, our system significantly reduces the processing time of these two months' worth of apps to less than four days! We believe these results show that RiskRanker can scale to handle the current rate at which new apps are being submitted to the various Android markets.

More importantly, among these 3,281 unique suspicious apps, we successfully uncover 322 (or 9.81%) zero-day malware samples[4] that belong to 11 distinct families. (The first- and second-order risk analyses contribute to identifying 40 and 282 zero-day malware instances, respectively.) In addition, from the same dataset, we also identify a further 396 (12.07%) malware samples from 18 known malware families. As a result, from our two-month dataset's apps, RiskRanker successfully detects 718 (21.88% of 3,281 suspicious apps) malware samples representing 29 different families.

In summary, this paper makes the following contributions:

- To uncover zero-day Android malware in existing Android markets, we propose a proactive scheme (with two-order risk analysis) to assess the security risks introduced by these apps. To the best of our knowledge, RiskRanker is the first system that performs such a large-scale security risk analysis for zero-day malware detection.

- We have implemented a RiskRanker prototype and used it to examine 118,318 apps collected in a two-month period – September and October 2011 – from multiple Android markets. Using RiskRanker, processing this large number of apps takes less then four days. Among 3,281 suspicious apps it reports, we find 718 malicious apps (from 29 malware families), 322 of them being zero-day (in 11 different families). These findings demonstrate the scalability and effectiveness of our scheme.

The rest of this paper is organized as follows: we first describe our system design in Section 2 and then present our prototyping and evaluation results in Section 3. After that, we discuss possible limitations and further improvements in Section 4. Finally, we discuss related work in Section 5 and conclude our work in Section 6.

2. DESIGN

RiskRanker is designed to scalably and accurately sift through a large number of apps from existing Android markets to uncover zero-day malware. By assessing and prioritizing potential risks from these untrusted apps, we aim to use the system to significantly narrow down the search space to a manageable size, which naturally raises the challenging requirements of *scalability*, *efficiency*, and *accuracy*. More specifically, our system must scale to handle hundreds of thousands of apps in a timely and resource-efficient manner. The system also needs to be efficient to winnow its input down to a list that is short enough to verify manually, and accurate enough to not miss malicious apps.

Figure 1 shows the overall architecture of RiskRanker. To meet the above design goals, we assess and translate

[2]Because of that, we in this paper mainly focus on the first two categories, i.e., *high-risk* and *medium-risk*.

[3]In this paper, we consider apps that have the same SHA1 value to be identical.

[4]In this paper, we consider a malicious app to be zero-day if it has not been reported before and cannot be detected by anti-virus software at the time of discovery.

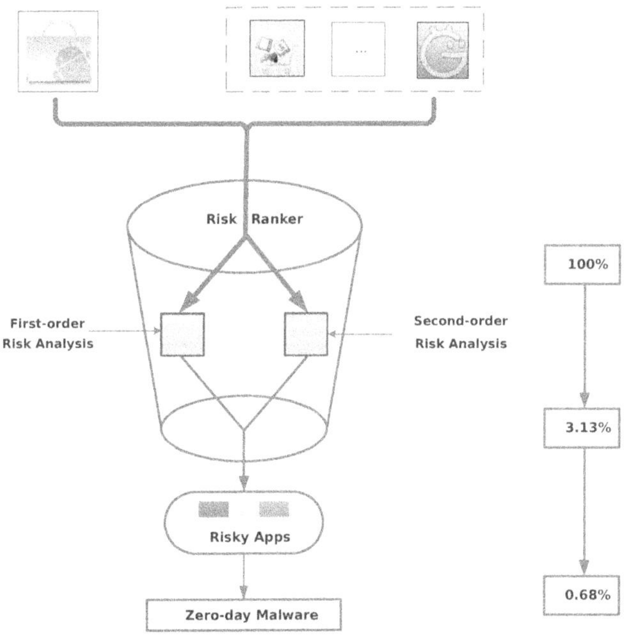

Figure 1: The RiskRanker architecture

potential security risks into corresponding detection modules of two orders of complexity. In particular, we analyze each app from an Android market with a set of analysis modules designed to detect behaviors that we classify as high- or medium-risk: The *first-order* modules handle non-obfuscated apps by evaluating the risks in a straightforward manner; The *second-order* modules capture certain behaviors (e.g., encryption and dynamic code loading) that are in themselves not of concern, but that in conjunction with others may form malicious patterns and be instrumental to detect stealthy malware.

These analysis modules ultimately produce output that includes a severity rating and related evidence to verify the behavioral pattern in each reported app. This output is then sorted by severity to produce a prioritized list of suspicious apps that merit further analysis. In the remainder of this section, we will detail these analysis modules.

2.1 First-Order Analysis

The first-order analysis modules are designed to scalably sift through untrusted apps and expose those high- and medium-risk apps.

Detecting high-risk apps RiskRanker flags any app as high-risk if it carries attack code that will exploit platform-level vulnerabilities in the OS kernel or privileged daemons to obtain superuser privileges. In Android, normal apps are typically constrained by the Linux process boundary. They are allowed to communicate with each other and may choose to contain native code (for improved performance with native speed execution). However, such a design also exposes the underlying OS kernel as well as other privileged daemons to malicious exploitation. If successful, such exploitation will allow an untrusted app to completely bypass the built-in security mechanisms, allowing it unfettered access to the device. Therefore, these exploits pose one of the greatest threats to users.

In order to detect these platform-level exploits, we distill each known vulnerability into a corresponding vulnerability-

Table 1: An overview of existing platform-level exploits in Android

Exploit Name	Vulnerable Program	Malware with the Exploit
`Asroot` [9]	`Linux Kernel`	`Asroot`
`Exploid` [8]	`init`	`DroidDream, zHash` `DroidKungFu`
`GingerBreak` [29]	`vold`	`GingerMaster`
`KillingInThe` `NameOf` [2]	`ashmem`	-

`RATC` [12]
`Zimperlich` [30]
	`adbd`
	`zygote`
		`DroidDream`
		`BaseBridge`
		`DroidKungFu`
		`DroidDeluxe`
		`DroidCoupon`

`zergRush` [20]	`libsysutils`	-

specific signature [52] to capture its essential characteristics, which will be exhibited when the vulnerability is being exploited. For example, `Exploid` [8] takes advantage of a vulnerability in the privileged `init` daemon by not verifying the source of an incoming `NETLINK` message (which contains commands for execution). Accordingly, we extract the following two preconditions as its signature: (1) it sends a message via the socket interface; (2) the message is in a particular format such as "`ACTION=add%cDEVPATH=`", which is used to trigger the vulnerability.

In Table 1, we show the list of known platform-level exploits in Android as well as the representative malware families that make use of these exploits. Notice that if the `RATC` exploit is launched by an app, it actually exploits a bug in `Zygote` instead of the `adbd` daemon. In this case, it behaves like the `Zimperlich` exploit, so we use `RATC` to represent both of these exploits.

To reduce the number of apps for exploit detection, we pre-process each app to detect the presence of native code. For each occurrence, we verify it with these signatures to spot the presence of any of these root exploits. If present, we flag the app as high-risk and take a note of the related vulnerability. Among our dataset of 104,874 distinct apps, we found that 9.42% contain native code.

Detecting medium-risk apps RiskRanker reports as medium-risk those behaviors that could result in the user being charged money surreptitiously or that upload undeniably private information to a remote server. For example, Android has a group of permissions named `android.permission-group.COST_MONEY`, which is defined as "permissions that can be used to make the user spend money without their direct involvement." [17] If abused, these features can be used to construct malware such as SMS Trojans, which send text messages to premium phone numbers that result in charges being placed on the user's phone bill. Such malware is popular due to the direct return it provides to malware authors, but these same features do have legitimate uses, typically in the form of instant-messaging, reminder, or social-networking apps.

To distinguish between such legitimate and malicious uses of potentially costly functions, we leverage the associated semantics defined in Android. Specifically, Android apps make extensive use of callbacks, and each such callback is invoked by the framework under well-defined conditions. As an example, a button on the screen, when pressed by the user, ultimately will lead to `onClick(...)` callback. Malware that intends to charge the user without their knowledge is unlikely to do so via such a callback handler – as such handlers are triggered by user interaction. Similarly, malware that transmits sensitive data is unlikely to prompt the user before doing so. Accordingly, in our search for risky behaviors, we look for code paths that can cost the user money *without* implying such user interaction.

To find such code paths, we perform static analysis on the reverse engineered Dalvik bytecode contained in each app. In particular, we elect to use a conservative program analysis, based on symbolic execution, to determine which callbacks can result in a call to a method of interest. This process involves both control- and data-flow analysis techniques in a bid to unambiguously identify the code path. While data-flow analysis presents some challenges on Android, the properties of Dalvik's bytecode assist us with the control-flow portion of our work.

To elaborate, Dalvik (similar to Java) requires that the control-flow graph of an app should be reliably determined in advance. When Dalvik code is loaded for execution, a static verifier is run to ensure that all references in the code can be resolved. To facilitate this process, the bytecode itself does not contain operations that can lead to ambiguity: for example, there are no computed jumps or field accesses via memory addresses. At any given point in a Dalvik program's execution, the type of all its registers are known and there are a limited number of subsequent instructions that can be branched to. Furthermore, all method and field references are by name, and method execution must always start at the first instruction in the method.

We combine these factors to remove much of the ambiguity that makes the static analysis of Dalvik bytecode potentially unreliable. However, the requirements of the static verifier and our RiskRanker system only overlap but so far. Specifically, the Dalvik's static verifier only aims to ensure that the code being loaded is well-formed, and its notions of well-formedness concern only methods and classes. The verifier therefore only is designed to perform intra-method dataflow analysis, to determine the type of each virtual register at each point in the program. However, Android apps have a very complicated lifecycle, one that often features concurrency and does not require that entry points be called in any strict sequence. Given the sheer number of developers and developer styles any large-scale static analysis system will encounter, inter-method dataflow analysis will necessarily introduce ambiguity for RiskRanker. This ambiguity presents four challenges that merit further discussion.

The first source of control-flow graph ambiguity involves the class hierarchy. When a method signature specifies that a parameter is an object, this can lead to unnecessary edges from that method in the call graph. For example, all objects must ultimately descend from the `java.lang.Object` class, which has a `toString()` method. Consider what happens when a method takes a `java.lang.Object` as a parameter, and subsequently calls that parameter's `toString()` method: the method call could resolve to the `toString()` method of literally *any* class! To mitigate this problem, we employ a form of points-to analysis. Type information within a method is guaranteed to be consistent by the static verifier; therefore, when considering a potential code path, we propagate this intra-method type information across method invocations, thereby reducing the set of methods such an ambiguous method call could resolve to.

Algorithm 1: Android app analysis

Input: method call of interest, set of entry points
Output: code path, constraints

states = (method call site, no constraints)

foreach *state* ∈ *states* **do**
  remove *state* from *states*
  foreach *predecessor of state* **do**
    if *predecessor is an entry point of interest* **then**
      report the code path and constraint information
    if *predecessor is a branch comparison* **then**
      states+ = (*predecessor*, *existing constraints* + *comparison constraints*)
    else
      states+ = (*predecessor*, *existing constraints*)

Second, the possible values that each term in a comparison can take must be known to determine which paths through a program are feasible. Since the verifier does not check this, it is perfectly legal to compile an Android app that has large tracts of dead (unreachable) code, so long as that code is composed of legal methods and classes. In RiskRanker, we address this issue by treating the comparisons encountered along a code path as constraints; the problem then becomes one of constraint satisfaction, where feasible paths have satisfiable constraints. Note that concurrency can still lead to certain ambiguity, as values tested in a comparison may be defined in other threads. As these values are not defined along the current code path, they result in satisfiable constraints if any consistent value exists that would result in the correct branches being taken along the code path. However, it is possible that this consistent value may not exist in practice, resulting in some false positives. We note that this problem is, in the general case, not decidable.

In addition, there are two last potential sources of ambiguity: native code and the reflection language feature. Fortunately, both can be flagged easily, as their APIs and usages are well-defined. Native code is obviously of interest to other analysis modules already, and so is itself a cause for concern. Reflection, on the other hand, simply allows for a programmer to bypass the static verifier in order to call a method by name – where that method's name is given as a string object, rather than in bytecode. Fortunately, this confluence of the data-flow and code-flow graphs of an app is only problematic if the arguments originate outside the current method body (recall that data-flow within a method is well-defined). Therefore, it is possible to ignore a large number of reflection calls, as many such calls use constant arguments in practice. Reflection calls that rely on data defined elsewhere in the current code path can be resolved similarly by stitching together the information that is known from the intra-method dataflow analysis performed in the course of static verification. Unfortunately, it is possible that some data used for reflection originates outside the scope of the current thread. In this case, we can again employ points-to analysis to determine the set of possible values this data can take, and thus the set of methods or fields the reflection call could be referencing.

Mindful of the challenges outlined above, we construct the control-flow graph, locate each method call of interest (e.g., those can be potentially abused for background sending of SMS messages), and perform reachability analysis. Our approach is summarized in Algorithm 1: For each method call of interest, we subsequently traverse the control-flow graph in reverse, looking for a callback method that does not imply user interaction. If such a method is found, we report the call sequence that leads to the potentially risky method call, which can then be verified. Due to the stability of these method call chains, we can further use white-listing to eliminate known-safe paths from repeated analysis (i.e., those paths that arise from a commonly-included library need only be verified once).

Our experience indicates that this algorithm is generic and can be readily extended to handle information leaks that disclose information that directly concerns the user's identity, such as their recent calls or the list of accounts. In particular, the general approach we take essentially employs *backwards slicing* to determine where the arguments to a network call originated from. If this slicing leads to a method call that obtains personal information, we flag the app as having a potential information leak and report the corresponding execution path. In other words, we essentially treat the propagation of dangerous information as a form of the type-resolution problem we addressed earlier. Intra-method dataflow is unambiguous on Android, and inter-method dataflow can be pieced together by propagating interesting data through the call chain. However, the type of fields used to store sensitive information must also be inferred, and such fields might be defined in other threads. We solve it by applying *forward slicing* to see what fields depend on the data returned by a sensitive call. This field information is then retained and referenced by the backwards slicing process used to determine where the arguments to a network operation originate from.

2.2 Second-Order Analysis

Our first-order analysis modules are mainly designed to handle non-obfuscated apps and may fall short for stealthy malware that intensively encrypts or dynamically changes its payload. To mitigate these weaknesses, we accordingly develop second-order analysis modules to collect and correlate various signs or patterns of behavior common among malware yet not among legitimate apps.

Pre-processing The first step in our second-order analysis is to capture certain distinct behavior which may not be malicious in itself but is commonly abused by existing malware. One example is the inclusion of a secondary (child) app inside a host app. The secondary app typically exists as an internal `.apk` or `.jar` file containing its executable code and is saved in the host app's `assets` or `res` directory that can be programmatically accessed. Note that the inclusion of these secondary apps is not inherently suspicious. In fact, it is a common practice that some popular ad and mobile payment service providers require the host apps to embed their ad and mobile payment frameworks inside. However, if the host app is malicious, the same mechanism can be leveraged to dynamically load and execute the embedded (malicious) child app. Motivations for this behavior are many. The child app can request additional privileges or persist after the host app has been uninstalled. Another example is that many legitimate apps use the Java encryption APIs to encrypt their communications and data. However, these same methods can be misused by malware to encrypt their malicious native code, frustrating our high-risk analysis module's efforts to detect embedded root exploits.

To recognize these abusable behaviors, we collect and aggregate various information for our second-order analysis. In our current prototype, RiskRanker is designed to automatically collect the following information: (1) the location of the secondary app; (2) background dynamic code loading and related execution path(s); (3) programmed access to internal `assets/res` directories; (4) use of encryption and decryption methods; and (5) native code execution (e.g., `Runtime.exec(...)`) and JNI accesses.

Encrypted native code execution The detection of high-risk apps in our first-order analysis sought to detect the presence of platform-level exploits in an app by using a set of signatures we developed. However, such an approach may be thwarted if malware authors attempt to encrypt the exploit code. In fact, many known Android malware store their exploit code in encrypted form within the `assets` or `res` directories, from which this encrypted code is read, decrypted and executed at runtime. Fortunately, these related behaviors are collected during our pre-processing phrase and can be correlated for their detection.

Specifically, in a normal scenario, native binaries are supposed to be contained within an app's `lib` directory. The `assets` and `res` directories are designed to contain art assets, user-interface descriptions and the like, but can also contain arbitrary data. The Android framework provides two classes, `android.content.res.AssetManager` and `android.content.res.Resources`, that gate access to this data. While many accesses to such data are innocent enough, accessing these systems in a code path that also contains the encryption and execution methods should raise red flags, as storing native binaries in such non-standard circumstances may signal that the app has something to hide.

In addition, Android contains the Java `javax.crypto` package, which provides a number of encryption and hashing functions in an easy-to-use, standardized form. While we recognize that malware could use their own decryption methods, or bundle third-party crypto libraries, the convenience this package provides appears alluring to malware authors. Consequently, in our current prototype, we opt to check for the use of its APIs.[5]

Moreover, in possible execution paths, we also look for related JNI calls as well as other methods (e.g., `Runtime.exec(...)`) to invoke native code. This is based on the observation that after the encrypted native binaries have been loaded and decrypted, they will be executed.

In RiskRanker, we combine the above three pieces of information together to detect encrypted native code execution. If an app programmatically accesses the `assets` or `res` directories, the encryption APIs, and calls `Runtime.exec(...)` in one execution path, then we assume that the app is executing an encrypted native binary stored in the `assets` or `res` directories. We classify such apps as potentially high-risk,

[5]A more sophisticated system could employ heuristics to detect the presence of in-app cryptographic methods [53].

Table 2: Overall results from RiskRanker

Family	# Samples	Zero-day?	First-Order Risk Analysis		Second-Order Risk Analysis	
			High-Risk	Medium-Risk	Encrypted Native Code Execution	Unsafe Dalvik Code Loading
Androidbox	13			√		
AnserverBot	185	√		√		√
BaseBridge	7		√			
BeanBot	6	√		√		
CoinPirate	1			√		
DogWars	1			√		
DroidCoupon	1	√	√			
DroidDream	2		√			
DroidDreamLight	30			√		
DroidFun	1	√		√		
DroidKungFu1	3				√	
DroidKungFu2	1		√		√	
DroidKungFu3	213				√	
DroidKungFu4	96	√			√	
DroidKungFuSapp	2	√			√	
DroidLive	10	√		√		
DroidStop	7	√		√		
FakePlayer	1			√		
Fjcon	9	√		√		
Geinimi	24		√	√		
GingerMaster	2		√			
GoldDream	21			√		
Kmin	48			√		
Pjapps	17			√		
Pushme	1			√		
RogueLemon	2	√		√		
RuPaidMarket	3			√		
TigerBot	3	√		√		
YZHC	8			√		
Total	718	11	6	20	5	1

just as with the earlier high-risk app detection in our first-order analysis. This module effectively leads to the discovery of 315 malware samples in five malware families, including two zero-day malware families (Section 3).

Unsafe Dalvik code loading Our next second-order detection module captures the unsafe behavior of dynamically loading untrusted Dalvik code. Typically, the bytecode that runs in a Dalvik virtual machine is from the `classes.dex` file embedded in the app or the framework itself. For extensibility, Android provides a mechanism that can be used to load and execute bytecode from an arbitrary source at runtime. Specifically, an app can leverage the `DexClassLoader` feature [11] to load classes from embedded `.jar` and `.apk` files.

The dynamic loading of new class files could potentially change the code to run and thus reduce the effectiveness of our earlier static-analysis efforts (as our analysis does not have access to all of the code that is about to run). On the other hand, we cannot consider all apps with dynamic loading behavior malicious, because this behavior can be used legitimately. For example, apps can use this mechanism to update their functionality without reinstalling the app itself. In fact, among the 118,318 apps we analyzed, 3.90% of them are using `DexClassLoader`. In our prototype, we further combine the dynamic code loading behavior with the existence of a secondary package. This can significantly reduce the number of apps for further analysis to only 424 apps. This combination leads to the discovery of the zero-day `AnserverBot` [21] malware, which accounts for 184 distinct samples.

3. PROTOTYPING & EVALUATION

We have implemented our RiskRanker prototype in Linux using Java and Python. The first- and second-order risk analysis modules are mainly implemented in Python, except for the medium-risk detection module, which is written in Java. In total, there are $3.6K$ lines of Python and $8.7K$ of Java code.

Our current prototype contains a high-risk root exploit detection module that is sensitive to seven kinds of known root exploits (Table 1). It also contains a medium-risk detection module that scans for four different kinds of behavior: sending background SMS messages, making background phone calls, uploading call logs and uploading received SMS messages. Of the medium-risk metrics, we find that background SMS sending behavior is the most useful, and will principally discuss results arising from it.[6]

[6]It is important to note that some malware samples such as

Table 3: Malware discovery in high-risk apps

	Apps with native code	High-risk apps	Actual malware
# of apps	9,877	24	14
Percentage	9.42%	0.02%	0.01%

In our prototype, to quickly index and look up various information associated with the collected apps, we use a MySQL database. In particular, before running our system, we will pre-process each app and extract related information into the database (e.g., where the app was downloaded from, whether it contains native code, etc.).

To demonstrate the effectiveness and accuracy of our prototype, we collected 118,318 apps over a two-month period – September and October 2011 – from 15 different Android markets, including the official one and 14 other alternative markets. As the same app may be present in multiple Android markets, we have in total 104,874 distinct apps: 52,208 (49.78%) of them came from the official Android Market and the remaining 52,666 (50.22%) from other marketplaces. Our prototype was deployed on a local cluster of 5 nodes, each containing 8 cores and $8GB$ of memory. The actual run of our system shows that it can handle approximately 3,500 apps per hour, which means that it took about 30 hours to process all of the apps in our sample.

Among the collected 104,874 distinct apps, RiskRanker successfully uncovers 718 malicious apps in 29 malware families, including 322 zero-day malware that belong to 11 distinct new families. In Table 2, we show a detailed breakdown of our results. Specifically, the first-order risk analysis reveals 220 malware samples from 25 families. The second-order risk analysis leads to 6 malware families with 499 samples. One malware family (`DroidKungFu2`) has a sample detected by both methods, while another (`AnserverBot`) has its host app detected by the second-order techniques and its child app by the first-order techniques. These results show the effectiveness of RiskRanker.

In the following, we examine each risk analysis module and present related results.

3.1 First-Order Analysis

Detecting high-risk apps As mentioned earlier, we consider apps with root exploits to be high-risk apps. In order to detect them, we first locate all apps with native code and then compare their native code with the signatures of known root exploits. More specifically, we check the properties of each file in the app to determine whether it is an `ELF` binary. If so, we then scan this `ELF` file with our predefined root exploit signatures. If there is a match, our system reports it for manual verification.

In Table 3, we report the detection results of high-risk apps. Overall, 9,877 (or 9.42%) of collected apps contain native code. Among them, we find that 24 embed root exploits. So far, we only observe three exploits being used in our dataset: `Exploid`, `RATC` and `GingerBreak`, with `RATC` being the most popular – it accounts for more than 60% of the detected samples.

Further analysis shows that among these 24 high-risk apps, 14 are actually malware from 6 distinct families. Figure 2 shows the breakdown of samples from each of these malware families: 50% of them are `BaseBridge`. From the results, we also successfully uncover two zero-day malware, i.e., `GingerMaster` [14] and `DroidCoupon` [23].

`GoldDream` were only detected due to other behaviors (e.g., uploading SMS messages to a remote server).

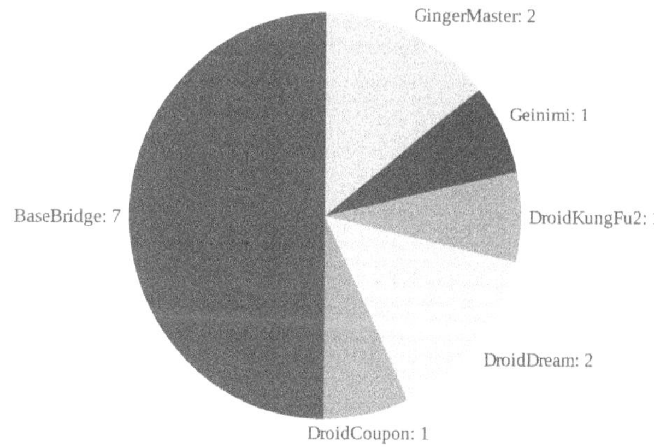

Figure 2: The breakdown of detected malware with root exploits

In particular, `GingerMaster` is the first malware discovered in the wild that makes use of a root exploit capable of rooting Android 2.3 devices. Similar to many others we detected in this study, `GingerMaster` repackaged its exploit code into a popular legitimate app (in this case, one that displays photographs of models). The embedded exploit code, once installed, will be triggered to root the phone and subsequently leverage the elevated privilege to download and install other apps from a remote server without the user's knowledge. The `DroidCoupon` malware similarly piggybacks on popular coupon apps by enclosing the `RATC` root exploit. If successful, it will also attempt to fetch additional apps from a remote server and install them silently. To help conceal their natures, both `GingerMaster` and `DroidCoupon` employ obfuscation techniques. For example, the root exploits used in these two malware families are saved as picture files, having `.png` as the filename suffix. `DroidCoupon` also disguises command strings and URLs as integer arrays.

Interestingly, our system spots one `Geinimi` [22] sample, which "normally" does not have a root exploit in its payload. In this case, the `Geinimi` sample is a repackaged version of `com.corner23.android.universalandroot`, a legitimate jail-breaking app with root exploits. Because of that, it is also detected by our system.

Besides these 14 malware, there are 10 additional apps that appear to have legitimate reasons to contain a root exploit: for example, `z4root` (4 samples) advertises itself as a one-click solution to root devices, while `universalandroot` (2 samples), `itfunzsupertools` (2 samples) and `holentech` (1 sample) provide very similar functionality. The last one is a security app, `com.lbe.security` (`SHA1: 34d90bbed455b370349 0a606e71d0232894c1ca4`), which contains the `GingerBreak` root exploit. In this particular case, we cannot determine whether this app should be considered malicious. Our sample contains several security apps, however, none of the others required such functionality, so this practice is certainly unusual.

Medium-risk apps Medium-risk app detection involves constructing the whole program function call graph and then checking whether any reachable paths exist between a set of source methods and a set of sink methods. For background

Table 4: Malware discovery in medium-risk apps

Family	# Samples	Zero-day?
Androidbox	13	
AnserverBot	1	√
BeanBot	6	√
CoinPirate	1	
DogWars	1	
DroidDreamLight	30	
DroidFun	1	√
DroidLive	10	√
DroidStop	7	√
FakePlayer	1	
Fjcon	9	√
Geinimi	23	
GoldDream	21	
Kmin	48	
Pjapps	17	
Pushme	1	
RogueLemon	2	√
RuPaidMarket	3	
TigerBot	3	√
YZHC	8	
Total	206	8

SMS sending behavior, we use `sendTextMessage/sendMultipartTextMessage(...)` in class `SmsManager` as sink methods. With respect to source methods, we use a comprehensive list of entry points and callback methods that do not involve user interaction. If we find such a path, we then mark the app as a medium-risk and save the detected call chain path in the MySQL database.

In total, our system reports 2,437 apps that exhibit background SMS sending behavior. Because there can be multiple paths from source to sink methods in one app, the total number of paths (3,406) is larger than the number of apps. Also, while it is possible to go through these 2,437 apps one by one, we find that we can still significantly reduce the number of apps that need to be analyzed. In particular, instead of analyzing each app, we opt to analyze each unique path our system detects (as such paths represent the actual behavior of concern). Moreover, as the same path can exist in multiple apps, the number of unique paths is less than the number of apps. For example, the path from the `run()` method of the `com.GoldDream.zj.zjService$1` class to the `sendTextMessage(...)` method exists in 5 different apps, all of which belong to the `GoldDream` malware family. In total, of these 3,406 paths, only 1,223 paths are unique.

Subsequently, we analyzed these 1,223 distinct paths individually. Our experience indicates that this analysis can be done by one person under two days. When a potentially-malicious path was identified by this analyst, we forwarded it to another for confirmation. Ultimately, we discovered 94 distinct malicious paths. We then marked all the related apps corresponding to these malicious paths as malware.

Overall, we discovered 206 infected apps among 2,437 medium-risk apps, representing 20 families, including 8 zero-day malware families. We tabulate these results in Table 4. The discovery of new samples from known malware families indicate that some of these families are still actively spreading in the wild. For example, the `Pjapps` malware was first discovered in February 2011. More than half a year later, it is still present in some alternative marketplaces.

Table 5: Malware discovery in second-order risk analysis

Family	# Samples	Zero-day?
DroidKungFu1 [26]	3	
DroidKungFu2 [25]	1	
DroidKungFu3 [24]	213	
DroidKungFu4 [16]	96	√
DroidKungFuSapp [18]	2	√
AnserverBot [21]	184	√
Total	499	3

Of these new discoveries, a few merit special mention. For example, the `AnserverBot` sample we detected is actually its child app, which contains the background SMS sending behavior. The `DroidStop` malware performs its activities at an interesting point in its host app's lifecycle. When the app is no longer visible to the user, DroidStop sends a SMS to a particular premium-rate number (2623588217) inside the USA. This last malware was once published on the official Android Market, but has now been removed by Google.

3.2 Second-Order Analysis

In our second-order risk analysis, we combine various information to locate apps that feature encrypted native code and unsafe Dalvik code loading behaviors.

Encrypted native code execution As discussed before, we combine three pieces of information together to detect apps that execute encrypted native binaries in the background. In order to collect this information, for each app, we first get all the paths from the app's entry points to the methods that represent the constituent behaviors that make up this greater pattern. More specifically, these methods include decryption methods (from the `javax.crypto` package), methods that access the assets or resources directories, and the `Runtime.exec(...)` method. The paths we detect to each method of interest are saved into a MySQL database. We then examine the entry point for each path in a given app; if all three types of path exist that start from the same entry point, we mark the app as potentially risky.

We found that 5,495 apps contain all three types of path. However, only 328 of those apps contain all three paths originating from the same entry method. We then analyzed these 328 apps and discovered 315 malware samples from 5 different malware families, including 2 zero day malware families; the remaining 13 apps are benign. In Table 5, we show the malware reported by our second-order risk analysis of encrypted native code execution (with the exception of `AnserverBot` which was detected due to unsafe Dalvik code loading). The results are encouraging as all known `DroidKungFu` variants are successfully detected. By capturing the intrinsic characteristics of the `DroidKungFu` malware, we expect that RiskRanker will be able to capture new `DroidKungFu` variants in the future.

Unsafe Dalvik code loading Besides the detection of encrypted native code execution, our second-order analysis also captures the unsafe practice of dynamic Dalvik code loading. In combination with our child app detection logic, we can effectively identify those apps that could load new code at runtime. In our prototype, we first recognize all apps

Table 6: The missed malware samples

Malware Name	#	Malware Name	#
ADRD	1	Gone60	2
BaseBridge	2	jSMSHider	1
DroidDream	1	BatteryDoctor	1
DroidDreamLight	2	TapSnake	1
FakeNetflix	1	-	-

that are capable of dynamic Dalvik code loading (e.g., by looking for references to the `DexClassLoader` class' methods). Among the resulting apps, we further detect the presence of a child app. If such an app is found, we immediately flag its parent for further analysis.

In total, we found that 4,257 apps feature dynamic code loading capability and 1,655 apps contain a child package. Combining these two categories, we find 492 apps in common. After analyzing these apps, we found that some of the uses of `DexClassLoader` originate from embedded libraries, rather than the app itself. For example, one particular library from iBuildApp [15] uses `DexClassLoader` methods, and is contained in 135 apps. Furthermore, one ad library [4] and the Adobe AIR [3] platform extensively use the `DexClassLoader` feature. Accordingly, we choose to white-list these libraries to reduce the number of apps that need to be analyzed.

After white-listing these libraries, we found one particular app that dynamically loads its child package `anserverb.db` from the `assets` directory. Upon analyzing this app, we found it is actually zero-day malware – which we named `AnserverBot` – and accordingly released a security alert about it. It turns out that this malware is one of the most sophisticated pieces of Android malware we have discovered so far. Specifically, it aggressively employs several sophisticated techniques to evade detection and analysis, including Plankton-like [27] dynamic code loading, Java reflection-based method invocation, heavy use of code obfuscation and data encryption, self-verification of signatures, as well as run-time detection and removal of installed mobile security software. The detailed analysis can be find in our report [5]. In total, we detected 184 `AnserverBot` samples.

3.3 False Negative Measurement

Our results so far demonstrate the effectiveness of RiskRanker in uncovering new (zero-day) Android malware. In the following, we further measure the false negatives of our system. To this end, we download malware samples from the public contagio dump repository [10] and use them as the ground truth. Note that the malware samples in this repository are contributed by volunteers and our study shows that there are some duplicates. After removing these duplicates, there are 133 distinct samples from 31 different families.

Our system reports 121 of the 133 samples we collected from the contagio repository. The remaining 12 malware samples are summarized in Table 6. Upon analysis, we discover that the samples that our system fails to detect all fall outside the scope of our current analysis in one of three ways. First off, some malware families engage in malicious behaviors that our system does not attempt to detect. For example, `FakeNetflix` impersonates a Netflix app in a bid to collect the user's account and password information; such social-engineering attacks are very difficult for an automated system to distinguish from legitimate app functionality. `BatteryDoctor`, `Gone60`, and `TapSnake` are spyware that harvest personal information and human judgment is still necessary to distinguish them from many other legitimate apps in the marketplace that may collect user information. The `ADRD` sample engages in click fraud by retrieving a list of ad URLs from a remote server and then visiting them in the background. Second, the missed samples do not share the same malicious payload as others in the same family. For example, our system missed `DroidDreamLight` samples that lack the background SMS sending behavior seen in other samples in the same family. Also, the missed `BaseBridge` samples do not contain the root exploits common to that family. Third, there are some samples that are, in a sense, guilty by association. For example, `jSMSHider` installs a malicious child app, while the missed `DroidDream` "sample" is in fact the malware's innocuous child app. Neither of these samples themselves engage in actual malicious behavior, but the apps associated with them do.

3.4 Malware Distribution Breakdown

We further analyze the distribution of the malware we detected among the Android markets we studied. Sadly, we find that malware can be detected in all of the markets we examined, including the official Android Market. In the worst case, 220 samples were detected in a single alternative market, representing 3.06% of all apps collected from that market. Four alternative markets contain more than 90 malware samples apiece. Specifically, if we sum up the malware samples from these four alternative markets, they together account for 85.98% of all malware samples we discovered. We stress that while no sizable market can be expected to be perfectly secure, some markets clearly do better than others at policing their contents. For example, in the official Android Market, only 2 apps out of 52,208 were malicious. Although we do not know the exact reason why Google's market is so much cleaner than some of the others, we believe that it is likely due to the adoption of Google Bouncer [6] service. Unfortunately, there is limited public information on how Bouncer works.

4. DISCUSSION

While our prototype demonstrates promising results, it is still a proof-of-concept system, and therefore has several important limitations. In this section, we will examine these limitations and suggest possible areas of improvement.

To begin with, our root-exploit detection scheme depends on signatures, which obviously implies that it can detect only known exploits and may also miss encrypted or obfuscated exploits. Our second-order risk analysis alleviates the situation and at present appears to work very well. However, if such techniques start to hinder the deployment of malware in app markets, malware authors might respond by attacking some of the assumptions that make this method work so well. For example, our prototype only considers the `javax.crypto` libraries for convenient encryption detection; nothing prevents an attacker from implementing their own in-app encryption or decryption scheme. Similarly, instead of packaging their native code in the assets or resource directories, such code could be included as large constant arrays in the Dalvik binary itself. While it is possible to counter all of these strategies, it is obvious that the situation can and likely will grow more complicated as malware matures on the platform.

Similarly, our dynamic Dalvik code execution scheme has limitations. Child apps are not the only sources of Dalvik code; such code can be downloaded from the Internet or stored as raw data contained in a constant array, for example. Nothing prevents an attacker from encrypting or otherwise obfuscating a child app, either, in much the same way that native binaries are hidden today. In the long term, these concerns may best be addressed by incorporating some dynamic analysis techniques, such as fuzzing, into systems like RiskRanker. Also, while static analysis may determine that dynamic code loading is taking place, it may not be very effective at identifying precisely what dynamic code is being loaded.

Our medium-risk modules are similarly imperfect. We test for only four distinct behaviors, for example, and our dataflow analyses do not account for code lying outside the execution path very well. While these are general concerns that can be remedied by applying more sophisticated techniques drawn from the existing program analysis literature, likely at the cost of additional processor time, we do recognize one interesting practical matter that came out of our experiences collecting the data for this work. A fraction of Android apps are obfuscated to frustrate casual analysis. When we identified 1,223 unique paths using our medium-risk analyses, 59 of those paths had their class names obfuscated. Obfuscation of this sort leads to two challenges. First, slightly different samples of the same app can have wildly different reported code paths that in fact represent the same behavior, because obfuscation is generally very sensitive to small changes in the input app. For example, the entry function `com.package.Class.run()` may become `a.b.c.run()` in one sample, yet `a.c.d.run()` in another. Secondly, known-safe app paths cannot be white-listed to save analyst effort over time; a known-safe ad library contained in a large number of apps, for example, may still consume analyst-hours of effort when obfuscated apps that contain it are uploaded to a market. Obfuscation is actually recommended by Google [19], so this problem is here to stay. The same issues apply to the second-order encrypted native binary execution analysis, as well. To resolve these issues, it would be better to report paths based on their semantic meaning – what they actually *do* – rather than simply their "names." Unfortunately, obfuscation can also rearrange (in limited ways) the opcodes along an execution path, so this is not a trivial problem to solve. We leave such semantic path description techniques to future work.

Finally, it is important to note that the search for malware is not always black and white. Many risky apps threaten user security and privacy, but are not necessarily malware. Sometimes, these apps are simply written in a potentially dangerous way, but are published by a reputable company; for example, Adobe AIR uses dynamic code execution, which a more aggressive system would flag. Other apps take pains to justify potentially dangerous actions to the user. For example, in the Chinese app markets, some apps display a user agreement when they are first launched, stating that the app will use a premium-rate SMS number to bill the user on some recurring basis. Some of these apps add this functionality to repackaged versions of existing apps, and so may be considered a soft kind of malware. It is impossible to determine that such apps are in fact malware, however, when they are considered in isolation. We also see many instances of classical gray-area malware in the app markets, particularly spyware. Many apps collect more information than they need to function and blithely send it to external parties – it is simply very easy to get such information on mobile platforms. However, it is an open question on where the line should be drawn on such information leaks, and how much apps need to disclose to the user about how their information is being used.

5. RELATED WORK

Recently, smartphone security has been an area of active research. Many researchers have identified areas of concern and proposed solutions to problems on smartphone operating systems. For example, TaintDroid [38] and PiOS [37] were developed to identify information leaks on smartphone platforms, where TaintDroid used dynamic analysis techniques on Android and PiOS applied program slicing to iOS binaries. A raft of follow-up work then sought to address these information leaks by altering the frameworks involved: Apex [48], MockDroid [32], TISSA [57] and AppFence [46] all offer extensions to the Android framework to provide finer-grained control over an app's access to potentially sensitive resources. Most of these efforts are aimed at addressing this problem on the user's phone; RiskRanker, on the other hand, attempts to identify such risky behaviors at the app market. Of the systems listed previously, RiskRanker has the most in common with PiOS, in that both employ program slicing to better understand app behavior. However, PiOS targets iOS binaries for information leaks while RiskRanker looks for potential security risks (including platform-level root exploits and background SMS message-sending) for zero-day Android malware detection.

Another line of research deals with the confused-deputy problem [45] on Android, where inter-process communication channels can inadvertently expose privileged functionality to unprivileged callers. ComDroid [34] and Woodpecker [44] both employ static analysis techniques to detect such attacks in third-party and firmware apps, respectively. To deal with this problem, QUIRE [36] and Felt *et al.* [42] offer extensions to the Android IPC model that allow the ultimate implementor of a privileged feature to check the IPC call chain to ensure unprivileged apps can not launch confused-deputy attacks unnoticed. Similarly, Bugiel *et al.* [33] use a run-time monitor to regulate communications between apps. RiskRanker at present does not check for confused-deputy attacks that target the IPC layer, but its design was informed by our experiences developing the Woodpecker system. We consider these efforts to be complementary to RiskRanker, because one of the lessons learned from these systems is that confused-deputy attacks do not target well-understood, standardized framework components. For example, Schrittwieser *et al.* [51] find that recent Internet-based messaging apps contain security flaws that can allow attackers to hijack accounts, spoof sender-IDs, or even enumerate subscribers. RiskRanker could be extended to identify such vulnerabilities in a similar fashion to ComDroid or Woodpecker, but it would be more difficult to automate the discovery of malware that attempts to *exploit* such vulnerabilities.

Other systems attempt to use or extend the Android permission system to defend against malware. For example, Kirin [40] blocks the installation of apps that request particularly dangerous combinations of permissions, while Saint [49] lets app developers specify permission policies that constrain permission assignment at install-time and use at run-time. RiskRanker conceptually has some similarities with these systems, in that its second-order analyses aim to identify patterns of seemingly innocent API uses that can be indicators of malware. However, Kirin is concerned about the end user and Saint the app developer, while RiskRanker targets app markets themselves. Another interesting system in this category is Stowaway [41], which aims to identify instances of app over-privilege, where an app requests more permissions than it uses. Such over-privilege itself could be used as an input to RiskRanker in the future.

Besides the above defenses, some work has been proposed to apply common security techniques from the desktop to mobile devices. For example, L4Android [47] and Cells [31] apply virtualization techniques to the mobile space, allowing for multiple virtual cellphones to be run on one device, isolated from one another. Meanwhile, MoCFI [35] enforces control-flow integrity in iOS apps without requiring access to their source code or cooperation from their developers. These approaches naturally complement RiskRanker by providing defense-in-depth, but again are designed to be deployed on endpoint devices.

Some work has focused on malware and the overall market health. Felt *et al.* [50] surveyed 46 malware samples on three different smartphone platforms, discussing the incentives that motivated their creation and possible defenses against them. However, this work did not discuss how to discover this malware to begin with. Enck *et al.* [39] studied the 1,100 top free apps from the official Android Market to understand their general security and privacy characteristics. Our work has a much stronger emphasis on malware detection than privacy leak detection. MalGenome [55] aims to systematically characterize existing Android malware from various aspects, including their installation methods, activation mechanisms as well as the nature of carried malicious payloads. Note that it did not focus on the methodology for discovering Android malware. However, the insights behind it is instrumental for RiskRanker to look for certain malicious patterns. Finally, we have developed two related systems in the past. The first, DroidRanger [56], aims to mainly detect known malware in the existing app markets. It requires malware samples to extract behavioral signatures for detection. RiskRanker, on the other hand, is designed to actively detect zero-day malware without relying on malware specimens. Also, while our second-order risk analysis shares a similar spirit with the two basic heuristics in DroidRanger, their differences are fundamentally rooted in their distinct goals: DroidRanger was designed to measure the overall health of an app market, whereas RiskRanker is designed to uncover more sophisticated zero-day malware by assessing and ranking potential risks in each app. The second related system, DroidMOSS [54], is designed to detect repackaged apps using fuzzy hashing. RiskRanker does not have any notion of whether an app has been repackaged, although many malware samples are bundled with a repackaged version of a legitimate app; likewise, DroidMOSS does not aim to detect malware. AdRisk is another recent system [43] that is proposed to systematically identify potential risks posed by existing in-app ad libraries. Although the ad libraries identified by AdRisk do not seem to have malicious intent, they contain potential risks ranging from leaking user's private information to executing untrusted code from the Internet. This potential to do harm places such libraries in a gray area.

6. CONCLUSION

In this paper, we present a proactive scheme to scalably and accurately sift through a large number of apps in existing Android markets to spot zero-day malware. Specifically, our scheme assesses the potential security risks from untrusted apps by analyzing whether dangerous behaviors are exhibited by these apps (with two-order risk analysis). We have implemented a prototype of RiskRanker and evaluate it using 118,318 apps from a variety of Android markets to demonstrate its effectiveness and accuracy: among the apps in the sample, our system successfully discovered 718 malware samples in 29 families, including 322 zero-day specimens from 11 distinct families.

Acknowledgment

We would like to thank our shepherd, Patrick McDaniel, and the anonymous reviewers for their insightful comments that greatly helped improve the presentation of this paper. We also want to thank Zhi Wang, Wu Zhou, Deepa Srinivasan, Minh Q. Tran, Chiachih Wu and Lei Wu for numerous helpful discussions.

7. REFERENCES

[1] 260,000 Android users infected with malware. `http://www.infosecurity-magazine.com/view/16526/260000-android-users-infected-with-malware/`.

[2] adb trickery #2. `http://c-skills.blogspot.com/2011/01/adb-trickery-again.html`.

[3] Adobe AIR 3. `http://www.adobe.com/products/air.html`.

[4] AdTouch. `http://www.adtouchnetwork.com/adtouch/sdk/SDK.html`.

[5] An Analysis of the AnserverBot Trojan. `http://www.csc.ncsu.edu/faculty/jiang/pubs/AnserverBot_Analysis.pdf`.

[6] Android and Security. `http://googlemobile.blogspot.com/2012/02/android-and-security.html`.

[7] Android Market Statistic. `http://www.androlib.com/appstats.aspx`.

[8] android trickery. `http://c-skills.blogspot.com/2010/07/android-trickery.html`.

[9] Asroot. `http://milw0rm.com/sploits/android-root-20090816.tar.gz`.

[10] Contagio mobile malware mini dump. `http://contagiominidump.blogspot.com/`.

[11] DexClassLoader. `http://developer.android.com/reference/dalvik/system/DexClassLoader.html`.

[12] Droid2. `http://c-skills.blogspot.com/2010/08/droid2.html`.

[13] Gartner Says Sales of Mobile Devices Grew 5.6 Percent in Third Quarter of 2011; Smartphone Sales Increased 42 Percent. `http://www.gartner.com/it/page.jsp?id=1848514`.

[14] GingerMaster: First Android Malware Utilizing a Root Exploit on Android 2.3 (Gingerbread). `http://www.csc.ncsu.edu/faculty/jiang/GingerMaster/`.

[15] iBuildApp. `http://ibuildapp.com/`.

[16] LeNa (Legacy Native) Teardown_Lookout Mobile Security. http://blog.mylookout.com/wp-content/uploads/2011/10/LeNa-Legacy-Native-Teardown_Lookout-Mobile-Security1.pdf.

[17] Manifest.permission_group definitions. http://developer.android.com/reference/android/Manifest.permission_group.html.

[18] New DroidKungFu Variant – DroidKungFuSapp – Emerges! http://www.csc.ncsu.edu/faculty/jiang/DroidKungFuSapp/.

[19] ProGuard. http://developer.android.com/guide/developing/tools/proguard.html.

[20] Revolutionary - zergRush local root 2.2/2.3. http://forum.xda-developers.com/showthread.php?t=1296916.

[21] Security Alert: AnserverBot, New Sophisticated Android Bot Found in Alternative Android Markets. http://www.csc.ncsu.edu/faculty/jiang/AnserverBot/.

[22] Security Alert: Geinimi, Sophisticated New Android Trojan Found in Wild. http://blog.mylookout.com/2010/12/geinimi_trojan/.

[23] Security Alert: New Android Malware – DroidCoupon – Found in Alternative Android Markets. http://www.csc.ncsu.edu/faculty/jiang/DroidCoupon/.

[24] Security Alert: New DroidKungFu Variant – AGAIN! – Found in Alternative Android Markets. http://www.csc.ncsu.edu/faculty/jiang/DroidKungFu3/.

[25] Security Alert: New DroidKungFu Variants Found in Alternative Chinese Android Markets. http://www.csc.ncsu.edu/faculty/jiang/DroidKungFu2/.

[26] Security Alert: New Sophisticated Android Malware DroidKungFu Found in Alternative Chinese App Markets. http://www.csc.ncsu.edu/faculty/jiang/DroidKungFu.html.

[27] Security Alert: New Stealthy Android Spyware – Plankton – Found in Official Android Market. http://www.csc.ncsu.edu/faculty/jiang/Plankton/.

[28] Smartphone shipments tripled since '08. Dumb phones are flat. http://tech.fortune.cnn.com/2011/11/01/smartphone-shipments-tripled-since-08-dumb-phones-are-flat/.

[29] yummy yummy, GingerBreak! http://c-skills.blogspot.com/2011/04/yummy-yummy-gingerbreak.html.

[30] Zimperlich sources. http://c-skills.blogspot.com/2011/02/zimperlich-sources.html.

[31] Andrus, J., Dall, C., Van't Hof, A., Laadan, O., and Nieh, J. Cells: A Virtual Mobile Smartphone Architecture. In *Proceedings of the 23rd ACM Symposium on Operating Systems Principles* (2011), SOSP '11.

[32] Beresford, A. R., Rice, A., Skehin, N., and Sohan, R. MockDroid: Trading Privacy for Application Functionality on Smartphones. In *Proceedings of the 12th International Workshop on Mobile Computing System and Applications* (2011), HotMobile '11.

[33] Bugiel, S., Davi, L., Dmitrienko, A., Fischer, T., Sadeghi, A.-R., and Shastry, B. Towards Taming Privilege-Escalation Attacks on Android. In *Proceedings of the 19th Annual Symposium on Network and Distributed System Security* (2012), NDSS '12.

[34] Chin, E., Felt, A. P., Greenwood, K., and Wagner, D. Analyzing Inter-Application Communication in Android. In *Proceedings of the 9th Annual International Conference on Mobile Systems, Applications, and Services* (2011), MobiSys 2011.

[35] Davi, L., Dmitrienko, A., Egele, M., Fischer, T., Holz, T., Hund, R., Nurnberger, S., and Sadeghi, A.-R. MoCFI: A Framework to Mitigate Control-Flow Attacks on Smartphones. In *Proceedings of the 19th Annual Symposium on Network and Distributed System Security* (2012), NDSS '12.

[36] Dietz, M., Shekhar, S., Pisetsky, Y., Shu, A., and Wallach, D. S. QUIRE: Lightweight Provenance for Smart Phone Operating Systems. In *Proceedings of the 20th USENIX Security Symposium* (2011), USENIX Security '11.

[37] Egele, M., Kruegel, C., Kirda, E., and Vigna, G. PiOS: Detecting Privacy Leaks in iOS Applications. In *Proceedings of the 18th Annual Symposium on Network and Distributed System Security* (2011), NDSS '11.

[38] Enck, W., Gilbert, P., Chun, B.-g., Cox, L. P., Jung, J., McDaniel, P., and Sheth, A. N. TaintDroid: An Information-Flow Tracking System for Realtime Privacy Monitoring on Smartphones. In *Proceedings of the 9th USENIX Symposium on Operating Systems Design and Implementation* (2010), USENIX OSDI '10.

[39] Enck, W., Octeau, D., McDaniel, P., and Chaudhuri, S. A Study of Android Application Security. In *Proceedings of the 20th USENIX Security Symposium* (2011), USENIX Security '11.

[40] Enck, W., Ongtang, M., and McDaniel, P. On Lightweight Mobile Phone Application Certification. In *Proceedings of the 16th ACM Conference on Computer and Communications Security* (2009), CCS '09.

[41] Felt, A. P., Chin, E., Hanna, S., Song, D., and Wagner, D. Android Permissions Demystified. In *Proceedings of the 18th ACM Conference on Computer and Communications Security* (2011), CCS '11.

[42] Felt, A. P., Wang, H. J., Moshchuk, A., Hanna, S., and Chin, E. Permission Re-Delegation: Attacks and Defenses. In *Proceedings of the 20th USENIX Security Symposium* (2011), USENIX Security '11.

[43] Grace, M., Zhou, W., Jiang, X., and Sadeghi, A.-R. Unsafe Exposure Analysis of Mobile In-App Advertisements. In *Proceedings of the 5th ACM Conference on Security and Privacy in Wireless and Mobile Networks* (2012), ACM WiSec '12.

[44] Grace, M., Zhou, Y., Wang, Z., and Jiang, X. Systematic Detection of Capability Leaks in Stock Android Smartphones. In *Proceedings of the 19th Annual Symposium on Network and Distributed System Security* (2012), NDSS '12.

[45] Hardy, N. The Confused Deputy: (or why capabilities might have been invented). *ACM SIGOPS Operating Systems Review 22* (October 1998).

[46] Hornyack, P., Han, S., Jung, J., Schechter, S., and Wetherall, D. These Aren't the Droids You're Looking For: Retrofitting Android to Protect Data

from Imperious Applications. In *Proceedings of the 18th ACM Conference on Computer and Communications Security* (2011), CCS '11.

[47] Lange, M., Liebergeld, S., Lackorzynski, A., Warg, A., and Peter, M. L4Android: A Generic Operating System Framework for Secure Smartphones. In *Proceedings of the 1st Workshop on Security and Privacy in Smartphones and Mobile Devices* (2011), CCS-SPSM '11.

[48] Nauman, M., Khan, S., and Zhang, X. Apex: Extending Android Permission Model and Enforcement with User-Defined Runtime Constraints. In *Proceedings of the 5th ACM Symposium on Information, Computer and Communications Security* (2010), ASIACCS '10.

[49] Ongtang, M., McLaughlin, S., Enck, W., and McDaniel, P. Semantically Rich Application-Centric Security in Android. In *Proceedings of the 2009 Annual Computer Security Applications Conference* (2009), ACSAC '09.

[50] Porter Felt, A., Finifter, M., Chin, E., Hanna, S., and Wagner, D. A Survey of Mobile Malware In The Wild. In *Proceedings of the 1st Workshop on Security and Privacy in Smartphones and Mobile Devices* (2011), CCS-SPSM '11.

[51] Schrittwieser, S., Fruhwirt, P., Kieseberg, P., Leithner, M., Mulazzani, M., Huber, M., and Weippl, E. Guess Who's Texting You? Evaluating the Security of Smartphone Messaging Applications. In *Proceedings of the 19th Annual Symposium on Network and Distributed System Security* (2012), NDSS '12.

[52] Wang, H. J., Guo, C., Simon, D. R., and Zugenmaier, A. Shield: Vulnerability-Driven Network Filters for Preventing Known Vulnerability Exploits. In *Proceedings of the ACM SIGCOMM 2004 Conference* (2004), ACM SIGCOMM '04.

[53] Wang, Z., Jiang, X., Cui, W., Wang, X., and Grace, M. ReFormat: Automatic Reverse Engineering of Encrypted Messages. In *Proceedings of the 14th European Symposium on Research in Computer Security* (September 2009), ESORICS '09.

[54] Zhou, W., Zhou, Y., Jiang, X., and Ning, P. DroidMOSS: Detecting Repackaged Smartphone Applications in Third-Party Android Marketplaces. In *Proceedings of the 2nd ACM Conference on Data and Application Security and Privacy* (2012), CODASPY '12.

[55] Zhou, Y., and Jiang, X. Dissecting Android Malware: Characterization and Evolution. In *Proceedings of the 33rd IEEE Symposium on Security and Privacy* (2012), IEEE Oakland '12.

[56] Zhou, Y., Wang, Z., Zhou, W., and Jiang, X. Hey, You, Get off of My Market: Detecting Malicious Apps in Official and Alternative Android Markets. In *Proceedings of the 19th Annual Symposium on Network and Distributed System Security* (2012), NDSS '12.

[57] Zhou, Y., Zhang, X., Jiang, X., and Freeh, V. W. Taming Information-Stealing Smartphone Applications (on Android). In *Proceedings of the 4th International Conference on Trust and Trustworthy Computing* (2011), TRUST '11.

Opaak: Using Mobile Phones to Limit Anonymous Identities Online

Gabriel Maganis*, Elaine Shi†, Hao Chen*, Dawn Song†
University of California, Davis* University of California, Berkeley†

ABSTRACT

Trust and anonymity are both desirable properties on the Internet. However, online services and users often have to make the trade off between trust and anonymity due to the lack of usable frameworks for achieving them both. We propose Opaak, a practical anonymous authentication framework. Opaak enables its users to establish identities with different online services while ensuring that these identities cannot be linked with each other or their real identity. In addition, Opaak allows online service providers to control the rate at which users utilize their services while preserving their anonymity. Hence, allowing the service providers to prevent abuse in the form of spam or Sybil attacks, which are prevalent in such online services that offer anonymity. Opaak leverages the mobile phone as a scarce resource combined with anonymous credentials in order to provide these features. We target two kinds of applications for Opaak and identify their requirements in order to achieve both trust and anonymity. We develop efficient protocols for these applications based on anonymous credentials. In addition, we design an architecture that facilitates integration with existing mobile and web applications and allows application developers to transparently utilize our protocols. We implement a prototype on Android and evaluate its performance to demonstrate the practicality of our approach.

Categories and Subject Descriptors

K.6.5 [**Management of Computing and Information Systems**]: Security and Protection—*Authentication*

General Terms

Security

Keywords

mobile applications, privacy, anonymous credentials, spam, Sybil attacks.

1. INTRODUCTION

Trust and anonymity are often conflicting goals for Internet users. One one hand, users wish to evaluate the trustworthiness of online reviews and recommendations and wish to only establish relations with other trustworthy users. On the other hand, users often wish to remain anonymous and protect their online privacy. For example, users do not want their activities being tracked by the websites they visit or by online advertisement companies.

Our online privacy depends on how we manage our identities. A common approach to remaining anonymous online is to use pseudonyms. However, pseudonymous identities come at the expense of trust i.e., it is difficult to determine whether a pseudonym has been created by a legitimate person or for malicious purposes such as spam. One approach to mitigate this problem is to employ trusted authorities to certify or vouch for the validity of these pseudonyms. Indeed, Douceur showed that Sybil attacks are always possible without a trusted authority to certify the uniqueness of identities in the system (except under extreme and unrealistic assumptions) [23]. Furthermore, a survey showed that trusted certification has been cited as the most common approach to preventing Sybil attacks [28]. However, a trusted certification approach usually comes at the expense of anonymity.

Meanwhile, the mobile phone has become the most personal and most frequently used computing device for people today. As a result, recent trends such as including near field communication (NFC) capabilities indicate a move towards the vision for the mobile phone to become one's universal authentication device i.e., the "mobile wallet" [9]. Mobile phones are also considered to generally be more secure computing platforms than desktop computers [12]. This is partly due to the fact that mobile phones have a more controlled computing environment, where third-party applications are installed through a central application vetting authority (e.g., Apple's App Store). Most importantly, mobile phones can serve as a natural form of scarce resource that can defend against Sybil attacks i.e., people can only finance a limited number of mobile phones.

For these reasons, websites currently employ mechanisms such as SMS verification before users can utilize the services they offer. For example, Craigslist [1] may require a mobile phone number from a user. Craigslist sends a verification code to the number and the user must submit this code before they can post their ad. However, such a mechanism also comes at the expense of anonymity i.e., Craigslist learns the user's mobile phone number.

Fortunately, there is a cryptographic mechanism that can help called *anonymous credentials* [15, 17, 18]. Anonymous credential schemes allow users to demonstrate possession of a credential granted by a trusted authority without revealing anything other than the fact that they own such a credential. We can use this tool to enable a trusted certification approach together with the scarcity of mobile phones in order to provide users with anonymity while preventing Sybil attacks at the same time. In fact, anonymous credentials have formed the basis for privacy-preserving digital identity management frameworks [16]. However, such frameworks have very broad goals and focus on demonstrating users' attributes (e.g., age) to online services. Proving such attributes in a privacy-preserving manner require complex algorithms (e.g., proofs such as [4, 31]) which can become impractical on the mobile phone.

In this paper, we propose a practical anonymous authentication framework, Opaak (OPen Anonymous Authentication frameworK), that provides user's with privacy and anonymity online without the expense of trust. We devise efficient protocols based on anonymous credentials that can be deployed on the mobile platform by targeting specific but practical kinds of applications.

1.1 Motivating Applications

Anonymous, unlinkable accounts across different websites. Single sign-on authentication frameworks such as OpenID [6], allow a user to authenticate to different websites using the same credentials. However, the user's identity is revealed to the websites. Even if the user chooses fictitious identities, these websites can link these identities to the same user. Moreover, the websites where the user has created accounts at can be tracked by the OpenID identity provider.

With Opaak, users can create anonymous accounts across different websites and they cannot be linked with each other. In addition, the Opaak anonymous identity provider does not learn which websites where the user has created these accounts.

Anonymous, unlinkable posts with rate limiting. User opinions — such as votes, ratings, reviews, and recommendations — are important contents in online communities, such as IMDB, Yelp, and Amazon. However, privacy concerns may discourage users from contributing. For example, a user may hesitate to rate a politically sensitive movie, or to review a doctor specializing in HIV treatment. To preserve anonymity, the user could use a different pseudonym for each of her posts, but this would reduce the trustworthiness of her posts, as malicious users could launch spam or Sybil attacks.

Opaak allow users to anonymously contribute votes or ratings in such online services. It guarantees that no two posts from the same user are linkable. To prevent malicious users from manipulating the ratings by aggressively spamming the recommendation system, Opaak allows the site administrator to limit the rate of user posts e.g., a user can contribute at most k posts within a time period T.

1.2 Contributions

We develop Opaak, an anonymous authentication framework that takes advantage of the mobile phone as a natural form of scarce resource in order to limit anonymous identities on the Internet. More specifically, this paper makes the following contributions:

- We demonstrate that applications based on full-fledged anonymous credentials can be deployed on the smartphones with very practical performance. We define two types of applications for Opaak and identify each of their requirements in order to enable trust and anonymity simultaneously. To the best of our knowledge, we are among the first to do this along with [37].
- Opaak emphasizes the use of mobile phones as a scarce resource to solve the Sybil problem on the Internet. Although we may not be the first to make this observation, we are the first to demonstrate how to leverage this fact while simultaneously respecting user privacy.
- We develop a set of protocols related to k-times anonymous authentication (k-TAA) schemes that are efficient and practical on the mobile platform. There are numerous existing k-times anonymous authentication schemes however, they are not suitable for the mobile platform precisely because of range proofs and the state-of-the-art algorithms for doing this do not scale well. We show that we can achieve rate limiting for our targeted applications without resorting to range proofs.
- We design an architecture for deploying Opaak that facilitates integration with existing web applications. Our proxy architecture allows application developers who may not be cryptography experts to utilize Opaak transparently. We believe our architecture is more lightweight compared to existing anonymous credential based systems as a result of our focus on smartphone deployment and our well-defined set of target applications.
- Finally, we implement a prototype of Opaak on Android, evaluate its performance and show that it is feasible for real-world deployment.

2. APPLICATIONS AND REQUIREMENTS

Opaak has three main participants namely the *users*, *relying parties* and *anonymous identity providers*. Relying parties offer some form of service that users wish to use (e.g., Wikipedia [8], Craigslist [1]). People who wish to post advertisements on Craigslist would be considered users. An anonymous identity provider (AIP) is a trusted third party, which vouches for the identity of a user. For example, a cellular service provider might serve the role of an anonymous identity provider and certify that a user is one of its subscribers.

Figure 1 shows a high-level overview of the interactions between the participants. First, users join the system by registering with an AIP and thus acquiring an anonymous credential. The anonymous credential forms the basis for users to create cryptographic proofs that enable them to anonymously authenticate to relying parties so they can use the services they offer. At the same time, relying parties are able to verify properties about the user that allow them to prevent abuse (e.g., spam) while respecting the user's privacy.

Our main goal is to design an authentication framework with privacy-preserving and single sign-on (SSO) properties. However, since anonymity necessarily introduces an avenue for abuse, we must go a step further than existing SSO frameworks (e.g., OpenID [6]) and provide mechanisms to

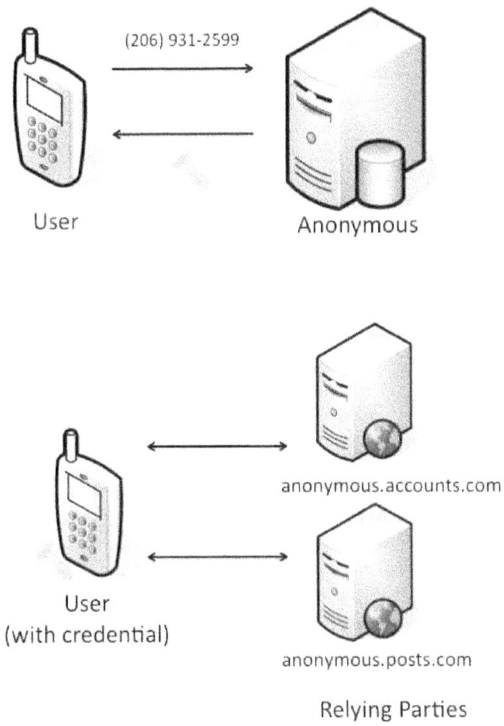

Figure 1: Overview of Opaak. A user is issued a credential by the anonymous identity provider upon presenting their phone number (top). The user can then anonymously utilize services offered by relying parties like posting messages (bottom).

protect relying parties from spam and Sybil attacks. To describe our goals, we logically separate them using two kinds of applications.

First, Opaak supports *anonymous single sign-on authentication*, where users can register anonymous accounts at relying parties then login to them afterwards. An example would be a service like Wikipedia where users would like to edit entries anonymously but at the same time, Wikipedia would like to maintain some form of accountability for edits from the users.

Second, Opaak supports *anonymous message boards* where users can post messages, rate a product, vote on a poll etc. These types of applications offer services but don't necessarily need the users to create an account with them. An example would be a service such as Craigslist where users can post advertisements. As shown by some of the mechanisms they have employed such as mobile phone number and e-mail address verification, spam and Sybil attacks are a concern for Craigslist.

We note that the two applications are not mutually exclusive. A relying party can adopt both anonymous SSO authentication protocols where a user has one or more unlinkable accounts, as well as the anonymous message board protocols where a user may anonymously post messages without being logged in.

In addition to the goals we detail below, we also make it our goal to make these applications accessible to developers who may not be experts in cryptography. Our framework must be designed such that developers can transparently utilize our protocols and easily integrate them into their applications.

2.1 Anonymous Single Sign-On Requirements

Some examples of existing SSO frameworks include OpenID and Facebook Connect [3]. In these frameworks, there are one or more identity providers such as Google or Facebook by which users consolidate their identity or login name. When users wish to authenticate with a relying party, instead of having a different usernames and passwords at each relying party, the user can simply login to Google's OpenID service, or use Facebook Connect. Hence, the identity provider (Google or Facebook) vouches for the user's identity to the relying party. Unfortunately, these SSO frameworks were not designed with user privacy in mind.

2.1.1 Privacy issues

Today's SSO frameworks raise important privacy concerns both for users and relying parties. A user's accounts with different relying parties can be linked with each other, and sometimes even linkable to their real identities. Consider users who use the Facebook Connect feature to login to different relying parties, their accounts and activities at these relying parties could be easily linked to their Facebook identity. For example, relying parties could collude to find out which users they have in common. Furthermore, since the majority of users use their real names on Facebook, these accounts can then be linked to their real identities.

The identity provider can track which relying parties a user is receiving services from. This presents a privacy concern both for the user and for the relying party. Clearly, users sometimes wish to remain anonymous when they visit certain kinds of relying parties (e.g., pornographic websites, or alcohol self-help groups). The relying party may also wish to hide from the identity provider how much traffic or users they have, as this can be sensitive business information.

Therefore, we believe that there is a need for an anonymous SSO framework. Such a framework should offer the following privacy and security guarantees.

2.1.2 Privacy requirements

Unlinkability across different relying parties. We wish to guarantee complete unlinkability for a user's accounts at different relying parties. Given an account at one relying party, there should be no way to determine whether an account at another different relying party belongs to the same user or not. In this way, a user's account at an alcohol self-help group website cannot be linked to his professional profile on sites such as LinkedIn [5].

Transactional privacy from the identity provider. The identity provider should not learn which relying parties a user receives service from or when a user logs in to a certain website. This protects the privacy of both the user and the relying party.

Unlinkability between different accounts with the same relying party. If a user registers more than one account at the same relying party, these accounts should be completely unlinkable. For example, if a user wishes to register two accounts with Wikipedia for editing, one under his real name, and another under a fake name, these two accounts should not be linkable to the same user.

2.1.3 Misuse prevention requirements

While enabling anonymity, Opaak must also prevent abuse in the form of Sybil attacks [23]. These attacks occur when a user is able to create multiple accounts with a relying party thus creating Sybil identities with that relying party. More specifically, we identify our requirements for preventing misuse below.

Limited number of anonymous identities per user. When a user joins the system and registers with an AIP, the AIP must request that the user demonstrate possession of a scarce resource, such as a valid phone number or a credit card number. In this way, a user can only have as much anonymous identities in the system as the resources they actually possess and effectively limiting the capability of a user to create Sybil identities.

Limited number of accounts with the same relying party. Relying parties may wish to limit how many accounts each user can have. Currently, relying parties achieve this through various approaches. Some common approaches include the CAPTCHA [39], the verification of phone numbers via SMS, or the verification of credit card numbers.

We recognize that in some cases, multiple accounts might be desirable like in the Wikipedia example above. In Opaak, we do not prevent users from creating multiple accounts but seek to do it in a controlled fashion with strong guarantees about identities in the system. Opaak allows each user to have up to k accounts at a relying party, where each relying party can choose its own value of k. In addition, if an account is found to be misusing or abusing the service, the relying party can simply freeze that account.

2.2 Anonymous Message Board Requirements

The second type of applications we wish to support are anonymous message board like applications. In these types of applications, users perform actions at a relying party where accounts are not necessary. For example, a user may wish to post a message anonymously, anonymously rate a product or cast a private vote.

2.2.1 Privacy requirements

Unlinkability for any two actions. In anonymous message board type of applications, we wish to ensure that any two actions (e.g., messages posted, ratings, or votes) of the same user, at the same or different relying parties, are completely unlinkable. Given an action at one relying party, there should be no way for the same relying party or another relying party to determine whether another action was performed by the same user or not.

Transactional privacy from the identity provider. Similar to the anonymous SSO application, we also wish to achieve transactional privacy against the identity provider, so that the identity provider does not learn the relying parties utilized by a user.

2.2.2 Misuse prevention requirements

Similar to the anonymous SSO application, our second type of application must also prevent abuse in the form of Sybil attacks. In addition, Opaak must include a mechanism to throttle the usage of services offered by the relying parties to prevent spam.

Limited number of anonymous identities per user. We also wish to defend against Sybil attacks and ensure that each user can only have a limited number of anonymous identities. As discussed above, this can be achieved by requiring that the user demonstrate possession of some scarce resource such as a valid phone number or credit card, when the user registers with the AIP.

Spam prevention. Anonymity is seemingly contradictory to the notion of spam prevention. If the relying party does not learn the identity of the user, how can it throttle the rate at which a user casts votes or posts messages? This question has been answered by earlier work which demonstrates how to realize k-times anonymous authentication [15,33]. Opaak also provides spam prevention and rate throttling features through its *rate limiting pseudonyms*. We elaborate on this technique in Section 3.

2.3 Non-Goals

We consider the following goals to be orthogonal to our efforts, and outside the scope of this paper.

We do not consider network-based attacks on the anonymity of users such as correlating the IP addresses from which requests come from. When requests are generated from mobile phones, today's cellular infrastructure provides some natural protection against IP correlation attacks, as many mobile phones share the same public IP address of the Internet gateway at their cellular provider. Furthermore, existing mechanisms to mitigate these attacks such as Tor [22] can be easily composed with our system so that relying parties do not trivially see the originating IPs of user requests.

Device fingerprinting is another potential approach to identify the origin of network packets. This class of attacks is very challenging to defend against especially when they are based on physical properties of a device (e.g., clock skew [27]). We do not consider such attacks.

3. OPAAK PROTOCOLS

In this section, we describe the protocols we developed in order to realize our target applications.

3.1 Preliminary

First, we briefly describe zero-knowledge proofs and the CL-signature scheme. For a more comprehensive and formal description of these cryptographic techniques, we refer the reader to the specification document for the Idemix cryptographic library [26].

3.1.1 Zero-knowledge Proofs

A zero-knowledge proof of knowledge (ZKPK) is a two-party protocol where a prover convinces a verifier knowledge of a secret value without disclosing the secret value itself. More concretely, consider a prover Alice with a secret key x and public key $y = g^x$. A verifier Bob only knows the values (g, y) but through a ZKPK, Alice can convince Bob that she knows x such that $y = g^x$ without revealing x itself.

In describing our protocols, we use the notation by Camenisch and Stadler [19]. Specifically, to denote the ZKPK of integers α and β such that $y = g^\alpha h^\beta$ and $u < \alpha < v$ holds, where y, g, h are elements of a mathematical group

$G = \langle g \rangle = \langle h \rangle$ we would write the following:

$$PK\{(\alpha, \beta) : y = g^{\alpha} h^{\beta} \wedge (u < \alpha < v)\}$$

We use the various ZKPK techniques available in the Idemix library implemented using a common three-move zero-knowledge protocol [32]. As the name suggests, these protocols require three rounds of communication. Fortunately, they can be made non-interactive (reduced to one round) by means of the Fiat-Shamir heuristic [24], and we denote these with SPK for "signature proof of knowledge" instead of PK. In addition, there are techniques implemented in the Idemix library which allows us to compose multiple ZKPK into a single proof for efficiency. We omit the details of these techniques in the description of our protocols for simplicity.

3.1.2 CL-signatures

The signature scheme proposed by Camenisch and Lysyanskaya [18] is a powerful technique that allows a user to prove possession of a signature to a verifier *without revealing* the underlying message or even the signature itself using efficient ZKPK. In the scheme, the public key of a signer is of the form (n, R, S, Z), where n is a special RSA modulus of length l_n and $R, S, Z \in QR_n$ are generators of QR_n, the group of quadratic residues modulo n. A CL-signature over a message m is of the form (A, e, v). The value A is computed such that $A^e \equiv R^m S^v Z \pmod{n}$ where v is a random number and e is a random prime number with bitlengths l_v and l_e, respectively.

The scheme includes the signing and verification algorithms i.e., CL-SIGN and CL-VERIFY, respectively. In Opaak, we utilize the variant of the CL-signature protocols where the signer and the verifier do not learn the value of the message m. Instead, the signers and verifiers can use ZKPK to ascertain m. In addition, the scheme also includes a CL-SHOW algorithm. This algorithm allows the signature holder to prove to a verifier that it has obtained a signature from a signer without revealing the message or the signature itself i.e., the *proof of possession* mentioned above. Note that implicit in the CL-SHOW algorithm is verification of the signature with CL-VERIFY. We denote a (non-interactive) proof of possession of a CL-signature over a message m as follows:

$$SPK\{(e, m, v) : A^e \equiv R^m S^v Z \bmod n\}$$

3.1.3 Rate Limiting Pseudonyms

At the foundation of the Opaak protocols lies our scheme for forming authentication tokens, which we call *rate limiting pseudonyms*. Simply put, a rate limiting pseudonym (rnym) is globally unique identifier that must belong to a single user i.e., a user can create multiple rnym but an rnym cannot be the same for two users (or even the same user, given different parameters). In the following sections, we show how this simple property enables us to develop protocols that enable anonymity while preventing spam and Sybil attacks at the same time. We discuss the details of how an rnym is formed below.

First, we define some public parameters of the scheme. We denote the group of non-negative integers $\mathbb{Z}_{\Gamma}^*$ with order $\Gamma - 1$. The public parameters consists of (Γ, ρ, g), where g is a random generator $\langle g \rangle$, which is the subgroup of $\mathbb{Z}_{\Gamma}^*$ of prime order ρ. Accordingly, the bitlengths of Γ and ρ are given by l_{Γ} and l_{ρ}. We also employ a cryptographic hash function H mapping $\{0,1\}^* \rightarrow \mathbb{Z}_{\Gamma}$.

Now, given a user's secret key uk and an arbitrary string s, we can form an rnym as follows:

$$g_{\mathsf{rnym}} = H(s)^{(\Gamma-1)/\rho}$$
$$\mathsf{rnym} = g_{\mathsf{rnym}}^{\mathsf{uk}} \bmod \Gamma$$

Rate limiting pseudonyms have the following useful properties we take advantage of: (1) an rnym is unique with respect to uk and the string s hence, two rnym formed with different (uk, s) pair, cannot be linked with each other, (2) two rnym formed with the same uk and s pair will always be the same. (3) following from these properties, other users who do not possess uk can form the set of rnym of the user who holds uk as a secret.

Next, we introduce the notion of *rate limiting strings* (rls). Rate limiting strings are strings which encode some form of counting, the very basic of which could be "1", "2", or "3" etc. In our approach, we use a concatenation of the domain name of a relying party, time period identifiers and counters to form the rate limiting string, depending on our target application. For the relying party at *anonymous.posts.com*, during time period T, a user posting for the i-th time would form the following rls = "$anonymous.posts.com||T||i$" to be allowed to post a message. Indeed, an rnym as defined in the equation above is the same as the "domain pseudonym" feature of the Idemix library [26], where s corresponds to the domain name of a website. However, by defining the meaning of the input to the hash function H more specifically and using rate limiting strings instead, we are able to create the protocols necessary for our target applications.

This way of forming identifiers is very similar to the authentication tokens or tags used in anonymous blacklisting schemes [34–36] and the first k-times anonymous authentication (k-TAA) scheme by Teranishi [33]. The key difference is that in our protocols, we do not hide the base g and how it was formed i.e., it is a public parameter. Hiding the base g would introduce more complex ZKPK as in the anonymous blacklisting and k-TAA schemes. Indeed, not hiding the base weakens our definition of anonymity slightly and the keen reader might have noticed that an rnym is not fully anonymous. We discuss this issue further in Section 6.

3.2 Setup

Recall that Opaak has three main participants which include the users, relying parties and the anonymous identity provider (AIP). In the setup phase, the participants generate cryptographic keys and agree on system parameters.

First, the system parameters are generated and agreed upon by all participants. The AIP generates the public group parameters (Γ, ρ, g) necessary for the rnym scheme. In addition, it is necessary to define the bitlengths $l_n, l_e, l_v, l_{\Gamma}, l_{\rho}$, which determine the security of the system. In the case of multiple AIPs, the federation must agree on these values. For a more comprehensive list of these parameters, we refer the reader to the Idemix specification [26].

Next, the participants generate their respective cryptographic keys. The AIP generates its keypair (sk, pk) and publishes its public key necessary for the CL-signature scheme pk $= (n, R, S, Z)$. Users generate a master secret uk upon joining the system perhaps via downloading the credential service smartphone application of an AIP. Relying parties do not require cryptographic keys and must simply obtain (securely) a copy of the system parameters described above.

3.3 Registration

The registration protocol is executed between a user Alice and an AIP as she joins the system. This process is analogous to the part of the login procedure in conventional SSO frameworks when a user logs in to the identity provider instead of the relying party. We note that unlike existing SSO frameworks, this procedure is only done once in Opaak. In fact, this is key to achieving transactional privacy.

As described in Section 2, we would like to limit the number of anonymous identities that a user can assume in Opaak. This both protects against Sybil attacks and assures relying parties that the users they are serving correspond to real human beings, or at least a close approximation. To achieve this, Opaak requires users to demonstrate possession of some scarce resource (ResourceName). Some examples of the scarce resource include the user's mobile phone number or the user's International Mobile Subscriber Identity (IMSI).

The protocol enforces that each user (with one unit of resource) can only register up to k times with an AIP. Of course, ideally $k = 1$ but we allow for flexibility depending on the policies an AIP would like to implement.

1. Alice initiates the protocol by sending the AIP a registration request and ResourceName.
2. The AIP maintains a database of ResourceName it has seen. If the ResourceName submitted by Alice has been $> k$ times, then it terminates the protocol and returns `failure` to Alice.
3. The AIP must verify if ResourceName is valid. For example, with mobile phone numbers, the AIP could send an SMS message with a verification code. If the verification fails, then it terminates the protocol and returns `failure` to Alice.
4. Next, the AIP adds ResourceName to its database or updates its count.
5. At this point, the AIP has some assurance that Alice is a real person via ResourceName and proceeds to grant Alice a credential. Now, Alice and the AIP execute the CL-SIGN algorithm. At the successful termination of this protocol, Alice obtains a CL-signature over her secret key uk of the form (A, e, v) such that $A^e \equiv R^{\mathsf{uk}} S^v Z \bmod n$. For the rest of the paper, we may refer to this CL-signature as Alice's anonymous credential or credential for brevity.
6. Alice saves (A, e, v) and keeps it secret.

3.4 Anonymous SSO: Registering accounts

Most relying parties require users to create an account on their site before they can give them access to their services. This registration protocol is executed between a user Alice and a relying party whenever Alice wishes to create a new account. We emphasize that the AIP is not involved in this protocol unlike existing SSO frameworks. In fact, the AIP is not involved in the any of the remaining protocols to be described i.e., it is only needed when Alice obtains her anonymous credential.

Similar to the AIP, relying parties may have different policies regarding users creating multiple accounts however, Opaak seeks to prevent Sybil attacks while simultaneously enabling anonymity. Taking this into consideration, the protocol ensures that each user can only create up to k anonymous accounts with a the relying party i.e., a malicious user who tries to create more than k accounts can be detected.

The relying party defines its parameters *dom* for its domain name and k for the maximum allowable anonymous accounts per user.

1. Alice downloads the parameters *dom* and k.
2. Alice chooses a fresh i such that $1 \le i \le k$ i.e., an i value that she has not used to to create an account with yet. This means that Alice must keep track of the i values that she has used so far.
3. Alice forms the corresponding rnym as follows:

$$g_{\mathsf{rnym}} = H(dom||i)^{(\Gamma-1)/\rho}$$
$$\mathsf{rnym} = g_{\mathsf{rnym}}^{\mathsf{uk}} \bmod \Gamma$$

4. Alice sends an account creation request to the relying party along with an optional FriendlyName.
5. Alice creates a proof of possession of her credential from the AIP and a ZKPK of her master secret used to form rnym as follows:

$$SPK\{(e, \mathsf{uk}, v) : A^e \equiv R^{\mathsf{uk}} S^v Z \bmod n \wedge g_{\mathsf{rnym}}^{\mathsf{uk}} \bmod \Gamma\}$$

Alice sends the proof to the relying party. As a result of this proof, the relying party learns rnym, and the values *dom* and i used to form it.

6. The relying party verifies the proof and checks that the parameters are valid i.e., *dom* is the correct domain name and $1 \le i \le k$. If the proof verification fails or the parameters are invalid, then the protocol is terminated and the relying party returns `failure` to Alice.
7. The relying party now checks the database it maintains of rnym it has already seen. If the rnym is valid, then it adds it to the database together with FriendlyName associated with it and returns `success` to Alice. Otherwise, the relying party returns `failure`.

3.4.1 Friendly names

An rnym is actually a very large integer value with a long string representation and might not be user friendly. For this reason, Opaak allows users to specify human-readable, friendly names (i.e., username) for their accounts. For example, a Wikipedia user may want to be known under a pen name. Similarly, a friendly name can also be an email address, or the user's real name. The relying party saves the mapping between each account's rnym and friendly name. Of course, this has direct privacy implications on the anonymity of the account. We discuss the privacy issues related to enabling friendly names in Section 6.

3.5 Anonymous SSO: Logging in

This authentication protocol is executed between a user Alice and a relying party. When Alice wishes to login to an account she has created with the relying party (using the previous protocol), she proves possession of her anonymous credential and shows a valid rnym along with a ZKPK of her master secret used to form the rnym. We describe the step by step process below.

The relying party defines its parameters *dom* for its domain name and k for the maximum allowable anonymous accounts per user.

1. Alice downloads the parameters *dom* and k.
2. Alice keeps track of the i values corresponding to the accounts she has created with the relying party. She looks up the correct i optionally corresponding to FriendlyName.

3. Alice forms the valid rnym as follows:

$$g_{\mathsf{rnym}} = H(dom||i)^{(\Gamma-1)/\rho}$$
$$\mathsf{rnym} = g_{\mathsf{rnym}}^{\mathsf{uk}} \bmod \Gamma$$

4. Alice creates a proof of possession of her credential from the AIP and a ZKPK of her master secret used to form rnym as follows:

$$SPK\{(e, \mathsf{uk}, v) : A^e \equiv R^{\mathsf{uk}} S^v Z \bmod n \wedge g_{\mathsf{rnym}}^{\mathsf{uk}} \bmod \Gamma\}$$

Alice sends this proof to the relying party along with a login request and the correct FriendlyName. As a result of this proof, the relying party learns rnym, and the values *dom* and *i* used to form it.

5. The relying party verifies the proof and checks that the parameters are valid i.e., *dom* is the correct domain name and $1 \leq i \leq k$. If the proof verification fails or the parameters are invalid, then the protocol is terminated and the relying party returns `failure` to Alice.
6. The relying party has authenticated Alice and must simply look up the rnym entry in its database (and optionally Alice's FriendlyName). If rnym is not in the database then Alice has made an error and has tried to authenticate with an *i* value that she has not used to create an account with the relying party. The relying party may proceed with the previous registration protocol and add the rnym entry to its database or simply terminate the protocol and return `failure` to Alice.
7. After the relying party has retrieved the account information pertaining to rnym, then the relying party's application protocols can proceed as regularly. For example, the relying party could send back a unique token to establish an HTTP session with Alice.

3.6 Periodic k-Times Anonymous Authentication

The second class of applications targeted by Opaak are those where the relying party offers services where users don't necessary need to create accounts are their site such as posting messages, reviews, ratings, or casting votes. Enabling anonymity for these kinds of applications means that we must prevent spamming behavior from malicious users as well. Below we describe how Opaak reconciles between the seemingly conflicting goals of anonymity and spam control.

The relying party publishes its parameters *dom* for its domain name and *k* for the maximum allowable requests per time period *T*. For example, if Craigslist would like to allow users only one advertisement post per hour, then $k = 1$ and the time period identifier *T* would be updated every hour. The only requirement for *T* is that it is a unique identifier the simplest example of which is a timestamp. The protocol ensures that a user Alice can only create one valid rnym per time period *T*.

1. Alice downloads the relying party's parameters *dom*, *k* and current time epoch identifier *T*. Note that a *k* is only valid for the current time epoch i.e., a relying party can change its policy at every time epoch if it wishes. For example, if the $k = 1$ for the current time period, the relying party can easily change it to $k = 2$ for the next time period without any problems whatsoever. This means that Alice cannot cache these parameters.
2. Alice must keep track of the *i* values she has used so far for the current time period. Alice chooses a fresh *i* such that $1 \leq i \leq k$ i.e., an *i* value that she has not used to issue a request so far.
3. Alice forms a valid rnym as follows:

$$g_{\mathsf{rnym}} = H(dom||i||T)^{(\Gamma-1)/\rho}$$
$$\mathsf{rnym} = g_{\mathsf{rnym}}^{\mathsf{uk}} \bmod \Gamma$$

4. Alice creates a proof of possession of her credential from the AIP and a ZKPK of her master secret used to form rnym as follows:

$$SPK\{(e, \mathsf{uk}, v) : A^e \equiv R^{\mathsf{uk}} S^v Z \bmod n \wedge g_{\mathsf{rnym}}^{\mathsf{uk}} \bmod \Gamma\}$$

Alice sends this proof to the relying party along with her request (e.g., a message for an anonymous message board application). As a result of this proof, the relying party learns rnym, and the values *dom*, *i* and *T* used to form it.

5. The relying party verifies the proof and checks that the parameters are valid i.e., *dom* is the correct domain name, $1 \leq i \leq k$ and *T* is the current time epoch. If the verification fails, then the protocol is terminated and the relying party returns `failure` to Alice.
6. The relying party searches its database for the current time epoch and checks whether it has seen rnym before. If rnym is not in database then it can proceed to grant Alice's request (e.g., add the message to it's message board) and return a `success` status to Alice. Otherwise, Alice is attempting to reuse the rnym and the protocol terminates and a `failure` status is returned to Alice.

Our technique can be thought of as a combination of the schemes proposed by Teranishi et. al. [33] and Camenisch et. al. [15]. However, it is more efficient since we don't require complex ZKPK (e.g., proving a number lies within a given range [4,31]). Moreover, the penalty for misbehavior is less severe i.e., we do not de-anonymize (by revealing the public key of) a misbehaving user. We discuss the privacy implication of this tradeoff for efficiency in Section 6.

4. OPAAK ARCHITECTURE

4.1 Components

Figure 2 shows the architecture of Opaak and an overview of how the components interact with each other. There are three major components (each corresponding to the participants in the system): a relying party (RP) application, a mobile phone (users), and the AIP application. An RP application (or app) is composed of a mobile phone app and a web app. Our approach allows RP apps to utilize our protocols transparently. The Opaak components serve as a proxy for the requests carried out between the RP mobile phone app and the RP web app. In other words, we augment the existing communication between the RP app components with the Opaak protocols by sitting in between the RP phone app and the RP web app. This approach lends itself to compatibility with current systems and incremental deployment since existing RP apps would require minimal modification in order to use our protocols.

We now describe the function of each component. First, the AIP application is a web service consisting of the Registration Service, which executes the AIP side of the reg-

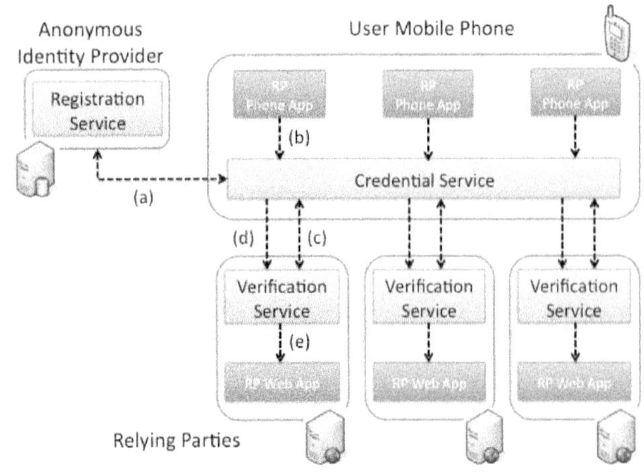

Figure 2: Architecture components. (a) User registers with an AIP through the Credential Service. (b) Phone app sends a request. (c, d) Credential Service runs Opaak protocols with the Verification Service. (e) Verification Service forwards the request to the web app.

istration protocol. It is also responsible for publishing the public parameters of the system (e.g., the RSA modulus n). Next, the mobile phone is responsible for interacting with the other components on behalf of the user. It consists of RP mobile phone apps which can utilize the Opaak protocols transparently through the Credential Service. The Credential Service is the main software component on the phone running all the protocols on behalf of the user. It registers the user with an AIP and obtains an anonymous credential by communicating with the Registration Service. It is also responsible for controlling access to the Opaak cryptographic material and keeping them privy to the other components. When using relying party services, it takes the requests from RP phone apps and executes the protocols with the Verification Service on the RP web app side. Finally, the Verification Service is responsible for maintaining the database of rnym and forwarding requests (upon successful execution of the protocols) to the RP web app. In the case of the anonymous SSO, the rnym could be exported since the RP app would need to use them as identifiers for its own database.

As a concrete example, consider a simple online message board RP app. It has a mobile phone app that takes an input message from the user, sends it as an HTTP POST request to a web app, which then takes care of displaying it on a webpage. From Figure 2, the RP app can be easily modified to allow anonymous posts (e.g., one post every five minutes) as follows:

(a) To participate, a user first registers with an AIP Opaak through the Credential Service. At the end of the protocol, the Credential Service saves the user's master secret and anonymous credential. This is only done once, for example, a cellular service provider serving as an AIP could distribute phones with the registration protocol already completed.
(b) The RP phone app (instead of sending the POST request directly to the RP web app) sends the message to the Credential Service. Of course, the RP phone app must let the Credential Service know the Internet location of the Verification Service. For example, the request to the Credential Service might look like the following: Request("Post this message!", `http://anonymous.posts.com/verify`).
(c) The RP web app publishes $k = 1$ and the current time period identifier T (updated every five minutes) through the Verification Service. The Credential Service authenticates with the Verification Service using the periodic k-times anonymous authentication protocol described in Section 3.
(d) Upon successful execution of the protocol, the Credential Service can proceed to send the message from the RP phone app to the Verification Service.
(e) Finally, the Verification Service forwards the message to the RP web app so it can proceed as originally intended.

We note that the RP phone app does not need to know about the first step, (a). It simply discovers the Credential Service on the phone and uses it.

4.2 Practical Considerations

In this section, we describe several practical and usability considerations that result from our mobile phone based architecture.

4.2.1 Credential Access Control

The Credential Service on the phone must control access to the user's master secret and anonymous credential. We can leverage several features of modern mobile operating systems (OS) such as Android. We describe them below assuming a deployment of the Credential Service as an Android app.

First, Android apps execute in a sandbox and are isolated from other apps on the phone. By default, only the Android app itself can access its resources such as files. We can rely on Android to enforce this mechanism so that no other third party RP phone app can read or write the files containing the user's master secret and anonymous credential.

Second, Android allows apps to communicate with each other through well-defined interfaces i.e., Android apps can call other Android apps. Moreover, it allows passing parameters to callee applications through generic "bundles". This feature actually lends itself well to our proxy architecture. RP phone apps can encapsulate their web requests in bundles and pass them along to the Credential Service to be forwarded to the RP web app.

Third, Android apps can declare custom permissions that they could require from other applications. For example, a Credential Service Android app can declare a custom permission `android.USE_CREDENTIAL_SERVICE` and require that other apps that would like to call it to declare this permission in their manifest. Android can enforce this permission check during installation of the RP phone app hence, only apps that the phone owner has allowed to use this custom permission would be able to use the Credential Service.

4.2.2 Registration using SMS messages

Recall that an AIP must require users to demonstrate possession of scarce resource such as a valid mobile phone number as a way to defend against Sybil attacks. To make the registration process more usable on a mobile phone, we propose the following method which minimizes user involvement.

Recall that the user and AIP execute the CL-SIGN algorithm during registration. One possible method is for the AIP to piggyback the mobile phone number verification onto the CL-signing protocol. Without going into much detail, in the last step of this protocol, the AIP sends the user a part of the CL-signature necessary for the user to execute further steps of computation before the actual CL-signature is obtained. The AIP can split this last part of the CL-signature C into two random shares, a small number s, and a larger share t, such that $C = s + t$. The smaller s will be sent over an SMS message, while t is sent directly over a data connection where the Credential Service receives it directly. We assume that the Credential Service is granted permission to read SMS messages, such that it can read the share s directly from the received SMS message, without having to prompt the user. In this way, the Credential Service can compute $C = s + t$, and then to proceed to computing the final CL-signature.

Alternatively, the AIP can simply send a verification code over SMS which the Credential Service reads and sends back to the AIP for verification. However, this approach introduces one more round trip than the above.

4.2.3 *Defense against Device Theft*

When storing the credential on the mobile phone, a major concern is device theft. Absent of any protection, if a legitimate user's device is captured by a malicious party, the malicious party will be able use the credential and impersonate the user when logging onto websites. The following methods can help mitigate the device theft problem.

First, we can employ conventional mechanisms such as master passwords. We may require that the user enter a master password on the phone during any authentication request. In addition, to prevent a malicious party from trivially reading credential from the phone's storage, we can also encrypt the credential under a secret key generated from the user's master password.

Second, we can employ credential revocation mechanisms. The Idemix library supports protocols for updating an existing credential i.e., the user simply re-runs parts of the CL-SIGN algorithm. This feature can be used to implement credential revocation. We can add an expiration date to the basic anonymous credential used by Opaak. When the date expires, the credential is rendered invalid and the user must re-run the registration protocol with the AIP. Clearly, there is trade-off between mitigating the problems of device theft effectively and efficiency. The smaller the time window for expiration is, the better for defending against device theft. Otherwise, it would require more resources on the AIP's part to enable frequent re-runs the registration protocol to re-validate the user's credential. Indeed, one of the advantages of our approach is that the user only needs to register with the AIP once.

5. IMPLEMENTATION AND EVALUATION

To evaluate the practicality of our approach, we implemented a prototype which includes both mobile and web applications. We used the Identity Mixer cryptographic library version 2.3.2 for Java to implement the ZKPK and CL-signature cryptographic operations. To implement our web applications, we used the Django framework. However, to carry out the actual cryptographic operations on the server side, we simply wrote Java programs that were called by Django applications. A more ideal solution would be to leverage projects like Jython but as a first step, we opted for less engineering overhead but even our unoptimized implementation performs well.

5.1 Protocol Parameters

We developed a Java program to generate the necessary parameters discussed in Section 3. First, we set the bitlengths for the `rnym` parameters to be $l_\Gamma = 1632$ and $l_\rho = 256$ and the bitlengths for the CL-signature parameters to be $l_n = 2048, l_e = 120$ and $l_v = 2724$. We follow the recommendations for strong security parameters in the Idemix specification. The program computes the group parameters (Γ, ρ, g) and generates the the public-private key pair for an AIP where the public key is (n, R, S, Z).

5.2 Setup and Registration

We developed a web application that implements the Registration Service for an AIP. Our AIP assumes all users requesting a credential are legitimate i.e., it does not require the user to demonstrate possession of a scarce resource such as a phone number. We believe this is reasonable for an experimental prototype. However, in a real world setting, this is absolutely necessary, otherwise Opaak does not have a basis for its guarantees against spam and Sybil attacks.

5.3 Anonymous Posts Website

We developed an example relying party, which lets users anonymously posts messages to a public forum using our mobile phone application. The website demonstrates the periodic k-times anonymous authentication protocol described in Section 3. It allows up to 1000 posts per hour i.e., $k = 1000$ and the time epoch identifier T is updated every hour. It can be viewed live at `http://trident.cs.ucdavis.edu/website/posts`.

The website includes a `posts` and a `verify` web application. The `posts` application is a standard database driven website which displays messages posted by users. The `verify` application unsurprisingly implements the Verification Service. Upon successful execution of the protocols with the Credential Service (running on the user's phone), the message string is inserted into the `post` application's database by the `verify` application.

5.4 Anonymous Post Android Application

We developed a mobile phone application on the Android 2.2 platform, which implements the Credential Service. The Android app can be downloaded at `http://trident.cs.ucdavis.edu/app/MobiCredAndroid.apk`.

The app has two Android activities corresponding to the two main responsibilities of the Credential Service. First, it includes an activity to run the registration protocol with the AIP registration service application we developed. This activity also downloads the necessary files containing the group parameters and AIP public key described previously. Second, an activity that runs the periodic k-times anonymous authentication protocol. It retrieves the T and k parameters from the above website, forms a valid `rnym` for the current monitoring period then submits the message entered by a user to the Verification Service via HTTP POST. Note that instead of developing a separate example RP phone app, we simply include the message post functionality into the

Credential Service in order to facilitate our benchmark experiments.

5.5 Performance

	Min	Median	Mean	Max
Setup (GP)	1.122	23.549	57.266	586.902
Setup (AKP)	2.041	30.362	39.662	191.088
Register	3.541	4.459	4.490	8.654
Post	2.076	2.238	2.553	7.385

Table 1: Wall clock times (in seconds) to perform 100 trials of the setup phase operations, registration protocol, and message post operation.

We ran several benchmarks to measure the performance of our implementation. The system that hosted our experiments for the setup operations was a quad-core 2.67 GHz Intel Xeon processor with 6GB of RAM. The system that hosted our website was a standard VMWare Workstation virtual machine hosted by the previous system and configured with 1 2.67 GHz processor, 1GB of RAM and a bridged network adapter. Both systems were connected to the Internet via ethernet on a university network. The smartphone that hosted our Android applications was a Motorola Droid running Android 2.2.

Table 1 gives the wall-clock time in seconds (calculated via the `java.util.Date.getTime` API call). For the setup operations, GP refers to the time to generate (Γ, ρ, g) while AKP refers to the time to generate the AIP's keypair where the public part is (n, R, S, Z).

For the register and post message operations, the time given is measured from the user's perspective and includes all the operations on the mobile phone, network requests, and all the operations on the server. More precisely, for the registration protocol, the time is measured starting from the first web request and ending after the credential has been saved to a file on the device. Similarly, for posting a message, the time is measured starting from the initial web request, including the proof computations on the mobile phone, the verification and database storage operations on the server, then ending after the user has received a status code from the server indicating a successfully posted message. Each operation was timed for 100 repetitions and given in the table are the min, median, mean and max times.

5.6 Discussion

Notice that the max execution time for GP is almost ten minutes long. Moreover, the standard deviation for the GP times is 107.104 while for the AKP times is 35.190. Compare these to the register and post operations which are both under one. The wide variation is due to the algorithms used to generate the random numbers required in the protocols.

For example, the GP operation involves generating the random generator g, and the algorithm for this is to simply keep generating random numbers until a number satisfying the constraints is found i.e., the order is prime and lies within a given interval. In addition, the AKP operation involves generating the "safe RSA modulus" n in the public key i.e., n is the product of two safe primes. This means that $n = pq$ where $p = 2p' + 1$, $q = 2q' + 1$ and p', q' are prime numbers as well. Here, the algorithm is the same, generate random numbers and loop until the constraints are satisfied. Indeed, there are optimizations to speedup the GP and AKP operations by taking advantage of multi-core processors. However, these operations are only performed once for the lifetime of the system hence we deemed the running times from our experiments reasonable and did not explore the optimizations further.

6. SECURITY ANALYSIS

The security of Opaak relies heavily on the security of CL-signatures and the proofs of knowledge used. In this section, we discuss the resistance of our approach to various attacks with respect to our goals described in Section 2. Recall that we are not concerned with network based attacks i.e., we are concerned with misbehaving participants in the system.

6.1 Linking User Identities and Tracking

6.1.1 *Identifying Different Users*

Recall from Section 3 that an rnym is of the form $\mathsf{rnym} = g^{\mathsf{uk}} \bmod \Gamma$ where $g = H(s)^{(\Gamma-1)/\rho}$ and s is an arbitrary string. Ideally, full anonymity means that given two different rnym, a relying party *cannot determine whether it belongs to the same user or not.* As previously mentioned, having a public base (g) has a privacy implication. However, in Opaak, two different rnym with the same input to the hash function means that they *must belong to two different users.* More concretely, consider users Alice and Bob with master secret keys uk_a and uk_b, respectively. If they try to create their first accounts at a relying party with domain name *dom*, then Alice would form $\mathsf{rnym}_a = g^{\mathsf{uk}_a} \bmod \Gamma$ and Bob would form $\mathsf{rnym}_b = g^{\mathsf{uk}_b} \bmod \Gamma$ where $g = H(dom||0)^{(\Gamma-1)/\rho}$.

Indeed, this is a slightly weakened definition of anonymity. A consequence of this is that a relying party would be able to count the number of unique users it has seen. We emphasize that the above example does not break our definition of linkability i.e., the relying party knows that two users have created accounts but it cannot identify which user. Moreover, if Alice chooses to create a second account where $g = H(dom||1)^{(\Gamma-1)/\rho}$, the relying party cannot tell whether it is Alice, or Bob. Hence, even if relying parties collude, the most information they can glean is limited to comparing the number of unique users they have seen.

Hiding the value of g_{rnym} as some existing schemes have done would require us to utilize ZKPK techniques [4, 31] that can hamper performance significantly in practice. We believe this tradeoff is reasonable in order to keep our protocols efficient on the mobile platform.

6.1.2 *Offline Identity Providers*

Our main concern with AIPs is that they attempt to track the user's online transactions with different relying parties. Unlike existing SSO frameworks, the user's transactions need not be brokered through the AIP in Opaak. This means that the AIP learns nothing about the transaction. Furthermore, the AIP can be offline and is never involved during transactions between users and relying parties.

During registration, the AIP can learn some identifying information from users because they must present some form of scarce resource such as a mobile phone number. However, the AIP is limited to knowing that it has certified such a phone number and nothing else.

6.1.3 *Optional Friendly Names*

If a user does not associate a FriendlyName with an account, then that account is an anonymous account. However, if a user associates the same FriendlyName to her accounts at different relying parties, then these accounts can be linked. Similarly, a user might willingly reveal her real identity to the relying party e.g., posting a message containing their real name, or using the same profile picture on different websites. To benefit from the Opaak's unlinkable accounts, a user must choose a different FriendlyName at an alcohol self-help group than her professional LinkedIn account. A user should also refrain from sending any data that may expose their identity to the alcohol self-help group website.

6.1.4 *Collusion*

Relying parties, users and AIPs might collude to attempt to de-anonymize an rnym they have seen, or link a user's proofs of possession of CL-signatures, with the goal of tracking the relying parties it has visited. Provided that the registration protocol was carried out honestly between the user and an AIP, then AIPs cannot do this. This is also true even in the case that an AIP is also a relying party. This is because of the property of CL-signatures that two proofs of possessions cannot be linked with each other.

At most, colluding participants could compare the number of unique rnym a relying party has seen. It is not possible for two relying parties to link the users they have seen hence, colluding participants could not even count the total number of unique users in the system.

6.2 User Impersonation

One of our concerns is that users might try impersonate other users in the system. They might do this by forging a CL-signature or by taking over another user's mobile phone. Provided that the registration protocol was carried out honestly between the user and an AIP, then this is not possible. CL-signatures are resistant to forgery even if a user colludes with other participants in the system. On the other hand, a malicious user could simply take over another user's mobile phone to either use it directly or perhaps copy the stored master secret. We describe ways to mitigate device theft in Section 4.2.

6.3 Spam and Sybil Attacks

As it enables user anonymity online, Opaak's main concern come in the form of spam and Sybil attacks. Users could attempt to spam service requests to relying parties. They might also attempt to register to an AIP multiple times and obtain multiple credentials.

Provided that the AIP requires the user to demonstrate possession of a scarce resource, then the protocols outlined in Section 3 prevent a user from creating an uncontrolled number of accounts with a relying party or spamming requests to a relying party by using multiple credentials. Furthermore, the periodic k-times anonymous authentication protocol which makes sure that a user can only create k valid rnym for a specified time period per credential they possess.

7. RELATED WORK

A number of digital identity management (DIM) frameworks have been proposed with the most notable ones being OpenID [6] and Microsoft's CardSpace [20]. OpenID and CardSpace focus on allowing users to consolidate and manage their digital identities and don't offer users anonymity from relying parties. In contrast, Opaak's main goal is to provide users anonymity from relying parties (as well as the identity providers) while simultaneously preventing spam and Sybil attacks. On the other hand, a suite of cryptographic techniques called U-Prove technology [2] developed by Stefan Brands has been integrated into CardSpace to offer similar capabilities as PRIME. Presumably, Opaak can also use U-Prove to achieve its goals, however we opt to focus on Idemix which provides better support and documentation. Similarly, PseudoID [21] augments the OpenID framework with blind signatures that allow users to anonymously authenticate with identity providers. The addition of a blind signature service (BSS) gives PseudoID users transactional privacy. However, identity providers can still log the relying parties users are visiting since users (although it doesn't know which user) are still redirected by relying parties to identity providers. Opaak eliminates this communication flow completely by using anonymous credentials and the AIP is never involved when a user authenticates with a relying party. Furthermore, Opaak enables relying parties to rate limit users while PseudoID must completely rely on the BSS to prevent Sybil attacks.

There are DIM frameworks that consider user privacy as a design goal and these include Privacy and Identity Management for Europe (PRIME) [16] (now PrimeLife [7]), VeryIDX [30], and PriMan [38] PRIME is an ongoing effort involving a consortium of companies and academia aiming to develop a comprehensive framework for privacy for the web and its applications. Indeed, the Identity Mixer cryptographic library [26] was developed as part of the PRIME project and is utilized by the PRIME core to enable anonymous authentication in applications. While both Opaak and PRIME utilize anonymous credentials, the PRIME architecture [10] has much more components as a result of its wide scope of design goals. Opaak has a more lightweight architecture as we focus on smartphone deployment and a well-defined set of target applications. In addition, our architecture allows developers to use our techniques more transparently. PRIME also focuses on managing the identity attributes (e.g., age, sex, marital status) of users and allowing them to selectively disclose these attributes to relying parties [11]. While Opaak could easily integrate such selective disclosure features as well, it opts to focus on giving relying parties the ability to control spam and Sybil attacks as these can be done with efficient protocols. Zero knowledge proofs of such identity attributes (e.g., proving an age value lies in a given interval like "≥ 18") can include techniques that do not scale well [4, 31]. VeryIDX also places an emphasis on proving identity attributes (e.g., credit card number, social security number) to relying parties. They do this with their "aggregate zero knowledge proof" protocol based on an aggregate signature scheme by Boneh et. al.. Similar to Opaak, they also realize the increasing ubiquity of mobile phones and store the secrets on the phone itself. As a result, they utilize a combination of cryptographic techniques (secret sharing) and secure hardware to protect the secrets on the phone. In contrast, Opaak focuses on controlling how third party applications access the credentials and leverage the isolation mechanisms and permission framework available with a modern smartphone OS such as Android. PriMan claims to be a user-centric identity middleware framework. PriMan takes into consideration third party application developers in their architecture design which reconciles different approaches to digital credentials and allows developers to easily switch from one approach to another.

Several applications based on anonymous credentials have also been described in literature. Blanton designed a system that adds anonymous access to online subscription services [14]. During the subscribe phase, users and service providers agree on a subscription type and an expiration date on which the service provider issues a CL-signature. Subsequently, users access the service anonymously until the expiration date by proving possession of the CL-signature. To prevent users from abusing the system, the access phase requires the user to obtain a new signature on a fresh access token where the old access token is stored by the service provider so that it can check whether tokens are being reused. Similar to Opaak, the anonymous ePoll system developed by Verhaeghe et. al. utilize the Idemix library on an Android phone [37]. Anonymous voting applications can be supported by Opaak as well however, their system goes further and provides features such as using verifiable encryption to provide voters with receipts that voters can use to prove participation. Furthermore, they conduct an evaluation and show how range proofs can affect performance negatively and that they do not scale well. This is precisely the reason why Opaak makes the tradeoff for privacy as discussed in Section 6. While Opaak and the ePoll system above show that full-fledged anonymous credentials can be utilized on commodity smartphones, Bichsel et. al. show that it can be done on severely resource constrained standard Java smart cards [13].

A number of cryptographic schemes have been proposed that have similar goals to rate limiting pseudonyms. The original k-times anonymous authentication scheme developed by Teranishi et. al. uses authentication tokens formed very similarly as rate limiting pseudonyms. They allow users to form up to k valid tokens which can allows them to authenticate with a relying party. It is based on group signatures instead of CL-signatures and could potentially be used to implement our protocols. More recently, a periodic n-times authentication scheme proposed by Camenisch et. al. [15] allows users to show up to n tokens anonymously and unlinkably to a verifier. Exceeding n means a user gets de-anonymized i.e., the verifier can compute the user's public key from (reused) tokens. In contrast, our protocols do not have such harsh punishment for misbehavior and only detects it so that the user's request can be denied. Both these schemes could potentially be modified to fit our scenario however, they both require complex ZKPK (e.g., proving a number lies in a given interval) that we opt to avoid in our scheme.

Anonymous blacklisting schemes are protocols that allow relying parties to maintain a blacklist of misbehaving users while preserving user privacy. There are the BLAC family of protocols [34–36] and the Nymble family of protocols [25,29]. In the BLAC protocols, users authenticate to relying parties by downloading the relying party's current blacklist and creating a ZKPK that they are not in the blacklist. Relying parties maintain their own blacklist containing "tickets" which are transcripts of previous authentications from users. Tickets are formed such that during authentication, users prove that they could not have created any of the tickets currently on the blacklist. In the Nymble protocols, users form "nymbles" based on their credentials and relying parties file complaints on a nymble to a Nymble Manager which can then include all other nymbles that a user may form on the blacklist. Since our protocols were tailored with specific applications in mind, they are more efficient than these blacklisting schemes. Moreover, it is unclear how they can be adapted for a periodic k times anonymous authentication scheme while maintaining the same security guarantees.

8. CONCLUSION

In this paper, we proposed Opaak, a framework for anonymous authentication that enables users to participate in online services anonymously while making sure that relying parties have the ability to prevent spam and Sybil attacks. It targets two kinds of applications, an anonymous single sign-on application, and anonymous message board types of applications. Opaak provides efficient protocols for these applications based on anonymous credentials. In addition, we designed a proxy based architecture that allows Opaak to be integrated easily with existing mobile and web applications. Our architecture also allows applications developers who may not be well-versed in cryptography to utilize our protocols transparently. We implemented a prototype on an Android phone, and demonstrated that it can perform anonymous authentications in a matter of seconds. With our work, we hope to encourage application developers to adopt systems like Opaak and motivate the deployment of anonymous credential based systems in the current Internet landscape.

9. ACKNOWLEDGEMENTS

The authors would like to thank the anonymous reviewers and our shepherd Yih-Chun Hu for their insightful comments. Special thanks to John Bethencourt for help with initial work on this project, Patrik Bichsel for advice on Idemix, and Man Ho Allen Au for advice on k-TAA schemes. This material is based upon work partially supported by the National Science Foundation (NSF) under Grant No. 0644450 and 1018964 and by the Air Force Office of Scientific Research (AFOSR) under MURI Grant No. 22178970-4170 and No. FA9550-08-1-0352. Any opinions, findings, and conclusions or recommendations expressed in this material are those of the authors and do not necessarily reflect the views of the AFOSR and NSF.

10. REFERENCES

[1] Craigslist. `http://www.craigslist.org`.
[2] Credentica. `http://www.credentica.com/`.
[3] Facebook connect. `http://developers.facebook.com/docs/guides/web`.
[4] Four squares. `http://schorn.ch/lagrange.html`.
[5] Linkedin. `http://www.linkedin.com`.
[6] OpenID website. `http://openid.net`.
[7] PrimeLife. `http://www.primelife.eu/`.
[8] Wikipedia. `http://www.wikipedia.org`.
[9] R. Anderson. Can We Fix the Security Economics of Federated Authentication? In J. A. Malcolm, editor, *International Workshop on Security Protocols*, 2011.
[10] C. Andersson, J. Camenisch, S. Crane, S. Fischer-Hubner, R. Leenes, S. Pearsorr, J. Pettersson, and D. Sommer. Trust in PRIME.

International Symposium on Signal Processing and Information Technology, 0:552–559, 2005.

[11] E. Bangerter, J. Camenisch, and A. Lysyanskaya. A Cryptographic Framework for the Controlled Release of Certified Data. In *Security Protocols Workshop*, 2004.

[12] M. Becher, F. C. Freiling, J. Hoffmann, T. Holz, S. Uellenbeck, and C. Wolf. Mobile Security Catching Up? Revealing the Nuts and Bolts of the Security of Mobile Devices. In *IEEE Symposium on Security and Privacy*, 2011.

[13] P. Bichsel, J. Camenisch, T. Groß, and V. Shoup. Anonymous Credentials on a Standard Java Card. In *CCS*, 2009.

[14] M. Blanton. Online subscriptions with anonymous access. In *ASIACCS*, pages 217–227, 2008.

[15] J. Camenisch, S. Hohenberger, M. Kohlweiss, A. Lysyanskaya, and M. Meyerovich. How to win the clonewars: Efficient Periodic n-times Anonymous Authentication. In *CCS*, 2006.

[16] J. Camenisch, R. Leenes, and D. Sommer, editors. *Digital Privacy: PRIME - Privacy and Identity Management for Europe*, volume 6545. Springer, 2011.

[17] J. Camenisch and A. Lysyanskaya. An efficient system for non-transferable anonymous credentials with optional anonymity revocation. In *EUROCRYPT*, pages 93–118, 2001.

[18] J. Camenisch and A. Lysyanskaya. A signature scheme with efficient protocols. In *Security in Communication Networks*, 2002.

[19] J. Camenisch and M. Stadler. Efficient group signature schemes for large groups. In B. Kaliski, editor, *Advances in Cryptology âĂŤ CRYPTO '97*, volume 1294 of *Lecture Notes in Computer Science*, pages 410–424. 1997.

[20] D. Chappell. Cardspace. `http://msdn.microsoft.com/en-us/library/aa480189.aspx`.

[21] A. Dey1 and S. Weis. PseudoID: Enhancing privacy for federated login. In *HotPETS*, 2010.

[22] R. Dingledine, N. Mathewson, and P. Syverson. Tor: The second-generation onion router. In *USENIX Security Symposium*, pages 303–320, 2004.

[23] J. R. Douceur. The sybil attack. In *IPTPS*, pages 251–260, 2002.

[24] A. Fiat and A. Shamir. How to prove yourself: Practical solutions to identification and signature problems. In A. Odlyzko, editor, *Advances in Cryptology âĂŤ CRYPTOâĂŹ 86*, volume 263 of *Lecture Notes in Computer Science*, pages 186–194. 1987.

[25] R. Henry, K. Henry, and I. Goldberg. Making a Nymbler Nymble Using VERBS. In *PETS*, pages 111–129, 2010.

[26] IBM Research Zurich Security Team. Specification of the Identity Mixer Cryptographic Library v. 2.3.2. `https://prime.inf.tu-dresden.de/idemix/`, December 2010.

[27] T. Kohno, A. Broido, and K. C. Claffy. Remote Physical Device Fingerprinting. In *IEEE Symposium on Security and Privacy*, pages 211–225, 2005.

[28] B. N. Levine, C. Shields, and N. B. Margolin. A Survey of Solutions to the Sybil Attack. Tech report 2006-052, University of Massachusetts Amherst, 2006.

[29] Z. Lin and N. Hopper. Jack: Scalable Accumulator-based Nymble System. In *WPES*, pages 53–62, 2010.

[30] F. Paci, E. Bertino, S. Kerr, A. C. Squicciarini, and J. Woo. An Overview of VeryIDX - A Privacy-Preserving Digital Identity Management System for Mobile Devices. *Journal of Software*, 4(7):696–706, 2009.

[31] M. O. Rabin and J. O. Shallit. Randomized algorithms in number theory. *Communications on Pure and Applied Mathematics*, 39, 1986.

[32] C.-P. Schnorr. Efficient signature generation by smart cards. *Journal of Cryptology*, pages 161–174, 1991.

[33] I. Teranishi, J. Furukawa, and K. Sako. k-Times Anonymous Authentication (Extended Abstract). In *ASIACRYPT*, pages 308–322, 2004.

[34] P. P. Tsang, M. H. Au, A. Kapadia, and S. W. Smith. PEREA: Towards Practical TTP-free Revocation in Anonymous Authentication. In *CCS*, pages 333–344.

[35] P. P. Tsang, M. H. Au, A. Kapadia, and S. W. Smith. Blacklistable anonymous credentials: Blocking misbehaving users without TTPs. In *CCS*, pages 72–81, 2007.

[36] P. P. Tsang, M. H. Au, A. Kapadia, and S. W. Smith. Blacklistable Anonymous Credentials: Blocking Misbehaving Users without TTPs. In *CCS*, pages 72–81, 2007.

[37] P. Verhaeghe, K. Verslype, J. Lapon, V. Naessens, and B. D. Decker. A mobile and reliable anonymous epoll infrastructure. In *Mobisec*, 2010.

[38] K. Verslype, P. Verhaeghe, J. Lapon, V. Naessens, and B. De Decker. Priman: a privacy-preserving identity framework. In *Data and Applications Security and Privacy XXIV*, pages 327–334, 2010.

[39] L. von Ahn, M. Blum, N. J. Hopper, and J. Langford. CAPTCHA: Using Hard AI Problems for Security. In *EUROCRYPT*, pages 294–311, 2003.

Plug-n-Trust: Practical Trusted Sensing for mHealth

Jacob Sorber, Minho Shin[†], Ron Peterson, David Kotz
Institute for Security, Technology, and Society, Dartmouth College, Hanover, NH, USA
[†]Dept. of Computer Engineering, Myongji University, South Korea

ABSTRACT

Mobile computing and sensing technologies present exciting opportunities for healthcare. Prescription wireless sensors worn by patients can automatically deliver medical data to care providers, dramatically improving their ability to diagnose, monitor, and manage a range of medical conditions. Using the mobile phones that patients already carry to provide connectivity between sensors and providers is essential to keeping costs low and deployments simple. Unfortunately, software-based attacks against phones are also on the rise, and successful attacks on privacy-sensitive and safety-critical applications can have significant consequences for patients.

In this paper, we describe Plug-n-Trust (PnT), a novel approach to protecting both the confidentiality and integrity of safety-critical medical sensing and data processing on vulnerable mobile phones. With PnT, a plug-in smart card provides a trusted computing environment, keeping data safe even on a compromised mobile phone. By design, PnT is simple to use and deploy, while providing a flexible programming interface amenable to a wide range of applications. We describe our implementation, designed for Java-based smart cards and Android phones, in which we use a split-computation model with a novel path hashing technique to verify proper behavior without exposing confidential data. Our experimental evaluation demonstrates that PnT achieves its security goals while incurring acceptable overhead.

Categories and Subject Descriptors

C.3 [**Special-purpose and application-based systems**]: Smartcards; J.3 [**Life and Medical Sciences**]: Health; K.6.5 [**Management of Computing and Information Systems**]: Security and Protection

General Terms

Security, Performance

Keywords

trusted computing, smartcards, smartphones, mHealth

1. INTRODUCTION

Healthcare information technology has potential to improve healthcare quality, improve efficiency, and reduce cost, and is currently on the cusp of major innovations and widespread deployment around the world. There are especially exciting opportunities for the application of mobile-sensing technologies to healthcare. Judicious use of this *mHealth* technology can enable physicians to remotely monitor their patients' health and improve the quality of healthcare, enable patients to manage their health more easily, and reduce the cost of care by allowing patients to spend less time in the hospital or make fewer visits to their doctor [35].

There are many applications of mHealth sensing for clinical purposes (such as long-term care for patients with diabetes [46] or hypertension [3]) and non-clinical purposes (including elder care [6], lifestyle coaching for people seeking to change unhealthy behavior [8], and fitness monitoring for athletes [10]). Regardless of the nature of the person being monitored – an outpatient being monitored remotely by their doctor, a resident of an assisted-living facility, a family member, or an athlete – for the purposes of this paper we always refer to the subject of sensing as the "patient."

Although healthcare information technology poses many security and privacy challenges, including some specific to mHealth [27], in this paper we focus on the *mobile* edge of the mHealth ecosystem, where all of the above settings share the same security and privacy challenges. These mHealth sensing applications collect personal health-related data that is inherently sensitive, and its unintended disclosure would violate the patient's privacy and possibly cause social or economic harm to the patient. Furthermore, since this data may be used to diagnose a patient's condition or adjust a patient's treatment, any data corruption caused by an adversary can have an adverse effect on the patient's health.

Nearly all mHealth sensors require a gateway to forward data from low-power sensors to back-end services, such as an Electronic Health Record (EHR) maintained by a hospital, a Personal Health Record (PHR) set up by the patient, or a vendor-managed portal that provides the patient access to her data. What device could serve as the gateway? For many of today's mHealth sensors, the gateway is a proprietary, stationary device located in the patient's home; this solution limits data collection to those times when the patient is at home. To enable mobility, the patient could carry a proprietary mobile gateway; this solution requires the patient to carry (and remember to charge) yet another device. A third option is to leverage a device that many patients already carry: their mobile phone. Such an approach would improve data collection (because the gateway would almost always be present), and would reduce cost, by obviating the need for a dedicated gateway device.

Unfortunately, it is risky to use mobile phones to collect and process privacy-sensitive and safety-critical medical data. The software running on today's mobile phones is increasingly complex – comparable to desktops and laptops – and vulnerable to malware and other software-based attacks. As phones are used in more sensitive processes (like banking or location-based services), they present an attractive target for attackers. Without reducing this risk, these attacks present a significant obstacle standing in the way of the widespread adoption and deployment of mobile healthcare systems.

We aim to provide strong security and privacy guarantees for mHealth-sensing applications that are based on wearable sensors and off-the-shelf mobile phones. At a high level, there are two aspects to this problem. First, we must ensure the desired confidentiality and integrity properties, within the sensing devices and the mobile phone. Second, we must ensure these properties in the communications between the phone and the sensors. We address the first aspect in this paper; we addressed the second aspect in an earlier paper [29].

Our approach, which we call Plug-n-Trust, is to create a tiny trusted component, implemented on a smart card, that literally "plugs in" to the patient's mobile phone. The system is designed to allow a health-related app to run on the phone, collecting data from a body-area network of sensor devices. Even if the phone has been compromised by malware, the phone will not leak sensitive sensor data, nor be able to tamper with the results reported to the back-end services used by the patient or his caregivers.

There are other possible approaches, which we discuss in more detail in Section 6. Trusted hypervisors [9, 19] try to provide a secure sandbox for sensitive applications, but require the user to replace (or underlay) their phone OS with a hypervisor—not easy for any user and not possible on some phones. Trusted hardware (such as a TPM [41]) provides a secure root of trust on which one may build a trusted software stack that can isolate sensing applications from other less-trusted applications. Such approaches require trust and agreement among many parties, most of whom have nothing to do with mHealth, including the handset manufacturer, the TPM chip maker, the hypervisor developer, and the OS developer. Furthermore, although TPM hardware is common in many desktop and laptop computers, the technology has yet to be broadly leveraged, and has not yet appeared in common mobile phones.

Ultimately, to be practical, the trusted computing base (TCB) on which mHealth rests should require trust between as few parties as possible and these trust relationships should be relevant to the application. For medical applications, it makes sense for the patient to trust the doctor or hospital, and the manufacturers of the sensing devices, but requiring them to trust the handset-maker or its OS vendor is unnecessarily risky. In Plug-n-Trust, as we show, the mobile portion of the trusted computing base is small and robust, and easily deployed: the physician (or other trusted caregiver) simply provides a smart card to the patient, who then plugs it into her phone.

We make three contributions in this paper. First, we describe the design of Plug-n-Trust (PnT), a practical approach to ensuring both the confidentiality and integrity of both the sensing and processing of medical data on untrusted smart phones. Second, we demonstrate the feasibility and limitations of PnT using a prototype implementation on Java-based smart cards and Android mobile phones. Third, we provide a careful security analysis of PnT, demonstrating that it meets the desired properties.

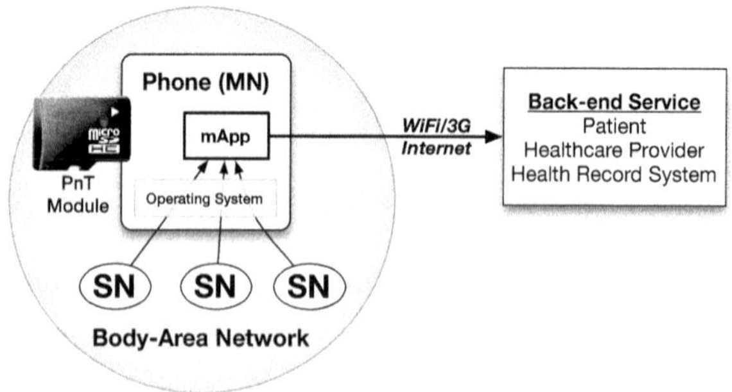

Figure 1: System model

Our discussion of Plug-n-Trust focuses primarily on enabling trusted computing simply, requiring little or no support from smart-phone vendors and OS providers; however, in Section 6.2 we discuss how PnT's secure computing environment can also be adapted to a vendor-deployed solution using built-in secure elements like TrustZone [42].

2. SECURITY MODEL

In this section, we describe the architecture of mHealth monitoring systems that we aim to protect, the adversary model and threat model, the security goals that we aim to achieve, and our trust assumptions. These assumptions form the foundation for the design presented in the next section, and for the security analysis in Section 5.

Our overall goal is to provide tools for vulnerable off-the-shelf mobile devices to host privacy- and safety-critical mHealth applications with minimal risk to patient privacy and security, in the event of an attack.

2.1 System model

Figure 1 provides a high-level view of our mHealth-sensing system. The mobile portion of the system consists of a *mobile node (MN)*, typically a mobile phone, and one or more *sensor nodes (SNs)*. The SNs communicate with the MN via a body-area wireless network, using a protocol like Bluetooth, Zigbee, or our secure protocol [29]. The MN communicates via the Internet to various back-end *services*, such as EHRs, PHRs, or vendor-managed portals that support customers of their mHealth device.

Inside the MN, an mHealth application (*mApp*) is responsible for collecting sensor data from the SNs, processing or aggregating the information in some application-specific manner, and then uploading (some of) the resulting data to the back-end services. Some mHealth sensors (like EKG or EEG sensors) collect data at a high rate, and some mHealth applications dynamically adjust sensing activity based on context. It may not be feasible to upload all of the raw encrypted data for processing on the back-end server. This approach may require too much bandwidth, with accompanying energy and monetary costs; with PnT we enable local operations on the data, reducing bandwidth needs and enabling disconnected operation. Thus, to support application-specific computations on the sensor data when the MN may itself be compromised, we include a plug-in smart card, the "PnT module" in the figure. This card provides a hardware root of trust, and in our implementation is a commercially-available microSD card that simply snaps into the microSD slot in off-the-shelf mobile phones.

Hardware assumptions

H1. *Crypto.* Each SN has the capability to encrypt and hash its sensor readings, with standard algorithms like AES and SHA [13, 45]. This capability is becoming increasingly feasible in mote-class devices as hardware support for encryption is present in an increasing number of low-power microcontrollers (e.g. TI CC430).

H2. *Clock.* Each SN has an embedded real-time clock that is sufficiently accurate to timestamp its data.

H3. *Platform.* The MN is a general-purpose mobile platform, such as a smart phone, with at least two wireless network interfaces: one for body-area communications with the SNs (comparable to Bluetooth or Zigbee), and one for Internet communications (e.g., Wi-Fi or 3G).

H4. *Internal sensors.* The MN may have internal sensors, such as a clock, camera, or GPS, but we must assume that these are compromised when the MN is compromised and thus PnT is unable to use them.

2.2 Adversary and threat model

A wide range of attacks have been demonstrated against computing devices and networks, ranging from passive network sniffing to hardware-based attacks that require physical possession of the device. We consider secure communication between the mobile phone and the back-end service, as well as securing the back-end service itself, to be relatively routine problems that have been addressed in a variety of ways. In this paper, therefore, we focus on the security of the MN. (We are less concerned with the SNs, which are single-purpose embedded devices with a small attack surface, and are not connected to the Internet. Although some attacks have been demonstrated [23], mobile phones are much easier to attack.)

Mobile phones are at risk from many kinds of compromise. Software-based attacks are particularly dangerous, since software vulnerabilities are universally common, and attacks can be launched remotely on many devices at once, at a small cost to the attacker. Continuous efforts are being made to protect mobile platforms against intrusion (e.g., MTM [12] or virtualization [9]); however, attacks will continue to evolve, and some attacks will inevitably succeed. All system components from the applications down to the kernel, hypervisor, and device drivers may potentially be compromised. These attackers may have the ability to read system memory, access communication channels, and arbitrarily modify system behavior – including the processing of medical sensor data.

The patient should be concerned about several threats.

T1. The adversary may recover medical sensor data (or data derived from it).

T2. The adversary may modify sensor data (or data derived from sensor data), or inject false data, without it being detected.

T3. The adversary may prevent the system from collecting or reporting sensor data (DoS) without it being apparent to the patient and back-end server.[1]

[1]Although DoS attacks are a real threat, we do not address DoS attacks in this paper. See Section 7.1 for discussion.

We assume a strong adversary with some limitations, relevant to the threats above:

A1. *Computationally bounded.* The adversary cannot break cryptographic primitives such as AES and SHA.

A2. *No compromise of SN.* The adversary does not have the capability to compromise either the software or the hardware of the SN, at least, without it being immediately evident to the patient.

A3. *Physical attacks.* The adversary is not capable of physical side-channel attacks such as power, heat, sound, and electromagnetic clues to learn about the information stored within the card.

A4. *Local adversary.* The adversary cannot compromise any of the back-end services, but can compromise the software running locally on the MN.

A5. *Secure body-area network.* The adversary is unable to compromise patient confidentiality or anonymity, or data integrity, by observing wireless body-area network traffic; solutions exist [29].

2.3 Security properties

In the face of a powerful adversary, and to avoid the threats listed above, we focus on the providing the following security and privacy properties in our design of a more secure approach to mHealth sensing.

SP1. *Data Confidentiality.* Sensor data and any data derived from sensor data remains confidential to all entities other than the system's trusted components (sensor node, the smart card, and the back-end server).

SP2. *Sensor Data Integrity.* The sensor data remains intact during its processing and delivery to the back-end server, and any injection of incorrect data or past data into the system is identified.

SP3. *Derived Data Integrity.* Derived data is obtained from sensor data by a series of computations defined by the doctor. Derived data remains intact in transit, and any attempt to deviate from the predefined derivation logic is identified.

2.4 Trust model

Any trustworthy system is built on certain assumptions about who trusts whom, and in what way. In our efforts to achieve the above goals, we make the following trust assumptions about the patient who uses the sensors, the healthcare provider (doctor, hospital, or other entity that consumes the data), and the companies that manufacture the sensors and smart cards.

TR1. The patient and the healthcare provider trust the SN manufacturer to produce calibrated sensors that operate correctly, so that the patient and the health provider can trust the sensor to provide the right reading.

TR2. The patient and the healthcare provider trust the card manufacturer to correctly implement PnT within the card; thus, it protects confidentiality and integrity according to the above goals, and in face of the above adversaries.

TR3. The patient trusts the healthcare provider (or other data consumer) to keep sensor data confidential after it has been delivered, and not to reveal encryption keys.

TR4. The healthcare provider trusts the patient to not be malicious and to not tamper with the hardware or software of the sensor node, or the PnT card.

The manufacturer has no stake in the system, so the manufacturer assumes nothing about other principals.

3. Plug-n-Trust

Plug-n-Trust (or PnT) is a novel approach that enables confidential and trustworthy processing of safety-critical medical sensor data. As its name implies, PnT uses a smart card that *plugs* into a phone's microSD slot and serves as a *tiny trusted third party* [24] to create a safe and trusted computing environment for critical data processing on an untrusted mobile phone. In short, sensitive applications rely on the card to perform sensitive processing, and the card provides verifiable proof that computations were conducted properly.

In addition to the security goals of ensuring confidentiality and integrity, PnT is designed to be usable, deployable, flexible, and minimal, as the following paragraphs explain:

Usable: Requiring users to remember secrets (e.g., passwords), configure permissions, and annotate content often makes systems more difficult to use, and may lead users to bypass or disable security features altogether [11]. Instead, PnT only requires a user to plug the trusted smart card into her phone. After that the system operates automatically in the background, without any additional interaction.

Easily Deployable: In addition to being difficult to use, many technically feasible security systems require too much infrastructure or agreement between too many different parties (e.g., hardware manufacturers, software providers, and network operators) to be deployed. For example, trusted software stacks whose validity is verified by a hardware Trusted Platform Module (TPM) [41] require a complex trust relationship between the TPM manufacturer, the platform manufacturer, the operating-system vendor, and the application developer, and in some cases, the cellular-network provider (who often controls aspects of the platform or its operating system). Furthermore, TPM programming remains highly challenging, so TPMs have only been used in limited settings [37]. New approaches, such as Logical Attestation [39], may ease the use of TPMs if smart-phone vendors were to adapt the Nexus operating system to small-scale platforms.

In contrast, PnT can be easily deployed because it only requires trust relationships between the parties that are directly involved (e.g., patient, provider, and card manufacturer), independent of those who are not (e.g., cell-phone service providers, phone manufacturers, operating system developers, or app stores).

Flexible: Applications change. Providers may want to adjust sensing and processing just like they adjust doses of medication. PnT is flexible by design. Applications can be updated, modified, and tuned. Application designers need only describe to the back-end server how data is processed, so that its integrity can be verified. (Specifically, they provide a compact representation of the sequence of mathematical operations on the sensor data, but need not provide the source code for the whole sensing application.)

Minimal: Minimizing a system's Trusted Computing Base (TCB) – the set of code that must remain intact to ensure security – is essential to building secure systems [38]. In existing systems the TCB is either the OS kernel, a hypervisor [19], or a secure coprocessor [24]. In PnT the TCB (on the phone) runs only on the plug-in smart card. This hardware separation from all other code running on the system, with a simple API, provides an extremely small attack surface.

Functional: The PnT card's storage space can also be used to store other media.

3.1 Plug-in smart card

For decades, smart cards have been used to protect encryption keys and other sensitive information in applications ranging from financial payment systems to cell phones (SIMs). Early cards were used primarily for identification and had limited storage; however, recent increases in both storage and computation resources (including hardware support for cryptography and secure random-number generation) make these tiny tamper-resistant devices capable candidates for more sophisticated tasks.

In Plug-n-Trust, we use a smart card in a microSD form factor [17] as a pluggable trusted third party or secure coprocessor that represents the patient's interests in a potentially unsafe mobile phone. Phones that do not have a microSD card slot could run PnT on a SIM card instead [33]. While not as powerful as traditional secure coprocessors (like the IBM 4758, used in the FaeriePlay prototype [24]), smart cards are much more portable and more affordable.

Smart cards also provide PnT users with portability. They are supported by many major mobile-phone operating systems (Android, Windows Mobile, BlackBerry, PocketLinux, and Symbian), and even more importantly they allow a patient's security to be independent of a single mobile device or service provider. When a patient buys a new phone, the user simply transfers the card to the new phone, leaving no sensitive information behind and requiring no configuration, other than installing an app—which may be loaded from the smart card—on the new phone. If the card is lost or stolen, its tamper-resistant design makes it more difficult for the attacker to access private information (see Section 7.1 for more detail). Note that we do not assume tamper-proof cards, for which tampering is impossible; tamper resistance, like all security mechanisms raises the bar for the attacker.

The scant resources (a few kB for data and code) prevent us from loading entire applications onto the card. Instead, our goal is to do as little on the smart card as possible. The application's processing of sensitive data must happen in the card, but program logic for sensor discovery, data collection, and data delivery are performed on the mobile phone. While this design choice is born from necessity, the benefits of a small code base and an extremely simple interface between the application and smart card (smaller attack surface) make PnT easier to secure.

3.2 Bootstrapping Plug-n-Trust

PnT uses encryption to protect sensitive data when it is transmitted between a sensor and the smart card and again during delivery to the back-end server. This encryption requires secret keys that are used to ensure both confidentiality of the data and authenticity of the system components. Specifically, PnT requires each sensor s to have a key, K_s, which is known to the smart card. The smart card has a key, K_c, which is also known to the back-end system.

A critical challenge with any cryptographic system is, of course, key distribution. We need a secure way to share secret keys between the PnT card and the sensor nodes, without trusting the MN and its operating system. One major advantage of PnT is that the card can be removed from the phone, and paired directly with the SNs. We imagine a range of possible approaches, depending on how the sensors and smart card are deployed. Both sensors and smart card may be provided by the care provider (who also administers the back-end server), allowing keys to be pre-installed before the patient receives the devices. More likely, the patient may receive the sensors from a trusted proxy of the provider, such as a pharmacy. In that case, the pharmacist would insert the patient's PnT card into a slot on the pharmacist's own trusted device, which then may use any number of existing techniques to develop a shared secret between the card and the SN. Alternately, the card might come with a pairing device, much like that in the pharmacy example, which the patient can keep at home and use for introducing new sensors to the card. (Later, if the SN is lost or discarded, the card needs to revoke the relevant keys; the card can be informed via one of these trusted devices, or in a secure message sent from the provider via the MN.) Finally, Go-Trust plans to ship smart cards that support Near-Field Communication (NFC), allowing the card to communicate directly with NFC-capable SNs [21].

Notably, PnT is not tied to a single crypto-system, allowing implementers considerable flexibility. An implementation could easily use either symmetric (AES, DES) or asymmetric (RSA, ECC) encryption. This decision has well-known trade-offs. Asymmetric encryption provides greater flexibility for key distribution and bootstrapping, but requires more computation and larger keys. Our implementation uses AES-256 (in CBC mode), a symmetric algorithm, which is efficient for low-power sensor devices.

3.3 Plug-n-Trust basic operation

In PnT, data is encrypted before it leaves the sensor, remains encrypted while stored on the mobile phone, and only is decrypted (and modifiable) in the card, as shown in Figure 2. The card exposes a simple API that facilitates communication between the application and the card. We assume that a sensing application is a series of *tasks*, and the mApp signals the beginning of a new task with the command:

$$\texttt{start}()$$

which resets internal data structures on the card.

Moving data in and out of the card

Raw data messages arrive from sensor s encrypted with the sensor's key, K_s. In addition to one or more raw data readings (*data*), the message also includes additional metadata—the ID of the originating sensor (s), the time of collection (t), a sequence number (n), an error flag (e), and a message authentication code (MAC) computed over all of the above (s,t,n,e, and *data*) after encryption. The HMAC algorithm uses a key, K_s^m, derived from K_s—to protect the integrity of the data and its processing.

$$m = \{s, t, n, e, data\}_{K_s}, HMAC(\{s, t, n, e, data\}_{K_s^m})$$

The application transfers the encrypted data message, m,

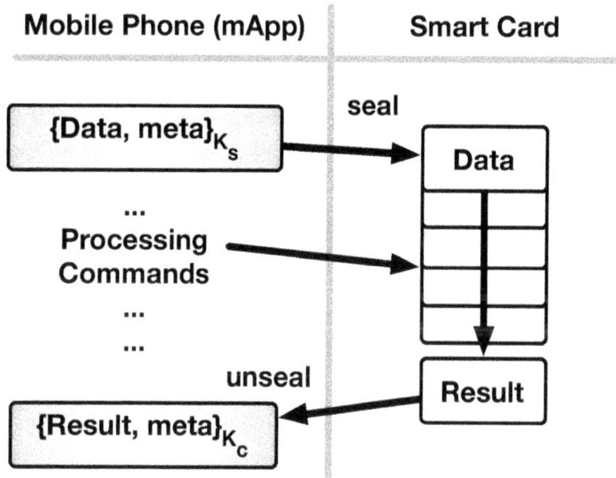

Figure 2: In Plug-n-Trust, encrypted sensor data is passed (sealed) into the smart card, processed according to application-supplied commands, and the encrypted result is then removed (unsealed) from the card.

as well as the ID of the originating sensor, s, into the card using the `seal` command. Passing in the sensor ID, s, is not strictly necessary, since the smart card could try to decrypt the message with all of its known keys; however, as the number of keys increases this quickly becomes inefficient. After the data and metadata are stored in the card, `seal` returns a data-independent reference, r_m, that the application can use in subsequent processing operations on the data.

$$\texttt{seal}(s, m) \Rightarrow r_m$$

Sensor readings can be sealed into the card individually, or as vectors of sequential readings, which is useful for reducing the amount of communication between the application and the card. For simplicity, our design assumes that sensor readings are integer values; floating-point support would be an easy extension to the architecture.

Once processing is complete, the application can use `unseal` to retrieve the computational result and associated metadata used by the server to validate the computation (see Section 3.5). The card encrypts all of this data before releasing it to the phone.

$$\texttt{unseal}(r_m) \Rightarrow \{result, metadata\}_{K_c}$$

Arithmetic operations

$$\begin{array}{ll} \texttt{add}(r_1, r_2) \Rightarrow r_3 & \texttt{addc}(r_1, c) \Rightarrow r_2 \\ \texttt{sub}(r_1, r_2) \Rightarrow r_3 & \texttt{subc}(r_1, c) \Rightarrow r_2 \\ \texttt{mult}(r_1, r_2) \Rightarrow r_3 & \texttt{multc}(r_1, c) \Rightarrow r_2 \\ \texttt{div}(r_1, r_2) \Rightarrow r_3 & \texttt{divc}(r_1, c) \Rightarrow r_2 \end{array}$$

After data has been sealed into the card, PnT provides a variety of commands for processing the data. Standard arithmetic operations are provided that add, subtract, multiply, and divide the values associated with two references, r_1 and r_2, and return a reference, r_3, to the result.

The references can refer to scalars or vectors, with straightforward semantics. If both references refer to scalar values, then the operation is applied to both and the result is also a scalar. If both are vectors, then the requested operation is applied in pairwise fashion to corresponding elements in each vector, resulting in a vector of the same length. If the vectors are of different lengths—which is likely an uncommon case—then some entries will be ignored and the length of the result is $min(length(r_1), length(r_2))$. Finally, if one value is a scalar and the other a vector, then the scalar value is applied to each of the vector's entries.

Corresponding commands are also provided for arithmetic with a fixed constant. For example, $\texttt{divc}(r, 2)$ divides an in-card scalar or vector referred to by r by 2.

Summarizing data

$$\texttt{sum}(r_1) \Rightarrow r_2 \qquad \texttt{max}(r_1) \Rightarrow r_2$$
$$\texttt{prod}(r_1) \Rightarrow r_2 \qquad \texttt{min}(r_1) \Rightarrow r_2$$
$$\texttt{len}(r_1) \Rightarrow r_2$$

A set of commands are also provided that reduce a vector of elements to a scalar. These commands allow an application to compute the sum or product of a vector, or determine the maximum or minimum element in the vector. These commands only make sense for vectors, and if applied to a scalar the result will just be a copy of the original argument, except `len()`, which outputs 1.

Logical comparisons

$$\texttt{gt}(r_1, r_2) \Rightarrow r_3 \qquad \texttt{gtc}(r_1, c) \Rightarrow r_2$$
$$\texttt{lt}(r_1, r_2) \Rightarrow r_3 \qquad \texttt{ltc}(r_1, c) \Rightarrow r_2$$
$$\texttt{eq}(r_1, r_2) \Rightarrow r_3 \qquad \texttt{eqc}(r_1, c) \Rightarrow r_2$$

In addition to arithmetic computations, the ability to compare sensor readings to each other, or to a constant value, is critical to many applications that process sensor readings.

$$\texttt{and}(r_1, r_2) \Rightarrow r_3 \qquad \texttt{or}(r_1, r_2) \Rightarrow r_3 \qquad \texttt{not}(r_1) \Rightarrow r_2$$

The Boolean results of those comparisons can also be modified using the standard **and**, **or**, and **not** Boolean operators. Both comparison and Boolean commands can be applied to both scalars and vectors. When vectors are used, the desired operation is applied to each element of the vector, and the result is a vector.

Conditional execution

$$\texttt{if}(r_1, r_2, r_3) \Rightarrow r_4$$

Finally, without the ability to support data-dependent processing, PnT would not be useful to a significant number of applications. The `if` command makes conditional execution possible as demonstrated in the following example, which computes the arithmetic mean of the elements of vector, d, and uses conditional execution to avoid dividing by zero.

$$\texttt{seal}(s, \{d\}_{K_s}) \Rightarrow r_1 \tag{1}$$
$$\texttt{sum}(r_1) \Rightarrow r_2 \tag{2}$$
$$\texttt{len}(r_1) \Rightarrow r_3 \tag{3}$$
$$\texttt{eqc}(r_3, 0) \Rightarrow r_4 \tag{4}$$
$$\texttt{div}(r_2, r_3) \Rightarrow r_5 \tag{5}$$
$$\texttt{if}(r_4, r_3, r_5) \Rightarrow r_6 \qquad (r_6 \Leftarrow r_3 \text{ or } r_5) \tag{6}$$
$$\texttt{unseal}(r_6) \Rightarrow \{\text{result}, \text{metadata}\}_{K_c} \tag{7}$$

The `if` command on line 6 accepts three arguments: a Boolean value stored at reference, r_4, which resulted from the comparison on line 4; and two possible results. The second possible result is r_3, which equals zero in the error condition. The second, r_5, is the result of the division on line 5. If a divide by zero occurred, then the division fails silently, the value in r_5 is undefined, and its error flag is set. According to the typical C-style ternary operator semantics, the value pointed to by r_6 will be the value of r_3 if r_4 is true, and the value of r_5 otherwise. Note that conditional execution in PnT requires both sides of a branch to be executed, which is wasteful in many cases, but necessary to avoid leaking information about the values of the data.

The other major control-flow construct, loops, can be implemented within the mApp. That is, the mApp can loop as many times as needed, sending commands to the card. Data-dependent loop conditions can be implemented (albeit inefficiently) using the `if` command (see Section 7.1).

Additional commands

When selecting commands for PnT, we focused on supporting common processing tasks used in mHealth applications, like summarizing data (statistics such as mean, median, variance, and covariance) and checking data ranges and thresholds (is the patient's blood sugar within a safe range?).

As shown in our evaluation section, our current set of operations also supports Actigraphy [4], which is a well-known metric for measuring activity level based on accelerometer data, often used for diagnosing sleep disorders. There are three ways to compute Actigraphy. The most popular and sophisticated method is Proportional Integral Mode (PIM), which computes the area below the curve; we integrate using the trapezoidal integration rule. Other methods include Zero Crossing Mode (ZCM), which computes the number of intersections between zero axis and the curve, and Time Above Threshold (TAT), which measures the length of time that the signal is above a certain threshold. Our proposed operation set easily supports both ZCM and TAT as well.

We discuss the potential for other, more complex computations in Section 7.2.

3.4 Handling errors and hiding variation

Errors, exceptions, and conditional execution may all potentially leak information about the data being processed. Returning a divide-by-zero error, for example, reveals that the divisor is equal to zero. Likewise, a naive implementation of `if` might allow an attacker to learn which block was executed, if one took longer than the other. Since we assume that the attacker issues arbitrary commands, errors and conditionals could provide the tools an attacker would need to discover the value of any data element stored in the card. In our design of Plug-n-Trust, we deal with these cases in three ways:

Fail silently: When data-dependent errors occur, like divide-by-zero, PnT fails silently. A result is returned as though the operation succeeded, and internally, the data's associated error bit is set. The error will be apparent to the back-end server when the data is received, but not to the attacker.

Take the long way: PnT avoids information leakage by removing timing variation from its execution. For example, in `if` operations, *both* possible results must be computed. This maintains the confidentiality of the data, at the cost of some wasted computation. In practice, we have found this wasted computation to be minimal, though pathological cases exist that would be impractical.

Stick to one size: When unsealing, PnT always returns a vector padded out to its maximum size, which in our implementation must be a multiple of the AES block size

(16 bytes); currently our vectors are all 32 bytes long. The adversary cannot learn anything about data values from the size of the returned data.

3.5 Detecting invalid processing

The architecture, as described thus far, is sufficient to keep sensor data and processing confidential. Of course, ensuring confidentiality alone is not sufficient. The attacker, who can issue arbitrary commands, may attempt to modify the data-processing results to cause harm without knowing any of the sensor values. To protect the patient against invalid processing, the back-end system (acting on behalf of the care provider) should be able to verify whether the expected data processing has occurred.

A naive approach might keep a history of all operations that are performed on each piece of sensor data, and ship that history with the data when it is unsealed. Whoever uses the data can then verify that the history is valid. This would be simple to implement, but completely impractical. A key reason for processing data locally is to reduce bandwidth requirements—a benefit that would be defeated by storing complete histories.

Instead, for each data element, d, PnT represents its processing history (or path) using a compact incremental hash, H_d. These hashes are similar in flavor to "path identifiers" used by path-profiling algorithms (e.g., Ball-Larus [7]); however, path-profiling algorithms only prevent path ID collisions within a program while PnT must protect against external hash collisions. The back-end server knows enough about the mApp to know the proper sequence of operations, the set of sensor IDs, and the expected sequence of operands. It can thus compute an expected path hash, and compare it with the path hash reported from the card, to verify whether the correct data was correctly processed.

Figure 3 shows an example of path hashes; in this computation of $(y_1x_1 + 5x_2)$ two readings x_1 and x_2 are collected from s_x (in that order), and y_1 is collected from s_y. If the adversary changes the operations, or rearranges the operands, the path hash will be different and allow the back-end to detect the attack. PnT computes a hash value at each step, incorporating information about both the op-code and the operands; the hash function H can be any cryptographically secure hash function (our implementation uses SHA-256). For arithmetic commands, such as `mult` or `add`, the path hash is computed by applying H to the concatenation of the op-code Op_{cmd} and a representation of the arguments: for a constant c, we hash in c; for a reference r, we hash in H_r.

At the beginning, the `start()` operator notifies the card of a new task. When operator $\texttt{seal}(s, \{s, t, n, e, data\}_{K_s})$ is called, where n is the sequence number, and t is the sensing time, the card does the following: If it is the first seal command for sensor s (since the `start`), the card stores the sequence number n in an internal variable $seq(s)$. For all `seal` operations, the card computes the hash value for the operation as $H(Op_{seal} \,|\, s \,|\, n - seq(s))$ where Op_{seal} is the op-code for sealing. We call $(n - seq(s))$ the *relative sequence number*. For example, the first sealings for sensors s_x and s_y (the second and third lines in Figure 3) sets each relative sequence number to 0, while the subsequent sealing (the fifth line) results in 1. With the relative sequence numbers embedded in the path hash, the back-end server can check the correct order of the data (see Section 5.2): if, for example,

Figure 4: Our prototype, configured for power measurement.

the adversary swaps x_1 and x_2, H_{r_1} and H_{r_4} would change and thus so would the overall H_{r_6}.

Upon receiving unsealed data, the back-end server precomputes the hashes it expects to receive (based on the desired data processing), and compares those hashes to those included in the data. Hashes that do not match in the back-end's computation are invalid, indicating an attack.

For the back-end server to verify freshness of the computation, it must be given some knowledge of the time during which the data was collected. Thus, we maintain for each reference a time range (t_{min}, t_{max}) that represents the timestamps of the data used to compute its value, and each operator's output is assigned a time range that is the union of the time ranges of its inputs. Figure 3 shows the final time range reported as part of the `unseal` result.

3.6 Managing card resources

Smart cards have limited resources; for example, the G&D card we use has 160 Kbytes of ROM, 4608 bytes of RAM and 72 Kbytes of EEPROM. The EEPROM is shared between program and data and when EEPROM is exhausted the smart card will start returning errors. A simple resource manager could keep track of data storage to handle errors of this nature more gracefully. Further, there is no automatic garbage collection so the programmer must remember to free references when done with them. When data is unsealed all intermediate values could instead be automatically freed. See Section 7 for a discussion of automatic path extraction and automatic reference freeing.

4. IMPLEMENTATION AND EVALUATION

To evaluate the potential for our approach in real applications, we implemented PnT using current mobile-phone and smart-card technology. Our test hardware, shown in Figure 4, includes a G&D smart card [17] and two smart phones – the G1 (shown) and Nexus One Android phones.

PnT's on-card component is implemented as an applet for the JavaCard v2.2.1 platform. The G1 and Nexus One phones

Operation:	Path Hash:	Time range:	Sequence numbers:
`start()`			
$\texttt{seal}(s_x, \{x_1, n_{x_1}, t_1\}_{K_{s_x}}) \Rightarrow r_1$	$H_{r_1} = H(Op_{seal} \mid s_x \mid 0)$	$time(r_1) = (t_1, t_1)$	$seq(s_x) = n_{x_1}$
$\texttt{seal}(s_y, \{y_1, n_{y_1}, t_2\}_{K_{s_y}}) \Rightarrow r_2$	$H_{r_2} = H(Op_{seal} \mid s_y \mid 0)$	$time(r_2) = (t_2, t_2)$	$seq(s_y) = n_{y_1}$
$\texttt{mult}(r_1, r_2) \Rightarrow r_3$	$H_{r_3} = H(Op_{mult} \mid H_{r_1} \mid H_{r_2})$	$time(r_3) = (t_1, t_2)$	
$\texttt{seal}(s_x, \{x_2, n_{x_2}, t_3\}_{K_{s_x}}) \Rightarrow r_4$	$H_{r_4} = H(Op_{seal} \mid s_x \mid 1)$	$time(r_4) = (t_3, t_3)$	$n_{x_2} - seq(s_x) = 1$
$\texttt{multc}(r_4, 5) \Rightarrow r_5$	$H_{r_5} = H(Op_{multc} \mid H_{r_4} \mid 5)$	$time(r_5) = (t_3, t_3)$	
$\texttt{add}(r_3, r_5) \Rightarrow r_6$	$H_{r_6} = H(Op_{add} \mid H_{r_3} \mid H_{r_5})$	$time(r_6) = (t_1, t_3)$	
$\texttt{unseal}(r_6) \Rightarrow \{time(r_6), value(r_6), H_{r_6}\}_{K_c}$			

Figure 3: An example procedure for computing path hash and generating the message to be sent to the back-end server. The task is to compute $y_1x_1 + 5x_2$. Assume that $t_1 < t_2 < t_3$ and $n_{x_1} + 1 = n_{x_2}$. For clarity we do not include the error codes (e) in the inputs and outputs or the computed HMACs for `seal` and `unseal`.

run Android versions 1.6 (Donut) and 2.3.5_r1 (Gingerbread) respectively, both modified with seek-for-android [36], a free implementation of PC/SC ("Personal Computer/Smart Card", a specification for smart-card integration into computing environments) for Android phones.

As examples, we implemented several computations that are commonly used in processing health-related sensor data: *Mean & variance*, which compute basic statistics (e.g., arithmetic mean, variance, data range) that help to remove noise and quantify volatility; *Zone detection*, which help to monitor heart-rate or blood-glucose levels, because providers often want to know if values fall outside a particular zone rather than the specific data values; and *Sleep actigraphy*, in which accelerometer readings are integrated over time to measure a patient's sleep patterns.

In this section, we evaluate the overhead and performance limitations of PnT, using our current implementation. Specifically, we focus on energy consumption and processing speed, using today's smart-card technology, and identify key focus areas for future improvement.

4.1 Energy overhead

To assess the impact of PnT on mobile device battery performance, we measured the quiescent power consumption as well as the per-command and per-application energy cost of our PnT prototype. We measured energy consumption using a Monsoon Power Monitor[2] and a microSD card extender, with custom modifications to facilitate power monitoring, as shown in Figure 4. For the sake of comparison, we use this same setup to measure the idle power and active energy characteristics of a standard Transcend microSD flash card plugged into the same phones.

The measured idle power draw of the G&D smart card is roughly 9mW—an order of magnitude higher than that of a standard flash card; however, this would still require 4 days to drain just 10% of the G1 phone's battery. For low-frequency applications where the smart card is used infrequently, the card can also be powered down between uses.

The per-task energy consumptions are shown in Table 1 for several PnT tasks, our applications, and 1 MB reads and writes to the flash card. While we only show results for a subset of PnT's processing commands, the time and energy requirements of the other data-processing operations are comparable. The impact of PnT on device lifetime will depend on how often it is used as well as other unrelated activities on the phone; however, even if PnT were constantly processing data, it would drain only 20% of the phone's battery over the course of a day.

[2]http://www.msoon.com

4.2 Processing performance

While the energy costs for processing are not a significant limitation, processing speed is still a challenge. Table 1 also shows the computation time of the PnT commands and applications, the maximum data rate (samples/s) that each can sustain, and the percentage of time that is spent by each in communication between the phone and the card. For tasks that must compute an HMAC, the percentage of time spent computing the HMAC is also shown. The max data rate is ten times one over the computation time, since ten samples were processed per seal. The percentage of time is the measured communication ping time divided by the computation time (for the apps the ping time was multiplied by the number of operations). These performance results are shown for the Nexus One phone. The results for the G1 phone running Donut were identical, except it had an additional communication latency of roughly 30 ms for each message sent to the card. These results show that, at least for this generation of smart cards, PnT is suitable primarily for applications with low-to-medium sampling rate. This poor performance stems from three key factors: slow communication, lack of HMAC support, and the card's built-in persistence mechanisms.

To date, smart cards have been used primarily as secure key stores that perform cryptographic functions for e-commerce applications with low data-rate requirements. Consequently, they have been optimized for fast cryptographic operations, but not for fast communication—a fact that results in PnT spending 16–93% of a given task's time in communication. Moving PnT to a newer phone (Nexus One) and a newer version of Android (Gingerbread) improved communication performance by nearly 50%, and we expect that overhead can be dramatically reduced further by optimizing the software that communicates with the card.

A second performance challenge is that our current cards do not provide support for HMAC computations. Our software HMAC implementation currently accounts for more than 50% of the `seal` and `unseal` operations and 12–36% of the total runtime in our test applications. Based on the next-generation JavaCard specifications, we expect newer cards to provide hardware support for HMAC in the near future.

The third reason for slow performance in PnT stems from the fact that today's smart-card hardware and software systems are designed for both powered and contactless appli-

Operation	Time (ms)	Max rate (samples/s)	Energy (mJ)	Overhead (% time) Comm	HMAC
`seal`	226.2	44.2	13.1	16.5%	50.3%
`unseal`	182.6	54.8	10.6	20.5%	58.0%
`free`	39.9	250.9	2.3	93.7%	
`add`	142.9	70.0	8.3	26.1%	
`multc`	111.3	89.8	6.4	33.5%	
`div`	117.9	84.8	6.8	31.7%	
`prod`	101.9	98.1	5.9	36.6%	
`if`	112.6	88.8	6.5	33.2%	
Sleep Actigraphy	600.9	16.6	34.7	31.1%	36.6%
Mean & Variance	1585.9	6.3	91.7	44.7%	12.1%
Zone Detection	2101.3	4.8	121.5	31.7%	30.8%
Flash card (read,1MB)	49.8		6.325		
Flash card (write,1MB)	184.6		22.702		

Table 1: Timing/energy results for individual PnT operations on a Nexus One smart phone using 10-sample vectors, and for comparison (in the bottom two rows) for read and write operations on a generic flash-memory card. The maximum sustainable sensing rate (sample/s) is also shown. For PnT, the key limiting factors are the slow communication speeds of current smart cards and slow HMAC performance, which combined result in 59-69% of the total latency for our test applications. As smart cards find increasing use in higher-rate applications (like sensor data processing), we expect hardware performance to improve dramatically.

cations. To survive transient power failures, the JavaCard environments store objects and variables in EEPROM by default, which has much longer access times than RAM, especially for write operations (note that `seal` takes nearly twice as long as `unseal`). JavaCard v2.2.1 allows simple arrays to be declared as *transient* and stored in RAM; however, many EEPROM accesses cannot be avoided (access to objects and their non-array attributes). The potential for improvement can be seen in the `free` operation—PnT's only EEPROM-free operation—which has more than 93% communication overhead. The next generation of smart cards will allow greater use of transient variables.

Finally, some smart-card companies have recently started selling high-performance microSD products that can encrypt voice and video traffic in real-time. These cards are not currently programmable by third-party developers. While communication, HMAC support, and mandatory EEPROM accesses represent significant limitations to PnT's current processing performance, these limitations are neither fundamental nor likely to remain for long. When resolved, we expect PnT to be able to support many more applications with much higher data rates.

5. SECURITY ANALYSIS

To assess the security of PnT, we describe how PnT guarantees the security properties set forth in Section 2.3, preventing adversaries from succeeding in their attacks, some of which were highlighted in Section 2.2. In particular, we discuss how PnT achieves data confidentiality and integrity.

5.1 Data confidentiality

PnT protects the patient's privacy by enforcing our first security property **SP1**:*Data Confidentiality*. The adversary may attempt to break this property either directly or indirectly. In a direct attack, the adversary tries to directly access the sensor data or any derived information, i.e., the final or intermediate results of the data processing. PnT prevents this attack by encrypting all sensor data whenever it leaves trusted components: sensor nodes, the plug-in card, and the back-end server. These components are assumed to protect data privacy and integrity (assumptions A2 and TR1 for the SN; A3 and TR2 for the card; A4 and TR3 for the server). The adversary cannot succeed in this attack unless she can break the chosen cryptosystem (e.g., AES). The intermediate results are never exposed outside the card; only data-independent references to the results are revealed.

In an indirect attack, the adversary tries to learn about the sensor data and derived information by observing the size or timing of responses by the plug-in card. The card returns only two types of values: references (which are small, opaque fixed-size objects), and unsealed results (which are encrypted, fixed-sized vectors). Thus, the size of intermediate and final results provide no information; what can the adversary learn from the computation time? Obviously, more-complex computations take longer. However, this information is useless because the adversary already knows what computation will be performed. For the basic operators (arithmetic, Boolean, and conditional operators) each operation takes a fixed amount of time; the adversary learns nothing about the data values. The `if` statement may seem promising; if the adversary can learn which branch (then/else) was chosen, it can infer properties of the data used in the conditional expression. PnT prevents this inference by performing both sides of the branches, and only retaining the result of the correct branch.

Since most smart cards are designed to resist physical attacks, we assume that physical side-channel attacks are outside the capability of the adversary (A5).

5.2 Integrity

The provision of security property **S2** (*Sensor Data Integrity*) is straightforward because the encryption, path hash, MAC, and timestamp protect data against modification, forgery, or replay (by assumptions H1, H2, A1, and A5). The adversary cannot break the data integrity unless she can break the chosen cryptographic primitives (A1) or the clock within the sensor (H2, A2).

Security property **S3** (*Derived Data Integrity*) is more challenging due to its larger attack surface. We next discuss possible attack methods and justify how PnT counteracts.

Wrong sensor type: When processing sensor data (body temperature, for example) the adversarial mApp may attempt to feed the processing routine with the wrong type of sensor data (say, systolic blood pressure). It does so by giving the wrong arguments to the `seal` operator: supplying the wrong sensor ID or sensor data. If the wrong sensor ID is given to the `seal` command, the card will fail to decrypt the data and will return an error. If the correct sensor ID is provided, but the sensor is the wrong type for the operations, the computation will proceed. Because the sensor ID is included in the computation of the path hash, however, the back-end server can immediately detect the use of an incorrect sensor.

Stale sensor data: The adversary may provide a set of sensor readings received in the past, rather than fresh readings. The back-end server can detect this replayed or delayed data immediately because the sensor timestamp is incorporated into the (t_{min}, t_{max}) range reported in the package exported by the card, as described in Section 3.

Wrongly ordered data: The adversary may provide fresh data, from the right sensor, but reorder, repeat, or omit some of the readings. When sensor readings are re-ordered, or repeated, or omitted, the computation may result in different values. PnT prevents this attack by adding the relative sequence number of each datum into the path hash. So, whenever a re-ordering or omission attack occurs, the path hash will not match what is expected.

Wrong constants: The adversary can easily change the processing algorithm by replacing one constant with another, for example, to call `div(r1,4)` instead of `div(r1,2)`. Since the constant value is included in the path-hash computation, this attack causes a mismatching hash value, detected at the back-end server.

Wrong procedure: The adversary may manipulate the procedure, change the order of operations, insert new operations, remove operations, and so on. These changes affect the hash value and the back-end server can detect them.

Hash collision attack: The most sophisticated attack is by an attacker who knows how all the above attacks can fail. Here, the attacker strives to find a computation path that has the identical hash value as the correct path, and that produces a different result, hoping the result will be harmful to the patient. She cannot succeed in this attack without, in effect, breaking the underlying hash function (A1). The point of cryptographic hashes, like SHA-256, is to make it difficult to find such collisions.

6. RELATED WORK

Plug-n-Trust builds on and is inspired by a large body of related work, providing support for trustworthy sensing, secure multiparty computation, and virtualization.

6.1 Trustworthy mobile sensing

Several systems have worked towards providing mobile sensing systems that can be trusted to provide users with reliable data. A common approach is to have wireless sensors, with or without hardware support for cryptography, sign their own sensor readings [13,34,45]. Encryption can be used with this sign-and-send model to protect privacy, and data can be trusted, since it cannot be modified until it is delivered. Unfortunately, this can result in a large volume of data that can strain bandwidth limits and increase patient cost.

Another proposed approach [19] allows local processing of sensor data within a trusted OS and hypervisor backed by a TPM [41], without trusting the application. While this allows greater flexibility than the simpler sign-and-send systems, it also presents a much larger attack surface, and other challenges faced by TPM-based and hypervisor-based systems, which we describe in the following sections.

6.2 Trusted platforms

The Trusted Computing Group [41] has proposed using an on-board hardware root of trust to establish a trusted software stack (or Trusted Platform) using binary attestation. Others [2,12,15,19] have extended this idea to mobile devices (e.g., Mobile Trusted Module (MTM) [31]).

To date, in spite of the widespread availability of TPMs in desktop and laptop computers, trusted platforms have failed to find widespread use. MTMs have not been embraced by industry, who have chosen instead to use legacy security solutions. Debates about Digital Rights Management (DRM), a primary driver for both TPMs and MTMs, have highlighted another challenge, determining what entities should have control over the TPM and its trusted software. Shubina, et al. [37] conclude that error handling for TPM primitives is a significant source of confusion for programmers, which has further limited the use of TPMs. Furthermore, flaws have been noted in the MTM specification [26] and attempts to address them are ongoing.

ARM's TrustZone [42] architecture also hopes to improve the situation, by integrating MTM-like secure processing and general purpose processing into a single chip. In addition to tighter integration, TrustZone supports additional features, including secure I/O to compatible peripherals (e.g., a secure display or keyboard). The future of both TrustZone and MTMs is unclear, but so far, neither has been widely adopted.

If MTMs do, at some point, become part of the mobile computing landscape, approaches like Flicker [30] provide an alternative to PnT; however, trusted computations in Flicker must occur while the OS is suspended, leading to competition that could render the mobile device unresponsive and unusable when under high load.

Additionally, PnT could be implemented using TrustZone or an MTM, rather than a smart card, for trusted processing. These alternatives, especially TrustZone, may provide better performance than current smart cards, while complicating the bootstrapping of the system. The MTM's root of trust would likely reside with the smart-phone manufacturer and/or service provider, whom the patient would have to trust with her private medical data.

The new Google Wallet technology [22] incorporates a secure coprocessor (the NXP PN65K) into the smart phone to support secure payment transactions over Near Field Communication (NFC). The PN65K provides storage, computation, and communication that is separate from the rest of the phone, and Android allows access only from the trusted Wallet app. It is not clear whether this chip could support other secure operations, such as health-data processing, but it seems unlikely that the Wallet business partners would be willing to open this payment-focused technology to other applications. In any case, as above, any phone-embedded solution brings the handset manufacturer and network operator into the trust relationships that should be narrowly focused on the patient and the health provider.

6.3 Virtual machines

Another popular approach to securing computation is to run applications in a virtual machine (VM), where sandboxing can limit the duration and severity of some attacks. If the hypervisor is able to detect attacks, VMs can also simplify recovery by reloading a fresh VM image [9]. This approach is being advanced by a number of mobile-phone industry initiatives, including a partnership between VMWare and LG Electronics [44]; however, virtualization is not yet widely available for mobile phones.

If (or when) mobile phone-based virtualization becomes commonplace, sensitive or safety-critical sensor data could be processed by a trusted hypervisor, instead of a smart card, at the cost of diminished security. Virtualization is complicated and even small hypervisors [19,38] must deal with much more than security and privacy. Hypervisor complexity presents an unnecessarily large attack surface for attackers to search for vulnerabilities, and the consequences of a successful attack on a hypervisor are no less severe than a successful attack on a traditional operating system [43]. Protecting vulnerable software with additional layers of vulnerable software may provide enough security for some applications, but not highly-sensitive or safety-critical computation.

Green Hills Software claims to have developed a formally-verified "hacker-proof" hypervisor [25], that they market to military and government customers. We were not able to verify these claims; however, if these security guarantees prove to be true *and* cell-phone manufacturers were willing to virtualize their current OS on top of the Green Hills VM, this approach may provide an alternative to PnT.

Traditional VM-based approaches also have performance and usability disadvantages. Software-only approaches require a trusted software stack and inherit all of the challenges of the trusted platforms described in Section 6.2. A trusted VM also competes with other applications for resources, while PnT provides its own computing resources by replacing an existing microSD card with a combination microSD card and secure processor.

Additional variants have been proposed that improve performance. The "Divide" system from Enterproid [14] provides "dual-persona" capability to Android phones – a promising approach, which might be extended to support multiple persona, e.g., work, play, finance, and health. However, Divide is nonetheless an Android app and is thus vulnerable to rootkits and other mechanisms that compromise the underlying operating system.

The "Cells" approach goes further, allowing a smart phone to support multiple *virtual phones (VPs)*, each of which can run the full range of Android applications [5]. With much lower overhead than a VM or hypervisor approach, Cells would allow one VP to run mHealth applications and keep them isolated from other activity on the phone. On the other hand, the TCB is still pretty large and the effectiveness of separation and containment still depends on the security of the kernel and the Cells implementation. In that regard, relative to PnT, it has the same drawbacks as other virtualization methods.

6.4 Secure multiparty computation

Another closely-related approach to keeping sensitive computations private is to use a secure multiparty computing (SMC) system like FaeriePlay [24]. In addition to keeping data confidential, SMC systems typically employ garbled Boolean circuits (which also hide the computation being executed). While attractive for mHealth-related applications, evaluating programs as Boolean circuits comes at a high performance cost, limits the kinds of computations that an application can perform, and is difficult to verify.

Existing SMC systems overcome the verifiability challenges by either evaluating the entire program circuit, one gate at a time, inside a secure coprocessor like the IBM 4758 [24] – which is impractical for today's smart cards – or by using a homomorphic encryption scheme [18]; some partially homomorphic encryption schemes are more practical [32], perhaps even for use on mobile phones; however, it is not clear whether these techniques are expressive enough to support general-purpose computation. If approaches to SMC become more practical in the future, they will provide another attractive solution to this problem.

6.5 Control flow integrity

Finally, the path-hashing approach used in PnT is inspired by the field of Control Flow Integrity (CFI). CFI systems [1, 16] use techniques like code rewriting and shadow call stacks to ensure that executing programs execute within a defined control flow graph. While PnT shares many of the same goals with existing CFI approaches, all rely on the assumption that the attacker cannot modify application code. Since we assume a more powerful attacker who can modify application code, we have chosen to make control flow verifiable rather than preventing abuses.

7. DISCUSSION AND FUTURE WORK

This section discusses the limitations of Plug-n-Trust as well as the extensions that are left as future work.

7.1 Limitations

Hiding computation type

While PnT protects the integrity and confidentiality of sensed data and its processing, it does not hide the nature of the data processing, which might also reveal the patient's health condition. Some applications may be coded in a general way that limits the amount of leaked information, and garbled circuits (as mentioned in Section 6.4) could be used if and when SMC techniques become more efficient. Meanwhile, code obfuscation techniques [28] may provide a lower-cost alternative to make it significantly more difficult for an adversary to identify the nature of PnT's computations.

Data accessibility

One current limitation of PnT is the inability to *display data* to the user. With an attacker that has full control over the mobile phone, releasing data in a form that can be displayed to the user also makes it available to the attacker, violating our confidentiality goals. This limitation applies to all other solutions unless the OS can be trusted to be free from malware. In theory, smart cards with a secure input or output channel (e.g., a button or light) can secure limited communications with the user [20]. (TrustZone [42] is capable of securing I/O and hence can display data without an attacker intercepting it, but has other limitations discussed above.) New smart cards from Go-Trust support Near-Field Communication (NFC), which may allow the card to communicate directly with nearby trusted peripherals [21].

The lack of data accessibility may also make debugging more difficult. Path hashing makes it possible to detect that a problem has occurred, but the specific attack or error may not be clear. This is a common challenge in secure systems, and techniques that embed more information about a specific error are needed to ease this tension.

Data-dependent operation

Another challenge is supporting *data-dependent communication*. An application that, in the interest of efficiency, only communicates when certain events occur (like an irregular heart beat), will unavoidably leak information: the presence of a message reveals information about the data. A potential compromise between efficiency and confidentiality would support data-dependent events masked by interspersed false events. Even with false event messages, this approach would likely be vulnerable to a variety of statistical attacks. This challenge is fundamental, because it represents a traffic-analysis attack on the data stream between the card and the back-end and there is no easy solution whenever the OS or phone cannot be trusted.

The analog of data-dependent communication is *data-dependent loop bounds*. PnT currently does not provide any in-card support for loops, though applications are free to execute commands in loops. When those loops depend on the value of confidential data, the number of iterations reveals information about the data, and could be used in PnT to discover the value of any arbitrary data element. Data-dependent loops with a bounded number of iterations can be implemented in PnT by iterating the maximum number of times, and using a conditional to turn the unneeded iterations into no-ops. This approach is potentially inefficient.

Another analog of data-dependent communication is *data-dependent sensor management*. In some applications, the results of data processing or aggregation within the MN may lead to a need to change the configuration of one or more SNs, e.g., to turn it on or off, to change the sampling rate, or to adjust the gain on a sensor. As with the above examples, feeding such commands back through the mApp would leak information into the untrusted MN. It would be possible, though expensive, for the card to return a vector of encrypted commands (one to be sent to each SN) each time it completed processing a batch of sensor data. Such a plan needs further investigation.

Cost

A search for current (late 2011) prices found microSD smart cards selling for around \$60-\$70 per card. These cards are new to the market and not in widespread use yet, while more widely deployed (non-microSD) smart cards cost only \$1-\$10. Mass production will have a similar effect on microSD smart cards; their price already dropped 80% over the past year.

Denial-of-Service (DoS)

An attacker that has compromised the phone can deny the application service in a wide variety of ways—to refuse delivery of messages, deny power to the card, and so forth—and there is nothing the card can do to prevent it. Instead of preventing DoS attacks, we hope to make them apparent, so that the attack can be remedied.

A variety of approaches could be used to detect the presence of a DoS attack. In most areas, mobile phones enjoy better than 90% network coverage and no data delivered for a long time might be a sign of DoS. If applications intentionally wait to deliver data, PnT could also send heart-beat messages. In any event, it is impossible for PnT to discern the difference between a DoS attack and a device failure. When a potential DoS attack is detected, notifying the patient will likely be necessary to determine the nature of the problem.

Theft or loss

An attacker may steal the smart card, the phone, or sensor nodes to learn about the patient's medical condition or to inject invalid data. Although PnT still protects the patient's privacy in face of theft (see Section 5.1), the attacker may carry the sensors, the phone, and the smart card and monitor her own medical condition, and submit the data to the back-end server, impersonating the owner of the smart card. In this case, reporting lost cards to the care provider will be the only effective measure.

To better counteract theft, the smart card should be able to authenticate the user. Since the phone can be malicious, the smart card and the patient cannot rely on the phone for mediating this authentication process. A direct method for the authentication involves some biometric, but at this time it is unfeasible to equip the card with a biometric sensor (e.g., fingerprint scanner) and necessary software. One could employ a physiology-based authentication scheme [40], which uses sensor data as a biometric to ensure authenticity.

Tolerance to data loss

Using sequence numbers and timestamps, PnT ensures detection of any re-ordering, repeating, or omission of sensor readings (within a given time window). While this feature is essential for critical monitoring applications, there are cases when some data loss should be tolerated. Since packet loss is not unusual in wireless communication, discarding all the sensor data because of one missing datum can significantly degrade data availability. Therefore, it is beneficial if PnT can support a computation task that is tolerant of a few missing data elements. We plan to extend our sequence-ensuring mechanism so that the computation is accepted if the number of missing data elements is below a certain threshold.

7.2 Extensions

There are several other extensions that we hope to address.

Complex computation support

PnT can easily be extended to support real numbers as well as integers. It can also be extended to support built-in operators, such as FFT or other signal-processing functions. We plan to add more mathematical operations such as exponentiation, logarithms, trigonometric functions, and set operations, and the ability to handle multi-datatype vectors.

However, in spite of these extensions, PnT is unlikely to fit all applications. Some computationally heavy and data intensive computations (like some machine-learning algorithms) may be too complex to perform in a small smart-card, and too complex to verify via path hashing. These analyses may be done by the back-end server.

Medical actuator support

Another important extension will support medical actuators, like insulin pumps, without requiring the back-end service to be in the loop. For this to work, the actuator devices must be paired with the card (to exchange keys), and the device should be able to verify the result derived by the PnT card as the back-end server does, that is, by checking the path hash. Path hashes may be pre-computed and pre-installed on the actuator device, reducing the computational cost that verification would place on a resource-constrained platform.

Language and compiler support

Currently, applications interact with PnT by specifying individual PnT commands; the result looks something like assembly language. We plan to improve this situation by providing higher-level programming interfaces that allow system designers to describe how data should be processed without worrying about sealing and unsealing data, unrolling `if` statements, or forgetting to free allocated references.

Internal sensors

Finally, we plan to make the wealth of sensor data provided by a phone's internal sensors accessible to PnT. Without additional hardware, these sensors cannot be trusted if the phone's OS has been compromised; however, many attacks compromise only the application. We plan to support multiple levels of trust in PnT, allowing data from less-trusted sensors to be processed by more risk-tolerant applications.

Multiple-task support

In practice, the patient's condition may require multiple monitoring tasks to run simultaneously. For example, the mobile node may need to analyze ECG data continuously while blood-pressure level is monitored intermittently. We plan to extend PnT to support multiple monitoring tasks by allowing the SD card to maintain multiple contexts of computation, by introducing a task-reference by which the applications can specify the context of each operation.

8. SUMMARY

The connectivity and processing provided by patient-carried smart phones is critical to the success of many mHealth applications; however, as private and safety-critical data processing moves onto commodity smart-phones that are vulnerable to software-based attacks, security and privacy concerns must be addressed.

In this paper, we address this challenge in three ways: First, we describe the design of Plug-n-Trust (PnT), a novel and practical approach to protecting both the confidentiality and integrity of safety-critical medical sensing and data processing on vulnerable mobile phones. Second, our implementation and experiments show that PnT is feasible—using current Java-based smart cards and Android phones—for applications with low-to-medium data-rate requirements, and we identify opportunities for dramatic performance improvements that will allow PnT to support more data-intensive sensing applications. None of PnT's performance challenges are fundamental, and all will be eliminated as smart-card technology evolves. Third, we provide a security analysis demonstrating that PnT meets the desired security goals.

PnT is applicable to a wide range of applications, not limited to healthcare, and is amenable to a range of deployment scenarios that include secure elements built into phones as well as the plug-in card model we have proposed.

Acknowledgements

This research results from a research program at the Institute for Security, Technology, and Society at Dartmouth College, supported by the National Science Foundation under Grant Award Number 0910842 and by the Department of Health and Human Services (SHARP program) under award number 90TR0003-01. The views and conclusions contained in this document are those of the authors and should not be interpreted as necessarily representing the official policies, either expressed or implied, of the sponsors.

We also thank the anonymous reviewers, our shepherd Urs Hengartner, and our colleagues at Dartmouth for their valuable feedback.

9. REFERENCES

[1] M. Abadi, M. Budiu, U. Erlingsson, and J. Ligatti. Control-flow integrity principles, implementations, and applications. *ACM Transactions on Information Systems Security*, 13(1), Nov. 2009. DOI 10.1145/1609956.1609960.

[2] O. Aciicmez, A. Latifi, J.-P. Seifert, and X. Zhang. A trusted mobile phone prototype. In *IEEE Consumer Communications and Networking Conference (CCNC)*, pages 1208–1209, Jan. 2008. DOI 10.1109/ccnc08.2007.270.

[3] S. Agarwal and C. T. Lau. Remote health monitoring using mobile phones and web services. *Telemedicine and e-Health*, 16(5):603–607, June 2010. DOI 10.1089/tmj.2009.0165.

[4] S. Ancoli-Israel, R. Cole, C. Alessi, M. Chambers, W. Moorcroft, and C. P. Pollak. The role of actigraphy in the study of sleep and circadian rhythms. *American Academy of Sleep Medicine Review Paper*, 26(3):342–392, 2003.

[5] J. Andrus, C. Dall, A. V. Hof, O. Laadan, and J. Nieh. Cells: a virtual mobile smartphone architecture. In *Proceedings of the ACM Symposium on Operating Systems Principles (SOSP)*, pages 173–187. ACM, Nov. 2011. DOI 10.1145/2043556.2043574.

[6] A. Arcelus, R. Goubran, H. Sveistrup, M. Bilodeau, and F. Knoefel. Context-aware smart home monitoring through pressure measurement sequences. In *Proceedings of IEEE International Workshop on Medical Measurement and Applications (MEMEA)*, pages 32–37, Apr. 2010. DOI 10.1109/MEMEA.2010.5480223.

[7] T. Ball and J. R. Larus. Optimally profiling and tracing programs. *ACM Transactions on Programming Languages Systems*, 16:1319–1360, July 1994. DOI http://doi.acm.org/10.1145/183432.183527.

[8] F. Buttussi and L. Chittaro. Smarter phones for healthier lifestyles: An adaptive fitness game. *IEEE Pervasive Computing*, 9(4):51–57, Oct. 2010. DOI 10.1109/MPRV.2010.52.

[9] L. P. Cox and P. M. Chen. Pocket hypervisors: Opportunities and challenges. In *Proceedings of the IEEE Workshop on Mobile Computing Systems and Applications (HotMobile)*, pages 46–50. IEEE, 2007. DOI 10.1109/HotMobile.2007.20.

[10] S. Coyle, F. Benito-Lopez, R. Byrne, and D. Diamond. On-body chemical sensors for monitoring sweat. In *Wearable and Autonomous Biomedical Devices and Systems for Smart Environment*, volume 75 of *Lecture Notes in Electrical Engineering*, pages 177–193. Springer, 2010. DOI 10.1007/978-3-642-15687-8_9.

[11] L. Cranor and S. Garfinkel. *Security and Usability*. O'Reilly Media, Inc., 2005.

[12] K. Dietrich and J. Winter. Towards customizable, application specific mobile trusted modules. In *Proceedings*

of the ACM Workshop on Scalable Trusted Computing (STC)*, pages 31–40. ACM, 2010. DOI 10.1145/1867635.1867642.

[13] A. Dua, N. Bulusu, W.-c. Feng, and W. Hu. Towards trustworthy participatory sensing. In *USENIX Conference on Hot Topics in Security*. USENIX, Aug. 2009. Online at http://www.usenix.org/event/hotsec09/tech/full_papers/dua.pdf.

[14] Enterproid.com. Enterproid: The Divide platform. Online at http://www.enterproid.com/features.php, visited Dec. 2011.

[15] J.-E. Ekberg, Nokia. Mobile Trusted Module (MTM) – an introduction. Online paper, 2007. Online at http://research.nokia.com/files/tr/NRC-TR-2007-015.pdf.

[16] U. Erlingsson, M. Abadi, M. Vrable, M. Budiu, and G. C. Necula. XFI: software guards for system address spaces. In *Proceedings of the Symposium on Operating Systems Design and Implementation (OSDI)*, pages 75–88. USENIX, Nov. 2006. Online at http://www.usenix.org/event/osdi06/tech/full_papers/erlingsson/erlingsson.pdf.

[17] Giesecke and Devrient GmbH. Online at http://www.gi-de.com/, visited Mar. 2011.

[18] R. Gennaro, C. Gentry, and B. Parno. Non-interactive verifiable computing: Outsourcing computation to untrusted workers. In T. Rabin, editor, *Advances in Cryptology (CRYPTO)*, volume 6223, pages 465–482. Springer, 2010. DOI 10.1007/978-3-642-14623-7_25.

[19] P. Gilbert, L. P. Cox, J. Jung, and D. Wetherall. Toward trustworthy mobile sensing. In *Proceedings of the Workshop on Mobile Computing Systems & Applications (HotMobile)*, pages 31–36. ACM, Feb. 2010. DOI 10.1145/1734583.1734592.

[20] H. Gobioff, S. Smith, J. D. Tygar, and B. Yee. Smart cards in hostile environments. In *Proceedings of the USENIX Workshop on Electronic Commerce*, pages 23–28, 1996. Online at http://www.cs.dartmouth.edu/~sws/pubs/gsty96.pdf.

[21] Go-Trust NFC smart cards. Online at http://www.go-trust.com/products/swp-secure-microsd/, visited Dec. 2011.

[22] Google. Google wallet. Online at http://www.google.com/wallet/, visited Dec. 2011.

[23] D. Halperin, T. S. Heydt-Benjamin, B. Ransford, S. S. Clark, B. Defend, W. Morgan, K. Fu, T. Kohno, and W. H. Maisel. Pacemakers and implantable cardiac defibrillators: Software radio attacks and zero-power defenses. In *Proceedings of the IEEE Symposium on Security and Privacy (S&P)*, pages 129–142. IEEE Press, May 2008. DOI 10.1109/SP.2008.31.

[24] A. Iliev and S. W. Smith. Small, Stupid, and Scalable: Secure Computing with Faerieplay. In *Proceedings of the ACM Workshop on Scalable Trusted Computing (STC)*, 2010.

[25] D. Kleidermacher. Next generation secure mobile devices. *Information Quarterly*, 7(4), 2008.

[26] R. Korthaus, A. R. Sadeghi, C. Stüble, and J. Zhan. A practical property-based bootstrap architecture. In *Proceedings of the ACM Workshop on Scalable Trusted Computing (STC)*, pages 29–38. ACM, 2009. DOI 10.1145/1655108.1655114.

[27] D. Kotz. A threat taxonomy for mHealth privacy. In *Proceedings of the Workshop on Networked Healthcare Technology (NetHealth)*. IEEE Press, Jan. 2011. DOI 10.1109/COMSNETS.2011.5716518.

[28] D. Low. Protecting Java code via code obfuscation. *Crossroads*, 4(3):21–23, Apr. 1998. DOI 10.1145/332084.332092.

[29] S. Mare, J. Sorber, M. Shin, C. Cornelius, and D. Kotz. Adapt-lite: Privacy-aware, secure, and efficient mhealth sensing. In *Proceedings of the Workshop on Privacy in the Electronic Society (WPES)*, Oct. 2011. DOI 10.1145/2046556.2046574.

[30] J. M. McCune, B. J. Parno, A. Perrig, M. K. Reiter, and H. Isozaki. Flicker: an execution infrastructure for TCB minimization. *SIGOPS Operating Systems Review*, 42:315–328, Apr. 2008. DOI 10.1145/1352592.1352625.

[31] Mobile Phone Work Group. Mobile Trusted Module FAQ. Online at http://www.trustedcomputinggroup.org/resources/mobile_trusted_module_faq, visited Dec. 2010.

[32] M. Naehrig, K. Lauter, and V. Vaikuntanathan. Can homomorphic encryption be practical? In *Proceedings of the ACM Workshop on Cloud Computing Security*, pages 113–124. ACM, Oct. 2011. DOI 10.1145/2046660.2046682.

[33] SIMalliance. SIMaLliance Web Site. Online at http://www.simalliance.org, visited Dec. 2011.

[34] S. Saroiu and A. Wolman. I am a sensor, and I approve this message. In *Proceedings of the Eleventh Workshop on Mobile Computing Systems & Applications (HotMobile)*, HotMobile '10, pages 37–42. ACM, 2010. DOI 10.1145/1734583.1734593.

[35] L. A. Saxon, D. L. Hayes, F. R. Gilliam, P. A. Heidenreich, J. Day, M. Seth, T. E. Meyer, P. W. Jones, and J. P. Boehmer. Long-term outcome after ICD and CRT implantation and influence of remote device follow-up: The ALTITUDE survival study. *Circulation*, 122(23):2359–2367, Dec. 2010. DOI 10.1161/CIRCULATIONAHA.110.960633.

[36] Open source. Seek-for-android. Online at https://code.google.com/p/seek-for-android/wiki/PCSCLite, visited Apr. 2011.

[37] A. Shubina, S. Bratus, W. Ingersol, and S. W. Smith. The diversity of TPMs and its effects on development: a case study of integrating the TPM into OpenSolaris. In *Proceedings of the ACM Workshop on Scalable Trusted Computing (STC)*, pages 85–90. ACM, 2010. DOI 10.1145/1867635.1867649.

[38] L. Singaravelu, C. Pu, H. HÃd'rtig, and C. Helmuth. Reducing TCB complexity for security-sensitive applications: Three case studies. In *Proceedings of European Conference on Computer Systems*, pages 161–174, 2006.

[39] E. G. Sirer, W. de Bruijn, P. Reynolds, A. Shieh, K. Walsh, D. Williams, and F. B. Schneider. Logical attestation: an authorization architecture for trustworthy computing. In *Proceedings of the ACM Symposium on Operating Systems Principles (SOSP)*, pages 249–264. ACM, Nov. 2011. DOI 10.1145/2043556.2043580.

[40] J. Sriram, M. Shin, T. Choudhury, and D. Kotz. Activity-aware ECG-based patient authentication for remote health monitoring. In *Proceedings of the International Conference on Multimodal Interfaces and Workshop on Machine Learning for Multi-modal Interaction (ICMI-MLMI)*, pages 297–304. ACM Press, Nov. 2009. DOI 10.1145/1647314.1647378.

[41] Trusted Computing Group. TPM Main Specification Level 2 Version 1.2, Rev 103. Online at http://www.trustedcomputinggroup.org/resources/tpm_main_specification, visited Dec. 2010.

[42] TrustZone. ARM TrustZone web site. Online at http://www.arm.com/products/processors/technologies/trustzone.php, visited Dec. 2010.

[43] S. Vaughan-Nichols. Virtualization sparks security concerns. *Computer*, 41(8):13 –15, Aug. 2008. DOI 10.1109/MC.2008.276.

[44] VMWare, Inc. and LG Electronics MobileComm U.S.A, Inc. VMWare News Release, Dec. 7, 2010. Online at http://www.vmware.com/company/news/releases/vmware-lge-partnership.html, visited Dec. 2010.

[45] A. Wolman, S. Saroiu, and V. Bahl. Using trusted sensors to monitor patients' habits. In *USENIX Workshop on Health Security and Privacy*. USENIX Association, Aug. 2010. Online at http://research.microsoft.com/en-us/um/people/alecw/healthsec-2010.pdf.

[46] F. Zhou, H.-I. Yang, J. Álamo, J. Wong, and C. Chang. Mobile personal health care system for patients with diabetes. In Y. Lee, Z. Bien, M. Mokhtari, J. Kim, M. Park, J. Kim, H. Lee, and I. Khalil, editors, *Aging Friendly Technology for Health and Independence*, volume 6159 of *Lecture Notes in Computer Science*, chapter 12, pages 94–101. Springer, 2010. DOI 10.1007/978-3-642-13778-5_12.

TapPrints: Your Finger Taps Have Fingerprints

Emiliano Miluzzo†, Alexander Varshavsky†, Suhrid Balakrishnan†
Romit Roy Choudhury§

†AT&T Labs – Research, Florham Park, NJ, USA
§Duke University, Durham, NC, USA

ABSTRACT

This paper shows that the location of screen taps on modern smartphones and tablets can be identified from accelerometer and gyroscope readings. Our findings have serious implications, as we demonstrate that an attacker can launch a background process on commodity smartphones and tablets, and silently monitor the user's inputs, such as keyboard presses and icon taps. While precise tap detection is nontrivial, requiring machine learning algorithms to identify fingerprints of closely spaced keys, sensitive sensors on modern devices aid the process. We present *TapPrints*, a framework for inferring the location of taps on mobile device touchscreens using motion sensor data combined with machine learning analysis. By running tests on two different off-the-shelf smartphones and a tablet computer we show that identifying tap locations on the screen and inferring English letters could be done with up to 90% and 80% accuracy, respectively. By optimizing the core tap detection capability with additional information, such as contextual priors, we are able to further magnify the core threat.

Categories and Subject Descriptors

I.5 [**Pattern Recognition**]: Applications

General Terms

Algorithms, Design, Experimentation, Security, Measurement, Performance

Keywords

Smartphone Sensing, Tap Detection with Motion Sensors, Mobile Device Security

1. INTRODUCTION

The proliferation of sensors on mobile devices, combined with advances in data analytics, has enabled new directions in personal computing. The ability to continuously sense people-centric data, and distill semantic insights from them, is a powerful tool driving a wide range of application domains. These domains range from remote patient monitoring, to localization, to fine grained context-awareness [16, 3, 2, 24]. While this ability to zoom into behavioral patterns and activities will only get richer over time, it is not surprising that there will be side effects. The research community and industry are already facing the ramifications of exposing location information; rumor has it that insurance companies are on the lookout to mine an individual's dietary pattern to guide insurance cost and coverage. While these kind of side effects will continue to surface over time, this paper exposes one that may be particularly serious. Briefly, we find that sensor information from mobile devices – specifically from the accelerometer and gyroscope – can be adequate to infer the locations of touch-screen taps. Interpreted differently, a background process running on a mobile device may be able to silently monitor the accelerometer and gyroscope signals, and infer what the user is typing on the software keyboard. Needless to say, the ramifications can be monumental.

This finding may not be surprising when one contemplates about it. Modern mobile devices, like smartphones and tablets, are being upgraded with sensitive motion sensors, mostly to support the rapidly growing gaming industry. Thus, when a user taps on different parts of the screen, it is entirely feasible that these taps produce an identifiable pattern on the 3-axis accelerometer and gyroscope. Figure 1 illustrates the intuition.

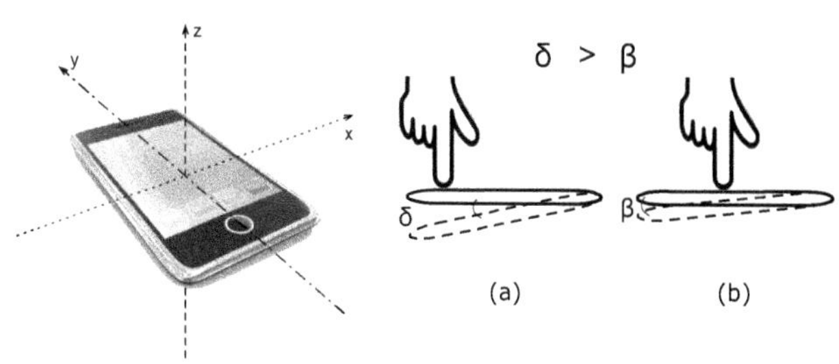

Figure 1: Smartphones equipped with a 3-axis accelerometer and gyroscope. A tap on the device nearer the edge (a) results in a larger deflection of the device than a tap near the middle (b).

The rotation of the device is larger when the tap occurs at the edge of the device. The linear acceleration of the device may also be different, depending on how the finger pressure is absorbed by the phone's metal/glass structure and casing. Of course, sensor noise and other factors will indeed pollute these signals, and in reality the problem of identifying a particular key-press from sensor data alone is hard. Nevertheless, theoretically, it should be conceivable that finger taps could have a fingerprint. This paper intends to verify this hypothesis.

In pursuit of this goal, we design a system called *TapPrints*, and implement it on two smartphone platforms (Google Nexus S and Apple iPhone 4), as well as on a tablet computer (Samsung Galaxy Tab 10.1). We collect labeled accelerometer and gyroscope data of about 40,000 taps, from 10 users over a period of 4 weeks. We apply machine learning algorithms to model the sensor readings from each tap, and test it for the English alphabet (26 letters) and icons arranged on the screen. The difficulty in this problem arises not only because the letters are close to each other, but also because a correct classification means discriminating one out of 26 classes. Nevertheless, our results show that English letters on the software keyboard can be detected with up to 80% accuracy. Taps on locations different from the keyboard, such as app icons, are easier to detect, and TapPrints can attain up to 90% accuracy. We believe these accuracies are already adequate to extract users' passwords; even if an attacker is unable to recognize a password in one attempt, we will show that 3 or 4 attempts will often be enough.

TapPrints should not be deemed as the upper bound of what is feasible; it is rather an early step towards exposing a latent security threat in modern mobile devices. There are many opportunities to raise the inference accuracy beyond what we report in this paper. For example, (1) by considering the sequence of tapped letters and applying sophisticated spell-checking [23], an attacker may be able to recognize words with higher accuracy. (2) Knowing that the user has tapped on the email app can enable the attacker to narrow down the search space to email addresses in the contact list. (3) Moreover, knowing that the email "send" button was pressed can offer confidence that the character "@" was typed, and perhaps ".com" or ".edu" – all these confidences can be valuable to train the attacker's classification models, helping her to become stronger over time. While these observations can be alarming, we will also discuss potential ideas to cope with this problem, such as adjusting the keyboard structure, developing special rubber cases that absorb device motion, or perhaps expediting the paradigm shift towards swiping-based keyboards.

The rest of this paper is organized as follows. Section 2 motivates our work and introduces the threat model in the current smartphone programming ecosystem. Section 3 follows with a detailed presentation of the TapPrints inference methodology. The system performance evaluation is presented in Section 4. A discussion on possible solutions that mitigate the vulnerability exposed by TapPrints is presented in Section 5, while the related work can be found in Section 6. We conclude with some final remarks in Section 7.

2. MOTIVATION AND THREAT MODEL

Given their popularity, sensor-rich smartphones and tablets are increasingly becoming targets of attacks that compromise users' privacy [8]. The malicious software may even take the form of a useful application that can be downloaded from a popular application store, such as the Apple App Store or the Google Android Market. Unless the access to sensor or personal data is restricted by the system or by the user, malicious software can use official APIs to access the data and upload it to third party servers without the user's consent [13].

The only sensor that requires explicit user access permission on both the Android and the iPhone platforms today is the location sensor. Indeed, most users perceive their location as being private and few would be happy to see their location tracks publicly available. In contrast, accessing the accelerometer and gyroscope sensor data does not require explicit user permission on any of the major mobile platforms. These sensors, initially introduced to drive the device's user interface, are mostly used for gaming and are widely perceived as innocuous. For example, Android OS allows background running services with accelerometer and gyroscope sensor access without restrictions. Moreover, there is work aimed at the standardization of Javascript access to a device's accelerometer and gyroscope sensors in order for any web application to perform, for example, website layout adaptation [4]. In this work, we show that accelerometer and gyroscope sensor data can create a serious privacy breach. Specifically, we demonstrate that it is possible to infer where people tap on the screen and what people type by applying machine learning analysis to the stream of data from these two motion sensors. The reason we focus on the accelerometer and gyroscope sensors is that they are able to capture tiny device vibrations and angle rotations, respectively, with good precision.

Our threat model makes three assumptions. First, we assume that the malicious software has access to the stream of accelerometer and gyroscope data. The software can: i) be secretly installed on the user's device through a remotely accessible backdoor or physical access; ii) take the form of a seemingly legitimate application; iii) or be embedded in a harmlessly looking web page (if the Javascript access to a device's motion sensors is ratified). Instead of continuously eavesdropping on the user, the malicious software could monitor the user's activity periodically or react to external events, such as the reception or transmission of a text message, or the use of the soft-keyboard. Second, we assume the malicious software has a way to send the observed data to a remote server either directly or through another application. With the growing computation capabilities of mobile devices, at some point it will be possible to run the learning algorithms entirely on the mobile device, thus requiring little communication with backend servers. Third, we assume that an attacker is able to obtain or learn a tap classification model. For best accuracy, this model should include taps from the specific user the attacker is eavesdropping upon. However, as we show in Section 4.4, the attacker could also train a tap classifier using data from a small number of other people and use it to effectively infer taps from unaware users at a large scale while collecting their motion sensor data.

The goal of this paper is to raise awareness around potential misuses of accelerometer and gyroscope sensor data, possibly requiring revisiting the policies governing sensor data access, for example, blocking the access to the motion sensors while the device's soft-keyboard is up, which prevents malicious software from correlating taps with sensor data.

3. TAPPRINTS INFERENCE ENGINE

In this section, we describe the TapPrints inference engine. Recall, TapPrints works by learning a classification model (an ensemble of such models, more accurately) of how sensor data from taps translates to particular key presses. We begin by describing the methodology for collecting the labeled data. Next, we detail the feature extraction process and move on to the TapPrints learning algorithm.

3.1 Labeled Data Acquisition

To collect labeled data, we develop custom applications for the iPhone and Android platforms. Each application allows users to type sequences of letters and to randomly tap icons on the screen while using the official Android and iOS APIs for motion sensor data access. We log four streams of records: (A) timestamped accelerometer and gyroscope readings; (B) timestamps of the beginning and the end of a tap; (C) the coordinates of each tap on the screen; and (D) the typed letters or the IDs of the tapped icons. We store the labeled data locally on the device during the data collection process and retrieve it later for the off-line learning and classification analysis.

Figure 2 shows a sample of the raw accelerometer data after a user has tapped on the screen – the timing of each tap marked by the digital pulses on the top line. As we can see, each tap generates perturbations in the accelerometer sensor readings on the three axes, particularly visible along the z-axis, which is perpendicular to the phone's display. Gyroscope data exhibits similar behavior and is not shown.

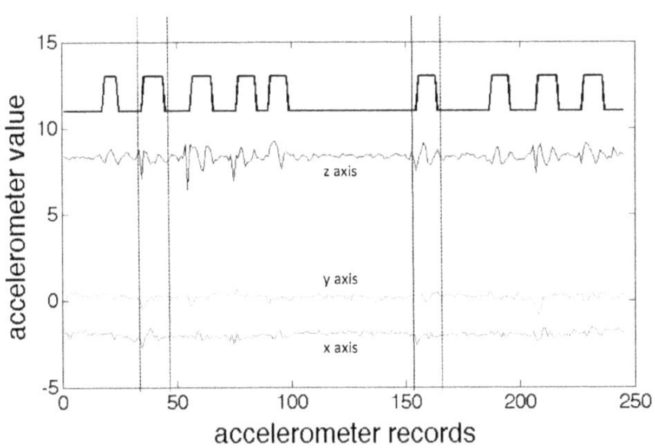

Figure 2: The square wave in the top line identifies the occurrence of a tap. We have also highlighted two particular taps by marking their boundaries with dashed vertical lines. Notice that the accelerometer sensor readings (on the z-axis in particular) show very distinct patterns during taps. Similar patterns can also be observed in the gyroscope data.

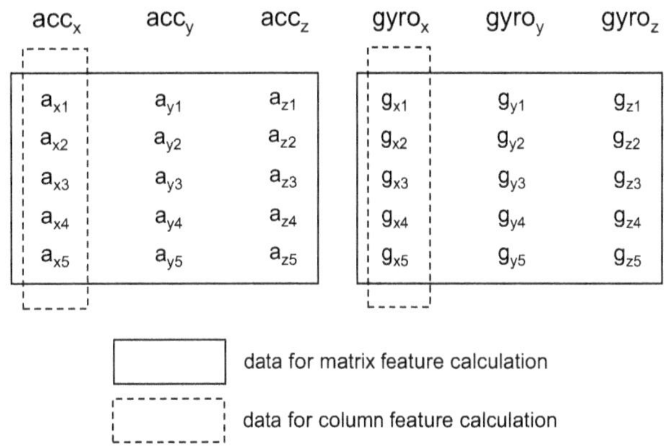

Figure 3: The solid box contains data used to extract matrix features, while the dashed box contains data used to extract column features.

3.2 Feature Extraction

In this section, we discuss the feature vector used by the TapPrints learning algorithm. To extract features effectively, we require the number of sensor records per tap to be the same across all taps. Unfortunately, this is not always the case for two reasons. First, taps may have different durations. The longer the tap, the more sensor records are available. The typical tap duration is between 100ms and 200ms. Second, high CPU loads or internal operating system constraints may reduce the number of sensor readings being delivered to the application. To equalize the sensor records to a fixed number for each tap, we use cubic spline interpolation [21].

In total, the TapPrints feature vector contains 273 features, extracted from both the time and frequency domains. The features are designed to capture as much information as possible from the underlying physical event of a tap. Some features are related to the amount of energy of the tap, others measure the rate of change of the sensor data vectors by computing, for example, the first order numerical derivative of the readings. We next describe our time and frequency domain features in more detail.

Time Domain Features. Since taps on different locations of the screen generate different sensor signatures (as evident in Figure 2) we design features that are able to capture the properties of the sensor readings generated by a tap.

At a high level, we extract two types of features: column features and matrix features (see Figure 3). Column features are extracted from the individual components of each sensor axis. Our column features include cubic spline interpolation, moments (mean and higher order), extreme values (min/max), skewness (to measure the asymmetry of the vector components) and kurtosis (to measure the 'peaked'-ness of the vector components).

Matrix features capture the correlation between the three-axis accelerometer and gyroscope vectors. These features consist of various matrix norms – the 1-norm (maximum absolute column sum of the matrix), the Infinity norm (maximum absolute row sum of the matrix), and Frobenius norm (square root of the squared sum of the entries in the matrix). We also extract features from the vector of the squared ℓ_2 norm of the rows in the matrix – we get one value per

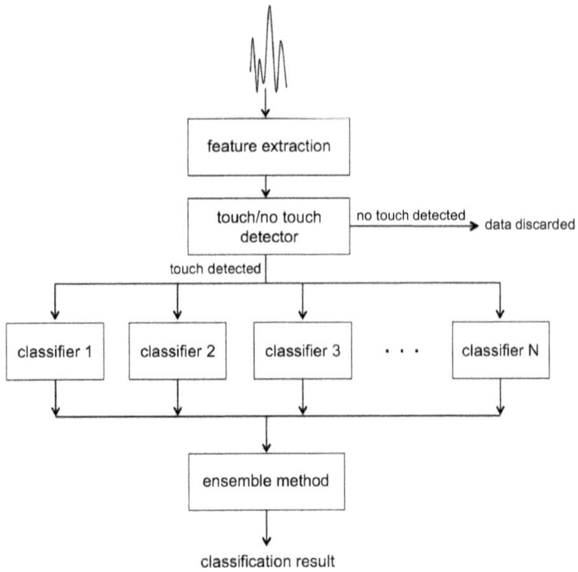

Figure 4: We adopt an ensemble classification technique, more robust and often more accurate than any single classifier. A bank of N different classifiers, with different error characteristics, concurrently produce a label given the same input feature vector. The final classification is the result of the ensemble algorithm's decision, currently, a simple winner-takes-all majority voting algorithm.

row/record in the tap event. From this vector we extract the mean and min/max values.

We also compute the first order numerical derivatives of the sensor components and compute features based on the magnitude of the acceleration and angular rotation vectors, which is an estimate of the energy associated with a tap.

To capture the correlation between the accelerometer and gyroscope vectors we compute two sets of features: the angle between these vectors and their rate of change. We also compute the Pearson correlation coefficients between all pairs of the individual sensor components to take into account the correlation between the accelerometer and gyroscope vector components.

Frequency Domain Features. To capture the tap characteristics in the frequency domain we compute the Fast Fourier Transform (FFT) of all 6 individual components of the accelerometer and gyroscope three-axis time series vectors. In addition, we calculate features from the power spectrum of the FFT values. These frequency domain features are useful to characterize the different oscillation and vibration patterns of the phone in response to taps.

3.3 Learning

For the learning phase of TapPrints we adopt an ensemble classification approach, which is generally more robust and often more accurate than any single classifier [10, 5, 9, 15].

3.4 Ensemble Classification

Once we extract the features from the labeled data, we use them to train various classification algorithms. We refer to these various classification algorithms as base learners. Since all the TapPrints prediction tasks, such as inferring a letter/icon-position, are multi-class classification problems, we choose various multi-class classification techniques as the base learners in our ensemble. Once these N base learners have been trained on the labeled data, features from the unlabeled taps are fed into the bank of classifiers concurrently at test time, and an ensemble technique is used to produce the final prediction for the label (see Figure 4).

One of the key factors in the amount of gain in accuracy achieved by using an ensemble is the amount of correlation between the various base learners [5]. The goal is to use a very diverse set of classifiers so that the classification errors made by the base learners are as uncorrelated as possible. Intuitively, uncorrelated base learners would add independent/orthogonal information about the classification task, helping to efficiently construct the final ensemble solution. An equally important factor in successful ensembles is the accuracy of each individual classifier. Clearly, for a strong ensemble, reasonably strong base learners are a must.

We construct our ensemble classifier with the following set of diverse and powerful multi-class base learners:

k-Nearest Neighbor (kNN). We use a set of kNN classifiers, each parametrized by the neighborhood size parameter, k. Roughly speaking, k controls the amount of smoothness we expect in the feature space with respect to the class boundaries. k also depends on the amount of labeled training data you have, and the dimensionality of the feature space. Large feature spaces warrant larger k values, or much larger amounts of training data.

Multinomial Logistic Regression. We include a family of multinomial logistic regression classifiers [20] (a generalization of the logistic regression binary classification technique to the case where the number of labels is greater than two) by varying the amount of regularization. Setting the amount of regularization essentially allows us to control the expressivity/capacity of the model, and thus provides us a parameter that we can tune to prevent overfitting.

Support Vector Machines (SVM). We also include a family of both linear and non-linear SVM classifiers [26]. A set of linear SVM classifiers is obtained by varying the regularization coefficient. We also obtain a family of non-linear/kernel SVM classifiers where we only use a radial basis function (rbf) kernel with varying regularization coefficient and the kernel bandwidth. Kernel classification is especially useful in situations where the function we are trying to fit has strongly non-linear decision boundaries, and the kernel bandwidth (rbf kernel) allows us to set the smoothness of the variations allowable in the function.

Random Forests. Our bank of classifiers also includes random forest classifiers [5], which are an ensemble classification technique themselves, composed of decision trees. Random forests grow a large number of diverse tree classifiers by randomly selecting a subset of the features and examples for each tree to learn from. The final classifier then uses voting.

Bagged Decision Trees. Another ensemble classifier, a set of bagged decision trees, is also composed of a collection of decision trees, but unlike random forests, each tree operates on the whole feature set. Diversity is instead created by bagging the examples, so each tree is learned on a bootstrap sample of the original data. We use several such bagged decision tree classifiers for our ensemble by varying the number of trees in the model.

With this choice of base learners we cover an extremely broad set of learning techniques. We cover parametric methods (logistic regression, SVM), and non-parametric methods (kNN, trees). We cover both linear (linear SVM, logistic regression), and non-linear (trees, non-linear SVM) techniques. Finally, we also include some powerful ensemble methods (random forests, bagged decision trees) in our mix of base classifiers. This diversity is what gives our ensemble the best chance to show significant gains in predictive accuracy over any single method.

3.5 Ensemble Method

The ensemble method is what makes our final prediction. It takes as input the base learner predictions/scores and it produces in response the final prediction of the class label. The idea is for the ensemble technique to find consensus among the different classifiers in the ensemble. This increases robustness, and noise-tolerance and thus helps to minimize spurious misclassifications. We employ one of the best studied techniques for ensemble methods, namely, winner-takes-all followed by majority voting. Essentially, we choose 20 different base classifiers and train them independently. At test time, for each classifier, we compute its predicted label for the test example. This is the winner-takes-all aspect, as we do not use the raw scores from the base classifiers, but rather just the predicted label, or vote. Then the final ensemble label output is the label with the most predicted votes (over the base classifier votes). Though simple, this is an effective ensembling technique [10, 5]. More intricate ensemble techniques can use the scores of the base learners or learn models to combine the base learner predictions [9, 11].

4. EVALUATION

In this section, we evaluate the icon and letter tap inference performance of TapPrints.

Our current TapPrints infrastructure consists of a client and a server. The client software runs on sensor-rich smartphones and tablets, collecting and storing accelerometer and gyroscope readings locally as the user taps on the screen. In order to evaluate the performance for different display form factors and mobile operating systems, we use a Samsung Galaxy Tab 10.1 Android tablet, a Samsung Nexus S Android smartphone, and an Apple iPhone 4. The server runs the inference algorithms, which consist of a combination of MATLAB routines and machine learning packages. In the future, we plan to run less computation-intensive machine learning algorithms directly on the phone to quantify the tradeoff between TapPrints' energy cost and accuracy.

We configure the sensors on the clients to generate sensor readings at the highest possible rate for best accuracy. Although both the iPhone and the Nexus S can theoretically generate sensor readings at about 100Hz, we find that in practice the frequency is lower. On the iPhone, the sampling frequency of the sensors is slightly lower than 100Hz, while on the Nexus S a sampling rate of 100Hz for both the accelerometer and gyroscope sensors causes frequent application reboots. The second highest sampling rate for the gyroscope is therefore selected on the Nexus S. We remove the gravity component from the accelerometer data and use only the linear acceleration, independent from the device orientation. We use a five-fold cross validation approach for all the experiments.

We next describe the data collection methodology before moving into the evaluation results of the TapPrints system.

4.1 Data Collection

We recruit 10 volunteers to collect taps and key presses using the methodology described in Section 3.1. Due to the long duration of our experiments, we vary the number of volunteers per experiment to accomodate their preferences. In total, we collect labeled data for more than 40,000 taps and key presses over the course of four weeks. Volunteers conduct four types of experiments:

Icon Taps. Users tap icons appearing in random order on the screen. The icons are arranged in a 5-by-4 grid and both the size and location of the icons match the size and location of application icons on both iOS and Android application launcher screens. Figure 5 shows the approximate size and location of the icons.

Sequential Letters. Users type each letter of the alphabet sequentially multiple times (50 repetitions per letter).

Pangrams. Users type 19 different pangrams, one after the other. A pangram is a sentence that includes every letter of the English alphabet at least once.

Repeated Pangram. Users type a single pangram (one not included in the Pangram experiment) 20 times. This experiment emulates the use of pass phrases.

4.2 Typing Modalities

To quantify the impact of the typing modalities on the inference accuracy while using a phone, the Pangrams and Sequential Letters experiments are performed in the following three configurations: (i) phone in portrait mode, tapping with the thumb of the holding hand; (ii) phone in portrait mode in one hand, tapping with the index finger of the other hand; (iii) and phone in landscape mode held with both hands, tapping with the thumbs. For the tablet, we consider only the most common typing configuration, which is the landscape mode. For the Icon Tap experiment, we collected data in portrait mode only. Unless otherwise noted, all the experiments, except for the Icon Taps case, are performed in landscape mode. The participants typed while sitting or standing, which are the two most common typing scenarios.

4.3 Detecting Taps

Before classifying individual taps, we need to be able to identify them in the sensor data stream. To this end, we develop a tap detection classifier. We first extract the features discussed in Section 3 from fixed length moving windows on the sensor data stream (with a 50% overlap between windows). We then train a bagged decision tree classifier. The resulting classifier achieves nearly 100% accuracy, generating virtually no false positives or false negatives. A close look at Figure 2 explains this result. Distinguishing between taps and non-taps is an easy classification problem as the motion sensor readings during a tap differ considerably from non-tap sensor data.

4.4 Icon Tap Experiment

The goal of the icon tap experiment is to show that it is possible to infer the location of taps across a smartphone and tablet's display with high accuracy. By knowing where taps occur, an application running the TapPrints algorithm would theoretically be able to profile detailed usage patterns

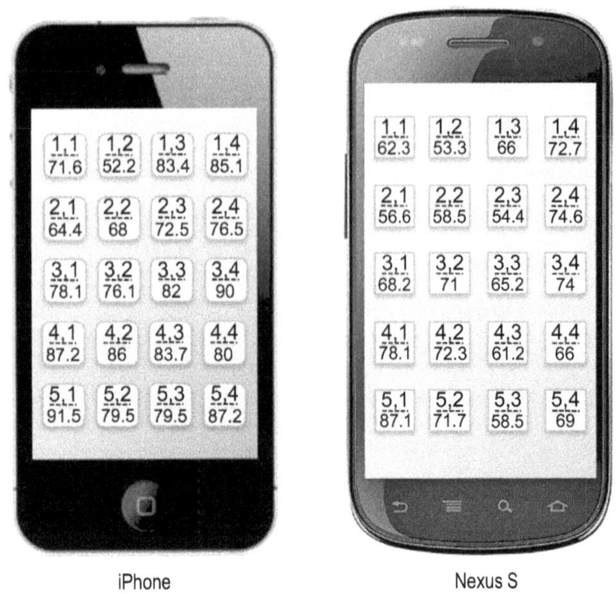

Figure 5: Results from the icon tap experiment: the top numbers on each icon are the row and column index of the icon in the matrix arrangement, while the bottom number is the average icon tap inference accuracy across 5 users. For example, iPhone's icon 1,1 inference accuracy is 71.6%.

of a mobile device. For example, TapPrints could detect the applications being launched by a user and the buttons being pressed within it.

Figure 5 shows the tap inference accuracy averaged across 5 users for both the iPhone and the Nexus S phones. Icons are indexed by the row and column number of their position on the screen. The figure shows that TapPrints achieves an average accuracy of 78.7% on the iPhone and of 67.1% on the Nexus S. Note that the probability of correctly inferring an icon by a random guess is, on average, only $\frac{1}{20} = 5\%$. We relate the better accuracy on the iPhone to the higher gyroscope sampling rate (recall that the gyroscope on the Nexus S is set to a lower sampling rate). The other reason might be a difference in the sensor's sensitivity or placement on the internal board.

Interestingly, some tap locations can be inferred more precisely than others, e.g., iPhone's icon (5,1) with 91.5% accuracy vs. icon (1,2) with just 52.2% accuracy. The confusion matrix in Table 1 shows, for each icon, the confusion probabilities with the four most confused icons. For example, taps on icon (1,1) are inferred correctly 71.64% of the time while being mostly misclassified as taps on icons (2,1), (1,2), (2,2), and (1,3) with 16.42%, 5.97%, 3.73%, and 0.75% probability, respectively. Given that misclassification errors localize with high probability in the proximity of the tapped icon, it is possible to limit the inference space to only the surrounding icons, as reported in the confusion matrix. This principle of error locality is particularly powerful in the letter inference case, as we discuss in Section 4.5.

Impact of the Amount of Training Data. Figure 6 shows the average icon tap inference accuracy as a function

Table 1: Confusion matrix for the iPhone icon tap inference. The first column to the left shows the average icon inference accuracy. The 2nd, 3rd, 4th, and 5th columns show the confusion percentage with, respectively, the 1st, 2nd, 3rd, and 4th most confused icons. Icons are indexed by the row and column number of their position on the screen.

Avg acc (%)	1st conf	2nd conf	3rd conf	4th conf
1,1: 71.64	**2,1**: 16.42	**1,2**: 5.97	**2,2**: 3.73	**1,3**: 0.75
1,2: 52.27	**1,1**: 17.42	**2,2**: 15.15	**1,3**: 7.58	**3,3**: 3.03
1,3: 83.46	**2,3**: 9.02	**1,2**: 3.01	**1,4**: 2.26	**2,2**: 0.75
1,4: 85.19	**2,4**: 5.93	**2,3**: 5.19	**1,3**: 1.48	**3,3**: 0.74
2,1: 64.44	**1,1**: 20.74	**3,1**: 6.67	**2,2**: 5.93	**1,2**: 1.48
2,2: 67.91	**1,2**: 11.19	**2,1**: 7.46	**1,3**: 5.22	**3,2**: 2.99
2,3: 72.59	**1,3**: 15.56	**3,3**: 5.93	**1,4**: 2.96	**2,4**: 0.74
2,4: 76.52	**1,4**: 9.09	**3,3**: 6.06	**3,4**: 6.06	**2,3**: 2.27
3,1: 78.20	**2,1**: 9.77	**4,1**: 5.26	**3,2**: 4.51	**1,1**: 1.50
3,2: 76.12	**2,2**: 8.96	**3,1**: 5.22	**2,3**: 2.24	**2,4**: 1.49
3,3: 82.09	**2,3**: 9.70	**1,4**: 3.73	**3,4**: 2.99	**2,4**: 0.75
3,4: 90.23	**2,4**: 6.77	**4,3**: 1.50	**3,3**: 0.75	**4,2**: 0.75
4,1: 87.22	**3,1**: 4.51	**4,2**: 3.01	**5,1**: 3.01	**3,2**: 2.26
4,2: 85.82	**4,1**: 4.48	**5,1**: 2.99	**5,2**: 2.99	**3,2**: 2.24
4,3: 83.70	**3,3**: 4.44	**4,4**: 3.70	**3,4**: 2.96	**5,3**: 2.96
4,4: 80.00	**5,4**: 11.85	**4,3**: 5.19	**3,4**: 1.48	**5,3**: 1.48
5,1: 91.54	**5,2**: 4.62	**4,1**: 2.31	**4,2**: 1.54	-
5,2: 79.55	**5,3**: 6.06	**4,2**: 5.30	**4,3**: 4.55	**5,1**: 4.55
5,3: 79.55	**5,4**: 9.85	**5,2**: 5.30	**4,3**: 3.79	**2,4**: 0.76
5,4: 87.22	**5,3**: 6.02	**4,3**: 3.76	**4,4**: 3.01	-

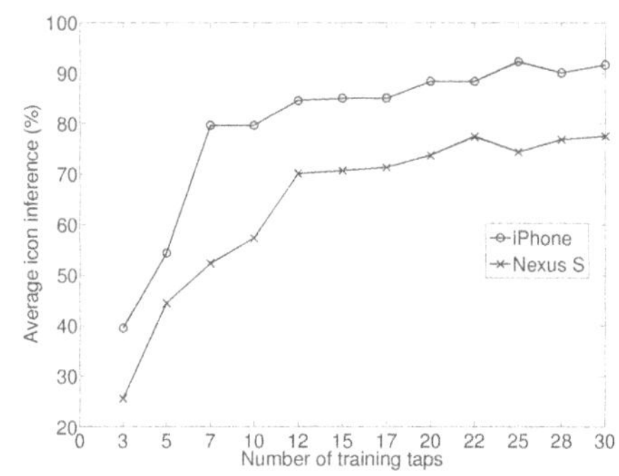

Figure 6: The average icon inference accuracy increases with the number of tap training samples. Even a small number of samples, e.g., 12, suffices to achieve 70% accuracy or higher.

of the number of taps available for training the classifier. The accuracy stabilizes at around 20 training taps per icon, with only 12 taps per icon needed to achieve 70% and 85% accuracy on the Nexus S and the iPhone, respectively. This result has important implications because a malicious application would require only few taps from an unaware user to successfully train an icon tap classifier. The malicious software could, for example, ask the user to tap on properly positioned buttons on the screen under the guise of a game.

Borrowing Strength Across Users. In the results we have shown so far, our classifiers are built using training data from all the users and tested on different data from the same users. However, a more threatening scenario arises if we can show that it is feasible to train a classifier with data from a set of users to infer taps from people not involved in the training phase. If this is possible, we prove an important point: an attacker could use data from a small number

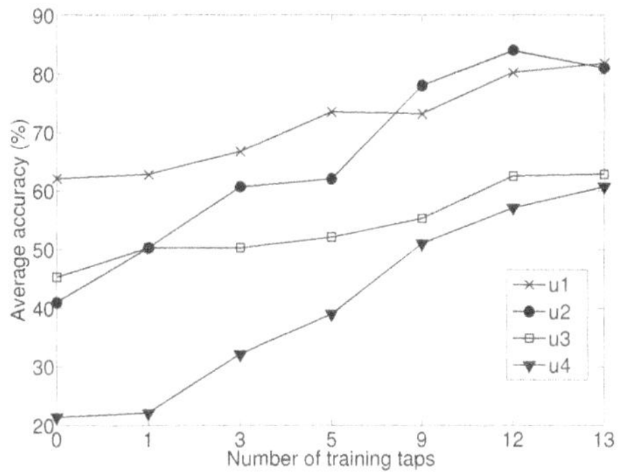

Figure 7: Borrowing strength from other users' training data. Each user's model is built with other's users data and tested on some of the user's data. The accuracy increases with adding user's data to the model.

of people to pre-train a classification model that, when deployed in the wild, is still able to effectively infer taps from new users. Here we show that indeed an attacker can use a pre-trained classifier to achieve reasonable inference accuracy on new users' taps. We also show that the classifier can improve its accuracy as more data from the new user becomes available.

To evaluate TapPrints' accuracy on users whose data isn't employed for training the classifier, we perform the following experiment. Given a pool of N users, we train the icon tap classifier using data from N-1 users and test it on the remaining user. Then, we incrementally add training data from the left-out user, recompute the model, and measure the resulting classification accuracy. The procedure is repeated for all users. The results of the experiment are shown in Figure 7 for four different users.

The figure shows that a pre-trained classifier with no data from the user being tested achieves a tap classification accuracy of at least 40% for users 1, 2, and 3 (refer to y-axis values at x-axis of 0). As training samples from the left-out users are added and the model is recomputed, the classification accuracy quickly increases. This is because the original classification model adapts to this new user by including her own data. After only 13 additional taps per icon it is possible to achieve an average classification accuracy comparable to the one from user-specific classifiers, which are 80.6%, 90%, 67.1%, and 71.7% for users 1, 2, 3, and 4, respectively.

This important result demonstrates TapPrints scalability and adaptability across users. It shows that malicious software, with a pre-trained classifier from a small number of people, could be rapidly and successfully deployed to infer other users' taps. The pre-trained model could also quickly boost its accuracy by silently collecting additional data from the new unaware users.

4.5 Letter Tapping Experiment

Letter tap inference is more challenging than icon tap inference because of the smaller distance between letters on the keyboard and the larger number of classes to distinguish between. Note that the inference probability of one of the 26 English alphabet letters by random guessing is, on average, $\frac{1}{26} = 3.8\%$. We now present the results of the TapPrints' letter inference experiment.

Table 2: Confusion matrix for the tablet experiment, where letters are typed in landscape mode. The first column to the left shows the average letter inference accuracy. The 2nd, 3rd, 4th, and 5th columns show the confusion percentage with, respectively, the 1st, 2nd, 3rd, and 4th most confused letters.

Avg acc (%)	1st conf	2nd conf	3rd conf	4th conf
a: 82.12	s: 9.70	e: 5.15	z: 1.52	d: 0.61
b: 56.74	n: 21.99	v: 10.64	g: 3.55	h: 2.13
c: 72.32	x: 7.91	v: 6.21	d: 2.82	f: 2.82
d: 57.73	s: 18.04	e: 10.31	f: 4.12	t: 3.09
e: 90.81	r: 5.55	w: 1.04	s: 0.87	a: 0.69
f: 46.90	d: 19.31	g: 13.79	e: 5.52	r: 5.52
g: 62.35	h: 9.41	f: 7.65	e: 4.12	t: 4.12
h: 64.48	j: 9.84	g: 7.10	n: 3.83	u: 3.83
i: 83.05	u: 7.75	o: 6.30	l: 0.97	j: 0.48
j: 48.78	h: 13.41	k: 11.59	u: 9.76	m: 5.49
k: 46.30	l: 18.52	i: 14.81	j: 8.64	u: 4.32
l: 82.72	o: 6.81	i: 4.19	k: 4.19	j: 0.52
m: 63.98	n: 22.98	j: 3.73	h: 2.48	l: 2.48
n: 65.98	m: 14.95	b: 6.19	h: 5.15	j: 2.58
o: 76.47	i: 15.22	p: 7.27	r: 0.69	l: 0.35
p: 69.15	o: 27.66	l: 1.60	i: 1.06	r: 0.53
q: 52.48	e: 30.50	a: 10.64	w: 4.26	l: 0.71
r: 54.13	e: 35.78	t: 8.56	a: 0.61	u: 0.61
s: 53.01	a: 30.92	e: 7.23	d: 6.83	x: 0.80
t: 71.02	r: 18.37	y: 4.08	e: 3.67	g: 1.22
u: 66.55	i: 24.65	y: 5.63	j: 0.70	k: 0.70
v: 60.00	c: 16.43	b: 13.57	g: 3.57	n: 2.86
w: 10.19	e: 67.52	q: 14.01	a: 4.46	s: 1.91
x: 47.30	c: 19.59	z: 18.92	a: 6.76	s: 4.05
y: 51.35	u: 24.86	t: 16.22	i: 3.24	r: 2.16
z: 38.19	a: 27.08	x: 20.14	s: 9.03	c: 2.08

4.5.1 *Tablet*

We start by showing the TapPrints' inference accuracy for the Pangrams experiment on the tablet. The average per-letter inference accuracy is shown in the first column to the left of the confusion matrix in Table 2. The second, third, fourth, and fifth columns show the misclassification error with the first, second, third, and fourth top most confused letters, respectively.

The overall mean inference accuracy is 65.11%, a much larger value than the 3.8% accuracy from random guessing. Moreover, given the locality of errors, a letter is misclassified with high probability only with nearby letters in the keyboard layout, as shown in Table 2. For example, the letter 'B' is confused with high probability only with the closest letters on the keyboard, i.e., 'N', 'V', 'G', and 'H'. The fact that misclassification errors tend to cluster in the region of the tap can be leveraged to deploy targeted dictionary attacks. Briefly, should a misclassification occur, it is possible to narrow down the exact letter with high probability by just focusing on a small subset of letters – the most confused ones – rather than on the entire keyboard.

From the results in Table 2, TapPrints' letter inference accuracy varies from 90.81% for the letter 'E' to 10.19% for the letter 'W'. The reason for this difference is the relative frequency of individual letters in the pangrams we use to train the models. As we show in Section 4.5.2, the amount of training taps per letter plays a major role in the quality of the classifier. The letter 'W', for example, with fewer occurrences in the pangrams than other letters (vowels for

true text:	six	big	devils	from	japan	quickly	forgot	how	to	waltz
after 1 rep:	dux	**bog**	aecols	**drom**	hapah	**quockly**	**foegot**	**hoe**	**to**	**ealtz**
after 4 reps:	**six**	**big**	sebils	**drom**	**japan**	**quickly** o	**foegot**	**hoe**	**to**	**ealtz**
after 15 reps:	**six**	**big**	**debils**	**feom**	**japan**	**quickly**	**foegot** g	**hoe**	**to**	**ealtz**

Figure 8: Example of error correction applied to repetitions of a same pangram on the tablet. The actual text is shown on the top row, while the 1st, 2nd, and 3rd row show the result of the inference after, respectively, one, four, and 15 repetitions of the pangram. The accuracy improves at each repetition by applying a majority voting scheme to each letter. When it is not possible to express a majority vote, possible choices are shown. Correctly inferred letters are represented in bold.

example), is subject to lower classification accuracy due to less training data and to its proximity to 'E', which occurs extremely frequently in the pangrams. As a result, given our training data, vowels tend to have higher accuracies than consonants because of their larger frequency in the English vocabulary. We further examine this issue in Section 4.5.2, where an equal amount of training data is collected for each letter, which results in high inference accuracy for all the letters.

Effect of Majority Voting Scheme. Next, we show that it is possible to further boost the letter inference accuracy by exploiting repetitions. Imagine, for instance, a banking application where a user needs to type a passcode or password every time she logs in. By monitoring the application being launched (e.g., the browser) and the sequence of key presses, an attacker might be alerted that some sensitive information is about to be entered by the user. Since users repeat the action of typing sensitive information (e.g., passwords, pass phrases) multiple times, the question is: Could an attacker leverage this repetitive pattern to retrieve the typed content? We emulate this scenario with the Repeated Pangram experiment where we ask users to repeatedly type the same pangram 20 times.

We show that when a letter is typed multiple times we can smooth individual letter inference errors by combining sequences of predictions. An example of such a smoothing approach is a majority-voting scheme on the individual predictions. As an example, N repeated taps of the letter 'E' would result in N predictions for these taps. If the majority of the labels is the letter 'E', only then would TapPrints conclude that the (repeatedly) typed letter is an 'E'. This voting scheme is robust since it determines 'E' as the correct final outcome even if there are few misclassification errors in the sequence of predictions.

Figure 8 shows the results of the Repeated Pangram experiment. The top line shows the actual typed pangram, while the following lines show the text inferred by TapPrints after one, four, and 15 repetitions, respectively. TapPrints applies the majority voting scheme to each letter after each repetition. The bold letters highlight a correct inference result. When a tie occurs and majority vote is impossible, the candidate labels are stacked on top of each other (e.g., 'f' and 'g' after the 15th repetition).

The figure shows that TapPrints augmented with a ma-

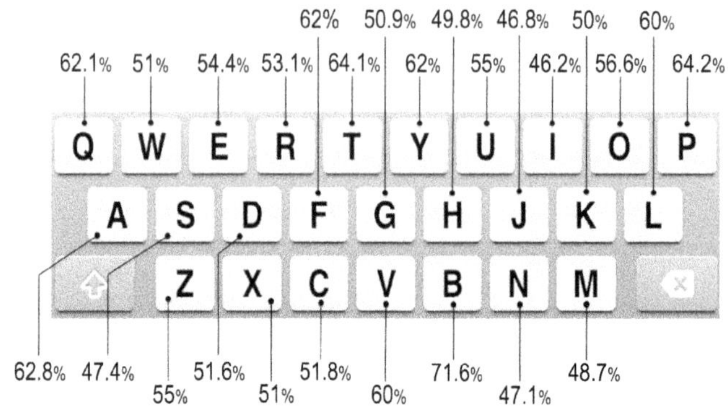

Figure 9: Average letter inference accuracy from sequentially tapping each letter multiple times on a smartphone in landscape mode.

jority voting scheme is able to correctly capture many of the typed words, or large portions of them, after only four repetitions. After 15 repetitions the number of correctly inferred letters is even larger. With the support of a dictionary over partially inferred words, further accuracy gains are possible as shown in [19]. This result proves that it is possible to infer typed words by mining motion sensor data and that the accuracy increases by exploiting the repetitive typing patterns of users as they go about their normal interaction with their mobile devices.

To estimate the number of iterations needed to infer each letter using the majority voting scheme we use a simulation approach. Essentially, we simulate our inference about a particular letter typed repeatedly, apply our majority voting scheme, and estimate how many repetitions on average are required for a reasonably sure chance of identifying the letter. For clarity, we discuss the simulation for the letter 'A', but we repeat this procedure for every letter of the alphabet. To start with, suppose we have the data of a sequence of 100 instances of a user typing the letter 'A'. Using the model we trained, we simulate what we would infer for each 'A' instance (independently) using the probabilities from the confusion matrix of the Pangrams experiment (Table 2 shows a subset of the full confusion matrix). The result is a sequence of 100 inferred letters, which are mostly 'A's, but with errors as we expect from the confusion matrix.

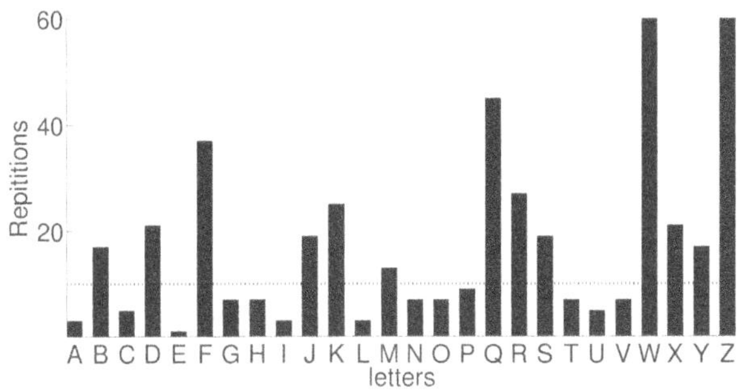

Figure 10: Number of typing repetitions required by the voting algorithm to correctly infer each letter with a probability of at least 90%. Few repetitions, less than 10, are necessary to correctly infer most of the letters.

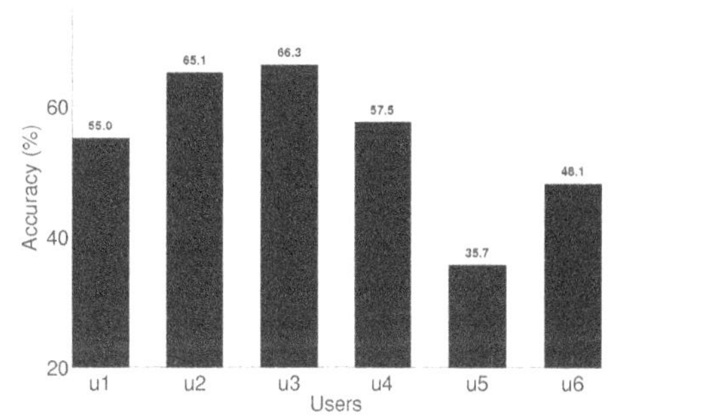

Figure 11: Average per-user inference accuracy for sequential letter typing.

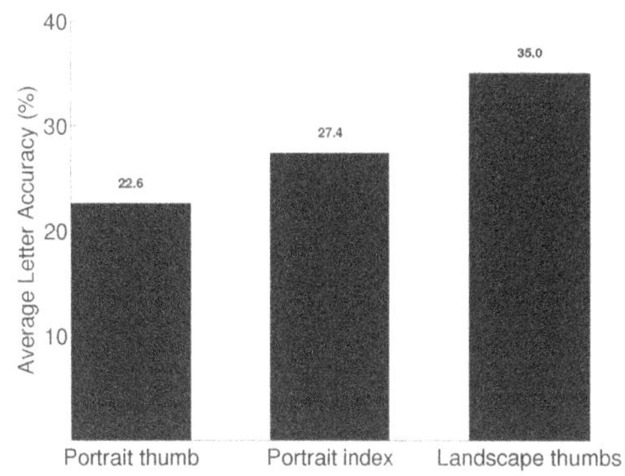

Figure 12: Average letter inference accuracy for the three typing configurations. The inference is more accurate in landscape mode because of the more firm grip on the device, which reduces the injection of noise in the sensor stream.

Now, for all sub-sequences of length k (the first k labels in the sequence), we apply our majority voting procedure to the inferred labels, and the result is either a correct identification of the letter 'A' (a hit) or not. We generate 100, 000 such sequences of length 100, and for any sub-sequence of length k, we can estimate the probability that the majority voting scheme will correctly infer an 'A': the number of sequences where a sub-sequence of length k is a hit divided by 100, 000. Finally, we estimate a particular value of k, which is the minimum sub-sequence length that allows us to correctly classify an 'A' with at least 90% probability (across the 100, 000 sequences). This is the likely number of repetitions we will need to correctly classify a letter with a probability of 90%.

We plot the results of our simulation in Figure 10. The results show that, indeed, most of the letters can be correctly inferred after less than 10 repetitions, some require between 10 and 20 and only two more than 20. Letters 'W' and 'Z' are never inferred correctly due to their individual low inference accuracy (see Table 2).

The impact of these results is profound. With millions of users electing tablets and smartphones as one of their means to access banking services, web, and email, malicious software can effectively and quickly capture users' sensitive information by just tapping into the accelerometer and gyroscope sensor streams on these devices.

4.5.2 *Smartphones*

In what follows, we discuss the performance of letter inference on smartphones. For our experiments we use a Samsung Nexus S and an Apple iPhone 4. However, given the similarity of the results for the two platforms (see Section 4.6), we focus on the Nexus S analysis only.

The letter inference task on smartphones is particularly challenging due to the small soft-keyboard form factor. The size of a letter button is about a quarter the size of an icon, with a separation between letters of about a dozen pixels. Yet, we show that TapPrints is able to infer letters, on average, 10 times more accurately than by random guessing. We also show, as we do with the tablet, that it is possible to boost the inference accuracy by smoothing individual letter errors using majority voting over sequences of letter presses.

Sequential Letters. To establish a performance baseline, we first measure TapPrints' accuracy in a controlled setting. Users are asked to type each letter repeatedly for a fixed number of times. There are two advantages to collecting data in this manner. First, we minimize the effect of rocking and tilting the phone, a common side effect of typing complete words and reaching out to letters in different parts of the screen. Secondly, we can collect the same amount of training data for each letter.

Our volunteers repeat this experiment in all the three typing modalities (see Section 4.2). Intuition leads us to believe that the wider keyboard layout in landscape mode results in higher inference accuracy for landscape typing. Surprisingly, we find little or no difference in the accuracy between the portrait and landscape modalities, with an average accuracy of about 55% for both. The reason being that a phone held with two hands is firmer than a phone held with one hand, with the result of less significant sensor reading perturbations.

Figure 9 shows the average per-letter accuracy in landscape mode, with each letter being marked with the corresponding inference accuracy. For the majority of the letters, the accuracy is greater than 50%, which makes it likely for the majority voting scheme to classify the letters correctly

Table 3: Confusion matrix for the Nexus S experiment for landscape typing. The first column to the left shows the average letter inference accuracy. The 2nd, 3rd, 4th, and 5th columns show the confusion percentage with, respectively, the 1st, 2nd, 3rd, and 4th most confused letters.

Avg acc (%)	1st conf	2nd conf	3rd conf	4th conf
a: 56.87	e: 9.76	s: 9.11	q: 8.22	z: 5.26
b: 27.11	n: 14.29	j: 11.66	h: 11.37	v: 8.75
c: 41.15	e: 9.51	x: 8.19	f: 6.86	v: 6.19
d: 17.19	e: 27.27	s: 14.62	r: 9.88	x: 5.73
e: 60.32	r: 11.64	s: 5.42	w: 5.42	d: 4.17
f: 23.71	r: 19.07	e: 12.81	c: 9.54	t: 9.54
g: 18.04	t: 14.43	h: 12.11	v: 8.76	y: 6.96
h: 32.23	u: 13.69	i: 10.82	j: 9.93	g: 6.62
i: 52.43	u: 12.87	o: 12.20	j: 4.67	k: 3.05
j: 29.78	i: 14.60	h: 10.06	n: 9.27	k: 7.69
k: 27.27	o: 19.14	i: 12.44	m: 9.09	j: 8.37
l: 39.37	p: 16.11	o: 11.41	k: 10.07	m: 8.95
m: 33.42	k: 13.90	n: 12.03	l: 9.36	j: 8.02
n: 35.75	j: 11.31	m: 10.86	b: 7.92	i: 7.92
o: 37.53	i: 26.83	p: 7.99	k: 7.05	l: 5.01
p: 51.50	o: 20.73	l: 13.46	i: 2.99	e: 1.50
q: 37.60	a: 32.11	e: 8.88	w: 7.05	s: 4.18
r: 33.46	e: 31.40	t: 11.76	f: 5.43	d: 3.49
s: 30.48	e: 21.31	a: 14.62	d: 8.40	z: 7.47
t: 41.27	r: 19.18	e: 7.53	y: 6.51	g: 5.65
u: 41.74	i: 24.93	y: 9.14	h: 7.96	j: 3.69
v: 32.20	g: 10.45	b: 10.17	h: 9.89	c: 9.32
w: 14.43	e: 45.62	s: 9.79	a: 9.54	q: 7.99
x: 22.65	c: 14.09	e: 13.54	d: 11.88	s: 10.50
y: 37.88	u: 21.94	h: 8.31	i: 7.62	t: 7.39
z: 29.33	s: 20.53	a: 19.20	x: 6.93	e: 6.67

after only three repetitions (since two out of three times a letter is likely to be inferred correctly by the classifier).

The average per-user accuracy for the six users who complete this experiment is shown in Figure 11. While for most users the average accuracy is above 50% or close to it, user 5's accuracy is lower (35.7%). This is because user 5 tends to exert very light taps on each letter, causing only little perturbations in the sensor readings. The consequence is a poorly trained and imprecise classifier. This result suggests that soft and easy typing reduces the likelihood of correct key press inference.

Pangrams. We now examine a more realistic use-case, which is typing actual words and sentences. Users are tasked to repeatedly type 19 different pangrams for a number of times (see Section 4.1) in the three different typing modalities discussed in Section 4.2. Users are asked to type as they naturally would.

Figure 12 shows the average inference accuracy for the three typing modalities. Interestingly, in contrast to sequential typing, there is a difference in the average accuracy values between the modalities, with portrait thumb, portrait index, and landscape thumbs achieving, on average, 23%, 27%, and 35%, respectively. Thumb typing in portrait orientation leads to the lowest accuracy because, in this mode, the phone is subject to large tilts and vibrations from the thumb reaching out to various letters across the keyboard. As a result, sensors generate noisier data, making the decoupling of noise from informative tap sensor readings challenging. In this realistic typing scenario, TapPrints best performance is achieved in landscape mode, where a firmer grip of the phone reduces the impact of spurious events on the sensor data. We chose to continue with this orientation for

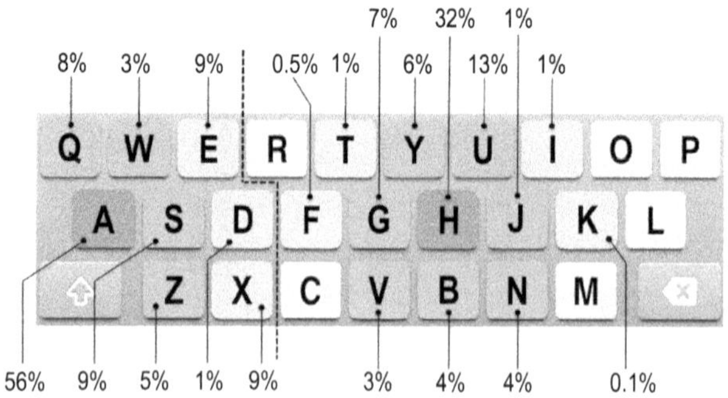

Figure 14: Confusion probabilities for 'A' and 'H', two representative letters at the edge and in the middle. The orange and yellow color code represents, respectively, 1-off and 2-off letters. The inference accuracy of letter 'A' (more sensitive to the phone tilt) is higher than of letter 'H'. Misclassification probabilities of 'A' and 'H' with letters, respectively, to the left and to the right of the dashed line, are also reported.

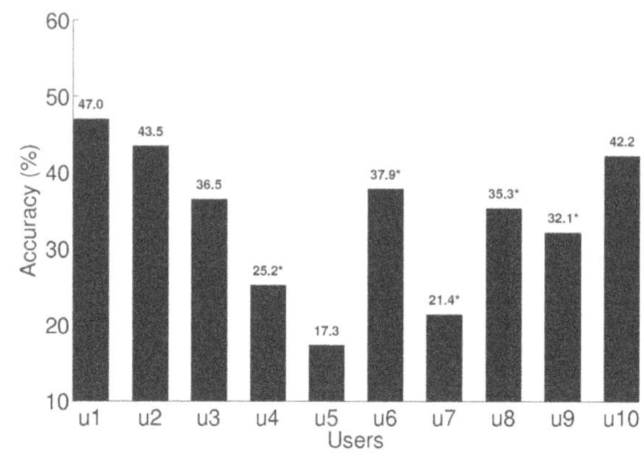

Figure 15: Average per-user inference accuracy in the pangram typing scenario with landscape orientation.

our experiments because it represents the most vulnerable typing modality.

The confusion matrix, with aggregate values across all users, is shown in Table 3. The top four most confused letters are also reported. Not surprisingly, misclassification errors follow the keyboard layout, since confusion is most likely to occur between neighboring letters than between letters far apart. A visually compelling representation of this property is shown in Figure 14, where two letters, one at the edge ('A') and one in the middle ('H'), are shown along with the 1-off and 2-off top confused letters. The probability of confusion with 1-off neighboring letters is, on average, ten times larger than the same probability with 2-off letters. Again, by exploiting this property, an attacker can narrow down the inference of a letter by just focusing on the topmost confused letters rather than the entire keyboard.

The average per-user accuracy for the realistic typing scenario is shown in Figure 15. Values marked with an asterisk are associated with users providing fewer training samples. The reason for the low accuracy for user 5 is the very gentle

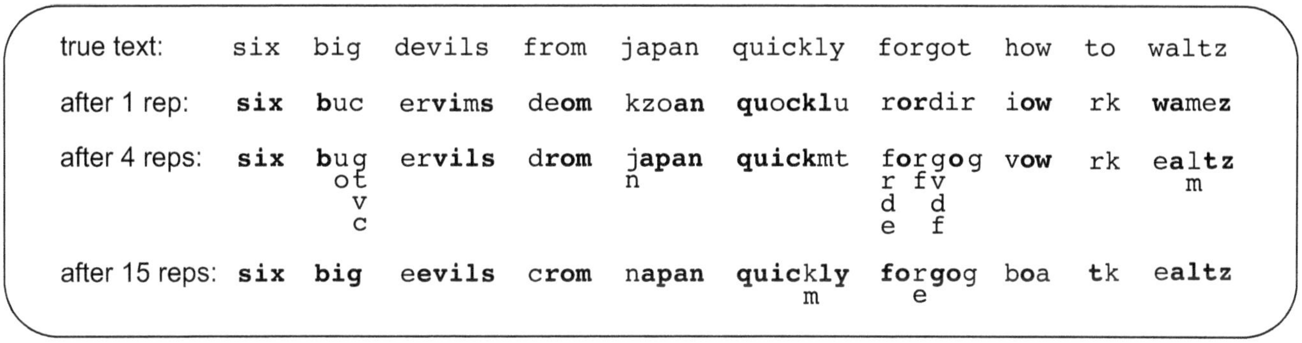

Figure 13: Example of error correction using repetitions of the same pangram on the Nexus S (iPhone performance is similar). Majority voting is again applied to the letters inferred after each pangram repetition. The accuracy is lower than the tablet case due to the lower per-letter accuracy inference on the phone compared to the tablet.

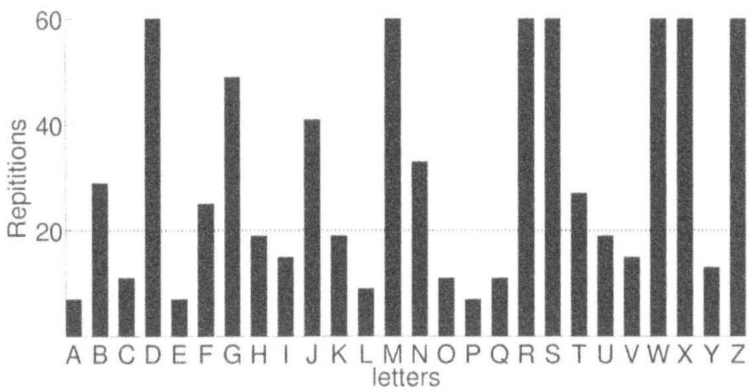

Figure 16: Number of typing repetitions required by the voting algorithm to correctly infer each letter with a probability of at least 90%.

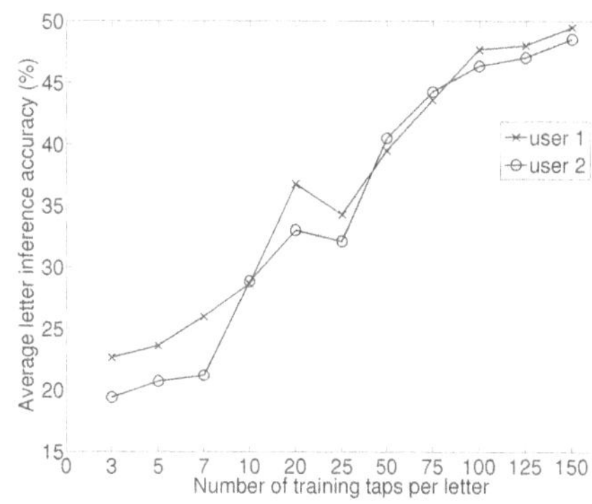

Figure 17: The average letter inference accuracy increases with the number of training samples. More training taps per letter than the icon case are required because of the close locations of letters on the keyboard.

tapping force that she exerts on the display, which creates not enough perturbations in the sensor readings to properly model the taps.

Effect of Majority Voting Scheme. To show the impact of majority voting, we perform again the Repeated Pangrams and simulation experiments discussed in Section 4.5.1. The confusion matrix in Table 3 is exploited in the simulation experiment. The result of the Repeated Pangrams experiment is shown in Figure 13, where the actual typed sentence is on top while the following lines show the inferred sentence after one, four, and fifteen repetitions, respectively. The majority voting scheme is able to correctly infer the majority of the letters after four repetitions only, providing hints on the candidate letters when no majority voting is possible. Some letters (mostly consonants) are harder to infer because of the relatively small number of training samples for these letters.

Figure 16 shows the repetition simulation result, with the number of repetitions needed to correctly infer each letter with a probability of at least 90%. The result highlights that the majority of the letters can be correctly inferred with a probability of at least 90% in less than 20 repetitions. Some letters, lacking enough training data from our realistic experiment, require more repetitions or, as for the case of 'W', 'X', 'D', and 'Z' are never correctly inferred even after 60 repetitions. However, as we have shown in Section 4.5.2, each letter can be inferred correctly when proper amount of training data per letter is collected.

Impact of the Amount of Training Data. Figure 17 shows the average per-letter inference accuracy as a function of the number of training samples. The figure shows that the inference accuracy increases with the number of training samples, a result consistent with the icon tap inference accuracy pattern discussed in Section 4.4. The only difference is the larger amount of training samples needed by the letters' learning model compared to the icons' model. We conjecture that the reason is twofold: the tiny spatial separation between letters and the smaller letter button size. When combined, these factors require more training data to effectively model letter tap differences.

4.6 Comparison across Devices

Figure 18 shows the average letter inference accuracy across different devices for the Pangrams experiment. We also include the average accuracy from typing on a tablet enclosed in an off-the-shelf padded leather case. While, on average, there is no noticeable difference between the TapPrints performance on the iPhone and Nexus S, we observe an accu-

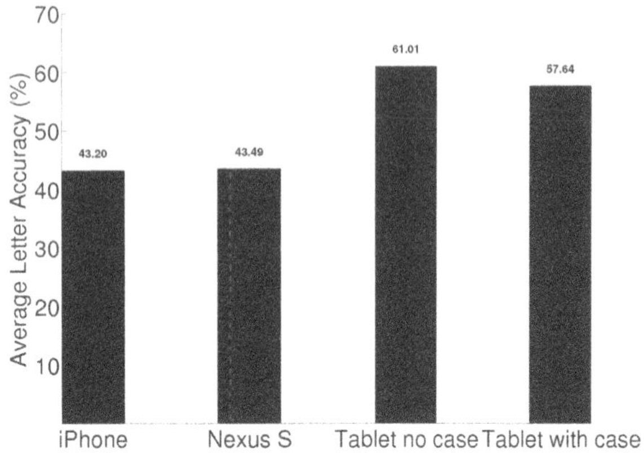

Figure 18: Comparison of the average letter inference accuracy for different devices: iPhone 4, Nexus S, Samsung Galaxy Tablet 10.1, and Samsung Galaxy Tablet 10.1 in a leather case.

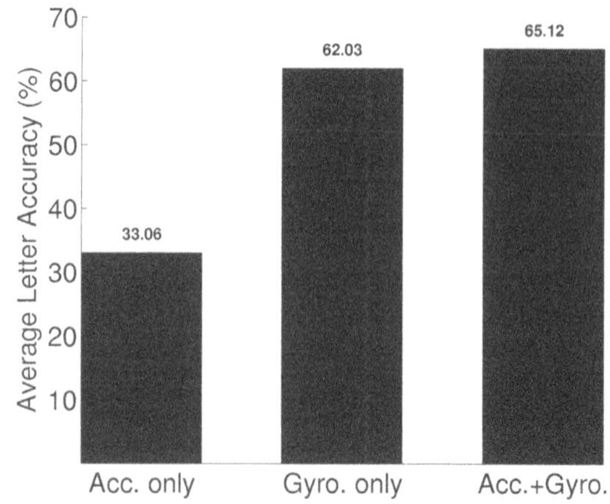

Figure 19: Impact of the sensing modality on the tap inference accuracy: the gyroscope is the most effective sensor.

racy drop when the tablet is placed in the leather case (for one of the users we observe a drop of about 15%). After a closer look at the tablet's accelerometer and gyroscope data from using the case, we notice less pronounced sensor perturbations in response to each tap. From this preliminary experiment it seems that dampening the sensor perturbations using padded cases is one of the possible solutions to reduce the effectiveness of attacks that exploit motion sensor data streams for key press inference.

4.7 Which is the Most Effective Sensor?

So far, we have shown TapPrints' performance exploiting both accelerometer and gyroscope data for the learning models. But, what is the contribution of each individual sensor to the inference accuracy? Or, in other words, which is the most effective sensor? To answer these questions, we train the classifiers for the letter tap experiment using only accelerometer data, only gyroscope data, and data from both the sensors combined. The results from the tablet experiment are shown in Figure 19 (the results for smartphones exhibit similar trends). Interestingly, the dominant contribution is from the gyroscope, achieving an average accuracy of over 60% when operating alone. The accelerometer adds only a 3% on average to the classification accuracy. Using just the accelerometer, the achievable accuracy is only about 33% on average. We conjecture that this is because angular variations – picked up by the gyroscope – from different letter taps are more distinct than vibrations – picked up by the accelerometer. A mobile device is a rigid body, and while vibrations from a tap tend to equally spread across the body, the linear velocities with respect to the device's three rotation axis are different even for small variations of the distance between a tap location and the axis.

4.8 Sensor Sampling Rate

Here we show the effect of varying the sensing sampling rate on the letter inference accuracy. Intuitively, the lower the sampling rate, the worse the tap inference performance because fewer sensor samples are available to capture each tap's sensor signal and to model it effectively. Figure 20

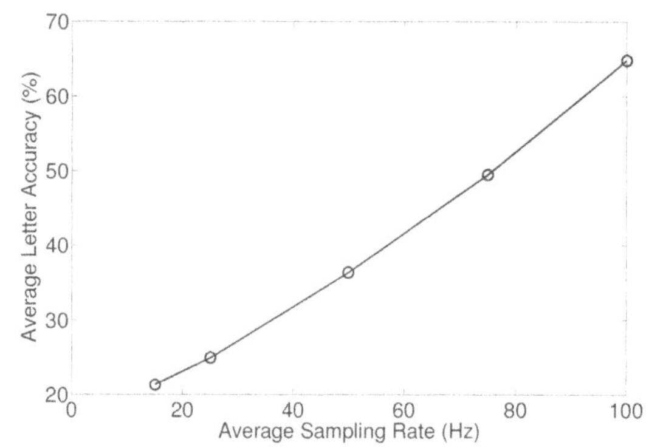

Figure 20: The inference accuracy increases with the sampling rate.

shows this effect by plotting the average letter inference accuracy on the tablet as a function of the sampling rate. The results from our two smartphones exhibit similar trends. Increasing the sampling rate from 20Hz to 100Hz improves the classification accuracy from 22% to 65%. Interestingly, each sampling rate increase produces large classification accuracies boosts. This leads to two conclusions. First, the classification accuracy may further increase with higher sensor sampling rates. Second, it is possible to reduce the effect of the attack by reducing the sampling rate, for instance, when the device's keyboard is up.

5. DISCUSSION

Although TapPrints should be only deemed as an early step towards the use of mobile devices' motion sensors for tap and letter inference, it exposes a latent security threat: An attacker could silently tap into the motion sensor data stream using a background running application with little effect on the device performance – the impact on the battery is minimal given the low power requirements of the accelerometer and gyroscope [22].

We show that the inference accuracy can increase by leveraging the correlated sensing data patterns from tapping a same letter or icon multiple times. More sophisticated attacks could combine TapPrints with targeted dictionary mining techniques, which exploit the information from the confusion matrix to narrow down the inference of a letter by just searching the dictionary space associated with the top-most confused letters [1].

We now discuss possible solutions that could mitigate the threats exposed by TapPrints. One approach is the modification of the mobile devices' operating systems to pause the motion sensors when a user is typing on the keyboard. In this way, it wouldn't be possible to correlate the sensor data with keyboard taps. Less obvious are the implications of limiting the sensor data access to prevent the inference of taps on locations other than the keyboard, i.e., icons and buttons. In this case, the reduction of the motion sensors sampling rate (e.g., down to 5 Hz or less) for any application running in the background is an option. As shown in Section 4.8, lower sensing rates prevent the sensors from picking up the very rapid sensor data perturbations generated by taps. However, any application legitimately using the motion sensors for activity, context, and gesture recognition, for example, while in the background [22] would suffer tremendously. Alternatively, the use of the sensors could be subject to special agreements between application developers and smartphone vendors or application distribution providers to hold developers accountable for any misconduct in the use of the sensor data. There might be also a need to re-think the future standard specifications for remote Javascript access to mobile devices' motion sensor data [4]. It might also be prudent to have users grant permission to motion sensor data access, as it is already done for the location engine. Special rubber cases that absorb the device motion or swiping-based keyboards could be a further alternative.

TapPrints raises important issues about the use and programming practices of the ever growing smartphone and tablet ecosystem. We have proposed just a few possible answers to the many research questions evolving from our findings and the involvement of the broader research community is necessary to address these issues.

6. RELATED WORK

Sensor-rich smartphones and tablets are increasingly becoming target of attacks that compromise users' privacy [12]. Schlegel et al. [25] demonstrate that it is possible to use the smartphone's microphone to retrieve sensitive data, such as credit card numbers and PINs from interacting with tone and speech-based phone menu systems. Cai et al. [8] survey existing smartphone-sniffing attacks, including the ones that use GPS, microphone, and camera. In contrast, our work deals with the inference of tapped icons and letters using just the phone's accelerometer and gyroscope readings.

Inferring keystrokes on a traditional desktop keyboard using side channels has received a lot of attention. Researchers exploit electromagnetic waves [18], acoustic signals [27], timing events [14], and specialized software [28] to intercept the typed content with high accuracy. Marquardt et al. [19] show that it is possible to use an iPhone's accelerometer to infer key presses on a regular keyboard when the phone lays two inches away. The difference between our work and previous projects is that we solely rely on motion sensors, easily accessible through off-the-shelf APIs on mainstream mobile devices. Moreover, touchscreen tapping doesn't generate distinct sounds or specific electromagnetic signals, making previous techniques not applicable to mobile device typing.

In parallel to our research, Cai et al. [6] and Owusu et al. [7] present, respectively, TouchLogger and ACCessory, both relying on the accelerometer for key press inference on soft-keyboards. TouchLogger is an application to infer digits from a number-only soft-keypad (with relative large buttons that are easier to detect), while ACCessory infers taps of keys and of areas arranged in a 60-region grid. Our work differs from these efforts in a number of important ways. While both TouchLogger and ACCessory are tested on just one Android smartphone model and in landscape mode only, TapPrints is evaluated across several platforms including different operating systems (iOS and Android) and form factors (smartphones and tablets). In contrast to the use of just the accelerometer in TouchLogger and ACCessory, we show a combined approach that uses both the accelerometer and gyroscope for achieving better accuracy. In addition, we present a more comprehensive evaluation, across multiple realistic scenarios, with a larger number of users, and with three different typing modalities.

7. CONCLUSION

In this paper, we presented TapPrints, a framework to infer where people tap and what people type on a smartphone or tablet display based on accelerometer and gyroscope sensor readings. Using about 40,000 labeled taps collected from ten participants on three different platforms and with the support of machine learning analysis, we showed that TapPrints is able to achieve up to 90% and 80% accuracy for, respectively, inferring tap locations across the display and letters. Although our findings are initial, we demonstrated that motion sensors, normally used for activity recognition and gaming, could potentially be used to compromise users' privacy. We hope that our research will accelerate follow up efforts to address the vulnerabilities stemming from unrestricted access to motion sensors.

8. ACKNOWLEDGMENTS

We would like to thank the volunteers for their participation in our experiments. We would also like to thank our shepherd Yang Li and the anonymous reviewers for their valuable comments.

9. REFERENCES

[1] S. Agrawal, I. Constandache, S. Gaonkar, R. Roy Choudhury, K. Caves, and F. DeRuyter. Using Mobile Phones to Write in Air. In *Proceedings of the 9th international conference on Mobile systems, applications, and services*, pages 15–28. ACM, 2011.

[2] M. Azizyan, I. Constandache, and R. Roy Choudhury. Surroundsense: Mobile Phone Localization via Ambience Fingerprinting. In *Proceedings of the 15th annual international conference on Mobile computing and networking*, pages 261–272. ACM, 2009.

[3] R. Becker, R. Cáceres, K. Hanson, J. Loh, S. Urbanek, A. Varshavsky, and C. Volinsky. A Tale of One City: Using Cellular Network Data for Urban Planning. *IEEE Pervasive Computing, Vol. 10, No. 4, October-December 2011*, 2011.

[4] S. Block and A. Popescu. Device Orientation Event Specification. W3C, Draft 12 July 2011.

[5] L. Breiman. Random Forests. In *Machine Learning*, volume 45(1), 2001.

[6] L. Cai and H. Chen. Touchlogger: Inferring Keystrokes on Touch Screen from Smartphone Motion. In *Proceedings of the 6th USENIX conference on Hot topics in security (HotSec'11). USENIX Association, Berkeley, CA, USA*, pages 9–9, 2011.

[7] E. Owusu, J. Han, S. Das, A. Perrig and J. Zhang. ACCessory: Password Inference using Accelerometers on Smartphones. In *Proceedings of the 13th Workshop on Mobile Computing Systems and Applications (HotMobile'12). San Diego, CA, USA*, 20121.

[8] L. Cai, S. Machiraju, and H. Chen. Defending Against Sensor-Sniffing Attacks on Mobile Phones. In *Proceedings of the 1st ACM workshop on Networking, systems, and applications for mobile handhelds*, 2009.

[9] R. Caruana, A. Niculescu-Mizil, G. Crew, and A. Ksikes. Ensemble Selection from Libraries of Models. In *Proceedings of the twenty-first international conference on Machine learning*, ICML '04, pages 18–, New York, NY, USA, 2004. ACM.

[10] T. G. Dietterich. Ensemble Methods in Machine Learning. In *Multiple Classifier Systems*, pages 1–15, 2000.

[11] P. Domingos. Bayesian Averaging of Classifiers and the Overfitting Problem. In *In Proceedings 17th International Conference on Machine Learning*, pages 223–230. Morgan Kaufmann, 2000.

[12] M. Egele, C. Kruegel, E. Kirda, and G. Vigna. Pios: Detecting Privacy Leaks in iOS Applications. In *Proceedings of the Network and Distributed System Security Symposium*, 2011.

[13] W. Enck, P. Gilbert, B. Chun, L. Cox, J. Jung, P. McDaniel, and A. Sheth. Taintdroid: an Information-Flow Tracking System for Realtime Privacy Monitoring on Smartphones. In *Proceedings of the 9th USENIX conference on Operating systems design and implementation*, pages 1–6. USENIX Association, 2010.

[14] D. Foo Kune and Y. Kim. Timing Attacks on Pin Input Devices. In *Proceedings of the 17th ACM conference on Computer and communications security (CCS '10)*, 2010.

[15] M. Jahrer, A. Töscher, and R. Legenstein. Combining Predictions for Accurate Recommender Systems. In *Proceedings of the 16th ACM SIGKDD international conference on Knowledge discovery and data mining*, KDD '10, pages 693–702, New York, NY, USA, 2010. ACM.

[16] N. Lane, E. Miluzzo, H. Lu, D. Peebles, T. Choudhury, and A. Campbell. A Survey of Mobile Phone Sensing. *Communications Magazine, IEEE*, 48(9):140–150, 2010.

[17] H. Lu, J. Yang, Z. Liu, N. Lane, T. Choudhury, and A. Campbell. The Jigsaw Continuous Sensing Engine for Mobile Phone Applications. In *Proceedings of the 8th ACM Conference on Embedded Networked Sensor Systems*, pages 71–84. ACM, 2010.

[18] V. M. and P. S. Compromising Electromagnetic Emanations of Wired and Wireless Keyboards. In *Proceedings of the 18th conference on USENIX security symposium*, 2009.

[19] P. Marquardt, A. Verma, H. Carter, and P. Traynor. (sp)iphone: Decoding Vibrations from Nearby Keyboards Using Mobile Phone Accelerometers. In *Proceedings of the 18th ACM conference on Computer and communications security*, pages 551–562. ACM, 2011.

[20] P. McCullagh and J. A. Nelder. *Generalized Linear Models (Second edition)*. London: Chapman & Hall, 1989.

[21] S. McKinley and M. Levine. Cubic Spline Interpolation. *College of the Redwoods*, 1998.

[22] E. Miluzzo, N. Lane, K. Fodor, R. Peterson, H. Lu, M. Musolesi, S. Eisenman, X. Zheng, and A. Campbell. Sensing Meets Mobile Social Networks: the Design, Implementation and Evaluation of the CenceMe Application. In *Proceedings of the 6th ACM conference on Embedded network sensor systems*, pages 337–350. ACM, 2008.

[23] B. Pinkas and T. Sander. Securing Passwords Against Dictionary Attacks. In *Proceedings of the 9th ACM Conference on Computer and Communications Security*, pages 161–170. ACM, 2002.

[24] M. Poh, K. Kim, A. Goessling, N. Swenson, and R. Picard. Cardiovascular Monitoring Using Earphones and a Mobile Device. *Pervasive Computing, IEEE*, (99):1–1, 2011.

[25] R. Schlegel, K. Zhang, X. Zhou, M. Intwala, A. Kapadia, and X. Wang. Soundcomber: a Stealthy and Context-Aware SoundTrojan for Smartphones. In *Proceedings of the 18th Annual Network and Distributed System Security Symposium (NDSS '11)*, 2011.

[26] B. Schoelkopf, C. Burges, and A. Smola. *Advances in Kernel Methods - Support Vector Learning*. MIT Press, 1998.

[27] L. Zhuang, F. Zhou, and J. D. Tygar. Keyboard Acoustic Emanations Revisited. *ACM Trans. Inf. Syst. Secur.*, 2009.

[28] K. Killourhy and R. Maxion. Comparing Anomaly-Detection Algorithms for Keystroke Dynamics. In *Dependable Systems & Networks, 2009. DSN'09. IEEE/IFIP International Conference on*, pages 125–134. IEEE, 2009.

Medusa: A Programming Framework for Crowd-Sensing Applications *

Moo-Ryong Ra* Bin Liu* Tom La Porta† Ramesh Govindan*

*Department of Computer Science
University of Southern California
{mra, binliu, ramesh}@usc.edu

†Department of Computer Science and Engineering
Pennsylvania State University
tlp@cse.psu.edu

ABSTRACT

The ubiquity of smartphones and their on-board sensing capabilities motivates *crowd-sensing*, a capability that harnesses the power of crowds to collect sensor data from a large number of mobile phone users. Unlike previous work on wireless sensing, crowd-sensing poses several novel requirements: support for humans-in-the-loop to trigger sensing actions or review results, the need for incentives, as well as privacy and security. Beyond existing crowd-sourcing systems, crowd-sensing exploits sensing and processing capabilities of mobile devices. In this paper, we design and implement Medusa, a novel programming framework for crowd-sensing that satisfies these requirements. Medusa provides high-level abstractions for specifying the steps required to complete a crowd-sensing task, and employs a distributed runtime system that coordinates the execution of these tasks between smartphones and a cluster on the cloud. We have implemented ten crowd-sensing tasks on a prototype of Medusa. We find that Medusa task descriptions are two orders of magnitude smaller than standalone systems required to implement those crowd-sensing tasks, and the runtime has low overhead and is robust to dynamics and resource attacks.

Categories and Subject Descriptors

D.3.3 [**Programming Languages**]: [Language Constructs and Features – Frameworks]

General Terms

Design, Experimentation, Human Factors, Reliability, Security

Keywords

Crowd-sensing, Macroprogramming, Programming Framework, Smartphone, Worker Mediation

*The first author, Moo-Ryong Ra, was supported by Annenberg Graduate Fellowship. This research was partially sponsored by the Army Research Laboratory under Cooperative Agreement Number W911NF-09-2-0053 and by the USC METRANS Transportation Center.

MobiSys'12, June 25–29, 2012, Low Wood Bay, Lake District, UK.

1. INTRODUCTION

The ubiquity of smartphones and other mobile devices, and the plethora of sensors available on them, have inspired innovative research that will, over time, lead to sophisticated context-aware applications and systems. Some of this research has explored ways in which mobile phone users can *contribute* sensor data towards enabling self-reflection, environmental awareness, or other social causes.

In this paper, we consider a specific form of acquiring sensor data from multiple smartphones that we call *crowd-sensing*. Crowd-sensing is a capability by which a *requestor* can recruit smartphone users (or *workers*) to provide sensor data to be used towards a specific goal or as part of a social or technical experiment (e.g., *tasks* such as forensic analysis, documenting public spaces, or collaboratively constructing statistical models). Workers' smartphones collect sensor data, and may process the data in arbitrary ways before sending the data to the requestor.

Crowd-sensing is a form of networked wireless sensing, but is different from prior research on networked sensing for two reasons: each smartphone is owned by an individual, and some sensing actions on the smartphone may require human intervention (e.g., pointing a camera at a specific target, or initiating audio capture at the appropriate moment). Crowd-sensing is different from crowd-sourcing in one important respect: a crowd-sourcing system like the Amazon Mechanical Turk (AMT [1]) permits requestors to task participants with human intelligence tasks like recognition or translation, while a crowd-sensing system enables requestors to task participants with acquiring processed sensor data.

For these reasons, any software system designed to support crowd-sensing is qualitatively different from systems for wireless sensing or crowd-sourcing. Crowd-sensing systems must support ways in which workers can be incentivized to contribute sensor data, since participating in crowd-sensing may incur real monetary cost (e.g., bandwidth usage). In some cases, it may be necessary to support "reverse" incentives, where workers pay for the privilege of participating in the requestor's task. They must also support *worker-mediation*, by which (human) workers can mediate in the sensor data collection workflows. Specifically, humans may trigger sensing actions, and annotate collected sensor data. In addition, a crowd-sensing system must support privacy of sensor data contributions, and worker anonymity (in cases where workers desire anonymity). Identifying the requirements for crowd-sensing is our first important contribution (Section 2).

In this paper, we tackle the challenge of *programming* crowd-sensing tasks. A high-level programming language can simplify the burden of initiating and managing crowd-sensing tasks, especially for non-technical requestors who we assume will constitute the majority of crowd-sensing users. Our second contribution is the design of a high-level programming language called *MedScript* and an associated runtime system called *Medusa* for crowd-sensing (Section 4). MedScript provides abstractions called *stages* for intermediate steps in a crowd-sensing task, and for control flow between stages (called *connectors*). Unlike most programming languages, it provides the programmer language-level constructs for incorporating workers into the sensing workflow. The Medusa runtime is architected (Section 3) as a partitioned system with a cloud component and a smartphone component. For robustness, it minimizes the task execution state maintained on the smartphone, and for privacy, it ensures that any data that leaves the phone must be approved by the (human) worker. Finally, it uses AMT to recruit workers and manage monetary incentives.

Our third contribution is an evaluation of Medusa on a complete prototype (Section 5). We use our prototype to illustrate that Medusa satisfies many of the requirements discussed above. We demonstrate that Medusa can be used to compactly express several novel crowd-sensing tasks, as well as sensing tasks previously proposed in the literature. For three different tasks, Medusa's task descriptions are almost two orders of magnitude more compact than stand-alone implementations of those tasks. Medusa's overhead is small, and it provides effective sandboxing against unauthorized access to smartphone resources as well as to excessive resource usage.

Medusa represents a convergence of two strands of research (Section 6): participatory sensing systems like [16] and crowd-sourcing frameworks like [1]. Unlike the former, Medusa explicitly supports incentives and worker-mediation; unlike the latter, Medusa enables requestors to gather raw or processed sensor data from mobile devices in a substantially automated way.

2. CROWD-SENSING: MOTIVATION AND CHALLENGES

Motivation. Consider the following scenario. A social science researcher, Alice, is interested in studying the dynamics of "Occupy" movements across the globe. Her methodology is to obtain ground truth videos about life in "Occupy" encampments across the United States and overseas. This methodology of obtaining observations of human life in urban spaces was pioneered by William Whyte [20] and is accepted practice in public planning.

However, it is logistically difficult for Alice to visit each of these locations to conduct research, so she resorts to an idea that we call *crowd-sensing*, which leverages the prevalence of mobile devices with sensors and the power of crowds. She recruits volunteers at each one of the encampments by offering them small monetary incentives to provide her with videos of daily life at the camps. Some volunteers take one or more videos on a mobile device, and upload summaries of each video (e.g, a few frames or a short clip) to an Internet-connected service. Other volunteers upload summaries of videos they may have taken before being recruited. Alice obtains a large corpus of summaries, which she makes available to other volunteers to *curate*, selecting only the relevant subset that may be useful for research. She then asks the volunteers whose summaries have been chosen to upload the full videos, and pays them for their participation. She analyzes the curated set of videos, and draws conclusions about similarities and differences between "Occupy" communities across the globe.

In the scenario above, we say that Alice is interested in conducting a *task*, that of collecting relevant videos for her research. A task is initiated by a *requestor*. Each task consists of several *stages*. In our example, the stages were, in order: recruiting volunteers, volunteers taking videos on their smartphones, uploading summaries, curating summaries using another set of volunteers, and uploading the full videos. Volunteers who participate in tasks are called *workers*; workers have smartphones[1] that are used to perform one or more stage actions.

In this paper, we consider the architectural requirements, the design of a high-level programming framework and an associated runtime system for crowd-sensing. In the absence of such a programming framework, Alice may need to manually recruit volunteers (e.g., by advertising in mass media or using her social network), follow-up with the volunteers to upload summaries and videos, manually recruit volunteers for curation (e.g., students in her class), implement software that simplifies the curation process, and ensure that payments are made to all volunteers.

Ideally, our programming framework should abstract all these details, and allow Alice in our scenario to compactly express the stages in her task. The associated runtime should automate stage execution, relieving the requestor from the burden of manually managing stages.

There are many other examples of crowd-sensing, beyond video documentation:

Spot Reporter A journalist is under deadline pressure to write an article about a developing forest fire and with its impact on nearby communities. He recruits volunteers with smartphones to send pictures of the extent of fire damage, interview local residents, and report on the efficacy of the first responders.

Auditioning A television station invites budding actors to submit videos of their acting skills, in order to select actors for a second stage of in-person auditioning for an upcoming television show.

Collaborative Learning A software company is developing a lifestyle monitoring mobile app, which lets users self-reflect on their activities. It would like to recruit volunteers to submit labeled samples of accelerometer readings; these are used to build a machine-learning classifier of human activity which will be embedded in the mobile app.

Forensic Analysis Security officials at a stadium are investigating an outbreak of violence and would like to determine people who were at the scene of the outbreak when the violence occurred. They request volunteers, who may have taken pictures during the event, to send snapshots of faces in the stadium.

We describe these tasks in detail later in Section 5.

Requirements. A practical programming framework that supports crowd-sensing tasks must satisfy several challenging requirements. Some of these requirements govern the expressivity of the programming language, while others constrain the underlying runtime system. We discuss these in order.

Expressivity Requirements. Requestors must be able to specify *worker-mediation*, i.e., a worker may need to perform an action to complete a stage such as initiating or stopping the recording of the video or audio clip, labeling data or approving data for upload (see below). However, not all stages will require workers to initiate sensor data collection; requestors may want to design stages that *access stored sensor data*, such as in the forensic analysis task.

Stages must support not just sensing, but also *in-network processing*; this helps conserve network bandwidth for users with data usage limits or reduces upload latency by reducing the volume of

[1] Henceforth, we will use the term smartphones, although our designs apply equally well to other mobile devices such as tablets.

information that needs to be transmitted on a constrained connection. For example, in our video documentation example, video summaries were first extracted from the video in order to weed out irrelevant videos. Similarly, feature vectors may be extracted in the collaborative learning task, or faces in the forensic analysis task. In addition, the programming language must be *extensible*, in order to support new sensors or new in-network processing algorithms.

Tasks may have *timeliness* requirements and any contribution received after the deadline are discarded; in general, we anticipate task deadlines to be on the order of hours or days after task initiation, so traditional quality-of-service concerns may not apply. Tasks may require *crowd-curation*; in our scenario, Alice relies on this to pre-select videos likely to be most relevant to her research objectives.

Requestors must be able to specify monetary *incentives* for workers. In our scenario, Alice promises to pay participants for uploading videos. Some tasks may also require *reverse-incentives*, where the workers pay the requestors for the privilege of participating in the task. In our auditioning example, a requestor may charge a small fee in order to disincentivize frivolous participants.

Runtime Requirements. Workers must be able to sign up for *multiple concurrent tasks*, and requestors should be able to initiate multiple tasks concurrently. Thus, a smartphone user may wish to participate in collaborative learning, but may also sign up to be a spot reporter. Stages in a task may be executed at different times by different workers; thus, stage execution is *not necessarily synchronized* across workers. In the video documentation task, one worker may upload their summaries several hours after another worker has uploaded his or her video. Task execution must be *robust* to intermittent disconnections or longer-term outages on smartphones.

The runtime should preserve subject *anonymity* with respect to requestors, and should contain mechanisms for ensuring data *privacy*. It should use best practices in *secure communication*, and ensure the *safety* of execution on worker smartphones. We discuss later what kinds of anonymity, privacy, security, and safety are achievable in our setting. Finally, the runtime must respect user-specified *resource usage policies* on the smartphone for crowd-sensing; for example, if generating a video summary is likely to consume significant battery power, the framework must terminate the stage or take other actions discussed later.

Summary. This is a fairly complex web of requirements, but we believe they are necessary for crowd-sensing. Worker-mediation and crowd-curation leverage human intelligence to ensure high-quality sensor data. Support for concurrent tasks and asynchrony enables high throughput of tasks. Monetary incentives, as well as anonymity, privacy and security features as well as resource controls incentivize worker participation. Ensuring task execution robustness in the runtime relieves programmers of the burden of dealing with failures.

In the next section, we describe an architecture that meets these requirements.

3. MEDUSA SYSTEM ARCHITECTURE

In this section, we first describe the architectural principles that guide the design of our crowd-sensing programming system, called Medusa. We then illustrate our design (Figure 1) by describing how the video documentation crowd-sensing task is executed by Medusa.

3.1 Architectural Principles

Medusa is a high-level language for crowd-sensing. In Medusa, programmers specify crowd-sensing tasks as a sequence of stages.

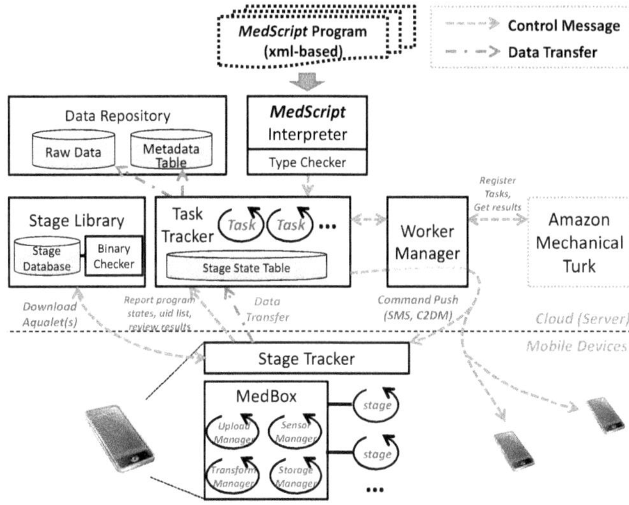

Figure 1: System Architecture

Thus, Alice would program her video documentation task using a description that looks approximately like this:

```
Recruit -> TakeVideo -> ExtractSummary
  -> UploadSummary -> Curate -> UploadVideo
```

This describes the sequence of steps in the task. The Medusa runtime executes these tasks. The design of the Medusa runtime is guided by three architectural decisions that simplify overall system design.

Principle #1: Partitioned Services. Medusa should be implemented as a partitioned system that uses a collection of services both on the cloud and on worker smartphones. This constraint follows from the following observation. Some of the requirements are more easily and robustly accomplished on an always Internet-connected cloud server or cluster: task initiation, volunteer recruitment, result storage, and monetary transactions. Others, such as sensing and in-network processing, are better suited for execution on the smartphone.

Principle #2: Dumb Smartphones. Medusa should minimize the amount of task execution state that is maintained on smartphones. This design principle precludes, for example, large segments of a task from being completely executed on the smartphone without the cloud being aware of execution progress. We impose this constraint to enable robust task execution in the presence of intermittent connectivity failures, as well as long-term phone outages caused, for example, by workers turning off their smartphones.

Principle #3: Opt-in Data Transfers. Medusa should automatically require a user's permission before transferring any data from a smartphone to the cloud. Data privacy is a significant concern in crowd-sensing applications. Our principle ensures that, at the very least, users have the option to opt-out of data contributions. Before workers opt-in, they may view a requestor's privacy policy.

Discussion. Other designs of the runtime are possible. For example, it might be possible to design a purely peer-to-peer crowd-sensing system, and it might also be possible to empower smartphones to exclusively execute crowd-sensing tasks. In our judgement, our architectural principles enable us to achieve all of the requirements outlined in the previous section without significantly complicating system design. Furthermore, these principles do not precisely determine exactly what functionality to a place on the

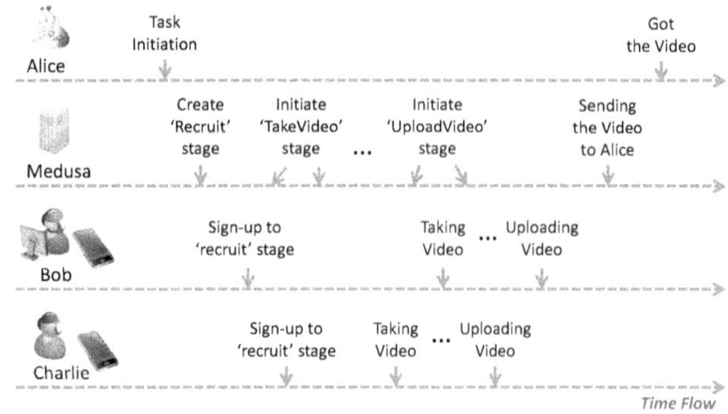

Figure 2: Illustration of video documentation task

cloud or on the phone: our design is one instantiation that adheres to these principles.

Some of these principles may appear to conflict each other. For example, if the smartphone components are designed to be "dumb", then intermediate data products during task execution must be communicated to the cloud, which may require significant user opt-in interactions and hamper usability. We describe later how we address this tension.

Finally, our opt-in requirement can impact usability by requiring user intervention whenever data leaves the phone. In the future, we may consider relaxing this constraint in two ways. First, users may be willing to sacrifice privacy for convenience or monetary reward. In the future, we may consider policies that allow users to opt-out of approving data transfers for tasks which promise significant monetary incentives, for example. An understanding of these kinds of exceptions can only come from experience using the system. Second, some of the in-network processing algorithms may be designed to be privacy-preserving (e.g., by using differential privacy [8]), so user assent may not be necessary for data generated by these algorithms.

3.2 How Medusa Works

In keeping with our *partitioned services* principle, Medusa is structured as a collection of services running on the cloud and on the phone (Figure 1). These services coordinate to perform crowd-sensing tasks. We give a high-level overview of Medusa by taking our running example of Alice's video documentation task.

In Medusa, Alice writes her video documentation task in MedScript, a high-level language designed for non-experts, which provides stages as abstractions and allows programmers to express control flow between stages. She submits the program to the MedScript interpreter, which is implemented as a cloud service. Figure 2 illustrates the sequence of actions that follows the submission of the task to Medusa.

The interpreter parses the program and creates an intermediate representation which is passed on to a Task Tracker. This latter component is the core of Medusa, coordinating task execution with other components as well as with workers' smartphones. For the `Recruit` stage, Task Tracker contacts Worker Manager (a back-end service), which initiates the recruitment of workers. As we shall describe later, Worker Manager uses Amazon Mechanical Turk [1] for recruitment and a few other tasks. When workers agree to perform the task, these notifications eventually reach Task Tracker through Worker Manager; different workers may agree to perform the task at different times (e.g., Bob and Charlie in Figure 2). More generally, workers may perform different stages at different times. Monetary incentives are specified at the time of recruitment.

Once a worker has been recruited, the Task Tracker initiates the next stage, `TakeVideo`, on that worker's smartphone by sending a message to the Stage Tracker, one instance of which runs on every phone. The `TakeVideo` stage is a downloadable piece of code from an extensible Stage Library, a library of pre-designed stages that requestors can use to design crowd-sending tasks. Each such stage executes on the phone in a sandboxed environment called MedBox. The `TakeVideo` stage requires human intervention – the worker needs to open the camera application and take a video. To remind the worker of this pending action, the stage implementation uses the underlying system's notification mechanism to alert the worker. Once the video has been recorded, Stage Tracker notifies the Task Tracker of the completion of that stage, and awaits instructions regarding the next stage. This is in keeping with our *dumb smartphones* principle. The video itself is stored on the phone.

The Task Tracker then notifies the Stage Tracker that it should run the `ExtractSummary` stage. This stage extracts a small summary video comprised of a few sample frames from the original video, then uploads it to the Task Tracker. Before the upload, the user is required to preview and approve the contribution, in keeping with our *opt-in data transfers* principle.

Execution of subsequent stages follows the same pattern: the Task Tracker initiates the `curate` stage while a completely different set of volunteers rates the videos. Finally, the selected videos are uploaded to the cloud. Only when all stages have completed is Alice notified (Figure 2).

Medusa maintains persistent state of data contributions in its Data Repository as well as the state of stage execution for each worker in its Stage State Table. As we shall discuss later, this enables robust task execution.

In the next section, we discuss each of these components in greater detail.

4. MEDUSA DESIGN

Medusa achieves its design objectives using a cloud runtime system which consists of several components, together with a runtime system on the smartphone. Before we describe these two subsystems, we begin with a description of the MedScript programming language.

4.1 The MedScript Programming Language

The MedScript programming language is used to describe crowd-sensing tasks, so each task described in Section 2 would be associated with a corresponding MedScript. More precisely, a MedScript defines a *task instance*, specifying the sequence of actions performed by a single worker towards that crowd-sensing task. In general (as described in the next section), requestors may ask the Medusa runtime to run multiple instances of a MedScript task at different workers, and a single worker may also run multiple task instances. Moreover, a single worker may concurrently run instances of different tasks (possibly) initiated by different requestors.

Listing 1 shows the complete listing of the video documentation task. We use this program listing to illustrate various features of the MedScript language.

MedScript consists of two high-level abstractions: *stages* and *connectors*. A stage (e.g., line 8) describes either a sensing or computation action, or one that requires human-mediation, and connectors (e.g., line 65) express control flow between stages. In our current realization, MedScript is an XML-based domain-specific language; we have left it to future work to explore a visual programming front-end to MedScript.

```
1  <xml>
     <app>
3      <name>Video-Documentation</name>
       <rrid>[User's Requestor ID]</rrid>
5      <rrkey>[User's Requestor Key]</rrkey>
       <deadline>21:00:00 12/16/2011</deadline>
7
       <stage>
9        <name>Recruit</name> <type>HIT</type>
         <binary>recruit</binary>
11       <config>
           <stmt>Video Documentation App. Demonstration</stmt>
13         <expiration>21:00:00 12/16/2011</expiration>
           <reward>.05</reward>
15         <output>W_WID</output>
         </config>
17     </stage>
       <stage>
19       <name>TakeVideo</name> <type>SPC</type>
         <binary>mediagen</binary>
21       <trigger>user-initiated</trigger> <review>none</review>
         <config>
23         <params>-t video</params>
           <output>VIDEO</output>
25       </config>
       </stage>
27     <stage>
         <name>ExtractSummary</name> <type>SPC</type>
29       <binary>videosummary</binary>
         <trigger>immediate</trigger> <review>none</review>
31       <config>
           <input>VIDEO</input>
33         <output>SUMMARY</output>
         </config>
35     </stage>
       <stage>
37       <name>UploadSummary</name> <type>SPC</type>
         <binary>uploaddata</binary>
39       <trigger>immediate</trigger>
         <config>
41         <input>SUMMARY</input>
         </config>
43     </stage>
       <stage>
45       <name>Curate</name> <type>HIT</type>
         <binary>vote</binary> <wid>all</wid>
47       <config>
           <stmt>Judging from Video Summaries</stmt>
49         <expiration>21:00:00 12/16/2011</expiration>
           <reward>.01</reward>
51         <numusers>2</numusers>
           <input>SUMMARY</input>
53         <output>SSMASK</output>
         </config>
55     </stage>
       <stage>
57       <name>UploadVideo</name> <type>SPC</type>
         <binary>uploaddata</binary>
59       <trigger>immediate</trigger>
         <config>
61         <input>SSMASK, VIDEO</input>
         </config>
63     </stage>

65     <connector>
         <src>Recruit</src>
67       <dst> <success>TakeVideo</success> <failure>Recruit</failure> </dst>
       </connector>
69     <connector>
         <src>TakeVideo</src>
71       <dst> <success>ExtractSummary</success> <failure>Recruit</failure> </dst>
       </connector>
73     <connector>
         <src>ExtractSummary</src>
75       <dst> <success>UploadSummary</success> <failure>Recruit</failure> </dst>
       </connector>
77     <connector>
         <src>UploadSummary</src>
79       <dst> <success>Curate</success> <failure>Recruit</failure> </dst>
       </connector>
81     <connector>
         <src>Curate</src>
83       <dst> <success>UploadVideo</success> <failure>Recruit</failure> </dst>
       </connector>
85   </app>
   </xml>
```

Listing 1: Video Documentation

Stages. A stage is a labeled elemental instruction in MedScript. Thus, `TakeVideo` is a label for the stage described starting on line 18. Stage labels are used in connector descriptions (see below).

Medusa defines two types of stages: a *sensing-processing-communication* stage or (SPC stage), and a *human-intelligence task* stage (or HIT stage). An SPC stage extracts sensor data from either a physical sensor (such as a camera, GPS, or accelerometer) or a logical sensor (such as a system log or a trace of network measurements), processes sensor data to perform some kind of summarization or recognition, or communicates sensor data to the requestor. A HIT stage exclusively requires human review and/or input and does not involve sensing or autonomous data processing. Thus, for example, `TakeVideo` or `ExtractSummary` or `UploadVideo` (lines 18, 27, and 56) are all examples of SPC stages, while `Recruit` and `Curate` (lines 8 and 44) are examples of HIT stages. Stages may take parameters; for example, the `reward` parameter for the `Recruit` stage specifies a monetary reward (line 14).

Each stage has an associated stage binary, referenced by the `binary` XML tag (e.g., line 10). Medusa maintains an extensible library of stage binaries in its Stage Library. Stage binaries may be re-used across tasks, and also within the same task: both the `UploadSummary` and `UploadVideo` stages use the `uploaddata` binary.

Connectors. Each stage can have one of two outcomes: *success* or *failure*. A stage may fail for many reasons: if a stage does not encounter failure, it is deemed to be successful. For each stage, *connectors* specify *control-flow* from that stage to the next stage upon either success or failure. Thus, in our example, if `TakeVideo` is successful (lines 69-72), control flow transfers to `ExtractSummary`, else control flow reverts to the `Recruit` stage.

In general, for each outcome (success or failure), there is at most one target next stage. If a target is not specified for an outcome, the Medusa runtime terminates the corresponding task instance.

Medusa allows a special control-flow construct called the *fork-join*. From one of the outcomes of the stage (usually the success outcome), a programmer may define multiple connectors to different target stages. This indicates that the target stages are executed *concurrently*. This construct is useful, for example, to concurrently read several sensors in order to correlate their readings. We give a detailed example of this construct later. Medusa enforces, through compile-time checks, that *both* the success and failure outcomes of these target stages are connected to (or *join*) a single stage, one which usually combines the results of the target stages in some way.

Data Model and Stage Inputs and Outputs. Each stage in Medusa produces exactly one output. This output is explicitly assigned by the programmer to a *named variable*. In our example, named variables include `VIDEO` and `SUMMARY`. The scope of a named variable is the task instance, so that any stage in an instance may access the named variable. If two instances execute on the same device, their variable namespaces are distinct.

Each stage can take zero or more inputs. Inputs to stages are defined by variables. The `ExtractSummary` stage takes `VIDEO` as input (line 32). Each named variable has an associated type (see below), and the MedScript compiler performs static type checking based on the known types of stage outputs.

In general, a named variable's value is a bag of one or more *homogeneously-typed* items, and the named variable is said to have the corresponding type. Supported types in Medusa include integer, floating-point, and many predefined *sensor data types*. Each sensor data type represents a contiguous sequence of sensor data, and examples include: a video clip, an audio clip, a contiguous sequence of accelerometer readings, and so on.

Medusa does not define an assignment operator, so variables can only be instantiated from stage outputs. For example, `VIDEO` is instantiated as the output of `TakeVideo` (line 24) and `SUMMARY` as the output of `ExtractSummary` (line 33). Both of these variables are bags of video clips: the former contains videos taken by the user and the latter the summary videos generated by `ExtractSummary`.

Stages and User-Mediation. One of the distinguishing aspects of Medusa is its support for worker-mediation. Workers can be in-the-loop in four ways.

First, the HIT stages `Recruit` and `Curate` require user intervention. Workers have to agree to perform a task instance during the `Recruit` stage. Requestors can specify monetary incentives as well as a deadline for recruitment. The `Curate` stage permits humans to review sensor data submissions and select submissions relevant to the task at hand. This takes as input a named variable containing one or more data items (in our example `SUMMARY`, line 52), and produces as output a bit mask (`SMASK`, line 53), that indicates selected submissions. This bitmask can be used in later stages to determine, for example, which data items to upload.

Second, only one stage binary, `uploaddata`, can transfer data from the phone to the cloud and, before data is uploaded, the worker is presented with a *preview* of the data and is given an opportunity to opt-in to the submission. Optionally, at this stage, the requestor may specify a link to a privacy policy that describes how workers' submissions will be used. If a worker chooses to opt out of the upload, the `uploaddata` stage fails.

Third, some stages, such as the `TakeVideo` stage, may require explicit worker action in order to *initiate* sensing. This form of worker-mediation leverages worker intelligence to determine the appropriate time, location, and other environmental conditions to take sensor data samples. This is indicated by a `trigger` parameter to a stage (e.g., line 21).

Fourth, requestors may provide humans with the option to *annotate* the output of any stage. Once a stage has produced an output, users are presented with a prompt to label the output by either selecting from a list of choices or by adding freeform text annotations. These annotations are stored as metadata with the data items. This capability is useful in our collaborative learning application, where users submit labeled training samples that, for example, indicate activities (like walking, running, sitting).

Task Parameters. In addition to stage parameters, programmers may specify *task parameters*, usually before the stage and connector definitions. In our example, the requestor specifies a deadline for the task. If a task instance is not completed before its deadline, it is deleted. Other parameters can include requestor credentials, described below.

Failure Semantics. As described above, each stage has two possible outcomes: success or failure. Stages can fail for several reasons. The smartphone may be turned off by its owner or its battery may die. The deadline on a stage expires because a worker failed either to initiate a sensing action or to annotate a sensor result within the deadline. The merge or join stage after a fork may fail, because one of the forked stages failed. Finally, stages may fail because components of the runtime fail.

Deadline and runtime failures result in task instances being aborted. However, other failures may result in one or more stages being re-tried; the requestor can specify, using appropriate connectors, which stages should be retried upon failure.

4.2 Medusa Cloud Runtime

The Medusa cloud runtime consists of several components described below.

MedScript Interpreter. Requestors submit their XML task descriptions to the MedScript interpreter and indicate how many instances of tasks should be run by Medusa. For example, if Alice wishes to receive about 50 videos in the hope that about 10-15 of them are of sufficient quality for her research, she can indicate to the interpreter that she wishes to spawn 50 instances. Requestors can add additional instances if their initial estimates proved to be incorrect.

The interpreter performs compile-time checks on the submitted task descriptions: static type checks, checks on restrictions for fork-join connectors, and sanity checks on named variables. It then converts the task description into an intermediate representation and stores this in the Stage State Table, together with the number of instances to be run. Finally, it notifies the Task Tracker, which is responsible for coordinating the stage execution for all instances of this task.

Task Tracker. In Medusa, each submitted task is associated with a fresh instance of a Task Tracker. The Task Tracker spawns multiple worker instances and keeps track of their execution. Recall that different stages may execute at different times at different workers. The Task Tracker creates *instance state* entries that store the current state of execution of each instance in the Stage State Table, a persistent store.

For HIT stages, the Task Tracker uses the Worker Manager to initiate those stages. Thus, if Alice wishes to invoke 50 instances of her video documentation task, the Task Tracker first instantiates 50 instances of the `Recruit` stage using Worker Manager. When a worker signs up for one of the task instances, the Worker Manager notifies the Task Tracker, which then proceeds to instantiate the `TakeVideo` stage on the worker's smartphone.

To do this, the Task Tracker notifies the Stage Tracker on the worker's smartphone to begin executing the corresponding stage. We describe the notification mechanism later in this section. When that stage is completed on the smartphone, the stage tracker returns a list of *references* (pointers) to data items in that stage's output variable; the data items themselves are stored on the smartphone. These references are stored as part of the instance state; if the smartphone is turned off immediately after the `TakeVideo` stage, the Task Tracker has all the information necessary to execute the subsequent stage when the smartphone restarts.

More generally, the Task Tracker notifies the Stage Tracker instructing it to execute an SPC stage on the phone. It passes the references corresponding to input variables for that stage. At the end of the execution, the Stage Tracker sends the Task Tracker all references for data items in the output of that stage.

Thus, in Medusa, the Task Tracker coordinates the execution of every stage and maintains instance state information in persistent storage. A single worker may concurrently sign up for multiple instances of the same task, and/or instances of many different tasks. If a Task Tracker fails, it can be transparently restarted and can resume stage execution at each worker. Between the time when the Task Tracker fails, and when it resumes, some stages may fail because the Stage Tracker may be unable to contact the Task Tracker; these stages will usually be re-executed, depending upon the control flow specified by the requestor.

The Task Tracker keeps track of the task deadline; when this expires, the Task Tracker terminates the corresponding worker instance by updating the program state appropriately. The Task Tracker also keeps internal timeouts for stage execution, and may decide to retry stages whose timeout expires. This ensures robustness to smartphone outages and connectivity failures, as well as to humans who may not have initiated sensing actions on time or completed an annotation of sensor data.

The Task Tracker does not know the identities of the requestor or the worker, instead referring to them in its internal data structures using an opaque machine-generated ID. Below, we describe how these IDs are mapped to identities of individual workers or of the requestor. This design ensures that knowledge about identities of participants is localized to a very small part of the system (described later), thereby reducing the vulnerability footprint.

The Task Tracker thus supports concurrent execution of worker instances, tracks unsynchronized stage execution across different workers, enforces timeliness of stage execution, as well as ensures robustness.

Worker Manager. The Worker Manager supports HIT stage execution, monetary transactions, crowd-curation, and worker smartphone notifications. To achieve these objectives, Worker Manager uses the Amazon Mechanical Turk (AMT [1]) system as a backend. AMT was designed to support crowd-sourcing of human intelligence tasks, such as translation, behavioral surveys, and so forth.

When the Task Tracker encounters a `Recruit` stage, it invokes the Worker Manager and passes to it the monetary incentive specified in the stage description, together with the number of desired instances. The Worker Manager posts AMT tasks (using the AMT API) seeking workers who are interested in signing up for the crowd-sensing task. Usually, requestors will include an optional URL as a parameter to the `Recruit` stage so that potential workers can see a description of the work involved. Workers may use a browser on their smartphone to sign up on AMT for this task instance. When a worker signs up, AMT notifies the Worker Manager, which in turn notifies the Task Tracker.

In order to post a task on AMT, the Worker Manager needs to be entrusted with an Amazon-issued ID and key for the requestor. These requestor credentials are specified as task parameters. However, the ID and key do not reveal the identity of the requestor. Thus, while requestor identity is not exposed, limited trust must be placed in the Worker Manager not to misuse IDs and keys.

When a task instance completes (e.g., when a worker uploads a selected video in our video documentation example), the Task Tracker contacts Worker Manager, which, in turn, presents the uploaded data to the requestor and seeks approval. Once the requestor approves the submission, money is transferred from the requestor's AMT account to the worker's AMT account.

Crowd-curation works in a similar manner. The Worker Manager spawns an AMT "voting" task inviting volunteers to vote on submissions in return for a monetary incentive. When the voting is complete, AMT notifies the Worker Manager, which in turn notifies the Task Tracker.

The Worker Manager also supports reverse-incentives, so workers can pay requestors for the privilege of contributing sensor data. Our auditioning example requires this, where the reverse incentive is designed to dissuade frivolous participation. The Worker Manager achieves this by posting a task requiring a null contribution on AMT with roles reversed: the worker as the "requestor" and the requestor as the only permitted "worker". As soon as the requestor signs up and the worker "accepts" the contribution, the requisite payment is transferred from the workers AMT account to that of the requestor.

Our use of AMT requires workers and requestors to have AMT accounts and to manage their balances on these accounts. Both workers and requestors have an incentive to do this: requestors are able to complete tasks using the system that they might not otherwise have been able to, and workers receive micro-payments for completing sensing tasks that require little cognitive load (since many of the steps can be automated).

Moreover, AMT preserves *subject anonymity* so that worker identities are not revealed to requestor (unless workers choose to). In Medusa, this subject anonymity is also preserved within the rest of the system (the runtimes on the smartphone and cloud) by identifying requestors and workers only using the opaque AMT IDs. However, since Medusa supports crowd-sensing, the sensing data submitted by a worker may reveal the identity of the worker. By opting-in to an upload of the sensor data, workers may give up subject anonymity; they can always choose to opt-out, or carefully initiate sensor data collection (e.g., taking videos whose content does not reveal the sender), should they wish to retain anonymity.

Finally, because workers are anonymous to the rest of the system, and AMT is the only component that knows the identity of the worker, the Task Tracker uses AMT (through the Worker Manager) to notify workers' smartphones of SPC stage initiation. This notification is done by using AMT to send an SMS message to the smartphone, which is intercepted by the Stage Tracker. That component initiates stage execution on the phone. When an SPC stage completes, the Stage Tracker directly transfers program states to the Task Tracker without AMT involvement. However, this transfer contains enough identifying information to uniquely identify the task instance.

Stage Name	Description
probedata	Searches for data items matching specified predicates.
uploaddata	Uploads data corresponding to specified input variable.
mediagen	Invokes external program to sense a data item (e.g., video clip or audio clip).
videosummary	Generates a motion jpeg summary from a original high-qualify video data.
facedetect	Extracts faces from videos or image inputs
netstats	Scans Bluetooth and Wifi interfaces. Users can set filters to get selective information.
vcollect	Collects samples directly from a sensor (e.g., from accelerometer or microphone).
vfeature	Extracts feature vectors from input raw sensor data.
gpsrawcollect	Extracts sequence of location samples at specified rate.
combiner	Temporally merges two sensor data streams.
cfeature	Extracts feature vectors from temporally merged streams.

Table 1: Stage Binaries in Stage Library.

Stage Library. Stage Library is a reusable and extensible library of SPC stage binary implementations. For example, the binaries for the `TakeVideo`, `ExtractSummary`, and `UploadVideo` stages (namely `mediagen`, `videosummary` and `uploaddata`) are stored here. These binaries are stored on the cloud, but when a stage is initiated on a smartphone, the Stage Tracker is responsible for downloading the binaries. Stage binaries may be cached on smartphones, and caches are periodically purged so that the latest versions of the stage binaries may be downloaded. Table 1 gives some examples of stages that we have implemented.

The Stage Library ensures Medusa's extensibility. When a requestor needs a specific functionality to be implemented as a stage binary, they may provide this implementation themselves or (more likely) make a feature request to the Medusa system developers. A larger-scale deployment, which is left to future work, will provide us with insights into how often a new stage needs to be added to Stage Library. We anticipate that there will likely be many such additions to the Stage Library initially, but, with time, requestors will be able to reuse stages frequently. To support the Stage Library's extensibility, stages are stored on the cloud and downloaded on demand to the phone.

Before stage binaries are admitted into the Stage Library, they are *vetted* for safety. As we describe below, stages are only allowed to invoke a restricted set of functionality on the smartphone. For example, stages cannot access the underlying file system on the phone or transmit data to arbitrary websites. Medusa ensures this by using static program analysis to determine the set of APIs invoked by stage binaries; it conservatively rejects any stage binary that accesses disallowed APIs or uses dynamic API binding.

In this way, Medusa achieves safety and extensibility of stages that implement sensing, in-network processing, and communication functions.

4.3 Medusa Runtime on the Smartphone

The Medusa runtime on the smartphone consists of the two components described below.

Stage Tracker. As described above, the Stage Tracker receives stage initiation notifications by intercepting SMS messages. Each SMS message contains an XML description of the stage; the Stage Tracker parses this description, downloads the stage binary if necessary, and initiates the execution of the stage. The SMS message also contains the values of named variables instantiated during previous stage computations; recall that these contain pointers to data items stored on the phone. The Stage Tracker is responsible for setting up the namespace (containing these named variables and their values) before the stage executes. It is also responsible for implementing the triggers necessary to initiate stage execution (e.g., to start recording a video clip), as well as the human-mediation required for annotations.

Once a stage's execution is complete, stage state is transferred back to the Stage State Table. As discussed before, this includes only references to data items, not actual data (the one exception is the `upload` stage, which may transfer data but only after the worker explicitly opts in). Task Tracker polls the Stage State Table periodically to determine stage completion.

If a stage execution fails for any reason (e.g., resource constraints or implementation bugs in a stage), the Stage Tracker returns a failure notification to the Task Tracker. If the Task Tracker does not hear back from the Stage Tracker within a timeout period, it tries to contact the Stage Tracker before determining whether the stage has failed or not.

By design, the Stage Tracker does not know (or care) to which instance of which task a given stage belongs. Furthermore, the Stage Tracker can initiate multiple concurrent stages. The Stage Tracker itself is stateless, and transfers all intermediate program state to the Task Tracker which maintains all instance state. If a Stage Tracker fails, it can be transparently restarted: any ongoing stage executions can complete successfully, but a stage that completed before the Stage Tracker was restarted may be marked as having failed and may then be re-tried.

In this manner, the Stage Tracker robustly supports multiple concurrent task instances.

MedBox. On the smartphone, stages executed in a sandboxed environment called MedBox. MedBox exports capabilities necessary for crowd-sensing, and monitors resource usage by stages.

MedBox provides libraries that allow stages to access sensors, or that provide functions to manipulate sensitive data, as well as to transfer sensor data. Moreover, it also provides a library to access restricted storage on the phone where sensor data is stored. Stages can query this storage system using metadata attributes like location and time, allowing them to extract previously generated sensor data. Finally, MedBox supports triggers for stage execution, including triggers based on time or location. For example, a worker may be notified that she is near the location of an "Occupy" site. Notifications of these triggers may be delivered to users using any of the methods available in the underlying operating system. Stages are restricted by cloud-side static analysis to only access these libraries (as described above).

Finally, Medusa allows smartphone owners to specify limits on usage of system components such as CPU, network, or memory on a stage or task instance. MedBox tracks the usage of each stage and aborts the stage if it exceeds its allocated resources. In the future, we envision extending the expressivity of these policies to permit privacy controls (e.g., opt-out for certain kinds of uploads) or reward-related stage-specific limits (e.g., tighter limits for stages where the monetary incentives are lower).

System Components	Requirements
MedScript Interpreter	Timeliness enforcement, robustness
Task Tracker	Supports multiple users, un-synchronized user operation, timeliness enforcement, robustness
Worker Manager	Incentives, crowd-curation, anonymity, security, worker-mediation
Stage Library	In-network processing, sandboxing, extensibility
Stage Tracker	Multiple concurrent stages, robustness, worker-mediation, privacy
MedusaBox	Stateless execution, sandboxing, concurrent stages, resource monitoring, access to stored data

Table 2: Medusa System Components and Properties

Summary. Medusa's various components together satisfy the complex web of crowd-sensing requirements discussed in Section 2. Table 2 summarizes which components satisfy which requirements; a requirement may be satisfied by multiple components.

5. EVALUATION

In this section, we demonstrate four properties of Medusa: the expressivity of MedScript, the ability of the Medusa runtime to support concurrent task instance execution, the scalability of the runtime, and the robustness of Medusa.

Our experiments are conducted on a prototype of Medusa. The MedScript interpreter and Task Tracker are written using Python. Apache and PHP are used for accessing the MySQL based Stage State Table, as well as Stage Library. Worker Manager runs Tomcat with a Java Servlet environment and accesses AMT through a Java SDK[2]. Finally, the Stage Library uses the `dex2jar` utility to convert Dalvik bytecode to Java, and the `javap` program from the Sun JDK for static program analysis.

Our smartphone software runs on Android v2.3.6. Stage Tracker is implemented as an Android service, and intercepts SMS and MMS messages[3], and uses secure-HTTP based connections to the Stage Library and the Task Tracker. MedBox is implemented as a collection of Java threads.

To evaluate our system, we use commodity hardware both for the server and client side. In all our experiments, workers use either the Google Nexus One or the Nexus S. On the server, our cloud-based service is emulated by a Dell PowerEdge 2950, which has a dual-core Intel Xeon E5405 2.00GHz processor and 6MB built-in cache. Finally, we use data service from two GSM carriers in the United States to evaluate our push mechanism via SMS messaging.

5.1 Language Expressivity

To demonstrate the expressivity of MedScript, we have implemented ten qualitatively different crowd-sensing tasks (Table 3). Each task demonstrates different facets of Medusa, ranging from the use of different sensors to the use of different facilities provided by MedScript. Some have been proposed for participatory

[2] `http://mturksdk-java.sourceforge.net/`

[3] Our notifications may be longer than that allowed by SMS. Some carriers break up long initiation XML messages into multiple SMSs, others convert the SMS into an MMS.

Application	LOC*	Sensors	Properties
Video Documentation	86	Camera(Video)	In-network processing, Crowd curation
Collaborative Learning	62	Accelerometer, Audio	Different sensors
Auditioning	86	Camera(Video)	Reverse incentive mechanism, Crowd curation
Forensic Analysis	86	GPS, Audio	Access to stored data
Spot Reporter	45	Camera(Video), Audio	Text/Voice tagging
Road Monitoring	90	Accelerometer	Fork-join construct, App from PRISM [7]
Citizen Journalist	45	GPS	Multiple triggers, App from PRISM [7]
Party Thermometer	62	GPS, Audio	Fork constraint, App from PRISM [7]
WiFi and Bluetooth Scanner	45	Network Sensors	Apps from AnonySense [6]

Table 3: Implemented Crowd-Sensing Apps. (*Line Of Code)

sensing in related work: we have implemented these in Medusa to demonstrate the power of the system.

Novel Crowd-Sensing Tasks. Many of the tasks in Table 3 are novel and are enabled by Medusa.

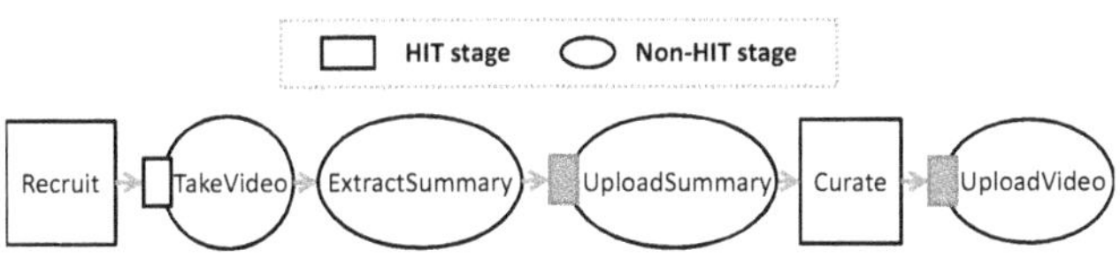

Figure 3: Video Documentation

Video Documentation. Although we have explained the code for the video documentation task in Listing 1, we discuss a few of the subtleties of this task here. First, for this and for other tasks discussed below, we illustrate the task program as a control flow diagram shown in Figure 3 (lack of space precludes a full program listing for the other tasks discussed below). HIT stages are represented by squares and SPC stages by ovals. Only the success connectors are shown, and the failure connectors are limited for brevity.

This task demonstrates some of the Medusa's key features: in-network processing for extracting video summaries, crowd-curation, and opt-in data transfers (both when summaries are uploaded as well as when full videos are uploaded). Some stages are worker-initiated (e.g. `TakeVideo`), while others are automatically triggered when the previous stage completes (e.g., `ExtractSummary`). Worker-initiated stages have an unfilled rectangle before them, automatically triggered stages do not. A user may take multiple videos, so that multiple summaries can be uploaded for curation. Each upload stage requires worker opt-in (denoted by the filled rectangle before the stage). Only a subset of these summaries may be selected; the `Curate` stage returns a bitmask, which is used to determine the set of full videos that need to be uploaded. Finally, this task has multiple data upload stages: MedScript does not restrict uploads to the end of the control flow.

Collaborative Learning. The Collaborative Learning (Figure 4) task allows, for example, software vendors to solicit training samples for constructing robust machine learning classifiers (Section 2). Compared with video documentation, this task uses a different set of sensors (either an accelerometer or a microphone) and a different in-network processing algorithm (feature extraction). This task requires *annotation*, denoted by an unfilled rectangle *after* the `GetRawData` stage: workers are required to label the collected samples (e.g., as a sample of "walking", or "running" etc.). If the worker collects multiple samples, she has to label each sample.

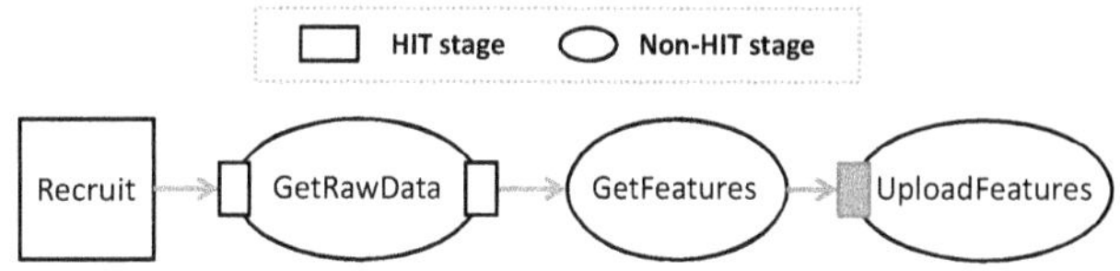

Figure 4: Collaborative Learning

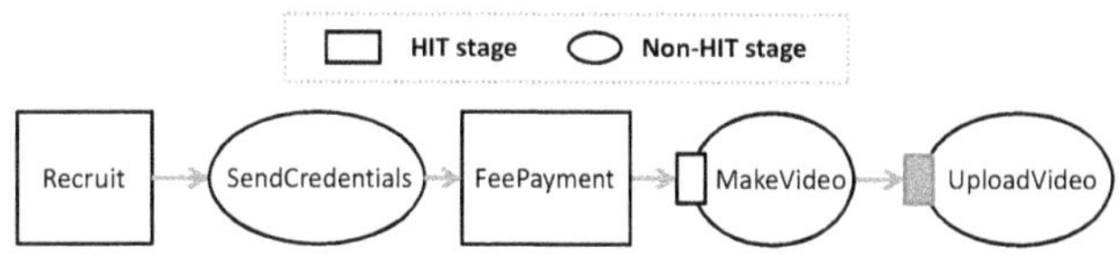

Figure 5: Auditioning (Full listing in Appendix A.)

Auditioning. Our auditioning task allows, for example, TV show producers to solicit videos from aspiring actors (Figure 5). The novel feature of this task is the support for reverse incentives: workers are required to *pay* the requestor for the privilege of submitting a video for auditioning. Reverse incentives are implemented by two stages. First, `SendCredentials` is an SPC stage that sends the worker's AMT credentials (stored on the phone) to the Medusa cloud runtime. This step is necessary to set up the payment, which is implemented as a HIT stage `FeePayment`. This stage is conceptually similar to `Recruit`, except that the Medusa runtime sets up AMT as explained in Section 4. Once the requestor accepts and completes this task, payment is transferred; the worker can subsequently take a video and upload it. Uploaded videos are sent to the requestor (the TV producers in our example); we assume that selected participants are notified out-of-band (e.g., invited to an on-site auditioning).

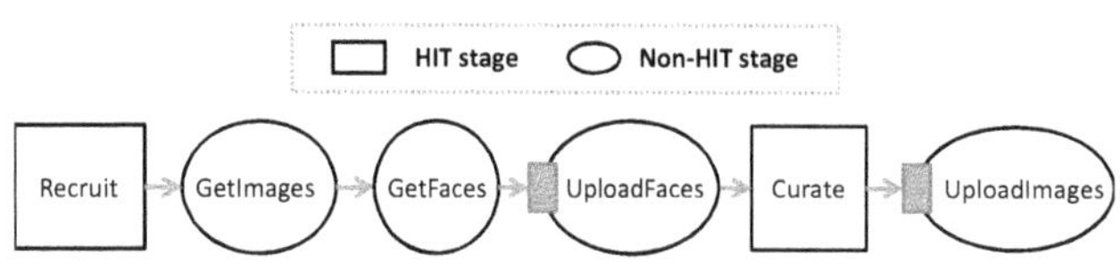

Figure 6: Forensic Analysis

Forensic Analysis. Our forensic analysis task permits security officials to obtain images of people who were at a specified location at a specified time when an incident occurred (Figure 6). The novel aspect of this task is access to historical sensor data stored on the smartphone. This access is accomplished through the `GetImages` stage, which, given a spatio-temporal predicate, obtains images whose meta-data matches the specified predicate. The subsequent `GetFaces` stage extracts faces from each image. Only the extracted faces are uploaded for curation. A security officer might visually inspect the uploaded faces to select faces corresponding to potential suspects. The images corresponding to the selected faces are then uploaded.

Previously-Proposed Sensing Applications. Prior work has explored several participatory sensing applications, and Medusa can be used to program these. We have implemented these applications as crowd-sensing tasks, each of which begins with a `Recruit` stage using which workers can sign up for the task.

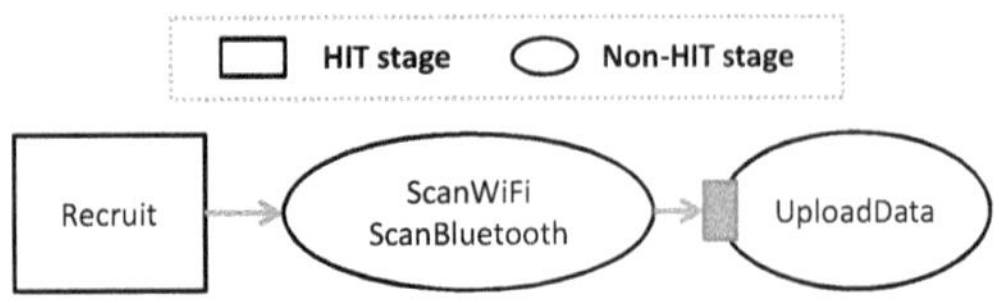

Figure 7: WiFi/Bluetooth Scanner

WiFi/Bluetooth Scanner [6] enables requestors to collect Wifi and Bluetooth neighborhoods from workers. This is implemented using a `netstats` stage, which models this network neighborhood information as outputs of a sensor. Requestors can use other stages to preprocess (e.g., filter) the data before upload.

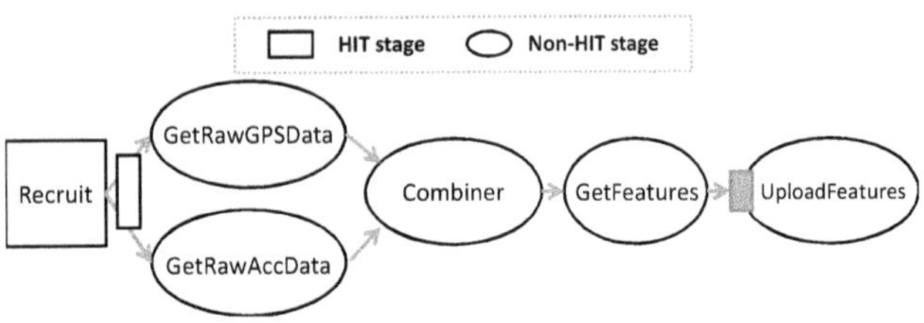

Figure 8: Road-Bump Monitoring

Road-Bump Monitoring [15], can be implemented using the task shown in Figure 8. This task collects GPS and accelerometer data, correlates data from the two sensors, and extracts features that are used to determine the location of bumps on roads. Its novelty is the use of the *fork-join* construct, which allows independent stages to concurrently sample GPS and the accelerometer.

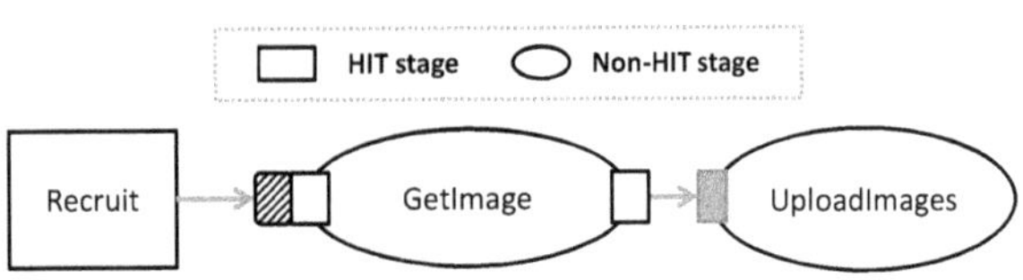

Figure 9: Citizen Journalist (Full listing in Appendix A.)

Citizen Journalist discussed in [10], asks workers to take an image at a specified location and label the image before uploading. The corresponding crowd-sensing task, shown in Figure 9, is unique in that one of the stages supports *two* triggers for initiating the stage: a location-based trigger which notifies the worker (shown as a hashed rectangle), who is then expected to initiate the taking of a photo.

Finally, we have also implemented a crowd-sensing task for the Party Thermometer application [7]. This task uses a location trigger to autonomously collect ambient sound, generate feature vectors, and upload them (Figure 10).

Quantifying Expressivity. Our Medusa task descriptions are very compact (Table 3): no task exceeds 100 lines. We also use code size to compare the complexity of three crowd-sensing tasks with standalone implementations of these tasks. Our comparison is only approximate because the implementations being compared are not precisely functionally equivalent.

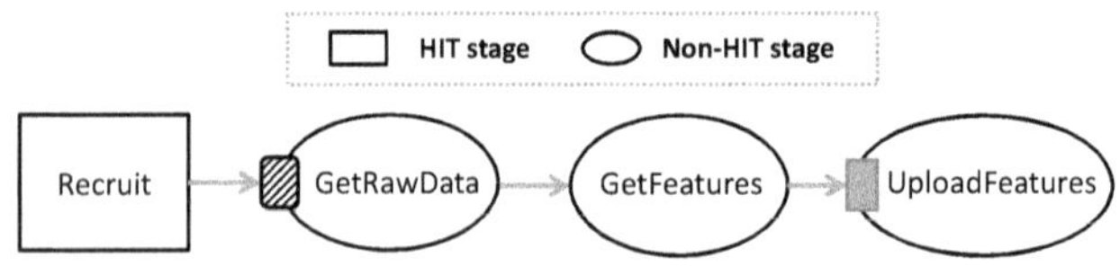

Figure 10: Party Thermometer

	Medusa	Standalone
Video Documentation	86	8,763
Collaborative Learning	62	7,076
Spot Reporter	45	17,238

Table 4: Line of Code Comparison

Our work on Medusa has been inspired by our earlier standalone prototypes for video documentation and collaborative learning. The earlier prototypes did not include support for worker-mediation or incentives but had additional user interface elements for software configuration and management. In addition, a research group with whom we collaborate has implemented a spot reporter application, which tasks human agents to send audio, video, or text reports from the field. Our implementation of this task requires three stages, one to recruit agents, another to take the video (together with annotation) and the third to upload the result.

Table 4 quantifies the difference in code complexity using lines of code as a metric. Even given the differences in implementations, Medusa versions are about two orders of magnitude more compact than the standalone versions! Medusa's runtime implements many functions common to these standalone applications, thereby reducing the programmer's burden.

5.2 Concurrent Task Instances

The dynamics of task execution in Medusa can be quite complex, since computational and sensing stages can be interleaved with worker mediation. Thus, execution dynamics can be strongly influenced by human actions. At the same time, Medusa task throughput can be enhanced by its support for concurrent task instance execution.

To illustrate some of these properties of Medusa, we conducted an experiment where we instantiated one task instance for each of the ten prototype applications discussed above. Then, we asked four volunteers to sign up for a subset of the task instances. Each volunteer was given a phone and assigned an AMT worker account, but were otherwise unconstrained in how and when to complete the tasks.

Figure 11 depicts a timeline of the progress of the experiment. All task instances were completed within about 30 min. by the four workers. The upper panel shows the completion time of the task instances: these range from 6 min. to 30 min., with most instances completing within about 10 mins. The lone outlier is Citizen Journalist, which was location-triggered and could only be initiated when the worker had reached the specified location.

The timeline also depicts several interesting features of Medusa. Task instances can have differing numbers of SPC stages, indicated by the vertical bars along each timeline. Workers can concurrently sign up for multiple task instances; for example, worker I executes three task instances simultaneously. Workers sign up for tasks at

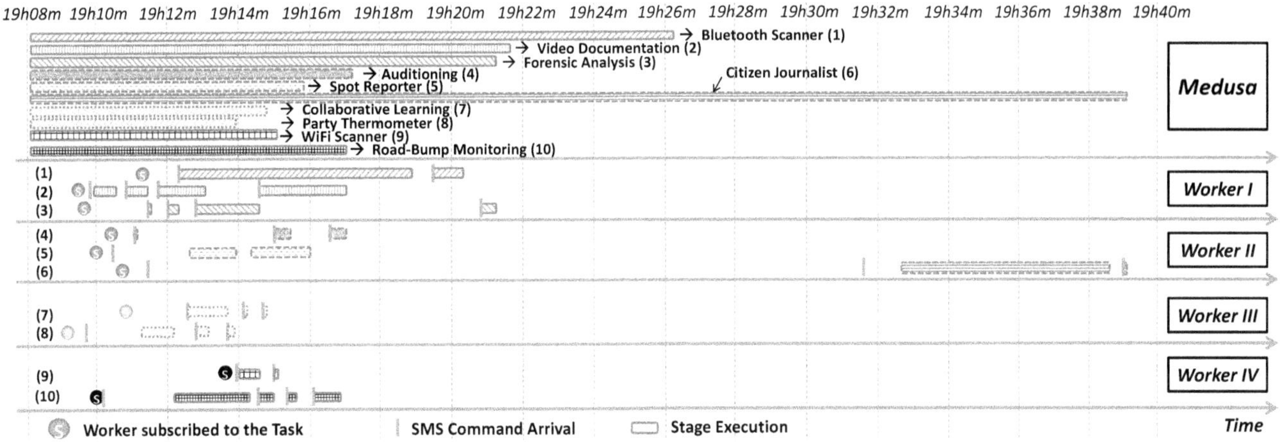

Figure 11: Concurrent Execution of Multiple MedScript Programs.

Component	Avg	Max	Min
Task Interpretation and Type Checking	26.64	26.89	26.45
Task Tracker Initiation Delay	0.84	0.90	0.77
Task Tracker latency amonst stages	0.87	0.94	0.83
[HIT] Delay upon the task registration request	4.83	12.03	3.61
[HIT] Waiting time for the registration response	2177.99	2824.52	1858.45
[HIT] Task execution delay	31039.83	50060.72	20028.91
[SPC] Delay on the request to AMT for commanding workers	1.30	2.56	0.77
[SPC] Messaging request confirmation from AMT	732.5	833.54	700.61
[SPC] Delay on SMS/MMS command message delivery	27000	78000	17000
[SPC] Task execution delay	3038.54	7039.54	1029.20
Total processing time	34.47	43.31	32.42
Total waiting time	63988.86	138758.3	40617.16
Total execution time	64023.33	138801.6	40649.58

Table 5: Delay break-down on the server (Unit: msec)

different times (between 1 min and 6 min into the experiment). Moreover, stage execution times for different tasks vary widely. For example, Video Documentation and Bluetooth Scanner have much longer stages than Spot Reporter. For some stages, stage execution commences immediately after Stage Tracker receives a command from Task Tracker; for others, waiting for worker initiation delays stage execution. Medusa seamlessly handles all these execution dynamics in a manner transparent to the task requestor.

5.3 Scalability and Overhead

To quantify the scalability and overhead, we instrumented our prototype to measure the time taken to perform several individual steps involved in task instance execution, both on the cloud runtime and the phone runtime. We then executed a single task instance on a Nexus One phone, where the task consisted of two stages: a `Recruit` (which requires human-mediation) followed by a null SPC stage (i.e., one that returns immediately and requires no worker-mediation). We present results from 10 trials.

Medusa Server Overhead. Table 5 shows server-side delays. Task

Component	Avg	Max	Min
Retrieve and parse SMS/MMS command message	38.2	58	26
Stage binary initiation time	402.7	506	374
Stage runner latency	6.2	14	3
Stage termination latency	12.7	14	3
Total overhead imposed	459.8	598	409

Table 6: Delay break-down on the phone (Unit: msec)

interpretation and type checking together take 26.64ms on average, Task Tracker initialization takes much less, 0.84ms. For the `Recruit` stage, a HIT request requires only 4.83ms; however, the time to register the task with AMT and receive a response requires 2.18s. To run an SPC stage, Task Tracker must prepare an XML command and send it to Worker Manager so that it can send a request for AMT to send SMS/MMS messages to the phone. Those two steps together incur 732.5ms of delay.

The largest component of delay in our system is the time that it takes to deliver an SMS message to the phone: between 20 to 50 *seconds*. This is because our implementation uses email-to-SMS gateway-ing; a more optimized implementation might use direct SMS/MMS delivery, reducing latency. That said, latency of task execution is not a significant concern for the kinds of tasks we consider: with humans in the loop, we anticipate crowd-sensing tasks to have deadlines on the order of hours from initiation. Indeed, the second largest component of delay in our system is the time that it takes for a human to sign up for the task.

The SPC stage execution time, measured from the perspective of the cloud runtime as the time from when the Task Tracker received confirmation message delivery from Worker Manager, to the time stage detection is completed is about 3 seconds. A large component of this latency is network delay, but another contributor is the polling period in the Task Tracker. Recall that Task Tracker polls the Stage State Table to determine if a state has completed. This polling time adds to the latency; in the future, we plan to alter our system so that Stage Tracker directly notifies Task Tracker of stage completion, which then saves state in the table, eliminating this latency.

Overall, we are encouraged by these results. The dominant delays in the system arise from notification using SMS and from waiting for human input. Actual processing overheads are only 34.47ms on average, which permit a throughput of 1740 task instances per minute on a single server. Moreover, this component is highly par-

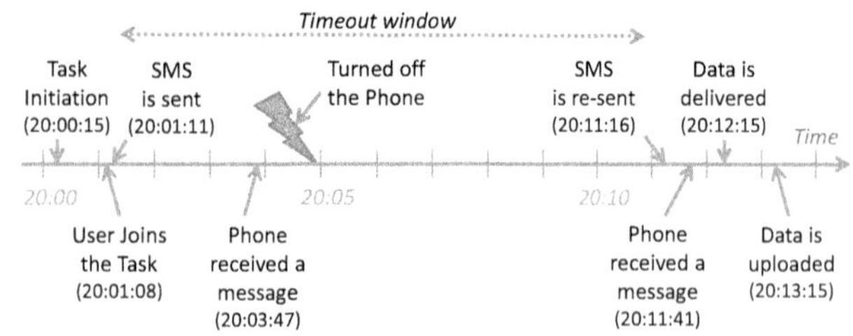

Figure 12: Failure recovery: turn off the phone in the middle.

allelizable, since each task instance is relatively independent and can be assigned a separate Task Tracker. Thus, if task throughput ever becomes an issue, it would be possible to leverage the cloud's elastic computing resources to scale the system.

Medusa Runtime Overhead on the Phone. Table 6 presents the break-down of delay on the Medusa runtime on a Nexus One phone. Retrieving a stage initiation message from Task Tracker involves merging multiple SMS messages and parsing the resulting XML data object. This takes 38.2ms of delay on average. The major component of delay on the phone is the time to instantiate the stage implementation binary (in our experiments, the stage implementation was cached on the phone), a step which takes 402.7ms. This overhead is imposed by Android; our stage implementations are Android package files and the Android runtime must unpack this and load the class files into the virtual machine. Stage Tracker then performs some bookkeeping functions before invoking the stage, which takes 6.2ms. When the stage finishes its execution, Stage Tracker performs additional bookkeeping before notifying the cloud runtime, and this takes 12.7ms. The total delay imposed by Medusa platform is only 459.8ms per stage, of which the dominant component is the stage binary initialization. On the timescale of expected task completion (e.g., several 10s of minutes in our experiments above), this represents negligible overhead. Furthermore, we are aware of methods such as binary alignment which will allow us to optimize stage loading, but have left it for future work.

5.4 Robustness

Finally, we briefly demonstrate Medusa's robustness.

Failure Recovery. The Medusa runtime maintains internal timeouts to recover from transient errors caused by messaging failure, battery exhaustion, or user actions (e.g., accidentally turning off the phone). To demonstrate failure recovery, we conducted an experiment where we turned off the phone during the scanning stage of the WiFi scanner task. Then we turned on the phone again after a couple of minutes. Figure 12 shows the sequence of events. Medusa waits for a pre-defined timeout (10 minutes in our implementation) before restarting the scanner stage. This stage then completes execution, eventually resulting in task completion. This simple retry mechanism works well across any transient failure mainly because stage execution state is maintained on the cloud.

Static Analysis on Stage Binary. To demonstrate the efficacy of using static analysis for sandboxing, we recruited three volunteers (none are authors of the paper) to add malicious code to existing stage implementations. We then tested whether our analysis was able to deter such attacks, which included opening HTTP connections, invoking sleep operations, accessing the SD card, and so forth. Of the nine modifications made by the volunteers, our static analyzer caught 7 correctly (Table 7). The two attacks that it did not catch targeted resource usage directly: infinite recursion,

Code Modification	Result
Make one function sleep for a long time	REJECT
Write a file to SD card	REJECT
Time string format translation	REJECT
Open HTTP connection	REJECT
Delete all files in SD card	REJECT
Vector operation	ACCEPT
Throws exceptions	REJECT
System.exit(-1)	REJECT
Recursive calls	ACCEPT
Fills the heap until memory is exhausted	ACCEPT

Table 7: The defense capability of static analyzer against arbitrary code modification.

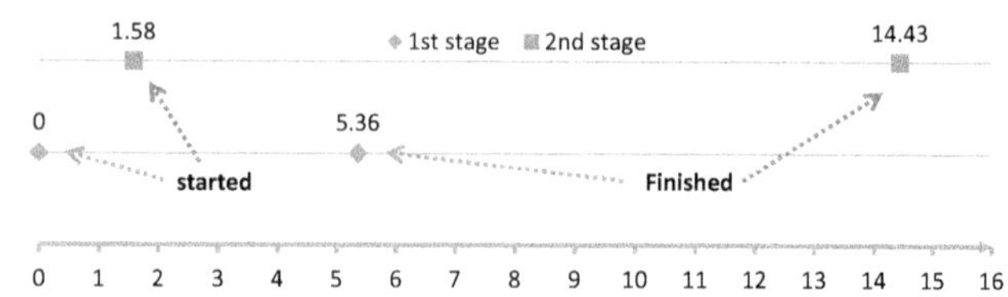

Figure 13: Robustness: forcing a limit on computation.

and attempts to exhaust the heap. These attacks require dynamic tracking: Medusa also has a dynamic resource tracking component, which we demonstrate next.

Limiting Resource Usage. Medusa allows smartphone owners to place limits on the computation used by a single stage (on a per-stage basis) and the amount of data transferred per task instance. Using these controls, users can implement policies which, for example, apportion more resources to tasks with higher monetary reward.

To illustrate how these limits are enforced, we created a single long-running stage, and concurrently ran two instances of that stage with limits of 5 and 10 seconds respectively. MedBox maintains a watchdog timer, which, in our implementation, checks stage resource usage every three seconds. As shown in Figure 13, it successfully terminated our stages at 5.36 seconds and 12.85 seconds; given the granularity of our timer, stages may get bounded additional resources beyond their limits. This kind of mechanism can protect against CPU resource exhaustion attacks.

Medusa can also protect against excessive network usage per task instance. Since all data transfers must use MedBox's data transfer library, that library can account for data transfers on a per task basis. We conducted an experiment in which a task instance attempted to upload 60 image files and was allocated a 3MB limit: MedBox successfully terminated the task (by notifying the Task Tracker) just before it reached the the 3MB limit (the transfer library transfers data in 200KB chunks, and checks whether the limit would be exceeded before transferring data) as depicted in Figure 14.

6. RELATED WORK

To our knowledge, no other prior work describes a programming language and runtime system for crowd sensing. Crowd-sensing, as we have defined it, combines two strands of research: participatory sensing and crowd sourcing. We now describe how these strands relate to Medusa.

Participatory sensing systems like PEIR [16] and SoundSense

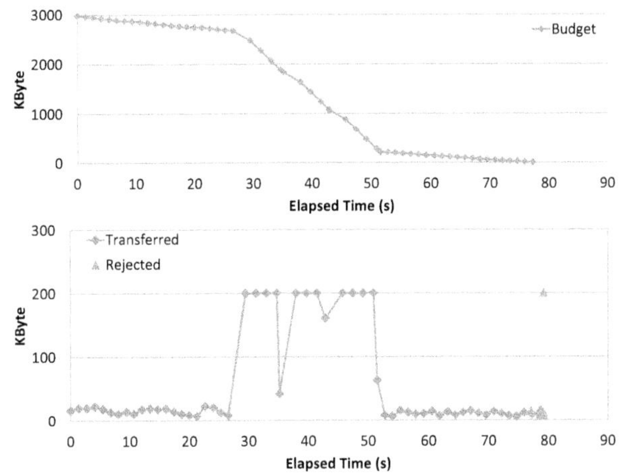

Figure 14: Failure recovery: harnessing network data usage.

[13] do not incorporate human-mediation or incentives into sensing. Moreover, they do not support programmable data collection from multiple sensors or an in-network processing library. Some pieces of work in this area, however, share some goals with Medusa. Campaignr [2] is an early effort on programming sensor data collection on a single phone for participatory sensing and uses an XML description language to specify data collection tasks and parameters.

AnonySense [6] is a privacy-aware tasking system for sensor data collection and in-network processing, while PRISM [7] proposes a procedural programming language for collecting sensor data from a large number of mobile phones. More recently, Ravindranath et al. [18] have explored tasking smartphones crowds and provide complex data processing primitives and profile-based compile time partitioning. Unlike these systems, Medusa incorporates incentives and reverse incentives, human-mediation, and support for curation into its programming framework.

Several systems have integrated worker-mediation into their workflow. Most of the systems have been inspired by, or directly employ, Amazon Mechanical Turk (AMT [1]). CrowdDB [9] extends relational databases to support human-mediation in SQL query workflows. CrowdSearch [22] exploits AMT to crowd-source video search. TurKit [12] enhances the original programming model of AMT by allowing repeated execution of existing tasks, and enhances the robustness of the AMT programming model. Additionally, several studies have explored who uses AMT and in what ways [11, 5]. Medusa is qualitatively different from this body of work in that it enables users to contribute processed sensor data, a capability that enables many interesting applications as described above.

Complementary to our work is PhoneGap [3], a system that enables developers to program mobile applications using HTML5 and JavaScript, then automatically generates platform-specific native binaries. Thus, PhoneGap focuses on single source multi-platform code development, and not on crowd-sensing.

Finally, many of our applications have been inspired by prior work. Conceptual descriptions were provided in [17] for video documentation and in [14, 4] for collaborative learning. However, these papers did not discuss incentives or worker-mediation. Road-bump monitoring was described in [15], and citizen journalist in [10], but the focus of research in these papers (sensing algorithms, energy, location discovery) was very different from ours.

7. CONCLUSIONS

Medusa is a programming system for crowd-sensing. The MedScript programming language provides high-level abstractions for stages in crowd-sensing tasks, and for control flow through the stages. Medusa supports specifying various forms of worker mediation in the sensing workflow. A runtime, partitioned between the cloud and smartphones, achieves a complex web of requirements ranging from support for incentives to user-specified controls on smartphone resource usage. Medusa task descriptions are concise, and the runtime system has low overhead.

Finally, our work suggests several future directions, including: gaining deployment experience with more crowd-sensing tasks; reducing the overhead of the Medusa runtime and the latency of its SMS-based notification; replacing the XML-based specification language with a visual paradigm using OpenBlocks [19] and AppInventor [21]; using Medusa a distributed programming framework that crowd-sources computations to mobile devices.

Acknowledgements

We would like to thank our shepherd, Landon Cox, and the anonymous referees, for their insightful suggestions for improving the technical content and presentation of the paper.

8. REFERENCES

[1] Amazon mechanical turk, `https://www.mturk.com/`.
[2] Campaignr:configurable mobile sensing, `http://urban.cens.ucla.edu/technology/campaignr/`.
[3] Phonegap, `http://phonegap.com/`.
[4] X. Bao and R. R. Choudhury. Movi: mobile phone based video highlights via collaborative sensing. In *Proc. ACM MOBISYS*, 2010.
[5] L. Chilton, J. Horton, R. Miller, and S. Azenkot. Task search in a human computation market. In *Proc. HCOMP*, 2010.
[6] C. Cornelius, A. Kapadia, D. Kotz, D. Peebles, M. Shin, and N. Triandopoulos. Anonysense: Privacyaware people-centric sensing. In *Proc. ACM MOBISYS*, 2008.
[7] T. Das, P. Mohan, V. N. Padmanabhan, R. Ramjee, and A. Sharma. Prism: Platform for remote sensing using smartphones. In *Proc. ACM MOBISYS*, 2010.
[8] C. Dwork. Differential privacy. In *ICALP*, pages 1–12. Springer, 2006.
[9] M. J. Franklin, D. Kossmann, T. Kraska, S. Ramesh, and R. Xin. CrowdDB: answering queries with crowdsourcing. In *Proc. ACM SIGMOD*, 2011.
[10] S. Gaonkar, J. Li, R. R. Choudhury, L. Cox, and A. Schmidt. "Micro-Blog: Sharing and Querying Content Through Mobile Phones and Social Participation". In *Proc. ACM MOBISYS*, 2008.
[11] G. Little, L. Chilton, M. Goldman, and R. Miller. Exploring iterative and parallel human computation processes. In *Proc. HCOMP*, 2010.
[12] G. Little, L. Chilton, M. Goldman, and R. Miller. Turkit: Human computation algorithms on mechanical turk. In *Proc. ACM UIST*, 2010.
[13] H. Lu, W. Pan, N. D. Lane, T. Choudhury, and A. Campbell. Soundsense: scalable sound sensing for people-centric applications on mobile phones. In *Proc. ACM MOBISYS*, 2009.
[14] E. Miluzzo, C. Cornelius, A. Ramaswamy, T. Choudhury, Z. Liu, and A. Campbell. Darwin phones: the evolution of

sensing and inference on mobile phones. In *Proc. ACM MOBISYS*, 2010.

[15] P. Mohan, V. N. Padmanabhan, and R. Ramjee. "Nericell: rich monitoring of road and traffic conditions using mobile smartphones". In *Proc. ACM SENSYS*, 2008.

[16] M. Mun, S. Reddy, K. Shilton, N. Yau, J. Burke, D. Estrin, M. Hansen, E. Howard, R. West, and P. Boda. Peir, the personal environmental impact report, as a platform for participatory sensing systems research. In *Proc. ACM MOBISYS*, 2009.

[17] M.-R. Ra, J. Paek, A. B. Sharma, R. Govindan, M. H. Krieger, and M. J. Neely. Energy-delay tradeoffs in smartphone applications. In *Proc. ACM MOBISYS*, 2010.

[18] L. S. Ravindranath, A. Thiagarajan, H. Balakrishnan, and S. Madden. Code In The Air: Simplifying Sensing and Coordination Tasks on Smartphones. In *Proc. HotMobile*, 2012.

[19] R. V. Roque. OpenBlocks: an extendable framework for graphical block programming systems. Master's thesis, Massachusetts Institute of Technology. Dept. of Electrical Engineering and Computer Science., 2007.

[20] W. H. Whyte. The Social Life of Small Urban Spaces. Project for Urban Spaces, 1980.

[21] D. Wolber. App inventor and real-world motivation. In *Proc. ACM SIGCSE*, 2011.

[22] T. Yan, V. Kumar, and D. Ganesan. Crowdsearch: Exploiting crowds for accurate real-time image search on mobile phones. In *Proc. ACM MOBISYS*, 2010.

APPENDIX

A. MORE MEDSCRIPT PROGRAMS

A.1 Citizen Journalist App

```
<xml>
  <app>
    <name>Citizen-Journalst </name>
    <rrid >[User's Requestor ID]</rrid >
    <rrkey >[User's Requestor Key]</rrkey >

    <stage >
      <name>Hiring </name> <type >HIT</type >
      <binary >recruit </binary >
      <config >
        <stmt>Citizen Journalist Demonstration </stmt>
        <expiration >21:00:00 12/16/2011</expiration >
        <reward >.05 </reward >
        <output >W_WID</output >
      </config >
    </stage >
    <stage >
      <name>TakePicture </name> <type >SPC</type >
      <binary >mediagen </binary >
      <trigger >location=34.252339|-118.277907|40, user-initiated </trigger >
      <review >textdesc </review >
      <config >
        <params>-t image </params>
        <output >IMAGE</output >
      </config >
    </stage >
    <stage >
      <name>UploadData </name> <type >SPC</type >
      <binary >uploaddata </binary >
      <trigger >none </trigger >
      <config >
        <input >IMAGE</input >
      </config >
    </stage >

    <connector >
      <src >Hiring </src >
      <dst> <success >TakePicture </success > <failure >Hiring </failure > </dst>
    </connector >
    <connector >
      <src >TakePicture </src >
      <dst> <success >UploadData </success > <failure >Hiring </failure > </dst>
    </connector >
  </app>
</xml>
```

Listing 2: Citizen Journalist

A.2 Auditioning App

```
<xml>
  <app>
    <name>Auditioning </name>
    <rrid >[User's Requestor ID]</rrid >
    <rrkey >[User's Requestor Key]</rrkey >
    <rwid >[User's Worker ID]</rwid>
    <deadline >21:00:00 12/16/2011</deadline >

    <stage >
      <name>Recruit </name> <type >HIT</type >
      <binary >recruit </binary >
      <config >
        <stmt>Recruiting for Audition App.</stmt>
        <expiration >21:00:00 12/16/2011</expiration >
        <reward >.05</reward >
        <output >W_WID</output >
      </config >
    </stage >
    <stage >
      <name>SyncENV </name> <type >SPC</type >
      <binary >helloworld </binary >
      <trigger >none </trigger >
    </stage >
    <stage >
      <name>FeePayment </name> <type >HIT</type >
      <rid >W_RID</rid > <rkey >W_RKEY</rkey > <wid>R_WID</wid>
      <binary >recruit </binary >
      <config >
        <stmt>Audition Fee Payment.(100 dollars)</stmt>
        <expiration >21:00:00 12/16/2011</expiration >
        <reward >100</reward >
      </config >
    </stage >
    <stage >
      <name>MakeVideo </name> <type >SPC</type >
      <binary >mediagen </binary >
      <trigger >user-initiated </trigger >
      <config >
        <params>-t video </params>
        <output >VIDEO</output >
      </config >
    </stage >
    <stage >
      <name>UploadVideo </name> <type >SPC</type >
      <binary >uploaddata </binary >
      <trigger >user-initiated </trigger > <review >yesno </review >
      <config >
        <input >VIDEO</input >
      </config >
    </stage >
    <stage >
      <name>Evaluation </name> <type >HIT</type >
      <rid >W_RID</rid > <rkey >W_RKEY</rkey > <wid>R_WID</wid>
      <binary >vote </binary >
      <config >
        <stmt>Evaluation Press. If you like the video, press Yes. Otherwise No
            .</stmt>
        <expiration >21:00:00 12/16/2011</expiration >
        <reward >.01</reward >
        <numusers >1</numusers >
        <input >VIDEO</input >
        <output >RESULT</output >
      </config >
    </stage >

    <connector >
      <src >Recruit </src >
      <dst >
        <success >SyncENV </success >
        <failure >Recruit </failure >
      </dst >
    </connector >
    <connector >
      <src >SyncENV </src >
      <dst >
        <success >FeePayment </success >
        <failure >Recruit </failure >
      </dst >
    </connector >
    <connector >
      <src >FeePayment </src >
      <dst >
        <success >MakeVideo </success >
        <failure >Recruit </failure >
      </dst >
    </connector >
    <connector >
      <src >MakeVideo </src >
      <dst >
        <success >UploadVideo </success >
        <failure >UploadVideo </failure >
      </dst >
    </connector >
    <connector >
      <src >UploadVideo </src >
      <dst >
        <success >Evaluation </success >
        <failure >Recruit </failure >
      </dst >
    </connector >
  </app>
</xml>
```

Listing 3: Auditioning

Open Data Kit Sensors: A Sensor Integration Framework for Android at the Application-Level

Waylon Brunette, Rita Sodt, Rohit Chaudhri, Mayank Goel,
Michael Falcone, Jaylen VanOrden, Gaetano Borriello
Department of Computer Science & Engineering
University of Washington
Box 352350
Seattle, WA 98195-2350
{wrb, rsodt, rohitc, mayank, mfalcone, dutchsct, gaetano}@cse.uw.edu

ABSTRACT

Smartphones can now connect to a variety of external sensors over wired and wireless channels. However, ensuring proper device interaction can be burdensome, especially when a single application needs to integrate with a number of sensors using different communication channels and data formats. This paper presents a framework to simplify the interface between a variety of external sensors and consumer Android devices. The framework simplifies both application and driver development with abstractions that separate responsibilities between the user application, sensor framework, and device driver. These abstractions facilitate a componentized framework that allows developers to focus on writing minimal pieces of sensor-specific code enabling an ecosystem of reusable sensor drivers. The paper explores three alternative architectures for application-level drivers to understand trade-offs in performance, device portability, simplicity, and deployment ease. We explore these tradeoffs in the context of four sensing applications designed to support our work in the developing world. They highlight a range of sensor usage models for our application-level driver framework that vary data types, configuration methods, communication channels, and sampling rates to demonstrate the framework's effectiveness.

Categories and Subject Descriptors

D.2.11 Software Architectures

General Terms

Design, Experimentation, Performance

Keywords

Mobile computing, drivers, smartphones, ICTD, sensing, Bluetooth, USB, Open Data Kit.

1. INTRODUCTION

Market penetration of smartphones as a computing and communications platform has increased significantly in recent years. Basic feature phones are gradually being replaced by relatively inexpensive smartphones in developing countries. For example, in Kenya the Android based Huawei Ideos is sold for approximately USD 80[27]. Researchers and practitioners in the information and communication technologies for development (ICTD) community are increasingly leveraging smartphones to improve information management in under-resourced environments. Our work is motivated by the platform shift from traditional PCs and standalone sensing appliances to mobile devices (*e.g.*, smartphones, tablets) coupled with cloud services to create mobile information systems. There is an unprecedented opportunity to integrate consumer mobile devices with external sensors enabling the collection of data directly on these devices. However, unlike traditional personal computing devices, the new consumer devices are locked by service providers or manufacturers, and most end-users do not have the administrative rights, technical ability, or organizational capacity to modify or customize the operating system. As a result, relying on conventional in-kernel device driver frameworks to integrate external sensors with consumer smartphones is impractical. Our project explores ways to package software so that non-technical users can access external sensors from a locked mobile device running a stock version of the Android operating system. The framework assumes the consumer device is 'locked down' and an end-user only has the skills to install applications from a standard app marketplace such as Google Play (Google's Android app store).

The ICTD community has begun investigating using phone-based sensing to perform in-situ and remote monitoring [3, 5, 9]. Even though capturing sensor data directly eliminates many of the errors that plague traditional data collection techniques, such as manual form-filling, it is still not widely used in developing regions because of the high level of technical expertise required to develop a mobile sensing application. The technical challenges include managing the details of different physical communication channels, processing sensor-specific data, developing a user interface and designing application control logic. Unfortunately, this level of expertise is usually not readily available in developing regions or even in developed regions on projects undertaken by resource-limited organizations (such as non-profits and community groups). Because of these complications, we hypothesize that including sensors in mobile data collection poses several technical barriers that, if reduced, would enable more applications to leverage sensors for data collection across varied domains. The Open Data Kit (ODK) Sensors framework aims to lower these barriers by simplifying the deployment of smartphone applications that use external sensors. More specifically, the goal of this work is four-fold:

1. Create a modular framework for adding new sensors by abstracting away management of discovery, communication channels, and data buffers. Integrating a new sensor should require adding only its data handling and configuration primitives.
2. Provide a high-degree of isolation between applications and sensor-specific code. Applications should continue to function even if sensor-specific code is buggy or a sensor becomes inoperative.
3. Understand the tradeoffs of several architectural approaches, especially modularity and performance.
4. Facilitate the integration of new sensors into applications by making it possible to download new sensor capabilities from an application market rather than requiring modifications to the OS configuration.

The ODK Sensors framework provides a single sensing interface for both built-in and external sensors. Having a single interface is appropriate for lightly trained technical workers because it hides a large number of the details involved in developing sensing applications. The framework also provides a simple, high-performing, and flexible abstraction on which to develop and deploy user-level device drivers on Android. While a device driver abstraction is a standard concept, the framework includes features that make development of device drivers easier by handling sensor state (*e.g.,* connection, buffered data, threading) and only requiring driver developers to implement sensor-specific commands and data processing. To evaluate and demonstrate the efficacy of the framework we implemented four applications that are exemplars of different classes of sensor data collection. Three of these applications were previously deployed in developing regions and were ported to the framework leading to significant code simplifications. We discuss these applications and how they leverage the ODK Sensors framework in detail in our previous work [4]. Here we use the applications to demonstrate the benefits of the ODK Sensors framework and compare their minimum performance requirements to the framework's throughput.

This paper examines the architectural implications of three alternative framework architectures that utilize different inter-process communication mechanisms. By comparing peak throughput across the: three framework versions, communication channels, and applications, we show that framework throughput is not the limiting factor of the sensing system. Therefore, our design choices are biased towards making it easier to create mobile sensing applications by focusing on how programming and deployment barriers can be reduced rather than on the relatively small differences in performance.

2. ODK and ODK SENSORS

Open Data Kit (ODK) [11] is a successful suite of mobile tools that exploit the rich interaction and high-performance computing capabilities of smartphones to improve information collection, distribution, and decision-support. ODK focuses on deployment contexts where conventional computing solutions (*i.e.,* informed by concerns of the developed world) are often inappropriate due to constraints such as affordability, infrastructure, institutional capacity, and technical support. ODK Sensors expands ODK by creating a framework to ease the augmentation of a mobile consumer device with sensing capabilities. The ODK Sensors framework supports a variety of external sensors that vary by the type of data they collect, the communication channel over which they interact with the smartphone, and the rate at which they generate data. It provides a unified interface for sensing on Android devices by combining both built-in and external sensors into a single interface. While this design maximizes the variety of sensors available through a uniform interface, the gains in ease of development are more significant for external sensors as more programming is required to interface these as compared to built-in sensors. ODK Sensors focuses on ease-of-use, in general, with a particular focus on appropriateness to our target contexts.

The framework reduces the complexity of building sensor-based mobile applications by providing abstractions that encapsulate communication channels in addition to delineating user-application functionality from sensor communication. The framework has three constituencies: Application Users, Application Developers, and Sensor Driver Developers. A typical Application User is assumed to be the least technically proficient of the three and is only expected to be able to use applications on an Android device. An Application Developer is expected to know how to create new Android applications (design UIs and implement application domain logic), but is not expected to have detailed knowledge of the specifics of sensor control or how the sensors represent and communicate their data. A Sensor Driver Developer is the only constituent expected to understand the low-level protocol used by a specific sensor for configuration and data packaging, but is not expected to deal with communication channel setup or multiplexing. The delineation of application logic from framework logic leads to a clean separation of developer roles and allows an application developer to focus on higher-level application specific concepts while a driver developer focuses on creating sensor-specific drivers.

The goal of the ODK Sensors project is to shift as much responsibility as possible to the framework developers to simplify the creation of sensing application while maintaining a high-level of flexibility for integrating new sensor types. By creating a framework to isolate these three development roles we hope to make it easier to create sensing applications by isolating development tasks that can be fulfilled independently by people with the appropriate levels of technical skill. To encourage new driver development, the framework assumes as much responsibility as possible for aspects common to many sensors, including management of connection state and threads. Additionally, decomposing the system into modules enables more effective testing and code reuse, thereby improving overall system robustness which is particularly important for ICTD deployment settings (since once a system is deployed in remote locations updating it in the field becomes logistically difficult in terms of costs, time, and complexity).

For the framework to successfully enable an ecosystem of external sensors it must be:

1. easy to create sensor drivers, that is, minimizing the knowledge and amount of code required to create a driver,
2. easy to integrate/reuse external sensors in a wide variety of applications,
3. easy to deploy the framework and device drivers, shielding an end user from the technical details of the sensing infrastructure,
4. easy to upgrade the framework and sensor drivers,
5. hard for bad driver code to damage the framework since Sensor Driver Developers may not be expert Android developers,

6. easy for an Application User to discover available sensors through a streamlined user interface, and
7. easy to manage communication channel details such as proper handling of dropped connections.

ODK Sensors attempts to meet all the above requirements by creating an environment of reusable components for the development of mobile sensing applications.

3. RELATED WORK

The variety and number of mobile applications has increased due to the popularity of smartphones and app stores. Despite this proliferation, there are still only a limited number of applications that make use of external sensing devices. This is in part due to the programming challenges of implementing communication between smartphones and external sensors and in part due to resource constraints that prevent adoption of these applications in under-resourced environments. This leads to two main areas of research: (1) reducing programming barriers [8, 21] and (2) making mobile sensing applications more efficient [22, 23, 26, 29]. Within these two areas, some related work focuses on on-device sensors, while other work seeks to expand communication to sensors not built into the phone itself. There is also a significant body of research in device driver design that examines tradeoffs of reliability, ease of use, and performance with user-level versus kernel-level drivers or a combination of the two [10, 17, 19, 20, 24, 25].

The concept of user-level drivers (or application-level) is not new; the L3 system incorporated user-level drivers in 1988 [20]. Leslie *et al.* [19] built user-level device drivers into Linux without significant performance degradation, even for high-bandwidth devices such as Ethernet, by implementing a framework that used shared data structures, batched work, and optimized event notification. Microdrivers [10] developed a program to split existing drivers into kernel-level and user-level parts by leaving critical path code in the kernel (*e.g.* data handling, I/O) and moving the rest of the driver code to a user-mode process. Similarly, Decaf Drivers [24] implemented ways to convert Linux kernel drivers to Java programs running in user mode. These systems demonstrated good performance, despite not using native kernel drivers. While ODK Sensors was influenced by these projects, it focuses on creating user-level drivers for locked consumer devices running Android. Therefore, unlike these projects we do not alter the kernel to provide the communication link between the OS and the user-level driver. Instead, ODK Sensors' communication managers run as user-level threads and use Android's APIs to handle sending and receiving data from the sensor and then forward the bytes to the appropriate device driver for processing.

The migration towards user-level drivers is in part motivated by the desire to make systems more fault tolerant and reliable in the face of driver-error. Maverick [25], a web-based system, provides security by using device drivers and frameworks that run as user-level web applications to support interacting with multiple USB devices. Alternatively, Carburizer [16] detects and tolerates interrupt-related bugs to proactively manage device failures for improved reliability in the presence of faulty devices. Like Maverick, ODK Sensors leverages user-level drivers to provide reliability and security; however, ODK Sensors runs each driver as a separate application causing each driver to be isolated in its own virtual machine.

Other frameworks similar to ODK Sensors have been proposed, but they seek to interface primarily with built-in sensors. Zhuang *et al.* [29] introduced an adaptive location-sensing framework that improves the energy efficiency of location-based applications through suppression or substitution of location requests from built-in GPS sensors. It seeks to increase energy efficiency of the system, which differs from our goal of lowering programming barriers.

Dandelion [21] supports building applications distributed across a Maemo Linux smartphone and wireless body sensors by providing abstractions that shield application developers from hardware specific code. Dandelion envisions a scenario where sensor vendors provide a runtime to enable a platform-agnostic programming abstraction called a 'senselet' written by application developers to run on the sensor itself. The ODK Sensors framework also shields application developers from sensor-specific hardware; however, the framework provides abstractions at a different level as sensor drivers execute on the smartphone and leverage the framework's communication channel abstractions and sensor state management. The initial processing of sensor data occurs on the Android device in the sensor driver (removing this concern from the scope of application developers), whereas Dandelion requires data processing in the 'senselet' on the sensor that must be written by the application developer in this limited sensor environment. Additionally, ODK Sensors does not require sensor vendors to include a runtime enabling the framework to support any standard sensor that communicates via a supported communication channel. The Reflex [22] project (a fork of Dandelion) is a suite of runtime and compilation techniques that conceals the heterogeneous distributed nature of the system and reduces power consumption by offloading data processing to lower-power co-processors. While Reflex focuses on energy efficiency and performance in mobile-sensing applications, ODK Sensors focuses on lowering programming barriers for application developers and supporting different data and application types. The ODK Sensors driver executes within the framework rather than on a separate co-processor. LittleRock [23] and Turducken [26] have similar goals as Reflex, and present other architectures that offload continuous sensor data processing to dedicated low-power processors.

Gadgeteer [28] is a rapid prototyping platform that eases development with embedded hardware devices through the use of modular hardware components and object-oriented programming in C#. While Gadgeteer and ODK Sensors are both focused on making it easier for users to integrate with different external sensors, Gadgeteer achieves this by simplifying how different hardware pieces talk to each other, whereas ODK Sensors aims to make it easy for the mobile application developer to leverage a variety of sensors in their application without significant programming knowledge about the specific sensor.

IOIO [15] is a development board designed to work with Android phones through a USB connection. It abstracts the communication between external hardware and software running on the smartphone, enabling Android applications to directly control hardware attached to the IOIO board. It is different from ODK Sensors because IOIO provides an abstraction to Android applications at the level of I/O pins of the IOIO board. ODK Sensors currently interfaces with Arduino boards to enable low-power sensing, by decoupling the interface board from the Android application. An independently operating sensor board enables sensing to occur at lower power allowing the Android device to remain in a sleep state longer. Amarino [18] is another toolkit that connects Android phones with Arduino microcontrollers via Bluetooth. It helps developers in easily sending information about phone's internal events such as phone calls, SMSs, and on-device sensor data over Bluetooth. It is similar to the IOIO Board but with communication over Bluetooth rather than USB.

AndWellness [12] lets researchers customize surveys to collect data from sensors on phones carried by study participants. It shares our framework's goal of lowering barriers for building sensing applications. However, unlike ODK Sensors, this application focuses primarily on customizing surveys and front-end visualizations of real-time data. Our work aims to lower barriers for the application and sensor driver developer and in the future interface with applications such as ODK Collect [11] or ODK Tables [13], which will help make it easier to develop information services directly by end-users.

PRISM [8], like ODK Sensors, is sensing application middleware whose aim is to provide reusable components and eliminate redundant efforts regarding distributed operation, security and privacy. PRISM has been evaluated for a variety of applications, all of which interface exclusively with sensors built into the phone. PRISM also has a focus on deploying these applications at scale. In contrast, our framework focuses on interaction with external sensors by abstracting away the communication layer to make programming easier.

These related projects address many problems that are common to mobile sensing applications. We aim to further lower development barriers by simplifying the process of connecting a smartphone to an external sensor through the creation of a framework with tailored abstractions that facilitate the integration of new and varied types of sensor data into mobile applications.

4. FRAMEWORK

The ODK Sensors framework simplifies the development of sensor-based mobile applications by creating a common abstraction point that enables all sensors to be accessed through a unified interface. Creating a single-interface reduces complexity since all external sensors as well as Android's built-in sensors are exposed through a common interface regardless of the communication medium used. The interface encapsulates communication and delineates user-application code from sensor-specific driver code, freeing application developers from understanding the specifics of the underlying communication between an Android device and an external sensor. From a user's perspective the overall architecture for ODK Sensors consists of three apps: the *User-Application App*, the *ODK Sensors Framework App*, and the *Sensor Driver Apps*. For the purposes of this paper an Android application (software downloaded from a market) is referred to as an "app", whereas the word "application" is used to refer to usage/deployment examples. The *ODK Sensors Framework App* is responsible for managing low-level, channel-specific communications and providing abstractions to isolate sensor driver code. The *User-Application App* communicates with sensors through the unifying framework API (explained in Section 4.1). Figure 1 shows an end-user view of ODK Sensors on a smartphone. In the figure, two apps (*User-Application App* and *ODK Sensors Framework App*), a *USB Bridge*, and a temperature probe are used to monitor the flash heat pasteurization of milk to eliminate contaminants (*e.g.* HIV) in breast milk (described in section 5).

We chose Android as the target platform for the ODK Sensors framework because it is open source and has extensive support for background processes and includes several built-in constructs for inter-application communication (IPC) between Android Applications. Examples these Android constructs are detailed in Table 1. For instance, a *Broadcast Receiver* uses non-blocking message passing to communicate between applications; whereas, a *Service* construct communicates through a blocking inter-process communication mechanism. The ODK Sensors framework uses the Android Interface Definition Language (AIDL) to specify the programming interface that both the client and service use to communicate. From the AIDL, Android generates IPC code to decompose objects into primitives that the operating system can marshal as *parcels* across process boundaries to provide blocking IPC functionality. These constructs enable a comparison between a single-threaded asynchronous model (*Broadcast Receiver*) and a multi-threaded synchronous model (*Service*) for inter-application communication. Additionally, Android supports multiple communication APIs that facilitate connecting to a wide variety of external sensors. While APIs for Bluetooth and Wi-Fi radios have long been available on Android devices, only recently has Android added USB support through the Android Accessory Protocol (AAP) [1] enabling compliant devices to connect to external hardware over USB. The AAP requires the external device to act as the USB Master while the Android device is the Slave. To include a wide variety of external sensors in the framework, we utilize a *USB Bridge* to connect sensors that use diverse digital I/O protocols (*e.g.* I^2C, SPI) to the phone's USB port via an interface board. For our initial prototype, we used an Arduino board [2] to act as the *USB Bridge*.

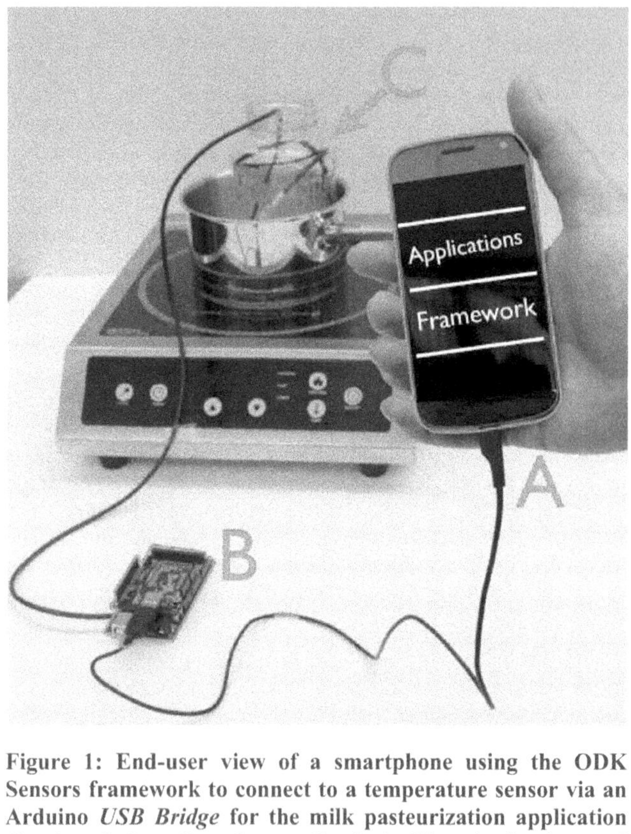

Figure 1: End-user view of a smartphone using the ODK Sensors framework to connect to a temperature sensor via an Arduino *USB Bridge* for the milk pasteurization application (Section 5 describes the application). The *Application* and *Framework* are Android apps that are installed on the mobile device (A). The mobile device is connected to an Arduino interfacing board (B) over USB; forming a *USB Bridge* to the temperature sensor (C) using the Arduino board's I/O ports (B). The *Application* uses the *Framework* to get data from the temperature sensor over USB using the ODK Sensors API.

Table 1: List of Android constructs used in the three framework implementations. The constructs listed are used for application metadata, packaging, storing, or sending data between applications and forming the basis of inter-process communication with sensor drivers in V2 and V3 (Section 4.2)

Android Construct	Description
Service	Services run in main thread of hosting process. Requests from other processes are handled concurrently by a threadpool.
Broadcast Receiver	Asynchronous broadcast receiver. Receives broadcasts sent via Intents from other processes.
Intent	Asynchronous messages that represent an operation to be performed, such as broadcasts, start services, start application, *etc*.
Bundle	Mapping from String values to parcel-able types. A *parcel* is a data container used for IPC.
Manifest	Provides essential information about the application (permissions, package name, etc.) to the Android OS.
Content Provider	Stores and retrieves data that is accessible by all applications.

The ODK Sensors framework (shown in Figure 2) presents a common interface to all top-level user applications via the Service Interface and Content Provider. User-Application apps only need to implement the application-specific logic that handles processed sensor data received from the framework. For each call to the service, the *Sensor Manager* dispatches the commands to the appropriate sensor object that, in turn, utilizes a sensor driver to perform specific low-level tasks. The framework supports multiple communication modalities by providing abstractions called *Channel Managers* that encapsulate complexities specific to each communication channel. ODK Sensors supports multiple data types, sample sizes, sampling frequencies, and sensor configurations by utilizing *Sensor Driver* abstractions that encapsulate sensor-specific data processing. These abstractions enable applications to interface with sensors using higher-level key-value pair constructs that are not constrained to be fixed-size arrays or values of a specific type. This enables developers to focus on the application logic instead of sensor-specific logic.

The framework's communication subsystems provide abstractions for lower-level, channel-specific communication protocols that make it easier for a driver developer to interface with an external sensor. The framework encapsulates communication channel specifics within the respective channel managers to hide them from application and sensor driver developers. For instance, Bluetooth-enabled sensors need to be discovered and paired with the smartphone and a socket needs to be set up for communication. ODK Sensors automatically manages this entire process for the application developer. The current implementation supports communications over Bluetooth and USB; in the future, we plan to add additional channel managers to support other communication methods such as NFC and Wi-Fi. To create a single unified sensing interface, the ODK Sensors framework also exposes all 11 built-in Android sensors (for Android 2.3 and greater) creating a single integration point for sensing.

A *Sensor Driver* handles the particular messaging protocol that configures and/or requests data from an external sensor by issuing commands to the appropriate *Channel Manager*. During data collection, the *Communication Manager* passes all raw data received from the sensor to the appropriate sensor driver via the *Sensor Manager*. Sensor drivers receive and process data encoded in formats specific to their respective sensors and generate configuration commands as required by the sensor. The sensor driver parses this sensor-specific data, transforming it into key-value pairs that can be easily consumed by top-level user applications. Likewise, the sensor's configuration parameters are specified as key-value pairs by user applications and passed to the appropriate *Sensor Driver* by the *Sensor Manager*. The driver encodes these key-value pairs according to the sensor's messaging protocol and the encoded data is sent to the sensor via the *Channel Manager*. By having channel managers handle communication details, the driver developer no longer needs to be aware of channel-specific protocols. Instead they can simply implement the interface described in Section 4.2 and convert raw sensor data coming from the communication channel into key-value pairs and vice versa for sensor configuration data.

In addition to allowing for easy reuse, the *Sensor Driver* design shields applications from changes in the communication protocol, configuration, or data type. Shielding applications from driver or channel changes leads to more robust systems that are easier to maintain. The sensor driver abstraction also enables multiple apps to interface with the same type of sensor by reusing an existing *Sensor Driver*. The framework's sensor drivers are designed as stateless processors of data to shield driver developers from tasks required to enable interaction with multiple identical sensors simultaneously (*e.g.*, channel management, threads, buffers).

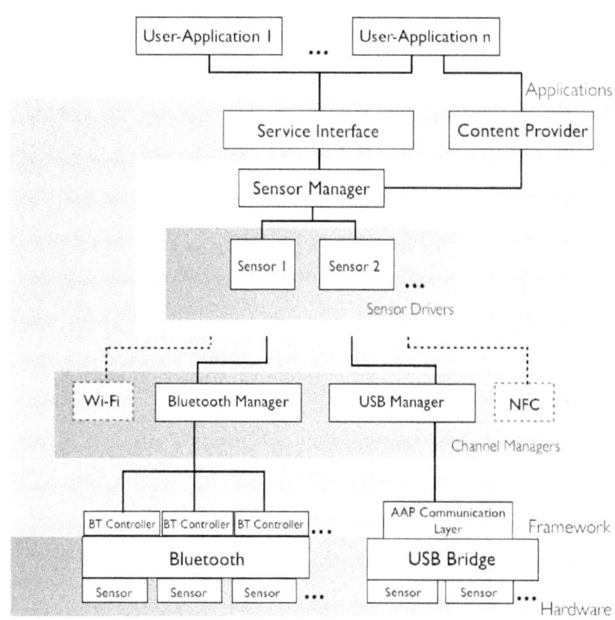

Figure 2: Architecture overview of ODK Sensors system. Sensor Manager maintains references to all sensors and the corresponding sensor drivers. Channel Managers manage the connections over the communication channels (e.g., USB, Bluetooth). The Service Interface and Content Provider provide the access point for applications to interact with the framework. Note: All architectures use the same data flows; however, in V2 and V3 sensor drivers move to separate apps as described in Section 4.2.

To better understand which Android IPC constructs to use to create our application-level (or user-level) sensor driver framework we implemented three different versions of the framework. The *Sensor Manager* and *Channel Managers* communicate with sensor drivers either with method calls inside the framework (V1), with IPCs through a Service Interface (V2), or with broadcasts to Broadcast Receivers (V3). To establish a baseline to compare against, version 'V1' of the framework was created as a single app. The device drivers in the framework are stateless processors of data, thus allowing the driver programmer to keep code simple by avoiding managing multiple sensors within the driver or having to run multiple apps for each sensor type. The details of the interfaces presented to developers and framework's communication subsystems are discussed in the following subsections.

4.1 Framework Interface

The ODK Sensors framework interface abstracts many sensor specific details and does not expose any channel specifics. The *Service Interface* creates a common interface for user applications to leverage both built-in sensors and external sensors connected over the communication channels through methods shown in the following code snippet:

```
interface ODKSensorService {
  boolean sensorConnect(in String id, boolean
      useContentProvider);
  void configure(in String id, in Bundle config);
  boolean startSensor(in String id);
  boolean stopSensor(in String id);
  List<Bundle> getSensorData(in String id, long
      maxNumReadings);
  boolean isConnected(in String id);
  boolean isBusy(in String id);
  boolean hasSensor(in String id);
}
```

The *Service Interface* requires the sensor's ID to control a specific sensor in the framework. If applications do not know the Sensor ID, then the framework provides an interactive discovery process for users, freeing the application from implementing their own sensor discovery interface. To launch the ODK Sensors interactive sensor discovery UI, the application simply sends an Intent to the framework as described in Section 4.2.4. Once the application has the sensor ID, it can connect to the sensor and begin retrieving sensor data by periodically calling the `getSensorData` method of the Service Interface. This method returns sensor readings to the application in a mapping of key-value pairs created by the *Sensor Driver* processing raw sensor data. As an example, a temperature sensor's driver parses data according to the messaging protocol of the sensor to extract multiple (or single) temperature readings in degrees Celsius and passes a list of Android *bundles* that map key = "temperature" to value = "value in °C" to the application. The framework expects the application to retrieve data in a timely fashion to clear the memory that is buffering the data. In cases where the application does not plan to read the data immediately, the framework provides applications a second mechanism for retrieving sensor data stored in a local database through a *Content Provider*. Applications specify whether the framework should put the data in a database to be accessed by a Content Provider when calling `sensorConnect`. An application can query the Content Provider whenever it wants to get data from a sensor that the framework considers owned by the application.

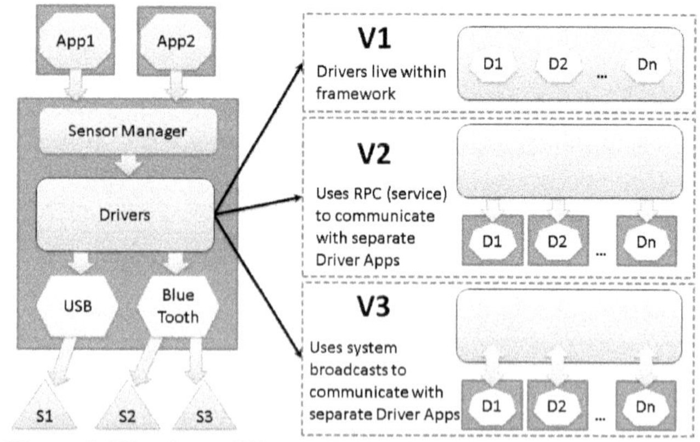

Figure 3: The three different framework implementations that vary with respect to how the framework communicates with the sensor drivers. Each dark rectangle represents a separate application. The design in which the drivers live within the framework is referred to as V1. The V2 design uses remote service calls to communicate with separate driver apps outside the framework. V3 uses system broadcasts to communicate with these separate driver apps.

4.2 Sensor Drivers

Sensor drivers are designed to abstract sensor specific control code from more general sensor management code. Certain concepts are common to all sensors, such as initialization, configuration, and taking readings, but the framework does not need to, nor should it know, specifically, how each sensor accomplishes these tasks. Sensor drivers enable the framework to communicate with sensors while maintaining the necessary abstractions that keep the framework modular and extensible. The same driver interface is used for all sensors – both built-in and external sensors – as the driver interface abstraction encapsulates sensor-specific data transfer and processing, and hides sensor specifics such as data types, frequency of collection, data size, and various configuration parameters. The driver abstractions enable the framework to reuse core functionality such as sensor configuration, connection, communication handshakes, buffering data, multiplexing, etc. for multiple types of applications. Simultaneous integration of different sensors involves complex tasks such as concurrent Bluetooth and USB setup that requires multiple data sockets and threads to buffer and process the data from these connections. The framework hides these complexities thereby significantly simplifying the job of both the application-level developer and the sensor driver developer.

Moving sensor drivers into separate Android apps that implement a common driver interface improves the framework modularity and extensibility. Separating drivers from the framework enables drivers for new sensors to be downloaded and installed onto an Android device from any Android marketplace or website like any other Android app. To understand the various architectural trade-offs of moving sensor drivers to external Android apps, three different framework architectures were created to compare two different Android IPC mechanisms. As depicted in Figure 3, one design keeps sensor drivers within the framework (V1) serving as a baseline and two designs move each driver to its own external Android app (V2 & V3). To understand the best way to communicate with the external drivers one framework used a blocking Binder IPC call (V2) while the other used a non-blocking message passing structure (V3). Both of these versions enable end-users to dynamically add drivers to the ODK Sensors framework as needed. User applications remained the same across the three implementations, as their interface with the framework did not change. In each of the three framework implementations every driver must implement the following interface:

```
public interface Driver {
  byte[] configureCmd(Bundle config);
  byte[] getSensorDataCmd();
  SensorDataParseResponse getSensorData(long
       maxNumReadings, List<SensorDataPacket>
       rawSensorData, byte [] remainingData);
  byte[] startCmd();
  byte[] stopCmd();
}
```

Drivers implement this interface to specify how to parse raw sensor data and what, if any, sensor-specific messages to send for configuration, getting data, and starting or stopping the sensor. The device drivers in the framework are stateless processors of data, which means the programmer does not need to manage multiple sensor instances. Methods such as `getSensorData` require the sensor driver to maintain access to buffered data it has previously processed as well as the raw data from the sensor it has not yet processed. To facilitate this `getSensorData` returns a `SensorDataParseResponse` which is simply a list of key-value pairs that have been fully parsed as well as any leftover bytes from the input stream. We eliminate state from the driver itself and instead have the framework handle all sensor state, including buffered data, and provide it to the driver at the appropriate time when the application has requested data. Implementing an external sensor driver for each of the three architectures is slightly different as described in the following subsections. In the case of the built-in sensors, their drivers are included with the framework even though it is possible to implement them as separate apps since communication is handled internally by the Android OS. The commands are Android-specific, not communication channel specific so it seems unnecessary to move them outside of the framework.

4.2.1 V1: Drivers within the Framework

In the first version, V1, we place all the sensor drivers inside the framework thus creating a single Android app. At runtime, the framework accesses individual sensor drivers from an in-memory map. In this architecture, the drivers, by executing within the framework, create a tight coupling that should have the best performance and provide a baseline against which to compare the other versions. However, this design is not ideal for our target users to dynamically add new drivers as it requires a recompilation or the use of a class loader. Using a class loader is not ideal for a population of non-technical users because it requires placing files into proper access-controlled directories on the device, thereby enabling the framework to dynamically load classes. Our goal is to build a system that uses established Android distribution and communication mechanisms to dynamically deliver and add drivers to the framework.

4.2.2 V2: Driver Communication via Services

The V2 version of the framework creates separate Android apps for each sensor driver. Each of these Apps implements an Android *Service* that defines the `Driver` Interface using AIDL, enabling the framework to communicate with the sensor driver via Android's Binder IPC. Android provides an AIDL to define the programming interface between the client and server and automatically generates stub Binder IPC classes. When the client calls these methods, the system copies the payload from the client into kernel memory, which is memory-mapped into the server's address space. The server-side procedure then handles the call. The sensor driver application includes driver-specific metadata that is used by the framework for driver discovery (described in Section 4.2.4). If an appropriate driver application is installed on the device, a generic sensor object is constructed to act as a proxy between the driver and framework by binding to the `Driver` interface that is presented as an Android Service. The framework maintains a reference to the specified generic sensor object that communicates with the driver using Android's Binder IPC. Each driver proxy acts as a thin wrapper for the driver and contains a reference to the appropriate channel manager. Communication with the sensor forwards commands returned from the driver so that they are transmitted over the appropriate channel. Since the driver application can be reused by multiple instances of the same type of sensor it is important that it be stateless. Therefore, the driver's remote function calls are designed to transfer the required state to the driver on each service call and allow the driver to return state information along with the parsed data. An example of state information stored by the framework is the excess bytes from the input data stream, enabling buffering of the input stream until enough data is received to parse and produce a full message.

4.2.3 V3: Driver Communication via Broadcasts

The third and final version of the framework, V3, also implements sensor drivers as separate Android applications but uses message passing with *Broadcast Receivers* for IPC (instead of synchronous blocking IPC). Each driver specifies a unique broadcast address that is discovered by the framework during the driver discovery phase (described in Section 4.2.4). The framework communicates with the driver by sending messages to the driver's unique broadcast address. Included in the broadcast message sent to the driver is a unique broadcast address for the driver to use to send a response back to the framework. The additional information included in each of the *Intents* (*i.e.* the data exchanged) is part of the API between the framework and drivers. By instantiating a broadcast receiver for each instance of a sensor, the framework can multiplex responses for each individual sensor. This API provides the same functionality as the interface implemented by sensor drivers in V1 and V2. Similarly as for V2, the driver does not maintain state, allowing the driver application to be reused by multiple instances of the same sensor type. The message passing interface is designed to communicate state between the framework and the sensor driver, enabling the sensor driver to cache incomplete information between processing sensor readings.

This architecture decouples the drivers from the thread of control allowing for a non-blocking message passing framework but incurs the overhead of using Android *Broadcasts* for data communication. The framework is better shielded from buggy drivers due to the decoupling enabled by the blocking semantics. However, we acknowledge that broadcasts can be intercepted by another Android application, which is a security risk. This will be addressed in future work by encrypting data sent between different processes.

4.2.4 Sensor & Driver Discovery

Each version of the framework semi-automates the sensor discovery process by providing a pre-built UI that application developers can launch when they need the user to select one of the currently available sensors. When the framework gets a request for an unknown sensor, it informs the user's application to launch the framework's built-in sensor discovery system to find and pair the appropriate device driver for the desired sensor. The user simply needs to select the sensor along with the corresponding driver from the list presented. The framework displays a list of available drivers for the relevant communication channel, as well as a list of sensors that are already physically connected to or in Bluetooth wireless range of the device. In V2 and V3, the framework automatically discovers installed driver applications by searching Android *manifest* files. Once the user has mapped the sensor with the appropriate sensor driver the framework will do the rest. After the user selects the sensor they want to use, the ID is returned to the application. This ID enables the application to control the sensor with the methods presented in the ODK Sensors Service Interface. Once a sensor has been discovered it is stored in the *Sensor Manager* database enabling the application to skip the discovery step in the future.

An advantage of having sensor drivers as separate applications is that it simplifies driver distribution. Ideally, we envision a scenario where manufacturers post their drivers on their own website or on an Android market (such as Google Play). By taking advantage of Google's standard application distribution system the framework is designed to lower barriers for novice users by using familiar application delivery channels. Additionally, if the device manufacturer needs to update their driver they simply need to post the updated application to market and the built-in Android application update system will take care of the rest. Another advantage of separating the device drivers into separate Android apps is it gives manufacturers who want to keep their protocols proprietary an opportunity to create drivers that can be easily used to create applications while protecting their protocols.

5. APPLICATIONS

We developed a few example applications to demonstrate reuse, flexibility, and extensibility of the framework. These applications exemplify the three basic dimensions of variation in sensing applications: *communication channel*, *sensor configuration,* and *data collection style*; all of which must be supported flexibly by the framework. First, the channel used to communicate with the phone can vary across sensors and applications (*e.g.,* USB, Bluetooth, NFC). Second, sensors have different configuration requirements, which may include various parameters or settings that need to be specified such as sampling rate, trigger conditions or alerts, identifiers, and calibration. Third, the data needed by an application can change in format, size, and frequency of collection. The framework aims to support any combination of communication, configuration, and data type transfer between phone and sensor. A top-level user application retains the same interaction with a sensor driver in the framework even if there are changes in communication protocol, configuration, or data type. In the event of such changes, the sensor driver only requires minimal adjustments for parsing a new type of data or specifying a new channel manager.

Developing these applications for the ODK Sensors framework helped to evaluate whether we could reuse the provided abstractions for different use cases, verifying the framework and sensor driver's interfaces. The applications exercise both the wired and wireless subsystems of the framework and, in our experience, are exemplary of the four commonly used modes of data collection in sensor-based systems (Table 2):

- Single Reading: The user requests data from the sensor and chooses to record data points by taking a single reading from a real time stream of data. A clinician's application that connects tools (*e.g.,* blood pressure, pulse oxymetry) to a phone is an example of this use case.
- Real-Time Time-Series: The user of a data collection node has an active session with the sensor and observes a stream of samples from the sensor in real-time. Monitoring the temperature curve of the milk pasteurization procedure [7] is an example of this use case. The temperature curve is also saved so that it can be reviewed later.
- Snapshot Time-Series: Sensors are deployed to autonomously monitor certain phenomena. They aggregate readings internally over a period of time and may report some to a remote location periodically (e.g., alert to detect a specific condition). Temperature and electrical current sensors deployed to monitor vaccine refrigerators are examples of this use case [5].
- Historical Time-Series: Sensors are deployed to autonomously monitor certain phenomena (*e.g.,* movement of an object such as a water can over a period of months). However, unlike a Snapshot Time-Series, data retrieval is not automatic and requires someone to be within range of the sensor to offload the sensor's stored data. The WaterTime monitoring [6] application exemplifies this approach and is dependent on field calibration for configuring the sensor with sampling rates and identifiers.

Table 2: Variations in the sample sensing applications.

Application	Comm. Channel	Configuration	Data Style
Medical	Bluetooth	Calibrate	Single Reading
MilkBank	USB	Sampling Rate	Real-Time Time-Series
Vaccine	USB	Alerts Sampling Rate Snapshot Size	Snapshot Time-Series
WaterTime	Bluetooth	Identifier Calibrate	Historical Time-Series

Three of these applications have already been deployed in developing regions as part of pilot studies. Rewriting these applications to leverage the ODK Sensors framework has simplified them significantly by separating out the application specific logic from the communication logic. Figure 4 shows how these applications interact within the V1 framework.

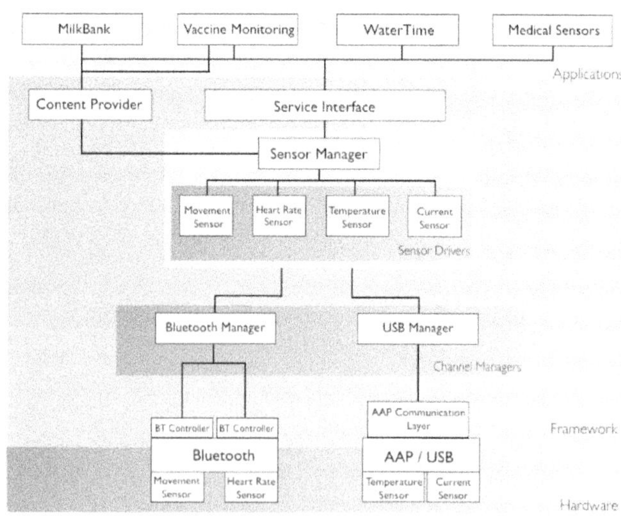

Figure 4: Example instance of V1 framework architecture incorporating the four applications discussed.

6. EXPERIMENTS

The experiments were designed to explore the tradeoffs between the three frameworks in terms of performance, power consumption, and ease of programming. Each of the architectures was benchmarked based on the same set of applications, sensor drivers, and communication channels. The code of the framework and drivers is identical for V1, V2, and V3 with the exception of the protocol the framework uses to communicate with sensor drivers. The protocol changes required four Java files to be different. User applications were not affected across the three implementations, as their interface with the framework did not change.

6.1 Performance

Performance experiments were conducted using several different Android devices to understand the impact of hardware variability. The Samsung Galaxy Tab and Samsung Nexus S were used to represent high-tier Android devices, the HTC Nexus One and Motorola Droid were used to represent mid-tier devices, and the Huawei IDEOS was used to represent low-tier Android devices likely to be common in developing regions. An Arduino Mega2560 with USB Host capability connected to a Motorola Xoom tablet was used to characterize the *USB Bridge*.

6.1.1 Sensor Application Throughput

First we measured the throughput of the four real-world applications described in Section 5. Each test consisted of 60 seconds of continuous data collection. The results in Table 3 show that the rate at which a sensor application gets sensor data is often on the order of one sample every 1 to 2 seconds. In fact, three out of four of the applications collect a data sample no more than twice per second; this rate is actually limited by the sensor itself. The Vaccine and MilkBank applications use a digital temperature sensor that requires a 750ms delay in sampling the ADC, while the heart rate sensor used in the Medical application transmits a packet every 2 seconds. The framework throughput is substantially higher than the saturation point for these three applications. However, in the WaterTime application, which requires a one-time bulk reading of historical data, the sensor sends its collected data as fast as possible over Bluetooth and achieves a throughput of 51 packets per second. This is close to the maximum throughput achieved on the Bluetooth channel using the framework (see Table 4).

Table 3: Observed data throughput from sensor to application on four real-world applications. Packet size is not consistent across applications as the size of data samples varies.

Application	Throughput (pkts/sec)
WaterTime (Bluetooth)	51.0
Medical (Bluetooth)	1.5
Vaccine (USB)	1.0
MilkBank (USB)	1.0

6.1.2 Communication Channel Throughput

Next, we systematically tested the saturation point of the Bluetooth and USB communication channels with a stress test that sent fake data as fast as possible over the channels. To do this we programmed a "spammer" sensor on two Arduino microprocessors; one that emulates a Bluetooth-based sensor and another that emulates a USB-based sensor. Each "spammer" sensor reacts to a "start" signal by executing a loop to send 1 byte data packets as fast as possible until it receives a "stop" signal. In our tests we allowed the "spammer" sensor to send a rapid, constant stream of data for ten minutes on each of our three framework architectures on Bluetooth and on USB. The maximum throughput that could be achieved with full saturation of the communication channels is shown in Table 4.

Table 4: Throughput (pkts/sec) on Bluetooth & USB Channels

Framework	Bluetooth	USB
Drivers in Framework	58.0	42.5
Drivers w/ Service	58.7	41.5
Drivers w/ Broadcasts	49.0	42.0

The throughput results were surprising as the USB channel had lower throughput than the Bluetooth channel. Investigating this issue further revealed that the USB Host driver on the Arduino adds some delays that impact the channel performance. We found that increasing the payload size did not significantly reduce the packets per second rate. After increasing the message size to 1KB per packet and disabling our reliability system, the *USB Bridge* gave a channel bandwidth of over 40KB per second. We found that increasing the payload size beyond 1KB was problematic and was primarily limited by the Arduino's memory size. This is acceptable for now because our real-world USB applications require throughput that is significantly lower than what we can already achieve with our current *USB Bridge*. However, in the future, we will explore alternative options for the *USB Bridge* to achieve higher performance to support more demanding sensing applications, such as those streaming high-resolution camera data.

6.1.3 Framework Throughput

Finally, we evaluated the performance of each framework by establishing an emulated communication channel for fake sensors to use with varying send rates and packet sizes. These experiments tested the throughput of the three framework versions by eliminating the limits imposed by real communication channels. For a baseline understanding, we evaluated the framework's throughput by varying packet size and the delay between packets (Table 5). To understand the effects of multiple sensors, we ran tests that varied the number of sensors that communicated simultaneously through the framework (Table 6). Finally, we verified that the framework's performance did not significantly degrade when operating on different classes of Android devices (high-tier, mid-tier and low-tier) (Table 7). This is important because users in developing regions will likely have a diverse set of devices.

We measured framework throughput (packets/second) by sending data at varying rates with varying sizes as shown in Table 5. The test results reported the number of packets received on average per second for each of the frameworks V1, V2, and V3 on a Nexus One. Each test ran for three minutes on five different Nexus One phones. The results of the tests were averaged. The send delay values began at 1ms and were increased by doubling the delay to a max of 128ms, while the packet size started at 1 byte and was increased by an order of magnitude for each test up to a maximum of 100,000 bytes. We varied these parameters to understand any limiting factors inherent to the different IPC mechanisms. Differences in performance of the various architectures become negligible as the send delay becomes the dominant limiting factor. None of the three frameworks were able to complete the most strenuous tests of 100,000 byte packets being sent with only a 1ms delay because of memory errors invalidating the results (indicated by "Error" in the table). Generally speaking, the throughput values of the framework are similar since the cycles spent in IPC are a small part of the total framework execution time. In the cases where the send delay was not the dominant factor, V1 appeared to perform the best since it

Table 5: Throughput results (pkts/sec) for a Nexus One when varying the packet size and inter-packet generation delay for the three framework versions. The top section of the table contains V1 results, the middle section contains V2 results, and the bottom section contains V3 results. The 'Error' value indicates a test run was unable to be completed because of a memory error. The Max column contains the theoretical maximum throughput.

V1	(bytes) 1	10	100	1K	10K	100K	Max
(ms) 1	798.1	796.9	790.2	747.9	Error	Error	1000.0
2	441.0	440.8	438.9	423.6	Error	Error	500.0
4	234.6	234.5	234.8	229.6	Error	Error	250.0
8	121.4	121.4	121.3	119.7	111.5	Error	125.0
16	61.6	61.6	61.5	60.9	58.1	Error	62.5
32	31.0	31.0	31.0	30.9	30.1	Error	31.3
64	15.5	15.5	15.5	15.5	15.3	13.8	15.6
128	7.8	7.8	7.8	7.8	7.7	7.3	7.8
V2	(bytes) 1	10	100	1K	10K	100K	Max
(ms) 1	785.5	786.8	781.1	Error	Error	Error	1000.0
2	437.6	438.0	436.0	421.8	Error	Error	500.0
4	233.7	233.8	233.0	229.0	Error	Error	250.0
8	120.6	120.6	120.3	119.6	Error	Error	125.0
16	61.4	61.4	61.3	61.0	58.8	Error	62.5
32	31.0	31.0	31.0	30.9	30.1	Error	31.3
64	15.5	15.5	15.5	15.5	15.3	Error	15.6
128	7.8	7.8	7.8	7.8	7.7	7.3	7.8
V3	(bytes) 1	10	100	1K	10K	100K	Max
(ms) 1	771.5	768.0	760.5	Error	Error	Error	1000.0
2	432.2	431.5	426.5	419.2	Error	Error	500.0
4	231.9	231.9	231.2	225.2	Error	Error	250.0
8	119.2	119.1	119.2	118.4	Error	Error	125.0
16	60.9	60.8	60.8	60.4	Error	Error	62.5
32	30.8	30.9	30.8	30.7	30.1	Error	31.3
64	15.5	15.5	15.5	15.4	15.3	Error	15.6
128	7.7	7.7	7.7	7.7	7.7	Error	7.8

does not incur any IPC overhead; whereas, V3 had lower performance since it communicates with drivers uses message-passing.

The primary difference between V2 and V3 is the IPC method. Our results confirm our expectation that the broadcast intent system has more overhead than a synchronous Binder IPC call. This finding correlates with other research that showed the Android system implements the *Intent* call as two IPC-style calls – one from the sender to the system and one from the system to the receiver [14]. While the 'double call' for V3 causes slower performance, the broadcast communication is still only a small amount of the overall work. Therefore, while the effects are visible, they are generally negligible except at faster data generation rates. Our results also matched the findings that there is a drop-off in performance for large packets. The Android system allocates a default kernel-side buffer 4kB in size for each process that will receive data from Binder IPC calls. When the payload size exceeds 4KB, the system allocates an additional, temporary buffer to transfer the data. These larger IPC calls incur additional allocation overhead and were shown to be inefficient as the Android system memory-maps each additional page separately, instead of mapping them all at once [14]. These factors result in a drop-off in performance for payload sizes over 4KB, which is reflected in our results for packets 10KB bytes or larger.

To compare how framework throughput (packets/second) changes when multiple sensors are simultaneously generating data, we ran tests with multiple fake sensors generating 100 byte data packets every 64ms. In this test case, the drivers simply copy the incoming bytes into a parsed sensor reading. The results in Table 6 show that the frameworks perform close to the theoretical max when there are a few sensors running simultaneously. Differences in framework throughput start to emerge as the number of sensors increases. However, we consider 12 sensors to be adequate for all the realistic applications we have considered to date.

Table 6: Framework throughput (pkts/sec) of multiple sensors sending 100 byte packets every 64ms on a Nexus One

Num Sensors	V1	V2	V3	Max
6	93.3	93.2	93.1	93.8
12	186.5	186.4	185.6	187.5
24	371.9	371.8	365.8	375.0
48	737.6	735.2	686.4	750.0

To understand how performance would change on various Android platforms, we measured V2's performance on a few different models of Android devices. Similar to the tests for Table 6, we used 100 byte packets sent every 64ms and varied the number of sensors simultaneously moving data through the driver. As expected, the high-end devices (Samsung Nexus S and Galaxy Tab) had the best performance while the IDEOS had the lowest performance with 48 sensors concurrently running. Overall, the throughput of the V2 framework for 12 sensors is similar across the four devices, confirming that the framework should be portable to a variety of Android devices. While the results presented in Table 7 show all of the devices were able to support multiple sensors with a send rate of 1 packet every 64ms, other tests showed that faster send rates eventually caused errors on all devices. Devices with smaller amounts of RAM and smaller default heap sizes seemed to have more issues. For instance, when delay between packets was reduced to 32ms the IDEOS experienced out of memory errors with 48 sensors, while the Nexus S did not experience errors until 96 sensors were simultaneously sending data. Again, only minor differences were seen when considering realistic numbers of sensors.

Table 7: Throughput (pkts/sec) of V2 framework on different Android devices (100 byte packets sent every 64ms)

Num Sensors	Galaxy Tab	Nexus S	Droid	IDEOS
3	46.4	46.6	44.4	46.3
6	92.8	93.2	92.1	92.2
12	185.0	186.2	184.0	183.8
24	368.6	372.5	366.4	360.6
48	733.7	741.0	720.4	677.4

6.2 Power

The power consumption of each of the three frameworks is negligible with respect to the power use of a device with active Bluetooth or an illuminated screen. We ran 12-hour tests where each framework processed 100 packets/second on a Nexus One. At the beginning of each test, the battery was charged to 100% and each test resulted in an approximate 3% drain of the battery. In these tests, the screen, Bluetooth, Wi-Fi, and cellular radios were off and the only applications running were the battery monitoring application and the sensor framework. These tests show there were only negligible differences between the frameworks with regards to power consumption for expected operating conditions. The tests also show that the power consumption of the framework is negligible in comparison to other phone components.

6.3 Ease of Use/Programming

The ODK Sensors framework provides a reusable code base that makes it easier to create sensing applications by providing a common interface for all sensors (both built-in and external) that abstracts sensor communication details. By decomposing a sensing application into discrete and reusable building blocks, we enable a plug-and-play capability where users with different technical skill levels can contribute in different ways. The system is designed to leverage app marketplaces, which provide a standard interface for users to find and download a variety of applications, making it easy to enhance their Android device to use external sensors. Both application users and developers can leverage marketplaces for distributing and locating sensing applications and sensor drivers that can be reused for a variety of mobile-sensing needs. An application user who wants to collect data from sensors can simply download the ODK Sensors framework, a sensing application (written by an application developer) and the sensor driver(s) (written by driver developers) that corresponds to the sensor(s) they require. An application developer can re-use any existing sensor driver, allowing them to write a sensing application by simply downloading the appropriate driver and interacting with it using the framework's interface. A driver developer's responsibilities are limited to implementing the sensor-specific communication requirements (*e.g.* configuration, start/stop commands, *etc.*). The driver developer can focus on the sensor-specific issues and ignore communication channel specifics of the various wired and wireless channels. A final measure of ease of programming is that the system is portable. In ODK Sensors, all Android devices are interchangeable within the framework as long as the requisite apps and drivers have been installed.

6.3.1 *Application User*

The application user is expected to be able to obtain and use basic Android applications. To make device driver distribution easy, the drivers in V2 and V3 are designed to be separate applications enabling device manufacturers to post them on their website or an Android marketplace. By using marketplaces such as Google Play, manufacturers take advantage of Android's application distribution system making it easy for users to find and keep their driver updated as the user's device will automatically receive updates (similar to Windows and Mac updates). The framework tries to minimize the work of the end-user by semi-automating the sensor discovery process. The framework automatically discovers all installed sensor drivers on the Android device and populates a list of available devices on each communication channel. The user simply selects the sensing device and is prompted with a filtered list of possible corresponding drivers. The framework's interface also guides the user through any channel-specific pairing tasks (such as on Bluetooth) required to interact with the sensor.

6.3.2 *Application Developer*

The framework simplifies the application developer's role by providing a single interface to communicate with all external and built-in sensors. This significantly reduces the number of software packages an application developer needs to understand and use in order to communicate with multiple types of sensors over varied communication channels. To further minimize developer responsibilities, ODK Sensors provides a base class for application developers that encapsulates the calls to framework's *Service Interface* to further simplify development by removing the need for the developer to understand Android IPC specifics. Additionally, the framework reduces application developer responsibilities by providing a sensor discovery and driver detection user interface. If an application needs the user to discover a sensor it can send an *Intent* to ODK Sensors to launch the framework's discovery UI. The UI guides the user through sensor discovery, pairing, registering, and driver selection then returns the selected sensor ID to the application, thus eliminating the need for the application developer to re-implement common functionality. If necessary, a specialized user interface could be developed, but at higher development cost.

To quantify the simplification of development we ported four existing standalone sensing applications to leverage the ODK Sensors framework (Section 5). The reduction in the lines of code (summarized in Table 8) in the ported applications provides a basic measure of how development was simplified. We expect standalone applications to have larger codebases because they must implement additional logic to manage communications and handle sensor-specific data parsing.

Table 8: Compares the lines of code needed to create a standalone sensing application vs. an application that uses the ODK Sensors framework. The Sensor Driver column reports the lines of code in the framework's sensor drivers.

Application	Standalone App	App using Framework	Sensor Driver
WaterTime	1350	956	139
Medical	933	246	80
MilkBank	1325	316	105
Built-in Accelerometer	546	538	29

The ported applications leverage the framework for channel and connection management, while sensor-specific data processing is delegated to sensor drivers. Table 9 shows the additional Android modules needed to implement a standalone application that interacts with sensors over Bluetooth, while Table 10 shows the Android modules needed to communicate with a sensor over a *USB Bridge*. Developers leveraging the ODK Sensors interface do not need to understand these Android constructs as the framework hides the internals of these modules from application developers.

Table 9: Additional modules used in a Bluetooth-based sensing application written without using ODK Sensors.

Module	Purpose
Permissions	Allow access to Bluetooth in Manifest
Bluetooth Adapter	Perform fundamental Bluetooth tasks like device discovery and creation of BluetoothDevice and BluetoothServerSocket
Bluetooth Device	Representation of Bluetooth device to create a connection or query information from it
Bluetooth ServerSocket	Listen for connections and create BluetoothSocket to manage the connection
Bluetooth Socket	RFCOMM socket like a TCP socket to allow streaming transport over Bluetooth
InputStream	For input to BluetoothSocket
OutputStream	For output from BluetoothSocket
IOException	Deal with exception on I/O from BluetoothSocket
UUID	Used to initiate an RFCOMM communication
UI Components	To make a UI that will allow a user to discover, pair and connect with a device
Handler	To receive updates from a Bluetooth I/O thread
Message	Contains data to be sent and handled with the Handler described above when Bluetooth device state changes or data is read / written.
IntentFilter	Used with BroadcastReceiver
Broadcast Receiver	Listen on BluetoothDevice state changes for when a device is discovered

Table 10: Additional modules used in a USB-based sensing application written without using ODK Sensors.

Module	Purpose
Permissions	Set up application to use USB in Manifest
UsbAccessory	Allows you to enumerate and communicate with connected USB accessories
UsbManager	Represents a USB accessory and contains methods to access its identifying information
ParcelFile-Descriptor	Descriptor that can be passed between processes.
FileDescriptor	Descriptor that can be passed between processes.
IOException	Deal with I/O error from USB
FileInputStream	Read from USB
FileOutputStream	Write to USB
PendingIntent	Intents that can be passed to and run by another application.
IntentFilter	Used with BroadcastReceiver
Broadcast Receiver	Used to discover when a USB accessory has been attached

6.3.3 Sensor Driver Developer

The framework shields device driver developers from tasks required for the app to interact with multiple sensors, such as management of channels, threads, buffers, and sensor states. The device drivers in the framework are designed to be stateless processors of data enabling the developer to keep their code simple. A slightly more experienced developer will need to be responsible for the sensor driver implementation because they will need to interpret sensor data sheets to determine how to manage sensor-specific communications, configuration parameters, and data formats. Table 8 lists how many lines of code were needed to create drivers for the application-specific sensors discussed in Section 5. To make driver implementation simple, driver developers can use a base class provided by the framework development team to handle the communication between the driver and the framework. Since the communication version information is contained within the provided base class, the framework will automatically communicate via the correct protocol when discovering the driver, thereby enabling drivers with varying versions of communication base classes to exist simultaneously on the same Android device. This eliminates the requirement of upgrading all driver applications to the newer protocol version simultaneously to updating the framework. Additionally, upgrading the driver to the latest framework communication protocol should be as simple as swapping in the latest binary containing the base class (assuming no changes to the driver interface functions). Driver developers also benefit from the framework's system design of deploying drivers through an Android marketplace. When a device manufacturer updates their driver, they simply need to post the updated application to market and the built-in Android application update system will push out the update to devices that previously installed the driver application.

7. DISCUSSION & FUTURE WORK

To build the next generation of information systems for developing regions, tools are needed to simplify the creation of mobile sensing applications. Barriers to connecting external sensors to mobile devices need to be reduced to enable a wide range of developers with varying programming skills to easily leverage external sensors. While the technical skills required to create and deploy mobile sensing applications are not uncommon in technically astute developer communities, in developing regions the lack of educational resources and technical expertise creates barriers to the development of sensing applications. ODK Sensors' single abstraction for both built-in and external sensors helps developers with basic Android coding skills by reducing the number of Android constructs a developer needs to understand. Additionally, the framework's interface creates a clean separation point between application code and sensor-specific code. This decoupling naturally lends itself to a reusable framework where sensor-specific code is placed into modular drivers. Thus, driver developers can keep their code simple acting as a stateless processor of data. This modular approach also enables different brands/models of sensors to be easily changed by replacing drivers without necessarily requiring application code changes. To further simplify development, the framework provides abstract base classes to make connecting to the framework easier by handling all the inter-process communication. Providing abstract base classes enables easy upgrades to both applications and drivers because the framework developers handle any protocol changes within the base class. The long-term goal of the framework is to enable a market of reusable application components and drivers that can be easily integrated by non-technical users to create sensor-based applications. Deploying mobile sensing applications in under-resourced environments is also challenging because in many circumstances the end-user will likely need to load the sensing software onto consumer devices that may be locked by their service providers. Therefore, for ODK Sensors to be easily deployable by end-users with limited technical abilities the framework and user-level drivers need to fit in Android's app distribution model.

The concept of user-level drivers is well studied and known to have many advantages such as ease of development, portability, and maintainability, but generally suffers in terms of performance. The experiments in Section 6 show the performance of all three frameworks is adequate compared to overall system throughput as limited by the communication channels (Bluetooth and USB). Since framework performance does not significantly impact the system, other factors that lower programming and deployment barriers quickly become more important when choosing the optimal framework architecture. One advantage of the V2 and V3 design is the separation of sensor drivers from the framework thus providing a sandbox environment to minimize any negative effects of misbehaving third-party driver code from the framework. In this respect, V2 and V3 are better framework choices because sensor drivers are isolated as separate Android apps and run in their own virtual machines (VMs). A disadvantage of V2's design is its IPC mechanism acts as a blocking call that can be potentially dangerous if the driver causes the framework to block forever. However, since each sensor is isolated in its own framework thread, the effect of a misbehaving driver on the framework will be minimal as the framework can handle such drivers with a timeout or exception. The main disadvantage to the V3 design is that it is inherently less secure than V2. As any program can register to receive broadcasts making it easy for rogue programs to eavesdrop or inject data. This security problem could be addressed by encrypting data sent between the different processes; however, V3 has other limitations such as timing issues caused by the fact that broadcast messages cannot be received until after the framework's *onCreate* method completes. This method is called when an application binds to the ODK Sensors framework; however, during construction no communication between drivers and framework is possible until after construction is complete because no messages can yet be received by the framework. Unfortunately, once framework construction completes, applications are capable of sending messages to sensors before the framework's driver connections are properly established causing timing issues.

With sensors becoming increasingly popular on mobile devices, the possibility of multiple applications simultaneously leveraging the same sensor arises and leads to contention for sensor control. Problems can occur when different applications using the same sensor issue conflicting start and stop commands. Additionally, it becomes more difficult for the sensor framework to know how long it should keep a copy of the sensed data cached so that it can deliver the data to all waiting applications. For example, if two applications are simultaneously polling a temperature sensor at different rates, then the framework must maintain data until both applications receive their copy. However, this can be challenging if one application decides it no longer wants the data and sends a stop command while the other application is still waiting. There is a need for techniques to resolve conflicts in resource sharing, leader selection, resource ownership, *etc.* Currently, the framework avoids these issues by only allowing one application at a time to own an external sensor. In the future, we plan to explore models for electing a 'leader' application based on the application that is most dependent on the current stream of data and would be adversely affected by changes to the sensor.

We also plan to continue to simplify sensor integration by expanding the types of sensors that can be connected to the ODK Sensors framework through a *USB Bridge* by including support for a variety of interfacing boards (*e.g.* IOIO, Phidgets). Additionally, as Android devices with USB Master or USB On-The-Go become common, we plan to enhance the *USB Manager* to act both as a USB Slave and USB Master. We are also working on channel managers to manage access to the new class of NFC-enabled low-power sensors that will soon be available.

8. CONCLUSION

The platform shift from traditional PCs to mobile devices with cloud services creates a need and opportunity to integrate these devices (*e.g.*, smartphones, tablets) with external sensors and deploy applications in new settings. To address this, we created an application-level driver framework that enables convenient reuse of sensor-specific code between applications by logically separating the high-level application from the underlying sensor driver. The focus of ODK Sensors is on enabling the integration of data from a variety of sensors over both wired and wireless communication channels. It simplifies application development by creating a single interface that can control virtually any kind of sensor (both external and built-in) and reduces the amount of code needed to access a sensor. Applications that leverage the framework to communicate with external sensors can be implemented in fewer lines of code (Table 8) by removing sensor communication code - on the order of ten fewer Android Java modules (Tables 9 and 10). Additionally, the sensor framework automatically multiplexes the communication channels allowing different types of sensors to be used simultaneously by an application. For example, an application can easily use two USB and three Bluetooth sensors simultaneously to record several phenomena at once. The ODK Sensors framework is designed to flexibly meet any application's needs regardless of data type, data collection rate and size, sensor configuration requirements, or communication channel.

After testing the three framework implementations, it was clear that performance was not the most important factor to consider when selecting the final design, as most sensing applications sample data at a significantly lower rate than the framework's maximum throughput. The performance analysis showed that the system bottleneck is the throughput of the Bluetooth and USB communication channels rather than framework throughput. Since the three frameworks performed similarly, other factors were examined before deciding which is optimal. The V2 framework offered the best tradeoff in terms of programming ease, deployment ease, and performance. The separate driver app design makes it easier for end-users to dynamically add new drivers and V2 has better performance than V3. Performance may become more important in the future to accommodate applications that use high bandwidth sensors such as external cameras for medical devices.

This work is part of the larger Open Data Kit [11] project that seeks to develop a modular set of tools to magnify human resources through appropriately designed technology. One of ODK's strengths is creating information systems that collect a wide variety of data types (*e.g.*, location, images, audio, video, and barcodes) that are difficult to record on paper forms. By lowering the barriers to add external sensing components, we hope to expand mobile data collection applications to include an even richer set of data types. ODK Sensors increases the variation of input data types possible by simplifying access to sensing resources through the creation of a single interface that makes external sensors as easy to integrate as built-in sensors. By creating a framework designed to follow ODK's modular components philosophy, we aim to expand the tool suite to allow end-users to easily augment their Android device with external sensing options. The component philosophy enables easy reuse of sensor drivers that will hopefully lead to an ecosystem of drivers further promoting the creation of novel mobile sensing applications. Using standard Android app distribution channels (e.g. Google Play) will make it easy for users to download functionality enhancements (application-level device drivers) to their unmodified Android OS. This simple method of deployment will hopefully lead to the creation of new sensing-based mobile data collection applications that improve information services in under-resourced contexts that typically lack a rich technology infrastructure (both physically and in terms of expertise).

9. ACKNOWLEDGMENTS

We thank Steven Gribble for his helpful advice on this project. We also thank Yaw Anokwa, Brian DeRenzi, and Mitch Sundt for their insightful feedback on this paper. Finally, we thank our anonymous reviewers, and our shepherd Landon Cox, for their guidance. The material in this paper is based upon work supported by NSF Research Grant No. IIS-1111433 and a NSF Graduate Research Fellowship under Grant No. DGE-0718124.

10. REFERENCES

[1] Android Open Accessory Development Kit. http://developer.android.com/guide/topics/usb/adk.html. Accessed April 2012.

[2] Arduino. http://www.arduino.cc/ Accessed April 2012.

[3] A. Bhardwaj, P. Arjunan, A. Singh, V. Naik, and P. Singh. MELOS: a low-cost and low-energy generic sensing attachment for mobile phones. In *Proc. of the 5th ACM Workshop on Networked Systems for Developing Regions*, 27-32, June 2011.

[4] R. Chaudhri, W. Brunette, M. Goel, R. Sodt, J. VanOrden, M. Falcone, and G. Borriello. Open data kit sensors: mobile data collection with wired and wireless sensors. In *Proc. of*

the 2nd ACM Symposium on Computing for Development, 9:1-10, March 2012.

[5] R. Chaudhri, E. O'Rourke, S. McGuire, G. Borriello, and R. Anderson. FoneAstra: enabling remote monitoring of vaccine cold-chains using commodity mobile phones. In *Proc. of the First ACM Symposium on Computing for Development*, 14:1-9, Dec. 2010.

[6] R. Chaudhri, R. Sodt, K. Lieberg, J. Chilton, G. Borriello, J. Cook, and Y. Masuda. Low-power Sensors and Smartphones for Tracking Water Collection in Rural Ethiopia. *IEEE Pervasive Computing*, (to appear), March 2012.

[7] R. Chaudhri, D. Vlachos, J. Kaza, J. Palludan, N. Bilbao, T. Martin, G. Borriello, B. Kolko, and K. Israel-Ballard. A system for safe flash-heat pasteurization of human breast milk. *In Proc. of the 5th ACM Workshop on Networked Systems for Developing Regions*, 9-14, June 2011.

[8] T. Das, P. Mohan, V. N. Padmanabhan, R. Ramjee, and A. Sharma. PRISM: platform for remote sensing using smartphones. *In Proc. of the 8th Int. Conf. on Mobile Systems, Applications, and Services*, 63-76, June 2010.

[9] N. Dell, S. Venkatachalam, D. Stevens, P. Yager, and G. Borriello. Towards a point-of-care diagnostic system: automated analysis of immunoassay test data on a cell phone. In *Procc of the 5th ACM workshop on Networked Systems for Developing Regions*, 3-8, June 2011.

[10] V. Ganapathy, M. J. Renzelmann, A. Balakrishnan, M. M. Swift, and S. Jha. The design and implementation of microdrivers. *SIGOPS Operating Systems Review*, 42(2):168-178, March 2008.

[11] C. Hartung, Y. Anokwa, W. Brunette, A. Lerer, C. Tseng, and G. Borriello. Open Data Kit: Building Information Services for Developing Regions. *ICTD 2010*, Dec. 2010.

[12] J. Hicks, N. Ramanathan, D. Kim, M. Monibi, J. Selsky, M. Hansen, and D. Estrin. AndWellness: an open mobile system for activity and experience sampling. *Wireless Health 2010*, 34-43, Oct. 2010.

[13] Y. Hong, H. K. Worden, and G. Borriello. ODK Tables: data organization and information services on a smartphone. In *Proc. of the 5th ACM Workshop on Networked Systems for Developing Regions*, 33-38, June 2011.

[14] C. Hsieh, H. Falaki, N. Ramanathan, H. Tangmunarunkit, D. Estrin. Performance Optimization of Android IPC for Continuous Sensing Applications. *CENS Technical Report #104*. April 2012.

[15] IOIO for Android. Android Development Tools. http://www.sparkfun.com/products/10748. Accessed April 2012.

[16] A. Kadav, M. J. Renzelmann, and M. M. Swift. Tolerating hardware device failures in software. In *Proc. of the ACM SIGOPS 22nd Symposium on Operating Systems Principles*, 59-72, Oct. 2009.

[17] A. Kadav and M. M. Swift. Understanding modern device drivers. In *Proc. of the 17h Int. Conf. on Architectural Support for Programming Languages and Operating Systems*, 87-98, March 2012

[18] B. Kaufmann and L. Buechley. Amarino: a toolkit for the rapid prototyping of mobile ubiquitous computing. In *Proc. of the 12th Int. Conf. on Human Computer Interaction with Mobile Devices and Services*, 291-298, Sept. 2010.

[19] B. Leslie, P. Chubb, N. Fitzroy-Dale, S. Götz, C. Gray, L. Macpherson, D. Potts, Y. Shen, K. Elphinstone, and G. Heiser. User-Level Device Drivers: Achieved Performance. *Journal of Computer Science and Technology*, 20(5):654-664, Sept. 2005.

[20] J. Liedtke, U. Bartling, U. Beyer, D. Heinrichs, R. Ruland, and G. Szalay. Two Years of Experience with a (Mu)-Kernel Based OS. *SIGOPS Operating Systems Review*, 25(2):51-62, April 1991.

[21] F. X. Lin, A. Rahmati, and L. Zhong. Dandelion: a framework for transparently programming phone-centered wireless body sensor applications for health. *Wireless Health 2010*, 74-83, Oct. 2010.

[22] F. X. Lin, Z. Wang, R. LiKamWa, and L. Zhong. Reflex: using low-power processors in smartphones without knowing them. *In Proc. of the 17th Int. Conf. on Architectural Support for Programming Languages and Operating Systems*, 13-24, March 2012.

[23] B. Priyantha, D. Lymberopoulos, and J. Liu. Enabling energy efficient continuous sensing on mobile phones with LittleRock. *In Proc. of the 9th ACM/IEEE Int. Conf. on Information Processing in Sensor Networks*, 420-421, April 2010.

[24] M. J. Renzelmann and M. M. Swift. Decaf: moving device drivers to a modern language. In *Proc. of the 2009 Conference on USENIX*, 14-14, June 2009.

[25] D. W. Richardson and S. D. Gribble. Maverick: providing web applications with safe and flexible access to local devices. In *Proc. of the 2nd USENIX conference on Web Application Development*, 12-12, Oct. 2011.

[26] J. Sorber, N. Banerjee, M. D. Corner, and S. Rollins. Turducken: hierarchical power management for mobile devices. In *Proc. of the 3rd Int. Conf. on Mobile Systems, Applications, and Services*, 261-274, June 2005.

[27] D. Talbot. Android Marches on East Africa. *Technology Review*, http://www.technologyreview.com/communications/37877, June 2011. Accessed April 2012.

[28] N. Villar, J. Scott, and S. Hodges. Prototyping with microsoft .net gadgeteer. In *Proc. of the 5th Int. Conf on Tangible, Embedded, and Embodied Interaction*, 377-380, Jan. 2011.

[29] Z. Zhuang, K.-H. Kim, and J. P. Singh. Improving energy efficiency of location sensing on smartphones. In *Proc. of the 8th Int. Con. on Mobile Systems, Applications, and Services*, 315-330, June 2010.

Software Abstractions for Trusted Sensors

He Liu[†], Stefan Saroiu, Alec Wolman, Himanshu Raj
Microsoft Research, [†]University of California San Diego

ABSTRACT

With the proliferation of e-commerce, e-wallet, and e-health smartphone applications, the need for trusted mobile applications is greater than ever. Unlike their desktop counterparts, many mobile applications rely heavily on sensor inputs. As a result, trust often requires authenticity and integrity of sensor readings. For example, applications may need trusted readings from sensors such as a GPS, camera, or microphone. Recent research has started to recognize the need for "trusted sensors", yet providing the right programming abstractions and system support for building mobile trusted applications is an open problem.

This paper proposes two software abstractions for offering trusted sensors to mobile applications. We present the design and implementation of these abstractions on both x86 and ARM platforms. We implement a trusted GPS sensor on both platforms, and we provide a privacy control for trusted location using differential privacy. Our evaluation shows that implementing these abstractions comes with moderate overhead on both x86 and ARM platforms. We find these software abstractions to be versatile and practical – using them we implement one novel enterprise mobile application.

Categories and Subject Descriptors

D.4.6 [**Security and Protection**]: Security Kernels

Keywords

Mobile Computing, Sensors, Trusted Platform Module, ARM TrustZone, Differential Privacy

1. INTRODUCTION

Sensor readings gathered from a smartphone have started to have high value. For example, location information is harvested to build maps of Wi-Fi access points at global scale [29]. Photos and videos taken with a smartphone are used by news media to determine abuse and human rights violations by various groups and governments [33, 17]. Blood sugar is being monitored by wireless sensors that control insulin pumps [19]. Location (e.g. geo-fencing) is now being used in mobile payment schemes [30].

As this value is rising, cloud services and mobile applications require "trusted sensors" – the ability to produce sensor readings that instill a high degree of confidence about their integrity and authenticity. Today, it is trivial to fabricate sensor readings by simply making up GPS locations, camera shots, or health readings. As long as fabricating sensor readings can be done by compromising a smartphone's software stack (e.g., deploying a piece of malware), the familiar spectrum of security miscreants will rise to exploit the value of sensor readings. Criminals will try to exploit financial transactions, steal health information for later resale, and prevent the use of sensors data for investigations or prosecutions. Recognizing the need for attesting the authenticity of sensor readings, the research community has started to describe the huge potential of such technology [2, 5, 25, 34] and to propose frameworks able to verify the authenticity of sensor readings captured and modified on a smartphone [6].

Designing the "right" software abstractions for trusted sensors is challenging and non-intuitive. For example, one way to detect tampering is to simply sign the sensor readings. Although this approach meets the definition of trusted sensors (i.e., it protects the integrity of sensor data from malicious applications), it is limited because it does not enable a common scenario where sensor readings are processed on the mobile device before uploading to the cloud [6]. For example, cropping a photo before uploading it to Facebook would invalidate the photo's signature. Signed readings is too rudimentary of an abstraction for trusted sensors because one cannot distinguish malicious tampering from legitimate application needs. Second, simply signing a sensor's reading without semantically understanding the sensor's state is insufficient and can even be insecure. For example, some GPS devices can be put in a "simulated mode" in which they simulate locations different than the actual physical location of the sensor. Finally, mobile applications might want to be able to reveal secrets based on a policy whose input is sensor readings. For example, a geo-fencing security application might want to implement access control based on GPS locations. Unfortunately, simply signing GPS locations does not meet the needs of such applications.

This paper's goal is to present two software abstractions designed to expose trusted sensors to mobile applications and cloud services. The first abstraction is called *sensor attestation* and its role is to protect the sensor reading's integrity and authenticity. This is done by attesting the code producing the reading as well as the sensor configuration (i.e., the sensor's state) when the reading was made. The second abstraction is called *sensor seal*. Sensor seal takes as input a secret, encrypts it, and binds it to a sensor policy. When an application calls sensor unseal, a sensor reading is produced and the secret is revealed only if the reading obeys the policy specified at seal time. For example, a geo-fencing application might seal authentication credentials (i.e., a "secret") to a virtual perimeter. Unseal is successful only if the location readings reveal that the smartphone is within the virtual perimeter. As their names suggest, these abstractions have been inspired from the two primitives used in trusted computing, *software attestation* and *sealed storage*. Like their trusted computing counterparts, these two primitives combined are sufficient for exposing trusted sensors to mobile applications and cloud services.

We have implemented these two abstractions on both x86 and ARM SoC platforms because we believe trusted sensors are needed for both laptops (x86) and smartphones and tablets (ARM). The two implementations are similar only at a very high-level – they both implement a runtime environment protected from the OS and the applications running on behalf of the user of the device. Inside these environments, we build a software stack that implements the two trusted sensor primitives – attestation and seal – and offers them to the untrusted OS and applications. The mechanisms used to build these isolated environments are different between the two architectures. On x86, we leverage trusted computing hardware and the Credo research hypervisor [21] that offers strong isolation properties similar to CloudVisor [36], to build guest VMs that can protect data integrity even if the management VM becomes compromised. On ARM, we leverage the ARM TrustZone extensions that can protect sensor readings' integrity from all other software running on the platform. The hardware support on ARM makes our system much simpler than its x86 counterpart.

Currently, both our implementations offer attestation and seal for GPS only. To demonstrate the benefits of trusted GPS, we implemented a new mobile application, called *TrustedDrive*. With TrustedDrive, a user can protect a storage partition by defining a geo-location policy. The storage partition is mounted only if the current location returned by the trusted GPS satisfies the original policy. Such an application is useful in an enterprise environment because it ensures that sensitive partitions are available only if employees are on premise.

Trusted sensors also raise privacy concerns for two reasons: 1) the digital signature that accompanies the sensor reading may identify who signed the reading, and 2) the content of the sensor reading may reveal sensitive information about the device's owner. In this paper, we describe approaches to deal with both problems, and we implement a differential privacy layer to address the latter problem. Our first approach is based on cryptography constructs that change the nature of the signature that accompanies a sensor reading when sent to the cloud. These mechanisms can provide user anonymity and sensor data non-transferability (e.g., while the cloud can verify a sensor reading's authenticity, the cloud cannot convince others of the reading's authenticity).

Our second approach aims to protect the information revealed by the sensor data before it is signed. For this we leverage differential privacy [3, 4], a technique that provides a mathematical framework for injecting "noise" into a query answer and measuring the amount of privacy loss in answering the query. We show how differential privacy can be added as a "privacy layer" to a trusted GPS implementation. This layer handles incoming queries for GPS coordinates and answers them in a differentially private-manner. The answers remain trustworthy because they continue to carry a trusted sensor attestation. Finally, because differential privacy introduces errors in the sensor readings, such errors may affect the correctness of mobile applications that depend on the GPS. We evaluate the magnitude of such errors using a previously collected trace of GPS locations [28], and show that certain classes of mobile applications will not be affected by the noise introduced by differential privacy.

The contributions of our paper are:

- We design two trusted computing abstractions that fit the needs of a growing class of mobile applications – those that require increased assurance about the sensor data produced by a mobile device.

- We demonstrate how to implement these abstractions in a virtualized system. Using a serial port GPS device as an example, we show how to hand over ownership of a sensor device from the management VM to a separate VM whose responsibility is to provide the trusted sensor stack.

- We demonstrate how to implement these abstractions using trusted computing hardware extensions found in ARM-based systems. In particular, we show how to leverage ARM TrustZone to provide a lightweight trusted sensor stack with a small trusted computing base (TCB).

- We take some initial steps toward overcoming the privacy challenges posed by trusted sensors. We demonstrate how differential privacy can be used with a GPS sensor to offer systematic a way of reasoning about how to add noise to sensor readings to improve privacy.

2. THE NEED FOR TRUSTED SENSORS

Previous work [2, 5, 6], including our own workshop papers [25, 34], describes many mobile applications that benefit from the presence of trusted sensors, and the attacks that rise due to the lack of them [31]. Rather than re-enumerating all trusted mobile applications in this paper, instead we classify them in a higher-level taxonomy. This classification helps to build a better intuition of why these two software abstractions meet the needs of these trusted applications.

1. Applications that collect proofs. These applications need to collect sensor readings and present them as "proofs" either immediately after they are collected, or at some later time. Sometimes these proofs are used to demonstrate the authenticity of content, as is the case with validating that photos have not been photoshopped. In other cases, such proofs are used to demonstrate a particular action or behavioral pattern of the user, as is the case with offering store discounts to loyal customers based on location proofs.

2. Participatory sensing/crowdsourcing applications. These applications upload sensor readings to the cloud. The cloud often combines readings from multiple users in a process referred to as *crowdsourcing*, to build an aggregate view of data. This aggregate view is then offered as a cloud service to all users.

This class of applications is subject to database manipulation attacks [31], where an attacker uploads fake sensor readings to the cloud. With sensor attestations, it is much harder for attackers to fabricate or alter sensor readings without being detected.

3. Applications that use authentication. These applications need to perform a security-sensitive operation based on a sensor reading. For example, a user might get access to a secret file or password only when present at a specific location. Sometimes the secret might be revealed by the cloud, whereas in other cases the secret might be revealed by the local smartphone. In both cases, the secret is revealed based on the reading values of a trusted sensor.

The sensor seal abstraction makes it easy to build such applications. The secret data can be first sealed according to a particular sensor-based policy. Any application that uses authentication can just issue an unseal operation to receive the secret data.

3. THREAT MODEL

A mobile user can install malware (whether accidentally or intentionally) on their mobile device, compromising the general purpose operating system. While such malware can tamper with sensor attestations, we enable detection of such tampering through signature verification. Malware may also modify all the software that runs on the machine (even the persistent copy of the trusted software), but such modifications will be detectable by using software attestation [32]. To provide these guarantees, we require that the trusted computing base (TCB) of our system cannot be compromised. If our system's TCB ever becomes compromised, we cannot provide any guarantees. On x86, the system TCB consists of the trusted sensors software stack plus the Credo hypervisor. On ARM, this consists of all code that runs in the ARM TrustZone secure world. Remember that, on both x86 and ARM, the general purpose OS, system services, and all third party applications are not part of the TCB.

Several classes of attacks are out of scope for our current system. As mentioned in the previous paragraph, all attacks that compromise our system's TCB are out of scope. Another class is side-channel attacks that attempt to infer the secrets (e.g., the signing key) of our trusted environment through covert channels. Another class is tampering with the trusted computing hardware, such as the TPM. TPMs were not designed to protect against physical attacks; the TPM spec does not require tamper-proof manufacturing [32]. Finally, it is possible to manipulate the physical environment to create false sensor readings. For example, keeping a lighter next to a temperature sensor will produce artificially high temperature readings.

As with all trusted computing systems, in addition to relying on the software TCB being exploit free, we must also trust the manufacturer of the trusted computing hardware. There are two aspects to this trust: 1) relying on a correct implementation of the TPM chip or TrustZone feature, and 2) relying on a secure provisioning process. To provision a TPM chip, the manufacturer injects a unique identity, known as the endorsement key-pair, into the TPM chip. The manufacturer then signs a certificate indicating that the public key of the endorsement key-pair is an authentic TPM chip. The security guarantees rely on the manufacturer's signing key remaining secret, and the private key of the endorsement key-pair remaining secret.

4. TRUSTED SENSORS ABSTRACTIONS

Trusted sensors are an important building block to building trustworthy systems in the mobile landscape. Although there is no precise definition, the term "trusted computing" often refers to systems that build upon hardware primitives to provide code integrity protection and confidentiality for their secret data. While the meaning of code integrity is well-understood, it is important to elucidate the meaning of data confidentiality. Typically, trusted computing systems are capable of protecting a piece of data in such a way that only a specific, pre-determined piece of code can access it. The combination of these two properties allows a system to guarantee that trusted code will run unmodified and will protect its secret data from all untrusted code.

Based on these insights, we borrow abstractions from trusted computing and map them to mobile sensing. The remainder of this section describes our abstractions and how they meet the needs of trusted mobile applications.

4.1 Abstraction #1: Sensor Attestation

A sensor attestation protects the sensor reading's integrity and demonstrates its authenticity. To offer these two properties, the sensor reading is signed. The key used for signing is a Trusted Platform Module (TPM)'s Attestation Identity Key (AIK) [32]. The private portion of the AIK is non-migratable and protected by the TPM. The underlying platform also binds the *same* AIK to the software configuration of the platform's TCB, and can use this AIK to sign *remote attestations*. The combination of binding the AIK to a trusted configuration and sharing the *same* AIK to sign both sensor and remote attestations provides a notion of authenticity: *this platform with this specific configuration produced this sensor reading.*

There is one additional practical requirement related to sensor attestation. The sensor attestation must also incorporate a notion of the sensor's state to ensure that software can correctly interpret how the reading was produced. In Section 5, we elaborate on this requirement.

4.2 Abstraction #2: Sensor Seal

Sensor seal protects a secret by encrypting and binding it to a *policy* that uses sensor readings. Unseal reads the sensors and evaluates a policy predicate to determine whether the sensor readings satisfy the policy. If so, the secret is decrypted. The encryption is performed using a secret storage key, similar to the storage root key (SRK) used by the TPMs [32]. As with the AIK, the private portion of the SRK is non-migratable and protected by the trusted computing hardware.

The policies used for sealing can be quite complex and span the readings of multiple sensors. We did not want to restrict application developers in the types of policies they can use. As a result, in our implementation, policies can be specified using a full-featured scripting language (we use Python in our prototype).

4.3 Local Processing of Sensor Readings

In some cases, sensor readings need to be processed on the local device before they can be uploaded to the cloud. For example, photos might be needed to be shrunk, cropped, or re-encoded at a lower resolution to reduce the bandwidth and energy costs of cloud upload [6]. Such legitimate pre-processing needs will affect the integrity of the sensor reading uploaded to the cloud.

To preserve the authenticity of these processed sensor readings, the local application code manipulating them must be able to protect its integrity at runtime. This code will first validate the sensor attestation, then process the sensor reading, and finally upload it to the cloud, all without its code being modified by an attacker. Without code integrity, the sensor readings being uploaded to the cloud could be modified by malware. Many systems for offering code integrity have been previously proposed [27, 14, 13, 26, 36] and they can all be integrated with our sensor attestations to offer end-to-end trust guarantees. In this paper, our x86 implementation supports local sensor processing because of our use of Credo hypervisor [21], as described in Sections 6 and 7.

5. TRUSTED SENSORS API

This section describes the API we use to provide the two software abstractions for trusted sensors. The abstractions are implemented by a runtime environment with strong isolation from the rest of the system, including the OS. Even in the presence of a compromised OS, applications can use our API to obtain sensor readings that protect their integrity and authenticity. The implementation of this API is described in Sections 7 and 8; this section focuses on describing the API design and how mobile applications can use it.

5.1 Sensor Attestation API

At a high-level, a sensor attestation is a sensor reading signed with an AIK [32]. To validate the reading's integrity, a verifier checks the signature using the public part of the AIK. To validate its authenticity, a verifier requests a remote attestation from the platform, using a standard remote attestation verification protocol. This last step attests the software configuration producing the sensor reading.

```
enum flash_mode_t {flash_off, flash_on, flash_auto};
enum autofocus_mode_t {focus_normal, focus_macro};
enum white_balance_t {automatic, incandescent,
                      fluorescent, daylight, cloudy};
enum scale_t {min, low, medium, high, max};
enum iso_t {isoauto, iso50, iso100, iso200,
            iso400, iso800};
enum metering_t {matrix, center_weighted, spot};

typedef struct camera_config {
    int resolution_width;
    int resolution_height;
    flash_mode_t flash_mode;
    autofocus_mode_t autofocus_mode;
    white_balance_t whitebalance;
    scale_t contrast;
    scale_t saturation;
    scale_t sharpness;
    int exposure_value;  // from -2 to +2
    iso_t iso;
    metering_t metering;
    bool wide_dynamic_range;
    bool anti_shake;
} camera_config_t;

CapturePhoto(camera_config_t *camera_config, // out
    int *image_size,                          // out
    unsigned char **image_bytes,              // out
    rsa_sig_t *signature)                     // out
```

Figure 1: Trusted Camera API. ***The settings available for this camera device are based on the Samsung Focus Windows Phone.***

We now describe how to design the API for a trusted sensor. The goal of a trusted sensor is simply to associate a digital signature with a sensor reading, to provide both authenticity and integrity for the sensor reading. Thus, the trusted sensor API must ensure that each call that returns sensor data also provides an additional out parameter for the signature, and this signature should cover all of the output data. We demonstrate the process using three example APIs: 1) for trusted Wi-Fi scanning; 2) for a trusted GPS sensor; and 3) for a trusted camera. We use three simple guidelines in turning a traditional sensor API into a trusted sensor API.

1. The developer must identify the different types of readings that a sensor can produce and the device-specific state needed to correctly interpret each sensor reading. All this device-specific state must be exposed through the trusted sensor API. For example, Figure 1 shows the API for a trusted camera. In this example, all of the camera settings that affect the resulting image, such as the flash mode and the exposure, must be explicitly represented in the API. The camera API allows a caller to read all of the relevant camera configuration state.

2. All API calls should be separated into either read or write operations – there should be no calls that mix both reads and writes into a single call. While sensor devices are typically thought of as read-only devices, most sensors support write operations in practice. A camera exposes operations that control its configuration state. These write operations are used to configure the sensor in ways that affect how the sensor readings should be interpreted.

As another example, many crowdsourced location systems use Wi-Fi scanning to determine the approximate location of a mobile device. In this example, the Wi-Fi radio acts as a sensor device, and many configuration parameters can affect the scan results. The list of channels the radio listens on, the listening duration, and the 802.11 band that the radio supports all affect which beacons will be heard during the scan. Figure 2 shows an example of the trusted sensor API that supports Wi-Fi passive scanning.

3. The signature provided with all read operations should cover

```
enum band_80211_t {80211a, 80211b, 80211g, 80211n};

typedef struct beacon_info {
   char *ap_bssid,
   char *ap_ssid,
   int  mean_rssi,
} beacon_info_t;

typedef struct sender_info {
   char *sender_mac,
   int channel,
   int mean_rssi,
} sender_info_t;

GetWifiPassiveScan(band_80211_t *band,   // out
     int *num_channels,                  // out
     int **channel_list,                 // out
     int *scan_delay_ms,                 // out
     int *num_beacons;                   // out
     beacon_info_t **beacons,            // out
     int *num_senders;                   // out
     sender_info_t **senders,            // out
     rsa_sig_t *signature)               // out
```

Figure 2: Trusted Wi-Fi Scanning API.

both the sensor's data and its configuration state that affects the sensor reading. One alternative that we considered but decided against was to have a separate API call to fetch the current configuration state along with a digital signature. The advantage of using one signature is that it ensures that the configuration is always sent along with the sensor reading. This ensures that configuration state needed to interpret the semantics of the sensor reading is always available to the software that needs it.

The API does not need to expose control of all device configuration state. The Wi-Fi and camera devices allow calls to modify the device state (elided from our examples), but the GPS example shown in Figure 3 does not. The GPS device does not allow selecting which satellites will be used to produce a location fix, but it does allow for reading which satellites were used in producing a given location fix.

5.2 Sensor Seal API

The API for sensor seal and unseal is shown in Figure 4. Seal takes as inputs a secret and a Python script specifying a sealing policy. For seal, the script's source code is concatenated with the secret and encrypted into a blob returned to the caller. To unseal, the caller passes in the encrypted blob. The blob is first decrypted using a storage key, similar to the TPM's storage root key (SRK), that never leaves the TCB of the system. Decryption produces both the secret and the Python script. The policy script predicate uses the API described above to obtain sensor readings and attestations. The policy script contains a predicate function that obtains the appropriate sensor readings, checks them against the policy, and if those checks pass the predicate returns true. If the predicate returns true, the system then decrypts the sealed blob and returns the secret to the caller.

6. HIGH-LEVEL SYSTEM ARCHITECTURE

The code implementing trusted sensors must be protected from the OS and the applications running on the mobile device. The code must be run in an isolated environment that offers (1) code integrity to ensure that malware cannot modify the trusted sensors' software stack, and (2) data confidentiality to protect the signing and encryption keys. This code is part of the system's TCB and

```
enum fix_t {no_fix, 2d_fix, 3d_fix};
enum fix_quality_t {invalid, gps_fix, dgps_fix,
                    pps_fix, rtk, float_rtk,
                    manual, simulation};

typedef struct sat_info {
    int sat_prn,
    int elevation_degrees,
    int azimuth_degrees,
    int snr,
} sat_info_t;

// combine output of GPGGA and GPGSA sentences to
// produce this data.
GetCurrentLocationFix(time *utc_time,          // out
    float *latitude,                           // out
    float *longitude,                          // out
    float *altitude_above_mean_sea_level,      // out
    fix_quality_t *fix_quality,                // out
    fix_t *fix_type;                           // out
    int *num_satellites,                       // out
    float *height_of_geoid,                    // out
    float *dilution_of_position,               // out
    float *vertical_dop,                       // out
    float *horizontal_dop,                     // out
    rsa_sig_t *signature)                      // out

// use output of GPGSV sentences to produce this data
GetCurrentSatInfo(int *num_sats_in_view,     // out
    sat_info_t **sat_info,                   // out
    rsa_sig_t *signature)                    // out
```

Figure 3: Trusted GPS API.

```
SensorSeal(int  secret_length,    // in
           char *secret,          // in
           int  script_length,    // in
           char *script,          // in
           int  seal_length,      // out
           char *sealed_blob)     // out

SensorUnseal(int  seal_length,      // in
             char *sealed_blob,     // in
             int  secret_length,    // out
             char *secret)          // out
```

Figure 4: Sensor Seal/Unseal API.

implements the trusted sensor APIs, the drivers needed to access the sensors, and a policy interpreter for the Python scripts needed for sensor seal.

Figure 5 illustrates the high-level architecture of our trusted sensors system. The TCB is presented at the bottom, while the OS and mobile applications are running at the top of the stack. The middle layer is used to run application code that needs to locally transform the sensor readings before uploading them to the cloud. This local processing needs to be isolated from the OS to preserve the integrity of the readings once processed. This layer is not within our system's TCB because it contains app-specific code and bugs in that code do not affect the security of our trusted sensor abstractions. However, the security of this layer is important to each application that needs to perform app-specific post-processing of sensor readings. As a result, our system needs to provide the ability to produce attestations that can be used to demonstrate the integrity of the application code used to perform this post-processing. The implementation of this layer is not a contribution of this paper, even though our system on the x86 platform does inherit this functionality from the Credo hypervisor. Instead, our system's focus is on the bottom layer of the figure – we present an implementation of the trusted sensor abstractions that remains secure even in the face of an OS compromise.

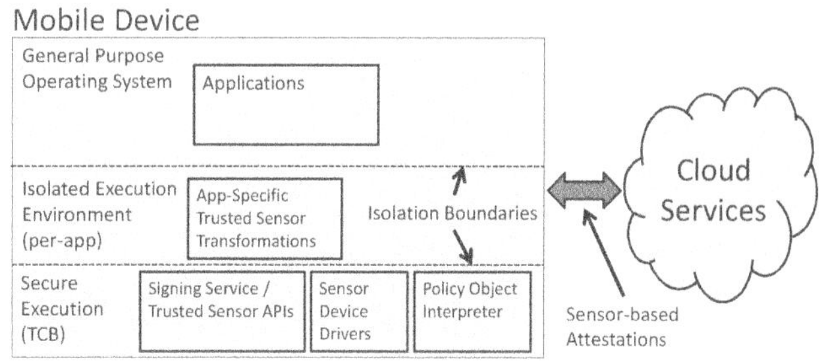

Figure 5: High-level Architecture of Trusted Sensors.

6.1 Design goals

Based on the description of the abstractions and their APIs, our system architecture has five goals.

1. Protect the integrity of the software producing the sensor attestation. Without integrity, our system cannot guarantee the properties of sensor attestation.
2. Protect the confidentiality of the attestation signing key. If the signing key for attestations were to be leaked, attackers can create their own attestations, compromising our system.
3. Protect the integrity of the sensor. It is paramount that the sensor produces correct readings and that it is not subject to erroneous configurations by untrusted software.
4. Provide support for seal/unseal for applications. To unseal successfully, the platform must check the sensor reading against a policy supplied by an application. Such checks can be complex and need to be expressed in a Turing-complete language.
5. The platform must offer adequate forms of privacy protection for mobile users' sensor data.

6.2 Design alternatives

Throughout the design of our systems (both x86 and ARM), we faced several design choices. In the remainder of this section, we describe three such choices we made and the design alternatives. The choices we made were not obvious at the time, and we believe that the alternatives are also interesting points in the trusted sensors design space.

6.2.1 *Hardware vs. Software*

Trusted sensors can be built both in software and in hardware, although each option presents a different set of challenges and drawbacks. A software-only implementation uses a small piece of trusted code to obtain a sensor reading and to sign that reading with a secret key to detect any tampering with the data. To be secure, this design must meet two requirements. First, the integrity of this code and the confidentiality of its secret key must be protected from all other software running on the machine. Second, the system must protect the integrity of the sensor device during the reading. Otherwise, malicious software could reprogram the sensor to cause its reading to become meaningless. Making such systems secure is challenging because the protection mechanisms must have a small TCB. For example, one option would be to rely on the general purpose OS to produce signed sensor readings. We believe such a design is insecure because OSes have large TCBs and cannot offer code integrity and data confidentiality. However, more secure designs have been proposed; some rely on hypervisors[27, 13, 36] to offer code integrity and data confidentiality, whereas others rely on secure co-processors[14]. None of this previous work was aimed at building trusted sensors and does not meet the sensor integrity requirement.

The other alternative is building trusted sensors in hardware. For this, sensor manufacturers need to equip their sensors with additional logic to perform digital signatures. Such designs are even more secure than software-only designs because their software TCB is significantly smaller. The main drawback of building trusted sensors in hardware is the added cost – building sensors is a low margin business and adding logic to compute signatures may increase the sensor's cost. Thus, building hardware trusted sensors at large-scale has a high barrier to deployment. To avoid this barrier, we have chosen a software-only design for the systems presented in this paper.

6.2.2 *Isolation mechanisms*

The software stack for producing trusted sensors must run in an isolated environment to protect its code and data. Such an isolation mechanism must not rely on an operating system, because OSes have too large of a trusted computing base to provide adequate security. In practice, there are two mechanisms to build such isolated environments – using hypervisors, and using hardware support for trusted computing, such as Intel's Trusted eXecution Technology (TXT) or ARM TrustZone.

We regard using trusted computing hardware as a more secure alternative over hypervisors. Building isolation mechanisms with these features can be done with a much smaller TCB than that of a hypervisor. Unfortunately, a pure TXT environment on x86 platforms has serious performance shortcomings in practice. These limitations are well-known [14] and led others to combine the use of TXT with a hypervisor to build isolation on x86 platforms [27, 13]. In particular, the TXT environment is used to perform a measured launch of a hypervisor whose trustworthiness can be attested by the TPM. The hypervisor then provides an isolation environment whose trust is rooted in the hypervisor.

We use a similar combination of a hypervisor and Intel TXT for our isolation mechanism on x86. On ARM however, we used the ARM TrustZones, as this mechanism does not suffer from the performance limitations of TXT. As a result, our ARM-based trusted sensors system has a much smaller TCB than its x86 counterpart.

6.2.3 *Privacy mechanisms*

Attaching a digital signature to every sensor reading has the potential to impact users' privacy, because one must provide the corresponding public key to any entity that wants to validate the authenticity of the sensor reading. At a high level, we see two general approaches to addressing these privacy concerns: 1) using cryptographic techniques to reintroduce anonymity, and 2) using differential privacy to provide statistical information about a sensor data set. Each approach provides a very different kind of privacy – the former hides the identity of the system that generates the sensor readings, whereas the latter reduces the accuracy of individual sensor readings to protect users' privacy.

The first and simplest option is to allow selective disabling of signing the sensor readings. To accomplish this, the system must allow users to remove the attestation from the corresponding sensor reading. With our implementation this is trivial because our trusted sensing API provides the attestation as an extra field in the sensor readings. Users can simply configure their systems to return *null* in place of an attestation. While this approach is simple, users who choose this approach will lose the benefits of trusted sensors.

Another option is to use cryptographic protocols that provide anonymity and information non-transferability. Our previous work [25] advocated using group signatures [1] for anonymity combined with zero-knowledge protocols [7] for non-transferability. While these approaches continue to be viable, several reasons made us not pursue a complete implementation of these approaches. First, group signatures require setting up a trusted group manager in charge of managing group members. Compromising the group manager leads to anonymity compromises of group members. In practice, setting up a trusted group manager appears expensive and technically difficult. Second, providing non-transferability has a high performance cost because these algorithms require high computational power. In previous work, we implemented a zero-knowledge protocol whose performance overhead was about 900 ms on a Pentium 1 GHz CPU. Adding an overhead of one second to each sensor reading may be too much for certain applications.

The third option (and the one we take) is to use differential privacy – a mathematical approach to measuring the amount of privacy loss given a query and a "noise" constant. Users can design policies that specify *privacy budgets* for each application or for the entire system. Applications run their queries on a set of sensor readings until the privacy budget is exhausted. The query results are noisy – they are inaccurate so that the amount of privacy loss due to the query stays within the budget. Section 10 describes how we use differential privacy for a trusted GPS sensor.

7. IMPLEMENTATION ON X86 PLATFORMS

On x86, we use Credo to provide an isolated environment for producing trusted sensor attestations. Credo is a research hypervisor, based on Microsoft's Hyper-V, which provides guest VMs additional protections from the Root VM. First, we provide a short primer on Credo (more details can be found in [21]). Next, we describe our modifications to Credo to enable trusted sensors.

7.1 Background on Credo

Credo is a hypervisor that leverages the TPM to establish trust in hypervisor that is launched at boot time. The Credo hypervisor also offers a new type of guest VM, called an *emancipated VM*. Emancipated VMs differ from traditional guest VMs in two respects: 1) the hypervisor provides stronger isolation, and 2) Credo enables measurement of the software state of the emancipated VM. The combination of these features ensures that as long as the hypervisor and the TPM remain uncompromised, then: 1) an emancipated VM can attest its configuration to external applications and services, and 2) an emancipated VM can persist secrets in untrusted storage that can only be decrypted by that same emancipated VM. Unlike most commodity hypervisors, the trusted computing base (TCB) of Credo's emancipated VMs is only the hypervisor; the Root VM (similar to Xen's Dom0) cannot compromise the integrity and confidentiality of an emancipated guest VM.

To verify the trustworthiness of the Credo hypervisor and to protect itself from attacks in the OS pre-boot environment, Credo uses the dynamic root of trust measurement (DRTM) features available on both Intel and AMD CPUs.

To support the emancipated guest VM features, Credo performs the following two actions when booting an emancipated guest VM. First, it isolates the memory and the virtual CPU state of the guest VM from all other VMs. Credo restricts the Root VM from accessing the guest VM's memory pages and from making any intercepts that could change the guest's virtual CPU state. Second, when resuming an emancipated guest VM image, Credo records a measurement to attest the software configuration of the emanci-

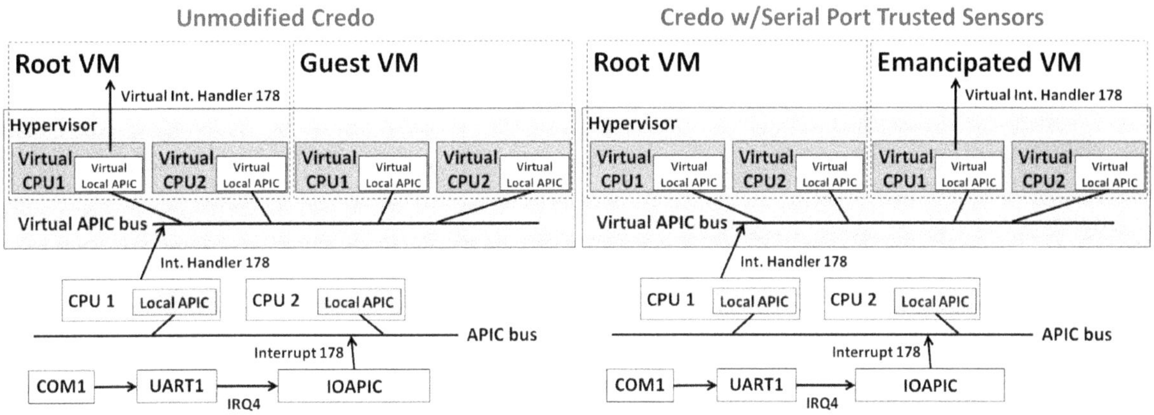

Figure 6: Interrupt Handling Architecture for the Serial Port on x86. *On the left, the figure shows the standard interrupt handling for Credo (or any other generic hypervisor). On the right, the figure shows our modification to Credo to add support for mapping the serial port to emancipated guest VM.*

pated guest VM. This enables software running inside this emancipated guest VM to seal a secret to the measurement of emancipated guest VM's software configuration. Credo then ensures that only the guest VM whose current measurement matches the specified measurement that came with the seal operation can decrypt this secret. A similar mechanism can be used by external cloud services: they can also create secrets in way that only specific emancipated VMs with a specific software stack can decrypt them.

7.2 Modifying Credo For Trusted Sensors

Credo provides a building block for two of the design requirements for trusted sensors – code integrity and data confidentiality. We enable trusted sensors with Credo by placing the trusted sensor software stack in an emancipated VM, thereby protecting it from the Root VM which runs the general purpose OS and applications. However, Credo leaves the Root VM in control of the physical I/O devices, and uses encryption to protect data before handing it off to the Root VM on its way to an I/O device. This ensures that the Root VM is no longer part of the system's TCB. However, we need direct access to sensor I/O devices from the emancipated VM that hosts the trusted sensor software stack. To support this, we modify Credo to support mapping certain I/O devices directly into an emancipated guest VM. In particular, for our implementation we enable one form of I/O support: our current implementation uses a GPS sensor that attaches through a serial port, as a result we need to enable serial port I/O from an emancipated guest VM.

7.3 Enabling Serial Port Trusted Sensors on Credo

We begin our implementation description by presenting how the serial port is handled by Credo. Note that for the features described is this discussion, Credo is identical to Hyper-V. Also, the architecture described here is very similar to other commodity hypervisors such as VMware or Xen. The goal of our modifications to Hyper-V/Credo is to take away ownership of the serial port from the Root VM and enable the emancipated guest VM to own the serial port. To accomplish this, we first enable mapping the serial port into an emancipated guest VM (rather than to the Root VM), and then we remove the Root VM's ability to communicate with that serial port.

7.3.1 Serial Port Interrupt Handling in Credo

X86 hardware uses an universal asynchronous receiver/transmitter (UART) to control the serial (COM) port. Whenever data is available on the COM port, the UART sends an interrupt request (IRQ4) to the I/O advanced programmable interrupt controller (IO APIC). The IO APIC translates this interrupt into IRQ 178, which it then sends on the APIC bus to which all CPUs are connected. All interrupts sent on the APIC bus are handled by the local APICs, interrupt controllers within each CPU core which handle delivery of the interrupt to the CPU. Each local APIC is programmed to control which IRQ messages should be delivered to the local CPU. The hypervisor configures one CPU, designated at boot time as the master, to handle all interrupts from I/O devices that arrive on the APIC bus. This master CPU also happens to be the CPU that runs the Root VM, and therefore the full virtualization stack. While the hypervisor is handling an interrupt, lower priority interrupts are masked (blocked) until the current handler completes. When the hypervisor interrupt handler finishes, it clears the interrupt signaling that the CPU is ready to receive additional interrupts.

When the local APIC interrupts the CPU to execute the appropriate interrupt handler, the hypervisor transforms this hardware interrupt into a message sent on the *virtual APIC bus* inside the hypervisor. Each virtual core has a virtual local APIC which receives these message and passes them on to the corresponding virtual core. For each VM, the hypervisor maintains a mapping of which virtual processors are hosted on which physical cores. Interrupts destined for a particular VM are always sent to the first virtual processor for that VM. *Synthetic interrupts* sent on the virtual APIC bus are implemented in two ways: if the destination virtual processor is hosted on the same physical core, the corresponding virtual local APIC can be manipulated directly, and if not then the interrupt is sent using an inter-processor interrupt (IPI) sent over the physical APIC bus. Upon receipt of an interrupt destined for the local virtual processor, the virtual local APIC delivers the interrupt to the hypervisor, which then dispatches it to the first virtual processor of the destination guest VM. By default, all I/O devices, including the serial port, are always owned by Root VM. On the left, the figure 6 shows a diagram of the standard serial port interrupt handling in Credo.

7.3.2 Interrupt Handling for Trusted Sensors

We modified the interrupt handling in Credo to enable mapping the serial port into an emancipated guest VM, rather than the Root VM. The reason for this is that we want to take away ownership of the serial port from the Root VM, and enable an emancipated guest VM to own the serial port. To accomplish this, we modified the virtual local APIC implementation in the Credo hypervisor. When an interrupt arrives, the hypervisor looks up which physical processor hosts the virtual processor 1 for the emancipated guest VM. It then uses the synthetic interrupt mechanism described in the previous section to deliver the interrupt to the correct virtual processor. The hypervisor then forwards the serial port interrupt up into the emancipated VM whose serial port driver handles it. In summary, we simply implement a bypass mechanism whereby interrupts for trusted sensor devices are delivered to the emancipated VM rather than the Root VM. The modifications to the hypervisor to enable this are quite simple, requiring less than 100 lines of code to implement.

7.3.3 Handling Serial Port I/O in Credo

The Root VM uses port-mapped I/O to configure the serial port and to perform actual I/O operations on the serial port. To enable port-mapped I/O from VMs, the hypervisor maintains an IO intercept bitmap that determines which of the port-mapped I/O addresses are trapped for each VM. These intercepts occur when the a VM executes any CPU instruction (such as IN or OUT) that accesses a port. By default, all port-mapped IO addresses are trapped for normal guest VMs, but for the Root VM only the address range corresponding to keyboard controller is trapped. As a result, the Root VM can use IN and OUT instructions to directly access the serial port UART.

7.3.4 Serial Port I/O for Trusted Sensors

To disable access to the serial port from the Root VM, we modify the IO intercept bitmap in the hypervisor so that the COM port IO addresses (0x3F8 to 0x3FF) are added to the Root VM's intercept bitmap. When the hypervisor traps these instructions, we modify the hypervisor so that all reads on these ports return 0xFF, and all writes are simply discarded. We chose this behavior because it matches the behavior of a serial port when no device is attached, and as a result this tells the Root VM that no device is present.

The final step is to enable the emancipated guest VM to access the serial port. To enable this, we modify the hypervisor to emulate all the IN, OUT, and OUTS instructions. We did not need to emulate the INS instruction because the serial port driver does not use that instruction. We chose to emulate, rather than modifying the guest VM IO intercept bitmap because it made our implementation simpler and because the overhead of emulation is not a problem at the low bitrates supported by the serial port.

7.4 Building Sensor Attestation and Seal

We use an emancipated guest VM to host the software that controls the sensor device and produces signed readings. Inside this VM, we run a 64-bit version of the Windows Preinstallation Environment (WinPE) created with the Windows Automated Installation Kit. WinPE is a minimal but fully functional Windows 7 environment with the full set of standard Windows 7 libraries. Because the software that runs in the emancipated guest VM is part of our TCB, it is critical to make the size of these components as small as possible. While WinPE has a much smaller TCB than a full installation of Windows 7, the TCB is still relatively large. In the future we could replace this with a much smaller software stack that is similar to the one we use for our ARM TrustZone implementation described in Section 8. The initial image for the emancipated VM that contains the trusted sensor software stack (and the measurement of that image) is created in a trusted environment to ensure that malware does not enter into the emancipated VM.

When the emancipated VM boots for the very first time, it creates a new RSA-1024 key-pair for signing, and uses the TPM's AIK to produce a *quote* for the public key of that key-pair. The reason for this step, rather than just directly using the TPM's AIK, is that it improves the performance of our sensor attestation operations. This is because RSA crypto operations on the main CPU are much faster than using the TPM for these operations. Further, the TPM quote includes relevant platform state which can be used to remotely attest that the Credo hypervisor has been launched using TXT, and that the VM is executing in emancipated mode. We also create a new symmetric storage key (an AES key) that will be used to implement sensor seal and unseal. We use the TPM's *seal* operation to protect this storage key and the private half of the signing key-pair. Finally, we save these items together on external storage: the sealed blob of keys, the TPM's AIK certificate, and the quote of the key-pair's public key.

Each time the emancipated VM boots, we initialize the network, and start the trusted sensors service. To support sensor seal and unseal, we also initialize a python interpreter inside the emancipated VM. Next, the trusted sensor service uses the TPM to unseal the blob of keys. Because the TPM seal operation includes the platform state (in a manner similar to the TPM quote), when the unseal is successful this ensures that: 1) the Credo hypervisor is running securely, 2) the VM is running emancipated, and 3) *this* emancipated VM performed the original seal operation. Finally, our trusted sensors service starts listening on a TCP port for requests to read a trusted sensor.

To handle incoming requests, our trusted sensor service reads from the specified sensor hardware, signs the readings with the private signing key, and replies with the signed data. We sign a SHA-1 hash of the data using the private RSA key. While the private signing key is kept confidential, the public key, the TPM's AIK certificate, and the TPM quote of the public key, are all exported to callers. This enables external programs to verify the integrity and authenticity of sensor readings produced by this VM.

8. IMPLEMENTATION ON ARM PLATFORMS

ARM TrustZone is ARM's hardware support for trusted computing. It is a set of security extensions found in many recent ARM processors (including Cortex A8 and Cortex A9). ARM TrustZone provides two virtual processors backed by hardware access control. The software stack can switch between the two states, referred to as "worlds". One world is called *secure world* (SW), and the other *normal world* (NW). With TrustZone, the entire OS and all applications run in the normal world, whereas a small trusted kernel runs in the secure world. The secure world provides code and data integrity and confidentiality because untrusted code running in the normal world cannot access the protected resources, such as memory pages and peripheral registers of the secure world.

An ARM platform first boots into the secure world. Here, the system firmware provisions the entire runtime environment of the secure world. Once provisioning is complete, the secure world yields to the normal world where the firmware loads up the bootloader which then loads the OS. The normal world must use a special ARM instruction called *smc* (secure monitor call), to call back into the secure world. When the CPU executes the *smc* instruction, the hardware switches into the secure monitor, which performs a secure context switch into the secure world.

Hardware interrupts can trap directly into the secure monitor code, which enables flexible routing of those interrupts to either world. This allows sensors (or other I/O devices) to map all their interrupts to the secure world and to protect the integrity of those sensors from untrusted code. However, this step is not sufficient

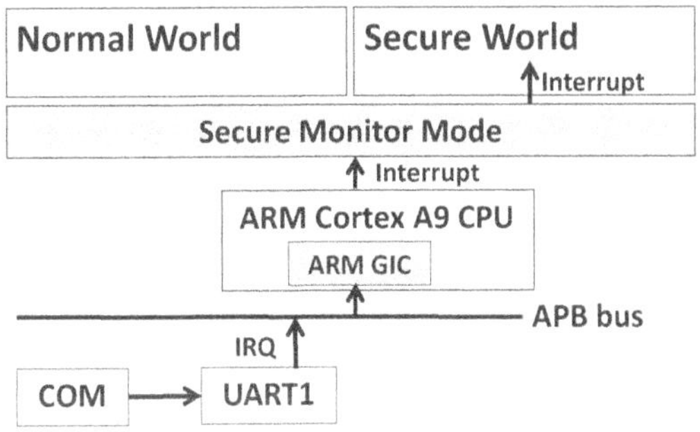

Figure 7: Architecture of handling the serial port on ARM. ***Overview of how the serial port interrupts are handled on the ARM platform.***

to fully protect I/O devices, because untrusted code executing in the normal world could potentially reprogram the device to disable it from generating interrupts, even if those interrupts are routed to the secure world. This is possible only if the normal world can memory-map the device and therefore access its control registers.

In this context, there are two approaches to secure access to a sensor device. One approach uses hardware configuration: the AXI-to-APB bridge reads an input signal that indicates which devices attached to the APB bus are only accessible from secure world, and these devices cannot be memory-mapped by the normal world. For the other approach, the device can be memory-mapped into both worlds. Thus, when the secure world starts executing, it must reset the device to a known good state, and then it must not relinquish control to the normal world until the sensor reading is complete.

8.1 Trusted GPS on ARM TrustZone

Before we describe the Trusted GPS implementation, we provide a brief description of the software stack we developed for the ARM TrustZone secure world, which we call the Secure-world Execution Environment (SEE). SEE is composed of three parts - a kernel, a runtime environment, and secure services. Trusted GPS is implemented as one secure service inside SEE.

Our kernel, which includes the secure monitor mentioned above, implements context switching between normal world and secure world, and dispatching of secure service requests from the NW to the appropriate service. Service requests are made via the *smc* instruction. The kernel also implements interrupt handling – for non-secure interrupts, the context is switched back to NW where they are handled by the NW operating system. This mechanism is also used to make the kernel and secure services preemptible. However, any time a secure service is preempted it returns to the NW with a special *continue* status. The NW is responsible for retrying the smc instruction to let the service continue execution. Since the SEE currently supports only a single context, the NW must serialize access to the SW among multiple secure service calls - it must ensure that a single secure service call finishes execution before allowing a new secure service call to start.

The SEE provides minimal runtime support in form of libraries to build secure services - in particular, it provides: **platform support**, such as a dynamic memory allocator, a clock, and secure storage; and **crypto support**, realized via a port of the OpenSSL library to the SEE.

Each secure service is uniquely identified using a 32-bit service identifier (SID). All secure services implement well defined entry points that are uniquely identified by 32-bit method identifiers (MID). These methods are used as callback routines by the dispatcher. Every secure service call from the NW identifies the SID and the MID as parameters. These parameters are passed in CPU registers r0 and r1, respectively. A secure service call can pass two additional service specific parameters using registers r2 and r3. The return code from the service call is returned in register r0. This calling convention matches the C calling convention for ARM, and allows the NW to wrap the smc assembly instruction with the following signature.

```
int SecureCall (int SID, int MID, int param0, int param1)
```

Besides registers, secure services use shared memory to communicate with the caller in the NW. To facilitate this communication, all of the NW memory is mapped in the SW with identity mapping (virtual address is same as physical address).

The trusted GPS secure service implements the trusted GPS API described earlier. For both functions, param0 provides the physical address of a shared memory buffer, and param1 provides the size of this buffer. The *GetCurrentLocationFix* secure service call returns a signed marshalled LocationFix structure in the buffer; the *SatelliteInfo* secure service call returns a signed marshalled SatInfo structure in the buffer.

Our SEE implementation is based on the Tianocore UEFI firmware [10]. We added the SEE functionality (the kernel and secure services) to the UEFI initialization "Security" phase that executes in the SW, and we implemented a trampoline to execute the remaining the UEFI phases (the Pre-EFI phase, the Driver Execution Environment (DXE), and the UEFI applications) in the NW. An example of a UEFI application is a bootloader that boots the normal world OS. Once the system starts executing in the NW, it can only interact with the SW by using the smc instruction.

Our SEE implementation provides an embedded model of secure service development, where services and the SEE kernel are compiled together to form a single binary executable image for the firmware. This firmware image is copied to the flash storage via a platform specific firmware utility. To ensure SEE integrity, the SEE firmware image is signed. Upon platform reboot, the first stage firmware loader verifies this signature before starting the SEE.

Figure 7 shows an overview of how serial port I/O and interrupts for the trusted GPS sensor are handled on the ARM platform. As with the x86 implementation, our GPS device is connected to the ARM system through a UART. Because our ARM development board does not setup the UART device as a secure-world only peripheral, this means that software running in the NW can access the UART's control registers and disable it generating interrupts. Therefore, we cannot rely on interrupts being delivered to the secure world. As a result, to protect the UART we must use the second approach we described in the previous section: our trusted GPS secure service uses polling to obtain a complete GPS reading, during which time the SW never yields to the NW.

9. EVALUATION

In this section, we present performance results from the prototype trusted sensor systems we have built. This includes our prototype running on an x86 system with a TPM chip, and our prototype running on an ARM SoC using ARM TrustZone. We also present performance results from the Trusted Drive application that we have built. Overall, our goal is simply to demonstrate that the performance overhead of the trusted sensor mechanisms do not introduce overhead that would be a significant barrier to deployment.

9.1 Experiment Platforms

We perform our x86 experiments on a system with an Intel Core2Duo E6850 CPU running at 3.0 GHz, with 4 GB of RAM.

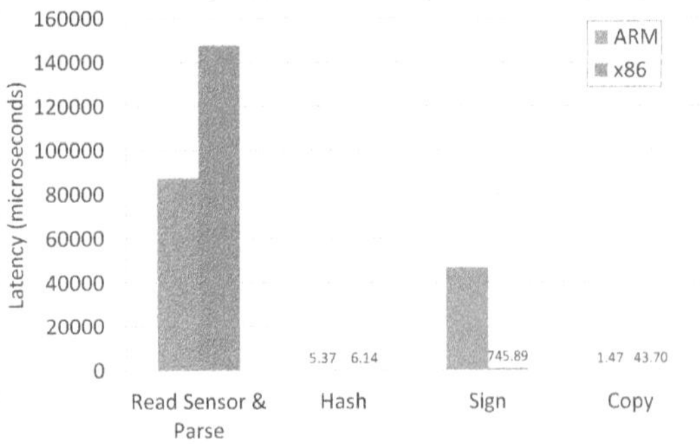

Figure 8: Latency Breakdown on ARM & x86 Platforms. ***Shows the latency of reading and parsing the sensor output, then hashing, signing, and copying the results.***

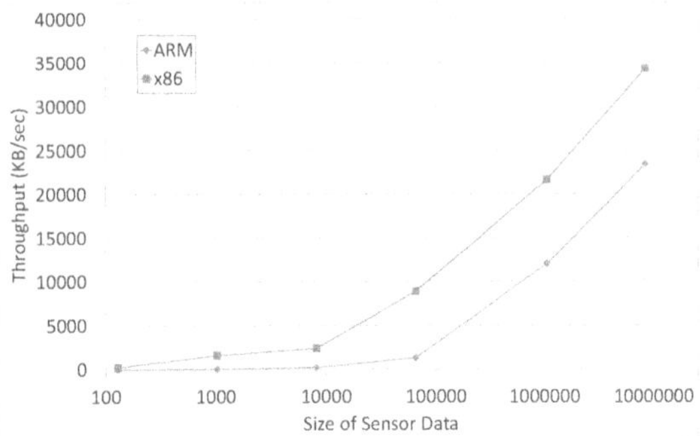

Figure 9: Sensor Attestation Throughput on ARM & x86 Platforms. ***Shows the throughput of sensor attestation for different sizes of sensor readings.***

Our Intel machine uses a Broadcom TPM 1.2 chip. Our ARM SoC platform prototype is built on an ARM SoC development board. This system contains 4 ARM Cortex A9 cores running at 1.2 GHz, with 2 GB of RAM. Our secure execution environment in the ARM TrustZone implementation uses only a single Cortex A9 core running in the TrustZone secure world, with 32 MB of RAM dedicated to the secure world. All of the individual points plotted in Figures 8 through 11 are the mean of 10 individual experiments.

9.2 Trusted Sensor Overhead

We look at three different aspects of trusted sensor overhead: the latency of each step in sensor attestation, the relationship between the size of the sensor reading and the performance overhead of signing, and the throughput of sensor-based seal and unseal.

In Figure 8, we break down the latency overhead of sensor attestation into the overhead of reading and parsing the sensor output and the overhead of hashing, signing, and copying the data back to the application. For these experiments, the reading is a 48 byte reading from the GPS sensor. We perform this experiment on both x86 and ARM, and our results demonstrate that the signing overhead is reasonable compared to the overhead of reading the sensor. Note that the overhead of signing on the ARM platform is surprisingly large (46 ms), but this is due to an unoptimized RSA implementation and the fact that our L2 cache controller is currently disabled in the SEE environment. We tested an optimized implementation of RSA in the general purpose OS on the same hardware, and it improved signing performance by more than a factor of 4.

We also observe that the overhead of reading the GPS device is noticeably larger on the x86 platform. This is because of our use of virtualization on the x86 platform. This additional overhead arises from instruction emulation that occurs for all port-mapped I/O, as well as higher interrupt latency.

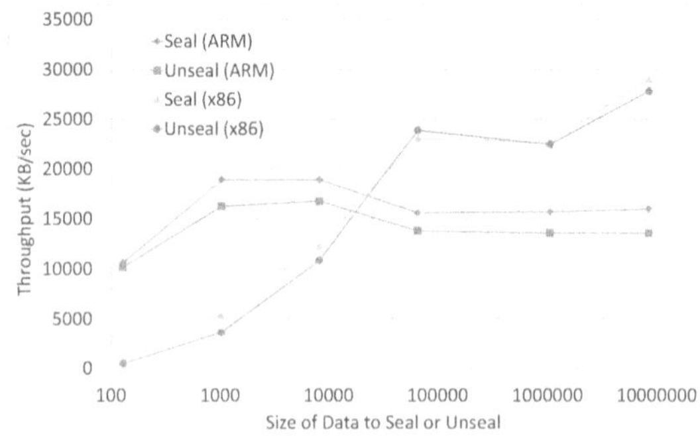

Figure 10: Sensor Seal & Unseal Throughput on ARM & x86 Platforms. ***Shows the throughput of sensor seal and unseal for different sizes of data.***

In Figure 9, we show the overhead of hashing the sensor reading, signing the hash, and copying the data as a function of the size of the sensor reading. In this graph, we omit the cost of reading the GPS sensor because we want to demonstrate the overhead of trusted sensors for high data rate sensors. Our GPS is a low data rate sensor that uses a UART interface which does not support DMA transfers. Such an interface is inappropriate for high data rate sensors. On both platforms, we see that the performance scales well because only the hashing and copying steps incur additional overhead as the size of the data grows.

Finally, in Figure 10, we show the performance overhead of sensor-based seal and unseal. The results in this graph show different components for our x86 and ARM implementation. On x86, as described in Section 7.4, our benchmark application runs in the Root VM and it communicates with the trusted sensor environment in the emancipated guest VM using a TCP connection. On ARM, as described in Section 8, our benchmark application runs in the normal world, and it communicates with the trusted sensor environment using a shared memory interface.

On both platforms, the overhead includes the performance overhead of copying combined with either AES encryption or decryption. Our results are already reasonably fast, and newer x86 platforms include AES CPU instructions which will improve performance even more. ARM SoC platforms often include custom hardware accelerators for AES, but our currently implementation does not yet take advantage of such improvements. In Figure 10, the ARM performance results show higher throughput when the size of the data is less than the L1 cache size of 32 KB, and slightly lower throughput for sizes greater than 32 KB. The x86 performance results show increasing throughput (due to TCP slow start) from 128 bytes up to 64 KB, and then the throughput flattens out at sizes larger than 64 KB.

9.3 Trusted Drive Application

The trusted drive is a trusted sensor application that manages an encrypted disk drive. The goal of this application is to avoid accidental disclosure of sensitive data, such as customers' personal information. To accomplish this, the trusted drive application ensures that the drive is only accessible (mounted) when the user's mobile device is in a trusted location, such as when the user is on the campus of a large enterprise. This provides a geo-fenced file system: the file-system is automatically mounted when the user is

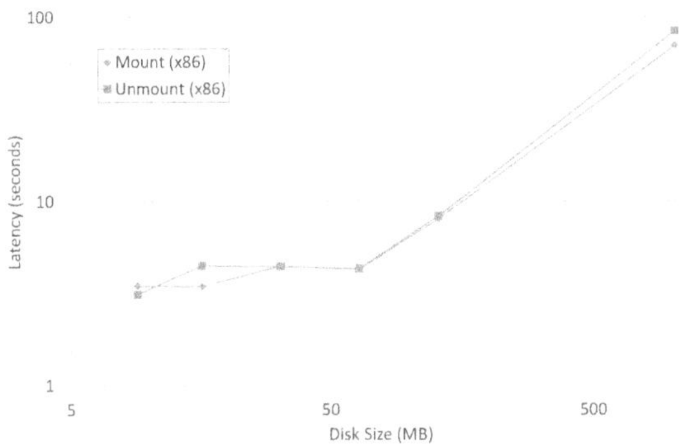

Figure 11: Trusted Drive Latency on x86 Platforms. ***Shows the latency of using the trusted drive application to mount (unseal) and unmount (seal) a trusted drive, as a function of the size of the drive.***

inside the fence, and automatically unmounted when the user ventures outside the fence.

Our scenario for this application makes two assumptions. First, we assume that users are much more likely to acquire malware outside of the trusted location. If malware infects the system while the trusted drive is mounted, our application offers little protection because the general purpose operating system can read data from the decrytped drive. Second, we assume that the mobile device implements an automatic anti-malware check whenever the device re-enters the trusted location. This ensures that the disk is not remounted if malware has infected the system.

To implement trusted drive, we use our trusted GPS sensor and the sensor seal abstraction. To ensure that the drive can only be accessed when the sensor reading shows the correct location, it utilizes the sensor seal and unseal abstraction to encrypt and decrypt the trusted drive. A sensor policy script is used to specify the set of locations where the drive should be enabled. This policy script is evaluated within the emancipated VM, and the drive will be unsealed if the script determines that the location falls within the specified boundaries.

The trusted drive application consists of a service that runs in the Root VM and manages mounting and unmounting the trusted drive. When the drive is not mounted, the trusted drive service attempts to mount the drive when the user attempts to access the drive. To mount the drive, the app calls unseal which will decrypt the drive as long as the policy allows it based on the current location. When the drive is mounted, the service periodically checks the sensor readings, and immediately unmounts and seals the drive when the user leaves the trusted location. Figure 11 shows the performance of mounting and unmounting the trusted drive as a function of the drive size. Our results show that all mount and unmount operations complete in less than 100 seconds.

10. DIFFERENTIAL PRIVACY

Attaching signatures to sensor data has the potential to negatively impact users' privacy, because the sensor readings produced by a device are signed by the same entity. Previous research [8] has shown that manipulating sensor readings (e.g., making a GPS location more coarse-grained) can be used to offset some of this privacy loss. With trusted sensors, any manipulation of sensor readings must occur *before* the readings are signed. In this section, we examine the potential of implementing a differential privacy layer inside the trusted sensors stack. Differential privacy offers a principled approach to manipulating sensor readings in a way that allows the system to measure how much information is revealed, and therefore how much privacy is lost, in answering a given query. Our initial work on differential privacy is limited to the GPS location sensor. We do not address how to apply differential privacy to all sensors: similar techniques may work well for an accelerometer, but it is not obvious how to apply differential privacy to the sensor readings produced by a smartphone camera.

10.1 Brief Primer

Differential privacy [3, 4] provides an intuitive formalization of privacy. Given a dataset and a query, differential privacy measures how much information is revealed by answering the query. Information is revealed when an attacker who knows the query answer is more likely to guess the existence of a data item in the dataset. Any query answered on the dataset leaks some information, however certain queries leak more information than others.

The amount of privacy loss is controlled by injecting *noise* in the query answer. Differential privacy frameworks expose a *noise knob* – if set to "high", the query answer has low privacy loss, but it is also more inaccurate, and vice-versa. Noise is generated dynamically for each query answer; if the same query is repeated, the answer changes from one run to another based on the random noise. Answering the same query twice increases the amount of privacy lost. To see why this is true, consider answering the same query repeatedly. Eventually, the attacker could infer the true query answer by looking at the distribution of query answers and factoring out the noise. Caching provides a practical way to answer repeated queries without losing more privacy. With caching, the query is not re-executed on the dataset, and as a result there is no additional privacy loss by returning the cached result. An attacker is no more likely to guess anything about the original dataset through caching.

10.2 Setting Up Differential Privacy in a System

Differential privacy is used as a *privacy layer* on top of the sensitive data. This layer's role is to manage the data produced by a sensor and answer incoming queries using differential privacy. Mobile applications must now issue queries to this privacy layer when making sensor readings. For example, a mobile application might ask for the user's average GPS location over the last hour. Whenever a query is issued to the system, this layer measures the degree of privacy lost by answering the query and decides whether to execute the query on the data. In this context, an implementation of a differential privacy layer requires three parameters.

The first parameter is P, the privacy budget of the entire dataset. The privacy lost by each query answer is deducted from this privacy budget. Once it reaches zero, the system refuses to answer any additional queries on this particular dataset. The data *owner* must set the value of P.

The second parameter is ϵ, the noise knob of a query. If set to "low", the query might disclose a lot of information and thus be rejected by the privacy layer because there is not enough budget. By changing ϵ to high, the query may be answered because there is enough privacy budget left for it. The issuer of the query controls setting this parameter. There is nothing security sensitive about controlling this parameter; it merely reflects how "interested" the query issuer is in a more "correct" query answer in exchange for spending more of the budget.

The final parameter is Δ, the *epoch* duration. After the privacy budget of a dataset has been exhausted, no future queries can be answered (except for the cached queries), putting the system into live-lock. A system in which new data is continually produced, as is the case for sensors, can overcome this situation by partitioning the data into epochs and assigning a separate privacy budget to each epoch.

10.3 Differential Privacy for Trusted Sensors

Differential privacy can be added to the sensor attestation abstraction. Although differential privacy makes the sensor readings "noisy", the readings' integrity remains protected by our system. This means that applications receive trusted sensor readings whose values are altered *due to differential privacy alone, and not due to malware.* With differential privacy, upon an application call, the trusted API cannot return raw sensor readings anymore because they are blocked by the privacy layer. Instead, applications can submit queries over the sensor readings which are then evaluated and executed by the differential privacy layer. Query results are returned together with attestations that validate their integrity. The privacy layer implementation must be part of the system's TCB. For illustration, our privacy layer is implemented as part of the Policy Object Interpreter module shown in Figure 5.

Our differential privacy layer is based on the Privacy Integrated Queries (PINQ) [15] language for writing differentially-private queries. PINQ supports a number of LINQ-based aggregators *Count, Sum, Average, Median* and transformations *Where, Select, Distinct, GroupBy, Join, Concat, Intersect,* and *Partition*. Any query expressed in terms of these LINQ-based aggregators can be answered by our privacy layer. Although less complete than the full SQL language, the combination of these aggregators and transformations is quite versatile.

The differentially private trusted sensor API is extended to provide an interface for PINQ queries. For example, suppose an application wants to know how often a user was at a particular geo-fenced location, say latitude between "-95.2" and "-95.1" and longitude between "29.29" and "29.30". This computation can be expressed as a PINQ query as shown below. Parameters in the body of the query, such as GetCurrentLocationFix in this example, correspond to the types of the output parameters defined by the original trusted API. In this example, the amount of the privacy budget to spend on this query is *epsilon*.

```
locFixes = new PINQueryable<GetCurrentLocationFix>();
locFixes.Where(fix => fix.latitude > -95.2 &&
                      fix.latitude < -95.1 &&
                      fix.longitude > 29.29 &&
                      fix.longitude < 29.3)
        .Count(epsilon);
```

10.4 Evaluation

Applying differential privacy to mobile sensor applications creates a trade-off between privacy and accuracy loss. Here, we look at whether mobile applications can continue to offer their functionality when their sensor inputs are subject to differential privacy. We examine two mobile applications that use GPS location. Our first application is centered on people's commute and measures the distance mobile users travel during a day. The second application applies clustering to users' locations, a common building block for context-based systems [35]. We use a real GPS dataset to drive these applications, both with and without differential privacy, and we compare the accuracy losses.

10.4.1 Methodology

The GPS traces used were gathered by a different research group [28]. They were obtained from 34 iPhone 3GS users over the course of one year for 24 of them, and half a year for the remaining 10. The GPS reads were gathered every 15 minutes as long as the phone was not turned off (the logging process was continuously run in the background). Each experiment was run with three different settings for ϵ of 0.1, 1.0, and 10.0, settings commonly used in other systems [16]. A setting lower than 0.1 is considered very strong privacy-wise, whereas a setting higher than 10 is considered weak. We are not advocating specific levels of differential privacy as sufficient, but are instead interested in understanding the trade-off between accuracy and privacy.

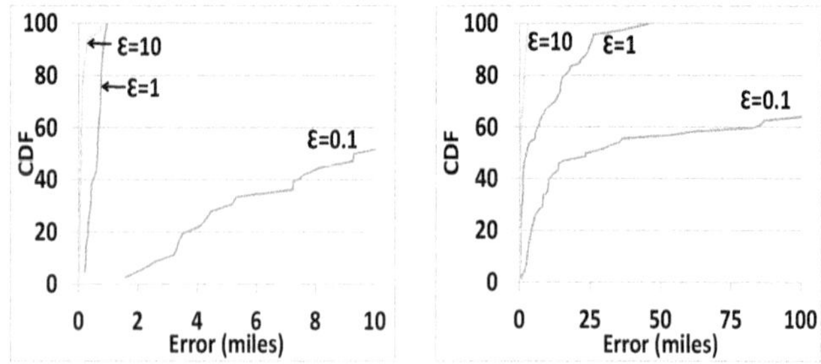

Figure 12: CDFs of error introduced by differential privacy. ***On the left, the graph shows the error for the commuting distance application; on the right, the graph shows the error for inferring places of interest.***

10.4.2 Results

For each user, we measure their daily commute by summing up the distances between each successive GPS reading in the trace, and computing the average daily commuting distance. We ignore all days with less than two GPS readings because we cannot compute a distance. On the left, Figure 12 shows the cumulative distribution function of the error in the average commute distance of all users measured in miles as a function of ϵ. With high levels of noise ($\epsilon = 0.1$), half of the users' commutes have an error higher than 10 miles. However, even with a moderate setting of noise levels ($\epsilon = 1$), the error never becomes greater than one mile.

To infer places of interest, we applied 2-means clustering to each user's GPS location history. Each cluster is likely to be centered around a place of interest for that user, such as their home, workplace, or school. We then compute the error in the location of the cluster center introduced by running this algorithm with differential privacy. On the right, Figure 12 shows the cumulative distribution function of these errors measured in miles as a function of ϵ. With high levels of noise ($\epsilon = 0.1$), 40% of cluster centers have an error of more than 80 miles, whereas with $\epsilon = 1$, the error introduced by differential privacy is always less than 3 miles. With a setting of $\epsilon = 1$, the error is within 16 miles 80% of the time, with a worst-case error of 46 miles. Such errors are tolerable if the application is trying to infer the city or town for the place of interest. On the other hand, if a more precise location is needed, then a lower level of noise must be used ($\epsilon = 10$).

Our preliminary results indicate that differential privacy has potential for protecting users' privacy with a trusted location sensor. Moderate noise does not appear to be a hindrance for the distance-based mobile application, and for coarse-grained inference of places-of-interest. Our analysis, however, is only the first step in understanding the potential of differential privacy, and the key question is how to apply it to a broad range of sensors.

11. RELATED WORK

1. Trusted sensors. Much recent work motivates the need of mobile applications to obtain sensor readings with a high degree of integrity and authenticity. Some sensors, such as GPS or Wi-Fi scanning, produce their readings by contacting an infrastructure, such as GPS satellites or Wi-Fi access points. In these cases, one alternative is for the infrastructure to sign the sensor readings, a technique sometimes referred to as "location proofs" [11, 24]. On the surface, such an approach seems to not require any changes on the client-side; as long as the infrastructure can sign its readings, mobile devices can collect them as proofs. However, such a design makes these proofs easily transferable from one device to another. A transferable proof has little value because it is unclear which device obtained the proof. Making proofs non-transferable requires devices to run client-side software that manages device identities; the infrastructure can then issue the proof to a specific device identity. Also, the presence of identities escalates the privacy concerns [12, 20].

The other approach is to rely on client-side mechanisms to increase the security of sensor readings. Most of this previous work only motivates the need for trusted sensors without describing an implementation [2, 5, 25, 34]. The piece of work closest to ours is YouProve [6]. YouProve generates a fidelity certificate for a sensor reading that has been manipulated by an untrusted application. A fidelity certificate describes how "close" the post-processed sensor reading is to the original one. One of YouProve's main challenges is coming up with a robust and complete set of techniques to characterize the loss of fidelity when applying any arbitrary transformation to a sensor reading. Furthermore, the set of techniques used for measuring fidelity loss is different from one sensor to another. Instead, our abstractions attest the code manipulating the readings and leave it up to the verifier to decide what pieces of code it trusts. Such a verification model is more closely aligned with the techniques used by trusted computing.

2. Trusted computing systems. Another area of related work is building systems that offer code integrity and data confidentiality for applications. Many of these systems can be extended to incorporate sensor I/O in their TCB. With such extensions, these systems could build and offer our two trusted sensor abstractions. Although we could have used a number of these systems in our x86 implementation, we chose to use Credo for several reasons. First, Credo offers an entire virtual machine as a secure runtime environment rather than just a hypervisor (e.g., TrustVisor [13]), a barebones TXT environment (e.g., Flicker [14]), or a new operating system (e.g., Nexus [26]). This makes programming in Credo much simpler. Credo's properties are similar to those of CloudVisor [36], but CloudVisor does not appear to be available for download.

On ARM, the Trusted Language Runtime (TLR) [23] runs parts of a mobile application inside the ARM TrustZone. We did not use TLR because its implementation is only partly done, and it has no support for I/O. Another related project implemented remote attestation on an Android platform [18]; however, this system does not appear to support running code in an isolated environment.

3. Differential privacy. Differential privacy is relatively new [3, 4] and a few recent systems have started to adopt it [22, 16]. Despite its theoretical guarantees, recently it has been shown that current implementations of differential privacy [22, 15] are subject to side-channel attacks [9]. While it is believed to be difficult for these attacks to reveal secrets, it appears possible to consume more than the pre-allocated privacy budget.

12. CONCLUSIONS

This paper presents two software abstractions for offering sensor readings to trusted mobile applications. With these abstractions, mobile applications can verify the integrity and authenticity of data produced by sensors. The paper presents two implementations of these abstractions one for x86, and one for ARM. Each of these implementations leverage the trusted computing mechanisms appropriate for each hardware platform, TPMs and hypervisors for x86, and ARM TrustZone for ARM. Finally, the paper presents a performance evaluation of these two implementations, and starts examining the potential of using differential privacy for trusted mobile applications.

13. ACKNOWLEDGEMENTS

Thanks to our shepherd, Lakshmi Subramanian, and the anonymous MobiSys reviewers for their helpful feedback and suggestions to improve this paper. We also wish to thank Landon Cox and Bryan Parno for their comments on earlier drafts of this paper. Thanks to Lin Zhong for making the Rice location traces available to us, and thanks to Stefan Savage and Geoff Voelker for supporting Lonnie's work on this project after his MSR internship was finished.

14. REFERENCES

[1] D. Chaum and E. van Heyst. Group Signatures. In *Proceedings of the 10th EUROCRYPT*, 1991.

[2] A. Dua, N. Bulusu, and W. Feng. Towards Trustworthy Participatory Sensing. In *Proceedings of the 4th HotSec*, August 2009.

[3] C. Dwork. Differential Privacy. In *Proceedings of the 33rd ICALP*, 2006.

[4] C. Dwork, F. McSherry, K. Nissim, and A. Smith. Calibrating Noise to Sensitivity in Private Data Analysis. In *Proceedings of the 3rd IACR Theory of Cryptography Conference*, March 2006.

[5] P. Gilbert, L. Cox, J. Jung, and D. Wetherall. Toward Trustworthy Mobile Sensing. In *Proceedings of the 11th HotMobile*, 2010.

[6] P. Gilbert, J. Jung, K. Lee, H. Qin, D. Sharekey, A. Sheth, and L. Cox. YouProve: Authenticity and Fidelity in Mobile Sensing. In *Proceedings of the 9th SenSys*, November 2011.

[7] S. Goldwasser, S. Micali, and C. Rackoff. The knowledge complexity of interactive proof-systems. In *Proceedings of the 17th STOC*, May 1985.

[8] M. Gruteser and D. Grunwald. Anonymous Usage of Location-Based Services through Spatial and Temporal Cloaking. In *Proceedings of the 1st MobiSys*, May 2003.

[9] A. Haeberlen, B. C. Pierce, and A. Narayan. Differential privacy under fire. In *Proceedings of the 20th USENIX Security Symposium*, August 2011.

[10] Intel. Intel UEFI implementation codenamed Tianocore. `http://tianocore.sourceforge.net`, last accessed December 2011.

[11] V. Lenders, E. Koukoumidis, P. Zhang, and M. Martonosi. Location-based Trust for Mobile User-generated Content: Applications, Challenges and Implementations. In *Proceedings of the 9th HotMobile*, 2008.

[12] W. Luo and U. Hengartner. Proving Your Location without Giving up Your Privacy. In *Proceedings of the 10th HotMobile*, 2009.

[13] J. M. McCune, Y. Li, N. Qu, Z. Zhou, A. Datta, V. Gligor, and A. Perrig. TrustVisor: Efficient TCB Reduction and Attestation. In *Proceedings of IEEE Symposium on Security and Privacy*, May 2010.

[14] J. M. McCune, B. Parno, A. Perrig, M. K. Reiter, and H. Isozaki. Flicker: An Execution Infrastructure for TCB Minimization. In *Proceedings of EuroSys*, 2008.

[15] F. McSherry. PINQ Library. http://research.microsoft.com/en-us/projects/pinq/.

[16] F. McSherry and R. Mahajan. Differentially-Private Network Trace Analysis. In *Proceedings of SIGCOMM*, 2010.

[17] MediaWatch. Beware the "trusted" source. http://www.abc.net.au/mediawatch/transcripts/s3218415.htm, 2011.

[18] M. Nauman, S. Khan, X. Zhang, and J.-P. Seifert. Beyond Kernel-level Integrity Measurement: Enabling Remote Attestation for the Android Platform. In *Proceedings of the 3rd TRUST conference*, June 2010.

[19] N. Paul, T. Kohno, and D. C. Klonoff. A Review of the Security of Insulin Pump Infusion Systems. *Journal of Diabetes Science and Technology*, 5(6):1557–1562, November 2011.

[20] K. P. N. Puttaswamy and B. Y. Zhao. Preserving Privacy in Location-based Mobile Social Applications. In *Proceedings of the 10th HotMobile*, 2009.

[21] H. Raj, D. Robinson, T. Tariq, P. England, S. Saroiu, and A. Wolman. Credo: Trusted Computing for Guest VMs with a Commodity Hypervisor. Technical Report MSR-TR-2011-130, Microsoft Research, 2011.

[22] I. Roy, S. T. V. Setty, A. Kilzer, V. Shmatikov, and E. Witchel. Airavat: Security and Privacy for MapReduce. In *Proceedings of the 7th NSDI*, 2010.

[23] N. Santos, H. Raj, S. Saroiu, and A. Wolman. Trusted Language Runtime (TLR): Enabling Trusted Applications on Smartphones. In *Proceedings of the 12th HotMobile*, 2011.

[24] S. Saroiu and A. Wolman. Enabling New Mobile Applications with Location Proofs. In *Proceedings of the 10th HotMobile*, 2009.

[25] S. Saroiu and A. Wolman. I am a Sensor, and I Approve This Message. In *Proceedings of the 11th HotMobile*, 2010.

[26] F. B. Schneider, K. Walsh, and E. G. Sirer. Nexus Authorization Logic (NAL): Design Rationale and Applications. *ACM Transactions on Information and System Security*, 14(1), May 2011.

[27] A. Seshadri, M. Luk, N. Qu, and A. Perrig. SecVisor: A Tiny Hypervisor to Provide Lifetime Kernel Code Integrity for Commodity OSes. In *Proceedings of the 21st SOSP*, October 2007.

[28] C. Shepard, A. Rahmati, C. Tossell, L. Zhong, and P. Kortum. LiveLab: measuring wireless networks and smartphone users in the field. *ACM SIGMETRICS Performance Evaluation Review*, 38(3), December 2010.

[29] Skyhook Inc. http://www.skyhookwireless.com, 2011.

[30] Slate. The End of the Credit Card? http://www.slate.com/articles/technology/technology/2011/11/card_cae_the_new_payments_app_that_could_make_cash_and_plastic_.single.html, 2011.

[31] N. O. Tippenhauer, K. Rasmussen, C. Pöpper, and S. Capkun. Attacks on Public WLAN-based Positioning Systems. In *Proceedings of the 7th Mobisys*, June 2009.

[32] Trusted Computing Group. Trusted Platform Module Main Specification, Part 1: Design Principles, Part 2: TPM Structures, Part 3: Commands. Revision 116, March 2011. http://www.trustedcomputinggroup.com/resources/tpm_main_specification.

[33] Washington Times. Guard at Hanging Blamed for Covert Video of Hussein. http://www.washingtonpost.com/wp-dyn/content/article/2007/01/03/AR2007010300358.html, 2007.

[34] A. Wolman, S. Saroiu, and V. Bahl. Using Trusted Sensors to Monitor Patients' Habits. In *Proceedings of the 1st HealthSec*, August 2010.

[35] T. Yan, D. Chu, D. Ganesan, A. Kansal, and J. Liu. Fast App Launching for Mobile Devices Using Predictive User Context. In *Proceedings of the 10th MobiSys*, June 2012.

[36] F. Zhang, J. Chen, H. Chen, and B. Zang. CloudVisor: Retrofitting Protection of Virtual Machines in Multi-tenant Cloud with Nested Virtualization. In *Proceedings of the 23rd SOSP*, 2011.

How Long to Wait?: Predicting Bus Arrival Time with Mobile Phone based Participatory Sensing

Pengfei Zhou, Yuanqing Zheng, Mo Li
Nanyang Technological University, Singapore
pfzhou@ntu.edu.sg, yuanqing1@e.ntu.edu.sg, limo@ntu.edu.sg

ABSTRACT

The bus arrival time is primary information to most city transport travelers. Excessively long waiting time at bus stops often discourages the travelers and makes them reluctant to take buses. In this paper, we present a bus arrival time prediction system based on bus passengers' participatory sensing. With commodity mobile phones, the bus passengers' surrounding environmental context is effectively collected and utilized to estimate the bus traveling routes and predict bus arrival time at various bus stops. The proposed system solely relies on the collaborative effort of the participating users and is independent from the bus operating companies, so it can be easily adopted to support universal bus service systems without requesting support from particular bus operating companies. Instead of referring to GPS enabled location information, we resort to more generally available and energy efficient sensing resources, including cell tower signals, movement statuses, audio recordings, etc., which bring less burden to the participatory party and encourage their participation. We develop a prototype system with different types of Android based mobile phones and comprehensively experiment over a 7 week period. The evaluation results suggest that the proposed system achieves outstanding prediction accuracy compared with those bus company initiated and GPS supported solutions. At the same time, the proposed solution is more generally available and energy friendly.

Categories and Subject Descriptors

C.2.4 [**Computer Communication Networks**]: Distributed Systems – *Distributed Applications*; C.3.3 [**Special-Purpose and Application-based Systems**]: Real-time and embedded systems; H.5.3 [**Information Interfaces and Presentation**]: Groups and Organization Interfaces – *Collaborative Computing*

General Terms

Design, Implementation, Experimentation, Measurement, Performance

Keywords

Bus arrival time prediction, Participatory sensing, Mobile phones

1. INTRODUCTION

Public transport, especially the bus transport, has been well developed in many parts of the world. The bus transport services reduce the private car usage and fuel consumption, and alleviate traffic congestion. As one of the most comprehensive and affordable means of public transport, in 2011 the bus system serves over 3.3 million bus rides every day on average in Singapore with around 5 million residents [1].

When traveling with buses, the travelers usually want to know the accurate arrival time of the bus. Excessively long waiting time at bus stops may drive away the anxious travelers and make them reluctant to take buses. Nowadays, most bus operating companies have been providing their timetables on the web freely available for the travelers. The bus timetables, however, only provide very limited information (e.g., operating hours, time intervals, etc.), which are typically not timely updated. Other than those official timetables, many public services (e.g., Google Maps) are provided for travelers. Although such services offer useful information, they are far from satisfactory to the bus travelers. For example, the schedule of a bus may be delayed due to many unpredictable factors (e.g., traffic conditions, harsh weather situation, etc). The accurate arrival time of next bus will allow travelers to take alternative transport choices instead, and thus mitigate their anxiety and improve their experience. Towards this aim, many commercial bus information providers offer the realtime bus arrival time to the public [20]. Providing such services, however, usually requires the cooperation of the bus operating companies (e.g., installing special location tracking devices on the buses), and incurs substantial cost.

In this paper, we present a novel bus arrival time prediction system based on crowd-participatory sensing. We interviewed bus passengers on acquiring the bus arrival time. Most passengers indicate that they want to instantly track the arrival time of the next buses and they are willing to contribute their location information on buses to help to establish a system to estimate the arrival time at various bus stops for the community. This motivates us to design a crowd-participated service to bridge those who want to know bus arrival time (querying users) to those who are on the bus and able to share their instant bus route information (sharing users). To achieve such a goal, we let the bus passengers themselves cooperatively sense the bus route information using commodity mobile phones. In particular, the sharing passengers may anonymously upload their sensing data collected on buses to a processing server, which intelligently processes the data and distributes useful information to those querying users.

Our bus arrival time prediction system comprises three major components: (1) Sharing users: using commodity mobile phones as well as various build-in sensors to sense and report the lightweight cellular signals and the surrounding environment to a backend server; (2) Querying users: querying the bus arrival time for a particular bus route with mobile phones; (3) Backend server: collecting the instantly reported information from the sharing users, and intellectually processing such information so as to monitor the bus routes and predict the bus arrival time. No GPS or explicit location services are invoked to acquire physical location inputs.

Such a crowd-participated approach for bus arrival time prediction possesses the following several advantages compared with conventional approaches. First, through directly bridging the sharing and querying users in the participatory framework, we build our system independent of the bus operating companies or other third-party service providers, allowing easy and inexpensive adoption of the proposed approach over other application instances. Second, based on the commodity mobile phones, our system obviates the need for special hardware or extra vehicle devices, which substantially reduces the deployment cost. Compared with conventional approaches (e.g., GPS supported ones [12, 29]), our approach is less demanding and much more energy-friendly, encouraging a broader number of participating passengers. Third, through automatically detecting ambient environments and generating bus route related reports, our approach does not require the explicit human inputs from the participants, which facilitates the involvement of participatory parties.

Implementing such a participatory sensing based system, however, entails substantial challenges. (1) Bus detection: since the sharing users may travel with diverse means of transport, we need to first let their mobile phones accurately detect whether or not the current user is on a bus and automatically collect useful data only on the bus. Without accurate bus detection, mobile phones may collect irrelevant information to the bus routes, leading to unnecessary energy consumption or even inaccuracy in prediction results. (2) Bus classification: we need to carefully classify the bus route information from the mixed reports of participatory users. Without users' manual indication, such automatic classification is non-trivial. (3) Information assembling: One sharing user may not stay on one bus to collect adequate time period of information. Insufficient amount of uploaded information may result in inaccuracy in predicting the bus route. An effective information assembling strategy is required to solve the jigsaw puzzle of combining pieces of incomplete information from multiple users to picture the intact bus route status.

In this paper, we develop practical solutions to cope with such challenges. In particular, we extract unique identifiable fingerprints of public transit buses and utilize the microphone on mobile phones to detect the audio indication signals of bus IC card reader. We further leverage the accelerometer of the phone to distinguish the travel pattern of buses to other transport means. Thus we trigger the data collection and transmission only when necessary (§3.3). We let the mobile phone instantly sense and report the nearby celltower IDs. We then propose an efficient and robust top-k celltower set sequence matching method to classify the reported celltower sequences and associate with different bus routes. We intellectually identify passengers on the same bus and propose a celltower sequence concatenation approach to assemble their celltower sequences so as to improve the sequence matching accuracy (§3.4). Finally, based on accumulated information, we are then able to utilize both historical knowledge and the realtime traffic conditions to accurately predict the bus arrival time of various routes (§3.5).

We consolidate the above techniques and implement a prototype system with the Android platform using two types of mobile phones (Samsung Galaxy S2 i9100 and HTC Desire). Through our 7-week experimental study, the mobile phone scheme can accurately detect buses with 98% detection accuracy and classifies the bus routes with up to 90% accuracy. As a result, the prototype system predicts bus arrival time with average error around 80 seconds. Such a result is encouraging compared with current commercial bus information providers in Singapore.

In the following of this paper, we first introduce the background and motivation in §2. In §3, we detail the challenges of our system and describe our technical solutions. The evaluation results are presented in §4. The limitations and possible improvements are discussed in §5 followed by the description of related works in §6. We summarize this paper in §7.

2. BACKGROUND AND MOTIVATION

The bus companies usually provide free bus timetables on the web. Such bus timetables, however, only provide very limited information (e.g., operating hours, time intervals, etc.), which are typically not timely updated according to instant traffic conditions. Although many commercial bus information providers offer the realtime bus arrival information, the service usually comes with substantial cost. With a fleet of thousands of buses, the installment of in-vehicle GPS systems incurs tens of millions of dollars [29]. The network infrastructure to deliver the transit service raises the deployment cost even higher, which would eventually translate to increased expenditure of passengers.

For those reasons, current research works [12, 29] explore new approaches independent of bus companies to acquire transit information. The common rationale of such approaches is to continuously and accurately track the absolute physical location of the buses, which typically uses GPS for localization. Although many GPS-enabled mobile phones are available on the market, a good number of mobile phones are still shipped without GPS modules [31]. Those typical limitations of the localization based schemes motivate alternative approaches without using GPS signal or other localization methods. Besides, GPS module consumes substantial amount of energy, significantly reducing the lifetime of power-constrained mobile phones [31]. Due to the high power consumption, many mobile phone users usually turn

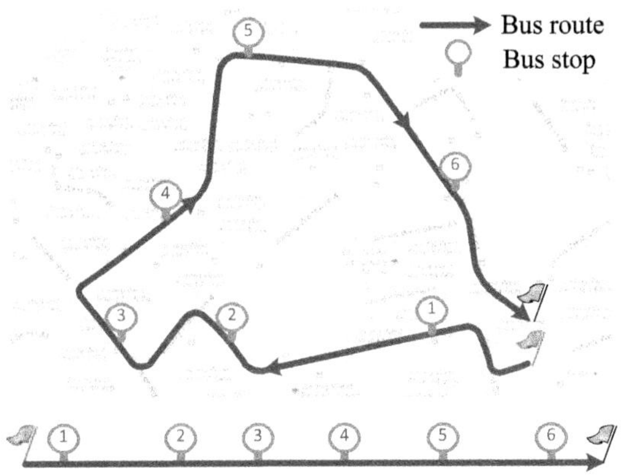

Figure 1: Absolute localization is unnecessary for arrival time prediction

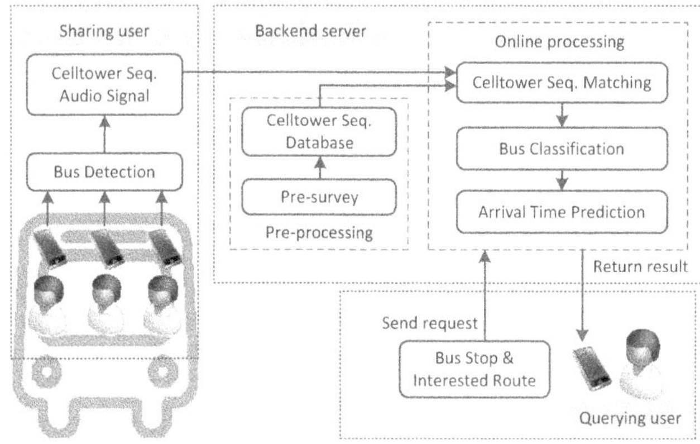

Figure 2: System architecture

off GPS modules to save battery power. The mobile phones in vehicles may perform poorly when they are placed without line-of-sight paths to GPS satellites [9].

To fill this gap, we propose to implement a crowd-participated bus arrival time prediction system utilizing cellular signals. Independent of any bus companies, the system bridges the gap between the querying users who want to know the bus arrival time to the sharing users willing to offer them real-time bus information. Unifying the participatory users, our design aims to realize the common welfare of the passengers.

To encourage more participants, no explicit location services are invoked so as to save the requirement of special hardware support for localization. Compared with the high energy consumption of GPS modules, the marginal energy consumption of collecting celltower signals is negligible on mobile phones. Our system therefore utilizes the celltower signals without reducing battery lifetime on sharing passengers' mobile phones. Our design obviate the need for accurate bus localization. As a matter of fact, since the public transport buses travel on certain bus routes (1D routes on 2D space), the knowledge of the current position on the route (1D knowledge) and the average velocity of the bus suffices to predict its arrival time at a bus stop. As shown in Figure 1, for instance, say the bus is currently at bus stop 1, and a querying user wants to know its arrival time at bus stop 6. Accurate prediction of the arrival time requires the distance between bus stop 1 and 6 along the 1D bus route (but not on the 2D map) and the average velocity of the bus. In general, the physical positions of the bus and the bus route on the 2D maps are not strictly necessary. In our system, instead of pursuing the accurate 2D physical locations, we logically map the bus routes to a space featured by sequences of nearby cellular towers. We classify and track the bus statuses in such a logical space so as to predict the bus arrival time on the real routes.

We leverage various lightweight sensors (e.g., microphone, accelerometer, etc.) on mobile phones to enable automatic and intelligent data collection and transmission. Although we can make use of a basket of instantly available sensor resources (e.g., magnetometer, gyroscope, camera, proximity sensors, etc.), we mainly focus on energy-friendly and widely available sensing signals (e.g., celltower and audio signals). The purpose is to make the solution lightweight and pervasively available to attract more participants.

3. SYSTEM DESIGN

Though the idea is intuitive, the design of such a system in practice entails substantial challenges. In this section, we describe the major components of the system design. We illustrate the challenges in the design and implementation, and present several techniques to cope with them.

3.1 System overview

Figure 2 sketches the architecture of our system. There are 3 major components.

Querying user. As depicted in Figure 2 (right bottom), a querying user queries the bus arrival time by sending the request to the backend server. The querying user indicates the interest bus route and bus stop to receive the predicted bus arrival time.

Sharing user. The sharing user on the other hand contributes the mobile phone sensing information to the system. After a sharing user gets on a bus, the data collection module starts to collect a sequence of nearby celltower IDs. The collected data is transmitted to the server via cellular networks. Since the sharing user may travel with different means of transport, the mobile phone needs to first detect whether the current user is on a bus or not. As shown in Figure 2 (left side), the mobile phone periodically samples the surrounding environment and extracts identifiable features of transit buses. Once the mobile phone confirms it is on the bus, it starts sampling the celltower sequences and sends the sequences to the backend server. Ideally, the mobile phone of the sharing user automatically performs the data collection and transmission without the manual input from the sharing user.

Backend server. We shift most of the computation burden to the backend server where the uploaded information from sharing users is processed and the requests from querying users are addressed. Two stages are involved in this component.

In order to bootstrap the system, we need to survey the corresponding bus routes in the offline pre-processing stage. We construct a basic database that associates particular bus routes to celltower sequence signatures. Since we do not require the absolute physical location reference, we mainly wardrive the bus routes and record the sequences of observed celltower IDs, which significantly reduces the initial construction overhead.

The backend server processes the celltower sequences and

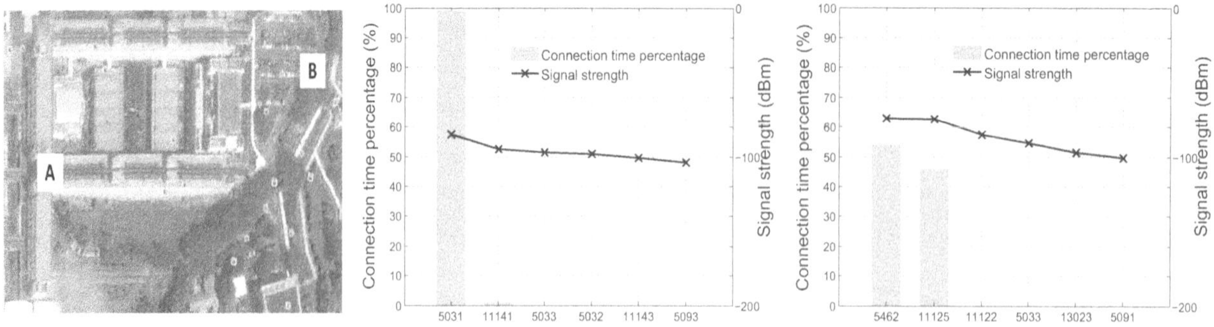

(a) Celltower coverage (b) Connection at position A (c) Connection at position B

Figure 3: Celltower connection time and received signal strength

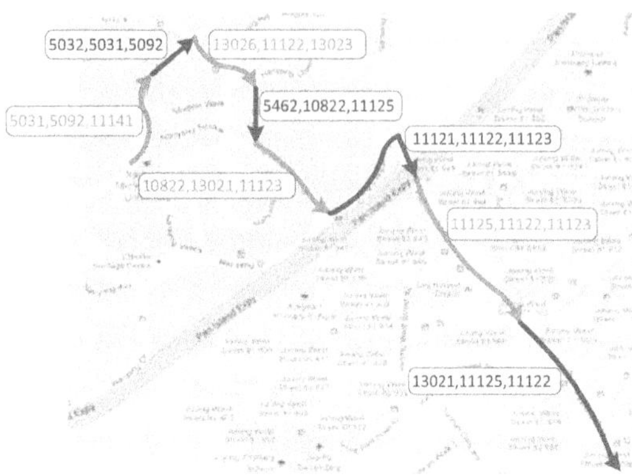

Figure 4: Celltower sequence set along a bus route

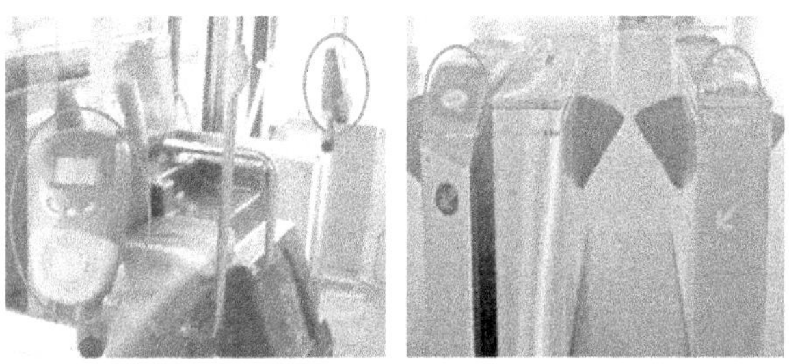

(a) On buses (b) At rapid train station entrances

Figure 5: Transit IC card readers

audio signals from sharing users on the buses in the online processing stage. Receiving the uploaded information, the backend server first distinguishes the bus route that the sharing user is currently traveling with. The backend server classifies the uploaded bus routes primarily with the reported celltower sequence information. The bus arrival time on various bus stops is then derived based on the current bus route statuses.

3.2 Pre-processing celltower data

The backend server needs to maintain a database that stores sequences of celltower IDs that are experienced along different bus routes. Wardriving along one bus route, the mobile phone normally captures several celltower signals at one time, and connects to the celltower with the strongest signal strength. We find in our experiments that even if a passenger travels by the same place, the connected celltower might be different from time to time due to varying celltower signal strength. To improve the robustness of our system, instead of using the associated celltower, we record a set of celltower IDs that the mobile phone can detect. To validate such a point, we do an initial experiment. We measure the celltower coverage at two positions A and B within the university campus, which are approximately 300 meters apart (Figure 3(a) depicts the two positions on the map).

Figure 3(b) and 3(c) report the celltower that the mobile phone can detect, as well as their average signal strength and connection time at A and B, respectively. We find that position A and position B are both covered by 6 celltowers with divergent signal strength. In Figure 3(b), we find that at position A the mobile phone is connected to the celltower 5031 over 99% of the time, while its signal strength remains consistently the strongest during the 10-hour measurement. In Figure 3(c), the mobile phone at position B observes two celltowers with comparable signal strength. We find that the mobile phone is more likely to connect to the celltower with stronger signal strength, and also may connect to the celltower with the second strongest signal strength. Nevertheless, during our 7-week experiments, we consistently observe that mobile phones almost always connect to the top-*3* strongest celltowers. Therefore, in practice we choose the set of the top-*3* strongest celltowers as the signature for route segments.

Figure 4 illustrates the celltower sequence collected on our campus bus traveling from our school to a rapid train station off the campus. The whole route of the bus is divided into several concatenated sub-route segments according to the change of the top-*3* celltower set. They are marked alternately in red and black in the figure. For example, the mobile phone initially connects to celltower 5031 in the first sub-route and the top-*3* celltower set is $\{5031, 5092, 11141\}$. Later the mobile phone is handed over to celltower 5032 and the celltower set becomes $\{5032, 5031, 5092\}$ in the second sub-route. We subsequently record the top-*3* celltower in each sub-route.

Such a sequence of celltower ID sets identifies a bus route in our database. By wardriving along different bus routes, we can easily construct a database of celltower sequences associated to particular bus routes.

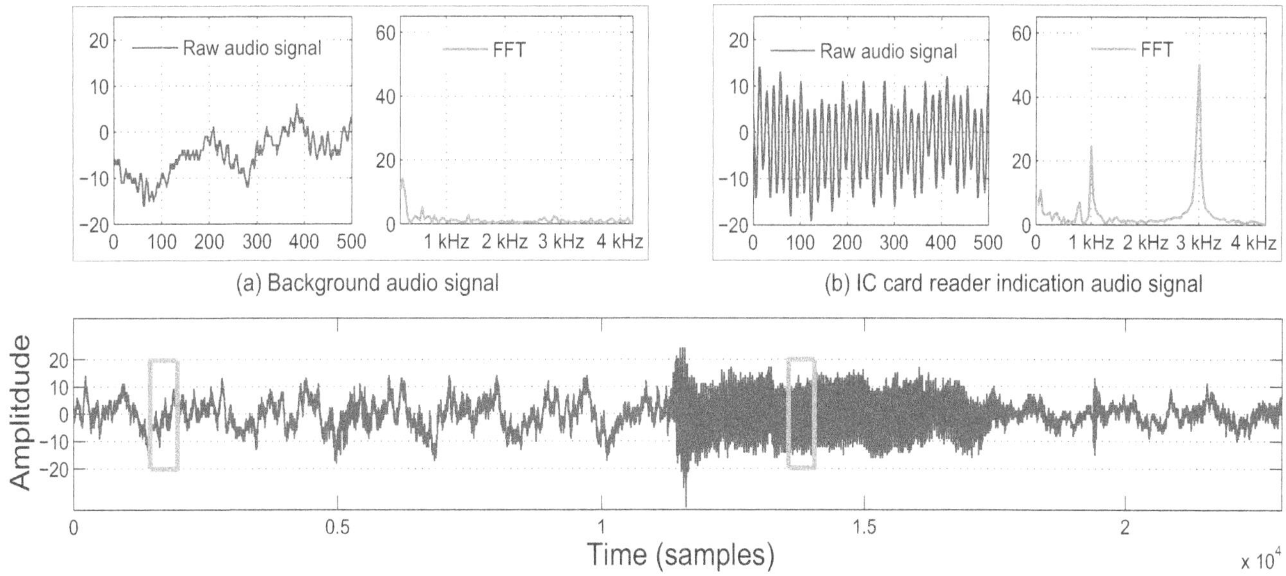

Figure 6: Bus detection using audio indication signal

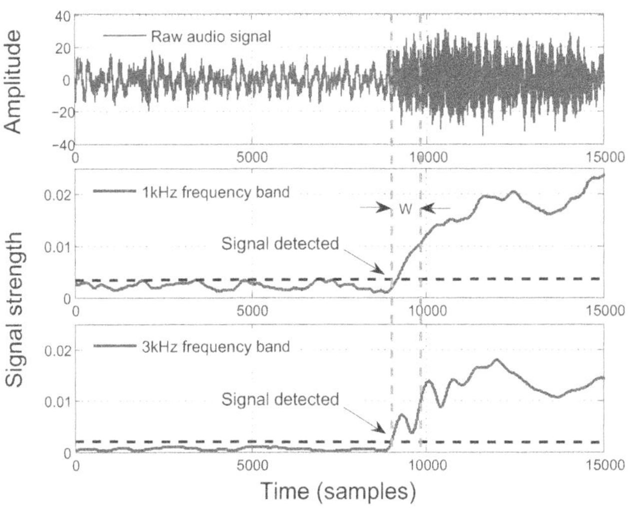

Figure 7: Detecting audio beeps in the frequency domain

3.3 Bus detection: Am I on a bus?

During the on-line processing stage, we use the mobile phones of sharing passengers on the bus to record the cell-tower sequences and transmit the data to the backend server. As aforementioned, the mobile phone should intelligently detect whether it is on a public transit bus or not and start to collect the data only when the mobile phone is on a bus. Some works [18, 21] study the problem of activity recognition and context awareness using various sensors. Such approaches, however, cannot be used to distinguish different transport modes (e.g., public transit buses and non-public buses). In this section, we explore multi-sensing resources to detect the bus environment and distinguish it from other transport modes. We seek a lightweight detection approach in terms of both energy consumption and computation complexity.

3.3.1 Audio detection

Nowadays, IC cards are commonly used for paying transit fees in many areas (e.g., EZ-Link cards in Singapore [2], Octopus cards in Hong Kong [3], Oyster cards in London [4], etc). On a public bus in Singapore, several card readers are deployed for collecting the fees (as depicted in Figure 5(a)). When a passenger taps the transit card on the reader, the reader will send a short beep audio response to indicate the successful payment. In our system, we choose to let the mobile phone detect the beep audio response of the card reader, since such distinct beeps are not widely used in other means of transportation such as non-public buses and taxis.

In order to exploit the unique beeps of IC card readers, in our initial experiment we record an audio clip on the bus at the audio sampling rate of 44.1kHz with Samsung Galaxy S2 i9100 mobile phone. Such a sampling rate is more than sufficient to capture the beep signals [26]. Figure 6 (bottom) plots the raw audio signal in the time domain, where the IC card reader starts beeping approximately from $11000th$ sample and lasts to $18000th$ sample. We crop the section of the beep audio signal and depict the section in Figure 6(b). After we convert the time domain signal to the frequency domain through 512pt Fast Fourier Transform (FFT) (Figure 6(b)), we observe clear peaks at 1kHz and 3kHz frequency bands. For comparison we depict the audio clip as well where no beep signal is sent. Both time domain and the frequency domain signals are plotted in Figure 6(a). We find no peaks at 1kHz and 3kHz frequency bands.

With the knowledge of the frequency range of the dual-tone beep signal sent by the IC card reader, in our system we can lower down the audio sampling rate of the mobile phone to 8kHz (8000 samples/s) which is sufficient to capture the beep signals with maximum frequency of 3kHz. We find that in practice 128pt FFT suffices to detect the IC card reader on the bus with tractable computation complexity on commodity mobile phones. We use the standard sliding window averaging technique with window size $w = 32$ samples to filter out the noises in both 1kHz and 3kHz frequency

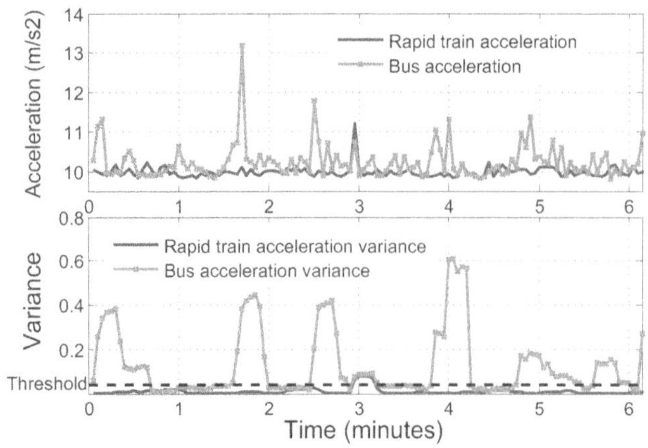

Figure 8: Accelerometer readings on rapid train and bus

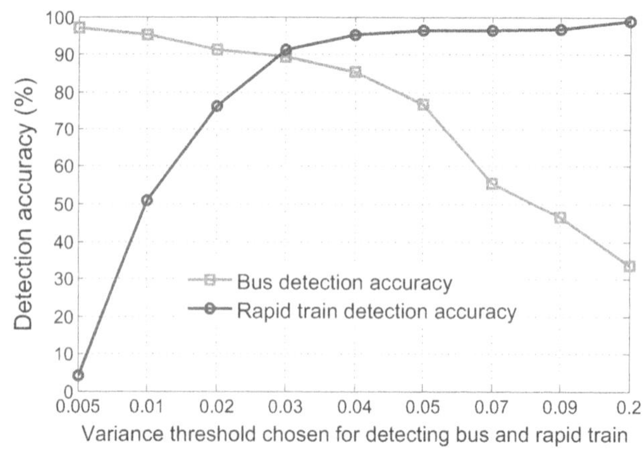

Figure 9: Variance threshold to distinguish rapid trains from buses

bands. We use an empirical threshold of three standard deviation (i.e., 99.7% confidence level of noise) to detect beep signals. If the received audio signal strengths in 1kHz and 3kHz frequency bands both exceed the threshold, the mobile phone confirms the detection of the bus. Figure 7 depicts the beep signal detection process. When the IC card reader starts beeping, the signal strengths in both 1kHz and 3kHz frequency bands jump significantly and therefore can be detected.

We test the audio indication based bus detection method with various scenarios, and the experiments show encouraging results for bus detection (§4.2.1). As the dual-tone responsive signal is universally used in almost all public transit buses in Singapore, we can use it as an identifiable signature to distinguish the buses from other vehicles. Therefore, we use the dual-tone as the acoustic trigger for the successive celltower data collection and transmission of the mobile phones of sharing users. We can easily adopt similar techniques [22] to detect certain audio indications to identify the public transports as well in other areas (e.g., the bell ringing tunes in Hong Kong buses).

3.3.2 Accelerometer detection: Bus v.s. Rapid train

In Singapore, however, transit IC cards are used in rapid train stations as well where the IC card readers in the entrances may send the same beep audio signal (Figure 5(b)). In practice, we find that solely relying on the audio detection the mobile phones may falsely trigger the celltower ID collection when they go with the rapid trains. Since the train routes have substantial above-ground segments that overlap with bus routes, simply using celltower signals does not effectively differentiate the two transit means. We expect to leverage the accelerometer sensor on the mobile phone to reduce such false detection.

Intuitively, the rapid trains are moving at relatively stable speeds with few abrupt stops or sharp turns. On the contrary, the buses are typically moving with many sharp turns and frequent acceleration and deceleration. We collect the accelerometer data at a moderate sampling rate of 20Hz. The raw accelerometer readings are first made orientation-independent by computing the L_2-norm (or magnitude) of the raw data [28]. Figure 8 (top) plots the accelerometer readings on a rapid train and a public transit bus which suggest that the accelerometer reading on the bus fluctuates much frequently with larger magnitudes. We explore such features of accelerometer readings to distinguish the buses from the rapid trains.

We measure the statistics of the accelerometer readings during 12.5 seconds (250 samples) to reduce the impact of noise, such as average and variance of the acceleration. Figure 8 (bottom) plots the variance of the accelerometer readings on the rapid train and the public transit bus, respectively. According to the figure, the variance on the bus is significantly larger than that on the train. Therefore, we distinguish the buses from the trains using the variance of accelerometer readings by setting a proper threshold.

We confirm the detection of buses if the measured acceleration variance is above the threshold, and the detection of rapid trains otherwise. In Figure 9, we vary the threshold from 0.005 to 0.2 and plot the detection accuracy. If the threshold is small, most buses will be correctly detected, while many trains will be misdetected as buses as well, which may lead to noisy inputs to the backend server and energy waste of mobile phones in collecting celltower IDs. On the other hand, if threshold is too big, most rapid trains will be filtered out, while we will miss the detection of many actual buses, which may lose the opportunities in collecting useful celltower information on the buses. We select an empirical threshold 0.03 to balance the false negative and false positive.

In practice, we find that accelerometer based detection can distinguish the buses from the trains with an accuracy of approximately 90% (§4.2.2). The error rate of falsely detecting rapid trains as buses is even smaller. The detection error of falsely classifying public buses into rapid trains is mainly due to the abnormality of the bus routes (e.g., long straight routes) especially during non-peak hours. Such a detection error is tolerable in the bus classification stage, where the backend server has information redundancy to handle the noisy reports.

3.4 Bus classification

When a sharing user gets on the bus, the mobile phone samples a sequence of celltower IDs and reports the information to the backend server. The backend server aggregates the inputs from massive mobile phones and classifies the in-

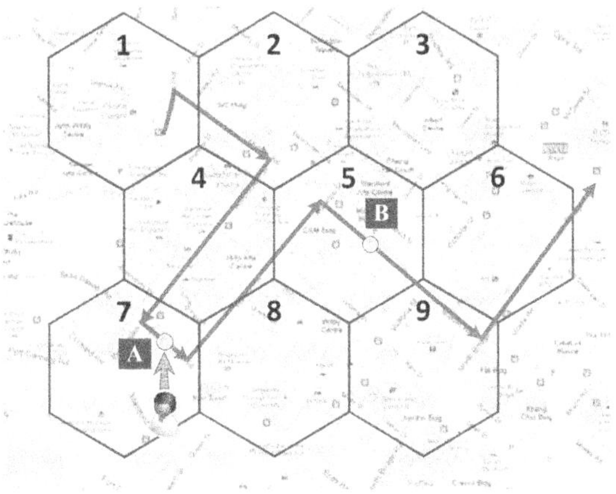

Figure 10: Celltower sequence matching

Database seq.	1 2 4	7 8 4 5	9 6
Uploaded seq.		7 8 4 5	
Matched seq.		7 8 4 5	

Table 1: Celltower sequence matching

puts into different bus routes. The statuses of the bus routes are then updated accordingly.

3.4.1 Celltower sequence matching

We match the received celltower sequences to those signature sequences store in the database. Figure 10 shows an illustrative example where a sharing passenger gets on the bus at location A. The backend server will receive a celltower sequence of $\langle 7, 8, 4, 5 \rangle$ when the sharing user reaches location B. Say that the celltower sequence of the bus route stored in the database is $\langle 1, 2, 4, 7, 8, 4, 5, 9, 6 \rangle$, then the sequence $\langle 7, 8, 4, 5 \rangle$ matches the particular bus route as a sub-segment as shown in Table 1.

In practical scenarios, the sequence matching problem becomes more complicated due to the varying celltower signal strength. Recall that for each sub-route we record the top-*3* celltower IDs instead of the connected celltower ID in the pre-processing period. We let each mobile phone send back the sequence of celltowers that the mobile phone has connected to. In the matching process on the server, we accordingly devise a top-k celltower sequence matching scheme by modifying the Smith-Waterman algorithm [33]. Smith-Waterman is a dynamic programming algorithm for performing local sequence alignment which has been widely used in bioscience (e.g., to determine similar regions between two nucleotide or protein sequences).

We make concrete modifications on the original algorithm to support the top-k celltower sequence matching. We weigh a matching of a celltower ID with a top-k set according to the celltower signal strength. Say that in a top-k set $S = \{c_1, c_2, \dots, c_k\}$ ordered by signal strength (i.e., $s_i \geq s_j, 1 \leq i \leq j \leq k$), where c_i and s_i denote celltower i and its signal strength, respectively.

We denote the uploaded celltower sequence from a sharing user as $Seq_{\texttt{upload}} = \langle u_1 u_2 \dots u_m \rangle$ where m is the sequence length. We also denote a celltower set sequence in database as $Seq_{\texttt{database}} = \langle S_1 S_2 \dots S_n \rangle$ where n is the set sequence length. If $u_i = c_w \in S_j$, u_i and S_j are considered matching with each other, and mismatching otherwise. We assign a score $f(s_w)$ for a match, where $f(s_w)$ is a positive non-decreasing scoring function and w is the rank of signal strength. In practice, we use $f(s_w) = 0.5^{w-1}$ as the scoring function according to the signal strength order in the set. The penalty cost for mismatches is set to be an empirical value of -0.5 which balances the robustness and accuracy in practice.

We choose top-*3* celltower IDs with strongest celltower signal strength to form a set based on our initial observations (§3.2). The distinctive advantage of the proposed classification algorithm is its robustness to the variation of celltower signal strength. Table 2 shows a celltower set sequence matching instance. In the example, the uploaded celltower sequence is $Seq_{\texttt{upload}} = \langle 1, 8, 10, 15, 16 \rangle$, and the celltower ID set is shown in the first three rows sorted in decreasing order of the associated celltower signal strength.

After running the sequence matching algorithm across all bus route sequences in the database, the backend server selects the bus route with the highest score. If the highest matching score is smaller or the sequence length is shorter than our empirical thresholds, the backend server postpones

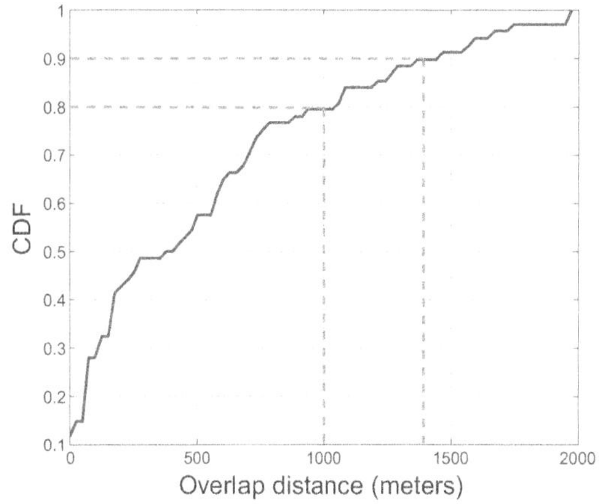

Figure 11: CDF of the overlapped route length

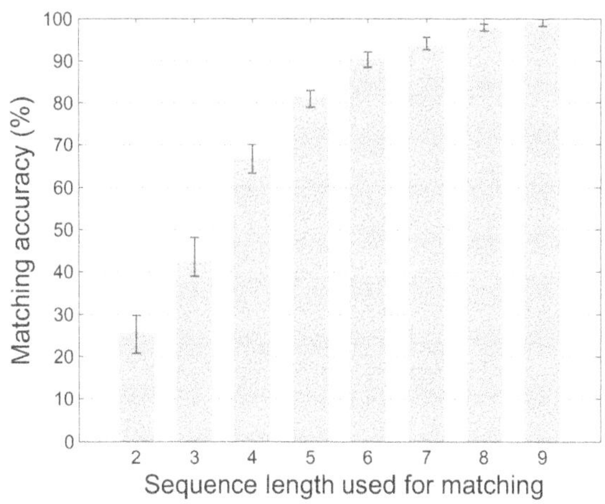

Figure 12: Matching accuracy with varying sequence length

Database celltower set seq.	19 20 21	1 2 3	4 5 6	7 8 9	10 11 12	13 14 15	16 17 18	22 23 24	Σ
Uploaded seq.		1	–	8	10	15	16		
Score	0	+1	-0.5	+0.5	+1	+0.25	+1	0	3.25

Table 2: Top-*3* set sequence matching

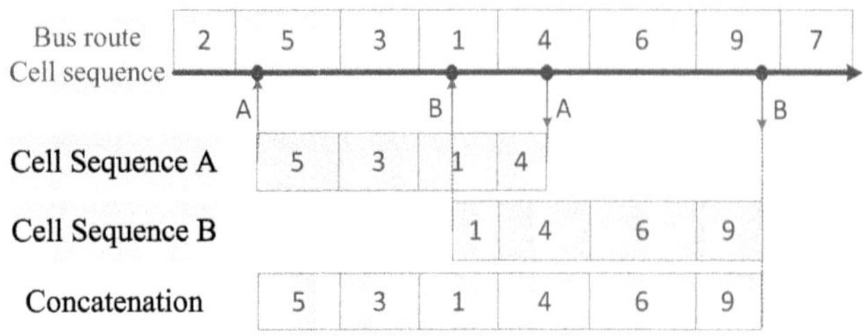

Figure 13: Celltower sequence concatenation

the updates to avoid errors. Intuitively, the small highest matching score would be due to mistriggering of sharing phones uploading celltower sequence not from interested bus routes (e.g., rapid trains, private cars, etc). Too short celltower sequence may not be informative since the misclassification rate of such short sequence is high and thus the backend server postpones the classification and the updating process until the sequence excesses the empirical threshold (which will be elaborated later).

One problem of the celltower sequence matching is that some bus routes may overlap with each other. The mobile phones on the overlapped road segments are likely to observe similar celltower sequences. Since many buses typically arrive at and depart from several major transit centers, such overlapping road segments among different bus routes are common.

We survey 50 bus routes in Singapore and measure their overlapped road segments using Google Maps. Figure 11 plots the distribution of the lengths of overlapped road segments, which suggests that over 90% of the overlapped route segments are shorter than 1400 meters, and over 80% are less than 1000 meters. Considering that the coverage range of each celltower in urban area is about 300-900 meters, we set the empirical threshold of celltower sequence length to 7.

Figure 12 plots the celltower sequence matching accuracy in classifying the bus routes. We vary the length of uploaded celltower sequence from 2 to 9. We find that the matching accuracy is low when the celltower sequence length is small (e.g, <4) largely because of the problem of route overlap. We observe that when the celltower sequence length reaches 6, the accuracy increases substantially to around 90%. When the celltower sequence length is larger than 8, the experimental results are reasonably accurate and robust.

3.4.2 *Celltower sequence concatenation: Solving jigsaw puzzles*

In many practical scenarios, the length of the celltower sequence obtained by a single sharing user, however, may be insufficient for accurate bus route classification. An intuitive idea is that we can concatenate several celltower sequences

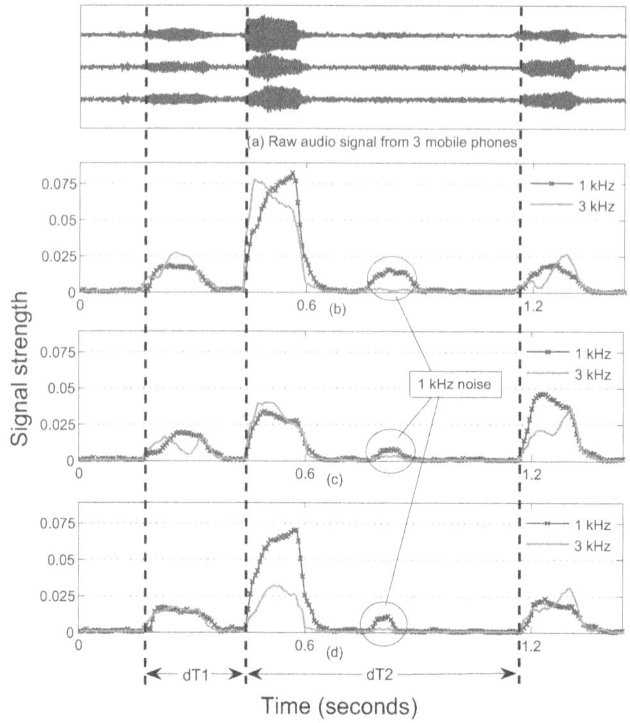

Figure 14: Time intervals of audio indication signals

of different sharing users on the same bus to form a longer celltower sequence. In Figure 13, both celltower sequences of sharing user A and B are short, while by concatenating the two celltower sequences the backend server may obtain an adequately long celltower sequence which can be used for more accurate bus classification. A simple way of concatenating the celltower sequences is to let the mobile phones of sharing passengers locally communicate with each other (e.g., over Bluetooth) [24]. This approach, however, mandates location exposure among sharing passengers and might raise privacy concerns. We thereby shift such a job to the backend server.

Recall that the mobile phone needs to collect audio signals for bus detection (§3.3.1). Here, we reuse such information to detect whether the sharing passengers are on the same bus for celltower sequence concatenation. At each bus stop, normally several passengers enter a bus and multiple beeps of the IC card readers can be detected. The time intervals between the consecutive beep signals fingerprint each bus in the time domain. Figure 14 depicts an instance of the audio signals captured by three different mobile phones on the same bus. We depict the raw audio signals in Figure 14(a), and corresponding frequency domain signals in Figure 14(b)-(d). Compared with the time domain signal, the frequency domain signal is robust against the background noise (e.g., though signal strength increases are observed in 1kHz frequency band around 0.8s, the signal strengths in 3kHz frequency band remain low). We can see that in the frequency domain the signals are highly cross-correlated and thus can be used to determine whether the phones are on the same bus. Specifically, the time intervals observed by three mobile phones are all approximately dT1 and dT2 in Figure 14.

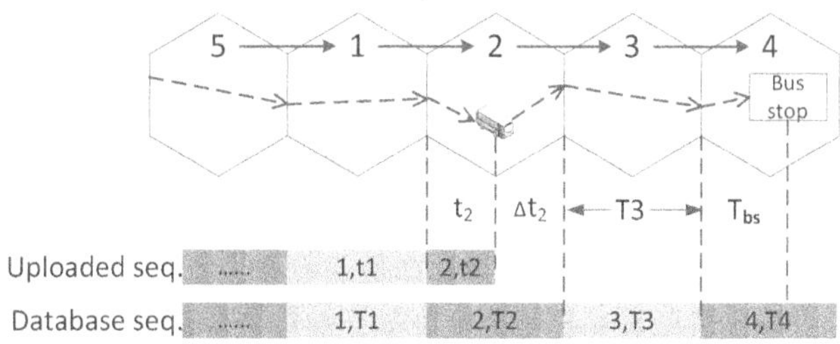

Figure 15: Bus arrival time prediction

We therefore use the time intervals between the detected beeps to determine whether multiple mobile phones are on the same bus. In our system, the mobile phones of sharing users keep sampling the audio signal and record the time intervals between the detected beeps. Such beep interval information is reported along with the celltower sequences to the backend server. Receiving the uploaded sensing data from sharing passengers, the backend server detects and groups the sharing passengers on the same bus by comparing both celltower sequences and the time intervals of the beep signals. The backend server concatenates the pieces of celltower sequences from the same bus and forms a longer celltower sequence.

3.5 Arrival time prediction

After the celltower sequence matching, the backend server classifies the uploaded information according to different bus routes. When receiving the request from querying users the backend server looks up the latest bus route status, and calculates the arrival time at the particular bus stop.

Figure 15 illustrates the calculation of bus arrival time prediction. The server needs to estimate the time for the bus to travel from its current location to the queried bus stop. Suppose that the sharing user on the bus is in the coverage of celltower 2, the backend server estimates its arrival time at the bus stop according to both historical data as well as the latest bus route status. The server first computes the dwelling time of the bus at the current cell (i.e., cell 2 in this example) denoted as t_2. The server also computes the traveling time of the bus in the cell that the bus stop is located denoted as t_{bs}. The historical dwelling time of the bus at cell 3 is denoted as T_3. The arrival time of the bus at the queried stop is then estimated as follows,

$$T = T_2 - t_2 + T_3 + t_{bs}$$

Without loss of generality, we denote the dwelling time in cell i as T_i, $1 \leq i \leq n$, the bus's current cell number as k, and the queried bus stop's cell number as q. The server can estimate the arrival time of the bus as follows,

$$T = \sum_{i=k}^{q-1} T_i - t_k + t_q$$

The server periodically updates the prediction time according to the latest route report from the sharing users and repsonds to querying users. The querying users may indicate desired updating rates and the numbers of successive bus runs to receive the timely updates.

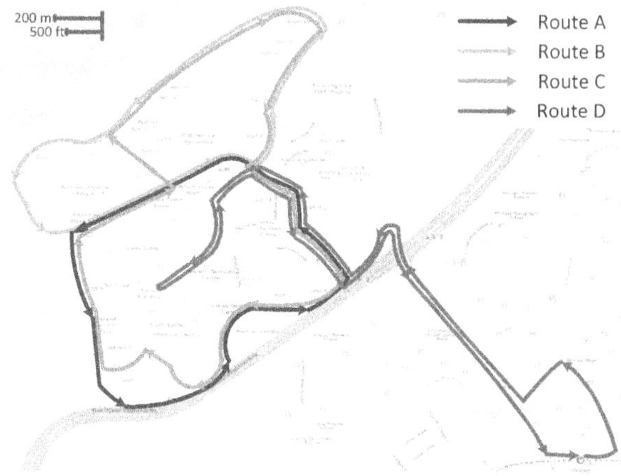

Figure 16: Campus shuttle bus routes

Route	Length	Avg. vel.	Stop	Seq. Length
A	4.0km	22.1km/h	11	14-15
B	3.8km	21.2km/h	9	9-10
C	5.5km	20.6km/h	13	16-17
D	5.8km	18.3km/h	9	20-22

Table 3: Campus bus route length, average velocity, number of bus stops, and celltower sequence length

Route	A	B	C	D
A	-	1.4km	3.4km	1.9km
B	1.4km	-	2.1km	0km
C	3.4km	2.1km	-	1.9km
D	1.9km	0km	1.9km	-

Table 4: The lengths of shared bus routes

4. IMPLEMENTATION AND EVALUATION

We implement a prototype system on the Android platform with different types of mobile phones, and collect the real data over a 7-week period. We first present the experiment environment and methodology (§4.1). We test and evaluate each component (bus detection in §4.2, and bus classification in §4.3) and present the overall performance of bus arrival time prediction in §4.4. The following details the experiment methodology and findings.

4.1 Experimental methodology

Mobile phones. We implement the mobile phone applications with the Android platform using Samsung Galaxy S2 i9100 and HTC Desire. Both types of mobile phones are equipped with accelerometers and support 16-bit 44.1kHz audio signal sampling from microphones. The Samsung Galaxy S2 i9100 has a 1GB RAM and Dual-core 1.2GHz Cortex-A9 processor, while the HTC Desire has a 768MB RAM and 1GHz Scorpion processor. For most of our experiments, we base on the SingTel GSM networks in Singapore.

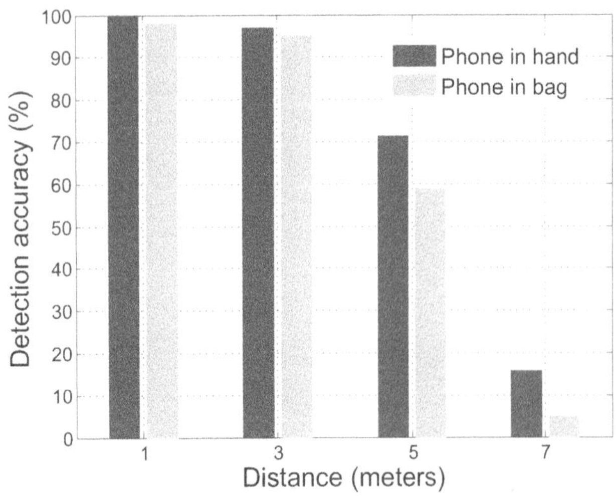

Figure 17: Bus detection accuracy

Scenario	DR	FPR	Accuracy
Mobile phone in hand			
1m	100%	3%	98%
3m	97%		97%
5m	71%		84%
7m	15%		56%
Mobile phone in bag			
1m	98%	1%	98%
3m	95%		97%
5m	59%		79%
7m	5%		52%

Table 5: Bus detection accuracy. Detection rate (DR), false positive rate (FPR) and accuracy under various scenarios

Backend server. We implement the backend server in Java running on the DELL Precision T3500 workstation with 4GB memory and Intel Xeon W3540 processor. The bus arrival time prediction service can be implemented in a computing cloud for dynamic and scalable resource provisioning as well [15].

Experiment environment. Public bus transit system serves millions of bus rides every day covering most parts of Singapore. The public bus transit system is supervised by Land Transport Authority (LTA) of Singapore and is commercially operated mainly by two major public transport providers, SBS Transit and SMRT Corporation [5, 20]. Many other transit means coexist with the public bus system. Mass Rapid Transit (MRT) trains form the backbone of the railway system. There are also tens of thousands of taxicabs operated by commercial companies and by individual taxi owners [10]. IC cards are widely used for paying transit fees. Several card readers are deployed for collecting the fees on SBS and SMRT public buses and at entrance gates of MRT stations.

We experiment in both campus shuttle buses and public transport buses (SBS Transit bus service in Singapore). As shown in Figure 16, there are 4 shuttle bus routes (i.e., Route A-D) in our campus. The shuttle buses serve from 08:00 to 23:00 with time intervals varying from 5 to 20 minutes. The bus route lengths span approximately from 3.8km to 5.8km with celltower set sequence lengths varying from 9 to 22. The average velocity of the buses is about 20km/h. Table 3 gives the details of the bus routes. The shuttle bus routes have overlapped road segments as depicted in Figure 16. The campus bus C travels in clockwise direction, while buses A, B, and D move in counterclockwise direction. We see that Route A and Route C have substantial overlapped segments. Table 4 summarizes the shared route segments between each pair of bus routes, which span from 0km to 3.4km. We see that around 85% (3.4km/4km) of Route A overlaps with Route C.

We experiment on SBS Transit bus route 179 and 241 as well. For comparison study, we also collect celltower sequences and accelerometer readings in East-West and the North-South MRT Lines in Singapore.

4.2 Bus detection performance

4.2.1 Audio detection accuracy

We collect more than 200 beep signals on different *public transit buses* during our 7-week experiments. We set the audio sampling rate to be 8kHz, and we use 128-pt FFT to detect the IC card reader. We test the bus detection method by varying the distances between the IC card reader and the mobile phones (approximately 1 meter to 7 meters). We also consider the scenarios where mobile phones may be held in hand and inside bags. We report the average detection accuracy of single beeps in different circumstances. In Figure 17, we see that the detection rate is over 95% when mobile phones are in close vicinity to the IC card reader (e.g., within 3 meters) even when they are placed in bags. With mobile phones placed 5 meters away from the reader, the detection accuracies are about 58% held in hand, and 71% placed in bags, respectively. As the distance increases further (e.g, >7 meters), the detection accuracy drops substantially. In addition, we list the detection rate, false positive rate, and accuracy of bus detection method in Table 5.

The experiment results suggest that the audio based method effectively detects the beep signal on the bus when the distance between the IC card reader and the mobile phone is within 3 meters. Considering that the entrance gate of the bus is about 1.4 meters wide, when a sharing user enters a bus, the mobile phone would be less than 1 meter away from the IC card reader (normally within 0.5 meters).

4.2.2 Bus vs. MRT train

We next evaluate the accelerometer based bus detection method that is used to distinguish the buses from the MRT trains. Figure 18 plots the accuracy in detecting the buses. We find that accelerometer based method can distinguish the buses from the MRT trains with an accuracy of over 90% on average. We analyzed the main reason for falsely detecting public buses as MRT trains, and find that it happens mostly when the buses are driving along long straight routes late during night time. The accelerometer readings may be relatively stable and very similar to those on the MRT trains.

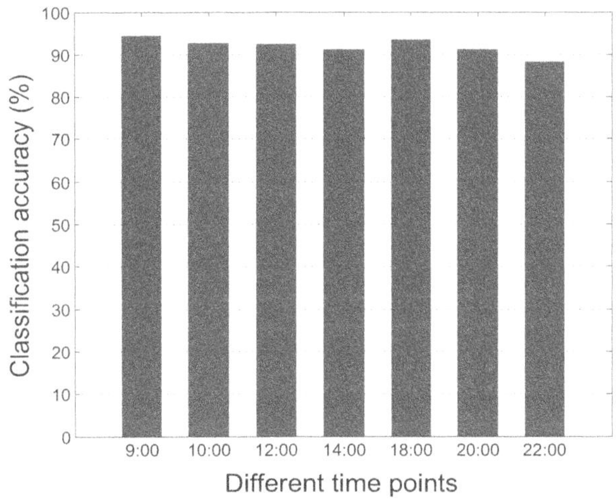

Figure 18: Bus vs. MRT using accelerometer

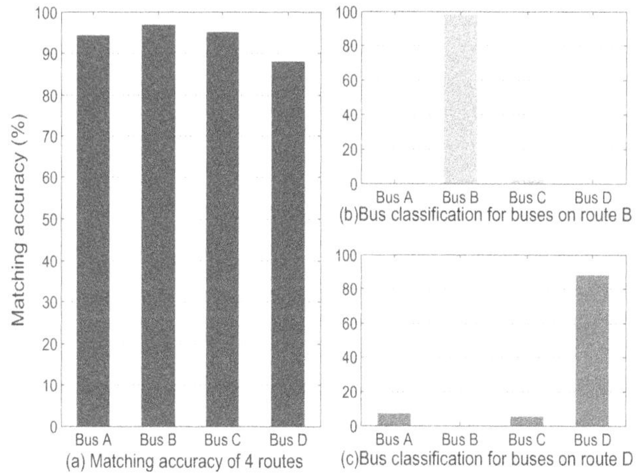

Figure 19: Bus classification accuracy

4.3 Bus classification performance

We present the evaluation results for our bus classification algorithms. In our prototype system, we collect the celltower sequences on the 4 campus bus routes and store them in the database. The campus buses do not have IC card readers, so we use the GNUradio to produce and play the dual-tone (1kHz and 3kHz) beeps. Mobile phones start to collect data after detecting the beeping signals on buses. For the public transit buses (e.g., SBS transit and SMRT Corporation buses), the mobile phones can directly detect their IC card readers. The data collection process spans over a period of 7 weeks. We collect 20 runs for each shuttle bus route for the bus route classification. As the cellular networks are likely to be updated incrementally, most celltowers along the bus routes typically remain consistent during the experiment period.

We implement the celltower sequence matching with the top-*3* celltower sequence matching algorithm. In Figure 19(a), we plot the bus classification results for the 4 campus bus routes. According to the experiment results, the bus classification accuracy is approximately 90% with the highest accuracy of 96% for Bus B and the lowest of 87% for Bus D. Although 85% of Route A is overlapped with Route C, the bus classification accuracy for Bus A and C are still around 94%. The main reason is that Bus A and C travel in the opposite directions. Since Route D shares a large portion of overlapped road segments with Route A and Route C, and buses travel in the same direction on the shared road segments, buses along Route D might be misclassified to Route A or Route C. Figure 19(c) depicts the classification ratio of buses along Route D. We can find that 7% of the buses are misclassified to Route A and 6% are misclassified to Route C. Although Route B has many overlapped road segments with Route A and C, the buses travel in the opposite directions on those road segments. (Figure 19(b)) depicts the classification ratio of buses along Route B. We find that only 3% of the buses are misclassified to Route C. Overall, the bus classification accuracy is satisfactory, considering the high overlap ratio of the four routes in the campus (the city-wide public bus routes are far less overlapped, e.g., SBS 179 and 241).

4.4 Arrival time prediction

We present the final bus arrival time prediction results based on above estimations. We collect the campus bus traces using a high accurate vehicle GPS navigator as the benchmarks. In the same buses, we collect celltower sequences using two mobile phones and stored the sequence in memory stick for our later trace-driven study.

In the trace-driven study, we generate queries at different campus bus stops according to poisson arrival process, and compare the predicted arrival time with the actual arrival time of the campus buses to compute the average of the absolute prediction error. Figure 20(a) shows the CDF of the absolute error of arrival time prediction results. The median prediction errors vary approximately from 40s for Bus B to 60s for Bus D. The 90th percentiles are approximately from 75s for Bus B to 115s for Bus D, respectively. Generally, the average estimation error increases as the length of bus route increases.

Figure 20(b) plots the average error against the distance between the sharing user and the querying user, where we approximate the distance using the number of bus stops. We observe that as the bus moves closer to the querying user, the prediction error becomes smaller. The error of Bus D increases faster than those of Bus A, B, and C.

We experiment with commercial bus system as well. For comparison, we also query the arrival time of public transit buses provided by LTA of Singapore. The public buses are periodically tracked with on-bus localization devices and respond to the queries for the bus information. People can send an SMS to query the bus arrival time indicating the interested bus route and stop. In the experiment we test the arrival time prediction on SBS bus route 179 and 241. We compute the prediction error by comparing the predicted results with the actual arrival time of the buses. Both prediction errors of LTA and our system are measured and we plot the CDF of the prediction results in Figure 20(c). According to the results, the average prediction error of our system is approximately 80 seconds, while the prediction result of LTA is around 150 seconds. Such a comparison result is surprising, as we expect more accurate prediction result from the commercial system of LTA where a rich set of resources including on-bus GPS sensors are proactively used.

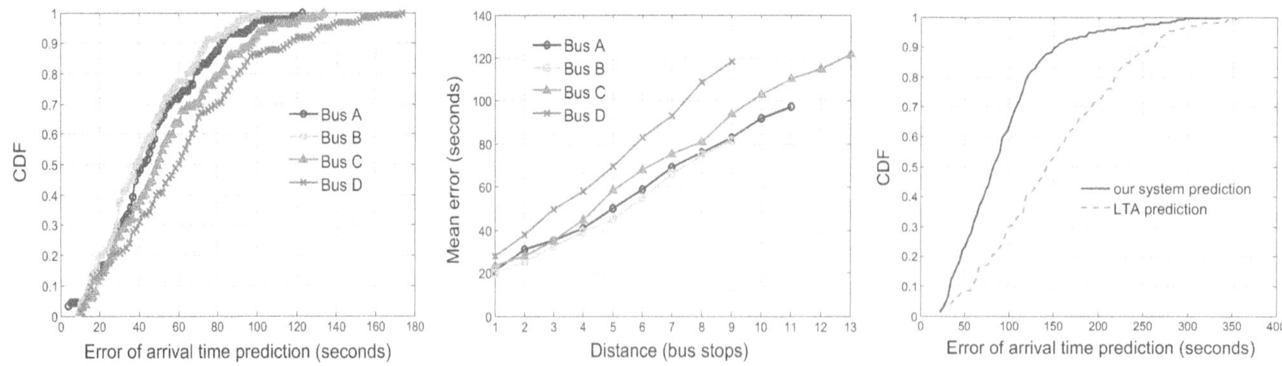

(a) Bus arrival time prediction error (b) Bus arrival time prediction (c) Our system v.s. LTA

Figure 20: Arrival time prediction performance

Sensors	Samsung i9100	HTC Desire
No sensor	18.2	15.3
Accelerometer 20Hz	18.0	15.2
Microphone 8kHz+FFT	17.5	14.9
Celltower 1Hz	17.8	15.0
GPS 1Hz	9.7	6.4

Table 6: Battery duration for different sensor settings (in hours)

We suspect that the deployed system of LTA is intentionally made inaccurate (e.g., using caching to reduce computation and communication cost), yet we cannot further dig into such a commercially running system for more details.

4.5 System overhead

Mobile phone. The computation complexity of the algorithms on mobile phones is bounded by the length of audio signals and accelerometer signals needed for the bus detection. In order to maintain the sample resolution and remove the noise, we extract the audio signal with sliding widows with the window size of 32. We record the audio signal at the sampling rate of 8kHz, and use $n = 128$pt FFT to convert the time domain audio signals to frequency domain signals. The major computational complexity is attributed to performing FFT on mobile phones which is $O(n \log n)$. Current mobile phones can finish the computation task in realtime. For example, it takes approximately 1.25ms and 1.8ms on average to finish to 128pt FFT on Samsung Galaxy S2 i9100 and HTC Desire, respectively.

We measure the power consumption of continuously sampling microphone, accelerometer, GPS, and cellular signals. Table 6 illustrates the measured battery lifetime when the mobile phones continuously trigger different sensors. The experiments were performed with the screen set to minimum brightness. We report the average results over 10 independent measurements. The battery duration was quite similar for sampling accelerometer at 20Hz, sampling audio signal at 8kHz with 128pt FFT, and sampling no sensors. Sampling the celltower signal consumes limited extra battery power as well. On the other hand the battery lifetime is substantially reduced when the GPS module in the phone is enabled.

Backend server. The computation overhead of backend server is mainly bounded by the bus classification algorithm, i.e., the uploaded celltower sequence length l, the celltower set sequence length k, and the number of celltower set sequences in the database N. The computation complexity of sequence matching using dynamic programming is $O(lk)$, and as we need to compare with N candidate sequences in database the overall computation complexity is $O(lkN)$. Since in practice both m and n are usually small (e.g., $\max\{l,k\}$ is around 40 according to our experiments), the computation complexity increases almost linearly to the number of candidate celltower sequences in the database.

5. LIMITATIONS AND ON-GOING WORK

Alternative referencing points. In practical implementation, we observe that the number of celltowers that a sharing user can capture on a bus influences the bus classification accuracy. It takes a few minutes for the passenger on a bus to observe several celltowers to form a reliable sequence for bus classifications. We are currently studying to utilize the ambient radio signals and extract useful information (e.g., WiFi points) to complement the celltower IDs as fingerprints. As a matter of fact, there are relatively stable WiFi points along different bus routes. Similar to using GPS modules, such a method consumes extra power on the mobile phones, though the WiFi module draws much less power. Our preliminary measurement indeed finds many reliable WiFi hotspots in our campus which may extend the celltower fingerprints. We are also studying how such complementary information can be utilized in an energy efficient manner.

Number of passengers. The number of sharing passengers affects the prediction accuracy in our system. When there is no sharing passengers on a bus, the backend server would miss the bus, which affects the prediction results. This common issue of crowd-sourced solutions is largely influenced by the penetration rate and popularity of the services. One may actively promote the service to reach a critical penetration rate so as to ensure that at least one sharing user is on the bus willing to report the bus status. At the initial stage, we may equip the bus driver with the mobile phone clients so that at least one sharing user (i.e., the bus driver) can update the bus status to the backend server.

First few bus stops. The bus classification method needs a sufficiently long celltower sequences for accurate bus route classification, and consequently the arrival time at the first few bus stops would not be timely updated. As discussed in [12], being less affected by unpredictable traffic conditions, the arrival time at the first few bus stops is stable and predictable according to the historical data. Therefore, the backend server may bias the arrival time prediction for those bus stops towards the official bus schedules and real-time traffic conditions on the bus routes.

Overlapped routes. By concatenating several celltower sequences from the same bus, the backend server may obtain a longer celltower sequence for bus classification. Although such a longer sequence alleviates the overlapped bus route issue, occasionally our bus classification algorithm still cannot confidently classify the bus routes and have to postpone responses to querying passengers. Since many buses typically arrive at and depart from several major transit centers, such route overlapping could be especially common in downtown area. In our experiments, we find that there are distinct differences of the bus speeds between the campus shuttle buses and the public transit buses even when the buses travel the same overlapped sub-routes. The buses with more frequent bus stops tend to drive at lower speed than those with fewer bus stops. We are currently exploring how to effectively utilize the differences of the bus speed to distinguish the buses with overlapped routes. The problem of efficiently and accurately classifying different bus routes sharing substantial portion of overlapped segments (e.g., in downtown areas) remains challenging.

6. RELATED WORK

Phone-based transit tracking. Our work is mostly related to recent works on the transit tracking systems [12, 20, 29]. EasyTracker [12] presents an automatic system for low-cost, real-time transit tracking, mapping and arrival time prediction using GPS traces collected by in-vehicle smartphones. Thiagarajan *et al.* [29] present a grassroots solution for transit tracking utilizing accelerometer data and GPS modules on participating mobile phones. Our work differs from them in that it predicts the bus arrival time based on celltower sequence information shared by participatory users. To encourage more participants, no explicit location services (e.g., GPS-based localization) are invoked so as to reduce the overhead of using such special hardware for localization.

EEMSS [32] presents an energy efficient sensor management framework which uses minimum number of sensors on mobile devices to monitor user states. Nericell [23] uses onboard sensors to efficiently monitor the road surface quality and traffic conditions. VTrack [31] predicts road traffic time based on a sequence of WiFi-based positioning samples using an HMM-based algorithm for map matching. Ravindranath *et al.* [27] use various sensor hints to improve wireless protocols. CTrack [30] presents energy-efficient trajectory mapping using celltower fingerprints and utilizes various sensors on mobile phones to improve the mapping accuracy. Balan *et al.* [10] present a realtime trip information system to predict taxi fares and trip time. SignalGuru [19] presents a software service that predicts traffic signals' future schedule which enables green light optimal speed advisory by leveraging opportunistic sensing on windshield-mount smartphones. Yang *et al.* [35] present a driver detection system that distinguishes a driver and a passenger leveraging car speakers and mobile phone microphones.

Celltower sequence matching. StarTrack [7] provides a comprehensive set of APIs for mobile application development. Applying new data structures, [17] enhances StarTrack in efficiency, robustness, scalability, and ease of use. CAPS [25] determines a highly mobile user's position using a cell-ID sequences matching technique which reduces GPS usages and saves energy on mobile phones. Unlike those proposals, our work does not aim to position the mobile users though similar in spirit to these existing works in utilizing the celltower sequences.

Participatory sensing. Many recent works develop participatory platforms for people-centric mobile computing applications [6, 13]. Micro-blog [16] presents a participatory sensing application which connects sharing parties and querying parties to allow geo-tagged multimedia sharing. MoVi [11] studies the problem of social activity coverage where participants collaboratively sense ambience and capture social moments through mobile phones. SoundSense [22] classifies ambient sounds to achieve context recognition. SurroundSence [8] utilizes various sensors on mobile phones to collect identifiable fingerprints signals for logical localization. Escort [14] obtains cues from social encounters and leverages an audio beacon infrastructure to guide a user to a desired person. WILL [34] designs an indoor logical localization technique leveraging user mobility and WiFi infrastructure while avoiding site survey. Although targeted at totally different applications and problems, the common rationale behind these works and our design is that the absolute physical location of users though sometimes sufficient not always necessary to accomplish particular tasks.

7. CONCLUSIONS

In this paper, we present a crowd-participated bus arrival time prediction system using commodity mobile phones. Our system efficiently utilizes lightweight onboard sensors which encourages and attracts participatory users. Primarily relying on inexpensive and widely available cellular signals, the proposed system provides cost-efficient solutions to the problem. We comprehensively evaluate the system through a prototype system deployed on the Android platform with two types of mobile phones. Over a 7-week experiment period, the evaluation results demonstrate that our system can accurately predict the bus arrival time. Being independent of any support from transit agencies and location services, the proposed scheme provides a flexible framework for participatory contribution of the community.

8. ACKNOWLEDGEMENT

We would like to thank our shepherd, Rajesh Krishna Balan, as well as the anonymous reviewers for providing constructive feedbacks and valuable input for improving the quality of this paper. We acknowledge the support from NTU SUG M4080103.020 and NAP M4080738.020.

9. REFERENCES

[1] Bus transport in Singapore. http://en.wikipedia.org/wiki/Bus_transport_in_Singapore.
[2] EZ-Link. http://www.ezlink.com.sg.
[3] Octupus. http://www.octopus.com.hk/home/en.
[4] Oyster. https://oyster.tfl.gov.uk/oyster.
[5] PublicTransport@SG. http://www.publictransport.sg/.

[6] T. Abdelzaher, Y. Anokwa, P. Boda, J. Burke, D. Estrin, L. Guibas, A. Kansal, S. Madden, and J. Reich. Mobiscopes for Human Spaces. *IEEE Pervasive Computing*, vol. 6(issue 2): pages 20–29, Apr. 2007.

[7] G. Ananthanarayanan, M. Haridasan, I. Mohomed, D. Terry, and C. A. Thekkath. Startrack: a framework for enabling track-based applications. In Proceedings of ACM MobiSys, pages 207–220, 2009.

[8] M. Azizyan, I. Constandache, and R. Roy Choudhury. Surroundsense: mobile phone localization via ambience fingerprinting. In Proceedings of ACM MobiCom, pages 261–272, 2009.

[9] P. Bahl and V. N. Padmanabhan. RADAR: an in-building RF-based user location and tracking system. In Proceedings of IEEE INFOCOM, pages 775–784, 2000.

[10] R. K. Balan, K. X. Nguyen, and L. Jiang. Real-time trip information service for a large taxi fleet. In Proceedings of ACM MobiSys, pages 99–112, 2011.

[11] X. Bao and R. Roy Choudhury. Movi: mobile phone based video highlights via collaborative sensing. In Proceedings of ACM MobiSys, pages 357–370, 2010.

[12] J. Biagioni, T. Gerlich, T. Merrifield, and J. Eriksson. Easytracker: automatic transit tracking, mapping, and arrival time prediction using smartphones. In Proceedings of ACM SenSys, pages 1–14, 2011.

[13] J. Burke, D. Estrin, M. Hansen, A. Parker, N. Ramanathan, S. Reddy, and M. B. Srivastava. Participatory sensing. In Workshop on World-Sensor-Web (WSW): Mobile Device Centric Sensor Networks and Applications, pages 117–134, 2006.

[14] I. Constandache, X. Bao, M. Azizyan, and R. R. Choudhury. Did you see bob?: human localization using mobile phones. In Proceedings of ACM MobiCom, pages 149–160, 2010.

[15] E. Cuervo, A. Balasubramanian, D.-k. Cho, A. Wolman, S. Saroiu, R. Chandra, and P. Bahl. Maui: making smartphones last longer with code offload. In Proceedings of ACM MobiSys, pages 49–62, 2010.

[16] S. Gaonkar, J. Li, R. R. Choudhury, L. Cox, and A. Schmidt. Micro-blog: sharing and querying content through mobile phones and social participation. In Proceedings of ACM MobiSys, pages 174–186, 2008.

[17] M. Haridasan, I. Mohomed, D. Terry, C. A. Thekkath, and L. Zhang. Startrack next generation: a scalable infrastructure for track-based applications. In Proceedings of USENIX OSDI, 2010.

[18] M. Keally, G. Zhou, G. Xing, J. Wu, and A. Pyles. Pbn: towards practical activity recognition using smartphone-based body sensor networks. In Proceedings of ACM SenSys, pages 246–259, 2011.

[19] E. Koukoumidis, L.-S. Peh, and M. R. Martonosi. Signalguru: leveraging mobile phones for collaborative traffic signal schedule advisory. In Proceedings of ACM MobiSys, pages 127–140, 2011.

[20] F. Li, Y. Yu, H. Lin, and W. Min. Public bus arrival time prediction based on traffic information management system. In Proceedings of IEEE International Conference on Service Operations and Logistics, and Informatics (SOLI), pages 336–341, 2011.

[21] Y. Liu, L. Chen, J. Pei, Q. Chen, and Y. Zhao. Mining frequent trajectory patterns for activity monitoring using radio frequency tag arrays. In Proceedings of IEEE PerCom, 2007.

[22] H. Lu, W. Pan, N. D. Lane, T. Choudhury, and A. T. Campbell. Soundsense: scalable sound sensing for people-centric applications on mobile phones. In Proceedings of ACM MobiSys, pages 165–178, 2009.

[23] P. Mohan, V. N. Padmanabhan, and R. Ramjee. Nericell: rich monitoring of road and traffic conditions using mobile smartphones. In Proceedings of ACM SenSys, pages 323–336, 2008.

[24] J. Paek, J. Kim, and R. Govindan. Energy-efficient rate-adaptive gps-based positioning for smartphones. In Proceedings of ACM MobiSys, pages 299–314, 2010.

[25] J. Paek, K.-H. Kim, J. P. Singh, and R. Govindan. Energy-efficient positioning for smartphones using cell-id sequence matching. In Proceedings of ACM MobiSys, pages 293–306, 2011.

[26] C. Peng, G. Shen, Y. Zhang, Y. Li, and K. Tan. Beepbeep: a high accuracy acoustic ranging system using cots mobile devices. In Proceedings of ACM SenSys, pages 1–14, 2007.

[27] L. Ravindranath, C. Newport, H. Balakrishnan, and S. Madden. Improving wireless network performance using sensor hints. In Proceedings of USENIX NSDI, 2011.

[28] S. Reddy, M. Mun, J. Burke, D. Estrin, M. Hansen, and M. Srivastava. Using mobile phones to determine transportation modes. *ACM Transactions on Sensor Networks*, vol. 6(issue 2): pages 1–27, March 2010.

[29] A. Thiagarajan, J. Biagioni, T. Gerlich, and J. Eriksson. Cooperative transit tracking using smart-phones. In Proceedings of ACM SenSys, pages 85–98, 2010.

[30] A. Thiagarajan, L. Ravindranath, H. Balakrishnan, S. Madden, and L. Girod. Accurate, low-energy trajectory mapping for mobile devices. In Proceedings of USENIX NSDI, 2011.

[31] A. Thiagarajan, L. Ravindranath, K. LaCurts, S. Madden, H. Balakrishnan, S. Toledo, and J. Eriksson. Vtrack: accurate, energy-aware road traffic delay estimation using mobile phones. In Proceedings of ACM SenSys, pages 85–98, 2009.

[32] Y. Wang, J. Lin, M. Annavaram, Q. A. Jacobson, J. Hong, B. Krishnamachari, and N. Sadeh. A framework of energy efficient mobile sensing for automatic user state recognition. In Proceedings of ACM MobiSys, pages 179–192, 2009.

[33] M. S. Waterman and T. F. Smith. Identification of common molecular subsequences. *Journal of Molecular Biology*, 147:195–197, 1981.

[34] C. Wu, Z. Yang, Y. Liu, and W. Xi. WILL: Wireless indoor localization without site survey. In Proceedings of IEEE INFOCOM, 2012.

[35] J. Yang, S. Sidhom, G. Chandrasekaran, T. Vu, H. Liu, N. Cecan, Y. Chen, M. Gruteser, and R. P. Martin. Detecting driver phone use leveraging car speakers. In Proceedings of ACM MobiCom, pages 97–108, 2011.

Beyond Deployments and Testbeds: Experiences with Public Usage on Vehicular WiFi Hotspots

Joshua Hare, Lance Hartung, and Suman Banerjee
Dept. of Computer Sciences, University of Wisconsin-Madison
Madison, WI, USA
{hare, hartung, suman}@cs.wisc.edu

ABSTRACT

We describe our experiences with deploying a vehicular Internet access service on public transit buses. Our system, called WiRover, has been running on these buses since April 2010 for about 18 months till date providing a WiFi hotspot to which bus passengers can connect. As of Dec. 1, 2011 we have observed 17,567 unique client devices connect to the WiRover system. These devices have downloaded more than 337.53 GB and uploaded 48.19 GB of content through our system. Since the initial deployment, the buses have logged more than 9,331 hours of operation and have traveled over much of the northern Midwest of the United States. Through this paper we provide different insights acquired in deploying and running a continuous service with real users, present various user and usage characteristics in these systems, discuss various design and management strategies for these networks, and explore different network traffic optimizations possible.

Categories and Subject Descriptors

C.2.1 [**Network Architecture and Design**]: Wireless Communications; C.2.3 [**Network Operations**]: Public Networks, Network Management

General Terms

Experimentation, Measurement, Performance

Keywords

Network Diversity, Vehicular Connectivity, Wireless, WiRover

1. INTRODUCTION

Internet connectivity for moving vehicles and their passengers is an important and challenging problem domain. It is now becoming common for public transportation buses to offer WiFi services to passengers [25]. The growth of WiFi connectivity in mass transportation is causing vehicular networking to become ubiquitous (famously Amtrak trains [1], Greyhound buses, and Google buses [24]). Despite the rapid growth in vehicular Internet connectivity, little is known about the network performance, usage characteristics, and operational challenges of providing such connectivity in vehicular environments. For instance, given the widespread availability of data plans on our mobile phones, would users even care to access the Internet through such services? If users do indeed use these services, how much can common

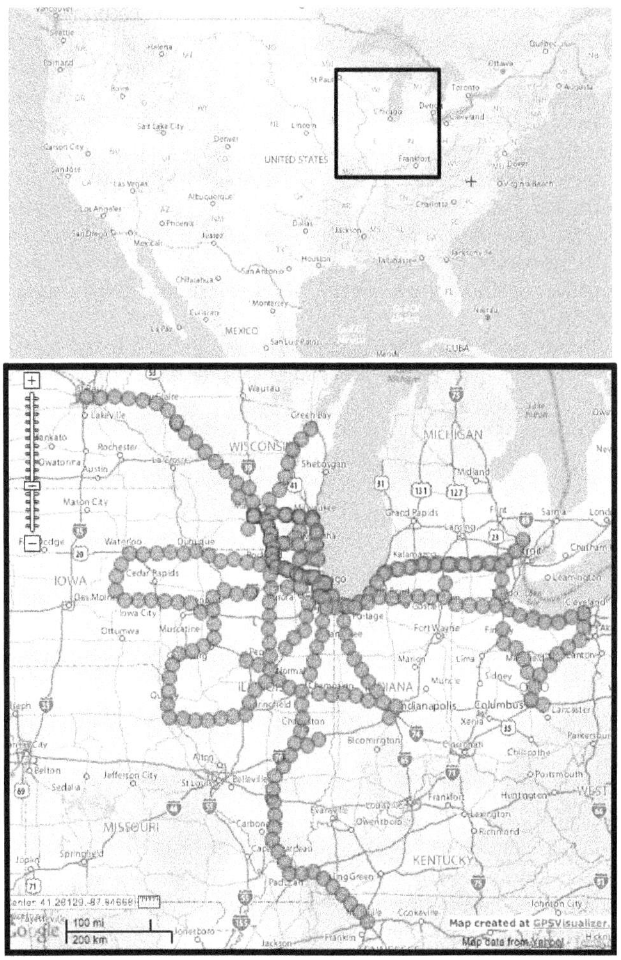

Figure 1: The 5 public buses on which we have installed our WiRover system have traveled over much of the northern Midwest region of the United States (the map covers 988763 sq km).

MobiSys'12, June 25–29, 2012, Low Wood Bay, Lake District, UK.

caching and compression techniques help if deployed on a single public bus carrying a few tens of users? How should one deal with outages of individual cellular networks when providing Internet connectivity to such public vehicles?

Over the last few years, we have designed and refined a vehicular Internet connectivity system, we call *WiRover* and have installed this system on a diverse set of vehicles based out of Wisconsin, USA. More specifically, our WiRover system has been installed on more than 10 different public buses, ambulances, and personal vehicles. In this work, we focus on 5 particular vehicles which happen to be passenger buses, and share our operational experience and learnings from deploying and operating an Internet service on these buses for more than 18 months. These 5 buses consist of 2 city transit buses and 3 coach buses, which have covered much of the northern Midwest of the United States, as shown in Figure 1. The WiRover system provides bus passengers with a free Internet connection by creating a WiFi hotspot within the bus. Passengers connect to this free service to obtain Internet services. In exchange for this free service, our terms of service allow us to monitor network traffic characteristics (anonymized, and in aggregate) for research purposes. The goal of our study has been to use our deployment to understand various issues and challenges in providing a continuous and robust service to all users in public transit systems.

When we first deployed our system we were unsure as to how many smart phone users might connect to our system. After all, each smart phone user already has a mobile cellular data plan and may not need to use the services of WiRover. Hence, to be successful as a popular vehicular WiFi hotspot, we needed to ensure that the experience through WiRover is superior to that of the individual's own smart phone. Further, our system needed to be robust and highly available so that users can begin to depend on it in their daily commutes or occasional long distance bus rides.

We believe that we have been quite successful in being able to attract and retain users. This is particularly apparent when we observe the large number of various smart phone users who continuously access Internet services through our WiRover system, despite having data plans readily available on their mobile phones.

We believe this is due to three reasons. First, WiRover provides a faster Internet download experience for users through a number of optimizations built into the WiRover system, particularly the ability to aggregate the bandwidth from multiple networks. Figure 2, shows that WiRover is able to provide an average speedup of 1.37 over a single interface. Second, it provides reliable connectivity to users by utilizing its multi-network design. The two cellular networks that we use commonly in each bus (let us call them NetA and NetB) , individually have an availability of about 77.96% and 80.88% respectively across the entire area under study. The lower numbers reflect poorer coverage on certain rural highways and cities, and also reflect periods of time when these individual networks might be too slow to provide a useful Internet service. However, in our typical dual-network solution mounted on the buses, the two networks together increase overall availability to 94.37%. Finally, we also feel that providing a free service is enough to attract curious passengers.

Ensuring a reliable Internet connection has not been without its challenges. Hardware failures are challenging to debug since access to the hardware platform first requires access to the vehicle. Software bugs may remain dormant for days or months, making them difficult to identify and reproduce. Throughout this paper, we share the challenges we have faced and the mechanisms we have adopted to overcome the challenges of running similar deployments.

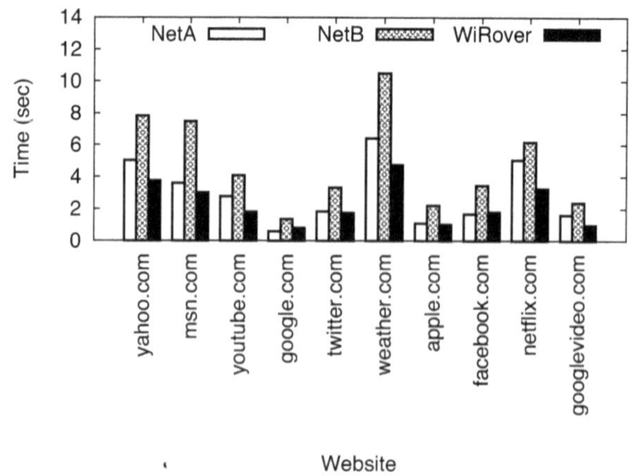

Figure 2: The website download times to ten of the most popular visited websites. WiRover provides an average speedup of 1.37 over a single NetA interface.

1.1 Contributions

The vehicular environment introduces many challenges that are not present in stationary networks. Users connect to our WiFi hotspot just as they would at a coffee shop and some users may have the expectation that their experience will be similar in both environments. However, in the vehicular environment it is necessary to use wireless back-haul networks and the fluctuating performance and link failures of these networks must be contended with.

In the past there have been many notable projects that presented innovative systems to improve Internet connectivity into vehicles to study different topics of interest, e.g., delay-tolerant networking (DieselNet) [23, 26, 3], to survey status of different WiFi hotspots across a metropolitan area (CarTel) [7, 9], to collect road condition information through vehicle-mounted sensors (Pothole Patrol) [10], to collect various types of wireless measurements across a wide-area (WiScape) [22], and to design techniques to improve achievable throughputs to moving vehicles (ViFi, Wiffler, Pluribus) [5, 4, 14] as well as others [8, 12, 15, 20]. This paper does not focus on any of these individual problem domains, including fine-tuning algorithms that improve end-to-end throughputs to moving vehicles. Instead the goal of this paper is to share our experiences in running a vehicular Internet access system on multiple city and long distance buses for a period of more than 18 months *as a service with thousands of real users depending and continuously using it* and the ensuing lessons learned from this effort. While all the above cited efforts worked significantly with vehicular deployments, some as very long running efforts, we believe that none of these prior efforts provided a continuous Internet access system to the multitude of bus passengers on the move that we are providing daily and are reporting in this

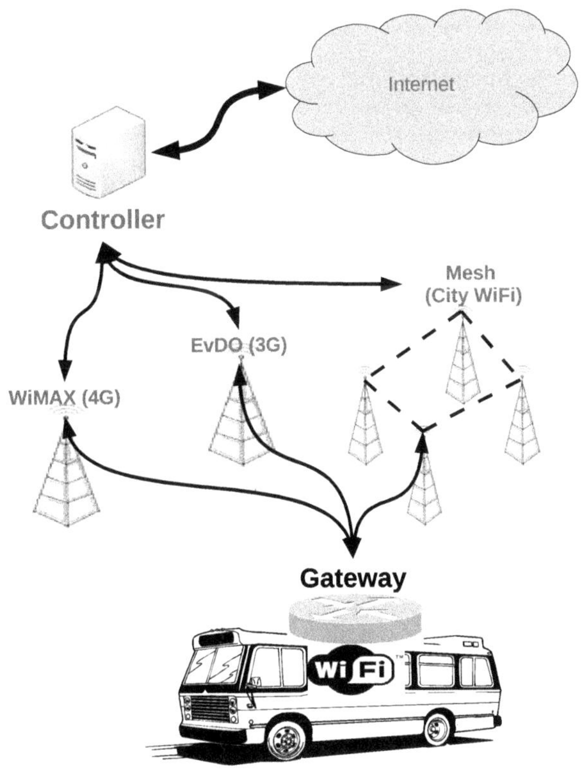

Figure 3: The WiRover system can connect to any wireless back-haul channel to provide Internet connectivity locally to bus passengers.

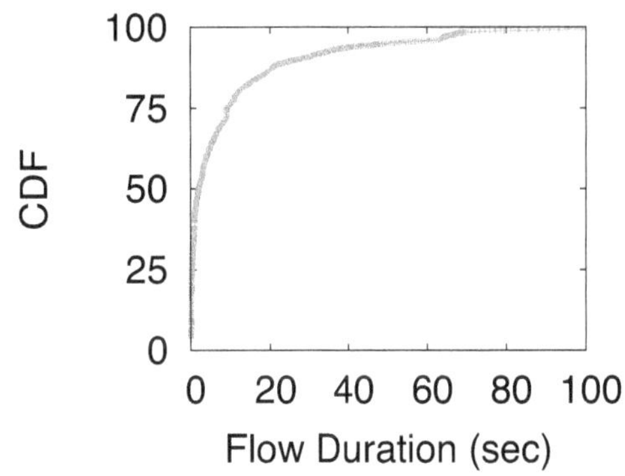

Figure 4: The CDF of flow durations as observed over the past 18 months show that more than 85% are less than 20 seconds.

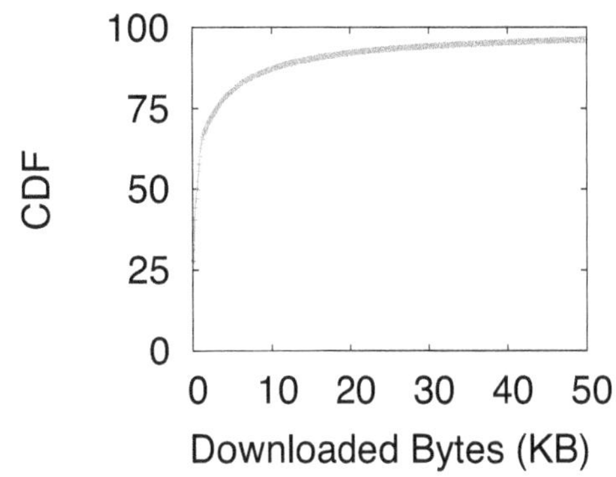

Figure 5: The CDF of flow size as observed over the past 18 months show that more than 62% of flows download less then 1 KB.

paper. While building vehicular testbeds is always hard, we have found that running a continuous service operating on these vehicles on which users can depend upon is a bigger challenge. While small failures that impact individual nodes in a testbed can be addressed in an opportunistic fashion, and often allow experimenters to ignore such nodes for small periods of time, failures in our system lead to unhappy users and often required us to quickly detect and repair such occurrences.

Summarizing, this paper makes the following contributions:

- Experiences in designing and implementing a simple yet sustainable vehicular Internet service, deploying and operating it for more than 18 months across both local and long distance public transit buses, and in managing this system to preserve uptime (Section 2 and 3).
- Characteristics of user behavior and usage, when accessing Internet services from these vehicles, including some differences in behaviors across the passengers of city transit and long distance buses (Section 4).
- Some potential optimizations to improving user experience and their potential usefulness in our system, especially in the applicability of techniques such as Packet Caches for reducing traffic across "expensive" cellular links. (Section 5).
- Exploration of the various types of failures experienced by our system, including those due to spotty and variable network coverage, their impact on user traffic, and our approaches to mitigate or manage them (Section 6).
- Discussion of some open research challenges while running our deployment over the past 18 months (Section 8).

2. DESIGN

Our WiRover system builds on certain basic design tenets of the MAR system [21]. The MAR system proposed a vehicular gateway which connected to multiple cellular networks to increase system bandwidth through bandwidth aggregation. These multiple cellular networks further increased the system's diversity and thus increased the reliability. We comment on some of our design choices and differences from the MAR system in this section, including some of our software scaffolding to implement a long running vehicular Internet service. They include: (i) our encapsulation structure and a simple flow scheduling approach, (ii) flow migration techniques in order to mitigate network failures and outages, (iii) a continuous active and passive network monitoring subsystem, and (iv) related traffic optimizations.

Encapsulation and simple flow scheduling: As shown in Figure 3, our WiRover system consists of a gateway node and a controller node (the proxy server). The gateway creates a single encapsulation tunnel across all the available wireless back-haul networks between the gateway and controller nodes. Traffic is modified through network address translation (NAT) at both the gateway and the controller nodes. While this encapsulation structure allows for partitioning of packets of an individual flow across multiple cellular paths, we chose a simpler alternative (as in MAR) of mapping individual flows to a single cellular path only, based on dynamic weights assigned to each path. (The only exception to this rule was during failures and slowdowns in individual cellular paths as described in the next section.) This design choice was pragmatic since in typical busy periods such flow mapping was adequate to extract all advantages of bandwidth aggregation in our system. Further, our flow logs collected over the last 18 months show that most flows are small and short-lived (Figure 4 and Figure 5), and the complexities of flow partitioning can easily outweigh the advantages in such cases.

Flow migration during link failures: Each cellular network has its own spatio-temporal fluctuations occasionally leading to significant slowdowns and outages. Since our flows are mapped entirely to a single cellular path, we chose to implement a simple flow migration scheme when such slowdowns or outages are detected. These events are detected on a cellular path triggering future packets for all flows in the failed path to be mapped to a new path. The receiver node (controller or gateway depending on the direction of the flow) detects the onset of a flow re-map and performs transient re-order management for that flow, i.e., out-of-order packets from a re-mapped flow are held for a dynamic period of time (the period is governed by the latency and bandwidth characteristics) to minimize end-to-end re-ordering of these packets.

Network performance measurement subsystem: The measurement subsystem monitors the performance of the different cellular back-haul paths. If the performance on a particular network drops below configurable thresholds (such as, average throughput below 20 Kbps, and at least $4\times$ lower than at least one other alternate path, or recent loss rates in excess of 50%), then the measurement subsystem can mark that cellular path as having failed. Of course, if a network interface card itself reports being disconnected from the network, that cellular path is also marked as failed. Flows currently mapped to these paths are then migrated to other alternate paths. To keep these measurements current, our measurement subsystem estimates achieved throughputs, latencies, and loss rates using an exponentially weighted moving average (EWMA) scheme. Samples for these metrics are obtained in two ways: (i) by passively observing ongoing user traffic mapped to the different cellular paths and (ii) in absence of recent observations, by generating periodic active measurements. The frequency of active measurements are also based on various statistics of historical measurements collected at different locations and at different times, insights for which were developed in our prior WiScape work [22].[1] We have observed that in busy periods of user activity, active measurements are rarely required.

Traffic optimizations: We have also experimented with the deployment of various caching and compression schemes in the gateway and the controller respectively. Such optimization can be extremely important to a system that utilizes expensive, low-bandwidth wireless network technologies, such as our cellular paths.

Remote management of a vehicular Internet service: Detecting problems and failures efficiently and swiftly is, perhaps, the most important challenge of operating a continuously running vehicular Internet service. Without specific safeguards, many problems that occur on the buses plying on local and long distant routes can go undetected for days and weeks. For instance, the mere fact that we can "ping" a gateway might hide the fact that the WiFi component has failed to boot up. Further, when problems are detected we need a systematic method to track down the relevant vehicle and be able to fix them in an expedited manner. The following are some of the current strategies in place based on our experiences.

- *GPS-based continuous tracking at controller:* Each gateway periodically reports its current location to our controller. Any time a gateway disappears from our controller's logs for an extended period of time, our system first checks its last known location. If the last known location is one of the designated garages for the vehicle, then we assume that the bus probably parked in the garage and went offline in a graceful manner. Often, individual buses can be garaged for a few days as they undergo servicing and repairs. If the last known location was not one of the garages, then it is a signal for likely problems encountered. In such cases, we have to contact the bus operator expeditiously to enquire about the whereabouts of the vehicle.

- *Forced reboots and automated log offloads:* In certain cases, a gateway might be unable to connect to the controller. This can happen due to poor connectivity across all cellular paths, e.g., it happens due to shielding effects in certain locations in the bus depots. Other times, it could be manifestation of software bugs that have not yet been fixed, or have been introduced in a new revision. If the gateway is unable to connect to the controller for a configurable period, we force a system re-boot. However, to understand the nature of the problem, we attempt to log all critical level events in the local hard disk in the gateway, and periodically offload these logs through the cellular interfaces when connectivity is available. This allows us to build detailed records of performance failures for future analysis.

- *Tracking usage activity:* If a gateway correctly reports to the controller, then we monitor the total user traffic that is transiting through our system and the number of users currently observed to be connected. If the number of users is zero for extended periods of time, despite the vehicle being observed to be moving, we match it against known routes. If it is on at least some designated route, then such a scenario can indicate a problem with the WiFi subsystem. We have SSH access into each gateway, and in many cases we can reset the WiFi interface if it has gone into some bad state. In case such repairs do not seem to work (in one instance, we observed that the gateway's hard disk was completely full which caused some associated malfunction), we use the GPS to determine when the buses are likely to pass by stops near our university campus. In our early months, we used to

[1]The WiScape project used a hardware platform similar to WiRover to collect measurements, however, the WiScape project did not provide WiFi service to passengers.

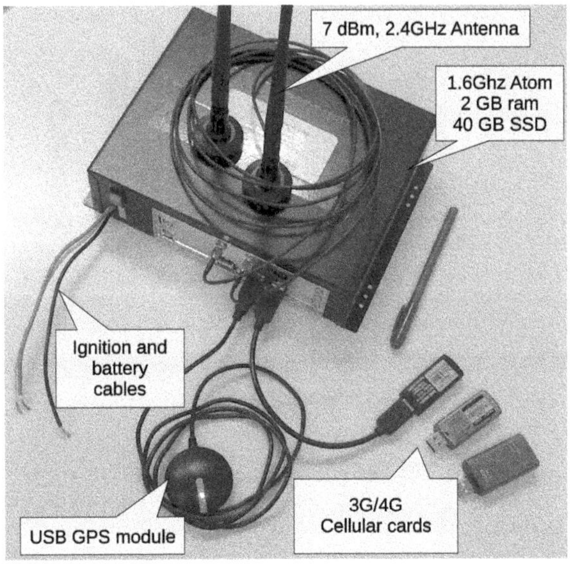

Figure 6: Our current version of our hardware setup consists of an off-the-shelf hardware platform outfitted with a power supply designed for vehicular settings and multiple wireless network modems as well as a GPS receiver.

board the bus during the stop durations and try some quick tests and fixes to revive the malfunctioning units.

- *Keeping backup units handy:* In some cases we have less than 5 minutes to fix the gateway unit before the bus continues on its routes, and we quickly realized that trying to fix nodes during these brief stops is quite hard. For this reason, in the recent months, we always keep a number of fully-configured backup gateway units handy. These days, we simply replace the malfunctioning unit with a backup unit, and bring the malfunctioning one into our lab for more careful evaluation and testing. This approach has significant streamlined our repair process.

- *User reports:* Finally, we routinely get a significant volume of comments, suggestions, and even bug reports from our user population. Many of these have helped in solving ongoing problems in the early months of operation.

We believe that a combination of all these approaches has significantly enhanced the robustness and the performance of our system over our continuing period of operation.

3. IMPLEMENTATION

Over the past year, our particular implementation of the WiRover system has evolved to address multiple issues as they have arisen. The WiRover code base runs on standard off-the-shelf hardware platforms with Linux as a combination of user-level processes and some specific kernel modifications. Our initial hardware setup consisted of a power inverter, a low cost netbook, and a Linksys wireless router.

3.1 Vehicular Power and Gateway Setup

This hardware setup was designed such that the system would initialize when the vehicle was turned on and would *gracefully* shutdown when the vehicle was turned off. In the early days of our deployment, it was most convenient to use a power inverter, a low-cost netbook (serving as the gateway), and a separate wireless router unit. A netbook platform was chosen so that it could gracefully stop the WiRover processes including archiving all unsaved data by using its own battery, once power supply to it was cut off. This design prevented daily crashes of the software at each power down event.

April 27, 2010	Deployed 1st Metro bus (M942).
Aug 12, 2010	Deployed 2nd Metro bus (M007).
Sept 17, 2010	Deployed 1st Coach bus (C001).
June 15, 2011	Deployed 2nd and 3rd Coach buses (C002-003)).

Table 1: Over the past year and a half we have deployed the WiRover system on 5 buses.

When the vehicle ignition was turned on, power was supplied to the inverter which in turn powered the netbook and the wireless router. The netbook was configured to boot-up using the Wake-On-LAN (WOL) mode, and once the wireless router started up, it was configured to wake up the netbook by sending WOL packets. This is a setup that had worked well for our prior *testbed-only* operations [17].

In our new installations for our vehicular Internet service, we initially encountered numerous failures with this simple setup. After some investigation we realized that unlike our previous deployment in [17] where their system was installed on a hybrid powered bus, our system was running on traditional diesel powered buses. The key difference between these two buses is the fact that diesel powered buses rely heavily on an electric powered motor to initially start the diesel engine. During this vehicle startup period of time the voltage level of the bus drops as this electric motor is consuming a large amount of power to start the engine. This drop in voltage was significant enough that the bus' voltage levels would drop below the voltage required for the power inverter to operate.

This would in turn cause the power to the wireless router to cut out in the middle of its startup process and *reset the wireless router's configuration settings.* Eventually all of the configuration settings that we had entered were reset and the wireless router no longer transmitted the WOL packets to wake up the laptop.[2] This served as a caution to check against treating the power supply issue casually, since if a power failure occurs once the bus is on the road, it becomes terribly inconvenient to determine the problems and fix them.

Current Hardware Solution: In light of our discovery, we moved to a more resilient hardware setup that could operate well on all types of buses. We chose to use a power supply that is designed for vehicular environments and has enough capacitance to overcome the voltage drop experienced during an engine turn over.

The current gateway nodes (Figure 6) consist of an embedded computer running an Atom 1.6 GHz processor with 2 GB of RAM and a 40 GB solid state hard drive. We use solid

[2]As an aside, we had multiple discussions with some researchers of the ViFi project, who complained of similar issues with WOL packets not suitably working with their deployment in a reliable way, and we were unaware as to why they worked for us in [17] and not for them.

state hard drives due to their resilience to vibrations, which thus reduces the likelihood of drive failures. The hardware platform further supports an *integrated WiFi card* which is capable of operating in AP mode which we leverage to provide WiFi connectivity to bus passengers. This also made our system quite compact.

For wireless backhaul and connectivity, our gateway currently uses two USB-based cellular modems (one NetA and one NetB), as well as a WiMAX modem. The WiMAX modem connects to our own experimental WiMAX base stations located on the roof of the CS building on campus. The buses can use this WiMAX network only when it is passing through specific parts of our campus in range of this base station.

The controller is a rack-mounted 2.66 GHz Quad core CPU with 4 GB of RAM located in our lab, and hosts the WiRover controller software along with various MySQL databases to store various performance logs and measurement data.

3.2 Vehicular Deployment

Our public vehicular Internet service is currently installed on 5 buses (at different times over the last 18 months as shown in Table 1), all of which share a single controller node. The buses are operated by two separate organizations: 2 buses are operated by Madison Metro Transit and will simply be referred to as Metro where the other 3 buses are operated by Van Galder Bus Company and will be referred to as Coach buses. There are three different types of buses: standard diesel Metro bus (M942), hybrid Metro bus (M007), and 3 standard Coach buses (C001-003). Our ability to install hardware on more buses is currently limited by the monthly cost of cellular data services. (The bus operators have expressed willingness to pay for this service and discussions on this are still underway.)

For the Metro buses, each bus is randomly assigned a bus route at the start of the day. Throughout the day the bus is likely to serve multiple different routes. For the Coach buses, each bus predominately serves a Madison, WI to Chicago, IL route but often times these buses are on charter routes and have covered much of the northern Midwest of the United States.

3.3 Basic System Benchmarks

Prior to presenting some of the detailed performance and usage characteristics of our system, there are two system overheads that are worth pointing out both of which relate to the use of a centralized controller.

Routing overheads: Our encapsulation tunnel forces all traffic to pass through our centralized controller. While it is possible to allow user traffic to flow directly to the Internet without touching the controller, it provides us two advantages: (i) flow migration in instances of individual cellular network failures, and (ii) ability to observe, manage, and configure all traffic from a convenient central location.

While this structure implies additional latencies on the path for users, given the high cellular latencies, the additional overhead is not overwhelming. Figure 7 shows the results of ICMP ping measurements (we found ICMP latencies to be representative of UDP and TCP latencies) to some popular web destinations that do not filter such packets, when using the cellular network with the least path latency

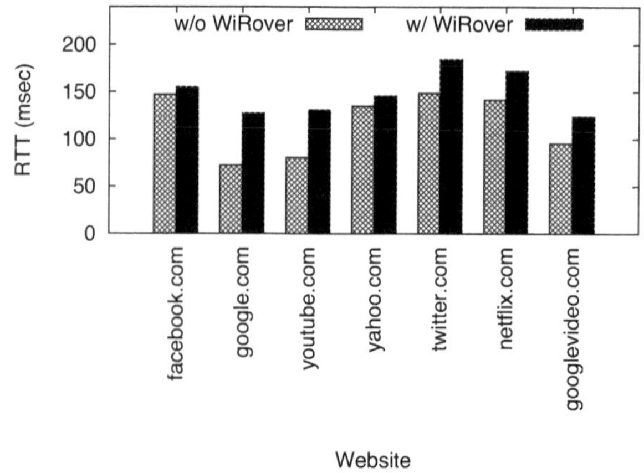

Figure 7: ICMP RTT measurements to some of the most popular web destinations when using the WiRover system as compared to each cellular network.

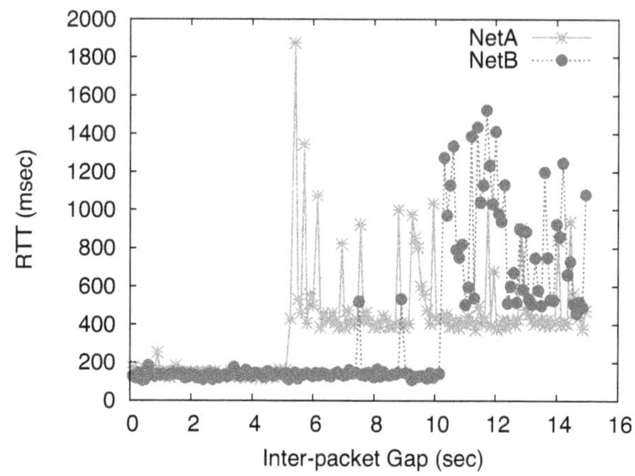

Figure 8: As the inter-packet gap is increased the cellular network re-installs network state increasing the packet delay.

in each case. We can see that the tunnel adds between 5 and 50 ms of additional latency for these ping packets. For most destinations the relative increase in latency is modest. The Google-controlled domains seem worst affected in a relative sense, possibly due to an efficient placement of the servers near the cellular networks.

Software overheads: Our benchmark of the WiRover system shows that the total software processing delay that is added to network packets is on the average of 320 μsec.

Tunnel overhead and fragmentation: The network tunnel adds 28 bytes of (IP and UDP) headers to data packets, which can increase the overall packet length beyond the maximum transmission unit (MTU) of the backhaul cellular networks. This reduces bandwidth efficiency on the end-to-end path. The bigger concern of this addition is the potential of packet fragmentation to packets that are already near the MTU.

Traditionally, such packet fragmentation can be avoided by running MTU discovery from the client device. However, we have found that most devices do not run MTU discovery and despite our efforts to provide the proper MTU values to clients via the dynamic host configuration protocol (DHCP) request some client devices choose to disregard this information. In such cases and as a last resort the WiRover system will overwrite the maximum segment size (MSS) field in TCP SYN packets which will reduce the maximum packet size for TCP packets. But packet fragmentation is still theoretically possible in our system for non-TCP packets, although we have not seen any occurrences.

NAT punching and other measurement issues: Various cellular networks also independently NAT traffic, and to ensure that our downlink path stays active and current, NAT punching was required. In absence of this, our system might inadvertently experience false failures. This is achieved through our periodic measurement subsystem. One additional issue addressed by our measurement subsystem shows up primarily during collecting latency estimates. It is well known that traffic inactivity in a certain cellular path can cause cellular base stations to remove bandwidth resources for such devices (in our case gateways). Arrival of new traffic requires synchronization between the cellular modem and the cellular base station [19], which causes additional delays for the first few new packets. Latency measurements on each path needs to be cognizant of these additional latencies. Figure 8, shows the round-trip time (RTT) delay for the second of two UDP ping packets as the inter-packet time gap is increased. The plot shows that packet RTTs in both networks go up significantly after some inactivity — the inactivity period is about 5 seconds for NetA and about 10 seconds for NetB.

4. USAGE CHARACTERISTICS

The WiRover system allows us to monitor usage and user characteristics of vehicular Internet services in the wild. These characteristics provide a valuable insight into how user's utilize such services, some differences in user behavior depending on travel routes, and as well as traffic characteristics that can be exploited for system optimizations. Further, the success of the system can also be monitored by presence and activity of repeat users who use our system. In this section, we first comment on usage volumes, including repeat users, the types of devices we see connect to our system, the popularity and types of traffic destinations, and draw specific conclusions from them.

4.1 Repeat Usage

As of Dec. 1, 2011 our WiRover system has provided Internet connectivity to 17,567 unique WiFi devices. However, it is important to understand how many of these devices have connected to our system multiple times. To accomplish this we define a *session* as a block of time where the device sends at least one byte of network traffic every 2 hours on a particular bus. If a time gap of greater than 2 hours is detected the data that the device transmits is grouped into two different sessions. Our notion of a session tries to capture all usage in a single bus ride as a single unit of activity. Since it is possible for a particular passenger to ride the same bus multiple times in one day, we have found that a 2 hour time period is sufficient to disambiguate the majority of these instances while still detecting passengers with sporadic usage on longer bus trips.

Figure 10, shows the growth of new devices and sessions over the duration of our deployment. The number of new devices grew more rapidly after July 2011 since only a short time before that we installed gateway nodes on 2 additional Coach buses. The number of repeat devices displays the total number of devices that have connected to the system at least twice since our data collection began. This demonstrates that despite the total number of repeat devices today is small (about 30%) of the total number of devices, repeat usage comprise a more significant fraction of our sessions (more than 55%).

We further find that 2427 of the 5079 repeat users are only observed on the Metro buses while only 761 repeat users are only observed on the Coach buses (1891 repeat users are observed on both). Despite the fact that we generally see a higher number of total users on the Coach buses (typically we see twice as many total users on the Coach buses), we observe that the Metro buses have almost twice as many repeat users as the Coach bus. This is likely because many Metro passengers ride the bus on a daily basis as they commute to and from work, where it is less likely many Coach passengers ride the bus with similar frequency.

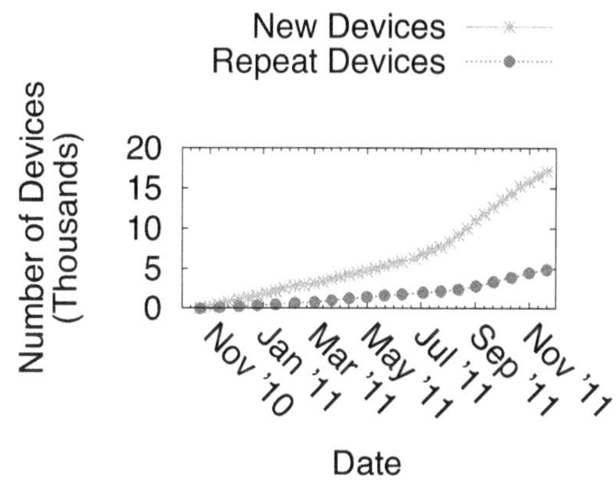

Figure 9: The growth of devices observed over time.

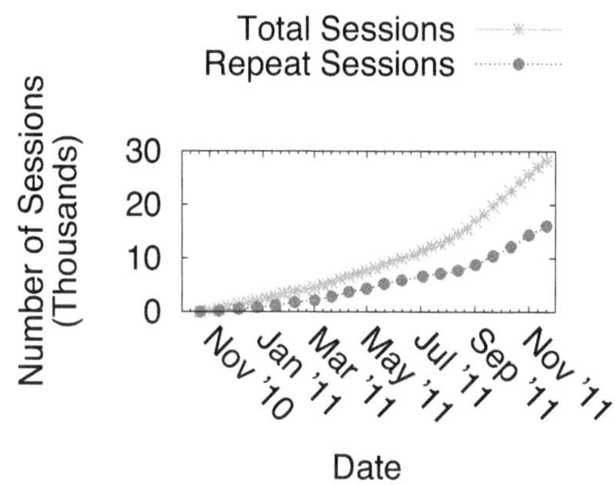

Figure 10: The growth of sessions observed over time.

4.2 Usage Characteristics

The usage characteristics for a particular device's session may be dependent on a number of factors such as the user's trip duration, the time of day, or even the type of device the

Device Platform	Metro buses Percent. of Sessions (%)	Avg Duration (Minutes)	Avg TX/RX Bytes	Coach buses Percent. of Sessions (%)	Avg Duration (Minutes)	Avg TX/RX Bytes
iPod	47.28	20.0	248 KB / 2.75 MB	44.44	83.0	825 KB / 9.56 MB
iPhone	36.69	16.6	189 KB / 1.62 MB	34.48	87.0	970 KB / 9.86 MB
Linux ARM	8.98	16.0	356 KB / 2.66 MB	8.45	93.2	1.29 MB / 15.2 MB
Apple Laptop	1.78	15.6	984 KB / 6.12 MB	1.67	48.8	5.42 MB / 55.3 MB
Windows Laptop	1.78	17.9	3.42 MB / 13.6 MB	1.67	48.1	3.67 MB / 27.5 MB
iPad	1.51	20.4	365 KB / 2.18 MB	1.42	76.1	1.94 MB / 17.0 MB
BlackBerry	0.80	5.83	26.4 KB / 338 KB	0.76	55.3	291 KB / 1.41 MB
Win Mobile	0.37	12.1	501 KB / 5.50 MB	0.35	54.6	199 KB / 1.07 MB
PSP	0.30	13.6	293 KB / 2.44 MB	0.28	64.3	297 KB / 6.34 MB
Nokia	0.29	8.0	89.2 KB / 384 KB	0.25	21.8	232 KB / 1.07 MB
Linux Laptop	0.23	9.24	749 KB / 4.54 MB	0.22	40.1	1.77 MB / 9.81 MB
Nintendo 3DS				0.02	97.6	1.04 MB / 10.7 MB

Table 2: Comparison of average usage characteristics for various device types. The overall average session duration for Metro buses is 18.13 minutes and was 53.23 minutes for Coach buses. The longest user session was observed on one of the Coach buses and was 645 minutes long (10.75 hours).

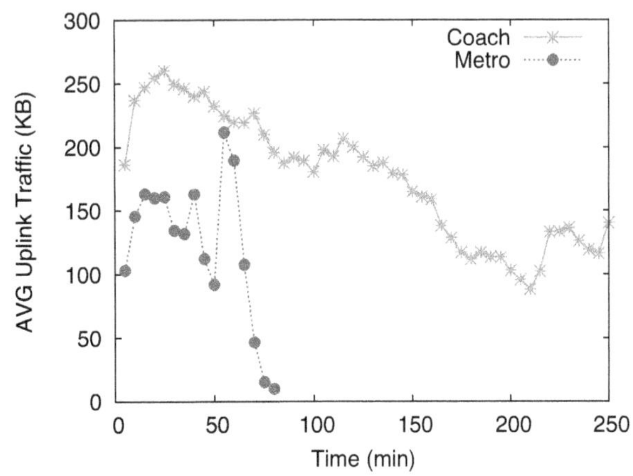

Figure 11: Average volume of uplink traffic over a user's session time.

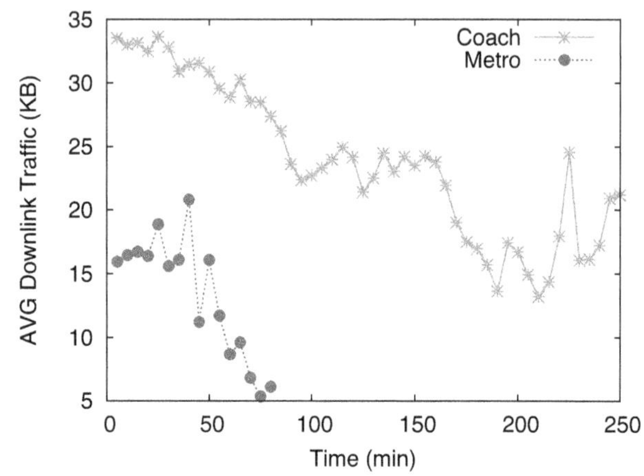

Figure 12: Average volume of downlink traffic over a user's session time.

user is interacting with. Table 2 shows the average session duration as well as the average number of bytes sent and received by various device types over the WiRover system. As can be seen in the table, there are differences in not only the device type but also based on the type of bus the passenger is on. Passengers on the Metro buses typically have shorter trip durations and this can be seen from the shorter average session durations and lower usage characteristics.

When we look at the average usage characteristics per session over time we notice that on average most usage is observed at the beginning of the session as shown in Figure 11 and Figure 12. For the Metro buses there is a sharp decrease in the average downlink traffic toward the end of user's session whereas usage on the Coach buses has a more graceful decrease. This may be due to the fact that users on the Metro buses use their devices right up to the point that they leave the bus. However, due to the longer durations of Coach bus trips, users may run out of battery life on their devices and scale back their usage over the trip duration.

Smart Phone Usage: Despite many smart phone users having dedicated cellular data plans we still find a large number of smart phone devices connecting to the system. As seen in Figure 13, smart phone devices consistently outnumber every other device category. We believe it is due to both the superior performance of the WiRover system (Figure 2), as well as the free WiFi-based service that help users conserve their data usage. Over the past few months we have seen over 3000 distinct devices across all of the buses with October 2011 yielding the highest single month total of 3787 distinct devices.

Other devices: We also observed a small number of unexpected devices that connected to our network. Examples include a Sony PlayStation Portable and a Nintendo 3DS. There were also a small number of feature phones that were observed to connect to our network, the maximum number of which never exceed 15 in a month.

4.3 Traffic Destinations

The user generated traffic that traverses our system is predominately HTTP traffic as is shown by Table 3. This is

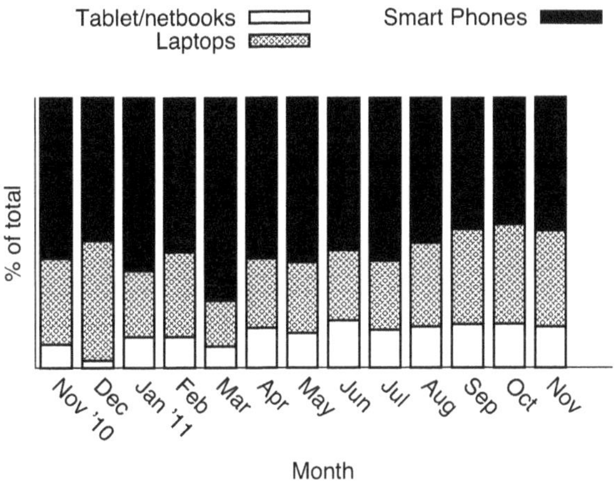

Figure 13: Distribution of various device type observed each month. Feature phones have been omitted from these results as they constitute less than 1% of devices for any given month.

Proto/Port	Bytes Received	Bytes Sent	Description
TCP 80	218 GB	19 GB	HTTP
TCP 443	17 GB	5.1 GB	HTTPs
TCP 1935	7.3 GB	227 MB	Flash RTMP
TCP 993	2.1 GB	307 MB	IMAPS
UDP 4500	870 MB	182 MB	IPsec NAT Trav.
TCP 22	832 MB	26 MB	SSH
TCP 995	829 MB	48 MB	POP3S
TCP 182	706 MB	15 MB	
UDP 443	653 MB	260 MB	
UDP 53	525 MB	331 MB	DNS

Table 3: The top 10 TCP and UDP ports sorted by the total number of received bytes.

not entirely unexpected and opens the possibility of employing caching, compression, and various other data reduction techniques as we will explore in Section 5.

Not unexpectedly, users seem to utilize our service for *mostly entertainment purposes.* Table 4 lists the top-15 websites by the volume of traffic that our users generate from each destination (most of these destinations are also our top 15 when sorted by the number of flows). For comparison we have also included the destination's rank as calculated by comScore (Nov 2011, US Top 50 list), a market research firm that monitors participants' network traffic and Alexa.com (Nov 2011, US rank) which ranks based on data collected from users that install its browser toolbar. Apple provides the highest amount of downlink traffic from a single domain, much of which includes their iTunes service. There are also a lot of other traffic from Apple that are specific to various mobile applications on different devices. There are further streaming media content flows from YouTube, Pandora, Megavideo, Netflix, and other such sites. In the uplink side, the highest amount of content flows into Facebook, indicating many of our users are uploading personal content utilizing our service.

Further, the top destinations exhibit significant differences between the Coach buses (plying inter-state routes) and the

Destination	Bytes Received	Bytes Sent	Alexa Rank	comScore Rank
apple.com	15 GB	623 MB	21	12
youtube.com	9.5 GB	361 MB	3	NA
googlevideo.com	7.7 GB	224 MB	NA	NA
limelight networks.com	6.6 GB	238 MB	185893	NA
facebook.com	6.2 GB	1.1 GB	2	4
pandora.com	4.3 GB	144 MB	68	NA
edgesuite.net	3.7 GB	112 MB	2547	NA
google.com	3.0 GB	419 MB	1	1
megavideo.com	2.8 GB	85 MB	471	NA
netflix.com	1.6 GB	79 MB	23	41
mzstatic.com	970 MB	69 MB	30078	NA
windows update.com	688 MB	25 MB	NA	NA
symatec liveupdate.com	657 MB	20 MB	NA	NA
tumblr.com	641 MB	39 MB	25	NA
yahoo.com	599 MB	92 MB	4	2

Table 4: The top 15 web destinations sorted by the total number of received bytes. For comparison we provide comScore's Nov 2011 Top 50 rankings as well as the Nov 2011 Alexa.com page rank.

Metro buses (plying city routes in Madison, WI). For instance, some of the top destinations in Metro buses, that do not figure in the Coach buses, include the National Public Radio (npr.org), the Onion (an American satire news publication with strong ties to Madison, WI), and cityofmadison.com, all of which have relevant local context in Madison, WI.

5. DATA REDUCTION TECHNIQUES

A significant cost of a vehicular Internet service stem from the monthly charges that need to be paid for cellular data plans. In our current deployment, we limit each bus to using two cellular modems to limit cost. Given the interest of cellular operators to impose various forms of bandwidth caps on cellular usage, it is critical to design efficient mechanisms that reduce the volume of uplink and downlink traffic in the system.

Given the dominance of HTTP traffic (Table 3), it is natural to expect that various forms of caching and compression techniques to be greatly beneficial. In this section, to explore these options, and also evaluate the usefulness of other data redundancy elimination techniques within packets, as proposed in recent literature [2].

5.1 Compression

When an application requests a web page it often specifies that it is capable of decompressing zipped content files. When applications make requests it is often to the benefit of the server to compress the content to both reduce bandwidth consumption as well as improve the download speed for the end user. However, not all servers compress all the content even when applications request compressed content. In such a situation a compression proxy could compress the content before it is forwarded to the end user.

For our WiRover system a compression proxy could be

added to the controller node to compress the downlink HTTP content before it is transferred over the cellular downlinks to achieve potential savings. To analyze how beneficial a compression proxy would be to the WiRover system we collected detailed packet traces at the gateways of two of our Coach buses. The traces were collected over multiple days and total over 10 hours of traffic. The relatively small amount of this collection stems from the difficulty of offloading traces from the gateways. We had to do this in brief periods of time when we had access to the corresponding buses (at a stop near our lab), or over the expensive cellular links. (Note that trace collection for compression analysis at the controller would have been simple but would not have been very helpful, due to traffic being modified via NAT at the gateway, and we would lose end client-specific information if logged at the controller.)

What was compressed: We then used these traces to analyze HTTP flows and found that 5,839 out of the 17,959 HTTP flows (corresponding to 188 MB of content) were already compressed by content servers. The compressed 5,839 flows contributed nearly 18 MB of content out of these 188 MB. When we uncompressed the compressed content, we realized that the original content volume was 79 MB which has been compressed down to 18 MB. Overall, the Internet-based servers had served 249 MB of HTTP content which was compressed down selectively to 188 MB — *a total savings of about 24.5%.*

We found that most of the compressed HTTP flows contained HTML, CSS, XML, javascript, or other text based content. Interestingly, we found that some servers had compressed images which when uncompressed were actually smaller than the compressed data size. This is due to the fact that compressed files must include a dictionary of replacement values that are used during the decoding process. We observed that 564 flows displayed such behaviors but accounted only about 13 KB of waste which is dwarfed by the (79 - 18 =) 61 MB that were saved thanks to content server compression.

What was not compressed: We also found that there were 10,350 flows which were served uncompressed when the client devices had advertised compression support.

To better understand the potential gains that compression could have provided in these situation we took each HTTP payload and ran the standard gzip compression, at the highest compression levels, and measured the file sizes before and after. The uncompressed flows accounted for a total of 170 MB of data and compression would have reduced that number to about 153 MB, a reduction of over 17 MB, or about 10%. Given that the highest monthly data consumption on the buses today routinely exceed 10 GB, *we could anticipate a wasted opportunity of eliminating of more than 1 GB of data consumption at this rate.*

For these uncompressed flows we analyze the content type for each flow and find that the majority of uncompressed flows contain images, audio files, or video files. Such content is often viewed as incompressible since the data is already stored in an optimized format, however, with enough volume even savings of only a few bytes per flow can add up to significant savings. Table 5 shows the top 10 uncompressed content types sorted by potential savings. The "text/plain" content type was only observed from netflix.com related servers and the "application/octet-stream" was only observed from pandora.com servers. Both of these content types display high volumes of traffic with relatively low compression ratios compared to text based content such as "text/html" or "application/atom+xml" both of which have smaller total bytes but have higher compression ratios.

Content Type	Total Bytes	Potential Savings	Total Flows
image/jpeg	23 MB	4.1 MB	463
image/png	12 MB	2.5 MB	351
text/plain	23 MB	2.2 MB	164
application/ octet-stream	39 MB	1.4 MB	29
application/ javascript	1.7 MB	1.4 MB	8
video/mp4	9.6 MB	720 KB	27
application/ x-rhapsodycontentaac	23 MB	539 KB	8
text/html	533 KB	401 KB	3
application/atom+xml	437 KB	395 KB	2
video/x-flv	15 MB	388 KB	4

Table 5: The top 10 HTTP content types sorted by the number of bytes that would have been saved if the content had been compressed using gzip.

Period	1	2	3	4	5
Total (MB)	47	57	56	59	59
Savings (MB)	8.5	8.3	5.0	16	6.1
Savings (%)	18.1	14.6	8.9	27.1	10.3
Period	**6**	**7**	**8**	**9**	**10**
Total (MB)	50	57	55	58	56
Savings (MB)	11	13	12	6.8	18
Savings (%)	22.0	22.8	21.8	11.7	32.1

Table 6: Savings in a 10-hour trace using Packet Caches between the controller and the gateway.

5.2 Packet-level redundancy elimination

An interesting optimization that could be added into the WiRover system would be to eliminate redundancies at the packet level using techniques such as Packet Caches [2]. That proposed technique creates fingerprint hashes of matching substrings of packet payloads and stores them at the receive side. If the same substrings are found in future packets (quickly detected by comparing hashes), then the matching substring can be eliminated from the data packet and be replaced by an appropriate identifier. One advantage that this method has over HTTP compression and caching is that it operates over all traffic flows and is not limited to HTTP traffic. Further, given that the vast majority of our traffic is HTTP based this method is also able to compress the text based HTTP *headers* that are found in each HTTP flow. Note, that server-based content compression techniques that operate on HTTP flows do not compress HTTP headers, which can leave out significant opportunities when the HTTP payloads are small (especially with many occurrences of 1 pixel by 1 pixel images in content these days).

Table 6 shows the possible compression achievable by using Packet Caches deployed across the controller and gateway pair for the 10-hour trace, partitioned into approxi-

Type	Hit Count	Total Bytes
image/jpeg	6408	77 MB
application/javascript	4028	38 MB
image/png	5302	30 MB
application/zip	2532	26 MB
text/css	2522	21 MB

Table 7: The top 5 HTTP content types sorted by the total number of bytes served locally by the web cache.

Failure Type	Count
Hardware failures	4
No connectivity on boot-up	403
Individual cellular connectivity failure	3,624

Table 8: The number of failures by type.

mately 1 hour periods. The achievable savings range between 8.9% and 32.1% for the most part.

5.3 Caching

To further reduce resource consumption on the wireless back-haul networks a web cache could be installed on each gateway node to serve cached content locally. To test the effectiveness of web caching on our WiRover system we ran a squid proxy (using the default configuration) for a little more than two months (July 28, 2011 to Oct. 10, 2011) on a single Coach bus. Over this period of time 36,270 cache hits were observed out of 265,867 total events. The total volume of traffic that was served locally due to the web cache amounted to over 286 MB out of a total of 7.8 GB.

Table 7, shows the top 5 HTTP content when sorted by the number of bytes served locally. As is seen the majority of gains come from image files. For these image files we find that the 6408 hits correspond to 4395 unique URLs. Of these unique URLs only 981 recorded only a single hit while 428 and 181 recorded 2 and 3 hits respectively. The recording the most hits (32 in total) was a logo icon from www.facebook.com.

Caching Considerations: Without knowledge of the client device's IP address it is difficult to determine whether these cache hits are the result of a single user downloading the same content multiple times or if there are multiple users accessing the same content. However, the WiRover system would benefit in either case.

It is further unclear to what extent browser caching benefits a systems such as ours. Since we do not have access to client devices this question is out of the scope of this particular piece of work.

6. FAILURE ANALYSIS

To ensure our system is providing a reliable service it is imperative that we monitor system failures and repair them quickly. System failures can range from uncontrollable events such as a lack of cellular network coverage to hardware specific failures that require replacing the gateway unit on a bus. Some of these failures can be extremely difficult to detect and diagnose and have put in place significant checks and balances to quickly detect and address them (see Section 2).

Provider	Total Failures	Metro	Coach	Avg Link Uptime (min)
NetA	2,807	270	2,537	147.6
NetB	817	301	516	457.0
Total	3,624	571	3,053	

Table 9: The number of individual cellular connectivity failures by provider.

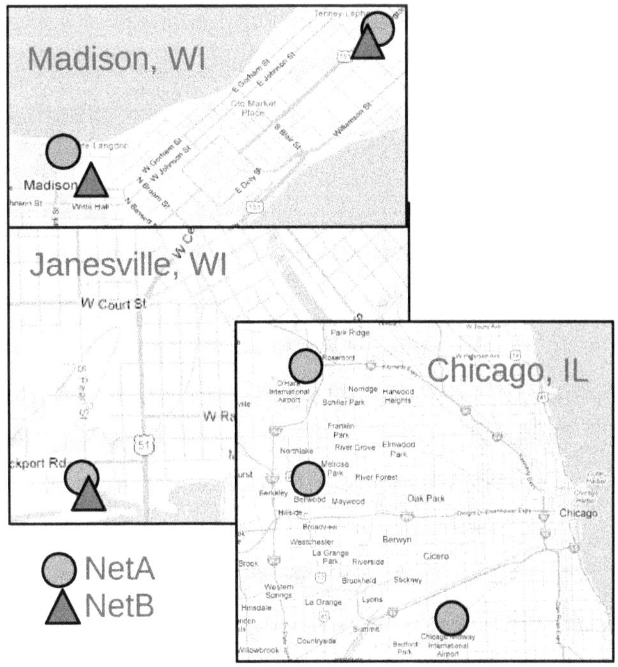

Figure 14: Locations with the highest density of link failures for each network.

Table 8 groups the various system failures we have encountered over the past 18 months. There were four occurrences of hardware related failures. Our first hardware failure was the result of an improper vehicle power supply (Section 3). The other three failures have been caused by a combination of hard disk failures and system clock skew. However, we routinely observe transient failure scenarios that are in some way related to connectivity issues. There were 403 recorded instances when a gateway booted-up, but none of the cellular interfaces were able to lock onto their corresponding cellular base stations. Our system is configured to re-boot itself after a 3-minute period elapses in this state. We have seen many occurrences of this event primarily in the bus garages, where various forms of RF shielding from the environment may cause this sort of disruption. In addition, there were 3,624 instances where one of the functioning cellular interfaces suddenly lost connectivity (as reported by the cellular interface card).

6.1 Causes of Link Failures

Table 9 shows the number of individual failures experiences by the different cellular interfaces. The table also shows a comparison of the number of link failures observed on each type of bus as the Metro buses only cover an urban area while the Coach buses also travel in rural areas.

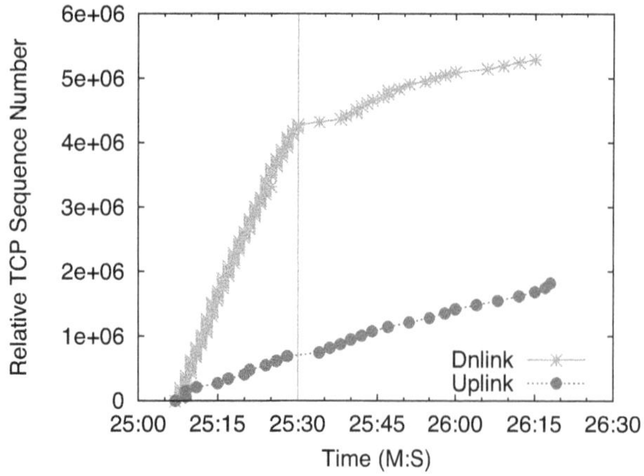

Figure 15: TCP sequence numbers for the downlink and uplink iperf streams. One of the cellular modems was unplugged at the time indicated by the vertical line.

By clustering link failures by their GPS location we are able to determine locations that have a history of causing connectivity failures in individual cellular interfaces. Figure 14, provides the locations where the most failures occur in a 1 km radius area. Not surprisingly, 2 of the 3 locations for which NetB is represented are the locations of the garages for both Metro buses and Coach buses. Historical data such as this could then be coupled with real-time measurements to improve our understanding of what causes each such failure.

6.2 Link Failure Prediction

We have observed that cellular interface cards actually take a long time to report that it has lost connectivity to the networks. Sometimes it may be as much as 10-15 seconds after the link starts to lose packets sent along it. In fact, 10-15 seconds prior to the interface reporting a failure, we observe an increased occurrence of packet losses from that network. This observation has been routinely observed in the link failure scenarios described above. It implies that one can use an increased occurrence of packet losses to predict the onset of a connectivity loss for an interface relatively easily.

An alternative to this design could be to use the received signal strength indication (RSSI) metric to estimate the quality of cellular connectivity and likely onset of a disconnection period as proposed in the MAR system. However, the APIs in our cellular code division multiple access (CDMA) modems cannot be used to query RSSI values while the modem has an established data connection to a cellular network. Hence, we believe a loss rate estimation technique might be a fast and efficient way to predict onset of link failures and to initiate flow migrations.

6.3 Flow Migration

Often times, link failures are caused by factors that are not controllable by our WiRover system. For example, if the bus enters an area with little to no signal strength then a link failure is likely to occur. Our goal is to ensure that our system is capable of mitigating such link failures. This is an important goal for user retention.

Cellular link failures can be frustrating for users in the vehicular environment but often times they cannot be avoided. For our WiRover system to provide users with a superior experience our system must then mitigate the effect of link failures. One mechanism the WiRover system uses to mitigate link failures is to migrate network flows from the failed network to an active network. However, it is important to understand the effect link migrations have on network traffic.

To best understand the effects of link migrations we show an example of how such flow migration operates over our WiRover system. This was a bidirectional TCP flow operating between a client device and an Internet-based server. We plot the TCP sequence numbers for both the uplink and downlink parts of the flow in Figure 15. The vertical line in the figure indicates the time at which connectivity to one of the cellular networks was lost. As can be seen, the downlink flow experiences some packet loss, and a brief stall in performance, but quickly migrates to the other interface and resumes to send packets on the new cellular link. The uplink direction, in this example, was not originally assigned to the failed cellular network, and hence did not get affected.

7. RELATED WORK

Providing Internet connectivity into a vehicle is not a new idea. Early work studied the challenges faced when attempting to connect to a wireless basestation from a vehicle using WiFi based technologies [18, 11, 13]. Multiple other works have attempted to connect to open WiFi access points to provide a reliable Internet connect into a vehicle [7, 15, 9, 12, 16, 8]. In particular Cabernet [9], proposed multiple optimizations to the process of connecting to a WiFi AP which in turned increased the amount of time they were able to maintain a connection while driving by.

The MAR system [21] avoided many of the challenges of vehicular Internet connectivity by using cellular technologies instead of WiFi based technologies. The MAR system implemented a multihomed designed which used multiple 2.5G and 3G cellular links from different cellular providers. Such a design allowed them to exploit network diversity and aggregate bandwidth to improve the reliability and performance of their system. The authors proposed multiple traffic scheduling mechanisms and implemented a "per-TCP connection state scheduler."

The PluriBus project [14] based their design from the MAR system but implemented their system with a combination of 3G cellular and WiMAX modems. The authors observed high loss rates on their WiMAX links and developed evolutionary codes to mitigate those losses.

A recent project (Wiffler) [4] explored the feasibility of offloading cellular traffic to WiFi networks. Other projects [6, 3, 5] have studied the performance of various application classes such as web search and interactive application from vehicular settings.

Unlike all such prior work, our effort focused on providing a continuous and reliable Internet connectivity service to these public transit passengers. This paper shares our experience in running such a service which continues to challenge us far beyond than rolling out and deploying an experimental testbed.

8. OPEN RESEARCH CHALLENGES

Our toils over the past 18 months have exposed challenges that we had not anticipated based on prior work in vehicular environments.

Management: We have found that managing a vehicular deployment is particularly challenging. The ability to track and monitor the health of mobile devices is challenging in itself. However, such a deployment generates troves of data that must be collected and analyzed which requires automated mechanisms to collect both raw usage data and user feedback data.

Traffic Optimizations: Our desire to expand our deployment is hampered by the financial burden of our cellular based connectivity. Recent works have suggested offloading data to less expensive technologies [4] but in rural areas other technologies may not be readily available. As such other traffic optimizations are needed to help reduce the network load on cellular back-haul links.

Link Failure Prediction: We have found that wireless link failures are extremely common in the vehicular environment. The ability to predict link failures could greatly improve the reliability of the system.

9. CONCLUSION

Deploying a system in the wild has provided many challenges over the past year. However, with the large number of users that have connected to our system we are continually working to improve the performance of the system so that we can support more and more users. We hope that in the long-term our WiRoversystem will provide valuable insight into the vehicular networking environment. However, in the short-term we are happy to provide a service to the community while we further our research interests.

Overall, our more than 18 months of operation has provided us with various insights on how performance in these vehicular Internet access systems can be tuned. While many other vehicular Internet services do exist today, little is known about their actual performance or potential pitfalls. In fact, when talking to one of our bus operators (Van Galder Bus) who have had experiences with a few other such systems at different times, *we were told that many existing systems have very unpredictable performance.* Performance can be quite spotty, and failures are unexpected, with the vendor being totally unaware of the duration, the period, and most importantly, the cause of such failure. While resetting the node on the vehicle often fixes the problem, without a systematic study, such problems are bound to recur. We hope that a study such as ours can help shed light on some of the issues one can expect in such vehicular systems.

Both our partner bus operators have been fairly satisfied with the performance of the WiRover system so far. Our users have voted with their feet (or their bytes), as we continue to see a growing number of repeat users. Most interestingly, there were a large number of smart phone users, who in spite of their existing phone data plans, form the highest user population in our system. Our study has provided us (and hopefully, others) with many insights on how to improve this system even further, and we hope to implement them in the months to come.

10. ACKNOWLEDGMENTS

We would like to thank Rob Madden for his significant implementation contributions on an early version of the system as well as Wes Miller and Sam Roth for efforts to create our management tools. We thank the Madison Metro Transit and the Van Galder Bus Company for partnering with our research efforts and for their feedback and suggestions. All authors are supported in part by the following grants of the US National Science Foundation: CNS-1040648, CNS-0916955, CNS-0855201, CNS-0747177, CNS-1064944, and CNS-1059306.

11. REFERENCES

[1] Amtrak to take free onboard wifi nationwide. `http://arstechnica.com/tech-policy/news/2010/06/amtrak-to-take-free-onboard-wifi-nationwide.ars`.

[2] Ashok Anand, Archit Gupta, Aditya Akella, Srinivasan Seshan, and Scott Shenker. Packet caches on routers: the implications of universal redundant traffic elimination. In *SIGCOMM*, 2008.

[3] Aruna Balasubramanian, Brian Neil Levine, and Arun Venkataramani. Enhancing interactive web applications in hybrid networks. In *MobiCom*, 2008.

[4] Aruna Balasubramanian, Ratul Mahajan, and Arun Venkataramani. Augmenting mobile 3G using WiFi. In *MobiSys*, 2010.

[5] Aruna Balasubramanian, Ratul Mahajan, Arun Venkataramani, Brian Neil Levine, and John Zahorjan. Interactive WiFi connectivity for moving vehicles. In *SIGCOMM*, 2008.

[6] Aruna Balasubramanian, Yun Zhou, W. Bruce Croft, Brian Neil Levine, and Aruna Venkataramani. Web search from a bus. In *CHANTS*, 2007.

[7] Vladimir Bychkovsky, Bret Hull, Allen Miu, Hari Balakrishnan, and Samuel Madden. A measurement study of vehicular Internet access using in situ Wi-Fi networks. In *Mobicom*, 2006.

[8] Pralhad Deshpande, Anand Kashyap, Chul Sung, and Samir R. Das. Predictive methods for improved vehicular wifi access. In *MobiSys*, 2009.

[9] Jakob Eriksson, Hari Balakrishnan, and Samuel Madden. Cabernet: vehicular content delivery using wifi. In *MobiCom*, 2008.

[10] Jakob Eriksson, Lewis Girod, Bret Hull, Ryan Newton, Samuel Madden, and Hari Balakrishnan. The pothole patrol: using a mobile sensor network for road surface monitoring. In *MobiSys*, 2008.

[11] Richard Gass, James Scott, and Christophe Diot. Measurements of in-motion 802.11 networking. In *WMCSA*, 2006.

[12] Anastasios Giannoulis, Marco Fiore, and Edward W. Knightly. Supporting vehicular mobility in urban multi-hop wireless networks. In *MobiSys*, 2008.

[13] David Hadaller, Srinivasan Keshav, Tim Brecht, and Shubham Agarwal. Vehicular opportunistic communication under the microscope. In *MobiSys*, 2007.

[14] Ratul Mahajan, Jitendra Padhye, Sharad Agarwal, and Brian Zill. E PluriBus Unum: High performance connectivity on buses: Technical Report: MSR-TR-2008-147, October 2008. `http://research.microsoft.com/apps/pubs/default.aspx?id=70650`.

[15] Ratul Mahajan, John Zahorjan, and Brian Zill. Understanding wifi-based connectivity from moving vehicles. In *IMC*, 2007.

[16] Anthony J. Nicholson and Brian D. Noble. Breadcrumbs: forecasting mobile connectivity. In *MobiCom*, 2008.

[17] Justin Ormont, Jordan Walker, Suman Banerjee, Ashwin Sridharan, Mukund Seshadri, and Sridhar Machiraju. A city-wide vehicular infrastructure for wide-area wireless experimentation. In *Mobile Computing and Networking*, 2008.

[18] J. Ott and D. Kutscher. Drive-thru internet: IEEE 802.11b for "automobile" users. In *INFOCOM*, 2004.

[19] Feng Qian, Zhaoguang Wang, Alexandre Gerber, Zhuoqing Morley Mao, Subhabrata Sen, and Oliver Spatscheck. Characterizing radio resource allocation for 3g networks. In *IMC*, 2010.

[20] Asfandyar Qureshi and John Guttag. Horde: separating network striping policy from mechanism. In *MobiSys*, 2005.

[21] Pablo Rodriguez, Rajiv Chakravorty, Julian Chesterfield, Ian Pratt, and Suman Banerjee. Mar: A commuter router infrastructure for the mobile Internet. In *MobiSys*, 2004.

[22] Sayandeep Sen, Jongwon Yoon, Joshua Hare, Justin Ormont, and Suman Banerjee. Can they hear me now?: A case for a client-assisted approach to monitoring wide-area wireless networks. In *IMC*, 2011.

[23] Hamed Soroush, Nilanjan Banerjee, Aruna Balasubramanian, Mark D. Corner, Brial Neil Levine, and Brian Lynn. Dome: A diverse outdoor mobile testbed. In *HotPlanet*, 2009.

[24] The New York Times. Google's buses help its workers beat the rush. `http://www.nytimes.com/2007/03/10/technology/10google.html`.

[25] USA Today. More cities offer wi-fi on buses. `http://www.usatoday.com/tech/wireless/2008-04-10-wifi\_N.htm`.

[26] Xiaolan Zhang, Jim Kurose, Brian Neil Levine, Don Towsley, and Honggang Zhang. Study of a bus-based disruption-tolerant network: mobility modeling and impact on routing. In *MobiCom*, 2007.

WIISARD: A Measurement Study of Network Properties and Protocol Reliability during an Emergency Response

Octav Chipara[1], William G. Griswold[2], Anders N. Plymoth[3], Ricky Huang[3], Fang Liu[3], Per Johansson[3], Ramesh Rao[3], Theodore C. Chan[4], Colleen Buono[4]
[1]Department of Computer Science, University of Iowa
[2] Department of Computer Science, University of California San Diego
[3] CalIT2, University of California San Diego
[4] Division of Emergency Medicine, Department of Medicine, University of California San Diego

ABSTRACT

This paper describes the design, deployment, and empirical evaluation of WIISARD – a novel emergency response system that provides reliable communication in dynamic wireless environments without extensive communication infrastructure. The main contribution of this paper is an in-depth empirical study of network properties that emerge during a drill in which WIISARD is deployed with minimal infrastructure support. The drill involves 19 first responders and 41 victims. The properties of links established among first responders vary between phases of the drill and depend upon the responder's role in the drill. The rescue phase – in which responders are highly mobile as they triage victims – poses significant challenges to reliable communication. During this phase, the contacts between responders are short-lived; however, they are reestablished within minutes. Once a contact between responders is established, the quality of the link between those responders is usually high. The connectivity graph observed during the rescue phase is usually connected and has a small diameter although there are times when it has a large diameter or it is partitioned. While mobility increases network dynamics, we also observe that the mobility patterns characteristic of the emergency response workflow can be leveraged to disseminate data efficiently through data muling. WIISARD employs a gossip-based protocol and supports data dissemination through local communication and data muling to achieve 98% reliability during the drill exercise. These results indicate the feasibility of providing reliable communication in emergency response with minimal infrastructure in spite of network dynamics.

Categories and Subject Descriptors

C.2.4 [**Computer-communication Networks**]: Distributed Systems

Keywords

Emergency response, Delay tolerant networking, Mobility, Reliability

1. INTRODUCTION

Every year, natural and human-caused disasters in the form of hurricanes, earthquakes, and infrastructure failures put the health of millions at risk. The key to successful emergency responses is to achieve *situational awareness* through effective communication: information about the event, casualties, and available resources must be exchanged among first responders in a *reliable* and *timely* manner. A breakdown in communication can result in ineffective use of the limited resources or jeopardize the safety of responders.

A typical emergency response involves four overlapping phases: staging, rescue, treatment, and transport. During the staging phase responders arrive on scene and establish command and control. In the rescue phase, the responders triage victims and provide minimal medical care to stabilize their condition. As the scale of the disaster increases, opportunities arise during the treatment phase to provide further, more comprehensive, medical treatment. In the transport phase, the victims are transported to hospitals based on the severity of their injuries.

Traditionally, information during emergency responses has been collected and exchanged either in writing or verbally using hand radios. While written and verbal communication is adequate for incidents that involve a few victims (e.g., car accidents), their effectiveness deteriorates rapidly as the incidents increase in scale. In large-scale incidents, information captured on paper can be lost, while verbal communication often introduces inaccuracies [24]. Moreover, neither form of communication is effective in sharing information rapidly among numerous first responders.

The aim of the WIISARD project is to develop a reliable communication infrastructure for emergency response by taking advantage of mobile computing technology. Capturing information in digital form and exchanging it wirelessly ought to reduce inaccuracies, limit accidental data loss, and facilitate the timely and effortless dissemination of information among first responders. Yet, to meet this goal, we must overcome three key challenges: (1) minimize the reliance on network infrastructure during emergency responses, (2) cope with a dynamic radio environment subject to interference, and (3) support communication among mobile users.

Existing emergency response systems may be divided into two classes based on the networking technology they use. Cellular networks are used increasingly to deliver patient information during their transport to hospitals [29,33]. However, cellular networks are not a reliable solution for emergency response systems as indicated by recent disasters (e.g., Katrina) where the cellular infrastructure either failed or was overloaded. A promising alternative is to adopt mesh networking technology: first responders would bring a number of wireless nodes to the scene that, when deployed, would self-organize in a mesh network that facilitates communication at the disaster scene. Typically, these systems would use either ad-hoc routing (e.g., OLSR, AODV) or Delay Tolerant Networking (DTN) protocols to share information.

The initial design of WIISARD [6] employed a client-server architecture and AODV-based routing [38]. Unfortunately, this design performed inconsistently across deployments. This experience is consistent with the results reported by other researchers [12] and highlights the acute need to develop an empirical characterization of the network properties observed during emergency responses in order to provide a sound basis for understanding the challenges of reliable communication in such a setting. While numerous emergency response systems have been designed and deployed, these systems seldom provide a detailed study of network properties and reliability.

This paper makes the following contributions: (1) We present one of the first empirical studies of network properties that emerge during a drill exercise. The novel aspect of our drill is that it evaluates the effectiveness of using DTN to support reliable communication during emergency responses when minimal infrastructure is deployed. Based on these results, we provide insights regarding the challenges of reliable communication in emergency response and discuss approaches to cope with these challenges. (2) WIISARD achieved 98.25% reliability in spite of significant network dynamics due to responder mobility and network partitions by using a peer-to-peer architecture in conjunction with a gossip-based communication protocol. (3) In a wider context, our work contributes to the growing body of empirical studies of the properties of DTN applications and human mobility patterns. In contrast to these previous studies, a unique aspect of emergency response systems is that responders *cooperate* to rescue victims according to a pre-established workflow (see Section 3). Therefore, we observe more complex network properties that vary with the role of the responders and among the phases of the drill.

We deployed WIISARD as part of a drill organized at University of California, San Diego (UCSD) on August 10th, 2011. The drill involved 19 first responders who rescued 41 victims. The collected data indicates that each stage of the drill has different network properties. The staging and treatment phases exhibited good and stable connectivity. In contrast, the rescue phase posed significant challenges to reliable communication. During this phase, the contacts between providers, which are indicative of the long-term link quality, tended to be short-lived; however, these contacts were reestablished within minutes. Once a contact between responders was established, the associated link was usually characterized by good short-term link quality, as indicated by a median packet reception rate that exceeds 70%.

The connectivity graph of responders during the rescue phase was typically dense and had a small diameter (less than 3 hops). However, due to limited infrastructure, at times the graph became sparse and had a large diameter (7 hops). Even worse, for 26% of the rescue phase, the connectivity graph was partitioned. The mobility of responders was a major source of variability in network properties. However, mobility also had a beneficial effect in that it improved data dissemination through data muling. In fact, due to the inherent mobility patterns of the emergency response workflow, a responder encountered all other responders within seven minutes. This indicates the potential of improving reliability through data muling.

The redesigned WIISARD system that we describe in this paper has a peer-to-peer architecture and data is disseminated through a gossip-based protocol called **W**IISARD **C**ommunication **P**rotocol (WCP). WCP uses local communication to handle variations in link quality and caches data aggressively to support data dissemination via data muling. WCP achieved 98% reliability in spite of significant variations in network properties discussed above. Thus, it is feasible to achieving high reliability in emergency responses even with minimal infrastructure.

The remainder of the paper is organized as follows. The next section discusses related work. Background information regarding emergency responses is provided in Section 3. WIISARD's architecture and software components are described in Section 4. The drill exercise is presented in Section 5. We analyze the network properties that emerge during the drill in Section 6. WIISARD's reliability is characterized in Section 7. We discuss the implications of the observed network properties on emergency response systems in Section 8. Conclusions are provided in Section 9.

2. RELATED WORK

In this section, we review the prior work on emergency response systems and place our work in the wider context of DTN empirical studies.

2.1 Emergency Response Systems

Numerous emergency response systems take advantage of mobile computing technology to improve communication accuracy and timeliness. These systems employ various wireless technologies: Wi-Fi [1,10,20,24,39], 802.15.4 [12,20,26], and cellular networks [29, 33]. We focus on systems that use Wi-Fi mesh networks, as they are closely related to our work, and discuss how they address the challenges of reliable communication in emergency responses.

Early emergency response systems adopted client-server architectures due to their simplicity [4, 17, 18, 24, 39]. An important limitation of this architecture is that clients that are within communication range cannot communicate unless they maintain connections to the server, potentially over multiple hops. To remove this limitation, several systems opted for more flexible network architectures that support either multi-cast [10] or publish-subscribe [20] primitives. However, similar to the client-server systems, these systems still require the construction and maintenance of multi-hop end-to-end routes. Emergency responses are highly dynamic as the wireless channel fluctuates due to the movement of people, vehicles, and equipment (see Section 5). Moreover, since infrastructure is often limited, network partitions are common. These factors make it difficult (if not impossible) to maintain end-to-end routes and, as a result, such systems often suffer from poor reliability in realistic deployments.

To address user mobility and network partitions, recent systems [3, 7, 19, 21] adopt decentralized peer-to-peer architectures and employ delay tolerant networking (DTN) techniques. For example, DistressNet [21] has a hierarchical network architecture that uses 802.15.4 to monitor the vital signs of victims/responders and 802.11 for long-range communication between responders. DistressNet copes with network partitions through DTN techniques. Dong et al. present the Emergency Delay Tolerant Network architecture and investigate the use of Session Initiation Protocol (SIP) based communication [19]. WIISARD also takes advantage of DTN techniques to improve data dissemination through data muling. However, in contrast to the above techniques [3, 19, 21], which have been evaluated only through simulations, WIISARD is evaluated through an in-situ deployment.

In spite of the numerous emergency response systems that have been developed, they are seldom evaluated with real users and in real-world deployments. However, there are a few notable exceptions. As part of the AID-N project [20], the functional requirements of emergency response systems were elicited through surveys of first responders. The usability of the AID-N system was evaluated on a 5-point Likert scale. Surveys indicate that the first responders found AID-N to be more effective in tracking victims than standard paper triage tags. The US Army (BMIST system [30]) and Navy (TACMED-CS [37]) have developed mobile systems to provide access to electronic patient records. These systems are designed to store changes to patient records locally and synchronize them when network connectivity is available. Both systems are currently in use. In our own prior work, we studied the impact of (an earlier version of) WIISARD on the emergency response workflow through a drill exercise [24]. WIISARD significantly reduced the rate of missing/duplicate patient identifiers. None of the above empirical studies provide an analysis of network properties and reliability during emergency responses.

In this paper, we present one of the first empirical studies of network properties during emergency responses. The presented results are complementary with those obtained as part of the Code Blue project. CodeBlue [26] was an earlier system aimed at monitoring patient vital signs during a disaster drill using wireless sensors. Unlike WIISARD, CodeBlue employed an ad hoc multicast routing protocol based on ADMR [13], requiring sensor nodes to maintain routing tables and link quality state throughout the network. The CodeBlue disaster drill study [12] demonstrated the challenges of using such complex, stateful protocols in highly dynamic environments with high mobility and poor link quality In contrast, the focus of this paper is to assess the feasibility of using DTN protocols to support reliable communication in emergency responses when minimal infrastructure is deployed.

2.2 Delay Tolerant Networking

DTN techniques have been proposed for delivering packets when contemporaneous end-to-end paths do not exist. DTN protocols may be classified based on their assumptions of user mobility. The most general DTN protocols (e.g., [8, 25]) do not make any assumptions regarding the mobility of users. These protocols aim at maximizing the likelihood of packet delivery and limiting the number of duplicate packet transmissions by deciding which packets are forwarded during node encounters or dropped when a node's storage capacity is reached. Since little is known about the mobility of responders, the WCP protocol used by WIISARD falls into this category. WCP has similarities to Trickle [25]: in both protocols nodes exchange metadata to determine the state that must be updated and suppress packet retransmissions based on overhearing. The novelty of this paper is not WCP but rather the characterization of network properties and mobility patterns that emerge during emergency responses.

DTN protocols that take advantage of the mobility patterns of users have been proposed (e.g., [15, 27]). Fundamental to these protocols is an accurate characterization of human mobility. Initial models of human mobility focused on random walk models, which did not accurately capture many of the properties of human mobility (for a review refer to [9]). Recently, more realistic models of human mobility have been developed based on empirical mobility traces collected using GPS [35] or based on Bluetooth contacts [11]. The macro scale human mobility has been studied using association traces of mobile laptops/PDAs and 802.11 access points (e.g., traces collected at UCSD [28] and Dartmouth [22]) and based on student class schedules [36]. These traces indicate that some popular locations are significantly more likely to be visited leading to networks in which user densities vary widely. Moreover, it has been observed that the distribution of inter-contact times of users is heavy tailed, a characteristic that cannot be reproduced by random walk models [11, 35]. However, a common modeling assumption is that the mobility patterns of users are identical and network properties are time-invariant [8, 11, 34, 35]. Exceptions do exist. For example, mobility models have been enhanced to include social relationships of users [15, 31]. These models have been shown to support efficient message delivery in DTNs [15]. Similarly, in recent study, Hsu et al. considers the time variant nature of mobility patterns.

A characteristic of the traces considered in these studies is that they studied populations of users that do not cooperate to complete tasks. In contrast, the results presented in this paper consider the cooperative task of rescuing victims during an emergency response according to the workflow discussed in Section 3. This leads to a richer set of mobility patterns and network properties. Specifically, we will show that network properties vary with both the role of the responders and the phases of the drill. This differentiates our results from previous studies where user behavior is considered to be uniform and network properties stationary. Moreover, our work also differs from the efforts of integrating social relationships into motion models since, in emergency response, the organizational structure is dictated by the response workflow rather than socialization behavior. Due to these properties, most of the existing models of mobility are not applicable. In fact, the analysis presented in this paper will be guided by an understanding of the emergency response workflow.

3. BACKGROUND

Most US emergency agencies follow the Incident Command System (ICS) protocol [2]. Next, we summarize the relevant concepts and terminology of the ICS.

The geographic layout of a drill is divided into *zones* of different risk to victims and responders (see Figure 2). The *hot zone* includes the incident location and, thus, it is considered to be unsafe. The time victims and responders spend in the hot zone should be minimized. The *warm* zone has a medium risk while the *cold zone* is considered to be safe.

A typical *emergency response workflow* involves the following teams and associated responsibilities. The entry team locates, triages, and evacuates the victims from the hot to the cold zone. In the cold zone, the medical team retriages victims, performs more detailed medical exams, and provides medical care. The transport team manages the arrival of ambulances and the loading of victims to hospitals. The med-com team tracks available hospital resources and allocates victims to hospital beds based on the victim's injuries and resource availability. Patients flow from one team to the other until they are transported to hospitals. For the workflow to scale successfully to mass casualty incidents, the teams are intended to work independently. However, for the system to work efficiently, communication plays a crucial role in ensuring effective hand-offs as victims are transferred between teams and enhancing global communication awareness as captured by victim counts, victim statuses, and resource availability.

The typical implementation of ICS results in functions such as the command, triage, treatment, and transport having defined locations for the duration of the response. Also, access into and out of the hot zone is typically managed with one or more entry points for safety reasons. Consequently, as responders move victims to the cold zone, they are guided through the hot zone checkpoint and all arrive in the triage area. In this way, the independent actions of the teams result in an orderly, coordinated response. As described in the following sections, the mobility patterns characteristic of the emergency response workflow have a profound impact on network properties.

4. WIISARD SYSTEM

This section presents the system architecture, hardware, and software components of WIISARD. Here we focus on the key design decisions we made to support reliable communication in emergency responses.

4.1 System Architecture

WIISARD supports the emergency response workflow by tracking victims during the response and by providing role-tailored interfaces to manage the electronic medical records of patients, the assignment of patients to ambulances, and the availability of hospital resources. WIISARD consists of *triage devices*, *mid-tier devices*, and *command centers*. The entry and medical teams use the *triage devices* to triage patients and manage their electronic medical records. The workflow of the transport and med-com teams is supported via *mid-tier devices*. The *mid-tier devices* implement capabilities for managing ambulances and hospital resources in addition to the functionality already provided by *triage devices*. The incident commander uses a *command center* to get an overview of the progress of the response by monitoring the locations of responders and accessing statistics including victim counts, victim statuses, and resource availability.

WIISARD supports a simple broadcast model: the information generated by a responder is disseminated to all other responders. The evaluation criterion is whether the information is disseminated *reliably*. To this end, WIISARD employs a peer-to-peer architecture in which information is cached aggressively and disseminated using the gossip-based communication protocol described in Section 4.3.3. All WIISARD devices operate in 802.11 ad-hoc mode.

This architecture has three notable features. First, WIISARD does not require the deployment of additional communication infrastructure: the devices carried by responders should be sufficient to support timely and reliable communication. This decision is motivated by our experiences with the previous version of WIISARD, which required the deployment of mesh nodes to provide coverage prior to the start of drill exercises. The deployment effort was onerous, requiring more than an hour to deploy, configure, and test the mesh nodes. Even worse, once responders and equipment arrived on-scene, the network properties changed dramatically, requiring us to move the mesh nodes. As it took time to notice the network failures and move the nodes, failures could be prolonged. Obviously, such a deployment process would not be viable in real emergencies. It is important to note that WIISARD can take advantage of additional infrastructure to improve connectivity, however, *infrastructure nodes provide a value-added service* rather than being required to support communication.

Second, we opt for a simple broadcast primitive coupled with aggressive caching rather than the more complex communication primitives adopted in other systems (e.g., [10,20, 21,24]). This approach is justified by the fact that a responder may need to access to the information of any patient at any time. A benefit of caching all data is that it facilitates data dissemination through data muling as responders move about the scene. We hypothesize (and verify in Section 5) that data muling plays a significant role in data dissemination. It is important to understand that caching all data is a viable option because WIISARD is a *low-data rate application*. WIISARD focuses on managing patient records along with ambulance and hospital information, which are generated through manual input. Moreover, even our most resource-constrained devices – the mobile phones – can store all the generated data.

Finally, WIISARD minimizes the impact of network partitions and mobility by using a gossip-based communication protocol. The gossip-based protocol relies on *local* communication to disseminate data. This has important advantages over traditional routing protocols that require the construction and maintenance of end-to-end multi-hop routes that are subject to significant temporal dynamics.

4.2 Hardware Components

WIISARD uses hardware that is heterogeneous across three relevant dimensions: display size, operating system, and packet transmission power.

Triage device: The triage device is a Nokia N900 mobile phone that runs Maemo OS – a derivative of the Linux operating system. The small form factor and weight make the triage devices a good choice for the mobile entry and medical teams. The Nokia N900 uses a high-end OMAP 3430 ARM Cortex A8, has 32GB flash storage, and Wi-Fi card with a transmission power of 10mW. The GPS available on Nokia N900 phones is used to track the locations of responders.

Mid-tier/command center: The mid-tier device is a TabletPC while the command center is a large-screen laptop. Their larger touch screens allow for complex user interfaces to manage ambulance and hospital resources. Both devices run Windows XP operating system, are resource rich, and transmit packets at 100mW.

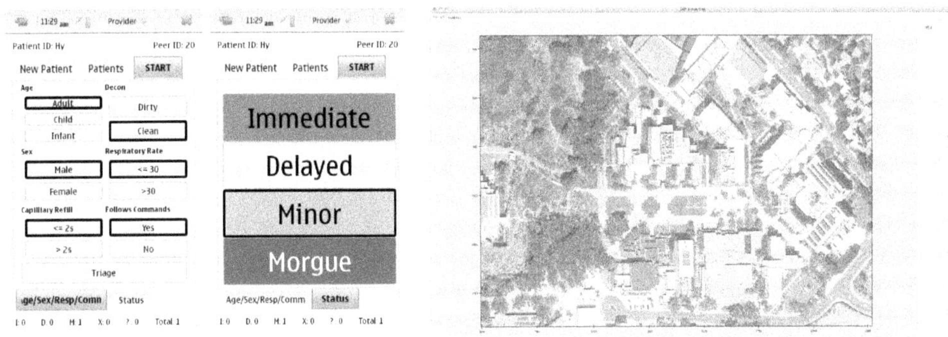

(a) START Triage (b) Command Center

Figure 1: START Triage and Command Center components

Mesh boxes: Mesh boxes may be deployed to improve network connectivity. Each mesh box contains an ALIX x86 board with a 500 MHz AMD CPU and 256 MB RAM, with a 4 GB compact flash card acting as hard drive. Each board also has two mini PCI 802.11abg cards and transmits packets at 100mW. The mesh boxes run Linux Ubuntu 10.04 LTS.

Patient tracking: Traditionally, victims are tracked during emergency responses using paper triage tags that capture the severity of their injuries. To facilitate electronic tracking, we evaluated bar code and RFID technologies. We opted for RFID technology due to the difficulties of using barcodes during sunny days. RFID tags are taped onto the traditional paper triage tags. A NephSystem N330 RFID reader is used to read the short-range RFID tags. The reader is connected to any WIISARD device using Bluetooth.

4.3 Software Components

WIISARD is a portable system that supports both ARM and Intel processors and may be deployed on multiple operating systems including Linux, Windows XP, and Mac OS X. The majority of WIISARD is implemented in Python to ensure cross-platform portability. Only the WIISARD communication protocol is implemented in C++, due to performance considerations. WIISARD has three key components: user interfaces, object storage, and networking.

4.3.1 User Interfaces

The WIISARD user interfaces are developed using the QT toolkit that is accessed via PyQt bindings. The interface design is the result of multiple iterations and interactions with first responders. Particular attention was given to adapting each interface to match the screen size of each device. WIISARD provides the following user interfaces:

Triage: The triage user interface implements the Simple TriAge Rapid Treatment (START) protocol [23] (see Figure 1(a)). The triage of a victim starts by scanning the RFID tag located on their triage tag. The RFID reader transmits the tag ID to the mobile phone over Bluetooth. The START triage interface is designed to enable fast triage by minimizing user input. Typically, a victim is triaged within a minute. Victims may be retriaged when necessary.

Transport: A transport supervisor is responsible for assigning patients to ambulances and designating destination hospitals. This decision is usually based on the nature of victim injuries, need for specialized hospital facilities (e.g., burn center), and hospital bed availability. Traditionally, this information is maintained using paper worksheets. WIISARD improves this process by implementing a transport interface that allows the officer to quickly identify a victim and assign them to ambulances and hospitals. Victims are identified either based on their injury status or by scanning their RFID tag. This enables responders to quickly bring up victim information. The more complex transport user interface is effectively supported by the larger screen size of the tablet devices compared to that of the mobile phones.

Command Center: The command center is capable of accessing all patient information, creates summaries about the progress of the response, and displays the locations of first responders and victims (see Figure 1(b)). WIISARD uses the GPS capabilities of mobile phones to track not only the location of the first responders that carry the mobile phones but also the location of the victims. The location of victims is inferred based on the location of the providers: when a provider scans a victim's RFID, the victim's location is updated based on the provider's current location.

4.3.2 Object Storage

The data generated by responders is modeled as objects that are persisted on disk using SqlLite. Each object has a globally unique identifier. In the case of patient data, their RFID tag is used as the unique identifier; the identifiers of the other objects are randomly generated Universally Unique IDentifiers (see RFC 4122).

WIISARD distinguishes between different versions of the same object through time stamps. Obviously, this requires all peers to be time synchronized. WIISARD achieves time synchronization either through NTP or GPS time. Version conflicts are resolved by keeping the object with the latest timestamp. While this does not eliminate the potential for version conflicts, such conflicts would seldom occur in practice because of two factors. First, it is unlikely for first responders at different locations to update the same record simultaneously. In fact, most of the time the responders are within physical proximity of the victim to update the records, as they need to scan their triage tag. Second, since data is entered manually, small clock synchronization errors have minimal impact on distinguishing between recent and old version of an object.

The object storage component implements two key functions. First, the object storage mediates the interactions between the network component and the user interface components through a standard model-view-control design pattern. A user interface component registers callbacks through which it is notified of object updates. Second, WIISARD is designed to recover from application crashes. WIISARD implements the following check-point policy: the data generated locally is committed to the database as soon as possible to ensure its persistence, while data received from other nodes is buffered before being committed to disk. The policy does not introduce high overhead as the users create or modify objects infrequently.

4.3.3 WIISARD Communication Protocol

The initial prototype of WIISARD used a client-server architecture and an AODV-based routing protocol [14, 24]. This initial prototype suffered from inconsistent network reliability across deployments. The poor reliability is partly explained by the need to construct and maintain multi-hop paths in a highly dynamic network environment. Moreover, network partitions disconnect the clients and server, preventing communication.

A pragmatic approach to improving reliability is to forgo the client-server architecture that requires the use of end-to-end routes and opt for a peer-to-peer architecture in which data is disseminated through a gossip-based protocol that relies on local communication. This approach is a good fit with the communication requirements of emergency responses, where information must be shared across all responders. A gossip protocol works by having each peer "gossip" the information it receives until all peers within the network share this information. The fundamental challenge of gossip protocols is to avoid the *broadcast-storm problem*: a peer requests a piece of information and its neighbors rush to send it, resulting in packet collisions and high overhead.

To meet this challenge, we developed the **W**IISARD **C**ommunication **P**rotocol (WCP), a reliable and efficient gossip protocol. A preliminary evaluation of WCP is presented in [14]. WCP divides the user-generated data into `Blocks` such that each `Block` fits in a packet. To improve response time and minimize memory utilization, WCP uses a caching scheme to maintain recently referenced blocks in memory while committing the rest to the object storage, as described in Section 4.3.2.

WCP works as follows. Peer n divides its local time into periods of length P and each period has W equally sized slots. Note that each peer operates independently without requiring time synchronization. In the beginning of each period, n transmits a `Beacon` that summarizes the blocks stored in its local storage component as version vectors [32]. Upon receiving a `Beacon` from a neighbor m, n inspects its object store to determine if it has any `Blocks` that m does not. Peer n selects random slots to transmit the `Blocks` that m is missing within the W slots following the reception of m's `Beacon`. Accordingly, in response to a `Beacon`, up to W `Blocks` may be transmitted. By picking the slot at random, the source of the packet is randomized over multiple requests. This avoids selecting a peer with poor link quality from transmitting in multiple rounds.

WCP uses two optimizations to reduce the number of duplicate packets. First, a peer n does not transmit the same `Block` more than once within W slots. This reduces the overhead when multiple neighbors are missing the same block. Peer n may fulfill many outstanding requests using a single transmission by taking advantage of the broadcast nature of wireless communication. Second, peer n cancels the transmission of a `Block` when it overhears the `Block` multiple times (twice in our deployment). Such an optimization is

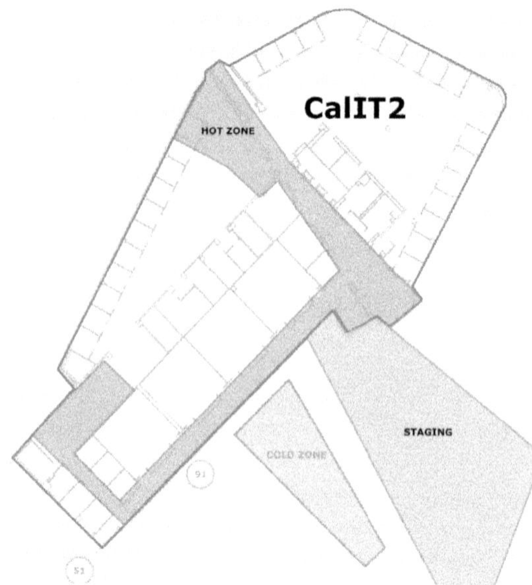

Figure 2: Physical layout of the drill

Variable	Statistic
Responders	16 responders / 3 supervisors
Victims	41 victims
Deployed devices	15 triage devices
	3 mid-tiers
	1 command center
	2 mesh boxes
Phases of the drill	staging (11:00:00 – 11:06:00)
	rescue (11:06:00 – 11:32:00)
	treatment (11:32:00 – 11:37:00)

Table 1: Deployment statistics

particularly effective in dense connectivity graphs in which many peers may receive the broadcast `Block`. This optimization was initially proposed as part of Trickle [25].

WCP makes minimal assumptions about network connectivity. WCP relies on *soft state* that is refreshed periodically through `Beacons`. In fact, the only assumption WCP makes is that network connectivity remains stable within a beacon period P. The beacon period represents a tradeoff between the protocol overhead and the assumed stability of the network. In our deployment, P is set to 5 seconds. Moreover, since WCP caches all data, it can also disseminate data through data muling as opportunistic connections are established with other peers. The use of local communication, soft state, and data muling enables WCP to achieve a network reliability of 98% during the drill exercise presented in Section 5. This is in spite of a highly dynamic environment characterized by variable link quality, mobile users, and network partitions.

5. DRILL EXERCISE

We deployed WIISARD as part of a drill organized at University of California, San Diego (UCSD) on August 10th, 2011. Drill exercises are a common and effective method of training and evaluating the readiness of emergency responders. The WIISARD team had previously participated in several drills administered by San Diego County emergency agencies, and our emergency doctors had been part of the

County team that designs and stages drills, enabling us to create a realistic drill.

The drill scenario called for a major earthquake that injured and trapped multiple victims on the fifth floor of Atkinson Hall. The mission of the first responders was to assess the criticality of injuries sustained by victims, remove them from the building, and provide necessary medical treatment.

The exercise involved 19 responders and 41 victims. The first responders were fire fighters from stations located in the vicinity of UCSD, plus two ambulance crews from Pacific Ambulance. The victims were volunteers mainly from the San Diego and UCSD chapters of the Community Emergency Response Team (CERT). Victims were provided with scenario cards that described the extent of their injuries. Scenarios were designed by our medical collaborators to cover a range of injuries consistent with those that may be sustained in earthquakes. To increase the realism of the drill, moulage was applied and victims were encouraged to act out their injuries: scream, act disoriented, complain, and ask for preferential treatment. Moreover, some of the office furniture on the fifth-floor of Atkinson Hall was positioned to simulate the impact of an earthquake.

Figure 2 shows the geographic layout of the drill. The hot zone is located on the fifth floor of Atkinson Hall. The cold zone is located in front of the building. The emergency response proceeded in three phases: *staging*, *rescue*, and *treatment*. The duration of each phase is captured in Table 1. During the *staging phase*, the first responders arrive on scene, and establish command and control. We instrumented the entry team and the transport and medical supervisors. During the analysis we will refer to the members of the entry team as ***responders*** and to the transport and medical supervisors as ***supervisors***. In the *rescue phase*, the entry team moved in the building to locate, triage, and evacuate the victims to the cold zone. Meanwhile, most of the supervisors remained in the cold zone. Once all victims were transported to the cold zone, the *treatment phase* started. During this phase the entry/medical teams retriaged victims and provided additional details regarding their injuries.

The deployment included all components of the WIISARD system: 16 triage devices, three mid-tiers, and a Command Center (see Table 1). As our goal was to create a realistic deployment scenario in which minimal infrastructure is used, we only include two mesh boxes. The mesh boxes were deployed outdoors as indicated by the circled numbers in Figure 2. The mesh boxes had minimal impact on the observed results: they improved connectivity during treatment. However, the boxes had no impact either during staging when they were off or during rescue when they received no data from within the building. We supplied the first responders with triage tags augmented to each include an RFID tag. As previously discussed, the triage tags are usually placed around the victims' necks to track them during the response.

To create a detailed communication record, each peer recorded the packets it transmitted and received. Since all peers communicated via a shared multicast address, each peer records the messages transmitted by all other peers within its communication range. The minimum transmission rate is 0.2 packet per second as WCP transmits beacons every 5 seconds. The WIISARD devices were time synchronized using NTP in the beginning of the drill and whenever connections to the NTP server (deployed on the command center) were established. We performed a retrospective analysis to verify that the peers were time synchronized by comparing the timestamps when different peers received the same packet. The analysis revealed a time synchronization error below a second.

6. NETWORK PROPERTIES

In this section we characterize the network properties that emerge during emergency responses, based on in-situ measurements from the drill exercise. Specifically, our study focuses on the following questions:

1. What are the underlying network properties of emergency responses, including distribution of link failures and density/diameter of connectivity graph?
2. How do network properties vary across the phases of the drill and with the roles responders have in the drill?
3. What is the impact of mobility on network properties?

Answering these questions provides a sound basis for understanding the challenges of reliable communication in emergency response and will guide the selection appropriate network architectures and protocols to overcome these challenges. Our analysis focuses on understanding the challenges of reliable communication when minimal infrastructure is deployed. We characterize both the properties of links and those of the connectivity graph. A common thread in our analysis is that the emergent network properties are the result of the emergency response workflow and team structure.

6.1 Link Properties

We investigate the reliability of links established among responders during the drill. We start by considering whether link properties change across the phases the drill or with the responder's roles in the drill. Data analysis indicates that the rescue phase posed the most significant challenges to reliable communication. Next, we characterize the long-term and the short-term variations in link quality during the rescue phase. The analysis of link properties at different time scales is motivated by the fact that different mechanisms are employed to handle short-term and long-term variations in link quality.

Mobility is a primary cause of long-term link variability: links are established and broken as the responders rescue victims. To quantify the long-term variations in links, we define *contacts* between pairs of peers. A contact (n, m) means that a peer n is in contact with peer m, while the time difference between consecutive packets that n receives from m is less than a minute. We use *contact length* and *inter-contact time* to assess the temporal properties of contacts. The *contact length* and *inter-contact time* are computed based on both beacon and data packets.

Of interest are the short-term variations in link quality during contacts. Many factors contribute to short-term variations in link quality, including interference, changes in antenna orientation, and movement over short distances. We quantify short-term link quality by measuring the packet reception rates (PRRs) during peer contacts. PRR is computed based on the beacons transmitted by WCP.

6.1.1 Link Properties Vary between Drill Phases

Figure 3(a) plots the number of distinct contacts of responders during a one-minute window. The background

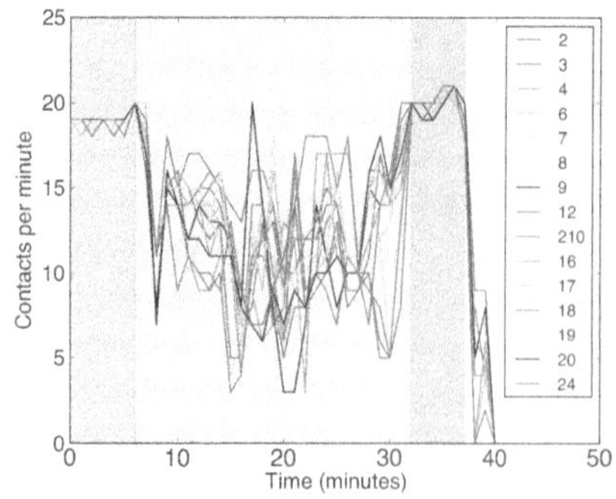

(a) Number of contacts per minute

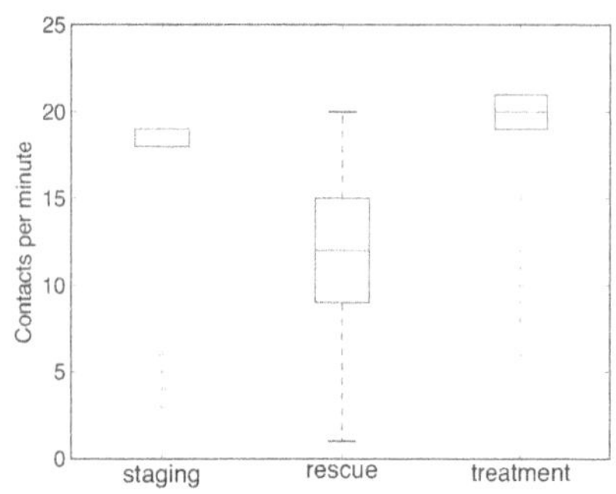

(b) Distribution of contacts for responders

Figure 3: Number of contacts during the drill

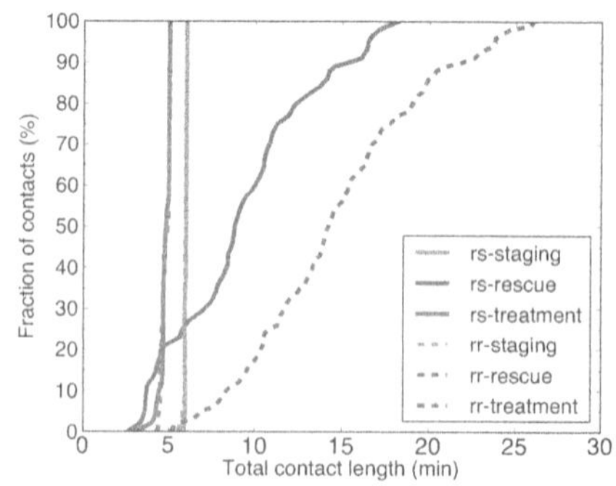

Figure 4: CDF of total contact lengths among rescuers (`rr`) and between rescuers and supervisors (`rs`) during each phase of the drill

color indicates the phase of the drill: staging, rescue, or treatment. During the staging and treatment phases, responders had a similar average number of contacts. According to Figure 3(b), the median number of contacts during staging and treatment was 19 and 20, respectively. The staging phase had little variability: the range was 18 – 19 contacts per minute. This is the result of the close proximity of responders as they planned their response. During treatment we observe a wider range of 6 – 21 contacts per minute. The higher maximum number of contacts was due to the deployment of two mesh boxes at the end of the staging phase. The increased variability in the number of contacts may be attributed to the larger area in which the responders were distributed and increased mobility as responders re-triaged and treated patients.

The number of contacts during the rescue phase (median 12, range 1 – 21) differs significantly from either the staging or treatment phases. The lower median value and higher variability are the compounded result of the indoor environment where walls significantly attenuated links as well as the increased mobility of the responders during the rescue phase. This result indicates that network properties change significantly among the different phases of the drill.

Result: *The long-term link properties vary across the phases of the drill. The rescue phase exhibits high variability in number of contacts per minute due to mobility.*

6.1.2 Link Properties depend on Responder's Roles

An emergency response system must disseminate data both among responders and between responders and supervisors. We define the link groups – `rr` and `rs` – to include all pairs of links that may be established among responders (`rr`) and between responders and supervisors (`rs`), respectively. To determine whether link properties depend on the responder's role in the drill, we compare the properties of links in `rr` and `rs`.

To this end, we define the *total contact length* of peers (n, m) as the sum of their contact lengths during each phase. The CDF of the total contact lengths during each phase is plotted in Figure 4.

During staging and treatment, there were small differences in the distribution of total contact lengths between the `rr` and `rs` groups. More importantly, a large fraction of the links was established for more than 90% of their respective phase durations. The sharp increase in the CDF for `rr` and `rs` during staging and treatment indicate the end of the phase after 5 and 6 minutes, respectively. This indicates stable connectivity among responders and between responders and supervisors.

During the rescue phase, there was good connectivity among responders. All contacts among responders (`rr`) had a total contact length of at least 5 minutes. Moreover, half of the contacts in `rr` had total contact lengths longer than 14 minutes. In sharp contrast, the connectivity between responders and supervisors was poor: 22% of the contacts in `rs` had a total contact length less than 5 minutes (there were no such short contacts in `rr`). Even worse, there is no contact between responders and supervisors (`rs` group) that has a total length longer than 17 minutes (i.e., 69% of the rescue phase) indicating that the two groups were disconnected for part of the rescue phase.

This data indicates that there was good connectivity both in the `rr` and `rs` groups during treatment and staging. However, the responders and supervisors (`rs`) were disconnected for part of the rescue phase. A likely explanation is that the supervisors stationed themselves in the cold zone, while the rescuers that were part of the entry team proceeded into the hot zone to rescue the victims. This shows that it may be more difficult to disseminate data from responders to their supervisors than it is among the responders.

Result: *The long-term link properties depend on the responder's roles in the drill. During the rescue phase, the connectivity between responders and supervisors may be intermittent as responders enter the hot zone and supervisors remain in the cold zone.*

6.1.3 *Long-term Variations in Link Quality*

Next, we characterize the long-term variations of links as measured by the contact metrics. The presented data shows that the rescue phase poses the most significant challenges to reliable communication. Accordingly, we will focus the remainder of our analysis on the network properties observed during this phase.

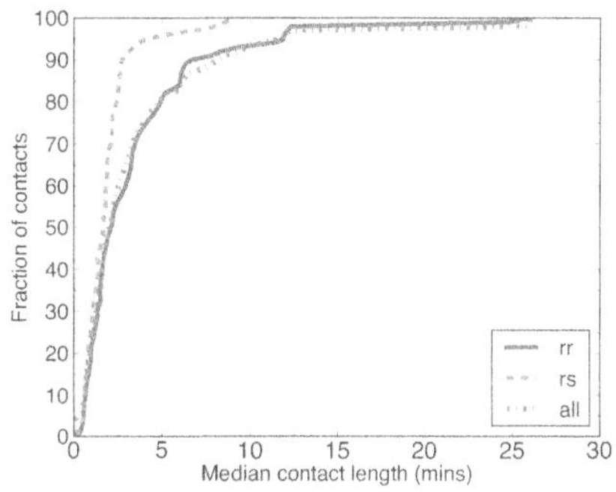

(a) CDF of median contact lengths

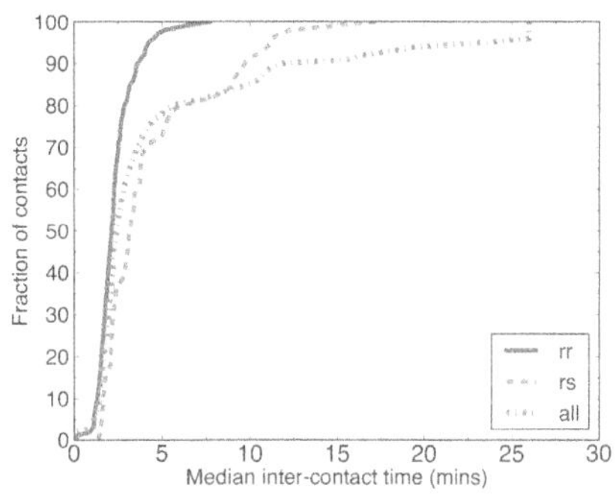

(b) CDF of median inter-contact times

Figure 5: CDF of median contract lengths and median inter-contact during rescue phase

Figure 5(a) plots the CDF of median contact-lengths for the `rr` and `rs` groups during rescue. Half of the links had a median contact-length of 1.94 minutes. The short median contact length is indicative of a dynamic network in which link failures are common. Moreover, the contacts among responders were longer than those between responders and supervisors. For example, 90% of the contacts in `rs` were shorter than 2.75 minutes, compared to only 58% of the contacts in `rr`. Another interesting property of the distribution is its long tail, particularly for the `rr` group of contacts. The long tail is due to a small subset of contacts that were stable for prolonged periods of time. In fact, a small fraction of the links was stable for the entire 26 minutes of the rescue. This is consistent with our empirical observation that responders tend to work in small groups. It also stands in contrast to prior empirical DTN studies in which this behavior was attributed to popular locations rather than teamwork.

Figure 5(b) plots the CDF of inter-contact lengths. Half of the contacts had a median inter-contact length of about 2.15 and 3.15 minutes for the `rr` and `rs` groups, respectively. Overall, contacts between responders were reestablished significantly faster than those occurring between responders and supervisors. In fact, 90% of the contacts in `rr` were reestablished within 2.8 minutes compared to 6.85 minutes for contacts in `rs`. As previously discussed, this is the result of the supervisors typically being located in the cold zone while the responders entered the hot zone.

Result: *Contacts between responders are typically short, however, they are often reestablished within minutes. Moreover, a subset of responders establishes contacts that are stable for prolonged periods of time.*

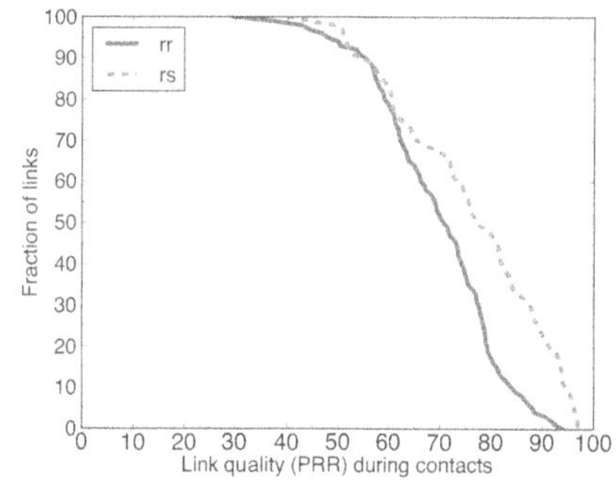

Figure 6: Complementary CDF of link quality as measured by PRR while peers are in contact

6.1.4 *Short-term Variations in Link Quality*

Figure 6 plots the complementary CDF of the observed PRR while peers are in contact. This result captures the variability in link quality that can be attributed to environmental factors including wall attenuation, interference, and changes in antenna orientation due to body movement. The figure indicates that once a contact was established, the link quality tended to be relatively high. For example, the median PRR of links in `rr` and `rs` exceeded 70% and 77%, respectively. A consequence of the good PPR observed while contacts were established is that retransmissions were effective in combating short-term variations in link quality.

Result: *The short-term link quality while contacts are established is high.*

6.2 Connectivity Properties

This section investigates the connectivity of the responders during the rescue phase. The analysis focuses on the key factors that affect the reliability and latency of data dissemination including: the prevalence of network partitions, the density and diameter of the connectivity graph, and the potential of using data muling to disseminate data. The connectivity graph was computed by merging the time synchronized logs of peers.

6.2.1 *Network Partitions*

For the majority of the rescue phase, all fifteen responders were located inside Atkinson Hall triaging patients. Due to the small physical area that was in play (see Figure 2), we expected that the network would remain connected as responders moved on the floor. Figure 7(a) plots the number of connected components in the communication graph. While for much of the rescue phase the graph was connected (i.e., it had a single connected component), to our surprise, the network was partitioned for 26.3% of the phase. Three factors contribute to this result. First, the responders operated indoors where walls limit signal propagation. Second, the transmission power of phones was 10 mW (compared to 100mW for the other devices). Finally, responders tended to operate in groups that were not distributed uniformly throughout the building.

Partitions can have a profound impact on network performance by preventing end-to-end paths to be established. Figure 7(b) plots the fraction of responder pairs for which there is no path at a time instant. During the rescue phase, 67% of responder pairs were affected by partitions. Thus, traditional routing protocols that require continuous connectivity may not be suitable for emergency responses.

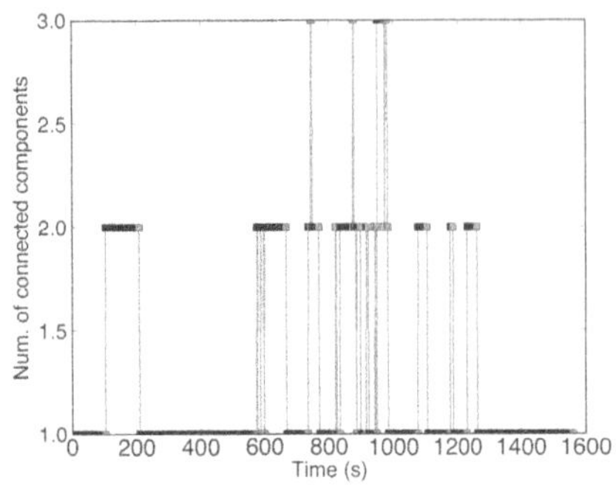

(a) Num. of connected components

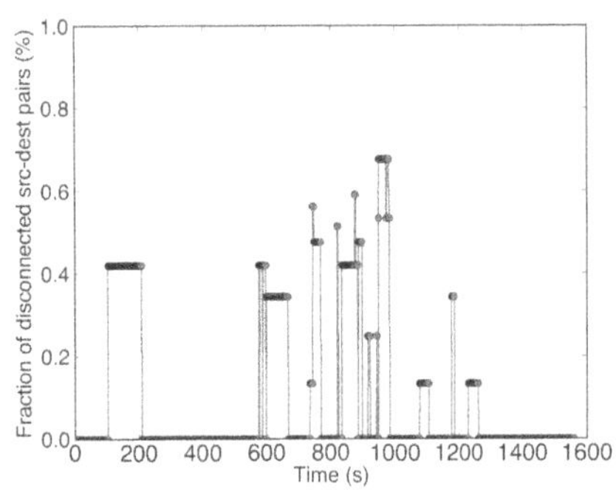

(b) Fraction of disconnected peers

Figure 7: Impact of network partitions

Result: *Network partitions are common and prevent many pairs of peers from communicating.*

6.2.2 Path Lengths and Density of the Connectivity Graph

Figure 8(a) plots the median and maximum path length (i.e., diameter) of all paths between responders during the rescue phase. The distribution of these metrics over time is plotted in Figure 8(b). For 73% of the rescue phase, the diameter of the graph was at most three while the median path length was one. During this time, the connectivity graph was dense and had a low diameter. Such graphs indicate low communication latencies and effective data dissemination, as a small number of broadcasts were sufficient to relay data to all peers. However, for the remainder of the time, the communication graph was rather sparse: the diameter was as large as seven with a median path length as high as three. These results indicate graphs with well-connected components (small median path length) that were connected through a few links (large diameters). The diversity of the graph structures observed during the rescue phase underline the need to develop communication algorithms that operate well in both dense and sparse communication graphs.

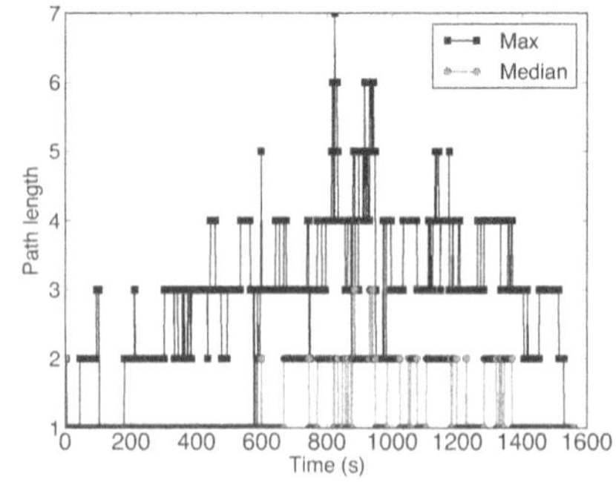

(a) Path lengths over time

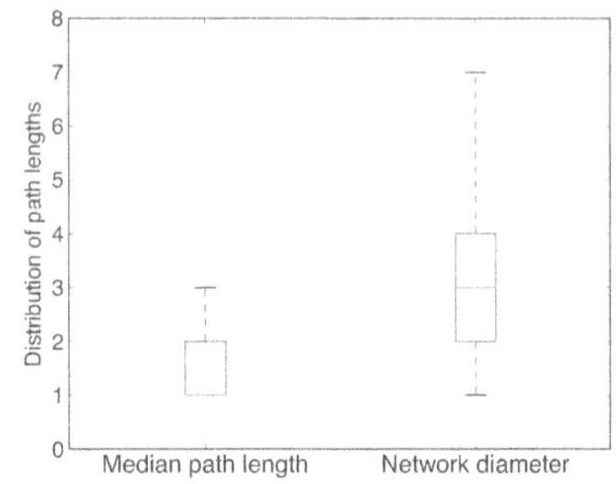

(b) Distribution of path lengths over time

Figure 8: Path lengths statistics

Result: *Reliable communication must be supported in both dense and sparse graphs.*

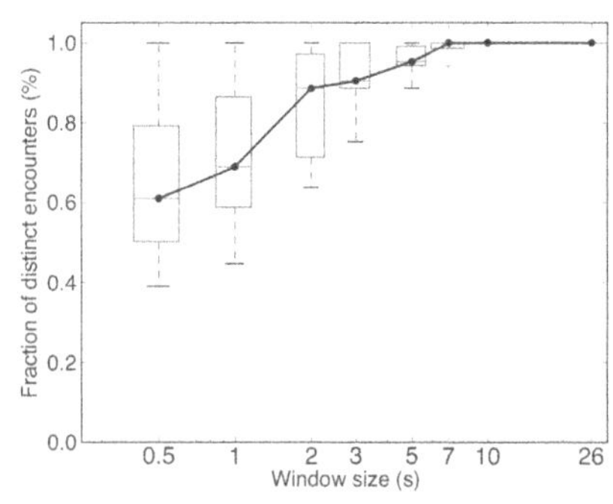

Figure 9: Fraction of encountered nodes over for different windows sizes during rescue phase

6.2.3 The Benefits of Mobility

DTN techniques enable protocols such as WCP to take advantage of mobility to improve data dissemination. To quantify the potential of delivering data through data muling, for each responder we compute the fraction of other responders they encountered during windows of different sizes. Figure 9 plots the distribution of encounters for each responder for a given window size. As the window size increases, a responder tends to encounter an increasing number of the other responders. The number of encountered peers increases at an exponential rate over time (note that the x-axis is logarithmic). This result is captured both in the increasing median values as well as the rapidly decreasing variance. In fact, within seven minutes, a responder encountered all other responders. This shows that if we are willing to tolerate a transmission latency of seven minutes, data among responders can be disseminated solely through data muling.

Previous studies [15] that considered the encounter patterns of nodes within WLAN traces of users show that it is unlikely for a user to meet a large fraction of the other users. The fundamental difference between the prior studies and our results is that we are specifically looking at the encounter patterns of first responders that work cooperatively to rescue victims, rather than considering populations of unrelated users. This indicates that disconnected routing protocols should be a good choice for delivering data among first responders during emergency responses.

Result: *Data dissemination may be improved through data muling as mobile responders rescue victims.*

7. NETWORK PERFORMANCE OF WIISARD

The previous section characterizes the link and network properties that are observed during drills. These results highlight the significant challenges that must be overcome to support reliable communication during emergency responses. As discussed in Section 4.3.3, WCP addresses these challenges through a simple, yet robust gossip-based protocol. In contrast to traditional routing protocols that construct and maintain end-to-end routes, WCP relies on local message exchanges to tolerate variations in link quality and takes advantage of data muling to disseminate data.

During the deployment, victim records were generated during triage/retriage and location records were generated when valid GPS coordinates were acquired. Figure 7 plots the number of generated reports by each responder that used a mobile phone. As responders spent a similar time outdoors, they transmit similar numbers of location reports. In contrast, the number of victim records varied significantly indicating that responders triaged an uneven number of victims. Figure 10(b) plots the total number of reports transmitted (including retransmissions) by all peers during each minute of the drill. On average, a total of 13 packets per second including a single record were transmitted during the drill. However, the sending rate varied significantly observing a maximum aggregated transmission rate of 33 packets per second. The increases in workload are correlated with WIISARD recovering from network partitions. Overall, the data indicates that the system is below its network capacity, indicating the potential to scale to larger numbers of responders. Moreover, the sending rate may be significantly reduced by transmitting multiple records in a single packet, since the typical size of a record is about 200 bytes, far below the maximum payload of 802.11 packets.

The goal of WIISARD is to deliver the data reports generated by a node to all other nodes. We say that a report was delivered reliably, if all nodes in the network receive a copy of the data report. Note that even if a single node does not receive the report, we count this as a failure of the protocol. Even with this strict requirement, WCP performed

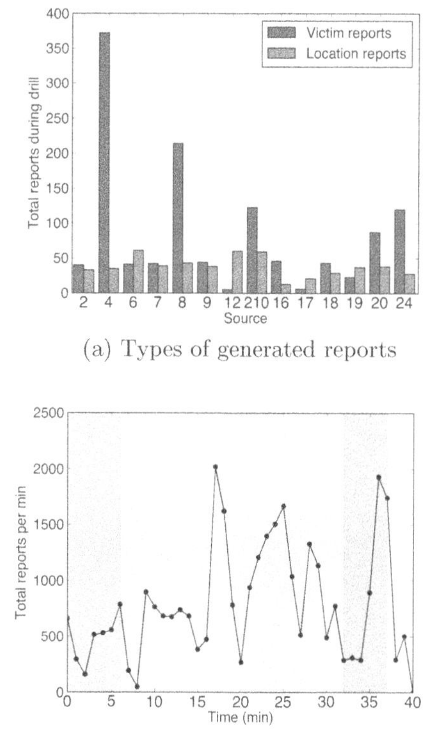

(a) Types of generated reports

(b) Total transmitted reports over time

Figure 10: Characterization of transmission rate

well: WCP delivered 98.25% of the generated data reports to all nodes. Most of the dropped data reports occurred towards the end of the trace, when the most recently captured patient data might not have had time to propagate through the network like data captured earlier.

Figure 11(a) plots the reliability of delivering the data reports created by each node. In accordance to the overall reliability of 98.25%, most of the data reports generated by a node were successfully disseminated. The clear exception is peer 17 for which only 80% of the data reports it created were disseminated to all other nodes. In part, this result is skewed due to the small number of data reports generated by peer 17. We note that no data reports were generated on the mid-tier or command center; these devices were used primarily to view victim statistics.

Figure 11(b) plots the CDF of dissemination latency for the subset of data reports that were delivered to all peers. The median and 90th percentile of the distribution were 30 seconds and 2.2 minutes, respectively. However, the distribution also has a long tail: the worst-case latency was 9.1 minutes. This indicates that while most of the time data was delivered within 2.2 to most responders, due to network partitions, some responders experienced prolonged delays.

8. DISCUSSION

Impact of Emergency Response Workflow: The presented study highlights the profound impact of the emergency response workflow on the mobility patterns and network properties observed during the drill. As a result, emergency response differs from the previously studied DTN applications in the following key aspects. First, link properties

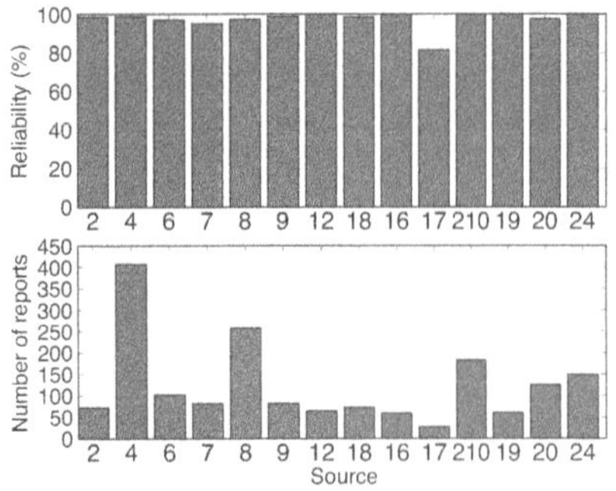

(a) Dissemination reliability

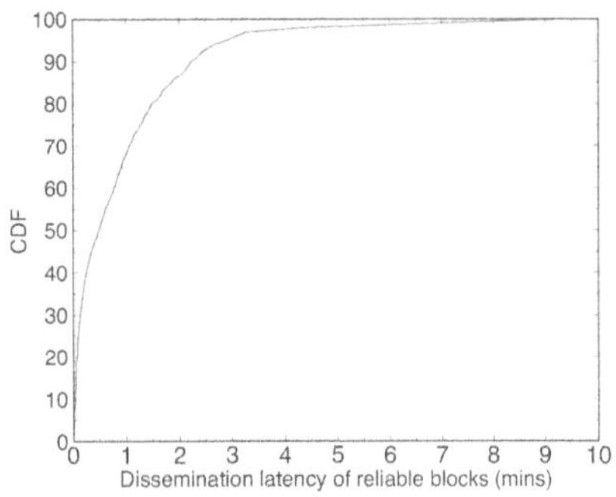

(b) CDF of dissemination latency

Figure 11: Dissemination reliability and latency

vary across the phases of the drill and with the roles of the responders. In contrast, prior DTN studies often assume that the mobility patterns of users are statistically similar and that network properties are stationary [8,11,34,35]. Second, the team structure – rather than popular locations [8,34] or social relationships [15,31] – are responsible for the long tail of the cumulative distribution of median contact lengths. Finally, the emergency response workflow offers unexpected opportunities for disseminating data through data muling: when (1) teams physically transfer patients from one to another and (2) as teams enter/exit the hot zone in which access is managed through a single or few entry points. This shows that emergency response has a more complex network properties that can be understood starting from the emergency response workflow.

Generalizability: Emergency response belongs to a larger class of applications that involve mobile entities that perform cooperative tasks according to well-established workflows. A typical workflow may dictate team structure, physical movement, and best practices for operation. In the case of emergency response, these aspects are defined as part of the ICS (as discussed in Section 3). Other applications that pertain to this class include military operations, participatory sensing tasks (e.g., surveys of earthquake-damaged buildings), or the operation of robotic teams/swarms (e.g., crop pollination [16]). Next, we discuss the impact of the empirical properties on emergency response. However, we expect that some of the observed challenges and associated solutions may be applicable to this application class.

Networking Architecture: The presented results provide a strong case against the use of traditional mesh routing protocols in emergency response systems. First, mesh routing requires end-to-end routes to be maintained in spite of network dynamics. Our empirical data clearly indicates the presence of significant network dynamics both within and across the phases of the drill. Second, mesh routing protocols have not been designed to cope with network partitions, which are common even in the relatively confined area in which the drill occurred. These two factors explain the poor performance that we observed with the initial WIISARD system (10% data reliability) and are consistent with those obtained as part of the Code Blue Project [26] (20% data reliability according to [12]). In contrast, WIISARD achieves 98.25% data delivery during the drill. This shows the superiority of DTN techniques over mesh routing protocols for emergency response systems.

State Consistency: Brewer's CAP theorem [5] states that a distributed system cannot simultaneously provide consistency, availability, and tolerate partitions. WIISARD is an "always available" system: it provides responders with the most up-to-date information available. Therefore, in systems such as WIISARD, there is a fundamental trade-off between consistency and tolerance to network partitions. WIISARD supports an eventual consistency model: WIISARD cannot guarantee consistency when the network is partitioned, however, WIISARD will automatically update its records to their latest versions as partitions merge. In WIISARD, inconsistent state can give rise to a situation where a responder or incident commander has an incorrect view of the status of a patient, such as whether the patient has been assigned a triage level.

An important property of WIISARD is that it requires different degrees of consistency based on the tasks and roles of providers. The primary function of WIISARD is to support cooperative triage and treatment of patients by the members of a response team. It is essential for these members to have consistent views of patient records. In practice, it is easy to achieve consistency among members of the same response team since they are usually within physical proximity and, as a result, they are part of the same connected graph component. Since according to the emergency response workflow patients are physically handed off from one response team to another, this ensures the presence of opportunities of synchronizing patient records between teams through data muling. An additional measure that may be taken to ensure consistency of patient records between teams is to provide each patient with a device that stores and forwards their patient records.

The secondary function of WIISARD is to provide incident commanders with statistics regarding patient counts and the distribution of injury severity, which requires less strict consistency guarantees. In our drill exercise we observed maximum delays of 9.1 minutes to deliver data to incident commanders. In its current implementation WIISARD does not provide differentiated service based on the consistency guarantees required by responders. We expect that such traffic differentiation could improve the performance of WIISARD.

Limiting Network Partitions: There are a number of opportunities to improve the performance of WIISARD. A non-technological approach to minimizing the observed latencies is to modify the response workflow to start the evacuation of victims earlier. This will ensure that data can be delivered from the hot zone to the cold zone through data muling. Technological alternatives focus on minimizing the likelihood of observing network partitions. A pragmatic approach is to deploy additional infrastructure to bridge the cold and hot zones since our study indicates that DTN techniques can already provide reliable communication within each zone. Alternatively, a long-range wireless technology, carried by responders or mounted on their vehicles, may be employed to a similar effect. However, as long-range wireless solutions usually have low bandwidth, it is important to prioritize patient information. The development of tools for network deployment and the evaluation of the effectiveness of combining long-range and short-range wireless technology remain research questions in emergency response systems.

Testing and Simulations: A fundamental challenge to developing robust emergency response system is testing: there are only limited opportunities to deploy a system as part of drill exercises to test their performance. Months of software testing typically go into preparing for a drill exercise to ensure correct operation stifling innovation in this area. For example, without an understanding of mobility patterns of responders, we had to opt for a gossip-based protocol that does not account for user mobility. The presented results highlight that the mobility patterns of responders tend to be structured. We plan on developing protocols that take advantage of these properties to improve delivery performance, particularly focusing on reducing the latency of packet delivery. Moreover, the insights presented here, combined with the collected data can be the basis for developing a realistic simulation environment for evaluating emergency response systems.

9. CONCLUSIONS

This paper presented the design, deployment, and empirical evaluation of WIISARD during a drill exercise. The drill exercise is designed to assess the feasibility of achieving reliable communication with limited communication infrastructure. The drill involved 19 instrumented first responders that triaged 41 simulated victims. Our work makes the following key contributions.

1. We provide a detailed characterization of the network properties that emerge during an emergency response. We show that network properties are highly variable and depend on both the drill phase and the responder's role during the emergency response. The contacts between responders tend to be short, however, they are often reestablished within minutes. The quality of links during contacts tends to be high. The connectivity graph usually has small diameter and is dense. However, due to limited infrastructure, at times it becomes sparse and even partitioned. Network partitions can be effectively mitigated by taking advantage of the frequent contacts between responders, which are an important characteristic of the emergency response workflow.

2. In spite of these challenges, WIISARD achieves a reliability of 98% during the drill. This shows the effectiveness of using gossip-based communication and taking advantage of data muling.

3. The empirical results also point towards several ways in which emergency response systems may be improved, including support supporting differentiated consistency guarantees, leveraging the structured motion patterns of responders to reduce communication latencies, and the development of a sound testing environment for such systems. The presented study highlights the profound impact of the emergency response workflow on the mobility patterns and network properties observed during the drill.

Acknowledgments

We express profound thanks to the first responders that participated in our emergency response drill and allowed us to work with them to evaluate WIISARD. We also thank the other members of the WIISARD team who played critical roles in preparation for exercises and administration of the project. Finally, we would like to thank our anonymous reviewers for their feedback and our shepherd Matt Welsh.

This work was supported by NIH/NLM grant number R01-LM009522 and by NSF through grants CNS-1144664 (NeTS) and CNS-1144757 (NeTS).

10. REFERENCES

[1] Emergency Patient Tracking System (EPTS). `http://www.raytheon.com/capabilities/products/epts/`.

[2] G. A. Bigley and K. H. Roberts. The incident command system: High-reliability organizing for complex and volatile task environments. *The Academy of Management Journal*, Vol. 44(No. 6):pp. 1281–1299, 2001.

[3] D. Bradler, B. Schiller, and E. Aitenbichler. Towards a distributed crisis response communication system. In *Proceedings of ISCRAM*, 2009.

[4] B. Braunstein, T. Trimble, R. Mishra, B. S. Manoj, R. Rao, and L. Lenert. Feasibility of using distributed wireless mesh networks for medical emergency response. In *Proceedings of AMIA*, 2006.

[5] E. A. Brewer. Towards robust distributed systems (Invited Talk). Principles of Distributed Computing, 2000.

[6] S. W. Brown, W. G. Griswold, B. Demchak, and L. A. Lenert. Middleware for reliable mobile medical workflow support in disaster settings. In *Proceedings of AMIA*, pages 309–313, 2006.

[7] R. Bruno and M. Conti. Opportunistic networking overlays for ICT services in crisis management. *Proceedings of ISCRAM*, 2008.

[8] J. Burgess, B. Gallagher, D. Jensen, and B. Levine. Maxprop: Routing for vehicle-based disruption-tolerant networks. In *Proceedings of INFOCOM*, 2006.

[9] T. Camp, J. Boleng, and V. Davies. A survey of mobility models for ad hoc network research. *Wireless communications and mobile computing*, 2(5):483–502, 2002.

[10] R. Carella and S. McGrath. ARTEMIS personal area networks for emergency remote triage and information management. *Proceedings of ISCRAM*, 2006.

[11] A. Chaintreau, P. Hui, J. Crowcroft, C. Diot, R. Gass, and J. Scott. Impact of human mobility on opportunistic forwarding algorithms. *IEEE Transactions on Mobile Computing*, 6:606–620, 2007.

[12] B. Chen, G. Peterson, G. Mainland, and M. Welsh. Livenet: Using passive monitoring to reconstruct

sensor network dynamics. *Distributed Computing in Sensor Systems*, pages 79–98, 2008.

[13] B.-r. Chen, K.-K. Muniswamy-Reddy, and M. Welsh. Ad-hoc multicast routing on resource-limited sensor nodes. In *Proceedings of REALMAN*, pages 87–94, 2006.

[14] O. Chipara, A. N. Plymoth, F. Liu, R. Huang, B. Evans, P. Johansson, R. Rao, and W. G. Griswold. Achieving reliable communication in dynamic emergency responses. In *Proceedings of AMIA*, 2011.

[15] E. Daly and M. Haahr. Social Network Analysis for Information Flow in Disconnected Delay-Tolerant MANETs. *IEEE Transactions on Mobile Computing*, 8(5):606–621, 2009.

[16] K. Dantu, B. Kate, J. Waterman, P. Bailis, and M. Welsh. Programming micro-aerial vehicle swarms with karma. In *Proceedings of SenSys*, Nov. 2011.

[17] R. Dilmaghani and R. Rao. An Ad Hoc Network Infrastructure: Communication and Information Sharing for Emergency Response. In *Proceedings of WIMOB*, 2008.

[18] R. Dilmaghani and R. Rao. A wireless mesh infrastructure deployment with application for emergency scenarios. In *Proceedings of ISCRAM*, 2008.

[19] F. Dong, Y. Hu, M. Tong, and X. Ran. Supporting Emergency Service by Retasking Delay-Tolerant Network Architecture. In *Proceedings of MSN*, 2009.

[20] T. Gao, T. Massey, L. Selavo, D. Crawford, B.-r. Chen, K. Lorincz, V. Shnayder, L. Hauenstein, F. Dabiri, J. Jeng, A. Chanmugam, D. White, M. Sarrafzadeh, and M. Welsh. The Advanced Health and Disaster Aid Network: A Light-Weight Wireless Medical System for Triage. *IEEE Transactions on Biomedical Circuits and Systems*, 1(3):203–216, 2007.

[21] S. George, W. Zhou, H. Chenji, M. Won, Y. O. Lee, A. Pazarloglou, R. Stoleru, and P. Barooah. DistressNet: a wireless ad hoc and sensor network architecture for situation management in disaster response. *IEEE Communications Magazine*, 48(3):128–136, 2010.

[22] T. Henderson, D. Kotz, and I. Abyzov. The changing usage of a mature campus-wide wireless network. In *Proceedings of MobiCom*, 2004.

[23] C. A. Kahn, C. H. Schultz, K. T. Miller, and C. L. Anderson. Does start triage work? an outcomes assessment after a disaster. *Annals of Emergency Medicine*, 54(3):424 – 430.e1, 2009.

[24] L. A. Lenert, D. Kirsh, W. G. Griswold, C. Buono, J. Lyon, R. Rao, and T. C. Chan. Design and evaluation of a wireless electronic health records system for field care in mass casualty settings. *Journal of American Medical Informatics Association*, 18(6):842–852, 2011.

[25] P. Levis, N. Patel, D. Culler, and S. Shenker. Trickle: a self-regulating algorithm for code propagation and maintenance in wireless sensor networks. In *Proceedings of NSDI*, 2004.

[26] K. Lorincz, D. Malan, T. Fulford-Jones, A. Nawoj, A. Clavel, V. Shnayder, G. Mainland, M. Welsh, and S. Moulton. Sensor networks for emergency response: challenges and opportunities. *Pervasive Computing, IEEE*, 3(4):16–23, 2004.

[27] L. McNamara, C. Mascolo, and L. Capra. Media sharing based on colocation prediction in urban transport. In *Proceedings of MobiCom*, 2008.

[28] M. McNett and G. M. Voelker. Access and mobility of wireless PDA users. *ACM SIGMOBILE Mobile Computing and Communications Review*, 9(2):40–55, Apr. 2005.

[29] S. F. Midkiff and C. W. Bostian. Mikobos - a mobile information and communication system for emergency response. In *Proceedings of ISCRAM*.

[30] T. J. Morris, J. Pajak, F. Havlik, J. Kenyon, and D. Calcagni. Battlefield Medical Information System-Tactical (BMIST): The application of mobile computing technologies to support health surveillance in the Department of Defense. *Journal of Telemedicine and E-Health*, 12(4):409–416, 2006.

[31] M. Musolesi and C. Mascolo. Designing mobility models based on social network theory. *ACM SIGMOBILE Mobile Computing and Communications Review*, 11(3):59–70, 2007.

[32] J. Parker, D.S., G. Popek, G. Rudisin, A. Stoughton, B. Walker, E. Walton, J. Chow, D. Edwards, S. Kiser, and C. Kline. Detection of mutual inconsistency in distributed systems. *IEEE Transactions on Software Engineering*, SE-9(3):240 – 247, May 1983.

[33] S. Pavlopoulos, E. Kyriacou, A. Berler, S. Dembeyiotis, and D. Koutsouris. A novel emergency telemedicine system based on wireless communication technology-ambulance. *IEEE Transactions on Information Technology in Biomedicine*, 2(4):261 –267, 1998.

[34] M. Piorkowski, N. Sarafijanovic-Djukic, and M. Grossglauser. A parsimonious model of mobile partitioned networks with clustering. *Communication Systems and Networks and Workshops, 2009. COMSNETS 2009. First International*, pages 1–10, 2009.

[35] I. Rhee, M. Shin, S. Hong, K. Lee, S. J. Kim, and S. Chong. On the Levy-Walk Nature of Human Mobility. *IEEE/ACM Transactions on Networking*, 19(3):630–643, 2011.

[36] V. Srinivasan, M. Motani, and W. T. Ooi. Analysis and implications of student contact patterns derived from campus schedules. In *Proceedings of MobiCom*, 2006.

[37] D. Williams and Naval Health Research Center (U. S.). *Tactical Medical Coordination System : (TacMedCS)*. Naval Health Research Center, San Diego, CA, 2007.

[38] W. Zhao. A new mac layer routing protocol for infrastructure wireless mesh networks. Master's thesis, Technical University of Denmark, 2008.

[39] X. Zhao, A. Rafiq, R. Hummel, D.-Y. Fei, and R. C. Merrell. Integration of information technology, wireless networks, and personal digital assistants for triage and casualty. *Journal of Telemedicine and E-Health*, 12(4):466–474, 2006.

DozyAP: Power-Efficient Wi-Fi Tethering

Hao Han[1,2], Yunxin Liu[1], Guobin Shen[1], Yongguang Zhang[1], Qun Li[2]
[1]Microsoft Research Asia, Beijing, China
[2]College of William and Mary, Williamsburg, VA, USA

ABSTRACT

Wi-Fi tethering (i.e., sharing the Internet connection of a mobile phone via its Wi-Fi interface) is a useful functionality and is widely supported on commercial smartphones. Yet existing Wi-Fi tethering schemes consume excessive power: they keep the Wi-Fi interface in a high power state regardless if there is ongoing traffic or not. In this paper we propose *DozyAP* to improve the power efficiency of Wi-Fi tethering. Based on measurements in typical applications, we identify many opportunities that a tethering phone could sleep to save power. We design a simple yet reliable sleep protocol to coordinate the sleep schedule of the tethering phone with its clients without requiring tight time synchronization. Furthermore, we develop a two-stage, sleep interval adaptation algorithm to automatically adapt the sleep intervals to ongoing traffic patterns of various applications. DozyAP does not require any changes to the 802.11 protocol and is incrementally deployable through software updates. We have implemented DozyAP on commercial smartphones. Experimental results show that, while retaining comparable user experiences, our implementation can allow the Wi-Fi interface to sleep for up to 88% of the total time in several different applications, and reduce the system power consumption by up to 33% under the restricted programmability of current Wi-Fi hardware.

Categories and Subject Descriptors

C.2.1 [Computer-communication Networks]: Network Architecture and Design — Wireless communication

General Terms

Algorithm, Design, Experimentation, Measurement

Keywords

802.11, Energy Efficiency, SoftAP, Wi-Fi Tethering

1. INTRODUCTION

Wi-Fi tethering, also known as a "mobile hotspot", means sharing the Internet connection (e.g., a 3G connection) of an Internet-capable mobile phone with other devices over Wi-Fi. As shown in Figure 1, a Wi-Fi tethering mobile phone acts as a mobile software access point (SoftAP). Other devices can connect to the mobile SoftAP through their Wi-Fi interfaces. The mobile SoftAP routes the data packets between its 3G interface and its Wi-Fi interface. Consequently, all the devices connected to the mobile SoftAP are able to access the Internet.

Figure 1: Wi-Fi tethering.

Wi-Fi tethering is highly desired. For example, even before the Android platform provided built-in support on Wi-Fi tethering, there were already some third-party Wi-Fi tethering tools on Android Market [8]. Two of them, called "Barnacle Wi-Fi Tether" and "Wireless Tether for Root Users", are very popular, each with more than one million installs. There are two main reasons why Wi-Fi tethering is desirable. First, cellular data networks provide ubiquitous Internet access but the coverage of Wi-Fi networks is spotty. Second, it is common for people to own multiple mobile devices but likely they do not have a dedicated cellular data plan for every device. Hence, it is desirable to share a data plan among multiple devices, e.g., sharing the 3G connection of an iPhone with a Wi-Fi only iPad. In response to this common user desire, Wi-Fi tethering is now widely supported, as a built-in feature on most smartphone platforms, including iPhones (iOS 4.3+), Android phones (Android 2.2+) and Windows phones (Windows Phone 7.5).

However, existing Wi-Fi tethering schemes significantly increase the power consumption of smartphones. When operating in the SoftAP mode, the Wi-Fi interface of a smartphone is always put in the high power state and never sleeps even when there is no data traffic going on. This increases the power consumption by one order of magnitude and reduces the battery life from days to hours (more details in Section 2.1). To save power, Windows Phone automatically turns off Wi-Fi tethering if the Wi-Fi network is inactive for a time threshold of several minutes. However, this simple scheme has two drawbacks. First, the Wi-Fi interface still operates in the high power state for all the idle intervals less than the threshold, leading to waste of energy. Second, it harms usability. If a user does not generate any traffic for a time period longer than the threshold (e.g., while reading a long news article) and then starts to use the network again (e.g., by clicking another news link), the user will have to go back to the smartphone and manually re-enable Wi-Fi tethering, which results in poor user experience.

The IEEE 802.11 standard defines a Power Saving Mode (PSM) [10] that allows the Wi-Fi interface of a device to stay in a lower-power state to save power. However, PSM is designed for client devices only. An AP should never enter PSM according to the IEEE 802.11 standard. It is not an issue for APs in traditional scenarios because they are externally powered. However, such old design wisdom does not work for the battery-powered SoftAPs in Wi-Fi tethering.

The IEEE 802.11 standard also defines a power saving mechanism for stations operating in ad hoc mode. In ad hoc mode all the stations equally share the responsibility for beacon generation. One station may sleep when another station generates beacons. However, in Wi-Fi tethering, a SoftAP is the single gateway to the Internet and can hardly sleep. Ad hoc mode is much less used than the infrastructure mode (i.e., AP mode) in practice. Interestingly, despite that the Wi-Fi hardware does support ad hoc mode, the OS on many mobile devices, including Windows phones, Android phones and iPhones, hides it [30], preventing a device from connecting to an ad hoc network. Thus, in this paper we focus on Wi-Fi tethering in the infrastructure mode.

We propose and design *DozyAP*, a system to reduce power consumption of Wi-Fi tethering on smartphones while still retaining a good user experience. The key idea of DozyAP is to put the Wi-Fi interface of a SoftAP into sleep mode to save power when possible. We measure the traffic pattern of various online applications used in Wi-Fi tethering. We find that the Wi-Fi network stays in the idle state for a large portion of the total application time (more details in Section 2.2), which means there are many opportunities to reduce the power consumption. With DozyAP, a SoftAP can automatically sleep to save power when the network is idle and wake up on demand if the network becomes active.

Putting a SoftAP to sleep imposes two challenges. First, without a careful design, it may cause packet loss. Existing Wi-Fi clients assume that APs are always available for receiving packets, so whenever a client receives an outgoing packet from applications, it will immediately send the packet to its AP. However, if the AP is in the sleep mode, this packet will be lost, even after the retrials that occur at the low layers of the network stack. Second, putting an AP to sleep will introduce increased network latency and may impair user experiences if the extra latency is user perceivable.

DozyAP addresses the first challenge with a sleep request-response protocol with which a SoftAP and its clients learn and agree on a valid sleep schedule of the SoftAP. To avoid possible packet loss, a client will transmit packets only when the SoftAP is active and buffer outgoing packets otherwise. To address the second challenge, we design an adaptive sleep scheme and limit the maximum sleep duration. Consequently, DozyAP is able to reduce power consumption of Wi-Fi tethering with negligible impact on the network performance. DozyAP does not require any changes to the 802.11 protocol and is incrementally deployable via software updates to mobile devices.

We have implemented the DozyAP system on existing commercial smartphones and evaluated its performance using various applications and the traces from real users. Evaluation results show that DozyAP can put the Wi-Fi interface of a SoftAP to sleep for up to 88% of the total time in several different applications. Due to the restricted programmability of current Wi-Fi hardware on smartphones, forcing a SoftAP to go to sleep or wake up consumes considerable overhead. Thus, DozyAP only saves power by up to 33% while increasing network latency by less than 5.1%.

To the best of our knowledge, we are the first to study power efficiency in Wi-Fi tethering for SoftAP. The main contributions of this paper are:

- We study the characteristics of existing Wi-Fi tethering implementations and present our findings. We show that current Wi-Fi tethering is power hungry and wastes energy unnecessarily. We analyze the traffic patterns of various applications and identify many opportunities to optimize the power consumption of Wi-Fi tethering.
- We propose DozyAP to improve power efficiency of Wi-Fi tethering. We design a simple yet reliable sleep protocol to schedule a mobile SoftAP for sleep without requiring tight time synchronization between the SoftAP and its clients. We develop a two-stage adaptive sleep algorithm to allow a mobile SoftAP to automatically adapt to the traffic load for the best sleep schedule.
- We implement DozyAP system on commercial smartphones and evaluate its performance through experiments with real applications and simulations based on real user traces. Evaluation results show that DozyAP is able to significantly reduce power consumption of Wi-Fi tethering and retain comparable user experience at the same time.

The rest of the paper is organized as follows. In Section 2 we report our findings on existing Wi-Fi tethering, focusing on the power consumption and the traffic patterns of various applications. Based on the findings, in Section 3 we design DozyAP to schedule a mobile SoftAP for sleep and present the design details. We describe our implementation in Section 4 and evaluate it in Section 5. We discuss limitations of DozyAP and future work in Section 6, survey the related work in Section 7 and conclude in Section 8.

2. UNDERSTANDING WI-FI TETHERING

In this section we report our findings on the characteristics of Wi-Fi tethering through real measurements on existing commercial smartphones. We have focused on two characteristics: the power consumption of existing Wi-Fi tethering implementations and the traffic pattern of various online applications used in Wi-Fi tethering. We also provide some background on Wi-Fi power management to set up the context of our DozyAP design.

2.1 Power Consumption

We first measure the power consumption of existing commercial smartphones regarding the Wi-Fi tethering. We impose *no traffic* but simply turn on/off the Wi-Fi tethering, i.e., the Wi-Fi interface and the 3G interface were kept on but idle. We used a Nexus One phone (Android 2.3.6), a HTC HD7 Windows Phone (Windows Phone 7.5) and an iPhone 4 (iOS 4.3.5) for experiments. For the Nexus One and the HTC HD7, we measured the power consumption of the whole system using a Monsoon Power Monitor [4]. However, it is not possible to use the Monsoon Power Monitor to measure the power consumption of the iPhone 4 without damaging the phone. Instead, we used MyWi 5.0 [5], a very popular third-party Wi-Fi tethering tool on iOS that is able to tell the draining current of the battery, to measure the power consumption of the iPhone 4 when Wi-Fi tethering is enabled. In all the experiments, the display was turned off.

With Wi-Fi tethering disabled, the power consumption was pretty low, because the Wi-Fi and 3G interfaces were in sleep for most of the time. The average power consumption was only 20mW for the Nexus One and 30mW for the HTC HD7, respectively. For the iPhone4, MyWi 5.0 read a draining current of 6mA, equivalently, a power consumption of 22mW.

With Wi-Fi tethering enabled, the power consumption of the smartphones increased significantly. Figure 2 shows the results for the Nexus One and the HTC HD7 smartphones. We can see that both smartphones operated in a high power state constantly even though there was no traffic at all. There are periodic spikes in the plots, caused by periodic Wi-Fi beacon transmissions. On average the power consumption was 270mW for the Nexus One and 302mW for the HTC HD7. These results indicate that Wi-Fi tethering leads to severe drop of the battery lifetime: from 70 hours to 5.2 hours for the Nexus One and from 43 hours to 4.3 hours for the HTC HD7, given their respective battery capacity of 1400mAh and 1300mAh. For the iPhone 4, MyWi 5.0 read a draining current of 90mA, equivalently a power consumption of 333mW. While this software approach may not be as accurate as using the Monsoon Power Monitor, the result still clearly indicates that Wi-Fi tethering on the iPhone 4 has similar power consumption as on the Nexus One and the HTC HD7.

The above results demonstrate that existing Wi-Fi tethering schemes on all the three mobile platforms are power hungry. They consume an *order of magnitude more power than necessary* when there is no ongoing traffic, i.e., in idle network state. In next subsection we will show that such *idle* cases are indeed frequent in various typical Internet access scenarios.

Intuitively, the Wi-Fi interface should be put to sleep when the Wi-Fi network is idle. As the battery is a very scarce resource on smartphones, this calls for a power-efficient Wi-Fi tethering solution and motivates us to conduct the work in this paper.

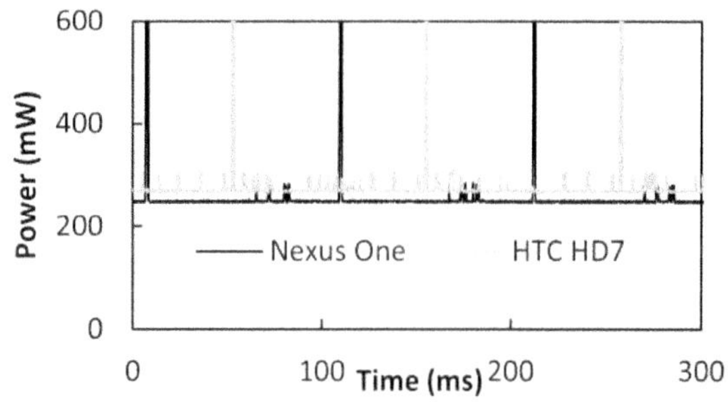

Figure 2: Measured power consumption of Wi-Fi tethering on Nexus One and HTC HD7 in idle case.

2.2 Traffic Patterns

Next, we study how frequently the Wi-Fi network is actually in an idle state and how long the idle state typically lasts. We enabled Wi-Fi tethering on a Nexus One smartphone with a China Unicom 3G connection, and connected a Wi-Fi client to the mobile SoftAP. On the client, we launched various applications that access the Internet and used those applications normally. In the meantime, we used a Lenovo T61 laptop running Linux 2.6.32 as a Wi-Fi sniffer to capture all the frames exchanged between the client and the SoftAP. We studied two different clients: a Nexus One smartphone and a Wi-Fi version iPad 2. The following seven applications were measured.

(1) News reading. We went to http://news.baidu.com, and read five news articles one by one. **(2) Online book reading**. We went to http://www.zangdimima.org and read three chapters of a popular book. **(3) Video streaming**. We went to http://www.youku.com (a popular video website in China like YouTube in US) and watched a 76-second long video clip. **(4) Search**. We used Google Search to search for "tethering" and read the first three results one by one. **(5) Map**. We used Bing Map to search and browse three locations in Beijing: Tiananmen Square, Peking University and Microsoft R&D Asia. **(6) Email**. We used Gmail to read three new emails and replied to one of them. **(7) RSS Reader**. We used Google reader to download and read five pieces of news.

Note that some websites detect the type of client devices and return different content for different device types. For example, when the Nexus One smartphone is used, Baidu News automatically redirects to its mobile version that returns less complex webpages than the normal version. Similarly, Youku streams low bitrate video clips to the Nexus One smartphone but high bitrate ones of the same videos to the iPad 2. Consequently, the same application may behave differently on different devices.

For each application, we study the traffic patterns by analyzing the packet inter-arrival time of all the captured packets. Figure 3 shows the results of the Nexus One. Due to the space limitation, we omit the curves of the iPad 2 which has similar results. We first study the distribution of packet in-

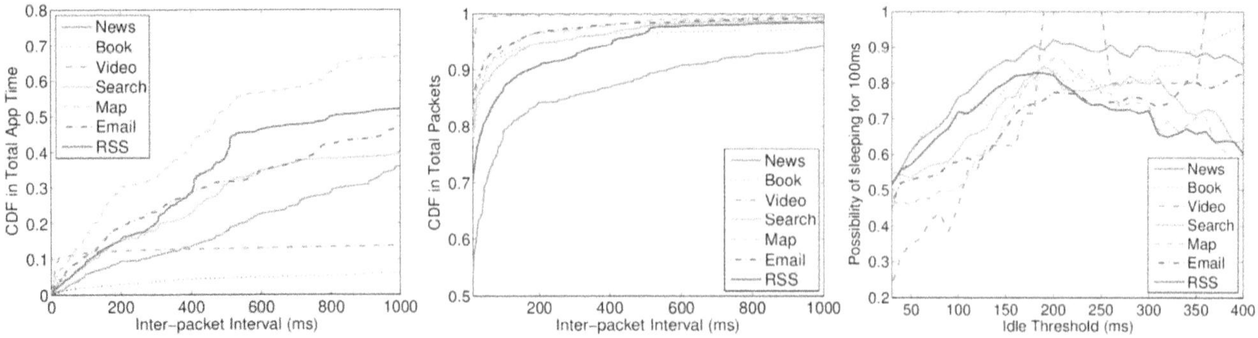

Figure 3. Traffic patterns. From left to right: CDF of packet inter-arrival interval in total application time; CDF of packet inter-arrival interval in total packets; Probability of sleeping for 100ms after an idle threshold.

ter-arrival intervals in the total application time which is the period from the first packet to the last one. The left figure in Figure 3 shows the Cumulative Distribution Function (CDF) for all the applications, where the y-axis depicts the percentage of packets with inter-packet intervals less than or equal to a specific value in the x-axis to the total application time. To make the curves easy to read, we only show the data for the time intervals less than one second. We can see that the intervals under 200ms only take less than 30% of the total application time for all the applications on the Nexus One. For the iPad 2, the corresponding number is 35%. For some applications, these intervals consume as low as 20% or even less than 10% on the Nexus One or the iPad 2. If we consider the network "idle" during the packet inter-arrival intervals larger than 200ms, then we can say that the Wi-Fi network was idle for 70%-90% of the total application time. This shows that these applications only spent a small portion of time for the Internet access and their network traffic is very sparse and bursty.

There are two main reasons for the above findings. First, all the applications consist of two phases: a *content fetching* phase and a *content consuming* phase. Once users download some content from a remote server (e.g., a Web server), they need to spend time to consume the content (e.g., reading the text). The content consuming time may vary from seconds to tens of seconds to even minutes. During such a time, the network is mostly idle. In the email case, replying to emails and composing new ones also result in significant network idle time. Secondly, the bandwidth of 3G is much lower than that of Wi-Fi. According to 3GTest [18], 3G typically offers 500Kbps – 1Mbps downlink throughput for US carriers but the Wi-Fi offers much higher data rates (54Mbps for 802.11a/g and 300Mbps for 802.11n). Furthermore, 3G has much higher RTTs, ranging from 200ms to 500ms [18], than that of Wi-Fi. Consequently, the Wi-Fi interface of a SoftAP in Wi-Fi tethering often has to wait for data to be received from or transmitted over 3G. Such a waiting period will put the Wi-Fi interface in an "idle" state.

While the results are somehow as expected for those interactive applications, we are surprised to see that similar patterns were observed in the video streaming case. Even for the iPad 2 on which a high bitrate video clip was continuously played back, the packet inter-arrival intervals larger than 200ms took more than 60% of the total streaming time. After carefully checking the captured trace, we found that it used a large video buffer when streaming video clips. It aggressively downloaded video content until the video buffer was full. Then it stopped video downloading. The downloaded bits were constantly consumed and drained from the buffer. Once the buffer level became lower than a threshold, the aggressive downloading was resumed again.

The large percentage of the Wi-Fi idle time in these applications demonstrates that there are many opportunities to reduce the power consumption of Wi-Fi tethering. During the large network idle intervals, the Wi-Fi interface of a mobile SoftAP *should* sleep to save power. More specifically, there are two kinds of network idle intervals that we exploit in this paper. The first one is the long network idle intervals resulting from the user content consuming behavior. The second one is the relatively shorter network idle intervals that occur during the content downloading. The latter case is mainly caused by the RTTs of 3G: after a client sends a request packet to a remote server, it has to wait for at least a RTT of 3G to get the first response packet from the server. For example, to access a Web server, we can typically see two such network idle intervals: one for the DNS name lookup for the server and the other for making a TCP connection to the server.

We further study how putting a SoftAP to sleep can affect the network performance. The middle figure in Figure 3 shows the CDF of packet inter-arrival interval in total packets, where the y-axis depicts the percentage of the packets whose inter-packet interval is less than or equal to a specific value in the x-axis to the total number of packets. We can see that the inter-packet intervals under 150ms cover more than 80% of all the packets for all the applications. For some applications the number is as high as 90% or even 95%. This means that if we use an idle threshold of 150ms to decide whether to put the SoftAP to sleep or not, most of the packets will not be affected. The right figure in Figure 3 further shows the probability that the SoftAP can successfully sleep for extra 100ms after waiting for different idle thresholds. Again, if we use a threshold of 150ms, the probability is higher than 60% for all the applications.

All the above findings demonstrate that a mobile SoftAP *should* and *can* sleep to save power in Wi-Fi tethering, which provides the foundation for DozyAP design.

2.3 Background: Wi-Fi Power Saving

The IEEE 802.11 standard defines PSM to save power of Wi-Fi client devices [10]. In PSM, the Wi-Fi interface of a client stays in a lower-power state to save power and cannot receive or transmit any data. If an AP has some packets for a PSM client, the AP buffers the packet and sets the Traffic Indication Map (TIM) in its periodic beacons which are broadcasted typically every 100ms. A PSM client periodically (e.g., every three beacon intervals) wakes up to listen to the beacons of the AP. If the client detects a TIM for itself, it sends a separate PS-Poll frame to receive each buffered packet. Otherwise it goes to sleep immediately. When the AP sends a buffered packet to the client, a MORE bit in the data frame is set if the AP has more packets for the client. This allows the client to decide when to stop sending PS-Poll messages.

On the Nexus One, the above static PSM scheme is called "PM_MAX". PM_MAX allows a client to sleep as much as possible by allowing the client to sleep immediately if the AP does not have any packet for it. However, it leads to long network latency and hence low network efficiency because a separate PS-Poll message is required for every packet. This makes it less suitable for interactive applications. Therefore, on the Nexus One, another power saving scheme called "PM_FAST" is used. In PM_FAST, a client stays in active unless its Wi-Fi interface is idle for a threshold of 200ms. Then it sends a Null-Data frame with power management flag set to 1 to tell its AP that it will sleep soon. If such a frame is acknowledged, the client is able to go to sleep since all packets destined for it will be buffered at AP. Otherwise, the client cannot go to sleep. Many other Wi-Fi devices today also implement a similar scheme known as adaptive PSM [19]. PM_FAST is designed for fast system response and PM_MAX is more suitable for background services. By default Nexus One smartphones use PM_FAST if the screen is on and switch to PM_MAX if the screen is turned off.

3. DOZYAP DESIGN

Guided by the findings in Section 2, we design the DozyAP system that aims to reduce the power consumption of Wi-Fi tethering by putting the Wi-Fi interface of a mobile SoftAP into sleep mode whenever possible. Below we present the detailed design of DozyAP and the rationale of the design decisions. We start with a single client and describe the extension to support multiple clients later on.

3.1 Scheduling a SoftAP for Sleep

We design a simple "*sleep request-response*" protocol to enable a mobile SoftAP to safely sleep in Wi-Fi tethering,

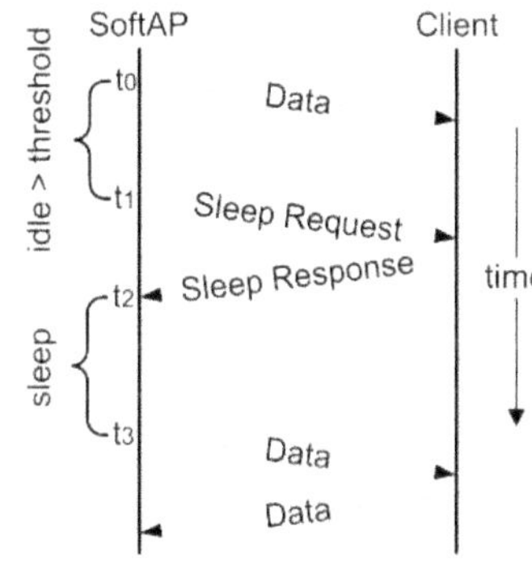

Figure 4: Interactions of a SoftAP and a client using the sleep request-response protocol.

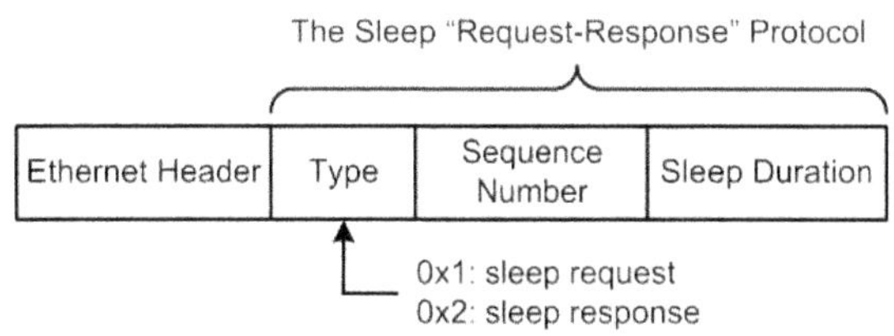

Figure 5: Packet format of the sleep protocol.

according to its own best schedule. While the SoftAP can sleep at will, it can only do so when the client agrees, to avoid possible packet loss. Therefore, before entering sleep mode, a SoftAP sends a "sleep request" to its client. If the client sends back a "sleep response" to accept the sleep request, the SoftAP then enters the sleep mode. Otherwise, it will continue to stay in the active state. Figure 4 shows a typical interaction procedure between a SoftAP and a client. At time t_1, the SoftAP decides to sleep and enters sleep mode at time t_2 after receiving the client's agreement. When the sleep times out at time t_3, the SoftAP wakes up and continue to communicate with the client.

Packet format. Both the sleep request and the sleep response are transmitted as a normal Wi-Fi unicast data packet. This design does not require any modification on existing Wi-Fi standard and is easy to implement. Figure 5 shows the packet format. The sleep protocol is implemented directly on top of the underlying link layer without TCP/IP headers in the middle to reduce the overhead. The sleep protocol packets have three fields. The "Type" field indicates the packet type: "0x1" means *sleep request* and "0x2" means *sleep response*. The "Sequence Number" field is a unique ID to identify a sleep request-response pair. It starts from zero and increases by one for every new sleep request. The "Time Duration" field specifies how long (in milliseconds) the SoftAP requests to sleep. All the sequence numbers and time durations are decided by the SoftAP. When the client accepts a sleep request, it simply copies the sequence number and time duration from the sleep request packet into its sleep response packet. Sleep response packets are used only for accepting a sleep request. If the client does not agree the SoftAP to sleep, it simply chooses not to send out the sleep response. There is only one case that the client will decline the sleep request of the SoftAP: it has more data packets to transmit. In that case, the client will send a data packet, instead of the sleep response packet, to the SoftAP. The SoftAP then learns that the client has declined the sleep request and thus stays active. This design reduces the overhead of the sleep protocol because a sleep response packet is transmitted only when it is necessary.

State machine. Figure 6 shows the state machine of a SoftAP and a client. A SoftAP has three states: *Normal*, *Pre-sleep* and *Sleep*. In the *Normal* state, the SoftAP is active and can transmit and receive packets normally. When the Wi-Fi interface of the SoftAP is idle for a time period larger than a pre-defined threshold, the SoftAP sends a sleep request packet to its client with a sequence number *seq* and a time duration *dur*. Then it enters the *Pre-sleep* state and waits for a sleep response. If it receives the right sleep response with the same sequence number *seq* and time duration *dur*, it will put its Wi-Fi interface into sleep mode and enters the *Sleep* state; if it receives any packet other than the expected sleep response, it will go back to the *Normal* state and invalidate the sleep request. In the *Sleep* state, the Wi-Fi interface of the SoftAP is turned to sleep to save power. Thus, the SoftAP cannot receive any packets over Wi-Fi. If it receives any data from its 3G interface, it will not send the received data to the client over Wi-Fi. Instead, it buffers them during the whole period of *Sleep* state. When the sleep timeout expires, if the SoftAP has any buffered data, it wakes up its Wi-Fi interface, switches to *Normal* state and transmits the buffered data to the client. Otherwise, it moves back to the *Pre-sleep* state, sends out another sleep request with a new sequence number *new_seq* and a new time duration *new_dur* and waits for the next sleep response.

The state machine of a client has only two states: *Normal* and *Block*. In the *Normal* state, the client communicates with the SoftAP as normal. It may use any Wi-Fi power saving schemes, such as PM_MAX and PM_FAST. If it receives a sleep request from the SoftAP and agrees, i.e., it does not have any packets to transmit, it tentatively sets its Wi-Fi power saving scheme to PM_MAX, sends back a sleep response to the SoftAP and enters the *Block* state. Note that by switching to PM_MAX, the firmware automatically sends a "Null-Data" packet *before* we send out the sleep response. The "Null-Data" packet tells the SoftAP that the client will sleep immediately. Doing this way, the client can go to sleep as quickly as possible (i.e., immediately after the sleep response packet) to save power and the SoftAP knows that the client is sleeping. Otherwise, if the client uses PM_FAST scheme, after sending out the sleep response, it stays in active until it sends out a "Null-Data" packet to tell the SoftAP that it will go to sleep. However, as the SoftAP has already entered the *Sleep* state after receiving the sleep response, it cannot receive the "Null-Data" packet following the sleep response. As a result, the client cannot receive an ACK from the SoftAP for the "Null-Data" packet and cannot go to sleep. Note that PM_MAX is the

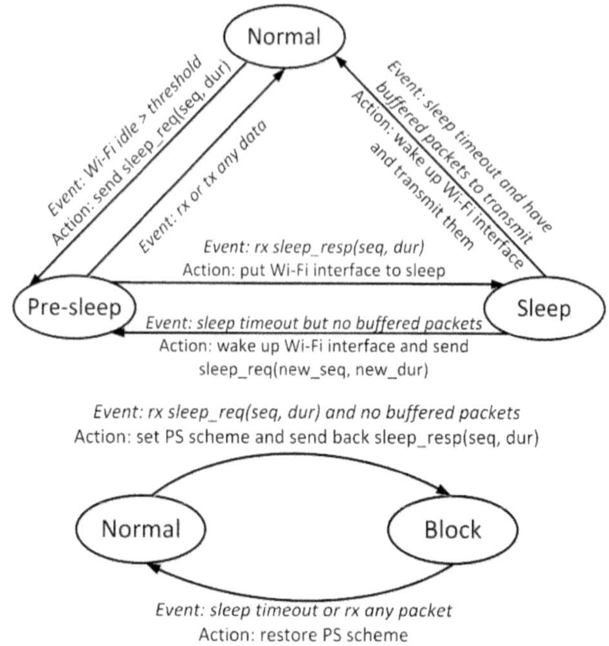

Figure 6: State machine for a SoftAP (top) and a client (bottom).

default scheme defined in PSM of IEEE 802.11 standard and is supported by all the client devices. Thus, this design does not depend on any specific devices.

In the *Block* state, the Wi-Fi interface of the client is in power saving mode and the client knows that the SoftAP is sleeping. Thus, it blocks all the packet transmissions by buffering all the packets from applications. If the sleep schedule times out or the client receives a data packet from the SoftAP, it restores the previous power saving scheme (e.g., back to PM_FAST) and moves back to the *Normal* state. At this time, both the SoftAP and the client are active and can communicate as normal (receiving or transmitting any data packet will cause the SoftAP moves to *Normal* state from *Pre-sleep* state). Otherwise, the SoftAP will send out a new sleep request, and both the SoftAP and the client can enter the *Sleep* or *Block* state again.

3.2 Synchronization

One advantage of the sleep request-response protocol is that it does not require tight time synchronization between the SoftAP and the client. If the SoftAP and the client can synchronize their time perfectly, they can coordinate their sleep scheduling to avoid packet loss without transmitting any extra sleep request and response packets. However, this is hard to achieve in practice. While very fine-grained hardware timestamps (e.g., at microsecond granularity) already exists at link layer, such timestamps are segregated inside firmware and are not available to the Wi-Fi driver and applications. It is possible to do time synchronization by explicitly exchanging packets with timing information between the SoftAP and the client. Such time synchronization must be done periodically due to clock drift, which increas-

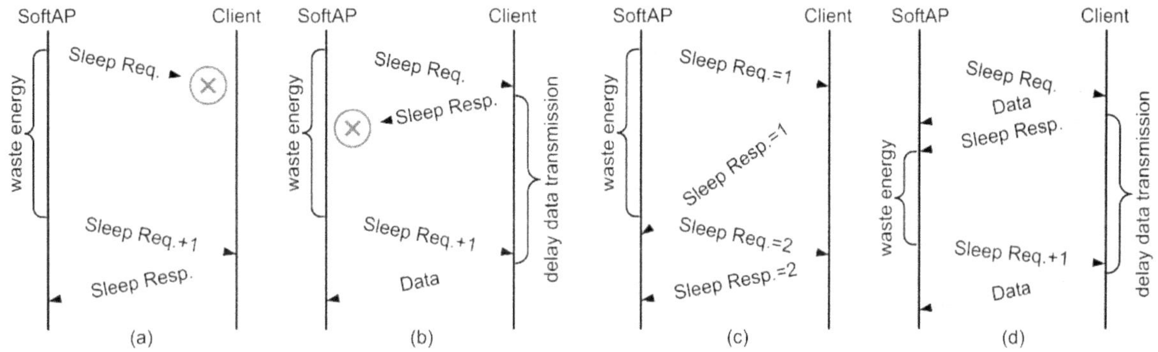

Figure 7: Abnormal cases: (a) a sleep request is lost; (b) a sleep response is lost; (c) a sleep response is delayed; (d) a data packet is delayed.

es power consumption. Due to these considerations, we intentionally avoided the time synchronization approach.

Interestingly, the proposed sleep protocol can achieve loose synchronization between a SoftAP and a client, with a desirable property: *the client will never conclude that the SoftAP is awake while it is sleeping.* Therefore, our approach will not lead to packet loss that would arise from wrong attempts of sending packets while the SoftAP is actually in sleep mode. In normal case, this is obvious because the SoftAP will sleep only after it receives a sleep response from the client. However, due to the uncertainty and complexity of wireless communication, the sleep protocol may not work as smoothly as expected. Below we analyze several possible abnormal cases and their consequences, as illustrated in Figure 7.

Packet loss. First, sleep request or response packets may be lost during their transmission. For example, a sleep request may be lost. In this case, the SoftAP will stay in active. If later on the client or the SoftAP has data to transmit, they start to communicate as normal. Or the SoftAP will send out a new sleep request after the network remains idle for a period longer than the pre-defined idle threshold, as shown in Figure 7a. The worst effect of losing a sleep request is that the SoftAP would waste some energy for staying in unnecessary active state between two successive sleep requests. Similarly, if a sleep response is lost, the SoftAP also has to stay in active until the next sleep request. However, in this case, as the client has concluded that the SoftAP is in sleep, it will stay in the *Block* state and start to buffer packets. Thus, it may further incur extra delay up to the idle threshold to the client, as shown in Figure 7b.

Packet out-of-order. Second, packet transmission may be delayed due to the hardware queuing and wireless contention. As a result, there is a slight chance that packets may not arrive at their destinations in the expected order. For example, Figure 7c shows the case that a sleep response is delayed by the client's hardware. The SoftAP receives sleep response 1 after sleep request 2 is sent out. In this case, the SoftAP just ignores sleep response 1 but it has to stay in active between the two sleep requests. Figure 7d shows a more complex case. The client has already passed a packet to the firmware and the packet is waiting for transmission in the hardware queue. At this moment the client receives a sleep request from the SoftAP. As the client does not have more data to transmit, it replies a sleep response. However, once the SoftAP receives the data packet, it resets its idle timer, stays in active and ignores the sleep response. Consequently, the SoftAP and the client are out of sync: the SoftAP stays in active, wasting energy, but the client assumes the SoftAP is in sleep and delays its packets transmission.

Based on the above analysis, we can see that those abnormal cases would at most cause some overhead in energy and transmission latency, but they would not break the desired synchronization property of our sleep protocol. A client will never try to send packets when the SoftAP is actually in sleep. The SoftAP and the client may run out of sync

```
// Parameters:
thresh, min, max, init, step, cur, pre, thresh_l, long;

ACTIVE:
measure the Wi-Fi network idle time;

SHORT_SLEEP:
if (Wi-Fi network idle time > thresh)
    first = true;
    sleep for a time period of init;
while (1)
    cur = first? init : (cur + step); first = false;
    if receive or transmit a packet
        if (cur < init)
            init = max(init - step, min);
        pre = cur; goto ACTIVE;
    if ((cur > init + step) && (pre > init + step))
        init = min(init + step, max);
    if (cur >= thresh_l)
        pre = cur; goto LONG_SLEEP;
    sleep for a time period of step;

LONG_SLEEP:
while (1)
    sleep for a time period of long;
    if receive or transmit a packet;
        goto ACTIVE;
```

Figure 8: The two-stage adaptive sleep algorithm.

temporally, but will always resume sync after the subsequent sync response. This demonstrates the robustness of our sleep protocol.

3.3 Adaptive Sleeping

We design an adaptive sleep scheduling algorithm for a SoftAP to adapt to the traffic pattern and also the 3G network property. Our adaptive sleep algorithm consists of two stages, namely a *short sleep* stage and a *long sleep* stage, that are designed to exploit the two distinctive phases (i.e., the content fetching phase and the content consuming phase) of interactive applications, respectively.

Sleep algorithm. Figure 8 shows how the two-stage adaptive sleep algorithm works. The basic idea is to probe the optimal sleep interval such that the SoftAP can wake up shortly before a packet arrives. Starting with an initial conservative sleep interval, the sleep interval is gradually increased, at a conservative pace, until a packet has arrived during the last sleeping. Then the initial sleep interval is updated according to the probing history. All the successive sleep slots are collectively called a sleep cycle.

More concretely, when the Wi-Fi interface remains idle for a time period of *thresh*, the SoftAP will enter the short sleep stage. It first sleeps for a time period of *init* which equals to *min* initially. When the SoftAP wakes up, it either goes back to the ACTIVE mode if there are pending outgoing or incoming packets, or continues to sleep for a fixed interval of *step*. Depending on the real packet arrival pattern, the length of the sleep cycle may become longer and longer between two subsequent wakeups. The sleep period can be expressed as "*init* + N * *step*" where "N" is the number of continuous sleep slots after the first sleep slot of *init*.

As waking the Wi-Fi hardware up introduces certain energy overhead [25], it is desirable to reduce the number of unnecessary wakeups. This calls for a good *init* value that can let the SoftAP sleep as long as possible while still being able to wake up in time, i.e., to avoid or shorten the probing process. We determine the *init* value by exploiting the sleep history. We use parameters *cur* and *pre* to track the gross length of all successful sleep slots in the current sleep cycle and that in the previous sleep cycle, respectively. That is, we have *cur* equals to "*init* + (N-1) * *step*" because a sleep cycle is always ended up by a false sleep slot during which a packet has arrived and been buffered. Parameter *pre* is simply a running record of the previous *cur*. Based on the values of *cur* and *pre*, we adjust the value of *init* with a simple algorithm INIT_UPDATE as follows: if both *cur* and *pre* are greater than current *init* plus *step*, we increase *init* by *step* for the next sleep cycle. If *cur* is less than current *init* minus *step*, we decrease *init* by *step*. To avoid excessive latency that may be caused by an overly greedy *init* value, we cap it to the value of *max*. In SHORT_SLEEP stage, if the SoftAP has continuously been in sleep for a time period of *thresh_l*, it will go to LONG_SLEEP stage. In LONG_SLEEP stage, the SoftAP simply sleeps for a time period of *long* periodically until it has to quit from sleep to communicate with the client.

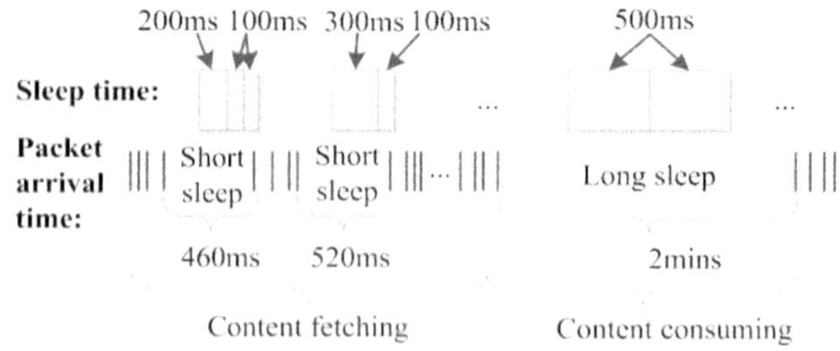

Figure 9: Short sleep and long sleep example.

Example. Figure 9 illustrates the algorithm with a concrete example. Some details such as the time for waking up the Wi-Fi interface between continuous sleep slots and the time taken to get permission from clients after each wake-up are omitted for sake of easier reading. Assume current value of *init* is 200ms, the value of *step* is 100ms and the value of *thresh* is 150ms. For the 460ms inter-packet interval in Figure 9, the SoftAP starts to sleep after 150ms. The SoftAP will first sleep for 200ms, followed by two 100ms sleep slots. Suppose then the value of *init* is qualified to increase to 300ms, for the 520ms inter-packet interval, the SoftAP will first sleep for 300ms followed by one more 100ms sleep slot. In the content consuming period, after sleeping for a time period of *thresh_l*, the SoftAP enters the long sleep stage and periodically sleeps for a time period of *long* (500ms).

The above algorithm is specially designed for the traffic patterns of typical applications in Wi-Fi tethering as shown in Section 2.2. The short sleep stage is designed for the SoftAP to sleep between the time when the client sends out a request to a remote server and the time when the first response packet from the remote server is received. That duration is roughly a RTT of the 3G connection (typically hundreds of milliseconds [18]). The purpose of *init* parameter is exactly to estimate the 3G's RTT in an elegant way, based on the length of last two sleep cycles. Note that our algorithm is conservative in the sense that it tries to reduce the energy consumption under minimal impairment to user experience, i.e., extra latency incurred. We decrease the value of *init* quickly, by considering only the length of the current sleep cycle, but increase the value of *init* slowly by considering the length of both the current and the previous sleep cycles. In addition, we use the parameter *thresh* to prevent the SoftAP from entering the short sleep stage during burst data transmission period (e.g., multiple response packets from a remote server for the same client request such as fetching a picture. In Section 4 we describe the parameter values used in our implementation.

In summary, our sleep algorithm automatically adapts to the traffic pattern of applications and achieves a good balance between power saving and network performance.

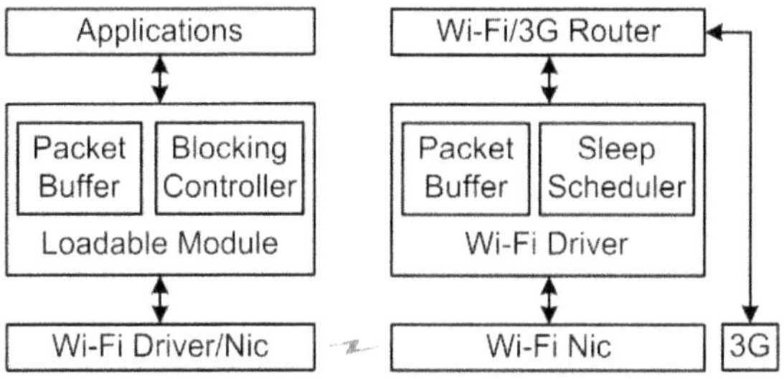

Figure 10: Implementation architecture of the client part (left) and the SoftAP part (right).

3.4 Supporting Multiple Clients

DozyAP can support multiple clients by repeatedly applying the sleep request-response protocol to each client. A client goes to sleep once it agrees to the AP's sleep request. A SoftAP can sleep only if it receives the sleep responses from all the clients. It is possible that some clients replied to a sleep request but other clients did not so that the SoftAP has to stay awake. However, this is not wrong because some clients may have data to send and the SoftAP must serve those clients. It is expected that the SoftAP sleeps less and consumes more power in the multi-client case. However, extending DozyAP to support multiple clients will not break the synchronization property of the sleep protocol: no client will send a packet when the SoftAP is in sleep. Note that we considered the possibility of broadcasting the sleep requests as it can obviously reduce the overhead of the sleep protocol. However, we do not take this approach for several reasons. First, broadcast packets are less reliable because they are transmitted without link layer retransmissions. In particular, the clients in PSM likely cannot receive them. Second, broadcast packets are transmitted at lowest data rate and thus take more air time than normal packets. Third, the sleep responses are still sent in unicast packets. Last but not the least, Wi-Fi tethering typically happens with very few simultaneous clients, often only one or two clients. Therefore, the improvement of using broadcast is expected to be very small in multi-client case, and for single client case, it is worse than using unicast.

Another minor issue with the multi-client case is the beacon. In our design a SoftAP does not send out beacons in the sleep mode. Thus, a new client cannot join the Wi-Fi network when the SoftAP is in sleep. However, the SoftAP sends out periodic beacons when it is active. Even in long sleep stage, it still wakes up periodically and can send out beacons. Consequently, a new client is still able to find the SoftAP but may experience slightly longer latency. As this only happens when a new client joins the network, we think it is acceptable.

4. IMPLEMENTATION

We have implemented a DozyAP prototype on Nexus One smartphones running Android 2.3.6 (the latest official version for Nexus One), with a Wi-Fi chipset of Broadcom BCM4329 802.11 a/b/g/n [2].

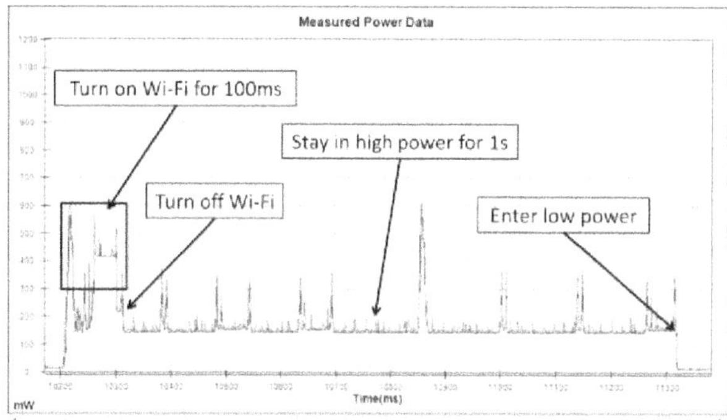

Figure 11: Power consumption of turning the Wi-Fi interface on/off in Wi-Fi tethering.

The overall architecture consists of two parts: the SoftAP part and the client part, as shown in Figure 10. The SoftAP part is directly modified from the open source Wi-Fi driver in which we embedded the sleep request-response protocol and the two-stage adaptive sleep algorithm. When the SoftAP is in sleep state, all the packets received from 3G interface are buffered. The client part is implemented as a loadable module where a packet buffer is implemented, together with a blocking controller to decide if and when application packets must be buffered. In our prototype we use a special Ethernet type of 0xfffff (a reserved value that should not be used in products) for the packets of the sleep request-response protocol. In real deployment, other approaches can be used to implement the sleep protocol, e.g., using dedicated IP packets rather than the special Ethernet type. We use a *netfilter* to intercept all the outgoing packets and to detect the packets of sleep requests and response. Implementing the client part as a loadable module does not require any modifications to the source code of the client OS. This makes it easy to deploy DozyAP on different types of client devices.

Putting a mobile SoftAP to sleep. One practical difficulty we met is how to put a mobile SoftAP to sleep. On smartphones (Nexus One, and other types of smartphones), most Wi-Fi MAC layer functionalities are implemented in the firmware running on the Wi-Fi chipset, not in the CPU-hosted Wi-Fi driver. When Wi-Fi tethering is enabled, the firmware of the Android One smartphone keeps the Wi-Fi hardware always on and in a high power state. There is no interface available to change the power states. All that we can do in the driver is to turn on/off the Wi-Fi interface. Consequently, in our implementation, we simply turn off and on the Wi-Fi interface when the SoftAP decides to sleep and wakeup, respectively. By modifying the source of the driver, we hide the fact that the Wi-Fi interface is turned off. Thus, applications and the OS can work as normal as if the Wi-Fi interface is always on.

Energy overhead of turning on/off the Wi-Fi. It costs some extra energy to turning on/off the Wi-Fi interface. We measured such energy overhead on a Nexus One smartphone and Figure 11 shows the measurement results. Initially, the Wi-Fi interface was off. Then we turned the Wi-Fi interface on for 100ms and turned it off again. We can observe two artifacts: first, when the Wi-Fi interface is turned on, the system jumps to a high power state of 600mW and stays in an average power state of 400mW which is higher than the normal power consumption of Wi-Fi tethering in idle case (270mW as shown in Figure 2). Second, when the off command is issued, the power consumption reduced immediately, but remains at a power level as high as 150mW for about one second before entering a very low power state of 10mW. This finding is similar to what the authors reported in [25]. They pointed out that when the Wi-Fi interface goes to sleep, it first enters a "light sleep" state and then enters a "deep sleep" state after some time. We cannot control this behavior but it significantly affects how much power we can save: the SoftAP never enters the "deep sleep" state because we have limited the maximum sleep time to 500ms. We want to point out that this is a *platform-specific limitation* caused by the restricted programmability over the Wi-Fi hardware. Our design itself does not impose any limitation. If the smartphone can enter the "deep sleep" state more quickly, our approach can save much more power.

Parameter values. We determine the parameter values in the adaptive sleep algorithm (refer to Figure 8) based on the findings in Section 2 and the measured power data. We set *thresh* to 150ms for triggering a SoftAP to enter sleep mode. With this setting, the SoftAP can handle over 80% of the total packets in its short awake time and over 60% cases it can successfully sleep for 100ms if the SoftAP decides to. Due to the energy overhead of turning on the Wi-Fi interface as shown in Figure 11, there is no obvious benefit if the SoftAP sleeps less than 100ms. Thus, we set *min* to 100ms. Considering the RTT of 3G and to limit the maximum extra latency, we set *max* to 500ms. The value of *init* varies between 100ms and 500ms. We set *thresh_l* to 3s for switching to the long sleep stage where the SoftAP periodically sleeps for 500ms (i.e., *long* equals to 500ms).

5. EVALUATION

We evaluate the performance of DozyAP by answering the following questions. 1) How much power can DozyAP save for a mobile soft AP in various applications? 2) What is the impact of DozyAP on client side power consumption? 3) How much extra latency does DozyAP introduce? 4) How much power can be saved in multi-client case?

5.1 Experiment Setup

Hardware devices. We use one Nexus One smartphone as a SoftAP with a China Unicom 3G connection (WCDMA), and another Nexus One smartphone as a client to run applications. Both smartphones run Android 2.3.6. We use a Monsoon Power Monitor [4] to measure the power consumption. We repeat every experiment for five times and present the average results.

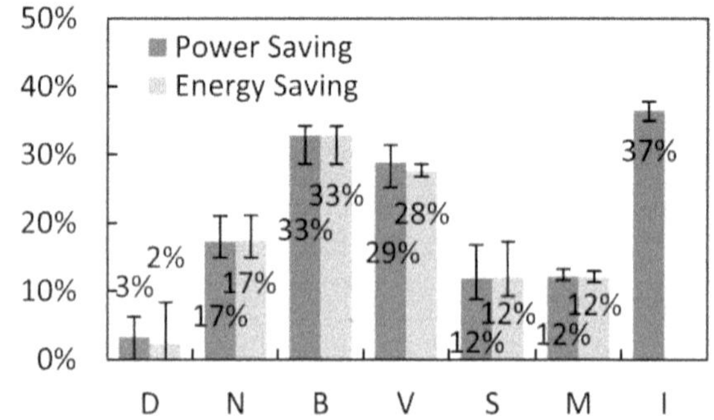

Figure 12: Power saving and energy saving of the SoftAP in idle (I), busy download (D) and the five applications of news reading (N), book reading (B), video streaming (V), search (S) and map (M).

Applications and methodology. We use five of the applications described in Section 2, including news reading, book reading, video streaming, search, and map. To make the experiments repeatable, except video streaming, we analyze the captured trace of the applications to find out all the HTTP requests contained in the traces. Then we write a test program in Java to send out those HTTP requests with the exact same order and timing as the traces. The program uses the WebView class in the WebKit package [1]. Thus, we can easily repeat every experiment. For video streaming, we manually play the same video clip.

Traces. To evaluate DozyAP with more diverse and realistic traffic patterns, we asked the authors of MoodSense [21] for the traces collected from real users. In MoodSense, the authors conducted a two-month field study with 25 iPhone users and collected their network traffic everyday using tcpdump [7]. We select the traces of the top eight most active 3G users. For each of them, we further select the trace of the day when the user generated the largest 3G traffic volume. We use the eight-day traces to evaluate the performance of DozyAP.

5.2 Power Consumption

Average power. We first measure the power consumption of a mobile SoftAP with DozyAP and without DozyAP. Besides the five applications, we also measure two extreme cases: *idle* case and busy *download* case. In the idle case, we measure the power consumption of the SoftAP with one client associated but without any network traffic. In the busy download case, we measure the power consumption of downloading a 1MB file from a Web server. The dark bars in Figure 12 show the average power saving of DozyAP. Without explicit mention, the error bars depict the minimum and maximum values in all the experiments. We can see that DozyAP can reduce the average power by 12.2% to 32.8% for the five applications. In the idle case, it can save power by 36.5%. Even for the busy download case, the average power can be reduced by 3.3%. It is worth noting that the power saving percentage is calculated in *the total power consumption of the whole system*, including the power con-

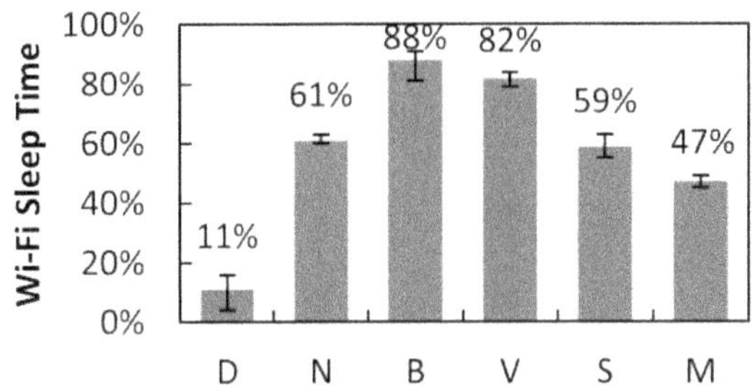

Figure 13: Wi-Fi interface sleep time of the SoftAP in busy download and the five applications.

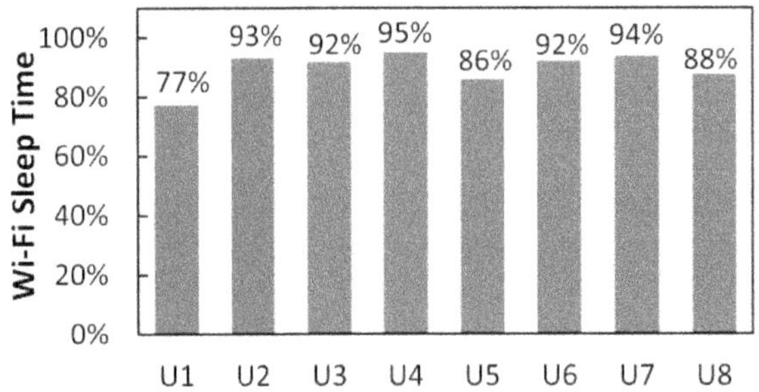

Figure 14: Wi-Fi interface sleep time calculated based on the traces of eight real users.

sumed by CPU and 3G. 3G consumes significant power when transmitting and receiving data. If we only consider the power consumption of Wi-Fi, the power saving percentage will be even higher in busy download and the five applications.

Total energy. As DozyAP buffers packets and delays their transmission, it may lead to longer application time comparing with the case without DozyAP. Thus, we measure the total energy for the busy download case and the five applications. Total energy does not make sense for the idle case. The light bars in Figure 12 show the results. We can see that DozyAP never increase the total energy. Instead, it can save the total energy by 12.2% to 32.9% for the five applications, which is almost the same as the result of average power. As we will show in Section 5.3, DozyAP introduces very little network latency which has negligible impact on total energy. Even in the busy download case, DozyAP can save the total energy by 2.3%.

Wi-Fi interface sleep time. As we point out in Section 4, with the current commercially available smartphones, forcing the SoftAP to go to sleep or wakeup can be only achieved by turning off/on the Wi-Fi interface. That results in significant overhead (see Figure 11). If we can have more control on the power states of the Wi-Fi hardware (e.g., if we can directly modify the firmware or if we have a MadWifi [9] style driver which implements most MAC layer functions in driver rather than in firmware), DozyAP should be able to save significantly more power. Therefore, we measure how much time DozyAP can put the Wi-Fi interface of a mobile SoftAP to sleep. Figure 13 shows the results. We can see that the Wi-Fi interface of a SoftAP can

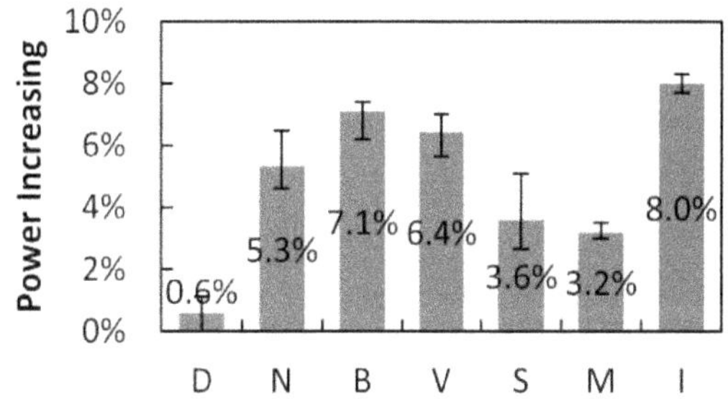

Figure 15: Power increasing of the Nexus One client in idle, busy download and the five applications.

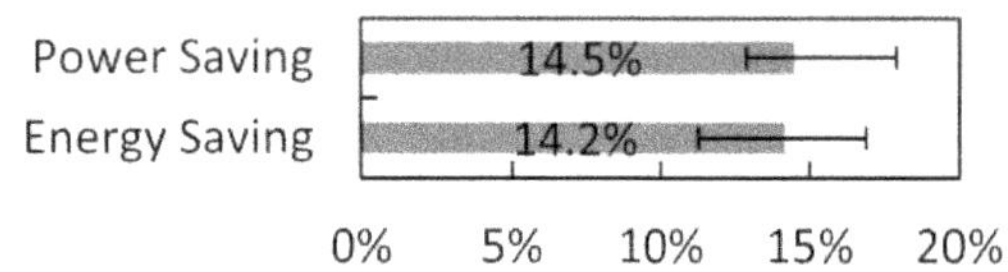

Figure 16: Power saving and energy saving of the SoftAP with two applications running on single client.

stay in sleep mode for 47%-88% of the total time in the five applications. Even in the busy download case, we can turn the Wi-Fi interface to sleep for 11% of the total time. These results demonstrate the potential of DozyAP to significantly reduce the power consumption of Wi-Fi tethering. Given proper control over the Wi-Fi hardware, more energy is expected to be saved from sleeping.

We also evaluate the Wi-Fi interface sleep time with the real traces of the eight users in MoodSense [21]. To do it, we wrote a program to analyze the packet inter-arrival time of the traces and calculate the Wi-Fi interface sleep time as if these traces have happened in Wi-Fi tethering. To make the calculation reasonable, we ignore all the inter-packet arrival intervals larger than 5 minutes. That is, for any intervals larger than 5 minutes, we treat it as if the user stopped using the phone and turned Wi-Fi tethering off. This treatment is conservative because a user may spend more than 5 minutes to read a long news article or Wi-Fi tethering might not be turned off even the user stopped using the phone for 5 minutes. Figure 14 shows the calculated results. We can see that DozyAP is able to allow the Wi-Fi interface of a SoftAP stay in sleep mode for 77%-95% of the time, for the mixed, multi-application real user traffic. The numbers in Figure 14 are higher than the ones in Figure 13. The reason is because the experiments in Figure 13 focused on single application usage only. In practice users may use multiple applications one by one. Switching from one application to another leads to more network idle time.

Power consumption of a client. We also measure the power consumption of the Nexus One client in the idle case, busy download and the five applications. Figure 15 shows the results. We can see that DozyAP increases the power consumption of the client by less than 7.1% for the five applications. The reason is because that the client needs to wake up to receive the sleep requests from the SoftAP and

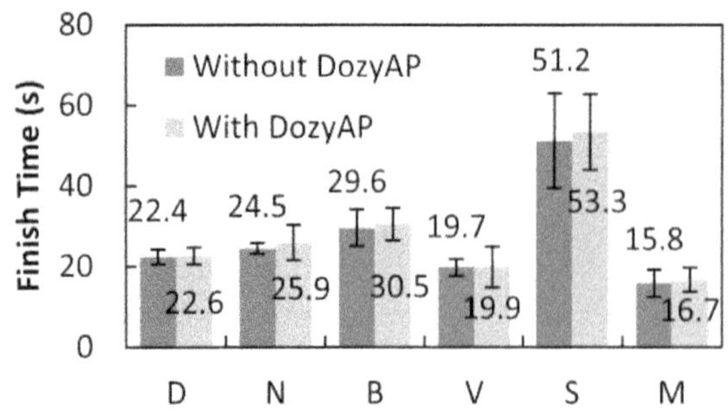

Figure 17: Finish time of busy download and the five applications.

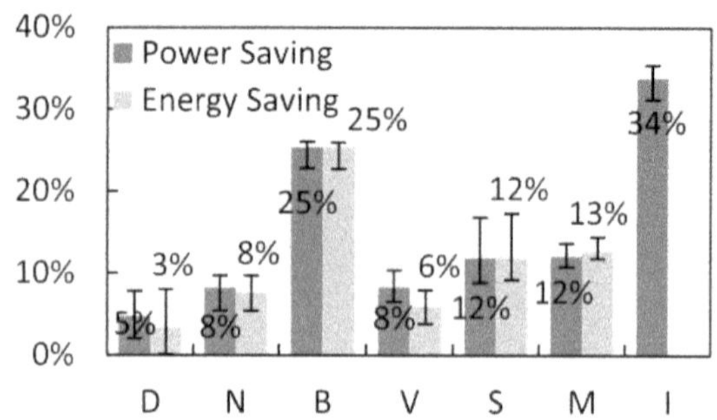

Figure 18: Power saving and energy saving of the SoftAP with two clients.

send back the sleep responses when the network is idle. Thus, the idle case introduces the highest overhead but it is still only 8%. Comparing to the large power saving of the SoftAP, this small overhead is acceptable.

Multiple applications on single client. In some cases, multiple applications may run on a single client simultaneously. We evaluate DozyAP in a typical scenario where a user is reading news in the foreground meanwhile listening online music in the background. To do it, we first started *Douban FM* (which is popular app in China like *Last.fm*). Once the music began to load, we started the news reading program (the same as before) immediately. Figure 16 shows the power saving and energy saving of the SoftAP. The average result over ten experiments is 14.5% and 14.2% for power saving and energy saving, respectively.

5.3 Latency

DozyAP incurs extra network latency because it delays packet transmissions when a SoftAP is in sleep mode. If the extra latency is user perceivable, it may impair user experience. As all the five applications are about fetching remote Web content, users care about the *page loading time* which is the period from the time when a user sends out a webpage request to the time when the webpage is fetched and rendered by the browser. The page loading time metric is widely used to evaluate the performance of browsers and Web servers. We evaluate the *finish time* of loading content which is the sum of the page load time of all the webpage requests in an application. The WebView object used in our test program is able to tell when a webpage is loaded. In the experiments, we sent out all the webpage requests of an application one by one *without any time interval* and calculated the total finish time.

Figure 17 shows the average result and the variance in busy download and the five applications. We can see that DozyAP introduces very small extra network latency, ranging from 0.9% to 5.1%. Such small extra latency is hardly perceivable by users because of two reasons. First, as the 3G network has limited throughput and large RTT, it takes several hundred milliseconds to even seconds to load a webpage. Second, the time variance of the page loading time is pretty large, up to several seconds. That is, even without DozyAP, users already experience long page loading time with large variance. Therefore, the small latency increase of less than 5.1% is very hard to detect.

5.4 Multiple Clients

We evaluate the performance of DozyAP when two clients are associated with a SoftAP. One client is a Nexus One smartphone and the other is a Kindle Fire tablet. Each client ran the same programs simultaneously. Figure 18 shows the average energy saving and power saving in busy download, the five applications and the idle case. As expected, the most power and energy savings are lower than the ones in single client scenario. However, the saving in download case does not drop as much as other applications. It is because in the download case the 3G connection was always fully utilized. No matter a client or multiple clients were downloading, the 3G connection was always the bottleneck so that the sleep time for the SoftAP would not drop much. Another finding is that the saving for video streaming has a significant drop (from about 28% to less than 10%). The reason is that two clients were competing in streaming video so that both of them needed more time to finish. Thus, the SoftAP had less opportunity to sleep.

It is worth noting that, in the rare case where many clients share the same SoftAP, the energy gain of DozyAP becomes less and the overhead on clients may increase significantly. DozyAP can disable and enable the sleep protocol on the fly, depending on the number of clients. In our implementation, we disable the sleep protocol if the SoftAP has more than two clients and enable the sleep protocol if there are only one or two clients.

6. DISCUSSIONS AND FUTURE WORK

DozyAP requires installing a loadable module on a client, which restricts the number of clients it can support. It may be hard or impossible to upgrade the software of dumb Wi-Fi client devices, e.g., music players and e-readers. However, to access the Internet, most people use "smart" devices including smartphones, tablets and laptops. All these devices are programmable and upgradable. The loadable module in our implementation can immediately work on all the Android-based smartphones and tablets. It can also be easily ported to the other devices running a Linux-style OS kernel including iOS. For Windows and Windows Phone based devices, a similar approach can be used as well. One can implement the client part in a loadable Network Driver Interface Specification (NDIS) [6] driver without modifying the source code of the OS. Therefore, DozyAP is able to support a wide range of diverse Wi-Fi client devices.

DozyAP takes advantage of the speed discrepancy between cellular and Wi-Fi. One may argue that such an advantage will not exist when 4G is deployed. However, the speed of Wi-Fi increases fast too. With 11n and 11ac, there is still a big gap between cellular and Wi-Fi. In addition, our solution benefits not only from such a speed discrepancy, but also from the long content consuming time of users.

Our implementation uses fixed parameter values derived from the measurement results, which can be improved. For example, one may use a dynamic approach to tune the parameters to better adapt to the network conditions. Even though we use fixed values, we take a conservative way, e.g., the sleep time starts from a small value of 100ms. As shown in Figure 8, the tuning procedure of parameter *init* is also conservative.

More power can be saved through transmission power adaptation. The built-in Wi-Fi tethering on existing smartphones always uses the highest transmission power. It wastes energy because a SoftAP is often close to its clients in Wi-Fi tethering. We plan to design a scheme to automatically adjust the transmission power based on the network conditions (e.g., RSSI and packet loss).

We also plan to further take advantage of the bandwidth discrepancy between 3G and Wi-Fi to *create* more opportunities for a SoftAP to sleep. The basic idea is shaping the traffic between 3G and Wi-Fi. For downlink traffic, the SoftAP can buffer the packets received from 3G and send them to the client over Wi-Fi in batch. For uplink traffic, if the 3G connection is congested, the SoftAP can ask the client to stop sending more data. Thus, the Wi-Fi interface of both the SoftAP and the client can sleep longer.

7. RELATED WORK

Wi-Fi power saving. There has been a lot of research effort devoted to power saving in Wi-Fi [11-13, 15, 17, 19, 23, 25-28], focusing on improving the existing PSM in general or targeting at specific applications or usage scenarios. To name some recent work, Catnap [15] exploits the bandwidth discrepancy between Wi-Fi and broadband to save energy for mobile devices. NAPman [28] employs an energy-aware scheduling algorithm to reduce energy consumption by eliminating unnecessary retransmissions. SleepWell [25] coordinates the activity circles of multiple APs to allow client devices to sleep longer. All these solutions are for Wi-Fi clients only. DozyAP is complementary, focusing on the power efficiency of APs. Putting an AP to sleep is more challenging than putting a client to sleep because client devices expect that their AP is always on. To avoid packet loss, a SoftAP in DozyAP must coordinate its sleep schedule with its clients, which is different from existing work.

There is little work on power saving of APs. In [20, 32], the authors propose to extend the IEEE 802.11 standard to support power saving access points for multi-hop solar/battery powered applications. Without building any real systems, they focus on protocol analysis and simulation, assuming Network Allocation Vector (NAV) can be used. Our work focuses on system design and implementation. We build real systems on commercial smartphones and do evaluation with real experiments. In addition, the NAV-based approach cannot work on existing smartphones because NAV is only visible in firmware. Cool-Tether [29] considers an alternative way to address the mobile hotspot problem that involves reversing the role of the phone and the client. However, it significantly increases the power consumption of the client and does not support multiple clients.

Traffic-driven design. Adapting to traffic load for better sleeping is not a new idea [11, 27]. Traffic patterns in different applications and scenarios have also been studied in some papers and the similar observations are identified (e.g., the large portion of network idle time) [19, 23]. DozyAP builds on top of the basic techniques and applies them to Wi-Fi tethering scenario. Furthermore, DozyAP can be improved by leveraging existing literature, e.g., by traffic shaping [13, 15, 26] and sleeping in short intervals [23].

Sleep scheduling. Sleep/wake scheduling has been extensively studied in Bluetooth domain, e.g., [16, 22] and sensor network domain, e.g., [14, 24, 31]. However, those approaches usually focus on MAC layer design, resulting in a new MAC protocol, and often require time synchronization. DozyAP employs a simple application-level protocol to coordinate the sleep schedule of a SoftAP with its client, without requiring time synchronization or any modifications on existing IEEE 802.11 protocol. Thus, DozyAP is easy to deploy on existing smartphones.

Dedicated Wi-Fi tethering devices. MiFi [3] is a dedicated mobile Wi-Fi hotspot device. However, such a device also stays in a high power state even without any ongoing traffic. We measured a Huawei E5830 MiFi device and found the average power consumption was as high as 420mw in idle case. We believe MiFi devices can benefit from DozyAP design if they are programmable.

In addition, MyWi Ondemand [5] makes Wi-Fi tethering easy to use for an iPhone and an iPad paired over Bluetooth. When a user leaves a Wi-Fi network and uses her iPad, Wi-Fi tethering can be automatically enabled between her iPad and her iPhone. MyWi Ondemand provides a convenient way to decide when to enable and disable Wi-Fi tethering but does not save power when Wi-Fi tethering is enabled.

8. CONCLUSIONS

In this paper we have studied the power efficiency of Wi-Fi tethering. We show that Wi-Fi tethering on existing smartphones is power hungry and wastes energy unnecessarily, but there are many opportunities to save power by putting a mobile SoftAP to sleep. We propose DozyAP system to improve the power efficiency of Wi-Fi tethering. DozyAP employs a lightweight yet reliable sleep request-response protocol for a mobile SoftAP to coordinate its sleep schedule with its clients without requiring tight time synchronization. Based on our findings on the traffic patterns of typical applications used in Wi-Fi tethering, we design a two-stage adaptive sleep algorithm to allow a mobile SoftAP to automatically adapt to the on-going traffic load for the best power saving. We have implemented DozyAP system on commercial smartphones. Experimental results demonstrate that DozyAP is able to significantly reduce the power consumption of Wi-Fi tethering without impairing the user experience.

ACKNOWLEDGEMENTS

We sincerely thank our shepherd, Dr. Kameswari Chebrolu, as well as anonymous reviewers, whose comments and feedback helped improve this paper. We also thank Dr. Chiu C. Tan (Temple University), Ahmad Rahmati (Rice University), and our colleagues at Wireless and Networking Group, Microsoft Research Asia for fruitful discussions and valuable suggestions. The W&M team was supported in part by NSF grants CNS-1117412 and CAREER Award CNS-0747108.

REFERENCES

[1] Android WebKit package and WebView class, http://developer.android.com/reference/android/webkit/package-summary.html

[2] Broadcom, BCM4329. http://www. broadcom.com.

[3] MiFi. http://en.wikipedia.org/wiki/MiFi.

[4] Monsoon Power Monitor. http://www.msoon.com/LabEquipment/PowerMonitor/.

[5] MyWi and MyWi Ondemand. http://intelliborn.com/mywi.html.

[6] Network Driver Interface Specification (NDIS), http://msdn.microsoft.com/en-us/library/ff559102.aspx.

[7] Tcpdump. http://www.tcpdump.org/.

[8] The Android Market, https://market.android.com/.

[9] The MadWifi project. http://madwifi-project.org/.

[10] Wireless LAN Medium Access Control (MAC) and Physical Layer (PHY) specifications. *IEEE Std 802.11*, 2007.

[11] M. Anand, E. Nightingale, and J. Flinn. Self-Tuning Wireless Network Power Management. In *MobiCom*, 2003.

[12] D. Bertozzi, L. Benini, and B. Ricco. Power Aware Network Interface Management for Streaming Multimedia. In *IEEE WCNC*, 2002.

[13] S. Chandra and A. Vahdat. Application-specific network management for energy-aware streaming of popular multimedia formats. In *USENIX ATC*, 2002.

[14] T. Dam and K. Langendoen. An Adaptive Energy-Efficient MAC Protocol for Wireless Sensor Networks. In *SenSys*, 2003.

[15] F. Dogar, P. Steenkiste and K. Papagiannaki. Catnap: Exploit High Bandwidth Wireless Interfaces to Save Energy for Mobile Devices. In *Mobisys*, 2010.

[16] S. Garg, M. Kalia and R. Shorey. MAC Scheduling Policies for Power Optimization in Bluetooth: A Master Driven TDD Wireless System. In *IEEE VTC*, 2000.

[17] Y. He and R. Yuan. A Novel Scheduled Power Saving Mechanism for 802.11 Wireless LANs. *IEEE Transactions on Mobile Computing*, 8(10):1368–1383, 2009.

[18] J. Huang, Q. Xu, B. Tiwana, Z. M. Mao, M. Zhang and P. Bahl. Anatomizing Application Performance Differences on Smartphones. In *MobiSys*, 2010.

[19] R. Krashinsky and H. Balakrishnan. Minimizing energy for wireless web access with bounded slowdown. In *MobiCom*, 2002.

[20] Y. Li, T. D. Todd and D. Zhao. Access Point Power Saving in Solar/Battery Powered IEEE 802.11 ESS Mesh Networks. In *IEEE QShine*, 2005.

[21] R. LiKamWa, Y. Liu, N. D. Lane and L. Zhong. Can Your Smartphone Infer Your Mood?. In *PhoneSense* workshop, 2011.

[22] T. Lin and Y. Tseng. An Adaptive Sniff Scheduling Scheme for Power Saving in Bluetooth. In *IEEE Wireless Communications*, 9(6):92-103, 2002.

[23] J. Liu and L. Zhong. Micro Power Management of Active 802.11 Interfaces. In *MobiSys*, 2008.

[24] G. Lu, N. Sadagopan and B. Krishnamachari. Delay Efficient Sleep Scheduling in Wireless Sensor Networks. In *Infocom*, 2005.

[25] J. Manweiler and R. R. Choudhury. Avoiding the Rush Hours: WiFi Energy Mangement via Traffic Isolation. In *Mobisys*, 2011.

[26] C. Poellabauer and K. Schwan. Energy-aware traffic shaping for wireless real-time applications. In *RTAS*, 2004.

[27] D. Qiao and K. Shin. Smart Power-Saving Mode for IEEE 802.11 Wireless LANs. In *Infocom*, 2005.

[28] E. Rozner, V. Navda, R. Ramjee, and S. Rayanchu. NAPman: Network-Assisted Power Management for WiFi Devices. In *MobiSys*, 2010.

[29] A. Sharma, V. Navda, R. Ramjee, V. N. Padmanabhan, and E. M. Belding. Cool-tether: energy efficient on-the-fly Wi-Fi hotspots using mobile phones. In *CoNEXT*, pages 109-120, 2009.

[30] H. Wirtz, R. Backhaus, R. Hummen and K. Wehrle. Establishing Mobile Ad-Hoc Networks in 802.11 Infrastructure Mode. In *WiNTECH* demo session, 2011.

[31] Y. Wu, S. Fahmy and N.B. Shroff. Optimal Sleep/Wake Scheduling for Time-Synchronized Sensor Networks with Qos Guarantees. In *IEEE/ACM Transactions on Networking*, 17(5):1508-1521, 2009.

[32] F. Zhang, T. D. Todd, D. Zhao and V. Kezys. Power Saving Access Points for IEEE 802.11 Wireless Network Infrastructure. *IEEE Transactions on Mobile Computing*, Vol. 5, No. 2, 2006.

CiAO/IP: A Highly Configurable Aspect-Oriented IP Stack

Christoph Borchert[a] Daniel Lohmann[b] Olaf Spinczyk[a]
christoph.borchert@tu-dortmund.de lohmann@cs.fau.de
olaf.spinczyk@tu-dortmund.de
[a]Technische Universität Dortmund
[b]Friedrich-Alexander-Universität Erlangen-Nürnberg

ABSTRACT

Internet protocols are constantly gaining relevance for the domain of mobile and embedded systems. However, building complex network protocol stacks for small resource-constrained devices is more than just porting a reference implementation. Due to the cost pressure in this area especially the memory footprint has to be minimized. Therefore, embedded TCP/IP implementations tend to be statically configurable with respect to the concrete application scenario. This paper describes our software engineering approach for building CiAO/IP – a tailorable TCP/IP stack for small embedded systems, which pushes the limits of static configurability while retaining source code maintainability. Our evaluation results show that CiAO/IP thereby outperforms both *lwIP* and *uIP* in terms of code size (up to 90% less than *uIP*), throughput (up to 20% higher than *lwIP*), energy consumption (at least 40% lower than *uIP*) and, most importantly, tailorability.

Categories and Subject Descriptors

D.4.7 [**Operating Systems**]: Organization and Design—*Real-time systems and embedded systems*; C.2.2 [**Computer-Communication Networks**]: Network Protocols—*TCP/IP*; D.2.2 [**Software Engineering**]: Design Tools and Techniques—*Modules and Interfaces*; D.3.3 [**Programming Languages**]: Language Constructs and Features

Keywords

AOP, Aspect-Oriented Programming, AspectC++, Operating Systems, Embedded Systems, TCP/IP, Internet Protocol, Network Protocol Stacks

1. INTRODUCTION

Communication has become an integral part of today's computer systems. Every personal computer is equipped with a built-in network interface and there is a clear trend towards smart (computerized) devices, which very often have the ability to interact with the Internet, such as modern TVs. Hence, small and, in the future, even smaller devices need to implement the TCP/IP protocol suite, which is the de-facto standard for data communication and by definition used in the Internet in order to achieve *interoperability*.

However, there is still a huge gap between vision and reality: Science and industry are thinking about *Service-Oriented Architectures (SOA) for embedded devices* [3], small devices being connected with the Internet using a lightweight middleware such as *Universal Plug'n'Play (UPnP)*[1] or its successor *Devices Profile for Web Services (DPWS)*[2]. Both are based on TCP/IP. On the other hand, the resource contraints of embedded microcontroller units often forbid to use the software-intensive Internet protocols, which in practice leads to hybrid systems and a "reinvention of the wheel".

Whether a full-featured TCP/IP stack is well suited for deeply embedded systems with tight resource constraints, has already been debated in great detail [11, 13]. The purpose of this paper is *not* to give the final answer. However, we can show *how* the TCP/IP protocol suite could be used in nearly any device, despite of the enormous diversity in hardware platforms and application requirements.

Today, many TCP/IP implementations are available. Most of them were designed as a component of a PC or server operating system – such as BSD, Linux, or Windows – and are quite resource intensive.

Common examples for TCP/IP stacks for resource-constrained embedded systems are *micro-IP (uIP)* and *lightweight-IP (lwIP)* [8]. uIP proved that a full TCP/IP implementation fits even into a small 8-bit microcontroller unit by leaving out all optional features, wheras lwIP provides better performance at the cost of higher memory consumption. It is remarkable that both, uIP and lwIP, stem from the same author and have been presented together in [8]. This underlines that a "one size fits all" TCP/IP implementation is not well suited for embedded systems. However, is a "two sizes fit all" solution so much better?

1.1 Contribution and Outline

In this paper we present the design approach of the CiAO/IP[3] stack and a comparison of CiAO/IP with *uIP* and *lwIP*. Our approach is to construct CiAO/IP as a highly configurable *Software Product Line* [26] that provides for very tailored stack implementations for embedded applications. In order to keep the highly configurable code maintainable, *Aspect-Oriented Software Development* [18] has been exercised not only for the implementation but already during the design phase.

In summary, our contribution is to show that *applying* aspect-oriented design and implementation techniques to a new domain – networked embedded systems – is promising from a code mainte-

[1] `http://www.upnp.org/`

[2] `http://docs.oasis-open.org/ws-dd/ns/dpws/2009/01/`

[3] The source code is available under the terms of the GNU GPL at: `http://ess.cs.tu-dortmund.de/Software/CiAO-IP`

MobiSys'12, June 25–29, 2012, Low Wood Bay, Lake District, UK.

nance perspective without incurring energy and performance overhead.

Moreover, the fine-grained configurability of CiAO/IP allows developers to tailor the IP stack at compile time to their actual usage scenario, so that memory footprint, performance, and energy consumption are even improved over the state-of-the-art. We regard that knowledge of *how to build* network stacks for small-footprint devices as a useful contribution that deserves to be communicated more broadly.

The outline of this paper is as follows:

- We show that configurability is an important requirement on system software for networked embedded systems, which is not sufficiently fulfilled by other state-of-the-art IP stacks in this domain (Section 2). Despite the tension between source code maintainability and fine-grained configurability, the paper shows that CiAO/IP achieves *both* objectives (Section 6).

- We apply our unique design methodology, which is based on several years of experience with the aspect-oriented operating system CiAO[4] [22], to the domain of the Internet Protocol stack. Starting from requirements with explicit variability we step-by-step derive an aspect-oriented software design. Reusable programming idioms help to turn the design into code (Section 3 and Section 4).

- We show how the methodology led to our CiAO/IP design and implementation. Thereby, we document the structure of CiAO/IP, which is the first aspect-oriented TCP/IP stack we are aware of.

- By comparing CiAO/IP with *uIP* and *lwIP* we prove the implementation's scalability (8 to 64-bit systems) and high configurability (in terms of functional features), which we could achieve without any drawbacks in terms of code size, performance, and energy consumption (Section 5).

2. CONFIGURABILITY OF IP STACKS

The development of an IP stack for small mobile and embedded devices is more than just porting a reference implementation.

2.1 Diversity of Platforms and Requirements

The first challenge is that two embedded systems rarely look the same. The microcontroller market is heterogeneous and offers many entirely different devices. Simple control tasks can be performed by small 8-bit CPUs, which are still dominating the market in terms of shipped units [34]. At the other end of the spectrum, complex signal-processing often requires high-performance 32-bit or even 64-bit devices. The memory sizes are as diverse as their layouts, but share the commonality that they define hard constraints that a software developer must fulfill. When it comes down to networking, the number of possible interfaces seems to be vast. Both wired and wireless technologies are available in many different kinds, such as Ethernet, wireless LAN, GSM, and IEEE 802.15.x.

System software that should support and cope with all this variability has to be exceptionally *flexible* and *portable*, and must be *minimalistic* in order to fit into the tiny memories of even the most miniaturized devices – often only a few KiB of RAM!

The second challenge is that the requirements on system software vary a lot in different embedded systems. Depending on the application scenario, the operating system has to support best-effort scheduling for concurrent tasks, while other scenarios require strict spacial and temporal isolation of each task in order to meet certain dependability properties. The requirements on network stacks are diverse as well: Multimedia applications, such as *Voice over IP*, call for low latencies and guaranteed quality of service. Video streaming, on the other hand, requires high throughput and is not tied to low latency constraints. Furthermore, systems differ in the protocols they use. Information exchange such as E-Mail requires a reliable transport protocol that confirms data reception, whereas best-effort data delivery is sufficient for sensor applications. In short: the functionality required from the network stack strongly depends on the concrete application.

The big challenge from a software engineer's point of view is to overcome this multi-dimensional diversity and to design software that fits into all those situations in an optimal way. Static (compile-time) *configurability* is the key to support a broad variety of hardware platforms and application requirements without sacrificing efficiency.

The CiAO/IP stack aims at a very high degree of configurability to tailor the stack for the chosen hardware platform and application scenario. The goal is to be able to leave everything off that is *not* needed – in order to achieve minimal resource consumption.

2.2 State of the Art

Table 1 gives an overview on the features provided by our implementation of CiAO/IP and those of uIP and lwIP[5] [8].

A configurable feature that can be omitted if not needed, is denoted by ✓. This kind of feature contributes to functional scalability. A fixed feature, which is always present and not removable by the end-user, hinders scalability and is denoted by $*$. Unimplemented features of each IP stack are shown as whitespace.

uIP has by far the most reduced feature set. It supports solely a single networking interface, manages a single buffer for exactly one packet and provides no support for an operating system. The variability of uIP is very limited, too, for instance the protocols ICMP and TCP are fixed[6]. Furthermore, uIP does not implement a sliding window for TCP, so that poor throughput can be expected. Thus, uIP is only suited for application scenarios on deeply embedded devices that require TCP connectivity .

lwIP provides rich functionality in terms of features. As shown in Table 1, lwIP implements almost every feature that we consider for this comparison. Nevertheless, the configurability of networking protocols is coarse-grained. For example, the feature TCP includes client *and* server functionality (often only one side is needed), a sliding window (again, only for *both* Tx and Rx) and a bunch of common optimization algorithms, such as Silly Window Syndrome (SWS) avoidance, Round-Trip Time estimation and congestion control. None of these optional protocol features is actually optional in the lwIP implementation – TCP can only be included and excluded at a whole. Because of this fixed set of rich functionality, which might be welcome on powerful devices, lwIP does not scale to devices with tighter resource constraints. The large number of fixed features sacrifices scalability.

The CiAO/IP implementation differs substantially from the preceding ones: In CiAO/IP *every* feature is optional and, thus, configurable by the application developer. We abstain from fixed features in favor of scalability. Thereby, the developer has full control and does not have to pay for features that are not needed. The overall functionality of CiAO/IP is comparable to lwIP, but configurability is much more fine-grained.

[4] CiAO is Aspect-Oriented

[5] versions 1.0 and 1.32 respectively

[6] Contiki 2.5 (`http://www.contiki-os.org/`) includes an updated version of uIP that provides configurability of TCP and ICMP by `#ifdef` directives.

Feature	uIP	lwIP	CiAO/IP
Multiple Networking Interfaces		*	✓
Checksum Offloading per Interface			✓
Multiple Connections (Concurrency)	*	*	✓
Buffers per Connection (Isolation)			✓
Operating System Support		✓	✓
IPv4	✓	✓	✓
IPv4 Tx	*	*	✓
IPv4 Rx	*	*	✓
IPv4 Fragment Reassembly	✓	✓	✓
IPv4 Fragmentation		✓	
IPv6	(✓)[a]	(✓)[a]	(✓)[b]
ARP	✓	✓	✓
ARP Reply	*	*	✓
ARP Request	*	*	✓
ARP Cache Timeout	*	*	✓
Static ARP Cache Entries			✓
ICMP	*	✓	✓
UDP	✓	✓	✓
UDP Tx	*	*	✓
UDP Rx	*	*	✓
UDP Checksumming	✓	✓	✓
TCP	*	✓	✓
Client (Connect)	✓	*	✓
Server (Listen)	*	*	✓
Sliding Window (Tx)		*	✓
Sliding Window (Rx)		*	✓
Avoid Silly Window Syndrome (Tx)		*	✓
Avoid Silly Window Syndrome (Rx)		*	✓
Round-Trip Time Estimation		*	✓
Congestion Control (Slow-Start)		*	✓
TCP Urgent Data	✓	*	
Limit Excessive Retransmissions	*	*	✓
MSS Option	*	*	✓
Timestamp Option		✓	

[a] provided by a different code base, no dual stack

[b] under development

Table 1: Features provided by uIP, lwIP and CiAO/IP

2.3 A Software Engineering Challenge

Traditionally, configurability in system software has been expressed by the use of the C preprocessor. It supports textual substitution by macro expansion and conditional compilation by means of `#ifdef` directives. The preprocessor transforms the source code before it is parsed by the C compiler. A typical idiom is to represent all configurable features as macros in common header file, which controls the source code adaptation process of the system.

Expressing software variability in that way has been heavily criticized as error-prone, unreadable and unmaintainable [29]. A recent study of Linux kernel code shows that `#ifdef`-based configuration led to hundreds of bugs [33]. There are two main problems:

1. Dependencies between features are represented on the source-code level: This leads to a mix of abstraction levels within the code, because for each particular feature the *high-level* requirements and dependencies become intermixed with *low-level* implementation details.

2. The implementation of features often "crosscuts" various modules (functions, files, etc.): For many configurable features, the implementation is "scattered" over many modules and "tangled" with the implementation of other features. The consequence is that configurable features lead to code that is bloated with lots of macros and `#ifdef` directives. The whole code base becomes hard to understand and maintain.

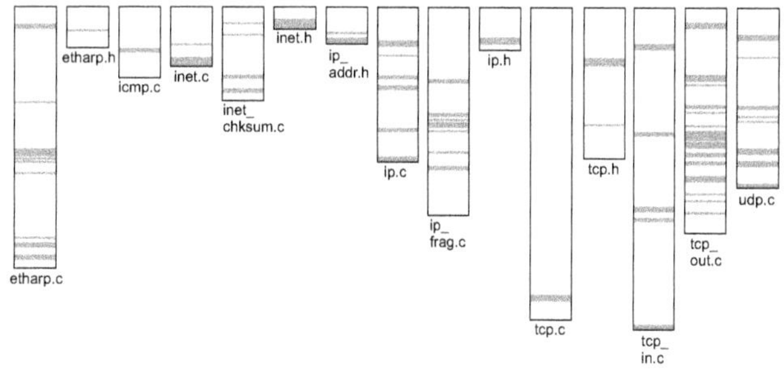

Figure 1: Scattered implementation of byte order conversion in lwIP. Each bar represents a source code file of lwIP; a horizontal line indicates the use of a conversion macro (`htons, ntohs,...`) in the respective source line.

Figure 1 illustrates these problems on the example of byte ordering in lwIP: In lwIP (as in uIP, Linux, and BSD) the conversion of 16-bit and 32-bit words to the network byte order is performed by dedicated preprocessor macros (`htons, ntohs, htonl, ntohl`). Access to network protocol headers is coupled to these macros – their use is compulsory to ensure data correctness. Hence, these macros appear 210 times in 15 files of the analyzed lwIP configuration. This implementation of the byte order conversion is tedious and error-prone, as a single forgotten macro call can already lead to data corruption in terms of swapped bytes.

The conclusion is that traditional preprocessor-based configuration does not scale: Software with lots of configurable features almost automatically leads to a maintenance nightmare [33]. This is a software engineering challenge. The problems can only be solved by using a configurability-oriented design approach combined with suitable programming language abstractions.

3. DESIGN APPROACH

The CiAO/IP stack is a generic platform for IP networking software. It supports the automatic *derivation* of custom IP implementations that fulfill application-specific constraints. An application developer just has to specify the relevant constraints by selecting software features in a graphical user interface. After that, he instantly gets a highly optimized IP implementation that meets his requirements. Even architectural properties, such as simultaneous support for different link layers, are configurable and do not cause any overhead if not needed.

CiAO/IP has been developed as an aspect-oriented software product line. A software product line is "a *family* of systems in a domain, rather than a single system" [15]. Due to systematic *reuse*, the software is better tested and, thus, less error-prone. From an economic perspective, the effort for maintaining a software product-line is much less than for several individual products.

One of the key ideas from the product-line engineering community is to strictly separate the so called *problem space* from the *solution space*. This addresses and solves the first problem identified in Section 2.3, namely the mix of abstraction levels in the source code. Typically a problem-space model of a product line describes the common and variable features and their dependencies in an abstract and problem-oriented manner. The problem-space model is independent from the product line's implementation. The solution space is the set of code artifacts that form the implementation.

Aspect-Oriented Programming is a paradigm that provides programming language means to modularize the implementation of crosscutting concerns [18]. This allows us to avoid the second problem from Section 2.3. As a result the highly configurable code of CiAO/IP is completely free of `#ifdef` directives.

In the following, we will first explain feature modeling, which is the methodology that we used to model the problem space of the CiAO/IP product line. Then we briefly introduce the concept of Aspect-Oriented Programming.

3.1 Feature Modeling

Feature modeling [15] is a mature technique for formalizing variability. The idea is to identify and document requirements in terms of *features*. A feature is a requirement of a specification or an informal demand of a stakeholder. Each feature can be either mandatory or optional – the latter case is where variability comes into play. Furthermore, features may have restrictions and dependencies among each other. A feature model encompasses all requirements, both mandatory and optional, that are imposed for the product being developed. The success of a fine-grained configurable software product-line relies mainly on high-quality feature models: It is crucial to have detailed information about the expected variability in advance in order to guide the subsequent design and implementation process.

We believe that the methodology of feature modeling is well suited for analyzing the complex requirements that IP networking software must fulfill. As a starting point for a feature model of the Internet TCP/IP protocol suite, we use the official specification published by the Internet Engineering Task Force (IETF). There are several RFC documents which define requirements for each protocol. In RFC1122 [6], the most important requirements for Internet hosts are summed up and categorized into three different kinds: *MUST*, *SHOULD*, and *MAY*. These capitalized verbs determine the significance of each particular requirement; in terms of a software product line *MUST* requirements map to mandatory features, whereas *SHOULD* and *MAY* requirements describe optional features.

Figure 2 shows a simplified feature diagram of the Internet Protocol. A feature diagram graphically represents the variability of a feature model by using a tree structure. Each node represents a feature that depends on all of its ancestors. A filled circle at the lower end of an edge indicates a mandatory feature (c.f. RFC *MUST*), whereas a nonfilled circle describes an optional one (c.f. RFC *SHOULD* or *MAY*). Cumulative features, of which a least one must be present, are grouped together by a filled arch at the upper ends of their edges. As shown in Figure 2, the Internet Protocol can be implemented as version 4 (IPv4) as well as 6 (IPv6). For each version, there is a distinction between sending and receiving, and each particular version requires at least one of them. Additionally, the Internet Protocol is extensible via *IPv4/IPv6 Options*, which are in fact optional features.

The development of feature models for each protocol of the TCP/IP protocol suite has been necessary for the sequel of this paper. A more complex example is given in Figure 3, which contains a still heavily simplified feature diagram of TCP, the most complex protocol we dealt with.

Compared to the RFC, the feature diagram is far more fine-grained. For instance, the feature of a *Sliding Window* is considered optional, because the minimal size of that window is not specified, which makes it effectively optional. A window size of exactly one segment is equivalent to a nonexisting Sliding Window.

Hence, feature modeling is a challenging discipline, which relies on the knowledge and farsightedness of experts in the analyzed domain in order to grasp all hidden features, which are not explicitly

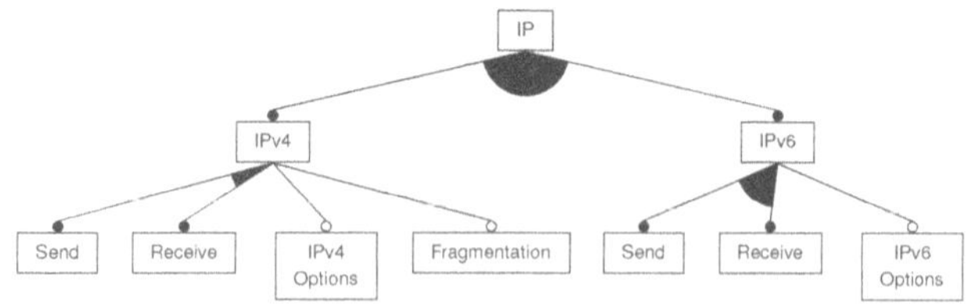

Figure 2: Feature Diagram of the Internet Protocol

stated in the specification, but relevant to stakeholders. On the other hand, no expertise in programming languages or specific paradigms is needed. This part can be left to solution space experts.

3.2 Aspect-Oriented Programming

Traditionally, the developer of an Internet Protocol stack decides upfront which features to implement, so that a particular system consists of a fixed subset of the RFC1122 requirements. Our idea is to postpone this decision as far as possible to preserve this freedom for the actual user of the protocol stack.

The design of a software architecture that adheres to the variability of feature models is a nontrivial task. We propose *Aspect-Oriented Programming (AOP)* [18] as a solution for this problem in order to avoid the already discussed pitfalls of the C preprocessor. AOP supports the decomposition of a system into orthogonal modules by featuring *implicit invocation*. In AOP, an *aspect* contains one or more pieces of *advice*, that intercept the control flow and extend the underlying types. A piece of advice targets several *join points*, which are either locations in the dynamic control flow or part of the static program structure, where arbitrary code can be invoked. Join points are described declaratively via *pointcut expressions* in a textual form.

AspectC++ [30], which has been developed by our group over the last 10 years, extends the C++ programming language by AOP mechanisms. It consists of an *aspect weaver* that processes ordinary C++ code and aspect code, and weaves the latter type of code into the C++ files at the relevant locations (i.e., join points). Thus, the advice code is *inlined* into existing C++ code and produces no overhead compared to an implementation by hand (see [23]). In constrast to C preprocessor directives, the AspectC++ language elements are fully integrated into the syntax and semantics of C++.

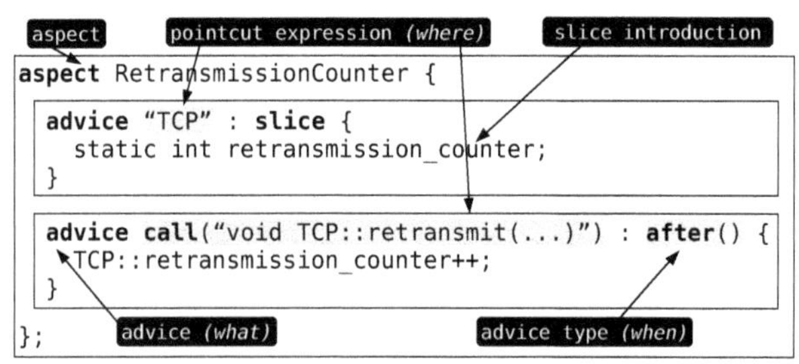

Figure 4: Syntax of AspectC++

Figure 4 outlines the essential syntactic elements of AspectC++ by taking the example of a retransmission counter for TCP. The given aspect implements the counter in a modular way by two pieces of advice. The first piece of advice extends the class `TCP` by an `int` member to store the counter's value. This way, the static program structure is modified by introducing additional class members, which is called a *slice introduction*. The second piece of advice targets the function `TCP::retransmit(...)` and implicitly

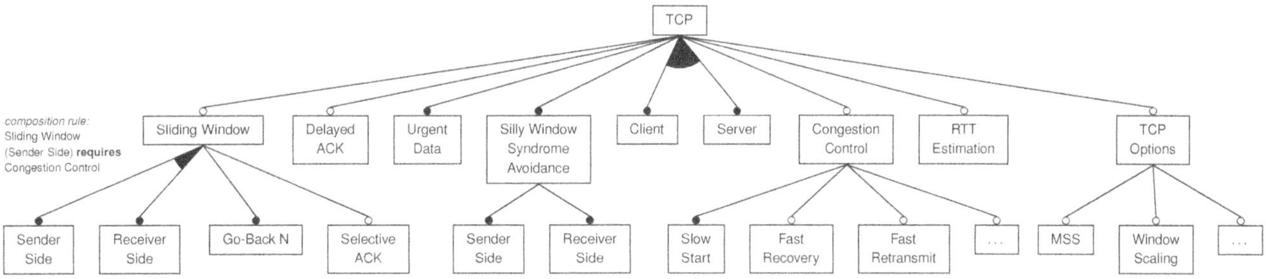

Figure 3: Simplified Feature Diagram of TCP

increases the counter *after* each *call* to that function. The ellipsis of the function's arguments in the pointcut expression ensures that in case of an overloaded function all overloads are matched regardless their arguments.

Furthermore, multiple pointcut expressions can be combined using the algebra of sets. Besides *after* advice, *before* and *around* (the latter replaces the original behavior of the join point) advice can be used to specify when advice code shall be invoked.

In the following section, we show that AOP provides a rich toolkit for the design and implementation of a highly modular networking stack. Even crosscutting features, which are conceptually scattered over many different software modules, can be developed as separate modules by using AOP. We provide a design method for deriving a mapping of features to aspects and ordinary C++ classes and answer the emerging question: Which features should be turned into aspects?

4. INSIDE THE IP STACK

The previous section outlined basic methods we use for the design and implementation of fine-grained configurable networking software. Feature modeling reveals all features that have to be considered during the following design process. Before the actual design of the software system, the interactions of all features and their impact on the system must be analyzed. The system designer needs to gather this information in order to design the fundamental software architecture. Section 4.1 covers the necessary analysis that results in a reference software architecture, which is elaborated in Section 4.4. In between, general design principles and idioms are introduced that guide the design process.

4.1 Concern Impact Analysis

Concern Impact Analysis, which originates from the design of automotive operating systems [21], is a method for the investigation of interrelation between features. Both, concerns that are part of the system's specification (e.g., RFCs) and internal concerns serve as input for the analysis. Internal concerns are inherent properties of the system that are not documented in the specification. For example, the network byte order is not discussed in RFC1122 although it is crucial for interoperability. The starting point for the following analysis is a set of both explicit concerns (i.e., features) and internal concerns. Since it is hardly possible to be aware of all internal concerns in advance, missing ones can be added iteratively. Networking software in general consists of

1. an *API*, that provides accessible system services to applications,

2. *connection state* to distinguish between different connections and their actual usage,

	Protocols				Options			Internal	
	ARP	IP	UDP	TCP	UDP Checksum	TCP Sliding Window	TCP MSS Option	Checksum Algorithm	Byte Order
`send()`		⊕	⊛	⊛	◐	●	◐		◐
`receive()`		⊕	⊛	⊛	◑	●	◑		◑
`connect()`				⊕		◐	◐		
`listen()`				⊕		◑	◑		
`close()`				⊕					
`bind()`			⊕						
`unbind()`			⊕						
`set_ipv4_addr()`		⊕							◐
`set_dst_port()`			⊕	⊕					◐
`set_src_port()`			⊕	⊕					◐
Packet Buffers	⊛	⊛	⊛	⊛		⊛			
ARP Cache	⊕								⊛
IP Addresses		⊕							⊛
Port Numbers			⊕	⊕					⊛
TCP States				⊕		⊛	⊛		
Receive IRQ	◑	◑	◑	◑					
Timeout	◑			◑		◑			
TCP/UDP ⇄ IP					◑			◑	
IP ⇄ Data Link	◐							◐	
Initialization		◑							

Table 2: Impact of configurable concerns on configurable IP networking software. Impact of concerns (columns) on the API, internal states, and events (rows).

3. and external events or rather internal *state transitions* that alter the system's behavior.

These three ingredients are of special interest because they emphasize places where various configurable concerns can interfere. An arrangement of all identified configurable concerns over these three ingredients leads to a comprehensive *crosscut table matrix* that reveals critical join points.

Table 2 shows an excerpt of the crosscut table matrix that we prepared for IP networking software. The columns list a few of the configurable concerns, which are themselves grouped into sets of networking *protocols*, *options* that extend protocols, and *internal* concerns. The aforementioned three categories (API, state, and transitions) form the rows of Table 2. Each category is separated by a horizontal line and covers a few relevant examples.

After the columns and rows are identified, the crosscut table matrix has to be populated. The influence of each concern on each item (row) is analyzed and classified into mainly three different types of impact:

1. **Introduction** of system services or states (static structure). This kind of impact implies the introduction of basic functionality that is added to the system and typically becomes visible

in the system's API. This fundamental type of influence provides join points for other concerns, denoted by a ⊕ sign in the crosscut table matrix.

2. **Extension** of existing system services or states (static structure). Some concerns modify already existing API services or existing system states. This less invasive type of impact, compared to the preceding one, is visually denoted by a ⊛.

3. **Modification** of the run-time behavior (dynamic structure). Concerns may have impact on the execution of API functions, the occurrence of external events or internal state transitions. The actual modification can take place before / after / around the event, as denoted by ◐ / ◑ / ●.

The crosscut table matrix provides a comprehensive overview of the concerns present in the system and their (subtle) relationships. It enables the developer to make informed decisions in the subsequent architecture design process – in which the concerns ultimately have to be mapped to classes and aspects.

4.2 Mapping to Classes and Aspects

As a driver for this process, the crosscut table matrix can be read in two directions:

A **horizontal analysis** focuses on a technical element, that is a state or service provided by the API for which it yields the contributing and dependent concerns. Once a state or service is introduced by a concern, each modifying concern (denoted by a ⊛) explicitly "*uses*" the introducing one. For example, the concerns *UDP* and *TCP* directly "use" *IP* – they built on the respective `send()` and `receive()` primitives, as observable in the first two rows of Table 2. With this information, the system designer is able to develop a *concern hierarchy* [21] by arranging the concerns according to their "uses" relationships. Figure 5 summarizes the concern hierarchy that encompasses the subset of concerns used in the preceding crosscut table matrix[7].

A **vertical analysis** provides more detailed information about a conceptual concern. Basic concerns can be identified by at least one introduction (⊕), for instance *ARP* in the first column. A new state, namely the *ARP Cache*, is introduced by this concern. Crosscutting concerns can easily be spotted by encountering multiple modifications (◐ / ◑ / ●) for a given concern, such as the *Byte Order* in the last column, which crosscuts several API functions and states.

The information from both analyses can then be used to achieve an initial mapping of concerns to classes and aspects by some simple rules of thumb:

- A *basic concern*, such as *ARP*, *IP*, *UDP*, and *TCP*, is mapped to a class.
- If a basic concern "uses" another basic concern, it becomes a derived class. For example, *UDP* and *TCP* both will be derived from *IP* (see Figure 5).
- A crosscutting concern, such as *Byte Order* or *Checksum Algorithm*, becomes an aspect.
- A configuration option, such as *TCP Sliding Window* or *UDP Checksum*, is also often crosscutting and thus becomes an aspect.

Of course, these rules of thumb only provide an initial software structure, that has to be refined in the subsequent design process, in which usually additional aspects and classes are employed to implement the binding and configurability of the identified concerns. In order to ease the following fine-grained design process, we propose design principles that aid the system designer to make "right" design decisions.

[7]Once again, all concerns for IP networking software do not fit into a single comprehensive diagram.

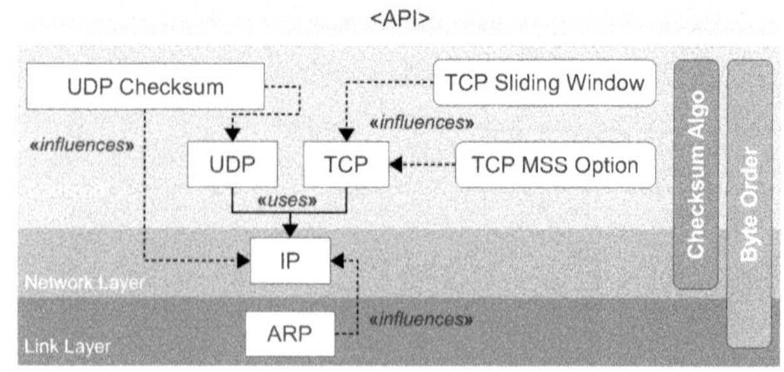

Figure 5: Layered concern hierarchy. Rectangles denote basic concerns, rounded boxes indicate crosscutting concerns. Concerns have "*uses*" and "*influence*" relationships, the latter denoted by a dashed arrow.

4.3 Design Principles

Based on our experience with the automotive operating system CiAO [22, 32], the following design process is guided by four reusable principles that lead to an aspect-aware system design. The corresponding AspectC++ idioms are beneficial for a clean system design and fine-grained configurability that does not induce run-time overhead.

4.3.1 *Loose Coupling*

The principle of *loose coupling* describes the relation between two software modules. Consider a networking device driver and the IP networking protocol implementation. Both modules should be exchangeable – the device driver should work with different protocol stacks and a protocol stack should be able to use various device drivers. Using AOP, loose coupling between these modules can be established easily:

The device driver exposes an *explicit join point*, which is an empty function without arguments named `ready()`, that is called after the device driver is initialized. Aspects can hook into this function and bind the device driver module to the IP stack on the initialization event (see last row of Table 2). The *binding aspect* is a recurring aspect role whenever several software modules, possibly developed by independent working groups, need to be integrated in a noninvasive manner.

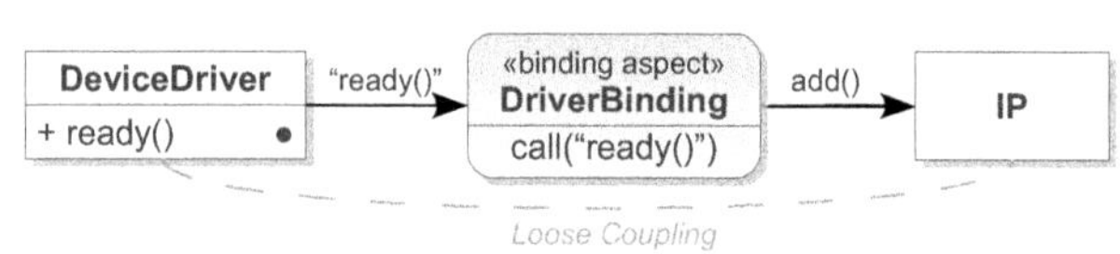

Figure 6: The principle of *loose coupling*. A *binding aspect* establishes a relation between two otherwise unrelated modules. The invocation of `ready()` is translated to an `add()` call, which registers the `DeviceDriver` at the `IP` stack.

4.3.2 *Minimal Extensions*

The principle of *minimal extensions* aims at incremental system design. Software modules are designed to provide only the minimal functionality that is required for a working system – no more, no less. Additional features are only introduced by *extension aspects*.

For instance, the TCP module itself does neither support a sliding window for transmitting nor for receiving. In Section 4.2, these features were already identified as crosscutting concerns and mapped to aspects. Thus, the sliding window for transmitting and for receiving are both designed as minimal extensions to the TCP module. As shown in Figure 7, the sliding window crosscuts the module TCP as well as both receive and transmit buffering.

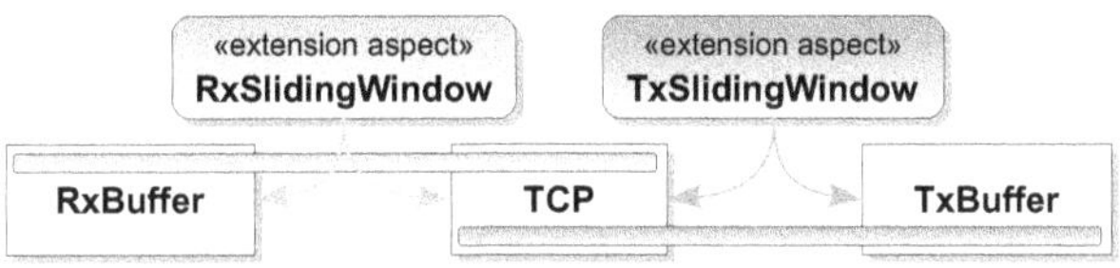

Figure 7: The principle of *minimal extensions*. *Extension aspects* introduce optional concerns by means of *slices*. Existing software modules are extended by additional member variables and functions in order to fulfill their duty.

4.3.3 Visible Transitions

The aspect-aware design enables the substitution of internal system policies. Thus, all relevant transitions need to be accessible by aspects. The system designer has to ensure that control-flow transitions are visible as unambiguous join points, that are preferably distinguishable at compile time. A fine-grained class hierarchy enclosed in expressive C++ namespaces provides a rich set of join points in the sense of *visible transitions*.

Often, a highly crosscutting concern represents a system policy. For example, the byte order conversion, which is examined in the last column of Table 2, is an exchangeable policy. On machines that internally operate at the network byte order, no conversion has to be performed. Otherwise, a conversion of every multi-byte entry of each protocol header takes place. We designed all protocol header structures as C++ classes, whose attributes are exclusively accessed throughout `get()` and `set()` functions. Therefore, aspects can alter the result and arguments of these functions and perform the byte order conversion directly in place. Figure 8 shows such an aspect for `get()` functions of 16-bit words. The source code for 32-bit words and `set()` functions is similar. Other software modules do not have to be aware of network byte ordering at all, because this concern is entirely encapsulated by a single *policy aspect*.

actual conversion function | pointcut expression *(where)*

```
aspect LittleEndian {
  static uint16_t ntohs(uint16_t n);

  pointcut NetworkToHost() = "% IP_Header::get%()" ||
                             "% TCP_Header::get%()";

  advice call( NetworkToHost() ) &&
         result(res) : after(uint16_t res) {
    *tjp->result() = ntohs(res);
  }
};
```

pointer to result *value* | binding to result *type*

Figure 8: Byte order conversion by *policy aspect*. Conversion logic is encapsulated in the *static* function `ntohs`, which is implicitly invoked by the advice. The pointcut declaration describes *where* the advice shall take effect, that is all `get()` functions exposed by protocol header structures (*visible transitions*). `tjp` (this join point) provides access to context information, and `tjp->result()` yields a typed pointer to the return value.

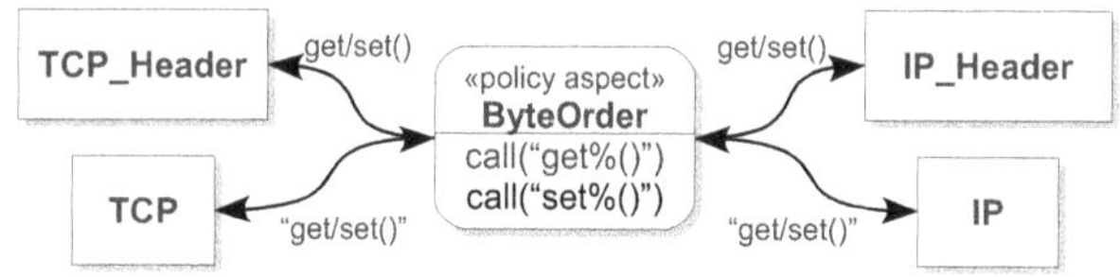

Figure 9: The principle of *visible transistions*. A fine-grained call structure makes system internal transitions explicit. Thus, system policies are designed to be exchangeable *policy aspects*, that alter the control-flow at relevant locations.

4.3.4 Upcall Dispatcher Hierarchy

A layered system defines a "uses" relation among its layers – typically top-down (see Figure 5). Each module is aware of all subordinate layers and directly depends on the ones it uses. The other way around, a module that is used by superordinate modules must not have knowledge of them. Otherwise, circular dependencies occur and configurability is lost.

Upcalls, which are invocations in the opposite direction of the "uses" relation, cause difficulties since they manifest in a tight coupling between layers due to circular dependencies. Consider an IP module that receives an IP packet and propagates it to superordinate UDP and TCP modules, which are in turn configurable and thus not always present. There may be even more subscribers for the content of an IP packet, such as a configurable ICMP module. The problematic nature of upcalls becomes even worse if multiple layers are involved. For example, an Ethernet packet, that is dispatched from the bottom layer up to the top layer, forces circular dependencies between all traversed layers.

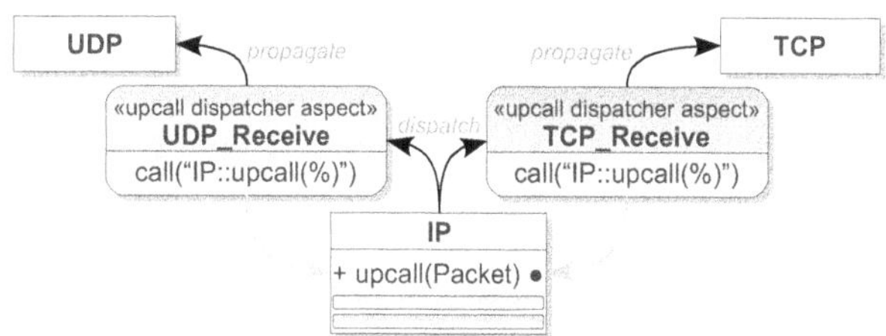

Figure 10: The principle of *upcall dispatcher hierarchy*. Bottom-up data propagation in a layered system is performed by independent *upcall dispatcher aspects*. Each aspect evaluates inherent dispatch conditions and eventually lifts data upwards.

The *upcall dispatcher hierarchy* circumvents this issue by strictly sticking to the inherent "uses" relation. Figure 10 illustrates this design principle based on the aforementioned example. The bottom-up propagation of data is implemented by explicit join points, which are empty functions (see Section 4.3.1), whose arguments contain the data to be propagated. In Figure 10, the module `IP` offers the public function `upcall(Packet)`, that is called by `IP` in order to dispatch a packet to superordinate modules. The actual dispatch is performed by independent *upcall dispatcher aspects*, which hook into the `upcall` function and evaluate whether the packet is to be accepted. Dispatch criteria are usually a check of the protocol's header identification field and a verification of the checksum. The dispatch process itself is distributed to several aspects, which provide configurability because they can be omitted easily.

A key benefit of upcall dispatcher aspects is that they are stackable. Consider a module `Ethernet`, that also provides an explicit join point `upcall(Ethernet_Frame)`. The upcall dispatching to superordinate modules, for instance `ARP` and `IP`, can be performed in the same way. Thus, circular dependencies between layers are avoided and the already developed "uses" relation is enforced by this design principle.

4.4 Architectural Design Decisions

The efficiency of a system is vitally determined by crucial design decisions during the development process. Several concerns are not covered in the system's specification (i.e., RFCs), although they define fundamental properties of the software architecture. These *hidden concerns* have to be identified, evaluated and carefully designed in order to narrow the design space. Thus, the unnecessary variability is eliminated that is known in advance to bear no convincing benefits. This is the major software engineering challenge at the very end of the design phase, which is involved with the overall performance of the system.

4.4.1 Memory Management

The most important concern with regard to performance and scalability is memory management [7]. Each network packet has to be buffered in data memory, and access to it must be very fast. On typical embedded systems, memory and especially data memory (RAM) is a scarce resource, so that an economical utilization is necessary. Furthermore, on such systems, a dynamic memory management (heap) is often not economical and thus not available.

In general, there are two possible designs for the memory management of an IP stack:

1. **Linear buffers**, large enough to hold a contiguous packet.
2. **Linked lists** of small buffers, each buffer containing a fragment of the packet.

Both designs are reasonable and used in popular networking systems – Linux applies linear buffers whereas BSD favors linked lists. The advantage of linear buffers is speed, because a contiguous packet often fits into a single cache line. Extra effort has to be paid to packet construction in the space of linear buffers. The size of all applied protocol headers has to be known in advance and an appropriate amount of memory has to be reserved in the front of each buffer. Linked lists simplify packet construction, because arbitrary protocol headers can be linked in front of the list. This flexibility comes at the price of poor cache locality due to the scattering of list elements across the data memory.

In order to gain maximal efficiency, we chose to use linear buffers with two configurable sizes. This design decision takes into account that the distribution of packet sizes tends to be bimodal [25]. Large packets deliver the payload while small packets acknowledge successful reception. Because the size of the payload heavily depends on the application, we leave the configuration of the buffer sizes to the application programmer. We designed the memory management to be part of a parametrized API, that is implemented as generic C++ templates.

Thus, the application programmer can tune the memory utilization for each particular connection by instantiating a template class of the API. For example, the memory management parameters for high-throughput connections should satisfy the *bandwidth-delay product*. In general, the optimal amount of memory is approximately $throughput_{desired} \times delay_{end-to-end}$ for a sliding window.

4.4.2 Concurrency

An implicit requirement for IP networking software is the concurrent handling of multiple connections. Full concurrency is achieved by threads that are scheduled by an operating system. Thus, IP networking software could be run inside the context of threads or inside the operating system itself.

Our design decision is to reuse the runtime context of threads in order to implement concurrency. This approach has several advantages: A *prioritization* of threads is enforced, because the networking routines of high-priority threads are preferred over low-priority ones. This is essential for real-time systems, where hard timing deadlines have to be met. As a consequence, the feature of prioritized connection comes for free if thread priorities are supported by the underlying operating system.

The second important advantage of this design is that *isolation* between different connections can easily be established. Each connection can constitute separate pools of buffers in the context of the actual thread, so that no interference between concurrent connections can occur. The isolation of memory regions is especially interesting for safety-critical applications. On the other hand, shared buffers use the memory resources more efficiently. Therefore, CiAO/IP implements buffer sharing as a configurable feature.

4.4.3 Synchronization

A structured coordination between multiple threads running IP networking software and the underlying operating system is crucial for the performance of the system. The actual synchronization primitives depend highly on the services provided by the operating system, and are even nonexistent if no operating system is used. Therefore, synchronization is an exchangeable system policy.

According to the design principles (see Section 4.3), we make synchronization explicit by providing unambiguous join points for visible transitions. Thus, policy aspects implement synchronization by using operating system services. For example, we integrated our IP stack into the automotive operating system CiAO. CiAO itself provides *events* and *alarms* for synchronization, both specified by the AUTOSAR [2] standard. Policy aspects invoke these services implicitly and allow a noninvasive integration, so that neither the operating system nor the IP stack have to be modified. Therefore, the integration of this IP Stack into other operating systems is straightforward.

5. EVALUATION

To evaluate the CiAO/IP stack, we have implemented most features specified by the relevant RFC documents on a wide range of platforms:

8-bit: A BTnode[8] sensor network platform, which uses an Atmel ATmega128L *AVR* microcontroller with 128KiB ROM and 4KiB RAM combined with a sub 1 GHz CC1000 radio.

16-bit: A TI EZ430-Chronos[9] development system, which comes in a sports watch. It combines a *MSP430* microcontroller with 32KiB ROM, 4KiB RAM and a sub 1 GHz CC1101 radio device.

32-bit: An *IA-32* PC with Intel Core 2 Quad 6600 CPU, 4GiB RAM and an Intel 82566DM Gbit Ethernet adapter.

64-bit: Same as above, but running in *AMD64* mode.

Table 1 in Section 2 has already shown that all features of CiAO/IP that we identified as "optional" during the domain analysis became configurable in the implementation. This section will prove that this high degree of variability does not come a cost. CiAO/IP clearly outperforms uIP and lwIP in terms of memory consumption (Section 5.1), performance (Section 5.2), and energy efficiency (Section 5.3).

[8] http://www.btnode.ethz.ch/

[9] http://processors.wiki.ti.com/index.php/EZ430-Chronos

5.1 Memory Consumption

Memory is a scarce resource on embedded devices. This is true for both program memory (ROM) as well as data memory (RAM). For instance, the aforementioned 16-bit MSP430 platform contains only 32KiB ROM and 4KiB RAM, which software must not exceed. Especially system software that offers services to the actual application and is itself not self-contained, has to be exceptionally small and specifically designed for the embedded systems domain. A counterexample is the Linux[10] kernel, whose basic feature *TCP/IP Networking* increases the uncompressed kernel size by about 234KiB on IA-32. Hence, we focus our evaluation on uIP, lwIP and CiAO/IP.

The requirements on data memory are dominated by buffering. Network packets are inherently dynamic content that has to be stored in the RAM. Besides buffers, networking software stores connection state in the RAM, which requires an infinitesimal amount of memory compared to a buffer for a single IP packet. The buffer sizes of uIP, lwIP and CiAO/IP are statically configurable and, thus, their RAM consumption is comparable.

However, CiAO/IP is more flexible as it offers to configure whether buffers for a particular connection should be allocated on the data memory section, on the dynamic heap (if present), or even on the runtime stack of the current application. This feature is especially useful for temporary connections that are active only for a short period of time (for instance, during a software update), and, hence, consume data memory only within this period. Buffers, that are globally allocated on the data memory or the heap at system startup, constantly take RAM.

Besides data memory, an efficient usage of program memory is crucial. The requirements to ROM directly depend on the functionality of the software: More features lead to higher ROM consumption. In the sequel, we discuss the issue of program memory in detail. All following values are obtained by using the *gcc*[11] with optimization for size (-Os).

5.1.1 The Cost of TCP

TCP is certainly the most complex part of the Internet protocol suite. As a consequence, an implementation of TCP requires more program memory than other protocols. Nevertheless, TCP is essential for interoperability since many applications rely on it, for example web services. A realistic scenario for TCP also includes IP, and on IA-32 and AMD64 additionally Ethernet and ARP, since almost every personal computer comes with an Ethernet adapter.

Figure 11(a) shows the memory consumption of a TCP client in the described scenario. We use the minimal possible feature selections of uIP, lwIP and CiAO/IP to evaluate the minimal costs for a TCP client. The stacked bar diagram outlines the costs for the IP stacks themselves (lower section) and the cost for the application code that has to be added for a working TCP client (upper hatched section). This is relevant, as the required application code differs substantially between the IP stacks: uIP, for instance, does not perform retransmissions of lost TCP segments itself – it is up to the application logic to detect packet loss and to retransmit missing packets. For lwIP, we use its memory-efficient event-driven, nonsequential API that, however, also requires more effort on the application side than the UNIX-like sockets of CiAO/IP. For all IP stacks, the application code also handles synchronization with network devices and timer events. The necessary synchronization primitives are provided by the underlying CiAO operating system.

Figure 11(b) shows the memory consumption of a TCP server in the same scenario. For both client and server, the cost of TCP varies among the different architectures. It is notable that uIP has been heavily optimized for 8-bit architectures by using 8-bit data types and arithmetic whenever possible [9]. This specialization saves about 50% of the TCP code size on AVR and MSP430 compared to CiAO/IP, whereas it leads to an increased code size on IA-32 and AMD64.

5.1.2 The Cost of UDP

UDP is the second essential transport protocol of the Internet protocol suite. Compared to TCP, UDP is almost a null protocol, which is reflected by its low memory consumption. Consider a node of a sensor network that periodically transmits sensor measurements. For such a scenario, UDP is sufficient.

Figure 11(c) summarizes the cost of transmitting UDP packets without optional checksumming[12]. Since TCP is a mandatory feature of uIP, there is a high overhead if used for UDP only. lwIP consumes even more program memory due to architectural properties, such as flexible packet buffers and generic driver interfaces. CiAO/IP is by far the most efficient implementation with regard to UDP. On the average over all architectures, CiAO/IP requires only twelve percent of the memory that uIP respectively lwIP take.

For receiving UDP datagrams the results are similar. Figure 11(d) outlines our results. The application code (hatched section of top the bars) for receiving is larger than for transmitting, because synchronization with receive interrupts is involved.

5.1.3 Feature Scalability

We have measured the code size induced by each configurable feature: Starting with the minimal possible feature selection of the particular IP stack we have incremented the number of active features one by one. The results are depicted in Figure 12.

The minimal code size of uIP (see Figure 12(a)) is rather high due to the mandatory TCP feature, ranging from 4253 byte on MSP430 to 6481 byte on AMD64. When all features of uIP are active, the code size almost doubles.

lwIP scales from a minimum of 6836 byte on IA-32 up to a maximum of 38.7kB on AVR. This spread indicates scalability to some extend, but the absolute values are quite high. For example, the sum of all considered lwIP features on MSP430 occupies 30KiB of its 32KiB internal ROM (94%).

In contrast, CiAO/IP scales from a minimum of 789 byte on AMD64 up to a maximim of 20.8kB on AVR. In between, each configurable feature increases the memory consumption by a very small amount. The sharp spike in this diagram is caused by the TCP feature, which adds an exceptional large portion of code due to its complexity.

5.2 Performance

The aforementioned 8-bit and 16-bit hardware platforms are ill-suited for performance benchmarking, since their gross networking throughput is limited to 76.8 kbit/s respectively 500 kbit/s by the radio hardware. Moreover, the net throughput of payload (*goodput*) is primarily determined by the link quality[13] and timing of TCP retransmissions. To analyze the influence of the different IP stacks on high-speed networking performance, we therefore focus on the Gbit Ethernet adapter of the IA-32 platform:

We have connected two PCs (Intel Core 2 Quad 6600 CPU, 4GiB RAM, Intel 82566DM Gbit Ethernet) via a crossover cable to create

[10]version 2.6.33

[11]version 4.4 (MSP430/IA-32/AMD64), version 4.3 (AVR)

[12]The values for IA-32 and AMD64 also encompass the cost for Ethernet and ARP.

[13]We measured a goodput of 1161 bit/s for CiAO/IP and 680 bit/s for uIP on the average in the wireless setup described in Section 5.3. Both systems were configured with exactly equivalent features.

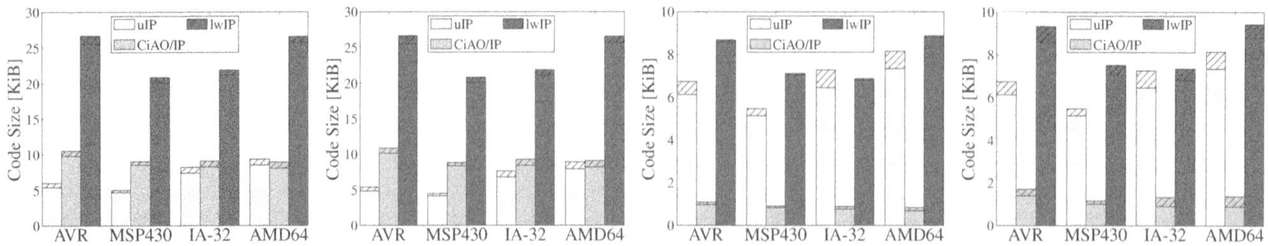

(a) TCP Client (b) TCP Server (c) UDP Tx w/o Checksum (d) UDP Rx w/o Checksum

Figure 11: Code size (ROM) for common use cases. Each use case includes IPv4. On IA-32 and AMD64, Ethernet and ARP is also included to get a working system. Figures (a) - (d) show the minimal memory consumption of each particular IP stack for the desired functionality. The lower sections of the stacked bars constitute the code size of the IP stacks themselves. The upper hatched sections indicate the cost for the application code that has to be added for full operation. Hardware platforms range from 8-bit (AVR) to 64-bit (AMD64).

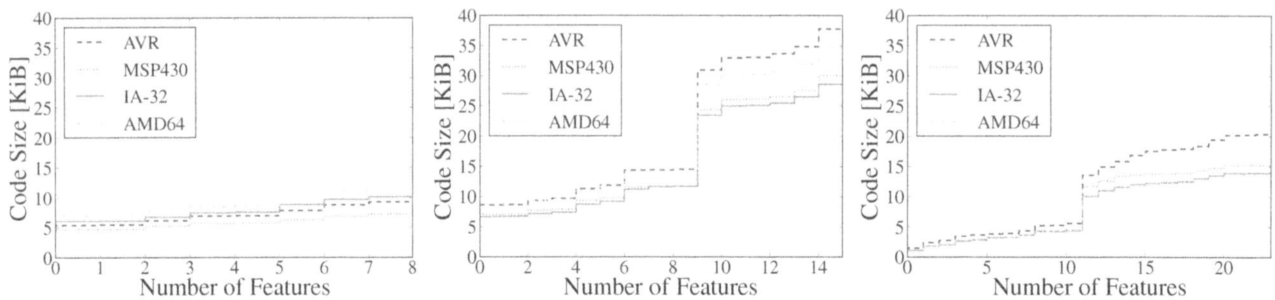

(a) **Feature increment of uIP:** TCP Server (IPv4) and ICMP, TCP Client, TCP Urgent Data, IPv4 Fragment Reassembly, UDP, UDP Checksumming, ARP, IPv4 Forwarding, NDP

(b) **Feature increment of lwIP:** UDP (IPv4), UDP Tx Checksumming, UDP Rx Checksumming, UDP Lite, IPv4 Forwarding, IPv4 Fragment Reassembly, IPv4 Fragmentation, ARP, ARP Queuing, VLAN, TCP, TCP Queuing, TCP Backlog, TCP Timestamps, ICMP, IGMP

(c) **Feature increment of CiAO/IP:** UDP Tx (IPv4), UDP Tx Cksum., UDP Rx (IPv4), UDP Rx Cksum., IPv4 Frag. Reass., Ethernet, Tx Cksum Offload., Rx Cksum Offload., ARP Reply, ARP Request, Static ARP Cache, ARP Timeout, TCP Client, TCP Server, Reset Invalid TCP Packets, Tx Slid. Window, Rx Slid. Window, SWS Tx, SWS Rx, MSS Opt., RTT Est., Cong. Control, Generic Buffer, Generic Rx Queue

Figure 12: Code size (ROM) vs. features. For each TCP/IP stack, the first measuring point shows the code size of the initial minimal feature selection. Each subsequent measuring point constitutes the addition of one more configurable feature. Thus, figures (a) - (c) outline an incremental feature selection and visually evince scalability.

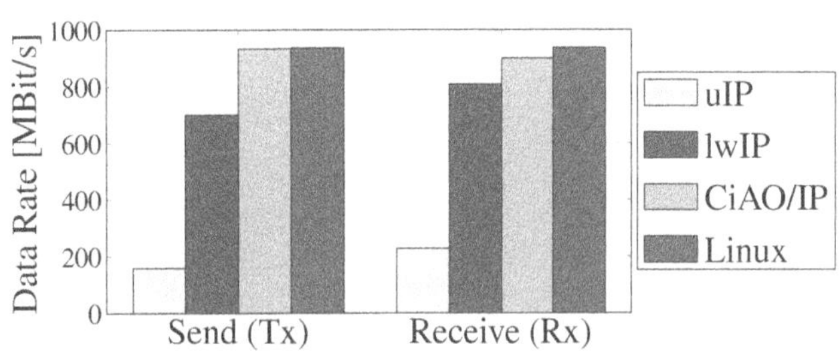

Figure 13: Goodput of a TCP connection. Depicted is the goodput between two local IA-32 machines, that are connected via gigabit Ethernet (crossover). One peer is always a Linux 2.6 box, whereas the machine under test runs the different IP stacks.

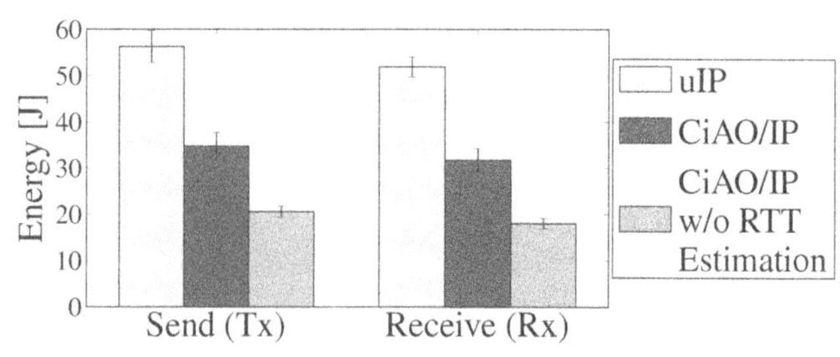

Figure 14: Energy consumption. Shown is the mean energy consumption and the sample standard deviation of unidirectional TCP data transfers (32 kB) between two wireless sensor networking nodes. *CiAO/IP w/o RTT Estimation* uses a fixed retransmission timeout (200 ms) for lost packets.

an actual loss-free environment. One box constantly runs Linux 2.6, whereas the other box runs the IP stack under test. We have established a single TCP connection between the two machines and measured the achieved goodput. Each IP stack is equipped with 100 KiB RAM for buffering and exactly equivalent feature configuration, except for uIP which solely supports a single buffer (1514 byte) due to the missing sliding window. For fairness, uIP is extended by *uip_split* [9] in order to reach maximal throughput. This uip_split extension avoids the default 250 ms delay between packet transmits. As uIP implements a stop-and-wait procedure, it otherwise would suffer from delayed acknowledgments of the Linux peer. Figure 13 shows the results of our measurements. We classify the results into sending and receiving, which describe the operational mode of the IP stack under test.

With Linux on both ends a maximal goodput of 937 Mbit/s is achieved, which here can be understood as the upper bound.

CiAO/IP, however, achieves almost the same transfer rates (934 Mbit/s for Tx, 900 Mbit/s for Rx).

lwIP clearly lacks behind. This is mostly caused by lwIP's linked packet buffer management (*pbufs*), which have to be copied into a contiguous memory region before delivery to the device driver. We intentionally do not use vectored I/O (scatter/gather) hardware accelerators that might overcome this design decision, since embedded devices generally do not provide such hardware support.

uIP is not competitive due to the missing sliding window, which results in a significant performance degradation.

5.3 Energy Consumption

Energy is probably the most scarce resource for battery-powered devices. We have chosen the aforementioned BTnode[14] sensornet platform for determining the impact of IP stacks on energy consumption. A typical duty for sensor nodes is a firmware update: a unidirectional, reliable data transfer of a single file. Thus, our setup for energy measurements[15] consists of a TCP client, that transmits 32 kB of data to a listening TCP server.

On the link layer, we use the *B-MAC* [27] protocol for collision avoidance and preamble sampling with an interval of 25 ms (*Low Power Listening*). The packet size is limited to 255 byte, and each IP stack is equipped with 300 byte for buffering. Due to a quite noisy radio channel, which leads to a high percentage of corrupt packets (approximately 50%), we disabled the exponential backoff algorithm of every TCP implementation, which otherwise would slow down the retransmission of lost packets unnecessarily. Unfortunately, we did not get lwIP working under these harsh conditions, so that we had to omit lwIP in the following measurements.

Figure 14 presents the mean energy consumption and the sample standard deviation of uIP and CiAO/IP. For fairness, we did not activate the sliding window feature of CiAO/IP, because uIP does not support it and would otherwise be disadvantaged. To sum up, both IP stacks were configured with exactly equivalent features. Consequently, no congestion control was performed, because the slow-start algorithm can only be applied to a sliding window.

However, even without this beneficial feature CiAO/IP consumes significantly less energy: A uIP sender consumes a mean of 56.3 Joule, whereas CiAO/IP takes a mean of 34.8 Joule for the same task. This is caused by the coarse granularity of uIP's round-trip time estimation, which can only be multiple of 500 ms. In contrast, CiAO/IP performs this estimation much more accurate in units of 1 ms. Thus, lost packets are detected earlier (up to 499 ms per lost packet). To prove this claim, we deactivated the round-trip time estimation of CiAO/IP (an optional feature in CiAO/IP) and fixed the retransmission timeout to 200 ms. This way, the energy consumption of the sender could be further reduced to 20.5 Joule. These observations are also valid for the receiver, whose energy consumption is determined by the time the transfer takes to complete.

[14]AVR microcontroller (8-bit) running at 7.37 MHz, CC1000 transceiver operating at 868 MHz and 19.2 kbit/s

[15]*Hitex PowerScale* with *ACM probes* used for energy measurements

5.4 Summary of Results

CiAO/IP outperforms uIP and lwIP in almost all scenarios. The only exception is the TCP client/server, where uIP requires on the AVR/MSP430 platform only half the code size of CiAO/IP – which is exactly the scenario uIP has been optimized for [8]. However, even in this configuration uIP consumes significantly more energy (1.6x) and delivers a much smaller throughput (0.59x). In all other configurations, our aspect-oriented CiAO/IP stack outperforms the C implementations of uIP and lwIP notably with respect to code size, throughput, energy, and feature-wise scalability.

6. DISCUSSION

The main goal of this work is to achieve feature-wise scalability of the layered TCP/IP protocols by a software engineering approach for static configurability. As shown in the previous section, CiAO/IP offers fine-grained configurability and thereby achieves excellent results with respect to code size, throughput and energy. In the following, we discuss the pros and cons of our approach with respect to more "soft" properties, including maintainability, configurability, portability, and optimization.

6.1 Maintainability

The complexity and maintainability of some piece of software is generally hard to quantify. However, source-code metrics, such as lines of code (LOC) can serve as an indicator: The minimalistic uIP consists of 2796 LOC[16] whereas CiAO/IP consists of 4943 LOC and a comparable subset of lwIP[17] encompasses 11987 LOC. Clearly, more features correspond to more LOC, but in general aspect-oriented software consists of less LOC than its C counterparts, because crosscutting concerns are modularized into aspects and, thus, avoid scattered fragments of similar code.

A good example for this is the macro-based implementation of the byte order concern illustrated in Figure 1. As pointed out in Section 2.3, every header access has to be surrounded by one of the byte ordering macros (`htons, ntohs, ...`). This is a tedious and error-prone design rule. Our aspect-oriented CiAO/IP stack circumvents this issue by a much better separation of concerns: Byte ordering is a policy, hence we implement the byte order conversion as a modular policy aspect (see Section 4.3.3) that implicitly affects the `get()` and `set()` functions of all classes that describe network protocol headers. The respective aspect consists of a single source code file with less than 50 lines of code. This reduces complexity, as the byte order in CiAO/IP is correct by construction – it is not possible to create an invalid packet by a forgotten call of a conversion macro. Therefore, AOP contributes to maintainability of source code by modularization and separation of concerns.

On the other hand, AOP may impair the independent development of modules of the system [19]. Since aspects do not expose an explicit interface, the interaction of the system's modules can become complicated, because conceptually, each aspect can potentially affect each module [31]. To obviate this drawback, our aspect-oriented

[16]effective lines of code (excluding empty lines and comments), obtained with *cloc*: `http://cloc.sourceforge.net/`

[17]version 1.32, without IPv6, DNS and DHCP

design principles (see Section 4.3) are also applied as a means to make such interactions explicit.

6.2 Configurability

The excellent configurability we achieved in CiAO/IP is primarily the result of a software product-line development process that considers configurability as a fundamental design goal from the very beginning. However, eventually the individual features have to be implemented in a way that makes it possible to "leave them off" if not needed. The advantage of AOP here is *loose coupling* by the advice concept: Essentially, advice inverts the direction in which control-flow relationships are specified. This facilitates the self-integration of optional features into the control flows of the base system. Furthermore, advice-based binding is inherently loose – if the addressed join point is not present, the binding is silently dropped. This property is useful for the implementation of *interacting* optional features, which are difficult to tackle with other decomposition approaches [17]. In CiAO/IP we thereby could keep a textbook-like strict layering in the design and implementation of the IP stack without scarifying efficiency.

6.3 Portability

Loose coupling is also beneficial with respect to portability to other platforms. The aspect-based integration of our IP stack into the operating system makes it easy to use CiAO/IP with other systems or to use it without any operating system. We are currently working on implementations for eCos [24] and FreeRTOS [4].

From the perspective of application developers, switching to CiAO/IP is relatively easy as CiAO/IP offers the well-understood BSD socket concept and does not, like uIP and lwIP, require the application engineer to implement and maintain stack-specific state machines on the application level.

6.4 Optimization

The main benefit of the high configurability offered by CiAO/IP is optimization for *nonfunctional properties*, such as memory footprint, throughput, and energy consumption. This is done by first activating only those network protocols that meet the *functional* requirements depending on the actual field of application, for example TCP and IPv4. Other protocols should be disabled. This way, the design space of valid feature selections is significantly narrowed. As outlined in Table 1, several optional features do not provide basic networking protocols by themselves, and, thus still remain *open* for this particular feature selection. The assignment of these open features, for instance the TCP sliding window and round-trip time estimation to either *on* or *off*, leads to a *design space exploration*.

We developed the *feedback approach* [28] for statistical reasoning on nonfunctional properties and assignment of open features. Information about previously configured feature selections have to be captured by run-time tests and the results feed into a database. Thus, nonfunctional properties for further feature selections that are not yet evaluated can be estimated, and a testing set for validation is generated automatically. This is a semi-automated process that relies on testing of specific feature selections until the database is saturated.

7. RELATED WORK

We originally developed our aspect-aware design method as well CiAO/IP for our CiAO operating system family for deeply embedded devices [22, 21]. CiAO implements the automotive AUTOSAR OS standard [2], a rapidly growing domain with respect to "mobile Internet" applications. Other operating system projects that aim to bridge deeply embedded systems and the Internet world are mostly from the domain of sensor networks: Recent versions of *TinyOS* [20] provide the BLIP IPv6 stack, which also provides configurability of UDP and TCP and is implemented with the component model offered by the NesC language [12]. BLIP has the distinction of operating solely at IPv6. All other systems we are aware of are based on either uIP or lwIP, both of which clearly dominate the domain: *Contiki* [10] uses uIP; *eCos* [24] supports lwIP and, alternatively, a BSD-derived stack; *FreeRTOS* [4] allows developers to choose between uIP (called FreeTCPIP) and, again, lwIP.

AOP as a modularization approach for configurable system software has also been applied in the domain of middleware [35, 14] and embedded databases [16]. Other modularization approaches in the domain are too numerous to be discussed here. However, several researchers favor *feature-oriented programming* [5] – and even the C preprocessor is experiencing a renaissance by the provisioning of better tool support towards a "virtual separation of concerns" [1].

8. CONCLUSIONS

Network protocol stacks for the domain of resource-constrained embedded systems have to fullfill a broad variety of functional requirements, while at the same being thrifty with respect to hardware resources, especially memory and energy. This calls for static configuration approaches to tailor the provided functionality to the actual application's needs.

We presented the CiAO/IP stack and its underlying design approach for embedded system software, which pushes the limits of static configurability to a new level. In our design approach, the domain (mainly given by RFCs) is analyzed with *software product line* methodology; the resulting features are then structured by a *concern impact analysis* and implemented with *aspect-oriented programming* as fine-grained and loosly coupled implementation artefacts.

By following this design approach in the development of CiAO/IP, we did not only achieve a clear and complete separation of concerns in the code (thus, maintainability and portability), but also excellent configurability and scalability of the resulting protocol stack: CiAO/IP outperforms *uIP* and *lwIP* in terms of code size (up to 84% / 88% less than uIP / lwIP for a UDP sender on an AVR), throughput (up to 587% / 33% higher than uIP / lwIP for a TCP sender on IA-32 with Gbit Ethernet) and energy consumption (up to 63% lower than uIP on AVR for a TCP sender).

We hope that our results encourage other developers of system software to follow the guidelines in this paper.

9. ACKNOWLEDGEMENTS

We wish to thank the anonymous reviewers for their helpful and encouraging comments. Special thanks go to Prabal Dutta, whose shepherding helped us to clarify the content of this paper.

This work was partly supported by the German Research Council (DFG) under grant no. SP 968/4-1, SP 968/5-1, and SFB 876 projects A1 and A4.

10. REFERENCES

[1] Sven Apel and Christian Kästner. Virtual separation of concerns - a second chance for preprocessors. *Journal of Object Technology*, 8(6):59–78, 2009.

[2] AUTOSAR. Specification of operating system (version 2.0.1). Technical report, Automotive Open System Architecture GbR, June 2006.

[3] D. Barisic, M. Krogmann, G. Stromberg, and P. Schramm. Making embedded software development more efficient with SOA. In *21st International Conference on Advanced*

Information Networking and Applications Workshops, 2007 (AINAW '07), volume 1, pages 941–946, May 2007.

[4] Richard Barry. *Using the FreeRTOS Real Time Kernel*. Real Time Engineers Ltd, 2010.

[5] Don Batory. Feature-oriented programming and the AHEAD tool suite. In *26th Int. Conf. on Software Engineering (ICSE '04)*, pages 702–703. IEEE, 2004.

[6] R. Braden. Requirements for Internet Hosts - Communication Layers. RFC 1122 (Standard), October 1989.

[7] David D. Clark, Van Jacobson, John Romkey, and Howard Salwen. An analysis of TCP processing overhead. *IEEE Communications Magazine*, 27:23–29, 1989.

[8] A. Dunkels. Full TCP/IP for 8-bit architectures. In *Proceedings of the 1st international conference on Mobile systems, applications and services*, pages 85–98. ACM, 2003.

[9] Adam Dunkels. *The uIP Embedded TCP/IP Stack. The uIP 1.0 Reference Manual*. Swedish Institute of Computer Science, 2006.

[10] Adam Dunkels, Björn Grönvall, and Thiemo Voigt. Contiki — a lightweight and flexible operating system for tiny networked sensors. In *Proceedings of the First IEEE Workshop on Embedded Networked Sensors (Emnets-I)*, Tampa, Florida, USA, November 2004.

[11] Deborah Estrin, Ramesh Govindan, John Heidemann, and Satish Kumar. Next century challenges: scalable coordination in sensor networks. In *Proceedings of the 5th annual ACM/IEEE international conference on Mobile computing and networking*, MobiCom '99, pages 263–270, New York, NY, USA, 1999. ACM.

[12] David Gay, Philip Levis, Robert von Behren, Matt Welsh, Eric Brewer, and David Culler. The nesC language: A holistic approach to networked embedded systems. In *ACM SIGPLAN Conf. on Programming Language Design and Implementation (PLDI '03)*, pages 1–11, San Diego, CA, USA, 2003. ACM.

[13] Jonathan W. Hui and David E. Culler. IP is dead, long live IP for wireless sensor networks. In *Proceedings of the 6th ACM conference on Embedded network sensor systems*, SenSys '08, pages 15–28, New York, NY, USA, 2008. ACM.

[14] Frank Hunleth and Ron Cytron. Footprint and feature management using aspect-oriented programming techniques. In *2002 Joint LCTES & SCOPES Conferences (LCTES/SCOPES '02)*, pages 38–45, Berlin, Germany, June 2002. ACM.

[15] Kyo Kang, Sholom Cohen, James Hess, William Novak, and A. Spencer Peterson. Feature-oriented domain analysis (FODA) feasibility study. Technical report, Carnegie Mellon University, Software Engineering Institute, Pittsburgh, PA, November 1990.

[16] Christian Kästner, Sven Apel, and Don Batory. A case study implementing features using AspectJ. In *11th Software Product Line Conf. (SPLC '07)*, pages 223–232. IEEE, 2007.

[17] Christian Kästner, Sven Apel, Syed Saif ur Rahman, Marko Rosenmüller, Don Batory, and Gunter Saake. On the impact of the optional feature problem: Analysis and case studies. In Dirk Muthig and John D. McGregor, editors, *13th Software Product Line Conf. (SPLC '09)*, Pittsburgh, PA, USA, 2009. Carnegie Mellon University.

[18] Gregor Kiczales, John Lamping, Anurag Mendhekar, Chris Maeda, Cristina Videira Lopes, Jean-Marc Loingtier, and John Irwin. Aspect-oriented programming. In Mehmet Aksit and Satoshi Matsuoka, editors, *11th Eur. Conf. on OOP (ECOOP '97)*, volume 1241 of *LNCS*, pages 220–242. Springer, June 1997.

[19] Gregor Kiczales and Mira Mezini. Aspect-oriented programming and modular reasoning. In Gruia-Catalin Roman, William G. Griswold, and Bashar Nuseibeh, editors, *27th Int. Conf. on Software Engineering (ICSE '05)*, pages 49–58, New York, NY, USA, 2005. ACM.

[20] Philip Levis, Sam Madden, Joseph Polastre, Robert Szewczyk, Kamin Whitehouse, Alec Woo, David Gay, Jason Hill, Matt Welsh, Eric Brewer, and David Culler. *TinyOS: An Operating System for Wireless Sensor Networks*. Ambient Intelligence. Springer, Heidelberg, Germany, 2005.

[21] Daniel Lohmann, Wanja Hofer, Wolfgang Schröder-Preikschat, and Olaf Spinczyk. Aspect-aware operating-system development. In Shigeru Chiba, editor, *10th Int. Conf. on Aspect-Oriented Software Development (AOSD '11)*, pages 69–80, New York, NY, USA, 2011. ACM.

[22] Daniel Lohmann, Wanja Hofer, Wolfgang Schröder-Preikschat, Jochen Streicher, and Olaf Spinczyk. CiAO: An aspect-oriented operating-system family for resource-constrained embedded systems. In *2009 USENIX ATC*, pages 215–228, Berkeley, CA, USA, June 2009. USENIX.

[23] Daniel Lohmann, Fabian Scheler, Reinhard Tartler, Olaf Spinczyk, and Wolfgang Schröder-Preikschat. A quantitative analysis of aspects in the eCos kernel. In Yolande Berbers and Willy Zwaenepoel, editors, *ACM SIGOPS/EuroSys Eur. Conf. on Computer Systems 2006 (EuroSys '06)*, pages 191–204, New York, NY, USA, April 2006. ACM.

[24] Anthony Massa. *Embedded Software Development with eCos*. New Riders, 2002.

[25] Bosko Milekic. Network buffer allocation in the FreeBSD operating system. In *Proceedings of BSDCan*, Ottawa, ON, Canada, May 2004.

[26] Linda Northrop and Paul Clements. *Software Product Lines: Practices and Patterns*. AW, 2001.

[27] Joseph Polastre, Jason Hill, and David Culler. Versatile low power media access for wireless sensor networks. In *Proceedings of the 2nd international conference on Embedded networked sensor systems*, SenSys '04, pages 95–107, New York, NY, USA, 2004. ACM.

[28] Julio Sincero, Wolfgang Schröder-Preikschat, and Olaf Spinczyk. Approaching Non-Functional Properties of Software Product Lines: Learning from Products. In IEEE Computer Society Press, editor, *Proceedings of the 17th Asia-Pacific Software Engineering Conference (APSEC 2010)*, pages 147–155, Los Alamitos, CA, USA, 2010.

[29] Henry Spencer and Gehoff Collyer. #ifdef considered harmful, or portability experience with C News. In *1992 USENIX ATC*, Berkeley, CA, USA, June 1992. USENIX.

[30] Olaf Spinczyk and Daniel Lohmann. The design and implementation of AspectC++. *Knowledge-Based Systems, Special Issue on Techniques to Produce Intelligent Secure Software*, 20(7):636–651, 2007.

[31] Friedrich Steimann. The paradoxical success of aspect-oriented programming. In *21st ACM Conf. on OOP, Systems, Languages, and Applications (OOPSLA '06)*, pages 481–497, New York, NY, USA, 2006. ACM.

[32] Jochen Streicher, Christoph Borchert, and Olaf Spinczyk. Upcall dispatcher aspects: Combining modularity with efficiency in the CiAO IP stack. In *1st AOSD W'shop on*

Modularity in Systems Software (AOSD-MISS '11), pages 23–27. ACM, March 2011.

[33] Reinhard Tartler, Daniel Lohmann, Julio Sincero, and Wolfgang Schröder-Preikschat. Feature consistency in compile-time-configurable system software: Facing the Linux 10,000 feature problem. In Christoph M. Kirsch and Gernot Heiser, editors, *ACM SIGOPS/EuroSys Eur. Conf. on Computer Systems 2011 (EuroSys '11)*, pages 47–60, New York, NY, USA, April 2011. ACM.

[34] Jim Turley. The two percent solution. *embedded.com*, December 2002. `http://www.embedded.com/story/OEG20021217S0039`, visited 2011-04-08.

[35] Charles Zhang and Hans-Arno Jacobsen. Quantifying aspects in middleware platforms. In *2nd Int. Conf. on Aspect-Oriented Software Development (AOSD '03)*, pages 130–139, New York, NY, USA, 2003. ACM.

Demo: ExerLink - Enabling Pervasive Social Exergames with Heterogeneous Exercise Devices

[1]Taiwoo Park, [1]Inseok Hwang, [2]Uichin Lee, [3]Sunghoon Ivan Lee, [1]Chungkuk Yoo, [1]Youngki Lee, [1]Hyukjae Jang, [1]Sungwon Peter Choe, [1]Souneil Park and [1]Junehwa Song

[1,2]KAIST, Republic of Korea [3]UCLA, Los Angeles, USA

[1]{twpark, inseok, ckyoo, youngki, hjjang, sungwon, spark, junesong}@nclab.kaist.ac.kr, [2]uclee@kaist.edu, [3]silee@cs.ucla.edu

ABSTRACT

We demonstrate a pervasive social exergame platform called *ExerLink* that converts exercise intensity to game inputs and intelligently balances intensity/delay variations for fair game play experiences. Also, we show the potential of using multiple exercise devices as game controllers and incorporating multiple heterogeneous controllers into a game. Specifically, we consider a class of exercise equipment used for repetitive, individual, and aerobic (RIA) exercises such as treadmill running, stationary cycling, hula hooping, and jump roping.

Categories and Subject Descriptors

C.3 [**Special-purpose and Application-based Systems**]: Real-time and embedded systems; H.5.2 [**Information Interfaces and Presentation**]: User Interfaces—*evaluation/methodology, input devices and strategies, prototyping*

Keywords:

Exergame, Exercise, Social, Pervasive, Heterogeneous

1. INTRODUCTION

We envision that pervasive social exergames supporting multiple exercise modalities will enable true ubiquitous social interactions, fostering social bonds and friendships. Further, such exergames will allow people with different exercising capabilities/preferences to gather from diverse situations and to play/exercise together. For instance, a homemaker running on a treadmill at home, a son who is a college student cycling in his school gym, the husband on a business trip using his jump rope, and a young daughter using a hula hoop at her friend's house can all meet virtually and exercise together over a multi-exercise game. They will be able to choose the right exercise appropriate for their given situation or preference, enjoying exercising together via social exergames.

However, realizing an interactive multi-player exergame using RIA exercise devices as game controllers gives rise to important challenges resulting from the differences between exercise modalities, e.g., rotating ropes and taking steps. First, converting exercise devices into game controllers requires exercise-specific considerations of sensor hardware and software support to acquire the exercise intensity metrics to be used as game inputs, e.g., exercise speeds. Traditionally such support has been provided only for specific types of exercise equipment such as stationary bikes. Second, it may cause significant imbalance in game performance among players who simultaneously utilize different types of controllers designed for different exercises. Little study has been done to understand the performance difference with heterogeneous devices and personal preferences/capabilities which is a unique problem in pervasive social exergames.

We demonstrate *ExerLink*, a pervasive social exergame platform

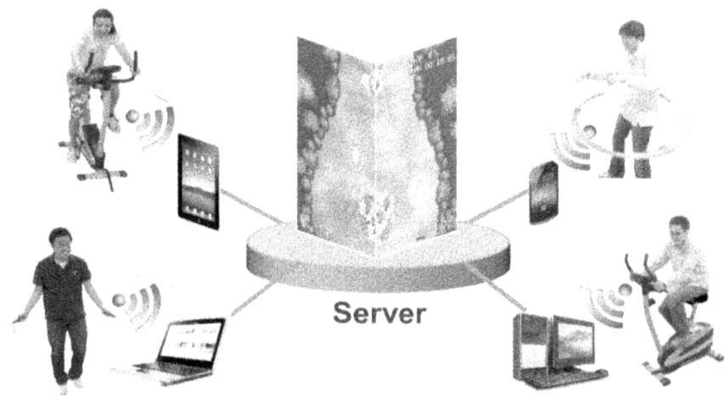

Figure 1 Team playing a game on ExerLink

that converts exercise intensity to game inputs and intelligently balances intensity/delay variations for fair game play experiences.

2. ARCHITECTURE OVERVIEW

ExerLink mediates social exergames and diverse RIA exercise devices and supports mobile devices including smartphones, tablets and Mobile Internet Devices (MIDs). ExerLink provides a run-time environment for game applications, and multiple exergames can run at the same time. The ExerLink platform processes exercise intensity information from devices and converts it into game input value. The conversion process considers heterogeneity of exercise devices and personal differences of physical abilities/preferences. The platform mediates communication between game applications and a game server considering different network delay among remote game players. It also provides a voice communication channel between game players to support social interactions in a remote play situation. Lastly, it tracks and summarizes amounts of exercises performed by players to constantly motivate the players.

3. DEMONSTRATION SETUP

We have implemented an ExerLink prototype on windows-based MIDs and laptops using Java. Also, we modified the existing RIA exercise devices to measure the exercise intensities in real-time. To this end, we implanted sensors and micro-controllers onto an off-the-shelf hula hoop and a jump rope. We will utilize those two near-mobile exercise devices for on-site demonstration. For the interactive treadmill and stationary cycle, their built-in sensing capabilities are exploited. We carefully augmented existing devices so as not to disrupt the players' natural exercise activities. To demonstrate the feasibility of a pervasive multi-exercise game, we utilize Swan Boat[1], an interactive and collaborative racing exergame that we have previously developed.

4. REFERENCES

[1] M. Ahn, et al., "Swan boat: pervasive social game to enhance treadmill running", In Proc. MM, 2010

MobiSys'12, June 25-29, 2012, Low Wood Bay, Lake District, UK.
ACM 978-1-4503-1301-8/12/06.

Demo: A Barcode Streaming System for Smartphones

Tian Hao, Ruogu Zhou, Guoliang Xing
Department of Computer Science and Engineering, Michigan State University, USA
{haotian, zhouruog, glxing}@cse.msu.edu

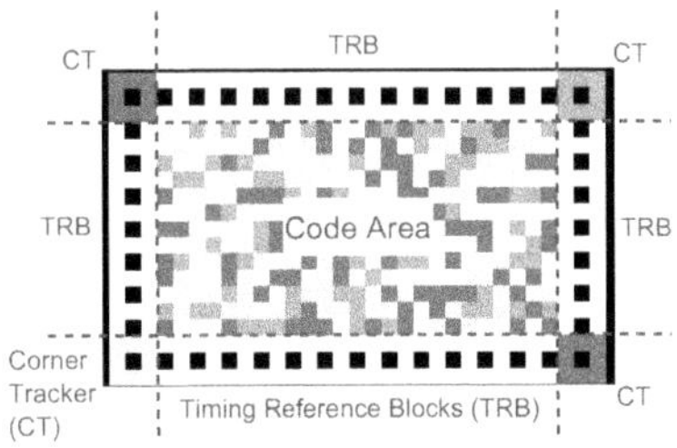

Figure 1: The design of the 2D color code.

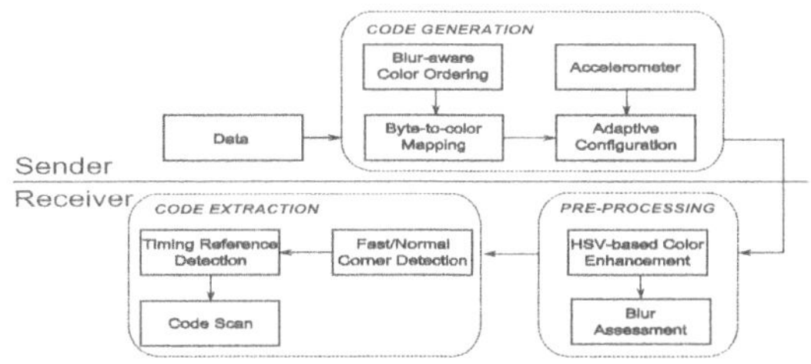

Figure 2: System architecture of COBRA

Figure 3: A Google Nexus S phone is communicating with a HTC Inspire phone using COBRA.

Categories and Subject Descriptors

C.2.1 [**COMPUTER-COMMUNICATION NETWORKS**]: Network Architecture and Design—*Wireless communication*

Keywords

Visible Light Communication, Color Barcode, Smartphone

1. INTRODUCTION

As a potential alternative for radio frequency (RF) technology, visible light communication (VLC) has received significant attention recently. In contrast to RF technologies, VLC can enable secure and interference-free wireless links due to the directional and short-range natures of visible light. We demonstrate COBRA [1], a special form of VLC for off-the-shelf smartphone systems, which can be potentially used in scenarios like contactless mobile advertisement/payment (e.g., in retail stores and museums) and data exchange between smartphones.

COBRA is designed to offer a secure, intuitive, and easy-to-use communication interface without the reliance on additional wireless interfaces or the Internet infrastructure. We develop a new 2D color barcode (as shown in Fig. 1) for COBRA that is optimized for streaming between small-size screen and low-speed camera of smartphones. As shown in Fig. 2, COBRA adapts the size and layout of code blocks in streamed barcodes to deal with the significant image blur in mobile environments, and adopts new image processing techniques such as fast corner detection to achieve real-time barcode stream decoding.

MobiSys'12, June 25–29, 2012, Low Wood Bay, Lake District, UK.
ACM 978-1-4503-1301-8/12/06.

2. SYSTEM DEMO

As shown in Fig. 3, two Android phones, a Google Nexus S smartphone and a HTC Inspire smartphone, are employed in the demonstration. During communication, the screen of the sender phone is faced towards the rear camera of the receiver phone. The user is required to select a file to transfer. Then the selected file will be encoded into a sequence of barcodes being displayed on the screen. The receiver phone will sample the camera and start decoding if barcodes are detected. The correctness of file transmission is verified with CRC check.

In this demonstration, we will simulate possible situations in mobile environment, including varying the distance, view angle and orientation between the sender and receiver. Moreover, we will also show how the sender adapts the size and layout of the code block according to the feedback from the accelerometer. To illustrate the performance of COBRA, a laptop is connected with the receiver to display the real-time statistic of the communication such as the decoding accuracy for each received barcode. During the demo, we will also invite the audience to test an Android app that exchanges photos and contacts between two phones using COBRA.

3. REFERENCES

[1] T. Hao, R. Zhou, and G. Xing. Cobra: Color barcode streaming for smartphone systems. In *MobiSys*. ACM, 2012.

Demo: SenseTogether- Cooperative Ambience Monitoring Platform with Continuity and Benefit Awareness

Youngki Lee, Younghyun Ju, Chulhong Min, Seungwoo Kang, Inseok Hwang, Junehwa Song
KAIST, Daejeon, Republic of Korea

Categories and Subject Descriptors
C.5.3[**Computer System Implementation**]: Microcomputers - *Portable devices, Personal computers*

Keywords
Cooperation, Ambience, Context, Sensing, Energy, CoMon

1. INTRODUCTION
In this demonstration, we present CoMon [1], a novel cooperative ambience monitoring platform. CoMon newly addresses a severe energy problem caused by continuous sensing and context processing, through opportunistic cooperation among nearby mobile users. For cooperative monitoring, CoMon automatically detects nearby cooperator candidates that are likely to remain in the vicinity for a long period of time. Then, it devises a cooperation plan that provides mutual and fair benefit to cooperators. Through continuity- and benefit-aware operation, CoMon enables applications to monitor the environment at much lower energy consumption. In this demo, we will show the key operations of CoMon and their effectiveness under various usage scenarios.

2. HARDWARE SETUP
We prototyped CoMon on Android phones and various types of sensor devices. Figure 1 shows example hardware devices for the demo. To demonstrate cooperative context monitoring of CoMon, we plan to use multiple Android phones as well as wearable sensor devices with various sensing modules (e.g., accelerometers, gyroscopes, ECG, CO_2 sensors, dust sensors, GPS, microphones).

3. DEMONSTRATION
We plan to demonstrate the key functionalities of CoMon.

Context monitoring: CoMon runs as a middleware on top of a smartphone OS. It provides sensing applications with APIs to specify the contexts of interest (e.g. location, activity) in a declarative query. We will first show diverse context types and sensing devices that CoMon supports, and the detailed operation of context monitoring.

Cooperative monitoring: As a key of CoMon, we will show the cooperative operation among multiple cooperators. Our demonstration will be performed in three phases: (1) cooperator detection, (2) cooperation planning, and (3) cooperative monitoring (See Figure 2 for screenshots).

In the first phase, CoMon finds nearby candidates via periodic Bluetooth scans. To establish stable cooperation channels, it selects the ones that are likely to stay together longer among the discovered ones, leveraging our heuristic

(a) Nexus One (b) BT-serial converter (c) Dust sensor (d) CO_2 sensor

Figure 1 Example Demo Setup (see [1] for details).

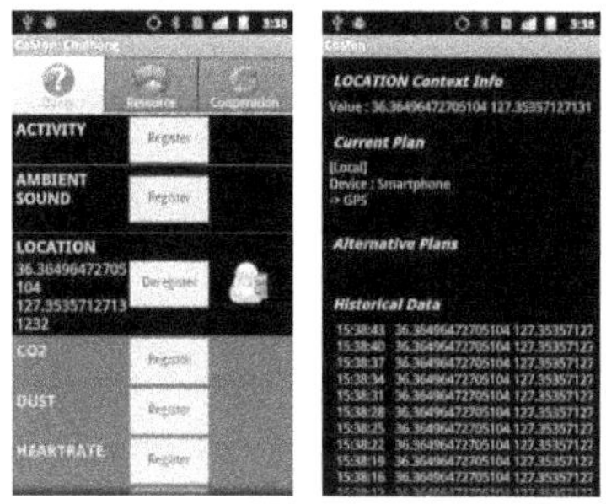

Figure 2 Screenshots showing CoMon operation

predictors. For the demonstration, we will pre-configure the smartphones to have different social relationships and encounter histories, such as friends, family and strangers. Then, we will show that CoMon discovers and selects the most proper ones among the nearby candidates.

In the second phase, CoMon conducts cooperation planning to decide which contexts to share and trade, so as to provide mutual and fair benefits for all cooperators. We will show that the planning decision is effectively made considering currently running applications, available sensor devices and their resources. For better understanding, we plan to visualize the planning process between cooperators.

Finally, we will demonstrate that once the cooperation channel is established, CoMon provides applications with monitoring results stably and seamlessly. We will also present how much each smartphone's energy consumption is saved through the cooperation, exposing system status.

Application deployment: In addition, we plan to show several proof-of-concept applications running on CoMon. We consider two example applications: *TripMemory* and *PollutionAlarm*. The former is the application that tracks a user's travelling path and logs her surrounding events extracted from ambient sound. The latter continuously monitors surrounding air quality and alarms users if severe. We will show that CoMon enables these applications to run at much reduced energy cost by leveraging opportunistic cooperation.

4. REFERENCES
[1] Lee, Y. et al., CoMon: Cooperative Ambience Monitoring Platform with Continuity and Benefit Awareness, In MobiSys, 2012.

MobiSys'12, June 25-29, 2012, Low Wood Bay, Lake District, UK.
ACM 978-1-4503-1301-8/12/06.

Demo– Satellites in Our Pockets: An Object Positioning System using Smartphones

Justin Manweiler
IBM T. J. Watson Research*
Hawthorne, NY, USA
jmanweiler@us.ibm.com

Puneet Jain
Duke University
Durham, NC, USA
puneet.jain@duke.edu

Romit Roy Choudhury
Duke University
Durham, NC, USA
romit.rc@duke.edu

ABSTRACT

We attempt to localize a distant object by looking at it through a smartphone. As an example use-case, while driving on a highway entering New York, we want to look at one of the skyscrapers through the smartphone camera, and compute its GPS location. While the problem would have been far more difficult five years back, the growing number of sensors on smartphones, combined with advances in computer vision, have opened up important opportunities. We harness these opportunities through a system called *Object Positioning System (OPS)* [1] that achieves reasonable localization accuracy. Our core technique uses computer vision to create an approximate 3D structure of the object and camera, and applies mobile phone sensors to scale and rotate the structure to its absolute configuration. Then, by solving (nonlinear) optimizations on the residual (scaling and rotation) error, we ultimately estimate the object's GPS position.

We present a demonstration of OPS, a system to appear in the *MobiSys 2012* main conference. The user is expected to bring the object of interest near the center of her viewfinder, and take as few as four photographs. GPS and compass readings are also recorded during the process. The position of the photographs can be seperated by a few steps from each other in any direction. OPS uses *Structure from Motion* to extract keypoints across the photographs to come up with a 3D structure, composed of the object and the camera locations. Finally, OPS minimizes errors in GPS and compass readings with help of 3D structure to converge on the object location.

Categories and Subject Descriptors

H.3.4 [**Information Storage and Retrieval**]: Systems and Software

Keywords

Augmented Reality, Localization, Structure from Motion

1. DEMONSTRATION

OPS is implemented in two parts, an OPS smartphone client and a back-end server application. We built and tested the OPS client on the Google NexusS phone, as a Java extension to the standard Android. Photographs are pre-processed locally on the phone to extract keypoints and feature descriptors. Along with keypoints and descriptors, the phone uploads all available sensor data from when each photograph was taken, to include GPS and compass.

We evaluated OPS in the most natural way possible. We took four photographs for each localization by moving few steps in no specified direction. Phone GPS and compass are subjected to noise and biases because of environmental factor. Computer vision techniques, such as structure from motion, can also break down in a variety of scenarios. The goal of this demo is to demonstrate how we addressed these challenges. We plan to ask participants to take four photographs of an outdoor object. We expect that participants will be able to see object position on the google earth, as shown in Fig. 1.

Figure 1: Sampled test; circle denotes object-of-interest (top), Google Earth view (bottom): stadium seats near goal post.

*This work was conducted at Duke University.

2. REFERENCES

[1] J. Manweiler, P. Jain, and R. Roy Choudhury. Satellites in our pockets: An object positioning system using smartphones. In *Proceedings of the 10th international conference on Mobile systems, applications, and services*. ACM, 2012.

Demo – MVChat: Flasher Detection for Mobile Video Chat

Lei Tian[†], Junho Ahn[†], Hanqiang Cheng[*], Xinyu Xing[§], Yuli Liang[†], Richard Han[†], Qin Lv[†], Shivakant Mishra[†], David Chu[‡], Xue Liu[*]

[†] Dept. of CS, University of Colorado Boulder [*] Dept. of CS, McGill University
[§] Dept. of CS, Georgia Institute of Technology [‡] Microsoft Research

[†] {lei.tian, junho.ahn, Yu-li.Liang, qin.lv, rick.han, shivakaht.mishra}@colorado.edu,
[*]{hanqiang.cheng, xueliu}@cs.mcgill.ca, [§]xinyu.xing@gatech.edu, [‡]davidchu@microsoft.com

1. ABSTRACT

Online video chat services such as Chatroulette [1] and Omegle [2] that randomly match pairs of users in video chat sessions have become increasingly popular, with over twenty thousand online users at anytime during a day. A key problem encountered in such systems is the presence of misbehaving users ("flashers") and obscene content. Our previous works [3] [4] prove that using some image recognition methods (skin-detection, dense SIFT) and machine learning algorithms could achieve significantly higher recall and better precision for flasher detection. Nowadays, with the rapid development of advanced mobile phones with both front and back cameras, we expect mobile video chat to become a popular extension of online video chat services. However, because of the computation-intensive features used by our previous solutions and mobile phones' hardware limitations such as memory size and CPU capacity, it is difficult to directly apply our previous works to mobile platforms. As smartphones are increasingly equipped with diverse sensing capabilities, we plan to utilize this multi-dimensional sensor information to extend flasher detection on mobile platform. This project explores how we can mine accelerometer and other mobile sensor data to infer some clues to optimize flasher detection accuracy while reducing the computation demands of flasher detection on the mobile device.

In this demo, we will show our Android-based mobile video chat system (MVChat) in operation, which extends video chat to the mobile domain, beyond merely online services such as Chatroulette and Omegle. We will show how our previous flasher detection algorithm is capable of detecting misbehaving user video streams in a mobile environment. We will further demonstrate the capability of our system for real-time collection of multi-dimensional mobile sensor data such as acceleration, light, gyroscope, GPS as well as audio and video snapshots. Moreover, we will demonstrate the analysis of mobile sensor data analysis in order to identify key characteristics for detecting flashers on the mobile platform.

Based on the Adobe AIR API, we establish our mobile video chat app by RTMFP (Real Time Media Flow Protocol) which is the same as the protocol used by most online video chat services such as Chatroulette and Omegle. It is thus possible to bind our solution with existing video chat services. Safety-permitting, we will show a demo of our client attached to an existing online video chat service.

Figure 1 is captured while using our app in a mobile video chat session.

Figure 1: Mobile Video Chat Session Snapshot

Categories and Subject Descriptors

H.5.1 [**Information Interfaces and Presentation**]: Multimedia Information Systems—*Video*

Keywords

Algorithms, Design, Experimentation, Human Factors

2. REFERENCES

[1] Chatroulette. `http://www.chatroulette.com/`.

[2] Omegle. `http://www.omegle.com/`.

[3] Hanqiang Cheng, Yu-Li Liang, Xinyu Xing, Xue Liu, Richard Han, Qin Lv, and Shivakant Mishra. Efficient misbehaving user detection in online video chat services. In *WSDM '12: Proceedings of the Fifth ACM International Conference on Web Search and Data Mining*, pages 23–32, 2012.

[4] Xinyu Xing, Yu-Li Liang, Hanqiang Cheng, Jianxun Dang, Sui Huang, Richard Han, Xue Liu, Qin Lv, and Shivakant Mishra. SafeVchat: Detecting obscene content and misbehaving users in online video chat services. In *WWW '12: Proceedings of the 20th International World Wide Web Conference*, pages 685–694, 2011.

MobiSys'12, June 25–29, 2012, Low Wood Bay, Lake District, UK.
ACM 978-1-4503-1301-8/12/06.

Demo – How Long to Wait?: Predicting Bus Arrival Time with Mobile Phone based Participatory Sensing

Pengfei Zhou, Yuanqing Zheng, Mo Li
Nanyang Technological University, Singapore
pfzhou@ntu.edu.sg, yuanqing1@e.ntu.edu.sg, limo@ntu.edu.sg

Categories and Subject Descriptors

C.2.4 [**Computer Communication Networks**]: Distributed Systems–*Distributed Applications*; H.5.3 [**Information Interfaces and Presentation**]: Groups and Organization Interfaces – *Collaborative Computing*

General Terms

Design, Experimentation, Measurement, Performance

1. INTRODUCTION

When traveling with buses, the travelers usually want to know the accurate bus arrival time. We present a novel bus arrival time prediction system based on crowd-participatory sensing. Our MobiSys'12 paper [3] provides the technical details. We briefly introduce its architecture as shown in Figure 1. There are three main components : (1) Sharing users: using commodity mobile phones together with the build-in sensors to sense the surrounding environment and report the lightweight cellular signals and cell tower sequences to a backend server; (2) Querying users: querying the bus arrival time for a particular bus route with mobile phones; (3) Backend server: collecting the instantly reported information from the sharing users, and processing such information so as to predict the bus arrival time.

Our system solely relies on the collaborative effort of the participating users. Independent from any bus companies, the system bridges the gap between the querying users who want to know the bus arrival time to the sharing users willing to offer the bus information. Instead of referring to GPS signal which is energy hungry and may perform poorly inside buses, we resort to more generally available and energy efficient sensing resources, including cell tower signals, movement statuses, audio recordings, etc., which bring less burden to the participatory party and encourage their participation.

Implementing such a system entails substantial challenges and we propose practical solutions to cope with them. (1)Bus detection: since the sharing users may travel with diverse means of transport, we extract unique identifiable fingerprints of public transit buses and utilize the microphone to detect the audio indication signals of bus IC card reader to identify the buses. (2) Bus classification: we use an efficient and robust top-k celltower set sequence matching method to classify the reported celltower sequences from the sharing users and associate them with different bus routes. (3) Information assembling: one sharing user may not stay on one bus to collect adequate time period of information. We identify passengers on the same bus based on the ambient sounds and propose a celltower sequence concatenation approach to assemble their celltower sequences so as to improve the sequence matching accuracy.

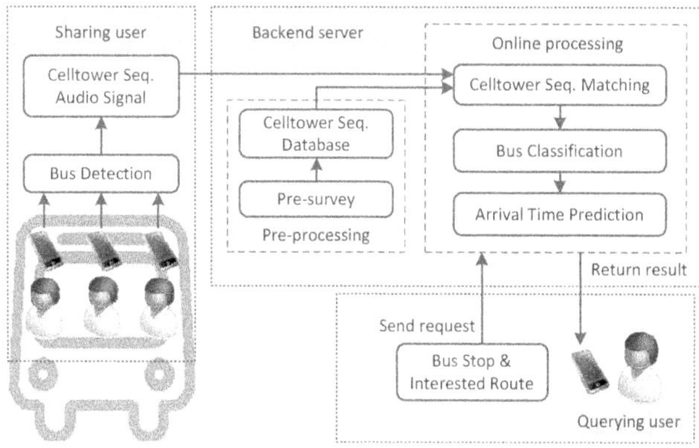

Figure 1: System architecture.

2. DEMONSTRATION

The goal of this demo is to demonstrate how the proposed system performs in practical situations. The demonstration system comprises a number of 5 Android mobile phones and a laptop as the backend server. We test and record the system operation in Singapore [1] on campus buses and several SBS public transit bus routes. The mobile phones will automatically detect the IC card indication audio signal, which is combined by 1kHz and 3kHz audio signals in Singapore, to identify the buses. We use accelerometer readings to distinguish buses from rapid trains. We will demonstrate the bus classification on the backend server using top-*3* celltower set sequence matching method and compare the prediction results to the actual arrival time of the buses.

We plan to extend our system and adapt to London bus transport system [2]. We will test the system in London and collect the experimental traces for demonstration on the conference.

3. REFERENCES

[1] Bus transport in Singapore. `http://en.wikipedia.org/wiki/Bus_transport_in_Singapore`.
[2] Bus transport in London. `http://en.wikipedia.org/wiki/Category:Bus_transport_in_London`.
[3] P. Zhou, Y. Zheng, and M. Li. How long to wait?: Predicting bus arrival time with mobile phone based participatory sensing. In Proceedings of ACM MobiSys, 2012.

MobiSys'12, June 25–29, 2012, Low Wood Bay, Lake District, UK.
ACM 978-1-4503-1301-8/12/06.

Demo: Smartlake: Lightweight Sensing and Optimizing Cleanup of Algae Blooms on Taihu Lake

Dong Li[1], Zijian Wang[1], Chenda Hou [1,2], Le Zhang[1,2], Ze Zhao[1,2], Li Cui[1],
[1]Institute of Computing Technology, Chinese Academy of Sciences, Beijing, China 100190
[2]Graduate School of Chinese Academy of Sciences, Beijing, China 100190
E-mail: {lidong, wangzijian, houchenda, zhangle, zhaoze, lcui}@ict.ac.cn

ABSTRACT
Facing the still unclear growth regularity of bl ooms, we propose rapid lightweight estimating method for the discovery of al gae blooms. An app lication-specific approximate algorithm for t imely dispatching salvaging boats is designed to minimize the total cleanup time.

Categories and Subject Descriptors
C.3 [Computer Systems Organization]: Special-Purpose and Application-Based Systems

Keywords
Algae BloomsSensing, Schedule Algorithm

1. INTRODUCTION
Nowadays, vicious algal blooms in the lakes poseda di rect threat to the water q uality and the public health of local people [1]. The most efficient method to regain the original environment is to physically clean them up from the water. SmartLake is the first practical auto sensing and cleanup system for al gae blooms in China on Tai hu Lake [2]. The goal is to catch the oc currences of alg ae blooms and automatically dispatch the salvaging boats to clean up t he algae scrums.However, the pe rsistentoccurrence of the algal blooms and t he mobility of sal vaging boats still pose great technical challenges for this system to work more efficiently.

2. SYSTEM DESIGN
Algae blooms estimation: The embedded sensing unit consists of a di gital camera and an ARM-linux based control and processing board. Picture s of water surface are used as data source. We tes ted the co lor features G-R, G-B and 2G-R-B to distinguish algae from water and found out that 2G-R-B resulted in t he best performance through thousands of pictures taken on Tai hu Lake. To deal with the fal se alarm, we re-implemented SVM on our embedded device to classify the algae and non-al gaepictures. Then we desi gned a lightweight algorithm for real -time in situ algae blooms identification with 89.3% accuracy.

The optimizing schedule for the cleanup of algae blooms: The consistent status and position variation of al gae blooms and salvagi ng boats require the system to make the cleanup schedule adaptively. We proposed a schedul e strategy to generate the boat assignment plan periodically with a configurable cycle. Each time a schedule plan need to be generated, a Generalized Assignment Problem (GAP with a set of cl eanup tasks and a set of boat s is formulated. An application-specific heuristic algorithm is designed to get the approximate solution. The ave rage cleanup time is recorded as the basis for adjusting the schedule cycle.

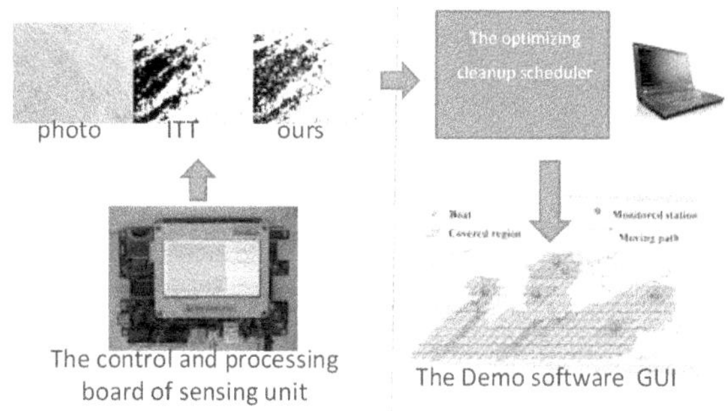

Figure 1 Demonstration Setup

3. DEMONSTRATION SETUP
We demonstrate the performance of SmarkLake using the ARM-based processing board and a laptop as shown in Figure 1. The conference participants can randomly select a picture taken on Tai hu Lake to see the analysis result of our method and compare the result with their judgments by eyes. The estimation result will trigger a simulating run of cleanup scheduler on the datasets of al gae blooms collected in 2010 and 2011. We will show t he dispatch process usi ng animations in a software GUI. Meanwhile the schedule plan generated by the nearest-first-dispatch greedy scheduler is also shown for comparison.

4. ACKNOWLEDGMENTS
This work i s supported by the National Ba sic Research Program of China (973 Program) under Grant No. 2011CB302803and the Beijing Natural Science Foundat ion (the energy-saving technology for internet of things based on the principle of locality) under Grant No.4112054.

5. Reference
[1] Lucie Guo, Doi ng battles with green monster of Taihu Lake, Science, 317: 1166, 2007.
[2] Dong Li, Cui Li, etc. A Cy ber Physical Networking System for monitoring and Cleaning up Alg ae Blooms with Agile sensor and Actuator Control Mechanism on Lake Tai , INFOCOM 2011 workshop CPNS.

Demo – MicroCast: Cooperative Video Streaming on Smartphones

Lorenzo Keller
School of I&C
EPFL, Lausanne, CH
lorenzo.keller@epfl.ch

Anh Le
CS Dept & CalIT2
UC Irvine
anh.le@uci.edu

Blerim Cici
Networked Systems & CalIT2
UC Irvine
bcici@uci.edu

Hulya Seferoglu
LIDS
MIT
hseferog@mit.edu

Christina Fragouli
School of I&C
EPFL, Lausanne, CH
christina.fragouli@epfl.ch

Athina Markopoulou
EECS Dept. & CalIT2
UC Irvine
athina@uci.edu

1. EXTENDED ABSTRACT

In this work, we are interested in a scenario where a group of smartphone users, within proximity of each other, are interested in watching the same video at the same time. The default operation today is that each user with a cellular connection downloads the video independently from the server. However, each phone's individual cellular connection may not be sufficient for providing high video quality.

We propose a novel cooperative scheme, called MicroCast, for video streaming to a group of smartphones within proximity of each other. Each phone utilizes simultaneously two network interfaces: one (cellular) to connect to the video server and download parts of the video and the other (WiFi) to connect to the rest of the group and exchange downloaded parts. Fig. 1 illustrates MicroCast scenario.

First, we propose a scheduling algorithm, MicroDownload, that decides which parts of the video each phone should download from the server, based on the phones' download rate. Second, we propose a novel all-to-all local dissemination scheme, MicroNC-P2, for sharing content among group members, which outperforms state-of-the-art peer-to-peer schemes in our setting. MicroNC-P2 is explicitly designed to exploit WiFi overhearing and network coding, based on a local broadcast framework, MicroBroadcast, which we developed specifically for Android phones. The full architecture of MicroCast is depicted in Fig. 2. To the best of our knowledge, MicroBroadcast is the first framework that allows smartphones to broadcast in high speed locally.

We perform performance evaluation on a testbed consisting of several Android phones, and we show that each component separately as well as the combined system as a whole significantly improve the performance compared to alternative approaches. Furthermore, this improvement comes at no significant battery cost. The detailed description of MicroCast can be found in [1]. A video demo and a poster of MicroCast can be found at [2] and [3]. More supporting materials are provided in [4].

Categories and Subject Descriptors

C.2.1 [**Computer-Communication Networks**]: Network Architecture and Design—*Wireless communication*; C.2.2 [**Computer-Communication Networks**]: Network Protocols—*Applications*

MobiSys'12, June 25–29, 2012, Low Wood Bay, Lake District, UK.
ACM 978-1-4503-1301-8/12/06.

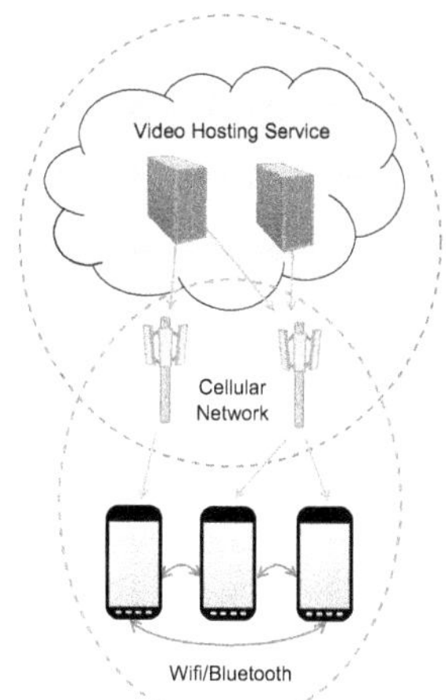

Figure 1: MicroCast Scenario.

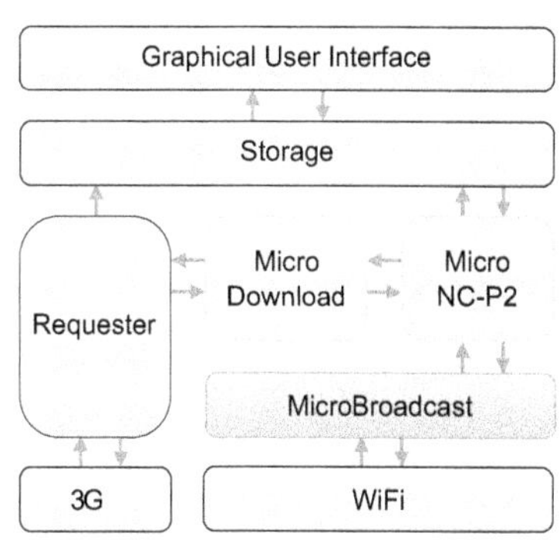

Figure 2: Architecture of MicroCast.

Keywords

Video Streaming, Smartphones, Network Coding

2. REFERENCES

[1] L. Keller, A. Le, B. Cici, H. Seferoglu, C. Fragouli, and A. Markopoulou, "Microcast: Cooperative video streaming on smartphones," MobiSys 2012.
[2] Video demo of MicroCast [Online] *http://www.youtube.com/watch?v=bEZCxLcjBw0.*
[3] MicroCast poster [Online] *http://goo.gl/L0xXt.*
[4] "Wireless network coding: from theory to practice," [Online] *http://odysseas.calit2.uci.edu/doku.php/public:muri09.*

Demo: A Programming Cloud of Smartphones

A. Konstantinidis
Dept. of Computer Science
University of Cyprus
akonstan@cs.ucy.ac.cy

C. Costa
Dept. of Computer Science
University of Cyprus
cs07cc6@cs.ucy.ac.cy

G. Larkou
Dept. of Computer Science
University of Cyprus
glarko01@cs.ucy.ac.cy

D. Zeinalipour-Yazti
Dept. of Computer Science
University of Cyprus
dzeina@cs.ucy.ac.cy

1. ABSTRACT

In this demonstration we present *SmartLab*[1], an exciting experimental testbed of approximately 40+ real Android Smartphones, plus emulated devices, deployed at the Department of Computer Science building at the University of Cyprus. SmartLab provides a public, permanent testbed for the development and testing of smartphone network applications via an intuitive web-based interface. Registered users can upload and install Android executables (APKs) on a number of Android smartphones, capture their output, reboot the devices, create concurrent interactive jobs using MonkeyRunner scripts, interact with the remote devices and many other exciting features. SmartLab aims to facilitate research in smartphone network programming environments, communication protocols, system design, and applications.

Categories and Subject Descriptors

H.4 [**Information Systems Applications**]: Miscellaneous

Keywords

Android Smartphones, Programming Cloud, System Design

2. SMARTLAB AND DEMONSTRATION

In this section, we overview the basic elements of our programming cloud, the basic steps of interaction as well as our demonstration scenario. SmartLab is inspired by both PlanetLab [1] and MoteLab [2].

Architecture: Our 40+ smartphones are located in an air-conditioned laboratory that features WiFi coverage and are remotely mounted to a VMWare VCenter located in the Computer Science building cluster-room. In particular, smartphone devices are connected with USB 2.0 cables, which provide sustained power, to a proxy PC that mounts our devices over TCP/IP to a virtualized Ubuntu Linux 10.10 servers.

Allocate: SmartLab supports real and emulated Android devices. For the former category, the user can browse through an intuitive web-based catalog in order to allocate devices for a certain amount of time (see web site). During sign up, a registered user is provided with credit hours that can be utilized to reserve *"1 smartphone for 1 hour"* (charges incur on a minute basis). Reservations are automatically released when the user budget is due. For the latter category, a user can allocate up to 5 emulated devices for an unlimited amount of time. SmartLab is free to the academic community, but the allocation scheme might be adjusted in the future depending on supply, demand and usage.

File-Transfer: A registered user is able to upload APKs and data to his allocated devices through an intuitive web-based drag-n-drop user interface. Subsequently, these files can be transferred to one or more devices through a simulated network share that is mounted on all devices rented by the user. Additionally, the user can perform any kind of interaction with the local file system of the allocated devices (i.e., both external and on-board flash).

Interact: SmartLab supports four (4) modes of user interaction: i) *Remote Control Terminals (RCT)*, which support our in-house implementation of an ajax-based web-based remote screen terminal for Android that can mimic user clicks and gestures such as sliding in order to unlock devices and conduct other functionalities; ii) *Remote Shells (RS)*, which supports our in-house implementation of an ajax-based web-based shell that can be utilized to issue a wide variety of known UNIX commands (e.g., `ls, ps, df, pwd, date`, etc.) to the Linux kernels that are located at the core of each Android device; iii) *Remote Scripting Environment (RSE)*, which allows users to author Android MonkeyRunner automation scripts (written in python) in order to quickly perform repetitive tasks on selected devices; and iv) *Remote Debug Tools (RDT)*, which provide web-based debugging extensions to the Android Debug Bridge (ADB) that is used during development.

We will start our demonstration out by overviewing the main elements of SmartLab. We shall then present the complete interaction workflow, i.e., allocate devices, transfer-files and initiate/control demonstration programs with our four modes of interaction (RCT, RS, RSE and RDT). We particularly plan to show how somebody can test both widely-known applications available through the market and interesting applications that we have developed in-house for crowd-sourced trajectory matching, optimal P2P search, distributed games, etc.

3. REFERENCES

[1] Peterson L. et. al., "A Blueprint for Introducing Disruptive Technology into the Internet," *HotNets'02*.

[2] Werner-Allen G. et. al., "MoteLab: a wireless sensor network testbed," *IPSN'05*.

[1]Video and System available: http://smartlab.cs.ucy.ac.cy/

MobiSys'12, June 25–29, 2012, Low Wood Bay, Lake District, UK.
ACM 978-1-4503-1301-8/12/06.

Demo: The Airplace Indoor Positioning Platform

Christos Laoudias
KIOS Research Center
University of Cyprus
laoudias@ucy.ac.cy

George Constantinou
Dept. of Computer Science
University of Cyprus
gconst02@cs.ucy.ac.cy

Marios Constantinides
Dept. of Computer Science
University of Cyprus
mconst02@cs.ucy.ac.cy

Silouanos Nicolaou
Dept. of Computer Science
University of Cyprus
snicol02@cs.ucy.ac.cy

Demetrios Zeinalipour-Yazti
Dept. of Computer Science
University of Cyprus
dzeina@cs.ucy.ac.cy

Christos G. Panayiotou
KIOS Research Center
University of Cyprus
christosp@ucy.ac.cy

ABSTRACT

In this demo paper, we present the *Airplace* indoor positioning platform developed for Android smartphones [1]. *Airplace* relies on existing WLAN infrastructure and exploits Received Signal Strength (RSS) values from neighboring Access Points (AP) to infer the unknown user location. Our system utilizes a number of RSS fingerprints collected a priori to build the so-called radiomap. Location is then estimated by finding the best match between the currently measured fingerprint and fingerprints in the radiomap [2].

We will demonstrate the real-time positioning capabilities of *Airplace* by allowing attendees to carry an Android tablet and viewing their position on a floorplan map, while walking around the demo area[1]. Our goal is to highlight the effectiveness of various algorithms found in the literature, as well as two state-of-the-art algorithms developed in-house [1].

The *Airplace* system consists of the *RSS Logger* and *Find Me* applications and the *Distribution Server*, while it follows a mobile-based network-assisted architecture to eliminate the communication overhead and respect user privacy. In a typical scenario, when a user walks inside a building a smartphone client conducts a single communication with our server to receive the RSS radiomap and is then able to position itself independently using the observed RSS values.

The *RSS Logger* application is developed around the Android RSS API for scanning and recording data samples in specific locations at predefined intervals; see Fig. 1a. These samples contain the MAC addresses and RSS levels (in dBm) of all neighboring WLAN APs, as well as the coordinates of the location where the user initiated the recording. The collected data are stored locally in log files and users may contribute their data to our system for building and updating the radiomap through crowdsourcing.

The *Find Me* application is a positioning client that downloads the radiomap from the server, thus enabling the user to self-locate independently thereafter. The interface is shown in Fig. 1b (left), where the user can set the preferences and select any of the available algorithms. Subsequently, the *Track Me* button can be switched on for tracking the user

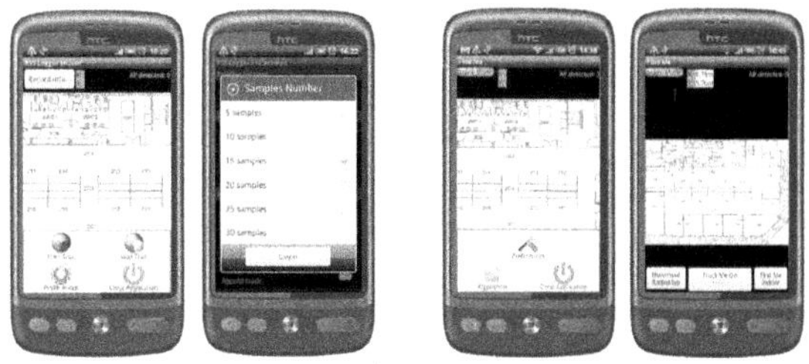

(a) (b)

Figure 1: *RSS Logger* (a) and *Find Me* (b).

while walking indoors. In this case, the current location estimate (green circle) is updated every one second, while the past locations are shown as red dots; see Fig. 1b (right).

Our *Distribution Server* is responsible for the construction and distribution of the radiomap. The server parses all available RSS log files, which may be contributed by several users, and merges the data in a single radiomap file.

For the demonstration we will use a Motorola Xoom tablet running Android 3.1 and featuring a 10.1′ screen that facilitates presentation. The *Distribution Server* will be running on a linux-based workstation and clients will use the tablet's built-in WLAN adapter to connect to the server, through the WLAN hotspots at the conference venue, for downloading the radiomap. First, our team will have collected adequate samples before the demo to guarantee good performance. Next, the participants may start positioning themselves with the *Find Me* application and they will be able to appreciate the potential of indoor location-oriented applications. To make the demo more appealing, a floorplan map of the demo area in .jpg format will be required for facilitating the collection of the data for the radiomap and displaying the location estimates during positioning.

Categories and Subject Descriptors

H.4 [**Information Systems Applications**]: Miscellaneous

Keywords

Indoor positioning, WLAN, Signal strength, Android

1. REFERENCES

[1] C. Laoudias et al., "The Airplace Indoor Positioning Platform for Android Smartphones," in *MDM*, 2012.

[2] M. Kjærgaard, "A taxonomy for radio location fingerprinting," in *LoCA*, 2007, pp. 139–156.

[1]A video demo of the *Airplace* system is available at: `http://www2.ucy.ac.cy/~laoudias/pages/platform.html`

MobiSys'12, June 25–29, 2012, Low Wood Bay, Lake District, UK.
ACM 978-1-4503-1301-8/12/06.

Demo: A Mobility Aware Cooperative Scheme for MAC Reconfiguration in Cognitive Wireless Networks

Xi Zhang, Junaid Ansari, Xuan Zhu and Petri Mähönen
{xzh, jan, zxu, pma}@inets.rwth-aachen.de
RWTH Aachen University, Institute for Networked Systems, Kackertstrasse 9, D-52072 Aachen, Germany

Categories and Subject Descriptors: C.2.1 [Network Architecture and Design]: Wireless Communication

Keywords: MAC, Adaptation, Cooperation

With increasing wireless communication applications, assigning dedicated spectral resource to each application is becoming impractical. Cognitive radios are emerging as a possible solution for enabling coexistence among networks and higher spectrum utilization by opportunistically using spectral holes through dynamic spectrum access schemes [1]. Since the spectral environment for a node can be highly dynamic and unstable due to mobility and spectrum sharing with other networks, medium access schemes must be flexible and adaptable in order to provide a stable and high Quality of Service (QoS).

Classically, protocols are implemented in a monolithic fashion which restricts reconfigurability required in a dynamic spectrum access paradigm and mobile environment. One common approach for reconfiguration is by appropriately setting the parameters of a particular protocol such as the selection of the modulation and coding schemes in LTE [2]. MultiMAC [3] switches among a few pre-defined standalone protocols based on QoS requirements. Although these approaches aim at providing flexibility, the design choice is limited to the subset of the tunable parameter values and pre-selected options. In our demonstration, we will use a toolchain for runtime protocol realization (TRUMP) [4] which is built based on Decomposable MAC Framework [5]. It enables flexible adaptation of MAC behaviour on-the-fly resulting from spectral and network dynamics, mobility and varying QoS application demands. Reconfiguration at individual nodes may lead to possibly incoherent MAC schemes in a neighbourhood. Therefore, we have designed and implemented a cooperative mechanism where coexistence of different MAC solutions is allowed in the same network without disruption in communication links. Furthermore, the cooperative scheme allows the neighbouring nodes to adapt their MAC behaviour in order to maximize the performance characteristics. This is achieved through a controlling mechanism which monitors the MAC behaviour of its neighbours and disseminates this information.

Demo This demonstration will provide an opportunity to the audience to interactively change the protocol configuration at individual nodes, spectral conditions and/or the network topology and correspondingly observe the resulting performance characteristics. The demonstration will show how our cooperative scheme enables MAC reconfiguration in a network neighbourhood. The demo will also show that our cooperative scheme allows nodes to reconfigure their MAC behaviour with a low controlling overhead while being able to communicate seamlessly even when a node joins or leaves the network. We have implemented our solution on WARP [6], a SDR platform with Virtex Pro II FPGA. In this demonstration, four WARP boards will be connected to a PC as shown in Figure 1. An interactive Graphical User Interface (GUI) displays live MAC behaviour and performance statistics. Different topologies can be formed by adjusting the positions of the WARP nodes on the GUI. The transmit power of the nodes will be changed accordingly to model the topology and node mobility. An interferer will be introduced to the network to model a dynamic spectral environment. The demonstration requires two channels at 2.4 GHz band to be dedicated for the usage since the interferer will be jamming the channels randomly.

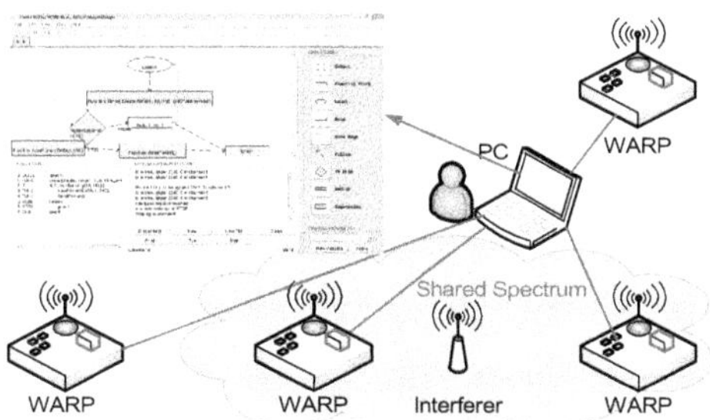

Figure 1: Demonstration scenario.

MobiSys'12, June 25–29, 2012, Low Wood Bay, Lake District, UK.
ACM 978-1-4503-1301-8/12/06.

Acknowledgment

We would like to thank partial financial support from RWTH Aachen University, DFG through UMIC research centre (Nucleus project) and EU through 2PARMA project.

1. REFERENCES

[1] Mitola J. III, "Cognitive radio: An integrated agent architecture for software defined radio" Ph.D. dissertation, KTH, Sweden, 2000

[2] 3rd Generation Partnership Project (3GPP). The LTE Protocol Specification, 3GPP Rel8 In http://www.3gpp.org/Release-8.

[3] C. Coerr *et al.*, "MultiMAC - an adaptive MAC framework for dynamic radio networking" In *Proc. of IEEE DySPAN*, 2005.

[4] X. Zhang, J. Ansari, G. Yang and P. Mähönen. "TRUMP: Efficient Realization of Protocols for Cognitive Radio Networks" In *Proc. of IEEE DySPAN*, 2011.

[5] J. Ansari *et al.*, "A Flexible MAC Development Framework for Cognitive Radio Systems" In *Proc. of IEEE WCNC*, 2011.

[6] A. Khattab *et al.*, "WARP: a flexible platform for clean-slate wireless medium access protocol design" In SIGMOBILE Mob. Comput. Commun. Rev, 2008.

Demo: Mobile Backhaul Bandwidth Optimization for Real-Time Content Using a Rule-based Proxy System

Sumedh Sathaye, Cesar Gonzales, Peter Westerink, Kang-Won Lee, Dinesh Verma
IBM T.J. Watson Research Center, Yorktown Heights, NY 10598
{sathaye, butron, peterw, kangwon, dverma}@us.ibm.com

Categories and Subject Descriptors
C.2.1 [**Computer Communication Network**]: Network Architecture and Design - *Network communications; Distributed networks; Wireless communication*

Keywords
Rule-Based; Mobile; Real-Time; Streaming; Hierarchical

1. PROBLEM STATEMENT
Mobile network infrastructures are facing ever increasing demand for data bandwidth due to the exponential rise in the number of users of smart devices and data intensive applications using video. The result is network congestion leading to poor quality of service. Caching frequently-accessed content close to end-users can help. However, there are many scenarios when caching is not possible such as when live, real-time content streams are accessed by many end-users simultaneously. Examples in the consumer space include: live web broadcasts of popular sports, catalogue launches, internet radio and TV. Examples in the enterprise space include live video conferencing, and other similar collaboration applications using mobile devices such as smart phones and tablets. IP multicasting is one approach to delivering real-time content to multiple users for which well-established standards exist, yet IP multicasting is not widely used in large networks because of a number of security concerns [1].

2. PROPOSED SOLUTION
We present a proxy system called SHERPA (Scalable, Hierarchical, Extensible, and Real-time Proxy Architecture) to help address this problem. SHERPA is designed to work at the edge of a mobile network (e.g. Wireless AP, a base station, or RNC). On the path between mobile clients and content servers, it can emulate multicasting in the backhaul network using unicast protocols. We demonstrate an implementation of SHERPA and show that it can scale well when deployed hierarchically within the network. In addition SHERPA can be used to customize streaming services via rules (with a set of rules for each real-time content service). SHERPA is also capable of multi-tenancy in that one or more services can be delivered concurrently from a single instance of the proxy.

3. SHERPA ARCHITECTURE
Connected to the mobile network edge equipment (as an embedded subsystem or a stand-alone system), SHERPA depends on the equipment to break-out IP traffic from/to mobile clients, but otherwise operates independently. The architecture of SHERPA is shown in Figure 1. SHERPA intercepts and transparently processes IP messages between clients and servers. Protocol-specific parsing/processing modules extract information such as protocol fields and data. A set of event-condition-action (ECA) rules fire upon four events: (1) Arrival of first mobile client request to new service such as CSPAN.org, (2) A subsequent request for that service, (3) A server response, and (4) Exit of a client from the service. Service rules are customized for a service, and are used to interpret protocol information such as session id, content flow id, and client join, leave, rejoin, channel change, exit etc. events. The rules invoke primitives in the SHERPA Core layer, which creates and manages state on behalf of each service connection. When a client requests a content stream already being served to another client, the core, under the control of service-specific rules, *automatically* serves the content locally from the SHERPA proxy, thus avoiding duplicate streams from the content server to the proxy. Similar automatic handling of events such as channel change, service join/exit is also supported. SHERPA Rules can be built-in, plugged-in, or written in languages such as JavaScript.

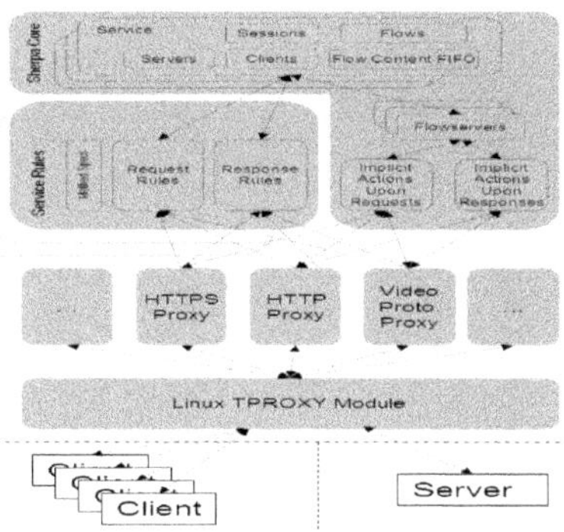

Figure 1: SHERPA Architecture

4. DEMO
This demo shows a SHERPA instance operating at a wireless AP, running 2 sets of rules, one for the C-SPAN web TV service, and another for IBM's SmartCloud Meetings web conferencing service (an example of an enterprise service that requires real-time delivery). When the attached N clients request the same content e.g. same C-SPAN channel, they are all served locally by SHERPA, thus reducing the mobile backhaul bandwidth required by a factor of (N-1). The demo requires only an Internet connection good enough to view C-SPAN web TV service live for a single client.

5. ACKNOWLEDGEMENT
Work partially funded by DARPA under contract no. Contract No. W911QX-10-C-0094.

6. REFERENCES
[1] "*IPv4 Multicast Security: A Network Perspective,*" by Tom Bachert, SANS Institute 2002. (Google: Bachert

Demo: Enabling Mobile Time-Dependent Pricing

Sangtae Ha
Dept. of Electrical Engineering
Princeton University
Princeton, NJ
sangtaeh@princeton.edu

Soumya Sen
Dept. of Electrical Engineering
Princeton University
Princeton, NJ
soumyas@princeton.edu

Carlee Joe-Wong
PACM
Princeton University
Princeton, NJ
cjoe@princeton.edu

Rudiger Rill
Dept. of Electrical Engineering
Princeton University
Princeton, NJ
ruediger.rill@gmail.com

Mung Chiang
Dept. of Electrical Engineering
Princeton University
Princeton, NJ
chiangm@princeton.edu

ABSTRACT

ISPs around the world have begun to offer new pricing plans for wireless data, such as usage-based pricing in the U.S., to mitigate recent growth in bandwidth demand. Time Dependent Pricing (TDP) represents a next step in this direction [1, 2]. With TDP, ISPs can shift traffic to off-peak periods, thus reducing their cost, while consumers save money by choosing the time of usage. TDP uses a feedback control loop between an ISP and its users to account for users' responses to offered prices in optimizing the future prices. We have implemented such a TDP system and are presently conducting a user trial at Princeton while planning larger trials with commercial ISPs. This demo will introduce the audience to our system's ISP- and user-side features. On the ISP side, we show the current network congestion, while on the user side, we show device UIs displaying the offered prices, the device usage history, and automated scheduling of applications to keep users within a specified budget.

Categories and Subject Descriptors

C.2.0 [**Computer-Communications Networks**]: General—*Data Communications, Pricing*; H.5.2 [**Information Interfaces and Presentation**]: User Interfaces—*GUI, Screen Design*

Keywords

Wireless pricing, Data plans, Economics

We will distribute iPads and iPhones to demo visitors, allowing real-time interaction with our GUI on these devices. We also show the network congestion condition from our trial. The prices shown on our devices are calculated by a pricing server located in our Princeton Edge Lab, which acts as the ISP within our trial. This server measures the aggregate traffic from all trial participants and updates its estimates of user response to prices, using these to compute the ISP's cost-minimizing prices for future time periods. For privacy reasons, the client-side UIs interact with the lab server only by exchanging price and usage information.

The client side consists of a UI installed on users' handheld devices, which serves two main purposes: self-education and automatic application scheduling. The education components allow users to view future prices, as well as the past prices and corresponding device usage (Figs. 1a and 1b). Users can also view the amount of bandwidth used by different apps, and in particular can see which apps use the most bandwidth (Fig. 1c). Finally, users can set their monthly budget on Fig. 1e's screen and turn on the "autopilot" mode, which automatically schedules applications. Users can specify delay tolerances for different applications, which are considered in the automated scheduling (Fig. 1d). Users can also manually schedule applications to future times (FIg. 1f) or to wait for WiFi access.

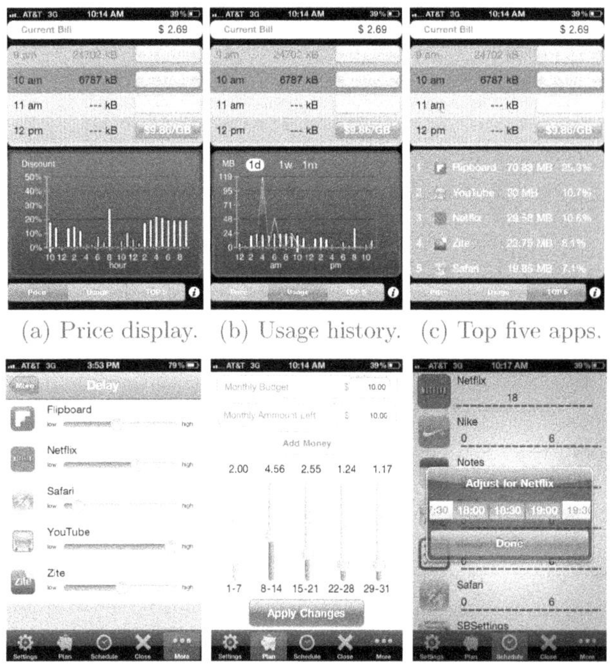

(a) Price display. (b) Usage history. (c) Top five apps.

(d) Delay indices. (e) Budget. (f) Scheduling.

Figure 1: Screenshots of the iOS app.

Acknowledgements

C.J.-W. was supported by the NDSEG fellowship.

1. REFERENCES

[1] TUBE Website. http://scenic.princeton.edu/tube/.

[2] JOE-WONG, C., HA, S., AND CHIANG, M. Time-Dependent Broadband Pricing: Feasibility and Benefits. *Proc. of ICDCS* (June 2011).

Demo: Code In The Air - Simplifying Tasking on Smartphones

Lenin Ravindranath, Tiffany Yu-Han Chen, Somak Das, Raluca Ifrim, Hari Balakrishnan, and Sam Madden
MIT Computer Science and Articial Intelligence Laboratory
Cambridge, MA, USA

ABSTRACT

Smartphones now are equipped with a variety of sensors including inertial sensors (accelerometers and gyroscopes) and multiple position sensors (GPS, WiFi, and cellular radios). These powerful capabilities have made smartphones an attractive platform for *tasking applications*. Tasking applications are rapid developing mobile applications which process data from multiple sensors continuously to determine user's context (such as location or activity) and take certain actions based on pre-defined conditions. Examples of such applications include location-based reminders, notifying when friends are nearby, changing the ring-mode of a phone automatically depending on the location, automatically tracking and storing movement tracks when driving, and inferring the number of steps walked each day. However, today, developing tasking applications is non-trivial for two reasons: poor abstractions and poor programming support [1].

We address these shortcomings through *Code In The Air* (CITA), a system that simplifies the development of tasking applications. CITA achieves this by providing an activity composition framework and a tasking execution framework that allows developers and end-users to easily write and compose tasks. Figure 1 shows the CITA architecture.

Categories and Subject Descriptors

C.3.3 [**Special Purpose and Application Based Systems**]: Real-time and embedded systems

Keywords

Tasking applications, Mobile programming,Mobile sensing

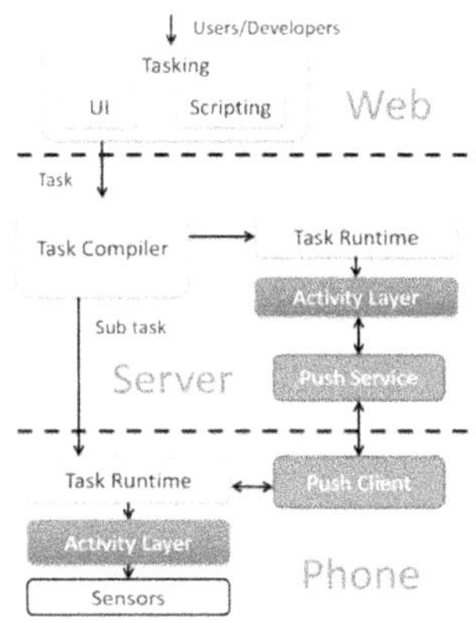

Figure 1: CITA Architecture

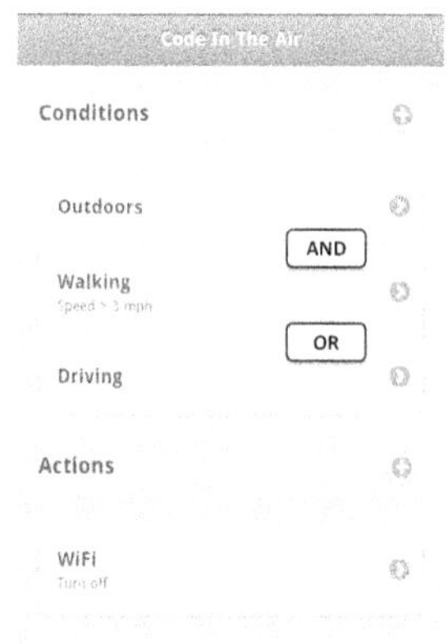

Figure 2: A screenshot of the end user interface.

1. DEMONSTRATION

During the demo, participants will be able to develop tasking applications by using CITA's programming model, and be able to use and interact with their applications in real time. We have implemented two programming platforms for participants to compose and create single- and multi-device tasking applications:

- **The end user interface.** The end user interface on the phone enables participants to compose their own tasks by "mixing and matching" available activity primitives or tasks. We have implemented various activity primitives such as *isWalking*, *isBiking*, *isRunning*, *isDriving*, *isOutdoors*, *enterPlace*, and *leavePlace*. Figure 2 shows a screenshot of using the user interface to compose a single-device application: "Disable WiFi when I am walking outdoors or driving."
- **The JavaScript programming model.** In CITA, participants can create tasks by writing only server-side JavaScript code. The execution framework will automatically divide these tasks into sub-tasks and distribute them among multiple devices or servers. The following CITA script expresses the multi-device application: "Alice would like to be alerted whenever Bob leaves his workplace":

```
p = getDevice('Bob','Phone');
q = getDevice('Alice','Phone');
p.registerActivityCallBack('leavePlace(
        work)','trueAction', null);
function trueAction(e){
        q.vibrate();
        q.alert('Bob left work');
}
```

We will also show how CITA's activity layer continuously processes data from multiple sensors to recognize user's context in an energy efficient way.

MobiSys'12, June 25–29, 2012, Low Wood Bay, Lake District, UK.
ACM 978-1-4503-1301-8/12/06.

2. REFERENCES

[1] L. Ravindranath, A. Thiagarajan, H. Balakrishnan, and S. Madden. Code In The Air: Simplifying Sensing and Coordination Tasks on Smartphones. In *HotMobile*, 2012.

Demo: CellChek: Demonstrating a Cost-effective Cell Phone-based Patient Monitoring and Advising System*

Mostafa Izz[1]
mostafaizz@eng.cu.edu.eg

Yossuf Khazbak[1]
yossuf.khazbak@gmail.com

Tamer Elbatt[1, 2]
telbatt@ieee.org

Moustafa Youssef[3]
moustafa.youssef@ejust.edu.eg

Categories and Subject Descriptors

J.3 [**Life and Medical Sciences**]: information systems; Health

Keywords

Health-care, Mobility, Cost-Effective

1. INTRODUCTION

Motivation: The ubiquity of mobile phones witnessed today along with the growing need for enhanced medical services worldwide provide a unique opportunity to save costs and introduce mHealth services, especially for low-income communities.

Objectives: In this demo, we exhibit the novel concept of cost-effective mobile healthcare which "optimally" leverages the multiple wireless interfaces onboard most mobile phones today. The demo, coined *CellChek*, is a proof-of-concept testbed that showcases a cost-effective, cell phone-based patient monitoring and advising system.

System Overview: *CellChek* leverages mobile phone sensors, as well as bluetooth-enabled on-body medical devices. The Wireless Interface Selection Algorithm (WISA) constitutes the core cost-saving mechanism which decides the **minimum cost** wireless interface for uploading the medical data to the healthcare provider. In addition, our research explores the benefits, and potential trade-offs, of leveraging "free" short range phone-to-phone communications (via Bluetooth/WiFi) for disseminating medical advisory messages to a target group of co-located individuals. This gives rise to a fundamental cost-delay trade-off that is modeled and formulated as a constrained optimization problem and solved in closed-form. The proposed *CellChek* demo is targeted towards enabling remote periodic checkups as well as detecting abnormal medical conditions for high-risk patients.

2. THE *CellChek* DEMO

Description: *CellCheck* demo is based on a Windows Mobile phone which hosts the major software components. It will demonstrate the following software modules and technologies: i) Establishing connection between the two medical devices in use, namely Pulse Oximeter[2] and Blood Pressure Monitor[3], and the smart phone application software using Bluetooth and ii) The WISA algorithm in the works deciding the minimum cost wireless interface, for uploading patients' data to the remote medical facility. In addition, the demo shows the usage of the internal mobile accelerometer for fall/activity detection. Finally, the demo exhibits the web interface of the medical facility [1] that is used to manage the medical profiles for the Application users.

[1] Cairo University [2] Nile University
[3] Egypt Japan University of Science and Technology (E-JUST)

*This work was funded in part by a research grant from Microsoft Research under the Mobile Healthcare for Africa Awards program.

MobiSys'12, June 25–29, 2012, Low Wood Bay, Lake District, UK.
ACM 978-1-4503-1301-8/12/06.

Figure 1: CellChek System Overview

The demo hosts three major research components:

I. As opposed to simply uploading the medical data using 3G cellular (Baseline), WISA examines the mobile phone wireless interfaces, in sequence, starting with the zero cost option(s) Thus, it first checks free WiFi, If not available, it compares the cost of using SMS/MMS (for plain text and images, respectively) to the cellular data plan, given the data size, modality and QoS. Towards this objective, we develop a Data Transfer Accounting (DTA) module for 3G cellular plans.

II. DTA module hinges on the proposed "generic" cost model that not only covers a variety of 3G cellular pricing plans offered by major mobile operators in Egypt (e.g., Vodafone), ranging from "Pay-per-MByte", "Pay-Per-Hour" to "Unlimited", but also captures the dynamics of plan usage for other purposes other than medical as well as the medical data transfer timing relative to plan commencement date. It relies on the pre-stored parameters of the plan in use, namely limited quota and time limit, among other parameters, to compute the cellular data transfer cost.

III. The medical facility server represents the *CellChek* backend at the healthcare provider (HCP) side and hosts a database for the registered patients, web services for communication between the server and the mobile devices and a website [1] that serves as the HCP end-user GUI for the purpose of monitoring remote patients' status updates. It is the central point for the data gathered from all system subscribers and can be used to remotely monitor the status of the patients 24/7, fire up alarms in case of emergencies, and send advisory information, as needed.

Finally, *CellChek* leverages the Accelerometer to detect sudden falls and fire up alarms at the medical facility. In case of an emergency situation, *CellChek* generates a medical profile that describes the patient's latest status and sends the data directly to the web server. If unable to reach the server (due to lack of Internet access), then the data is sent via SMS to a pre-specified emergency number for the medical personnel.

3. REFERENCES

[1] *CellChek* medical facility server website http://aspspider.info/cellcheck/web/default.aspx
[2] http://www.nonin.com/PulseOximetry/Fingertip/Onyx9560
[3] http://www.aandd.jp/products/medical/bluetooth/ua_767pbt.html

Demo: Mobile Interaction with Ads on Public Display Networks

Stefan Schneegaß, Florian Alt, Albrecht Schmidt
VIS, University of Stuttgart
{firstname.lastname}@vis.uni-stuttgart.de

ABSTRACT

In public places we can observe that many conventional displays are replaced by digital displays, a lot of them networked. These displays mainly show advertising in a similar way to television´s commercial break not exploiting the opportunities of the new medium [1]. Several approaches of interaction between mobile devices and public displays have been investigated over the last 15 years. In this demo we concentrate on challenges that are specific to public displays used for advertising. In particular we focus on how new approaches for interaction with content, means for content creation, and tools for follow-ups can be implemented based on mobile devices. With Digifieds we present a research system that has been used to explore different research questions and to showcase the potential of interactive advertising in public space.

Keywords

Digifieds, public displays, interaction, urban computing

Categories and Subject Descriptors

H.4.3 [**Information System Applications**]. Communications Applications— Bulletin Boards

1. RESEARCH CHALLENGES & GOALS

When designing and implementing public display networks, the research challenges span from technical issues related to display technologies and networking to challenges in HCI and user acceptance. The central goal is to create an infrastructure valuable to the users that has at the same time a valid business model. This can be achieved by creating systems mixing information and ads.

In order to do this, very basic challenges have to be addressed:

- **How to capture a user's attention?** In order to be of any value it is essential to make the user aware of displays and to communicate that there is something to see. Without mobile devices this is usually done by graphical means (e.g., eye-catching content). When using mobile devices these can also be used to draw the users' attention towards the display.
- **How to communicate interactivity?** Given current installation users do not expect interactivity; hence it is a challenge to show the users that there is an opportunity for interaction. This can be supported via mobile devices, e.g., by providing messages that offer direct means for interaction or by providing a generic mobile client as means for interaction.
- **How to support workflows across devices?** In many cases the interaction should not be limited to a single point in time. Supporting interactivity over time and with different devices provides new opportunities. Relating advertising content to search carried out on the mobile device or to activities is one option. Another challenge is to provide means for users to take content and interaction with them, e.g., interaction on the public display while waiting for the bus and then taking the content with them on the bus.

MobiSys'12, June 25–29, 2012, Low Wood Bay, Lake District, UK.
ACM 978-1-4503-1301-8/12/06.

Figure 1: Digifieds can use the mobile phone to attract attention and enable content exchange with public displays.

- **How to lower the bar for users to become advertisers?** Advertising on public displays is so far restricted to big players. By integrating mobile devices it may be feasible to make such systems more open, similar to public bulletin boards. How users can create and post content and personal ads as they are on the move is a resulting question.

2. ARCHITECTURE

The Digifieds platform used to carry out our research consists of four components: (1) a central server back-end for the data, (2) a web-based display client for visualizing information and direct interaction, (3) an Android mobile client, and (4) a web client.

To tackle our research challenges we aim at using the mobile client in the following ways. The *user's attention* can be raised through push-based notifications as they enter the display vicinity. *Interactivity* can be communicated through attract sequences showing how to interact. *Workflows across devices* are supported by Digifieds through enabling the exchange of content via QR and alphanumerical codes. Digifieds is targeted at individuals who want to advertise items through classified ads, as well as to retailers and event organizers to post professionally designed content.

3. DEMO SETUP

Our demonstrator consists of a display node that is connected to the Digifieds network. Visitors will be able to install our Android application, (available from the Android Market) or use one of our demo phones to try out the system and interaction techniques. By showcasing ways of raising the users' attention, communicating the interactivity of our application, and different interaction techniques we aim at triggering a discussion about how to best implement future advertising applications for public displays.

4. CONCLUSION

Advertisements on public displays need to become more attractive in the future in order to raise the interest of passersby. Attracting the user's attention, communicating interactivity, and supporting intuitive interaction techniques are essential. The Digifieds installation shows how the mobile phone can be used to achieve this.

5. REFERENCES

[1] Alt, F., Schmidt, A., Müller, J. Advertising on Public Display Networks. In: IEEE Computer (to appear), 2012.

Demo – Medusa: A Programming Framework for Crowd-Sensing Applications

Moo-Ryong Ra* Bin Liu* Tom La Porta† Ramesh Govindan*

*Department of Computer Science
University of Southern California
{mra, binliu, ramesh}@usc.edu

†Department of Computer Science and Engineering
Pennsylvania State University
tlp@cse.psu.edu

ABSTRACT

The ubiquity of smartphones and their on-board sensing capabilities motivates *crowd-sensing*, a capability that harnesses the power of crowds to collect sensor data from a large number of mobile phone users. Unlike previous work on wireless sensing, crowd-sensing poses several novel requirements: support for humans-in-the-loop to trigger sensing actions or review results, the need for incentives, as well as privacy and security. Beyond existing crowd-sourcing systems, crowd-sensing exploits sensing and processing capabilities of mobile devices. In this paper, we design and implement Medusa, a novel programming framework for crowd-sensing that satisfies these requirements. Medusa provides high-level abstractions for specifying the steps required to complete a crowd-sensing task, and employs a distributed runtime system that coordinates the execution of these tasks between smartphones and a cluster on the cloud. We have implemented ten crowd-sensing tasks on a prototype of Medusa. We find that Medusa task descriptions are two orders of magnitude smaller than standalone systems required to implement those crowd-sensing tasks, and the runtime has low overhead and is robust to dynamics and resource attacks.

Categories and Subject Descriptors

D.3.3 [**Programming Languages**]: [Language Constructs and Features – Frameworks]

General Terms

Design, Experimentation, Human Factors, Reliability, Security

Keywords

Crowd-sensing, Macroprogramming, Programming Framework, Smartphone, Worker Mediation

1. MEDUSA DESIGN

In Medusa, programmers specify crowd-sensing tasks as a sequence of stages. As an example, video documentation app. is depicted in Figure 2. The Medusa runtime executes the sequence of steps. The design of the Medusa runtime is guided by three architectural decisions that simplify overall system design. a) Partitioned Services: Medusa should be implemented as a partitioned system that uses a collection of services both on the cloud and on worker smartphones. b) Dumb Smartphones: Medusa should minimize the amount of task execution state that is maintained in smartphones. This design principle precludes, for example, large segments of a task from being completely executed on the smartphone without the cloud being aware of execution progress. c) Opt-in Data Transfers: Medusa should automatically require a user's permission before transferring any data from a smartphone to the cloud. Data privacy is a significant concern in crowd-sensing applications. Our principle ensures that, at the very least, users have the option to opt-out of data contributions. As a consequence, resulting system architecture is presented in Figure 1.

MobiSys'12, June 25–29, 2012, Low Wood Bay, Lake District, UK.
ACM 978-1-4503-1301-8/12/06.

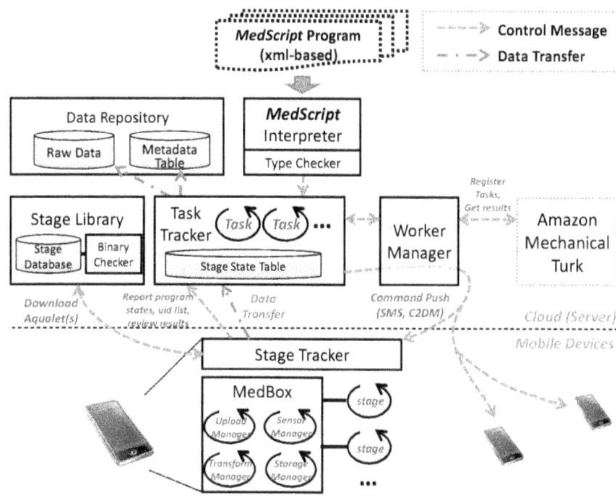

Figure 1: System Architecture

2. DEMO DETAILS

We have built Medusa prototype system using a commodity server machine and Android smartphones. Our demonstration will use video documentation app.(Figure 2) to illustrate crucial steps of the system such as a) executing MedScript program b) sending notification to users c) sensing and processing data on the smartphone d) review function to ensure data privacy. We anticipate that the audience can get realistic sense on the system's pratical usage scenarios.

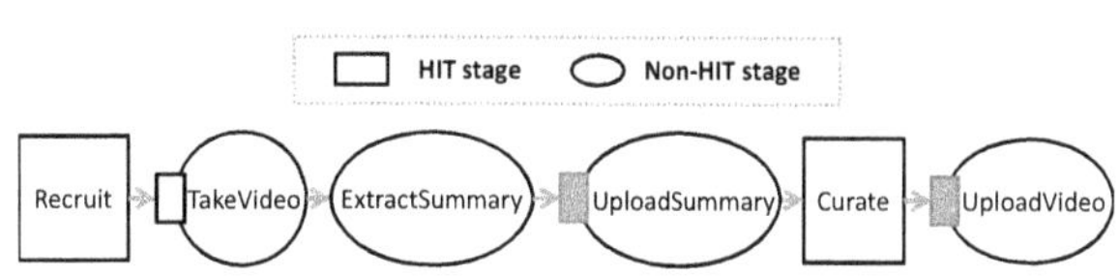

Figure 2: Video Documentation

3. DEMO REQUIREMENT

We need a reliable WiFi connection for Android smartphones as well as a laptop.

Demo: User Identification and Authentication with Capacitive Touch Communication

Tam Vu[†], Ashwin Ashok[†], Akash Baid[†], Marco Gruteser[†],
Richard Howard[†], Janne Lindqvist[†], Predrag Spasojevic[†], Jeffrey Walling[‡]
[†] WINLAB, Rutgers University, [‡] Dept. of Electrical and Computer Engineering,
{tamvu, aashok, baid, gruteser, reh, janne, spasojev}@winlab.rutgers.edu
{jeffrey.s.walling}@rutgers.edu

ABSTRACT

Today's identification and authentication mechanisms for touchscreen-enabled devices are cumbersome and do not support brief usage and device sharing. To address this challenge, this work explores a novel form of "wireless" communication that exploits the capacitive touchscreens which are now used in laptops, phones, and tablets, as a signal receiver. Using a custom built hardware token, in the form of a wearable ring, we show a proof-of-concept system that transmits a user identification code to the mobile device through the touchscreen. This mechanism works without any modification to the hardware or the firmware of the mobile device.

Categories and Subject Descriptors

C.2.1 [**Computer-Communication Networks**]: Network Architecture and Design—*Wireless Communication*

General Terms

Design, Measurement, Experimentation, Performance

Keywords

Signet Ring, Touchscreen Communication, User Identification, Capacitive Touchscreen

1. INTRODUCTION

Mobile devices provide us ubiquitous access to a vast array of media content and digital services, for example, access to our emails, photos, vehicle [4], house [1] and even accounting services such as bill-pay. As we increasingly rely on a variety of such mobile devices, we tend to quickly switch between them and temporarily share them with others [2]. We may let our children play games on our smartphones or share a tablet with colleagues or family members. Sometimes a device may be used by two or more people simultaneously, for example when playing a multi-player game. In all these situations, it would be of great benefit for the device to know who is interacting with it.

In this work, we show a proof-of-concept demonstration[1] of a novel form of "wireless" communication, that we term *capacitive touch communication* to address this challenge.

[1]Demo video is available at http://youtu.be/xmK1jH7iTuU

MobiSys'12, June 25–29, 2012, Low Wood Bay, Lake District, UK.
ACM 978-1-4503-1301-8/12/06.

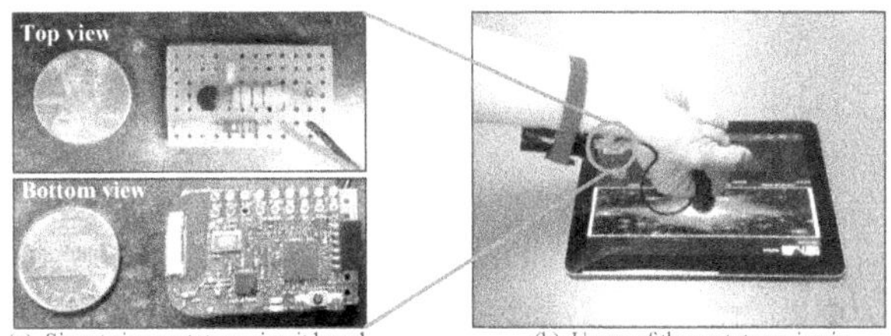

(a) Signet-ring prototype circuit board using off-the-self components

(b) Usage of the prototype ring in identification and authentication

Figure 1: Ring Prototype

The key idea is to exploit the pervasive capacitive touchscreen and touchpad input devices as receivers to identify bit-strings being transmitted from transmitters embeded in wearable artifacts, such as a ring.

These hardware tokens transmit electrical signals when it comes in contact with the screen either directly or through the finger. This signal spoofs the effect of human touch on the screen and thus results in touch events being registered to the device's operating system. Our software inside the device uses the number and timing information of those registered events for demodulating the transmitted data.

2. DEMO

We will show a custom-built hardware token, as in Figure 1-a, controlled by a battery-powered microcontroller TI-MSP430F2722 [3] that is programmed to carry a user's identification in the form of a bit sequence. The microcontroller generates modulated 3 Volt square waves at a frequency of 1 KHz. We will show how the user's identification embedded on the ring hardware is demodulated by the tablet, a Samsung Galaxy Tab 10.1, from the artificial touch events triggered by the square waves.

We will also demonstrate the capability of indirect communication in enabling a novel technique to differentiate two users simultaneously interacting with the same touchscreen in a shared-screen two player game such as Fruit Ninja. The two users (only one of them wears the ring) will play the game on the same touchscreen and our receiver software will differentiate each of the users.

3. REFERENCES

[1] Craftsman. Assurelink garage doors openers. http://bit.ly/HJyH2V.
[2] A. K. Karlson, A. B. Brush, and S. Schechter. Can I borrow your phone?: understanding concerns when sharing mobile phones. In *Proc. of CHI*, Apr. 2009.
[3] Msp430 ultra-low power microcontrollers. http://msp430.com.
[4] E. Sofge. Can a smartphone make your car smarter? *MSN Autos*, 2010.

Demo: Bluetooth TouchPoint

Gerald R. DeJean, Jeff Herron, Jie Liu, Darko Kirovski
Microsoft Research
One Microsoft Way, Redmond, WA 98052
{dejean; v-jefher; jie.liu; darkok}@microsoft.com

ABSTRACT

A new technology breakthrough which allows any standard Bluetooth mobile phone to access information services using the same selective and deliberate gesture envisioned for NFC. This technology is called Bluetooth Touchpoint, and it consists of reconfigurable coverage that combines NFC-like, close-proximity communications with the long-range, roaming solution of today's Bluetooth devices. Phase shifting technology combined with a unique antenna configuration makes this realization possible. Therefore, as the world moves toward contactless, wireless communication links, it is believed that the near ubiquity of Bluetooth in mobile devices makes this technology an excellent choice for delivering the benefits of NFC today without the wait, effort and cost associated with adopting NFC globally.

The main technology contribution is a new Bluetooth antenna design. This device is shown in Fig. 1. Instead of a single antenna, an array of antennas is integrated into the radio module. By carefully choosing the phase of radio signals sent at each antenna, constructive interference is created in certain special regions and cancel out the signals in other regions. Combining directionality with power control, the same antenna can be used to achieve three kinds of selectivity in device pairing, tapping, pointing, and approaching. Each is briefly explained below.

Tapping – With tapping, the user explicitly place one device against another to make the connection. The key advantage of tapping is the deliberateness. The user knows explicitly which counterpart she is interacting with. Examples include tapping a phone at a sales terminal for payment and scanning a bar code for object identification.

Pointing – In this mode of operation, the user points remotely to another device. The experience is best captured by remote control of home appliances. Unlike tapping, the users enjoy a better flexibility and range of the interaction range, without losing the deliberateness.

Approaching – Two devices can be paired as long as they are in the proximity of each other. This mode gives users the best mobility as long as it is within the communication range of another device. However, it also loses deliberateness. When multiple target devices are in the same space, users typically need to resolve them manually. For example, with WiFi links, a mobile device can see multiple access points, and the user must select which one to use.

Catagories and Subject Descriptors

B.0 [Hardware]: General

MobiSys'12, June 25–29, 2012, Low Wood Bay, Lake District, UK.
ACM 978-1-4503-1301-8/12/06.

General Terms

Design, Human Factors

Keywords

Directionality, PIFA, Bluetooth, NFC

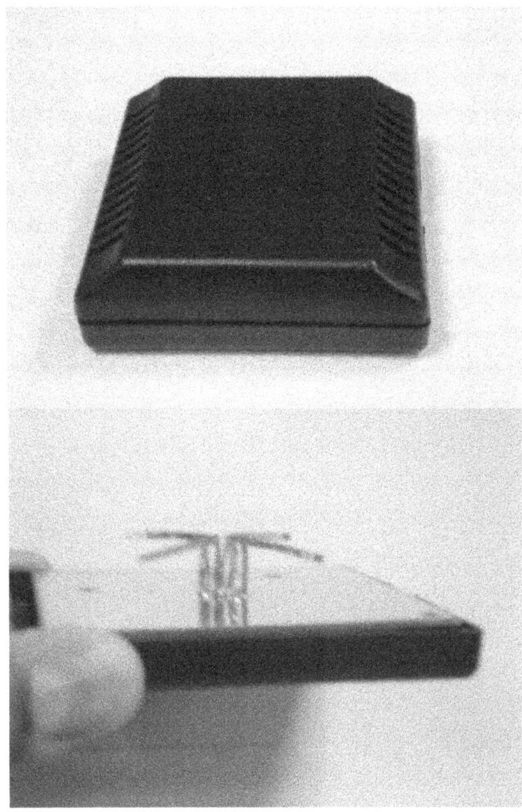

Figure 1. Prototype of Bluetooth TouchPoint appliance.

REFERENCES

[1] Jiang, X., Liang, C.-J., Chen, K., Zhang, B., Hsu, J., Liu, J., Cao, B., and Zhao, F. Design and Evaluation of a Wireless Magnetic-based Proximity Detection Platform for Indoor Applications. *Proc. IPSN 2012*, ACM Press (2012).

[2] Rulf, B., Robertshaw, G., Understanding Antennas for Radar, Communications, and Avionics. Van Nostrand Reinhold, NY, 1987.

[3] Kingsley, N., Papapolymerou, J. Organic "Wafer-Scale" Packaged Miniature 4-bit RF MEMS Phase Shifter. *Microwave Theory and Techniques* IEEE (2006).

Demo: Context Driven Advertisement Optimizer

Azeem J. Khan, V. Subbaraju and Archan Misra
Singapore Management University

Srinivasan Seshan
Carnegie Mellon University

Categories and Subject Descriptors

C.2.1 [**Computer-Communication Networks**]: Network Architecture and Design–Wireless Communication

Keywords

Mobile advertisements, Context-aware

1. INTRODUCTION

This proposal[1] describes an early demonstration of CAMEO[2], a framework that maintains uninterrupted operation of advertisement supported mobile applications, while reducing the amount of cellular data quota consumed by the delivery of advertisements. CAMEO is motivated by two recent trends: (1) Free "advertisement supported" mobile applications now dominate consumer marketplaces, and (2) Wireless telecom providers are increasingly adopting metered data plans. In such an environment, we have shown [1] that advertisement-supported applications are not truly 'free', as the advertisement traffic can represent a non-negligible fraction of overall data consumption. CAMEO both *a) preserves the delivery and display of mobile advertisements, and b) opens up new models of niche advertisement delivery*, while mitigating the advertisement-related consumption of precious metered wireless bandwidth.

2. SYSTEM ARCHITECTURE

CAMEO hinges on two observations: (1) Inexpensive or un-metered bandwidth (such as Wi-Fi) is often available opportunistically, and (2) Such bandwidth may be deployed solely for local advertising (e.g., in shopping malls). To support these goals, the CAMEO middleware (show in Fig. 1) requires a mobile application (that wants to display advertisements) to request a local proxy (residing on the mobile), which serves advertisements from a local (context aware) advertisement selector. The ads have been populated by a context driven ad-prefetcher which predicts likely future context (by mining past user behaviour) and then pre-fetches relevant advertisements during periods of cheap bandwidth and/or energy availability. Pre-fetching in CAMEO requires prediction of not just application request patterns but also broader user, device and even network context.

[1]This research is supported by the **Singapore National Research Foundation** under its International Research Centre Singapore Funding Initiative and administered by the IDM Programme Office.

[2]Context-Driven Advertisement Modulator and Optimizer

MobiSys'12, June 25–29, 2012, Low Wood Bay, Lake District, UK.
ACM 978-1-4503-1301-8/12/06.

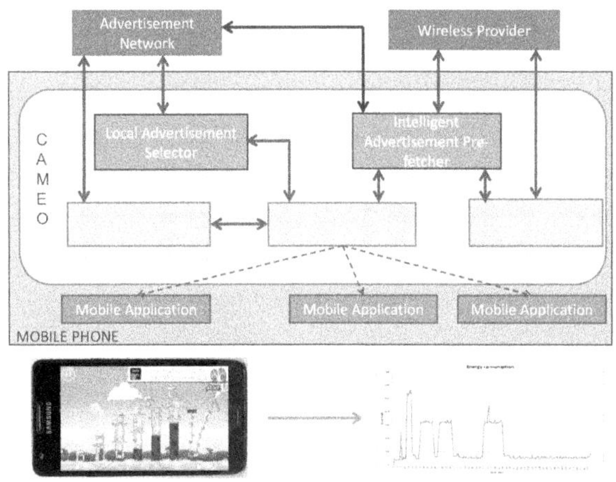

Figure 1: Proposed CAMEO demo

3. DEMONSTRATION DETAILS

In our demo, we shall demonstrate how *a)* CAMEO prefetches relevant advertisements from a real ad-network by predicting user context and *b)* the local phone proxy serves different advertisements based on the changing context.

Setup requirements: The demonstration will involve the use of one smartphone and one laptop, both provided by us. Ideally, we will require Wi-Fi Internet connectivity for the phone and the laptop, but we can modify the demonstration flow in case the Internet connectivity is absent.

Demo Scenario: The demonstration will let visitors use the smartphone to execute a designated application, such as Angry Birds. Prior to launching the game, we will select from a couple of geographical locations (e.g., UK and Singapore); this will cause real location & application specific advertisements to be downloaded and stored in the CAMEO proxy. The visitor can then continue playing the game, and will see the advertisements match the corresponding location chosen. For comparison purposes, we will also disable the CAMEO software and then show how the advertisements will instead be retrieved directly from the real ad-network. The demo will show that the user experience remains unchanged across both scenarios. Using appropriate real-time plots displayed on the laptop, we shall demonstrate how CAMEO helps a smartphone reduce both network cost and latency (conceptually illustrated in Figure 1).

4. REFERENCES

[1] A. J. Khan, V. Subbaraju, A. Misra, and S. Seshan. Mitigating the True Cost of Advertisement-Supported "Free" Mobile Applications. In *Proceedings of the Twelfth Workshop on Mobile Computing Systems &Applications*, HotMobile '12, pages 1–6, Mar 2012.

Demo: Introducing NextShare^{Mobile}, An Interactive Second Screen Application

William Knowles, Mu Mu, Will Bamford, Nicholas Race
Lancaster University, Bailrigg, Lancaster, LA1 4YW
{w.knowles, m.mu, w.bamford, n.race}@lancaster.ac.uk

Chris Needham
BBC Research & Development
MediaCity UK, Salford, M50 2LH
chris.needham@bbc.co.uk

ABSTRACT
The paper introduces NextShare^{Mobile}, an integrated second screen application. Design and evaluations are presented.

Categories and Subject Descriptors
H.5.2 User Interfaces (D.2.2, H.1.2, I.3.6)

General Terms
Design, Experimentation, Human Factors.

1. INTRODUCTION
The paper introduces NextShare^{Mobile}, an integrated second screen application. This work was produced as part of the P2P-Next project, which developed a next generation peer-to-peer content delivery platform into a consumer electronics device, called NextShare^{TV}. User feedback obtained from an initial Living Lab deployment of NextShare^{TV} suggested that many users were dissatisfied with the basic functionalities offered by the infrared remote control. This feedback was the predominant motivation for the development of NextShare^{Mobile} (Figure 1).

NextShare^{Mobile}, has been developed for Apple iOS devices, and designed to work in conjunction with the NextShare^{TV}. NextShare^{Mobile} supports functionalities of a conventional infrared remote control, through virtual keypads, rich content discovery with customisable EPG styles, keypad search within the application or on the primary screen with real-time updates, the granular management of playback sessions, and interactive applications.

Figure 1: Virtual RC, chapter selection, and quizzing

2. LIMO INTERACTIVITY
LIMO (Lightweight Interactive Media Objects) is a framework for timed metadata developed as part of the P2P-Next project, and integrated into NextShare^{Mobile}. In the LIMO framework, timed metadata can be used to build interactive applications associated with the media content, such as chapter navigation or quiz games. It may also be used to attach descriptive metadata relating to periods of time within the content, such as location or characters. In NextShare^{Mobile}, it is used to provide chapter navigation, quizzing, and subtitling of primary screen (Figure 2).

Figure 2: Interaction between devices

NextShare^{Mobile} connects to the NextShare^{TV} STB using Universal Plug and Play (UPnP). In UPnP terms, NextShare^{Mobile} acts as a wireless control point (client) with the STB acting as the controlled device (server) (Figure 3).

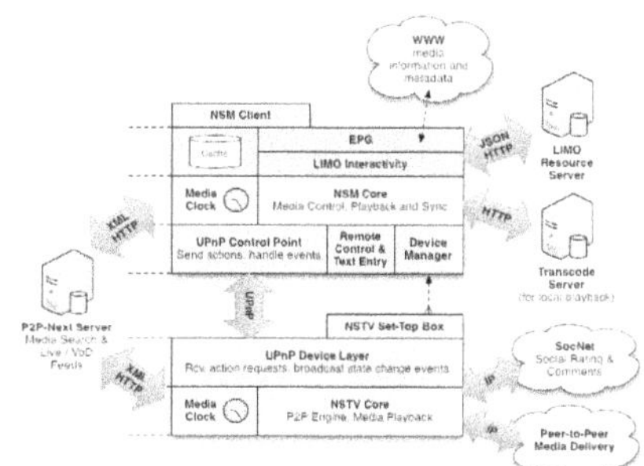

Figure 3: NextShare^{Mobile} Architecture

3. ROAMING SCREENS
Multiple NextShare^{Mobile} clients can control and stay synchronised to a single STB, and also move between STBes, transferring their current viewing context with them. Furthermore, the software supports local on-device playback of media, allowing, for example, a user to begin playback on the STB, continue watching on device, and then resuming playback on a STB in another room.

4. DEMONSTRATION REQUIREMENTS
The NextShare^{Mobile} demonstration requires an Ethernet connection, one television, and two power sockets.

5. ACKNOWLEDGEMENT
The work presented in this paper is supported by EC FP7 P2P-Next and UK FIRM project.

MobiSys'12, June 25–29, 2012, Low Wood Bay, Lake District, UK.
ACM 978-1-4503-1301-8/12/06.

Demo: Using Mobile Devices to Personalize Pervasive Displays

Sarah Clinch
Lancaster University
s.clinch@comp.lancs.ac.uk

Thomas Kubitza
Lancaster University and
University of Duisburg-Essen
thomas.kubitza@uni-due.de

Nigel Davies
Lancaster University
nigel@comp.lancs.ac.uk

Marc Langheinrich
University of Lugano (USI)
marc.langheinrich@usi.ch

Categories and Subject Descriptors

H.4 [**Information Systems Applications**]: Miscellaneous

Keywords

public displays, personalisation

1. INTRODUCTION

Personalising content on public displays can be achieved through many methods; recent work has often used the Bluetooth capabilities of mobile devices [4, 1]. However, power and privacy concerns may lead users to switch off the Bluetooth on their mobile devices, making them less useful for display personalisation. In this demonstration we show applications built using Tacita, an alternative method for allowing mobile users to make display personalisation requests.

2. DEMONSTRATION OVERVIEW

We previously demonstrated Tacita at HotMobile 2012 [3]. The system is comprised of four components: 1) an Android application that allows viewers to define content preferences; 2) a display component that schedules and renders content onto a display; 3) a set of Web applications designed for rendering on public displays; and 4) a map provider, which provides a database of display locations and available applications.

When the mobile application (Figure 1) determines that a user is near a public display that can provide applications matching their preferences it sends a personalisation request to the corresponding application. Each request specifies the display upon which to show the content plus any personalisation parameters. The application then sends a visualisation request to the display to ensure that the generated content is shown on the display.

This demonstration builds on previous work by expanding the set of public display applications and integrating with a novel display component, Yarely [2]. Yarely is Python-based display node software (handling media playback, sensing, and scheduling) designed specifically with the goal of supporting open display networks. The player is intended to support novel display applications that are dynamically scheduled, making it an ideal choice for Tacita.

Figure 1: Tacita Mobile App UI

3. REQUIREMENTS

The demonstration requires mains power, wireless Internet connectivity, wall space for projection, and table space to support a small projector and computer.

4. ACKNOWLEDGMENTS

The research leading to this demo has received funding from the European Union Seventh Framework Programme (FP7/2007-2013) under grant number 244011.

5. REFERENCES

[1] N. Davies, A. Friday, P. Newman, S. Rutlidge, and O. Storz. Using Bluetooth Device Names to Support Interaction in Smart Environments. In *Proc. of Mobisys '09*, 2009.

[2] N. Davies, M. Langheinrich, R. Jose, and A. Schmidt. Open display networks: A communications medium for the 21st century. *IEEE Computer*, 2012.

[3] T. Kubitza, S. Clinch, N. Davies, and M. Langheinrich. Using mobile devices to personalize pervasive displays. In *Demo. at HotMobile '12*, 2012.

[4] M. Sharifi, T. Payne, and E. David. Public display advertising based on bluetooth device presence. In *Proc. of MIRW '06*, 2006.

Demo: VisualComm - New Robust Channel of File Transfer in Mobile Communications

Zeyu Liu, Khiem Lam, Sirote Chaiwattanapong, Kyunghan Lee and Injong Rhee
North Carolina State University, Raleigh, NC, USA
zliu8, kvlam, schaiwa, klee8, rhee@ncsu.edu

ABSTRACT

There is a lack of method to communicate wirelessly without underlying networking infrastructure. We adopt the concept of camera-LCD communication and proposed this VisualComm system, a channel to communicate wirelessly through transmission and reception of a QR-code stream. VisualComm implements a series of techniques to overcome big challenges of frame missing, perspective distortion and viewing angle issue. In this demo, we show our VisualComm system can achieve real-time wireless file transfer with a reasonable bandwidth (of hundreds of Kbps) and good robustness.

Categories and Subject Descriptors

C.2.1 [**Network Architecture and Design**]: Wireless communication

Keywords

Mobile File Transfer, Information Dispersal, Robustness

1. BACKGROUND

A couple of works ([1] etc.) provided feasible solutions to video channel communication, however their works did not overcome at least one of the challenges of large portion of frame missing, real-time file transfer, or viewing angle issue.

2. VISUALCOMM IMPLEMENTATION

2.1 Framework of File Transfer System

To implement file transfer, files are fragmented into frames, and headers are added to each frame to enable file reconstruction. Contents in the frame headers include file ID, file type, frame ID, error correction byte and data padding size.

2.2 Robustness and Error Correction

In the real-time capturing of QR-code streams, frames may be lost or corrupted in the communication due to unsteady hands or misalignment of camera to the stream. To enhance the robustness and correct errors in practical use, we apply information dispersal algorithm (IDA) to the file fragmentation. IDA works in a way such that m original frames are encoded into n frames ($n \gg m$), but only m out of the n encoded frames are needed to reconstruct the data. With the help of IDA algorithm, user is guaranteed to receive the file even with a large number of frames missing.

MobiSys'12, June 25–29, 2012, Low Wood Bay, Lake District, UK.
ACM 978-1-4503-1301-8/12/06.

Figure 1: VisualComm demo setup illustration.

2.3 Viewing Angles and Perspective Distortion

Large viewing angles may blur the rear part of the QR-code, making it unreadable. In order to enhance the throughput of video communications in large viewing angles, we used a new approach of encoding multiple QR-code in one video frame, so that the reception of valid local information can be decoded without missing a frame opportunity.

Perspective distortion may impact the readability of the QR-code. Distortion correction is achieved by detecting the finder patterns at the four corners and scale the captured image back to readable QR-code.

2.4 Performance

Our VisualComm system delivers a reasonable bandwidth of hundreds of Kbps in real-time communication. With the help of our multi-coded-frame video display, 2/3 of the maximum bandwidth can be achieved with a 45-degree viewing angle and 1/4 with a 60-degree angle.

3. DEMO SETUP

Our demo will use the setup as shown in Figure 1, where the Android phone at the receiver side captures the QR-code frames displayed on the LCD screen, and decodes them into files. We will use different series of pre-encoded video clips to transfer a couple of files with different types and names, which can be opened on phone after the transfer finishes. Moreover, to illustrate the robustness of our VisualComm system, the mobile phone will be moved away from the screen for a short period of time, to mimic the careless hand. A multi-coded-frame video clip will also be used to show the ability that VisualComm deals with viewing angle issue.

4. REFERENCES

[1] Samuel David Perli, Nabeel Ahmed, Dina Katabi PixNet: interference-free wireless links using LCD-camera pairs *MOBICOM 2010: 137-148.*

Demo: An Experience Sampling Toolkit for the Transport Domain

Dave Gullick
Lancaster University
Lancaster, England
d.gullick@lancs.ac.uk

Prof. Nigel Davies
Lancaster University
Lancaster, England
nigel@comp.lancs.ac.uk

Mike Harding
Lancaster University
Lancaster, England
hardingm@comp.lancs.ac.uk

Categories and Subject Descriptors

H.1.2 [**Information Systems**]: User/Machine Systems; J.4 [**Social and Behavioral Sciences**]:

General Terms

Design, Experimentation, Human Factors, Measurement.

Keywords

Experience Sampling, Mobile Computing, Crowd-sourcing.

1. INTRODUCTION

There are a large number of projects that aim to use mobile computing technologies to help address transport related issues such as proving better travel information. These projects can often benefit from experience sampling in which mobile devices are used to collect in-situ user feedback on either the mobile computing application or the journey itself. In this demonstration we show elements of the Reflect toolkit, an experience sampling toolkit designed to help survey traveller experiences in-situ using mobile smartphones. The Reflect experience sampling toolkit differs from previous experience sampling research such as MyExperience [1] by specifically focusing on supporting behavioral researchers in the transportation research domain. It enables simple and rapid configuration and deployment of surveys on modern mobile platforms to crowd-source new forms of travel experience data.

1.1 Experience Sampling Toolkit

The Reflect toolkit consists of two distinct sets of components (Figure 1). A mobile client that can monitor mobile sensor data, dynamically validate sensor readings against pre-defined XML trigger condition scripts at runtime and display user interfaces to collect qualitative user data in-situ. Additionally, the mobile client enables users to manage their subscription to surveys and provide consent for data capture.

A web server application in the cloud has three sub-components that (1) provide a graphical user interface for researchers to create experiments with associated ethics and access rules, configure survey question actions, setup associated context trigger states and dynamically deploy survey configurations to participant devices in real-time, (2) a back-end database to store research surveys, question actions and trigger configurations and (3) a component that handles the generation of XML that describes survey context conditions and the associated actions that need to occur when a condition is validated.

1.2 Demonstration

Delegates to the conference can observe how the toolkit can be used to sample user feelings for a particular journey. Specifically, the demonstration shows (1) how a researcher can configure surveys on the web server, (2) how the toolkit generates XML to define condition triggers and coupled actions, (3) how surveys are dynamically synchronized to multiple client applications in real-time and (4) how the client is able to use a deployed survey configuration to recognize user contexts and render specific survey question interfaces.

Figure 1 Reflect Experience Sampler Framework

2. ACKNOWLEDGEMENTS

This work was conducted as part of the EPSRC funded Reflect Project.

3. REFERENCES

[1] Jon Froehlich, Mike Y. Chen, Sunny Consolvo, Beverly Harrison, and James A. Landay. 2007. MyExperience: a system for *in situ* tracing and capturing of user feedback on mobile phones. In *Proceedings of the 5th international conference on Mobile systems, applications and services* (MobiSys '07). ACM, New York, NY, USA, 57-70

Demo: Phone-to-Phone Mobile Motion Gaming on Commodity Phones

Zengbin Zhang
UC Santa Barbara
zengbin@cs.ucsb.edu

David Chu
Microsoft Research
davidchu@microsoft.com

Xiaomeng Chen
University of Science and Technology of China
monachen@mail.ustc.edu.cn

Thomas Moscibroda
Microsoft Research Asia
moscitho@microsoft.com

Categories and Subject Descriptors

C.3 [**Special Purpose and Application-Based Systems**]: Real-time and embedded systems; C.5.3 [**Computer System Implementation**]: Microcomputers—*Portable devices*

Keywords

acoustic localization, mobile gaming, mobile motion games, smartphones

1. INTRODUCTION

In this demo, we show examples of a novel class of mobile games called *Phone-to-Phone Mobile Motion Games* (MMG), in which *position, location, orientation, or movement of the phone is an integral part of game play.* Consider for example the following SwordFight game: two players wield their phones as imaginary swords, and try to attack each other. Relative positions and orientations of the two phones are combined to determine whether one is Attacking or Blocking [1].

MMG games represent a class of highly interactive games in which gameplay directly involves the user's physical activity – just as the Kinect and Wii do for console gaming. Such MMG games are a marked departure from the vast majority of today's mobile games. Rather than requiring players to conduct game actions via the screen, MMG games enable players to translate real-world motion into virtual gameplay. Yet MMG games require no external infrastructure and are played purely phone-to-phone.

The enabling technology for a game like SwordFight is the ability to conduct *very fast*, *accurate*, and *robust distance measurements* between the phones without any external infrastructure. Previous studies have shown that measurement lag lower than 100ms is needed to support action games without impacting user satisfaction. Additionally, the distance measurement should be accurate to within centimeters, and robust even when the players move fast or are in noisy environments.

Our system described in [1] utilizes acoustic sound for distance measurement. Previous studies have shown that acoustic localization can achieve accurate ranging in static scenarios with sufficient computation time. However, for MMG games, high-frequency and low-lag measurements are required to support correct game decisions. Furthermore, the players are often highly mobile with hand speeds as high as 2m/s. In the accompanying paper, we address these challenges in a systematic manner. First, we enable real-time distance measurement by an autocorrelation-based multi-stage algorithm, instead of using traditional yet computationally-intensive cross-correlation, all the while still maintaining high accuracy. Furthermore, we implement a pipelined streaming strategy to significantly improve the measurement frequency by overlapping different stages of the measurement. Finally, we introduce effective methods to tackle user mobility and environment factors, such as Doppler effect, multipath issues and ambient noise. Our evaluations have shown that the system is suitable for Mobile Motion Games even under high noise and mobility.

2. GAME DESIGN

In this demo we will demonstrate the following two games.

Game: SwordFight In SwordFight, the core game play works as described in the introduction. We use the accelerometer to tell whether the player is Attacking or Blocking. Additionally, players maintaining Attack or Block mode for too long suffer health exhaustion.

Game: ChaseCat ChaseCat is a round-based game. In each round, one player plays as the Cat while the other plays as the Dog. At the first half of the game, which we call as *the Chasing Phase*, the Dog will try to catch the Cat by getting her phone as close to the Cat's phone as possible, while the Cat will try to escape. Then, an audio cue will instruct both of them to Freeze. Distance measurements will be performed in *the Freeze Phase*, and the Dog wins if she's is within 30cm of the Cat. Otherwise, the Cat wins.

3. REFERENCES

[1] Z. Zhang, D. Chu, X. Chen, and T. Moscibroda. Swordfight: Enabling a new class of phone-to-phone action games on commodity phones. In *Proceedings of the 10th International Conference on Mobile Systems, Applications and Services (MobiSys)*, 2012.

MobiSys'12, June 25–29, 2012, Low Wood Bay, Lake District, UK.
ACM 978-1-4503-1301-8/12/06.

Demo: Unsupervised Indoor Localization

He Wang, Souvik Sen, Alex Mariakakis, Romit Roy Choudhury
Duke University
Durham, NC, USA
romit.rc@duke.edu

Ahmed Elgohary, Moustafa Farid, Moustafa Youssef
E-JUST
New Borg El-Arab City, Alexandria, Egypt
moustafa.youssef@ejust.edu.eg

ABSTRACT

We propose UnLoc [1], an unsupervised indoor localization scheme that bypasses the need for war-driving. Our key observation is that certain locations in an indoor environment present an identifiable signature on one or more sensing dimensions. An elevator, for instance, imposes a distinct pattern on a smartphone's accelerometer; a specific spot may experience an unusual magnetic fluctuation (refer to Figure 1). This form of urban sensing and activity recognition has already been demonstrated in literature [2, 3], but not yet applied in pure localization applications. We hypothesize that these kind of signatures naturally exist in the environment and can be envisioned as internal *landmarks* of a building. Mobile devices that "sense" these landmarks can recalibrate their locations, while dead-reckoning schemes can track them between landmarks. Neither war-driving nor floorplans are necessary - the system simultaneously computes the locations of users and landmarks, in a manner so that they converge reasonably quickly. We believe this is an unconventional approach to indoor localization, holding promise for real-world deployment.

Categories and Subject Descriptors

H.3.4 [**Information Storage and Retrieval**]: Systems and Software; C.2.4 [**Computer-Communication Networks**]: Distributed Systems

General Terms

Design, Experimentation, Performance

Keywords

Location, Mobile phones, Sensing, Landmarks, Recursion

1. DEMONSTRATION

We conducted real-life experiments in 3 different university buildings - Computer Science, Engineering, and North Gate shopping mall [1]. We covered approximately $1750m^2$, $3000m^2$, and $4000m^2$, respectively, in these buildings. Users walked around arbitrarily in the building for 1.5 hours, covering multiple floors; they carried 2 Android NexusS phones with UnLoc, one in the pocket and another in the hand with the screen facing up. The results of those experiments are shown in Figure 2.

We propose a live demonstration of the UnLoc system to show its localization capabilities. Prior to the day of the demonstration, we will bootstrap the system for the demonstration area; this entails having multiple people walk around various parts of the hotel carrying phones with UnLoc in order to generate a database of landmarks. On the day of the demonstration, we will invite participants to walk around the vicinity while holding Android NexusS phones installed with our system. They will be able to track their location on the screen of their phone through a floor map display. Meanwhile, other participants on site will be able to view a more detailed graphical display, showing the locations of both users and landmarks.

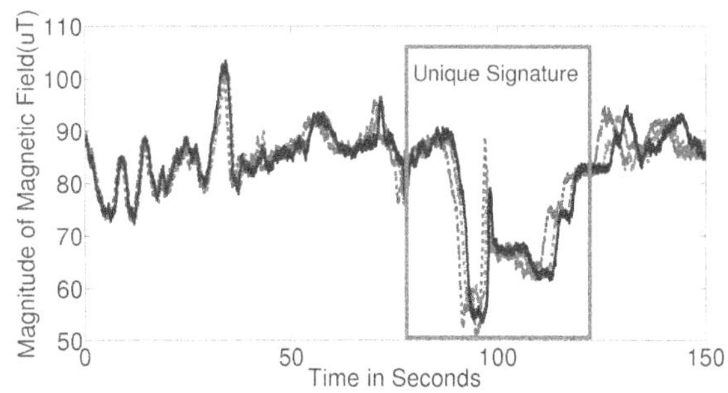

Figure 1: Magnetic signature near a networking lab.

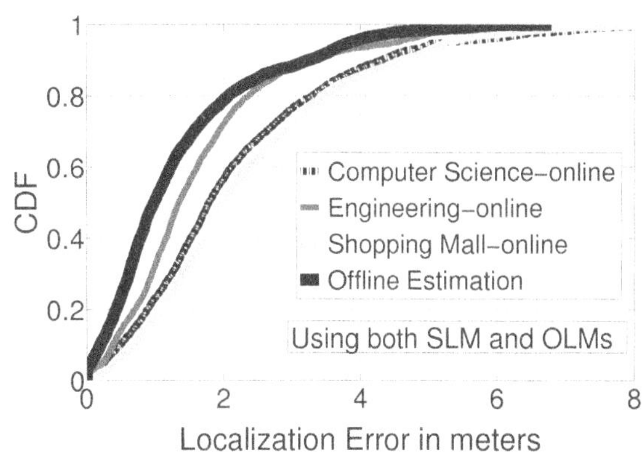

Figure 2: CDF of localization error using SLMs and OLMs

2. REFERENCES

[1] H. Wang et al. Unsupervised indoor localization. In *MobiSys*, 2012.

[2] J. Chung et al. Indoor location sensing using geo-magnetism. In *MobiSys*. ACM, 2011.

[3] L. Bao and S. Intille. Activity recognition from user-annotated acceleration data. *Pervasive Computing*, pages 1–27, 2004.

MobiSys'12, June 25–29, 2012, Low Wood Bay, Lake District, UK.
ACM 978-1-4503-1301-8/12/06.

Demo: NFC-based Sensor Data Caching

Jeremy Gummeson,
Pengyu Zhang, Deepak Ganesan
Department of Computer Science
University of Massachusetts, Amherst, MA 01003
{gummeson, pyzhang, dganesan}
@cs.umass.edu

Nissanka Bodhi Priyantha
Microsoft Research
Redmond, WA 98052
bodhip@microsoft.com

ABSTRACT

Near Field Communications (NFC) is an emerging technology that conveniently establishes radio communication by bringing two entities in close proximity of one another. Many use cases for devices equipped with this technology have been proposed ranging from payment systems to convenient data exchange.

In this demo, we implement a prototype NFC-based sensor device that periodically writes sensor data into a non-volatile memory; this data may be read out at a later time with an NFC-equipped smartphone or other device. The sensor features an ambient energy harvesting unit that allows sensing operations while disconnected from the phone; the phone is used as a supplemental harvesting source in addition to a way to offload collected sensor data.

Categories and Subject Descriptors

C.2 [**Computer-Communication Networks**]: Network Architecture and Design

Keywords

NFC, Energy Harvesting, Batteryless

1. DEMO DESCRIPTION

This demo shows the operation of an atonomous sensor data cache that provides sensing, computation, ambient energy harvesting, and NFC-based communication capabilities. The key differentiating technology used here is NFC; as opposed to other wireless standards (e.g. Bluetooth, 802.15.4), NFC allows every user equipped with an NFC-enabled smartphone to become a potential data mule that gathers accumulated sensor data.

As shown in Figure 1, our system architecture comprises an energy-efficient microcontroller, temperature sensor, EEPROM, capacitor-based energy buffer, a front-end circuit that handles NFC communications and harvesting, and a solar harvesting unit. The microcontroller's system role is to periodically query the temperature sensor attached to one of its ADC ports. The results, along with timestamps, are written into the EEPROM using an I2C serial interface. An analog front end circuit and harvesting unit provide 13.56MHz wireless energy harvesting and NFC read/write access to the EEPROM. The energy required for periodic sensor readings and EEPROM writes is provided by the capacitor-based energy buffer and attached solar harvesting unit.

MobiSys'12, June 25–29, 2012, Low Wood Bay, Lake District, UK.
ACM 978-1-4503-1301-8/12/06.

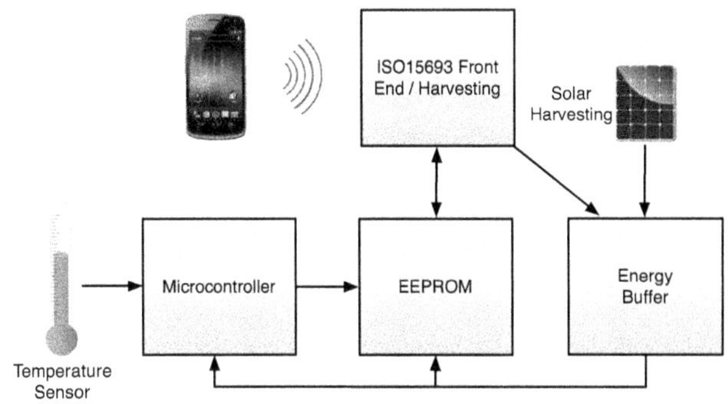

Figure 1: System Architecture

An Android application for a Samsung Galaxy Nexus smartphone is used to read the sensor data that accumulates in the EEPROM. The phone user initiates a tag read, which reads accumulated sensor data from the EEPROM and provides supplemental harvested energy; after the read completes, a graph of temperature readings vs time is displayed on the phone screen. This demo provides several novel contributions:

- We design and implement an energy harvesting (solar) based sensing system. Specifically, it exploits ambient energy harvesting for environmental monitoring (temperature sensing) and sensor data caching.
- We interface a low-power microcontroller with an analog front end that provides access to cached sensor data as well as RF-based energy harvesting.
- We design and implement a user program on a Galaxy Nexus Smartphone which communicates with our data cache sensor via NFC. The application provides a convenient interface for initiating reads and visualizing sensor data.

This proof-of-concept design demonstrates a new class of sensor device that bridges the gap between energy efficient sensor networks and crowd-sourced smartphone applications. This approach to sensing provides users the unique opportunity to directly interact with sensors deployed in the field and allows more exotic, disconnected deployment locations to be considered where always-connected networks may be difficult to maintain.

2. ACKNOWLEDGEMENTS

This work was funded by NSF grants CNS-0916577, CNS-0855128, and CNS-0923313.

Poster: Multi-track Map Matching

Adel Javanmard
Stanford University
adelj@stanford.edu

Maya Haridasan
Google Research
haridasan@google.com

Li Zhang
Microsoft Research
lzha@microsoft.com

Map matching is the procedure for determining the path of a user on a map from a sequence of location data (which we refer to as *track*). This process serves as a common preprocessing step for reasoning about traffic on the road network as well as for providing better location-based services [3, 5, 1]. Converting a track to a topological path on a map not only makes it easier to reason about paths, but also leads to reduced storage requirements and more efficient operations on the same. Some examples of such operations are comparison, querying and retrieval of tracks based on similarity to other tracks [1].

Due to its importance, many methods for map matching have been previously proposed [3, 4, 5], focusing on matching a single sufficiently dense and accurate sequence of locations. In this work, we consider the *multi-track map matching problem*, where we are given a number of tracks generated from trips through the same path, and we wish to recover the underlying path that generates these tracks. The problem is much more challenging than the single-track problem since each track contains a small number of samples (i.e. sampling intervals are large). This captures the realistic scenario where users repeatedly travel on regular routes, and samples are sparsely collected due to restraints in energy consumption on the mobile device. Another scenario is when a service collects users' location information only when the user actively uses such service. In such a scenario, the collected location data can be very sparse. However, since users typically travel on the same (or similar) routes repeatedly, one may compensate the sparseness of the data by combining location data from different trips.

The main challenge in multi-track map matching is that in combining data from multiple tracks, global ordering on all samples is not available, a necessary condition for applying existing single-track map matching algorithms. Instead, each track only gives us the order on a subset of locations. If we apply the map matching algorithm on each individual track, we would obtain paths with very poor quality given the low sampling rate on each path.

We propose two methods, the *iterative projection scheme* and the *graph Laplacian scheme*, to solve the multi-track problem by using a single-track map-matching subroutine. We also propose a boosting technique which may be applied to either approach to improve the accuracy of the estimated paths. In addition, in order to deal with variable sampling rates in single-track map matching subroutine, we revisit the single-track map matching problem and propose a method based on a particular regularized cost function that can be adapted for different sampling rates and measurement errors. The adaptation is in fact done through the regularization parameter λ in the cost function which balances two types of errors in measuring the quality of a path: the data error (fidelity of the path to the data points), and the model error (measuring "niceness" of the obtained path). By analyzing a simplified modeling of the problem, we characterize the optimal choice of λ in terms of the sampling rate and the variance of the measurement errors.

Finally, we evaluate the effectiveness of our proposed techniques for reconstructing tracks under several different configurations of sampling error and sampling rate. The evaluations are done on the dataset available in [2]. The data set contains tracks collected from real users in Seattle, WA, using commercially available consumer grade GPS device. Our results indicate that the proposed approaches lead to reasonable estimates of the route, significantly better than what would be achieved in case tracks were map matched individually.

MobiSys'12, June 25–29, 2012, Low Wood Bay, Lake District, UK.
ACM 978-1-4503-1301-8/12/06.

Categories and Subject Descriptors

I.5.1 [**Computing Methodologies**]: Pattern Recognition, Statistical Modeling

Keywords

GPS trajectory, map matching

1. REFERENCES

[1] M. Haridasan, I. Mohomed, D. Terry, C. A. Thekkath, and L. Zhang. Startrack next generation: A scalable infrastructure for track-based applications. In *OSDI*, pages 409–422, 2010.

[2] J. Krumm and E. Horvitz. The Microsoft multiperson location survey. Technical Report MSR-TR-2005-103, Microsoft Research, Redmond, WA, USA, 2005.

[3] J. Krumm, J. Letchner, and E. Horvitz. Map matching with travel time constraints. In *Society of Automotive Engineers (SAE) World Congress*, 2007.

[4] P. Newson and J. Krumm. Hidden markov map matching through noise and sparseness. In *GIS*, pages 336–343, 2009.

[5] A. Thiagarajan, L. Ravindranath, H. Balakrishnan, S. Madden, and L. Girod. Accurate, low-energy trajectory mapping for mobile devices. In *NSDI*, 2011.

Poster: Towards Mobile GPU-Accelerated Context Processing for Continuous Sensing Applications on Smartphones

Chulhong Min[1], Wookhyun Han[2], Inseok Hwang[1], SangJeong Lee[1], Youngki Lee[1], Insik Shin[3], Junehwa Song[1]
Computer Science Department, KAIST
{chulhong, inseok, peterlee, youngki, junesong}@nclab.kaist.ac.kr[1]
hanwook@cps.kaist.ac.kr[2] insik.shin@cs.kaist.ac.kr[3]

Categories and Subject Descriptors

K.8 [**Personal Computing**]: General; C.3 [**Special-Purpose and Application-based Systems**]: Real-time and embedded systems

Keywords

Acceleration, Context, GPU, Mobile sensing, Smartphone

1. MOTIVATION

Mobile sensing applications (MS apps) are coming into the spotlight as next generation applications on smartphones. These applications continuously monitor a user's situational contexts using mobile sensors and provide the user with proactive services. To prevent obesity and help weight control, for instance, the smartphone can track daily calories expended by physical activity. A number of such applications will be running simultaneously and contribute to the quality of user experiences through diverse services.

Despite their usefulness, many of these applications are not widely deployed yet. This is because they require a considerable amount of resources in a consistent manner. According to our measurement, indoor place tracking with ambient sound fingerprint continuously consumes about 50% of CPU utilization and 600mW of energy on Nexus One phone; note that GPS consumes about 450mW on average. With such high CPU overhead, the smartphone is able to support only a small number of the applications. The performance of other user-interactive applications would be degraded due to the CPU conflict. Also, such continuous energy consumption shortens the battery lifetime.

2. VISION

In this poster, we present the vision of mobile GPU-accelerated context processing for sensing applications on smartphones. The key idea is to leverage Graphics Processing Units (GPUs) on smartphones for context processing. By offloading a part of computation into GPUs, the GPU-accelerated context processing disburdens a smartphone's CPUs and thus achieves performance and energy benefits. A number of GPU-enhanced MS apps would be able to run in the background while little causing the foreground applications to slug or the smartphone battery to be depleted in the middle of a day.

The key design philosophies of GPUs share many analogies with the major workload characteristics of MS apps. GPUs are inherently designed to support highly parallel, iterative computation workload; continuous stream of voluminous data under low-latency and high bandwidth requirement. A typical example would be the real-time processing of a large dataset of textures and polygons through a well- defined pipeline consisting of Gouraud shading, anti-aliasing, etc. Similarly, many MS apps consist of pipelines with multi-step operations, each of which is an iterative signal processing operation such as FFT. Executing these operations often involves piece-wise computation on a large volume of windowed frames of raw data, in which we can expect a large potential of parallelism as in the common GPU tasks. More similarities lie in that many MS apps impose latency requirements, e.g., a few hundreds of milliseconds in hand gesture recognition. Moreover, the use of GPUs instead of CPUs often yields great energy-saving benefits, which is a critical issue in mobile systems.

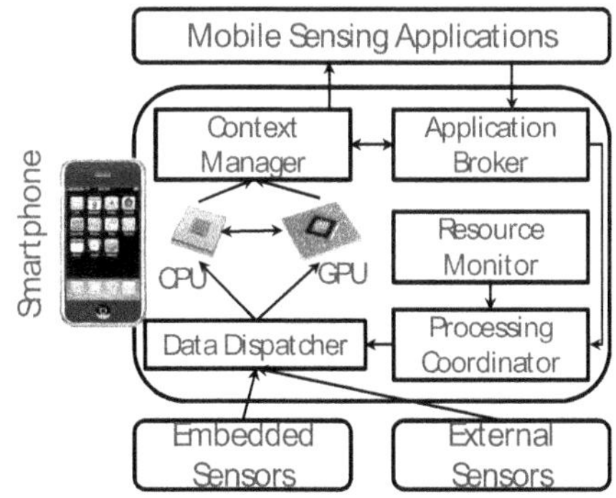

Figure 1. Overall architecture

Figure 1 shows the overall architecture. The *Application Broker* interacts with MS apps through a set of APIs we provide; the support for APIs is essential due to the lack of knowledge about underlying execution on GPUs. The *Processing Coordinator* carefully partitions the context processing computation into CPU and GPU parts in the way of achieving the maximum benefit. According to the decision, the *Data Dispatcher* properly dispatches the sensing data to CPU and GPU.

We believe that mobile GPU-accelerated context processing will be one of the key enabling forces to take the MS apps out of the lab and make them everyday reality. With them, people could enjoy diverse useful and proactive services without compromising the user experiences with their phones. It also would have substantial impact on mobile computing system as a first attempt to leverage new opportunities of mobile GPU for MS apps.

MobiSys'12, June 25–29, 2012, Low Wood Bay, Lake District, UK.
ACM 978-1-4503-1301-8/12/06

Poster: Evaluating Energy Consumption of Sensing Algorithms - Solely via Models?

Henrik Blunck, Mikkel Baun Kjærgaard, Christian Melchior
Department of Computer Science, Aarhus University, Denmark

ACM Classification Keywords
C.0 Systems Application Architecture

Keywords
Mobile Sensing, Power Consumption, Evaluation Methods

Automatic detection of significant places is an important concept in many emerging context-aware applications. Providing such detection energy-efficiently poses a crucial challenge on mobile devices with limited battery capacity [1]. Several solutions have been proposed to address the problem, but no comparison between them has yet been published running on mobile devices, making it hard for researchers and developers to provide suitable trade-off solutions between accuracy and energy efficiency in an informed manner. To close this gap, we investigate the impact of four common types of energy efficiency strategies on a fingerprint-based algorithm for place sensing based on WiFi measurements, namely SensLoc [2]. Using experimental prototypes on the Android platform, we show that in empirical evaluation only one of the strategies, namely sample rate management, affects the energy usage in a positive manner. The three other investigated strategies show either no effect or use more energy than the baseline implementation. This results stands in contrast to the conclusion of previous simulation based results [2]. Based on our experimental results, we question the viability of purely model-based validation of energy efficiency.

For our evaluation we implemented the SensLoc algorithm [2], including the following energy efficiency strategies: **Sample Rate Management (SRM)** trades accuracy for energy-efficiency by lowering the WiFi scanning rate. **Sensor Replacement (SR)** replaces a power hungry sensor with a more efficient one; for SensLoc we replace WiFi scanning with accelerometer-based motion detection, which exploits that devices do not change place, when stationary. **Distribution** offloads the computations of the SensLoc algorithm to a remote server over GSM / WiFi. **Code Optimized** requires the optimization of code with regards to Android performance guidelines in order to lower CPU and memory usage.

As the total power consumption is what matters to the end users, we use a holistic measurement approach, i.e. measuring power consumption using the battery information available from the operating system as this makes it possible to perform the tests in every day conditions of end users. The smartphone used for testing was a HTC Desire with a 1400 mAh battery. The phone was jail broken using Unrevoked in order to access battery information on the operating system level. Table 1 lists the experimental results w.r.t. measured battery drain per hour, and the respective percentual change compared to the baseline algorithm. The results indicate that power consumption is reduced neither via distribution (due to communication overheads) nor via sensor replacement (due to software overheads [3]), and that optimizing code has only an insignificant impact. Solely the sample rate management strategy shows a positive impact, reducing consumption by up to 14%, at the price, however, of reducing accuracy, as evaluated in prior work [2].

Table 1 - Change in power consumption for different energy saving strategies, when used with the SensLoc algorithm.

SensLoc + Strategy	Power per 1h [mAh]	Change
Distributed SRM-0.1 Hz	31.3	+6%
SR Accelerometer	31.3	+6%
SRM-0.1 Hz (Baseline)	**29.5**	-
Code Optimized	29.3	-0%
SRM-0.05 Hz	28.4	-4%
SRM-0.033 Hz	26.0	-12%
SRM-0.025 Hz	25.3	-14%

These results differ from prior simulation-based results [2] and therefore question the validity of simple model-based validation. Therefore, we recommend that systems are either evaluated through empirical measurements on mobile devices or that more advanced power models are developed, e.g., taking into account dynamic effects [1].

REFERENCES

1. Mikkel Baun Kjærgaard: Location-Based Services on Mobile Phones: Minimizing Power Consumption. IEEE Pervasive Computing 11(1): 67-73 (2012)
2. Donnie H. Kim, Younghun Kim, Deborah Estrin, Mani B. Srivastava: SensLoc: Sensing Everyday Places and Paths using Less Energy. SenSys 2010: 43-56
3. Bodhi Priyantha, Dimitrios Lymberopoulos, Jie Liu: LittleRock: Enabling Energy-Efficient Continuous Sensing on Mobile Phones. IEEE Pervasive Computing 10(2): 12-15 (2011)

MobiSys'12, June 25–29, 2012, Low Wood Bay, Lake District, UK.
ACM 978-1-4503-1301-8/12/06.

Poster: Saving Power for Mobile Phones with Partial Wi-Fi Scans

Songchun Fan
Duke University
Durham, NC, USA
songchun.fan@duke.edu

Mahanth Gowda
Duke University
Durham, NC, USA
mahanth.gowda@duke.edu

Romit Roy Choudhury
Duke University
Durham, NC, USA
romit.rc@duke.edu

ABSTRACT

The Wi-Fi interface is one of the major energy consuming components in smart phones. Wi-Fi scanning process contributes significantly to this. We present a solution based on partial scan, which produces full scan results, by performing only incomplete scan process. Using a decision tree algorithm, we propose such a prediction scheme, with the help of cached scan results.

Categories and Subject Descriptors

G.3 [**Probability and Statistics**]: Probabilistic algorithms; C.2.1 [**Computer-Communication Networks**]: Network Architecture and Design—*Wireless Communication*

Keywords

Wi-Fi, Scanning, Energy Efficiency, Desicion Tree

1. INTRODUCTION

We need to keep connected to the Internet, and our Wi-Fi-enabled mobile phones help us. Most of the smart phones today remember the connected Wi-Fi networks, and automatically connect to them when in their ranges. However, to seize the connection opportunities, our phones keep scanning Wi-Fi all the time, until the phone is able to connect to a network. This scanning process drains energy. Typically, a scan happens every 15 seconds, and each time it consumes about 20uAh energy, with an average power 57% more than the non-scan state, as shown in Figure 1. Many of the times the scans are unnecessary. Works such as [2] and [1] have optimized the scan interval and reduced the times of scans using cellular information or user mobility information retrieved from sensor readings.

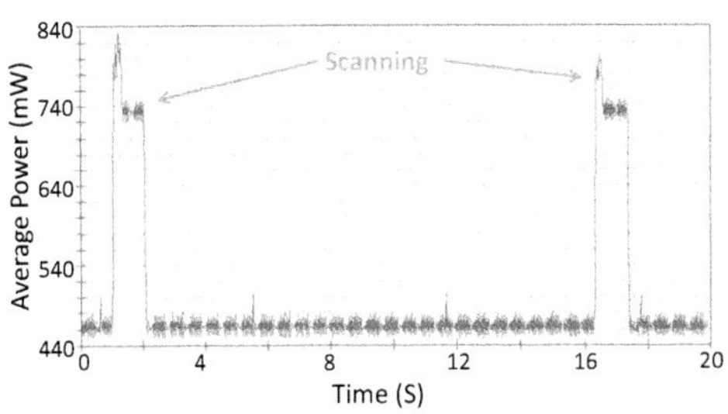

Figure 1: Scanning activity when the phone is screen-on.

MobiSys'12, June 25–29, 2012, Low Wood Bay, Lake District, UK.
ACM 978-1-4503-1301-8/12/06.

This poster attempts to solve the problem from another angle, by changing the scanning process itself. The key idea is as follows. We observe that not every Wi-Fi scan is necessary because often the Wi-Fi APs discovered from the scan are not open for free/public access. This part of the scan results are useless, and scanning them takes energy. If we can reproduce ideal scan results by scanning only a small subset of APs, we can save energy. The limited numbers of remembered networks in our phones show that we are more likely to connect to Wi-Fi networks only at certain places, such as home, cafe and office. These places always have a static set of networks, providing us the possibility to predict the rest of the scan results based on incomplete scan results.

Partial Scan - Our experiments are deployed on Nexus S phones, which use Broadcom DCM4329 chip as the network interface card (NIC). By default, the NIC listens on each of the channels (1-11) for tens of milliseconds. If during this process it hears beacons from APs on these channels, those APs are put into a result list. Then the NIC does the same process again, trying to catch some new beacons. This recurrence ends when the result list converges. We change this behavior such that the NIC stops scanning after obtaining partial results from the first round. This leads to a significantly lesser energy consumption compared to multiple (5 - 7) rounds of scan.

Prediction - If one round of partial scan contains an AP X, and if X was in the "previously connected" list, the NIC automatically connects to it. If not, we adopt a decision tree on the result set and predict the existence of an available AP in the rest of the results. If the possibility is high, we continue the rest of the full scan, otherwise, we stop this scan. For example, if a user once connected to "Starbucks" when she was in the range of "Target", later when her phone detects "Target", there will be a high chance that an available "Starbucks" network exists, even it is not in the current scan results. In this case, the phone continues the full scan. In cases where the phone does not see "Target", but see "WellsFargo", (and the user never connected to a network when "WellsFargo" shows up), the chance that any accessable network shows up is low, so the phone suppresses the rest of the scan and saves energy.

This poster shows an opportunity of utilizing partial scan results in Wi-Fi scanning. By cutting off each scan activity, we expect to extend the battery life.

2. REFERENCES

[1] K.-H. Kim, A. Min, D. Gupta, P. Mohapatra, and J. Singh. Improving energy efficiency of wi-fi sensing on smartphones. In *INFOCOM*, 2011.

[2] H. Wu, K. Tan, J. Liu, and Y. Zhang. Footprint: cellular assisted wi-fi ap discovery on mobile phones for energy saving. In *WiNTECH*, 2009.

Poster: Supporting Collaborative Sensing Applications

Xuan Bao
Duke University

Aman Kansal
Microsoft Research

Romit Roy Choudhury
Duke University

Paramvir Bahl
Microsoft Research

David Chu
Microsoft Research

Alec Wolman
Microsoft Research

ABSTRACT

Many context aware applications can benefit from using high-level sensing results with semantic meanings (e.g, busy/idle). This paper proposes a platform design that provides high-level "virtual sensor" abstractions and enables new virtual sensors to be bootstrapped from existing ones.

Categories and Subject Descriptors

C.2.4 [**Distributed Systems**]: Distributed Applications

Keywords

Context aware computing; User modeling

Introduction

The rich suite of sensors on mobile devices is enabling a new class of sensing applications that recognize user activities, their context, and environments, for several purposes. Different platform designs are proposed to exploit the power of crowds in large-scale sensing [1]. However, several challenges still remain to be the hurdles in bringing collaborative sensing apps to today's mobile app store.

Lack of High-Level "Virtual Sensor" Abstractions. Many applications can benefit from using high-level sensing results with semantic meanings (e.g., indoor/outdoor, alone/in-group, run/walk, busy/idle). These semantic meanings can be viewed as output from high-level "virtual sensors" that are built on top of real physical sensors For example, a virtual sensor differentiating running and walking can examine GPS and accelerometer readings and pass the classification results to upper-layer applications. The application can treat this process like using a new sensor without worrying about underlying techniques.

Difficulty in Establishing "Virtual Sensors". The inference algorithms used by virtual sensors often follow common machine learning principles. A large subset of the virtual sensors required for applications can be learned using data, through the use of classifiers based on known methods such as clustering, support vector machines, or Bayesian networks. Rather than requiring each application developer to acquire such expertise, we can provide a set of tools that allow developers to simply upload training data, and the inference models can be trained in the cloud using machine learning libraries in the cloud. The classifiers learnt can later be applied to sensor data acquired at mobile devices.

New "Virtual Sensors" on Top of Existing Ones. Higher-level virtual sensors can be built on top of existing lower-level instances. For example, a user being busy or idle can be inferred from other status information such as indoor/outdoor, alone/in-group. Suppose the latter two virtual sensors (indoor/outdoor, alone/in-group) are already established for existing applications. A new virtual sensor for busy/idle detection can certainly obtain these information through calling existing virtual sensors as shown in Figure 1. More broadly, the virtual sensor abstractions or even sensing data, contributed by existing apps, can be used to bootstrap new apps, particularly those that rely on collaborative sensing(e.g., pothole detection can benefit from knowing the user is driving and even reuse the data collected by a driver detection app [2]).

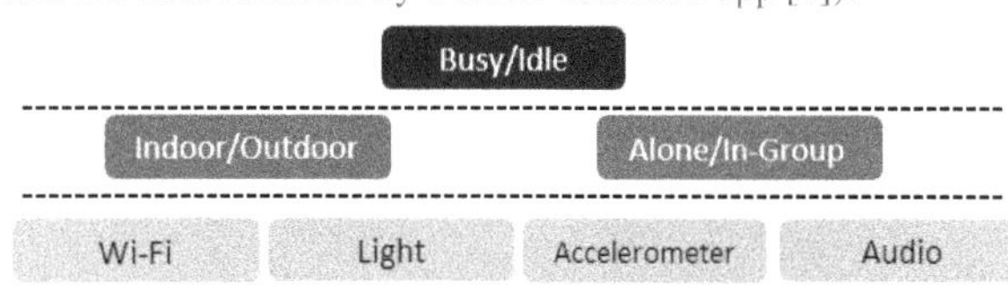

Figure 1: Virtual sensors example

Figure 2 shows the envisioned cloud-based cross-app platform. Cloud provides the necessary computation power for virtual sensor establishment. The cross-app design enables new applications to call inference results from existing ones.

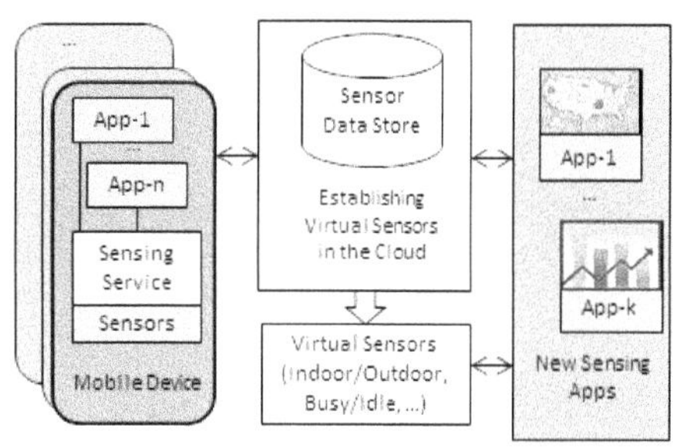

Figure 2: Overview of the Bootstrapping Module

All applications contribute sensing data to a central data store through our standard agent. Developer developing virtual sensors by provide labels to training data and the system automates the process of establishing inference models. New applications can call existing virtual sensors for inference results with semantic meanings. Applications that need new virtual sensors can build their logic on top of existing ones as well as reuse the collected sensing data.

1. REFERENCES

[1] Moo-Ryong Ra, Bin Liu, Tom La Porta, and Ramesh Govindan, "Medusa: A programming framework for crowd-sensing applications," *ACM MobiSys*, 2012.

[2] Jakob Eriksson, Lewis Girod, et al., "The pothole patrol: using a mobile sensor network for road surface monitoring," in *ACM MobiSys*, 2008.

MobiSys'12, June 25–29, 2012, Low Wood Bay, Lake District, UK.
ACM 978-1-4503-1301-8/12/06.

Poster: The Development of a Semantic Feature Analysis Based Mobile Application for Individuals with Aphasia

Conor Higgins
Interaction Design Centre
Department of Computer Science & Information Systems
University of Limerick
Limerick, Ireland
conor.higgins@ul.ie

Aine Kearns
Department of Speech & Language Therapy
University of Limerick
Limerick, Ireland
sue.franklin@ul.ie

Prof. Sue Franklin
Department of Speech & Language Therapy
University of Limerick
Limerick, Ireland
aine.kearns@ul.ie

ABSTRACT

We present a Semantic Feature Analysis based mobile application and paired web application, which have been developed in order to facilitate speech and language therapy with a Person with Aphasia (PWA). The goal of this research is to streamline current speech and language therapy practices, by enabling rapid data entry, aggregation and remote analysis. The mobile application has been developed in Objective C and is targeted at iPad users, though a smaller hand-held version may be developed in future. The paired web application has been built using standard Flex components, while its transition to a Javascript/HTML5 version is planned. The developed system has been trialled clinically on two occasions with a small number of participants, the intention of which was to investigate the potential for the delivery of Semantic Feature therapy programs in both domestic and clinical settings. These trials also offered the opportunity to gauge the reaction of the participants towards the presence of a novel mobile technology, as well as assessing the attitudes of carers and speech and language professionals towards the use of the device in question.

Categories and Subject Descriptors

K.4.2 [**Computers and Society**]: Social Issues—*Assistive technologies for persons with disabilities*; H.5.2 [**Information Interfaces and Presentation**]: User Interfaces—*User-centered design, Prototyping*

Keywords

Aphasia, Semantic Feature Analysis

1. INTRODUCTION

Aphasia is the loss of the ability to understand or convey language coherently, and is caused by damage to the brain, though it is most commonly associated with cerebrovascular accidents (CVAs), commonly known as strokes. The majority of PWAs experience word-finding difficulties known as Anomia, which is described as the "most consistent feature of aphasia" and as one of the most frustrating and distressing aspects of aphasia. Semantic Feature Analysis (SFA) is a rehabilitative approach to naming therapy, which provides an organised method of activating semantic networks to cue a target concept. According to models of lexical retrieval, when attempting to name an object its features are activated. During SFA therapy the PWA is guided through a framework, consisting of 6 semantic categories (Action, Association, Group, Location, Properties and Use), in which they provide features/words that are semantically related to the target concept. Responses are elicited from the PWA through a combination of prompts and cueing.

MobiSys'12, June 25–29, 2012, Low Wood Bay, Lake District, UK.
ACM 978-1-4503-1301-8/12/06.

2. SYSTEM EVALUATION

The developed system was subject to a clinical trial, whereby 2 Speech and Language Therapists (SLTs) were tasked with assessing the reaction and overall impression of the participants towards the application and device in general. While participants had expressive and in some cases comprehensive difficulties, the SLTs were capable of eliciting their responses using colloquial language and open-ended questions. SLTs met with and observed their respective participants a minimum of twice a week. Participants were additionally tasked with using the application without the presence of their assigned SLT. Following each week of therapy, participants and their carers were questioned in relation to the usability of the iPad and the SFA application, while general responses and events of significance were documented. Over the course of the study, it was deduced that while in general there were less difficulties reported as the study progressed, perhaps due to an increase in familiarity with the system, several complications were raised with respect to the practice of SFA in addition to the usability and suitability of touch screen devices to clinical therapy.

3. CONCLUSION

A Semantic Feature Analysis based mobile application has been presented, while an overview of its evaluation has been provided. The system's evaluation has demonstrated the attitudes of PWAs regarding the incorporation of novel technologies in speech and language therapy programs. It was observed that perhaps due to increased familiarity, over the course of the clinical study, the mobile and web applications were deemed more usable. However, the application's evaluation uncovered issues related to the understandability and potential suitability of semantic feature analysis to unsupervised speech and language therapy programs.

Using Local and Global Knowledge in Wireless Sensor Networks

Christopher Gwilliams, Alun Preece and Alex Hardisty
Cardiff University School of Computer Science and Informatics, Cardiff University, UK.
{C.Gwilliams, A.D.Preece, Alex.Hardisty}@cs.cardiff.ac.uk

Abstract

Wireless Sensor Networks (WSNs) produce large quantities of raw data from multiple sources, such as cameras, temperature or humidity sensors. We propose a network that uses the knowledge of domain experts that use the network, as well as previously sensed data, to classify incoming sensed data in near real-time. Using a field centre in Malaysia as a testbed, we are currently implementing a prototype network to classify images of animals that trigger wildlife cameras. Initial experiments to test the individual components of the network have yielded positive results and a prototype is planned for deployment in July of 2012.

Categories and Subject Descriptors

C.2.1 [**Network Architecture and Design**]: Network topology; C.2.1 [**Network Architecture and Design**]: Wireless communication

Keywords

sensor network, wireless, in-network processing

1. SCENARIO

Danau Girang is a field centre in the Kinabatangan Wildlife Reserve, Malaysia. It is operated by Cardiff University's School of Biosciences and owned by the Sabah Wildlife Department. Twenty-two wildlife cameras have been deployed to detect the movement of carnivores between parts of the rainforest, connected through corridors. Currently, research students are charged with the task of maintaining the cameras and manually processing the resulting images. Every two weeks, students must collect the SD cards and change the batteries of each camera. The images are then loaded onto a netbook and each picture is classified.

Using Danau Girang as a testbed, we aim to implement a prototype WSN that uses the knowledge of domain experts and the deployed environment of the WSN to classify sensed data before it reaches the user.

2. PROPOSED SOLUTION

There are existing systems that have been designed to merge Context Aware Systems (CAS) with WSNs to apply high level semantics to sensed data [3]. We aim to use a WSN to automate image retrieval that can, through the use of *local knowledge*, process sensed images within the network, thus providing users with classified images at the Field Centre in near real-time. We define local knowledge as knowledge of an area, that has been gained through experience or experimentation [2].

We have developed a WSN that consists of three tiers that aims to automate image retrieval and processing. The lower tier is the Data Collection (DC) nodes, attached to the wildlife cameras, that take images when motion is detected and route sensed data to the middle tier. The middle tier comprises Data Processing (DP) nodes tasked with serving a subset of Data Collection nodes and processing received data, classifying it via of image processing and a knowledge base of previous classifications. The upper tier comprises Data Aggregation (DA) nodes that stores all sensed data and acts as a gateway to the network itself. DA nodes mirror sensed data to online services, such as Flickr, when internet access is available. Data Aggregation nodes run GSN, a sensor middleware that integrates heterogeneous sensors through the use of virtual sensors [1].

3. EVALUATION AND IMPLEMENTATION

A visit to Danau Girang was made in 2011 to test individual components of the network, such as wireless range and impact of humidity on the hardware. A second visit is planned for 2012 to test the interoperability of each tier and implement a small-scale prototype network.

Acknowledgements The authors would like to thank Dr. Benoit Goossens, Cardiff University School of Biosciences.

4. REFERENCES

[1] K. Aberer, M. Hauswirth, and A. Salehi. A middleware for fast and flexible sensor network deployment. In *Proc 32nd Int Conf on Very large Databases*, VLDB'06, pages 1199–1202, 2006.

[2] L. Joshi, M. van Noordwijk, and F. Sinclair. Bringing local knowledge into perspective – a case of sustainable technology development in jungle rubber agro forests in jambi. In *Participatory Technology Development and Local Knowledge for Sustainable Land Use in Southeast Asia*, 2001.

[3] A. Wood, J. Stankovic, G. Virone, L. Selavo, Z. He, Q. Cao, T. Doan, Y. Wu, L. Fang, and R. Stoleru. Context-aware wireless sensor networks for assisted living and residential monitoring. *IEEE Network*, 22(4):26–33, July–Aug 2008.

MobiSys'12, June 25–29, 2012, Low Wood Bay, Lake District, UK.
ACM 978-1-4503-1301-8/12/06.

Poster: An Energy Profiler for Android Applications Used in the Real World

Hiroki Furusho
Dept. of AIT
Grad School of ISEE
Kyushu Univ.
Motooka 744, Nishi-ku,
Fukuoka, 819-0395 Japan
furusho
@f.ait.kyushu-u.ac.jp

Kenji Hasazumi
System LSI Research
Center, Kyushu Univ.
3-8-33 Momochihama,
Sawara-ku, Fukuoka, 814-
0001 Japan
nel
@slrc.kyushu-u.ac.jp

Takeshi Kamiyama
Hiroshi Inamura
Research Laboratories
NTT DOCOMO, Inc., 3-5,
Hikarinooka, Yokosuka,
Kanagawa, 239-8536 Japan
kamiyamat
@nttdocomo.co.jp,
inamura
@nttdocomo.com

Tsuneo Nakanishi
Akira Fukuda
Faculty of ISEE
Kyushu Univ.
Motooka 744, Nishi-ku,
Fukuoka, 819-0395 Japan
{tun,fukuda}
@f.ait.kyushu-u.ac.jp

ABSTRACT

Reducing the energy consumed in the use of smart phones has become a major challenge for application developers. While this problem can be addressed at various levels, it is important to reduce the energy consumption of individual applications which can vary greatly depending on the behavior of the application. Energy profiling methods are required in order to identify the points in which the applications are consuming excessive energy and to examine how to reduce overall energy consumption by applications.

The existing power estimating method[1,2] can estimate the energy consumption of entire terminal using data obtained from the OS (e.g. CPU-time, amount of access to file systems, traffic, etc.). It is, however, unable to identify points where excessive energy is consumed and to take into account all possible usage patterns and environments since developers are able to use the limited number of test cases in the profiling.

This paper proposes a novel method to analyze and visualize energy consumption of applications that run on Android OS under actual users' real usage. This method consists of four phases: 1) embedding logging codes to obtain hardware resource consumption from the OS for estimating the energy consumption in an application using AspectJ[3], 2) distributing the application to the users, 3) gathering that data from the application, and 4) creating visual of the data.

To explore the potential of the method, we profiled Crowdroid[4], an open-source twitter client, with four users and estimated the energy consumption for each activity with our method. An activity in the Android is an object that corresponds to the screen of the application. In order to calculate energy consumption from data we logged, we employ a model for estimation proposed by the existing power estimating method[1].

Figure 1 illustrates the cumulative energy consumption broke down into individual activities. The graph indicates that the second activity, which is most frequently used activity by the name of HomeTimelineActivity, consumes the biggest energy in the Crowdroid's activities.

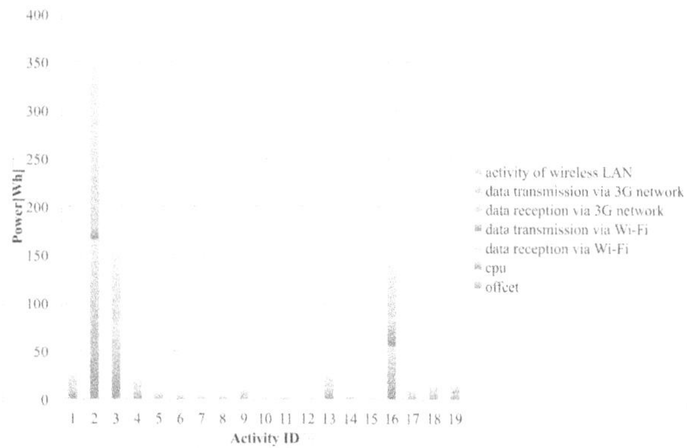

Figure 1. Cumulative energy consumption of each Activity

Categories and Subject Descriptors

D.2.5 [**Software Engineering**]: Testing and Debugging – *testing tools*

General Terms

Experimentation

Keywords

energy consumption

REFERENCES

[1] Yusuke KANEDA, Takumi OKUHIRA, Tohru ISHIHARA, Kenji HISAZUMI, Takeshi KAMIYAMA and Masaji KATAGIRI, 2010, A Run-Time Power Analysis Method using OS-Observable Parameters for Mobile Terminals. In *Int'l Conf. on Embedded Systems and Intelligent Technology*, Chiang Mai, Thailand, Febrary, 2010, 39.

[2] Lide Zhang, Birjodh Tiwana, Zhiyun Qian, Zhaoguang Wang, Robert P. Dick, Z. Morley Mao and Lei Yang, 2010, Accurate Online Power Estimation and Automatic Battery Behavior Based Power Model Generation for Smartphones. In *CODES+ISSS'10*, Arizona, USA, October 2010, ACM New York, USA, 105-114.

[3] The Eclipse Foundation. The AspectJ Project. http://www.eclipse.org/aspectj/

[4] Anhui OSS, Crowdroid, http://www.anhuioss.com/crowdroid/

Poster: A Smart Scheduling Mechanism for Energy Saving in Android System

Chang-Hung Hsieh, Yu-Yu Chen, Chih-Chieh Yang, Shih-Lung Chao, Hung-Yu Wei*
Dept. of Electrical Engineering, National Taiwan University
Taipei, Taiwan
hywei@cc.ee.ntu.edu.tw*

Categories and Subject Descriptors: C.2.1 [Network Architecture and Design]: Wireless communication

Keywords: Smart Scheduling Mechanism, Energy Saving, Android

1. INTRODUCTION

In this paper we provides a smart scheduling mechanism to address the issue of power efficiency on Android-based smartphones. We design an algorithm in the network kernel to align the applications to enable the scheduling ability for synchronizing the time of periodic applications. We first analyze the measurement of multiple applications and the result of delay test in Android kernel, based on which the algorithm manages the data traffic by dropping packets to defer the sync time of applications to shorten their time difference. By minimizing the synchronization time difference between applications, we can align the applications to achieve simultaneous synchronization. Additionally, we also consider applications of different periods to offer a synchronization solution of a fuller extent. To implement the idea, we design the algorithm for different application types. This algorithm is composed of several mechanisms: approaching, aligning, and maintaining.

Approaching: Assume that two applications have the same period of sync. If one application is delayed by the technique and the other application is not, their sync time will get closer next time. At first, these two applications have time difference t originally. If the sync of App 1 is deferred by MAX DELAY TIME, next round their difference will be shortened to t-MAX DELAY TIME. If we keep using the technique for the following rounds, they will have the time difference that is less than MAX DELAY TIME, getting chances to transmit together eventually. By then, the next mechanism will be adopted.

Aligning: In Fig. 1(a), these two applications are close, but they are still sent out separately in original system. In our proposed system, as shown in Fig. 1(b), the packet of App 1 can be buffered, so it can wait for a packet of App 2 as long as it can. When the packet of App 2 arrives in waiting period, these two packets can be sent out together. The system will consume less energy by reducing the amount of extra power in overhead. The black lines in Fig. 1 stand for the power level of wireless device. In original scheme, the two transmissions are separate, so the overhead is doubled.

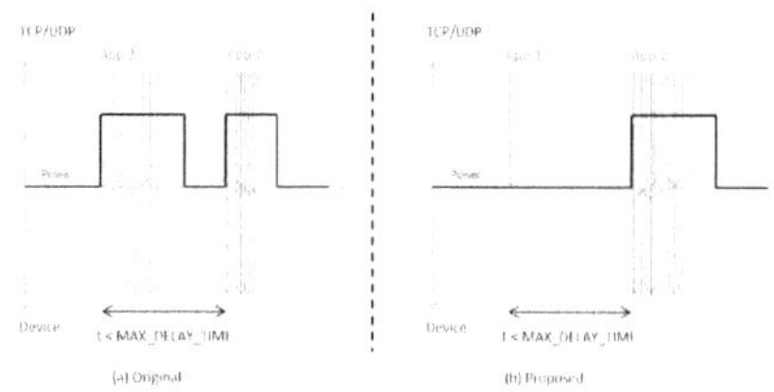

Figure 1: An example of aligning two near applications

Figure 2: An example of considering different periods and maintaining the alignment

In proposed scheme, the two transmissions are combined into single one, so the overhead is nearly the same as that in one transmission, halved compared with the original scheme.

Maintaining In Fig. 2, App 1 and App 2 are aligned at a moment, transmitting DNS packets at the same time. After T1, the DNS packet of App 1 arrives. The system first detects that the expected arrival of next DNS packet of App 2 is larger than MAX DELAY TIME, which means that there's no alignment action for the two applications. Next, the system calculates the next arrival of DNS packet of App 1 and the corresponding new time difference. In the example, the system finds out that without delaying the current DNS packet of App 1, the time difference of next App 1 sync and next App 2 sync is smaller than MAX DELAY TIME, which means that they will be aligned next time. As a result, the system sends out the current DNS packet of App 1 directly. By doing so, the property of alignment remains, enabling us to save power in the subsequent rounds.

Acknowledgment

This work was also supported by Ministry of Education and National Taiwan University under Grants AE00-00-04, and AE00-00.

This work was also supported by National Science Council, National Taiwan University and Intel Corporation under GrantsNSC 100-2911-I-002-001, and 101R7501

Poster: Cloud-enabled Content Search and Sharing System for Mobile Phones (MobiShare*)

Kuldeep Yadav, Vibhas Kumar, Shrey Jairath, Rushil Khurana,
Vinayak Naik, Amarjeet Singh
Indraprastha Institute of Information Technology (IIIT), Delhi
kuldeep@iiitd.ac.in

Categories and Subject Descriptors

C.2.1 [**Network Architecture and Design**]: Store and forward networks

Keywords

Mobile System, Content Sharing, Bluetooth, Cloud

1. INTRODUCTION AND MOTIVATION

Most of the users in developing countries use 2G (GPRS/ EDGE) connection which has very limited bandwidth and can not be used for large content transfers. As a result, content exchange happens through local communication technologies i.e. Bluetooth or manually swapping flash cards. Users typically search content by asking their friends and fix a meeting place to exchange that content. This model of content sharing is not scalable when number of users or shared items increase. Current research literature in mobile phone based delay tolerant networks (DTNs) are restricted to designing routing protocols or designing distributed content search and sharing evaluated using simulations. DTNs employs multi-hop routing which is not feasible in real-world due to resource constrained mobile devices, trust issues, and interfaces such as Bluetooth requiring human intervention for device pairing. Due to above described limitations, there is lack of a real system that can enhance pair-wise content sharing using local interfaces such as Bluetooth.

Here, we propose a mobile-cloud based system i.e *MobiShare* which uses a hybrid architecture to facilitate pair-wise content transfers. It uses two different communication channels, control channel facilitates content search and sharing with the help of Cloud whereas actual content transfer happens through local interfaces such as Bluetooth. The information routed through control channel is small in size compared to data channel and hence, can use a low bandwidth communication such as 2G. Notification mechanism of *MobiShare* allows mobile application to automatically switch ON the interface (Bluetooth or WiFi) whenever a content exchange opportunity arises to save energy.

2. SYSTEM OVERVIEW

The two core components of the system are mobile application and Cloud as show in Figure 1. User's social network is inferred from the phone contacts and Facebook friend list. User decides to share some content (songs, movies etc) with his friends, mobile application extracts meta-data of shared content and uploads it to Cloud. *MobiShare* removes the trust concerns of the users by allowing them to share only in their social network. Further, large degree of regularity in human mobility is predicted to find encounters between two users and help user in deciding appropriate content source. For location, we use low power consuming location interface such as Cell ID and opportunistically complement it by scanning WiFi APs and Bluetooth. Use of Cell IDs come up with some non-trivial challenges to solve i.e fluctuation of Cell IDs at same place, different operators etc. We solve all these challenges to build an accurate spatio-temporal mobility profile of a user from the location updates which are sent from mobile application to Cloud.

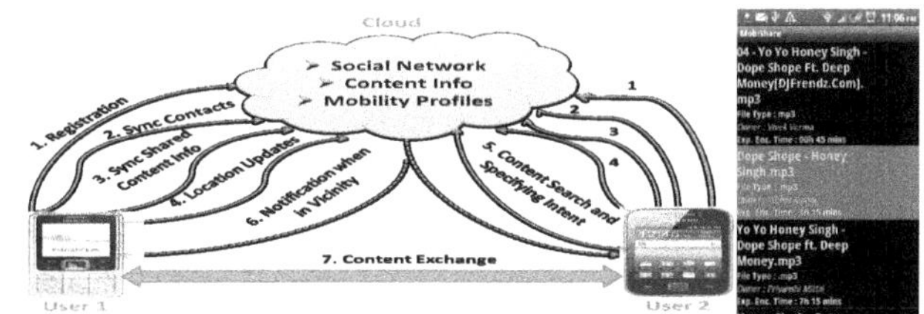

Figure 1: Left: Architecture of *MobiShare* system, Right: A snapshot of Android app's content search results

1. User 1 searches content "V" in mobile app. Cloud search in her social network. Assuming that "V" is owned by user 2 and user 3 both, it predicts an encounter time from the mobility profiles of 1, 2 and 1,3 and rank the search results based on encounter time which is shown to the user 1 (Refer Figure 1).
2. User 1 specifies her intent by selecting user(s) from she wants to take the content. Assuming it selected user 2, Cloud starts tracking location updates of 1 and2 to see whenever they come in proximity and then send notifications (using SMSes) to both users' phones.
3. On reception of notifications, mobile app starts neighbor discovery to find each other and exchange "V".

We have developed *MobiShare* mobile app for Android and Symbian (both S60 & S40) operating systems and setup a Cloud instance on our university's server. Proposed system is deployed with 14 different users for about two weeks. From our collected location data in two weeks, 40% of pairs have shown more than 20 interactions in 2 weeks, median interaction length is found to be approximately 30 minutes which can be utilized for content exchange among pairs. We have also observed that these pair-wise interactions are consistent across different weeks.

***http://www.iiitd.edu.in/muc/content/mobishare**

MobiSys '12, June 25–29, 2012, Low Wood Bay, Lake District, UK.
ACM 978-1-4503-1301-8/12/06.

Poster: Towards Position-based Routing for Mobile Environments

David Cairns
University of Stirling
dec@cs.stir.ac.uk

Marwan Fayed
University of Stirling
mmf@cs.stir.ac.uk

Categories and Subject Descriptors

C.2.1 [**Communication Networks**]: Wireless Communications; C.2.2 [**Communication Networks**]: Routing

Keywords

face-routing, wireless routing, mobile communication

ABSTRACT

Despite its simplicity there exists, as yet, no position-based routing that is suitable for dynamic networks: Existing schemes construct subgraphs or overlays that come at a cost that may be too high. In a network where nodes are mobile, such constructions risk becoming stale before they become stable, potentially invalidating routing decisions.

In this work we are investigating position-based routing for dynamic networks based on left- or right-hand rule (LHR). A node using LHR forwards messages to the neighbour that sits next in rotational order. When used to recover from greedy routing failures, LHR guarantees success if implemented over planar graphs; for this reason it is often called 'face-routing'. We note, however, that if planarity is violated then LHR is only guaranteed to eventually return to the point of origin. We have chosen LHR for two reasons. First, it is most prevalent in position-based routing literature and hence well-studied. Second, it is a simple rule requiring little-to-no overhead.

We have started by looking for the specific causes of failure in LHR with the hope that we might quickly recognise and avoid them in mobile environments [1]. First, in the unit disc graph where communication ranges are normalised to 1, we analyze the way in which intersecting links may interact to invoke sides of a 4-gon (Figure 1a). We have provably enumerated all possible configurations to three possible cases. Surprisingly, only one of these configurations is able to defeat LHR. This is because there are two distinct traversals, each of which has the potential to restrict exit points in the other. We call this the 'umbrella' configuration due to its appearance, as shown in Figure 1b.

Table 1: Ratio of umbrellas in sample networks.

Network Size (Density)	Node Distribution	Ratio U/I
1500 (7.5)	uniform	0.013
	norm	0.013
	skew	0.011

In determining how best to design a new routing protocol we first ask how likely the umbrella is to occur. Sample results are provided in Table 1. In all node densities and distributions tested, the proportion of intersections that form an umbrella configuration is a little more than 1%. In practice, there are additional conditions that must be true before it can defeat LHR. Referring to Figure 1b, one such condition is that no other path exists from node c to any of nodes a, b, d. As a consequence of such additional conditions, its impact is felt far less often than is implied by its frequency.

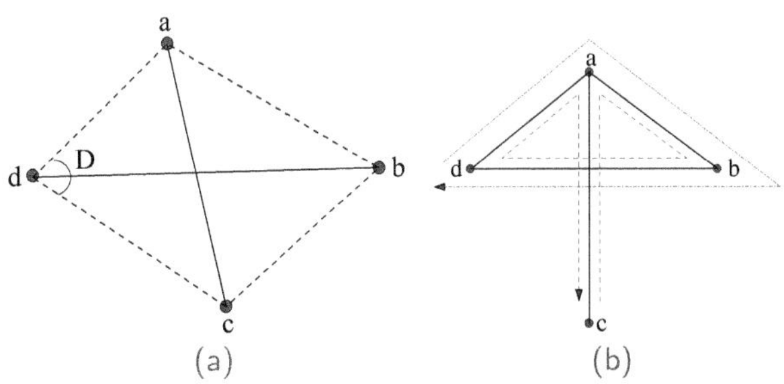

Figure 1: (a) Intersections invoke 4-gon edges; (b) Only 'umbrella' configurations defeat LHR.

Table 2: Measures from contrived graphs.

Path Length				Message Overhead			
GDSTR	CLDP	LCR	PLRP	GDSTR	CLDP	LCR	PLRP
5	9	7	7	624	245	0	0
4	6	6	11	792	227	227	6
5	10	15	15	774	298	0	0
9	8	8	29	1641	1082	1082	8

We have extended our analysis to include arbitrary graphs. We are able to show that, from the perspective of an LHR traversal, there are only five (5) possible intersecting configurations. Even with the relaxation on the graph, *the umbrella remains the only configuration that is able to defeat LHR.*

The infrequency of problem cases, in general, is supported by previous studies. We are currently developing a face-routing protocol for mobile networks called PLRP. In PLRP we are exploring different methods to deal with the umbrella configuration that avoid disturbing the network graph, or invoking global cooperation. Initial results are promising, though a trade-off is clear. In Table 2 we present sample results from contrived graphs that are designed to invoke worst-case behaviour. PLRP has almost no additional overhead when compared against well-known (and deserving) position-based schemes; the trade-off comes at a longer path length. We believe this may be a worthwhile trade-off in mobile environments.

[1] M. Fayed, D. Cairns, and H. T. Mouftah. An Analysis of Planarity in Face-Routing. TR-184, Computing Science and Maths, University of Stirling, Stirling, UK. 2010.

MobiSys'12, June 25–29, 2012, Low Wood Bay, Lake District, UK.
ACM 978-1-4503-1301-8/12/06.

Poster: A Framework to Sense Planet-Scale Vehicular Mobility using Online Traffic Webcameras

Gautam S. Thakur
CISE, University of Florida
Gainesville, USA
gsthakur@cise.ufl.edu

Pan Hui
Telekom Innovation
Laboratories, Berlin, Germany
pan.hui@telekom.de

Ahmed Helmy
CISE, University of Florida
Gainesville, USA
helmy@cise.ufl.edu

Categories and Subject Descriptors

H.2.8 [[**Database Management**]]: Database Applications-Data mining, Image databases, Spatial databases and GIS

General Terms

Experimentation, Human Factors, Measurement

Keywords

Urban Infrastructure, Vehicular Traffic, Social Networks

1. INTRODUCTION

Realistic modeling of vehicular mobility and sensing has been particularly challenging due to the lack of large libraries of measurements in the research community and the unavailability of Off-the-shelf analysis tools [1]. In this work, we propose *GlobalSense* - a novel framework for large-scale monitoring, analysis, and visualization of spatio-temporal vehicular mobility and sensing using the freely available online traffic webcams. We collect vehicular mobility traces from 2,700 traffic webcams in 10 different cities for six months and generate a mobility dataset of 7.5 Terabytes consisting of 125 million of images. We use background subtraction algorithm on these images to estimate traffic density.

The GlobalSense in Figure-1 is divided into four main components: (i) Data collection and pre-processing, (ii) Knowledge discovery, (iii) Modeling and analysis, and (iv) Graphical visualization. In data collection and pre-processing we download media feeds and process them to generate a quantifiable traffic value (density). The second and third components are designed for knowledge discovery, analysis, modeling and validating the data distribution using known statistical techniques. Finally, visualization helps to qualitatively identify hidden trends and traffic patterns.

2. MEASUREMENTS

We utilize the power of online traffic webcams as primary source of data collection, which are deployed on highways and critical traffic intersection of regions under study. At regular time interval, they capture still pictures of on-going road traffic and send them in the form of visual feeds to Department of Transportation(DoT). We download these images (from DoTs website) and store them in our media image storage server. After collecting these images we use background subtraction to generate traffic densities showing the recorded traffic. We view the connected global network of these webcams as a highly versatile platform, enabling us to visualize the traffic flow of the region and realistically model vehicular mobility that aid in traffic engineering purposes.

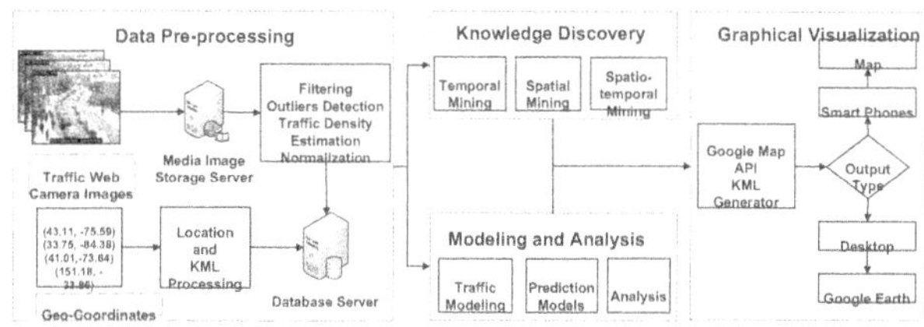

Figure 1: A modular framework for GlobalSense

3. PRELIMINARY RESULTS

In knowledge discovery we perform mining for spatio-temporal activity of traffic across all the cameras of a region. Our results [2] show that: (i) High correlation between driving time and distance indicate congestion-free traffic, (ii) Traffic follow certain patterns that are stable for a long time (42 days). (iii) Traffic conditions show high auto-correlation (80%) for 1-2 hour lag then decrease significantly to 25-30% for four hours lag. Next, we represent cities' roadways by network graphs in which nodes are camera locations and edges are urban streets that connect nodes. Such representation exhibit high centralities and small world properties with short path lengths and large clustering coefficient. Finally, modeling these densities against known theoretical distributions show less than 5% deviation for heavy-tailed models such as Weibull, also indicating self-similar process.

In graphical visualization the framework provides visual demonstration of traffic overlay on the maps. It also provides knobs to control and visualize the traffic in several different forms and thus help predict the traffic not only for the current but for a week time and several hours of days. We are using combination of GoogleEarth, OpenEarth, KML and Matlab to show the traffic on desktop machine. On handhelds, we are developing applications for both iOS and Android. We believe our framework will aid in the design and development of future vehicular networks and smart cities enabling ubiquitous computing in urban infrastructure.

4. REFERENCES

[1] J. Harri. Mobility models for vehicular ad hoc networks: a survey and taxonomy. *Communications, IEEE*, 2009.

[2] G. S. Thakur, P. Hui, and A. Helmy. Vehicular network modeling using public webcams. *CoRR*, 2011, `http://www.cise.ufl.edu/~gsthakur/globalsense`.

MobiSys'12, June 25–29, 2012, Low Wood Bay, Lake District, UK.
ACM 978-1-4503-1301-8/12/06.

Poster: A Hybrid Approach for Indoor Mobile Phone Localization

Valentin Radu, Jiwei Li, Lito Kriara, Mahesh K. Marina
The University of Edinburgh

Richard Mortier
University of Nottingham

ABSTRACT

We consider the mobile phone location tracking problem in indoor environments. We propose a hybrid approach for energy-efficient indoor mobile phone localization by combining two well-known techniques namely, pedestrian dead reckoning and WiFi fingerprinting. Our preliminary evaluation of the proposed approach shows that it achieves better location accuracy (median error around 2m) than either of the two techniques on which it is based.

Categories and Subject Descriptors

C.2.1 [**Computer-Communication Networks**]: Network Architecture and Design—*Wireless communication*

Keywords

Indoor localization, Mobile phones, Pedestrian dead reckoning, WiFi fingerprinting

1. MOTIVATION

Location tracking of mobile users and devices is needed for a variety of location-based applications, spanning different indoor environments (e.g., offices, shopping malls, airports). But research seeking practical and easily deployable indoor mobile location tracking solutions is still very active. Such solutions are increasing in demand in recent years with the emergence of smartphones with high-speed wireless Internet connectivity and their rapidly growing adoption. On the plus side, smartphones come with several built-in sensors such as accelerometers and compasses that can aid in mobile phone localization. In fact, researchers have already started exploiting phone sensors to realize *dead reckoning* based mobile location tracking systems, mainly for outdoor environments [1, 2]. While dead reckoning based schemes naturally enable continuous location tracking, they also have to deal with the well-known issue of error accumulation over time by having access to reliable anchor points once in a while to bound location errors. Moreover, dead reckoning in indoor environments is relatively more challenging. *WiFi fingerprinting* based localization approach is an attractive alternative as it can leverage the existing WiFi infrastructure (that is commonplace nowadays in most indoor environments) as well as exploit the presence of WiFi interfaces on smartphones. But the WiFi fingerprinting approach is not suitable for continuous location tracking of a mobile user because WiFi scanning operations are relatively quite power hungry (consume around 1W) and are also time consuming ($\sim$500ms).

MobiSys'12, June 25–29, 2012, Low Wood Bay, Lake District, UK.
ACM 978-1-4503-1301-8/12/06.

2. APPROACH

In this work, we propose a hybrid approach that combines the dead reckoning approach with the WiFi fingerprinting approach in such a way that they complement each other and retain their best aspects. Specifically, our approach primarily relies on dead reckoning based on phone accelerometer and compass while using WiFi fingerprinting from time to time to provide a new anchor point to correct any error accumulated while tracking the trajectory since the previous anchor point. Different from [2], the dead reckoning component of our system is fairly sophisticated including machine learning based activity recognition, speed estimation and map matching capabilities to better cope with the complex movement patterns (e.g., going up/down stairs) in indoor environments and increased likelihood of sensors being noisy (e.g., erroneous compass readings when in proximity of metal structures within buildings). The fingerprint database in the WiFi fingerprinting component is generated in an online manner through crowd-sourcing among system users, who are prompted periodically to identify their current location on the map which is used to tag a concurrently obtained WiFi fingerprint from the phone.

3. PRELIMINARY RESULTS

We have implemented our system on Android Nexus One phones and use Google App Engine as the back-end. Early results evaluating the location accuracy of our system were obtained through experimentation in the University of Edinburgh Informatics Forum building. These results show that pedestrian dead reckoning (PDR) on its own results in median location error of almost 10m despite its careful design because a single misclassification of activity (e.g., user taking the stairs when he is actually still walking) is sufficient to derail the estimated trajectory with PDR. We also find that combining PDR with WiFi fingerprinting does bring down the error drastically to a median error around 2m provided the frequency of using WiFi fingerprinting component is suitably chosen. The impact of WiFi scan frequency on location error is found to be complex in that scanning too frequently or too infrequently did not result in the best location tracking behavior.

4. ACKNOWLEDGMENTS

We thank Ciprian Balan for his contribution at the early stage of this work. We also thank Charles Sutton and Mike Hazas for helpful discussions.

5. REFERENCES

[1] I. Constandache et al. Towards Mobile Phone Localization without War-Driving. In *Proc. IEEE INFOCOM*, 2010.

[2] M. Youssef et al. GAC: Energy-Efficient Hybrid GPS-Accelerometer-Compass GSM Localization. In *Proc. IEEE GLOBECOM*, 2010.

Poster: A Construction of a Long Distance Communication Sensor Network Node using Arduino and Mad-SS Shield

Norihisa Segawa, Tomoaki Yamaguchi, Jun Sawamoto
Iwate Prefectural University, Takizawa, Iwate, Japan
mad-ss@dais.soft.iwate-pu.ac.jp

Masato Yazawa, Haruo Tamaki
Mathematical Assist Design Laboratory, Gunma, Japan
{yazawa,mad}@mail.wind.ne.jp

ABSTRACT

In this paper, we describe the construction of a long distance communication sensor network node for Arduino. We develop a Mad-SS Shield prototype system, which succeeds in about 5-km transmission with a 1 mW output.

ACM Classifiers: B.4.1 Data Communications Devices: Transmitters

Keywords: Arduino, Mad-SS, Direct Sequence of Spread Spectrum communication (DS-SS)

1. INTRODUCTION

In recent years, the technology using a sensor network is attaining an exalted of focus on the information collection in the real world, such as temperature and humidity. In order to experiment in a sensor network outdoors, the sensor node that can be treated easily is required.

In this paper, we propose a construction of a long distance communication sensor network using Arduino and Mad-SS shield.

2. MAD-SS

We developed the ultra low power, compact, long distance spread spectrum communication technology "Mad-SS" using in outdoor where no information infrastructure is available. This technology enables both the high sensitivity of SNR=-14.5db and 1.4 second synchronization time [1]. Mad-SS is based on direct sequence of spread spectrum communication (DS-SS). DS-SS has a problem for using sensor network. The problem is that synchronization acquisition of takes too much time to realization long-distance communication using weak radio wave. Mad-SS has the high-speed synchronization technology.

3. MAD-SS on Arduino

We develop a long distance sensor communication sensor node using Arduino and Mad-SS Shield.

Arduino is an open-source electronics prototyping platform based on flexible, easy-to-use hardware and software [2]. Arduino is a development system that consists of a microcomputer equipped with simple I/O. Arduino hardware is programmed using a Processing/Wiring language. Since a code can be described easily and various bases are moreover offered according to the use, Arduino is broadly used all over the world. Arduino Fio is a kind of the Arduino base which carries CPU (ATMega328P) and I/O, and a lithium polymer charge circuit in a size about 3 cm size long and about 7 cm wide.

To use Mad-SS for Arduino Fio, we developed MAD-SS shield (Fig. 2). The MAD-SS shield aims at about 10-km transmission with 1 mW of effective outputs with the 150 MHz band. Moreover, it has a serial communication port and is designed by the transmission speed of 1200 bps by 1 packet 31byte.

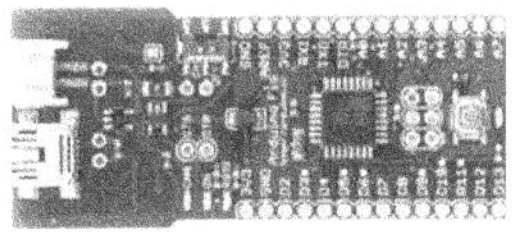

Figure 1. Arduino Fio

Figure 2. Mad-SS Shield

4. MAD-SS Application

Using Arduino and MAD-SS shield, it is easy to make the prototype of a long-distance communication sensor network [3]. Fig. 3 shows the prototype system that equipped Arduino Fio with a MAD-SS shield, a GPS module and a 2000mAh LiPo battery. Latitude and longitude information from a GPS module are changed into 10 bytes of packet, and is sent to the transmitter. The number of lines of this program is only 400 lines. Fig. 4 shows the screen that visualized the position information transmitted with this prototype by Google map. This prototype succeeds in about 5-km transmission with a 1 mW output. This is throughput of 10 times or more of an ordinary 2.4 GHz ZigBee sensor node.

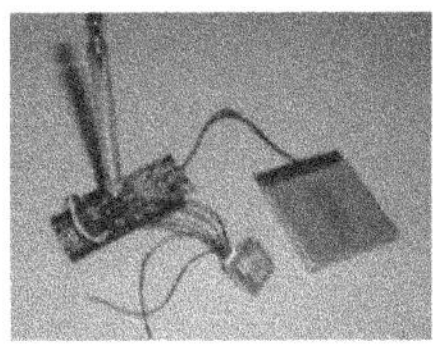

Figure 3. Prototype system

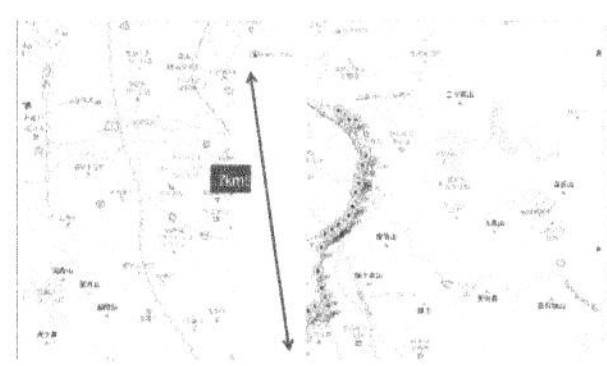

Figure 4. Visualization system

5. Conclusions

We describe a construction of a long distance sensor network using Arduino and a Mad-SS shield. In that, we describe its sensor node that generates electricity. Moreover, we show its sensor node that examines the activity of the animal and its wearable sensor node for the person who climbs the mountain.

Acknowledgments: This research was supported by Adaptable and Seamless Technology Transfer Program through Target-driven R&D, JST

References

[1] Mad-SS technology, http://www.madlabo.com/mad/product/ss/English/20090500esccEng.pdf

[2] Arduino, http://www.arduino.cc/

[3] T. Yamaguchi, K. Asakawa, M. Kashiwada, N.Segawa, J. Sawamoto, M. Yazawa, H. Tamaki, Proposal of the help support system using the Arduino & MAS-SS shield, Lecture Notes of IEICE, vol. 111, no. 470, LOIS2011-76, pp. 19-2(Japanese), 2012

Poster: SixthSense - Shaping Travel Activities Through Flexible Interpretations of Time

Mike Harding[1], Nigel Davies[1], Tom Cherrett[2], Janet Dickinson[3], Sarah Norgate[4] and Chris Speed[5]

[1]Lancaster University, [2]Southampton University, [3]Bournemouth University, [4]Salford University and [5]University of Edinburgh

Categories and Subject Descriptors
H.3.5 [**Online Information Systems**]: Data sharing

Keywords
Mobile Computing, Intelligent Travel Systems, Crowd-sourcing.

1. INTRODUCTION
During travel planning and journey execution our interpretation of time is shaped by static schedules (e.g. bus timetables, fixed meeting points and times), social norms and subjective time pressures that result from ill-informed assumptions about future travel activities. This rigid interpretation of time has contributed to unsustainable travel practices that impact negatively on our health, economy and the environment. In [1] we proposed SixthSense Transport, a mobile participatory sensing and travel visualization platform that aims to provide users with a more flexible view of time and travel through:

- Heightened awareness of potential transport options and opportunities.
- Shared visualizations of predicted future travel.
- An open model to capture, analysis and present travel activity data (e.g. journey traces, transport modes, transport network conditions).

1.1 School Walking Bus Scenario
The following use-case first presented in [1] describes the application of SixthSense in a common travel activity, delivering children to school.

Tom is doing the school run again and, as usual, he is late this morning. His school has subscribed to the SixthSense travel application and Tom is able to see the flows of cars, busses and walking-trains heading towards the school gates. Scrolling forwards in time in the app he can see that his intended plan of walking the kids to school is going to mean they miss the school bell. However, he can see that a 5 minute drive will enable him to intersect with one of the school's walking trains – enabling him to drop the kids off so they can benefit from walking at least part of the way to school while he can continue on his way to work.

The scenario above demonstrates how a heightened awareness of real-time and future travel patterns through the utilization of travel sensor data and collaborative visualizations can deliver new forms of planning and journey optimization. The user is not simply consulting a timetable that provides them with a set of fixed arrival and departure times to specific points or even relying on live traffic updates but rather they are able to see in real-time multi-modal travel options and to combine them in new, opportunistic ways.

1.2 SixthSense Platform
The SixthSense platform comprises four core components (1) a mobile participatory sensing component running on smartphone devices equipped with rich sensor capabilities to capture travel data, (2) an open server-side travel database that provides the basis for journey-learning prediction algorithms to infer future travel patterns, (3) interactive graphical visualizations of travel information and (4) a social networking framework to support the user creation of community orientated travel groups to enable users to share visualization predictions and personalized travel information.

1.3 Poster
In our poster we present a concrete SixthSense usage scenario and describe the design of an early prototype of the system.

2. ACKNOWLEDGEMENTS
This work was conducted as part of the EPSRC funded Sixth Sense project.

3. REFERENCE
[1] N. Davies, M. Lau, C. Speed, T. Cherrett, J. Dickinson, and S. Norgate. 2012. Sixth sense transport: challenges in supporting flexible time travel. In *Proceedings of the Twelfth Workshop on Mobile Computing Systems & Applications* (HotMobile '12). ACM, New York, NY, UK

MobiSys '12, June 25–29, 2012, Low Wood Bay, Lake District, UK.
ACM 978-1-4503-1301-8/12/06.

Author Index

www.ingramcontent.com/pod-product-compliance
Ingram Content Group UK Ltd.
Pitfield, Milton Keynes, MK11 3LW, UK
UKHW051416290926
15471UKWH00009B/1627

9 781450 318792